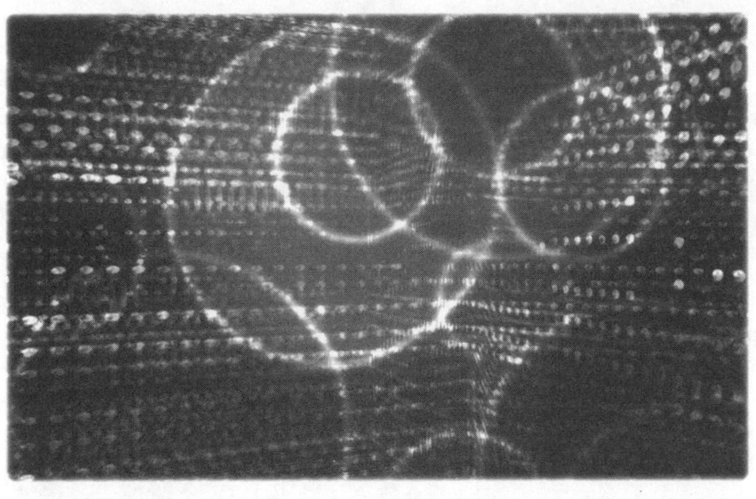

JAMSA'S C/C++/C# PROGRAMMER'S BIBLE

The Ultimate Guide to C/C++/C# Programming

SECOND EDITION

KRIS JAMSA, PHD, MBA

ONWORD PRESS
THOMSON LEARNING

Australia • Canada • Mexico • Singapore • Spain • United Kingdom • United States

Jamsa's C/C++/C# Programmer's Bible: The Ultimate Guide to C/C++/C# Programming, 2e

by Kris Jamsa, Ph.D., M.B.A.

Business Unit Director:
Alar Elken

Executive Editor:
Sandy Clark

Senior Acquisitions Editor:
Gregory L. Clayton

Senior Development Editor:
Michelle Ruelos Cannistraci

Executive Marketing Manager:
Maura Theriault

Marketing Coordinator:
Karen Smith

Channel Manager:
Fair Huntoon

Executive Production Manager:
Mary Ellen Black

Production Manager:
Larry Main

Production Coordinator:
Sharon Popson

Art/Design Coordinator:
David Arsenault

Editorial Assistant:
Jennifer M. Luck

Full Production Services:
SoundLightMind

Printed in the United States of America
2 3 4 5 XXX 05 04 02

For more information contact Delmar,
3 Columbia Circle, PO Box 15015,
Albany, NY 12212-5015.

Or find us on the World Wide Web at
http://www.delmar.com

ISBN 0-7668-4682-2

NOTICE TO THE READER

Contents

Getting Started with C

MACROS AND CONSTANTS

UNDERSTANDING STRINGS

FUNCTIONS

KEYBOARD OPERATIONS

MATH

FILES, DIRECTORIES, AND DISKS

ARRAYS, POINTERS, AND STRUCTURES

DOS AND BIOS SERVICES

MEMORY MANAGEMENT

DATE AND TIME

REDIRECTING I/O AND PROCESSING COMMAND-LINES

PROGRAMMING TOOLS

ADVANCED C

GETTING STARTED WITH C++

OBJECTS

COMMON CLASS FUNCTIONS

I/O WITH C++

INHERITANCE AND POLYMORPHISM

GENERIC FUNCTIONS AND TEMPLATES

EXCEPTION HANDLING AND TYPE PORTABILITY

CREATING SAMPLE REUSABLE CLASSES

STANDARD TEMPLATE LIBRARY

GETTING STARTED WITH WINDOWS PROGRAMMING

MESSAGES AND MENUS

DIALOG BOXES

WINDOWS MEMORY MANAGEMENT

PROCESSES AND THREADS

GRAPHICS DEVICE INTERFACE

BITMAPS, METAFILES, AND ICONS

WINDOWS I/O

THE .NET ENVIRONMENT

AN INTRODUCTION TO PROGRAMMING

1

Computer programs, also known as *software*, are made up of a series of instructions that the computer executes. When you create a program, you must specify the instructions that the computer must execute to perform the desired operations. The process of defining the instructions the computer is to execute is known as *programming*. When you create a program, you store the instructions in an ASCII file whose name usually contains the extension C for a C program, CPP for a C++ program, and CS for a C# program. For example, if you create a C program that performs payroll operations, you might name the file containing the program instructions *payroll.c*. When you create programs, you specify the desired instructions using a *programming language*. C, C++, and C# are only three of many programming languages. Many programmers use programming languages such as BASIC, Pascal, and FORTRAN. Different programming languages provide unique features and have their own strengths (and weaknesses). In any case, programming languages exist to let us define the instructions that we want the computer to execute. The instructions a computer executes are actually sets of 1s and 0s (binary digits) that represent electronic signals that occur inside the computer. To program the earliest computers (in the 1940s and 1950s), programmers had to understand how the computer interpreted different combinations of 1s and 0s because the programmers wrote all their programs using binary digits. As programs became larger, it became very impractical to make programmers work in terms of the computer's 1s and 0s. Instead, researchers created programming languages that let people express the computer instructions in a form more meaningful to humans. After programmers place their instructions in a file (called a *source file*), a second program (called a *compiler)*, converted the programming language instructions into the 1s and 0s (known as *machine code*) the computer understood. Files with the EXE and COM extensions contain the machine code (1s and 0s) the computer will execute. Figure 1 shows the process of compiling a source code file into an executable program.

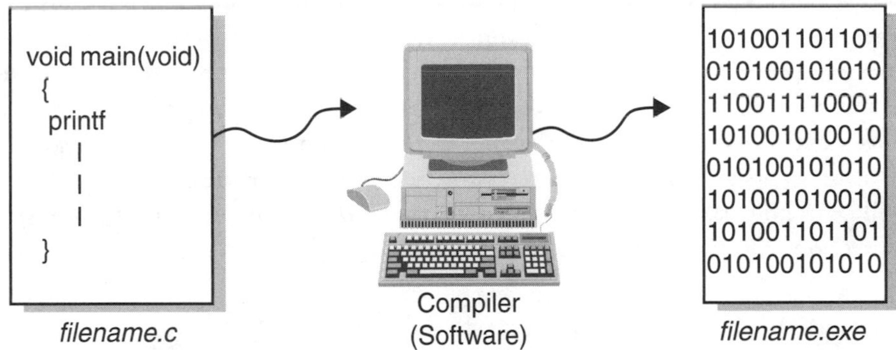

***Figure 1** A compiler converts source code instructions into machine code.*

After you create a source code file, you run a compiler to convert the instructions into a format the computer can execute. If you are using, for example, *Turbo C++ Lite™* (included on the CD-ROM that accompanies this book), you will use the Compile menu Compile to OBJ option to invoke the compiler (that is, instruct it to compile the source file). The following Tips walk you through the steps required to create and compile a C program.

CREATING AN ASCII SOURCE FILE

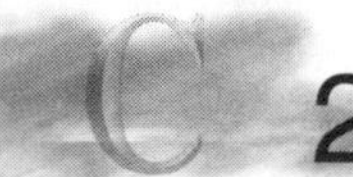

2

When you create a program, you must place the program statements that you want the computer to execute in a file called a *source file*. If you are not using *Turbo C++ Lite* or a full-featured compiler and editor, you should create your program files using an ASCII editor, such as the EDIT program that DOS provides. You should not create programs using a word processor (such as Microsoft *Word*® or Corel's *WordPerfect*®). As you know, word processors let you format documents by aligning margins, italicizing and underlining text, and so on. To perform these operations, word processors embed special characters within your documents. Although these characters are meaningful to your word processor, they will confuse the compiler that converts your source file to machine code, and this confusion will cause errors. When you create your source file, make sure you assign a meaningful name that accurately describes the program's function to the file. For example, you might name the source code for a billing program *billing.c*, and the source file for a game program *football.c*.

If, on the other hand, you are using a compiler that includes a built-in editor, you should create your programs within that editor. For example, if you are using *Turbo C++ Lite*, you will use the File menu New option to create a new program file. To create your first program within *Turbo C++ Lite*, perform the following steps:

1. Select the File menu New option. *Turbo C++ Lite* will create the *noname00.cpp* file.
2. Enter the following code into the *noname00.cpp* window:

```
#include <stdio.h>

void main(void)
 {
   printf ("Jamsa\'s C/C++/C# Programmer\'s Bible!");
 }
```

3. Select the File menu Save As option. *Turbo C++ Lite* will display the Save File As dialog box.
4. Within the Save File As dialog box, enter the name *first.c* and press ENTER. *Turbo C++ Lite* will save the *first.c* program file.

Although the *first.c* program contains six lines, only the *printf* statement actually performs any work. When you execute this program, *printf* will display the message *Jamsa's C/C++/C# Programmer's Bible!* on your screen. Every programming language (just like languages such as English and French) has a set of rules, called *syntax rules*, which you must follow when you use the language. When you create C programs, you must obey the syntax rules of the C programming languages. Examples of syntax rules include the parentheses that follow the name *main* and the semicolon at the end of the *printf* instruction. When you type your program, you must be very careful that you do not omit any of these elements. Double-check your typing to ensure you have successfully typed in the C program instructions exactly as they appear earlier. If the instructions are correct, save the contents of the file to your disk. In the next Tip you will learn how to compile your source file and to convert your C programming instructions to the machine language that your computer can understand and execute.

3 COMPILING YOUR C PROGRAM

In the previous Tip you created the C source file, *first.c*, which contains the *printf* statement that will display the message *Jamsa's C/C++/C# Programmer's Bible!* on your screen when you execute the program. A source file contains instructions in a format you can understand (or at least you will be able to understand after you learn C). An executable program, on the other hand, contains instructions expressed as 1s and 0s that the computer understands. The process of converting your C source file to machine code is known as *compiling*. Depending on the C compiler you are using, the command you must perform to compile your source file will differ. Assuming you are using Borland's *Turbo C++ Lite*, you can compile the program (*first.c*) that you created in Tip 2, using the following command sequence:

1. Select the Compile menu Build All option. *Turbo C++ Lite* will display the Compiling dialog box.

2. If the compiler successfully completes the compilation, it will prompt you to *Press any key*. If the C compiler does not create the file *first.exe*, but instead displays error messages on your screen, you have probably violated a C syntax rule, as the next Tip discusses.

3. If you successfully typed in the C statements as shown in Tip 2, the C compiler will create an executable file named *first.exe*. To execute the *first.exe* program, you can either select the Run menu Run option or press the CTRL+F9 keyboard shortcut.

When you execute the program, your screen will display the following output:

```
Jamsa's C/C++/C# Programmer's Bible!
C:\>
```

Note: *In some installations,* ***Turbo C++ Lite*** *will generate the output and return you immediately to the editing window. In such cases, select the File menu DOS Shell option to view the program's output.*

Understanding Syntax Errors 4

As you read in Tip 2, every programming language has a set of rules, called *syntax rules,* which you must obey as you specify your program statements. If you violate a syntax rule, your program will not successfully compile. Instead, the compiler will display error messages that specify the line of your program that contains the error and a brief description of the error. Using your editor, create the file *syntax.c*, which contains a syntax error. In the following example, the program fails to include an ending quote at the end of the message *Jamsa's C/C++/C# Programmer's Bible!*:

```
#include <stdio.h>

void main(void)
 {
   printf ("Jamsa\'s C/C++/C# Programmer\'s Bible!);
 }
```

When you compile this program, your compiler will display a syntax error message when it encounters line five. Depending on your compiler, the actual error message will differ. In the case of *Turbo C++ Lite*, your screen will display the following error messages:

```
Error syntax.c 5: Unterminated string or character constant in function main
Error syntax.c 6: Function call missing ) in function main()
Error syntax.c 6: Statement missing ; in function main()
```

Although the source code *syntax.c* only contains one error, the C compiler will display three error messages. The missing quote caused a series of cascading errors (one error leads to another) within the compiler. To correct syntax errors within your programs, perform the following steps:

1. Write down the line number of each error and a brief description.

2. Edit your source file, moving your cursor to the first line number the compiler displays.

3. Within the source file, correct the error and move the cursor to the next line number. Most editors will display the current line number to help you locate specific lines within the file.

In the case of the file *syntax.c*, edit the file and add the missing quote. Save the file to disk and use your compiler to compile it. After you correct the syntax error, the compiler will create the file *syntax.exe*. To execute *syntax.exe*, select the Run menu Run option. The program will run and will yield the output shown here:

```
Jamsa's C/C++/C# Programmer's Bible!
C:\>
```

5 THE STRUCTURE OF A TYPICAL C PROGRAM

In Tip 2 you created the source file *first.c*, which contained the following statements:

```
#include <stdio.h>

void main(void)
 {
   printf ("Jamsa\'s C/C++/C# Programmer\'s Bible!");
 }
```

These statements are similar to those you will find in most C programs. In many cases, a C source file may begin with one or more *#include* statements. The *#include* statement directs the C compiler to use a specific file's content. In the case of the file *first.c*, the *#include* statement directs the C compiler to use a file named *stdio.h*. The files an *#include* statement specifies are ASCII files that contain C source code. You can print or display each file's contents by following the steps discussed in Tip 13. Files that you name within an *#include* statement, which usually use the *h* extension, are called *include files* or *header files*. Most header files contain statements your programs commonly use, although you will learn later in this book about other uses for header files. When you direct the C compiler to include the file's contents, you do not have to type the statements into your programs yourself. After the *#include* statements, you will usually find a statement similar to the following:

```
void main(void)
```

Each C program you create will include a line similar to the *void main* statement. As you read in Tip 1, a C program contains a list of instructions you want the computer to execute. As the complexity of your programs increases, you will break them into small pieces that are easier for you (and for others who read your programs) to understand. The group of instructions you want the computer to execute first is called your *main program*. The statement *void main* identifies these statements (the main program) to the C compiler.

Clearly, because the C compiler will determine which statements form the main program and which statements are supplementary, you must have a way to tell the C compiler which instructions correspond to each section of your program. To assign program statements to a specific section of your program, place the statements within an opening brace ({) and a closing brace (}). The braces are part of the C syntax. For every opening brace you must have a brace that closes the group of statements.

6 ADDING STATEMENTS TO YOUR PROGRAM

As you have read, the program *first.c* used the *printf* statement to display a message on your screen. The following C program, *3_msgs.c*, uses three *printf* statements to display the same message. Each statement is contained within the program's opening and closing braces:

```
#include <stdio.h>

void main(void)
 {
   printf("Jamsa\'s ");
   printf("C/C++/C# Programmer\'s ");
   printf("Bible!");
 }
```

Note the space character within the *printf* statements. The space character is important because it ensures that the program will correctly display the text on your screen (by placing a space between words). As the number of statements in your programs increases, so too does the likelihood of syntax errors. Double-check your program to ensure that you correctly typed each statement, and then save the file to disk. When you compile and execute the *3_msgs* program, your screen will display the following output:

```
Jamsa\'s C/C++/C# Programmer\'s Bible!
C:\>
```

DISPLAYING OUTPUT ON A NEW LINE 7

Several of the previous programs have displayed the message *Jamsa's C/C++/C# Programmer's Bible!* on your screen display. As your programs become more complex, you might want the programs to display their output on two or more lines. In Tip 6, you created the program *3_msgs.c,* which used three *printf* statements to display a message on your screen:

```
printf("Jamsa\'s ");
printf("C/C++/C# Programmer\'s ");
printf("Bible!");
```

Unless you tell *printf* to do otherwise, *printf* will continue its output on the current line. The goal of the following program, *one_line.c,* is to display output on two successive lines:

```
#include <stdio.h>

void main(void)
 {
   printf("This is line one.");
   printf("This is the second line.");
 }
```

When you compile and execute the *one_line.c* program, your screen will display the following output:

```
This is line one.This is the second line.
C:\>
```

When you want *printf* to begin its output on a new line, you must include the special *newline character* (\n) within the text that you direct *printf* to display. When *printf* encounters the \n character, it will advance the cursor to the start of the next line. The following program, *two_line.c,* uses the newline character to display the second line of text on a new line as desired:

```
#include <stdio.h>

void main(void)
 {
   printf("This is line one.\n");
   printf("This is the second line.");
 }
```

When you compile and execute the program *two_line.c,* your screen will display the following output:

```
This is line one.
This is the second line.
C:\>
```

Many of the programs this book presents use the newline character. In fact, almost every program you write will normally use the newline character in one or more places.

C CONSIDERS UPPER- AND LOWERCASE LETTERS AS DIFFERENT 8

As you type your programs, you must keep in mind that C considers upper- and lowercase letters as different. As a rule, most C commands use lowercase, most C constants use all uppercase, and most C variables use mixed case. C programs make extensive use of lowercase letters. Because the following program, *uppererr.c,* uses the uppercase letter M in the name *Main* when C expects the name *main,* the program will not successfully compile:

```
#include <stdio.h>

void Main(void)
 {
   printf("This program does not compile.");
 }
```

When you compile the *uppererr.c* program, the *Turbo C++ Lite* compiler will display the following message:

```
Linker error: Undefined symbol _main in module TURBO_C\C0S.ASM
```

The relatively meaningless message that the *Turbo C++ Lite* compiler returns is the result of your spelling *Main* with a capital *M*. In this case, to correct the error you must simply change *Main* to *main*. After you change *Main* to *main*, recompile and execute the program.

9 Understanding Logic Errors (Bugs)

In Tip 4 you learned that if you violate one of the C language rules, the compiler will display a syntax error message and your program will not successfully compile. As your programs become more complex, there will be many times when the program successfully compiles, but does not correctly perform the task you wanted. For example, assume that you want the following program, *one_line.c*, to display its output on two lines:

```
#include <stdio.h>

void main(void)
 {
   printf("This is line one.");
   printf("This is the second line.");
 }
```

Because the program does not violate any of C's syntax rules, the program will successfully compile. When you execute the program, however, it will not display its output on two lines; instead it will display the output on one line, as shown here:

```
This is line one.This is the second line.
C:\>
```

When your program does not work as you desire, the program contains *logic errors,* or *bugs*. When your programs contain a logic error (and eventually your programs will), you must try to discover and correct the cause of the error. The process of removing logic errors from your program is called *debugging*. Later in this book you will learn several different techniques you can use to locate logic errors within your program. For now, however, the best way to locate such errors is to print a copy of your program and examine the program line-by-line until you locate the error. A line-by-line program examination is called *desk checking*. In the case of the program *one_line.c*, your desk checking should reveal that the first *printf* statement does not contain the newline character (\n).

10 Understanding the Program Development Process

When you create programs, you will usually follow the same steps. To begin, you will use an editor to create your source file. Next, you will compile the program. If the program contains syntax errors, you must edit the source file and correct the errors. After the program successfully compiles, you will try to run the program. If the program runs successfully and performs as you expect, you are done creating the program. On the other hand, if the program does not work as you expected, you must desk check the source code to locate the logic error (as discussed in Tip 9). After you correct the error, you must compile the source code to create a new executable file. You can then test the new program to ensure that it performs the task you want. Figure 10 illustrates the program development process.

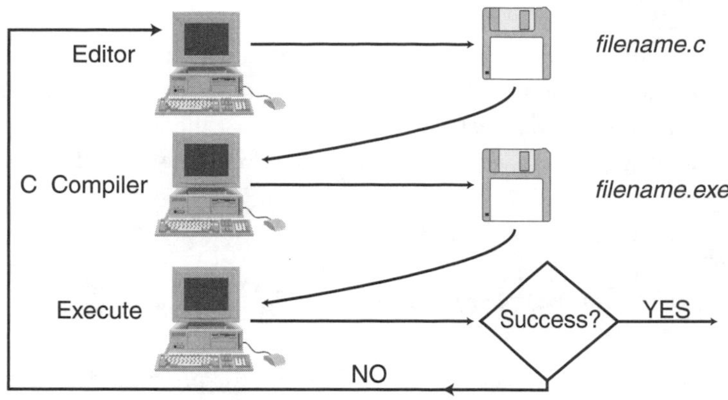

Figure 10 The program development process.

UNDERSTANDING THE FILE TYPES 11

When you create a C program, you place your statements in a source file that uses the C extension. If your program successfully compiles, the compiler will create an executable program file with the EXE extension. As you read in Tip 5, many programs use header files (which use the *h* extension) that contain commonly used statements. If you examine your directory after compiling a program, you will likely find one or more files with the *OBJ* extension. These files, called *object files,* contain instructions in the form of 1s and 0s that the computer understands. You cannot execute object files, however, because their contents are not quite complete.

The C compiler provides routines (such as *printf)* that perform commonly used operations and reduce the number of statements you must include in your programs. After the compiler examines your program's syntax, it creates an object file. In the case of the program *first.c*, the compiler would create an object file named *first.obj*. Next, a program called a *linker* combines the program statements in your object file with the functions (such as *printf*) the compiler provides to build the executable program. In most cases, when you invoke the compiler to examine your source code, the compiler will automatically invoke the linker for you if your program successfully compiles. Figure 11 illustrates the process of compiling and linking a program.

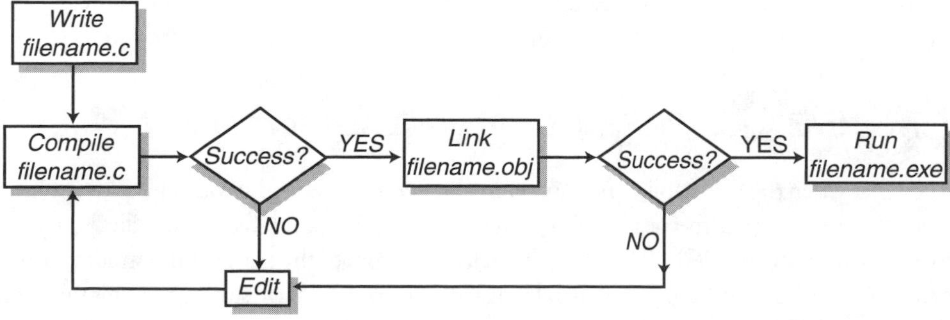

Figure 11 The process of compiling and linking a program.

12 Better Understanding the Linker

In Tip 11 you learned that when you compile your C program, a second program called a linker combines your program statements with predefined routines (which the compiler provides) to convert an object file to an executable program. As is the case with the compilation process, which can detect syntax errors, the linker process can also encounter errors. Consider, for example, the following program, *no_print.c*, which erroneously uses *print* instead of *printf*:

```
#include <stdio.h>

void main(void)
 {
   print("This program does not link");
 }
```

Because the *no_print.c* program does not violate any C syntax rules, the program will successfully compile, producing an OBJ file. However, the *Turbo C++ Lite* linker will display the following error message because of the undefined *print* statement:

```
Error: Function 'print' should have a prototype in function main()
```

Because the C compiler does not provide a function named *print*, the linker cannot create the executable program *no_print.exe*. Instead the linker will display the error message shown previously. To correct the error edit the file, changing *print* to *printf*, and recompile and link your program.

13 Understanding Header Files

Each program presented throughout this book uses one or more *#include* statements to direct the C compiler to use the statements a *header file* contains. A header file is an ASCII file whose contents you can print or display on your screen. If you examine the directory that contains your compiler (the directory *tclite* in the case of the *Turbo C++ Lite* compiler), you will find a subdirectory named *include*. The *include* subdirectory contains the compiler's header files. Take time now to locate your compiler's header files. You might even want to print the contents of a commonly used header file, such as *stdio.h*. You will find C programming statements within the *include* file. When the compiler encounters an *#include* statement in your program, the compiler compiles the code the header contains just as if you had typed the header file's contents into your program source code. Header files contain commonly used definitions and provide the compiler with information about compiler-provided functions, such as *printf*. For now, you may find a header file's contents difficult to understand. As you become more conversant in C, C++, and C#, however, you should print a copy of and examine each header file you use. The header files contain valuable information and provide you with programming techniques that will make you a better C programmer.

14 Helping the Compiler Locate Header Files

In Tip 13 you learned that when the C compiler encounters an *#include* statement, the compiler adds the header file's contents to your program, just as if you typed the header file's contents into your source file. Depending on your compiler, your environment entries may contain an INCLUDE entry that tells the compiler the name of the subdirectory that contains the header files. If when you compile a program your compiler displays an error message stating that it is unable to open a specific header file, first check the subdirectory that contains your compiler's header files to ensure that the file exists. If you find the file, issue the SET command at the DOS prompt, as shown here:

```
C:\> SET <ENTER>
COMSPEC=C:\DOS\COMMAND.COM
PATH=C:\DOS;C:\WINDOWS;C:\BORLANDC\BIN
PROMPT=$P$G
TEMP=C:\TEMP
```

If your environment does not contain an INCLUDE entry, check the documentation that accompanied your compiler to determine if your compiler requires such an entry. Usually, the compiler's installation will place within your *autoexec.bat* file a SET command that assigns the INCLUDE entry to the subdirectory that contains the header files, as shown here:

```
SET INCLUDE=C:\BORLANDC\INCLUDE
```

If your compiler uses the INCLUDE entry and your *autoexec.bat* file does not define the entry, you can create the entry yourself, placing it in your *autoexec.bat* file.

Note: *The **Turbo C++ Lite** program will only look for include files within its **include** subdirectory.*

SPEEDING UP COMPILATIONS

When you compile a source file, the C compiler might create one or more temporary files that exist only while the compiler and linker are working. Depending on your compiler, you may be able to use the TEMP environment entry to specify where the compiler creates these temporary files. If your computer has multiple hard drives, some of which have more available space than others (especially if your compiler runs within Windows and therefore uses virtual memory and swap files), you might consider assigning the TEMP entry to point to the drive with the most available space. This way, the compiler will create its temporary files on the very fast hard drive, which will speed up the compilation process. Assuming that your *D* drive has that extra space, you can place a SET command within your *autoexec.bat* file to assign the TEMP entry to the *D* drive, as shown here:

```
SET TEMP=D:
```

COMMENTING YOUR PROGRAMS

As a rule, each time you create a program you must ensure that you include in the program *comments* that explain the processing the program performs. In short, a comment is a message that helps you read and understand the program. As your programs increase in length, the programs become more difficult to understand. Because you may eventually create hundreds and possibly thousands of programs, you will not be able to remember the purpose of every statement within every program. If you include comments in your program, you will not have to remember each program's details. Instead, the program's comments will explain the processing.

Most newer C and C++/C# compilers provide two ways for you to place comments within your source file. First, you place two forward slash (//) characters together, as shown here:

```
// This is a comment
```

When the C compiler encounters the double slashes, it ignores the text that follows to the end of the current line. The following program, *comment.c*, illustrates the use of comments:

```
// Program: comment.c
// Written by: Kris Jamsa
// Date written: 6-05-01
// Purpose: Illustrates the use of comments in a C program.
#include <stdio.h>
void main(void)
 {
   printf("Jamsa\'s C/C++/C# Programmer\'s Bible!"); // Display a message
 }
```

In this example, you immediately know by reading these simple comments when, why, and who wrote the program. You should get in the habit of placing similar comments at the start of your programs. Should other programmers who must read or change the program have questions, they will quickly know the program's original author.

When the C compiler encounters the double slashes (//), it ignores the text on the rest of that line. Most newer C source files use the double slashes to designate a comment. If you are reading an older C program, you may encounter comments written in a second form. In the second acceptable comment form, the comment appears between a set of slashes and asterisks, as shown here:

```
/* This is a comment */
```

When the compiler encounters the opening comment symbol (/*) it ignores all text up to and including the closing comment symbol (*/). Using the /* *comment* */ format, a single comment can appear on two or more lines. The following program, *comment2.c*, illustrates the use of the /* *comment* */ format:

```
/* Program: COMMENT.C
   Written by: Kris Jamsa
   Date written: 6-05-01

   Purpose: Illustrates the use of comments in a C program. */

#include <stdio.h>

void main(void)
 {
   printf("Jamsa\'s C/C++/C# Programmer\'s Bible!");   /* Display a message */
 }
```

As you can see, the program's first comment contains five lines. When you use the /* *comment* */ format for your comments, make sure that every start comment symbol (/*) has a corresponding end symbol (*/). If the end symbol is missing, the C compiler will ignore much of your program, which will eventually result in syntax errors that will be difficult for you to detect.

Most C compilers will return a syntax error if you try to place one comment within another (nest comments), as shown here:

```
    /* This comment has /* a second */ comment inside */
```

17 IMPROVING YOUR PROGRAM READABILITY

In Tip 16 you learned how to use comments within your programs to improve their readability. Each time you create a program, assume that you or another programmer will eventually have to change the program in some way. Therefore, it is essential that you write your programs so they are easy to read. The following C program, *hardread.c*, will display a message on your screen:

```
#include <stdio.h>
void main(void){printf("Jamsa\'s C/C++/C# Programmer\'s Bible!");}
```

Although this program will compile and successfully display the desired message, the program is difficult to read, at best. A good program not only works, but is also easy to read and understand. The key to creating readable programs is to include comments that explain the program's processing and to use blank lines to improve the program's format. In later Tips you will learn the important role indentation plays in producing readable program code.

18 PAYING ATTENTION TO COMPILER WARNING MESSAGES

When your program contains one or more syntax errors, the C compiler will display error messages on your screen and will not create an executable program. As you create programs, there may be times when your compiler will display one or more *warning messages* on your screen, but will still create the executable program file. For example, the following C program, *no_stdio.c*, does not include the header file *stdio.h*:

```
void main(void)
 {
   printf("Jamsa\'s C/C++/C# Programmer\'s Bible!");
 }
```

When you compile this program, the *Turbo C++/C# Lite* compiler will display the following warning message:

```
Warning no_stdio.c 3: Function 'printf' should have a prototype in function main().
```

When the compiler displays a warning message, you should immediately determine the cause of the compiler complaint and correct it. Although the warnings might never cause an error during your program's execution, some warnings create the opportunity for errors that are very difficult to debug later. By taking time to locate and correct the cause of compiler warnings, you will learn much more about the inner workings of C and C++.

CONTROLLING COMPILER WARNINGS 19

In Tip 18, you have learned that you should pay attention to the warning messages that your compiler displays on your screen. To help you make better use of compiler warnings, many compilers let you set the message level you desire. Depending on your compiler, you may use a command-line switch to control the warning level or you may use *pragmas*, which Tip 145 explains. A pragma is a directive to the compiler. As you will learn, different compilers support different pragmas. For example, to disable the *Identifier is declared but never used* warning within *Turbo C++ Lite*, your code would include the following pragma:

```
#pragma warn -use
```

If you are not using *Turbo C++ Lite,* refer to the documentation that accompanies your compiler to determine if you can turn off specific warning messages.

USING COMMENTS TO EXCLUDE PROGRAM STATEMENTS 20

In Tip 16 you learned that you should use comments within your programs to improve your program's readability. As your programs become more complex, you may use comments to help you debug (remove errors) from your programs. When the C compiler encounters the double slashes (//), the compiler ignores all the remaining text on the current line following the double slashes. Likewise, when the compiler encounters the starting comment symbol (/*), the compiler ignores all the text that follows, up to and including the closing comment symbol (*/). As you test your programs, there may be times when you want to eliminate one or more statements from your program. One way to eliminate the program statements is to simply delete the statements from your source file. A second way to eliminate statements is to *comment them out.* The following program, *nooutput.c*, comments out all the *printf* statements:

```
#include <stdio.h>

void main(void)
 {
   // printf("This line does not appear");
   /* This is a comment
      printf("This line does not appear either");
   */
 }
```

Because both *printf* statements appear within comments, the compiler ignores both of them. As a result, no output appears when you execute the program. As your programs become more complex, using comments to disable statements will become very convenient.

As you learned in Tip 16, most C compilers will return one or more syntax errors if you try to place one comment within another (nested comments). When you use comments to disable statements, be careful that you do not inadvertently nest comments.

21 UNDERSTANDING THE IMPORTANCE OF NAMES

As you examine the Tips presented throughout this book, you will encounter variable names and functions whose names begin with an underscore, such as *_dos_getdrive* or *_chmod*. You usually only use such variables and functions within the DOS environment. If you are writing programs that will execute under DOS, Windows, Macintosh, UNIX, or possibly some other operating system, you should avoid using these functions because they will probably not be available under the other systems. Therefore, to move your program from DOS to another operating system, you will have to perform additional programming. Some functions may have two implementations, one with an underscore (*_chmod*), and one without (*chmod*). As a rule, use the function or variable that does not use the underscore, which is in this case *chmod*.

22 UNDERSTANDING THE SEMICOLON

As you examine C programs, you will find that the programs make extensive use of semicolons. The semicolon in C has special meaning. As you know, a program is a list of instructions that you want the computer to perform. When you specify those instructions in C, you use the semicolon to separate one statement from another. As your programs become more complex, you may find that a statement does not fit on one line. When the C compiler examines your program, it uses the semicolon to distinguish one statement from the next. The C language syntax defines the use of the semicolon. If you omit the semicolon, a syntax error will occur and the program will not successfully compile.

23 INTRODUCING VARIABLES

To perform useful work, programs must store information, such as a document that you edit over multiple computer sessions, within a file and also internally. As you know, each time you run a program, the operating system loads your program's instructions into the computer's memory. As the program runs, it stores values in memory locations. For example, assume that you have a program that prints a document. Each time you run the program, it will display a message asking you the name of the file, as well as the number of copies you want to print. As you type in this information, the program stores the values you enter in specific memory locations. To help your program track the memory locations in which it has placed data, each memory location has a unique *address*, such as location 0, 1, 2, 3, and so on. Because there can be billions of such addresses, keeping track of individual storage locations can become very difficult. To simplify storing information, programs define *variables*, which are names that the program associates with specific locations in memory. As the word variable implies, the *value* that the program stores in these locations can change or vary throughout the program's lifetime. Each variable has a specific *type*, which tells the computer how much memory the data the variable stores requires and which operations the program can perform on the data. Given the previous example of a program that prints a document, the program might use a variable named *filename* (which stores the name of the file you want to print) and one named *count* (which stores the number of copies you want to print). Within your program, you reference variables by name. Therefore, you should assign meaningful names to each variable. Within your C programs, you usually declare your variables immediately following *main*, before your program statements, as shown here:

```
void main(void)
 {
   // Variables go here
   printf("Jamsa\'s C/C++/C# Programmer\'s Bible!");
 }
```

The following program shows how you would declare three integer variables (variables that store counting numbers, such as 1, 2, and 3):

```
void main(void)
 {
   int age;        // The user's age in years
   int weight;     // The user's weight in pounds
   int height;     // The user's height in inches

   // Other program statements go here
 }
```

Each variable has a type that defines the amount of memory the variable requires, as well as the operations the program can perform on the data. To declare an integer variable, your C programs use the type *int*. After you declare a variable (that is, tell the program the variable's name and type), you can then assign a value to the variable (that is, store information).

ASSIGNING A VALUE TO A VARIABLE 24

A variable is a name that your program associates with a storage location in memory. After you declare a variable within your program, you can assign it a value. In C, you assign a value to a variable by using the equal sign (called the *assignment operator*). The following program declares three variables of type *int* and then assigns each variable a value:

```
void main(void)
 {
   int age;          // The user's age in years
   int weight;       // The user's weight in pounds
   int height;       // The user's height in inches

   age = 41;         // Assign the user's age
   weight = 165;     // Assign the user's weight
   height = 73;      // Assign the user's height

    // Other program statements
 }
```

UNDERSTANDING VARIABLE TYPES 25

When you declare variables within your programs, you must tell the C compiler the variable's name and type. A type defines the set of values the variable can store, as well as the set of operations that the program can perform on the data. C supports four basic types, each of which is listed in Table 25.

Type Name	Purpose
char	Stores a single character, such as a letter from A through Z
int	Stores counting numbers (called integers), such as 1, 2, and 3, as well as negative numbers
float	Stores single-precision floating-point numbers (with a decimal point), such as 3.14 or –54.1343
double	Stores a double-precision floating-point number (which is more precise than a single-precision floating-point number). You will use *double* for very large or very small numbers

Table 25 *The four basic types C supports.*

Many of the Tips presented throughout this book examine each of these types in detail. Most of the Tips in this book will use one or more variables of the basic types.

26 Declaring Multiple Variables of the Same Type

As you learned in Tip 24, when you declare a variable within your program, you must tell the C compiler the variable's name and type. The following statements declare three variables of type *int*:

```
int age;
int weight;
int height;
```

When you declare variables of the same type, C lets you list the variable names on one or more lines, with commas separating each variable name, as shown here:

```
int age, weight, height;
float salary, taxes;
```

27 Commenting Your Variables at Declaration

In C programs, comments help someone who is reading your program to better understand it. When you choose variable names, you should select names that meaningfully describe the value the variable will store. For example, consider the following declarations:

```
int age, weight, height;
int x, y, z;
```

Both declarations create three variables of type *int*. In the first declaration, however, you have an idea of how to use the variable simply by examining the variable's name. In addition to using meaningful names, you should also place a comment next to each variable declaration that further explains the variable, as shown here:

```
int age;          // The user's age in years
int weight;       // The user's weight in pounds
int height;       // The user's height in inches
```

28 Assigning Values to Variables at Declaration

After you declare a variable within your program, you can use the C *assignment operator* (the equal sign) to assign a value to the variable. C lets you assign a value to a variable within the variable's declaration. Programmers refer to the process of assigning a variable's first value as *initializing* the variable. The following statements, for example, declare and initialize three variables of type *int*:

```
int age = 41;           // The user's age in years
int weight = 165;       // The user's weight in pounds
int height = 73;        // The user's height in inches
```

29 Initializing Multiple Variables During Declaration

In Tip 26 you learned that C lets you declare two or more variables on the same line, as shown here:

```
int age, weight, height;
```

When you declare multiple variables on the same line, C lets you initialize one or more of the variables:

```
int age = 44, weight, height = 73;
```

In this example, C will initialize the variables *age* and *height*, and leave the variable *weight* uninitialized.

USING MEANINGFUL VARIABLE NAMES 30

When you declare variables in your programs, you should choose meaningful variable names that describe the variable's use. You can use a combination of upper- and lowercase letters in your variable names. As discussed in Tip 8, the C compiler distinguishes between upper- and lowercase letters. If you use upper- and lowercase letters in your variable names, you must always specify the same upper- and lowercase letter combinations. As you get started, you should probably stick to lowercase letters, because doing so reduces the possibility of entry errors from mixed cases. You must give a unique name to each variable you declare within your programs. In general, you can use an unlimited number of characters in a variable name. Your variable names can contain a combination of letters, numbers, and the underscore character—however, the names must start with a letter or underscore. The following statements illustrate some valid variable names:

```
int hours_worked;
float tax_rate;
float _6_month_rate;    // Starting _underscore is valid
```

C predefines several keywords that have special meaning to the C compiler. A *keyword* is a word that has meaning to the compiler without your giving it meaning. For example, *float, int*, and *char* are all keywords. As you create variable names, do not use these keywords. Tip 31 lists C keywords.

UNDERSTANDING C'S KEYWORDS 31

The C programming language defines several keywords that have special meaning to the compiler. As you choose variable names (and create your own functions), do not use these keywords. Table 31 lists C keywords.

C Keywords					
auto	*default*	*float*	*register*	*struct*	*volatile*
break	*do*	*for*	*return*	*switch*	*while*
case	*double*	*goto*	*short*	*typedef*	
char	*else*	*if*	*signed*	*union*	
const	*enum*	*int*	*sizeof*	*unsigned*	
continue	*extern*	*long*	*static*	*void*	

Table 31 The C keyword list.

UNDERSTANDING VARIABLES OF TYPE INT 32

A *variable* is a name that the C compiler associates with one or more memory locations. When you declare a variable within your program, you must specify the variable's type and name. A variable's *type* specifies the kind of values that variable can store and the set of operations that the program can perform on the data. C uses the type *int* to store integer values (positive and negative counting numbers). Older 16-bit C compilers normally allocate sixteen bits (two bytes) to store values of type *int*. Thus, a variable of type *int* can store values in the range –32,768 through 32,767. Newer compilers use 32 bits to store values of type *int*. Figure 32 shows how C represents an integer value:

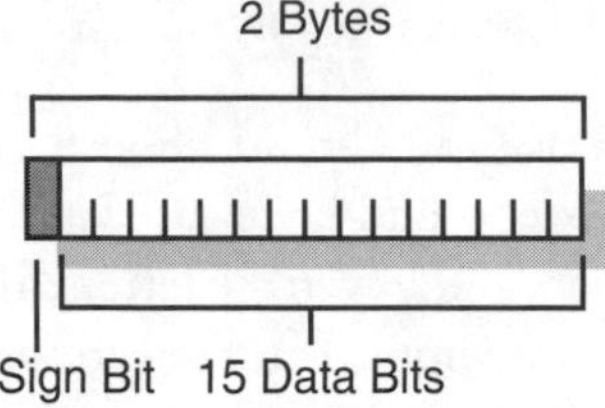

Figure 32 How C represents an integer value.

Values of type *int* are whole numbers; they do not include a fractional portion as do floating-point numbers. If you assign a floating-point value to a variable of type *int*, most C compilers will simply truncate the fractional portion. If you assign a variable of type *int* a value outside the range -32,768 through 32,767, an overflow condition will occur and the assigned value will be in error.

33 UNDERSTANDING VARIABLES OF TYPE CHAR

A *variable* is a name that the C compiler associates with one or more memory locations. When you declare a variable within your program, you must specify the variable's type and name. A variable's *type* specifies the kind of values that variable can store and the set of operations that the program can perform on the data. C uses the type *char* to store character (byte) values. The C compiler normally allocates eight bits (one byte) to store values of type *char*. A variable of type *char* can store whole number values in the range –128 through 127. Figure 33 shows how C represents a value of type *char*.

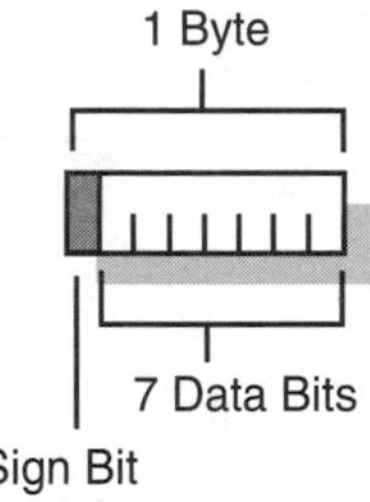

Figure 33 How C represents a value of type **char**.

Programs can assign a value to a variable of type *char* in one of two ways. First, the program can assign a character's ASCII value. For example, the letter A has the ASCII value 65:

```
char letter = 65;   // Assign letter the character A
```

Second, your program can use a character constant, which appears within single quotes, as shown here:

```
char letter = 'A';
```

Variables of type *char* only hold one letter at a time. To store multiple characters, you must declare a character string, which is discussed in this book's Strings section.

34 UNDERSTANDING VARIABLES OF TYPE FLOAT

A *variable* is a name that the C compiler associates with one or more memory locations. When you declare a variable within your program, you must specify the variable's type and name. A variable's *type* specifies the kind of values that variable can store and the set of operations that the program can perform on the data. C uses the type *float* to store floating-point values (positive and negative numbers that contain fractional portions). The C compiler normally allocates 32 bits (4 bytes) to store values of type *float*. A variable of type *float* can store values with six to seven digits of precision in the range 3.4E–38 through 3.4E+38.

C stores the value as a 23-bit *mantissa*, which contains the fractional number, an 8-bit exponent, which contains the power to which the computer raises the number when resolving its value, and a single sign bit which determines whether the value is positive or negative. In other words, if a variable contained the value 3.4E+38, the sign bit would be 0, which indicates the number is positive; the 23-bit mantissa would include a binary representation of 3.4, and the 8-bit exponent would include a binary representation of the 10^{38} exponent. Figure 34 illustrates how C represents a value of type *float*. Tip 337 explains mantissas and exponents in detail.

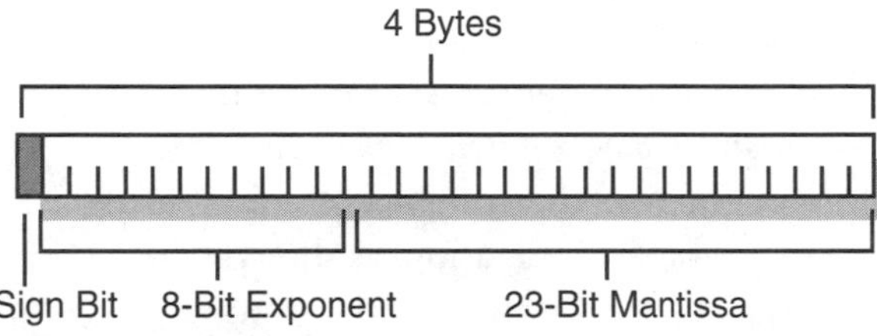

Figure 34 *How C represents a value of type* ***float****.*

Note: *This Tip and other, later Tips represent floating-point numbers in* ***scientific notation****. Simply put, scientific notation lets you represent any number as a single digit to the left of the decimal point, and unlimited number of digits to the decimal point's right, and an exponent representing 10 raised to that exponent's value. When you determine the number's actual value, you multiply the number (the* ***mantissa****) by the value* 10^x *(where* ***x*** *represents the* ***exponent****). For example, the number 3.1415967E+7 evaluates as 31415967.0 or 3.1415967 ** 10^7*.*

Understanding Variables of Type double 35

A *variable* is a name that the C compiler associates with one or more memory locations. When you declare a variable within your program, you must specify the variable's type and name. A variable's *type* specifies the kind of values that variables can store and the set of operations that the program can perform on the data. C uses the type *double* to store floating-point values (positive and negative numbers that contain fractional portions). The C compiler normally allocates 64 bits (8 bytes) to store values of type *double*. A variable of type *double* can store values with 14 to 15 digits of precision, in the range –1.7E-308 through 1.7E+308. Figure 35 illustrates how C represents a value of type *double*.

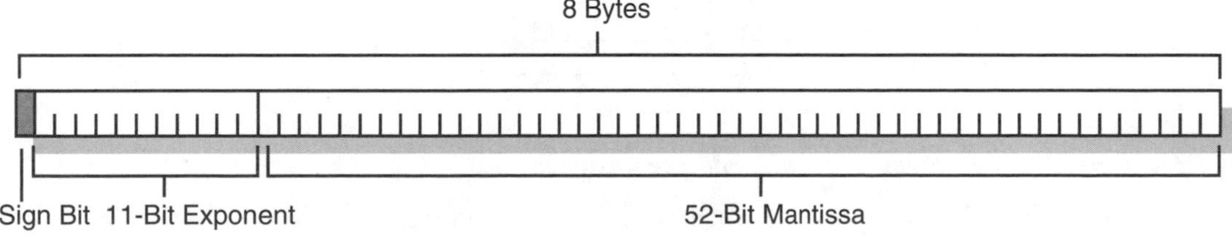

Figure 35 *How C represents a value of type* ***double****.*

Assigning Values to Floating-Point Values 36

A *floating-point* value is a value that contains a fractional part, such as 123.45. When you work with floating-point values within your programs, you can refer to the values using their decimal format, such as 123.45, or you can use the value's exponential format, 1.2345E2. Therefore, both the following statements assign the variable *radius* the same value:

```
radius = 123.45;
radius = 1.2345E2;
```

In a similar way, both the following statements assign the variable *radius* the same fractional value:

```
radius = 0.12345;
radius = 12.345E-2;
```

37 Understanding Type Modifiers

C provides four basic data types (*int, char, float,* and *double).* As you learned, each type defines a set of values the variable can store and a set of operations that the program can perform on the data. As you have learned, variables of type *int* can store values in the range –32,768 through 32,767. Likewise, variables of type *char* can store values in the range –128 through 127. To help you change the range of values that variables of type *int* and *char* can store, C provides a set of type modifiers—*unsigned, long, register, signed,* and *short*. A *type modifier* changes (modifies) the range of values a variable can store or the way the compiler stores a variable. To modify a type, place the type modifier in front of the type name in a variable declaration, as shown here:

```
unsigned int inventory_count;
register int counter;
long int very_large_number;
```

Several following Tips discuss these four type modifiers in detail.

38 Understanding the unsigned Type Modifier

A *type modifier* changes (modifies) the range of values a variable can store or the way the compiler stores a variable. As you have learned, variables of type *int* can store positive and negative values in the range –32,768 through 32,767. Within the representation of a value of type *int*, the value's most significant bit indicates the value's sign (positive or negative), as you learned in Tip 32. In some cases, your program may never need to store a negative value within a specific variable. The *unsigned* type modifier tells the compiler not to use the most significant bit as a sign bit, but instead to let the bit represent larger positive values. A variable of type *unsigned int* can store values in the range 0 through 65,535. Figure 38.1 illustrates how the C compiler stores an *unsigned int* variable. Again, newer C compilers normally use 32 bits to store unsigned values.

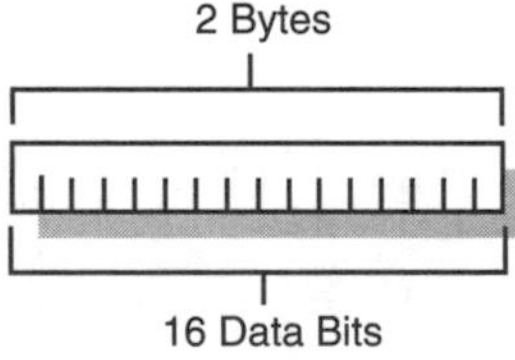

Figure 38.1 How the C compiler represents values of type ***unsigned int.***

As discussed in Tip 33, variables of type *char* can hold values in the range –128 through 127. When you use the *unsigned* type modifier with variables of type *char*, you can create variables that can store values in the range 0 through 255. Figure 38.2 illustrates how the C compiler represents an *unsigned char* variable.

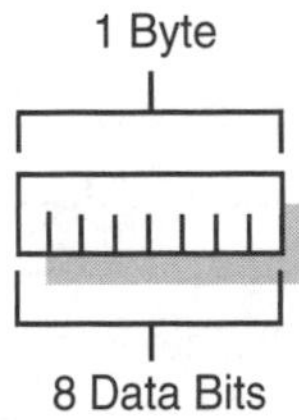

Figure 38.2 How the C compiler represents variables of type ***unsigned char.***

The following statements illustrate declarations of variables with type *unsigned int* or *unsigned char*:

```
void main(void)
 {
   unsigned int current_seconds;
   unsigned int status_indicator;
   unsigned char menu_border;     // Extended ASCII character
 }
```

UNDERSTANDING THE LONG TYPE MODIFIER 39

A type modifier changes (modifies) the range of values a variable can store or the way the compiler stores a variable. Variables of type *int* can store positive and negative values in the range -32,768 through 32,767. As previously shown in Tip 32, older C compilers represent values of type *int* using 16 bits, with the most significant bit indicating the value's sign. In many cases, your programs must store integer values that are larger (greater than 32,767) or smaller (less than –32,768) than the range of values a variable of type *int* can hold. The *long* type modifier tells the compiler to use 32 bits (four bytes) to represent the integer values. A variable of type *long int* can store values in the range –2,147,483,648 through 2,147,483,647. Figure 39 shows how the C compiler stores a *long int* variable.

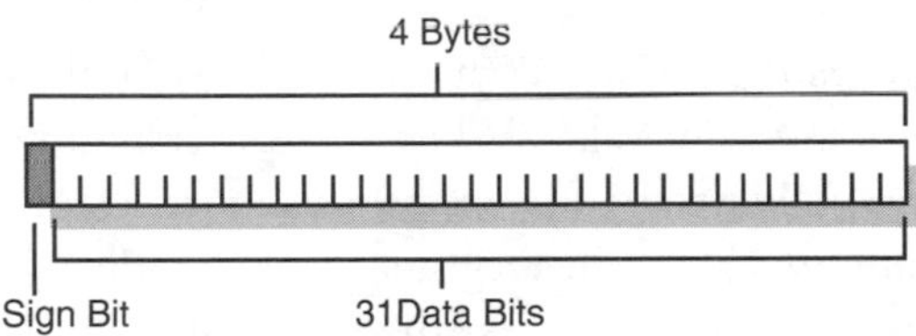

Figure 39 How the C compiler represents values of type ***long int.***

Note: *Many C++ compilers also support the type* ***long double****, which your programs can use to represent floating-point numbers with up to 80 digits of precision, rather than the standard 64 digits of precision. Values of type* ***long double*** *use 10 bytes of memory, with a 60-bit mantissa and a 19-bit exponent. The range for a long double value is 3.4E-4932 to 1.1E+4932. To determine whether your compiler supports* ***long double*** *declarations, check your compiler's documentation.*

COMBINING THE UNSIGNED AND LONG TYPE MODIFIERS 40

In Tip 38 you learned that the *unsigned* type modifier directs the C compiler not to interpret a value's most significant bit as a sign indicator, but rather, to use that bit to represent a larger value. Likewise, in Tip 39, you learned that the *long* type modifier directs the compiler to double the number of bits it uses to represent an integer value. In some cases, your programs may need to store very large positive values. By combining the *unsigned* and *long* type modifiers, you can direct the C compiler to allocate a 32-bit variable capable of values in the range 0 through 4,292,967,265. Figure 40 illustrates how the C compiler would represent an *unsigned long int* variable.

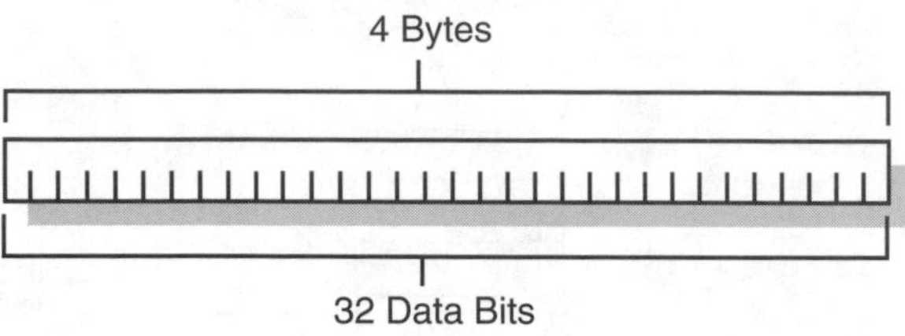

Figure 40 How the C compiler represents values of type ***unsigned long int.***

The following statements declare variables of type *unsigned long int*:

```
void main(void)
 {
   unsigned long int very_large_value;
   unsigned long int national_debt;
 }
```

41 Working with Large Values

As you have learned, variables of type *long int* can store values in the range –2,147,483,648 through 2,147,483,647. When you use large values in your programs, do not include commas. In addition, when you specify a long integer value, you should append an L to the constant as shown here:

```
long int big_number = 1234567L;
long int one_million = 1000000L;
```

If you include commas within your numbers, the C compiler will generate a syntax error.

42 Understanding the register Type Modifier

A variable is the name that your program associates with a memory location. When you declare a variable, the C compiler allocates memory to hold the variable's value. When your program must access the variable, slight overhead occurs (the computer consumes time) while the CPU accesses memory. Depending on the variable's use, you can sometimes direct the compiler to store the variable in a register (which resides within the CPU itself) to increase your program's performance. Because the compiler can access the value much faster when it resides in a register, your program will execute faster. The *register* type modifier directs the compiler to keep the variable in a register as often as possible. Because the CPU has a limited number of registers, the compiler cannot permanently assign the variable to a register. Instead, the compiler will try to keep the variable in a register as often as possible. The following statements show use of the *register* type modifier:

```
void main(void)
 {
   register int counter;
   register unsigned status_flags;
 }
```

You should use the *register* type modifier with variables that your program repeatedly accesses, such as a *loop* variable that the program accesses each time it loops.

43 Understanding the short Type Modifier

As discussed in Tip 32, older C compilers usually represent variables of type *int* using 16 bits. Therefore, the variables of type *int* can store values in the range –32,768 through 32,767. If you are using a 32-bit compiler, however, the compiler may represent an integer value using 32 bits, which means a variable of type *int* could store values in the range –2,147,483,648 through 2,147,483,647. If you store a value that is outside the range a variable of type *int* can store, an *overflow condition* occurs, and the assigned value is in error. (Tip 50 explains overflow in detail.) Programmers write some programs knowing that when an overflow occurs, the compiler consistently assigns the errant value (meaning that the errant value is always the same) to the overflowed value. In other words, the programmer writes the program to use overflow. Should you move a program that uses values of type *int* in this way (that is, that counts on the value overflowing) from a 16-bit to 32-bit environment, the overflow would no longer occur because the 32-bit integer can store a larger value. If you write a program based on overflow, which presumes that the compiler represents *int* variables with 16 bits, you can use the *short* type modifier to ensure that the compiler represents a variable using 16 bits. The following statements illustrate declarations of variables of type *short int*:

```
void main(void)
 {
   short int key_value;
   short int small_number;
 }
```

OMITTING INT FROM MODIFIED DECLARATIONS 44

Within this section you have learned about several C type modifiers, including *long*, *short*, and *unsigned*. The following statements illustrate how to use these three modifiers:

```
unsigned int status_flags;
short int small_value;
long int very_big_number;
```

When you use these three type modifiers, most compilers will let you omit the *int*, as shown here:

```
unsigned status_flags;
short small_value;
long very_big_number;
```

UNDERSTANDING THE SIGNED TYPE MODIFIER 45

As you learned in Tip 33, C compilers usually represent variables of type *char* using eight bits, with the most significant bit representing the value's sign. Therefore, variables of type *char* can store values in the range –128 through 127. In Tip 38, you learned you can use the *unsigned* qualifier to direct the C compiler to not interpret the sign bit, but instead to use the bit to represent a larger positive value. Using the *unsigned* type modifier, a variable of type *char* can store values in the range 0 through 255. If you are using a variable of type *char* and you assign the variable a value outside the range of valid values, overflow will occur and the value the computer assigns to the variable will not be the value you desire. In some cases, however, you will write programs with overflow in mind. If you plan to move such a program to a different compiler, which may represent variables of type *char* as unsigned, you can use the *signed* type modifier to ensure that the second compiler represents variables of type *char* using 7 bits for the data and 1 bit for the sign bit. The following statements show declarations of type *signed char*:

```
void main(void)
 {
   signed char byte_value;
   signed char menu_choice;
 }
```

46 MULTIPLE ASSIGNMENT OPERATIONS

As you have learned, C uses the equal sign (=) as its assignment operator. Normally, your C programs will assign values to variables on distinct lines, as shown here:

```
count = 0;
sum = 0;
value = 0;
```

When you want to assign the same value to multiple variables, C lets you perform all the assignments at one time, as shown here:

```
count = sum = value = 0;
```

When C encounters a multiple assignment operation, C assigns values from right to left. As a rule, only use multiple assignments to initialize variables. Using such operations for more complex operations will decrease your program's readability. For example, the following program assigns two variables the uppercase equivalent to the character the user types:

```
ltr_save = letter = toupper(getchar());
```

47 ASSIGNING ONE VARIABLE TYPE'S VALUE TO A DIFFERENT TYPE

A type defines the set of values a variable can store and the set of operations that your programs can perform on the data. C provides four basic data types (*int*, *float*, *char*, and *double*). Some cases may require you to assign the value of a variable of type *int* to a value of type *float*, or vice versa. As a general rule, you can successfully assign a value of type *int* to a variable of type *float*. When you assign the value of a variable of type *float* to a variable of type *int*, however, you must use caution. Most compilers will truncate the floating-point value, discarding the fractional portion. On the other hand, another compiler might round the value rather than truncate it (meaning that, if the fractional portion of the value is greater than .5, the two compilers will convert the value differently). If you want to ensure that your program performs floating-point value to integer value assignments consistently, you might consider using the *ceil* and *floor* functions, which this book's Math section presents.

48 CREATING YOUR OWN TYPES

A type defines the set of values a variable can store and the set of operations that your program can perform on the data. C provides four basic data types (*int*, *float*, *char*, and *double*). As you have learned, you can combine type modifiers to change the range of values a variable can store. As the number of variables your program declares increases, you may find it convenient to create your own variable name that provides a shorthand name for a commonly used type. For example, consider the following declarations of type *unsigned long int*:

```
unsigned long int seconds_since_january;
unsigned long int world_population_in_2000;
```

Using C's *typedef* statement, you can define the type name *ULINT*, which is identical to the type *unsigned long int*, as shown here:

```
typedef unsigned long int ULINT;
```

After you create the type name, you can use the name to define variables, as shown here:

```
ULINT seconds_since_january;
ULINT world_population_in_2000;
```

As your programs use more complex variable declarations, you may find that creating a new type name is very convenient, because type names can save you excess typing time and reduce the possibility of entry errors.

Note: *The code within this Tip defines ULINT in all capital letters because it is easier for another programmer to determine custom types if you represent the types differently than default types. You can make the type all capital letters, all lowercase letters, or a combination of both—it is your choice. However, you should be consistent in how you name custom types across multiple programs or multiple types within the same program.*

ASSIGNING A HEXADECIMAL OR OCTAL VALUE 49

Depending on your application, there may be times when you must work with *octal* (base 8) or *hexadecimal* (base 16) values. At such times, you tell the compiler you want to work with values that are not decimal values. If you precede a numeric value with a 0 (zero), such as 077, the C compiler treats the value as octal. Likewise, if you precede a value with 0x, such as 0xFF, the compiler treats the value as hexadecimal. The following statements illustrate how to use an octal and hexadecimal constant:

```
int octal_value = 0227;
int hex_value = 0xFF0;
```

UNDERSTANDING OVERFLOW 50

As you have learned, a variable's type defines the range of values a variable can store and the operations that a program can perform on the variable. Variables of type *int*, for example, can store values in the range –32,768 through 32,767. If you assign a value outside this range to a variable of type *int*, an *overflow* error will occur. As you have already learned, C uses 16 bits to represent variables of type *int*. The C compiler uses the most significant of the 16 bits to determine the variable's sign. If the most significant bit is 0, the value is positive. If the most significant bit is 1, the value is negative. C then uses 15 bits to represent the variable's value. To understand why overflow occurs, you must consider the value's bitwise implementation. Consider the following values:

```
      0    0000 0000 0000 0000
      1    0000 0000 0000 0001
      2    0000 0000 0000 0010
      3    0000 0000 0000 0011
      4    0000 0000 0000 0100

 32,765    0111 1111 1111 1101
 32,766    0111 1111 1111 1110
 32,767    0111 1111 1111 1111
```

If you add 1 to the value 32,767, you would expect the result to be 32,768. However, to C the value becomes –32,768, as shown here:

```
  32,767    0111 1111 1111 1111
   +   1    0000 0000 0000 0001
 -------  --------------------
 -32,768    1000 0000 0000 0000
```

The following program, *overflow.c*, illustrates how overflow occurs:

```
#include <stdio.h>

void main(void)
 {
   short positive = 32767, negative = -32768;
   short error_A = positive + 1;
   short error_B = negative - 1;

   printf("%d + 1 is %d\n", positive, error_A);
   printf("%d - 1 is %d\n", negative, error_B);
 }
```

When you compile and execute this program, your screen will display the following output:

```
32767 + 1 is -32768
-32768 - 1 is 32767
```

As you can see, adding a value to 32,767 yields a negative number, while subtracting a value from –32,768 produces a positive number. One problem that makes overflow difficult is that within your programs, you will often not notice the error because the C compiler does not return an error when the overflow occurs. In other words, the program continues to execute, despite the overflow. As a result, when you debug your programs, you may have a difficult time detecting errors that result from overflow.

Note: *If you use the* ***Turbo C++ Lite*** *compiler, or most other newer compilers, the compiler will warn you of the potential overflow problem. The* ***Turbo C++ Lite*** *compiler will display warnings of* ***Constant is long in function main*** *and* ***Conversion may lose significant digits in function main()****, but will nevertheless run the program (and overflow the variables). As a general rule, even if a compiler warning message does not stop the compiler from compiling the program, you should observe such messages closely and respond appropriately.*

51 Understanding Precision

As you have learned, computers represent numbers internally using combinations of 1s and 0s (binary digits). In previous Tips, you learned that because a type has a fixed number of bits, each type can only hold a specific range of values. If you assign a value outside the type's range, an overflow error occurs. Floating-point values can experience overflow and can suffer from insufficient precision. A value's *precision* defines its degree of accuracy. Values of type *float*, for example, provide six to seven significant digits. Assume, for example, you assign the value 1.234567890 to a variable of type *float*. Because type *float* only provides seven digits of significance, you can only count on values 1.23456 to be accurate. Values of type *double*, on the other hand, provide 14 to 15 significant digits. As a result, a value of type *double* could accurately store 1.234567890.

When you work with floating-point numbers, you must be aware that the computer represents values using a fixed number of bits. Therefore, it is impossible for the computer to always represent values exactly. For example, the computer may represent the value 0.4 as 0.3999999, or the value 0.1 as 0.099999, and so on. The following program, *precise.c*, illustrates the difference between double and single precision:

```
#include <stdio.h>

void main(void)
 {
   float accurate = 0.123456790987654321;
   double more_accurate = 0.1234567890987654321;

   printf("Value of float\t %21.19f\n", accurate);
   printf("Value of double\t %21.19f\n", more_accurate);
 }
```

When you compile and execute the *precise.c* program, your screen will display the following output:

```
Value of float  0.1234567890432815550
Value of double 0.1234567890987654380
C:\>
```

52 Assigning Quotes and Other Characters

As you work with variables of type *char*, or with character strings, there may be times when you must assign a single- or double-quote character to a variable. For example, to write *Jamsa's C/C++/C# Programmer's Bible*, you must include two single quotes within the string. In such cases, you must place the character within single quotes preceded by a backslash (\), as shown here:

```
char single_quote = '\'';
char double_quote = '\"';
```

In addition to the quote characters, your programs may often require that you assign one of the special characters listed in Table 52. To do so simply place the character's symbol immediately after the backslash character. In all cases, you must use lowercase letters to represent the special character.

Escape Character	Meaning
\a	ASCII bell character
\b	Backspace character
\f	Formfeed character
\n	Newline character
\r	Carriage return (no linefeed)
\t	Horizontal tab
\v	Vertical tab
\\	Backslash character
\'	Single quote
\"	Double quote
\?	Question mark
\nnn	ASCII value in octal
\xnnn	ASCII value in hexadecimal

Table 52 *C-defined escape characters.*

GETTING STARTED WITH PRINTF 53

Several tips presented throughout this book have used the *printf* function to display messages on the screen. When your program uses *printf,* the information you direct *printf* to print is called *printf's parameters* or *arguments.* The following statement uses *printf* to display the message *Jamsa's C/C++/C# Programmer's Bible!* on your screen display:

```
printf("Jamsa\'s C/C++/C# Programmer\'s Bible!");
```

In this case, the character string (the letters that appear within the double quotes) is *printf's* only parameter. When your programs begin to work with variables, you may want to use *printf* to display each variable's values. The *printf* function supports more than one parameter. The first parameter must always be a character string. You can make the parameters following the first character string be numbers, variables, expressions (such as 3 * 15), or even other character strings. When you want *printf* to display a value or variable, you must provide *printf* with information about the variable's type within the first parameter. In addition to specifying characters within the first parameter, you can include *format specifiers,* which tell *printf* how to print the other parameters. Such format specifiers take the form of a percent sign (%) followed by a letter. For example, to display an integer value, you use the *%d* (*d* for decimal value). Likewise, to print a floating-point value, you can use *%f.* The following *printf* statements illustrate how you might use format specifiers with *printf:*

```
printf("The users age is %d\n", age);
printf("The sales tax is %f\n", cost * 0.07);
printf("The user\'s age: %d weight: %d height: %d\n",
  age, weight, height);
```

As you can see, within *printf's* first parameter, you can specify one or more format specifiers. Note that the third statement does not fit on one line and continues to the next. When your statements cannot fit on one line, try to find a good place to wrap the line (such as immediately after a comma), and then indent the line that follows. The purpose of indentation is to improve your program's visual appeal and to make obvious to someone who is reading your program that the line contains a continuation of the previous line. Several tips that follow discuss in detail different *printf* format specifiers.

54 Displaying Values of Type int Using printf

The *printf* function supports format specifiers that provide *printf* with information about its parameter types (such as *int, float, char,* and so on). To display values of type *int* with *printf,* use the *%d* format specifier. The following program, *intout.c,* uses the *%d* format specifier to display values and variables of type *int*:

```
#include <stdio.h>

void main(void)
 {
   int age = 41;
   int height = 73;
   int weight = 165;

   printf("The user\'s age: %d weight: %d height: %d\n", age, weight, height);
   printf("%d plus %d equals %d\n", 1, 2, 1 + 2);
 }
```

When you compile and execute the *intout.c* program, your screen will display the following output:

```
The user's age: 41 weight: 165 height: 73
1 plus 2 equals 3
C:\>
```

Note: *Many C compilers treat the* **%i** *format specifier as identical to* **%d**. *If you are creating a new program, however, use the* **%d** *specifier, because the* **%i** *specifier is a legacy specifier and future compilers may not support it.*

55 Printing an Octal or Hexadecimal Integer Value

The *printf* function supports format specifiers that provide *printf* with information about its parameter types (such as *int, float, char,* and so on). Depending on your program, there may be times when you want to display an integer value in its octal (base 8) or hexadecimal (base 16) format. The *%o* (letter o, not zero) format specifier directs *printf* to display a value in octal. In a similar way, the *%x* and *%X* specifiers direct *printf* to display a value in hexadecimal format. The difference between *%x* and *%X* is that the latter displays hexadecimal values in uppercase. The following program, *oct_hex.c,* illustrates the use of the *%o, %x,* and *%X* format specifiers:

```
#include <stdio.h>

void main(void)
 {
   int value = 255;

   printf("The decimal value %d in octal is %o\n", value, value);
   printf("The decimal value %d in hexadecimal is %x\n", value, value);
   printf("The decimal value %d in hexadecimal is %X\n", value, value);
 }
```

When you compile and execute the *oct_hex.c* program, your screen will display the following output:

```
The decimal value 255 in octal is 377
The decimal value 255 in hexadecimal is ff
The decimal value 255 in hexadecimal is FF
C:\>
```

DISPLAYING VALUES OF TYPE UNSIGNED INT USING PRINTF 56

As you have learned, the *printf* function supports format specifiers that provide *printf* with information about its parameter types (such as *int, float, char*, and so on). To display values of type *unsigned int* with *printf*, you should use the *%u* format specifier. If you use *%d* instead of *%u*, *printf* will treat the value specified as type *int*, which will probably display the wrong result. The following program, *u_intout.c*, uses the *%u* format specifier, as well as *%d*, to display the value 42000. The *u_intout.c* program illustrates the type of error that can occur if you use the wrong format specifier:

```
#include <stdio.h>

void main(void)
 {
   unsigned int value = 42000;

   printf("Displaying 42000 as unsigned %u\n", value);
   printf("Displaying 42000 as int %d\n", value);
 }
```

If you compile and this program using a 16-bit compiler, your screen may display the following output:

```
Displaying 42000 as unsigned 42000
Displaying 42000 as int -23536
C:\>
```

Note: *When you compile this program under* ***Turbo C++ Lite****, the compiler will display two error messages, because the compiler sees the constant value 42,000 that the program tries to assign to the* ***unsigned int*** *variable* ***value*** *as a* ***long*** *number, rather than an* ***int****. In this case, because the program's purpose is to show the mistakes that can arise from* ***unsigned int*** *declarations, you should ignore the compiler warnings. Other 16-bit compilers will issue similar warnings.*

DISPLAYING VALUES OF TYPE LONG INT USING PRINTF 57

As you have learned, the *printf* function supports format specifiers that provide *printf* with information about its parameter types (such as *int, float, char*, and so on). To display values of type *long int* with *printf*, you should use the *%ld* format specifier. If you use *%d* instead of *%ld*, *printf* will treat the value specified as type *int*, quite probably displaying the wrong result. The following program, *longout.c*, uses the *%ld* format specifier, as well as *%d* to display the value 1000000L (note the L for a long value). The *longout.c* program illustrates the type of error that can occur should you use the wrong format specifier:

```
#include <stdio.h>

void main(void)
 {
   long int one_million = 1000000L;

   printf ("One million is %ld\n", one_million);
   printf ("One million is %d\n", one_million);
 }
```

Again, if you compile this program using a 16-bit compiler, your screen may display the following output:

```
One million is 1000000
One million is 16960
C:\>
```

58 DISPLAYING VALUES OF TYPE FLOAT USING PRINTF

The *printf* function supports format specifiers that provide *printf* with information about its parameter types (such as *int, float, char*, and so on). To display values of type *float* with *printf*, you should use the *%f* format specifier. The following program, *floatout.c*, uses the *%f* format specifier to display floating-point values:

```
#include <stdio.h>

void main(void)
 {
   float price = 525.75;
   float sales_tax = 0.06;

   printf("The item cost is %f\n", price);
   printf("Sales tax on the item is %f\n", price * sales_tax);
 }
```

When you compile and execute the *floatout.c* program, your screen will display the following output:

```
The item cost is 525.750000
Sales tax on the item is 31.544999
C:\>
```

As you can see, by default, the *%f* format specifier provides little output formatting. However, several Tips in this section present ways to format output using *printf*.

59 DISPLAYING VALUES OF TYPE CHAR USING PRINTF

As you have learned, the *printf* function supports format specifiers that provide *printf* with information about its parameter types (such as *int, float, char*, and so on). To display values of type *char* with *printf*, you should use the *%c* format specifier. The following program, *char_out.c*, uses the *%c* format specifier to display the letter A on your screen:

```
#include <stdio.h>

void main(void)
 {
   printf("The letter is %c\n", 'A');
   printf("The letter is %c\n", 65);
 }
```

As you can see, the *char_out.c* program will display the letter A using the character constant 'A', as well as the ASCII value 65. When you compile and execute the *char_out.c* program, your screen will display the following output:

```
The letter is A
The letter is A
C:\>
```

60 DISPLAYING FLOATING-POINT VALUES IN AN EXPONENTIAL FORMAT

As you have learned, the *printf* function supports format specifiers that provide *printf* with information about its parameter types (such as *int, float, char*, and so on). In Tip 58, you learned that by using the *%f* format specifier, you can display floating-point values. Depending on your program's requirements, there may be times when you will want to display values using an exponential format. To display a floating-point value in an exponential format, use the *%e* or *%E* format specifier. The difference between *%e* and *%E* is that the *%E* format specifier directs *printf* to use a capital E in the output. The following program, *exp_out.c*, uses the *%e* and *%E* format specifiers to display floating-point values in their exponential format:

```
#include <stdio.h>

void main(void)
 {
   float pi = 3.14159;
   float radius = 2.0031;

   printf("The circle's area is %e\n", 2 * pi * radius);
   printf("The circle's area is %E\n", 2 * pi * radius);
 }
```

When you compile and execute the *exp_out.c* program, your screen will display the following output:

```
The circle's area is 1.258584e+01
The circle's area is 1.258584E+01
C:\>
```

As you can see, by default, the *%e* and *%E* format specifiers provide little output formatting. However, several Tips in this section present ways to format output using *printf*.

DISPLAYING FLOATING-POINT VALUES 61

In Tip 58 you learned that using the *%f* format specifier, you can direct *printf* to display floating-point values using their decimal point format. Likewise, in Tip 60 you learned that you can use the *%e* and *%E* format specifiers to direct *printf* to display a floating-point value using an exponential format. In a similar way, *printf* supports the *%g* and *%G* format specifiers. When you use the *%g* and *%G* format specifiers, *printf* decides whether it should use the *%f* or *%e* format, depending on the technique that will display the output in the format most meaningful to the user. The following program, *flt_pt.c*, illustrates the use of the *%g* format specifier:

```
#include <stdio.h>

void main(void)
 {
   printf("Displaying 0.1234 yields %g\n", 0.1234);
   printf("Displaying 0.00001234 yields %g\n", 0.00001234);
 }
```

When you compile and execute the *flt_pt.c* program, your screen will display the following output:

```
Displaying 0.1234 yields 0.1234
Displaying 0.00001234 yields 1.234e-05
C:\>
```

DISPLAYING A CHARACTER STRING USING PRINTF

A *character string* is a sequence of zero or more characters. (The Strings section discusses character strings in more detail.) One of the most common operations your programs will perform is character string output. As you have learned, the *printf* function supports format specifiers that provide *printf* with information about its parameter types (such as *int, float, char*, and so on). To display a character string using *printf*, you should use the *%s* format specifier. The following program, *str_out.c*, uses the *%s* format specifier to display a character string:

```
#include <stdio.h>

void main(void)
 {
   char title[255] = "Jamsa\'s C/C++/C# Programmer\'s Bible";

   printf("The name of this book is %s\n", title);
 }
```

When you compile and execute the *str_out.c* program, your screen will display the following output:

```
The name of this book is Jamsa's C/C++/C# Programmer's Bible
C:\>
```

63 Displaying a Pointer Address Using printf

As you have learned, the *printf* function supports format specifiers that provide *printf* with information about its parameter types (such as *int, float, char*, and so on). You have also learned that a variable is a name that your program corresponds with a memory location. As your program's complexity increases, you will eventually work with memory addresses (called *pointers*). When you begin to work with pointers, there may be times when you must display a pointer address. To display a pointer address using *printf*, use the *%p* format specifier. The following program, *ptr_out.c*, uses the *%p* format specifier to display a memory address:

```
#include <stdio.h>

void main(void)
 {
   int value;

   printf("The address of the variable value is %p\n", &value);
 }
```

When you compile and execute the *ptr_out.c* program, your screen will display output similar to the following:

```
The address of the variable value is FFF4
C:\>
```

When you use the *%p* format specifier, the pointer's actual value and the format *printf* uses to display the value will differ from one operating system to another. The Pointers section discusses in detail the use of pointers.

64 Preceding a Value with a Plus or Minus Sign

As you have learned, *printf* supports various format specifiers that control how *printf* displays output. By default, when you use *printf* to display a negative value, it will precede the value with a minus sign. Depending on your program, there may be times when you want *printf* to display the sign for positive values as well. To direct *printf* to display a value's sign, simply include a plus sign immediately following the *%* in the format specifier. The following program, *showsign.c*, illustrates the use of the plus sign within the format specifier:

```
#include <stdio.h>

void main(void)
 {
   int neg_int = -5;
   int pos_int = 5;
   float neg_flt = -100.23;
   float pos_flt = 100.23;

   printf("The integer values are %+d and %+d\n", neg_int, pos_int);
   printf("The floating-point values are %+f %+f\n", neg_flt, pos_flt);
 }
```

When you compile and execute the *showsign.c* program, your screen will display the following:

```
The integer values are -5 and +5
The floating-point values are -100.230003 +100.230003
C:\>
```

FORMATTING AN INTEGER VALUE USING PRINTF 65

As you read in Tip 54, the *%d* format specifier directs *printf* to display an integer value. As your programs become more complex, you will want *printf* to format your data better. For example, assume you want to print out a table on your computer screen which is similar to the following output:

```
Salesman  Quantity
Jones     332
Smith     1200
Allen     3311
David     43
```

When you use the *%d* format specifer, you can direct *printf* to display a minimum number of characters. The following program, *int_fmt.c*, illustrates how you might format integer values using *%d*:

```
#include <stdio.h>

void main(void)
 {
   int value = 5;

   printf ("%1d\n", value);
   printf ("%2d\n", value);
   printf ("%3d\n", value);
   printf ("%4d\n", value);
 }
```

When you compile and execute the *int_fmt.c* program, your screen will display the following:

```
5
 5
  5
   5
C:\>
```

The digit you place after the % specifies the minimum number of characters *printf* will use to display an integer value. If, for example, you specify *%5d* and the value you want to display is 10, *printf* will precede the value with three spaces. Note that the value specifies the minimum number of characters the output will consume. If the value you want to display requires more characters than you specified, *printf* will use the number of characters *printf* requires to correctly display the value.

ZERO-PADDING INTEGER OUTPUT 66

In Tip 65 you learned how to format an integer value by placing the desired number of digits immediately after the % in the *%d* format specifier. If the integer value *printf* displays does not require the number of characters you specified, *printf* will precede the value with the necessary number of spaces. Depending on your program's purpose, there may be times when you want *printf* to precede the value with zeros (called *zero-padding*), as opposed to spaces. To direct *printf* to zero-pad a value, place a 0 (zero) immediately after the % in the format specifier, prior to the desired number of digits. The following program, *zero_pad.c*, illustrates zero-padding:

```
#include <stdio.h>

void main(void)
 {
   int value = 5;

   printf ("%01d\n", value);
```

```
   printf ("%02d\n", value);
   printf ("%03d\n", value);
   printf ("%04d\n", value);
 }
```

When you compile and execute the *zero_pad.c* program, your screen will display the following:

```
5
05
005
0005
C:\>
```

67 Displaying a Prefix Before Octal and Hexadecimal Values

In Tip 55 you learned how to use the *%o* format specifier to display octal values and the *%x* and *%X* format specifiers to display hexadecimal values. When your programs output such values, there may be times when you want to precede octal values with a zero (for example, *0777*), and hexadecimal values with *0x* (for example, *0xFF*).

To direct *printf* to precede an octal or hexadecimal value with the appropriate prefix, place a pound sign character (#) immediately after the *%* in the format specifier. The following program, *show_oh.c*, illustrates the use of the pound sign character (#) in the *printf* format specifier:

```
#include <stdio.h>

void main(void)
 {
   int value = 255;

   printf("The decimal value %d in octal is %#o\n", value, value);
   printf("The decimal value %d in hexadecimal is %#x\n", value, value);
   printf("The decimal value %d in hexadecimal is %#X\n", value, value);
 }
```

When you compile and execute the *show_oh.c* program, your screen will display the following output:

```
The decimal value 255 in octal is 0377
The decimal value 255 in hexadecimal is 0xff
The decimal value 255 in hexadecimal is 0xFF
C:\>
```

68 Formatting a Floating-Point Value Using printf

In Tip 65 you learned how to format an integer value by placing the desired number of digits immediately after the *%* in the *%d* format specifier. Using a similar technique, *printf* lets you format floating-point output. When you format a floating-point value, you specify two values. The first value tells *printf* the minimum number of characters you want to display. The second value tells *printf* the number of digits you want displayed to the right of the decimal point. The following program, *flt_fmt.c*, illustrates how to format floating-point values using *printf*:

```
#include <stdio.h>

void main(void)
 {
   float value = 1.23456;

   printf ("%8.1f\n", value);
   printf ("%8.3f\n", value);
   printf ("%8.5f\n", value);
 }
```

When you compile and execute the *flt_fmt.c* program, your screen will display the following output:

```
      1.2
    1.235
  1.23456
C:\>
```

FORMATTING EXPONENTIAL OUTPUT 69

In Tip 68 you learned how to use the *%f* format specifier to format floating-point values. Using similar formatting techniques, you can instruct *printf* to display floating-point output in an exponential format. The following program, *exp_fmt.c*, illustrates the formatted exponential output:

```
#include <stdio.h>

void main(void)
 {
   float value = 1.23456;

   printf ("%12.1e\n", value);
   printf ("%12.3e\n", value);
   printf ("%12.5e\n", value);
 }
```

When you compile and execute the *exp_fmt.c* program, your screen will display the following output:

```
     1.2e+00
   1.235e+00
 1.23456e+00
C:\>
```

LEFT-JUSTIFYING PRINTF'S OUTPUT 70

By default, when you output text using *printf*'s formatting characters, *printf* will display the text right-justified. Depending on your program, there may be times when you want *printf* to left-justify your output. To left-justify text, place a minus sign (–) immediately after the % in the format specifier. The following program, *leftjust.c*, illustrates the use of the minus sign to left-justify output:

```
#include <stdio.h>

void main(void)
 {
   int int_value = 5;
   float flt_value = 3.33;

   printf("Right justified %5d value\n", int_value);
   printf("Left justified %-5d value\n", int_value);
   printf("Right justified %7.2f value\n", flt_value);
   printf("Left justified %-7.2f value\n", flt_value);
 }
```

When you compile and execute the *leftjust.c* program, your screen will display the following output:

```
Right justified     5 value
Left justified 5     value
Right justified    3.33 value
Left justified 3.33    value
C:\>
```

71 COMBINING PRINTF FORMAT SPECIFIERS

Several Tips presented in this section have discussed various *printf* format specifiers. As you use *printf*'s format specifiers, there may be times when you want to take advantage of two or more format specifiers. For example, you may want to display a left-justified hexadecimal value, preceded with the characters *0x*. In such cases, simply place each of the specifiers after the *%*. The following program, *full_fmt.c*, illustrates the use of multiple format specifiers:

```
#include <stdio.h>

void main(void)
 {
   int int_value = 5;

   printf("Left justifed with sign %-+3d\n", int_value);
 }
```

When you compile and execute the *full_fmt.c* program, your screen will display the following output:

```
Left justified with sign +5
C:\>
```

72 WRAPPING A CHARACTER STRING TO THE NEXT LINE

When your programs use *printf*, there may be times when a character string within *printf* will not fit on the current line. In such cases, simply place a backslash (\) at the end of the line, which will continue the text at the start of the next line, as shown here:

```
printf("This line is very long and because it is so very long, it would not \
fit on the same line.");
```

Note: *If you wrap text to the next line, do not include spaces at the start of the new line's text. If spaces are present, the C compiler will include the spaces within the string.*

73 DISPLAYING NEAR AND FAR STRINGS

The Memory section of this book discusses *near* and *far* pointers in detail. Briefly, *near* and *far* pointers represent variable addresses within the program's memory space. Programs that run within older operating systems, such as MS-DOS, use *far* pointers to increase the range of memory addresses the program can use to store information. When your programs work with *far* string pointers, there may be times you will want to display the string's contents using *printf*. As you will learn in the Functions section, however, the compiler will generate an error if you pass a *far* pointer to a function that expects a *near* address. If you want to display the contents of a *far* string (whose start a *far* string pointer indicates) using *printf*, you must tell *printf* that you are using a *far* pointer. To do so, place an uppercase *F* (for *far*) immediately after the *%* in the format specifier, as shown here:

```
printf("%Fs\n", some_far_string);
```

Because *%Fs* tells *printf* you are using a *far* pointer, the function call is correct. In a similar way, you can tell *printf* that you are passing a *near* string by placing an uppercase *N* in the format specifier. However, because *printf* expects *near* strings by default, the format specifiers *%Ns* and *%s* have the same result. The following C program, *near_far.c*, illustrates the use of *%Fs* and *%Ns* within *printf*:

```
#include <stdio.h>

void main(void)
 {
```

```
   char *near_title = "Jamsa\'s C/C++/C# Programmer\'s Bible";
   char far *far_title = "Jamsa\'s C/C++/C# Programmer\'s Bible";

   printf("The book\'s title: %Ns\n", near_title);
   printf("The book\'s title: %Fs\n", far_title);
 }
```

***Note:** Many compilers, such as the **Visual C++** compiler, do not distinguish between **near** and **far** pointers. If you try to compile the **near_far.c** program under **Visual C++**, the compiler will return an error. To automatically update your programs to run under **Visual C++**, include the **windef.h** header file within your programs.*

WORKING WITH PRINTF'S ESCAPE CHARACTERS 74

When you work with character strings, there will be times when you will want to use special characters, such as the tab, carriage return, or linefeed characters. C defines several *escape characters* (that is, characters you precede with C's escape symbol, the backslash) to make it easy to include special characters within a string (such as characters you want *printf* to output). For example, several of the programs presented in this book have used the newline character (\n) to advance output to the start of the next line, as shown here:

```
printf("Line 1\nLine 2\Line 3\n");
```

Table 74 lists the escape characters you can use within your character strings (and hence the *printf* output).

Escape Character	Meaning
\a	ASCII bell character
\b	Backspace character
\f	Formfeed character
\n	Newline character
\r	Carriage return (no linefeed)
\t	Horizontal tab
\v	Vertical tab
\\	Backslash character
\'	Single quote
\"	Double quote
\?	Question mark
\nnn	ASCII value in octal
\xnnn	ASCII value in hexadecimal

Table 74 C-defined escape characters.

DETERMINING THE NUMBER OF CHARACTERS PRINTF HAS DISPLAYED 75

When your programs perform sophisticated screen formatting, there may be times when you want to know the number of characters *printf* has displayed. When you use the *%n* format specifier, *printf* will assign to a variable (passed by pointer) a count of the number of characters *printf* has displayed. The following program, *prt_cnt.c*, illustrates the use of the *%n* format specifier:

```
#include <stdio.h>

void main(void)
 {
```

```
   int first_count;
   int second_count;

   printf("Jamsa%n\'s C/C++/C# Programmer\'s Bible%n\n", &first_count, &second_count);
   printf("First count %d Second count %d\n", first_count, second_count);
 }
```

When you compile and execute the *prt_cnt.c* program, your screen will display the following:

```
Jamsa's C/C++/C# Programmer's Bible
First count 5 Second count 35
C:\>
```

76 USING PRINTF'S RETURN VALUE

In Tip 75 you learned how to use *printf*s *%n* format specifier to determine the number of characters *printf* has written. Using the *%n* format specifier is one way to ensure that *printf* has successfully displayed its output. In addition, when *printf* completes, it returns the total number of characters that *printf* wrote. If *printf* encounters an error, it will return the constant EOF (which, as you will learn, indicates the end of a file). The following program, *printfok.c*, uses *printf's* return value to ensure that *printf* was successful:

```
#include <stdio.h>

void main(void)
 {
   int result;

   result = printf("Jamsa\'s C/C++/C# Programmer\'s Bible!\n");
   if (result == EOF)
     fprintf(stderr, "Error within printf\n");
 }
```

If the user has redirected a program's output to a file or device (such as a printer), and redirected I/O experiences an error (such as *device off-line* or *disk full*), your programs can detect the error by testing *printf*'s return value.

77 USING THE ANSI DEVICE DRIVER

Several Tips presented throughout this book have made extensive use of *printf*'s output formatting capabilities. Although *printf* provides format specifiers that you use to control the number of digits displayed, to display output in octal or hexadecimal, or to left- or right-justify text, *printf* does not provide other format specifiers. *Printf* does not provide format specifiers that let you position the cursor to a specific row and column, clear the screen, or display output in colors. However, depending on the operating system you are using, you can probably perform such operations using the ANSI device driver. The ANSI driver supports different escape sequences that direct it to use specific colors, to position the cursor, and even to clear the screen. Programmers name such formatting statements *escape sequences* because they begin with the ASCII escape character (the value 27). If you are using DOS, install the ANSI driver by placing an entry such as the following within your *config.sys* file (and then rebooting):

```
DEVICE=C:\DOS\ANSI.SYS
```

After you install the ANSI driver, your programs can write escape sequences using *printf.*

Note: *If you are running Windows 95 or later on the same machine on which you compile programs, adding the ANSI driver to your system's* ***config.sys*** *file will not interfere with Windows operations.*

USING THE ANSI DRIVER TO CLEAR YOUR SCREEN DISPLAY 78

One of the most common operations each of your programs will perform when the program first begins execution is to clear the screen display. Unfortunately, the C run-time library does not provide a function that clears the screen display. To clear the screen display, use the ANSI driver discussed in Tip 77, and then invoke the following escape sequence to clear your screen display:

```
Esc[2j
```

An easy way to invoke the escape sequence is to use the octal representation of the escape character (\033), as shown here:

```
printf("\033[2J");
```

USING THE ANSI DRIVER TO DISPLAY SCREEN COLORS 79

Several Tips presented throughout this book have made extensive use of the *printf* function to display output. Although *printf* provides powerful format specifiers, *printf* does not provide the means for you to display output in color. However, if you are using the ANSI driver, as discussed in Tip 77, you can use the escape sequences listed in Table 79 to display output in color.

Escape Sequence	Color
Esc[30m	Black foreground color
Esc[31m	Red foreground color
Esc[32m	Green foreground color
Esc[33m	Orange foreground color
Esc[34m	Blue foreground color
Esc[35m	Magenta foreground color
Esc[36m	Cyan foreground color
Esc[37m	White foreground color
Esc[40m	Black background color
Esc[41m	Red background color
Esc[42m	Green background color
Esc[43m	Orange background color
Esc[44m	Blue background color
Esc[45m	Magenta background color
Esc[46m	Cyan background color
Esc[47m	White background color

***Table** 79 ANSI escape sequences you can use to set screen colors.*

The following *printf* statement selects the blue background color:

```
printf("\033[44m");
```

Similarly, the following *printf* statement selects red text on a white background:

```
printf("\033[47m\033[31m");
```

In the previous example, *printf* writes two escape sequences. The ANSI driver lets you specify the screen colors, which you separate with semicolons, as shown here:

```
printf("\033[47;31m");
```

80 Using the ANSI Driver to Position the Cursor

As you have learned, the ANSI driver supports escape sequences that let you clear your screen and display output in color. In addition, the ANSI driver provides escape sequences that let you position the cursor to specific row and column positions, which lets you display your output at specific screen locations. Table 80 shows the ANSI driver's cursor-positioning escape sequences.

Escape Sequence	Function	Example
Esc[x;yH	Set the cursor at row *x* and column *y*	*Esc[10;25H*
Esc[xA	Move the cursor up *x* rows	*Esc[1a*
Esc[xB	Move the cursor down *x* rows	*Esc[2b*
Esc[yC	Move the cursor right *y* columns	*Esc[10c*
Esc[yD	Move the cursor left *y* columns	*Esc[10d*
Esc[S	Store the current cursor position	*Esc[S*
Esc[U	Restore the cursor position	*Esc[U*
Esc[2j	Clear the screen, moving the cursor to the home position	*Esc[2j*
Esc[K	Clear to end of the current line	*Esc[K*

Table 80 ANSI driver cursor-positioning escape sequences you can use within your programs.

81 Performing Basic Math Operations in C

In all but the simplest programs, your programs will perform arithmetic operations such as addition, subtraction, multiplication, or division. To perform these basic math operations, use the operators described in Table 81.

Operator	Purpose
+	Addition
–	Subtraction
*	Multiplication
/	Division

Table 81 Basic arithmetic operators in C.

The following program, *math.c*, illustrates how you use C's basic arithmetic operators:

```
#include <stdio.h>

void main(void)
 {
   int seconds_in_an_hour;
   float average;

   seconds_in_an_hour = 60 * 60;
   average = (5 + 10 + 15 + 20) / 4;
   printf("The number of seconds in an hour %d\n", seconds_in_an_hour);
   printf("The average of 5, 10, 15, and 20 is %f\n", average);
   printf("The number of seconds in 48 minutes is %d\n",
          seconds_in_an_hour - 12 * 60);
 }
```

When you compile and execute the *math.c* program, your screen will display the following output:

```
The number of seconds in an hour 3600
The average of 5, 10, 15, and 20 is 12.000000
The number of seconds in 48 minutes is 2880
C:\>
```

UNDERSTANDING MODULO ARITHMETIC 82

In Tip 81 you learned that C uses the forward slash operator (/) for division. Depending on your application, there may be times when your program needs the remainder of an integer division. In such cases, use C's modulo (remainder) operator. The following program, *modulo.c*, illustrates how you use C's modulo operator:

```
#include <stdio.h>

void main(void)
 {
   int remainder;
   int result;

   result = 10 / 3;
   remainder = 10 % 3;
   printf("10 Divided by 3 is %d Remainder %d\n", result, remainder);
 }
```

When you compile and execute the *modulo.c* program, your screen will display the following output:

```
10 Divided by 3 is 3 Remainder 1
```

UNDERSTANDING OPERATOR PRECEDENCE AND ASSOCIATIVITY 83

In Tip 81 you learned that C uses the following operators: the plus sign (+) for addition; the hyphen (-) for subtraction; the asterisk (*) for multiplication; and the forward slash (/) for division. When your programs use these operators within arithmetic expressions, you should understand C's operator precedence, which specifies the order in which C performs arithmetic operations. For example, consider the following expression:

```
result = 5 + 2 * 3;
```

If you assume that C performs the operations from left to right (the addition before the multiplication), the expression's result is 21:

```
result = 5 + 2 * 3;
       = 7 * 3;
       = 21;
```

If, however, C performs the multiplication first, the result is 11:

```
result = 5 + 2 * 3;
       = 5 + 6;
       = 11;
```

To prevent the problem of indeterminate results, C defines an *operator precedence*. An operator precedence determines which operations C executes first. Table 83 illustrates C's operator precedence.

Operator Precedence (High to Low)										
()	[]	.	->							
++	--	+	–	*	&	!	~	(type) sizeof		
*	/	%								
+	-									
>>	<<									
==	!=									
&										
^										
|										
&&										
||										
? :										
=	+=	–=	*=	/=	%=	&=	^=	|=	<<=	>>=
,										

Table 83 C's operator precedence.

When you create an expression, C will execute the operations with the highest precedence first. If two operators have the same precedence, C performs the operations from left to right.

84 Forcing the Order of Operator Evaluation

As you learned in Tip 83, C performs operations in an expression based on each operator's precedence within the expression. In many cases, the order that C will use to evaluate operators is not the order you want. For example, consider the following expression—its goal is to calculate the average of three values:

```
average = 5 + 10 + 15 / 3;
```

Mathematically, the average of the three values 5, 10, and 15 is 10. However, if you let C evaluate the previous expression, C will assign the variable *average* the value 20, as shown here:

```
average = 5 + 10 + 15 / 3;
        = 5 + 10 + 5;
        = 15 + 5;
        = 20;
```

If you examine C's operator precedence table (which Tip 83 presents), you will find that C's *division* operator (/) has a higher precedence than C's *addition* operator (+). Therefore, you need a way to change the order in which C performs the operations. When C evaluates an expression, C will always perform operations that appear within parentheses before performing other operations. When you group the values you want to sum within parentheses, C will calculate the correct average, as shown here:

```
average = (5 + 10 + 15) / 3;
        = (15 + 15) / 3;
        = (30) / 3;
        = 10;
```

C performs operations within parentheses based on its operator precedence rules. If an expression contains multiple expressions within multiple parentheses sets, C performs the operations within the innermost parentheses first, as shown here:

```
result = ((5 + 3) * 2) - 3;
       = ((8) * 2) - 3;
       = (16) - 3;
       = 13;
```

UNDERSTANDING C'S INCREMENT OPERATOR 85

One very common operation programs perform is to increment a variable's current value by 1. For example, the following statement increments the value of *variable* count by 1:

```
variable = variable + 1;
```

Because increment operations are so common, C provides a shorthand notation you can use to increment variables within your programs, called the *increment operator*. The following statement uses the *increment* operator to add 1 to a value of *variable*:

```
variable++;
```

The following program, *0_to_100.c*, uses the *increment* operator to print the values 0 through 100:

```
#include <stdio.h>

void main(void)
 {
   int value = 0;

   while (value <= 100)
     {
       printf("%d\n", value);
       value++;
     }
 }
```

C provides both a *prefix* and *postfix increment* operator. The following statements both increment the variable *count* by 1:

```
count++;
++count;
```

The first statement uses C's *postfix increment* operator. The second statement uses the *prefix increment* operator. You should distinguish between the two operators because C treats *prefix* and *postfix* operators differently. When you use the *postfix increment* operator, C first uses the variable's value and then performs the *increment* operation. On the other hand, when you use the *prefix increment* operator, C first increments the variable's value and then uses the variable. To better understand the difference between the *prefix* and *postfix increment* operators, consider the following program, *prepost.c*, which uses both operators:

```
#include <stdio.h>

void main(void)
 {
   int value = 1;

   printf("Using postfix %d\n", value++);
   printf("Value after increment %d\n", value);
   value = 1;
   printf("Using prefix %d\n", ++value);
   printf("Value after increment %d\n", value);
 }
```

When you compile and execute the *prepost.c* program, your screen will display the following output:

```
Using postfix 1
Value after increment 2
Using prefix 2
Value after increment 2
C:\>
```

As you see, when you use the *postfix* operator, C first uses the variable's value (displaying the value 1) and then increments the variable (yielding 2). When you use the *prefix* operator, C first increments the variable (yielding 2) and then displays the already incremented value.

86 UNDERSTANDING C'S DECREMENT OPERATOR

Just as there will be many times when you want to increment a variable's value, so too will there be many times when you want to decrement a variable's current value by 1, as shown here:

```
variable = variable - 1;
```

Because decrement operations are so common, C provides a shorthand notation you can use to perform such operations—C's *decrement* operator. The following statement uses the *decrement* operator to subtract 1 from the value of *variable*:

```
variable--;
```

As was the case with C's *increment* operator, C provides both a *prefix* and a *postfix decrement* operator. The following statements both decrement the variable *count* by 1:

```
count--;
--count;
```

The first statement uses C's *postfix decrement* operator and the second statement uses the *prefix decrement* operator. You should distinguish between the two operators because C treats *prefix* and *postfix* operators differently. When you use the *postfix* operator, C first uses the variable's value and then performs the decrement operation. In the opposite manner, when you use the *prefix* operator, C first decrements the variable's value and then uses the variable. To better understand the difference between the *prefix* and *postfix increment* operators, consider the following program, *postpre.c*, which uses both operators:

```
#include <stdio.h>

void main(void)
 {
   int value = 1;

   printf("Using postfix %d\n", value--);
   printf("Value after decrement %d\n", value);
   value = 1;
   printf("Using prefix %d\n", --value);
   printf("Value after decrement %d\n", value);
 }
```

When you compile and execute the *postpre.c* program, your screen will display the following output:

```
Using postfix 1
Value after decrement 0
Using prefix 0
Value after decrement 0
C:\>
```

As you can see, when you use the *postfix decrement* operator, C first uses the variable's value (displaying the value 1) and then decrements the variable's value (yielding 0). When you use the *prefix decrement* operator, C first decrements the variable (yielding 0) and then displays the already decremented value.

Understanding a Bitwise OR Operation 87

As the complexity of your programs increases, you may find that you can increase a program's performance or reduce a program's memory requirement by using *bitwise operations*. Bitwise operations are operations that manipulate values one or more bits at a time. When you must manipulate a value a single bit at a time, you may take advantage of C's *bitwise OR* operator (|). The *bitwise OR* operator examines each bit within two values and yields a third value as a result. For example, assume that two variables contain the values 3 and 4, whose bits are, respectively, 00000011 and 00000100. The *bitwise OR* operator returns the value 7, as shown here:

```
3       00000011
4       00000100
        --------
7       00000111
```

In the value 3, bits 0 and 1 have a one value, and all the other bits have a zero value. In the value 4, bit 2 has a one value, and all the other bits have a zero value. The result of a *bitwise OR* operation will have a one value within each corresponding bit which has a one value in either of the original values. In this case, the result has a one value in bits 0, 1, and 2. The following program, *bit_or.c*, illustrates how you use C's *bitwise OR* operator:

```
#include <stdio.h>

void main(void)
 {
   printf("0 | 0 is %d\n", 0 | 0);
   printf("0 | 1 is %d\n", 0 | 1);
   printf("1 | 1 is %d\n", 1 | 1);
   printf("1 | 2 is %d\n", 1 | 2);
   printf("128 | 127 is %d\n", 128 | 127);
 }
```

When you compile and execute the *bit_or.c* program, your screen will display the following output:

```
0 | 0 is 0
0 | 1 is 1
1 | 1 is 1
1 | 2 is 3
128 | 127 is 255
C:\>
```

Understanding a Bitwise AND Operation 88

As you learned in Tip 87, you may find that you can increase your program's performance or reduce your program's memory requirements by using bitwise operations. Bitwise operations are operations that manipulate values one or more bits at a time. When manipulating data one bit at a time, you may take advantage of C's *bitwise AND* operator (&). The *bitwise AND* operator examines each bit within two values and yields a third value as a result. For example, assume that two variables contain the values 5 and 7, whose bits are, respectively, 00000101 and 00000111. The *bitwise AND* operator returns the value 5, as shown here:

```
5       00000101
7       00000111
        --------
5       00000101
```

If a bit within both terms has a one value, the *bitwise AND* operator sets to one the corresponding bit within the result. If a bit within either term contains a zero value, the *bitwise AND* operator sets to zero the corresponding bit within the result. In this case, bits 0 and 2 contain one values in both terms, so the result has one values in bits 0 and 2, and zero values in the remaining bits. The following program, *bit_and.c*, illustrates the use of C's *bitwise AND* operator:

```
#include <stdio.h>

void main(void)
 {
   printf("0 & 0 is %d\n", 0 & 0);
   printf("0 & 1 is %d\n", 0 & 1);
   printf("1 & 1 is %d\n", 1 & 1);
   printf("1 & 2 is %d\n", 1 & 2);
   printf("15 & 127 is %d\n", 15 & 127);
 }
```

When you compile and execute the *bit_and.c* program, your screen will display the following output:

```
0 & 0 is 0
0 & 1 is 0
1 & 1 is 1
1 & 2 is 0
15 & 127 is 15
C:\>
```

89 Understanding a Bitwise Exclusive OR Operation

As you have already learned, bitwise operations are operations that manipulate values one or more bits at a time. When you are manipulating data one bit at a time, there may be situations in which you must take advantage of C's *bitwise exclusive OR* operator (^), which examines the bits in two values and sets to 1 the bits of the result based on the truth table shown in Table 89.

X	Y	Result
0	0	0
0	1	1
1	0	1
1	1	0

Table 89 Bit results from a ***bitwise exclusive OR*** *operation.*

Assume that two variables contain the values 5 and 7, whose bits are, respectively, 00000101 and 00000111. The *bitwise exclusive OR* operator returns the value 2, as shown here:

```
5       00000101
7       00000111
        --------
2       00000010
```

The following program, *bit_xor.c*, illustrates the use of C's *bitwise exclusive OR* operator:

```
#include <stdio.h>

void main(void)
 {
   printf("0 ^ 0 is %d\n", 0 ^ 0);
   printf("0 ^ 1 is %d\n", 0 ^ 1);
   printf("1 ^ 1 is %d\n", 1 ^ 1);
```

```
   printf("1 ^ 2 is %d\n", 1 ^ 2);
   printf("15 ^ 127 is %d\n", 15 ^ 127);
 }
```

When you compile and execute the *bit_xor.c* program, your screen will display the following output:

```
0 ^ 0 is 0
0 ^ 1 is 1
1 ^ 1 is 0
1 ^ 2 is 3
15 ^ 127 is 112
C:\>
```

UNDERSTANDING A BITWISE INVERSE OPERATION 90

As you have learned, bitwise operations are operations that manipulate values one or more bits at a time. When you must manipulate values one bit at a time, you may take advantage of C's *bitwise inverse* operator (~). The *bitwise inverse* operator examines each bit within a value and yields a second value as a result. The *bitwise inverse* operation makes each bit which contains a one in the original value a zero in the resulting value, and makes each bit which contains a zero in the original a one in the resulting value. As an example, assume an unsigned character variable contains the value 15. The bitwise inverse of the operation would therefore return 240, as shown here:

```
15   00001111
240  11110000
```

As you can see, each bit which was set to one in the original value is zero in the result, and each bit which was set to zero in the original value is set to one in the result. The following program, *bit_inv.c*, illustrates the use of C's *bitwise inverse* operator:

```
#include <stdio.h>

void main(void)
 {
   int value = 0xFF;

   printf("The inverse of %X is %X\n", value, ~value);
 }
```

When you compile and execute the *bit_inv.c* program, your screen will display output similar to the following:

```
The inverse of FF is FF00
C:\>
```

APPLYING AN OPERATION TO A VARIABLE'S VALUE 91

As you perform arithmetic operations within your programs, you may find that you often assign a variable the result of an expression that includes the variable's current value. For example, consider the following statements:

```
total = total + 100;
count = count - 5;
half = half / 2;
```

For cases in which an assignment operation updates a variable with the result of an operation on the variable's current value, C provides a shorthand technique for expressing the operation. In short, you place the operator in front of the assignment operator. When you use C's shorthand assignment technique, the following statements will be equivalent to the three statements just shown:

```
total += 100;
count -= 5;
half /= 2;
```

When you use this shorthand technique, the following statements are equivalent:

```
variable += 10;         variable = variable + 10;
variable <<= 2;         variable = variable << 2;
variable &= 0xFF;variable = variable & 0xFF;
variable *= 1.05;variable = variable * 1.05;
```

92 UNDERSTANDING C'S CONDITIONAL OPERATOR

As you will learn, C's *if-else* statement examines a condition and performs one set of operations if the condition is true and another if the condition is false. In a similar way, C provides a *conditional* operator that examines a condition and, based on whether the condition evaluates to true or false, returns one of two values. The *conditional* operator's format is as follows:

```
(condition) ? trueResult: falseResult
```

To understand the *conditional* operator better, consider the following condition, which tests whether a test score is greater than or equal to 60. If the value is greater than or equal to 60, the statement assigns the variable *grade* a P, for pass. If the value is less than 60, the statement assigns the variable *grade* an F, for fail:

```
grade = (score >= 60) ? 'P': 'F';
```

The statement is similar to the following *if-else* statement:

```
if (score >= 60)
  grade = 'P';
else
  grade = 'F';
```

The following *printf* statement displays the string *"Pass"* or *"Fail,"* based on the test score:

```
printf("Score %d Result %s\n", score, (score >= 60) ? "Pass", "Fail");
```

When you use C's conditional operator to assign a condition's result to a variable, you can reduce the number of *if-else* statements you use within your programs.

93 UNDERSTANDING C'S SIZEOF OPERATOR

When your programs declare a variable, the C compiler allocates memory to store the variable's value. When you write programs that perform file *input/output* operations or allocate memory for dynamic lists, you will find it convenient to know the amount of memory your program has allocated for a specific variable. C's *sizeof* operator returns the number of bytes a variable or type requires. The following program, *sizeof.c*, illustrates the use of the *sizeof* operator:

```
#include <stdio.h>

void main(void)
 {
   printf("Variables of type int use %d bytes\n", sizeof(int));
   printf("Variables of type float use %d bytes\n", sizeof(float));
   printf("Variables of type double use %d bytes\n", sizeof(double));
   printf("Variables of type unsigned use %d bytes\n", sizeof(unsigned));
   printf("Variables of type long use %d bytes\n", sizeof(long));
 }
```

Depending on your compiler and system hardware, the *sizeof* program output may differ. When you use *Turbo C++ Lite*, your program will display the following (the output of a 32-bit compiler will differ):

```
Variables of type int use 2 bytes
Variables of type float use 4 bytes
Variables of type double use 8 bytes
Variables of type unsigned use 2 bytes
Variables of type long use 4 bytes
C:\>
```

PERFORMING A BITWISE SHIFT 94

When you work with values at the bit level, some common operations you will perform are *bitwise shifts*, either to the right (away from the most significant bit) or to the left (towards the most significant bit). To help your programs perform bitwise shifts, C provides two *bitwise shift* operators: one operator that shifts bits to the right (>>) and one that shifts bits to the left (<<). The following expression uses the *bitwise left-shift* operator to shift the values in the variable *flag* two positions to the left:

```
flag = flag << 2;
```

Assume that the variable *flag* contains the value 2, as shown here:

```
0000 0010
```

When you shift the value two places to the left, the result will be 8, as shown here:

```
0000 1000
```

When you shift values to the left, C zero-fills the lower bit positions. When you shift values to the right, however, the value that C places in the most significant bit position depends on the variable's type. If the variable is unsigned (that is, you declared it in the program as an *unsigned* type), C zero-fills the most significant bit during a *right-shift* operation. However, if the variable is of a signed type (in other words, if you did not declare the variable as *unsigned*), C uses the value 1 if the value is currently negative or 0 if the value is positive. The following program, *shiftem.c*, illustrates the use of C's *bitwise right-shift* and *bitwise left-shift* operators:

```
#include <stdio.h>

void main(void)
 {
   unsigned u_val = 1;
   signed int value = -1;

   printf("%u (unsigned) shifted left 2 times is %u\n", u_val, u_val << 2);
   printf("%u (unsigned) shifted right 2 times is %u\n", u_val, u_val >> 2);
   u_val = 65535;
   printf("%u (unsigned) shifted left 2 times is %u\n", u_val, u_val << 2);
   printf("%u (unsigned) shifted right 2 times is %u\n", u_val, u_val >> 2);
   printf("%d (signed) shifted left 2 times is %d\n", value, value << 2);
   printf("%d (signed) shifted right 2 times is %d\n", value, value >> 2);
 }
```

PERFORMING A BITWISE ROTATION 95

In Tip 94 you learned how to use C's *left-shift* and *right-shift* operators. When you perform a *left-shift* operation, C zero-fills the least significant bit. When you perform a *right-shift* operation, on the other hand, the value C places in the most significant bit position depends on the value's type and current value. As you work at the bit level, there may be times when you may want to simply rotate bits rather than shifting bits left or right. When you rotate bits to the left, the value's most significant bit becomes the least significant, while the other bits move one position to the right. When you rotate values to the right, the value's least significant bit becomes the most significant. To help you rotate bits, many C compilers provide the *_rotl* and *_rotr* functions, which rotate the bits comprising an unsigned value to the left and the right, as shown here:

```
#include <stdlib.h>

unsigned _rotl(unsigned value, int count);
unsigned _rotr(unsigned value, int count);
```

The *count* variable specifies the number of times you want to rotate the value. The following program, *rotate.c*, illustrates the use of the *_rotl* and *_rotr* functions:

```
#include <stdio.h>
#include <stdlib.h>

void main(void)
 {
   unsigned value = 1;
   printf("%u rotated right once is %u\n", value, _rotr(value, 1));
   value = 5;
   printf("%u rotated right twice is %u\n", value, _rotr(value, 2));
   value = 65534;
   printf("%u rotated left twice is %u\n", value, _rotl(value, 2));
 }
```

If you compile and execute the *rotate.c* program using a 16-bit compiler, your screen will display the following:

```
1 rotated right once is 32768
5 rotated right twice is 16385
65534 rotated left twice is 65531
C:\>
```

Note: *Many C compilers also provide the* ***_lrotl*** *and* ***_lrotr*** *functions, which rotate* ***unsigned long*** *integer values left or right.*

96 UNDERSTANDING CONDITIONAL OPERATORS

All the programs previously presented in this book began their execution with the first instruction in *main* and have executed each instruction that followed in order. As your programs become more complex, there will be times when the program must perform one set of instructions if a condition is true and, possibly, other instructions if the condition is not true. For example, your program might have different instructions for different days of the week. When a program performs (or does not perform) instructions based on a specific condition, the program is performing *conditional processing*. To perform conditional processing, the program evaluates a condition that results in true or false result. For example, the condition *Today is Monday* is either true or false. To help your programs perform conditional processing, C provides the *if*, *if-else*, and *switch* statements. Several of the Tips that follow discuss these statements in detail.

97 UNDERSTANDING ITERATIVE PROCESSING

All the programs previously presented in this book executed their instructions only one time. In some cases, a program may or may not have executed a set of instructions, based on the result of a tested condition. As your programs become more complex, there will be times when a program must repeat the same set of instructions a specific number of times or until the program meets a specific condition. For example, if you are writing a program that calculates student grades, the program must perform the same steps for each student in the class. Similarly, if a program displays a file's contents, the program will read and display each line of the file until the program finds the end-of-file marker. When programs repeat one or more statements until a given condition is met, the program is performing *iterative processing*. Each pass the program makes through the statements it is repeating is an *iteration*. To help your programs perform iterative processing, C provides the *for*, *while*, and *do while* statements. Several Tips presented in this book discuss the *for*, *while*, and *do while* statements in detail.

UNDERSTANDING HOW C REPRESENTS TRUE AND FALSE 98

Several Tips presented in this section have discussed C's *conditional* and *iterative* constructs, which perform one set of instructions if a condition is true and, possibly, another set of instructions if the condition is false. As you work with *conditional* and *iterative* constructs, it is important that you understand how C represents a true and a false value. C interprets any value that is not 0 as true. Likewise, the value 0 represents false. The following condition, therefore, will always evaluate to true:

```
if (1)
```

Many inexperienced C programmers write their conditions as follows:

```
if (expression != 0)  // Test if expression is true
```

When you want to test if a condition is true, simply include the expression as shown here:

```
if (expression)
```

When the expression evaluates to a non-zero (true) value, then C executes the statement that immediately follows the condition. When the expression evaluates to zero (false), C does not execute the statement that immediately follows the condition. Operators which work with true and false values are *Boolean* operators. The result of a Boolean expression is always a true or false value.

TESTING A CONDITION WITH IF 99

As your programs become more complex, they will often perform one set of statements when the condition is true and another set of statements when the condition is false. When your program must perform such *conditional processing*, you will use the C *if* statement. The *if* statement's format is as follows:

```
if (condition)
  statement;
```

The condition the *if* statement evaluates must appear within parentheses and is either true or false. When the condition is true, C performs the statement that immediately follows the condition. When the condition is false, your program does not perform the statement that follows the condition. As an example, the following *if* statement tests whether the variable *age* is greater than or equal to 21. If the condition is true, the program will execute the *printf* statement. If the condition is false, the program will not execute the *printf* statement and will continue its execution at the first statement that follows *printf* (the height assignment statement):

```
if (age >= 21)
  printf("The variable age is 21 or over\n");
height = 73;
```

UNDERSTANDING SIMPLE AND COMPOUND STATEMENTS 100

When your program performs conditional processing, there will be times when your program performs one or more statements when a condition is true and, possibly, several other statements if the condition is false. Likewise, when your program performs iterative processing, there will be times when your program repeats one statement, while at other times the program might repeat several statements. When you perform *conditional* and *iterative* processing, C classifies statements as either simple or compound. A *simple statement* is a single statement, such as a variable assignment or a call to *printf.* The following *if* statement invokes a simple statement *(printf)* when the condition is true:

```
if (condition)
  printf("The condition is true\n");
```

A *compound statement,* on the other hand, consists of one or more statements contained within right and left braces. The following *if* statement illustrates a compound statement:

```
if (condition)
  {
    age = 21;
    height = 73;
    weight = 165;
  }
```

When your program must perform multiple statements based on a condition, or when it must repeat several statements, you will use a compound statement and place the statements within the right and left braces.

101 TESTING FOR EQUALITY

As your programs become more complex, they will compare a variable's value to known conditions and determine which statements the program will execute next. To make such decisions, your programs will use the *if* or *switch* statements. As you learned in Tip 99, the format of the *if* statement is as follows:

```
if (condition)
  statement;
```

Most *if* statements will test whether a variable's value equals a second, specific value. For example, the following *if* statement tests whether the variable *age* contains the value 21:

```
if (age == 21)
  statement;
```

C uses the double equal sign (==) in tests for equality. When you write tests for equality, use the double equal sign (==) instead of the single equal sign (=) that C uses for an assignment. As you will learn in Tip 112, if you use the *assignment* operator (=), as opposed to the double equal sign, C will consider your condition correct syntax. Unfortunately, when the statement executes, C will not test to see if the variable equals the specified value. Instead, C will assign the specified value to the variable.

***Note:** Depending on your compiler warning level, the compiler might display a warning message about the assignment within the expected condition.*

Just as your programs must sometimes test values for equality, they must also sometimes test values for inequality. C uses the != symbol to test for inequality. The following statement tests whether the variable *age* is not equal to 21:

```
if (age != 21)
  statement;
```

The following program, *eql_neql.c,* uses the C tests for equality (==) and inequality (!=):

```
#include <stdio.h>

void main(void)
 {
   int age = 21;
   int height = 73;

   if (age == 21)
```

```
    printf("User\'s age is 21\n");
  if (age != 21)
    printf("User\'s age is not 21\n");
  if (height == 73)
    printf("User\'s height is 73\n");
  if (height != 73)
    printf("User\'s height is not 73\n");
}
```

When you compile and execute the *eql_neql.c* program, your screen will display the following output:

```
User's age is 21
User's height is 73
C:\>
```

To understand how to use the equality and inequality operators, experiment with the *eql_neql.c* program by changing the values of the *height* and *age* variables.

PERFORMING RELATIONAL TESTS 102

As your programs become more complex, there may be times when you must test whether a value is greater than another value, less than another value, greater than or equal to another value, or less than or equal to another value. To help you perform such tests, C provides a set of *relational operators*. Table 102 lists C's relational operators.

Operator	Function
>	*Greater-than* operator
<	*Less-than* operator
>=	*Greater-than-or-equal-to* operator
<=	*Less-than-or-equal-to* operator

***Table 102** C's relational operators.*

The following *if* statement uses C's *greater-than-or-equal-to* operator (>=) to test whether the integer variable *age* is over 20:

```
if (age >= 21)
  printf("The age is over 20\n");
```

PERFORMING A LOGICAL AND TO TEST TWO CONDITIONS 103

In Tip 99 you learned how to use C's *if* statement to test conditions within your programs. As your programs become more complex, your programs eventually will test for multiple conditions. For example, you may want an *if* statement to test whether a user has a dog and, if so, whether that dog is a Dalmatian. In cases when you want to test if two conditions are true, use C's *logical AND* operator. C represents the *logical AND* operator with two ampersands (&&), as shown in the following *if* statement:

```
if ((user_has_dog) && (dog == dalmatian))
  {
    // Statements
  }
```

When C encounters an *if* statement that uses the *logical AND* operator (&&), C evaluates the conditions from left to right. If you examine the parentheses, you will find that the previous *if* statement is in the following form:

```
if (condition)
```

In the following example, the condition is actually two conditions connected with the *logical AND* operator:

```
(user_has_dog) && (dog == dalmatian)
```

For the resulting condition to evaluate as true when your programs use the *logical AND* operator, both conditions must evaluate as true. If either condition is false, the resulting condition will evaluate as false.

Many Tips presented throughout this book will use the *logical AND* operator. In each case, to ensure that each expression evaluates with the correct operator precedence, the programs will place the conditions within parentheses.

***Note:** Do not confuse C's **logical AND** operator (&&) with C's **bitwise AND** operator (&). The **logical AND** operator evaluates two **Boolean** (true or false) expressions to produce a true or false result. The **bitwise AND** operator, on the other hand, manipulates bits (1s and 0s).*

104 PERFORMING A LOGICAL OR TO TEST TWO CONDITIONS

In Tip 99 you learned how to use C's *if* statement to test conditions within your programs. As your programs become more complex, you eventually will test for multiple conditions. For example, you may want an *if* statement to test if a user has a dog or if the user has a computer. In cases when you want to test whether either of two conditions is true (or if both are true), you can use C's *logical OR* operator. C represents the *logical OR* with two vertical bars (||), as shown here:

```
if ((user_has_dog) || (user_has_computer))
  {
    // Statements
  }
```

When C encounters an *if* statement that uses the *logical OR* operator (||), C evaluates the conditions from left to right. If you examine the parentheses, you will find that the previous *if* statement is in the following form:

```
if (condition)
```

In this particular example, the condition is actually two conditions connected by the *logical OR* operator, as shown here:

```
(user_has_dog) || (user_has_computer)
```

For the resulting condition to evaluate as true when you use the *logical OR* operator, only one of the two conditions must evaluate as true. If either condition (or both) is true, the resulting condition evaluates as true. If both conditions evaluate as false, the result is false.

Many Tips presented throughout this book use the *logical OR* (||) operator. In each case, to ensure that each expression evaluates with the correct operator precedence, the programs will place the conditions within parentheses.

***Note:** Do not confuse C's **logical OR** operator (||) with C's **bitwise OR** operator (|). The **logical OR** operator evaluates two **Boolean** (true or false) expressions to produce a true or false result. The **bitwise OR** operator, on the other hand, manipulates bits (1s and 0s).*

105 PERFORMING A LOGICAL NOT OPERATION

When your programs use the *if* statement to perform conditional processing, the *if* statement evaluates an expression that yields a true or false result. Depending on your program's processing, there may be times when you will only want the program to perform a set of statements when the condition evaluates as false. For example, assume that you want a program to test whether the user has a dog. If the user does not have a dog, the program should display a message telling the user to buy a Dalmatian. If the user has a dog, the program should not do anything. When you want your program to perform one or more statements when a condition is false, you should use C's *logical NOT* operator, which C represents using the exclamation mark (!). Consider the following *if* statement:

```
if (! user_has_dog)
   printf("You need to buy a dalmatian\n");
```

Conditions that use the *logical NOT* operator essentially say that when a certain condition is not true (in other words, when the condition evaluates as false), you should perform the *if* statement (or compound statements). Several Tips presented in this book use the *logical NOT* operator within conditions.

ASSIGNING THE RESULT OF A CONDITION 106

Several Tips in this section have presented different conditions that evaluate as true or false within an *if*, *while*, *for*, or other statement. In addition to letting you use conditions within C conditional and iterative control structures, C also lets you assign a condition's result to a variable. Assume, for example, that your program uses the result of the same condition more than once, as shown here:

```
if ((strlen(name) < 100) && (today == MONDAY))
  {
    // Statements
  }
else if ((strlen(name) < 100) && (today == TUESDAY))
  {
    // Statements
  }
else if (strlen(name) >= 100)
  {
    // Statements
  }
```

As you can see, the program uses the condition *(strlen(name) < 100)* three times. Each time the condition appears, the program invokes the *strlen* function. In the preceding statements, the program could (depending on *today's* value) invoke *strlen* three times. The following statements will assign the condition's result (true or false) to the variable *name_ok* and then repeatedly use the variable (as opposed to the condition). Using the variable rather than the condition, as shown here, will improve the program's performance:

```
name_ok = (strlen(name) < 100);

if (name_ok && (today == MONDAY))
  {
    // Statements
  }
else if (name_ok && (today == TUESDAY))
  {
    // Statements
  }
else if (! name_ok)
  {
    // Statements
  }
```

DECLARING VARIABLES WITHIN COMPOUND STATEMENTS 107

In Tip 100 you learned about the difference between simple and compound statements. As you have learned, a compound statement is one or more statements grouped within left and right braces. The following *while* loop, (which will read lines from a file and display the lines in uppercase letters), illustrates a compound statement:

```
while (fgets(line, sizeof(line), fp))
  {
    strupr(line);
    fputs(line, stdout);
  }
```

As your programs become more complex, sometimes the processing they perform within a compound statement will require that you use one or more variables whose values you use only within the loop (as might be the case with counter variables). When you use counter variables, for example, you usually declare those counter variables at the start of your program, immediately following the *main* statement. However, if you use a variable only within a compound statement, you can declare the variable at the start of the statement, as shown here:

```
if (condition)
  {
    int counter;
    float total;

    // Other statements
  }
```

In this case, the program declares two variables at the start of the compound statement. Within the compound statement, you can use these two variables just as though you defined the variables at the start of your program. You cannot, however, refer to these variables outside the compound statement's opening and closing braces. An advantage of declaring variables within the compound statement is that another programmer reading your program code will better understand how and when you use a variable. Several Tips presented later in this book focus on a variable's *scope*, or the locations within your program in which the program "knows" a variable. As a rule, you should limit the program's knowledge of a variable to only those locations that use the variable—in other words, you should limit the variable's scope. Declaring variables at the start of a compound statement, as this Tip describes, limits the variable's scope to the compound statement's starting and ending braces.

Note: *If you declare variables within a compound statement that have the same name as variables you defined outside the statement, the C compiler will use the newly declared variables within the compound statement and the original variables outside the statement.*

108 Using Indentation to Improve Readability

As you create your programs, one of the best ways you can improve program readability is to use indentation. Each time your program uses a brace (such as at the start of a compound statement), you should consider indenting your code two or more spaces. For example, consider the following program, *use_ind.c*:

```
#include <stdio.h>

void main(void)
 {
   int age = 10;
   int user_has_dog = 0;    // 0 is false

   if (age == 10)
     {
       printf("Dogs are important pets\n");
       if (! user_has_dog)
         printf("You should get a dalmatian\n");
     }
   printf("Happy is a dalmatian\n");
 }
```

You can examine only the indentation and still quickly get a feel for related program statements (for example, compound statements). Indentation is meaningless to the compiler. To the compiler, the following program, *no_ind.c*, is identical to the previous example:

```
#include <stdio.h>

void main(void)
{
int age = 10;
int user_has_dog = 0;    // 0 is false
if (age == 10)
{
printf("Dogs are important pets\n");
if (! user_has_dog)
printf("You should get a dalmatian\n");
}
printf("Happy is a dalmatian\n");
}
```

As you can see, the indentation makes the first program much easier for you, or another reader, to understand.

USING EXTENDED CTRL+BREAK CHECKING 109

When you create programs that use the *for*, *while*, and *do* loops for iteration and that run in the DOS environment, there may be times when you must press the CTRL+BREAK keyboard combination to end a program that is stuck in an infinite loop. By default, DOS checks for a CTRL+BREAK after each time it writes to the screen, the disk, or the printer, or if it reads a character from the keyboard. If your program does not perform these operations within the loop you want to break, you cannot use the CTRL+BREAK command to end the program's processing. When you use the DOS BREAK command, however, you can increase the number of operations that will, upon each operation's completion, instruct DOS to check for a CTRL+BREAK entry. Programmers call this additional testing *extended CTRL+BREAK checking*. The following BREAK command enables extended CTRL+BREAK checking:

```
C:\> BREAK ON <ENTER>
```

If you want DOS to automatically enable extended CTRL+BREAK checking as soon as the system starts, place a BREAK=ON entry in your *config.sys* file. Because DOS is performing more extended CTRL+BREAK checking, your overall system performance will decrease slightly. However, while you are first getting the feel for iterative processing, you might find that your ability to end a program using CTRL+BREAK is more important than the slight loss of performance.

TESTING FLOATING-POINT VALUES 110

Several Tips presented in this section have used the *if* and *while* statements to test a variable's value. For example, the following statements test several integer variables:

```
if (age == 21)
  // Statements

if (height > 73)
  // Statements
```

When you work with floating-point values, however, you must be careful when you test a variable's value. For example, the following statement tests a floating-point variable named *sales_tax*:

```
if (sales_tax == 0.065)
  // Statements
```

In Tip 51 you learned about floating-point precision and the fact that the computer must represent floating-point values using a fixed number of bits. It is impossible for the computer to represent all values exactly. In the case of the previous *if* statement, for example, the computer may represent the value 0.065 as 0.0649999. As a result, the *if* statement will never evaluate as true. To prevent such errors within your program, do not test for exact floating-point values. Instead, test for an acceptable range of values, as shown here:

```
if (fabs(sales_tax - 0.065) <= 0.0001)
  // Statements
```

In the previous example, because the difference between the value in the variable *sales_tax* and 0.065 is less than or equal to 0.0001, the program will consider the values as equal.

111 LOOPING FOREVER

As you have learned, C's *for*, *while*, and *do while* statements let you repeat one or more statements until they meet a given condition. Depending on your program, there may be times when you want the program to loop forever. For example, a program that detects radiation leaks at a nuclear reactor should always run. To make your program loop forever, simply place a non-zero constant within the loop, as shown here:

```
while (1)
```

Because you can use a non-zero value to force your programs to loop forever, you might want to define constants to improve your program's readability. For example, you might use the constant *FOREVER*, as shown here:

```
#define FOREVER 1
while (FOREVER)
```

To create a loop for the previous nuclear-reactor example, you might use the following:

```
#define MELT_DOWN 0
while (! MELT_DOWN)
```

112 TESTING AN ASSIGNMENT

As you have learned, C uses the single equal sign as the *assignment* operator and the double equal sign to test for equality, as shown here:

```
score = 100;
if (score == MAX)
  {
    // Statements
  }
```

In the preceding code fragment, the first statement assigns the value 100 to the variable *score*. Next, the *if* statement tests the variable's value. To help you reduce the number of statements in your program, C lets you test an assignment's result. For example, the following *if* statement combines the previous assignment and condition test:

```
if ((score = 100) == MAX)
  {
    // Statements
  }
```

C will first perform the expression the parentheses contain, assigning the value 100 to the variable *score*. Next, C will compare the value you assigned to the variable *score* to the constant *MAX*. If you remove the parentheses, as shown here, C will assign a different value and perform a different test:

```
if (score = 100 == MAX)
```

Without the parentheses, C will test whether the value 100 equals the constant *MAX*, and if so, it will assign the value 1 (true) to the variable *score*. If the value 100 does not equal *MAX*, the statement will assign the value 0 (false) to the variable *score*.

You will most commonly use assignment testing when you want to test the value a function (such as *fopen* or *getchar*) returns, as shown here:

```
if ((fp = fopen("CONFIG.SYS", "r")) == NULL)
  {
    // Statements
  }

if ((letter = getchar()) == 'A')
  {
    // Statements
  }
```

BEWARE OF IF-IF-ELSE STATEMENTS 113

When you use *if-else* statements, a sneaky logic error can cause problems if you do not keep track of which *else* statement corresponds to which *if* statement. For example, consider the following code fragment:

```
test_score = 100;
current_grade = 'B';

if (test_score >= 90)
  if (current_grade == 'A')
    printf("Another A for an A student\n");
else
  printf("Should have worked harder\n");
```

The first *if* statement tests whether a student's test score was greater than or equal to 90. If so, a second *if* statement tests whether the student already has an 'A' grade and, if so, prints a message. Based on the indentation, you would expect the *else* statement to display its message if the test score was less than 90. Unfortunately, that is not how the code fragment processes the conditions. When you place an *else* statement within your program, C associates the *else* with the first *else*-less *if* statement. Although the student's test score was 100, the previous code fragment will print the message telling the student that he or she should have worked harder. In other words, the fragment executes the statements shown here:

```
if (test_score >= 90)
  if (current_grade == 'A')
    printf("Another A for an A student\n");
  else
    printf("Should have worked harder\n");
```

To prevent C from associating the *else* statement with the wrong *if* statement, place the second *if* statement within braces, forming a *compound statement*, as shown here:

```
if (test_score >= 90)
  {
     if (current_grade == 'A')
       printf("Another A for an A student\n");
  }
else
  printf("Should have worked harder\n");
```

114 PERFORMING STATEMENTS A SPECIFIC NUMBER OF TIMES

One operation that your programs will commonly perform is to repeat a set of statements a specific number of times. For example, you might want to calculate test scores for 30 students, determine the highs and lows of 100 stock quotes, or even sound your computer's built-in speaker three times. To help your programs repeat one or more statements a specific number of times, C provides the *for* statement. You will implement the *for* statement as shown here:

```
for (starting_value; ending_condition; increment_value)
  statement;
```

When your program repeats (loops through) statements a specific number of times, you will normally use a variable, called the *control variable*, that counts the number of times you have performed the statements. The *for* statement contains four sections. The *starting_value* section assigns the control variable the variable's initial value, which is most often 0 or 1. The *ending_condition* section normally tests the control variable's value to determine if the program has performed the statements the desired number of times. The *increment_value* section normally adds the value 1 to the control variable each time the statements execute. Finally, the *for* loop's fourth section is the statement or statements you want to repeat. Because your program repeatedly performs the statement or statements you specified (loops back to the start of the statement), the *for* statement is often called a *for loop*. Consider the following *for* statement, which will display the numbers 1 through 10 on your screen:

```
for (counter = 1; counter <= 10; counter++)
  printf("%d\n", counter);
```

In the previous example, *counter* is the loop's control variable. First, the *for* statement assigns the value 1 to the variable. Second, the *for* loop immediately tests whether or not *counter's* value is less than or equal to 10 (the loop's *ending condition*). If *counter* is less than or equal to 10, the *for* loop immediately executes the next statement, which, in this example, is *printf*. After the program completes the *printf* statement, the *for* loop performs the expression you specified in the *increment_value* loop section. In this case, the *for* loop increments *counter's* value by 1. Then, the *for* loop immediately performs the *ending_value* test. If *counter's* value is less than or equal to 10, the loop continues. Therefore, the first time through the loop, the *printf* statement will display the value 1. The second time through the loop, *counter's* value is 2, then 3, and so on. After *printf* displays the value 10, the *increment_value* loop section increments *counter's* value, making it 11. When the *for* loop performs the *ending_value* test, *for* will find that *counter's* value is no longer less than or equal to 10, so the loop will end and your program will resume processing at the statement immediately following the *for* loop.

To better understand the *for* loop's processing, consider the following program, *for_test.c*:

```
#include <stdio.h>

void main(void)
 {
   int counter;

   for (counter = 1; counter <= 5; counter++)
     printf("%d ", counter);
   printf("\nStarting second loop\n");
   for (counter = 1; counter <= 10; counter++)
     printf("%d ", counter);
   printf("\nStarting third loop\n");
   for (counter = 100; counter <= 5; counter++)
     printf("%d ", counter);
 }
```

When you compile and execute the *for_test.c* program, your screen will display the following:

```
1 2 3 4 5
Starting second loop
1 2 3 4 5 6 7 8 9 10
```

```
Starting third loop
C:\>
```

As you can see, the first *for* loop displays the numbers 1 through 5. The second *for* loop displays the values 1 through 10. The third *for* loop does not display any values. If you look closely, you will see that the program initially assigns the loop's control variable the value 100. When the *for* statement tests the value, the *for* loop immediately meets the ending condition, so the loop does not execute.

All the examples presented in this Tip have used single statements in the *for* loop. If you must repeat more than one statement, place the statements in left and right braces, forming a *compound statement*, as shown here:

```
for (i = 1; i <= 10; i++)
  {
    // Statements
  }
```

PARTS OF THE FOR STATEMENT ARE OPTIONAL 115

In Tip 114 you learned that the *for* statement lets your program repeat one or more statements a specific number of times. As you learned, the *for* loop uses three sections within the *for* statement: an initialization, a test, and an increment (the loop's fourth section contains the statements the *for* loop repeats):

```
for (initialization; test; increment)
```

Depending on your program, sometimes you may not need to use each of the *for* statement's sections. For example, if you have already assigned the variable *count* the initial value 0, you can skip the loop's initialization section. Then to display the numbers 0 through 999, your loop would contain the following:

```
for (; count < 1000; count++)
  printf(" %d", count);
```

If you omit one of the *for* loop sections, however, you must include the corresponding semicolon. For example, the following *for* loop skips the initialization and increment sections:

```
for (; count < 1000; )
  printf(" %d", count++);
```

Likewise, the following *for* statement will loop forever:

```
for (;;)
  // Statement
```

Although the *for* statement provides these optional sections, your program will become more difficult to read if you omit them. As a rule, if you do not need to use all three parts of the *for* statement, you should use a different looping construct, such as the *while* statement.

DECREMENTING VALUES IN A FOR STATEMENT 116

As you have learned, a *for* statement lets you repeat one or more statements a specific number of times. Tips 114 and 115 presented several *for* statements. In each case, the *for* loop counted up, from 1 to 5, 1 to 10, and so on. The *for* statement also lets you decrement the control variable. For example, the following *for* loop counts down the numbers 10, 9, 8 and so on to 1:

```
for (counter = 10; counter >= 1; counter--)
  printf("%d ", counter);
```

As you can see, the preceding *for* statement is nearly the opposite of the *for* statements you saw in previous Tips. The loop initializes the control variable *counter* to a high value and then decrements *counter* by one each time the loop repeats.

The following program, *for_down.c,* uses the *for* statement to count down, first from 5 to 1, and then from 10 to 1:

```
#include <stdio.h>

void main(void)
 {
   int counter;

   for (counter = 5; counter >= 1; counter--)
     printf("%d ", counter);
   printf("\nStarting second loop\n");
   for (counter = 10; counter >= 1; counter--)
     printf("%d ", counter);
   printf("\nStarting third loop\n");
   for (counter = 0; counter >= 1; counter--)
     printf("%d ", counter);
 }
```

When you compile and execute the *for_down.c* program, your screen will display the following output:

```
5 4 3 2 1
Starting second loop
10 9 8 7 6 5 4 3 2 1
Starting third loop
C:\>
```

As you can see, the third loop does not display any values. In this example, the *for* statement initializes *counter* to a value that is less than the ending value of 1. Therefore, the loop immediately ends.

117 Controlling the for Loop Increment

As you have learned, the *for* loop lets your programs repeat one or more statements a specific number of times. In previous Tips, each *for* loop has either incremented or decremented the loop's control variable by one. C, however, lets you increment the variable by any amount you want. For example, the following *for* loop increments the control variable *counter* by 10 with each iteration of the loop:

```
for (counter = 0; counter <= 100; counter += 10)
  printf("%d\n", counter);
```

In a similar way, the previous *for* loops initialized the control variable to 1 or 0. Just as you can set the increment or decrement amount to whatever size you want, C, again, lets you initialize the variable to any value you want. The following program, *for_diff.c,* uses different increment and initialization values:

```
#include <stdio.h>

void main(void)
 {
   int counter;

   for (counter = -100; counter <= 100; counter += 5)
     printf("%d ", counter);
   printf("\nStarting second loop\n");
   for (counter = 100; counter >= -100; counter -= 25)
     printf("%d ", counter);
 }
```

USING FOR LOOPS WITH CHAR AND FLOAT VALUES 118

As you have learned, the *for* statement lets your programs repeat a set of statements a specific number of times. Each of the *for* statements presented in the previous Tips used only values of type *int*. You can, however, use character and floating-point values in your *for* loops. For example, the following *for* loop displays the letters of the alphabet:

```
for (letter = 'A'; letter <= 'Z'; letter++)
  printf("%c", letter);
```

Likewise, the following loop increments a floating-point value by 0.5:

```
for (percent = 0.0; percent <= 100.0; percent += 0.5)
  printf("%f\n", percent);
```

The following program, *for_more.c,* illustrates the use of letters and floating-point values in a *for* loop:

```
#include <stdio.h>

void main(void)
 {
   char letter;
   float percent;

   for (letter = 'A'; letter <= 'Z'; letter++)
     putchar(letter);
   for (letter = 'z'; letter >= 'a'; letter--)
     putchar(letter);
   putchar('\n');
   for (percent = 0.0; percent < 1.0; percent += 0.1)
     printf("%3.1f\n", percent);
 }
```

UNDERSTANDING A NULL LOOP 119

As you have learned, the *for* loop lets you repeat one or more statements until the loop's control variable meets a specific condition. In the past when programmers wanted their programs to pause briefly, perhaps to display a message, the programmers placed a "do-nothing" or *NULL loop* in their programs. For example, the following *for* loop does nothing 100 times:

```
for (counter = 1; counter <= 100; counter++)
  ;  // Do nothing
```

When you place a *NULL* loop in your program, C will perform the loop's initialization and then repeatedly test and increment the control variable until the control variable meets the ending condition. The loop's repeated testing consumes processor time, which causes the program to delay. If the program needs a longer delay, you can increase the ending condition:

```
for (counter = 1; counter <= 10000; counter++)
  ;  // Do nothing
```

Using delay techniques such as the *NULL* loop can cause problems, however. First, if the program is running on a 286, 386, or 486 computer, the length of the delay will differ simply because of the difference in speed between the different microprocessors. Second, if the program is running in a multitasking environment, such as Windows or Linux, "do-nothing" loops consume time that the processor could spend doing meaningful work for another program. If your programs must use such a delay, see the functions presented in the Date and Time section of this book.

120 UNDERSTANDING AN INFINITE LOOP

As you have learned, the *for* loop lets you repeat one or more statements a specific number of times. When the *for* loop meets its ending condition, your program will continue its execution at the statement that immediately follows. When you use *for* loops, you must ensure that the loop will meet its ending condition. Otherwise, the loop will continue to execute forever. Such unending loops are called *infinite* loops. In most cases, *infinite* loops occur as the result of a programming error. For example, consider the following loop:

```
for (i = 0; i < 100; i++)
  {
     printf("%d ", i);
     result = value * --i;    // cause of error
  }
```

As you can see, the loop's second statement decrements the value of the control variable *i*. Specifically, the loop decrements the value to –1 and then later increments the value to 0. As a result, the value never reaches 100, so the loop does not end. When your program enters an infinite loop, you might be able to press CTRL+C to end the program. The following program, *infinite.c,* illustrates an infinite loop:

```
#include <stdio.h>

void main(void)
 {
   int i;
   int result = 0;
   int value = 1;

   for (i = 0; i < 100; i++)
    {
      printf("%d ", i);
      result = value * --i;
    }
   printf("Result %d\n", result);
 }
```

When you compile and execute the *infinite.c* program, it will repeatedly display the value 0. To end the program, press CTRL+C.

121 USING C'S COMMA OPERATOR WITHIN A FOR LOOP

As you have learned, when you declare variables, C lets you declare multiple variables of the same type by separating the variable names with commas:

```
int age, height, weight;
```

In addition, C lets you separate variable initializations with commas, as shown here:

```
int age = 25, height = 73, weight = 160;
```

In a similar way, C lets you initialize and increment multiple variables in a *for* loop by separating the operations with commas. Consider the following loop, which works with the variables *i* and *j*:

```
for (i = 0, j = 100; i <= 100; i++, j++)
  printf("i = %d j = %d\n", i, j);
```

You will most commonly work with multiple variables in a *for* loop (also known as *nesting for loops*) in your programs that work with arrays. You will learn more about arrays in the Arrays and Structures section of this book. The following program, *for_2var.C*, illustrates the use of C's comma operator in a *for* loop:

```
#include <stdio.h>

void main(void)
 {
   int i, j;

   for (i = 0, j = 100; i <= 100; i++, j++)
     printf("i = %d j = %d\n", i, j);
 }
```

AVOID CHANGING THE CONTROL VARIABLE'S VALUE IN A FOR LOOP 122

As you have learned, the *for* statement lets you repeat one or more statements a specific number of times. To perform such processing, the *for* loop uses a *control variable*, which works as a counter. As a rule, you should not change the control variable's value in the *for* loop's statement. The only place the control variable's value should change is in the *for* loop's initialization and increment sections. When you change the control variable's value in the program statements, you run a greater risk of creating an *infinite* loop, and you make your programs more difficult to understand.

However, there may be times when you will want the loop to end or to skip the current iteration when the *control* variable equals a certain value. For such cases, use C's *break* or *continue* statements, which later Tips in this section discuss in detail.

REPEATING ONE OR MORE STATEMENTS USING A WHILE LOOP 123

As you have learned, the *for* statement lets you repeat one or more statements a specific number of times. In many cases, however, your programs must repeat one or more statements until the loop meets a specific condition that does not necessarily involve a count. For example, if you write a program that will display a file's contents on your screen, you will want the program to display each line of the file. In most cases, you will not know in advance how many lines the file contains. Therefore, you cannot use a *for* loop to display, for example, 100 lines. The file might contain more or fewer lines. Instead, you want the program to read and display lines until it reaches the end of the file. To do so, your programs can use the *while* loop. You will format the *while* loop as follows:

```
while (condition)
  statement;
```

When C encounters a *while* loop in your program, C will test the condition specified. If the condition is true, C will perform the statements contained in the loop. If the statement is false, C will continue your program's execution at the first statement that follows. A *while* loop can repeat a single statement or a compound statement you enclose between left and right braces, as shown here:

```
while (condition)
  {
    // Statements
  }
```

The following program, *wait_yn.c*, uses the *while* loop to loop repeatedly until you press the Y or N key in response to a yes or no question:

```
#include <stdio.h>
#include <ctype.h>
#include <conio.h>

void main(void)
 {
   char letter;                                  // Letter typed by the user

   printf("Do you want to continue? (Y/N): ");
```

```
   letter = getch();                                  // Get the letter
   letter = toupper(letter);                          // Convert letter to uppercase

   while ((letter != 'Y') && (letter != 'N'))
     {
       putch(7);                                      // Beep the speaker
       letter = getch();                              // Get the letter
       letter = toupper(letter);                      // Convert letter to uppercase
     }

   printf("\nYour response was %c\n", letter);
 }
```

First, the program will display the message the first *printf* statement contains. Second, the program will use *getch* to get the keystroke pressed. To simplify the loop's testing, the program will convert the letter to uppercase, so the loop will only need to test for Y or N. Third, the *while* loop will test the letter the user typed. If the letter is a Y or N, the condition will fail and the loop's statements will not execute. If the letter pressed is not Y or N, the loop's condition is true, and its statements will execute. In the loop, the program will beep the computer's built-in speaker to indicate an invalid character. Next, the program will get the new keystroke and convert the letter to uppercase. The loop will then repeat its test to determine if the user typed a Y or N. If not, the loop's statements will repeat. Otherwise, the program's execution will continue at the first statement that follows the loop.

124 UNDERSTANDING THE PARTS OF A WHILE LOOP

A *while* loop lets you execute one or more commands until the program meets the loop's condition. In Tip 114 you learned that a *for* loop actually contains four sections: an initialization, a test, a statement execution, and an increment. A *while* loop, on the other hand, contains only a test and the statements you want to repeat, as shown here:

```
while (condition)
  statement;
```

As you learned in Tip 120, an infinite loop is a loop whose ending condition is never met, and so the loop will continue executing forever. When you write programs that use *while* loops, you can reduce the possibility of an infinite loop by ensuring that your *while* loops perform the same four steps as those a *for* loop performs. To help you remember the four steps, you might want to use the acronym *ITEM,* as illustrated in Table 124.

Action	Description
Initialize	Initialize the loop's control variable
Test	Test the loop's control variable or condition
Execute	Execute the desired statements in the loop
Modify	Modify the control variable's value or perform an operation that will affect the condition you are testing

Table 124 The ITEM acronym's components.

Unlike the *for* loop, which lets you explicitly initialize and increment a control variable, a *while* loop requires that you include statements in the program which perform these steps for you. The following program, *item.c,* illustrates how your program performs these four steps. Unlike previous programs you have written, *item.c* uses a *while* loop to display the numbers 1 through 100:

```
#include <stdio.h>

void main(void)
 {
   int counter = 1;                        // Initialize the control variable
```

```
      while (counter <= 100)       // Test the control variable
        {
          printf("%d ", counter);// Execute the statements
          counter++;              // Modify the control variable
        }
   }
```

If you write a program that uses the *while* loop and the program experiences an infinite loop, one of the ITEM operations in your program is not correct.

REPEATING ONE OR MORE STATEMENTS USING DO 125

As you have learned, C's *while* statement lets you repeat one or more statements until a specific condition is met. Likewise, C's *for* statement lets you repeat one or more statements a specific number of times. In addition, C provides the *do* statement, which lets you execute one or more statements at least one time and then, if necessary, repeat statements. The *do* statement format is as follows:

```
do
  statement;
while (condition);
```

The *do* statement is ideal for situations that require you to perform one or more statements at least one time. For example, consider the following code fragment:

```
printf("Do you want to continue? (Y/N): ");
letter = getch();             // Get the letter
letter = toupper(letter);     // Convert letter to uppercase
while ((letter != 'Y') && (letter != 'N'))
  {
    putch(7);                       // Beep the speaker
    letter = getch();               // Get the letter
    letter = toupper(letter);       // Convert letter to uppercase
  }
```

As you can see, the code prompts the user for a keystroke, gets the keystroke, and converts the keystroke to uppercase. Depending on the key the user pressed, the fragment will start a *while* loop, which performs the same commands. Note how you can simplify the statements using the *do* statement, as shown in the following code fragment:

```
printf("Do you want to continue? (Y/N): ");
do
 {
   letter = getch();             // Get the letter
   letter = toupper(letter);     // Convert letter to uppercase
   if ((letter != 'Y') && (letter != 'N'))
     putch(7);   // Sound bell for invalid letter
 }
while ((letter != 'Y') && (letter != 'N'));
```

When C encounters a *do* statement in your program, C executes the statements between the words *do* and *while*. C then tests the condition the *while* clause specifies to determine whether or not the statements should repeat. Therefore, the statements a *do* loop specifies always execute at least one time. Programs often use the *do* loop to display and process menu options. The following program, *do_menu.c*, uses the *do* statement to display and process menu options until the user selects the Quit option:

```
#include <stdio.h>
#include <conio.h>
#include <ctype.h>
```

```
#include <stdlib.h>

void main(void)
 {
   char letter;

   do
     {
       printf("A Display directory listing\n");
       printf("B Change system time\n");
       printf("C Change system date\n");
       printf("Q Quit\n");
       printf("Choice: ");
       letter = getch();
       letter = toupper(letter);
       if (letter == 'A')
         system("DIR");
       else if (letter == 'B')
         system("TIME");
       else if (letter == 'C')
         system("DATE");
     }
   while (letter != 'Q');
 }
```

126 UNDERSTANDING C'S CONTINUE STATEMENT

As you have learned, the *for*, *while*, and *do* statements let your programs repeat one or more statements until a specific condition evaluates to either true or false. Depending on your program's purpose, there may be times when, based on a second specific condition, you will want your program to skip the current iteration. C's *continue* statement lets you do just that. If C encounters a *continue* statement in a *for* loop, C will immediately execute the loop's increment portion and then perform the ending condition test. If C encounters a *continue* statement in a *while* or *do* loop, then C will immediately perform the ending condition test. To better understand the *continue* statement, consider the following program, *odd_even.c*, which uses *continue* in a *for* and *while* loop type to display the odd and even values between 1 and 100:

```
#include <stdio.h>

void main(void)
 {
   int counter;

   printf("\nEven values\n");
   for (counter = 1; counter <= 100; counter++)
     {
       if (counter % 2)  // Odd
         continue;
       printf("%d ", counter);
     }
   printf("\nOdd values\n");
   counter = 0;
   while (counter <= 100)
     {
       counter++;
       if (! (counter % 2)) // Even
         continue;
       printf("%d ", counter);
     }
 }
```

The program uses the *modulo* (remainder) operator to determine if a value is even or odd. If you divide a value by 2 and get a remainder of 1, the value is odd. Likewise, if you get a remainder of 0, the value is even.

It is important to note that you can normally eliminate the need to use a *continue* statement by redesigning your program's use of *if* and *else* statements. For example, the following program *no_cont.c*, also displays even and odd values without having to use *continue*:

```
#include <stdio.h>

void main(void)
 {
   int counter;

   printf("\nEven values\n");
   for (counter = 1; counter <= 100; counter++)
     {
       if (!(counter % 2))   // Even
         printf("%d ", counter);
     }
   printf("\nOdd values\n");
   counter = 0;
   while (counter <= 100)
     {
       counter++;
       if (counter % 2) // Odd
         printf("%d ", counter);
     }
 }
```

Before you place a *continue* statement in your program, examine your code closely to determine if you can write the same statements without using *continue*. In most cases, you will find the resulting *continue*-less code easier to understand.

ENDING A LOOP USING C'S BREAK STATEMENT 127

As you have learned, the *for*, *while*, and *do* statements let your programs repeat one or more statements until a specific condition evaluates to either true or false. Depending on your program's purpose, there may be times when, based on a second specific condition, you will want the loop to end immediately, with your program continuing its processing at the statement that follows the loop. C's *break* statement lets you do just that. When C encounters a break in a loop, the loop's execution will immediately end. The next statement the program executes is the statement that immediately follows the loop. In the case of a *for* loop, C will not perform the loop's increment section—instead, the loop immediately stops. The following program, *usebreak.c*, illustrates the use of the *break* statement. The program loops through the numbers 1 through 100 and then 100 down to 1. Each time the loop reaches the value 50, the *break* statement immediately ends the loop:

```
#include <stdio.h>

void main(void)
 {
   int counter;

   for (counter = 1; counter <= 100; counter++)
    {
      if (counter == 50)
        break;
      printf("%d ", counter);
    }
   printf("\nNext loop\n");
   for (counter = 100; counter >= 1; counter--)
    {
      if (counter == 50)
        break;
```

```
        printf("%d ", counter);
     }
  }
```

As was the case with C's *continue* statement, you can usually rewrite your program's *if-else* and looping conditions to eliminate the need for the *break* statement in loops. In most cases, when you rewrite your program statements to eliminate *break*, your program will be much easier for the reader to understand. As a rule, limit your use of the *break* statement to C's *switch* statement.

128 BRANCHING WITH THE GOTO STATEMENT

If you have previously programmed in BASIC, FORTRAN, or in assembly language, you may be used to implementing *if-else* operations and loops using the GOTO statement. Like most programming languages, C provides a *goto* statement, which lets your program's execution branch go to a specific location, called a *label.* The *goto* statement's format is as follows:

```
goto label;
label:
```

The following C program, *goto_100.c*, uses the *goto* statement to display the numbers 1 through 100:

```
#include <stdio.h>

void main(void)
 {
   int count = 1;

   label:
     printf("%d ", count++);

     if (count <= 100)
       goto label;
 }
```

When you use the *goto* statement, the label must reside in the current function. In other words, you cannot use *goto* to branch from *main* to a label that appears in another function, or vice versa.

Because programmers have misused the *goto* statement in the past, you should restrict your use of *goto* whenever possible, and instead use constructs such as *if*, *if-else*, and *while*. In most cases, you can use these three constructs to rewrite a code fragment that uses *goto* and therefore produce more readable code.

129 TESTING MULTIPLE CONDITIONS

As you have learned, C's *if-else* statements let you test multiple conditions. For example, consider the following test of the variable *letter*:

```
letter = getch();
letter = toupper(letter);

if (letter == 'A')
   system("DIR");
else if (letter == 'B')
   system("TIME");
else if (letter == 'C')
   system("DATE");
```

In cases where you are testing the same variable for multiple possible values, C provides a *switch* statement, with the following format:

```
switch (expression) {
   case Constant_1: statement;
   case Constant_2: statement;
   case Constant_3: statement;
    :    :          :
 };
```

Instead of using the previous *if-else* statements, you could instead use *switch* as follows:

```
switch (letter) {
   case 'A': system("DIR");
             break;
   case 'B': system("TIME");
             break;
   case 'C': system("DATE");
             break;
  };
```

When C encounters a *switch* statement in your program, C evaluates the expression that follows to produce a result. C then compares the result to each of the constant values that you specify which follow the *case* keyword. If C finds a match, it executes the corresponding statements. The *break* statement separates corresponding statements from one case to another. You will normally place a *break* statement after the last statement that corresponds to an option. In Tip 130 you will learn the details governing the use of the *break* statement in *switch*. The following program, *swt_menu.c*, uses the *switch* statement to process a user's menu selection:

```
#include <stdio.h>
#include <conio.h>
#include <ctype.h>
#include <stdlib.h>

void main(void)
 {
   char letter;

   do {
     printf("A Display directory listing\n");
     printf("B Change system time\n");
     printf("C Change system date\n");
     printf("Q Quit\n");
     printf("Choice: ");
     letter = getch();
     letter = toupper(letter);
     switch (letter) {
       case 'A': system("DIR");
                 break;
       case 'B': system("TIME");
                 break;
       case 'C': system("DATE");
                 break;
     };
    }
   while (letter != 'Q');
 }
```

UNDERSTANDING BREAK WITHIN SWITCH 130

In Tip 129 you learned that C's *switch* statement lets you perform conditional processing. As you learned, you specify one or more possible matching cases using the *switch* statement. For each case, you specify the corresponding statements. At the end of the statements, you normally place a *break* statement to separate one *case* statement from another. If you omit the *break* statement, C will continue to execute all statements that follow, regardless of the case to which the statements belong. For example, consider the following *switch* statement:

```
switch (letter) {
  case 'A': system("DIR");
  case 'B': system("TIME");
  case 'C': system("DATE");
};
```

If the variable *letter* contains the letter A, C will match the first case, executing the DIR command. However, because no *break* statement follows, the program will also execute the TIME and DATE commands. If the variable *letter* contained the letter B, the program would execute the TIME and DATE commands. To prevent the execution of another case's statements, use the *break* statement, as shown here:

```
switch (letter) {
   case 'A': system("DIR");
             break;
   case 'B': system("TIME");
             break;
   case 'C': system("DATE");
             break;
 };
```

There may be times when you may want your programs to cascade through *case* options. For example, the following program, *vowels.c*, uses a *switch* statement to count the number of vowels in the alphabet:

```
#include <stdio.h>

void main(void)
 {
   char letter;
   int vowel_count = 0;

   for (letter = 'A'; letter <= 'Z'; letter++)
     switch (letter) {
       case 'A':
       case 'E':
       case 'I':
       case 'O':
       case 'U': vowel_count++;
     };
   printf("The number of vowels is %d\n", vowel_count);
 }
```

In this case, if the variable *letter* contains the A, E, I, or O, the match occurs and the C falls through to the statement that corresponds to the letter U, which increments the variable *vowel_count*. Because the *switch* statement contains no other cases following the letter U, the program does not include the *break* statement.

131 USING THE SWITCH STATEMENT'S DEFAULT CASE

As you have learned, C's *switch* statement lets you perform conditional processing. When you use the *switch* statement, you specify one or more cases that you want C to match, as shown here:

```
switch (letter) {
   case 'A': system("DIR");
             break;
   case 'B': system("TIME");
             break;
   case 'C': system("DATE");
             break;
 };
```

As you use the *switch* statement, you might find that sometimes you will want C to perform specific statements when the other cases do not match. To do so, you can include a *default* case with the *switch* statement, as shown here:

```
switch (expression) {
   case Constant_1: statement;
   case Constant_2: statement;
   case Constant_3: statement;
    :    :          :
   default: statement;
 };
```

If C does not match any of the case options that precede the default, C will execute the default statements. The following program, *con_vowl.c,* uses the *default* case to track the number of consonant letters in the alphabet:

```
#include <stdio.h>

void main(void)
 {
   char letter;
   int vowel_count = 0;
   int consonant_count = 0;

   for (letter = 'A'; letter <= 'Z'; letter++)
     switch (letter) {
       case 'A':
       case 'E':
       case 'I':
       case 'O':
       case 'U': vowel_count++;
                 break;
       default: consonant_count++;
     };
   printf("The number of vowels is %d\n", vowel_count);
   printf("The number of vowels is %d\n", consonant_count);
 }
```

DEFINING CONSTANTS IN YOUR PROGRAMS 132

As a rule, you can improve your program's readability and portability by replacing references to numbers, such as 512, with a more meaningful constant name. A *constant* is a name the C compiler associates with a value that does not change. To create a constant, you use the *#define* directive. For example, the following directive creates a constant named *LINE_SIZE*, and assigns to the constant the value 128:

```
#define LINE_SIZE 128
```

When the C preprocessor later encounters the *LINE_SIZE* constant name in your program, the preprocessor will replace the constant name with the constant's value. For example, consider the following character string declarations:

```
char line[128];
char text[128];

char current_line[LINE_SIZE];
char user_input[LINE_SIZE];
```

The first two declarations create character strings that contain 128-byte character strings. The second two declarations create character strings that are based on a constant named *LINE_SIZE.* When other programmers read your program code, one of the first questions they might ask is why you used 128 in your string declarations. In the case of the second declaration, however, the programmer knows that you declared all your strings in terms of a predefined *LINE_SIZE.* Within your programs, you might include loops similar to the following:

```
for (i = 0; i < 128; i++)
  // statements

for (i = 0; i < LINE_SIZE; i++)
  // statements
```

The second *for* loop makes your program more readable and easier to change. Assume, for example, that your program uses the value 128 throughout to refer to the string size. Should you later want to change the size to 256 characters, you must change every occurrence of the value 128 in your program—a time-consuming process. On the other hand, if you are using a constant such as *LINE_SIZE*, you only need to change the *#define* directive—a one-step process—as shown here:

```
#define LINE_SIZE 256
```

133 UNDERSTANDING MACRO AND CONSTANT EXPANSION

In Tip 132 you learned that your programs can use the *#define* directive to define a constant within your program. The following program, *sho_mac.c*, for example, uses three constants:

```
#define LINE 128
#define TITLE "Jamsa\'s C/C++/C# Programmer\'s Bible"
#define SECTION "Macros"

void main(void)
 {
   char book[LINE];
   char library_name[LINE];

   printf("This book's title is %s\n", TITLE);
   printf(SECTION);
 }
```

When you compile a C program, a *preprocessor* program runs first. The preprocessor's purpose is to include any specified header files and to expand macros and constants. Before the C compiler actually begins compiling your program, the preprocessor will substitute each constant name with the constant's value, as shown here:

```
void main(void)
 {
   char book[128];
   char library_name[128];

   printf("This book\'s title is %s\n", "Jamsa\'s C/C++/C# Programmer\'s Bible");
   printf("Macros");
 }
```

Because the preprocessor works with *#define*, *#include*, and other # statements, these statements are often called *preprocessor directives*.

134 NAMING CONSTANTS AND MACROS

As you have learned, a constant is a name the C compiler associates with a value that does not change. In Tip 144, you will learn about C macros. When you use constants and macros within your programs, you should use meaningful names that accurately describe their use. To help programmers who read your code differentiate between constants and variables, you should usually use uppercase letters for your constant and macro names. The following *#define* directives illustrate several macro definitions:

```
#define TRUE 1
#define FALSE 0
#define PI 3.1415
#define PROGRAMMER "Kris Jamsa"
```

As you can see, constants can contain *int*, *float*, or even *char* values.

USING THE __FILE__ PREPROCESSOR CONSTANT 135

When you work on a large project, sometimes you may want the preprocessor to know the name of the current source file. For example, you might use the filename within a processor directive that includes a message to the user stating that the program is still under development, as shown here:

```
The program PAYROLL.C is still under development and testing.
This is a BETA release only.
```

To help your programs perform such processing, the C preprocessor defines the *__FILE__* constant as equal to the name of the current source file. The following program, *filecnst.c*, illustrates the use of the *__FILE__* constant:

```
#include <stdio.h>

void main(void)
 {
   printf("The file %s is under Beta testing\n", __FILE__);
 }
```

When you compile and execute the *filecnst.c* program, your screen will display the following:

```
The file filecnst.c is under Beta testing
C:\>
```

Note: *While many preprocessor constants will change from compiler to compiler, the __FILE__ constant is consistent across most compilers.*

USING THE __LINE__ PREPROCESSOR CONSTANT 136

When you work on a large project, sometimes you may want the preprocessor to know, and potentially use, the current line number of the current source file. For example, if you are debugging a program, you might want the compiler to display messages from various points within your program, as shown here:

```
Successfully reached line 10
Successfully reached line 301
Successfully reached line 213
```

The following program, *linecnst.c*, illustrates the use of the *__LINE__* preprocessor constant:

```
#include <stdio.h>

void main(void)
 {
   printf("Successfully reached line %d\n", __LINE__);
   // Other statements here
   printf("Successfully reached line %d\n", __LINE__);
 }
```

When you compile and execute this program, your screen will display the following output:

```
Successfully reached line 5
Successfully reached line 7
C:\>
```

Note: *While many preprocessor constants will change from compiler to compiler, the __LINE__ constant is consistent across most compilers.*

137 CHANGING THE PREPROCESSOR'S LINE COUNT

In Tip 136 you learned how to use the preprocessor's *__LINE__* constant within your programs. When you use the *__LINE__* constant, sometimes you may want to change the preprocessor's current line number. For example, assume you are using *__LINE__* to help debug your program, as discussed in Tip 136. If you have narrowed down the error to a specific set of instructions, you might want the preprocessor to display line numbers relative to a specific location. To help you perform this processing, the C preprocessor provides the *#line* directive that lets you change the current line number. The following directive, for example, directs the preprocessor to set its line number to 100:

```
#line 100
```

You can also use the *#line* directive to change the name of the source code filename which the *__FILE__* constant will display:

```
#line 1 "FILENAME.C"
```

The following program, *chg_line.c*, illustrates how you use the *#line* directive:

```
#include <stdio.h>

void main(void)
 {
   printf("File %s: Successfully reached line %d\n", __FILE__, __LINE__);

   // Other statements here

#line 100 "FILENAME.C"

   printf("File %s: Successfully reached line %d\n", __FILE__, __LINE__);
 }
```

When you compile and execute the *chg_line.c* program, your screen will display the following output:

```
File chg_line.c: Successfully reached line 6
File FILENAME.C: Successfully reached line 102
C:\>
```

138 GENERATING AN UNCONDITIONAL PREPROCESSOR ERROR

As your programs become complex and use a large number of header files, sometimes you may not want the program to successfully compile if the program has not defined one or more constants. Likewise, if you are working with a group of programmers and you want them to be aware of a change you have made to the program, you can use the *#error* preprocessor directive to display an error message and end the compilation. The following directive, for example, ends the compilation, displaying a message to the user about the update:

```
#error The routine string_sort now uses far strings
```

Before the other programmers can successfully compile the program, they must remove the *#error* directive, therefore becoming aware of change.

139 OTHER PREPROCESSOR CONSTANTS

Several Tips in this section presented preprocessor constants most compilers support. Some compilers define many other preprocessor constants. The Microsoft *Visual C++* compiler, for example, uses dozens of reprocessor constants this book does not discuss. Refer to the documentation that accompanied your compiler to determine if your programs can take advantage of other preprocessor constants. Additionally, consult the on-line help documentation, under the *Predefined Macros* heading.

RECORDING THE PREPROCESSOR DATE AND TIME 140

As you work on large programs, you may want your preprocessor to work with the current date and time. For example, you might want the program to display a message that states the date and time that you last compiled the program, as shown here:

```
Beta Testing: PAYROLL.C Last compiled Dec 31 2001 12:00:00
```

To help you perform such processing, the C preprocessor assigns the constants *__DATE__* and *__TIME__* to the current date and time. The following program, *datetime.c*, illustrates how you might use the *__DATE__* and *__TIME__* constants:

```
#include <stdio.h>

void main(void)
 {
   printf("Beta Testing: Last compiled %s %s\n", __DATE__, __TIME__);
 }
```

TESTING FOR ANSI C COMPLIANCE 141

Although most C compilers are very similar, every compiler provides unique capabilities. To help you write programs that you can easily move from one system to another, the American National Standards Institute (ANSI) defines standards for operators, constructs, statements, and functions a compiler should support. Compilers that comply with these standards are called *ANSI C compilers*. As you build programs, sometimes you may want to determine whether or not you are using an ANSI compiler. To help you do so, ANSI C compilers define the constant *__STDC__* (for STandarD C). If the compiler is compiling for ANSI C compliance, the compiler will define the constant. Otherwise, the compiler does not define the constant. The following program, *chk_ansi.c*, uses the *__STDC__* constant to determine if the current compiler complies to the ANSI standards:

```
#include <stdio.h>

void main(void)
 {
   #ifdef __STDC__
     printf("ANSI C compliance\n");
   #else
     printf("Not in ANSI C mode\n");
   #endif
 }
```

Note: *Most compilers provide command-line switches or inline pragmas that direct it to use ANSI compliance. You will learn about both command-line switches and inline pragmas later in this book.*

TESTING FOR C++ VERSUS C 142

You can use several of the Tips this book presents for both C programming and C++ programming, while other Tips only apply to C++. As you create your own programs, there may be times when you want the preprocessor to determine if you are using C or C++ and process your statements accordingly. To help you perform such testing, many C++ compilers define the *__cplusplus* constant. If you use a standard C compiler, the constant will be undefined. The following program, *chk_cpp.c*, uses the *__cplusplus* constant to determine the compiler's current mode:

```
#include <stdio.h>

void main(void)
 {
   #ifdef __cplusplus
```

```
    printf("Using C++\n");
  #else
    printf("Using C\n");
  #endif
}
```

If you examine the header files the compiler provides, you will find many uses of the *__cplusplus* constant.

***Note:** Many C++ compilers provide command-line switches that direct them to compile using C++, as opposed to standard C.*

143 Undefining a Macro or Constant

Several Tips this section presented discussed constants and macros the preprocessor defines or a header file contains. Depending on your program, you may want the preprocessor to remove the definition of one or more of these constants from your programs. Alternately, you may want the preprocessor to redefine one or more of these constants. For example, the following macro redefines the macro *_toupper*, which is defined in the header file *ctype.h*:

```
#define _toupper(c) ((((c) >= 'a')&&((c)<='z')) ? (c)-'a' + 'A': c)
```

When you compile this program, many preprocessors will display a warning message stating that you have redefined the macro. To avoid display of this warning message, use the *#undef* directive to remove the macro's current definition before you redefine the macro, as shown here:

```
#undef _toupper
#define _toupper(c) ((((c)>='a')&&((c)<='z')) ? (c) - 'a' + 'A': c)
```

144 Comparing Macros and Functions

New C programmers often become confused as to when they should use macros or functions because of the similarities between the two. As you have learned, each time the preprocessor encounters a macro reference within your program, the preprocessor replaces the reference with the macro statements. Therefore, if your program uses a macro 15 times, the program will have 15 different copies of the macro placed in its statements. As a result, the executable program's size will grow. When your program uses a function, on the other hand, the program only contains one copy of code, which reduces the program's size. When the program uses the function, the program calls (branches to) the function's code. The disadvantage of using functions, however, is that each function call incurs additional processing that makes the function call take slightly longer to execute than a comparable macro. Therefore, if you want fast performance, use a macro. However, if program size concerns you more, use a function.

145 Understanding Compiler Pragmas

Several Tips in this section presented different preprocessor directives such as *#define*, *#include*, and *#undef*. Depending on your compiler, your preprocessor may support various compiler directives, called *pragmas*. The format of a pragma is as follows:

```
#pragma compiler_directive
```

For example, the *Turbo C++ Lite* compiler provides the *startup* and *exit* pragmas, which let you specify the functions that you want your program to automatically execute when the program starts or ends:

```
#pragma startup load_data
#pragma exit close_all_files
```

Note that the function you name within the *startup* pragma will actually execute *before main*, so you should not use the *startup* pragma too often. Depending on your compiler, the available pragmas will differ. Refer to the documentation that accompanies your compiler for a complete description of the pragmas available for your program use.

Note: *When you use the* ***startup*** *and* ***exit*** *pragmas, the function you name within the pragma must take no parameters and return no value; in other words, you must write the function in the following manner:*

```
void function(void)
```

LEARNING ABOUT PREDEFINED VALUES AND MACROS 146

Several Tips in this section discussed macros, constants, and various preprocessor directives. One of the most effective ways of learning how to better use macros, constants, and other preprocessor directives is to thoroughly examine how the C compiler uses these options. The C compiler places macros and constants within header files that reside in the compiler's *include* subdirectory. Many of the header files present ways to use various preprocessor directives. You should examine the contents of various header files to help you learn many ways to improve your programs by taking advantage of these preprocessor capabilities.

CREATING YOUR OWN HEADER FILES 147

As you know, the C compiler provides different header files that contain related macros, constants, and function prototypes. As the number of programs you create increases, you may find that many of your programs use the same constants and macros. Rather than repeatedly typing these macros and constants into your programs, you might consider creating your own header file and placing into the file the corresponding macros and constants. Assuming that you create a header file named *my_defs.h,* you can include the file at the start of your programs using the *#include* preprocessor directive, as shown here:

```
#include "my_defs.h"
```

When you include your macros and constants in a header file in this way, you can quickly change several programs by editing the header file and then recompiling the programs that include the file.

USING #INCLUDE <FILENAME.H> OR #INCLUDE "FILENAME.H" 148

All the programs presented throughout this book have included the header file *stdio.h,* as shown here:

```
#include <stdio.h>
```

In Tip 147 you learned how to create and include your own header file, *my_defs.h*. You can include *stdio.h* and *my_defs.h* within your programs with the following statements:

```
#include <stdio.h>
#include "my_defs.h"
```

As you examine the two *include* statements, note that left and right brackets <> enclose the header file *stdio.h,* while double quotes enclose *my_defs.h.* When you enclose a header filename within the left and right brackets, the C compiler will first search its header file directory for the specified file. If the compiler locates the file, the preprocessor will use the file. If the compiler does not find the file, the compiler will search the current directory or a directory you specify. When you enclose a header filename in double quotes, on the other hand, the compiler will only search the current directory for the file.

149 TESTING WHETHER A SYMBOL IS DEFINED

Several Tips in this section presented predefined C compiler symbols. In addition, some Tips have discussed how you can define your own constants and macros. Depending on your program, there may be times when you want the preprocessor to test whether the program has previously defined a symbol and, if so, process a specific set of statements. To help your program test whether or not the program has previously defined a symbol, the C preprocessor supports the *#ifdef* directive. The format of the *#ifdef* directive is as follows:

```
#ifdef symbol
  // statements
#endif
```

When the preprocessor encounters the *#ifdef* directive, the preprocessor tests whether or not the program has previously defined the specified symbol. If so, the preprocessor processes the statements that follow the directive up to the *#endif* statement. There may be times when you want the preprocessor to process statements if the program has not defined a symbol. In such cases, you can use the *#ifndef* directive. The following statements use *#ifndef* to direct the preprocessor to define the *_toupper* macro if a similar macro is not defined:

```
#ifndef _toupper
  #define _toupper(c) ((((c)>='a')&&((c)<='z')) ? (c) - 'a' + 'A': c)
#endif
```

150 PERFORMING IF-ELSE PREPROCESSING

In Tip 149, you learned how to use the *#ifdef*, *#ifndef*, and *#endif* statements to specify a set of statements you want the preprocessor to perform if a program has previously defined (*#ifdef*) or not defined (*#ifndef*) a symbol. There may be times when you want to take this processing one step further and include a set of statements you want the preprocessor to perform when the condition tested within the *#ifdef* statement is true, and a different set if the condition is false. To perform such processing, you can use the *#else* directive, shown here:

```
#ifdef symbol
  // Statements
#else
  // Other statements
#endif
```

For example, some compilers include unique preprocessor constants that indicate which compiler and version you are using to compile the program. You can use each compiler's unique constants to respond to different compilers. For example, the following code fragment will print *Microsoft* if the compiler is *Visual C++* and *Borland* if the compiler is Borland *C++*:

```
#ifdef _MSC_VER
  printf ("Microsoft");
#endif
#ifdef __BORLANDC__
  printf ("Borland");
#endif
```

151 PERFORMING MORE POWERFUL PREPROCESSOR CONDITION TESTING

In Tip 149 you learned how to use the *#ifdef* and *#ifndef* statements to direct the preprocessor to test whether a program has previously defined or not defined a symbol and then to process the statements that follow based on the test's result. In some cases, you may need the processor to test if several symbols are defined, not defined, or some combination of both. The following directives first test to see if the symbol *MY_LIBRARY* is defined. If your program has previously defined *MY_LIBRARY*, the preprocessor directives test whether your program has also previously defined the symbol *MY_FUNCTIONS*. If your program has not already defined *MY_FUNCTIONS*, the code directs the preprocessor to include the header file *my_code.h:*

```
#ifdef MY_LIBRARY
#ifndef MY_ROUTINES
#include "my_code.h"
#endif
#endif
```

Although the directives perform the desired processing, the nested conditionals make it potentially difficult for another programmer to follow your intent. As an alternative, your programs can use the *#if* directive with the *defined* operator to test if the program has previously defined the symbol, as shown here:

```
#if defined(symbol)
  // Statements
#endif
```

Your advantage in using the *#if* directive is that you can combine testing, unlike with the *#ifdef* or *#ifndef* directives. The following directive performs the same testing as the first example in this Tip:

```
#if defined(MY_LIBRARY) && !defined(MY_ROUTINES)
#include "my_code.h"
#endif
```

You can use *#if defined* to build conditions that use C's logical operators (including &&, ||, and !).

PERFORMING IF-ELSE AND ELSE-IF PREPROCESSING 152

In Tip 151 you learned how to use the *#if* preprocessor directive to test whether or not your program has previously defined a symbol. When you use the *#if* directive, there may be times when you want the preprocessor to process one set of statements when a symbol is defined and another set if the symbol is undefined (*conditional preprocessing*). You can perform conditional preprocessing using the *#else* directive:

```
#if defined(symbol)
  // Statements
#else
  // Statements
#endif
```

Taking the previous preprocessing example one step further, there may be times when you will want the preprocessor to test the status of other symbols when a specified condition fails. The following directives, for example, instruct the preprocessor to process one set of statements if the symbol *MY_LIBRARY* is defined; another set if *MY_LIBRARY* is not defined but *MY_ROUTINES* is defined; and a third set if neither symbol is defined:

```
#if defined(MY_LIBRARY)
  // Statements
#else if defined (MY_ROUTINES)
  // Statements
#else
  // Statements
#endif
```

As you can see, when you use *#if* and *#else* directives you significantly increase your control over the preprocessor.

Note: *Some compilers, including **Turbo C++ Lite**, support the **#elif** preprocessor directive, which performs the same processing as the **#else if** construction.*

153 DEFINING MACROS AND CONSTANTS THAT REQUIRE MULTIPLE LINES

Several Tips presented throughout this section have defined constants and macros. As your constants and macros become more complex, sometimes a definition will not fit on one line. When you must wrap a constant or macro definition to the next line, place a backslash character at the end of the line, as shown here:

```
#define very_long_character_string "This extremely long string constant\
requires two lines"

#define _toupper(c) ((((c) >= 'a') && ((c) <= 'z'))\
 ? (c) - 'a' + 'A': c)
```

154 CREATING YOUR OWN MACROS

As you have learned, macros provide a way for you to define constants that the preprocessor substitutes throughout your program before compilation begins. In addition, macros let you create function-like operations that work with *parameters*. Parameters are values that you pass to the macro. For example, the following macro, *SUM*, returns the sum of the two values that you pass to the macro:

```
#define SUM(x, y) ((x) + (y))
```

The following program, *show_sum.c*, uses the *SUM* macro to add several values:

```
#include <stdio.h>
#define SUM(x, y) ((x) + (y))

void main(void)
 {
   printf("Adding 3 + 5 = %d\n", SUM(3, 5));
   printf("Adding 3.4 + 3.1 = %f\n", SUM(3.4, 3.1));
   printf("Adding -100 + 1000 = %d\n", SUM(-100, 1000));
 }
```

Within the *SUM* macro's definition, the *x* and *y* represent macro parameters. When you pass two values to the macro, such as *SUM(3, 5)* the preprocessor substitutes the parameters into the macro, as shown in Figure 154.

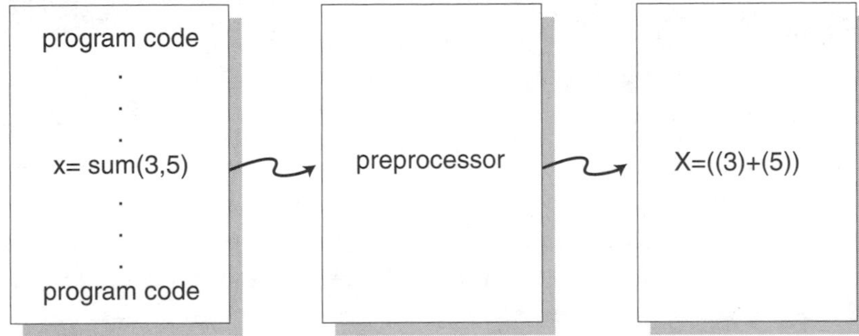

***Figure 154** The parameter substitution for SUM.*

In the program *show_sum.c*, the preprocessor's substitutions will result in the following code:

```
printf("Adding 3 + 5 = %d\n", ((3) + (5)));
printf("Adding 3.4 + 3.1 = %f\n", ((3.4) +  (3.1)));
printf("Adding -100 + 1000 = %d\n", ((-100) + (1000)));
```

DO NOT PLACE SEMICOLONS IN MACRO DEFINITIONS 155

When you examine the macro definition of the following *SUM* macro, note that the macro does not include a semicolon:

```
#define SUM(x, y) ((x) + (y))
```

If you include a semicolon within your macro, the preprocessor will place the semicolon at each macro occurrence throughout your program. Assume, for example, you placed a semicolon at the end of the *SUM* macro definition, as shown here:

```
#define SUM(x, y) ((x) + (y));
```

When the preprocessor expands the macro, the preprocessor will include the semicolon, as shown here:

```
   printf("Adding 3 + 5 = %d\n", ((3) + (5)););
   printf("Adding 3.4 + 3.1 = %f\n", ((3.4) +  (3.1)););
   printf("Adding -100 + 1000 = %d\n", ((-100) + (1000)););
```

Because the semicolon now occurs in the middle of the *printf* statement (indicating the statement's end), the compiler will generate errors.

***Note:** Unless you want the preprocessor to include a semicolon in the macro expansion, do not include a semicolon in the macro definition.*

CREATING MIN AND MAX MACROS 156

In Tip 154 you created the *SUM* macro, which added two values together. The following macros, *MIN* and *MAX*, return the minimum and maximum of two values:

```
#define MIN(x, y) (((x) < (y)) ? (x): (y))
#define MAX(x, y) (((x) > (y)) ? (x): (y))
```

The following program, *min_max.c*, illustrates how to use *MIN* and *MAX* macros:

```
#include <stdio.h>

#define MIN(x, y) (((x) < (y)) ? (x): (y))
#define MAX(x, y) (((x) > (y)) ? (x): (y))
void main(void)
 {
   printf("Maximum of 10.0 and 25.0 is %f\n", MAX(10.0, 25.0));
   printf("Minimum of 3.4 and 3.1 is %f\n", MIN(3.4, 3.1));
 }
```

When you execute the *min_max.c* program, the preprocessor's substitutions result in the following code:

```
printf("Maximum of 10.0 and 25.0 is %d\n", (((10.0) < (25.0)) ? (10.0): (25.0)));
printf("Minimum of 3.4 and 3.1 is %f\n", (((3.4) > (3.1)) ? (3.4): (3.1)));
```

CREATING SQUARE AND CUBE MACROS 157

As you have learned, C lets you define and pass values to macros. The last two macros you will examine in this section are the *SQUARE* and *CUBE* macros, which return, respectively, a value squared (x * x) and a value cubed (x * x * x) results:

```
#define SQUARE(x) ((x) * (x))
#define CUBE(x) ((x) * (x) * (x))
```

The following program, *sqr_cube.c*, illustrates how to use the *SQUARE* and *CUBE* macros:

```
#include <stdio.h>
#define SQUARE(x) ((x) * (x))
#define CUBE(x) ((x) * (x) * (x))

void main(void)
 {
   printf("The square of 2 is %d\n", SQUARE(2));
   printf("The cube of 100 is %f\n", CUBE(100.0));
 }
```

In the *sqr_cube.c* program, the preprocessor's substitutions result in the following code:

```
   printf("The square of 2 is %d\n", ((2) * (2)));
   printf("The cube of 100 is %f\n", ((100.0) * (100.0) * (100.0)));
```

Note: *To avoid overflow, the **sqr_cube.c** program uses the floating-point value 100.0 within the **CUBE** macro.*

158 BE AWARE OF SPACES IN MACRO DEFINITIONS

Several previous Tips have presented macros that support parameters. When you define macros that support parameters, you must be careful of white space in the macro definition. Do not place a space between the macro name and its parameters. For example, consider the following macro definition, *SQUARE*:

```
#define SQUARE (x) ((x) * (x))
```

When the preprocessor examines your program, the spaces between the macro name cause the preprocessor to assume that it should replace each occurrence of the name *SQUARE* with *(x) ((x) * (x))*, rather than with *((x) * x))*. As a result the macro will not evaluate correctly, and in most cases, the compiler will generate syntax error messages or warnings because of the substitution. To understand the preprocessor's macro substitution process better, experiment with the program *sqr_cube.c* (presented in Tip 157), by placing a space after each macro name.

159 UNDERSTANDING HOW TO USE PARENTHESES

Several of the previous Tips have presented macros to which your programs will pass values (parameters). If you take a close look at each macro's definitions, you will find that parentheses enclose the values:

```
#define SUM(x, y) ((x) + (y))
#define SQUARE(x) ((x) * (x))
#define CUBE(x) ((x) * (x) * (x))
#define MIN(x, y) (((x) < (y)) ? (x): (y))
#define MAX(x, y) (((x) > (y)) ? (x): (y))
```

Macro definitions enclose parameters within parentheses in order to support expressions. As an example, consider the following statement:

```
result = SQUARE(3 + 5);
```

The statement should assign the variable *result* the value 64 (8 * 8). Assume, for example, that you define the *SQUARE* macro as follows:

```
#define SQUARE(x) (x * x)
```

When the preprocessor substitutes the expression 3 + 5 for *x*, the substitution becomes the following:

```
result = (3 + 5 *  3 + 5);
```

Recall C's operator precedence and note that multiplication has higher precedence than does addition. Therefore, the program would calculate the expression as follows:

```
result = (3 + 5 * 3 + 5);
         = (3 + 15 + 5);
         = 23;
```

When you enclose each parameter within parentheses, however, you ensure that the preprocessor evaluates the expression correctly:

```
result = SQUARE(3 + 5);
         = ((3 + 5) * (3 + 5));
         = ((8) * (8));
         = (64);
         = 64;
```

***Note:** As a rule, you should always enclose your macro parameters within parentheses.*

MACROS ARE TYPELESS 160

In the Functions section of this book, you will learn how to create functions that perform specific operations. You will learn that C lets you pass parameters to your functions, just as you have passed values to macros. If your function performs an operation and returns a result, you must specify the result's type (such as *int, float*, and so on). For example, the following function, *add_values*, adds two integer values and returns a result of type *int:*

```
int add_values(int x, int y)
 {
   return(x + y);
 }
```

Within your program you can only use the *add_values* function to add two values of type *int*. If you try to add two floating-point values, an error will occur. As you have seen, macros let you work with values of any type. The *SUM* macro that you previously created, for example, supported values of both type *int* and type *float*:

```
printf("Adding 3 + 5 = %d\n", SUM(3, 5));
printf("Adding 3.4 + 3.1 = %f\n", SUM(3.4, 3.1));
```

When you use macros for simple arithmetic operations, you eliminate the need to duplicate functions simply because you want to work with values of different types. However, as you will learn in this book's Functions section, there are other tradeoffs to consider when deciding whether you are going to use macros or functions.

VISUALIZING A C STRING 161

Your computer requires a byte of memory to store a single ASCII character. As you have learned, a *string* is a sequence of ASCII characters. When you declare a string constant, C automatically assigns the *NULL* character. When your programs create their own strings by reading characters from the keyboard, your programs must assign the *NULL* character to the end of the string to indicate the string's end. Therefore, the best way for you to visualize a string is as a collection of bytes terminated by a *NULL* character, as shown in Figure 161.

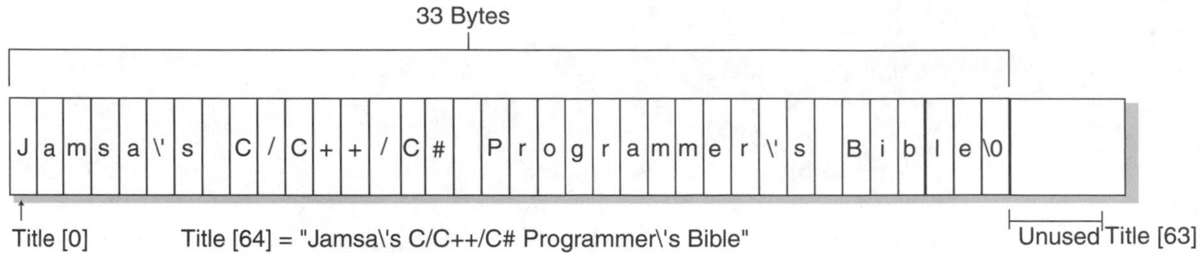

Figure 161 C stores strings in consecutive byte locations in memory.

When functions work with strings, the function usually only receives the location at which the string starts. After the function knows the string's start location, the function can traverse successive memory locations until the function encounters the *NULL* character (which indicates the string's end).

162 HOW THE COMPILER REPRESENTS A CHARACTER STRING

Several Tips presented throughout this book use character string constants enclosed by double quotes, as in the following example:

```
"Jamsa\'s C/C++/C# Programmer\'s Bible"
```

When you use a character string constant within your program, the C compiler automatically assigns the *NULL* character (\0) at the end of the string. Given the previous string constant, the C compiler will actually store the constant in memory, as shown in Figure 162.

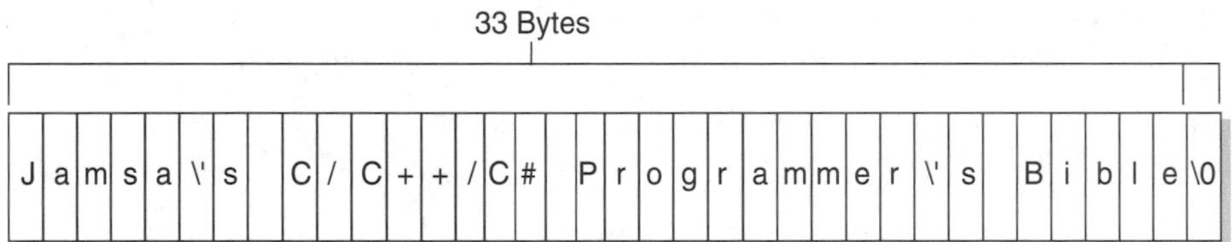

Figure 162 C automatically appends the NULL character to string constants.

163 HOW C STORES CHARACTER STRINGS

Many of the Tips this book presents make extensive use of character strings. For example, some programs use strings to read files and keyboard input and to perform other operations. In C, a character string is a *NULL*- or zero-terminated array of characters. To create a character string, you simply declare a character array, as shown here:

```
char string[256];
```

C will create a string capable of storing 256 characters, which C indexes from *string*[0] through *string*[255]. Because the string may contain less than 256 characters, C uses the *NULL* (ASCII 0) character to represent the string's last character. C does not typically place the *NULL* character after the last character in the string. Instead, functions such as *fgets* or *gets* place the *NULL* character at the string's end. As your programs manipulate strings, it is your responsibility to ensure that the *NULL* character is present. The following program, *buildabc.c*, defines a 256-character string and then assigns the uppercase letters of the alphabet to the first twenty-six digits of the string:

```
#include <stdio.h>

void main(void)
 {
   char string[256];
   int i;

   for (i = 0; i < 26; i++)
     string[i] = 'A' + i;
   string[i] = NULL;
   printf ("The string contains %s\n", string);
 }
```

The *buildabc.c* program uses the *for* loop to assign the letters A through Z to the string. The program then places the *NULL* character after the letter Z to indicate the string's end. The *printf* function will then display each character in the string up to the *NULL* character. The C functions that work with strings use the *NULL* character to determine the string's end. The following program, *a_thru_j.c,* also assigns the letters A through Z to a character string. However, the program then assigns the *NULL* character to *string*[10], which is the location that immediately follows the letter J. When *printf* displays the string's contents, it will stop at the letter J:

```
#include <stdio.h>

void main(void)
 {
   char string[256];
   int i;

   for (i = 0; i < 26; i++)
     string[i] = 'A' + i;
   string[10] = NULL;
   printf ("The string contains %s\n", string);
 }
```

Note: *As you work with strings, you must make sure you correctly include the **NULL** character to represent the end of the string.*

LEARNING HOW 'A' DIFFERS FROM "A" 164

As you learned in Tip 161, a character string is a sequence of zero or more ASCII characters that C typically terminates with *NULL* (an ASCII 0). When you work with characters within C, you can use a character's numeric ASCII value or you can place the character within single quotes, such as 'A'. On the other hand, when you use double quotes, such as "A," C creates a character string that contains the specified letter (or letters) and terminates the string with the *NULL* character. Figure 164 illustrates how C stores the constants 'A' and "A".

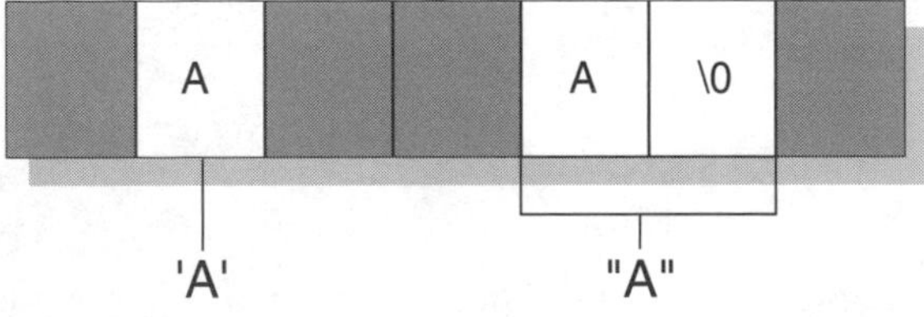

Figure 164 *How C stores the constants 'A' and "A".*

Because C stores them differently, character and string constants are not the same and you should be sure to treat the two sets of constants differently within your programs.

165 REPRESENTING A QUOTE WITHIN A STRING CONSTANT

As you have learned, to create a string constant your program must place the desired characters within double quotes:

```
"This is a string constant"
```

Depending on your programs, there may be times when a string constant will contain a double quote character. For example, assume you must represent the following string:

```
"Stop!" he said.
```

Because C uses the double quotes to define the string constants, you must have a way to tell the compiler that you want to include a quote within the string. To do so, use the escape sequence \", as shown here:

```
"\"Stop!\" he said."
```

The following program, *quotes.c,*, uses the \" escape sequence to place quotes within a string constant:

```
#include <stdio.h>

void main(void)
 {
   char string[] = "\"Stop!\" he said.";

   printf(string);
 }
```

166 DETERMINING THE LENGTH OF A STRING

In Tip 163 you learned that C functions usually use the *NULL* character to represent the end of a string. Functions such as *fgets* and *cgets* automatically assign the *NULL* character to indicate the end of a string. The following program, *show_str.c*, uses the *gets* function to read a character string from the keyboard. The program then uses a *for* loop to display the string's characters one at a time until the conditional *string[i] != NULL* evaluates to false:

```
#include <stdio.h>

void main(void)
 {
   char string[256];  // String input by user
   int i;             // Index into the string

   printf("Type a string of characters and press Enter:\n");
   gets(string);
   // Display each string character until NULL is found
   for (i = 0; string[i] != NULL; i++)
     putchar(string[i]);
   printf("\nThe number of characters in the string is %d\n", i);
 }
```

167 USING THE STRLEN FUNCTION

As you work with strings within your programs, you will perform many operations based on the number of characters in the string. To help you determine the number of characters in a string, most C compilers provide a *strlen* function, which returns the number of characters in a string. The format of the *strlen* function is as follows:

```
#include <string.h>

size_t strlen(const char string);
```

The following program, *strlen.c*, illustrates how to use the *strlen* function:

```
#include <stdio.h>
#include <string.h>

void main(void)
 {
   char book_title[] = "Jamsa\'s C/C++/C# Programmer\'s Bible";

   printf("%s contains %d characters\n", book_title, strlen(book_title));
 }
```

When you compile and execute the *strlen.c* program, your screen will display the following:

```
Jamsa's C/C++/C# Programmer's Bible contains 35 characters
C:\>
```

To better understand how the *strlen* function works, consider the following implementation. The function simply counts the characters in a string up to, but not including, the *NULL* character:

```
size_t strlen(const char string)
 {
    int i = 0;
    while (string[i])
        i++;
    return(i);
 }
```

168 COPYING ONE STRING'S CHARACTERS TO ANOTHER STRING

As your programs work with strings, there may be times when you must copy the contents of one character string to another string. To help you perform character string operations, most C compilers provide a *strcpy* function, which copies the characters in one string (the *source* parameter) to another string (the *destination* parameter):

```
#include <string.h>

char *strcpy(char *destination, const char *source);
```

The *strcpy* function returns a pointer to the beginning of the destination string. The following program, *strcpy.c*, illustrates how you will use the *strcpy* function within your programs:

```
#include <stdio.h>
#include <string.h>

void main(void)
 {
   char title[] = "Jamsa\'s C/C++/C# Programmer\'s Bible";
   char book[128];

   strcpy(book, title);
   printf("Book name %s\n", book);
 }
```

To better understand how the *strcpy* function works, consider the following implementation:

```
char *strcpy(char *destination, const char *source)
  {
    while (*destination++ = *source++)
        ;
    return(destination-1);
  }
```

The *strcpy* function simply copies letters from the source string to the destination, up to and including the *NULL* character.

169 APPENDING ONE STRING'S CONTENTS TO ANOTHER STRING

As your programs work with strings, there may be times when you must append one string's contents to another string. For example, if one string contains a subdirectory name and another contains a filename, you might append the filename to the subdirectory to create a complete pathname. C programmers often refer to the process of appending one string to another as *concatenating* strings. To help you append one string to another, most C compilers provide a function named *strcat*, which concatenates (appends) a source string to a target string, as shown here:

```
#include <string.h>

char *strcat (char target, const char *source);
```

The following program, *strcat.c*, illustrates how to use the *strcat* function:

```
#include <stdio.h>
#include <string.h>

void main(void)
  {
    char name[64] = "Triggerhill\'s I\'m so";
    strcat(name, " Happy");
    printf("Happy\'s full name is %s\n", name);
  }
```

When you compile and execute the *strcat.c* program, your screen will display the following:

```
Happy's full name is Triggerhill's I'm so Happy
C:\>
```

To better understand the *strcat* function, consider the following implementation:

```
char *strcat(char *target, const char *source)
 {
   char *original = target;

   while (*target)
     target++;      // Find the end of the string
   while (*target++ = *source++)
     ;
   return (original);
 }
```

As you can see, the *strcat* function loops through the destination string's characters until the function finds the *NULL* character. The *strcat* function then appends each character in the source string, up to and including the *NULL* character, to the destination string.

170 APPENDING N CHARACTERS TO A STRING

In Tip 169 you learned that the *strcat* function lets you append (concatenate) the characters in one string to another. In some cases, you will not want to append all the characters in a string, but rather only the first two, three, or *n* characters in the string. To help you append *n* characters to a string, most C compilers provide a function named *strncat*, which appends the first *n* characters of a source string to a destination string, as shown here:

```
#include <stding.h>

char *strncat(char *destination, const *source, size_t n);
```

If *n* specifies a number of characters greater than the number of characters in the source string, *strncat* will copy characters up to the end of the string and no more. The following program, *strncat.c*, illustrates how to use the *strncat* function:

```
#include <stdio.h>
#include <string.h>

void main(void)
 {
   char name[64] = "Bill";

   strncat(name, " and Hillary", 4);
   printf("Hasta la vista %s?\n", name);
 }
```

When you compile and execute the *strncat.c* program, your screen will display the following:

```
Hasta la vista Bill and?
C:\>
```

To help you better understand the *strncat* function, consider the following implementation:

```
char *strncat(char *destination, const char *source, int n)
 {
   char *original = destination;
   int i = 0;

   while (*destination)
     destination++;
   while ((i++ < n) && (*destination++ = *source++))
     ;
   if (i > n)
     *destination = NULL;
   return(original);
 }
```

TRANSFORMING ONE STRING TO ANOTHER STRING 171

Several of the Tips in this book have shown you ways to copy the contents of one string to another. The *strxfrm* function copies the contents of one string to another (up to the number of characters you specify within the *n* parameter) and then returns the length of the resultant string:

```
#include <string.h>

size_t strxfrm(char *target, char *source, size_t n);
```

The *target* parameter is a pointer to which the *strxfrm* function copies the source string. The *n* parameter specifies the maximum number of characters to copy. The following program, *strxfrm.c*, illustrates how to use the *strxfrm* function:

```
#include <stdio.h>
#include <string.h>

void main(void)
 {
   char buffer[64] = "Jamsa\'s C/C++/C# Programmer\'s Bible";
   char target[64];
   int length;

   length = strxfrm(target, buffer, sizeof(buffer));
   printf("Length %d Target %s Buffer %s\n", length, target, buffer);
 }
```

DO NOT OVERWRITE A STRING'S BOUNDS 172

Several of the Tips in this section have presented functions that copy or append characters from one string to another. When you perform character string operations, you must ensure that you do not overwrite a string's memory locations. As an example of problems with overwriting a string's bounds, consider the following declaration, which creates a character string capable of storing 10 characters:

```
char string[10];
```

If you assign more than 10 characters to the string, your operating system may not detect the error. Rather, the characters you intended to assign to the string may overwrite the memory locations that correspond to other variables. Not only is correcting an overwrite error very difficult, but an overwrite error might also cause both your program and the operating system to cease execution. As a rule, declare your strings slightly larger than you think you will need. Doing so lets you reduce the likelihood of overwriting a string. If your programs experience intermittent errors, examine your program code to determine if your program may be overwriting a character string.

173 DETERMINING WHETHER TWO STRINGS ARE THE SAME

As your programs work with strings, you will often compare two strings to determine if the strings are the same. To help you determine if two strings contain the same characters, you can use the *streql* function, shown here:

```
int streql(char *str1, char *str2)
 {
   while ((*str1 == *str2) && (*str1))
     {
       str1++;
       str2++;
     }
   return((*str1 == NULL) && (*str2 == NULL));
 }
```

The *streql* function will return the value 1 if the two strings are equal, and 0 if the strings are not equal. The following C program, *streql.c,* illustrates how to use the *streql* function:

```
#include <stdio.h>

void main(void)
 {
   printf("Testing Abc and Abc %d\n", streql("Abc", "Abc"));
   printf("Testing abc and Abc %d\n", streql("abc", "Abc"));
   printf("Testing abcd and abc %d\n", streql("abcd", "abc"));
 }
```

When you compile and execute the *streql.c* program, your screen will display the following output:

```
Testing Abc and Abc 1
Testing abc and Abc 0
Testing abcd and abc 0
C:\>
```

174 IGNORING CASE WHEN DETERMINING WHETHER STRINGS ARE EQUAL

In Tip 173 you created the function *streql,* which lets your programs quickly determine whether two strings are equal. When the *streql* function compares two strings, *streql* considers upper- and lowercase characters as distinct. There may be times when you will want to compare two strings without considering case. To compare strings without considering case, you can create the function *strieql,* as shown here:

```
#include <ctype.h>

int strieql(char *str1, char *str2)
 {
```

```
   while ((toupper(*str1) == toupper(*str2)) && (*str1))
     {
       str1++;
       str2++;
     }
   return((*str1 == NULL) && (*str2 == NULL));
 }
```

As you can see, the *strieql* function converts each character in each string to uppercase before comparing the two strings. The following program, *strieql.c*, illustrates how to use *strieql*:

```
#include <stdio.h>
#include <ctype.h>

void main(void)
 {
   printf("Testing Abc and Abc %d\n", strieql("Abc", "Abc"));
   printf("Testing abc and Abc %d\n", strieql("abc", "Abc"));
   printf("Testing abcd and abc %d\n", strieql("abcd", "abc"));
 }
```

When you compile and execute the *strieql.c* program, your screen will display the following:

```
Testing Abc and Abc 1
Testing abc and Abc 1
Testing abcd and abc 0
C:\>
```

CONVERTING A CHARACTER STRING TO UPPER- OR LOWERCASE 175

When your programs work with strings, there may be times when you may want to convert the string to uppercase. For example, when a user types in a filename or customer name, you may want the program to convert the entered string to uppercase to simplify string compare operations or to ensure that the program stores data in a consistent format. To help you perform these conversions, most C compilers provide the functions *strlwr* and *strupr*, as shown here:

```
#include <string.h>

char *strlwr(char *string);
char *strupr(char *string);
```

The following program, *strcase.c*, illustrates how to use the *strlwr* and *strupr* functions:

```
#include <stdio.h>
#include <string.h>

void main(void)
 {
   printf(strlwr("Jamsa\'s C/C++/C# Programmer\'s Bible!\n"));
   printf(strupr("Jamsa\'s C/C++/C# Programmer\'s Bible!\n"));
 }
```

To help you better understand these two functions, consider the following implementation of *strlwr*:

```
#include <ctype.h>

char *strlwr(char *string)
 {
   char *original = string;

   while (*string)
     {
       *string = tolower(*string);
       string++;
     }
```

```
   return(original);
 }
```

As you can see, both *strlwr* and *strupr* loop through the characters in a string, converting each character to either upper- or lowercase, depending upon the invoked function.

176 Obtaining the First Occurrence of a Character in a String

As your programs work with strings, there may be times when you will want to find the first (leftmost) occurrence of a specific character within a string. For example, if you are working with a string that contains a pathname, you might search the string for the first backslash (\) character. To help you search for the first occurrence of a string, most compilers provide a function named *strchr*, which returns a pointer to the first occurrence of a specific character within a string, as shown here:

```
#include <string.h>

char *strchr(const char *string, int character);
```

If *strchr* does not find the specified character within the string, *strchr* returns a pointer to the *NULL* character that marks the end of the string. The following program, *strchr.c*, illustrates how to use the *strchr* function:

```
#include <stdio.h>
#include <string.h>

void main(void)
 {
   char title[64] = "Jamsa\'s C/C++/C# Programmer\'s Bible!";
   char *ptr;

   ptr = strchr(title, 'C');
   if (*ptr)
     printf("First occurrence of C is at offset %d\n", ptr - title);
   else
     printf("Character not found\n");
 }
```

When you compile and execute the *strchr.c* program, your screen will display the following:

```
The first occurrence of C is at offset 5
C:\>
```

You should note that *strchr* does not contain an index to the first occurrence of a character; rather, *strchr* contains a pointer to the character. To help you better understand the *strchr* function, consider the following implementation:

```
char *strchr(const char *string, int letter)
 {
   while ((*string != letter) && (*string))
     string++;
   return((string);
 }
```

177 Returning an Index to the First Occurrence of a String

In Tip 176 you learned how to use the function *strchr* to obtain a pointer to the first occurrence of a character within a string. If you treat strings as arrays, as opposed to pointers, however, you probably prefer to work with an index to the character, rather than with a pointer. You can use the *strchr* function to obtain an index to the desired character by subtracting the string's starting address from the pointer *strchr* returns, as shown here:

```
char_ptr = strchr(string, character);
index = char_ptr - string;
```

If *strchr* does not find the character in the string, then the value *strchr* assigns to the index will be equal to the string's length. In addition to using *strchr* in the manner this Tip details, you can also use the function *str_index*, as shown here:

```
int str_index(const char *string, int letter)
 {
   char *original = string;
   while ((*string != letter) && (*string))
     string++;
   return(string - original);
 }
```

FINDING THE LAST OCCURRENCE OF A CHARACTER IN A STRING 178

As your programs work with strings, there may be times when you will want to find the last (rightmost) occurrence of a specific character within a string. For example, if you are working with a string that contains a pathname, you might search the string for the last backslash (\) character in order to find the location where the filename begins. To help you search for the last occurrence of a character within a string, most compilers provide a function named *strrchr*, which returns a pointer to the last occurrence of a specific character within a string, as shown here:

```
#include <string.h>
char *strrchr(const char *string, int character);
```

If *strrchr* does not find the character you specify within the string, *strrchr* returns a pointer to the *NULL* character that marks the end of the string. The following program, *strrchr.c*, illustrates how to use the *strrchr* function:

```
#include <stdio.h>
#include <string.h>

void main(void)
 {
   char title[64] = "Jamsa\'s C/C++/C# Programmer\'s Bible!";
   char *ptr;
   if (ptr = strrchr(title, 'C'))
     printf("Rightmost occurrence of C is at offset %d\n", ptr - title);
   else
     printf("Character not found\n");
 }
```

You should note that *strrchr* does not contain an index to the last occurrence of a character, but rather contains a pointer to the character. To help you better understand the *strrchr* function, consider the following implementation:

```
char *strrchr(const char *string, int letter)
 {
   char *ptr = NULL;
   while (*string)
     {
       if (*string == letter)
         ptr = string;
       string++;
     }
   return(ptr);
 }
```

179 RETURNING AN INDEX TO THE LAST OCCURRENCE OF A STRING

In Tip 178 you learned how to use the function *strrchr* to obtain a pointer to the last occurrence of a character within a string. If you treat a string as an array, as opposed to a pointer, however, you probably prefer to work with an index to the character rather than a pointer. If you use the *strrchr* function, you can obtain an index to the desired character by subtracting the string's starting address from the pointer *strrchr* returns:

```
char_ptr = strrchr(string, character);
index = char_ptr - string;
```

If *strrchr* does not find the character in the string, the value *strrchr* assigns to the index will be equal to the string's length. In addition to using *strrchr*, you can use the function *strr_index*, as shown here:

```
int strr_index(const char *string, int letter)
 {
   char *original = string;
   char *ptr = NULL;
   while (*string)
    {
     if (*string == letter)
       ptr = string;
     string++;
    }
   return((*ptr) ? ptr-original: string-original);
 }
```

180 WORKING WITH FAR STRINGS

As discussed in the Memory section of this book, *far* pointers let DOS programs access data that reside outside of the current 64Kb data segment. When you are working with *far* pointers, you must also use functions that expect their parameters to be *far* pointers. Unfortunately, none of the string manipulation routines this section presents anticipate *far* pointers to strings. Passing a *far* pointer to one of the string manipulation functions this section details will make an error occur. To support *far* pointers, however, many compilers provide *far*-pointer implementations of these functions. For example, to determine the length of a string a *far* pointer references, you might use the function *_fstrlen*, as shown here:

```
#include <string.h>
size_t _fstrlen(const char *string)
```

To determine which *far* functions your compiler supports, refer to your compiler documentation.

Note: *As you have learned previously,* ***Visual C++*** *does not support* ***far*** *declarations (either functions or pointers), so you can use the* ***strlen*** *function with* ***char*** *pointers of any size in* ***Visual C++***.

181 WRITING STRING FUNCTIONS FOR FAR STRINGS

In Tip 180 you learned that several compilers provide functions that support strings which *far* pointers reference. If your compiler does not provide such functions, you can create the *far* string functions yourself by modifying the functions in this section. As an example, the following function, *fstreql*, illustrates a *far* pointer-based implementation of *streql* (rather than the standard, local pointer-based implementation):

```
int fstreql(char far *str1, char far *str2)
 {
   while ((*str1 == *str2) && (*str1))
```

```
      {
        str1++;
        str2++;
      }
    return((*str1 == NULL) && (*str2 == NULL));
  }
```

Note: *As you have learned previously,* ***Visual C++*** *does not support* ***far*** *declarations, so you can use the* ***streql*** *function with* ***char*** *pointers of any size in* ***Visual C++****.*

COUNTING THE NUMBER OF CHARACTER OCCURRENCES IN A STRING 182

As your programs work with strings, there may be times when you will want to know the number of times a character occurs within a string. To help you count the number of times a character occurs within a string, your programs can use the *charcnt* function, as shown here:

```
int charcnt(const char *string, int letter)
 {
   int count = 0;

   while (*string)
     if (*string == letter)
       count++;
   return(count);
 }
```

REVERSING A STRING'S CONTENTS 183

As your programs perform different string operations, there may be times when you must reverse the order of characters within a string. To simplify such operations, most compilers provide a *strrev* function, as shown here:

```
#include <string.h>

char *strrev(char *string);
```

To better understand the *strrev* function, consider the following implementation:

```
char *strrev(char *string)
 {
   char *original = string;
   char *forward = string;
   char temp;

   while (*string)
     string++;
   while (forward < string)
    {
      temp = *(--string);
      *string = *forward;
      *forward++ = temp;
    }
  return(original);
 }
```

ASSIGNING A SPECIFIC CHARACTER TO AN ENTIRE STRING 184

As your programs work with strings, there may be times when you will want to set all the characters in a string to a specific character. For example, there may be times when you want to overwrite a string's current value before passing the string to a function. To simplify the overwriting of every character within a string, most C compilers provide a *strset* function, which assigns every character in the string a specified character, as shown here:

```
#include <string.h>

char *strset(char *string, int character);
```

The *strset* function assigns the specified character to each string location until the *strset* function encounters the *NULL* character. To better understand the *strset* function, consider the following implementation:

```
char *strset(char *string, int letter)
 {
   char *original = string;

   while (*string)
      *string++ = letter;
   return(original);
 }
```

As you can see, the function loops through the string assigning the specified character until the function finds the *NULL* character.

185 COMPARING TWO CHARACTER STRINGS

In Tip 173 you created the *streql* function, which let your programs test whether or not two character strings are equal. Depending on the processing your program must perform, there will be times (such as when your program performs a *sort* operation) when you must know if one string is greater than another. To help your programs perform operations that determine the value of various strings, most C compilers provide a function named *strcmp*, which compares two character strings, as shown here:

```
#include <string.h>

int strcmp(const char *str1, const *char str2);
```

If the strings are equal, *strcmp* returns the value 0. If the first string is greater than the second, *strcmp* returns a value less than 0. Likewise, if the second string is greater than the first, *strcmp* returns a value greater than 0. The following program, *strcmp.c*, illustrates how to use the *strcmp* function:

```
#include <stdio.h>
#include <string.h>

void main(void)
 {
   printf("Comparing Abc with Abc %d\n", strcmp("Abc", "Abc"));
   printf("Comparing abc with Abc %d\n", strcmp("abc", "Abc"));
   printf("Comparing abcd with abc %d\n", strcmp("abcd", "abc"));
   printf("Comparing Abc with Abcd %d\n", strcmp("Abc", "Abcd"));

   printf("Comparing abcd with abce %d\n", strcmp("abcd", "abce"));
   printf("Comparing Abce with Abcd %d\n", strcmp("Abce", "Abcd"));
 }
```

To better understand the *strcmp* function, consider the following implementation:

```
int strcmp(const char *s1, const char *s2)
 {
   while ((*s1 == *s2) && (*s1))
     {
       s1++;
       s2++;
     }
```

```
    if ((*s1 == *s2) && (! *s1))    // Same strings
      return(0);
    else if ((*s1) && (! *s2))      // Same but s1 longer
      return(-1);
    else if ((*s2) && (! *s1))      // Same but s2 longer
      return(1);
    else
      return((*s1 > *s2) ? -1: 1); // Different
  }
```

COMPARING THE FIRST N CHARACTERS OF TWO STRINGS 186

In Tip 185 you learned how to use the *strcmp* function to compare two strings. Depending on your program's function, there may be times when you may only want to compare the first *n* characters of two strings. To make comparing only *n* characters with two strings easier to perform, most C compilers provide a function named *strncmp*, as shown here:

```
#include <string.h>

int strncmp(const char *s1, const char *s2, size_t n);
```

Like *strcmp*, the *strncmp* function returns the value 0 if the strings are equal and a value less than or greater than 0, depending on whether the first or second string is greater. The following program, *strncmp.c*, illustrates how to use the *strncmp* function:

```
#include <stdio.h>
#include <string.h>

void main(void)
 {
   printf("Comparing 3 letters Abc with Abc %d\n", strncmp("Abc", "Abc", 3));
   printf("Comparing 3 letters abc with Abc %d\n", strncmp("abc", "Abc", 3));
   printf("Comparing 3 letters abcd with abc %d\n", strncmp("abcd", "abc", 3));
   printf("Comparing 5 letters Abc with Abcd %d\n", strncmp("Abc", "Abcd", 5));
   printf("Comparing 4 letters abcd with abce %d\n", strncmp("abcd", "abce", 4));
 }
```

To understand the *strncmp* function better, consider the following implementation:

```
int strncmp(const char *s1, const char *s2, int n)
 {
   int i = 0;

   while ((*s1 == *s2) && (*s1) && i < n)
     {
       s1++;
       s2++;
       i++;
     }
   if (i == n)                                  // Same strings
     return(0);
   else if ((*s1 == *s2) && (! *s1)) // Same strings
     return(0);
   else if ((*s1) && (! *s2))                   // Same but s1 longer
     return(-1);
   else if ((*s2) && (! *s1))                   // Same but s2 longer
     return(1);
   else
     return((*s1 > *s2) ? -1: 1);
 }
```

187 COMPARING STRINGS WITHOUT CONSIDERING CASE

In Tip 185 you learned how to use the *strcmp* function to compare two strings. Likewise, in Tip 186 you learned how to use the function *strncmp* to compare the first *n* characters of two strings. Both *strcmp* and *strncmp* consider upper- and lowercase letters as distinct. Depending on your program's processing, you may want the string comparison to ignore case. For such operations, most C compilers provide the functions *stricmp* and *strncmpi,* as shown here:

```
#include <string.h>

int stricmp(const char s1, const char s2);
int strncmpi(const char *s1, const char *s2, size_t n);
```

If your compiler does not support these functions, you can use the *strupr* function to convert the strings to uppercase and then pass the uppercase strings to *strcmp*. The following program, *cmpcase.c*, illustrates how to use the *stricmp* and *strncmpi* functions:

```
#include <stdio.h>
#include <string.h>

void main(void)
 {
   printf("Comparing Abc with Abc %d\n", stricmp("Abc", "Abc"));
   printf("Comparing abc with Abc %d\n", stricmp("abc", "Abc"));
   printf("Comparing 3 letters abcd with ABC %d\n", strncmpi("abcd", "ABC", 3));
   printf("Comparing 5 letters abc with Abcd %d\n", strncmpi("abc", "Abcd", 5));
 }
```

When you compile and execute the *cmpcase.c* program, your screen will display the following output:

```
Comparing ABC with ABC 0
Comparing abc with Abc 0
Comparing 3 letters abcd with ABC 0
Comparing 5 letters abc with Abcd -1
```

188 CONVERTING A CHARACTER STRING REPRESENTATION OF A NUMBER

When your programs work with strings, one of the most common operations you must perform is to convert an ASCII representation of a value to a numeric value. For example, if you prompt the user to input his or her salary, you might need to convert the character string input into a floating-point value. To help you convert ASCII values, most C compilers provide a set of run-time library functions that perform ASCII to numeric conversion. Table 188 briefly describes the standard ASCII conversion functions.

Function	Purpose
atof	Converts a character string representation of a floating-point value
atoi	Converts a character string representation of an integer value
atol	Converts a character string representation of a long integer value
strtod	Converts a character string representation of a double precision value
strtol	Converts a character string representation of a long value

***Table 188** Run-time library functions your programs can use to convert ASCII representations of a numeric value.*

The following program, *asciinum.c*, illustrates how to use the ASCII-to-numeric functions:

```
#include <stdio.h>
#include <stdlib.h>

void main(void)
```

```
{
  int int_result;
  float float_result;
  long long_result;

  int_result = atoi("1234");
  float_result = atof("12345.678");
  long_result = atol("1234567L");
  printf("%d %f %ld\n", int_result, float_result, long_result);
}
```

DUPLICATING A STRING'S CONTENTS 189

When your programs work with strings, sometimes you want to duplicate a string's contents quickly. If there are times when your program will need to copy the string and other times when your program might not, you might want the program to allocate the memory *dynamically* (during the program's execution) to hold the string copy as needed. To let your programs allocate memory during execution (dynamically) in order to create a character string copy, most C compilers provide the *strdup* function, as shown here:

```
#include <string.h>

char *strdup(const char *some_string);
```

When you invoke *strdup*, the function uses *malloc* to allocate memory and then copies the string's location to the memory location. When your program has finished using the string copy, it can release the memory using the *free* statement. The following program, *strdup.c*, illustrates how to use the *strdup* function:

```
#include <stdio.h>
#include <string.h>

void main(void)
 {
   char *title;

   if ((title = strdup("Jamsa\'s C/C++/C# Programmer\'s Bible")))
     printf("Title: %s\n", title);
   else
     printf("Error duplicating string");
 }
```

To better understand the *strdup* function, consider the following implementation:

```
#include <string.h>
#include <malloc.h>

char *strdup(const char *s1)
 {
   char *ptr;

   if ((ptr = malloc(strlen(s1)+1)))   // Allocate buffer
     strcpy(ptr, s1);
   return(ptr);
 }
```

FINDING A CHARACTER FROM A GIVEN SET'S FIRST OCCURRENCE 190

In Tip 176 you learned how to use the function *strchr* to find the first occurrence of a specific character. Depending on the function your program performs, there may be times when you may want to search a string for the first occurrence of any one character from a given character set. To help you search a string for any character in a set, most C compilers provide the *strspn* function, as shown here:

```
#include <string.h>
size_t strspn(const char *s1, const char *s2);
```

Within the *s1* string, the function returns the offset of the first character not contained within the *s2* string. The following program, *strspn.c*, illustrates how to use the *strspn* function:

```
#include <stdio.h>
#include <string.h>

void main(void)
 {
   printf("Searching for Abc in AbcDef %d\n", strspn("AbcDef", "Abc"));
   printf("Searching for cbA in AbcDef %d\n", strspn("AbcDef", "cbA"));
   printf("Searching for Def in AbcAbc %d\n", strspn("AbcAbc", "Def"));
 }
```

When you compile and execute the *strspn.c* program, your screen will display the following:

```
Searching for Abc in AbcDef 3
Searching for cbA in AbcDef 3
Searching for Def in AbcAbc 0
C:\>
```

To better understand *strspn*, consider the following implementation:

```
size_t strspn(const char *s1, const char *s2)
 {
   int i, j;

   for (i = 0; *s1; i++, s1++)
    {
      for (j = 0; s2[j]; j++)
       if (*s1 == s2[j])
         break;
      if (s2[j] == NULL)
        break;
    }
   return(i);
 }
```

191 Locating a Substring Within a String

As your programs work with strings, there will be times when you must search a string for a specific substring. To help you search a string for a substring, most C compilers provide a function named *strstr*, as shown here:

```
#include <string.h>
strstr(string, substring);
```

If the substring exists within the string, *strstr* returns a pointer to the first occurrence of the string. If *strstr* does not find the substring, the function returns *NULL*. The following program, *strstr.c*, illustrates how to use *strstr*:

```
#include <stdio.h>
#include <string.h>

void main(void)
 {
   printf("Looking for Abc in AbcDef %s\n",
    (strstr("AbcDef", "Abc")) ? "Found" : "Not found");
   printf("Looking for Abc in abcDef %s\n",
    (strstr("abcDef", "Abc")) ? "Found" : "Not found");
```

```
   printf("Looking for Abc in AbcAbc %s\n",
    (strstr("AbcAbc", "Abc")) ? "Found" : "Not found");
 }
```

To help you better understand *strstr*, consider the following implementation:

```
char *strstr(const char *s1, const char *s2)
 {
   int i, j, k;

   for (i = 0; s1[i]; i++)
     for (j = i, k = 0; s1[j] == s2[k]; j++, k++)
       if (! s2[k+1])
         return(s1 + i);
   return(NULL);
 }
```

COUNTING THE NUMBER OF SUBSTRING OCCURRENCES 192

In Tip 191 you learned how to use the function *strstr* to locate a substring within a string. You may sometimes want to know the number of times a substring appears within a string. The following function, *strstr_cnt*, lets you determine how many times a given substring occurs within a string:

```
int strstr_cnt(const char *string, const char *substring)
  {
    int i, j, k, count = 0;

    for (i = 0; string[i]; i++)
      for (j = i, k -= 0; string[j] == substring[k]; j++, k++)
         if (! substring[k + 1])
             count++;
    return(count);
  }
```

OBTAINING AN INDEX TO A SUBSTRING 193

In Tip 191 you learned how to use the function *strstr* to obtain a pointer to a substring within a string. If you treat character strings as arrays, instead of using pointers, there may be times when you will want to know the character index at which a substring begins within a string. Using the value *strstr* returns, you can subtract the string's address to produce an index:

```
index = strstr(string, substr) - string;
```

If *strstr* does not find the substring, the index value will be equal to the length of the string. In addition, your programs can use the function *substring_index* to obtain an index to a substring, as shown here:

```
int substring_index(const char *s1, const char *s2)
 {
   int i, j, k;

   for (i = 0; s1[i]; i++)
     for (j = i, k = 0; s1[j] == s2[k]; j++, k++)
       if (! s2[k+1])
         return(i);
   return(i);
 }
```

194 OBTAINING THE RIGHTMOST OCCURRENCE OF A SUBSTRING

In Tip 191 you used the function *strstr* to determine the first occurrence of a substring within a string. Depending on your program's function, there may be times when you will want to know the last (rightmost) occurrence of a substring within a string. The following function, *r_strstr,* returns a pointer to the rightmost occurrence of a substring within a string or the value *NULL* if the substring does not exist:

```
char *r_strstr(const char *s1, const char *s2)
 {
   int i, j, k, left = 0;

   for (i = 0; s1[i]; i++)
     for (j = i, k = 0; s1[j] == s2[k]; j++, k++)
       if (! s2[k+1])
         left = i;
   return((left) ? s1+left: NULL);
 }
```

195 DISPLAYING A STRING WITHOUT THE %S FORMAT SPECIFIER

Several of the Tips in this section have used the *%s* format specifier to display character strings. The following statement, for example, uses *printf* to display the contents of the character string variable named *title*:

```
printf("%s", title);
```

The first argument passed to the *printf* statement is a character string, which may contain one or more format specifiers. When your programs use *printf* to display only one character string, as the previous example shows, you can omit the character string that contains the format specifier and pass to *printf* the character string you want to display, as shown here:

```
printf(title);
```

As you can see, *printf's* first argument is nothing more than a character string that contains one or more special symbols.

196 REMOVING A SUBSTRING FROM WITHIN A STRING

In Tip 191 you used the *strstr* function to determine the starting location of a substring within a string. In many cases, your program must remove a substring from within a string. To do so, you can use the function *strstr_rem,* which removes the first occurrence of a substring, as shown here:

```
char *strstr_rem(char *string, char *substring)
  {
    int i, j, k, loc = -1;

    for (i = 0; string[i] && (loc == -1); i++)
      for (j = i; k = 0; str[j] == substring[k]; j++, k++)
        if (! substring[k + 1])
          loc = i;
    if (loc != -1)  // Substring was found
      {
       for (k = 0; substr[k]; k++)
         ;
       for (j = loc, i = loc + k, string[i]; j++, i++)
         string[j] = string[i];
       string[i] == NULL;
      }
   return(string);
 }
```

REPLACING ONE SUBSTRING WITH ANOTHER 197

In Tip 196 you used the function *strstr_rem* to remove a substring from within a string. In many cases, your programs must replace the first occurrence of one substring with another substring. You can do so with the following function, *strstr_rep*, as shown here:

```
#include <string.h>

char *strstr_rep(char *source, char *old, char *new)
 {
   char *original = source;
   char temp[256];
   int old_length = strlen(old);
   int i, j, k, location = -1;

   for (i = 0; source[i] && (location == -1); ++i)
     for (j = i; k = 0; source[j] == old[k]; j++, k++
       if (! old[k+1])
         location = i;
   if (location != -1)
    {
      for (j=0; j < location; j++)
        temp[j] = source[j];
      for (i=0; new[i]; i++, j++)
        temp[j] = new[i];
      for (k = location + old_length; source[k]; k++, j++)
        temp[j] = source[k];
      temp[j] = NULL;
      for (i = 0; source[i] = temp[i]; i++; // NULL Loop
    }
   return(original);
}
```

CONVERTING AN ASCII NUMERIC REPRESENTATION 198

When your programs work with character strings, your programs must often convert an ASCII representation of a value, such as 1.2345, to the corresponding *int, float, double, long,* or *unsigned* value. To help you perform such operations, C provides the functions defined in Table 198.

Function	Purpose
atof	Converts an ASCII representation of a value of type *float*
atoi	Converts an ASCII representation of a value of type *int*
atol	Converts an ASCII representation of a value of type *long int*

***Table 198** C functions which you can use to convert ASCII numeric representations.*

The formats of the functions Table 198 details are as follows:

```
#include <stdlib.h>

double atof(char *string);
int atoi(char *string);
int atol(char *string);
```

If a function is unable to convert the character string to a numeric value, the function returns 0. The following program, *ascii_to.c*, illustrates how to use the *ato* functions:

```
#include <stdio.h>
#include <stdlib.h>
```

```
void main(void)
 {
   int int_value;
   float flt_value;
   long long_value;

   int_value = atoi("12345");
   flt_value = atof("33.45");
   long_value =atol("12BAD");
   printf("int %d float %5.2f long %ld\n", int_value, flt_value, long_value);
 }
```

When you compile and execute the *ascii_to.c* program, your screen will display the following output:

```
int 12345 float 33.45 long 12
C:\>
```

Note the program's function call to *atol*. As you can see, when the function encounters the non-numeric value (the letter B), the function ends the conversion, returning the value that the function had already converted up to that point.

199 DETERMINING WHETHER A CHARACTER IS ALPHANUMERIC

An *alphanumeric* character is either a letter or a digit. In other words, an alphanumeric character is an uppercase letter from A through Z, a lowercase letter from a through z, or a digit from 0 through 9. To help your programs determine whether a character is alphanumeric, the header file *ctype.h* contains a macro named *isalnum*. The macro examines a letter and returns the value 0 if the character is not alphanumeric and a non-zero value for alphanumeric characters, as shown here:

```
if (isalnum(letter))
```

To better understand the macro *isalnum*, consider the following implementation:

```
#define isalnum(c) ((toupper((c)) >= 'A' && (toupper((c)) <= 'Z') || ((c) >= '0'
&& ((c) <= '9'))
```

200 DETERMINING WHETHER A CHARACTER IS A LETTER

As your programs work with characters within strings, there may be times your programs must test whether a character contains a letter of the alphabet (either upper- or lowercase). To help your programs determine whether a character is a letter of the alphabet, the header file *ctype.h* provides the macro *isalpha*. The macro examines a letter and returns the value 0 if the character does not contain a letter from uppercase A through Z or lowercase a through z. If the character contains a letter of the alphabet, then the macro returns a non-zero value:

```
if (isalpha(character))
```

To better understand the *isalpha* macro, consider the following implementation:

```
#define isalpha(c) (toupper((c)) >= 'A' && (toupper((c)) <= 'Z')
```

201 DETERMINING WHETHER A CHARACTER CONTAINS AN ASCII VALUE

An ASCII value is a value in the range 0 through 127. When your programs work with a string's characters, there may be times when you must determine whether a character contains an ASCII value. To help your programs determine an ASCII value, the header file *ctype.h* contains the macro *isascii*, which examines a letter and returns the value 0 if the character does not contain an ASCII character, and a non-zero value if the character does contain an ASCII value, as shown here:

```
if (isascii(character))
```

To better understand the *isascii* macro, consider the following implementation:

```
#define isascii(ltr) ((unsigned) (ltr) < 128)
```

As you can see, the *isascii* macro considers a value in the range 0 through 127 as ASCII.

DETERMINING WHETHER A CHARACTER IS A CONTROL CHARACTER 202

A *control character* is a value from ^A through ^Z or ^a through ^z. Different applications use control characters differently. For example, DOS uses the CTRL+Z character to represent the end of a file. Other word processors use control characters to represent boldface or italics. When you work with characters in a string, there may be times when you must determine whether a character is a control character. To help your programs perform such testing, the header file *ctype.h* contains the macro *iscntrl*, which returns a non-zero value for a control character and 0 if the letter is not a control character, as shown here:

```
if (iscntrl(character))
```

DETERMINING WHETHER A CHARACTER IS A DIGIT 203

A *digit* is an ASCII value from 0 through 9. When you work with strings, there may be times when you must determine if a character is a digit. To help you test whether a character is a digit, the header file *ctype.h* provides the macro *isdigit*. The *isdigit* macro examines a letter and returns the value 0 if the character is not a digit and a non-zero value for characters in the range 0 through 9, as shown here:

```
if (isdigit(letter))
```

To better understand the macro *isdigit*, consider the following implementation:

```
#define isdigit(c) ((c) >= '0' && (c) <= '9')
```

DETERMINING WHETHER A CHARACTER IS A GRAPHICS CHARACTER 204

A *graphics character* is a printable character (see *isprint),* excluding the space character (ASCII 32). When your programs perform character output operations, there may be times when you want to know whether a character is a graphics character. To help your programs perform such testing, the header file *ctype.h* provides the macro *isgraph*. The *isgraph* macro examines a letter and returns the value 0 if the character is not a graphic and a non-zero value for graphics characters:

```
if (isgraph(letter))
```

To better understand the *isgraph* macro, consider the following implementation:

```
#define isgraph(ltr) ((ltr) >= 33) && ((ltr) <= 127)
```

As you can see, a graphics character is any ASCII character in the range 33 to 127.

DETERMINING WHETHER A CHARACTER IS UPPER- OR LOWERCASE 205

As your programs work with characters within a string, there may be times when you must know whether a character is an upper- or lowercase letter. To help your programs test for case, the header file *ctype.h* provides the macros *islower* and *isupper*. These macros examine a character and return a 0 value for characters that are not lowercase *(islower)* or uppercase *(isupper)*, and a non-zero value otherwise:

```
if (islower(character))
if (isupper(character))
```

To better understand the *islower* and *isupper* macros, consider the following implementations:

```
#define islower(c) ((c) >= 'a' && (c) <= 'z')
#define isupper(c) ((c) >= 'A' && (c) <= 'Z')
```

206 DETERMINING WHETHER A CHARACTER IS PRINTABLE

When your programs perform character output, you might want to examine each character to ensure that you only output *printable characters*. A printable character is any character in the range 32 (the space character) through 127 (the DEL character). To help your programs test for a printable character, the header file *ctype.h* provides the macro *isprint*. The *isprint* macro returns a non-zero value for printable characters and a 0 value for characters that are not printable:

```
if (isprint(character))
```

To better understand the *isprint* macro, consider the following implementation:

```
#define isprint(ltr) ((ltr) >= 32) && ((ltr) <= 127)
```

As you can see, the *isprint* macro considers any ASCII character in the range 32 through 127 as a printable character.

207 DETERMINING WHETHER A CHARACTER IS A PUNCTUATION SYMBOL

Within a book, punctuation symbols include commas, semicolons, periods, question marks, and so on. Within C, however, a punctuation symbol is any graphics ASCII character that is not alphanumeric. As your programs work with characters in a string, there may be times when you must test whether a character contains a punctuation symbol. To help your programs test for punctuation symbols, the header file *ctype.h* defines the macro *ispunct*. This macro examines a character and returns a non-zero value for a character that contains a punctuation symbol and a 0 value for a character that does not contain a punctuation symbol:

```
if (ispunct(character))
```

To better understand the *ispunct* macro, consider the following implementation:

```
#define ispunct(c) (isgraph(c)) && ! isalphanum((c)))
```

208 DETERMINING WHETHER A CHARACTER CONTAINS WHITESPACE

The term *whitespace characters* includes the following characters: space, tab, carriage return, newline, vertical tab, and formfeed. When your programs perform character output, sometimes you must test whether or not a character contains a whitespace character. To help your programs test for whitespace, the header file *ctype.h* provides the macro *isspace*. This macro examines a character and returns a non-zero value for whitespace characters and a 0 value for non-whitespace characters:

```
if (isspace(character))
```

To better understand the *isspace* macro, consider the following implementation:

```
#define isspace(c)  (((c) == 32) || ((c) == 9) || ((c) == 13))
```

DETERMINING WHETHER A CHARACTER IS A HEXADECIMAL VALUE 209

A *hexadecimal value* is a digit in the range 0 through 9 or a letter from uppercase A through F or lowercase a through f. When your programs work with characters in a string, sometimes you must determine whether a character contains a hexadecimal digit. To help your programs test for hexadecimal digits, the header file *ctype.h* defines the macro *isxdigit*. This macro examines a character and returns a non-zero value if the character is a hexadecimal value and a 0 value if the character is not:

```
if (isxdigit(character))
```

To better understand the *isxdigit* macro, consider the following implementation:

```
#define isxdigit(c) (isnum((c)) || (toupper((c)) >= 'A' && toupper((c)) <= 'F'))
```

CONVERTING A CHARACTER TO UPPERCASE 210

As you work with character strings, a common operation your programs must perform is to convert a character from lowercase to uppercase. When you want to convert cases, your programs have two choices. Your programs can use the macro *_toupper*, which is defined in the header file *ctype.h*, or they can use the run-time library function *toupper*. The macro and function formats are as follows:

```
#include <ctype.h>

int _toupper(int character);
int toupper(int character);
```

Although the macro and function both convert a character to uppercase, the macro and function work differently. The macro *_toupper* does not test to make sure the character it is converting is lowercase. If you invoke the macro with a character that is not lowercase, the macro will cause an error. The function *toupper*, on the other hand, only converts lowercase letters and leaves all other characters unchanged. If you are sure that the character contains a lowercase letter, use the *_toupper* macro; this macro will execute faster than the function. If you are not sure if the character is lowercase, however, use the *toupper* function. The following program, *toupper.c*, illustrates the use of *_toupper* and *toupper*, as well as the errors that can occur when you use the macro with characters that are not lowercase:

```
#include  <stdio.h>
#include <ctype.h>

void main(void)
 {
   char string[] = "Jamsa\'s C/C++/C# Programmer\'s Bible";
   int i;

   for (i = 0; string[i]; i++)
     putchar(toupper(string[i]));
   putchar('\n');
   for (i = 0; string[i]; i++)
     putchar(_toupper(string[i]));
   putchar('\n');
 }
```

When you compile and execute this program, your screen will display the first string (using *toupper*) in correct uppercase letters. The second string, however, will contain non-standard characters (symbols, graphics, and so on) because *_toupper* tries to convert uppercase characters as it does lowercase characters.

211 Converting a Character to Lowercase

As you work with character strings, a common operation your programs must perform is to convert a character from uppercase to lowercase. When you want to convert cases, your programs have two choices. Your programs can use the macro *_tolower*, which is defined in the header file *ctype.h*, or your programs can use the run-time library function *tolower*. The macro and function formats are as follows:

```
#include <ctype.h>

int _tolower(int character);
int tolower(int character);
```

Although the macro and function both convert a character to lowercase, the macro and the function work differently. The macro *_tolower* does not test to make sure the character it converts is uppercase. If you invoke the macro with a character that is not uppercase, you will cause an error. The function *tolower*, on the other hand, only converts uppercase letters and leaves all other characters unchanged. If you are sure that the character contains an uppercase letter, then use the *_tolower* macro, which will execute faster than the function. If you are not sure if the character is uppercase, however, use the *tolower* function. The following program, *tolower.c*, illustrates the use of *_tolower* and *tolower*, as well as the errors that can occur if you use the macro with characters that are not uppercase:

```
#include <stdio.h>
#include <ctype.h>

void main(void)
 {
   char string[] = "Jamsa\'s C/C++/C# Programmer\'s Bible";
   int i;

   for (i = 0; string[i]; i++)
     putchar(tolower(string[i]));
   putchar('\n');
   for (i = 0; string[i]; i++)
     putchar(_tolower(string[i]));
   putchar('\n');
 }
```

When you compile and execute this program, your screen will display the first string (using *tolower*) in correct lowercase letters. The second string, however, will contain non-standard characters (symbols, graphics, and so on) because *_toupper* tries to convert uppercase characters as it does lowercase characters.

212 Working with ASCII Characters

When you work with character strings and different character functions, sometimes you must ensure that a character is a valid ASCII character; that is, that the value is in the range 0 through 127. To ensure a character is a valid ASCII character, you can use the *toascii* macro, which is defined in the header file *ctype.h*, as shown here:

```
#include <ctype.h>

int toascii(int character);
```

To better understand the *toascii* macro, consider the following implementation:

```
#define toascii(character) ((character) & 0x7F)
```

To improve its performance, the *toascii* macro performs a *bitwise AND* operation that clears the most significant bit of the character's byte value. The *bitwise AND* operation helps the macro ensure that the value falls in the range 0–127.

WRITING FORMATTED OUTPUT TO A STRING VARIABLE 213

As you know, the *printf* function lets you write formatted output to the screen display. Depending on your program's requirements, there may be times when you must work with a character string that contains formatted output. For example, say that your employees have a five-digit employee number and a three-character region identifier (such as Sea for Seattle). Assume that you store information about each employee in a file you name with a combination of these two values (such as SEA12345). The *sprintf* function lets you write formatted output to a character string. The format of the *sprintf* function is as follows:

```
#include <stdio.h>

int sprintf(char *string, const char *format [,arguments...]);
```

The following program, *sprintf.c*, uses the *sprintf* function to create an eight-character employee filename:

```
#include <stdio.h>

void main(void)
 {
   int employee_number = 12345;
   char region[] = "SEA";
   char filename[64];

   sprintf(filename, "%s%d", region, employee_number);
   printf("Employee filename: %s\n", filename);
 }
```

READING INPUT FROM A CHARACTER STRING 214

As you have learned, the *scanf* function lets you read formatted input from *stdin*. Depending on your program's processing, there may be times when a character string will contain fields you want to assign to specific variables. The *sscanf* function lets your programs read values from a string and assign the values to the specified variables. The format of the *sscanf* function is as follows:

```
#include <stdio.h>

int sscanf(const char *string, const char *format [,arguments]);
```

The arguments your program passes to the *sscanf* function must be pointers to variable addresses. If *sscanf* successfully assigns fields, it returns the number of fields it assigned. If *sscanf* does not assign fields, then it returns 0 or *EOF* if *sscanf* encountered the string's end. The following program, *sscanf.c*, illustrates the *sscanf* function:

```
#include <stdio.h>

void main(void)
 {
   int age;
   float salary;
   char string[] = "33 25000.00";

   sscanf(string, "%d %f\n", &age, &salary);
   printf("Age: %d Salary %f\n", age, salary);
 }
```

TOKENIZING STRINGS TO SAVE SPACE 215

Tokenizing strings is the process of using a unique value to represent a string. For example, assume that you have a program that works with a large number of character strings. Say the program contains a database of your customer accounts by city and state. Depending on how the program performs its processing, you might end up with many different tests, as the following example illustrates:

```
if (streql(city, "Seattle"))
  // Statement
else if (streql(city, "New York"))
  // Statement
else if (streql(city, "Chicago"))
  // Statement
```

Within each of your program's functions that perform repetitive *else if* testing, the program consumes a considerable amount of space for the string constants, as well as a considerable amount of time performing the string comparisons. Rather than using repetitive string calls, you can create a function called *tokenize_string*, which returns a unique token for each string. Within the example function, your program testing will become the following:

```
int city_token;

city_token = tokenize_string(city);
if (city_token == Seattle_token)
  // Statement
else if (city_token == NewYork_token)
  // Statement
else if (city_token == Chicago_token)
  // Statement
```

Using tokens in this way will help you eliminate the amount of data space the string consonants consume. Eliminating the string comparisons will also improve your program's performance.

216 INITIALIZING A STRING

In the Arrays and Pointers section of this book you will learn how to assign values to arrays while your program is declaring the array. C represents character strings as an array of bytes. When you declare a string, you will generally specify an initial value, as shown here:

```
char title[] = "Jamsa\'s C/C++/C# Programmer\'s Bible";

char section[64] = "Strings";
```

In the *title* string, the C compiler will allocate an array large enough to hold the specified characters (plus the *NULL* character). Because the string "Jamsa's C/C++/C# Programmer's Bible" is 32 characters long, the *title* string can hold 32 printable characters plus the *NULL* character. If you later assign more than 33 characters to the string, you will overwrite the memory that stores another variable's value. In the *section* string, the compiler will allocate a string capable of storing 64 characters. The compiler will assign to the first seven bytes of the string the letters in the word "Strings" and to the eighth byte the *NULL* character. The compiler usually initializes the remaining 56 characters to *NULL*.

217 UNDERSTANDING FUNCTIONS

Most programs presented throughout this book thus far use only the *main* function. As your programs become larger and more complex, you can simplify your work and improve your program's clarity by breaking the program into smaller pieces, called *functions*. For example, assume that you are creating an accounting program. You might have a function that performs the general ledger operations, a different function for accounts payable, a third for accounts receivable, and a fourth to generate a balance sheet. If you were place all your program's statements within *main*, your program would become very large and difficult to understand. As a program's size and complexity increases, so too does the chance of program errors. If you divide the program into smaller, more manageable blocks, you can avoid errors. A *function* is a named collection of statements that perform a specific task. For example, the following function, *hello_world*, uses *printf* to display a message:

```
void hello_world(void)
 {
   printf("Hello, world!\n");
 }
```

The keyword *void* tells C that the function does not return a value. In many cases, your functions will use *return* to return a calculation's result to the calling function. If the function does not use *return* to return a result, you should precede the function's name with *void*. The *void* that appears in parentheses tells C that the function does not use any parameters. A *parameter* is information the program passes to the function. When your programs invoke *printf*, for example, the information you specify within the parentheses are *parameters*. When a function does not use parameters, you should place the word *void* within the parentheses. To use a function, you simply specify the function's name followed by parentheses, much as you use *printf*. Programmers refer to using a function as a *function call*. The following program, *use_func.c*, uses the *hello_world* function:

```
#include <stdio.h>

void hello_world(void)
 {
   printf("Hello, world!\n");
 }

void main(void)
 {
   hello_world();
 }
```

When you execute this program, the *main* function executes first. As you can see, the only statement in *main* is the function call to the *hello_world* function. When C encounters the function call, C immediately transfers the program's execution to the function, beginning the program's execution with the first statement in the function. After the last statement in the function completes, C transfers the execution to the statement that immediately follows the function call. To better understand this process, change the *main* function within the *usefunc.c* program, as shown here:

```
void main(void)
 {
   printf("About to call function\n");
   hello_world();
   printf("Back from the function call\n");
 }
```

When you compile and execute the *use_func.c* program, your screen will display the following:

```
About to call function
Hello, world!
Back from the function call
C:\>
```

USING VARIABLES WITHIN FUNCTIONS 218

As you create useful functions, you will find that many functions require variables to generate valuable results. To use a variable within a function, you must first declare the variable, just as you do in *main*. For example, the following program, *three_hi.c*, calls the function *three_hellos*, which uses the variable *counter* in a *for* loop to display a message three times:

```
#include <stdio.h>

void three_hellos(void)
```

```
 {
   int counter; // Variable
   for (counter = 1; counter <= 3; counter++)
     printf("Hello, world!\n");
 }
void main(void)
 {
   three_hellos();
 }
```

When you declare variables within a function, the names you use for these variables are unique to the function. Therefore, if your program uses 10 different functions and each function uses a variable named *counter*, C considers each function's variable as distinct. If your function requires many variables, you should declare the variables at the function's start, just as you would within *main*.

219 UNDERSTANDING MAIN AS A FUNCTION

When you create a C program, C uses the function name *main* to determine the first statement the program will execute. Actually, *main* is a function, so if you have questions about the types of operations you can perform within your functions, the rule is fairly simple: *Anything you can do in main, you can do in a function.* Just as you can declare variables in *main*, you can also declare variables in your functions. You can also use constructs such as *if*, *while*, and *for* in your functions. Finally, one function can call (use) another. For example, the following program, *call_2.c*, uses two functions. When the program begins, *main* calls the function *three_hellos*, which in turn calls the function *hello_world* three times to display messages on your screen, as shown here:

```
#include <stdio.h>
void hello_world(void)
 {
   printf("Hello, world!\n");
 }
void three_hellos(void)
 {
   int counter;

   for (counter = 1; counter <= 3; counter++)
     hello_world();
 }
void main(void)
 {
   three_hellos();
 }
```

220 GETTING STARTED WITH PARAMETERS

A *parameter* is a value passed to a function. Most of the programs presented throughout this book have passed parameters to the *printf* function, as shown here:

```
printf("The value is %d\n", result);
```

As you use functions on a regular basis, you can pass parameters to a function to improve the function's usefulness. For example, consider the following construction of the *three_hellos* function, which calls the *hello_world* function three times:

```
void three_hellos(void)
 {
```

```
int counter;

   for (counter = 1; counter <= 3; counter++)
     hello_world();
 }
```

A more useful function might let you specify, as a parameter, the number of times you want the program to display the message. To use a parameter, your function must specify the parameter's name and type, as shown here:

```
void hello_count(int message_count)
```

In this case, the function *hello_count* supports one parameter of type *int* named *message_count*. When another function, such as *main*, wants to use *hello_count*, the function must specify the value C assigns to the parameter *message_count*:

```
hello_count(2);     // Displays the message twice
hello_count(100);   // Displays the message 100 times
hello_count(1);     // Displays the message 1 time
```

The following program, *useparam.c*, illustrates how you might use a function with a parameter:

```
#include <stdio.h>

void hello_world(void)
 {
   printf("Hello, world!\n");
 }

void hello_count(int message_count)
 {
   int counter;

   for (counter = 1; counter <= message_count; counter++)
     hello_world();
 }

void main(void)
 {
   printf("Display the message twice\n");
   hello_count(2);
   printf("Display the message five times\n");
   hello_count(5);
 }
```

As you can see, in *main*, the function call to *hello_count* includes the value that C is to assign to the *message_count* parameter.

Note: *When you pass a parameter to a function, the type of value you pass to the parameter (such as **int, float, char**, and so on), must match the parameter type. Depending on your C compiler, the compiler might detect parameter type mismatches. If your compiler does not detect parameter type mismatches, errors may emerge that are often difficult to detect and correct.*

USING MULTIPLE PARAMETERS 221

As you have learned, a *parameter* is a value you pass to a function. In general, you can pass an unrestricted number of parameters to a function. However, research has shown that when the number of parameters exceeds seven, the function becomes more difficult to understand and use correctly, therefore making the function more susceptible to error. When your function uses more than one parameter, you must specify each parameter's type and name and separate the parameters with commas, as shown here:

```
void some_function(int age, float salary, int job_number)
  {
    // Function statements
  }
```

When your program wants to call the function, you must specify values for each parameter, as shown here:

```
some_function(33, 40000.00, 534);
```

C, in turn, will assign the values to the parameters, as shown in Figure 221.

```
some_function (33, 40000.00, 534);

void some_function (int age, float salary, int job_number);
{
  // Function statements
}
```

Figure 221 Mapping parameter values.

222 UNDERSTANDING PARAMETER DECLARATIONS IN OLDER C PROGRAMS

When you create a function that uses parameters, you will usually specify each parameter's type and name, separated by commas, within the function header, as shown here:

```
void some_function(int age, float salary, int job_number)
 {
   // Function statements
 }
```

If you work with older C programs, you might find that the programmer declared the parameters as follows:

```
void some_function(age, salary, job_number)
  int age;
  float salary;
  int job_number;
 {
   // Function statements
 }
```

If you encounter such parameter declarations, you should understand that although the declaration format differs slightly, the purpose remains the same—to specify the parameter's type and name. If you feel tempted to update the function's format, make sure that your compiler fully supports the new format. Also, remember that the more changes you make to your program, the greater your chances are of introducing an error. As a general rule, "*If it ain't broke, don't fix it!*"

223 RETURNING A VALUE FROM A FUNCTION

As your functions become more complex, they will usually perform a calculation and return a result. To provide a result to the caller, a function must use the *return* statement, which you will implement as shown here:

```
return(result);
```

The type of value the function returns (*int*, *float*, *char*, and so on) determines the function's type. If a function returns a value of type *int*, for example, you must precede the function name with the type name, as shown here:

```
int some_function(int value)
 {
   // Function statements
 }
```

The following function, *i_cube*, returns the cube of the integer value the program specifies as its parameter. For example, if the calling function passes the value 5 to the function, *i_cube* will return the value 5 * 5 * 5 or 125:

```
int i_cube(int value)
 {
    return(value * value * value);
 }
```

As you can see, the function uses the *return* statement to pass back the calculation's result to the caller. The code within the calling function can assign the called function's result (also known as a *return value*) to a variable, or the code can use the return value within a third function, such as *printf*, as shown here:

```
result = i_cube(5);

printf("The cube of 5 is %d\n", i_cube(5));
```

The following program, *i_cube.c*, uses the *i_cube* function to determine several different cube values:

```
#include <stdio.h>

int i_cube(int value)
 {
   return(value * value * value);
 }

void main(void)
 {
   printf("The cube of 3 is %d\n", i_cube(3));
   printf("The cube of 5 is %d\n", i_cube(5));
   printf("The cube of 7 is %d\n", i_cube(7));
 }
```

The values you pass to a function must match the parameter types contained within that function's declaration. If you want to determine the cube of a floating-point value, for example, you would create a second function called *f_cube*, as shown here (note that the return value is also of type *float*):

```
float f_cube(float value)
 {
  return (value * value * value);
 }
```

UNDERSTANDING THE RETURN STATEMENT 224

As you have learned, for a function to provide its caller with a result, the function must use the *return* statement. When C encounters a *return* statement in a function, C immediately ends the function's execution, and returns the specified value to the calling function. The program does not execute any statements within the function that follow the *return* statement. Instead, the program resumes execution in the calling function.

As you examine other C programs, you might encounter functions that contain multiple *return* statements, each of which returns a value for a specific condition. For example, consider the function *compare_values*, shown here:

```
int compare_values(int first, int second)
  {
    if (first == second)
      return(0);
    else if (first > second)
      return(1);
    else if (first < second)
      return(2);
  }
```

The function *compare_values* examines two integer values and returns one of the values listed in Table 224.

Result	Meaning
0	The values are the same.
1	The first value is greater than the second value.
2	The second value is greater than the first value.

*Table 224 Values the **compare_values** function returns.*

As a rule, you should try to limit your functions to one *return* statement. As your functions become larger and more complex, having multiple *return* statements often will make the functions more difficult to understand. In most cases, you can rewrite your function so that it only uses one *return* statement, as shown here:

```
int compare_values(int first, int second)
  {
    int result;

    if (first == second)
      result = 0;
    else if (first > second)
      result = 1;
    else if (first < second)
      result = 2;

    return(result);
  }
```

In this case, because the function is so simple, you may have difficulty understanding the advantage you gain by using only one *return* statement. As your functions become more complex, however, the advantage might become more clear. You should note, however, that sometimes using more than one *return* statement produces more readable code than the single *return* alternative. You should write the most readable and modifiable code possible; if using multiple *return* statements achieves that goal, then use as many *return* statements as necessary.

225 UNDERSTANDING FUNCTION PROTOTYPES

If you take a close look at each preceding program, you will find that the calling functions always appear in the program source code following the functions they call. Most new C compilers require knowledge of a function's return and parameter types before the program calls the function. By placing the functions in front of their callers within your program code, you let the C compiler know the information it must have before it encounters the function call. As your programs become more complex, however, it might become impossible for you to always place the functions in the correct order. Therefore, C lets you place *function prototypes* in your program that describe a function's return and parameter types. For example, consider a program that uses the functions *i_cube* and *f_cube*, presented in Tip 223. Before the functions' first use, the program can include a prototype similar to the following:

```
int i_cube(int);         // Returns an int--one int parameter
float f_cube(float);     // Returns a float--one float parameter
```

As you can see, the function prototype specifies the function's return and parameter types. The following program, *useproto.c*, uses two function prototypes to eliminate the need for function ordering:

```
#include <stdio.h>

int i_cube(int);
float f_cube(float);

void main(void)
```

```
  {
    printf("The cube of 3 is %d\n", i_cube(3));
    printf("The cube of 3.7 is %f\n", f_cube(3.7));
  }

int i_cube(int value)
  {
    return(value * value * value);
  }

float f_cube(float value)
  {
    return(value * value * value);
  }
```

If you examine the *.h* header files, such as *stdio.h*, you will find that these files contain many function prototypes.

UNDERSTANDING THE RUN-TIME LIBRARY 226

As you write your own functions, you will often find that a function you created for one program meets the needs of a second program. The ability to reuse functions in more than one program can save you considerable programming and testing time. In the Tools section of this book, you will learn how to place your commonly-used functions within a library to make them easier to use in multiple programs. For now, however, you may need to cut and paste the function's statements from one source code file to another.

Before you spend a great deal of time writing a wide variety of all-purpose functions, make sure you examine the functions your compiler provides. Many compilers refer to these built-in functions as the *run-time library*. Most C compilers provide hundreds of run-time library functions with purposes ranging from opening and working with files to accessing disk or directory information to determining a character string's length. The hour or two you spend reading your compiler's run-time library documentation will save you many programming hours that you might have otherwise spent "reinventing the wheel."

UNDERSTANDING FORMAL AND ACTUAL PARAMETERS 227

As you read different books on C, you might encounter the terms *formal* and *actual* parameters. In short, formal parameters are the parameter names that appear in the function definition. For example, the names *age*, *salary*, and *job_number* are the formal parameters for the *job_information* function, as shown here:

```
void job_information(int age, float salary, int job_number)
  {
    // Function statements
  }
```

When a function calls another function, the values the calling function passes are the actual parameters. In the case of the following function invocation, the values 30, 42000.00, and 321 are the actual parameters:

```
job_information(30, 42000.00, 321);
```

The actual parameters you pass to a function can be constant values or variables. The value or variable's type must match that of the formal parameter. For example, the following code fragment illustrates how to use variables as actual parameters:

```
int workers_age = 30;
float workers_salary = 42000.00;
int job_number = 321;

job_information(workers_age, workers_salary, job_number);
```

When you invoke a function with variables as the actual parameters, the variable names used for the actual parameters have no relationship to the names of the formal parameters. Instead, C concerns itself only with the values the variables contain.

228 Resolving Name Conflicts

As you have learned, most C compilers provide an extensive library of functions you can call to perform specific tasks. For example, to obtain the absolute value of an integer expression, you can use the *abs* function. Likewise, to copy one string's contents to another, you can use the *strcpy* function. As you create your own functions, sometimes a function you define will have the same name as a run-time library function. For example, the following program, *mystrcpy.c*, creates and uses a function named *strcpy*:

```
#include <stdio.h>

char *strcpy(char *destination, const char *source)
 {
   char *start = destination;

   while (*destination++ = *source++)
     ;

   return(start);
 }

void main(void)
 {
   char title[64];

   strcpy(title, "Jamsa\'s C/C++/C# Programmer\'s Bible");
   printf(title);
 }
```

When a function name that you declare within your program conflicts with a run-time library function, C uses your program's function, not the run-time library function.

229 Functions That Do Not Return int

Several of the functions you saw earlier returned values of type *int*. When your function does not return a value type *int* (rather, it might return *float*, *double*, *char*, and so on), you must tell the compiler the function's return type. The following program, *show_avg.c*, uses the function *average_value* to determine the average of three values of type *int*. The function returns the average using a value of type *float*:

```
#include <stdio.h>

float average_value(int a, int b, int c)
 {
   return ((a + b + c) / 3.0);
 }

void main(void)
 {
   printf("The average of 100, 133, and 155 is %f\n", average_value(100, 133, 155));
 }
```

As you can see, the function header specifies the function's return type:

```
float average_value(int a, int b, int c)
```

Note: *If you do not specify a function's return type, the C compiler will assume that the function returns the type* ***int****.*

UNDERSTANDING LOCAL VARIABLES 230

C lets you declare variables within your functions. Such variables are called *local variables* because their names and values only have meaning within the function that contains the variable declaration. The following program, *localerr.c*, illustrates the concept of a local variable. The function *local_values* declares three variables, *a*, *b*, and *c*, and assigns to the variables the values 1, 2, and 3, respectively. The function *main* tries to print each variable's value. However, because the variable names are local to the function *local_values*, the compiler generates errors that state that the symbols *a*, *b*, and *c* are undefined, as shown here:

```
#include <stdio.h>

void local_values(void)
 {
   int a = 1, b = 2, c = 3;

   printf("a contains %d b contains %d c contains %d\n", a, b, c);
 }

void main(void)
 {
   printf("a contains %d b contains %d c contains %d\n", a, b, c);
 }
```

HOW FUNCTIONS USE THE STACK 231

The Memory section of this book describes the *stack*, which programs use to temporarily hold detailed information. The stack's primary purpose is to support function invocations. When your program invokes a function, C places the address of the instruction that follows the function invocation (called the *return address*) onto the stack. Next, C places the function's parameters, from right to left, onto the stack. Finally, if the function declares local variables, C allocates stack space that the function then uses to hold the variable's value. Figure 231 shows how C uses the stack for a simple function call.

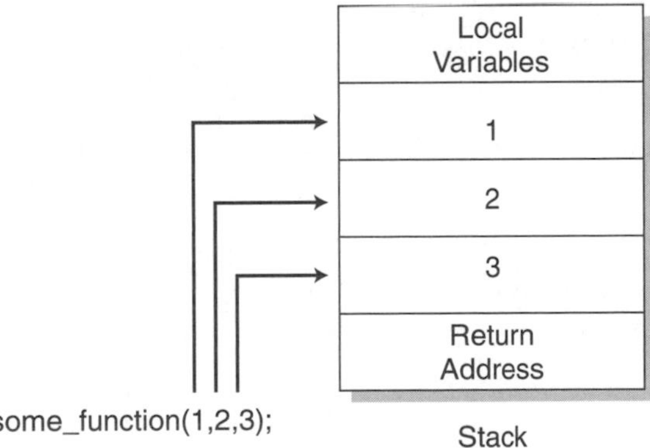

Figure 231 C uses the stack for a function call.

When the function ends, C discards the stack space that contained the local variables and parameters. Next, C uses the return value to determine the instruction that the program executes next. C removes the return value from the stack and places the address into the IP (instruction pointer) register.

232 UNDERSTANDING FUNCTION OVERHEAD

As you learned in Tip 231, when your program uses a function, C pushes the return address, parameters, and local variables onto the stack. When the function completes, C discards (pops) the stack space that contained the local variables and parameters and then uses the return value to resume executing the program at the correct location.

Although C's use of the stack is powerful because it lets the program invoke and pass information to functions, C's use also consumes processing time. Programmers call the amount of time the computer requires to push and pop stack information *function overhead*. To better understand function overhead's impact on your program's performance, consider the following program, *functovr.c*. The program first uses a loop to sum the values 1 through 100,000. Next, the program loops again, but instead uses a function to add the values, as shown here:

```
#include <stdio.h>
#include <time.h>

float add_em(long int a, float b)
 {
   float result;

   result = a + b;
   return(result);
 }

void main(void)
 {
   long int i;
   float result = 0;
   time_t start_time, stop_time;

   printf("Working...\n");
   time(&start_time);
   for (i = 1; i <= 100000000L; i++)
     result += i;
   time(&stop_time);
   printf("Using loop %d seconds\n", stop_time - start_time);
   printf("Working...\n");
   time(&start_time);
   for (i = 1; i <= 100000000L; i++)
     result = add_em(i, result);
   time(&stop_time);
   printf("Using function %d seconds\n", stop_time - start_time);
 }
```

On most systems, the function-based calculations might require almost twice as much processing time. When you use functions within your programs, therefore, you must consider the benefits the functions provide (such as ease of use, reuse of an existing function, reduction of testing, ease of understanding, and so on) versus the performance overhead they introduce.

233 UNDERSTANDING WHERE C PLACES LOCAL VARIABLES

As you have learned, C lets you declare variables within your functions. These variables are *local* to the function, which means only the function in which you declared the variables knows their values and existence. The following function, *use_abc*, declares three local variables named *a*, *b*, and *c*:

```
void use_abc(void)
 {
   int a, b, c;

   a = 3;
```

```
    b = a + 1;
    c = a + b;
    printf("a contains %d b contains %d c contains %d\n", a, b, c);
 }
```

Each time your program invokes the function, C allocates stack space to hold the local variables *a*, *b*, and *c*. When the function ends, C discards both the previously allocated stack space and the values that the local variables contained. Even if your function declares many local variables, C stores each variable's value on the stack.

DECLARING GLOBAL VARIABLES 234

In Tip 218 you learned that local variables are variables defined within a function whose names and existence are known only to the function. In addition to local variables, C also lets your programs use *global variables*, whose names, values, and existence are known throughout your program. In other words, all your C programs can use global variables. The following program, *global.c*, illustrates the use of three global variables, *a*, *b*, and *c*:

```
#include <stdio.h>
int a = 1, b = 2, c = 3;   // Global variables

void global_values(void)
 {
   printf("a contains %d b contains %d c contains %d\n", a, b, c);
 }

void main(void)
 {
   global_values();
   printf("a contains %d b contains %d c contains %d\n", a, b, c);
 }
```

When you compile and execute this program, the functions *global_values* and *main* both display the global variable values. Note that you declare the variables outside all the functions. When you declare global variables in this way, all your program's functions can use and change the global variable values simply by referring to the global variable name. Although global variables might at first appear convenient, misusing them can lead to errors that are very difficult to debug, as you will learn in Tip 235.

AVOID USING GLOBAL VARIABLES 235

In Tip 234 you learned how to declare global variables, which your program knows throughout all its functions. At first glance, using global variables seems to simplify your programming because it eliminates the need for function parameters and, more importantly, the need to understand *call by value* and *call by reference*. Unfortunately, however, global variables often create more errors than they fix. Because your code can change a global variable's value at virtually any location within your program, it is very difficult for another programmer who is reading your program to find each location in the program where the global variable changes. Therefore, other programmers might make changes to your program without fully understanding the effect the change has on a global variable. As a rule, functions should only change those variables passed to the functions as parameters. This lets programmers study the function prototypes to quickly determine which variables a function changes.

If you find that your program uses global variables, you may want to reconsider your program design. Your goal should be to eliminate (and definitely minimize) your use of global variables.

236 Resolving Global and Local Variable Name Conflicts

As you have learned, local variables are variables you declare within a function whose names are known only to that function. On the other hand, when you declare global variables outside all functions, every function throughout your program will know their names. If your program uses global variables, there may be times when a global variable's name is the same as that of a local variable your program declares within a function. For example, the following program, *conflict.c*, uses the global variables *a*, *b*, and *c*. The *conflict_a* function uses a local variable named *a* and the global variables *b* and *c*:

```
#include <stdio.h>
int a = 1, b = 2, c = 3;  // Global variables

void conflict_a(void)
 {
   int a = 100;

   printf("a contains %d b contains %d c contains %d\n", a, b, c);
 }

void main(void)
 {
   conflict_a();
   printf("a contains %d b contains %d c contains %d\n", a, b, c);
 }
```

When you compile and execute the *conflict.c* program, your screen will display the following:

```
a contains 100 b contains 2 c contains 3
a contains 1 b contains 2 c contains 3
C:\>
```

When global variable names and local variable names conflict, C will always use the local variable. As you can see, the changes the function *conflict_a* made to the variable *a* only appear within the function.

Note: *Although this program's purpose is to illustrate how C resolves name conflicts, it also illustrates the confusion that can occur when you use global variables. In this case, a programmer who is reading your code must pay close attention to determine that the function does not change the global variable* ***a****, but rather a local variable. Because the function combines the use of global and local variables, the code can become difficult to understand.*

237 Better Defining a Global Variable's Scope

In Tip 234 you learned that a global variable is a variable that all the functions throughout your program know. Depending on where you define a global variable, you can control which functions are actually able to reference the variable. In other words, you can control the global variable's *scope*. When your program declares a global variable, any functions that follow the variable declaration can reference that variable, up to the end of the source file. Functions that have definitions that appear prior to the global variable's definition cannot access the global variable. As an example, consider the following program, *gloscope.c*, which defines the global variable *title*:

```
#include <stdio.h>

void unknown_title(void)
 {
   printf("The book's title is %s\n", title);
 }

char title[] = "Jamsa\'s C/C++/C# Programmer\'s Bible";

void main(void)
 {
```

```
   printf("Title: %s\n", title);
 }
```

As you can see, the *unknown_title* function will try to display the variable *title*. However, because the global variable declaration occurs after the function definition, the global variable is unknown within the function. When you try to compile this program, your compiler will generate an error. To correct the error, move the global variable declaration to before the function.

UNDERSTANDING CALL BY VALUE 238

As you have learned, your programs pass information to functions using parameters. When you pass a parameter to a function, C uses a technique known as *call by value* to provide the function with a copy of the parameter's value. Using call by value, any changes the function makes to the parameter exist only within the function itself. When the function completes, the value of variables the calling function passed to the function are not changed within the calling function. For example, the following program, *nochange.c*, passes three parameters (the variables *a*, *b*, and *c*) to the function *display_and_change*. The function, in turn, will display the values, add 100 to the values, and then display the result. When the function ends, the program will display the values of the variables. Because C uses call by value, the function does not change the values of the variables within the calling function, as shown here:

```
#include <stdio.h>

void display_and_change(int first, int second, int third)
 {
   printf("Original function values %d %d %d\n", first, second, third);
   first += 100;
   second += 100;
   third += 100;
   printf("Ending function values %d %d %d\n", first, second, third);
 }

void main(void)
 {
   int a = 1, b = 2, c = 3;
   display_and_change(a, b, c);
   printf("Ending values in main %d %d %d\n", a, b ,c);
 }
```

When you compile and execute the *nochange.c* program, your screen will display the following:

```
Original function values 1 2 3
Ending function values 101 102 103
Ending values in main 1 2 3
C:\>
```

As you can see, the changes the function makes to the variables are only visible within the function itself. When the function ends, your variables within *main* are unchanged.

***Note:** When you use call by reference (which Tip 240 presents in detail), a function can change a parameter's value so that the change is visible outside of the function.*

PREVENTING PARAMETER VALUE CHANGE WITH CALL BY VALUE 239

In Tip 238 you learned that by default C uses call by value to pass parameters to functions. As a result, any changes to the parameter's value only occur within the function itself. When the function ends, the values of variables the program passed to the function are unchanged. As the Getting Started with C section of this book details, a variable is essentially a name assigned to a memory location. Every variable has two attributes of interest—its current value and its memory address. In the case of the program *nochange.c,* presented in Tip 238, the variables *a, b,* and *c* might use the memory addresses shown in Figure 239.1.

Address	Memory	Variable
1000	1	a
1002	2	b
1004	3	c

Figure 239.1 Variables store a value that resides in a specific memory location.

When you pass parameters to a function, C places the corresponding values onto the stack. In the case of the variables *a, b,* and *c,* the stack contains the values 1, 2, and 3. When the function accesses the variable's values, the function references the stack locations, as shown in Figure 239.2.

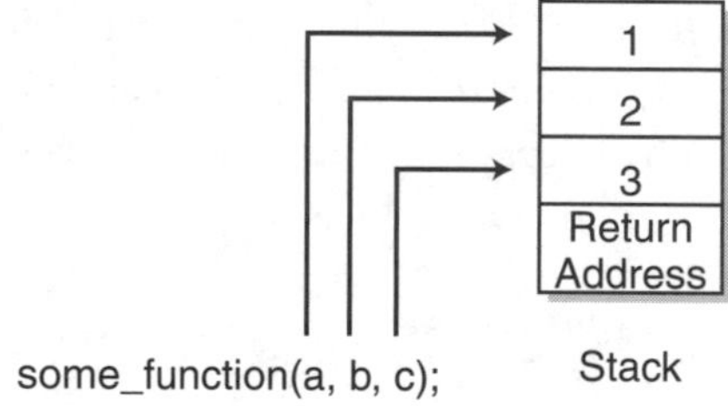

Figure 239.2 Functions reference parameter values stored on the stack.

Any changes the function makes to the parameter values actually change the stack values, as shown in Figure 239.3.

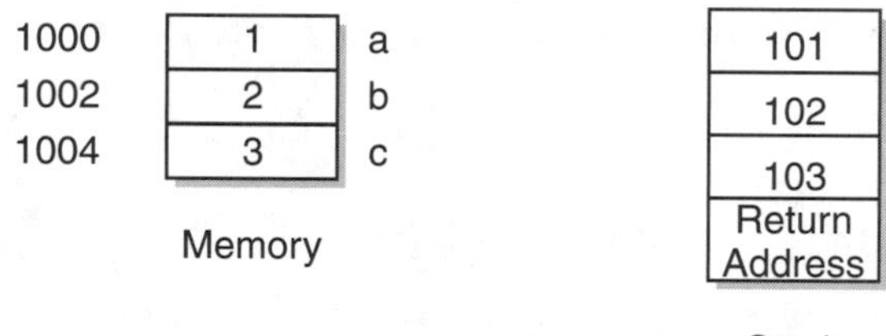

Figure 239.3 Changes functions make to parameter values affect only those values on the stack.

When the function ends, C discards the values on the stack as well as the changes the function has made to the stack locations. The function never references the memory locations that contain each variable's value, so your functions cannot make changes that exist after the function ends to any parameter the function receives using call by value.

240 UNDERSTANDING CALL BY REFERENCE

As you have learned, C passes parameters to functions using call by value by default. Using call by value, functions cannot change the value of a variable passed to a function. In most programs, however, your functions will change variables in one way or another. For example, a function that reads information from a file must place the information in a character string array. Likewise, a function such as *strupr* (presented in the Strings section of this book) must convert the letters in a character string to uppercase. When your functions change a parameter's value, your programs must pass the parameter to the function using *call by reference.* The difference between call by value and call by reference is that using call by value, functions receive a copy of a parameter's value. With call by reference, on the other hand, functions receive the variable's memory address. Therefore, the functions can make changes to the value stored at a specific memory location (in other words, the variable's value), which remain after the function ends. To use call by reference, your program must use *pointers*. The Pointers section of this book discusses pointers in detail. For now, however, think of a pointer simply as a memory address. To assign a variable's address to a pointer, use C's address operator (&). To later access the value in the memory location to which the pointer points, use C's redirection operator (*). Tips 241 and 242 discuss these operators in detail.

GETTING AN ADDRESS 241

A variable is essentially a name assigned to one or more memory locations. When your program runs, each variable resides in its own memory location. Your program locates variables in memory using the variable's *memory address*. To determine a variable's address, you use C's address operator (&). The following program, *showaddr.c,* for example, uses the address operator to display the addresses (in hexadecimal format) of the variables *a*, *b*, and *c*:

```
#include <stdio.h>

void main(void)
 {
   int a = 1, b = 2, c = 3;

   printf("The address of a is %x the value of a is %d\n", &a, a);
   printf("The address of b is %x the value of b is %d\n", &b, b);
   printf("The address of c is %x the value of c id %d\n", &c, c);
 }
```

When you compile and run this program, your program will display output similar to the following (the actual address values shown may differ):

```
The address of a is fff4 the value of a is 1
The address of b is fff2 the value of b is 2
The address of c is fff0 the value of c is 3
C:\>
```

When your programs later pass parameters to functions for variables whose value the function must change, your programs will pass the variables by reference (memory address), using the address operator, as shown here:

```
some_function(&a, &b, &c);
```

USING A VARIABLE'S ADDRESS 242

In Tip 241 you learned how to use C's address operator to obtain a variable's memory address. When you pass an address to a function, you must tell the C compiler that the function will be using a pointer (the memory address) of a variable, as opposed to the variable's value. To do so, you must declare a *pointer variable*. Declaring a pointer variable is very similar to a standard variable declaration in that you specify a type and variable name. The difference, however, is that an asterisk (*) precedes pointer variable names. The following declarations create pointer variables to values of type *int*, *float*, and *char*:

```
int *i_pointer;
float *f_pointer;
char *c_pointer;
```

After you declare a pointer variable, you must assign to the variable a memory address. The following statement, for example, assigns the address of the integer variable *a* to the pointer variable *i_pointer*:

```
i_pointer = &a;
```

Next, to use the value pointed to by the pointer variable, your programs must use C's redirection operator—the asterisk (*). For example, the following statement assigns the value 5 to the variable *a* (whose address is contained in the variable *i_pointer*):

```
*i_pointer = 5;
```

In a similar way, the following statement assigns to the variable *b* the value to which the variable *i_pointer* currently points:

```
b = *i_pointer;
```

When you want to use the value pointed to by a pointer variable, use the redirection operator (*). When you want to assign a variable's address to a pointer variable, you use the address operator (&). The following program, *use_addr.c*, illustrates the use of a pointer variable. The program assigns the pointer variable *i_pointer* the address of the variable *a*. The program then uses the pointer variable to change, display, and assign the variable's value:

```
#include <stdio.h>

void main(void)
 {
   int a = 1, b = 2;
   int *i_pointer;

   i_pointer = &a;                  // Assign an address
   *i_pointer = 5;                  // Change the value pointed to by i_pointer to 5

   // Display the value
   printf("Value pointed to by i_pointer %d the variable a %d\n", *i_pointer, a);
   b = *i_pointer;                  // Assign the value
   printf("Value of b is %d\n", b);
   printf("Value of i_pointer %x\n", i_pointer);
 }
```

Remember that a pointer is nothing more than a memory address. Your program must assign the value the pointer (the address) contains. In the program *use_addr.c*, the program assigned the pointer the address of the variable *a*. The program could have just as easily assigned the address of the variable *b*.

***Note:** When you use pointers, you must still keep in mind value types, such as **int**, **float**, and **char**. Your programs should only assign the address of integer values to integer pointers, and so on.*

243 CHANGING A PARAMETER'S VALUE

As you have learned, to change a parameter's value within a function, your programs must use call by reference, passing the variable's address. Within the function, you must use pointers. The following program, *chgparam.c*, uses pointers and addresses (call by reference) to display and then change the parameters the program passes to the *display_and_change* function:

```
#include <stdio.h>

void display_and_change(int *first, int *second, int *third)
 {
   printf("Original function values %d %d %d\n", *first, *second, *third);
   *first += 100;
   *second += 100;
   *third += 100;
   printf("Ending function values %d %d %d\n", *first, *second, *third);
 }

void main(void)
 {
   int a = 1, b = 2, c = 3;

   display_and_change(&a, &b, &c);
   printf("Ending values in main %d %d %d\n", a, b ,c);
 }
```

As you can see, when the program invokes the function, it passes as parameters the addresses of the variables *a*, *b*, and *c*. Within *display_and_change*, the function uses pointer variables and C's redirection operator to change and display the parameter's values. When you compile and execute the *display_and_change* program, your screen will display the following output:

```
Original function values 1 2 3
Ending function values 101 102 103
Ending values in main 101 102 103
C:\>
```

CHANGING ONLY SPECIFIC PARAMETERS 244

As you have learned, your functions can change a parameter's value using call by reference. Tip 243, for example, presented the *display_and_change* function, which used call by reference to change the value of each of its parameters. In many cases, however, your functions may change one parameter's value while leaving a second parameter's value unchanged. For example, the following program, *chgfirst.c*, uses the *change_first* function to assign to the *first* parameter the value of the *second* parameter:

```
#include <stdio.h>

void change_first(int *first, int second)
 {
   *first = second;  // Assign value of second to first
 }

void main(void)
 {
   int a = 0, b = 5;

   change_first(&a, b);
   printf("Value of a %d value of b %d\n", a, b);
 }
```

As you can see, the *change_first* function uses call by reference to change the first parameter's value, and call by value for the second parameter. When your functions use both techniques—and they will—you must keep in mind when to use pointers and when to directly reference the variable. As a rule, the parameters whose values you want to change will require call by reference. To better understand the impact of call by reference versus call by value, change the *change_first* function, as shown here:

```
void change_first(int *first, int second)
 {
  *first = second;  // Assign the value of second to first
  second = 100;
 }
```

When you compile and execute this program, you will see that the value of *first* has changed but the value of *second* has not. Because the parameter *second* is passed using call by value, the change to the parameter is not visible outside the function.

CALL BY REFERENCE STILL USES THE STACK 245

As you have learned, when C passes parameters to functions, C places the parameter's values on the stack. C uses the stack to hold parameters whether you are using call by value or call by reference. When you pass a parameter by value, C places the parameter's *value* on the stack. When you pass a parameter by reference, C places the parameter's *address* on the stack. Tip 244 presented the *chgfirst.c* program, which used the *change_first* function to assign the value of the function's second parameter to the function's first parameter. When the program invokes the function, C places the address of variable *a* and the value of variable *b* on the stack, as shown in Figure 245.

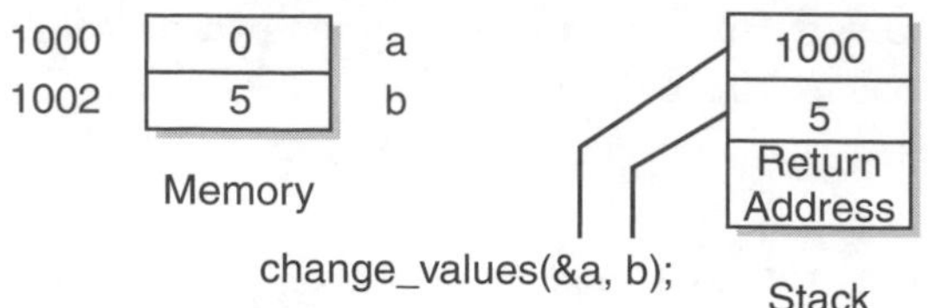

Figure 245 C places an address and a value on the stack.

Because the *change_values* function actually references the memory location that contains the value of variable *a,* the changes *change_values* makes to the variable exist after the function ends.

246 INTRODUCING FUNCTION VARIABLES THAT REMEMBER

In C, the variables you declare within functions are often called *automatic* because the C compiler automatically creates them when the function begins and then destroys them when the function ends. The variable's automatic life occurs because the compiler stores function variables temporarily on the stack. As a result, should a function assign a value to a variable during one invocation, the variable loses its value when the function completes. The next time you invoke the function, the variable's value is again undefined. Depending on the processing your function performs, sometimes you may want the function's variables to remember the last value they were assigned within the function.

For example, assume that you have written a function called *print_reportcard,* which prints a report card for every student in a school. Your function might use the variable *student_id* to hold the student identification number of the last student whose report card the function printed. In this way, without being told to, the function can begin with the next student. To make your function's local variables remember their values in this way, you must declare the variables using the keyword *static*, as shown here:

```
void print_reportcard(int printer_number)
  {
    static int student_id;
    // Other statements
  }
```

The following program, *static.c,* illustrates the use of a *static* variable within a function. The program, which uses the *print_reportcard* function, begins by assigning the variable *student_id* the value 100. Each time the program invokes the function, the function will display the variable's value and then will increment the value by 1, as shown here:

```
#include <stdio.h>

void print_reportcard(int printer_number)
 {
   static int student_id = 100;
   printf("Printing report card for student %d\n", student_id); student_id++;
   // Other statements here
 }

 void main(void)
  {
    print_reportcard(1);
    print_reportcard(1);
    print_reportcard(1);
  }
```

When you compile and execute the *static.c* program, your screen will display the following output:

```
Printing report card for student 100
Printing report card for student 101
```

```
Printing report card for student 102
C:\>
```

As you can see, the *student_id* variable retains its value from one invocation to the next.

Note: *When you declare static variables, the C compiler does not store the variables on the stack. Instead, the compiler places the variables within the data segment so their values can remain.*

UNDERSTANDING HOW C INITIALIZES STATIC VARIABLES 247

In Tip 246 you learned that the *static* keyword directs the compiler to retain a variable's value from one function invocation to the next. When your function declares a static variable, C lets you initialize the variable, as shown here:

```
void print_reportcard(int printer_number)
  {
    static int student_id = 100;  // Initialized once

    // Other statements
  }
```

When you declare a variable as *static*, the C compiler will initialize the variable to the value you specify. When you invoke the function later, C *will not perform* the initialization assignment again. This function variable initialization is different from the processing that C usually performs within a function. In the case of the following function, C will initialize the variable *count* every time the program calls the function:

```
void some_function(int age, char *name)
  {
    int count = 1;  // Initialized on every call

    // Other statements
  }
```

USING THE PASCAL CALLING SEQUENCE 248

As you create C programs, you might find that you would like to use a function that you have previously created in Pascal. Depending on your compiler, linker, and library type, you might still be able to call the Pascal function from your C program. The steps you must perform to do so, however, will depend on your compiler. Additionally, within your program code, you must include a function prototype at the start of your program that includes the *pascal* keyword, as shown here:

```
int pascal some_function(int score, int grade);
```

If you programmed in older Windows environments, you may have found many run-time library functions use the Pascal calling sequence. Functions that use the *pascal* keyword cannot support a variable number of arguments (as can *printf*).

UNDERSTANDING THE PASCAL KEYWORD'S EFFECT 249

You learned in Tip 248 that when your programs invoke a function, C passes parameters to the function using the stack. C places parameters on the stack from right to left. Figure 249.1 illustrates the stack contents for a C function call.

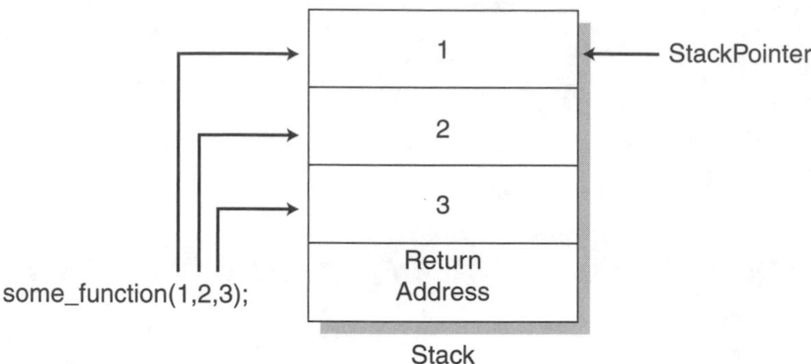

Figure 249.1 The stack contents for a C function call.

Pascal, on the other hand, pushes arguments onto the stack from left to right. Figure 249.2 illustrates the stack contents for a Pascal function call.

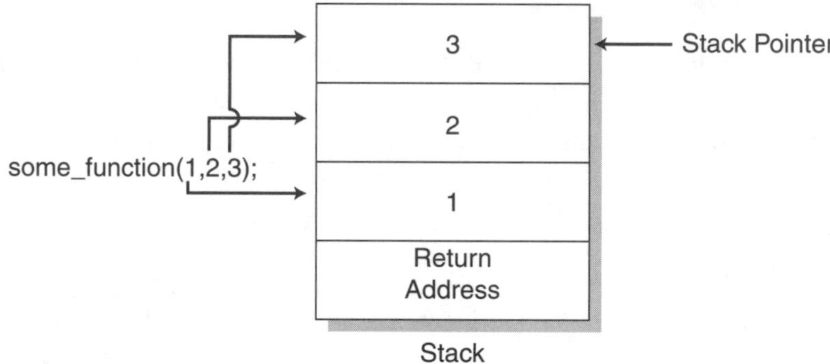

Figure 249.2 The stack contents for a Pascal function call.

If you are using a Pascal function from within your C program, use the *pascal* keyword to direct the C compiler to place the parameters on the stack from left to right, in the order that Pascal expects.

250 WRITING A MIXED LANGUAGE EXAMPLE

As you have learned, many C compilers let you invoke functions that were written in a different programming language. If you are calling a Pascal function from within your C program, for example, you can precede the function prototype using the *pascal* keyword. As you have learned, the *pascal* keyword directs the compiler to push parameters onto the stack from left to right. To illustrate the processing the *pascal* keyword performs, create the following function, *show_values*, and precede the function with the *pascal* keyword:

```
#include <stdio.h>

void pascal show_values(int a, int b, int c)
  {
     printf("a %d b %d c %d\n", a, b, c);
  }
```

Next, call the function using the following program code:

```
void main(void)
  {
     show_values(1, 2, 3);
     show_values(100, 200, 300);
  }
```

To experiment with the *show_values* function, remove the *pascal* keyword and note the change in the order of the parameter values C displays. Should your programs later call a Pascal routine, you must use the *pascal* keyword in the function prototype.

UNDERSTANDING THE CDECL KEYWORD 251

In Tip 250 you learned that if you use functions written in Pascal, you will use the *pascal* keyword to inform the compiler so that the compiler will place parameters onto the stack in the correct order. When you use functions written with multiple programming languages, you will probably want to include the keyword *cdecl* within your function prototypes to indicate C functions and to provide clarity to the reader. For example, the following function prototype informs the compiler that the function *change_values* uses the C calling structure:

```
int cdecl change_values(int *, int *, int *);
```

When the compiler encounters the *cdecl* keyword within a function header, the compiler will ensure that parameters passed to the function are placed on the stack from right to left. In addition, the compiler will ensure that the linker uses the C format for the function's name.

UNDERSTANDING RECURSION 252

C lets you divide your program into smaller pieces called functions. Using functions, your program becomes easier to understand, program, and test. In addition, you can often use the functions you create for one program within another program. As your programs execute, one function may call another, which calls another, which may, in turn, call several other functions. Within the series, each function performs a specific operation. As it turns out, C even lets a function call itself! A *recursive function* is a function that calls itself to perform a specific operation. The process of a function calling itself is named *recursion*. As the complexity of your programs and functions increases, you might find that you can easily define many operations in terms of themselves. When you use complex programs and functions, you might want to create a recursive function. Many programming books, for example, use the factorial problem to illustrate how recursion works. The factorial of the value 1 is 1. The factorial of the value 2 is 2*1. The factorial of the value 3 is 3*2*1. Likewise, the factorial of the value 4 is 4*3*2*1. The factorial process can essentially go on indefinitely. If you take a close look at the processing that the factorial performs, you will find that the factorial of 4, for example, is actually 4 times the factorial of 3 (3*2*1). Likewise, the factorial of 3 is actually 3 times the factorial of 2 (2*1). The factorial of 2 is 2 times the factorial of 1 (1). Table 252 illustrates the factorial processing.

Value	Calculation	Result	Factorial
1	1	1	1
2	2*1	2	2 * Factorial(1)
3	3*2*1	6	3 * Factorial(2)
4	4*3*2*1	24	4 * Factorial(3)
5	5*4*3*2*1	120	5 * Factorial(4)

Table 252 Factorial processing.

The following program, *fact.c*, creates the recursive function *factorial* and then uses the function to return the factorial values for the values 1 through 5:

```
#include <stdio.h>

int factorial(int value)
 {
```

```
    if (value == 1)
      return(1);
    else
      return(value * factorial(value-1));
  }
void main(void)
  {
    int i;
    for (i = 1; i <= 5; i++)
      printf("The factorial of %d is %d\n", i, factorial(i));
  }
```

As you can see, the *factorial* function returns a result that is based on the result of the function itself. Tip 253 examines the *factorial* function in detail.

253 UNDERSTANDING THE RECURSIVE FACTORIAL FUNCTION

In Tip 252 you learned that a recursive function is a function that calls itself to perform a specific task. Tip 252 presented the *factorial* function to illustrate recursion. The *factorial* function receives a specific parameter value. When the function begins, it first checks whether the value is 1, which by factorial definition is 1. If the value is 1, the function returns the value 1. If the value is not 1, the function returns the result of the value times the factorial of the value minus 1.

Assume, for example, that the program invokes the function with the value 3. The function will return the result of 3 * *factorial(3–1)*. When C encounters the function call within the *return* statement, C will invoke the function a second time—this time with the value of 3–1 or 2. Again, because the value is not 1, the function returns the result of the 2 * *factorial(2–1)*. On the third invocation of the function, the value is 1. As a result, the function returns the value 1 to the calling function, which, in turn, returns the result of 2*1 to its calling function. That calling function then returns the result of 3*2*1 to *its* calling function. Figure 253 illustrates the chain of recursive function invocations and return values for the factorial(3) function call.

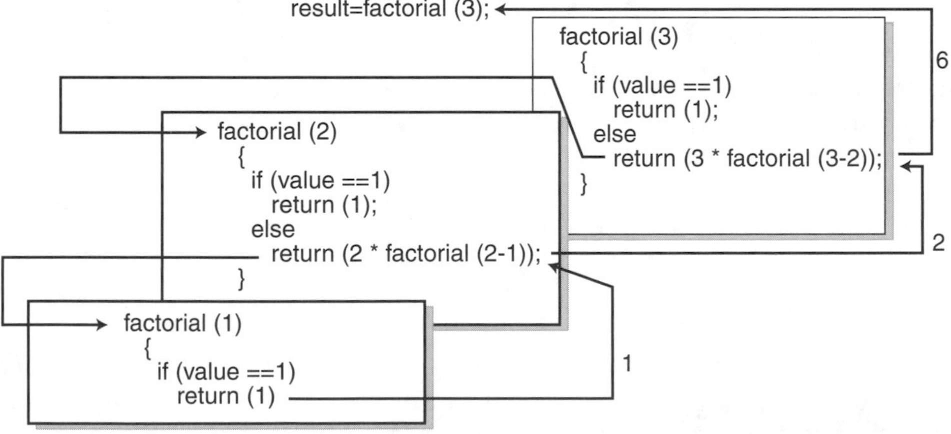

***Figure 253** The chain of function calls and value returns for the recursive **factorial** function.*

A recursive function is somewhat like a looping construc in that you must specify an ending condition. If you do not specify an ending condition, the function will never end. In the factorial problem, the ending condition is the factorial of 1, which is, by definition, 1.

PROGRAMMING ANOTHER RECURSIVE EXAMPLE 254

In Tip 252 you learned that a recursive function is a function that calls itself in order to perform specific processing. Tip 253, in turn, presented and explained the recursive *factorial* function. Because recursion can be a difficult concept, this Tip presents one more recursive function, *display_backward*, which will display a string's letters in reverse order. Given the letters ABCDE, the function will display the letters on your screen as EDCBA. The following program, *backward.c*, uses the *display_backward* function:

```
#include <stdio.h>

void display_backward(char *string)
 {
   if (*string)
    {
      display_backward(string+1);
      putchar(*string);
    }
 }

void main(void)
 {
   display_backward("ABCDE");
 }
```

DISPLAYING VALUES TO BETTER UNDERSTAND RECURSION 255

As you have learned, a *recursive* function is a function that calls itself to perform a specific operation. Tip 252 presented the recursive *factorial* function. To help you to better understand the recursion process, the program *showfact.c* includes *printf* statements within the *factorial* function that illustrate the function's recursive processing within the program:

```
#include <stdio.h>

int factorial(int value)
 {
   printf("In factorial with the value %d\n", value);
   if (value == 1)
    {
      printf("Returning the value 1\n");
      return(1);
    }
   else
    {
      printf("Returning %d * factorial(%d)\n", value, value-1);
      return(value * factorial(value-1));
    }
 }

void main(void)
 {
   printf("The factorial of 4 is %d\n", factorial(4));
 }
```

When you compile and execute the *showfact.c* program, your screen will display the following output:

```
In factorial with the value 4
Returning 4 * factorial(3)
In factorial with the value 3
Returning 3 * factorial(2)
In factorial with the value 2
Returning 2 * factorial(1)
In factorial with the value 1
Returning the value 1
The factorial of 4 is 24
C:\>
```

Inserting *printf* statements throughout your recursive functions will help you to better understand the processing the functions perform.

256 UNDERSTANDING DIRECT AND INDIRECT RECURSION

A recursive function is a function that calls itself to perform a specific operation. Several previous Tips in this section have presented recursive functions. When a function invokes itself to perform a task, the function performs a *direct recursion*. After you have examined a few recursive functions, you should be able to understand most functions that use direct recursion. A more difficult form of recursion, *indirect recursion*, occurs when a function (function A) calls another function (function B), which, in turn, calls the original function (function A). Because indirect recursion can result in code that is very difficult to understand, as a rule you should avoid using indirect recursion whenever possible.

257 DECIDING WHETHER TO USE RECURSION

A recursive function is a function that calls itself to perform a specific task. When you create functions, you can use recursion to create elegant solutions to many problems. However, you should avoid recursion whenever possible for two reasons. First, recursive functions can be difficult for novice programmers to understand. Second, as a rule, recursive functions are often considerably slower than their non-recursive counterparts. The following program, *no_recur.c*, invokes the non-recursive function *string_length* with the string "Jamsa's C/C++/C# Programmer's Bible" 10,000,000 times and then displays the amount of time required to perform the processing:

```
#include <stdio.h>
#include <time.h>

int string_length(const char *str)
 {
   int length = 0;

   while (*str++)
     length++;
   return(length);
 }

void main(void)
 {
   long int counter;

   time_t start_time, end_time;
   time(&start_time);
   for (counter = 0; counter < 10000000L; counter++)
     string_length("Jamsa\'s C/C++/C# Programmer\'s Bible");
   time(&end_time);
   printf("Processing time %d\n", end_time - start_time);
 }
```

Next, the program *ok_recur.c* uses a recursive implementation of the *string_length* function to perform the same processing:

```
#include <stdio.h>
#include <time.h>

int string_length(const char *str)
 {
   if (*str)
     return(1 + string_length(str+1));
   else
     return(0);
 }

void main(void)
 {
   long int counter;

   time_t start_time, end_time;
   time(&start_time);
   for (counter = 0; counter < 10000000L; counter++)
     string_length("Jamsa\'s C/C++/C# Programmer\'s Bible");
   time(&end_time);
   printf("Processing time %d\n", end_time - start_time);
 }
```

Experiment with these programs by, for example, changing the number of function calls to one or two million. As you will find, the non-recursive function executes considerably faster than its recursive counterpart. Therefore, when you design a recursive function, keep in mind that you may be adding significant overhead to your program's execution time.

UNDERSTANDING WHY RECURSIVE FUNCTIONS ARE SLOW

A recursive function is a function that calls itself to perform a specific task. As you learned in Tip 257, one reason to avoid using recursion is that recursive functions are usually considerably slower than their non-recursive counterparts. Recursive functions are slow because the function *call overhead* occurs with every invocation. As Tip 231 details, each time your program calls a function, the C compiler pushes onto the stack the address of the statement that immediately follows the function call (named the *return address*). Next, the compiler pushes the parameter values onto the stack. When the function completes, the computer's operating system pops the return address off the stack into the CPU's program counter. Although computers can perform these push and pop operations very quickly, the operations still require time.

As an example, assume that you invoke the recursive *factorial* function with the value 50. The function will then invoke itself 49 times. If each function call adds 10 milliseconds to your program, the function will be a half-second slower than a non-recursive counterpart, which only has the overhead of one function invocation. A half-second of overhead does not seem like much, yet assume that the program calls the function ten times. The half-second delay quickly turns into five seconds. If the program uses the function 100 times, the delay becomes 50 seconds, and so on. If you are writing a program that requires maximum performance, you should try to eliminate recursive functions whenever possible.

Note: *With newer, faster microprocessors, the operating system slowdown from recursive functions is not as important as it once was. However, the impact of recursive functions is still significant, and you should try to write effective, readable code without recursion whenever possible.*

259 UNDERSTANDING HOW TO REMOVE RECURSION

A recursive function is a function that calls itself to perform a specific task. As you have learned, you can improve your program's performance using non-recursive functions. As a rule, any function you are able to write recursively you can also write in terms of looping constructs, such as a *for* or *while* statement. The following program, *loopfact.c*, uses a *for* loop to implement the *factorial* function:

```
#include <stdio.h>

int factorial(int value)
 {
   int result = 1;
   int counter;

   for (counter = 2; counter <= value; counter++)
     result *= counter;
   return(result);
 }

void main(void)
 {
   int i;

   for (i = 1; i <= 5; i++)
     printf("Factorial of %d is %d\n", i, factorial(i));
 }
```

Whenever you eliminate recursion within your programs using a looping construct, you will generally improve your program's performance. However, keep in mind that users may more easily understand some operations your programs will perform when you implement the operations with recursion. Just as there are times when you must make tradeoffs between your program's speed and memory consumption, there may also be times when you must choose between readability and performance.

260 PASSING STRINGS TO FUNCTIONS

As you have learned, when you pass parameters to functions, C, by default, passes the parameters *by value*. Therefore, any changes that your function makes to the parameter do not exist outside of the function. To change a parameter's value, you must pass the parameter *by reference*. The exception to this rule is character strings. When you invoke a function with a character string, you are simply passing a byte array to the function. When C passes an array (any type of array—not just a string), C passes the array's starting address to the function. In other words, *C always uses call by reference for arrays*, so you do not need to use the *address* operator.

261 PASSING SPECIFIC ARRAY ELEMENTS

As you learned in Tip 260, C always passes arrays to functions using call by reference. As you work with character strings, there may be times when you want a function to work with specific array elements. For example, the following program, *halfcaps.c*, uses the *strupr* function to convert a section of a character string to uppercase:

```
#include <stdio.h>
#include <string.h>

void main(void)
 {
   char alphabet[] = "abcdefghijklmnopqrstuvwxyz";

   strupr(&alphabet[13]);
```

```
    printf(alphabet);
  }
```

The *strupr* expects the starting address of a *NULL*-terminated string to be a parameter. In this case, the program passes to *strupr* the address of the letter n, which the function call then follows with several *NULL*-terminated characters. By passing the address of a specific array element, your programs can use functions to manipulate specific array elements.

UNDERSTANDING CONST IN FORMAL PARAMETERS 262

If you examine the function prototypes for the string manipulation functions presented in the Strings section of this book, you will find that many of the parameter declarations place the keyword *const* before character string arguments, as shown here:

```
char *strcpy(char *destination, const char *source);
```

In the *strcpy* function's definition example, the *const* keyword specifies that the function code should not change the variable *source* within the function. Should your function code try to change the string's value, the compiler will generate an error. The following program, *chkconst.c*, uses the *const* keyword for the parameter *string*:

```
#include <stdio.h>

void no_change(const char *string)
 {
   while (*string)
     *string++ = toupper(*string);
 }

 void main(void)
  {
    char title[] = "Jamsa\'s C/C++/C# Programmer\'s Bible";

    no_change(title);
    printf(title);
  }
```

As you can see, the function *no_change* tries to convert the string's letters to uppercase. However, because the program uses the *const* keyword, the compiler will display an error message and the code will not successfully compile. You should use the *const* keyword before parameters that a function receives by reference when you do not want the parameter's value changed. Because C normally passes non-pointer parameters by value, parameters by value do not require the *const* keyword.

USING CONST WILL NOT PREVENT PARAMETER MODIFICATION 263

As you learned in Tip 262, the *const* keyword informs the compiler that the function should not change a specific parameter's value. Should a function try to modify such a parameter's value, the compiler will generate an error, and the program will not compile. However, you should note that just because the function's header specifies a parameter as a constant, it does not mean that function cannot change the parameter's value. The following program, *chgconst.c*, uses a pointer to the constant parameter *string* to convert the string's contents to uppercase:

```
#include <stdio.h>
#include <ctype.h>

void no_change(const char *string)
 {
   char *alias = string;

   while (*alias)
     *alias++ = toupper(*alias);
```

```
}
void main(void)
 {
   char title[] = "Jamsa\'s C/C++/C# Programmer\'s Bible";
   no_change(title);
   printf(title);
 }
```

When you compile and execute the *chgconst.c* program, the function *no_change* will convert the string's characters to uppercase. Because you used *pointer aliasing* (referring to a variable's memory locations using a different name), the compiler does not detect the parameter value change. Depending on your compiler type, the compiler may generate a warning message. If you are creating your own functions, do not use aliasing to change a parameter's value as does the *chgconst.c* program. If a parameter is truly a constant, its value should not change. The program within this Tip should teach you that the *const* keyword cannot actually prevent a parameter's value from changing.

264 UNDERSTANDING UNBOUNDED STRING DECLARATIONS

In C, a character string is an array of character values. You learned in the Strings section of this book that you specify the maximum number of characters the string will ever hold to create a string, as shown here:

```
char name[64];
char title[32];
char buffer[512];
```

When you pass a character string to a function, you actually pass the string's starting address. Because the *NULL* character terminates the string, C functions do not care how many characters the string contains. As a result, many functions declare character string parameters as unbounded arrays (arrays that do not specify a size), as shown here:

```
int strlen(char string[])
```

The declaration *char string[]* tells the compiler that the function will receive a pointer to a *NULL*-terminated string. The string might contain 64 characters, 1,024 characters, or maybe just the *NULL* character. The following program, *strarray.c*, uses an unbounded array to implement the *strlen* function:

```
#include <stdio.h>
int strlen(char str[])
 {
   int i = 0;
   while (str[i] !=NULL)
     i++;
   return(i);
 }
void main(void)
 {
   printf("Length of ABC is %d\n", strlen("ABC"));
   printf("Length of C/C++/C# Programmer\'s Bible is %d\n",
     strlen("C/C++/C# Programmer\'s Bible"));
   printf("Length of a NULL string is %d\n", strlen(""));
 }
```

When you compile and execute the *strarray.c* program, you will see that the function works for any size string. However, like most C functions that work with strings, the function will fail if the string is not terminated by the *NULL* character.

USING POINTERS VERSUS STRING DECLARATIONS 265

As you examine different C functions that manipulate strings, you may find the character strings declared as either unbounded arrays or as pointers, as shown here:

```
char *strcpy(char destination[], char source[]);
char *strcpy(char *destination, char *source);
```

Both declarations in the previous example inform the compiler that it is working with strings. Both are functionally identical, and both are correct. If you are creating your own functions, the format you choose should depend on how you reference the parameter within the function. If you treat the parameter as a pointer, use the *pointer-style* declaration. If you instead treat the parameter as an array, use the array. By treating the parameter in a consistent way, you will make your programs easier to understand.

HOW C USES THE STACK FOR STRING PARAMETERS 266

As you have learned, when your programs pass a parameter to functions, C places the parameter's value or address on the stack. When you pass a character string to a function, C places the string's starting address on the stack. For example, Tip 264 presented the program *strarray.c*, which passed several strings to the *strlen* function. Figure 266 illustrates the parameter value C places on the stack for the first function invocation.

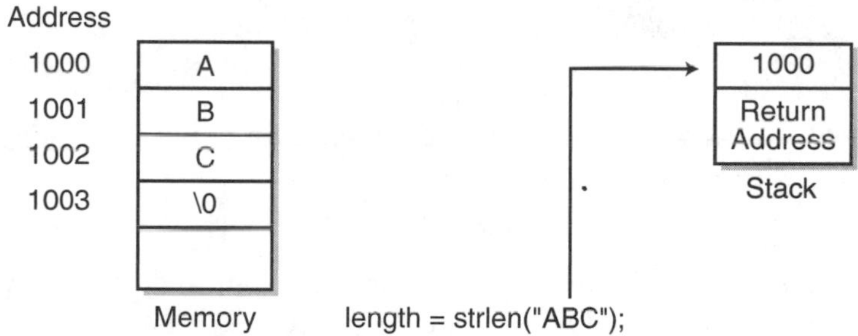

***Figure 266** How C passes string parameters to functions.*

As you can see, C does not place the string's characters onto the stack. Instead, C simply places the address of the *NULL*-terminated string on the stack. Because the function receives only an address (as opposed to an array of bytes), the function does not care how many characters the string contains.

UNDERSTANDING EXTERNAL VARIABLES 267

You can often use the functions you create for one program within another. To simplify function reuse, programmers often place functions in *object code libraries*. The Tools section of this book discusses the use of such libraries. In some cases, a library might define a global variable, such as the *_fmode*, *_psp*, or *errno* variables discussed throughout this book. When code outside of the current program defines a global variable and you want to use the global variable within your program, you must declare the variable using the *extern* keyword. The *extern* keyword tells the compiler that another program has declared the variable *externally* (outside of the current source file). If you examine the header file *dos.h*, for example, you will find several external variable declarations, including those shown here:

```
extern int const _Cdecl _8087;
extern int _Cdecl _argc;
extern char **_Cdecl _argv;
extern char **_Cdecl environ;
```

If you do not use the *extern* keyword, the compiler will assume that you are creating a variable with the name specified. When you include the *extern* keyword, on the other hand, the compiler will search for the global variable you specified.

268 PUTTING EXTERN TO USE

Tip 267 introduced the *extern* keyword, which you will use within your programs to tell the compiler to use a global variable that another program has declared outside of the current program. To better understand how the *extern* keyword works, compile the file *external.c*, which contains the declaration of the variable *tip_count* and the function *show_title*:

```
#include <stdio.h>

int tip_count = 1700;   // Global variable

void show_title(void)
 {
   printf("Jamsa\'s C/C++/C# Programmer\'s Bible");
 }
```

When you compile the *external.c* program, C will create the object file *external.obj*. The program *showext.c*, shown here, uses the external variable *tip_count* within the *external.obj* file:

```
#include <stdio.h>

void main(void)
 {
   extern int tip_count;

   printf("The number of tips is %d\n", tip_count);
 }
```

When you compile the *showext.c* program, perform the following steps within the *Turbo C++ Lite* compiler (if you are not using *Turbo C++ Lite,* check your compiler documentation):

1. Select the Project menu Open Project option.
2. Change to the directory which contains the *showext.c* program and enter the project's name as *showext*. Click your mouse on OK to create the project.
3. Select the Project menu Add Item option.
4. Add the *external.obj* file to the project.
5. Add the *showext.c* file to the project.
6. Select the Compile menu Build All option to build the file.

In this case, the *showext.c* program displays the value of the external variable *tip_count*. The program does not use the *show_title* function, although the program could have—simply by invoking *show_title*. However, the program's goal was to illustrate the use of the *extern* keyword.

Note: *To use externals within other compilers, refer to the compiler's on-line help documentation or the printed documentation that came with the compiler.*

269 UNDERSTANDING EXTERNAL STATIC

In Tip 267 you learned that the *extern* keyword tells the C compiler that you are referencing a global variable that a program in a different file defines. When the linker links your program modules, the linker will determine the variable's memory location. In Tip 268 you used the global variable *tip_count*, which was defined in the object file *external.obj*. Because the *showext.c* program referred to the variable using the *extern* keyword, the program could access the variable. Depending on your programs, there may be times when you use global variables in an object file that you do not want functions outside of the object file to access. In such cases, you simply precede the variable name with the *static* keyword:

```
static int variable_name;
```

The following file, *stextern.c*, declares two global variables, one named *tip_count* and one named *title*:

```
#include <stdio.h>

int tip_count = 1700;  // Global variable
static char title[] = "Jamsa\'s C/C++/C# Programmer\'s Bible";

void show_title(void)
 {
   printf(title);
 }
```

Compile the *stextern.c* file to create the object file *stextern.obj*. Next, create the following program, *nostatic.c*, which tries to use both global variables contained within the *stextern.obj* file:

```
#include <stdio.h>

void main(void)
 {
   extern int tip_count;
   extern char *title;
   void show_title(void);

   printf("The number of tips is %d\n", tip_count);
   printf("The book\'s title is %s\n", title);
   show_title();
 }
```

As you learned in Tip 268, to compile and link the program using *Turbo C++ Lite*, perform the following steps:

1. Select the Project menu Open Project option.
2. Change to the directory which contains the *nostatic.c* program and enter the project's name as *nostatic*. Click your mouse on OK to create the project.
3. Select the Project menu Add Item option.
4. Add the *extern2.obj* file to the project.
5. Add the *nostatic.c* file to the project.
6. Select the Compile menu Build All option to build the file.

When you compile and link the *nostatic.c* program, the linker should display a message stating that the compiler could not resolve the *title* variable. Because the *static* keyword precedes the declaration of the *title* variable, the variable is only known within the object file *extern2.obj*.

UNDERSTANDING THE VOLATILE KEYWORD 270

As the complexity of your programs increases, you might eventually write low-level functions and routines that access the PC's I/O ports or that service the PC's interrupt registers (also referred to simply as *interrupts*). When your programs perform such operations, using an interrupt or accessing a port might change your variables that correspond to specific memory locations or port addresses. Because both your program and many factors external to your program can change such variables, you must tell the compiler that the variable's value can change at any time. To inform the compiler that operations outside of the program might change a variable's value, use the *volatile* keyword, as shown here:

```
volatile int some_variable;
```

When the compiler encounters the *volatile* keyword, the compiler knows not to make assumptions about the variable's value at any time. For example, the compiler will not place the variable's value into a register for quick access. Doing so would run the risk that the register value is not the same as the variable's memory contents, which an interrupt (for example) might have changed after the variable's storage into the register, without the program's knowledge. Instead, when your program must access a variable's value, the compiler will specifically reference the variable's memory location.

Note: *You should generally declare* ***volatile*** *variables as global variables. In this way, programs and the outside operations reference memory locations contained within the program's data segment, as opposed to stack locations, which the program discards when the corresponding function ends.*

271 UNDERSTANDING THE CALL FRAME AND BASE POINTER

You have learned that when your program invokes a function, C pushes the return address and the function's parameters onto the stack. Within the stack, C refers to the saved function call information as a *call frame*. To help your functions quickly locate the call frame, C assigns the base pointer register (BP) to the address of the start of the frame.

C also places the function's local variables onto the stack (within the call frame). Figure 271 illustrates the contents of a simple call frame:

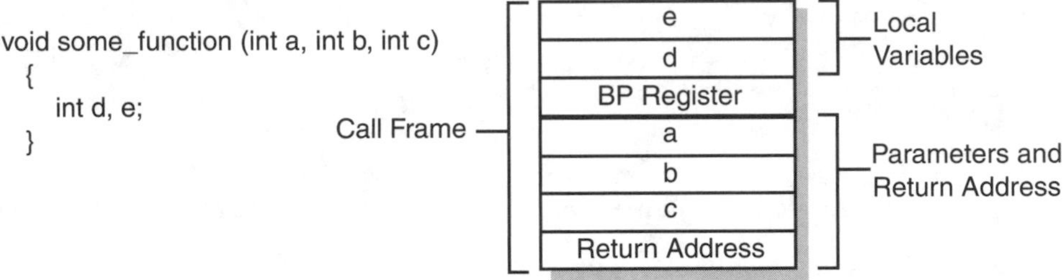

Figure 271 The information C places onto the stack for a function call constitutes a call frame.

When you write assembly language functions that you expect to call from within your C programs, you must understand the use and structure of the call frame so that your assembly language functions can access the parameter values stored within that call frame.

272 CALLING AN ASSEMBLY LANGUAGE FUNCTION

In Tip 236 you learned that your programs can call functions written in other programming languages, such as Pascal. In addition, your programs can call assembly language routines. The following assembly language routine, *swap_values*, exchanges the values of two variables passed to the function by reference (by address):

```
          .MODEL   small
          .CODE
          PUBLIC   _swap_values

_swap_values     PROC
          push     bp
          mov      bp,sp
          sub      sp,2
          push     si
          push     di

          mov      si,word ptr [bp+4]               ;Arg1
```

```
            mov     di,word ptr [bp+6]                  ;Arg2

            mov     ax,word ptr [si]
            mov     word ptr [bp-2],ax

            mov     ax,word ptr [di]
            mov     word ptr [si],ax

            mov     ax,word ptr [bp-2]
            mov     word ptr [di],ax

            pop     di
            pop     si
            mov     sp,bp
            pop     bp
quit:           ret

_swap_values    ENDP
            END
```

The companion CD-ROM that accompanies this book contains the *swap.asm* file. If you are using Borland C++, assemble the file to create the object file *swap.obj*, as shown here:

```
C:\> TASM SWAP.ASM <ENTER>
```

Next, create the following C program, *use_swap.c*, which uses the *swap_values* function:

```
#include <stdio.h>

void swap_values(int *, int *);

void main(void)
 {
   int a = 1, b = 2;

   printf("Original values a %d b %d\n", a, b);
   swap_values(&a, &b);
   printf("Swapped values a %d b %d\n", a, b);
 }
```

In this case, you wrote the function *swap_values* to support near pointers. If you change memory models, you must change the assembly language routine.

RETURNING A VALUE FROM AN ASSEMBLY LANGUAGE FUNCTION 273

In Tip 261 you learned how to call an assembly language function from within your C programs. In the *swap.asm* program, the function did not return a result. The following assembly language routine, *get_maximum*, however, returns the larger of two integer values:

```
            .MODEL small
            .CODE
            PUBLIC _get_maximum

_get_maximum        PROC
            push    bp
            mov     bp,sp

Arg1                equ     [bp+4]
Arg2                equ     [bp+6]

            mov     ax,Arg1         ;Move Arg1 into AX
            cmp     Arg2,ax         ;Compare Arg2 to Arg1
            jg      arg2_bigger     ;Jump if Arg2 is bigger
            jmp     finished

arg2_bigger:        mov     ax,Arg2
```

```
finished: pop    bp
          ret
_get_maximum     ENDP
          END
```

The companion CD-ROM that accompanies this book contains the *get_max.asm* file, which contains the *get_maximum* routine. As you can see, the assembly language routine places its result in the AX register. Later Tips explain the different registers in detail; for now, however, you can consider the AX register as similar to the BP register which the previous Tip explains. The following C program, *use_max.c*, invokes the assembly language function to determine the larger of two values:

```
#include <stdio.h>
extern int get_maximum(int, int);

void main(void)
 {
   int result;

   result = get_maximum(100, 200);
   printf("The larger value is %d\n", result);
 }
```

When the program calls the function, the C compiler will assign the value of the AX register as the function's result.

274 Introducing Functions that Do Not Return Values

As the number of functions you create increases, you will eventually create a function that does not return a value. As you have learned, the C compiler, unless told otherwise, assumes that a function returns the type *int*. If your function does not return a value, you should declare the function as type *void*, as shown here:

```
void my_function(int age, char *name);
```

Should the program later try to use the function's return value, as shown here, the compiler will generate an error:

```
result = my_function(32, "Jamsa");
```

275 Understanding Functions that Do Not Use Parameters

As the number of programs and functions that you create increases, you might eventually create a function that does not use any parameters. When you define the function (and the function prototype), you should use the *void* keyword to inform the compiler (and other programmers) that the function does not use parameters:

```
int my_function(void);
```

Should the program later try to invoke the function with parameters, the compiler will generate an error.

276 Understanding the auto Keyword

As you examine C programs, you might find variable declarations that use the keyword *auto*, as shown here:

```
auto int counter;
auto int flags;
```

The *auto* keyword informs the compiler that the variable is local to the function and that the compiler should automatically create and destroy the variable. The compiler creates automatic variables by allocating stack space.

Because variables are automatic by default, most programs omit the *auto* keyword. Within a function, the following variable declarations are identical:

```
auto int counter;
int counter;
```

UNDERSTANDING SCOPE 277

Within your programs, functions and variables have a *scope* that defines the areas within the program where their names have meaning. For example, consider the following program, *twocount.c*, which uses two variables named *count*:

```
#include <stdio.h>

void beeper(int beep_count)
 {
   int count;

   for (count = 1; count <= beep_count; count++)
     putchar(7);
 }

void main(void)
 {
   int count;

   for (count = 1; count <= 3; count++)
     {
       printf("About to beep %d times\n", count);
       beeper(count);
     }
 }
```

As you can see, the functions *beeper* and *main* both use variables named *count*. To C, however, both variables are distinct—each has a different scope. In the case of the function *beeper*, C only knows its *count* variable (that is, *count* has a defined scope) while the function is executing. Likewise, in the case of *main*, its *count* variable only has meaning while *main* is executing. As a result, the *for* loop that changes the *count* variable in the function *beeper* has no effect on the *count* variable within *main*.

When you discuss a variable's scope, you will often use the terms *local* and *global* variables. A local variable is one with scope restricted to a specific function. On the other hand, the entire program can know a global variable. In the case of the *twocount.c* program, each function defines each occurrence of the *count* variable as local.

UNDERSTANDING C'S CATEGORIES OF SCOPE 278

As you have learned, an identifier's *scope* (usually a variable or function name) is the part of the program within which the identifier has meaning (in other words, where the program can use the identifier). C defines four categories of scope: block, function, function prototype, and file. Additionally, C++ defines class scope. *Block scope* defines the bracketed region within which your program has defined a variable. Usually, block scope refers to a function. Local variables have block scope. As you learned in the Getting Started with C section of this book, however, you can declare variables after any open brace. The variable's scope exists up to the closing brace—meaning that a parameter with block scope may only have scope within an *if* conditional. Formal parameters have block scope, with the scope limited to the function defining the parameter. *Function scope* defines the region between a function's opening and closing brace. The only item with function scope is a label used by the *goto* statement. *Function prototype scope* specifies the region within the start and end of a function prototype. Identifiers that appear within a function prototype have meaning only within the function prototype, as shown here:

```
int some_function(int age, char *name);
```

File scope specifies a region from an identifier's declaration to the end of the source file. Global variables have file scope, which means only functions that physically follow the global variable declaration within the file can reference a global variable. In C++, *class scope* defines the named collection of methods and data structures that comprise the class.

279 UNDERSTANDING NAME SPACE AND IDENTIFIERS

As you have learned, *scope* defines the region of a program within which an identifier has meaning. Similarly, *name space* defines a region within which identifier names must be unique. In the simplest sense, an *identifier* is a name. C defines four classes of identifiers, as shown in the following list:

- *goto* label names: The label names a *goto* statement uses must be unique within a function.
- Structure, union, and enumeration tags: A *tag* is the name of a structure, union, or enumerated type. Tags must be unique within a block.
- Structure and union member names: The member names that appear within a structure or union must be unique. Different unions or structures can have the same member names.
- Variables, *typedef* identifiers, functions, and enumerated members: These identifiers must be unique within the scope (as explained in Tip 278) in which the identifier is defined.

280 UNDERSTANDING IDENTIFIER VISIBILITY

As you have learned, *scope* defines the program region within which an identifier has meaning. In a similar way, an identifier's *visibility* defines the region of code within which a program can access an identifier. Usually, an identifier's scope and visibility are the same. However, when your program declares an identifier with the same name within a block that appears inside of an existing identifier's scope, the compiler temporarily hides the outer identifier (in other words, the outer identifier loses visibility and the compiler does not recognize it). Consider the following program, *visible.c*, which uses two identifiers named *value*:

```
#include <stdio.h>

void main(void)
 {
   int value = 1500;

   if (value > 1499)
     {
       int value = 1;

       printf("Inner value is %d\n", value);
     }
   printf("Outer value is %d\n", value);
 }
```

When you compile and execute the *visible.c* program, your screen will display the following output:

```
Inner value is 1
Outer value is 1500
C:\>
```

When the program declares the variable *value* within the *if* statement, the variable declaration instructs the compiler to hide the outer occurrence of the variable with the same name. Outside of the block, however, the outer variable becomes visible to the compiler once again.

UNDERSTANDING DURATION 281

When you discuss variables, *duration* specifies the amount of time that an identifier possesses system-allocated memory. C supports three types of duration: *local*, *static*, and *dynamic*. Automatic variables created during a function invocation or variables defined within a block of statements have *local duration*. Your programs must always initialize local variables. If your program does not initialize a local variable, your program cannot predict the variable's contents. The compiler creates *static variables* as the program execution begins. Static variables usually correspond to global variables. Most C compilers initialize static variables to 0. The compiler allocates *dynamic variables* from the heap during the program's execution. In most cases, programs must initialize dynamic variables.

Note: *Some run-time library functions will initialize dynamic memory locations to 0 (zero), while some will not.*

FUNCTIONS THAT SUPPORT A VARYING NUMBER OF PARAMETERS 282

As you have learned, C maps the actual parameters that it passes to a function to the formal parameters defined in the function header. If the function expects three parameters, your function invocation should include three parameter values. If you consider functions such as *printf* or *scanf*, however, you will find that the functions support a varying number of parameters. For example, the following *printf* function calls are all valid:

```
printf("Jamsa\'s C/C++/C# Programmer\'s Bible");
printf("%d %d %d %d %d, 1, 2, 3, 4, 5);
printf("%f %s %s %d %x", salary, name, state, age, id);
```

As you will learn in Tip 283, you can use the macros *va_arg*, *va_end*, and *va_start* (defined in the header file *stdarg.h*) to direct your programs to create their own functions that support a varying number of parameters. The macros essentially pull parameters from the stack, one at a time, until the program reaches the last parameter. When you use these macros to get parameters, you must know each parameter's type. In the case of *printf*, the function uses the format specifiers (for example, *%d*, *%s*, and *%f*) to track the parameter types.

SUPPORTING A VARYING NUMBER OF PARAMETERS 283

In this Tip you will create a function called *add_values*, which adds up all the integer values the calling function passes to it. As shown here, the function supports a varying number of parameters. The value 0 within the function call indicates the last parameter (which does not affect the sum):

```
result = add_values(3,0);      // Returns 3
result = add_values(3,5,0);           // Returns 8
result = add_values(100,3,4,2,0);    // Returns 109
```

The following program, *addvalue.c*, contains and uses the *add_values* function:

```
#include <stdio.h>
#include <stdarg.h>

int add_values(int value, ...)
 {
   va_list argument_ptr;
   int result = 0;

   if (value != 0)
     {
       result += value;
       va_start(argument_ptr, value);
       while ((value = va_arg(argument_ptr, int)) != 0)
         result += value;
```

```
        va_end(argument_ptr);
      }
    return(result);
  }

 void main(void)
  {
    printf("Sum of 3 is %d\n", add_values(3, 0));
    printf("Sum of 3 + 5 is %d\n", add_values(3, 5, 0));
    printf("Sum of 3 + 5 + 8 is %d\n", add_values(3, 5, 8, 0));
    printf("Sum of 3 + 5 + 8 + 9 is %d\n", add_values(3, 5, 8 , 9, 0));
  }
```

The function *add_values* uses the *va_start* macro to assign a pointer (*argument_ptr*) to the first parameter on the stack. Next, the function uses the *va_arg* macro to get the values one at a time. The *va_arg* macro returns a value of the specified type and then increments the *argument_ptr* to point to the next argument. When the *argument_ptr* encounters the zero terminator, the function uses the *va_end* macro to assign a value to the *argument_ptr* that prevents the future use of the *argument_ptr* (until *va_start* reinitializes the *argument_ptr*). When you create functions that support a varying number of parameters, your functions must have a way to know the number of parameters and each parameter's type. In the case of *printf*, the format specifier defines the parameters and their types. In the case of *add_values*, the zero terminator marks the last parameter. Likewise, all the arguments passed to the function are the same type.

***Note:** Note the use of the ellipses (...) within the **add_values** function header to indicate a variable number of parameters.*

284 How VA_START, VA_ARG, AND VA_END WORK

In Tip 283 you learned that you can use the *va_start*, *va_arg*, and *va_end* macros, defined within the header file *stdarg.h*, to create functions that support a varying number of parameters. To better understand how these macros work, consider the following function call to *add_values*:

```
add_values(10, 20, 30, 0);
```

When the program makes the function call, the compiler will place the parameters onto the stack from right to left. Within the function, the *va_start* macro assigns a pointer to the first parameter, as shown in Figure 284.

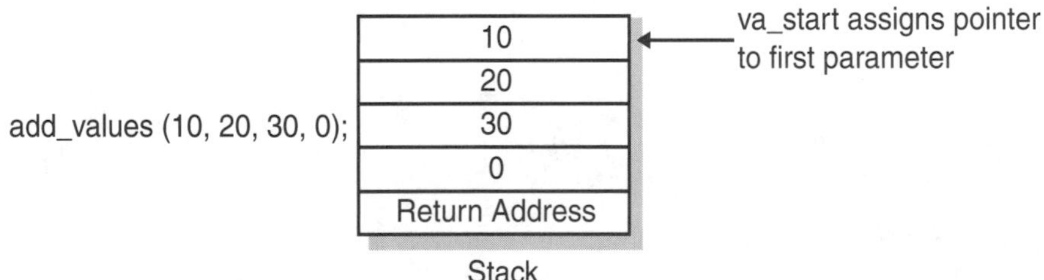

***Figure 284** Using **va_start** to assign a pointer to the first parameter.*

The *va_arg* macro returns the value pointed to by the argument pointer. To determine the value, the macro must know the parameter's type. A parameter of type *int*, for example, will use 16 bits, whereas a parameter of type *long* will use 32. After retrieving the parameter's value, the *va_arg* macro will increment the argument pointer so that it points to the next argument. To determine the number of bytes to add to the pointer, *va_arg* will again use the parameter's type. After the *va_arg* macro retrieves the last argument, the *va_end* macro will nullify the argument pointer's value.

CREATING FUNCTIONS THAT SUPPORT MANY PARAMETERS AND TYPES 285

In Tips 282 and 283 you learned how to create functions that support a variable number of parameters. Unfortunately, the *add_values* function you created supports only parameters of type *int*. The following program, *alltypes.c*, changes the *add_values* function to support values of all types. The function returns a value of type *float*. To help the function determine the parameter types, you pass to the function a format specifier similar to that which *printf* uses. For example, to add *three integer* values, use the following invocation:

```
result = add_values("%d %d %d", 1, 2, 3);
```

Likewise, to add three *floating-point* values, use the following invocation:

```
result = add_values("%f %f %f", 1.1, 2.2, 3.3);
```

Finally, to add *integer* and *floating-point* values, use the following invocation:

```
result = add_values("%f %d %f %d", 1.1, 2, 3.3, 4);
```

Using the format specifier, you eliminate the need to use the zero terminator. Additionally, the format specifier lets you determine how many bits each parameter uses, as shown here:

```
#include <stdio.h>
#include <stdarg.h>

double add_values(char *str, ...)
{
    va_list marker;
    double result = 0.0;

    va_start(marker, str);              // mark first additional argument
    while (*str)                        // examine each character in the string
      {
        if (*str == '%')                // if not a %_ format specifier, skip it
          {
            switch (*(++str))
              {
                case 'd': result += va_arg(marker, int);
                          break;
                case 'f': result += va_arg(marker, double);
                          break;
              }
          }
        str++;
      }
   va_end(marker);
   return(result);
 }

void main(void)
 {
   double result;

   printf("Result %f\n", add_values("%f", 3.3));
   printf("Result %f\n", add_values("%f %f", 1.1, 2.2));
   printf("Result %f\n", add_values("%f %d %f", 1.1, 1, 2.2));
   printf("Result %f\n", add_values("%f %d %f %d", 1.1, 1, 2.2, 3));
 }
```

286 Reading a Character from the Keyboard

Even the simplest C programs must often read characters from the keyboard. The character may correspond to a menu option, a Yes or No response, or even one of many letters in a name. Programs often perform character input operations using the *getchar* macro. You will implement the *getchar* macro as shown here:

```
#include <stdio.h>

int getchar(void);
```

If successful, *getchar* returns the ASCII value for the read character. If an error occurs or *getchar* encounters an end of file (usually for redirected input), *getchar* returns *EOF*. The following program, *getchar.c*, uses *getchar* to read a Yes or No response from the keyboard:

```
#include <stdio.h>
#include <ctype.h>

void main(void)
 {
   int letter;

   printf("Type Y or N to continue and press Enter\n");
   do
    {
      letter = toupper(getchar());
    }
   while ((letter != 'Y') && (letter != 'N'));
   printf("You typed %c\n", ((letter == 'Y') ? 'Y': 'N'));
 }
```

As you can see, the program uses a *do while* loop to repeatedly invoke *getchar* until the user types either Y or N on the keyboard.

***Note:** To support I/O redirection, C actually defines the **getchar** macro in terms of **stdin** (which corresponds to the keyboard by default).*

287 Displaying a Character of Output

In Tip 286 you learned how to use the *getchar* macro to read a character from the keyboard. In a similar way, C provides the *putchar* macro. The *putchar* macro writes a character to the screen (*stdout*). The format of the *putchar* macro is as follows:

```
#include <stdio.h>

int putchar(int letter);
```

If *putchar* succeeds, it returns the character written. If an error occurs, *putchar* returns *EOF*. The following program, *putchar.c*, uses *putchar* to display the letters of the alphabet:

```
#include <stdio.h>

void main(void)
 {
   int letter;

   for (letter = 'A'; letter <= 'Z'; letter++)
     putchar(letter);
 }
```

***Note:** Because C defines **putchar** in terms of **stdout**, you can use the DOS output redirection operators to redirect the output of the **putchar.c** program to a file or printer.*

UNDERSTANDING BUFFERED INPUT 288

When your program uses buffered input, the operating system does not actually pass the letters the user types to the program until the user presses ENTER. In this way, the user can change characters as he or she types, using the BACKSPACE key to erase characters as needed. When the user presses ENTER, all the characters typed are available to the program. The *getchar* macro uses buffered input. If you use *getchar* to read a single character response, *getchar* does not read a character until the user presses ENTER. If the user types multiple characters, all the characters are available to *getchar* within the input buffer. The following program, *bufferio.c*, illustrates buffered input. Run the program and then type in a line of text. The characters you type will not be available to the program until you press ENTER. After you press ENTER, however, the program will read and display characters until it encounters the *newline* character (which the operating system creates when you press ENTER), as shown here:

```
#include <stdio.h>

void main(void)
 {
   int letter;

   do
    {
      letter = getchar();
      putchar(letter);
    }
   while (letter != '\n');
 }
```

When you run the *bufferio.c* program, experiment with the letters you input, using the BACKSPACE key to rub out letters, and so on. As you will find, the letters passed to the program correspond with your final text.

ASSIGNING KEYBOARD INPUT TO A STRING 289

The Strings section of this book looks at several different ways to manipulate strings. When you perform keyboard input, one of the most common operations your programs will perform is to assign the characters resulting from the keyboard input to a string. The following program, *fillstr.c*, uses the *getchar* macro to assign letters to the *string* variable. To assign characters, the program simply loops, assigning characters to the string elements until the program encounters the *newline* character. The program then assigns the *NULL* character (end of string) marker to the current string position, as shown here:

```
#include <stdio.h>

void main(void)
 {
   char string[128];
   int index = 0;
   int letter;

   printf("Type in a string and press Enter\n");
   while ((letter = getchar()) != '\n')
     string[index++] = letter;
   string[index] = NULL;
   printf("The string was: %s\n", string);
 }
```

290 COMBINING GETCHAR AND PUTCHAR

As you have learned, *getchar* lets you read a letter from the keyboard (*stdin*), while *putchar* lets you display a letter on the screen (*stdout*). Depending on your program's function, there may be times when you want to read and display characters. The following *do while* loop, for example, will read and display characters up to and including the *newline* character:

```
do
 {
   letter = getchar();
   putchar(letter);
 }
while (letter != '\n');
```

Because *getchar* and *putchar* both work with integer values, you can combine the previous statements, as shown here:

```
do
  putchar(letter = getchar());
while (letter != '\n');
```

In this case, *getchar* will assign the character typed to the *letter* variable. The *putchar* macro, in turn, will display the value assigned to *letter*.

291 REMEMBER, GETCHAR AND PUTCHAR ARE MACROS

As you create your programs, remember that *getchar* and *putchar* are macros, not functions. Therefore, some compilers will not let you leave spaces between their names and parentheses, as shown here:

```
letter = getchar();

putchar(letter);
```

If you examine the header file *stdio.h*, you will find the macro definitions for *getchar* and *putchar*. The I/O Redirection section of this book explains the *getchar* and *putchar* macro definitions in detail.

292 READING A CHARACTER USING DIRECT I/O

You learned in Tip 288 that when you perform keyboard input, your programs can perform direct or buffered input. When your programs use direct input operations, the characters users type at the keyboard are immediately available to the program (in other words, the operating system does not buffer the characters). If the user presses the BACKSPACE key to erase a previous character, the program itself must handle the editing operation (erasing the previous character from the screen and removing the character from the buffer). The *getche* function lets your programs read a character from the keyboard using direct input. The format of the *getche* function is as follows:

```
#include <conio.h>

int getche(void);
```

The following program, *getche.c*, uses the *getche* function to read a Yes or No response from the keyboard:

```
#include <stdio.h>
#include <ctype.h>
#include <conio.h>

void main(void)
 {
   int letter;

   printf("Do you want to continue? (Y/N): ");
```

```
   do
     {
       letter = getche();
       letter = toupper(letter);
     }
   while ((letter != 'Y') && (letter != 'N'));
   if (letter == 'Y')
     printf("\nYour response was Yes\n");
   else
     printf("\nWhy not?\n");
 }
```

Unlike the program *getchar.c*, which requires the user to press ENTER to make the response available, the keys the user typed into the *getche.c* program are immediately available to the program.

DIRECT KEYBOARD INPUT WITHOUT CHARACTER DISPLAY 293

In Tip 292 you learned how to use the *getche* function to read characters from the keyboard as the user types the characters (using direct I/O). When you use *getche*, the program will display the letters the user types on the screen as the user types them. Depending on your program, there may be times when you want to read characters from the keyboard without displaying the characters on the screen. For example, if your program prompts the user for a password, the letters the user types should not appear on the screen for others to see. The *getch* function lets your programs read characters from the keyboard without displaying (echoing) the characters to the screen. The format of the *getch* function is as follows:

```
#include <conio.h>

int getch(void);
```

The following program, *getch.c*, uses the *getch* function to read characters from the keyboard. As the user types, the program uses *getch* to read each character, converts each character to uppercase, and then displays each character's uppercase equivalent on the screen. The following program, *getch.c,* shows how you can quickly implement such processing:

```
#include <stdio.h>
#include <conio.h>
#include <ctype.h>

void main(void)
 {
   int letter;

   printf("Type in a string of characters and press Enter\n");
   do
    {
      letter = getch();
      letter = toupper(letter);
      putch(letter);
    }
   while (letter != '\r');
 }
```

KNOWING WHEN TO USE '\R' AND '\N' 294

As you have learned, C uses the '\r' escape sequence to indicate a carriage return. Likewise, C uses '\n' to represent a *newline* (carriage return and linefeed). When your programs perform buffered input using *getchar*, C will convert the ENTER key to a carriage return and linefeed (*newline*) sequence. On the other hand, when you perform direct I/O using *getch* or *getche*, either function will return the ENTER key simply as a carriage return ('\r'). Therefore, you must test for the correct character within your programs, as shown here:

```
do
 {
   letter = getchar();
   putchar(letter);
 }
while (letter != '\n');
do
 {
   letter = getch();
   putchar(letter);
 }
while (letter != '\r');
```

295 PERFORMING DIRECT OUTPUT

As you have learned, the functions *getch* and *getche* let your programs read characters directly from the keyboard, bypassing C's buffered (file system-based) input streams. In a similar way, your programs can perform fast screen output using the *putch* function, as shown here:

```
#include <conio.h>

int putch(int letter);
```

If successful, *putch* returns the letter it displayed. If an error occurs, *putch* returns *EOF*. To perform fast output, the *putch* function communicates with the BIOS video services or directly accesses the PC's video memory. Functions such as *putchar*, on the other hand, use the file system, which in turn calls the BIOS. The *putch* function does not convert a linefeed character into a carriage return and linefeed sequence. The following program, *putch.c*, uses *putch* and *putchar* to display the letters of the alphabet 100,000 times. The program then displays the amount of time each function required, as shown here:

```
#include <stdio.h>
#include <conio.h>
#include <time.h>

void main(void)
 {
   int letter;
   int count;

   time_t start_time, stop_time;
   time(&start_time);
   for (count = 0; count < 100000L; count++)
    for (letter = 'A'; letter <= 'Z'; letter++)
     putchar(letter);
   time(&stop_time);
   printf("\n\nTime required for putchar %d seconds\n", stop_time-start_time);
   printf("Press any key...\n");
   getch();
   time(&start_time);
   for (count = 0; count < 100000L; count++)
     for (letter = 'A'; letter <= 'Z'; letter++)
       putch(letter);
   time(&stop_time);
   printf("\n\nTime required for putch %d seconds\n", stop_time-start_time);
 }
```

PLACING A KEYSTROKE BACK INTO THE KEYBOARD BUFFER 296

As you have learned, the *getch* function lets your programs read a character from the keyboard. Depending on how you write your program, there may be times when you read keystrokes up to a specific character and then process the keystrokes. When the processing completes, you read the remaining characters. When you write such code, there may be times when you want your program to "unread" a character. The *ungetch* function lets your programs "unread" a character. To do so, you will implement the *ungetch* function, as shown here:

```
#include <conio.h>

int ungetch(int character);
```

In addition, at times you may want to place a character into the keyboard buffer so that your program can reread the keystroke it just read. Using *ungetch,* your programs can do just that. The following program, *ungetch.c*, reads letters from the keyboard until it encounters a non-lowercase letter. The program then displays the letters and afterwards reads and displays any remaining characters on a different line:

```
#include <stdio.h>
#include <ctype.h>
#include <conio.h>

void main(void)
 {
   int letter;
   int done = 0;
   int uppercase_found = 0;

   do
    {
       letter = getch();
       if (islower(letter))
         putchar(letter);
       else
        {
          if (isupper(letter))
           {
              ungetch(letter);
              uppercase_found = 1;
              putchar('\n');
           }
          done = 1;
        }
    }
   while (! done);
   if (uppercase_found)
     do
      {
        letter = getch();
        putchar(letter);
      }
     while (letter != '\r');
 }
```

If you are reading characters using *getchar*, you can use the *ungetc* function to unread a character, as shown here:

```
ungetc(letter, stdin);
```

297 Fast Formatted Output Using CPRINTF

As you know, the *printf* function lets your programs perform formatted output. C actually defines the *printf* function in terms of the file handle *stdout*. As a result, you can redirect *printf's* output from the screen to a file or device. Because *printf* uses *stdout* to display characters, *printf* uses the C file system, which, in turn, uses the DOS functions. Each of the DOS functions, in turn, calls the BIOS. For faster formatted output, your programs can use the following function, *cprintf*, which works directly with the BIOS or your computer's video memory:

```
#include <conio.h>

int cprintf(const char *format[,arguments...]);
```

The following program, *cprintf.c*, writes the string "Jamsa's C/C++/C# Programmer's Bible" to your screen 1,700 times using *printf* and then *cprintf*. The program then displays the amount of time both functions required:

```
#include <stdio.h>
#include <conio.h>
#include <time.h>

void main(void)
 {
   int count;

   time_t start_time, stop_time;
   time(&start_time);
   for (count = 0; count < 1700; count++)
     printf("Jamsa\'s C/C++/C# Programmer\'s Bible\n");
   time(&stop_time);
   printf("\n\nTime required for printf %d seconds\n", stop_time-start_time);
   printf("Press any key...\n");
   getch();
   time(&start_time);
   for (count = 0; count < 1700; count++)
     cprintf("Jamsa\'s C/C++/C# Programmer\'s Bible\r\n");
   time(&stop_time);
   printf("\n\nTime required for cprintf %d seconds\n", stop_time-start_time);
 }
```

***Note:** The **cprintf** function does not convert the **newline** character into a carriage return linefeed sequence.*

298 Fast Formatted Input from the Keyboard

In Tip 297 you learned that the *cprintf* lets your programs bypass the file system to perform fast output to the screen display. In a similar way, the *cscanf* function lets your programs perform fast formatted input from the keyboard, as shown here:

```
#include <conio.h>

int cscanf(char *format[,arguments]);
```

The following program, *cscanf.c*, prompts you for three integer values. The program then reads the values using *cscanf*:

```
#include <conio.h>

void main(void)
 {
   int a, b, c;

   cprintf("Type 3 integer values and press Enter\r\n");
   cscanf("%d %d %d", &a, &b, &c);
```

```
   cprintf("The values entered were %d %d %d\r\n", a, b, c);
 }
```

WRITING A CHARACTER STRING 299

As you have learned, the *printf* function lets your programs write formatted output to the screen display. Using *printf*, your programs can write strings, integers, floating-point numbers, or combinations of different values to the screen. When your programs only need to write a character string, however, you may be able to improve your program's performance by using the *puts* function instead of *printf*, as shown here:

```
#include <stdio.h>

int puts(const char *string);
```

The *puts* function writes a *NULL*-terminated string to the screen (actually to *stdout*). If *puts* succeeds, *puts* returns a non-negative value. If an error occurs, *puts* returns *EOF*. The *puts* function automatically writes a *newline* character at the end of the string. The following program, *puts.c*, uses *printf* and *puts* to output the string "Jamsa's C/C++/C# Programmer's Bible" 1,700 times. The program displays the amount of time each function required:

```
#include <stdio.h>
#include <conio.h>
#include <time.h>

void main(void)
 {
   int count;

   time_t start_time, stop_time;
   time(&start_time);
   for (count = 0; count < 1700; count++)
     printf("Jamsa\'s C/C++/C# Programmer\'s Bible\n");
   time(&stop_time);
   printf("\n\nTime required for printf %d seconds\n", stop_time-start_time);
   printf("Press any key...\n");
   getch();
   time(&start_time);
   for (count = 0; count < 1700; count++)
     puts("Jamsa\'s C/C++/C# Programmer\'s Bible");
   time(&stop_time);
   printf("\n\nTime required for puts %d seconds\n", stop_time-start_time);
 }
```

Note: *Because the* ***puts*** *function automatically appends a* ***newline*** *character, the character string the program instructs* ***puts*** *to display does not include the newline character.*

FASTER STRING OUTPUT USING DIRECT I/O 300

In Tip 299 you learned that the *puts* function lets your programs quickly output a character string. However, because C defines the *puts* function in terms of *stdout* (so it can support redirection), the function must use the file system. For faster string output to the screen, your programs might want to use the *cputs* function, as shown here:

```
#include <conio.h>

int cputs(const char string);
```

Like *puts*, the *cputs* function outputs a *NULL*-terminated string. Unlike *puts*, however, *cputs* does not automatically append a *newline* character. The following program, *cputs.c*, uses the *puts* and *cputs* functions to display the string

"Jamsa's C/C++/C# Programmer's Bible" 1,700 times. The program displays the amount of time each function required to generate the output:

```
#include <stdio.h>
#include <conio.h>
#include <time.h>

void main(void)
 {
   int count;

   time_t start_time, stop_time;
   time(&start_time);
   for (count = 0; count < 1700; count++)
     puts("Jamsa\'s C/C++/C# Programmer\'s Bible");
   time(&stop_time);
   printf("\n\nTime required for puts %d seconds\n", stop_time-start_time);
   printf("Press any key...\n");
   getch();
   time(&start_time);
   for (count = 0; count < 1700; count++)
     cputs("Jamsa\'s C/C++/C# Programmer\'s Bible\r\n");
   time(&stop_time);
   printf("\n\nTime required for cputs %d seconds\n", stop_time-start_time);
 }
```

301 READING A CHARACTER STRING FROM THE KEYBOARD

In Tip 299 you learned that C provides the *puts* function, which you can use within your programs to write a character string to the screen display. In a similar way, your programs can use the *gets* function to read a character string from the keyboard, as shown here:

```
#include <stdio.h>

char *gets(char *string);
```

If *gets* is successful, it will return a pointer to the character string. If an error occurs or if *gets* encounters the end of file marker, *gets* will return the *NULL* value. The *gets* function reads characters up to and including the *newline* character. However, *gets* replaces the *newline* character with *NULL*. The following program, *gets.c*, uses the *gets* function to read a string of characters from the keyboard:

```
#include <stdio.h>

void main(void)
 {
   char string[256];

   printf("Type in a string of characters and press Enter\n");
   gets(string);
   printf("The string was %s\n", string);
 }
```

Note: *C actually defines the* ***gets*** *function in terms of* ***stdin*** *(which is by default the keyboard), which lets the function support I/O redirection.*

302 PERFORMING FASTER KEYBOARD STRING INPUT

In Tip 301 you learned how your programs can use the *gets* function to read a character string from the keyboard. Because C defines *gets* in terms of *stdin*, *gets* must use the file system to perform its input operations. If you do not need support for I/O redirection, you can use the *cgets* function to read characters from the keyboard, thereby improving your program's performance. You will implement *cgets* as shown here:

```
#include <conio.h>

char *cgets(char *string);
```

If *cgets* successfully reads characters from the keyboard, *cgets* will return a pointer to the string beginning at *string[2]*. If an error occurs, *cgets* will return *NULL*. The *cgets* function behaves differently from the *gets* function. Before you call *cgets* with a character string, you must first assign the maximum number of characters *cgets* will read to *string[0]*. When *cgets* returns, *string[1]* will contain a number count of characters *cgets* has read. The *NULL*-terminated character string actually begins at *string[2]*. The following program, *cgets.c*, illustrates how to use the *cgets* function:

```
#include <stdio.h>
#include <conio.h>

void main(void)
 {
   char buffer[256];

   buffer[0] = 253;   // Number of characters that can be read
   printf("Type in a string and press Enter\n");
   cgets(buffer);
   printf("\n\nThe number of characters read was %d\n", buffer[1]);
   printf("The string read: %s\n", &buffer[2]);
 }
```

To experiment with this program, reduce the number of characters *cgets* can read to 10. If the user tries to type more than 10 characters, the function will ignore the extra characters.

DISPLAYING OUTPUT IN COLOR 303

Using the *ansi.sys* device driver, your programs can display screen output in color. In addition, many C compilers provide text-based output functions that let you display color output. If you are using *Turbo C++ Lite,* Borland *C++* or Microsoft *C++*, the *outtext* (called *_outtext* for Microsoft *C++*) function lets you display colored output. If you are using *Turbo C++ Lite* or Borland *C++*, you can only use the *outtext* function in graphics mode. The Microsoft *_outtext* function, on the other hand, works in either text or graphics mode. If you must perform colored output, refer to the documentation that accompanied your compiler for specifics on these functions. As you will find, the compilers provide functions that set text position, colors, and graphics modes. If your compiler does not provide the functions the following tips present, turn to Tips 79 and 80. By writing ANSI escape sequences to the screen, you can easily implement functions that perform equivalent tasks.

CLEARING THE SCREEN DISPLAY 304

Most C compilers do not provide a function that lets you clear the screen display. If you are using *Turbo C++ Lite,* Borland C, or Microsoft C, however, you can use the *clrscr* function to clear a text mode window's contents, as shown here:

```
#include <conio.h>

void clrscr(void);
```

The following program, *clrscr.c*, uses the *clrscr* function to clear the screen display:

```
#include <conio.h>

void main(void)
 {
   clrscr();
 }
```

305 ERASING TO THE END OF THE CURRENT LINE

As your programs perform screen I/O, there may be times when you want to erase the contents of a line from the current cursor position to the end of the line. To do so, your programs can use the *clreol* function, as shown here:

```
#include <conio.h>
void clreol(void);
```

The *clreol* function erases the remaining contents of the current line beyond the cursor without moving the cursor.

306 DELETING THE CURRENT SCREEN LINE

As your programs perform screen-based I/O, there may be times when you want to delete the current line's contents, moving all the output that follows up one line. In such cases, your programs can use the *delline* function, as shown here:

```
#include <conio.h>
void delline(void);
```

The following program, *delline.c*, fills the screen with 24 lines of text. When you press ENTER, the program will use *delline* to erase lines 12, 13, and 14, as shown here:

```
#include <conio.h>

void main(void)
 {
   int line;

   clrscr();
   for (line = 1; line < 25; line++)
    cprintf("This is line %d\r\n", line);
   cprintf("Press a key to Continue: ");
   getch();
   gotoxy(1, 12);
   for (line = 12; line < 15; line++)
     delline();
   gotoxy(1, 25);
 }
```

307 POSITIONING THE CURSOR FOR SCREEN OUTPUT

As you have learned, you can use the *ansi.sys* device driver to position the cursor for screen output operations. If you are working in a DOS environment, many C compilers provide the *gotoxy* function, which lets you position the cursor at a specific column and row intersection, as shown here:

```
#include <conio.h>
void gotoxy(int column, int row);
```

The *column* parameter specifies a column position (x) from 1 through 80. The *row* parameter specifies a row position (y) from 1 through 25. If either value is invalid, the compiler will ignore the *gotoxy* operation. The following program, *gotoxy.c*, uses the *gotoxy* function to display screen output at specific locations:

```
#include <conio.h>

void main(void)
 {
   clrscr();
   gotoxy(1, 5);
```

```
   cprintf("Output at row 5 column 1\n");
   gotoxy(20, 10);
   cprintf("Output at row 10 column 20\n");
 }
```

DETERMINING THE ROW AND COLUMN POSITION 308

In Tip 307 you learned how to use *gotoxy* to place the cursor at a specific row and column position. In many cases, your programs will want to know the current cursor position before performing a screen I/O operation. The functions *wherex* and *wherey* return the cursor's column and row, as shown here:

```
#include <conio.h>

int wherex(void);
int wherey(void);
```

The following program, *wherexy.c*, clears the screen, writes three lines of output, and then uses the *wherex* and *wherey* functions to determine the current cursor position:

```
#include <conio.h>

void main(void)
 {
   int row, column;

   clrscr();
   cprintf("This is line 1\r\n");
   cprintf("Line 2 is a little longer\r\n");
   cprintf("This is the last line");
   row = wherey();
   column = wherex();
   cprintf("\r\nThe cursor position was row %d column %d\n", row, column);
 }
```

INSERTING A BLANK LINE ON THE SCREEN 309

As your programs perform screen-based I/O operations, there may be times when you want to insert a blank line on the screen so that you can insert text in the middle of existing text. To do so, your programs can use the *insline* function, as shown here:

```
#include <conio.h>

void insline(void);
```

When you invoke the *insline* function, all text below the current cursor position moves down one line. The line at the bottom of the screen will scroll off the window. The following program, *insline.c*, writes 25 lines of text on the screen. The program then uses the *insline* function to insert text at line 12, as shown here:

```
#include <conio.h>

void main(void)
 {
   int line;

   clrscr();
   for (line = 1; line < 25; line++)
    cprintf("This is line %d\r\n", line);
   cprintf("Press a key to Continue: ");
   getch();
   gotoxy(1, 12);
   insline();
   cprintf("This is new text!!!");
```

```
   gotoxy(1, 25);
 }
```

310 COPYING SCREEN TEXT TO A BUFFER

When your programs perform extensive screen I/O, there may be times when the program must copy the current screen contents to a buffer. To copy screen text, your programs can use the *gettext* function, as shown here:

```
#include <conio.h>

int gettext(int left, int top, int right, int bottom, void *buffer);
```

The *left* and *top* parameters specify the column and row positions of the upper-left corner of the screen region that you want to copy. Likewise, the *right* and *bottom* parameters specify the region's lower-right corner. The *gettext* function places the text and its attributes in the *buffer* parameter. The PC uses an attribute byte for every letter of text it displays on your screen. If you want to buffer 10 characters, for example, your buffer must be large enough to hold the 10 ASCII characters plus the 10 attribute bytes (20 bytes in length). The following program, *savescr.c*, saves the current text mode screen contents to the file *savescr.dat*:

```
#include <conio.h>
#include <io.h>
#include <fcntl.h>
#include <sys\stat.h>

void main(void)
 {
   char buffer[8000];
   int handle;

   if ((handle = creat("SAVESCR.DAT", S_IWRITE)) == -1)
    cprintf("Error opening SAVESCRN.DAT\r\n");
   else
     {
       gettext(1, 1, 80, 25, buffer);
       write(handle, buffer, sizeof(buffer));
       close(handle);
     }
 }
```

***Note:** In most cases, the current text attribute is 7. If you try to display the contents of the **savescr.dat** file using the **TYPE** command, your system will beep for every attribute value.*

311 WRITING A TEXT BUFFER AT A SPECIFIC SCREEN LOCATION

As you have learned, many DOS-based compilers provide functions that your programs can use to control video output. In Tip 310 you learned that your programs can use the *gettext* function to copy a range of screen characters (and their attributes) to a buffer. After you copy a text buffer, you can later copy it back to the screen using the *puttext* function, as shown here:

```
#include <conio.h>

int puttext(int left, int top, int right, int bottom, void *buffer);
```

The *left, top, right,* and *bottom* parameters specify the screen location to which you want the buffer's contents written. The *buffer* parameter contains the characters and attributes that *gettext* previously stored. The following program, *puttext.c*, moves the text "Jamsa's C/C++/C# Programmer's Bible" around your screen until you press any key:

```
#include <conio.h>
#include <io.h>
```

```
#include <fcntl.h>
#include <sys\stat.h>
#include <stdlib.h>
#include <dos.h>

void main(void)
 {
   char buffer[128];
   int row, column;

   clrscr();
   cprintf("Jamsa\'s C/C++/C# Programmer\'s Bible\r\n");
   gettext(1, 1, 23, 1, buffer);
   while (! kbhit())
    {
      clrscr();
      row = 1 + random(24);
      column = 1 + random(58);
      puttext(column, row, column+22, row, buffer);
      delay(2000);
    }
 }
```

312 Determining Text Mode Settings

As you have learned, many C compilers provide several text-based functions that your programs can use to control screen output operations. To help your programs determine the current screen settings, your programs can use the *gettextinfo* function, as shown here:

```
#include <conio.h>

void gettextinfo(struct text_info *data);
```

The data parameter is a pointer to a structure of type *text_info*, as shown here:

```
struct text_info
 {
   unsigned char winleft;          // Left column
   unsigned char wintop;           // Top row
   unsigned char winright;         // Right column
   unsigned char winbottom;        // Bottom row
   unsigned char attribute;        // Text attribute
   unsigned char normattr;         // Normal attribute
   unsigned char currmode;         // Current text mode
   unsigned char screenheight;     // In rows
   unsigned char screenwidth;      // In columns
   unsigned char curx;             // Cursor column
   unsigned char cury;             // Cursor row;
 };
```

The following program, *textinfo.c*, uses the *gettextinfo* function to display the current text settings:

```
#include <conio.h>

void main(void)
 {
   struct text_info text;

   gettextinfo(&text);
   cprintf("Screen coordinates %d,%d to %d,%d\r\n",
            text.wintop, text.winleft, text.winbottom, text.winright);
   cprintf("Text attribute %d Normal attribute %d\r\n", text.attribute,
            text.normattr);
```

```
   cprintf("Screen height %d width %d\r\n", text.screenheight, text.screenwidth);
   cprintf("Cursor position was row %d column %d\r\n", text.cury, text.curx);
 }
```

313 CONTROLLING SCREEN COLORS

As you have learned, your programs can use the *ansi.sys* device driver to display screen output in color. In addition, many DOS-based compilers provide the *textattr* function, which lets you select the foreground and background text colors:

```
#include <conio.h>

void textattr(int attribute);
```

The attribute parameter contains eight bits that specify the colors you desire. The least significant four bits specify the foreground color. The three bits that follow specify the background color, and the most significant bit controls blinking. To select a color, you must assign the desired color value to the correct bits. Table 313 specifies the color values.

Color Constant	Value	Use
BLACK	0	Foreground/background
BLUE	1	Foreground/background
GREEN	2	Foreground/background
CYAN	3	Foreground/background
RED	4	Foreground/background
MAGENTA	5	Foreground/background
BROWN	6	Foreground/background
LIGHTGRAY	7	Foreground/background
DARKGRAY	8	Foreground
LIGHTBLUE	9	Foreground
LIGHTGREEN	10	Foreground
LIGHTCYAN	11	Foreground
LIGHTRED	12	Foreground
LIGHTMAGENTA	13	Foreground
YELLOW	14	Foreground
WHITE	15	Foreground
BLINK	128	Foreground

***Table 313** Color attribute parameters.*

The following program, *textattr.c*, illustrates the available foreground colors:

```
#include <conio.h>

void main(void)
 {
  int color;

  for (color = 1; color < 16; color++)
    {
      textattr(color);
      cprintf("This is color %d\r\n", color);
    }
  textattr(128 + 15);
  cprintf("This is blinking\r\n");
 }
```

ASSIGNING BACKGROUND COLOR 314

As you learned in Tip 313, the *textattr* function lets your programs select foreground and background colors. To set the background color using *textattr*, your program must assign the color value you desire to bits 4 through 6 of the color value. To assign color value, your programs can use bitwise shift operations, or you can declare a structure with bit fields, as shown here:

```
struct TextColor {
  unsigned char foreground:4;
  unsigned char background:3;
  unsigned char blinking:1;
};
```

The following program, *setback.c,* uses the *TextColor* structure to set the current screen colors:

```
#include <conio.h>

void main(void)
 {
   union TextColor
     {
        struct
         {
            unsigned char foreground:4;
            unsigned char background:3;
            unsigned char blinking:1;
         } color_bits;
        unsigned char value;
     } colors;
   colors.color_bits.foreground = BLUE;
   colors.color_bits.background = RED;
   colors.color_bits.blinking = 1;
   textattr(colors.value);
   clrscr();
   cprintf("This is the new text color\n");
 }
```

SETTING THE FOREGROUND COLOR USING TEXTCOLOR 315

As you have learned, many DOS-based compilers provide the *textattr* function, which lets you select your desired foreground and background colors. To simplify the process of assigning a foreground color, you might want to use the *textcolor* function, as shown here:

```
#include <conio.h>

void textcolor(int foregroundcolor);
```

The *foregroundcolor* parameter will specify one of the color values listed in Table 315.

Color Constant	Value	Color Constant	Value
BLACK	0	*DARKGRAY*	8
BLUE	1	*LIGHTBLUE*	9
GREEN	2	*LIGHTGREEN*	10
CYAN	3	*LIGHTCYAN*	11
RED	4	*LIGHTRED*	12
MAGENTA	5	*LIGHTMAGENTA*	13
BROWN	6	*YELLOW*	14
LIGHTGRAY	7	*WHITE*	15
		BLINK	128

***Table 315** Valid foreground color values for **textcolor**.*

The following program, *txtcolor.c*, illustrates how to use the *textcolor* function to set the foreground color:

```
#include <conio.h>

void main(void)
 {
  int color;

  for (color = 1; color < 16; color++)
   {
     textcolor(color);
     cprintf("This is color %d\r\n", color);
   }
  textcolor(128 + 15);
  cprintf("This is blinking\r\n");
 }
```

316 SETTING THE BACKGROUND COLOR USING TEXTBACKGROUND

As you have learned, many DOS-based compilers provide the *textattr* function, which lets you select the foreground and background colors you want for text display. To simplify the process of assigning a background color, you can use the *textbackground* function, as shown here:

```
#include <conio.h>

void textbackground(int backgroundcolor);
```

The *backgroundcolor* parameter must specify one of the color values listed in Table 316.

Color Constant	Value	Color Constant	Value
BLACK	0	*RED*	4
BLUE	1	*MAGENTA*	5
GREEN	2	*BROWN*	6
CYAN	3	*LIGHTGRAY*	7

***Table 316** Valid background color values.*

The following program, *backgrnd.c*, uses the *textbackground* function to display the different background colors:

```
#include <conio.h>

void main(void)
 {
   int color;
```

```
   for (color = 0; color < 8; color++)
    {
     textbackground(color);
     cprintf("This is color %d\r\n", color);
     cprintf("Press any key to continue\r\n");
     getch();
    }
 }
```

CONTROLLING TEXT INTENSITY 317

As you have learned, many DOS-based compilers provide functions that let you control your screen output. When you use these functions to write text to the screen, you want to control the intensity (brightness) of information your programs write to the screen. To control intensity, you can use one of the following three functions to select the text output intensity:

```
#include <conio.h>

void highvideo(void);
void lowvideo(void);
void normvideo(void);
```

The functions control the intensity with which your screen will display text. The following program, *ntensity.c*, illustrates how to use these three functions:

```
#include <conio.h>

void main(void)
 {
   clrscr();
   highvideo();
   cprintf("This text is high video\r\n");
   lowvideo();
   cprintf("This text is low video\r\n");
   normvideo();
   cprintf("This text is normal video\r\n");
 }
```

DETERMINING THE CURRENT TEXT MODE 318

As you have learned, many DOS-based compilers provide functions that your programs can use to control text-based output. When your programs perform screen output, they must know, and possibly change, the PC's current text mode. For example, a program that expects 80 columns will display inconsistent results on a screen that is in 40-column mode. To help your programs change the current text mode, your programs can use the *textmode* function, as shown here:

```
#include <conio.h>

void textmode(int desired_mode);
```

The *desired_mode* parameter specifies the text mode you desire. Table 318 lists the valid text modes.

Constant	Value	Text Mode
LASTMODE	-1	Previous mode
BW40	0	Black and white 40 column
C40	1	Color 40 column
BW80	2	Black and white 80 column
C80	3	Color 80 column
MONO	7	Monochrome 80 column
C4350	64	EGA 43 line or VGA 50 line

***Table 318** Valid text mode operations.*

The following statement, for example, will select a 43 mode on an EGA monitor or 50 line mode on a VGA monitor:

```
textmode(C4350);
```

Note: *If you use **textmode** to change the current text mode, the change will remain in effect after your program ends.*

319 MOVING SCREEN TEXT FROM ONE LOCATION TO ANOTHER

As you have learned, many DOS-based compilers provide functions that let you control your screen's text output. If your program performs extensive screen output, there may be times when you want to copy or move the text that appears on one section of your screen to another section. To copy screen text, your programs can use the *movetext* function, as shown here:

```
#include <conio.h>

int movetext(int left, int top, int right, int bottom,
  int destination_left, int destination_top);
```

The *left, top, right,* and *bottom* parameters describe a box which encloses the region of text that you want to move. The *destination_left* and *destination_top* parameters specify the desired location of the box's upper-left corner. The following program, *movetext.c*, writes five lines of text to the screen and then asks you to press a key. When you do, the program will then copy the text to a new location, as shown here:

```
#include <conio.h>

void main(void)
 {
   int i;

   clrscr();
   for (i = 1; i <= 5; i++)
     cprintf("This is line %d\r\n", i);
   cprintf("Press any key\n\r");
   getch();
   movetext(1, 1, 30, 6, 45, 18);
   gotoxy(1, 24);
 }
```

To move the text to the new location, as opposed to just copying the text to the new location, you must delete the original text after the program completes the *movetext* operation.

320 DEFINING A TEXT WINDOW

As you have learned, many DOS-based compilers provide functions that your programs can use to better control screen output. By default, these functions write their output to the entire screen. Depending on your program's purpose, there may be times when you want to restrict the program's output to a specific screen region. To do so, your programs can use the *window* function, as shown here:

```
#include <conio.h>

void window(int left, int top, int right, int bottom);
```

The *left, top, right,* and *bottom* parameters define the upper-left and lower-right corners of a screen region within which you want to write output. The following program, *window.c*, restricts the program's output to the top-left quarter of the screen:

```
#include <conio.h>

void main(void)
 {
   int i, j;

   window(1, 1, 40, 12);
   for (i = 0; i < 15; i++)
    {
      for (j = 0; j < 50; j++)
       cprintf("%d", j);
      cprintf("\r\n");
    }
 }
```

When program output reaches the right edge of the window, the output wraps to the next line. After the program ends, output operations will have access to the entire screen.

321 Using the Absolute Value of an Integer Expression

The *absolute value* specifies the number-line distance of the value from 0. Absolute values are always positive. For example, the absolute value of 5 is 5. Likewise, the absolute value of –5 is 5. To help your programs determine an absolute value, C provides the *abs* function. The function *abs* returns the absolute value for an integer expression. You will construct the *abs* function as shown here:

```
#include <stdlib.h>

int abs(int expression);
```

The following program, *show_abs.c*, illustrates how to use the *abs* function:

```
#include <stdio.h>
#include <stdlib.h>

void main(void)
 {
   printf("The absolute value of %d is %d\n", 5, abs(5));
   printf("The absolute value of %d is %d\n", 0, abs(0));
   printf("The absolute value of %d is %d\n", -5, abs(-5));
 }
```

When you compile and execute the *show_abs.c* program, your screen will display the following:

```
The absolute value of 5 is 5
The absolute value of 0 is 0
The absolute value of -5 is 5
C:\>
```

Note: *Many C compilers also provide the function* ***labs****, which returns the absolute value for an expression of type* ***long int****.*

322 USING THE ARCCOSINE

The arccosine is the ratio between the hypotenuse of a right triangle and the leg adjacent to a given acute angle. In other words, the arccosine is the geometric inverse of an angle's cosine. In other words, if *y* is the cosine of some angle theta, theta is the arccosine of *y*. To help your programs determine the arccosine, C provides the *acos* function. The *acos* function returns an angle's arccosine (0 through pi) specified in radians (as type *double*), as shown here:

```
#include <math.h>

double acos(double expression);
```

If the specified expression is not in the range –1.0 through 1.0, *acos* will set the global variable *errno* to *EDOM* and display a *DOMAIN* error to *stderr*. The following program, *showacos.c*, illustrates how to use the *acos* function:

```
#include <stdio.h>
#include <math.h>

void main(void)
 {
   double radians;

   for (radians = -0.5; radians <= 0.5; radians += 0.2)
     printf("%f %f\n", radians, acos(radians));
 }
```

Note:** Many C compilers also provide the function* ***acosl, *which returns the arccosine value for a* ***long double*** *expression.*

323 USING THE ARCSINE

The arcsine is the ratio between the hypotenuse of a right triangle and the leg opposite a given acute angle. In other words, the arcsine is the geometric inverse of an angle's sine. If *y* is the sine of some angle theta, then theta is the arcsine of *y*. To help your programs determine the arcsine, C provides the *asin* function. The *asin* function returns an angle's arcsine (–pi/2 through pi/2), specified in radians (as type *double*), as shown here:

```
#include <math.h>

double asin(double expression);
```

If *expression* is not in the range –1.0 through 1.0, then *asin* will set the global variable *errno* to *NAN* and display a *DOMAIN* error to *stderr*. The following program, *showasin.c*, illustrates how to use the *asin* function:

```
#include <stdio.h>
#include <math.h>

void main(void)
 {
   double radians;

   for (radians = -0.5; radians <= 0.5; radians += 0.2)
     printf("%f %f\n", radians, asin(radians));
 }
```

Note:** Many C compilers also provide the function* ***asinl, *which returns the arcsine value for a* ***long double*** *expression.*

324 USING THE ARCTANGENT

The arctangent is the ratio between the leg adjacent to a given acute angle and the leg opposite that angle in a right triangle. In other words, the arctangent is the geometric inverse of an angle's tangent. If *y* is the tangent of some angle theta, theta is the arctangent of *y*. To help your programs determine the arctangent, C provides the *atan* function. The *atan* function returns an angle's arctangent (–pi/2 through pi/2), specified in radians (as type *double*), as shown here:

```
#include <math.h>

double atan(double expression);
```

The following program, *showatan.c*, illustrates how to use the *atan* function:

```
#include <stdio.h>
#include <math.h>

void main(void)
 {
   double radians;

   for (radians = -0.5; radians <= 0.5; radians += 0.2)
     printf("%f %f\n", radians, atan(radians));
 }
```

Note: *Many C compilers also provide the function* ***atanl****, which returns the arctangent value for a* ***long double*** *expression. Also, C provides the functions* ***atan2*** *and* ***atan2l****, which return the arctangent of* ***y/x****.*

OBTAINING A COMPLEX NUMBER'S ABSOLUTE VALUE 325

As you have learned, a complex number contains a real and imaginary part. C functions represent complex numbers as a structure with an *x* and *y* member, as shown here:

```
struct complex
 {
   double x, y;
 };
```

When you work with complex numbers, there may be times when you must calculate the number's absolute valuc (its positive distance from zero). To let your program calculate a complex number's absolute value, C provides the *cabs* function, as shown here:

```
#include <math.h>

double cabs(struct complex value);
```

The *cabs* function is similar to taking the square root of the sum of the square of each complex number. In the following example, the *cabs* function will return $(10^2 + 5^2)^{1/2}$. The following program, *showcabs.c*, illustrates how to use the C language *cabs* function:

```
#include <stdio.h>
#include <math.h>

void main(void)
 {
   struct complex complex_number;

   complex_number.x = 10;
   complex_number.y = 5;
   printf("Absolute value of 10,5 is %f\n", cabs(complex_number));
 }
```

When you compile and execute the *showcabs.c* program, your screen will display the following output:

```
Absolute value of 10,5 is 11.180340
C:\>
```

Note: *Many C compilers also provide the function* ***cabls****, which returns an absolute value for* ***long double*** *complex numbers. The companion CD-ROM that accompanies this book includes a substantially longer program to test for the absolute value of a complex number, because the C++ implementation of complex numbers is substantially different than C's. The* ***showcabs2.cpp*** *program compiles in both the C and C++ environments.*

326 Rounding Up a Floating-Point Value

When you work with floating-point numbers, there may be times when you must round up the value of a floating-point variable or expression to the next higher integer. For such cases, C provides the *ceil* function, as shown here:

```
#include <math.h>

double ceil(double value);
```

As you can see, *ceil* receives a parameter of type *double* and returns a value of type *double*. The following program, *showceil.c*, illustrates how to use the *ceil* function:

```
#include <stdio.h>
#include <math.h>

void main(void)
 {
   printf("The value %f ceil %f\n", 1.9, ceil(1.9));
   printf("The value %f ceil %f\n", 2.1, ceil(2.1));
 }
```

When you compile and execute the *showceil.c* program, your screen will display the following output:

```
The value 1.900000 ceil 2.000000
The value 2.100000 ceil 3.000000
C:\>
```

***Note:** Many C compilers also provide the function **ceill**, which rounds up a value of type **long double**.*

327 Using the Cosine of an Angle

For a triangle, the cosine of an angle is the ratio of the angle's adjacent edge to the hypotenuse. To help your programs determine the cosine, C provides the *cos* function. The *cos* function returns an angle's cosine, specified in radians (as type *double*), as shown here:

```
#include <math.h>

double cos(double expression);
```

The *cos* function returns a value in the range –1.0 through 1.0. The following program, *show_cos.c*, illustrates how to use the *cos* function:

```
#include <stdio.h>
#include <math.h>

void main(void)
 {
   printf("cosine of pi/2 is %6.4f\n", cos(3.14159/2.0));
   printf("cosine of pi is %6.4f\n", cos(3.14159));
 }
```

When you compile and execute the *show_cos.c* program, your screen will display the following output:

```
cosine of pi/2 is 0.0000
cosine of pi is -1.0000
C:\>
```

***Note:** Many C compilers also provide the function **cosl**, which returns the cosine value for a **long double** expression.*

USING THE HYPERBOLIC COSINE OF AN ANGLE 328

The hyperbolic cosine of an angle is the cosine of a "circular-like" angle defined using ratios of hyperbolic radians. To help your programs determine the hyperbolic cosine, C provides the *cosh* function. The *cosh* function returns a "circular-like" angle's hyperbolic cosine, specified in radians (as type *double*), as shown here:

```
#include <math.h>

double cosh(double expression);
```

If overflow occurs, *cosh* will return the value *HUGE_VAL* (or *_LHUGE_VAL* for *coshl*) and set the global variable *errno* to *ERANGE*. The following program, *showcosh.c*, illustrates how to use the *cosh* function:

```
#include <stdio.h>
#include <math.h>

void main(void)
 {
   double radians;

   for (radians = -0.5; radians <= 0.5; radians += 0.2)
     printf("%f %f\n", radians, cosh(radians));
 }
```

Note: *Many C compilers also provide the function* ***coshl****, which returns the hyperbolic cosine value for a* ***long double*** *expression.*

USING THE SINE OF AN ANGLE 329

In a triangle, the sine of an angle is the ratio of the angle's opposite edge to the hypotenuse. To help your programs determine the sine, C provides the *sin* function. The *sin* function returns an angle's sine, specified in radians (as type *double*), as shown here:

```
#include <math.h>

double sin(double expression);
```

The following program, *show_sin.c*, illustrates how to use the *sin* function:

```
#include <stdio.h>
#include <math.h>

void main(void)
 {
   double radians;

   for (radians = 0.0; radians < 3.1; radians += 0.1)
     printf("Sine of %f is %f\n", radians, sin(radians));
 }
```

Note: *Many C compilers also provide the function* ***sinl****, which returns the sine value for a* ***long double*** *expression.*

USING THE HYPERBOLIC SINE OF AN ANGLE 330

The hyperbolic sine of an angle is the sine of a "circular-like" angle defined using ratios of hyperbolic radians. To help your programs determine the hyperbolic sine, C provides the *sinh* function. The *sinh* function returns a "circular-like" angle's hyperbolic sine, specified in radians (as type *double*), as shown here:

```
#include <math.h>

double sinh(double expression);
```

If overflow occurs, *sinh* will return the value *HUGE_VAL* (or *_LHUGE_VAL* for *sinhl*) and set the global variable *errno* to *ERANGE*. The following program, *showsinh.c*, illustrates how to use the *sinh* function:

```
#include <stdio.h>
#include <stdlib.h>
#include <math.h>

void main(void)
 {
   double radians;
   double result;

   for (radians = 0.0; radians < 3.1; radians += 0.1)
     if (((result = sinh(radians)) == HUGE_VAL) && (errno == ERANGE))
       printf("Overflow error\n");
     else
       printf("Sine of %f is %f\n", radians, result);
 }
```

Note: *Many C compilers also provide the function* ***sinhl****, which returns the hyperbolic sine value for a* ***long double*** *expression.*

331 USING THE TANGENT OF AN ANGLE

In a triangle, the tangent of an angle is the ratio of the angle's opposite edge to the adjacent edge. To help your programs determine the tangent, C provides the *tan* function. The function returns an angle's tangent, specified in radians (as type *double*), as shown here:

```
#include <math.h>

double tan(double expression);
```

The following program, *show_tan.c*, illustrates how to use the *tan* function:

```
#include <stdio.h>
#include <math.h>

void main(void)
 {
   double pi = 3.14159265;

   printf("Tangent of pi is %f\n", tan(pi));
   printf("Tangent of pi/4 is %f\n", tan(pi / 4.0));
 }
```

When you compile and execute the *show_tan.c* program, your screen will display the following output:

```
Tangent of pi is -0.000000
Tangent of pi/4 is 1.000000
C:\>
```

Note: *Many C compilers also provide the function* ***tanl****, which returns the tangent value for a* ***long double*** *expression.*

332 USING THE HYPERBOLIC TANGENT OF AN ANGLE

The hyperbolic tangent of an angle is the tangent of a "circular-like" angle defined using ratios of hyperbolic radians. To help your programs determine the hyperbolic tangent, C provides the *tanh* function. The *tanh* function returns an angle's hyperbolic tangent, specified in radians (as type *double*), as shown here:

```
#include <math.h>

double tanh(double expression);
```

Note: *Many C compilers also provide* ***tanhl****, which returns the hyperbolic tangent value for a* ***long double*** *expression.*

PERFORMING INTEGER DIVISION 333

As you have learned, C provides the division (/) and modulo (%) operators that let your programs perform a division or determine the remainder of a division operation. Similarly, C provides the function *div*, which divides a numerator value by a denominator, returning a structure of type *div_t* that contains the quotient and remainder, as shown here:

```
struct div_t
 {
   int quot;
   int rem;
 } div_t;
```

The *div* function works with integer values, as shown here:

```
#include <stdlib.h>

div_t div(int numerator, int denominator);
```

The following program, *div_rem.c*, illustrates how to use the *div* function:

```
#include <stdio.h>
#include <stdlib.h>

void main(void)
{
  div_t result;

  result = div(11, 3);
  printf("11 divided by 3 is %d Remainder %d\n", result.quot, result.rem);
}
```

When you compile and execute the *div_rem.c* program, your screen will display the following output:

```
11 divided by 3 is 3 Remainder 2
C:\>
```

Note: *Many C compilers also provide the function* ***ldiv****, which returns the quotient and remainder for* ***long*** *values.*

WORKING WITH AN EXPONENTIAL 334

When your programs perform complex mathematical operations, your programs will often need to calculate the exponential of e^x. In such cases, your programs can use the *exp* function, which returns a value of type *double*, as shown here:

```
#include <math.h>

double exp(double x);
```

The following program, *show_exp.c*, illustrates how to use the *exp* function:

```
#include <stdio.h>
#include <math.h>

void main(void)
{
  double value;

  for (value = 0.0; value <= 1.0; value += 0.1)
    printf("exp(%f) is %f\n", value, exp(value));
}
```

Note: *Many C compilers also provide the function* ***expl****, which works with values of type* ***long double****.*

335 USING THE ABSOLUTE VALUE OF A FLOATING-POINT EXPRESSION

As you have learned, the *absolute value* specifies a value's number line distance from zero. Absolute values are always positive. For example, the absolute value of 2.5 is 2.5. Likewise, the absolute value of –2.5 is 2.5. When you work with absolute values, there may be times when you must calculate the absolute value of a floating-point expression. For such cases, C provides the *fabs* function. The *fabs* function returns the absolute value for a floating-point number, as shown here:

```
#include <math.h>

float fabs(float expression);
```

The following program, *showfabs.c*, illustrates how to use the *fabs* function:

```
#include <stdio.h>
#include <math.h>

void main(void)
 {
   float value;

   for (value = -1.0; value <= 1.0; value += 0.1)
     printf("Value %f fabs %f\n", value, fabs(value));
 }
```

***Note:** Many C compilers also provide the function* **fabsl**, *which returns the absolute value for an expression of type* **long double**.

336 USING THE FLOATING-POINT REMAINDER

In Tip 82 you learned how to use C's modulo operator (%) to get the remainder of an integer division. Depending on your program, there may be times when you want to know the remainder of a floating-point division. In such cases, your programs can use C's *fmod* function to divide two floating-point values. The *fmod* function will return the remainder as a floating-point value, as shown here:

```
#include <math.h>

double fmod(double x, double y);
```

As an example, if you invoke *fmod* with the values 10.0 and 3.0, *fmod* will return the value 1.0 (10 divided by 3 is 3 remainder 1). The following program, *showfmod.c*, illustrates how to use the *fmod* function:

```
#include <stdio.h>
#include <math.h>

void main(void)
 {
   double numerator = 10.0;
   double denominator = 3.0;

   printf("fmod(10, 3) is %f\n", fmod(numerator, denominator));
 }
```

When you compile and execute the *showfmod.c* program, your screen will display the following output:

```
fmod(10, 3) is 1.000000
C:\>
```

***Note:** Many C compilers also provide the function* **fmodl**, *which returns the fractional remainder of a* **long double** *value.*

USING A FLOATING-POINT VALUE'S MANTISSA AND EXPONENT 337

When your programs work with floating-point values, the computer stores the values using a mantissa (whose value is between 0.5 and 1.0) and an exponent, as shown in Figure 337.

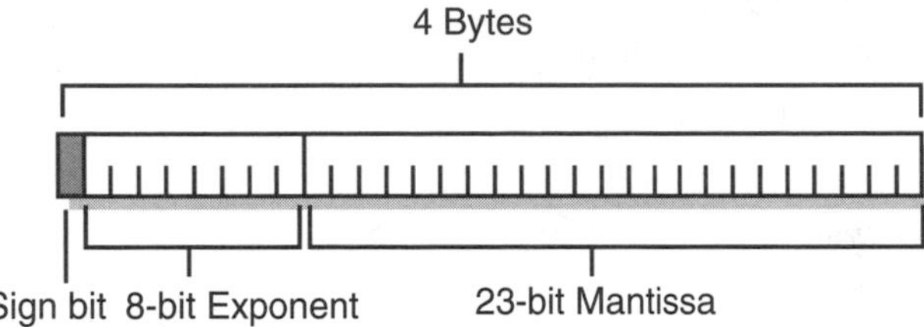

Figure 337 The computer stores floating-point values using a mantissa and exponent format.

To determine the stored value, the computer combines the mantissa and exponent, as shown here:

```
value = mantissa * (2 * exponent);
```

Usually, you do not need to be aware that the computer is using the mantissa and exponent. Depending on your program, however, there may be times when you want to know the mantissa and exponent values. For such cases, C provides the *frexp* function, which returns the mantissa and assigns the exponent to the variable *exponent*, which the calling function must pass to the *frexp* function by reference:

```
#include <math.h>

double frexp(double value, int *exponent);
```

The following program, *frexp.c*, illustrates how to use the *frexp* function:

```
#include <stdio.h>
#include <math.h>

void main(void)
 {
   double value = 1.2345;
   double mantissa;
   int exponent;

   mantissa = frexp(value, &exponent);
   printf("Mantissa %f Exponent %d Value %f\n",
     mantissa, exponent, mantissa * pow(2.0, 1.0 * exponent));
 }
```

When you compile and execute the *frexp.c* program, your screen will display the following output:

```
Mantissa 0.617250 Exponent 1 Value 1.234500
c:\
```

Note: *Many C compilers also provide the function* ***frexpl****, which returns the exponent and mantissa of a* ***long double*** *value.*

CALCULATING THE RESULT OF X * 2E 338

In Tip 334 you learned how to use C's *exp* function to obtain the result *ex*. Depending on your programs, there may be times when you must calculate *x* * 2e. In such situations, you can use C's *ldexp* function, as shown here:

```
#include <math.h>

double ldexp(double value, int exponent);
```

The following program, *ldexp.c*, illustrates how to use the *ldexp* function:

```
#include <stdio.h>
#include <math.h>

void main(void)
 {
   printf("3 *\' 2 raised to the 4\' is %f\n", ldexp(3.0, 4));
 }
```

When you compile and execute the *ldexp.c* program, your screen will display the following output:

```
3 * 2 raised to the 4 is 48.000000
C:\>
```

Note: *Many C compilers also provide the function* ***ldexpl*** *to support* ***long double*** *values.*

339 CALCULATING THE NATURAL LOGARITHM

The natural logarithm of a number is the power to which *e* must be raised to equal the given number. To help your programs determine the "natural log," C provides the *log* function, which returns the natural logarithm of a floating-point value:

```
#include <math.h>

double log(double value);
```

If the *value* parameter is less than 0, log will set the global variable *errno* to *ERANGE* and return the value *HUGE_VAL* (or *_LHUGE_VAL* for *logl*). The following program, *show_log.c*, illustrates how to use the *log* function:

```
#include <stdio.h>
#include <math.h>

void main(void)
 {
   printf("Natural log of 256.0 is %f\n", log(256.0));
 }
```

When you compile and execute the *show_log.c* program, your screen will display the following output:

```
Natural log of 256.0 is 5.545177
C:\>
```

Note: *Many C compilers also provide the function* ***logl****, which returns the natural logarithm of a* ***long double*** *expression.*

340 CALCULATING THE RESULT OF LOG10X

In Tip 339 you learned how to use C's *log* function to calculate a natural logarithm. As your programs perform mathematical operations, there may be times when you must determine the log to the base 10 of a value (commonly written as *log10x*). For such cases, C provides the *log10* function, as shown here:

```
#include <math.h>

double log10(double value);
```

If the *value* parameter is 0, *log10* will set the global variable *errno* to *EDOM* and return the value *HUGE_VAL* (or *_LHUGE_VAL* for *log10l*). The following program, *log_10.c*, illustrates how to use C's *log10* function:

```
#include <stdio.h>
#include <math.h>

void main(void)
 {
   printf("Log10 of 100 is %f\n", log10(100.0));
```

```
    printf("Log10 of 10000 is %f\n", log10(10000.0));
  }
```

When you compile and execute the *log_10.c* program, your screen will display the following output:

```
Log10 of 100 is 2.000000
Log10 of 10000 is 4.000000
C:\>
```

Note: *Many C compilers also provide the function* ***log10l****, which supports* ***long double*** *values.*

DETERMINING MAXIMUM AND MINIMUM VALUES 341

When your programs compare two numbers, there will be times when you will want to know the minimum or maximum of two values. For such cases, the header file *stdlib.h* provides the macros *min* and *max*. The following program, *min_max.c*, illustrates how to use these two macros:

```
#include <stdio.h>
#include <stdlib.h>

void main(void)
 {
   printf("Maximum of %f and %f is %f\n", 10.0, 25.0, max(10.0, 25.0));
   printf("Minimum of %f and %f is %f\n", 10.0, 25.0, min(10.0, 25.0));
 }
```

To better understand these two macros, consider the following implementations:

```
#define max(x,y)   (((x) > (y)) ? (x) : (y))
#define min(x,y)   (((x) < (y)) ? (x) : (y))
```

BREAKING A DOUBLE INTO ITS WHOLE AND REAL COMPONENTS 342

As you have learned, a floating-point value consists of two parts, an integer portion and a fractional portion. For example, given the number 12.345, the value 12 is the integer portion and 0.345 is the fractional portion. Depending on your program, there may be times when you will want to work with both a value's integer and fractional components, or with each component individually. For such cases, C provides the *modf* function, as shown here:

```
#include <math.h>

double modf(double value, double *integer_part);
```

The *modf* function returns the value's fractional portion and assigns the integer portion to the specified variable. The following program, *int_frac.c*, illustrates how to use the *modf* function:

```
#include <stdio.h>
#include <math.h>

void main(void)
 {
   double value = 1.2345;
   double int_part;
   double fraction;

   fraction = modf(value, &int_part);
   printf("Value %f Integer part %f Fraction %f\n", value, int_part, fraction);
 }
```

When you compile and execute the *int_frac.c* program, your screen will display the following output:

```
Value 1.234500 Integer part 1.000000 Fraction 0.234500
C:\>
```

***Note:** Many C compilers also provide the function **modfl**, which returns the integer and fractional parts of a **long double** expression.*

343 CALCULATING THE RESULT OF X^N

Raising a value to a given power is one of the most common mathematical operations your programs will perform. C provides the function *pow*, which returns the result of a value raised to a given power, as shown here:

```
#include <math.h>

double pow(double value, double power);
```

If evaluating the *value* raised to the given *power* results in overflow, *pow* will assign the global variable *errno* the value *ERANGE* and return *HUGE_VAL* (or *_LHUGE_VAL* for *powl*) to the calling function. If the calling function passes a *value* parameter to *pow* which is less than 0 and the power is not a whole number, then *pow* will set the global variable *errno* to *EDOM*. The following program, *show_pow.c*, illustrates how to use C's *pow* function:

```
#include <stdio.h>
#include <math.h>

void main(void)
 {
   int power;

   for (power = -2; power <= 2; power++)
     printf("10 raised to %d is %f\n", power, pow(10.0, power));
 }
```

When you compile and execute the *show_pow.c* program, your screen will display the following output:

```
10 raised to -2 is 0.010000
10 raised to -1 is 0.100000
10 raised to 0 is 1.000000
10 raised to 1 is 10.000000
10 raised to 2 is 100.000000
C:\>
```

***Note:** Many C compilers also provide the function **powl**, which supports values of type **long double**. Likewise, if you are working with complex values, the header file **complex.h** defines a function prototype for **pow** that works with complex numbers.*

344 CALCULATING THE RESULT OF 10^X

In Tip 343 you learned how to use the *pow* function to determine the result of a value raised to a given power. There may be times when your programs must calculate the result of 10^X. In such cases, you can use the *pow* function or, if your compiler supports it (as the *Turbo C++ Lite* compiler does), you can use C's *pow10*, as shown here:

```
#include <math.h>

double pow10(int power);
```

The following program, *pow10.c*, illustrates how to use the *pow10* function:

```
#include <stdio.h>
#include <math.h>

void main(void)
 {
   printf("10 raised to -1 is %f\n", pow10(-1));
   printf("10 raised to 0 is %f\n", pow10(0));
   printf("10 raised to 1 is %f\n", pow10(1));
```

```
    printf("10 raised to 2 is %f\n", pow10(2));
  }
```

When you compile and execute the *pow10.c* program, your screen will display the following output:

```
10 raised to -1 is 0.100000
10 raised to 0 is 1.000000
10 raised to 1 is 10.000000
10 raised to 2 is 100.000000
C:\>
```

Note: *Many C compilers also provide the function* ***pow10l****, which supports values of type* ***long double****.*

GENERATING A RANDOM NUMBER 345

Depending on your program, there may be times when you must generate one or more random numbers. For such cases, C provides two functions, *rand* and *random*, which both return integer random numbers, as shown here:

```
#include <stdlib.h>

int rand(void);
int random(int ceiling);
```

The *rand* function returns a random number in the range 0 through *RAND_MAX* (see *stdlib.h*). The *random* function returns a number in the range through *ceiling*, which the calling function passes to the *random* function. If your compiler does not support the random function, the file *random.c* on this book's CD-ROM provides a macro you can use in the function's place. The following program, *random.c*, shows how to use both functions:

```
#include <stdio.h>
#include <stdlib.h>

void main(void)
 {
   int i;

   printf("Values from rand\n");
   for (i = 0; i < 100; i++)
     printf("%d ", rand());

   printf("Values from random(100))\n");
   for (i = 0; i < 100; i++)
     printf("%d ", random(100));
 }
```

MAPPING RANDOM VALUES TO A SPECIFIC RANGE 346

In Tip 345, you learned that the C functions *rand* and *random* return random numbers. When your program generates random numbers, there may be times when your program must map the values to a specific range. If you are working with integer values, you can use the *random* function by using a parameter to specify the highest value in the range of random numbers. If you are working with floating-point values, however, such as values in the range 0.0 through 1.0, you can divide the number by a constant to derive a random floating-point number. To map a random integer series to a floating-point number series, simply divide the random number by the random number's upper bound, as shown here:

```
random(100)/100.0
```

The previous example yields a random value in the range 0.01-0.99. If your program requires more digits in the random floating-point number, you can generate a random number through 1000 and divide by 1000, as shown here:

```
random(1000)/1000.0
```

The previous example yields a random value in the range 0.001-0.999. If your program requires more precision in your random numbers, simply increase the maximum random integer size and the constant by which you divide *random's* result. The following program, *map_rand.c*, maps random numbers to the range 0.0 through 1.0 and integer values to the range –5 though 5:

```
#include <stdio.h>
#include <stdlib.h>

void main(void)
 {
   int i;

   printf("Values from random\n");
   for (i = 0; i < 10; i++)
     printf("%f\n", random(100)/100.0);
   printf("Values from random(-5) to random(5)\n");
   for (i = 0; i < 100; i++)
     printf("%d\n", random(10)-5);
 }
```

347 SEEDING THE RANDOM NUMBER GENERATOR

Tip 345 presented C's *rand* and *random* functions that you will use within your programs to generate random numbers. When you work with random numbers, there will be times when you will want to control the series of numbers that the random number generator creates (so that you can test your program's processing with the same set of numbers). There will also be times when you will want the generator to create actual numbers at random. The process of assigning the random number generator's starting number is called *seeding the generator*. To help you seed the random number generators, C provides two functions, *randomize* and *srand*, as shown here:

```
#include <stdlib.h>

void randomize(void);
void srand(unsigned seed);
```

The first function, *randomize*, uses the PC's clock to produce a random seed. On the other hand, the second function, *srand*, lets you specify the starting value of the random number generator. Your programs can use *srand* to control the range of numbers the random number generator creates. The following program, *randseed.c*, illustrates the *srand* and *randomize* functions (if your compiler does not support *randomize*, simply comment the corresponding lines):

```
#include <stdio.h>
#include <time.h>
#include <stdlib.h>

void main(void)
 {
   int i;

   srand(100);
   printf("Values from rand\n");
   for (i = 0; i < 5; i++)
     printf("%d ", rand());
   printf("\nSame 5 numbers\n");
   srand(100);
   for (i = 0; i < 5; i++)
     printf("%d ", rand());
   randomize();
   printf("\nDifferent 5 numbers\n");
   for (i = 0; i < 5; i++)
```

```
    printf("%d ", rand());
 }
```

CALCULATING A VALUE'S SQUARE ROOT 348

When your programs calculate mathematical expressions, they often must perform square root operations. To help your programs perform square root operations, C provides the *sqrt* function, as shown here:

```
#include <math.h>

double sqrt(double value);
```

The *sqrt* function only works with positive values. If your program invokes *sqrt* with a negative value, *sqrt* will set the global variable *errno* to *EDOM*. The following program, *sqrt.c*, illustrates how to use the *sqrt* function:

```
#include <stdio.h>
#include <math.h>

void main(void)
 {
   double value;

   for (value = 0.0; value < 10.0; value += 0.1)
     printf("Value %f sqrt %f\n", value, sqrt(value));
 }
```

Note: *Many C compilers also provide the function **sqrtl**, which returns the square root of a **long double** value.*

CREATING A CUSTOMIZED MATH ERROR HANDLER 349

Several functions presented in this section detect range and overflow errors. By default, when such errors occur, the functions invoke a special function named *matherr*, which performs additional processing, such as assigning the global variable *errno* a specific error number. As it turns out, if your programs define their own *matherr* function, the C math routines will invoke your custom handler. When the math routines invoke your *matherr* function, they will pass to *matherr* a pointer to a variable of type *exception*, as shown here:

```
struct exception
 {
   int type;
   char *function;
   double arg1, arg2, retval;
 };
```

The *type* member contains a constant that describe the error's type. Table 349 describes the error values.

Error Value	Meaning
DOMAIN	An argument is not in the domain of values the function supports
OVERFLOW	An argument produces a result that overflows the resulting type
SING	An argument produces a result in a singularity
TLOSS	An argument produces a result in which all the digits of precision are lost
UNDERFLOW	An argument produces a result that overflows the resulting type

***Table 349** C constants that describe mathematical errors.*

The *function* member contains the name of the routine that experienced the error. The members *arg1* and *arg2* contain the parameters the function experiencing the error passed to *matherr*, while *retval* contains a default return value (which you can assign). The following program, *matherr.c*, illustrates a custom error handler (the companion CD-ROM contains the _matherr.c program for compilers that use _matherr):

```
#include <stdio.h>
#include <math.h>

void main(void)
 {
   printf("Sqrt of -1 is %f\n", sqrt(-1.0));
 }

int matherr(struct exception *error)
 {
   switch (error->type)
    {
       case DOMAIN:     printf("Domain error\n");
                        break;
       case PLOSS:      printf("Partial precision loss error\n");
                        break;
       case OVERFLOW:   printf("Overflow error\n");
                        break;
       case SING:       printf("Error in singularity\n");
                        break;
       case TLOSS:      printf("Total precision loss error\n");
                        break;
       case UNDERFLOW:  printf("Underflow error\n");
                        break;
    };
   printf("Error occurred in %s values %f\n", error->name, error->arg1);
   error->retval = 1;
   return(1);
 }
```

Note: *The **matherr** function only catches domain and overflow errors. To detect divide-by-zero errors, use **signal**. Many C compilers also support the function **matherrl**, which supports arguments of type **long double**.*

350 DETERMINING THE CURRENT DISK DRIVE

If your programs work in the DOS environment, there will be many times when they must determine the current disk drive. For such cases, many compilers provide the *getdisk* function, as shown here:

```
#include <dir.h>

int getdisk(void);
```

The function returns a disk drive number, where 1 is drive A, 2 is drive B, and so on. The following *Turbo C Lite* program, *getdrive.c*, uses the *_dos_getdrive* function to display the current disk drive letter:

```
#include <stdio.h>
#include <dir.h>

void main(void)
 {
   printf("The current drive is %c\n", getdisk() + 'A');
 }
```

Note: *Because they exist to create programs for operating systems other than MS-DOS, many newer compilers do not support MS-DOS-based functions.*

SELECTING THE CURRENT DRIVE 351

In Tip 350 you learned how to use the *getdisk* function to determine the current disk drive in a DOS-based environment. Just as there may be times when your programs must determine the current disk drive, at other times your programs must select a specific disk drive. For such cases, some compilers provide the function *setdisk*, as shown here:

```
#include <dir.h>

int setdisk(int drive);
```

The *drive* parameter is an integer value that specifies the desired drive, where 0 is drive A, 1 is drive B, and so on. The function returns the number of disk drives present in the system. The following *Turbo C Lite* program, *select_c.c*, uses the *setdisk* function to select drive C as the current drive. The program also displays a count of the number of available drives (as set by the LASTDRIVE entry in the *config.sys* file):

```
#include <stdio.h>
#include <dir.h>

void main(void)
 {
   int drive_count;

   drive_count = setdisk(3);
   printf("The number of available drives is %d\n", drive_count);
 }
```

Note: *Many newer 32-bit compilers do not support the setdisk program.*

DETERMINING AVAILABLE DISK SPACE 352

When your programs store considerable amounts of information onto a disk—whether the disk is a floppy disk, hard disk, or other type—each program should keep track of the available disk space to reduce the possibility of running out of room during a critical disk operation. If you are working in a DOS-based system, your compiler may provide the *getdfree* function. The *getdfree* function returns a structure of type *dfree*, as shown here:

```
struct dfree
 {
   unsigned df_avail;              // Available clusters
   unsigned df_total;              // Total clusters
   unsigned df_bsec;               // Bytes per sector
   unsigned df_sclus;              // Sectors per cluster
 };
```

The format of the *getdfree* function is as follows:

```
#include <dos.h>

void getdfree(unsigned char drive, struct dfree *dtable);
```

The *drive* parameter specifies the desired drive, where 1 is drive A, 2 is drive B, and so on. The following *Turbo C Lite* program, *diskfree.c*, uses the *getdfree* function to obtain specifics about the current disk drive:

```
#include <stdio.h>
#include <dos.h>

void main(void)
 {
   struct dfree diskinfo;
   long disk_space;
```

```
    getdfree(3, &diskinfo);
    disk_space = (long) diskinfo.df_avail *
                 (long) diskinfo.df_bsec *
                 (long) diskinfo.df_sclus;
    printf("Available disk space %ld\n", disk_space);
  }
```

Note: *If you run this program on an operating system other than MS-DOS that uses a different file system, the program's output will not be correct.*

353 WATCHING OUT FOR DBLSPACE

Some Tips in this section show you ways to perform absolute disk read and write operations that work with a disk's sectors. Before your programs perform low-level disk I/O operations, make sure that the disk you are going to read is not a compressed disk with contents that *dblspace* or another third-party disk utility has previously compressed. Compressed disks store information on a sector-by-sector basis. If you write a compressed disk sector, you run considerable risk of corrupting the compressed disk—losing the information it contains. As a rule, most programs do not need to perform such low-level disk read and write operations. If you are writing a disk utility program such as *undelete*, make sure that you know how to test for and work with compressed disks before you begin.

354 READING FILE ALLOCATION TABLE INFORMATION

If you are working in a DOS-based system, the file allocation table tracks which parts of your disk are in use, which parts are damaged, and which parts are available (for file and program storage). If your programs perform low-level disk operations, there may be times when you must know information such as the disk's type, bytes per sector, number of sectors per cluster, and the number of clusters on the disk. In such cases, your compiler may provide the *getfat* or *getfatd* functions, as shown here:

```
#include <dos.h>

void getfat(unsigned char drive, struct fatinfo *fat);
void getfatd(struct fatinfo *fat);
```

The *getfat* function lets you specify the desired drive, whereas *getfatd* returns the information for the current drive. To specify a disk drive letter to the *getfat* function, specify a number value where 1 is drive A, 2 is drive B, 3 is drive C, and so on. The *getfat* and *getfatd* functions assign the information to a structure of type *fatinfo*, as shown here:

```
struct fatinfo
 {
   char fi_sclus;          // sectors per cluster
   char fi_fatid;          // disk type
   unsigned fi_nclus;      // clusters per disk
   int fi_bysec;           // bytes per sector
 };
```

The following *Turbo C Lite* program, *getfatd.c*, uses *getfatd* to display information about the current disk drive:

```
#include <stdio.h>
#include <dos.h>

void main(void)
 {
   struct fatinfo fat;

   getfatd(&fat);

   printf("Sectors per cluster %d\n", fat.fi_sclus);
   printf("Clusters per disk %u\n", fat.fi_nclus);
```

```
   printf("Bytes per cluster %d\n", fat.fi_bysec);
   printf("Disk type %x\n", fat.fi_fatid & 0xFF);
}
```

Note: *If you run this program on an operating system other than MS-DOS that uses a different file system, the program's output will not be correct.*

UNDERSTANDING THE DISK ID 355

In Tip 354 you used the *getfat* and *getfatd* functions to get information about the current disk drive. As you found, these functions returned a byte called the *fi_fatid*, which contains a representation of the DOS disk ID. Table 355 specifies the possible values for *fi_fatid*.

Value (Hex)	Disk Type
FOH	3 1/2 inch 1.44Mb or 2.88Mb
	Zip Disk
F8H	Hard Disk
	CD-ROM Drive
F9H	3 1/2 inch 720Kb or 5 1/4 inch 1.2Mb
FAH	5 1/4 inch 320Kb
FCH	5 1/4 inch 180Kb
FDH	5 1/4 inch 360Kb
FEH	5 1/4 inch 160Kb
FFH	5 1/4 inch 320Kb

Table 355 Disk ID values DOS returns.

PERFORMING AN ABSOLUTE SECTOR READ OR WRITE 356

If you work in a DOS-based environment, DOS lets you perform absolute disk read and write operations at the sector level. Normally, your programs use the DOS services to perform these operations. However, to make these operations easier to perform, many C compilers provide the *absread* and *abswrite* functions, as shown here:

```
#include <dos.h>

int absread(int drive, int number_of_sectors, long starting_sector, void *buffer);
int abswrite(int drive, int number_of_sectors, long starting_sector, void *buffer);
```

The *drive* parameter specifies the disk drive you want to read, where 0 is drive A, 1 is drive B, and so on. The *number_of_sectors* parameter specifies the number of sectors you want to read or write, beginning at the sector the *starting_sector* parameter specifies.

Finally, the *buffer* parameter is a pointer to the buffer into which information is read or from which the output is written. If the functions succeed, they return the value 0. If an error occurs, the functions return the value –1. The following *Turbo C Lite* program, *chk_disk.c*, reads every sector on drive C.

If the program experiences errors reading a sector, it will display the sector number:

```
#include <stdio.h>
#include <dos.h>
#include <alloc.h>
```

```
void main(void)
 {
   struct fatinfo fat;
   long sector, total_sectors;
   void *buffer;

   getfat(3, &fat);
   total_sectors = fat.fi_nclus * fat.fi_sclus;
   if ((buffer = malloc(fat.fi_bysec)) == NULL)
     printf("Error allocating sector buffer\n");
   else
     for (sector = 0; sector < total_sectors; sector++)
       if (absread(2, 1, sector, buffer) == -1)
        {
         printf("\n\007Error reading sector %ld press Enter\n", sector);
         getchar();
        }
       else
         printf("Reading sector %ld\r", sector);
 }
```

Note: *While you can perform absolute sector read and writes in Windows, the way Windows writes information to the disk makes absolute read and writes both dangerous and inconsistent. You should avoid absolute disk activities in Windows and process your disk reads and writes through the Windows Application Programming Interface (API).*

357 Performing BIOS-Based Disk I/O

When your programs perform file operations, they use the DOS system services to manipulate files. The DOS services, in turn, call other DOS services to read and write logical disk sectors. To perform the actual disk I/O operations, the DOS services then call BIOS disk services. If you write disk utility programs, for example, your programs may need to perform low-level disk I/O operations. In such cases, your compiler may provide the *biosdisk* function, as shown here:

```
#include <bios.h>

int biosdisk(int operation, int drive, int head, int track, int sector,
  int sector_count, void *buffer);
```

The *drive* parameter specifies the drive number, where 0 is drive A, 1 is drive B, and so on. For a hard disk, 0x80 is the first hard drive, 0x81 is the second drive, and so on. The *head*, *track*, *sector*, and *sector_count* parameters specify the physical disk sectors that you want to read or write. The *buffer* parameter is a pointer to the buffer into which *biosdisk* reads the data or from which *biosdisk* writes the data. Finally, the *operation* parameter specifies the desired function. Table 357.1 lists the valid operations.

Operation	Function
0	Reset the disk system
1	Return the status of the last disk operation
2	Read the specified number of sectors
3	Write the specified number of sectors
4	Verify the specified number of sectors
5	Format the specified track—buffer contains a table of bad locations
6	Format the specified track, setting bad sectors
7	Format the drive beginning at the specified track

Table 357.1 Valid ***biosdisk*** *operations.*

8	Return the drive parameters in the first four bytes of buffer
9	Initialize the drive
10	Perform a long read—512 sector bytes plus four extra
11	Perform a long write—512 sector bytes plus four extra
12	Perform a disk seek
13	Alternate disk reset
14	Read sector buffer
15	Write sector buffer
16	Test drive ready
17	Recalibrate the drive
18	Perform the controller RAM diagnostic
19	Perform the drive diagnostic
20	Perform the controller internal diagnostic

Table 357.1 Valid ***biosdisk*** *operations. (continued from previous page)*

If successful, the function returns the value 0. If an error occurs, the return value of the function specifies the error. Table 357.2 lists the error values.

Error Value	Error
0	Successful
1	Invalid command
2	Address mark not found
3	Write-protected disk
4	Sector not found
5	Hard disk reset failed
6	Disk change line
7	Drive parameter activity failed
8	DMA overrun
9	DMA across 64Kb boundary
10	Bad sector
11	Bad track
12	Unsupported track
16	CRC/ECC read error
17	CRC/ECC corrected data
32	Controller failure
64	Seek failed
128	No response
170	Hard disk not ready
187	Undefined error
204	Write fault
224	Status error
255	Sense operation failed

Table 357.2 Error status values ***biosdisk*** *returns.*

Note: *Many compilers also provide a function named* ***_bios_disk****, which performs processing identical to* ***biosdisk****, with the exception that your programs pass to the function a structure of type* ***diskinfo_t****, which contains the* ***drive****,* ***head****,* ***track****,* ***sector****, and* ***sector count*** *values.*

Note: *While you can use* ***bios_disk*** *to perform BIOS-based disk I/O under Windows, the methods Windows uses to write information to the disk makes BIOS-based disk I/O both dangerous and inconsistent. You should avoid BIOS-based disk I/O in Windows, and process your disk reads and writes through the Windows Application Programming Interface (API).*

358 Testing a Floppy Drive's Readiness

In Tip 357 you learned how to use the *biosdisk* function to invoke BIOS disk services. A useful operation the *biosdisk* function can perform is to test whether a floppy disk contains a disk and is ready for access. The following *Turbo C Lite* program, *test_a.c*, uses the *biosdisk* function to check the floppy disk:

```
#include <stdio.h>
#include <bios.h>

void main(void)
 {
   char buffer[8192];

   // Try reading head 1, track 1, sector 1
   if (biosdisk(2, 0, 1, 1, 1, 1, buffer))
     printf("Error accessing drive\n");
   else
     printf("Drive ready\n");
 }
```

Note: *Because the biosdisk function interacts with the BIOS, bypassing the operating system, many newer compilers do not support the function.*

359 Opening a File Using FOPEN

Many C programs you create will store and retrieve information in a file. Before your programs can read information from or write information to a file, the program must open the file. The *fopen* function lets your programs open a file. The format of *fopen* is as follows:

```
#include <stdio.h>

FILE *fopen(const char *filename, const char *mode);
```

The *filename* parameter is a character string that contains the name of the desired file, such as "*c:\datafile.dat*". The *mode* parameter specifies how you want to use the file—to read, write, or append. Table 359 describes the mode values *fopen* supports.

Mode	Meaning
a	Opens the file for append operations—if the file does not exist, the operating system creates the file
r	Opens an existing file for read operations
w	Opens a new file for output—if a file with the same name exists, the operating system overwrites the file
r+	Opens an existing file for reading and writing

Table 359 Mode values ***fopen*** *supports.*

Mode	Meaning
w+	Opens a new file for reading and writing—if a file with the same name exists, the operating system overwrites the file
a+	Opens a file for append and read operations—if the file does not exist, the operating system creates the file

Table 359 Mode values ***fopen*** *supports. (Continued from previous page.)*

The *fopen* function returns a pointer (called a *file pointer*) to a structure of type *FILE* that the header file *stdio.h* defines. Your program will use the file pointer for its input and output operations. If the *fopen* function cannot open the specified file, it returns the value *NULL*. Your programs should always test *fopen's* return value to make sure it successfully opened the file, as shown here:

```
if ((fp = fopen("FILENAME.EXT", "r")) != NULL)
  {
    // File successfully opened
  }
else
  {
    // Error opening the file
  }
```

Within your program, you must declare the file pointer variable as follows:

```
void main(void)
  {
     FILE *fp;  // Pointer to a structure of type FILE
```

Many programs open one file for input and another for output. In such cases, you would declare two file pointers, as shown here:

```
FILE *input, *output;
```

Many Tips in this section use *fopen* to open a file for read, write, or append operations.

UNDERSTANDING THE FILE STRUCTURE 360

As you have learned, when your programs perform file input and output operations, they normally declare file pointers using the *FILE* structure, as shown here:

```
FILE *input, *output;
```

If you examine the header file *stdio.h*, you will find the definition of the *FILE* structure. In the case of *Turbo C++ Lite*, the structure takes the following form:

```
typedef struct
 {
   short level;                 // fill/empty level of buffer
   unsigned flags;              // File status flags
   char fd;                     // File descriptor
   unsigned char hold;          // Ungetc char if no buffer
   short bsize;                 // Buffer size
   unsigned char *buffer;       // Data transfer buffer
   unsigned char *curp;         // Current active pointer
   unsigned istemp;             // Temporary file indicator
   short token;                 // Used for validity checking
 } FILE;                        // This is the FILE object
```

The *FILE* structure contains the low-level *file descriptor* the operating system uses to access the file, the file's buffer size and location, the character buffer *unget* uses, a flag that indicates whether the file is a temporary file, and other flag variables. In addition, the *FILE* structure stores the file pointer that keeps track of your current location within the file.

If you are working in the DOS environment, most compilers define a fixed size array (usually 20) of file pointers that hold the information for each file your program opens. If your program must open more than 20 files, you must refer to the documentation that accompanied your compiler for the steps you must perform to change the file pointer array size.

361 Closing an Open File

Just as your programs must open a file before they use it, your programs should also close the file when they no longer need it. Closing a file directs the operating system to flush all the disk buffers associated with the file and to free up system resources the file consumed, such as the file pointer data. The C *fclose* function closes the file associated with the specified file pointer, as shown here:

```
#include <stdio.h>

int fclose(FILE *file_pointer);
```

If *fclose* is successful, it will return the value 0. If an error occurs, *fclose* returns the constant *EOF*, as shown here:

```
if (fclose(fp) == EOF)
  printf("Error closing the data file\n");
```

As you examine C programs, you will find that most programs do not test *fclose*'s return status value, as shown here:

```
fclose(fp);
```

In most cases, should a file close operation experience an error, the program can do very little to correct the situation. However, if you are working with critical data files, you should display an error message to the user so the user can examine the file's contents.

***Note:** If you do not invoke the **fclose** function, C will close your open files when the program ends.*

362 Reading and Writing File Information One Character at a Time

When your programs perform file input and output operations, your programs can read and write data one character at a time or one line at a time. For character input and output operations, your programs can use the *fgetc* and *fputc* functions, whose formats are shown here:

```
#include <stdio.h>

int fgetc(FILE *input_pointer);
int fputc(int character, FILE *output_pointer);
```

The *fgetc* function reads the current character from the specified input file. If the file pointer has reached the end of the file, *fgetc* returns the constant *EOF*. The *fputc* function writes a character to the current file pointer location within the specified output file. If an error occurs, *fputc* returns the constant *EOF*. The following program, *confcopy.c*, uses *fgetc* and *fputc* to copy the contents of the root directory file *config.sys* to a file named *config.tst*:

```
#include <stdio.h>

void main(void)
 {
   FILE *input, *output;
   int letter;
```

```
   if ((input = fopen("\\CONFIG.SYS", "r")) == NULL)
     printf("Error opening \\CONFIG.SYS\n");
   else if ((output = fopen("\\CONFIG.TST", "w")) == NULL)
     printf("Error opening \\CONFIG.TST\n");
   else
     {
       // Read and write each character in the file
       while ((letter = fgetc(input)) != EOF)
         fputc(letter, output);
       fclose(input);      // Close the input file
       fclose(output);     // Close the output file
     }
 }
```

UNDERSTANDING THE FILE POINTER'S POSITION POINTER 363

Tip 360 presented the *FILE* structure. As you learned, one of the structure's fields holds a *position pointer* to the current location within the file. When you first open a file for read or write operations, the operating system sets the position pointer to the start of the file. Each time you read or write a character, the position pointer advances one character. If you read a line of text from the file, the position pointer advances to the start of the next line. Using the position pointer, the file input and output functions can always keep track of the current location within the file. When you open a file in append mode, the operating system sets the position pointer to the very end of the file. In later Tips you will learn how to change the position pointer to specific file locations using the *fseek* and *fsetpos* functions. Table 363 specifies the location at which *fopen* places the position pointer when you open the file in read, write, and append modes.

Open Mode	File Pointer Position
a	Immediately after the last character in the file
r	At the start of the file
w	At the start of the file

***Table 363** The file position pointer settings that result from a call to **fopen**.*

DETERMINING THE CURRENT FILE POSITION 364

In Tip 363 you learned how C tracks the current position in files open for input or output operations. Depending on your program, there may be times when you must determine the position pointer's value. In such cases, your programs can use the *ftell* function, as shown here:

```
#include <stdio.h>

long int ftell(FILE *file_pointer);
```

The *ftell* function returns a long integer value that specifies the byte offset of the current position in the specified file. The following program, *show_pos.c*, uses the *ftell* to display position pointer information. The program begins by opening the root directory file *config.sys* in read mode. The program then uses *ftell* to display the current position. Next, the program reads and displays the file's contents. After finding the end of the file, the program again uses *ftell* to display the current position, as shown here:

```
#include <stdio.h>

void main(void)
 {
   FILE *input;
```

```
    int letter;

    if ((input = fopen("\\CONFIG.SYS", "r")) == NULL)
      printf("Error opening \\CONFIG.SYS\n");
    else
      {
        printf("Current position is byte %d\n\n", ftell(input));
          // Read and write each character in the file
        while ((letter = fgetc(input)) != EOF)
          fputc(letter, stdout);
        printf("\nCurrent position is byte %d\n", ftell(input));
        fclose(input);      // Close the input file
      }
  }
```

365 UNDERSTANDING FILE STREAMS

Many books and magazines refer to C's file pointers as *pointers to file streams.* Unlike many other programming languages, C does not assume that files contain information in a specific format. Instead, C considers all files as nothing more than a collection of bytes. As you read a file, you read one byte after another, in other words, a *stream* of bytes. Your programs or functions, such as *fgets*, are left to interpret the bytes. For example, *fgets* considers the linefeed character as the end of one line and the start of another. The *fgets* function makes this character interpretation by itself. That is, C itself does not interpret the bytes. As you write programs and functions that manipulate files, think of the files as nothing more than a collection of bytes.

366 UNDERSTANDING FILE TRANSLATIONS

The C file manipulation functions, such as *fgets* and *fputs*, can interpret files in one of two ways: *text* and *binary* mode. By default, the *fgets* and *fputs* functions use text mode. In text mode, functions such as *fputs*, which write information to a file, convert the linefeed character to a carriage return linefeed combination. During an input operation, functions such as *fgets* convert the carriage return linefeed combination to a single linefeed character. In binary mode, on the other hand, the functions do not perform these character translations. To help you determine the current translation mode, many DOS- and Windows-based compilers provide the global variable *_fmode*, which contains one of the values listed in Table 366.

Constant	Description
O_TEXT	Text mode translations
O_BINARY	Binary mode translations

Table 366 The constant values for _fmode.

The following program, *fmode.c*, displays the current value of the *_fmode* variable:

```
#include <stdio.h>
#include <fcntl.h>  // Contains the _fmode declaration
#include <stdlib.h> // Some compilers define _fmode here
void main(void)
 {
   if (_fmode == O_TEXT)
     printf("Text mode translations\n");
   else
     printf("Binary mode translations\n");
 }
```

UNDERSTANDING THE CONFIG.SYS FILES= ENTRY 367

If you are working in a DOS-based environment, the FILES entry in the *config.sys* file specifies the number of files that the system can open at one time (in a Windows-based environment, Windows limits the number of open files based on the system's available memory, disk space, other resource usage, and so on). As briefly discussed in the Tip 371, DOS uses the first five file handles for *stdin*, *stdout*, *stderr*, *stdaux*, and *stdprn*. By default, DOS provides support for eight file handles. Because this number is too few for all but the simplest programs, most users increase the number of available handles to 20 or 30, as shown here:

```
FILES=30
```

The FILES entry defines the number of files DOS can open—not the number each program running under DOS can open. If you are running memory-resident programs, for example, the programs can have open files about which you are not aware. If you set the FILES entry to a large number of handles (DOS supports up to 255 handles), it does not mean that your C programs can open that many files. As it turns out, there are two problems with opening a large number of files within your C programs. First, most C compilers restrict the size of the file-pointer array to 20. Before you can open more than 20 files, you must change the array size. Second, as you will learn, DOS restricts the number of files a program can open to 20. Before you can open more than 20 files, you must use a DOS system service to request DOS to support more than 20 open files for the current program.

Note: *Tip 369 explains file handles.*

USING LOW-LEVEL AND HIGH-LEVEL FILE I/O 368

When your C programs work with files, they can perform two types of input and output operations: *low-level* and *high-level* file I/O. All the Tips presented to this point have used C's high-level (or stream-based) capabilities, such as *fopen*, *fgets*, and *fputs*. When you use C's high-level file I/O functions, they in turn, use operating system services that are based on *file handles*. The C run-time library provides low-level functions that your programs can use. Instead of working with a stream pointer, the low-level functions use *file descriptors*. Table 368 briefly describes several of C's most commonly used low-level functions.

Function Name	Purpose
close	Closes the file associated with the specified file handle, flushing the file's buffers
creat	Creates a file for output operations, returning a file handle
open	Opens an existing file for input or output, returning a file handle
read	Reads the specified number of bytes from the file associated with the file handle
write	Writes a specified number of bytes to the file associated with a given handle

Table 368 *C's common low-level file functions.*

When you write your programs, your choice of using low-level or high-level functions depends upon your personal preference. However, keep in mind that most programmers have a better understanding of C's high-level file manipulation functions. As a result, if you use the high-level functions such as *fopen* and *fgets*, more programmers will readily understand your program code.

Note: *The Windows I/O section of this book discusses low-level and high-level file I/O under Windows in detail.*

369 UNDERSTANDING FILE HANDLES

As you know, the FILES entry in the *config.sys* file lets you specify the number of file handles DOS supports. In short, a *file handle* is an integer value that uniquely defines an open file. When you use C's low-level file I/O functions, you will declare your program's file handles as type *int*, as shown here:

```
int input_handle, output_handle;
```

The functions *open* and *creat* return file descriptors or the value –1 if the function cannot open the file:

```
int new_file, old_file;

new_file = creat("FILENAME.NEW", S_IWRITE); // Create a new file for output
old_file = open("FILENAME.OLD", O_RDONLY); // Open an existing file for reading
```

DOS assigns each file you open or create a unique file handle. The handle's value is actually an index into the process file table, within which DOS keeps track of the program's open files.

370 UNDERSTANDING THE PROCESS FILE TABLE

When you run a program in the DOS environment, DOS keeps track of the program's open files using a *process file table*. Within the program segment prefix, DOS stores a far pointer to a table that describes the program's open files. Actually, the table contains entries into a second table, the *system file table*, within which DOS tracks all open files. Figure 370 illustrates the relationship between the file handle, process file table, and system file table.

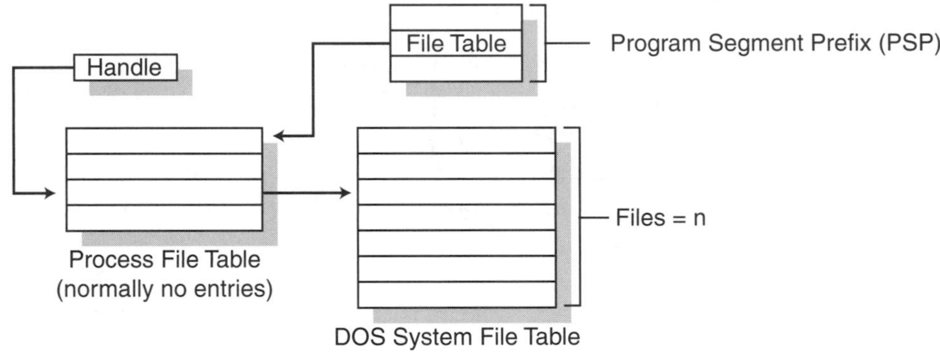

Figure 370 *The relationship between a file handle, process file table, and system file table.*

371 VIEWING THE PROCESS FILE TABLE ENTRIES UNDER MS-DOS

As Tip 370 describes, DOS keeps track of a program's open files using a process file table. At offset 18H, within the program segment prefix, is an array of integer values. The values that comprise this array specify indexes into the DOS system file table. If a value is not in use, the operating system sets it to FFH (decimal 255). The following *Turbo C Lite* program, *filetabl.c*, will display the values in the process file table. Remember, the table contains integer values that serve as indexes into the system file table:

```
#include <stdio.h>
#include <dos.h>
#include <stdlib.h>

void main(void)
  {
    struct fcbs
```

```
   {
     char drive;
     char filename[8];
     char extension[3];
     int current_block;
     int record_size;
   };

  typedef struct fcbs fcb;

  struct program_segment_prefix
   {
     char near *int20;
     char near *next_paragraph_segment;
     char reserved_1;
     char dos_dispatcher[5];
     char far *terminate_vector;
     char far *ctrlc_vector;
     char far *critical_error_vector;
     char near *parent_psp;
     unsigned char file_table[20];
     char near *environment_block_segment;
     char far *stack_storage;
     int  handles_available;
     char far *file_table_address;
     char far *shares_previous_psp;
     char reserved_2[20];
     char dos_int21_retf[3];
     char reserved_3[9];
     fcb fcb1;
     fcb fcb2;
     char reserved_4[4];
     char command_tail[128];
   } far *psp;

  int i;

  psp = (struct program_segment_prefix far *) ((long) _psp << 16);

  for (i = 0; i < 20; i++)
    printf("Entry %d contains %x\n", i, psp->file_table[i]);
}
```

When you compile and execute the *filetabl.c* program, you will find that the first five entries in the process file table are in use. These entries correspond to *stdin, stdout, stderr, stdaux,* and *stdprn*. Edit this program and open one or more files before displaying the file table entries, and you will find more entries within the process file table.

UNDERSTANDING THE SYSTEM FILE TABLE 372

File handles are index values into the process file table, which in turn points to the system file table. The system file table stores information about every file that either DOS, a device driver, a memory-resident program, or your program has open. Figure 372 illustrates the contents of the system file table.

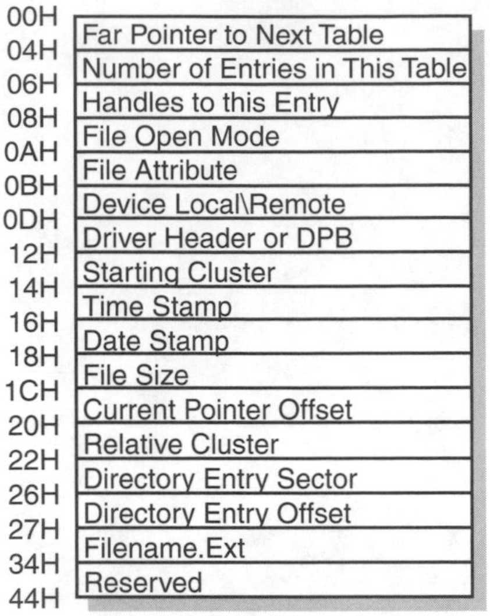

Offset	Field
00H	Far Pointer to Next Table
04H	Number of Entries in This Table
06H	Handles to this Entry
08H	File Open Mode
0AH	File Attribute
0BH	Device Local\Remote
0DH	Driver Header or DPB
12H	Starting Cluster
14H	Time Stamp
16H	Date Stamp
18H	File Size
1CH	Current Pointer Offset
20H	Relative Cluster
22H	Directory Entry Sector
26H	Directory Entry Offset
27H	Filename.Ext
34H	Reserved
44H	

Figure 372 The contents of the DOS system file table.

DOS actually divides the system table into two sections. The first section contains five entries. The second section provides enough space for the number of entries your FILES entry specifies in the *config.sys* file (minus five—the entries that reside in the table's first section).

373 Displaying the System File Table

DOS stores information about every open file within the system file table. Using the DOS list of lists, discussed in the DOS and BIOS section of this book, the *Turbo C Lite* program, *systable.c*, displays the system file table entries:

```
#include <stdio.h>
#include <dos.h>
#include <stdlib.h>

void main(void)
  {
    union REGS inregs, outregs;
    struct SREGS segs;
    int i, j;
    int structure_size;
    struct SystemTableEntry
      {
      struct SystemTableEntry far *next; // Next SFT entry
      unsigned file_count;               // Files in table
      unsigned handle_count;             // Handles to this file
      unsigned open_mode;                // File open mode
      char file_attribute;               // Attribute byte
      unsigned local_remote;             // Bit 15 set means remote
      unsigned far *DPD;                 // Drive parameter block
      unsigned starting_cluster;
      unsigned time_stamp;
      unsigned date_stamp;
      long file_size;
      long current_offset;
      unsigned relative_cluster;
      long directory_sector_number;
      char directory_entry_offset;
      char filename_ext[11];             // No period, space padded
                                         // Ignore SHARE fields for example
```

```
      } far *table_ptr, far *file;
      long far *system_table;

      // Get DOS version
    inregs.x.ax = 0x3001;
    intdos (&inregs, &outregs);
    if (outregs.h.al < 3)
      {
        printf ("This program requires DOS version 3 or later\n");
        exit (1);
      }
    else if (outregs.h.al == 3)
      structure_size = 0x35;
    else if (outregs.h.al >= 4)
      structure_size = 0x3B;
      // Get the list of lists pointer
    inregs.h.ah = 0x52;
    intdosx (&inregs, &outregs, &segs);
      // The pointer to the system file table is at offset 4
    system_table = MK_FP(segs.es, outregs.x.bx + 4);
    table_ptr = (struct SystemTableEntry far *) *system_table;
    do {
      printf ("%d entries in table\n", table_ptr->file_count);
      for (i = 0; i < table_ptr->file_count; i++)
        {
          file = MK_FP(FP_SEG(table_ptr), FP_OFF(table_ptr) +
            (i * structure_size));
          if (file->handle_count)
            {
              for (j = 0; j < 8; j++)
                if (file->filename_ext[j] != ' ')
                  putchar(file->filename_ext[j]);
                else
                  break;
              if (file->filename_ext[8] != ' ')
                putchar('.');
              for (j = 8; j < 11; j++)
                if (file->filename_ext[j] != ' ')
                  putchar(file->filename_ext[j]);
              printf ("  %ld bytes %x attribute %d references\n",
                file->file_size, file->file_attribute,
                file->handle_count);
            }
        }
      table_ptr = table_ptr->next;
    } while (FP_OFF(table_ptr) != 0xFFFF);
  }
```

When you run the *systable.c* program from the DOS prompt, its output probably is not very exciting. However, if you have Windows available, start Windows and use the MSDOS icon to open a DOS window. From within the DOS window, run the *systable* program. You might also want to edit the program and use *fopen* to open one or more files before displaying the system file table contents.

DERIVING FILE HANDLES FROM STREAM POINTERS 374

Tip 360 presented the *FILE* structure defined in the header file *stdio.h*. You have learned that when you perform high-level file operations using *fopen* or *fgets*, you declare stream pointers in terms of the *FILE* structure, as shown here:

```
FILE *input, *output;
```

The C functions later convert the stream pointers to file handles to perform the actual I/O operation. To better understand the relationship between stream pointers and file handles, consider the following program, *handles.c*, which opens the root directory file *config.sys* and then displays the file descriptor for the file, as well as the predefined file handles *stdin, stdout, stderr, stdaux*, and *stdprn*:

```
#include <stdio.h>

void main(void)
 {
   FILE *input;

   if ((input = fopen("\\CONFIG.SYS", "r")) == NULL)
     printf("Error opening \\CONFIG.SYS\n");
   else
     {
       printf("Handle for CONFIG.SYS %d\n", input->fd);
       printf("Handle for stdin %d\n", stdin->fd);
       printf("Handle for stdout %d\n", stdout->fd);
       printf("Handle for stderr %d\n", stderr->fd);
       printf("Handle for stdaux %d\n", stdaux->fd);
       printf("Handle for stdprn %d\n", stdprn->fd);
       fclose(input);
     }
 }
```

If your compiler displays errors regarding the FILE structure, see the *_handles.c* program on the CD-ROM.

375 PERFORMING FORMATTED FILE OUTPUT

Several Tips in this section present ways that your programs can write output to a file. In many cases, your programs must perform formatted file output. For example, if you are creating an inventory report, you will want to line up columns, work with text and numbers, and so on. In this book's Getting Started section, you learned how to use the *printf* function to perform formatted I/O to the screen display. In a similar way, C provides the *fprintf* function, which uses format specifiers to write formatted file output, as shown here:

```
#include <stdio.h>

int fprintf(FILE *file_pointer, const char *format_specifier, [argument[,...]]);
```

The following program, *fprintf.c*, uses *fprintf* to write formatted output to a file named *fprintf.dat*:

```
#include <stdio.h>

void main(void)
 {
   FILE *fp;

   int pages = 1000;
   float price = 54.95;

   if (fp = fopen("FPRINTF.DAT", "w"))
     {
       fprintf(fp, "Book Title: Jamsa\'s C/C++/C# Programmer\'s Bible\n");
       fprintf(fp, "Pages: %d\n", pages);
       fprintf(fp, "Price: $%5.2f\n", price);
       fclose(fp);
     }
   else
     printf("Error opening FPRINTF.DAT\n");
 }
```

RENAMING A FILE 376

As your programs work with files, there may be times when you must rename or move a file. For such cases, C provides the *rename* function. The format of the *rename* function is as follows:

```
#include <stdio.h>

int rename(const char *old_name, const char *new_name);
```

If *rename* successfully renames or moves a file, the function will return the value 0. If an error occurs, *rename* will return a non-zero value and assign to the global variable *errno* one of the error status values listed in Table 376.

Value	Meaning
EACCES	Access denied
ENOENT	File not found
EXDEV	Cannot move from one disk to another

*Table 376 The error status values for **rename**.*

The following program, *my_ren.c*, uses the *rename* function to create a program that can rename or move the file specified in the command line:

```
#include <stdio.h>

void main(int argc, char *argv[])
 {
  if (argc < 3)
    printf("Must specify a source and target filename\n");
  else if (rename(argv[1], argv[2]))
    printf("Error renaming file\n");
 }
```

Note: *Tip 1472 details how you will rename a file using the Windows API.*

DELETING A FILE 377

When your programs work with files, there will be many times when you must delete one or more files. In such cases, your programs can use C's *remove* function. The format of the *remove* function is as follows:

```
#include <stdio.h>

int remove(const char *filename);
```

If the function successfully removes the file, it returns the value 0. If an error occurs, *remove* returns the value –1 and assigns to the global value *errno* one of the values listed in Table 377.

Value	Meaning
EACCES	Access denied
ENOENT	File not found

*Table 377 The errors C's **remove** function returns.*

The following program, *my_del.c*, uses the *remove* function to delete all the files specified in the command line:

```
#include <stdio.h>

void main(int argc, char *argv[])
```

```
{
  while (*++argv)
    if (remove(*argv))
      printf("Error removing %s\n", *argv);
}
```

In addition to the *remove* function, most C compilers support the *unlink* function, which also deletes a file:

```
#include <io.h>

int unlink(const char *filename);
```

If *unlink* successfully deletes the file, it returns the value 0. If an error occurs, *unlink* returns the error status –1, assigning to the global variable *errno* the error status constants listed in Table 377. The following program, *unlink.c*, uses the *unlink* function to delete the files specified in the program's command line:

```
#include <stdio.h>

void main(int argc, char *argv[])
 {
   while (*++argv)
     if (unlink(*argv))
       printf("Error removing %s\n", *argv);
 }
```

Note: *Tip 1473 details how you will delete a file using the Windows API.*

378 DETERMINING HOW A PROGRAM CAN ACCESS A FILE

When your program works with files, there may be times when you must determine whether your program can access a specific file as required. The C function *access* checks whether a file exists as specified and whether you can open the file as required. The format of the *access* function is as follows:

```
#include <io.h>

int access(const char *filename, int access_mode);
```

The *access_mode* parameter specifies how your program needs to use the file, as shown in Table 378.1.

Value	Meaning
0	Checks if the file exists
2	Checks if the file can be written to
4	Checks if the file can be read
6	Checks if the program has read and write permission for the file

Table 378.1 *The values for the* ***access_mode*** *parameter.*

If the program can access the file as specified, *access* returns the value 0. If an error occurs, *access* returns the value –1 and assigns the global variable *errno* one of the error values listed in Table 378.2.

Value	Meaning
EACCES	Access denied
ENOENT	File not found

Table 378.2 *The error values for the* ***access*** *function.*

The following program, *access.c*, uses the *access* function to determine how your program can access the file specified in the program's command line:

```
#include <stdio.h>
#include <io.h>

void main (int argc, char *argv[])
 {
   int access_mode;

   access_mode = access(argv[1], 0);

   if (access_mode)
     printf("File %s does not exist\n");
   else
     {
       access_mode = access(argv[1], 2);
       if (access_mode)
         printf("File cannot be written\n");
       else
         printf("File can be written\n");
       access_mode = access(argv[1], 4);
       if (access_mode)
         printf("File cannot be read\n");
       else
         printf("File can be read\n");
       access_mode = access(argv[1], 6);
       if (access_mode)
         printf("File cannot be read/written\n");
       else
         printf("File can be read/written\n");
     }
 }
```

Note: *Tip 1462 details how you will use file attributes under Windows to determine how a program can access a file.*

SETTING A FILE'S ACCESS MODE 379

When your programs work with files, there will be times when you want to change a program's read and write access. For example, assume that you have an important data file. To protect the file when the program is not in use, you might set the file to read-only access. In this way, the user cannot accidentally delete the file. When the program starts, you can change the file to read and write access, as required. For such cases, your programs can use C's *chmod* function, as shown here:

```
#include <sys\stat.h>
#include <io.h>

int chmod(const char *filename, int access_mode);
```

The file header file *sys\stat.h* defines the access mode constants listed in Table 379.1.

Value	Meaning
S_IWRITE	Write permission is authorized
S_IREAD	Read permission is authorized

*Table 379.1 The access mode constants for **chmod**.*

To provide read and write access, perform a *bitwise OR* of the two constants (*S_IWRITE | S_IREAD*). If *chmod* successfully changes the file's attributes, it returns the value 0. If an error occurs, *chmod* returns the value –1 and sets the global variable *errno* to one of the error status values listed in Table 379.2.

Value	Meaning
ENOENT	File not found
EACCES	Permission denied

*Table 379.2 The error values **chmod** returns.*

The following program, *readonly.c*, sets the file specified in the command line to read-only access:

```
#include <stdio.h>
#include <sys\stat.h>
#include <io.h>

void main(int argc, char *argv[])
 {
   if (chmod(argv[1], S_IREAD))
     printf("Error setting %s\n", argv[1]);
 }
```

Note: *Tip 1463 details how you will use file attributes under Windows to change how a program can access a file.*

380 GAINING BETTER CONTROL OF FILE ATTRIBUTES

In Tip 379 you learned how to use C's *chmod* function to set a file's read and write attributes. When you use the DOS operating system, you can work with the attributes shown in Table 380.1, although your compiler may use different attribute names.

Value	Meaning
FA_ARCH	Archive attribute
FA_DIREC	Directory attribute
FA_HIDDEN	Hidden attribute
FA_LABEL	Disk volume label
FA_RDONLY	Read-only attribute
FA_SYSTEM	System attribute

Table 380.1 The attributes you can use with files within the DOS operating system.

Note: *Some compilers name these constants differently. Examine the include file **dos.h**, provided with your compiler, for the correct constant names.*

To help you work with these attributes, some C compilers provide the *_chmod* function, whose format is shown here (remember, parameters shown within brackets are optional):

```
#include <dos.h>
#include <io.h>

int _chmod(const char *filename, int operation [,int attribute]);
```

The operation tells *_chmod* if you want to set or get the attribute's setting. If the calling function sets the operation to 0, *_chmod* returns the file's current attributes. If the calling function sets the operation 1, *_chmod* sets the specified attribute. The left and right brackets, therefore, indicate that the *attributes* parameter is optional. If *_chmod* is successful, it returns the file's current attributes. If an error occurs, *_chmod* returns the value –1 and assigns the global variable *errno* one of the following values shown in Table 380.2.

Value	Meaning
ENOENT	File not found
EACCES	Permission denied

***Table 380.2** The errors **_chmod** returns.*

The following *Turbo C Lite* program, *tellattr.c*, uses *_chmod* to display a file's current attributes:

```
#include <stdio.h>
#include <dos.h>
#include <io.h>

void main(int argc, char *argv[])
 {
   int attributes;

   if ((attributes = _chmod(argv[1], 0)) == -1)
     printf("Error accessing %s\n", argv[1]);
   else
     {
        if (attributes & FA_ARCH)
          printf("Archive ");
        if (attributes & FA_DIREC)
          printf("Directory ");
        if (attributes & FA_HIDDEN)
          printf("Hidden ");
        if (attributes & FA_LABEL)
          printf("Volume label ");
        if (attributes & FA_RDONLY)
          printf("Readonly ");
        if (attributes & FA_SYSTEM)
          printf("System ");
     }
 }
```

Some older C compilers also provide the functions *_dos_getfileattr* and *_dos_setfileattr*, which let you get or set a file's DOS attributes, as shown here:

```
#include <dos.h>

int _dos_getfileattr(const char *filename, unsigned *attributes);
int _dos_setfileattr(const char *filename, unsigned attributes);
```

The *_dos_getfileattr* and *_dos_setfileattr* functions use the attribute constants detailed in Table 380.3.

Value	Meaning
_A_ARCH	Archive attribute
_A_HIDDEN	Hidden attribute
_A_NORMAL	Normal attribute
_A_RDONLY	Read-only attribute
_A_SUBDIR	Directory attribute
_A_SYSTEM	System attribute
_A_VOLID	Disk volume label

***Table 380.3** The attribute constants the **_dos_getfileattr** and **_dos_setfileattr** functions use.*

If the *_dos_getfileattr* and *_dos_setfileattr* functions succeed, the functions return the value 0. If an error occurs, the functions return the value –1 and assign the global variable *errno* the value *ENOENT* (file not found).

As a rule, your programs should only manipulate the archive, read-only, and hidden file attributes, reserving the other attributes for use by DOS. If you only change the read-only attribute, use the *chmod* function presented in Tip 379 to increase your program's portability.

Note: *Tip 1463 details how you will use file attributes under Windows to change how a program can access a file.*

381 TESTING FOR A FILE STREAM ERROR

When your programs perform file I/O operations, they should always test the return values of functions such as *fopen*, *fputs*, *fgets*, and so on to verify that the operations succeeded. To help your programs perform such testing, C provides the *ferror* macro, which examines an I/O stream for a read or write error. If an error has occurred, *ferror* returns a true value. If no error has occurred, *ferror* returns false, as shown here:

```
#include <stdio.h>

int ferror(FILE *stream);
```

After a file I/O error occurs, the *ferror* macro will remain true until your programs invoke the *clearerr* macro for the given stream:

```
#include <stdio.h>

void clearerr(FILE *stream);
```

The following program, *ferror.c*, reads and displays a file's contents to the screen. After each I/O operation, the program tests for an error. If an error occurs, the program ends, displaying an error message to *stderr*:

```
#include <stdio.h>
#include <stdlib.h>

void main(int argc, char *argv[])
 {
   FILE *fp;
   char line[256];

   if (fp = fopen(argv[1], "r"))
     {
       while (fgets(line, sizeof(line), fp))
         {
           if (ferror(fp))
             {
               fprintf(stderr, "Error reading from %s\n", argv[1]);
               exit(1);
             }
           else
             {
               fputs(line, stdout);
               if (ferror(fp))
                 {
                   fprintf(stderr, "Error writing to stdout\n");
                   exit(1);
                 }
             }
         }
     }
   else
     printf("Error opening %s\n", argv[1]);
 }
```

DETERMINING A FILE'S SIZE 382

As your programs perform file I/O operations, at times you must determine a file's size in bytes. For such cases, you can use C's *filelength* function. The *filelength* function returns a *long* value. You must pass the program a file handle, not a file pointer, as shown here:

```
#include <io.h>

long filelength(int file_handle);
```

If *filelength* succeeds, it returns the file size in bytes. If an error occurs, *filelength* returns the value –1 and sets the global variable *errno* to *EBADF* (bad file number). The following program, *filelen.c*, will display the size of a given file to the screen:

```
#include <stdio.h>
#include <io.h>
#include <fcntl.h>
#include <sys\stat.h>

void main(int argc, char *argv[])
 {
   int file_handle;
   long file_size;

   if ((file_handle = open(argv[1], O_RDONLY)) == -1)
     printf("Error opening the file %d\n", argv[1]);
   else
     {
       file_size = filelength(file_handle);
       printf("The file size in bytes is %ld\n", file_size);
       close(file_handle);
     }
 }
```

Note: *Tip 1463 details how you will determine a file's size using the Windows API.*

FLUSHING AN I/O STREAM 383

To improve your program performance, the C run-time library normally buffers your file output until it has a complete buffer (usually a disk sector) to write to disk, or until you close the file. In this way, the run-time library reduces the number of slow disk I/O operations. Unfortunately, when your programs use such a buffer, they leave the door open for the possibility of losing data. When your program performs a function such as *fputs* to write output and the function does not return an error, the program assumes the operating system has correctly recorded the data to the disk. In reality, however, the data still may reside in your computer's memory. If the user should turn off the computer, the user will lose the data. If you have a program for which you must ensure that all data writes to the disk, you can use *fflush* to direct the run-time library to write the data from its buffer in memory to disk. The format of the *fflush* function is as follows:

```
#include <stdio.h>

int fflush(FILE *file_stream);
```

If *fflush* succeeds, it returns the value 0. If an error occurs, *fflush* returns the constant *EOF*. The following statements illustrate how you can use *fflush* to empty the file buffer to disk following each output operation:

```
while (fgets(line, sizeof(line), input_file))
  {
    fputs(line, output_file);
```

```
    fflush(output_file);
  }
```

***Note:** When you use the **fflush** function, you direct the C run-time library to invoke an operating system service to write the data to disk. If the operating system performs its own buffering (called a **disk cache**), the operating system may place your data into its memory buffer, as opposed to disk. Depending on the disk-caching software, you may be able to invoke another system service to flush the output.*

384 CLOSING ALL OPEN FILES IN ONE STEP

As discussed in Tip 361, before your programs end you should use the *fclose* function to close your open files. Assume that you have a function that performs a critical operation. If the function experiences an error, the program should immediately end. Unfortunately, the function might not be aware of the open files. In such cases, your program can use C's *fcloseall* function to close all open files, as shown here:

```
#include <stdio.h>

int fcloseall(void);
```

If *fcloseall* succeeds, it returns the number of files it successfully closed. If an error occurs, *fcloseall* returns the *EOF* constant. The following statements illustrate how you might use *fcloseall*:

```
if (error_status == CRITICAL)
  {
    fprintf(stderr, "Critical device error\n");
    fcloseall();
    exit(1);
  }
```

385 GETTING A FILE STREAM'S FILE HANDLE

As discussed in Tip 360, when your programs perform file operations, they can perform high-level operations using file streams (*FILE *stream*). You can also use low-level file handles (*int handle*). As you have learned, several of C's run-time library functions require file handles. If your program uses file streams, you can close the file and reopen it using a file handle, or you can obtain a file handle using C's *fileno* function, as shown here:

```
#include <stdio.h>

int fileno(FILE *stream);
```

The following program, *fileno.c*, uses the *fileno* function to get the file handle for an open file stream:

```
#include <stdio.h>
#include <io.h>

void main(int argc, char *argv[])
 {
   FILE *stream;
   int handle;
   long file_length;

   if (stream = fopen(argv[1], "r"))
     {
       // Some statements
       handle = fileno(stream);
       file_length = filelength(handle);
       printf("The file length is %ld\n", file_length);
       fclose(stream);
     }
```

```
    else
      printf("Error opening %s\n", argv[1]);
  }
```

CREATING A TEMPORARY FILENAME USING P_TMPDIR 386

As your programs perform file I/O operations, your programs often must open one or more temporary files or write output to a nonexistent file on disk. In such cases, the difficulty then becomes determining a unique filename so that the program does not overwrite an existing file. To help your programs generate unique filenames, you can use the *tmpnam* function, as shown here:

```
#include <stdio.h>

char *tmpnam(char *buffer);
```

If your program passes a buffer to *tmpnam*, the function will assign the temporary name to the buffer. If you invoke *tmpnam* with *NULL*, *tmpnam* will allocate memory for the filename, returning to the program a pointer to the start of the filename. The *tmpnam* function examines the *P_tmpdir* entry in the *stdio.h* header file. If *P_tmpdir* is defined, *tmpnam* creates the unique filename in the corresponding directory. Otherwise, *tmpnam* will create the file in the current directory. Note that *tmpnam* does not actually create the file, but rather, it returns a filename that your program can use with *fopen* or *open*. The following program, *tmpnam.c*, illustrates the use of the *tmpnam* function:

```
#include <stdio.h>

void main(void)
 {
   char buffer[64];
   int counter;

   for (counter = 0; counter < 5; counter++)
     printf("Temporary filename %s\n", tmpnam(buffer));
 }
```

Note: *The companion CD-ROM that accompanies this book includes the program* ***mak_temp.cpp****, which creates a temporary file with the Windows API.*

CREATING A TEMPORARY FILENAME USING TMP OR TEMP 387

As your programs perform file I/O operations, your programs often must open one or more temporary files or write output to a nonexistent file on disk. In such cases, the difficulty then becomes determining a unique filename so that the program does not overwrite an existing file. To help your programs generate a unique filename, you can use the *tempnam* function, as shown here:

```
#include <stdio.h>

char *tempnam(char *buffer, char *prefix);
```

If your program passes a buffer to *tempnam*, the function assigns the temporary name to the buffer. If you invoke *tempnam* with *NULL*, *tempnam* allocates memory for the filename, returning to the program a pointer to the start of the filename. The *prefix* parameter lets you define a set of characters that you want *tempnam* to place at the start of each filename. The *tempnam* function examines the environment entries to determine if a TMP or TEMP entry exists. If TMP or TEMP is defined, *tempnam* will create the unique filename in the corresponding directory. Otherwise, *tempnam* creates the file in the current directory. Note that *tempnam* does not actually create the file, but rather, it returns a filename that your program can use with *fopen* or *open*. The following program, *tempnam.c*, illustrates the use of the *tempnam* function:

```
#include <stdio.h>

void main(void)
 {
   char buffer[64];
   int counter;

   printf("Temporary filename %s\n", tempnam(buffer, "Bible"));
 }
```

388 Creating a Truly Temporary File

In Tips 386 and 387 you learned how to use the *tmpnam* and *tempnam* functions to generate temporary filenames. As you learned, *tmpnam* and *tempnam* do not actually create a file, rather they simply return a filename not currently in use. In addition, C also provides a function named *tmpfile* that determines a unique filename and then opens the file, returning a file pointer to the program. You will implement the *tmpfile* function as shown here:

```
#include <stdio.h>

FILE *tmpfile(void);
```

If *tmpfile* succeeds, it opens the file in read and write mode, returning a file pointer. If an error occurs, *tmpfile* returns *NULL*. The file *tmpfile* returns is a temporary file. When your program ends (or calls *rmtmp*), the operating system deletes the file and discards its contents. The following statements illustrate how your program might use the *tmpfile* function:

```
FILE *temp_file;

if (temp_file = tmpfile())
  {
    // Temporary file successfully opened
    // Statements that use the file
  }
else
  printf("Error opening temporary file\n");
```

389 Removing Temporary Files

In Tip 388 you learned that the *tmpfile* function lets your programs create a temporary file with contents that exist only for the duration of the program's execution. Depending on your programs, you may want to discard temporary files before the program ends. In such cases, your program can use the *rmtmp* function, whose format is shown here:

```
#include <stdio.h>

int rmtmp(void);
```

If *rmtmp* succeeds, it returns the number of files it successfully closed and deleted.

390 Searching the Command Path for a File

When you work within the DOS environment, the PATH command defines the directories that DOS searches for EXE, COM, and BAT files when you execute an external command. Because the subdirectories defined in the PATH normally contain your most commonly used commands, there may be times when you want a program to search the PATH subdirectory entries for a data file. For such cases, some compilers provide the *searchpath* function. You invoke the function with the desired filename. If *searchpath* successfully locates the file, it returns a complete pathname to the file that your programs can use within *fopen*. If *searchpath* does not find the file, it returns *NULL*, as shown here:

```
#include <dir.h>

char *searchpath(const char *filename);
```

The following *Turbo C Lite* program, *srchpath.c*, illustrates the use of the *searchpath* function:

```
#include <stdio.h>
#include <dir.h>

void main (int argc, char *argv[])
 {
   char *path;

   if (path = searchpath(argv[1]))
     printf("Pathname: %s\n", path);
   else
     printf("File not found\n");
 }
```

Note: *The* ***searchpath*** *function searches the current directory for the specified file before searching the command path subdirectories.*

SEARCHING AN ENVIRONMENT ENTRY'S SUBDIRECTORY LIST FOR A FILE 391

In Tip 390 you used the *searchpath* function to search directories in the command path for a specified file. In a similar way, you may want to search the directories specified in a different environment entry for a file. For example, many C compilers define *LIB* and *INCLUDE* entries that specify the location of library files (with the *.lib* extension) and header files (with the *.h* extension). To search the directories the *LIB* and *INCLUDE* entries specify, you can use the *_searchenv* function, as shown here:

```
#include <dos.h>

char *_searchenv(const char *filename, const char *environment_entry, *pathname);
```

The *_searchenv* function searches the directories specified in the *environment_entry* for the specified filename. If *_searchenv* finds the filename, *_searchenv* assigns the file's pathname to the pathname character string buffer, returning a pointer to the pathname. If *_searchenv* does not find the file, *_searchenv* returns *NULL*. The following program, *srch_env.c*, uses the *_searchenv* function to search the subdirectories specified in the *LIB* entry for a specified file:

```
#include <stdio.h>
#include <stdlib.h>

void main (int argc, char *argv[])
 {
   char path[128];

   _searchenv(argv[1], "LIB", path);
   if (path[0])
     printf("Pathname: %s\n", path);
   else
     printf("File not found\n");
 }
```

Note: *The* ***_searchenv*** *function searches the current directory for the specified file before searching the environment entry's subdirectories.*

Note: *In Tips 1474 through 1476 you will learn how to use the Windows API to find files on your system.*

OPENING FILES IN THE TEMP DIRECTORY 392

As you know, many programs create their temporary files in the subdirectory that the TEMP environment entry specifies within the *config.sys* file. Within your programs, you can easily create your own files within the directory the

TEMP entry names, using the *getenv* function. The following statements illustrate how your programs can open a file named *tempdata.dat* within the temporary directory:

```
char pathname[_MAX_PATH];
strcpy(pathname, getenv("TEMP"));
if (pathname[0])
  strcat(pathname, "\\TEMPDATA.DAT");
else
  strcat(pathname, "TEMPDATA.DAT");
if (fp = fopen(pathname, "w"))
```

In this code fragment, if the TEMP entry exists, the program opens the file in the corresponding subdirectory. If there is no TEMP entry, the program opens the file in the current directory. Note that the code fragment assumes that the TEMP variable does not contain a value that ends with a backslash. Ideally, your programs will test TEMP's current value and process accordingly.

393 MINIMIZING FILE I/O OPERATIONS

Compared to the fast electronic speed of your computer's CPU and memory, the mechanical disk is very slow. As a result, you should try to minimize the number of disk I/O operations your programs must perform. With respect to file operations, the file that is open probably consumes the most time. Therefore, you should always examine your programs to make sure you do not open and close a file unnecessarily or repeatedly open a file from within a loop. For example, consider the following statements:

```
while (menu_choice != QUIT)
  {
    if (fp = fopen("DATABASE.DAT", "r"))
      {
        // Get customer name
        get_customer(name);
        // Search file for customer info
        search_customer_info(name, fp, data_buffer);
        fclose(fp);
      }
    else
      {
        file_open_error("Aborting...");
      }
    menu_choice = get_menu_choice();
  }
```

The statements repeatedly loop, getting customer information until the user selects the QUIT option. Note that the *fopen* function call occurs within the loop. Therefore, the program repeatedly performs the slow disk I/O operation. To improve the system's performance, the program should pull the *fopen* outside of the loop. If the *search_customer* function must start at the beginning of the file, the program can rewind the file, as shown here:

```
if (fp = fopen("DATABASE.DAT", "r"))
  file_open_error("Aborting...");
while (menu_choice != QUIT)
  {
    // Get customer name
    get_customer(name);
    rewind(fp);
    // Search file for customer info
    search_customer_info(name, fp, data_buffer);
    menu_choice = get_menu_choice();
```

```
   }
 fclose(fp);
```

WRITING CODE THAT USES BACKSLASHES IN DIRECTORY NAMES 394

Several Tips presented in this section work with directory names. For example, the *chdir* function lets your programs select a specific directory. When your program specifies a directory name as a constant value, make sure you use double backslashes (\\) within pathnames, as required. The following *chdir* function call, for example, tries to select the subdirectory DOS:

```
status = chdir("\DOS");
```

When you use a backslash character within a C string, remember that C treats the backslash character as a special symbol. When the C compiler encounters the backslash, it checks the character that follows to determine if it is a special symbol and, if so, replaces the character with the correct ASCII counterparts. If the character that follows the backslash is not a special symbol, the C compiler ignores the backslash character. Therefore, the previous *chdir* function would try to select the directory DOS as opposed to \\DOS. The correct function invocation in this case would be as follows:

```
status = chdir("\\DOS");
```

CHANGING THE CURRENT DIRECTORY 395

As your programs execute, there may be times when your programs must change the current directory. To help you perform such operations, most C compilers provide the *chdir* function. The *chdir* function is very similar to the DOS CHDIR command: if you invoke the function with a string that does not contain a disk drive letter, *chdir* looks for the directory on the current drive. The following function call, for example, selects the directory *data* on drive C:

```
status = chdir("C:\\DATA");   // Note the use of \\
```

In a similar way, the following command selects the directory *tclite* on the current drive:

```
status = chdir("\\TCLITE");
```

If the *chdir* function succeeds, it returns the value 0. If the directory does not exist, *chdir* returns the value –1 and sets the global variable *errno* to the constant *ENOENT*. The following *Turbo C Lite* program, *newchdir.c*, implements the DOS CHDIR command:

```
#include <stdio.h>
#include <stdlib.h>
#include <dir.h>
#include <errno.h>

void main(int argc, char *argv[])
  {
   char directory[MAXPATH];

   if (argc == 1)                         // Display the current directory
     {
       getcwd(directory, MAXPATH);
       puts(directory);
     }
   else if ((chdir(argv[1])) && (errno == ENOENT))
     puts("Invalid directory");
  }
```

Note: *Some compilers define the symbol _MAX_PATH in the **direct.h** (or **dir.h**) include file, as opposed to using MAXPATH. Some compilers name the functions _getcwd and _chdir. See the file _newchdir.c on the CD-ROM.*

396 CREATING A DIRECTORY

As your programs execute, they may need to create a directory. To help your programs do so, most C compilers provide a *mkdir* function. The *mkdir* function is very similar to the DOS MKDIR command. If you invoke the function with a string that does not contain a disk drive letter, *mkdir* creates the directory on the current drive. The following function call, for example, creates the directory DATA on drive C:

```
status = mkdir("C:\\DATA");  // Note the use of \\
```

In a similar way, the following command creates the directory TEMPDATA on the current drive, in the current directory:

```
status = mkdir("TEMPDATA");
```

If the *mkdir* function succeeds, it returns the value 0. If *mkdir* cannot create the directory, it returns the value –1.

Note: *Some compilers refer to the function as _mkdir.*

397 REMOVING A DIRECTORY

As your programs execute, they may need to create or remove a directory. To help your programs remove a directory, most C compilers provide a *rmdir* function. The *rmdir* function is very similar to the DOS RMDIR command. If you invoke the function with a string that does not contain a disk drive letter, *rmdir* creates the directory on the current drive. The following function call, for example, removes the directory DATA from drive C:

```
status = rmdir("C:\\DATA");  // Note the use of \\
```

In a similar way, the following command removes the directory TEMPDATA from the current drive and directory:

```
status = rmdir("TEMPDATA");
```

If the *rmdir* function succeeds, it returns the value 0. If the directory does not exist or *rmdir* cannot remove it, *rmdir* returns the value –1 and assigns the global variable *errno* one of the values listed in Table 397.

Value	Meaning
EACCES	Access denied
ENOENT	No such directory

*Table 397 The error values for **rmdir**.*

Note: *Some compilers refer to the function as _rmdir.*

398 REMOVING A DIRECTORY TREE

In MS-DOS version 6, Microsoft introduced the DELTREE command. DELTREE lets you, in one step, delete a directory, its files, and any subdirectories within the directory. If you do not use DOS version 6, you can create your own DELTREE command using the *Turbo C Lite* program *deltree.c*, as shown here:

```
#include <dos.h>
#include <stdio.h>
#include <stdlib.h>
#include <dir.h>
#include <alloc.h>
#include <string.h>

void main(int argc, char **argv)
 {
   void delete_tree(void);
```

```
   char buffer[128];
   char drive[MAXDRIVE], directory[MAXDIR], filename[MAXFILE], ext[MAXEXT];

   if (argc < 2)
     {
       printf ("Syntax error\n");
       exit(0);
     }
   fnsplit (argv[1], drive, directory, filename, ext);
   getcwd (buffer, sizeof(buffer));
   if (drive[0] == NULL)
     {
       fnsplit (buffer, drive, directory, filename, ext);
       strcpy (buffer, directory);
       strcat (buffer, filename);
       strcat (buffer, ext);
     }
   else
     {
       printf ("Do not specify drive letter\n");
       exit (1);
     }
   if (strcmpi(buffer, argv[1]) == 0)
     {
      printf ("Cannot delete current directory\n");
       exit (1);
     }
   getcwd (directory, 64);
   if (chdir (argv[1]))
     printf ("Invalid directory %s\n", argv[1]);
   else
     delete_tree();
   chdir (directory);
   rmdir (argv[1]);
 }

union REGS inregs, outregs;
struct SREGS segs;

void delete_tree(void)
 {
   struct ffblk fileinfo;
   int result;
   char far *farbuff;
   unsigned dta_seg, dta_ofs;

   result = findfirst("*.*", &fileinfo, 16);
   inregs.h.ah = 0x2f;
   intdosx (&inregs, &outregs, &segs);
   dta_seg = segs.es;
   dta_ofs = outregs.x.bx;
   while (! result)
    {
      if ((fileinfo.ff_attrib & 16) && (fileinfo.ff_name[0] != '.'))
       {
         inregs.h.ah = 0x1A;
         inregs.x.dx = FP_SEG(farbuff);
         segread(&segs);
         intdosx (&inregs, &outregs, &segs);
         chdir (fileinfo.ff_name);
         delete_tree();
         chdir ("..");
         inregs.h.ah = 0x1A;
         inregs.x.dx = dta_ofs;
         segs.ds = dta_seg;
```

```
          rmdir (fileinfo.ff_name);
        }
      else if (fileinfo.ff_name[0] != '.')
        {
          remove (fileinfo.ff_name);
        }
      result = findnext (&fileinfo);
    }
}
```

Note: *Because the program uses MS-DOS-based system services, many newer compilers will not support the corresponding functions.*

399 BUILDING A FULL PATHNAME

When your programs work with files and directories, you might need to know the file's complete (full) pathname. For example, if the current directory is *data* and the current drive is C, the full name of the file *report.dat* is *c:\data\report.dat*. To help you resolve a file's full name (that is, combine its components), some C compilers provide a function named *fnmerge*. The function uses five parameters: a buffer within which the function places the full pathname, the drive name, the directory name, the file name, and the extension, as shown here:

```
#include <dir.h>

void fnmerge (char *buffer, const char *drive, const char *dir,
              const char *filename, const char *extension);
```

If the value of the *buffer* parameter is *NULL*, *fnmerge* will allocate the memory used to hold the full pathname. If *fnmerge* successfully resolves the filename, it will return a pointer to the buffer. If an error occurs, the function returns *NULL*. The following *Turbo C Lite* program, *fullname.c*, illustrates the use of the *fnmerge* function:

```
#include <string.h>
#include <stdio.h>
#include <dir.h>

void main(void)
 {
    char s[MAXPATH];
    char drive[MAXDRIVE];
    char dir[MAXDIR];
    char file[MAXFILE];
    char ext[MAXEXT];

    getcwd(s,MAXPATH);
    strcat(s,"\\");
    fnsplit(s,drive,dir,file,ext);
    strcpy(file,"DATA");
    strcpy(ext,".TXT");
    fnmerge(s,drive,dir,file,ext);
    puts(s);
 }
```

Note: *Some compilers use the include file **direct.h**, as opposed to **dir.h**.*

400 PARSING A DIRECTORY PATH

As your programs work with files and directories, you may need to parse a pathname into a disk drive letter, subdirectory path, filename, and extension. To help you parse a pathname (that is, separate it into its components), some C compilers provide the *_splitpath* function. The format of the function call is as follows:

```
include <dir.h>

int fnsplit (const char *path, const char *drive, const char *directory,
             const char *filename, const char *ext);
```

The following *Turbo C Lite* program, *split.c*, illustrates the use of the *fnsplit* function:

```
#include <stdio.h>
#include <dir.h>
#include <stdlib.h>

void main(void)
  {
    char *path_1 = "C:\\SUBDIR\\FILENAME.EXT";
    char *path_2 = "SUBDIR\\FILENAME.EXT";
    char *path_3 = "FILENAME.EXE";
    char subdir[MAXDIR];
    char drive[MAXDRIVE];
    char filename[MAXFILE];
    char extension[MAXEXT];
    int flags;                           // holds the fnsplit return value

    flags = fnsplit(path_1, drive, subdir, filename, extension);
    printf ("Splitting %s\n", path_1);
    printf ("Drive %s Subdir %s Filename %s Extension %s\n",
      drive, subdir, filename, extension);
    flags = fnsplit(path_2, drive, subdir, filename, extension);
    printf ("Splitting %s\n", path_2);
    printf ("Drive %s Subdir %s Filename %s Extension %s\n",
      drive, subdir, filename, extension);
    flags = fnsplit(path_3, drive, subdir, filename, extension);
    printf ("Splitting %s\n", path_3);
    printf ("Drive %s Subdir %s Filename %s Extension %s\n",
      drive, subdir, filename, extension);
  }
```

Note the use of the constants to define the proper buffer sizes. When you compile and execute the *split.c* program, your screen will display the following:

```
Splitting C:\SUBDIR\FILENAME.EXE
Drive C: Subdir \SUBDIR\ Filename FILENAME Extension .EXE
Splitting \SUBDIR\FILENAME.EXE
Drive  Subdir \SUBDIR\ Filename FILENAME Extension .EXE
Splitting FILENAME.EXE
Drive  Subdir  Filename FILENAME Extension .EXE
C:\>
```

BUILDING A PATHNAME 401

As you work with files and directories within your programs, at times you may want to combine a disk drive letter, subdirectory, filename, and extension into a complete pathname. To help you perform such operations, some C compilers provide the *fnmerge* function. The format of the *fnmerge* function is as follows:

```
fnmerge(pathname, drive, subdir, filename, ext);
```

The following program, *makepath.c*, illustrates how to use the *fnmerge* function:

```
#include <stdio.h>
#include <stdlib.h>
#include <dir.h>

void main (void)
 {
```

```
    char pathname[MAXPATH];

    char *drive = "C:";
    char *subdir = "\\SUBDIR";
    char *filename = "FILENAME";
    char *extension = "EXT";

    fnmerge(pathname, drive, subdir, filename, extension);

    printf("The complete pathname is %s\n", pathname);
 }
```

When you compile and execute the *makepath.c* program, your screen will display the following output:

```
The complete pathname is C:\SUBDIR\FILENAME.EXT
C:\>
```

402 Opening and Closing a File Using Low-Level Functions

C supports high-level file I/O operations that work with file streams and low-level operations that work with byte ranges. When your programs perform low-level I/O, you can open an existing file using the *open* function. To close the file later, you use *close*, as shown here:

```
#include <fcntl.h>
#include <sys\stat.h>

int open(const char *path, int access_mode [,creation_mode]);
int close(int handle);
```

If *open* successfully opens the file, it returns a handle to the file. If an error occurs, *open* returns –1 and sets the global variable *errno* to one of the values listed in Table 402.1. Some compilers refer to the function as *_open*.

Value	Meaning
ENOENT	No such file or directory entry
EMFILE	Too many open files
EACCES	Access permission denied
EINVACC	Invalid access code

Table 402.1 Error status codes ***open*** *assigns to* ***errno****.*

The *path* parameter is a character string that contains the name of the desired file. The *access_mode* parameter specifies how you want to use the file. The *access_mode* value can be a combination (use a *bitwise OR*) of the values listed in Table 402.2.

Access Mode	Meaning
O_RDONLY	Read-only access
O_WRONLY	Write-only access
O_RDWR	Read and write access
O_NDELAY	Delay value UNIX uses
O_APPEND	Positions pointer for append operations
O_TRUNC	Truncates an existing file's contents
O_EXCL	If *O_CREAT* is specified and the file already exists, *open* returns an error
O_BINARY	Open file in binary mode
O_TEXT	Open file in text mode

Table 402.2 The possible values for the ***access_mode*** *parameter when you use it with* ***open****.*

By default, *open* will not create an output file if the file does not exist. If you want *open* to create files, you must include the *O_CREAT* flag along with the desired access modes (for example, *O_CREAT* | *O_TEXT*). If you specify *O_CREAT*, you can use the *creation_mode* parameter to specify the mode with which you want to create the file. The *creation_mode* parameter can use a combination of the values that Table 402.3 specifies.

Creation Mode	Meaning
S_IWRITE	Create for write operations
S_IREAD	Create for read operations

*Table 402.3 The possible values for the **creation_mode** parameter that **open** uses.*

The following statement illustrates how to use *open* to open the root directory file *config.sys* for read-only operations:

```
if ((handle = open("\\CONFIG.SYS", O_RDONLY)) == -1)
  printf("Error opening the file \\CONFIG.SYS\n");
else
  // Statements
```

If you want to open the file *output.dat* for write operations and you want *open* to create a file that does not yet exist, use *open* as follows:

```
if ((handle = open("\\CONFIG.SYS", O_RDONLY | O_CREAT, S_IWRITE)) == -1)
  printf("Error opening the file \\CONFIG.SYS\n");
else
  // Statements
```

When you finish using a file, you should close it using the *close* function, as shown here:

```
close(handle);
```

CREATING A FILE 403

In Tip 402 you learned that by default, the *open* function does not create a file if the file does not exist. As you also learned, however, you can direct *open* to create a file when you specify *O_CREAT* in the access mode. If you are using an older compiler, the *open* function might not support *O_CREAT*. As a result, you may need to use the *creat* function, as shown here:

```
#include <sys\stat.h>

int creat(const char *path, int creation_mode);
```

As before, the path parameter specifies the file you want to create. The *creation_mode* parameter can contain a combination of the values listed in Table 403.

Mode	Meaning
S_IWRITE	Create for write operations
S_IREAD	Create for read operations

*Table 403 The possible values for the **creation_mode** parameter.*

If *creat* succeeds, it will return a handle to the file. If an error occurs, *creat* will return the value –1 and assign an error status value to the global variable *errno*. The mode of translation (binary or text) that *creat* uses depends on the setting of the *_fmode* global variable. If a file with the specified name already exists, *creat* will truncate the file's contents. The following statement illustrates how to use *creat* to create the file *output.dat*:

```
if ((handle = creat("OUTPUT.DAT", S_IWRITE)) == -1)
  printf("Error creating file\n");
else
  // Statements
```

***Note:** If you want it to be obvious to another programmer that you are creating a file, you might want to use the **creat** function, as opposed to using **open** with the **O_CREAT** flag set.*

404 Performing Low-Level Read and Write Operations

When you use file handles to perform low-level file I/O operations, you open and close files using the *open* and *close* functions. In a similar way, you read and write files using the *read* and *write* functions, as shown here:

```
#include <io.h>

int read(int handle, void *buffer, unsigned length);
int write(int handle, void *buffer, unsigned length);
```

The *handle* parameter is the handle the *open* or *creat* functions return. The *buffer* parameter is either the data buffer into which the *read* function reads information or from which the *write* function writes data. The *length* parameter specifies the number of bytes *read* or *write* will transfer (the maximum is 65,534). If *read* succeeds, it returns the number of bytes read. If *read* encounters the end of the file, *read* returns 0. On an error, *read* returns –1 and sets the global variable *errno* to one of the values listed in Table 404.

Value	Meaning
EACCES	Invalid access
EBADF	Invalid file handle

***Table 404** The possible error values **read** returns.*

If *write* is successful, it returns the number of bytes written. If an error occurs, *write* returns the value –1 and assigns the global variable *errno* one of the values previously shown. The following loop illustrates how you might use *read* and *write* to copy the contents of one file to another:

```
while ((bytes_read = read(input, buffer, sizeof(buffer))
   write(output, buffer, bytes_read);
```

405 Testing for the End of a File

In Tip 404 you learned that the *read* function returns the value 0 when it encounters *EOF*. Depending on your program, at times you may want to test for the end of file before performing a specific operation. When you use file handles, the *eof* function returns the value 1 if the file pointer has reached the end of the file, 0 if the pointer is not at the end of the file, and –1 if the file handle is invalid:

```
#include <io.h.>

int eof(int handle);
```

The following statements modify the code shown in Tip 404 to use *eof* to test for the end of the input file:

```
while (! eof(input))
  {
     bytes_read = read(input, buffer, sizeof(buffer));
     write(output, buffer, bytes_read);
  }
```

406 Putting the Low-Level File Routines to Work

Several Tips in this section discuss C's low-level file I/O routines. To help you better understand each routine's use, consider the following program, *lowcopy.c*, which uses the *read* and *write* functions to copy the contents of the first file specified in the command line to the second:

```
#include <stdio.h>
#include <io.h>
#include <fcntl.h>
#include <sys\types.h>
#include <sys\stat.h>

void main(int argc, char *argv[])
 {
    int source, target; // file handles
    char buffer[1024];  // I/O buffer
    int bytes_read;

    if (argc < 3)
      fprintf(stderr, "Must specify source and target files\n");
    else if ((source = open(argv[1], O_BINARY | O_RDONLY)) == -1)
      fprintf(stderr, "Error opening %s\n", argv[1]);
    else if ((target = open(argv[2], O_WRONLY | O_BINARY | O_TRUNC |
             O_CREAT, S_IWRITE)) == -1)
       fprintf(stderr, "Error opening %s\n", argv[2]);
    else
      {
         while (!eof(source))
          {
            if ((bytes_read = read(source, buffer, sizeof(buffer)))<= 0)
             fprintf(stderr, "Error reading from source file");
            else if (write(target, buffer, bytes_read) != bytes_read)
             fprintf(stderr, "Error writing to target file");
          }
         close(source);
         close(target);
      }
 }
```

SPECIFYING THE MODE FOR A FILE-HANDLE TRANSLATION 407

As you have learned, C translates a file's contents using either binary or text translation. Unless you specify otherwise, C uses the setting in the *_fmode* global variable to determine the translation type, either *O_BINARY* or *O_TEXT*. When you open or create a file using C's low-level routines, you can specify the file's translation mode. In some cases, your program must specify the translation mode after you open the file. To specify the mode, you can use the *setmode* function, as shown here:

```
#include <fcntl.h>

int setmode(int handle, int translation_mode);
```

If *setmode* succeeds, it returns the previous translation mode. If an error occurs, *setmode* returns –1 and sets the global variable *errno* to *EINVAL* (invalid argument). The following statement, for example, sets the file associated with the handle *output* to text translation:

```
if ((old_mode = setmode(output, O_TEXT)) == -1)
  printf("Error changing file mode\n");
```

POSITIONING THE FILE POINTER USING LSEEK 408

As you work with C's low-level file I/O functions, you may want to position the file pointer to a specific location within the file before you perform a read or write operation. To do so, you can use the *lseek* function, as shown here:

```
#include <io.h>
long lseek(int handle, long offset, int relative_to)
```

The *handle* parameter specifies the file pointer you want to position. The *offset* and *relative_to* parameters combine to specify the desired position. The *offset* parameter contains the byte offset into the file. The *relative_to* parameter specifies the location in the file from which the *lseek* function should apply the *offset*. Table 408 specifies the values you can use for the *relative_to* parameter.

Constant	Meaning
SEEK_CUR	From the current file position
SEEK_SET	From the beginning of the file
SEEK_END	From the end of the file

Table 408 File positions from which the ***lseek*** *function can apply an offset.*

To position the file pointer at the end of a file, for example, you can use *lseek* as follows:

```
lseek(handle, 0, SEEK_END);   // At end of file
```

If successful, *lseek* will return the value 0. If an error occurs, *lseek* will return a non-zero value.

409 OPENING MORE THAN 20 FILES

As you have learned, a file handle is an integer value that identifies an open file. Actually, a file handle is an index into the process file table, which contains entries for up to 20 files. If your DOS-based program must open more than 20 files, the easiest way you can do so is to use the DOS file services. To begin, your program must request support for more than 20 files. You can do this by using the DOS *INT 21H function 67H* to increase the number of file handles. DOS will then allocate a table large enough to hold the number of handles specified (up to 255 minus the number of handles currently in use). Next, your program should open the files using the DOS services, as opposed to the C run-time library. In this way, your program can bypass the compiler's file limit. The following code fragment increases the number of file handles to 75:

```
inregs.h.ah = 0x67;
inregs.x.bx = 75;    // Number of handles
intdos(&inregs, &outregs);

if (outregs.x.ax)
  printf("Error allocating handles\n");
```

Note: *The number of available file handles is only an issue in a DOS environment or a DOS window. Windows determines the limit on the number of files that you can open at a single time based on your current memory, hard drive free space, and other Windows-specific considerations.*

410 USING DOS-BASED FILE SERVICES

As the DOS and BIOS section of this book details, DOS provides a collection of file services that let you open, read, write, and close files. To make these services easier to use from within C, many C compilers provide the functions listed in Table 410.

Function	Purpose
_dos_creat	Creates a file, returning a file handle
_dos_close	Closes a specified file
_dos_open	Opens a file, returning a file handle
_dos_read	Reads the specified number of bytes from a file
_dos_write	Writes the specified number of bytes to a file

***Table 410** Functions that use the DOS file system services.*

To help you better understand file services, consider the following program, *doscopy.c*, which copies the contents of the first file specified in the command line to the second file specified in the command line:

```
#include <stdio.h>
#include <dos.h>
#include <fcntl.h>

void main(int argc, char *argv[])
 {
   char buffer[1024];
   int input, output;                        // file handles
   unsigned bytes_read, bytes_written;       // actual number of bytes transferred

   if (argc < 3)
     fprintf(stderr, "Must specify source and target file\n");
   else if (_dos_open (argv[1], O_RDONLY, &input))
     fprintf(stderr, "Error opening source file\n");
   else if (_dos_creat (argv[2], 0, &output))
     fprintf(stderr, "Error opening target file\n");
   else
    {
      while (!_dos_read(input, buffer, sizeof(buffer), &bytes_read))
        {
          if (bytes_read == 0)
              break;
          _dos_write(output, buffer, bytes_read, &bytes_written);
        }
      _dos_close(input);
      _dos_close(output);
    }
 }
```

Note: *Although the DOS-based file routines are very similar to C's low-level file functions, you will increase your programs' portability by using C's* **open, read,** *and* **write** *functions, as opposed to the DOS-based functions. Most C compilers support C's low-level functions.*

Note: *When you program in Windows, you will use the Windows API functions rather than the DOS-based file routines to manage files. Tips 1450 through 1478 detail many Windows File API functions.*

OBTAINING A FILE'S DATE AND TIME STAMP 411

When you perform a directory listing, the DOS DIR command will display each file's name, extension, size, and the date and time the file was created or last changed. The date and time DOS stores for the file is called the file's *date and time stamp*. DOS only changes the date and time stamp when you make changes to the file. Some operating systems, on the other hand, track the date and time the file was created or last modified, as well as the date and time the file was last used (read). The operating systems refer to this second date and time stamp as the *last access time*. Depending on your program's purpose, there may be times when you must know a file's date and time stamp. Therefore, most compilers provide the *_dos_getftime* function, as shown here:

```
#include <dos.h>
unsigned _dos_getftime(int handle, unsigned *datefield, unsigned *timefield);
```

If the function successfully gets the file's date and time stamp, the function returns the value 0. If an error occurs, the function returns a non-zero value and assigns the global variable *errno* the value *EBADF* (invalid handle). The *handle* parameter is an open file handle to the desired file. The *datefield* and *timefield* parameters are pointers to unsigned integer values with bit meanings, as listed in Tables 411.1 and 411.2, respectively.

Date Bits	Meaning
0–4	Day from 1 through 31
5–8	Month from 1 through 12
9–15	Years since 1980

Table 411.1 The components of the ***datefield*** *parameter.*

Time Bits	Meaning
0–4	Seconds divided by 2 (1 through 30)
5–10	Minutes from 1 through 60
11–15	Hours from 1 through 12

Table 411.2 The components of the ***timefield*** *parameter.*

The following program, *filedt.c*, uses the *_dos_getftime* function to display the date and time stamp of the file specified in the command line:

```
#include <stdio.h>
#include <dos.h>
#include <fcntl.h>

void main(int argc, char *argv[])
 {
   unsigned date, time;
   int handle;

   if (_dos_open(argv[1], O_RDONLY, &handle))
      fprintf(stderr, "Error opening source file\n");
   else
    {
      if (_dos_getftime(handle, &date, &time))
         printf("Error getting date/time stamp\n");
      else
         printf("%s last modified %02d-%02d-%d %02d:%02d:%02d\n",
                argv[1],
                (date & 0x1E0) >> 5,   /* month */
                (date & 0x1F),         /* day */
                (date >> 9) + 1980,    /* year */
                (time >> 11),          /* hours */
                (time & 0x7E0) >> 5,   /* minutes */
                (time & 0x1F) * 2);    /* seconds */
     _dos_close(handle);
    }
 }
```

As you can see, the program uses C's bitwise operators to extract the *date* and *time* fields. You learned how to perform similar processing using structure bit fields in Tip 380.

Note: *Many newer compilers do not support the _dos file-manipulation functions.*

OBTAINING A FILE'S DATE AND TIME USING BIT FIELDS 412

In Tip 411 you used the function *_dos_getftime* to obtain a file's date and time stamp. As you learned, the *_dos_getftime* function encodes the *date* and *time* fields as bits within two unsigned values. To extract the field values, the program *filedt.c* uses C's bitwise operators. To make your program easier to understand, you might consider using *bit* fields within a structure. To do so, you can use the following program, *dtbits.c*:

```
#include <stdio.h>
#include <dos.h>
#include <fcntl.h>

void main(int argc, char *argv[])
 {
    struct Date
     {
       unsigned int day:5;
       unsigned int month:4;
       unsigned int years:7;
     } date;
    struct Time
     {
       unsigned seconds:5;
       unsigned minutes:6;
       unsigned hours:5;
     } time;
    int handle;

    if (_dos_open(argv[1], O_RDONLY, &handle))
       fprintf(stderr, "Error opening source file\n");
    else
     {
       if (_dos_getftime(handle, &date, &time))
          printf("Error getting date and time stamp\n");
       else
          printf("%s last modified %02d-%02d-%d %02d:%02d:%02d\n",
                 argv[1],
                 date.month,           // month
                 date.day,             // day
                 date.years + 1980,    // year
                 time.hours,           // hours
                 time.minutes,         // minutes
                 time.seconds * 2);    // seconds
       _dos_close(handle);
     }
 }
```

By using *bit* fields, the program eliminates the need for other programmers to understand the complicated bitwise operations that occurred in the *filedt.c* program.

Note: *Many newer compilers no longer support the _dos file-manipulation functions.*

SETTING A FILE'S DATE AND TIME STAMP 413

In Tips 411 and 412, you used the *_dos_getftime* function to obtain a file's date and time stamp. Depending on your program, you may need to set a file's date and time stamp. For such cases, many C compilers provide the *_dos_setftime* function, as shown here:

```
#include <dos.h>

unsigned _dos_setftime(int handle, unsigned date, unsigned time);
```

If the function succeeds, it returns the value 0. If an error occurs, the function returns a non-zero value. The *handle* parameter is a handle to an open file. The *date* and *time* parameters contain the bit-encoded date and time values (similar to those shown in Tip 411). The following program, *july4_01.c*, sets the date and time stamp of the file the command line specifies to noon, July 4, 2001:

```
#include <stdio.h>
#include <dos.h>
#include <fcntl.h>

void main(int argc, char *argv[])
 {
    union
     {
        struct Date
         {
            unsigned int day:5;
            unsigned int month:4;
            unsigned int years:7;
         } bits;
        unsigned value;
     } date;
    union
     {
        struct Time
         {
           unsigned seconds:5;
           unsigned minutes:6;
           unsigned hours:5;
         } bits;
        unsigned value;
     } time;
    int handle;

    if (_dos_open(argv[1], O_RDONLY, &handle))
        fprintf(stderr, "Error opening source file\n");
    else
     {
        date.bits.day = 4;
        date.bits.month = 7;
        date.bits.years = 21;  // 1980 + 21
        time.bits.hours = 12;
        time.bits.minutes = 0;
        time.bits.seconds = 0;
        if (_dos_setftime(handle, date.value, time.value))
          printf("Error setting date/time stamp\n");
        _dos_close(handle);
     }
 }
```

The *july4_01.c* program uses *bit* fields to simplify the assignment of the date and time bits. However, the *_dos_setftime* function requires parameters of type *unsigned int*. Because the bits must be viewed in two different ways, they are excellent candidates for a *union*. Tip 481 discusses unions in detail.

Note: *Tip 1465 details how you will set a file's date and time stamp within Windows.*

414 SETTING A FILE DATE AND TIME STAMP TO THE CURRENT DATE AND TIME

Several Tips in this book show ways to set a file's date and time stamp. When you want to set a file's date and time stamp to the current date and time, you can do so quickly with the *utime* function, as shown here:

```
#include <utime.h>

int utime(char *path, struct utimbuf *date_time);
```

The *path* parameter is a character string that specifies the name and directory of the file you want. The *date_time* parameter is a structure that contains the date and time the file was last changed and last accessed, as shown here:

```
struct utimbuf
 {
    time_t actime;     // Last access
    time_t modtime;    // Last modification
 };
```

If you are working in the DOS environment, DOS uses only the modification time. If you invoke the *utime* function with *date_time* set to *NULL*, the function sets the date and time stamp to the current date and time. If the function succeeds, it will return 0. If an error occurs, the function will return –1 and set the global variable *errno*. The following program, *utime.c*, uses the *utime* function to set the date and time stamp of the file specified to the current date and time:

```
#include <stdio.h>
#include <utime.h>

void main(int argc, char **argv)
 {
   if (utime(argv[1], (struct utimbuf *) NULL))
     printf("Error setting date and time\n");
   else
     printf("Date and time stamp set\n");
 }
```

Note: *Tip 1465 details how you will set a file's date and time stamp within Windows.*

READING AND WRITING DATA ONE WORD AT A TIME 415

As you have learned, the *getc* and *putc* functions let you read and write file information one byte at a time. Depending on your file's contents, at times you may want to read and write data one word at a time. To help you do so, most C compilers provide the *getw* and *putw* functions, as shown here:

```
#include <stdio.h>

int getw(FILE *stream);
int putw(int word, FILE *stream);
```

If *getw* succeeds, it will return the integer value read from the file. If an error occurs or *getw* encounters the end of file, *getw* will return *EOF*. If *putw* succeeds, it will return the integer value which *putw* wrote to the file. If an error occurs, *putw* will return *EOF*. The following program, *putwgetw.c*, uses the function *putw* to write the values 1 to 100 to a file. *Putwgetw.c* then opens the same file and reads the values using *getw*, as shown here:

```
#include <stdio.h>
#include <stdlib.h>

void main(void)
 {
   FILE *fp;
   int word;

   if ((fp = fopen("DATA.DAT", "wb")) == NULL)
     {
       printf("Error opening DATA.DAT for output\n");
       exit(1);
     }
   else
```

```
     {
       for (word = 1; word <= 100; word++)
         putw(word, fp);
       fclose(fp);
     }
   if ((fp = fopen("DATA.DAT", "rb")) == NULL)
     {
       printf("Error opening DATA.DAT for input\n");
       exit(1);
     }
   else
     {
       do
        {
          word = getw(fp);
          if ((word == EOF) && (feof(fp)))
            break;
          else
            printf("%d ", word);
        }
       while (1);
       fclose(fp);
     }
 }
```

416 CHANGING A FILE'S SIZE

As you work with files, there may be times when you must allocate a large amount of disk space for a file or when you want to truncate a file's size. For such cases, your programs can use the *chsize* function, as shown here:

```
#include <io.h>

int chsize(int handle, long size);
```

The *handle* parameter is the file handle that *open* or *creat* previously returned to the program. The *size* parameter specifies the desired file size. If *chsize* succeeds, it will return the value 0. If an error occurs, *chsize* will return the value –1 and set the global variable *errno* to one of the values listed in Table 416.

Value	Meaning
EACCES	Invalid access
EBADF	Invalid file handle
ENOSPC	Insufficient space (Unix)

Table 416 Error values ***chsize*** *returns.*

If you increase a file's size, then *chsize* will fill the new file space with *NULL* characters. The following program, *chsize.c*, creates a file named *100zeros.dat* and then uses the *chsize* function to zero-fill the file's first 100 bytes:

```
#include <stdio.h>
#include <io.h>
#include <fcntl.h>
#include <sys\types.h>
#include <sys\stat.h>

void main(void)
 {
    int handle;

    if ((handle = creat("100ZEROS.DAT", S_IWRITE)) == -1)
       fprintf(stderr, "Error opening 100ZEROS.DAT");
    else
```

```
    {
      if (chsize(handle, 100L))
        printf("Error changing file size\n");
      close(handle);
    }
}
```

CONTROLLING READ AND WRITE FILE-OPEN OPERATIONS 417

As you have learned, when you open a file, whether you are using *open*, *creat*, or *fopen*, you must specify whether you want to access the file in read, write, or read and write mode. The *umask* function lets you control how the program later opens files. The format of the *umask* function is as follows:

```
#include <io.h>

unsigned umask(unsigned access_mode);
```

The *access_mode* parameter specifies the modes you want to prevent files from using. Valid values for the *access_mode* parameter are shown in Table 417.

Access Mode	Meaning
S_IWRITE	Prevents write access
S_IREAD	Prevents read access
S_IWRITE | S_IREAD	Prevents read and write access

Table 417 The valid values for the ***access_mode*** *parameter of the* ***umask*** *function.*

As an example, if you want to prevent a program from opening files with write access, then you would use *umask* as follows:

```
old_mode = umask(S_IWRITE);
```

As shown here, the function returns the previous setting. The following program, *umask.c*, uses the *umask* function to set the access mode to *S_IWRITE*, which will clear the file's write-access bit (making the file read-only). The program then creates and writes output to the file *output.dat*. After the program closes the file, it tries to open *output.dat* for write access. Because *umask* previously set the file to read-only, the open operation fails, as shown here:

```
#include <stdio.h>
#include <io.h>
#include <fcntl.h>
#include <sys\stat.h>
#include <stdlib.h>

void main(void)
 {
   int output;
   int old_setting;

   old_setting = umask(S_IWRITE);
   if ((output = creat("OUTPUT.DAT", S_IWRITE)) == -1)
     {
       fprintf(stderr, "Error creating OUTPUT.DAT\n");
       exit(1);
     }
   else
     {
       if (write(output, "Test", 4) == -1)
         fprintf(stderr, "Cannot write to file\n");
       else
         printf("File successfully written to\n");
       close(output);
```

```
      }
    if ((output = open("OUTPUT.DAT", O_WRONLY)) == -1)
      fprintf(stderr, "Error opening OUTPUT.DAT for output\n");
    else
      printf("File successfully opened for write access\n");
  }
```

Note: *To remove the file* ***output.dat*** *from your disk, you must issue the command* ***ATTRIB -R output.dat*** *and then delete the file.*

418 ASSIGNING A FILE BUFFER

In the Keyboard section of this book you will learn that C provides I/O functions that perform buffered and direct I/O. For buffered I/O operations, data writes to or reads into a buffer before becoming available to your program. File operations, for example, use buffered I/O. When your programs perform direct I/O, on the other hand, the data is immediately available to your programs without being placed in an intermediate buffer. You can often use direct I/O to gain direct access to the keyboard. Usually, C automatically allocates a buffer for file streams. However, you can use the *setbuf* function to specify your own buffer, as shown here:

```
#include <stdio.h>

void setbuf(FILE *stream, char *buffer);
```

The *stream* parameter corresponds to the open file to which you want to assign the new buffer. The *buffer* parameter is a pointer to the desired buffer. If the *buffer* parameter contains *NULL*, the open file that *stream* specifies will not buffer the data. The following program, *setbuf.c*, uses the *setbuf* function to change the buffer that C assigns to the *stdout* file handle. The program then writes output to *stdout*. However, because the program is placing the data in a large buffer, the data will not appear on your screen until a three-second delay passes. The program then fills the buffer one character at a time, delaying ten milliseconds between characters. (If your compiler does not support *delay*, try using the *sleep* function.) When the buffer becomes full, it flushes (writes) to the screen, as shown here:

```
#include <stdio.h>
#include <dos.h>
#include <conio.h>

void main(void)
 {
   char buffer[512];
   int letter;

   setbuf(stdout, buffer);
   puts("First line of output");
   puts("Second line of output");
   puts("Third line of output");
   delay(3000);
   printf("About to fill buffer\n");
   fflush(stdout);
   for (letter = 0; letter < 513; letter++)
    {
     putchar('A');
     delay(10);
    }
 }
```

ALLOCATING A FILE BUFFER 419

In Tip 418 you learned how to use the *setbuf* function to assign a buffer to a file. When you use *setbuf*, you must specify the desired buffer. In a similar way, many C compilers provide the *setvbuf* function, which allocates a buffer (using *malloc*) of the desired size and then assigns the buffer to the specified file. In addition, *setbuf* lets you specify the buffering you desire, as shown here:

```
#include <stdio.h>

int setvbuf(FILE *stream, char *buffer, int buffer_type, size_t buffer_size);
```

The *stream* parameter is a pointer to an open file. The *buffer* parameter is a pointer to the buffer into which C buffers your data. If the *buffer* parameter is *NULL*, the *setvbuf* function will allocate the buffer for you. The *buffer_type* parameter lets you control the buffer type. Finally, the *buffer_size* parameter lets you specify a buffer size up to 32,767 bytes. If *setvbuf* succeeds, it returns 0. If an error occurs (such as insufficient memory), *setvbuf* returns a non-zero value. Table 419 lists the valid values for the *buffer_type* parameter.

Buffer Type	Buffering
_IOFBF	Full buffering. When the buffer is empty, the next read operation will try to fill the buffer. For output, the buffer must be full before *setvbuf* writes data to the disk.
_IOLBF	Line buffering. When the buffer is empty, the next read operation will try to fill the buffer. For output, *setvbuf* writes the buffer to disk when the buffer is full or when *setvbuf* encounters the *newline* character.
_IONBF	Unbuffered. The program will perform direct I/O.

Table 419 Valid buffering types ***setvbuf*** *uses.*

The following program, *setvbuf.c*, uses *setvbuf* to allocate an 8Kb buffer for full buffering:

```
#include <stdio.h>
#include <dos.h>
#include <conio.h>

void main(void)
 {
   char line[512];
   char *buffer;
   FILE *input;

   if ((input = fopen("\\AUTOEXEC.BAT", "r")) == NULL)
     printf("Error opening \\AUTOEXEC.BAT\n");
   else
    {
      if (setvbuf(input, buffer, _IOFBF, 8192))
        printf("Error changing file buffer\n");
      else
        while (fgets(line, sizeof(line), input))
          fputs(line, stdout);
      fclose(input);
    }
 }
```

CREATING A UNIQUE FILENAME USING MKTEMP 420

As you work with files, the ability to create a unique filename for temporary files is very important. Some of the Tips in this section demonstrate ways to create random filenames. Often, you will want to create a unique filename, but you will also want the filename to follow a specific format that will relate it to the application. For example, for an accounting program you might want all your filenames to begin with the letters *ACCNTG*. To control the creation of unique filenames, many compilers provide the *mktemp* function (some compilers call it *_mktemp*), as shown here:

```
#include <dir.h>

char *mktemp(char *template);
```

The *template* is a pointer to a character string that contains six characters followed by six Xs and a *NULL*. In the case of the accounting example, the template would be a pointer to "ACCNTGXXXXXX". The *mktemp* function replaces the Xs with two filename characters, a period, and three characters for the extension. If *mktemp* succeeds, it will return a pointer to the template string. If an error occurs, the function will return *NULL*. The following program, *mktemp.c*, illustrates how to use the *mktemp* function. If your compiler uses *_mktemp* instead of *mktemp*, see the file *_mktemp.c* on the CD-ROM:

```
#include <stdio.h>
#include <dir.h>

void main(void)
 {
   char name_a[13] = "ACCTNGXXXXXX";
   char name_b[13] = "COMPUTXXXXXX";
   char name_c[13] = "PCCHIPXXXXXX";

   if (mktemp(name_a))
     puts(name_a);
   if (mktemp(name_b))
     puts(name_b);
   if (mktemp(name_c))
     puts(name_c);
 }
```

When you compile and execute the *mktemp.c* program, your screen will display the following:

```
ACCTNGAA.AAA
COMPUTAA.AAA
PCCHIPAA.AAA
C:\>
```

421 READING AND WRITING STRUCTURES

The Structures section of this book presents many programs that work with structures. When your programs work with structures, there will be many times when your programs must store the structure data onto either a floppy disk or the computer's hard drive and then later read the data. As a rule, when you must read or write a structure, you can treat the structure as a long byte range. For example, the following *Turbo C Lite* program, *dtout.c*, uses C's *write* function to write the current system date and time to the file *datetime.dat*:

```
#include <stdio.h>
#include <dos.h>
#include <io.h>
#include <sys\stat.h>

void main(void)
 {
   struct date curr_date;
   struct time curr_time;
   int handle;

   getdate(&curr_date);
   gettime(&curr_time);
   if ((handle = creat("DATETIME.OUT", S_IWRITE)) == -1)
```

```
      fprintf(stderr, "Error opening file DATETIME.OUT\n");
    else
      {
        write(handle, &curr_date, sizeof(curr_date));
        write(handle, &curr_time, sizeof(curr_time));
        close(handle);
      }
  }
```

As you can see, to write the structure the program simply passes the structure's address. In a similar way, the following *Turbo C Lite* program, *dtin.c*, uses the *read* function to read the date and time structures:

```
#include <stdio.h>
#include <dos.h>
#include <io.h>
#include <fcntl.h>

void main(void)
 {
   struct date curr_date;
   struct time curr_time;
   int handle;

   if ((handle = open("DATETIME.OUT", O_RDONLY)) == -1)
     fprintf(stderr, "Error opening file DATETIME.OUT\n");
   else
     {
       read(handle, &curr_date, sizeof(curr_date));
       read(handle, &curr_time, sizeof(curr_time));
       close(handle);
       printf("Date: %02d-%02d-%02d\n", curr_date.da_mon,
              curr_date.da_day, curr_date.da_year);
       printf("Time: %02d:%02d\n", curr_time.ti_hour, curr_time.ti_min);
     }
 }
```

READING STRUCTURE DATA FROM A FILE STREAM 422

In Tip 421 you learned how to use C's *read* and *write* functions to perform file I/O operations that use structures. If your programs use file streams, as opposed to file handles, for file I/O you can perform similar processing using the *fread* and *fwrite* functions, as shown here:

```
#include <stdio.h>

size_t fread(void *buffer, size_t buffer_size, size_t element_count, FILE *stream);
size_t fwrite(void *buffer, size_t buffer_size, size_t element_count, FILE *stream);
```

The *buffer* parameter contains a pointer to the data you want to output. The *buffer_size* parameter specifies the data's size in bytes. The *element_count* parameter specifies the number of structures that you are writing, and the *stream* parameter is a pointer to an open file stream. If the functions succeed, they will return the number of items read or written. If an error occurs or either function encounters the end of file, both functions will return 0. The following program, *dtoutf.c*, uses the *fwrite* function to write the current date and time structures to a file:

```
#include <stdio.h>
#include <dos.h>

void main(void)
 {
   struct date curr_date;
   struct time curr_time;
   FILE *output;

   getdate(&curr_date);
```

```
   gettime(&curr_time);
   if ((output = fopen("DATETIME.OUT", "w")) == NULL)
     fprintf(stderr, "Error opening file DATETIME.OUT\n");
   else
     {
       fwrite(&curr_date, sizeof(curr_date), 1, output);
       fwrite(&curr_time, sizeof(curr_time), 1, output);
       fclose(output);
     }
 }
```

Likewise, the program *dtinf.c* uses the *fread* function to read the structure values, as shown here:

```
#include <stdio.h>
#include <dos.h>

void main(void)
 {
   struct date curr_date;
   struct time curr_time;
   FILE *input;

   if ((input = fopen("DATETIME.OUT", "r")) == NULL)
     fprintf(stderr, "Error opening file DATETIME.OUT\n");
   else
     {
       fread(&curr_date, sizeof(curr_date), 1, input);
       fread(&curr_time, sizeof(curr_time), 1, input);
       fclose(input);
       printf("Date: %02d-%02d-%02d\n", curr_date.da_mon,
              curr_date.da_day, curr_date.da_year);
       printf("Time: %02d:%02d\n", curr_time.ti_hour, curr_time.ti_min);
     }
 }
```

423 DUPLICATING A FILE HANDLE

Several Tips in this section present functions that work with file handles. Depending on your programs, there may be times when you want to duplicate a handle's value. For example, if your program performs critical I/O operations, you might want to duplicate a file handle and then close the new copied handle in order to flush the file's output to disk. Because the first file handle remains open, you do not have the overhead of reopening the file after the flush operation, as shown here:

```
#include <io.h>

int dup(int handle);
```

The *handle* parameter is the open file handle that you want to duplicate. If *dup* successfully duplicates the handle, it will return a non-negative value. If an error occurs, *dup* will return –1. The following program, *dup.c*, illustrates how you might use the *dup* function to flush a file's buffers:

```
#include <stdio.h>
#include <fcntl.h>
#include <io.h>
#include <sys\stat.h>

void main(void)
 {
   int handle;
   int duplicate_handle;
   char title[] = "Jamsa\'s C/C++/C# Programmer\'s Bible!";
   char section[] = "Files";
```

```
    if ((handle = open("OUTPUT.TST", O_WRONLY | O_CREAT, S_IWRITE)) == -1)
      printf("Error opening OUTPUT.TST\n");
    else
     {
       if ((duplicate_handle = dup(handle)) == -1)
         printf("Error duplicating handle\n");
       else
         {
           write(handle, title, sizeof(title));
           close(duplicate_handle);                    // Flush the buffer
           write(handle, section, sizeof(section));
           close(handle);
         }
     }
 }
```

FORCING A FILE HANDLE'S SETTING 424

In Tip 423 you learned how to use the *dup* command to make a duplicate copy of a file handle's contents. There may be times when you want to change an open file handle's setting and assign the value of a different handle. When performing change and assign operations with files, you can use *dup2*, as shown here:

```
#include <io.h>

int dup2(int source_handle, int target_handle);
```

The *target_handle* parameter is the file handle whose value you want to update. If the function successfully assigns the handle, it will return the value 0. If an error occurs, the function will return –1. The *source_handle* parameter is the file handle whose value you want to assign to the target. The following program, *dup2.c*, uses the *dup2* function to assign the value of the *stderr* function to *stdout*. In this way, users cannot redirect the program's output from the screen display:

```
#include <stdio.h>
#include <io.h>

void main(void)
 {
   dup2(2, 1);  // stdout is handle 1 stderr is handle 2
   printf("This message cannot be redirected!\n");
 }
```

ASSOCIATING A FILE HANDLE WITH A STREAM 425

Many Tips in this section present functions that work with either file streams or file handles. Depending on your program, there may be times when you are working with a file handle and want to use a function that corresponds to a file stream. In such cases, your programs can use the *fdopen* function to associate a file handle with a file stream, as shown here:

```
#include <stdio.h>

FILE *fdopen(int handle, char *access_mode);
```

The *handle* parameter is the handle of an open file that you want to associate with a file stream. The *access_mode* parameter is a pointer to a character string that specifies how you plan to use the file stream. The *access_mode* value must be one of the mode values you would usually use with *fopen*. If the function succeeds, it returns the stream pointer. If an error occurs, the function returns *NULL*. The following statement, for example, associates the file handle *input* with the file pointer *fpin* for read access:

```
if ((fp = fdopen(input, "r")) == NULL)
  printf("Error associating file\n");
```

```
else
  {
    gets(string, sizeof(string), fpin);
    fclose(fpin);
  }
```

426 UNDERSTANDING FILE SHARING

If you are working in a network environment and have installed the DOS SHARE command, you can write programs that let more than one program access different parts of the same file at the same time. For example, consider a program that lets multiple users assign seats in an airplane. When one user wants to assign a specific seat, the program locks that seat so another user will not also assign it. After the program assigns the seat, the user unlocks the seat.

When you share files in this way, you must first use the *sopen* function to open the file for sharing. Next, when your program wants to access a range of bytes in the file, the program tries to lock the data. If no one else is currently using (locking) the data, then the program's lock succeeds. After the program finishes with the data, it can unlock the range of bytes in the file.

When a program locks a range of bytes within a file, the program can assign a lock that will let other users access the data in specific ways. For example, the program might let another file read the locked range or it might let other programs read and write the same byte range. Several of the following Tips discuss C run-time library functions that support file sharing and locking.

427 OPENING A FILE FOR SHARED ACCESS

In Tip 426 you learned that you can use the DOS SHARE command to open files for multiple programs to use at the same time. To open a file for shared use, your programs must use the *sopen* function, as shown here:

```
#include <share.h>

int sopen(char *pathname, int access_mode, int share_flag[, int create_mode]);
```

The *pathname*, *access_mode*, and *create_mode* parameters are similar to those the *open* function uses. The *share_flag* parameter specifies how different programs can share the file. If *sopen* successfully opens the file, it will return a file handle. If an error occurs, *sopen* will return –1. Table 427 lists the valid values for the *share_flag* parameter.

Share Flag	Sharing Allowed
SH_COMPAT	Allows compatible sharing
SH_DENYRW	Prevents read and write access
SH_DENYWR	Prevents write access
SH_DENYRD	Prevents read access
SH_DENYNONE	Allows all access (read and write)
SH_DENYNO	Allows all access (read and write)

Table 427 Shared access modes ***sopen*** *supports.*

The following program, *sopen.c*, opens the file specified in the command line for shared read access. The file then waits for you to press a key before reading and displaying the file's contents, as shown here:

```
#include <stdio.h>
#include <share.h>
#include <io.h>
#include <fcntl.h>
```

```
void main(int argc, char *argv[])
 {
   int handle, bytes_read;
   char buffer[256];

   if ((handle = sopen(argv[1], O_RDONLY, SH_DENYWR)) == -1)
     printf("Error opening the file %s\n", argv[1]);
   else
     {
       printf("Press Enter to continue\n");
       getchar();
       while (bytes_read = read(handle, buffer, sizeof(buffer)))
         write(1, buffer, bytes_read);  // 1 is stdout
       close(handle);
     }
 }
```

To better understand how the *sopen.c* program works, invoke the SHARE command. Next, start Windows and create a DOS window within which you run the program using the filename *sopen.c* as the shared file. When the program prompts you to press a key, open a second DOS window and use TYPE to display the file's contents. As TYPE displays the *sopen.c* file's contents, two programs have the file open at the same time. Close the window and return to the first window. Press ENTER to display the file's contents. Experiment with the *sopen.c* program by chaining the shared modes. Repeat the process of trying to access the file using two programs.

LOCKING A FILE'S CONTENTS 428

As you have learned, when you share a file's contents, there may be times when you want to lock a range of bytes within a file to prevent another program from changing them. To lock a specific range of bytes within a file, your programs can use the *lock* function, as shown here:

```
#include <io.h>

int lock(int handle, long start_position, long byte_count);
```

The *handle* parameter is a handle that corresponds to a file that *sopen* opened for sharing. *The start_position* parameter specifies the starting offset of the range of bytes you want to lock within the file. *The byte_count* parameter specifies the number of bytes you want to lock. If the *lock* function successfully locks the range of bytes, it will return the value 0. If an error occurs, the function will return –1. You must have the DOS SHARE command installed for the *lock* function to work.

After you lock a range of bytes, other programs will try three times to read or write the locked range. If after the third try the program cannot read the data, then the *read* or *write* function will return an error. The following *Turbo C Lite* program, *lockauto.c*, locks the first five bytes of the root directory file *autoexec.bat* and then waits for you to press `a key`:

```
#include  <stdio.h>
#include <io.h>
#include <share.h>
#include <fcntl.h>

void main(void)
 {
   int handle;

   if ((handle = sopen("\\AUTOEXEC.BAT", O_RDONLY, SH_DENYNO)) == -1)
     printf("Error opening AUTOEXEC.BAT\n");
   else
    {
      lock(handle, 0L, 5L);
      printf("File locked--press Enter to continue\n");
```

```
      getchar();
      close(handle);
    }
 }
```

Next, the following program, *tryauto.c*, tries to read the file *autoexec.bat* one byte at a time. If an error occurs while it reads the file, then the program will display an error message, as shown here:

```
#include  <stdio.h>
#include <io.h>
#include <share.h>
#include <fntl.h>

void main(void)
 {
   int handle;
   int offset = 0;
   int bytes_read;
   char buffe[128];

   if ((handle = sopen("\\AUTOEXEC.BAT", O_BINARY | O_RDONLY, SH_DENYNO)) == -1)
     printf("Error opening AUTOEXEC.BAT\n");
   else
    {
      while (bytes_read = read(handle, buffer, 1))
       {
         if (bytes_read == -1)
           printf("Error reading offset %d\n", offset);
         else
           write(1, buffer, bytes_read);
         offset++;
         lseek(handle, offset, SEEK_SET);
       }
      close(handle);

       }
```

429 GAINING FINER FILE-LOCKING CONTROL

In Tip 428 you learned how to use the *lock* function to lock a range of bytes within a file. When you use the *lock* function, the operation either succeeds or immediately fails. If you want finer control of the lock operation, you can use the *locking* function, as shown here:

```
#include  <io.h>
#include  <sys\loking.h>
int  locking(int  handle,  int  lock_command,  long  bytcunt);
```

The *handle* parameter is the handle associated with the file you want to lock. The *lock_command* parameter specifies the desired locking operation. The *byte_count* parameter specifies the number of bytes you want to lock. The start of the region depends on the file's current position pointer. If you want to lock a specific region, you can first use the *lseek* function to position the file pointer. Table 429.1 specifies the possible values *for lock_command.*

Lock Command	Meaning
LK_LOCK	Locks the specified region. If the lock does not succeed, locking will try once every second for ten seconds to apply the lock.
LK_RLCK	Performs the same functions as *LK_LOCK.*

*Table 429.1 Commands the **locking** function uses.*

Lock Command	Meaning
LK_NBLCK	Locks the specified region. If the lock does not succeed, *locking* will immediately return an error.
LK_UNLCK	Unlocks a previously locked region.

*Table 429.1 Commands the **locking** function uses. (Continued from previous page.)*

If the *locking* function successfully locks the file, it will return the value 0. If an error occurs, the *locking* function will return the value –1 and set the global variable *errno* to one of the values specified in Table 429.2.

Error Status	Meaning
EBADF	Invalid file handle
EACCESS	File already locked or unlocked
EDEADLOCK	File cannot be locked after 10 tries
EINVAL	Invalid command specified

*Table 429.2 Error status values the **locking** function returns.*

The following program, *locking.c*, changes the program *lockauto.c*, presented in Tip 428, to use the *locking* function to lock the first five bytes of *autoexec.bat:*

```
#include <stdio.h>
#include <io.h>
#include <share.h>
#include <fcntl.h>
#include <sys\locking.h>

void main(void)
 {
   int handle;

   if ((handle = sopen("\\AUTOEXEC.BAT", O_RDONLY, SH_DENYNO)) == -1)
     printf("Error opening AUTOEXEC.BAT\n");
   else
    {
      printf("Trying to lock file\n");
      if (locking(handle, LK_LOCK, 5L))
         printf("Error locking file\n");
      else
       {
         printf("File locked--press Enter to continue\n");
         getchar();
         close(handle);
       }
    }
 }
```

As before, if you have Windows available, try running the *locking.c* program from within two DOS windows at the same time.

***Note:** Before you can use the **locking** function, you must install the DOS SHARE command.*

WORKING WITH DOS DIRECTORIES 430

Within your C programs, you can use the *findfirst* and *findnext* functions to work with files that match a specific wildcard combination (for example, "*.exe"). Because DOS does not treat directories as files, your programs cannot use the DOS services to "open" a directory and read its contents. If you understand how DOS lays out information on a disk, however, your programs can read from the DOS file allocation table and root directory and then read and track the sectors that contain a directory's entries. Disk utility commands (such as UNDELETE) and a directory sort tool perform these low-level disk I/O operations. Several of the following Tips illustrate how your programs can use these directory I/O functions. To simplify the task of reading a directory, some C compilers provide the functions listed in Table 430.

Function	Purpose
closedir	Closes a directory stream
opendir	Opens a directory stream for read operations
readdir	Reads the next entry in a directory stream
rewinddir	Moves the directory stream pointer back to the start of the directory

***Table 430** Directory I/O functions and their purposes.*

431 Opening a Directory

In Tip 430 you learned that some C compilers provide functions that let you open and read the names of files that reside in a specific directory. To open a directory for read operations, your programs can use the *opendir* function, as shown here:

```
#include <dirent.h>

DIR *opendir(char *directory_name);
```

The *directory_name* parameter is a pointer to a character string that contains the desired directory name. If the directory name is *NULL*, *opendir* opens the current directory. If the *opendir* function succeeds, it returns a pointer to a structure of type *DIR*. If an error occurs, the function returns *NULL*. The following statement, for example, illustrates how you would open the DOS directory for read operations:

```
struct DIR *input_directory;

if ((input_directory = opendir("\\DOS")) == NULL)
  printf("Error opening directory\n");
else
  // Statements
```

After you have performed your directory read operations, you should close the directory stream using the *closedir* function, as shown here:

```
#include <dirent.h>

void closedir(DIR *directory);
```

432 Reading a Directory Entry

In Tip 431, you learned how to use the *opendir* function to open a directory listing. After you open a directory, you can use the *readdir* function to read the name of the next entry in the directory list, as shown here:

```
#include <dirent.h>

struct dirent readdir(DIR *directory_pointer);
```

The *directory_pointer* parameter is the pointer which the *opendir* function returns. If *readdir* successfully reads a directory entry, it will return the entry read. If an error occurs or *readdir* reaches the end of the directory, the function will return *NULL*. The *readdir* function reads all the entries in the directory list, including the "." and ".." entries.

USING DIRECTORY SERVICES TO READ C:\WINDOWS 433

In Tip 431 you learned how to open and close a directory listing. In Tip 432 you learned how to use the *readdir* function to read the next entry in the directory list. The following program, *showdir.c*, uses the run-time library directory entries to open, read, and then close the directory specified in the command line:

```
#include <stdio.h>
#include <dirent.h>

void main(int argc, char *argv[])
 {
   DIR *directory_pointer;
   struct dirent *entry;

   if ((directory_pointer = opendir(argv[1])) == NULL)
     printf("Error opening %s\n", argv[1]);
   else
     {
       while (entry = readdir(directory_pointer))
         printf("%s\n", entry);
       closedir(directory_pointer);
     }
 }
```

The following command, for example, uses the *showdir.c* program to display the names of the files in the directory *c:\windows*:

```
C:\> SHOWDIR C:\WINDOWS  <ENTER>
```

REWINDING A DIRECTORY 434

In Tip 433 you learned that some C compilers provide run-time library functions that let you open and read the names of files in a specified directory. As you read directories, there may be times when you want to start reading files at the start of the directory list for a second time. One way to perform this operation is to close and then reopen the directory list. Alternately, your programs can use the *rewinddir* function, as shown here:

```
#include <dirent.h>

void rewinddir(DIR *directory_pointer);
```

The *directory_pointer* parameter is the pointer to the directory list that you want to reset. If you experiment with the *rewinddir* function, you will find that it is much faster to use than closing and reopening the directory list.

READING A DISK'S FILES RECURSIVELY 435

In Tip 433 you used the *showdir.c* program to display the files in a directory list. The following program, *allfiles.c*, uses the run-time library functions to display the names of every file on your disk. To do so, the program uses the recursive function *show_directory* to display filenames, as shown here:

```
#include <stdio.h>
#include <dirent.h>
#include <dos.h>
#include <io.h>
#include <direct.h>
#include <string.h>

void show_directory(char *directory_name)
```

```
{
  DIR *directory_pointer;
  struct dirent *entry;
  unsigned attributes;

  if ((directory_pointer = opendir(directory_name)) == NULL)
    printf("Error opening %s\n", directory_name);
  else
  {
     chdir(directory_name);
     while (entry = readdir(directory_pointer))
       {
         attributes = _chmod(entry, 0);
           // Check if entry is for a subdirectory and is not "." or ".."
         if ((attributes & FA_DIREC) &&
            (strncmp(entry, ".", 1) != 0))
          {
            printf("\n\n----%s----\n", entry);
            show_directory(entry);
          }
         else
          printf("%s\n", entry);
       }
    closedir(directory_pointer);
     chdir("..");
  }
}

void main(void)
 {
   char buffer[MAXPATH];

   // Save current directory so you can restore it later
   getcwd(buffer, sizeof(buffer));
   show_directory("\\");
   chdir(buffer);
 }
```

436 DETERMINING THE CURRENT FILE POSITION

You have learned previously how C tracks the current position in files that are open for input or output operations. Depending on your program, there may be times when you must determine the position pointer's value. If you are working with file streams, you can use the *ftell* function to determine the file pointer position. If you are working with file handles, however, your programs can use the *tell* function, as shown here:

```
#include <stdio.h>

long tell(int handle);
```

The *tell* function returns a long value that specifies the byte offset of the current position in the specified file. The following program, *tell.c*, uses the *tell* function to display position pointer information. The program begins by opening the root directory file *config.sys* in read mode. The program then uses *tell* to display the current position. Next, the program reads and displays the file's contents. After the program finds the end of file, the program again uses *tell* to display the current position, as shown here:

```
#include <stdio.h>
#include <io.h>
#include <fcntl.h>

void main(void)
 {
```

```
    int handle;
    char buffer[512];
    int bytes_read;

    if ((handle = open("\\CONFIG.SYS", O_RDONLY)) == -1)
      printf("Error opening \\CONFIG.SYS\n");
    else
      {
        printf("Current file position %ld\n", tell(handle));
        while (bytes_read = read(handle, buffer, sizeof(buffer)))
          write(1, buffer, bytes_read);
        printf("Current file position %ld\n", tell(handle));
        close(handle);
      }
 }
```

OPENING A SHARED FILE STREAM 437

Several Tips in this section present ways to share and lock files using file handles. If you usually work with file streams, your programs can use the *_fsopen* function, as shown here:

```
#include <stdio.h>
#include <share.h>

FILE * _fsopen(const char *filename, const *access_mode, int share_flag);
```

The *filename* and *access_mode* parameters contain character-string pointers to the desired filename and access mode that you would normally use with *fopen*. The *share_flag* specifies the sharing mode. If the function succeeds, it will return a file pointer. If an error occurs, the function will return *NULL*. Table 437 lists the valid values you can assign to *share_flag*.

Share Flag	Sharing Allowed
SH_COMPAT	Allows compatible sharing
SH_DENYRW	Prevents read and write acccss
SH_DENYWR	Prevents write access
SH_DENYRD	Prevents read access
SH_DENYNONE	Allows all access (read and write)
SH_DENYNO	Allows all access (read and write)

Table 437 *Valid values for the* ***access_mode*** *parameter.*

The following statements, for example, open the root directory file *autoexec.bat* for shared read operations:

```
if ((fp = _fsopen("\\AUTOEXEC.BAT", "r", SH_DENYWR)) == NULL)
  printf("Error opening \\AUTOEXEC.BAT\n");
else
  // Statements
```

CREATING A UNIQUE FILE IN A SPECIFIC DIRECTORY 438

Several Tips in this section show ways that your programs can create temporary files. If you usually work with file handles, you can use the function *creattemp*, which returns a handle, as shown here:

```
#include <dos.h>

int creattemp(char *path, int attribute);
```

The *path* parameter specifies the name of the directory within which you want to create the file. The name must end with two backslash characters ('\\'). The *creattemp* function will append the filename to the string to produce a complete pathname. The *attribute* parameter specifies the desired file attributes (or 0 for none). Table 438 lists the valid settings for the *attribute* parameter.

Constant	Description
FA_RDONLY	Read-only file
FA_HIDDEN	Hidden file
FA_SYSTEM	System file

Table 438 *The valid settings for the* ***attribute*** *parameter.*

If the function succeeds, it will return a file handle. If an error occurs, the function will return –1. The following program, *creattmp.c*, uses the *creattemp* function to create a unique file in the TEMP directory. If your compiler does not support the *creattemp* function, you can use the *mktemp* and **open** functions:

```
#include <stdio.h>
#include <dos.h>
#include <io.h>

void main(void)
 {
   char path[64] = "C:\\TEMP\\";
   int handle;

   if ((handle = creattemp(path, 0)) == -1)
     printf("Error creating file\n");
   else
    {
      printf("Complete path: %s\n", path);
      close(handle);
    }
 }
```

439 CREATING A NEW FILE

Several Tips in this section show ways to create files. In many cases, if you try to create a file and the name specified in the function already exists, the function will truncate the file's contents. However, you often might only want to create a file if a file with the same name does not already exist. For such cases, your programs can use the *creatnew* function, as shown here:

```
#include <dos.h>

int creatnew(const char *pathname, int attribute);
```

The *pathname* parameter specifies the complete path of the file you want to create. The *attribute* parameter specifies the desired file attributes (or 0 for none). Table 439.1 lists the possible settings for the *attribute* parameter.

Attribute	Meaning
FA_RDONLY	Read-only file
FA_HIDDEN	Hidden file
FA_SYSTEM	System file

Table 439.1 *The possible settings for the* ***attribute*** *parameter of the* ***creatnew*** *function.*

If *createnew* succeeds, it will return a file handle. If an error occurs, the function will return the value –1 and set the global variable *errno* to one of the values listed in Table 439.2.

Error	Meaning
EXISTS	File already exists
ENOENT	Path not found
EMFILE	Too many open files
EACCES	Access violation

Table 439.2 The error return values for ***creatnew****.*

The following program, *creatnew.c*, uses *creatnew* to create a file named *new.dat* in the current directory. If your compiler does not support the *creatnew* function, try using *_creat* as shown in *_creat.c* on the companion CD-ROM:

```
#include <stdio.h>
#include <dos.h>
#include <io.h>

void main(void)
 {
   int handle;

   if ((handle = creatnew("NEW.DAT", 0)) == -1)
     printf("Error creating NEW.DAT\n");
   else
    {
      printf("File successfully created\n");
      close(handle);
    }
 }
```

USING THE DOS SERVICES TO ACCESS A FILE 440

As you have learned, when your programs must access more than 20 files, you might want to use the DOS services, which will let you bypass the C run-time library routines. The following *Turbo C Lite* program, *copydos.c*, uses the DOS services to copy the contents of the first file specified in the command line to the second:

```
#include <stdio.h>
#include <dos.h>

void main(int argc, char **argv)
  {
    union REGS inregs, outregs;
    struct SREGS segs;
    char buffer[256];
    unsigned source_handle, target_handle;

    if (*argv[1] && *argv[2])
      {
        // Open the file to copy
        inregs.h.ah = 0x3D;
        inregs.h.al = 0;       // Open for read access
        inregs.x.dx = (unsigned) argv[1];
        segread (&segs);
        intdosx(&inregs, &outregs, &segs);
        if (outregs.x.cflag)
          printf ("Error opening source file %s\n", argv[1]);
        else
          {
             source_handle = outregs.x.ax;
             // Create the target file, truncating an
             // existing file with the same name
             inregs.h.ah = 0x3C;
             inregs.x.cx = 0;      // Open with normal attribute
```

```
               inregs.x.dx = (unsigned) argv[2];
               intdosx (&inregs, &outregs, &segs);
               if (outregs.x.cflag)
                 printf ("Error creating target file %s\n", argv[2]);
               else
                 {
                    target_handle = outregs.x.ax;
                    do {
                      // Read the source data
                      inregs.h.ah = 0x3F;
                      inregs.x.bx = source_handle;
                      inregs.x.cx = sizeof(buffer);
                      inregs.x.dx = (unsigned) buffer;
                      intdosx (&inregs, &outregs, &segs);
                      if (outregs.x.cflag)
                        {
                          printf ("Error reading source file\n");
                          break;
                        }
                      else if (outregs.x.ax)  // Not end of file
                        {
                          // Write the data
                          inregs.h.ah = 0x40;
                          inregs.x.bx = target_handle;
                          inregs.x.cx = outregs.x.ax;
                          inregs.x.dx = (unsigned) buffer;
                          intdosx (&inregs, &outregs, &segs);
                          if (outregs.x.cflag)
                            {
                              printf ("Error writing target file\n");
                              break;
                            }
                        }
                    } while (outregs.x.ax != 0);
                   // Close the files
                   inregs.h.ah = 0x3E;
                   inregs.x.bx = source_handle;
                   intdos (&inregs, &outregs);
                   inregs.x.bx = target_handle;
                   intdos (&inregs, &outregs);
                 }
          }
     }
   else
     printf ("Specify source and target filenames\n");
 }
```

441 FORCING A BINARY OR TEXT FILE OPEN

You have learned previously that many C compilers use the global variable *_fmode* to determine whether the program has opened files in text or binary mode. When you use the *fopen* function, you can control which mode *fopen* uses by placing the letter *t* or *b* immediately after the desired mode, as shown in Table 441.

Access Specifier	Access Mode
ab	Append access binary mode
at	Append access text mode
rb	Read access binary mode
rt	Read access text mode
wb	Write access binary mode
wt	Write access text mode

*Table 441 File mode specifiers for **fopen**.*

The following *fopen* statement, for example, opens the file *filename.ext* for read access in binary mode:

```
if ((fp = fopen("FILENAME.EXT", "rb")))
```

Reading Lines of Text 442

When your programs read text files, they will usually do so one line at a time. To read a line from a file, your programs can use the *fgets* function, whose format is shown here:

```
#include <stdio.h>
char *fgets(char string, int limit, FILE *stream);
```

The *string* parameter is the character buffer into which *fgets* reads the file data. Usually your programs will declare an array of 128 or 256 bytes to hold the data. The *limit* parameter specifies the number of characters the buffer can hold. When *fgets* reads characters from the file, *fgets* will read up to *limit*–1 (*limit* minus one) or to the first *newline* character (\n), whichever comes first. The function will then place a *NULL* character in the buffer to indicate the end of the string.

Many programs will use the *sizeof* function to specify the buffer size, such as *sizeof(string)*. Finally, the *stream* parameter specifies the file from which *fgets* must read the string. You must have previously opened the stream using *fopen* or used a predefined handle, such as *stdin*. If *fgets* successfully reads information from the file, *fgets* will return a pointer to the string. If an error occurs or if it reaches the end of file, *fgets* will return *NULL*.

Writing Lines of Text 443

You learned in Tip 442 that your programs will typically read from a file one line at a time. When writing to a file, your programs will typically write one line at a time. To write a string to a file, your programs can use the *fputs* function, as shown here:

```
#include <stdio.h>
int fputs(const char *string, FILE *stream);
```

The *fputs* function writes the characters in a specified string up to the *NULL* termination character. If *fputs* successfully writes the string, it will return a positive value to the calling function. If an error occurs, *fputs* will return the constant *EOF*.

Putting fgets and fputs to Use 444

In Tips 442 and 443 you learned that your programs can use the functions *fgets* and *fputs* to read and write file data. The following program, *textcopy.c*, uses *fgets* and *fputs* to copy the contents of the first file specified in the command line to the second file specified in the command line:

```
#include <stdio.h>
```

```
void main(int argc, char **argv)
 {
   FILE *input, *output;
   char string[256];

   if ((input = fopen(argv[1], "r")) == NULL)
     printf("Error opening %s\n", argv[1]);
   else if ((output = fopen(argv[2], "w")) == NULL)
     {
       printf("Error opening %s\n", argv[2]);
       fclose(input);
     }
   else
     {
       while (fgets(string, sizeof(string), input))
         fputs(string, output);
       fclose(input);
       fclose(output);
     }
 }
```

As you can see, the program opens an input file and output file and then reads and writes text until the *fgets* function encounters the end of file (*fgets* returns *NULL*). To copy the contents of the file *test.dat* to *test.sav*, for example, you would use the *textcopy.c* program as follows:

```
C:\> TEXTCOPY TEST.DAT TEST.SAV<ENTER>
```

445 FORCING BINARY FILE TRANSLATION

As you have learned, many compilers use the global variable *_fmode* to determine text or binary file access. In text mode, the C run-time library functions translate linefeed characters into carriage return linefeed combinations and vice versa. As you learned, by setting the *_fmode* variable to *O_TEXT* or *O_BINARY*, you can control the access mode. In addition, by placing a *t* or *b* within the access mode specified in *fopen*, you can set the access mode for text or binary mode access. The following *fopen* function call, for example, opens the file *filename.ext* for read access in binary mode:

```
if ((fp = fopen("FILENAME.EXT", "rb")) == NULL)
```

446 UNDERSTANDING WHY TEXTCOPY CANNOT COPY BINARY FILES

Tip 444 presented the program *textcopy.c*, which copied the contents of the first file specified in the command line to the second file. If you try to use *textcopy* to copy a binary file, such as an *exe* file, the copy operation will fail. When the *fgets* function reads a text file, *fgets* considers the CTRL+Z character (the ASCII character 26) as the end of the file. Because a binary file is likely to contain one or more occurrences of the value 26, *fgets* will end its copy operation at the first occurrence. If you want to copy an executable or other binary file, you must use C's low-level I/O routines.

447 TESTING FOR END OF FILE

As you have learned, when the *fgets* function encounters the end of a file, it returns *NULL*. Likewise, when *fgetc* reaches the end of a file, it returns EOF. There may be times when your programs must determine if a file pointer is at the end of the file before it performs a specific operation. In such cases, your programs can call the *feof* function, as shown here:

```
#include <stdio.h>

int feof(FILE *stream);
```

If the file pointer specified is at the end of the file, *feof* will return a non-zero value (true). If it has not yet reached the end of the file, the *feof* will return 0 (false). The following loop reads and displays the characters from the file that corresponds to the file pointer *input*:

```
while (! feof(input))
  fputc(fgetc(input), stdout);
```

Note: *After a function such as* ***fgetc*** *sets the end-of-file indicator for a file, it remains set until the program closes the file or calls the* ***rewind*** *function.*

UNGETTING A CHARACTER 448

Many programs, such as a compiler, for example, often read characters from a file one at a time until they find a specific character (a delimiter or token). After the program finds the character, the program performs specific processing. After the program completes the processing, it continues to read from the file. Depending on the structure of the file your program is reading, there may be times when you want the program to "unread" a character. In such cases, the program can use the *ungetc* function, whose format is shown here:

```
#include <stdio.h>

int ungetc(int character, FILE *stream);
```

The *ungetc* function places the character specified back into the file buffer. You can only "unget" one character. If you call *ungetc* two times in succession, the second character will overwrite the first character that you unget. The *ungetc* function places the specified character in the *FILE* structure *hold* member. In turn, the next file read operation will include the character.

READING FORMATTED FILE DATA 449

You have learned how to use the *fprintf* function to write formatted output to a file. In a similar way, the *fscanf* function lets you read formatted file data, just as the *scanf* function that you have learned about previously lets you read formatted data from the keyboard. The format of the *fscanf* function is as follows:

```
#include <stdio.h>

int fscanf(FILE *stream, const char *format[, variable_address, ...]);
```

The *stream* parameter is a pointer to the file from which you want *fscanf* to read. The *format* parameter specifies the data format—using the same control character as *scanf*. Finally, the *variable_address* parameter specifies an address into which you want the data read. The ellipsis (...) that follows the *variable_address* parameter indicates that you can have multiple addresses separated by commas.

When it completes, *fscanf* returns the number of fields it read. If *fscanf* encounters the end of file, it returns the constant *EOF*. The following program, *fscanf.c*, opens the file *data.dat* for output, writes formatted output to the file using *fprintf*, closes the file, and then reopens it for input, reading its contents with *fscanf*:

```
#include <stdio.h>

void main(void)
 {
   FILE *fp;

   int age;
   float salary;
   char name[64];

   if ((fp = fopen("DATA.DAT", "w")) == NULL)
```

```
      printf("Error opening DATA.DAT for output\n");
    else
      {
        fprintf(fp, "33 35000.0 Kris");
        fclose(fp);

        if ((fp = fopen("DATA.DAT", "r")) == NULL)
          printf("Error opening DATA.DAT for input\n");
        else
         {
           fscanf(fp, "%d %f %s", &age, &salary, name);
           printf("Age %d Salary %f Name %s\n", age, salary, name);
           fclose(fp);
         }
      }
  }
```

450 Positioning the File Pointer Based on its Current Location

You have learned that the file pointer contains a position pointer to track your current position within the file. When you know the format of your file, there may be times when you want to advance the position pointer to a specific location before you start reading the file. For example, the first 256 bytes of your file can contain header information that you do not want to read. In such cases, your programs can use the *fseek* function to position the file pointer, as shown here:

```
#include <stdio.h>

int fseek(FILE *stream, long offset, int relative_to);
```

The *stream* parameter specifies the file pointer you want to position. The *offset* and *relative_to* parameters combine to specify the desired position. The *offset* contains the byte offset into the file. The *relative_to* parameter specifies the location in the file from which *fseek* should apply the offset. Table 450 specifies the values you can use for the *relative_to* parameter.

Constant	Meaning
SEEK_CUR	From current file position
SEEK_SET	From the beginning of the file
SEEK_END	From the end of the file

*Table 450 File positions from which **fseek** can apply an offset.*

To position the file pointer immediately after the first 256 bytes of header information in a file, you would use *fseek,* as follows:

```
fseek(fp, 256, SEEK_SET);  // Offset 0 is the start
```

If successful, *fseek* will return the value 0. If an error occurs, *fseek* will return a non-zero value.

451 Getting File Handle Information

When you work with a file handle, there may be times when you must know specifics about the corresponding file, such as the disk drive that stores the file. In such cases, your programs can use the *fstat* function, which has the following format:

```
#include <sys\stat.h>

int fstat(int handle, struct stat *buffer);
```

The function assigns specifics about the file to a structure of type *stat* defined within the *include* file *stat.h,* as shown here:

```
struct stat
 {
   short st_dev;     // Drive number of disk
   short st_ino;     // Not used by DOS
   short st_mode;    // File open mode
   short st_nlink;   // Always 1
   short st_uid;     // User id--Not used
   short st_gid;     // Group id--Not used
   short st_rdev;    // Same as st_dev
   long st_size;     // File size in bytes
   long st_atime;    // Time file was last opened
   long st_mtime;    // Same as st_atime
   long st_ctime;    // Same as st_atime
 };
```

If *fstat* succeeds, it returns the value 0. If an error occurs, *fstat* returns the value –1 and sets the global variable *errno* to *EBADF* (for a bad file handle). The following program, *autoinfo.c*, uses the *fstat* function to display the date and time of the *autoexec.bat* file's last modification, as well as its size:

```
#include <stdio.h>
#include <io.h>
#include <fcntl.h>
#include <sys\stat.h>
#include <time.h>

void main(void)
  {
    int handle;
    struct stat buffer;

    if ((handle = open("\\AUTOEXEC.BAT", O_RDONLY)) == -1)
     printf("Error opening \\AUTOEXEC.BAT\n");
    else
      {
        if (fstat(handle, &buffer))
          printf("Error getting file information\n");
        else
          printf("AUTOEXEC.BAT is %ld bytes Last used %s\n",
            buffer.st_size, ctime(&buffer.st_atime));
        close(handle);
      }
  }
```

Note: *Tip 1465 details how you can obtain a file's date and time stamp from within Windows.*

REOPENING A FILE STREAM 452

As your programs work with files, at times you may want to override an open file pointer. For example, DOS does not provide a way to redirect the output of the *stderr* file handle from the command line. However, from within your program, you can override the destination of the *stderr* file pointer by reopening it using the *freopen* function:

```
#include <stdio.h>

FILE *freopen(const char *filename, const char, access_mode, FILE *stream);
```

The *freopen* function is similar to *fopen*, except that you pass to the function a file pointer whose value you want to overwrite. If the function succeeds, it returns a pointer to the original file stream. If an error occurs, *freopen* returns *NULL*. The following program, *nostderr.c*, for example, redirects *stderr* functions to the file *standard.err*, as opposed to the screen:

```
#include <stdio.h>

void main(void)
 {
   if (freopen("STANDARD.ERR", "w", stderr))
     fputs("stderr has been redirected", stderr);
   else
     printf("Error in reopen\n");
 }
```

453 UNDERSTANDING ARRAYS

As you have learned, a type describes the set of values a variable can hold and the set of operations that your programs can perform on the variable. Except for character strings, all the types you have examined so far can hold only one value. As your programs begin to perform more useful work, there will be times when you want a variable to hold many values. For example, the variable *scores* might keep track of 100 students' test scores. Likewise, the variable *salaries* might keep track of each company employee's salary. An *array* is a data structure that can store multiple values of the same type. For example, you can create an array that can hold 100 values of type *int* and a second array that can hold 25 values of type *float*.

Every value you assign to an array must be of the same type as the array's type. In this section you will learn how to create and work with arrays in your programs. After you work with one or two arrays, you will find that arrays are easy to understand. If you already feel comfortable with strings, you will soon feel just as comfortable working with arrays. Remember, a character string is simply an array of characters.

454 DECLARING AN ARRAY

In Tip 453 you learned that an *array* is a variable that can store multiple values of the same type. To declare an array, you must specify the desired type (such as *int*, *float*, or *double*), as well as the array size. To specify an array size, you place the number of values the array can store within brackets that follow the array name. The following declaration, for example, creates an array named *scores* that can store 100 test scores of type *int*:

```
int scores[100];
```

In a similar way, the following declaration creates an array of type *float* that contains 50 salaries:

```
float salaries[50];
```

When you declare an array, C allocates enough memory to hold all the elements. The first entry is at location 0. For example, in the arrays *scores* and *salaries*, the following statements assign the values 80 and 35,000 to the first array elements:

```
scores[0] = 80;
salaries[0] = 35000.0;
```

Because the first array element begins at offset 0, the array's last element occurs one location before the array's size. Given the previous arrays, *scores* and *salaries*, the following statements assign values to the last element of each array:

```
scores[99] = 75;

salaries[49] = 24000.0;
```

VISUALIZING AN ARRAY 455

As you have learned, an array is a variable that can store multiple values of the same type. To help you better understand how an array stores information, consider the following array declarations:

```
char string[64];
float salaries[50];
int scores[100];
long planets[13];
```

After you assign values to each array, the arrays will reside in memory in a manner similar to that shown in Figure 455.

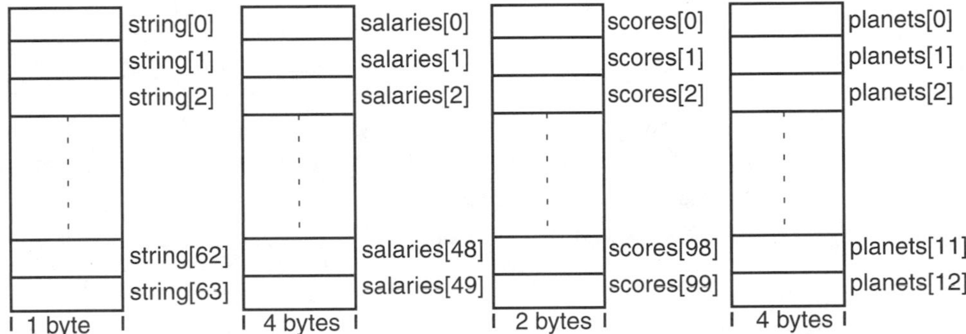

Figure 455 Storing values in arrays.

As you can see, each array's first value resides at offset 0. In the Getting Started with C section of this book, you learned that a variable is a name you assign to one or more memory locations. In an array, you may have a large number of memory locations that correspond to a single array.

UNDERSTANDING AN ARRAY'S STORAGE REQUIREMENTS 456

As you have learned, an array is a named collection of values of the same type. When you declare an array, the C compiler allocates enough memory to hold the number of values you specify. The actual amount of memory the compiler allocates depends on the array type. For example, an array of 100 elements of type *int* will usually require 100 * 2 or 200 bytes of memory. An array of 100 elements of type *float*, on the other hand, will require 100 * 4 bytes or 400 bytes. The following program, *arraysiz.c*, uses C's *sizeof* operator to display the amount of memory different array types require:

```
#include <stdio.h>

void main(void)
 {
   int scores[100];
   float salaries[100];
   char string[100];

   printf("Bytes used to hold int scores[100] is %d bytes\n", sizeof(scores));
   printf("Bytes used to hold int salaries[100] is %d bytes\n",
    sizeof(salaries));
   printf("Bytes used to hold char string[100] is %d bytes\n", sizeof(string));
 }
```

When you compile and execute the *arraysiz.c* program, your screen will display the following output:

```
Bytes used to hold int scores[100] is 200 bytes
Bytes used to hold float salaries[100] is 400 bytes
Bytes used to hold char string[100] is 100 bytes
C:\>
```

457 INITIALIZING AN ARRAY

Throughout this book, many of the programs have initialized character strings as follows:

```
char title[] = "Jamsa\'s C/C++/C# Programmer\'s Bible";
char section[64] = "Arrays";
```

In the first case, the C compiler will allocate 35 bytes to hold the string. In the second case, the compiler will allocate an array of 64 bytes, initializing the first seven characters to the letters "Arrays" and the *NULL* character. Most compilers will also initialize the remaining byte locations to *NULL*. When you declare arrays of other types, you can initialize the arrays in the same way. For example, the following statement initializes the integer array *scores* to the values 80, 70, 90, 85, and 80:

```
int scores[5] = {80, 70, 90, 85, 80};
```

When you assign initial values to an array, you must enclose the values within right and left braces ({}). In the previous case, the array size matches the number of values assigned to the array. The following statement, however, assigns four *floating-point* values to an array that can store 64 values:

```
float salaries[64] = {25000.0, 32000.0, 44000.0, 23000.0};
```

Depending on your compiler, it may assign 0 to the elements to which your program does not assign explicit values. However, as a rule, you should not assume that the compiler will initialize the other elements. Moreover, if you do not specify an array size, the compiler will allocate enough memory to hold only the values you specify. The following array declaration, for example, creates an array large enough to hold three values of type *long*:

```
long planets[] = {1234567L, 654321L, 1221311L};
```

458 ACCESSING ARRAY ELEMENTS

The values stored in an array are called *array elements*. To access an array element, you specify the array name and the element you want. The following program, *elements.c*, initializes the array *scores* and then uses *printf* to display the element values:

```
#include <stdio.h>

void main(void)
 {
   int scores[5] = {80, 70, 90, 85, 80};

   printf("Array Values\n");
   printf("scores[0] %d\n", scores[0]);
   printf("scores[1] %d\n", scores[1]);
   printf("scores[2] %d\n", scores[2]);
   printf("scores[3] %d\n", scores[3]);
   printf("scores[4] %d\n", scores[4]);
 }
```

When you compile and execute the *elements.c* program, your screen will display the following output:

```
Array Values
scores[0] = 80
scores[1] = 70
scores[2] = 90
scores[3] = 85
scores[4] = 80
C:\>
```

As you can see, to access a specific array element, you specify the element number you want within the left and right brackets that follow the array name.

LOOPING THROUGH ARRAY ELEMENTS 459

In Tip 458 you used the values 0 through 4 to display the elements of the array *scores*. When you reference many array elements, specifying numbers for each array element individually can become time consuming. As an alternative, your programs can use a variable to reference array elements. For example, assuming the variable *i* contains the value 2, the following statement would assign *array[2]* the value 80:

```
i = 2;
array[i] = 80;
```

The following program, *showarra.c*, uses the variable *i* and a *for* loop to display the elements of the array *scores*:

```
#include <stdio.h>
void main(void)
 {
   int scores[5] = {80, 70, 90, 85, 80};
   int i;
   printf("Array Values\n");
   for (i = 0; i < 5; i++)
     printf("scores[%d] %d\n", i, scores[i]);
 }
```

USING CONSTANTS TO DEFINE ARRAYS 460

As you have learned, when your programs work with arrays, you must specify the array size. For example, the following program, *5_values.c*, declares an array of five values and then uses a *for* loop to display the array's values:

```
#include <stdio.h>
void main(void)
 {
   int values[5] = {80, 70, 90, 85, 80};
   int i;
   for (i = 0; i < 5; i++)
     printf("values[%d] %d\n", i, values[i]);
 }
```

Assume, for example, that you later want to change the *5_values.c* program so that it supports 10 values. You must then change not only the array declaration, but also the *for* loop. The more changes you must make to a program, the greater your chance of error. As an alternative, your programs should declare arrays using constants. The following program, *5_const.c*, declares an array based on the constant *ARRAY_SIZE*. As you can see, the program not only uses the constant to declare the array, but also uses the constant as the ending condition for the *for* loop:

```
#include <stdio.h>
#define ARRAY_SIZE 5
void main(void)
 {
   int values[ARRAY_SIZE] = {80, 70, 90, 85, 80};
   int i;
   for (i = 0; i < ARRAY_SIZE; i++)
     printf("values[%d] %d\n", i, values[i]);
 }
```

If you must later change the array size, you can change the value assigned to the *ARRAY_SIZE* constant so that the program automatically updates the loops that control the array and the array size.

461 PASSING AN ARRAY TO A FUNCTION

As you have learned, an array is a variable that can store multiple values of the same type. Like all variables, your programs can pass arrays to functions. When you declare a function that works with an array parameter, you must tell the compiler. For example, the following program, *arrfunct.c*, uses the function *show_array* to display the values in an array. As you can see, the program passes to the function both the array and the number of elements the array contains, as shown here:

```
#include <stdio.h>

void show_array(int values[], int number_of_elements)
 {
   int i;

   for (i = 0; i < number_of_elements; i++)
    printf("%d\n", values[i]);
 }

void main(void)
 {
   int scores[5] = {70, 80, 90, 100, 90};

   show_array(scores, 5);
 }
```

When a function receives an array as a parameter, your program does not have to specify the array size in the parameter declaration. In the case of the function *show_values*, the brackets that follow the variable name *value* inform the compiler that the parameter is an array. Other than knowing that the parameter is an array, the compiler does not care about the size of the array your program passes to the function.

462 REVISITING ARRAYS AS FUNCTIONS

In Tip 461 you learned that when you declare the formal parameter for an array, you do not need to declare an array size. Instead, you can specify only the left and right brackets. The following program, *arrparam.c*, passes three different arrays (of different sizes) to the function *show_values*:

```
#include <stdio.h>

void show_array(int values[], int number_of_elements)
 {
   int i;

   printf("About to display %d values\n", number_of_elements);
   for (i = 0; i < number_of_elements; i++)
    printf("%d\n", values[i]);
 }

void main(void)
 {
   int scores[5] = {70, 80, 90, 100, 90};
   int count[10] = {1, 2, 3, 4, 5, 6, 7, 8, 9, 10};
   int small[2] = {-33, -44};

   show_array(scores, 5);
   show_array(count, 10);
   show_array(small, 2);
 }
```

When you compile and execute the *arrparam.c* program, your screen will display each array's values. As you have learned, the function does not care about the array size. However, note that the arrays the *arrparam.c* program passes to the function are all type *int*. If you tried to pass an array of type *float* to the function, the compiler would generate an error.

UNDERSTANDING HOW STRING ARRAYS DIFFER 463

Many Tips presented throughout this book have passed strings to functions. In most cases, the functions did not specify the string size. For example, the following statement uses the *strupr* function to convert a string to uppercase:

```
char title[64] = "Jamsa\'s C/C++/C# Programmer\'s Bible";
strupr(title);
```

As you have learned, in C, the *NULL* character represents the end of a character string. Therefore, functions can search the array elements for the *NULL* to determine where the array ends. Arrays of other types, such as *int*, *float*, or *long*, however, do not have an equivalent "ending" character. Therefore, you usually must pass to functions that work with arrays the number of elements the array contains.

PASSING ARRAYS ON THE STACK 464

Several previous Tips have discussed passing arrays as parameters to functions. When you pass an array to a function, C only places the address of the array's first element on the stack. Figure 464, for example, illustrates the array *scores* and a function call to *show_array* using *scores*.

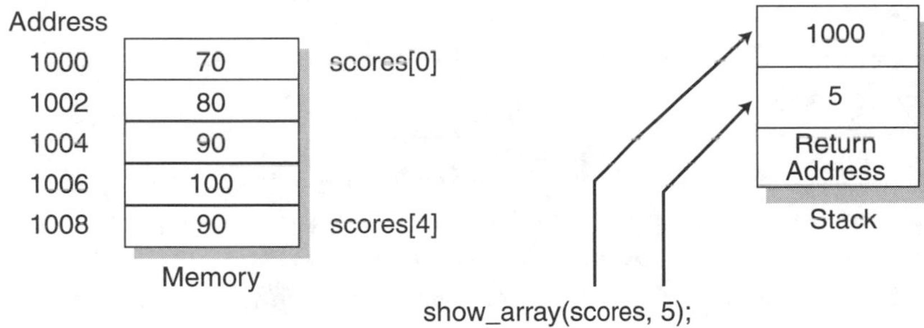

Figure 464 *When you pass an array parameter, C places the array's starting address on the stack.*

As you can see, C places only the array's starting address on the stack. Also, notice that the function receives no information from C regarding the size of the array.

DETERMINING HOW MANY ELEMENTS AN ARRAY CAN HOLD 465

As you have learned, depending on an array's type, the actual amount of memory an array can consume will differ. If you are working in the DOS environment, the amount of memory your arrays can consume will depend on the current memory model. In general, an array cannot consume more than 64Kb of space. The following program, *too_big.c*, for example, may fail to compile because the arrays consume too much memory:

```
void main(void)
 {
   char string[66000L];      // 66,000 bytes
   int values[33000L];       // 33,000 * 2 = 66,000 bytes
   float numbers[17000];     // 17,000 * 4 = 68,000 bytes
 }
```

***Note:** Because Windows uses the virtual memory model to manage memory, it does not place limits on array size to the extent that a DOS C program does. For example, you can declare strings (character arrays) in Windows as large as INT_MAX (2,147,483,647) characters in length. If, however, you try to declare an oversized variable within a DOS window, you will cause a stack fault and Windows will close the window.*

466 Using the Huge Memory Model for Big Arrays

If the amount of memory an array consumes exceeds 64Kb, you can direct many DOS-based compilers to use the *huge memory model* by treating the array as a pointer and including the word *huge* within the declaration, as shown here:

```
float huge values[17000];
```

The following program, *huge_flt.c*, creates a huge floating-point array:

```
#include <stdio.h>
#include <malloc.h>

void main (void)
  {
     int i;
     float huge *values;

     if ((values = (float huge *) halloc (17000, sizeof(float))) == NULL)
       printf ("Error allocating huge array\n");
     else
       {
         printf("Filling the array\n");
         for (i = 0; i < 17000; i++)
           values[i] = i * 1.0;
         for (i = 0; i < 17000; i++)
           printf ("%8.1f ", values[i]);
         hfree(values);
       }
  }
```

***Note:** Because Windows uses the virtual memory model to manage memory, it does not place limits on array size to the extent that a DOS C program does. For example, you can declare an **unsigned char** array in Windows as large as INT_MAX (2,147,483,647) without using the **huge** keyword. If, however, you try to declare an oversized variable within a DOS window without the **huge** keyword, you will cause a stack fault and Windows will close the window.*

467 The Tradeoff Between Arrays and Dynamic Memory

As you become more comfortable with C and how to use pointers within C, you may start to use arrays less often and instead allocate memory dynamically as you need it. There are several tradeoffs you must consider as you determine whether to use dynamic memory or an array. To begin, many users find arrays simpler to understand and use. As a result, your program itself might be easier for other programmers to follow. Second, because the compiler allocates space for arrays, your programs do not experience the run-time overhead associated with dynamic memory allocation. As a result, an array-based program might execute slightly faster.

As you have learned, however, when you declare an array, you must specify the array size. If you do not know the size you will need, you might have a tendency to allocate a larger array than necessary. As a result, you might waste memory. On the other hand, if the array size is too small, you must edit your program, change the array sizes, and recompile your program.

When you declare an array within your programs, keep in mind that you can perform identical processing by allocating memory dynamically. As you will learn in the Pointer section of this book, you can reference dynamically allocated memory using array indexes and eliminate the pointer confusion that often frustrates new C programmers. Because most operating systems let programs allocate memory very quickly, you might prefer the flexibility and improved memory management opportunities that dynamic memory allocation provides over arrays, despite the slight system overhead it incurs.

UNDERSTANDING MULTIDIMENSIONAL ARRAYS 468

As you have learned, an array is a variable that can store multiple values of the same type. In all the examples presented so far, the arrays have consisted of a single row of data. However, C also supports two-, three-, and multidimensional arrays. The best way to visualize a two-dimensional array is as a table with rows and columns. If an array contains three dimensions, visualize the array as several pages, each of which contains a two-dimensional table, as shown in Figure 468.

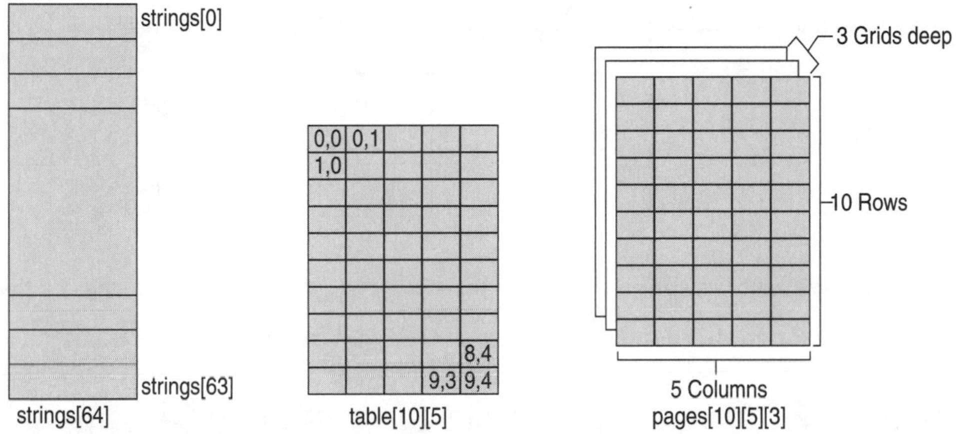

Figure 468 Logical model of multidimensional arrays.

The following array declarations create the arrays shown in Figure 468:

```
char strings[64];
int table[10][5];
float pages[10][5][3];
```

UNDERSTANDING ROWS AND COLUMNS 469

As you have learned, C supports multidimensional arrays that are similar to tables of values. When you work with a two-dimensional array, think of the array as a table of rows and columns. The table's rows go from left to right while the columns go up and down the page, as shown in Figure 469.

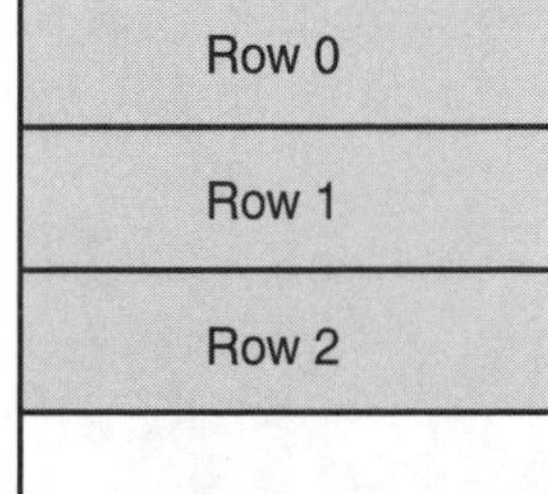

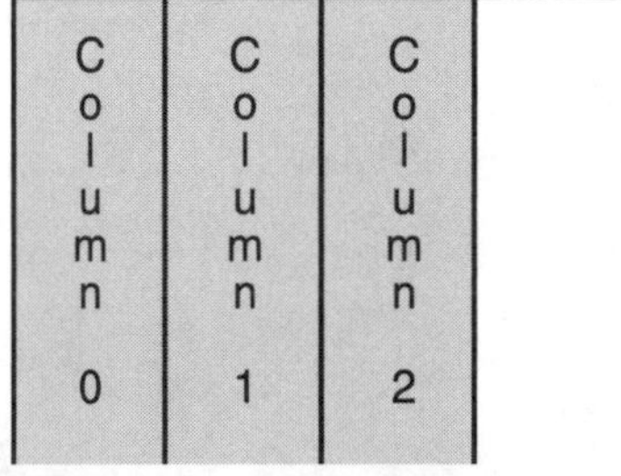

Figure 469 Rows and columns in a two-dimensional array.

When you declare a two-dimensional array, the first value you specify states the number of rows and the second value the number of columns:

```
int table [2][3];
```

470 Accessing Elements in a Two-Dimensional Array

As you have learned, you can best visualize a two-dimensional array as a table containing rows and columns. To reference a specific array element, you must specify the corresponding row and column position. Figure 470 illustrates statements that access specific elements within the array *table*.

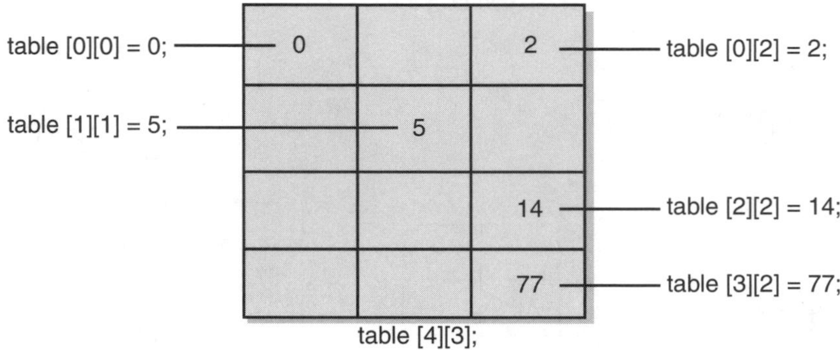

Figure 470 *To access elements in a two-dimensional array, you must specify the element's row and column position.*

As you can see, when you access a two-dimensional array, the row and column offsets begin at 0.

471 Initializing Elements in a Two-Dimensional Array

In Tip 457 you learned that to initialize array elements, you can place the element values within left and right braces after the array declaration. The following statement uses the same technique to initialize a two-dimensional array. However, in this case, the statement specifies the values for each array row within their own braces:

```
int table[2][3] = {{1, 2, 3},
                   {4, 5, 6}};
```

The C compiler will initialize the array elements as shown in Figure 471.

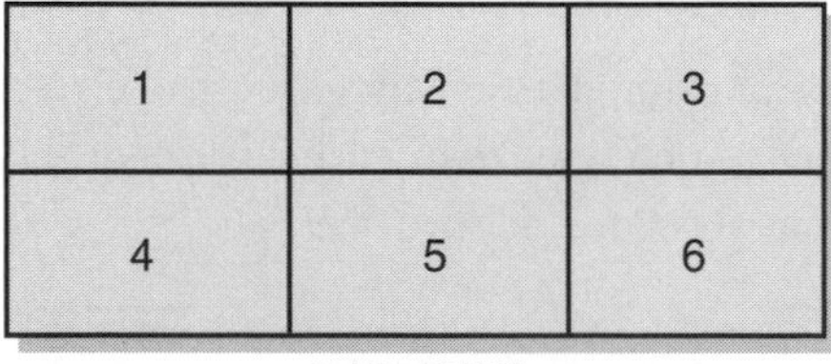

Figure 471 *Initializing the elements of a two-dimensional array.*

In a similar manner, the following statement initializes the elements of a larger array:

```
int sales[4][5] {{1, 2, 3, 4, 5},
                 {6, 7, 8, 9, 10},
                 {11, 12, 13, 14, 15},
                 {16, 17, 18, 19, 20}};
```

DETERMINING A MULTIDIMENSIONAL ARRAY'S MEMORY CONSUMPTION 472

In Tip 456 you learned that your programs can determine the amount of memory an array consumes by multiplying the number of elements in the array by the number of bytes required to represent the array's type (such as 2 for *int*, 4 for *float*, and so on). To determine the memory a multidimensional array consumes, you can perform the same calculation. To determine the number of elements in a multidimensional array, simply multiply the number of rows by the number of columns. The following expressions illustrate the memory amount different array declarations consume:

```
int a[5][10];          // 2 * 5 * 10 == 100 bytes
float b[5][8];         // 4 * 5 * 8 == 160 bytes
int c[3][4][5];        // 2 * 3 * 4 * 5 = 120 bytes
```

The following program, *md_size.c*, uses the *sizeof* operator to determine the number of bytes different array declarations consume:

```
#include <stdio.h>

void main(void)
 {
   int box[3][3];
   float year_sales[52][5];
   char pages[40][60][20];

   printf("Bytes to hold int box[3][3] %d bytes\n", sizeof(box));
   printf("Bytes to hold float year_sales[52][5] %d bytes\n"
          sizeof(year_sales));
   printf("Bytes to hold char pages[40][60][20] %ld bytes\n", sizeof(pages));
 }
```

When you compile and execute the *md_size.c* program, your screen will display the following:

```
Bytes to hold int box[3][3] 18 bytes
Bytes to hold float year_sales[52][5] 1040 bytes
Bytes to hold char pages[40][60][20] 48000 bytes
C:\>
```

LOOPING THROUGH A TWO-DIMENSIONAL ARRAY 473

In Tip 458 you learned how to use a variable to access elements in an array. When your programs work with two-dimensional arrays, you will normally use two variables to access array elements. The following program, *show_2d.c*, uses the variables *row* and *column* to display the values contained within the array *table*:

```
#include <stdio.h>

void main(void)
 {
   int row, column;
   float table[3][5] = {{1.0, 2.0, 3.0, 4.0, 5.0},
                        {6.0, 7.0, 8.0, 9.0, 10.0},
                        {11.0, 12.0, 13.0, 14.0, 15.0}};

   for (row = 0; row < 3; row++)
     for (column = 0; column < 5; column++)
       printf("table[%d][%d] = %f\n", row, column, table[row][column]);
 }
```

By nesting the *for* loops as shown, the program will display the elements contained in the array's first row (1.0 through 5.0). Next, the program will move to the second row and then the third row, displaying each element within each row in turn.

474 TRAVERSING A THREE-DIMENSIONAL ARRAY

In Tip 473 you learned how to traverse a two-dimensional array using two variables named *row* and *column*. The following program, *show_3d.c*, uses the variables *row*, *column*, and *table* to traverse a three-dimensional array:

```
#include <stdio.h>

void main(void)
 {
   int row, column, table;
   float values[2][3][5] = {
                              {{1.0, 2.0, 3.0, 4.0, 5.0},
                               {6.0, 7.0, 8.0, 9.0, 10.0},
                               {11.0, 12.0, 13.0, 14.0, 15.0}},

                            {{16.0, 17.0, 18.0, 19.0, 20.0},
                               {21.0, 22.0, 23.0, 24.0, 25.0},
                               {26.0, 27.0, 28.0, 29.0, 30.0}}
                             };

   for (row = 0; row < 2; row++)
     for (column = 0; column < 3; column++)
       for (table = 0; table < 5; table++)
         printf("values[%d][%d][%d] = %f\n", row, column, table,
                 values[row][column][table]);
 }
```

475 INITIALIZING MULTIDIMENSIONAL ARRAYS

In Tip 474 you learned how to display the contents of a three-dimensional array using three variables: *row, column*, and *table*. The *show_3d.c* program, presented in Tip 474, initialized the three-dimensional array values, as shown here:

```
   float values[2][3][5] = {
                              {{1.0, 2.0, 3.0, 4.0, 5.0},
                               {6.0, 7.0, 8.0, 9.0, 10.0},
                               {11.0, 12.0, 13.0, 14.0, 15.0}},

                            {{16.0, 17.0, 18.0, 19.0, 20.0},
                               {21.0, 22.0, 23.0, 24.0, 25.0},
                               {26.0, 27.0, 28.0, 29.0, 30.0}}
                             };
```

At first glance, initializing a multidimensional array can seem confusing. To better understand how to initialize such arrays, this Tip presents several sample initializations. As you examine the initializations, perform the initializations from right to left:

```
int a[1][2][3] =     {
                      { {1, 2, 3}, {4, 5, 6} }
                     };  // Array braces

int b[2][3][4] =     {
                      { {1, 2, 3, 4}, {5, 6, 7, 8}, {9, 10, 11, 12} },
                      { {13, 14, 15, 16}, {17, 18, 19, 20}, {21, 22, 23, 24} }
                     };  // Array braces

int c[3][2][4] =     {
                      { {1, 2, 3, 4}, {5, 6, 7, 8} },
                      { {9, 10, 11, 12}, {13, 14, 15, 16}},
                      { {17, 18, 19, 20}, {21, 22, 23, 24}}
                     };  // Array braces

int d[1][2][3][4] = {
                      {{{1, 2, 3, 4}, {5, 6, 7, 8}, {9, 10, 11, 12}},
```

```
                    {{13, 14, 15, 16}, {17, 18, 19, 20}, {21, 22, 23, 24}}}
                   }; // Array braces
```

Each array initialization gets a set of outer braces. Within the outer braces, you then define the different array elements within additional braces.

PASSING A TWO-DIMENSIONAL ARRAY TO A FUNCTION 476

As your programs work with multidimensional arrays, there will be times when you must write functions that work with the arrays. In Tip 461, you learned that when you pass arrays to a function, you do not need to specify the number of array elements. When you work with two-dimensional arrays, you do not need to specify the number of rows in the array, but you must specify the number of columns. The following program, *funct_2d.c*, uses the function *show_2d_array* to display the contents of several two-dimensional arrays:

```
#include <stdio.h>

void show_2d_array(int array[][10], int rows)
 {
   int i, j;

   for (i = 0; i < rows; i++)
     for (j = 0; j < 10; j++)
       printf("array[%d][%d] = %d\n", i, j, array[i][j]);
 }

void main(void)
 {
   int a[1][10] = {{1, 2, 3, 4, 5, 6, 7, 8, 9, 10}};
   int b[2][10] = {{1, 2, 3, 4, 5, 6, 7, 8, 9, 10},
                   {11, 12, 13, 14, 15, 16, 17, 18, 19, 20}};
   int c[3][10] = {{1, 2, 3, 4, 5, 6, 7, 8, 9, 10},
                   {11, 12, 13, 14, 15, 16, 17, 18, 19, 20},
                   {21, 22, 23, 24, 25, 26, 27, 28, 29, 30}};

   show_2d_array(a, 1);
   show_2d_array(b, 2);
   show_2d_array(c, 3);
 }
```

TREATING MULTIDIMENSIONAL ARRAYS AS ONE-DIMENSIONAL 477

In Tip 476 you learned that when you pass a two-dimensional array to a function and you want to access the array's row and column positions, you must specify the number of columns, as shown here:

```
void show_2d_array(int array[][10], int rows)
```

If you want to work with the elements of a multidimensional array, but you do not need to access the elements in their row or column positions, your functions can treat the multidimensional array as if it were one-dimensional. The following program, *sum_2d.c*, returns the sum of the values in a two-dimensional array:

```
#include <stdio.h>

long sum_array(int array[], int elements)
 {
   long sum = 0;
   int i;

   for (i = 0; i < elements; i++)
     sum += array[i];
   return(sum);
```

```
 }
void main(void)
 {
   int a[10] = {1, 2, 3, 4, 5, 6, 7, 8, 9, 10};
   int b[2][10] = {{1, 2, 3, 4, 5, 6, 7, 8, 9, 10},
                   {11, 12, 13, 14, 15, 16, 17, 18, 19, 20}};
   int c[3][10] = {{1, 2, 3, 4, 5, 6, 7, 8, 9, 10},
                   {11, 12, 13, 14, 15, 16, 17, 18, 19, 20},
                   {21, 22, 23, 24, 25, 26, 27, 28, 29, 30}};

   printf("Sum of first array elements %d\n", sum_array(a, 10));
   printf("Sum of second array elements %d\n", sum_array(b, 20));
   printf("Sum of third array elements %d\n", sum_array(c, 30));
 }
```

As you can see, the function *sum_array* supports one-, two-, or multidimensional arrays. To understand how *sum_array* works, you must first understand how C stores multidimensional arrays in memory. Tip 478 discusses how C stores multidimensional arrays in detail.

478 UNDERSTANDING HOW C STORES MULTIDIMENSIONAL ARRAYS

In Tip 454 you learned that when you declare an array, such as *int scores[100]*, C allocates enough memory to hold each array element. When you allocate a multidimensional array, the same is true. Although multidimensional arrays conceptually consist of rows, columns, and pages, to the compiler a multidimensional array is one long byte range. For example, assume that your program declares the following array:

```
int  table[3][5];
```

Figure 478 illustrates the array's conceptual appearance and actual memory use.

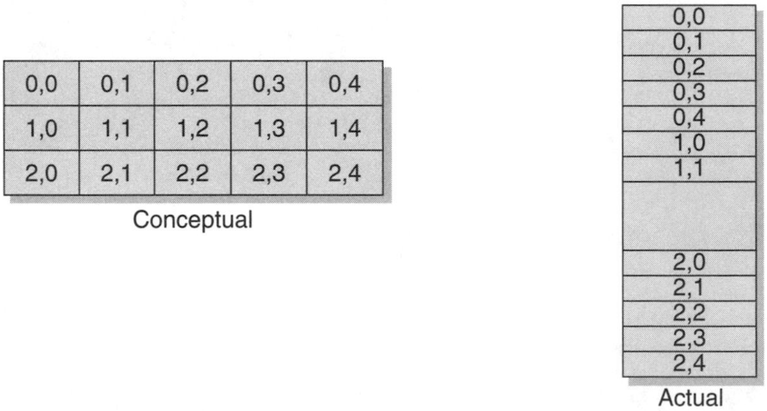

0,0	0,1	0,2	0,3	0,4
1,0	1,1	1,2	1,3	1,4
2,0	2,1	2,2	2,3	2,4

Conceptual

0,0
0,1
0,2
0,3
0,4
1,0
1,1

2,0
2,1
2,2
2,3
2,4

Actual

***Figure 478** Mapping a multidimensional array to memory.*

In Tip 477 you created a function that treated a multidimensional array as one-dimensional to add the values the array contained. Because the C compiler actually maps the multidimensional array to a one-dimensional memory range, treating the array as one-dimensional is valid.

479 UNDERSTANDING ROW-MAJOR VERSUS COLUMN-MAJOR ORDER

In Tip 478 you learned that the C compiler maps multidimensional arrays to one-dimensional memory. When the compiler maps a multidimensional array to memory, the compiler has two options. As shown in Figure 479, the compiler can place the array's row elements in memory before the column values, or the compiler can place the column elements first.

0,0	0,1	0,2
1,0	1,1	1,2

Row Major
0,0
0,1
0,2
1,0
1,1
1,2

Column Major
0,0
1,0
0,1
1,1
0,2
1,2

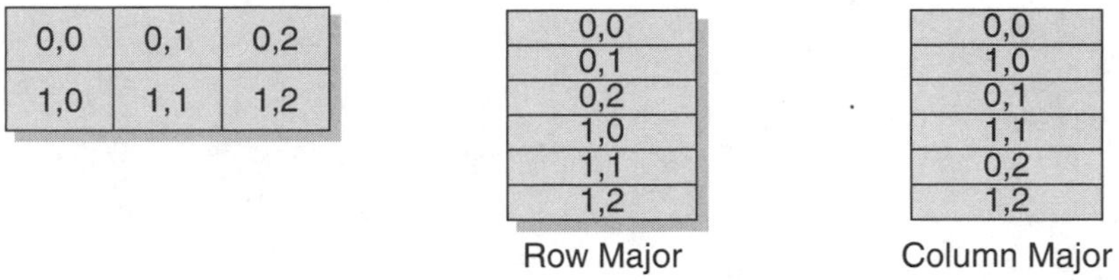

***Figure 479** Mapping array elements to memory.*

When the compiler places the array's row elements in memory before the column elements, the compiler is performing *row-major ordering*. Likewise, when the compiler places the column elements first, the compiler performs *column-major ordering*. C compilers store multidimensional arrays in row-major order.

ARRAYS OF STRUCTURES OF ARRAYS 480

Arrays and structures let you group related information. As you have learned, C lets you create arrays of structures or use arrays as structure members. In general, C does not place a limit on the depth to which your programs can go with respect to nested data structures. For example, the following declaration creates an array of 100 employee structures. Within each structure is an array of *Date* structures that correspond to the employee's hire date, first review, and last review:

```
struct Employee
 {
   char name[64];
   int age;
   char ssan[11];    // Social security number
   int pay_grade;
   float salary;
   unsigned employee_number;
   struct Date
    {
      int month;
      int day;
      int year;
    } emp_dates[3];
 } staff[100];
```

To access members and array elements, you simply work from left to right, starting from the outside and working inward. For example, the following statements assign an employee's hire date:

```
staff[10].emp_dates[0].month = 7;
staff[10].emp_dates[0].day = 7;
staff[10].emp_dates[0].year = 7;
```

Although nesting structures and arrays in this way can be very convenient, keep in mind that the deeper your programs nest such data structures, the more difficult the structure will become for other programmers to understand.

UNDERSTANDING UNIONS 481

As you have learned, structures let your programs store related information. Depending on your program's purpose, there may be times when the information you store in a structure will be only one of two values. For example, assume that your program tracks two special date values for each employee. For current employees, the program tracks the number of days the employee has worked. For an employee who no longer works for the company, the program tracks the employee's last employment date. One way to track such information is to use a structure, as shown here:

```
struct EmpDates
 {
   int days_worked;
   struct LastDate
    {
      int month;
      int day;
      int year;
    } last_day;
 };
```

Because the program will either use the *days_worked* or *last_day* members, the memory holding the unused value for each employee goes to waste. As an alternative, C lets your programs use a *union*, which allocates only the memory the union's largest member requires, as shown here:

```
union EmpDates
 {
   int days_worked;
   struct LastDate
    {
      int month;
      int day;
      int year;
    } last_day;
 };
```

To access the union members, you use the *dot* operator just as you would with a structure. Unlike the structure, however, the union can only store one member's value. Figure 481 illustrates how C allocates memory for the structure and member.

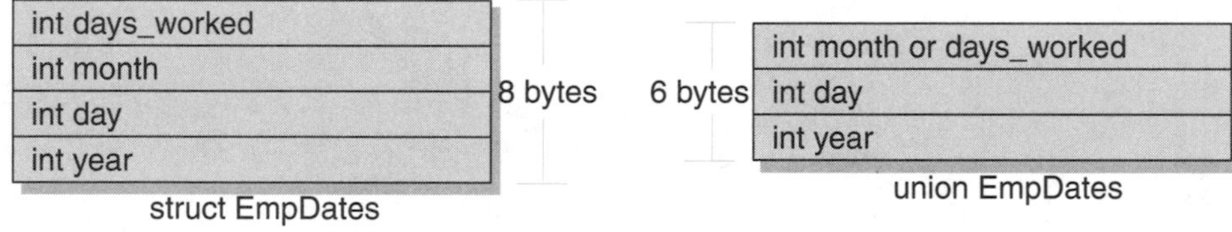

Figure 481 Allocating memory for a similar structure and union.

As you will learn, using unions not only saves memory, but also provides your programs with the ability to interpret memory values differently.

482 SAVING MEMORY WITH UNIONS

In Tip 481 you learned that C lets you store information within a *union*. When you use a union, C allocates the amount of memory required to hold the union's largest member. The following program, *unionsiz.c*, uses the *sizeof* operator to display the amount of memory different unions consume:

```
#include <stdio.h>

void main(void)
 {
   union EmployeeDates
    {
      int days_worked;
      struct Date
```

```
      {
        int month;
        int day;
        int year;
      } last_day;
    } emp_info;

  union Numbers
    {
      int a;
      float b;
      long c;
      double d;   // Largest--requires 8 bytes
    } value;

  printf("Size of EmployeeDates %d bytes\n", sizeof(emp_info));
  printf("Size of Numbers %d bytes\n", sizeof(value));
}
```

When you compile and execute the *unionsiz.c* program, your screen will display the following output:

```
Size of EmployeeDates 6 bytes
Size of Numbers 8 bytes
C:\>
```

USING REGS—A CLASSIC UNION 483

As you have learned, unions let your programs reduce their memory requirements and view information in different ways. In the DOS and BIOS section of this book, you will learn that to access the DOS and BIOS services, your programs usually assign parameters (at the assembly language level) to specific PC registers. To make the DOS and BIOS services available to your C programs, most C compilers provide access through run-time library routines that use a union of type *REGS*:

```
struct WORDREGS
 {
   unsigned int ax, bx, cx, dx, si, di, cflag, flags;
 };

struct BYTEREGS
 {
   unsigned char al, ah, bl, ah, cl, ch, dl, dh;
 };

union REGS
 {
   struct WORDREGS x;
   struct BYTEREGS h;
};
```

When your programs access one of the PC's general purpose registers (AX, BX, CX, and DX), the PC lets you refer to the register in a 16-bit (word) format. Alternatively, you can refer to the register's high and low bytes (AL, AH, BL, BH, CL, CH, DL, and DH). Because both methods refer to the same register, you have two ways of accessing the same storage location. Using a union, your programs have two ways to access the general purpose registers. Figure 483 illustrates how C stores variables of the *REGS* union in memory.

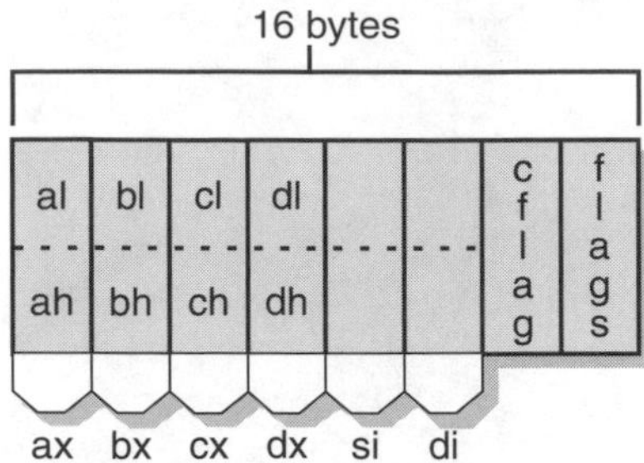

Figure 483 How C stores variables of the union REGS.

484 Putting the REGS Union to Use

In Tip 483 you learned that one of the most frequently used unions in DOS-based programs is the *REGS* union. The following *Turbo C Lite* program, *get_verx.c*, uses the *REGS* union to display the current DOS version, accessing the general purpose registers in their word form:

```
#include <stdio.h>
#include <dos.h>

void main(void)
 {
   union REGS inregs, outregs;

   inregs.x.ax = 0x3000;
   intdos(&inregs, &outregs);
   printf("Current version %d.%d\n", outregs.x.ax & 0xFF, outregs.x.ax >> 8);
 }
```

The following program, *get_verh.c*, uses the union's byte registers to display the current DOS version:

```
#include <stdio.h>
#include <dos.h>

void main(void)
 {
   union REGS inregs, outregs;

  inregs.h.ah = 0x30;
  inregs.h.al = 0;
  intdos(&inregs, &outregs);
   printf("Current version %d.%d\n", outregs.h.al, outregs.h.ah);
 }
```

485 Understanding Bit-Field Structures

Many functions in this book reduce the number of variables (and hence the amount of allocated memory) your programs must use by returning values whose bits have specific meanings. When a value's bits have specific meanings, your programs can use C's bitwise operators to extract the values (the specific bits). Assume, for example, that your program must track 100,000 dates. You can create a structure of type *Date* to track the dates, as shown here:

```
struct Date
 {
   int month; // 1 through 12
   int day;   // 1 through 31
   int year;  // last two digits
 };
```

As an alternative, your programs can use specific bits within an *unsigned int* value to hold the date fields, as shown in Figure 485.

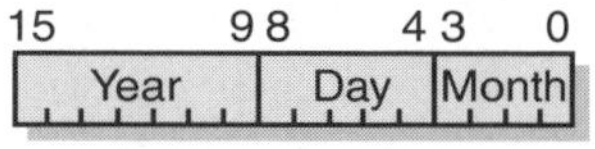

15 – 9	8 – 4	3 – 0
Year	Day	Month

***Figure 485** Using bits to represent a date.*

Then, each time your program must assign a date, it can perform the correct bitwise operations, as shown here:

```
unsigned date;

date = month;
date = date | (day << 4);
date = date | (year << 9);

printf("Month %d Day %d Year %d\n", date & 0xF, (date >> 4) & 0x1F, (date >> 9));
```

However, to make your programs easier to understand, C lets you create a *bit-field structure*. When you declare a bit-field structure, you define a structure that specifies the meaning of the corresponding bits:

```
struct Date
 {
   unsigned month:4;
   unsigned day:5;
   unsigned year:7;
 } date;
```

Your programs will then reference the bit-fields individually, as shown here:

```
date.month = 12;
date.day = 31;
date.year = 94;

printf("Month %d Day %d Year %d\n", date.month, date.day, date.year);
```

***Note:** When you declare a bit-field structure, the structure's members must each be **unsigned int** values.*

VISUALIZING A BIT-FIELD STRUCTURE 486

In Tip 485 you learned that C lets you represent bits within a value using a bit-field structure. When you declare a bit-field structure, C allocates enough bytes of memory to hold the structure's bits. If the structure does not use all the bits in the last byte, most C compilers will initialize the bits to 0. To help you better visualize how C stores a bit-field structure, Figure 486 illustrates how the C compiler will represent the bit-field structure, *Date*, as shown in the following code:

```
struct Date
 {
   unsigned month:4;
   unsigned day:5;
   unsigned year:7;
 } date;
```

15 – 9	8 – 4	3 – 0
Year	Day	Month

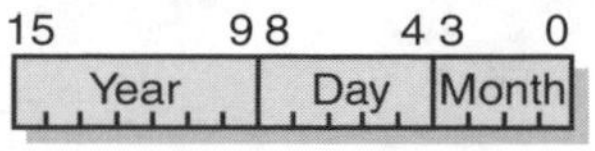

***Figure 486** How C represents the **Date** bit-field structure.*

487 UNDERSTANDING A BITWISE STRUCTURE'S RANGE OF VALUES

In Tip 486 you learned that C lets you represent bits within a value using a bit-field structure. When you create a bit-field structure, you must allocate enough bits to hold each member's desired value. To help you determine the number of bits you require, Table 487 specifies the range of values that a given number of bits can represent.

Size of Field	Range of Values
1	0–1
2	0–3
3	0–7
4	0–15
5	0–31
6	0–63
7	0–127
8	0–255
9	0–511
10	0–1023
11	0–2047
12	0–4095
13	0–8191
14	0–16383
15	0–32767
16	0–65535

Table 487 *The range of values your programs can represent with a given number of bits.*

488 SEARCHING AN ARRAY FOR A SPECIFIC VALUE

As you have learned, arrays let you store related values of the same type. At times, you may want to search an array for a specific value. There are two common ways to search an array: a *sequential search* and a *binary search*. To perform a sequential search, your program starts at the array's first element and searches one element at a time until the program finds the desired value or until the program reaches the last element in the array. For example, the following *while* loop illustrates how your programs might search an array for the value 1,500:

```
found = 0;
i = 0;

while ((i < ARRAY_ELEMENTS) && (! found))
  if (array[i] == 1500)
    found = true;
  else
    i++;

if (i < ARRAY_ELEMENTS)
  printf("Value found at element %d\n", i);
else
  printf("Value not found\n");
```

If you have previously sorted the values in an array from lowest to highest, your programs can perform a binary search, which you will learn more about in Tip 489.

UNDERSTANDING A BINARY SEARCH 489

As you have learned, one way of locating a value within an array is to search through every array element. Although such a sequential search is acceptable when your array size is small, looping through a large array can be time consuming. If your program has already sorted the values in the array from lowest to highest, it can use a *binary search* to locate the value. This type of search is called a binary search because with each operation the search divides the number of values it must examine by two.

The best way to conceptualize the binary search is to think of how you look up a word in the dictionary. Assume you want to find the word "Dalmatian." To begin, you may open the dictionary to the middle and examine the words on the page. Assuming that you open to the letter M, you know that "Dalmatian" appears before the current page, so you have just eliminated more than half the words in the dictionary. If you turn to the middle of the remaining pages, you will very likely find words that begin with the letter F. Again, you can discard half the possible choices, and continue your search in the pages that precede the current page. This time when you turn to the middle page, you will probably turn to the letter C. The word "Dalmatian" appears somewhere in the pages between C and F. When you select the middle page, you will likely be in the D words. By repeatedly discarding pages and selecting the middle page, you can quickly close in on the page containing the word "Dalmatian."

Note: *To perform a binary search, your program must sort the values in the array either from lowest to highest or from highest to lowest before you try the search.*

USING A BINARY SEARCH 490

As you learned in Tip 489, a binary search provides a quick way to search a sorted array for a specific value. The following program, *binary.c*, uses a binary search to search for several values in the array *count*, which contains the values 1 to 100. To help you better understand the processing the binary search performs, the function *binary_search* will print out messages that describe its processing:

```
#include <stdio.h>

int binary_search(int array[], int value, int size)
 {
   int found = 0;
   int high = size, low = 0, mid;

   mid = (high + low) / 2;
   printf("\n\nLooking for %d\n", value);
   while ((! found) && (high >= low))
     {
       printf("Low %d Mid %d High %d\n", low, mid, high);
       if (value == array[mid])
         found = 1;
       else if (value < array[mid])
         high = mid - 1;
       else
         low = mid + 1;
       mid = (high + low) / 2;
     }
   return((found) ? mid: -1);
 }

void main(void)
 {
   int array[100], i;

   for (i = 0; i < 100; i++)
     array[i] = i;
```

```
   printf("Result of search %d\n", binary_search(array, 33, 100));
   printf("Result of search %d\n", binary_search(array, 75, 100));
   printf("Result of search %d\n", binary_search(array, 1, 100));
   printf("Result of search %d\n", binary_search(array, 1001, 100));
 }
```

Compile and execute the *binary.c* program and observe the number of operations the search must perform to find each value. The program uses the variables *high*, *mid*, and *low* to keep track of the range of values it is currently searching.

491 SORTING AN ARRAY

As you have learned, arrays let you store related values of the same type. As your programs work with arrays, there will be times when your programs must sort an array's values, either from lowest to highest (ascending order) or from highest to lowest (descending order). Your programs can use several different sorting algorithms to sort arrays, including the *bubble sort*, the *selection sort*, the *Shell sort*, and the *quick sort*. Several Tips that follow discuss each of these sorting methods.

492 UNDERSTANDING THE BUBBLE SORT

The *bubble sort* algorithm is a simple array-sorting technique that is usually the first method most programmers learn. Because of its simplicity, the bubble sort is not very efficient and will consume more processor time than other sorting techniques. However, if you are sorting small arrays with 30 or fewer elements, using the bubble sort is fine. Assuming that you sort values from lowest to highest, the bubble sort loops through the values in an array, comparing and moving the largest array value to the top of the array (like a bubble in water rises to the surface). Figure 492 illustrates four iterations of a bubble sort.

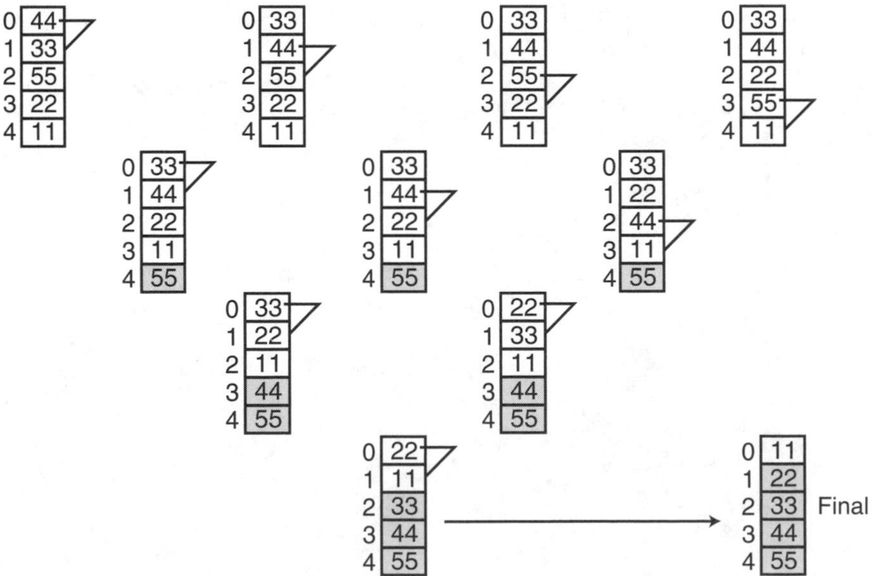

***Figure 492** Four iterations of a bubble sort.*

The first iteration moves the array's largest value to the top of the array. The second iteration moves the array's second largest value to the second-to-the-top position. The third iteration moves the third-largest value, and so on.

PUTTING A BUBBLE SORT TO USE 493

Tip 492 briefly illustrated how the bubble sort functions. The following program, *bubble.c*, uses the bubble sort to sort an array containing 30 random values:

```
#include <stdio.h>
#include <stdlib.h>

void bubble_sort(int array[], int size)
 {
   int temp, i, j;

   for (i = 0; i < size; i++)
    for (j = 0; j < size; j++)
      if (array[i] < array[j])
        {
          temp = array[i];
          array[i] = array[j];
          array[j] = temp;
        }
 }

void main(void)
 {
   int values[30], i;

   for (i = 0; i < 30; i++)
     values[i] = rand() % 100;
   bubble_sort(values, 30);
   for (i = 0; i < 30; i++)
     printf("%d ", values[i]);
 }
```

***Note:** The **bubble_sort** function sorts values from lowest to highest. To reverse the sort order, simply change the comparison to if (array[i] > array[j]).*

UNDERSTANDING THE SELECTION SORT 494

The *selection sort* is a simple sorting algorithm similar to the bubble sort Tip 492 presented. Like the bubble sort, your programs should only use the selection sort to sort small arrays (30 elements or fewer). The selection sort begins by selecting an array element (such as the first element). The sort then searches the entire array until it finds the minimum value. The sort places the minimum value in the element, selects the second element, and searches for the second smallest element. Figure 494 illustrates two iterations of the selection sort on an array of values.

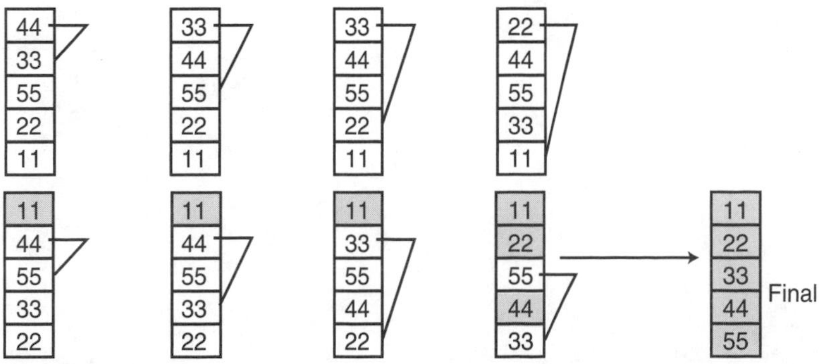

***Figure 494** Sorting values with the selection sort.*

495 PUTTING A SELECTION SORT TO USE

Tip 494 briefly illustrated the selection sort's functioning. The following program, *select.c*, uses the selection sort to sort an array containing 30 random values:

```
#include <stdio.h>
#include <stdlib.h>

void selection_sort(int array[], int size)
 {
   int temp, current, j;

   for (current = 0; current < size; current++)
    for (j = current + 1; j < size; j++)
      if (array[current] > array[j])
        {
          temp = array[current];
          array[current] = array[j];
          array[j] = temp;
        }
 }

void main(void)
 {
   int values[30], i;

   for (i = 0; i < 30; i++)
     values[i] = rand() % 100;
   selection_sort(values, 30);
   for (i = 0; i < 30; i++)
     printf("%d ", values[i]);
 }
```

Note: *The* ***selection_sort*** *function sorts values from lowest to highest. To reverse the sort order, simply change the comparison to* ***if (array[current] < array[j])****.*

496 UNDERSTANDING THE SHELL SORT

The *Shell sort* is named after its creator, Donald Shell. The Shell sort technique compares array elements separated by a specific distance (known as a *gap*) until the elements it compares with the current gap are in order. The Shell sort then divides the gap by two, and the process continues. When the gap is finally one and no changes occur, the Shell sort has completed its processing. Figure 496 illustrates how the Shell sort might sort an array.

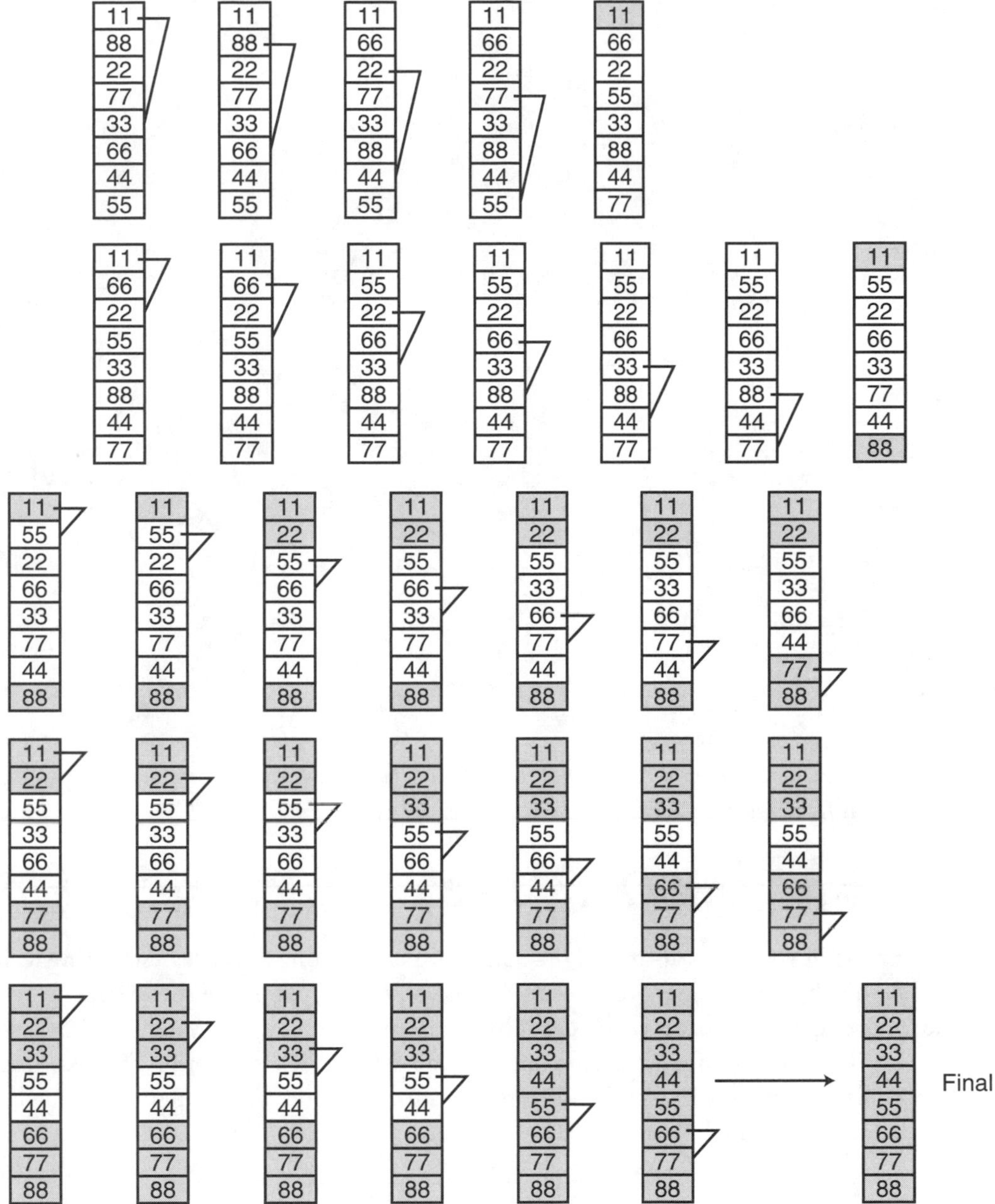

Figure 496 Sorting an array with the Shell sort.

PUTTING A SHELL SORT TO USE 497

Tip 496 briefly illustrated the Shell sort's functioning. The following program, *shell.c*, uses the Shell sort to sort an array containing 50 random values:

```
#include <stdio.h>
#include <stdlib.h>

void shell_sort(int array[], int size)
 {
   int temp, gap, i, exchange_occurred;

   gap = size / 2;
```

```
   do
    {
      do
       {
          exchange_occurred = 0;
          for (i = 0; i < size - gap; i++)
           if (array[i] > array[i + gap])
            {
              temp = array[i];
              array[i] = array[i + gap];
              array[i + gap] = temp;
              exchange_occurred = 1;
            }
       }
      while (exchange_occurred);
    }
   while (gap = gap / 2);
 }

void main(void)
 {
   int values[50], i;

   for (i = 0; i < 50; i++)
     values[i] = rand() % 100;
   shell_sort(values, 50);
   for (i = 0; i < 50; i++)
     printf("%d ", values[i]);
 }
```

Note: *The **shell_sort** function sorts values from lowest to highest. To reverse the sort order, simply change the comparison to **if (array[i] < array[i + gap])**.*

498 Understanding the Quick Sort

As the number of elements in your array increases, the *quick sort* becomes one of the fastest sorting techniques your programs can use. The quick sort considers your array as a list of values. When the sort begins, it selects the list's middle value as the *list separator*. The sort then divides the list into two lists, one with values that are less than the list separator and a second list whose values are greater than or equal to the list separator. The sort then recursively invokes itself with both lists. Each time the sort invokes itself, it further divides the elements into smaller lists. Figure 498 illustrates how the quick sort might sort an array of 10 values.

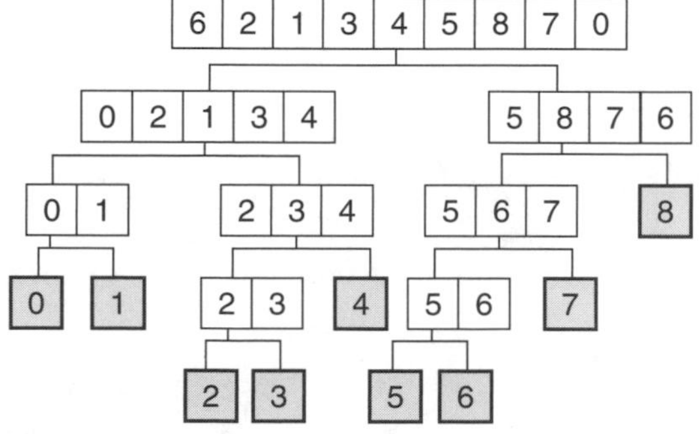

***Figure 498** Sorting values with the quick sort.*

PUTTING A QUICK SORT TO USE 499

Tip 498 briefly illustrated how the quick sort functions. The following program, *quick.c*, uses the quick sort to sort an array containing 100 random values:

```
#include <stdio.h>
#include <stdlib.h>

void quick_sort(int array[], int first, int last)
 {
   int temp, low, high, list_separator;

   low = first;
   high = last;
   list_separator = array[(first + last) / 2];
   do
    {
      while (array[low] < list_separator)
        low++;
      while (array[high] > list_separator)
        high--;
      if (low <= high)
       {
         temp = array[low];
         array[low++] = array[high];
         array[high--] = temp;
       }
    }
   while (low <= high);
   if (first < high)
     quick_sort(array, first, high);
   if (low < last)
     quick_sort(array, low, last);
 }

void main(void)
 {
   int values[100], i;

   for (i = 0; i < 100; i++)
     values[i] = rand() % 100;
   quick_sort(values, 0, 99);
   for (i = 0; i < 100; i++)
     printf("%d ", values[i]);
 }
```

Note: *The **quick_sort** function sorts values from lowest to highest. To reverse the sort order, change the comparisons in the two **while** statements, as shown here:*

```
while (array[low] > list_separator)
  low++;

while (array[high] < list_separator)
  high++;
```

PROBLEMS WITH PREVIOUS SORTING SOLUTIONS 500

Several previous Tips have shown different sorting techniques your programs can use to sort arrays. However, each of the Tips presented worked with arrays of type *int*. If your programs need to sort a different array type, you must create new functions. For example, to sort an array of type *float*, your programs must change the *quick_sort* function's header and variable declarations, as shown here:

```
void quick_sort(float array[], int first, int last)
 {
   float temp, list_separator;
   int low, high;
```

If you want to sort an array of *long* values later, you must create a different function. As you will learn, however, your programs can use the C run-time library *qsort* function to sort different array types. The *qsort* function uses memory indirection to sort values of all types.

501 SORTING AN ARRAY OF CHARACTER STRINGS

As you have learned, C lets you create an array of character strings, as shown here:

```
char *days[] = {"Monday", "Tuesday", "Wednesday" };
```

Just as there may be times when your programs must sort arrays of other types, the same is true for sorting character-string arrays. The following program, *str_sort.c*, uses a bubble sort to sort an array of character strings:

```
#include <stdio.h>
#include <stdlib.h>
#include <string.h>

void bubble_sort(char *array[], int size)
 {
   char *temp;
   int i, j;

   for (i = 0; i < size; i++)
    for (j = 0; j < size; j++)
      if (strcmp(array[i], array[j]) < 0)
        {
          temp = array[i];
          array[i] = array[j];
          array[j] = temp;
        }
  }

void main(void)
 {
   char *values[] = {"AAA", "CCC", "BBB", "EEE", "DDD"};
   int i;

   bubble_sort(values, 5);
   for (i = 0; i < 5; i++)
     printf("%s ", values[i]);
 }
```

When the function sorts the array of character strings, the function does not change the string contents to rearrange the array; rather, it arranges the character string pointers so the character strings are in order.

502 SEARCHING AN ARRAY WITH LFIND

As you have learned, a sequential search operation searches the elements of an array in order until it finds a specific value. To help your programs search arrays of any type, the C run-time library provides the *lfind* function:

```
#include <stdlib.h>

void *lfind(const void *element, void *base, size_t *number_of_entries,
            size_t element_width, int (*compare)(const void *, const void *));
```

As you can see, the function makes tremendous use of pointers. The *element* parameter is a pointer to the desired value. The *base* parameter is a pointer to the start of the array. The *number_of_entries* parameter is a pointer to the number of elements in the array. The *element_width* parameter specifies the number of bytes required for each array element. Finally, the *compare* parameter is a pointer to a second function that compares two array elements. Unlike the functions previously shown, which returned an array index to the desired value, the *lfind* function returns a pointer to the desired value or the value 0 if *lfind* did not find the element. The following program, *lfind.c*, uses the *lfind* function to search for a value of type *int* and a value of type *float*:

```
#include <stdlib.h>
#include <stdio.h>

int compare_int(int *a, int *b)
 {
   return(*a - *b);
 }

int compare_float(float *a, float *b)
 {
   return((*a == *b) ? 0: 1);
 }

void main(void)
 {
   int int_values[] = {1, 3, 2, 4, 5};
   float float_values[] = {1.1, 3.3, 2.2, 4.4, 5.5};
   int *int_ptr, int_value = 2, elements = 5;
   float *float_ptr, float_value = 33.3;

   int_ptr = (int *) lfind(&int_value, int_values, &elements, sizeof(int),
                     (int (*) (const void *, const void *)) compare_int);
   if (int_ptr)
     printf("Value %d found\n", int_value);
   else
     printf("Value %d not found\n", int_value);
   float_ptr = (float *) lfind(&float_value, float_values, &elements, sizeof(float),
                     (int (*) (const void *, const void *)) compare_float);
   if (float_ptr)
     printf("Value %3.1f found\n", float_value);
   else
     printf("Value %3.1f not found\n", float_value);
 }
```

Using pointers, the *lfind* function is able to eliminate the type conflicts that affected the searching and sorting functions previously discussed.

SEARCHING FOR VALUES WITH LSEARCH 503

In Tip 502 you learned how to use the *lfind* function to search an array of values for a specific element. If the function found the element, the function returned a pointer to the element. If the function did not find the element, the function returned 0. Depending on your programs, there may be times when you will want to add the element to the array if the function does not find it. In such cases, your programs can use the *lsearch* function, as shown here:

```
#include <stdlib.h>

void *lsearch(const void *element, void *base, size_t *number_of_entries,
            size_t element_width, int (*compare)(const void *, const void *));
```

The following program, *lsearch.c*, uses the *lsearch* function to search for the value 1500. If *lsearch.c* does not find the value, the *lsearch* function appends the value to the array:

```
#include <stdlib.h>
#include <stdio.h>

int compare_int(int *a, int *b)
 {
   return(*a - *b);
 }

void main(void)
 {
   int int_values[10] = {1, 3, 2, 4, 5};
   int *int_ptr, int_value = 1500, elements = 5, i;

   printf("Array contents before search\n");
   for (i = 0; i < elements; i++)
     printf("%d ", int_values[i]);
   int_ptr = (int *) lsearch(&int_value, int_values, &elements, sizeof(int),
                    (int (*) (const void *, const void *)) compare_int);
   printf("\nArray contents after search\n");
   for (i = 0; i < elements; i++)
     printf("%d ", int_values[i]);
 }
```

As you can see, when the function adds the array value it also updates the value parameter that specifies the number of array elements.

Note: *When your programs use the* ***lsearch*** *function, you must include additional space within the array into which you can append values.*

504 SEARCHING A SORTED ARRAY WITH BSEARCH

In Tip 489 you learned that a binary search locates a value in a sorted array by repeatedly reducing the number of array elements it will continue by a factor of two with each iteration. To help your programs perform binary search operations, the C run-time library provides the *bsearch* function:

```
#include <stdlib.h>

void *bsearch(const void *key, const void *base, size_t number_of_elements,
              size_t width, int (*compare)(const void *, const void *));
```

Like the *lsearch* function you learned about in Tip 503, the *bsearch* function makes extensive use of pointers. The *element* parameter is a pointer to the desired value. The *base* parameter is a pointer to the start of the array. The *number_of_entries* parameter specifies the number of elements in the array. The *element_width* parameter specifies the number of bytes required for each array element. Finally, the *compare* parameter is a pointer to a second function, which compares two array elements. Unlike the functions previously shown, which returned an array index to the desired value, the *bsearch* function returns a pointer to the desired value or the value 0 if the *bsearch* function did not find the element. The following program, *bsearch.c*, uses the *bsearch* function to search two different arrays, one for a value of type *int* and one for a value of type *float*:

```
#include <stdlib.h>
#include <stdio.h>

int compare_int(int *a, int *b)
 {
   return(*a - *b);
 }

int compare_float(float *a, float *b)
 {
   return((*a == *b) ? 0: 1);
 }
```

```
void main(void)
 {
   int int_values[] = {1, 3, 2, 4, 5};
   float float_values[] = {1.1, 3.3, 2.2, 4.4, 5.5};
   int *int_ptr, int_value = 2, elements = 5;
   float *float_ptr, float_value = 33.3;

   int_ptr = (int *) bsearch(&int_value, int_values, elements, sizeof(int),
                        (int (*) (const void *, const void *)) compare_int);
   if (int_ptr)
     printf("Value %d found\n", int_value);
   else
     printf("Value %d not found\n", int_value);
   float_ptr = (float *) bsearch(&float_value, float_values, elements, sizeof(float),
                        (int (*) (const void *, const void *)) compare_float);
   if (float_ptr)
     printf("Value %3.1f found\n", float_value);
   else
     printf("Value %3.1f not found\n", float_value);
 }
```

Note: *To use the* ***bsearch*** *function, the array values must be sorted from lowest to highest.*

SORTING ARRAYS WITH QSORT 505

In Tip 498 you learned that a quick sort operation sorts array elements by treating an array as a list. Because the quick sort repeatedly separates elements into smaller sorted lists, it is very efficient. To help your programs sort arrays of all types using a quick sort, the C run-time library provides the *qsort* function, as shown here:

```
#include <stdlib.h>

void *qsort(void *base, size_t number_of_entries, size_t element_width,
            int (*compare)(const void *, const void *));
```

Like the *lsearch* and *bsearch* functions you have already learned about, the *qsort* function makes extensive use of pointers. The *base* parameter is a pointer to the start of the array. The *number_of_entries* parameter specifies the number of elements in the array. The *element_width* parameter specifies the number of bytes required for each array element. Finally, the *compare* parameter is a pointer to a second function, which compares two array elements and returns a value, as shown here:

```
*a < *b          // Return value < 0
*a == *b         // Return 0
*a > *b          // Value > 0
```

The following program, *qsort.c*, uses the *qsort* function to search for a value of type *int* and a value of type *float*:

```
#include <stdlib.h>
#include <stdio.h>

int compare_int(int *a, int *b)
 {
   if (*a < *b)
     return(-1);
   else if (*a == *b)
     return(0);
   else
     return(1);
 }

int compare_float(float *a, float *b)
 {
```

```
    if (*a < *b)
      return(-1);
    else if (*a == *b)
      return(0);
    else
      return(1);
  }

void main(void)
  {
    int int_values[] = {51, 23, 2, 44, 45};
    float float_values[] = {21.1, 13.3, 22.2, 34.4, 15.5};
    int elements = 5, i;

    qsort(int_values, elements, sizeof(int),
          (int (*) (const void *, const void *)) compare_int);
    for (i = 0; i < elements; i++)
      printf("%d ", int_values[i]);
    putchar('\n');
    qsort(float_values, elements, sizeof(float),
          (int (*) (const void *, const void *)) compare_float);
    for (i = 0; i < elements; i++)
      printf("%4.1f ", float_values[i]);
  }
```

506 DETERMINING THE NUMBER OF ARRAY ELEMENTS

Several of the previous Tips have included the number of array elements as a function parameter. If the number of elements in your array might change, you can reduce the number of changes you must make to your program by using a constant value, such as the following:

```
#define NUM_ELEMENTS = 5
```

Alternately, your programs can use the *sizeof* operator to determine the number of elements in an array, as shown here:

```
elements = sizeof(array) / sizeof(array[0]);
```

The following program, *num_els.c*, uses the *sizeof* operator to derive the number of elements in two arrays and then display the number of elements in those arrays:

```
#include <stdio.h>

void main(void)
  {
    int int_values[] = {51, 23, 2, 44, 45};
    float float_values[] = {21.1, 13.3, 22.2, 34.4, 15.5};

    printf("Number of elements in int_values %d\n",
           sizeof(int_values) / sizeof(int_values[0]));
    printf("Number of elements in float_values %d\n",
           sizeof(float_values) / sizeof(float_values[0]));
  }
```

507 UNDERSTANDING POINTERS AS ADDRESSES

As you learned in the Getting Started with C section of this book, a *variable* is the name of a location in memory that is capable of storing a value of a specific type. Your program references each location in memory by using a unique *address*. A *pointer* is a variable or value that contains an address. The C programming language makes extensive use of pointers. When you pass arrays or strings to functions, the C compiler passes a pointer. Likewise, when a function must change a parameter's value, the program must pass to the function a pointer to the variable's memory address. Several of the Tips that follow examine pointers in detail.

DETERMINING A VARIABLE'S ADDRESS 508

A pointer is an address to a location in memory. When your program works with arrays (and strings), the program works with a pointer to the array's first element. When your programs need to determine a variable's address, your programs must use the C *address* operator ampersand (&). For example, the following program, *address.c*, uses the *address* operator to display the address of several different variables:

```
#include <stdio.h>

void main(void)
 {
   int count = 1;
   float salary = 40000.0;
   long distance = 1234567L;

   printf("Address of count is %x\n", &count);
   printf("Address of salary is %x\n", &salary);
   printf("Address of distance is %x\n", &distance);
 }
```

When you compile and execute the *address.c* program, your screen will display the following output:

```
Address of count is fff4
Address of salary is fff0
Address of distance is ffec
C:\>
```

UNDERSTANDING HOW C TREATS ARRAYS AS POINTERS 509

As you have already learned, the C compiler treats arrays as pointers. When your program passes an array to a function, for example, the compiler passes the array's starting address. The following program, *arrayadd.c*, displays the starting address of several different arrays:

```
#include <stdio.h>

void main(void)
 {
   int count[10];
   float salaries[5];
   long distances[10];

   printf("Address of the array count is %x\n", count);
   printf("Address of the array salaries is %x\n", salaries);
   printf("Address of the array distances is %x\n", distances);
 }
```

When you compile and execute the *arrayadd.c* program, your screen will display the following output:

```
Address of the array count is ffe2
Address of the array salaries is ffce
Address of the array distances is ffa6
C:\>
```

510 APPLYING THE ADDRESS OPERATOR (&) TO AN ARRAY

As you have learned, the C compiler treats an array as a pointer to the array's first element. In Tip 508, you learned that C uses the *address* operator (&) to return a variable's address. If you apply the *address* operator to an array, C will return the array's starting address. Therefore, applying the *address* operator to an array is redundant. The following program, *arraytwo.c*, displays an array's starting address, followed by the pointer that C's *address* operator returns:

```
#include <stdio.h>

void main(void)
 {
   int count[10];
   float salaries[5];
   long distances[10];

   printf("Address of the array count is %x &count is %x\n", count, &count);
   printf("Address of the array salaries is %x &salaries is %x\n",
          salaries, &salaries);
   printf("Address of the array distances is %x &distances is %x\n",
         distances, &distances);
 }
```

When you compile and execute the *arraytwo.c* program, your screen will display the following output:

```
Address of the array count is ffe2 &count is ffe2
Address of the array salaries is ffce &count is ffce
Address of the array distances is ffa6 &distances is ffa6
C:\>
```

511 DECLARING POINTER VARIABLES

As your programs become more complex, you will find that you work with pointers on a regular basis. In order to store pointers, your program must declare pointer variables. To declare a pointer you must specify the type of value to which the pointer points (such as *int, float, char* and so on), and an asterisk (*) before the variable name. For example, the following statement declares a pointer to a value of type *int*:

```
int *iptr;
```

Like any variable, you must assign a value to a pointer variable before you can use the pointer within your program. When you assign a value to a pointer, you really assign an address. Assuming that you have previously declared *int count;* the following statement assigns the address of the variable *count* to the pointer *iptr*:

```
iptr = &count;    // Assign count's address to iptr
```

The following program, *iptr.c*, declares the pointer variable *iptr* and assigns to the pointer the address of the variable *count*. The program then displays the pointer variable's value, along with the address of *count*:

```
#include <stdio.h>

void main(void)
 {
   int *iptr;        // Declare pointer variable
   int count = 1;

   iptr = &count;
   printf("Value of iptr %x Value of count %d Address of count %x\n",
          iptr, count, &count);
 }
```

When you compile and execute the *iptr.c* program, your screen will display the following output:

```
Value of iptr fff2 Value of count 1 Address of count fff2
C:\>
```

DEREFERENCING A POINTER 512

As you have learned, a pointer contains an address that points to a value of a specific type. Using the address a pointer contains, you can determine the value in memory to which the pointer points. *Dereferencing a pointer* is the process of accessing the value at a specific memory location. To dereference a pointer's value, you use the *asterisk* (*) *indirection* operator. For example, the following *printf* statement displays the value the integer pointer *iptr* points to:

```
printf("Value pointed to by iptr is %d\n", *iptr);
```

Likewise, the following statement assigns to the variable *count* the value the variable *iptr* points to:

```
count = *iptr;
```

Finally, the next statement assigns the value 7 to the memory location *iptr* points to:

```
*iptr = 7;
```

Note: *To use the value stored in the memory location pointed to by a pointer, you must dereference the pointer's value using the* ***asterisk (*) indirection*** *operator.*

USING POINTER VALUES 513

In Tip 510, you learned that you can assign an address to a pointer variable using the ampersand (&) *address* operator. In Tip 512 you learned that to access the value stored in memory at the memory location pointed to by a pointer, you must use the *asterisk (*) indirection* operator. The following program, *ptr_demo.c*, assigns the *int* pointer *iptr* the address of the variable *counter*. The program then displays the pointer's value and the value stored in the location at which the pointer points (the *counter's* value). The program then changes the value pointed to by the pointer, as shown here:

```
#include <stdio.h>

void main(void)
 {
   int counter = 10;
   int *iptr;                                      // Declare pointer value

   iptr = &counter;                                // Assign the address
   printf("Address in iptr %x Value at *iptr %d\n", iptr, *iptr);
   *iptr = 25;                                     // Change the value in memory
   printf("Value of counter %d\n", counter);
 }
```

USING POINTERS WITH FUNCTION PARAMETERS 514

The Functions section of this book examines in detail the process of passing parameters to functions. As you will learn, when you need to change the value of a parameter, you must pass to the function a pointer to a parameter. The following program, *swapvals.c*, uses the pointers to two parameters of type *int* to exchange (swap) the variable's values, as shown here:

```
#include <stdio.h>

void swap_values(int *a, int *b)
```

```
{
   int temp;

   temp = *a;   // Temporarily hold the value pointed to by a
   *a = *b;     // Assign b's value to a
   *b = temp;   // Assign a's value to b
 }

void main(void)
 {
   int one = 1, two = 2;

   swap_values(&one, &two);
   printf("one contains %d two contains %d\n", one, two);
 }
```

As you can see, within the function the statements dereference the pointers using the *indirection* operator (*). The program passes each variable's address to the function using the *address* operator (&).

515 UNDERSTANDING POINTER ARITHMETIC

A pointer is an address that points to a value of a specific type in memory. In the simplest sense, a pointer is a value that points to a specific memory location. If you add the value 1 to a pointer, the pointer will point to the next location in memory. If you add 5 to a pointer's value, the pointer will point to the memory location five locations ahead of the current address. However, pointer arithmetic is not quite as simple as you might guess. For example, assume that a pointer contains the address 1000. If you add 1 to the pointer, you would expect the result to be the address 1001. However, the resulting address depends on the pointer's type. For example, if you add 1 to a pointer variable of type *char* (which contains 1000), the resulting address is 1001. If you add 1 to a pointer of type *int* (which requires two bytes of memory), the resulting address is 1002. Likewise, if you add 1 to a pointer of type *float* (which requires four bytes), the resulting address is 1004. When you perform pointer arithmetic, keep in mind the pointer's type. In addition to adding values to pointers, your programs can subtract values or add and subtract two pointers. Several Tips in this section present various pointer arithmetic operations.

516 INCREMENTING AND DECREMENTING A POINTER

As your programs work with pointers, one of the most common operations they will perform is to increment and decrement a pointer's value to point to the next or previous memory location. The following program, *ptrarray.c*, assigns the starting address of an array of integer values to the pointer *iptr*. The program then increments the pointer's value to display the five elements the array contains:

```
#include <stdio.h>

void main(void)
 {
   int values[5] = {1, 2, 3, 4, 5};
   int counter;
   int *iptr;

   iptr = values;
   for (counter = 0; counter < 5; counter++)
     {
       printf("%d\n", *iptr);
       iptr++;
     }
 }
```

When you compile and execute the *ptrarray.c* program, your screen will display the values 1 through 5. The program initially assigns the pointer the array's starting address. The program then increments the pointer to point to each element.

COMBINING A POINTER REFERENCE AND INCREMENT 517

In Tip 516 you used the pointer *iptr* to display the contents of an array. To display the array contents, the pointer used a *for* loop, as shown here:

```
for (counter = 0; counter < 5; counter++)
 {
   printf("%d\n", *iptr);
   iptr++;
 }
```

As you can see, the *for* loop accesses the pointer's value on one line and then increments the pointer on the next. As you have learned, you can use C's *postfix increment* operator to use a variable's value and then increment the value. The following *for* loop uses the *postfix increment* operator to reference the value pointed to by the pointer variable and then increments the pointer's value:

```
for (counter = 0; counter < 5; counter++)
  printf("%d\n", *iptr++);
```

LOOPING THROUGH A STRING USING A POINTER 518

The Strings section of this book makes extensive use of pointers. As you have learned, a string is a *NULL*-terminated character array. The following program, *str_ptr.c*, uses the function *show_string* to display a character string using a pointer:

```
#include <stdio.h>

void show_string(char *string)
 {
   while (*string)
     putchar(*string++);
 }

void main(void)
 {
   show_string("Jamsa\'s C/C++/C# Programmer\'s Bible");
 }
```

As you can see, the *show_string* function declares the *string* variable as a pointer. Using the pointer, the function simply loops through the string's characters until it encounters the *NULL* character. To display the character, the *show_string* function first dereferences the pointer's address (getting the character). Then the function increments the pointer to point to the next character in the string.

USING FUNCTIONS THAT RETURN POINTERS 519

As you have learned, functions can return a value to your programs. The value that a function returns is always of the type declared in the function's prototype or header (such as *int*, *float*, or *char*). In addition to returning these basic types, functions can declare pointers to values. For example, the *fopen* function, which most C programs use to open a file stream, returns a pointer to a structure of type *FILE*, as shown here:

```
FILE *fopen(const char *pathname, const char *mode);
```

In a similar way, many of the functions presented in the Strings section of this book return pointers to character strings. As you examine the function prototypes presented in this book, note the functions that return a *pointer to a value*, as opposed to a value of a basic type.

520 CREATING A FUNCTION THAT RETURNS A POINTER

In Tip 519 you learned that many of C's run-time library functions return pointers. As your programs become more complex, you will create functions that return pointers to specific types. For example, the following program, *ptr_upr.c*, creates a function named *string_uppercase* that converts each character of a string to uppercase and then returns a pointer to the string:

```
#include <stdio.h>
#include <ctype.h>

char *string_uppercase(char *string)
 {
   char *starting_address;

   starting_address = string;
   while (*string)
     toupper(*string++);
   return(starting_address);
 }

void main(void)
 {
   char *title = "Jamsa\'s C/C++/C# Programmer\'s Bible";
   char *string;

   string = string_uppercase(title);
   printf("%s\n", string);
   printf("%s\n", string_uppercase("Arrays and Pointers"));
 }
```

As you can see, to create a function that returns a pointer, you simply place the asterisk before the function name, as shown here:

```
char *string_uppercase(char *string);
```

521 UNDERSTANDING AN ARRAY OF POINTERS

Several Tips presented in this section have discussed arrays in detail. So far, all the arrays have used C's basic types (such as *int*, *float*, or *char*); however, C does not restrict arrays to such simple types. Just as you can create functions that return pointers, so too can you create arrays of pointers. You will most commonly use arrays of pointers to hold character strings. As an example, the following declaration creates an array, named *weekdays*, that contains pointers to character strings:

```
char *weekdays[7] = {"Sunday", "Monday", "Tuesday", "Wednesday", "Thursday",
                     "Friday", "Saturday"};
```

If you examine the array's type from right to left, you will note that the array contains seven elements. The asterisk before the variable name specifies a pointer. If you combine the type name *char* that precedes the variable name, the declaration will become an *array of pointers to character strings* (in the previous example there are seven strings). One of the most widely used arrays of pointers to character strings is *argv*, which contains your program's command line, as this book's Command Line section details.

Note: *When you declare an array of pointers to character strings, the C compiler does not append a NULL entry to indicate the end of the array as it does for character strings.*

VISUALIZING AN ARRAY OF CHARACTER STRINGS 522

As you have learned, C treats an array as a pointer to the array's starting location in memory. In Tip 521 you created a character string array named *weekdays* that contained the days of the week. When you create an array of character strings, the C compiler stores pointers to the array's strings. Figure 522 illustrates how the C compiler would store the array *letters*, as shown here:

```
char *letters[4] = {"AAA", "BBB", "CCC", "DDD"};
```

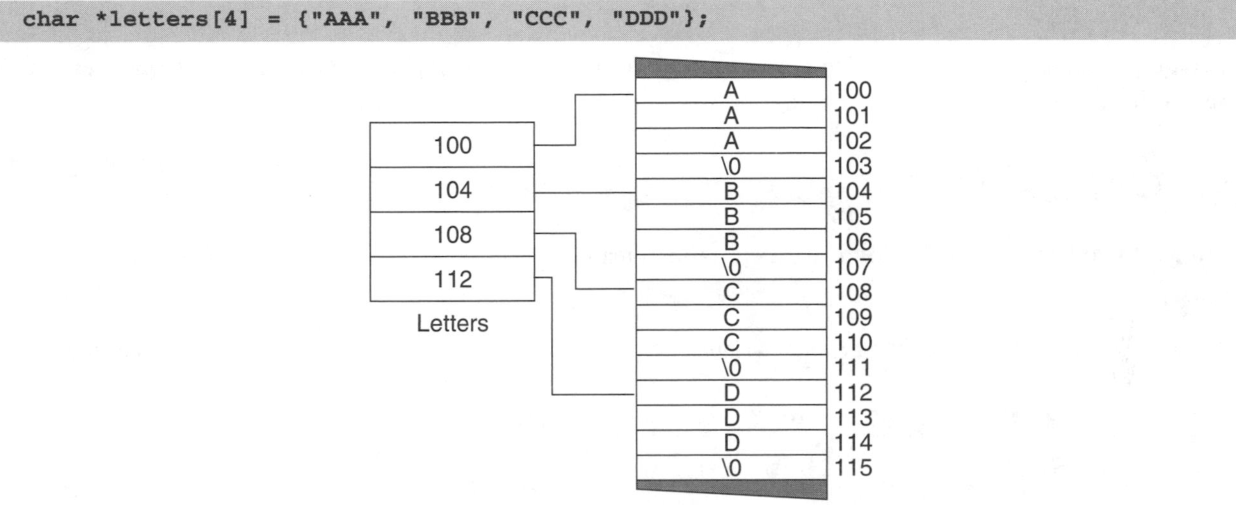

***Figure 522** C stores character string arrays as an array of pointers.*

Note: *When you declare an array of character strings, the C compiler does not append a NULL entry to indicate the end of the array as it does for character strings.*

LOOPING THROUGH AN ARRAY OF CHARACTER STRINGS 523

As you have learned, when you create an array of character strings, C stores pointers to each string within the array elements. The following program, *weekdays.c*, loops through the *weekdays* array, which contains pointers to strings that contain the names of the days of the week, as shown here:

```
#include <stdio.h>

void main(void)
 {
   char *weekdays[7] = {"Sunday", "Monday", "Tuesday", "Wednesday",
                        "Thursday", "Friday", "Saturday"};
   int i;

   for (i = 0; i < 7; i++)
     printf("weekdays[%d] contains %s\n", i, weekdays[i]);
 }
```

As you can see, the program simply loops through the array elements, using *printf's %s* format specifier.

TREATING A CHARACTER STRING ARRAY AS A POINTER 524

As you have learned, C treats an array as a pointer to the array's starting element in memory. Several Tips presented in the Strings section of this book access character string arrays using a pointer that is similar to the following:

```
char *string;
```

You have also learned that C lets you create arrays of character strings. The following declaration, for example, creates an array named *workdays* that can store the pointers to five character strings:

```
char *workdays[5];
```

Because the declaration creates an array, C lets you access the array using a pointer. To access the array using a pointer, you must declare a pointer variable that points to an array of character strings. In the case of the *workdays* array, the declaration of the reference pointer would become the following:

```
char **work_day_ptr;
```

The double asterisks, in this case, state that *work_day_ptr* is *a pointer to a pointer to a character string.* Several Tips presented in this book's Command Line section work with a pointer to a pointer to character strings.

525 USING A POINTER TO A POINTER TO CHARACTER STRINGS

In Tip 524, you learned that the following declaration creates a pointer to a pointer to a character string:

```
char **work_day_ptr;
```

The following C program, *workdays.c,* uses a pointer to a pointer to character strings to display the contents of the *workdays* array:

```
#include <stdio.h>

void main(void)
 {
   char *workdays[] = {"Monday", "Tuesday", "Wednesday", "Thursday", "Friday",
   char **work_day;

   work_day = workdays;
   while (*work_day)
     printf("%s\n", *work_day++);
 }
```

When the program begins, it assigns to the pointer *week_day* the starting address of the *workdays* array (the address of the string Monday). The program then loops until it encounters the *NULL* pointer (0, the ending condition).

Note: *When you declare an array of character strings, the C compiler does not append a* ***NULL*** *entry to indicate the end of the array as it does for character strings. Therefore, the declaration of the* ***workdays*** *array explicitly included a* ***NULL*** *value for which the program can test within the loop.*

526 DECLARING A STRING CONSTANT USING A POINTER

Several Tips presented throughout this book have initialized character strings at declaration, as shown here:

```
char title[] = "Jamsa\'s C/C++/C# Programmer\'s Bible";
```

When you declare an array with empty brackets, the C compiler allocates enough memory to hold the characters specified (and the *NULL* terminator), assigning to the variable *title* a pointer to the first character. Because the C compiler automatically allocates the necessary memory and then works with a pointer to the memory, your programs can use a character string pointer, as opposed to an array, as shown here:

```
char *title = "Jamsa\'s C/C++/C# Programmer\'s Bible";
```

UNDERSTANDING THE TYPE VOID POINTER 527

As you have learned, when you declare a pointer variable, you must specify the type of value to which the pointer points (such as *int*, *float*, or *char*). When you do so, the compiler can later correctly perform pointer arithmetic and add the correct offset values when you increment or decrement the pointer. In some cases, however, your programs will not manipulate a pointer's value in any way. Instead, your programs will only want to obtain a pointer to a memory location with which the program will determine the pointer's usage. In such instances, your programs can create a pointer to the type *void*, as shown here:

```
void *memory_pointer;
```

As you examine the C run-time library functions presented in the Memory section of this book, you will find that several of the functions return pointers to the type *void*. Such functions essentially tell you that they return a pointer to a memory location about which the compiler makes no assumptions as to the memory's contents or access.

CREATING POINTERS TO FUNCTIONS 528

As you have learned, C lets you create pointers to all data types (such as *int*, *char*, *float*, and even character strings). In addition, C lets your programs create and use pointers to functions. The most common use of pointers to functions is to let your programs pass a function as a parameter to another function. The following declarations create pointers to functions:

```
int (*min)();
int (*max)();
float (*average)();
```

Note the use of the parentheses around the variable names. If you removed the parentheses, the declarations would serve as function prototypes for functions that return pointers to a specific type, as shown here:

```
int *min();
int *max();
float *average();
```

When you read a variable declaration, begin with the innermost declaration that appears within parentheses and then work right to left:

```
int (*min)();
```

USING A POINTER TO A FUNCTION 529

In Tip 528 you learned that C lets you create pointers to functions. The most common use of a pointer to a function is to pass that function as a parameter to another function. Earlier in this section, you examined C's run-time library sorting and searching functions. As you learned, if you want to sort values from smallest to largest, you will pass a specific function to the run-time library routine. If you want to sort values from highest to lowest, you will pass a different function. The following program, *passfunc.c*, passes either the function *min* or *max* to the function *get_result*. Depending on the function the program passes, the value *get_result* returns will differ:

```
#include <stdio.h>

int get_result(int a, int b, int (*compare)())
 {
   return(compare(a, b));  // Invoke the function passed
 }
```

```
int max(int a, int b)
 {
   printf("In max\n");
   return((a > b) ? a: b);
 }
int min(int a, int b)
 {
   printf("In min\n");
   return((a < b) ? a: b);
 }
void main(void)
 {
   int result;
   result = get_result(1, 2, &max);
   printf("Max of 1 and 2 is %d\n", result);
   result = get_result(1, 2, &min);
   printf("Min of 1 and 2 is %d\n", result);
 }
```

530 USING A POINTER TO A POINTER TO A POINTER

As you have learned, C lets you create variables that are pointers to other pointers. In general, there is no limit on the number of *indirections* (pointers to pointers) your programs can use. However, for most C programmers, using more than a *pointer to a pointer* will result in considerable confusion and will make your program very difficult to understand. For example, the following program, *ptrtoptr.c*, uses three levels of pointers to a value of type *int*. Take your time with this program and draw the levels of indirection on a piece of paper until you understand the processing *ptrtoptr.c* performs:

```
#include <stdio.h>
int what_is_the_value(int ***ptr)
 {
   return(***ptr);
 }
void main(void)
 {
   int *level_1, **level_2, ***level_3, value = 1001;
   level_1 = &value;
   level_2 = &level_1;
   level_3 = &level_2;
   printf("The value is %d\n", what_is_the_value(level_3));
 }
```

531 UNDERSTANDING STRUCTURES

As you have learned, an array is a variable that lets your programs store multiple values of the same type. In other words, an array lets your programs group related information into a single variable, such as 100 test scores or 50 employee salaries. As your programs become more complex, there will be times when you want to group together related information that has differing types. For example, assume that you have a program that works with employee information. You might need to track the following information for each employee:

```
char name[64];
int age;
char ssan[11];    // Social security number
```

```
int pay_grade;
float salary;
unsigned employee_number;
```

Assume that you have several different functions in your program that work with employee information. Each time your program invokes a function, you must ensure that you have specified all the parameters, and in the correct order. As discussed in the Functions section of this book, the more parameters that your programs pass to functions, the harder your programs are to understand and the greater the chance of error. To reduce such complexity, your programs can create a *structure*, which groups the related information into one variable. For example, the following structure declaration creates a structure, named *Employee*, that contains the employee fields previously shown:

```
struct Employee
 {
   char name[64];
   int age;
   char ssan[11];    // Social security number
   int pay_grade;
   float salary;
   unsigned employee_number;
 };
```

As you will learn in the following Tips, this declaration creates a structure of type *Employee*.

A Structure Is a Template for Variable Declarations 532

In Tip 531 you learned that C lets you group related information into a structure. By itself, a structure definition does not create any variables. Instead, the definition specifies a template that your programs can later use to declare variables. Therefore, a structure definition does not allocate any memory. Rather, the compiler simply makes a note of the definition in case your program later declares a variable of the structure type.

A Structure Tag is the Structure's Name 533

In Tip 531 you learned that C lets you group related variables in a structure. Using the *struct* keyword, your programs can declare a structure, as shown here:

```
struct Employee
 {
   char name[64];
   int age;
   char ssan[11];    // Social security number
   int pay_grade;
   float salary;
   unsigned employee_number;
 };
```

In the previous example, the structure name is *Employee*. C programmers refer to the structure name as the structure's *tag*. As you will learn in Tip 534, your programs can use the structure tag to declare variables of a specific type. The following declaration creates a structure named *Shape*:

```
struct Shape
 {
   int type;      // 0 = circle, 1 = square, 2 = triangle
   int color;
   float radius;
   float area;
   float perimeter;
 };
```

534 DECLARING A STRUCTURE VARIABLE IN DIFFERENT WAYS

In Tip 531 you learned that C lets you group related information into a structure. As you have learned, a structure definition on its own does not create a usable variable. Rather, the definition simply serves as a template for future variable declarations. C provides two ways to declare variables of a specific structure type. First, assume that your program declares a structure of type *Employee*, as shown here:

```
struct Employee
 {
   char name[64];
   int age;
   char ssan[11];    // Social security number
   int pay_grade;
   float salary;
   unsigned employee_number;
 };
```

Following the structure definition, your programs can declare variables of type *Employee*, as shown here:

```
struct Employee employee_info;
struct Employee new_employee, terminated_employee;
```

Next, C lets you declare variables of a structure type following the structure definition, as shown here:

```
struct Employee
 {
   char name[64];
   int age;
   char ssan[11];    // Social security number
   int pay_grade;
   float salary;
   unsigned employee_number;
 } employee_info, new_employee, terminated_employee;
```

535 UNDERSTANDING STRUCTURE MEMBERS

As you have learned, C lets you group related information within structures. For example, the following statement creates a variable called *triangle* using the structure *Shape*:

```
struct Shape
 {
   int type;      // 0 = circle, 1 = square, 2 = triangle
   int color;
   float radius;
   float area;
   float perimeter;
 } triangle;
```

Each piece of information in the structure is a *member*. In the case of the *Shape* structure, there are five members: *type*, *color*, *radius*, *area*, and *perimeter*. To access a specific member, you use C's *dot* operator (.). As an example, the following statements assign values to different members of the *triangle variable.*

```
triangle.type = 2;
triangle.perimeter = 30.0;
triangle.area = 45.0;
```

VISUALIZING A STRUCTURE 536

As you have learned, C lets you group related information in structures. When you declare a variable of a specific structure type, C allocates enough memory to hold the values for each structure member. For example, if you declare a structure of type *Employee*, C will allocate memory as shown in Figure 536.

```
struct Employee{
    char name[64];
    int age;
    char ssan[11];
    int pay_grade;
    float salary;
    unsigned employee_number;
};
```

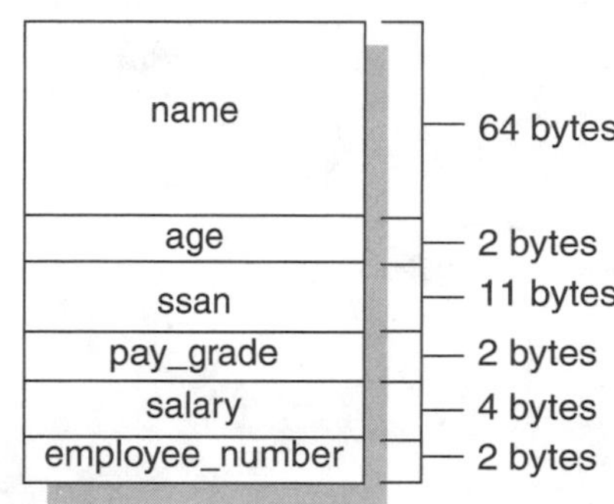

***Figure 536** A logical map of the memory C allocates to hold a structure.*

PUTTING A STRUCTURE TO USE 537

As you have learned, C lets you group related information in a structure. In the Date and Time section of this book, you will use the *getdate* function to determine the current system date. The function assigns the current date to members of a structure of type *date*, as shown here:

```
struct date
 {
   int da_year;   // Current year
   char da_day;   // Day of month
   char da_mon;   // Month of year
 };
```

The following *Turbo C Lite* program, *dosdate.c*, uses the *getdate* function to assign the date to the variable *curr_date*:

```
#include <stdio.h>
#include <dos.h>

void main(void)
 {
   struct date curr_date;

   getdate(&curr_date);
   printf("Current date: %d-%d-%d\n", curr_date.da_mon,
          curr_date.da_day, curr_date.da_year);
 }
```

Because the function must change the parameter's value, the program passes the structure variable to the function by reference (by address).

PASSING A STRUCTURE TO A FUNCTION 538

As you have learned, C lets you group related information within a structure. Like all variables, C lets you pass variables to a function. The following program, *strufunc.c*, passes a structure of type *Shape* to the function *show_structure*, which in turn displays each of the structure's members:

```
#include <stdio.h>
```

```
struct Shape
 {
   int type;
   int color;
   float radius;
   float area;
   float perimeter;
 };

void show_structure(struct Shape shape)
 {
   printf("shape.type %d\n", shape.type);
   printf("shape.color %d\n", shape.color);
   printf("shape.radius %f shape.area %f shape.perimeter %f\n",
          shape.radius, shape.area, shape.perimeter);
 }

void main(void)
 {
   struct Shape circle;

   circle.type = 0;
   circle.color = 1;
   circle.radius = 5.0;
   circle.area = 22.0 / 7.0 * circle.radius * circle.radius;
   circle.perimeter = 2.0 * 22.0 / 7.0 * circle.radius;
   show_structure(circle);
 }
```

539 CHANGING A STRUCTURE WITHIN A FUNCTION

In Tip 538 you learned that you can pass structures to functions just as you can pass variables of all types. To change a structure's members within a function, you must pass the structure by address (just as you would pass a variable whose value you want to change). The following program, *chgstruc.c*, invokes the function *change_structure,* which changes the values contained in a structure of type *Shape*:

```
#include <stdio.h>

struct Shape
 {
   int type;
   int color;
   float radius;
   float area;
   float perimeter;
 };

void change_structure(struct Shape *shape)
 {
   (*shape).type = 0;
   (*shape).color = 1;
   (*shape).radius = 5.0;
   (*shape).area = 22.0 / 7.0 * (*shape).radius * (*shape).radius;
   (*shape).perimeter = 2.0 * 22.0 / 7.0 * (*shape).radius;
 }

void main(void)
 {
   struct Shape circle;

   change_structure(&circle);
   printf("circle.type %d\n", circle.type);
   printf("circle.color %d\n", circle.color);
```

```
    printf("circle.radius %f circle.area %f circle.perimeter %f\n",
           circle.radius, circle.area, circle.perimeter);
  }
```

To change the structure's members, the program passes to the function a pointer to the structure. Within the function, the statements dereference the pointer's members using the *asterisk indirection* operator:

```
(*pointer).member = value;
```

UNDERSTANDING (*POINTER).MEMBER INDIRECTION 540

To change a structure member within a function, the program must pass a pointer to the structure. Within the function, the statements dereference the pointer using the *asterisk indirection* operator, as shown here:

```
(*pointer).member = value;
```

To resolve the pointer, C starts within the parentheses, first obtaining the location of the structure. Next, C adds to the address the offset of the specified member. If you omit the parentheses, C assumes that the member itself is a pointer and uses the asterisk indirection operator to resolve it, as shown here:

```
*pointer.member = value;
```

The parentheses-omitted syntax would be correct for a structure having a member that was a pointer, such as the following:

```
struct Planet
 {
   char name[48];
   int *some_pointer;
 } planet;
```

As you can see, the second member is a pointer to a value of type *int*. Assuming the program previously assigned the pointer to a memory location, the following statement places the value 5 in the memory location:

```
*planet.some_pointer = 5;
```

USING THE POINTER->MEMBER FORMAT 541

In Tip 540 you learned that to change a structure member within a function, the program must pass a pointer to the structure. To dereference the pointer within the function, C provides two formats. First, as you have seen, you can refer to a structure member in the following form:

```
(*pointer).member = value;
some_value = (*pointer).member;
```

Second, C lets you use the following format:

```
pointer->member = value;
some_value = pointer->member;
```

The following program, *chmember.c*, uses the second format within the *change_structure* function to reference members of a structure passed to the function by address:

```
#include <stdio.h>

struct Shape
 {
   int type;
   int color;
   float radius;
   float area;
   float perimeter;
```

```
};

void change_structure(struct Shape *shape)
 {
   shape->type = 0;
   shape->color = 1;
   shape->radius = 5.0;
   shape->area = 22.0 / 7.0 * shape->radius * shape->radius;
   shape->perimeter = 2.0 * 22.0 / 7.0 * shape->radius;
 }

void main(void)
 {
   struct Shape circle;

   change_structure(&circle);
   printf("circle.type %d\n", circle.type);
   printf("circle.color %d\n", circle.color);
   printf("circle.radius %f circle.area %f circle.perimeter %f\n",
          circle.radius, circle.area, circle.perimeter);
 }
```

542 USING A TAGLESS STRUCTURE

As you have learned, a structure *tag* is the structure's name. Using the tag, your programs can declare variables of a specific structure type. When you declare variables of structure types immediately following the structure definition, however, you do not necessarily have to specify a structure tag. For example, the following declaration creates two variables of a structure:

```
struct
 {
   int type;      // 0 = circle, 1 = square, 2 = triangle
   int color;
   float radius;
   float area;
   float perimeter;
 } triangle, circle;
```

If your program will not later refer to a structure by name (such as in a function prototype or formal parameters), then you can omit the structure tag, as just shown. However, when you include the tag, you provide other programmers who read your program with insight into the structure's purpose. When you include meaningful tag names, you may make your programs more readable.

543 UNDERSTANDING A STRUCTURE DEFINITION'S SCOPE

In the Functions section of this book, you learned that *scope* defines the region of the program within which an identifier (such as a variable or function) is known. When you define a structure, you must consider the structure's scope. If you examine the previous programs that work with structures within functions, you will find that the program defines the structure outside of and before the functions that use it. As a result, the structure definitions have a *global scope*, which lets all the functions that follow make reference to them. If, instead, the program had defined the structures within *main*, the only function that would be aware of the structure's existence would be *main*. If you need several of your program's functions to use a structure definition, you should define the structure outside of your functions at some point prior to all the functions that must access the structure.

INITIALIZING A STRUCTURE 544

As you have learned, C lets you initialize arrays when you declare them. In a similar way, your programs can also initialize a structure at declaration. The following program, *initstru.c*, declares and initializes a structure of type *Shape*:

```
#include <stdio.h>

void main(void)
 {
   struct Shape
    {
      int type;
      int color;
      float radius;
      float area;
      float perimeter;
    } circle = {0, 1, 5.0, 78.37, 31.42};

   printf("circle.type %d\n", circle.type);
   printf("circle.color %d\n", circle.color);
   printf("circle.radius %f circle.area %f circle.perimeter %f\n",
     circle.radius, circle.area, circle.perimeter);
 }
```

Because the program only uses the structure within *main*, the program defines the structure within *main*.

PERFORMING STRUCTURE I/O 545

Several Tips presented in this section have used *printf* to display the value of one or more structure members. When you perform screen or keyboard I/O operations that affect structure members, you must perform your I/O one member at a time. However, when you read or write structures from or to a file your programs can work with the entire structure. If your program uses file streams, you can use the *fwrite* and *fread* functions to write and read structures. The Files section of this book illustrates how to use *fwrite* and *fread* functions to perform structure I/O. To better understand this process, refer to the companion disk programs *dtoutf.c* and *dtinf.c*, located within the Tip0545 directory. If your programs use file handles, you can use the *read* and *write* functions to perform structure I/O. The companion disk files *dtout.c* and *dtin.c* (within the Tip0545 directory) illustrate how your programs can use *write* and *read* to perform structure I/O. Each of the I/O functions just discussed read or write a byte range. When C stores a structure in memory, the structure is really just a range of bytes. Therefore, to use a structure with these functions, simply pass a pointer to the structure, as shown in the sample programs.

USING A NESTED STRUCTURE 546

As you have learned, C lets you store related information within structures. Within a structure, you can include members of any type (*int*, *float*, and so on), as well as members that are structures themselves. For example, the following structure declaration includes a structure of type *Date*, which contains an employee's hiring date:

```
struct Employee
 {
   char name[64];
   int age;
   char ssan[11];    // Social security number
   struct Date
    {
      int day;
      int month;
      int year;
```

```
    } hire_date;
   int pay_grade;
   float salary;
   unsigned employee_number;
 } new_employee;
```

To access a member of a nested structure you use the *dot* operator; first to specify the nested structure and then to specify the desired member, as shown here:

```
new_employee.hire_date.month = 12;
```

547 Structures that Contain Arrays

As you have learned, structure members can be any type, including structures or arrays. When a structure member is an array, your programs reference the array member just as they would any array, except that the variable name and *dot* operator precede the array name. For example, the following program, *struarra.c*, initializes several structure fields, including an array. The program then loops through the array elements, displaying their value:

```
#include <stdio.h>

void main(void)
 {
   struct Date
    {
      char month_name[64];
      int  month;
      int  day;
      int  year;
    } current_date = { "July", 7, 4, 2001 };
   int i;

   for (i = 0; current_date.month_name[i]; i++)
     putchar(current_date.month_name[i]);
 }
```

548 Creating an Array of Structures

As you have learned, an array lets your programs store multiple values of the same type. Most of the arrays presented throughout this section have been of type *int*, *float*, or *char*. However, C also lets you declare arrays of a specific structure type. For example, the following declaration creates an array capable of storing information about 100 employees:

```
struct Employee
 {
   char name[64];
   int age;
   char ssan[11];    // Social security number
   int pay_grade;
   float salary;
   unsigned employee_number;
 } staff[100];
```

Assuming that the program has assigned values to each employee, the following *for* loop will display each employee's name and number:

```
for (emp = 0; emp < 100; emp++)
  printf("Employee: %s Number: %d\n", staff[emp].name,
    staff[emp].employee_number);
```

When you use an array of structures, you simply add the *dot* operator to each array element.

UNDERSTANDING DOS SYSTEM SERVICES 549

As you know, DOS was the original operating system for the IBM PC. DOS let users run programs and store information on disk. In addition, DOS provided services that let programs allocate memory, access devices such as the printer, and manage other system resources. To help your programs take advantage of capabilities built into DOS—such as determining the amount of free disk space, creating or selecting a directory, or even capturing keystrokes—DOS provides a set of services your programs can use. Unlike the functions the C run-time library provides, your programs do not access the DOS services using a simple function-call interface. Instead, programmers wrote the services so that other programmers can access the services at the assembly language level using registers and interrupts. However, as you will learn in this section, C actually makes it easy for your programs to take advantage of the DOS services without forcing you to use assembly language. In addition, the C run-time library often provides an interface to many of the DOS services through a function. Admittedly, very few users run DOS-based systems today. However, I have left this section of tips in the book, because they illustrate how easy C makes it for programmers to interact with the operating system. C's original popularity was as a systems programming language. When you create your programs, there may be times when you have the choice of using a C run-time library function or a DOS, Windows, or Linux service. As a rule, you should use the run-time library functions whenever possible, because using the run-time library functions will increase your program's portability. When you use the C run-time library function—as opposed to a DOS, Linux, or Windows-specific service—you do not need to change your program for it to run under a different operating sytem. Instead, you simply must recompile. If you have an unused system available, you should consider booting DOS and experimenting with the Tips this lesson presents. As you do, you will increase your understanding of C as well as how operating systems work behind the scenes. Although Windows lets you open an MS-DOS window, you should not run the applications the MS-DOS-based tips create under Windows. Many of the programs perform low-level operations that may return errant results under Windows.

UNDERSTANDING THE BIOS SERVICES 550

BIOS stands for *Basic Input/Output Services*. In short, the BIOS is a chip within your computer that contains the instructions your computer uses to write to the screen or printer, to read characters from the keyboard, or to read and write to your disk. As was the case with the DOS services, your programs can use the BIOS services to perform different operations. For example, you might use a BIOS service to determine the number of parallel or serial ports, the video display type, or the number of available disk drives. Like the DOS services, programmers designed the BIOS routines for assembly language programs to use. However, most C compilers provide run-time library functions that let your programs use these services without the need for assembly language. Many programmers confuse the DOS and BIOS services. As Figure 550.1 shows, the BIOS resides directly above your computer's hardware. The DOS services sit on top of the BIOS, and your programs sit on top of DOS.

***Figure 550.1** The relationship between the BIOS, DOS, and programs.*

However, there are times when DOS, and even your programs, can bypass the BIOS and directly access the computer's hardware. An application that must provide very fast video, for example, might bypass DOS and the BIOS to work directly with video memory. As a rule, however, only experienced programmers should bypass DOS and BIOS. DOS and the system BIOS perform considerable error-checking, which simplifies your programming tasks.

All varieties of Windows will call their own system services. However, just as with DOS system services, Windows system services eventually call the BIOS-level services to access the computer's hardware. However, while it is still possible to bypass Windows system services and call the BIOS-level services directly, it is generally not a good idea to do so, because of the Windows operating system design. Typically, you get the best results within Windows programs using the Windows API, and you should only call the BIOS services if it is absolutely necessary. Figure 550.2 shows the relationship between the BIOS, DOS, Windows, and programs.

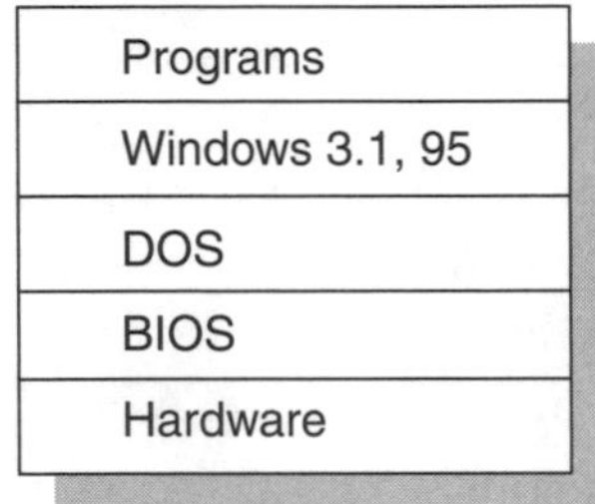

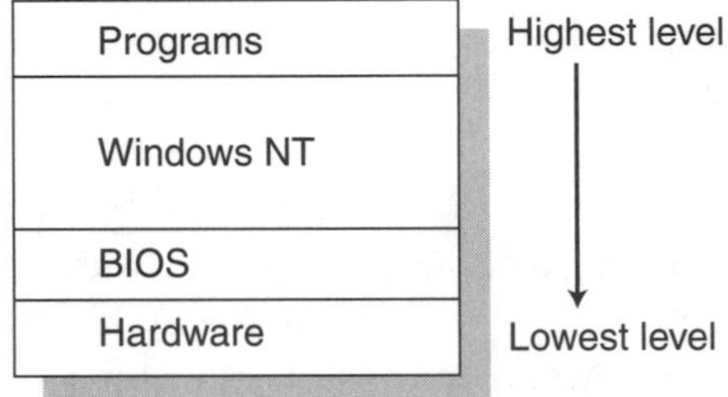

Figure 550.2 The relationship between the BIOS, DOS, Windows, and programs.

551 UNDERSTANDING REGISTERS

When your program executes, the program must reside in your computer's memory. Your computer's central processing unit (the CPU) will get your program's instruction and data as it needs to from memory. To improve performance, the CPU contains several temporary storage locations named *registers*. Because these registers reside within the CPU itself, the CPU can access each register's contents very quickly. In general, the CPU uses four types of registers: *segment, offset, general-purpose,* and *flags*. Table 551 briefly describes each register type's use.

Register Type	Use
Segment	Holds the starting memory address of a block of memory, such as the start of your program's code or data.
Offset	Holds the 16-byte offset into a block of memory, such as the location of a specific variable within your program's data segment.
General-Purpose	Temporarily stores program data.
Flags	Contains processor status and error information.

Table 551 The PC's register types.

The PC uses a segment and offset value to locate items in memory. When you use DOS services, you might want to assign the segment and offset addresses of one or more variables to different segment registers. The PC provides four general-purpose registers, named *AX, BX, CX,* and *DX.* Each general-purpose register can hold 16 bits of data (2 bytes). In some cases, you might only want to store one byte of information within a register. To help you do so, the PC lets you access each register's upper and lower bytes using the names shown in Figure 551.1.

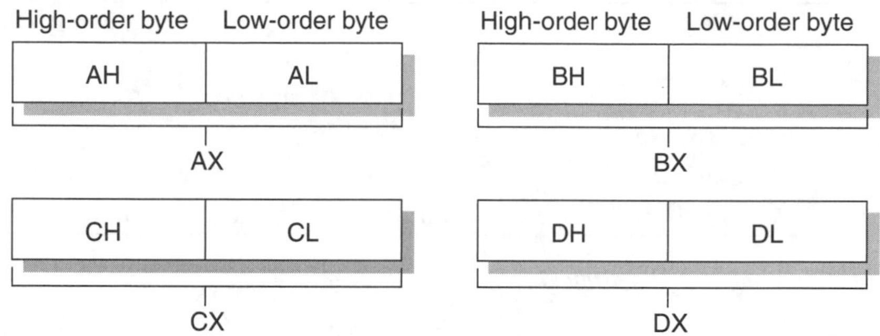

Figure 551.1 The PC's four general-purpose registers.

When you use the DOS and BIOS services, you will place parameters for whatever service your program is about to call within the general-purpose registers. When the service completes, DOS or the BIOS may place the service's result in one of the general-purpose registers. Finally, the flags register will store the CPU's state and possible error status values. When DOS and BIOS services complete, they often set or clear different bits within the flags register to indicate success or an error. Figure 551.2 illustrates the bits within the flags register (gray boxes represent unused bits).

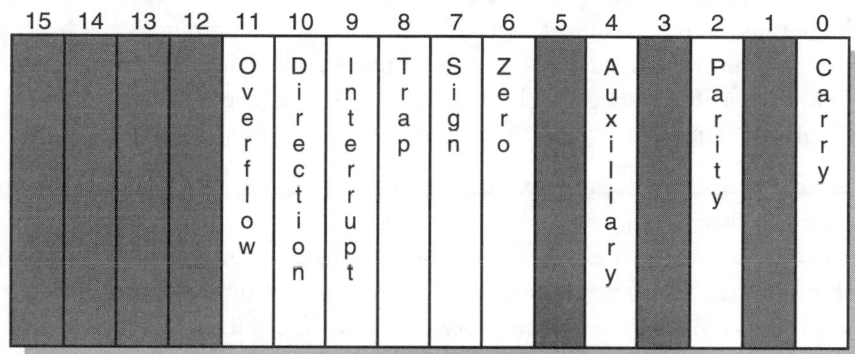

Figure 551.2 Bits within the PC's flags register.

Several of the Tips presented throughout this section discuss segments, offsets, and registers in greater detail.

UNDERSTANDING THE FLAGS REGISTER 552

As you have learned, the flags register contains CPU status and error information. After the PC completes different operations, such as add, subtract, or compare, it sets different bits in the flags register. Likewise, many BIOS and DOS services set the carry flag to indicate an error. Table 552 describes the bits the BIOS and DOS services use within the flags register.

Bit	Flag	Meaning
0	Carry	Indicates an arithmetic carry
2	Parity	Indicates an arithmetic operation resulted in an even number of bits set to 1
4	Auxiliary	Indicates an adjustment is required following a BCD (binary coded decimal) arithmetic operation

Table 552 Bits within the flags register.

Bit	Flag	Meaning
6	Zero	Indicates a zero result from a comparison or arithmetic operation
7	Sign	Indicates a negative result
8	Trap	Used for debugger trapping
10	Direction	Controls the direction of string instructions
11	Overflow	Indicates an arithmetic overflow

Table 552 Bits within the flags register. (Continued from previous page.)

When you use a DOS or BIOS service within your program, make sure that your program tests the flag bit the service sets to determine whether the service was successful or not.

553 Understanding Software Interrupts

An *interrupt* occurs when the CPU must temporarily stop what it is doing so that it can perform a different operation. When the operation finishes, the CPU can resume its original work as if it had never stopped. There are two types of interrupts: hardware and software interrupts. Devices connected to or inside your computer, such as the clock, disk drive, or keyboard, cause interrupts. When the original developers designed the PC, they provided support for up to 256 interrupts, numbered 0 through 255. Because your computer's hardware only needs a small number of these interrupts, many are available for software to use. The BIOS services, for example, make use of interrupts 5 and 10H through 1FH (16 decimal through 31 decimal). Likewise, DOS uses interrupts 21H through 28H (33 decimal through 40 decimal) and 2FH (47 decimal).

When you write programs in assembly language, you assign parameters to the PC registers and then invoke the interrupt that corresponds to the system service you want. For example, the BIOS uses interrupt 10H to access the video. To display a letter on the screen, for example, you assign the letter you want to the AL register, assign the value 9H to the AH (which directs the BIOS to perform a video write operation), assign the attribute you want (bold, blinking, normal, and so on) to the BX register, and then invoke INT 10H, as shown here:

```
MOV AL,41   ; A is the ASCII 41H
MOV AH,9    ; Request a video write
MOV BX,7    ; Character attribute
MOV CX,1    ; Number of characters to write
INT 10      ; Perform the video service
```

As you will learn, most DOS services use INT 21H. Fortunately, you do not have to work with assembly language within your C programs to invoke a service.

Note: *Although you can use software interrupts within your Windows programs, you should perform most activities within your Windows programs using the Windows API.*

554 Using the BIOS to Access the Printer

Several Tips presented throughout this book have written output to the printer using the *stdprn* file handle. Before your programs perform printer I/O, however, you might want the program to verify that the printer is on line and has paper. To do so, your programs can use the *biosprint* function. You will implement the *biosprint* function, as shown here:

```
#include <bios.h>

int biosprint(int command, int byte, int port_number);
```

The *command* parameter specifies one of the operations listed in Table 554.1.

Command	Meaning
0	Print the byte specified
1	Initialize the printer port
2	Read the printer status

*Table 554.1 The possible values for the **command** parameter.*

If you are printing a character, the *byte* specifies ASCII or extended ASCII value or the character you want. The *port_number* parameter specifies the printer port to which you want to print, where 0 is LPT1, 1 is LPT2, and so on.

The *biosprint* function returns an integer value in the range 0 through 255, whose bits are defined in Table 554.2.

Bit	Meaning if Set
0	Device timeout
3	I/O error
4	Printer selected
5	Out of paper
6	Device acknowledgment
7	Device not busy

*Table 554.2 Status bits **biosprint** returns.*

The following *Turbo C Lite* program, *printchk.c*, uses the *biosprint* function to test your printer status repeatedly until you press any key. Run the following program and experiment with your printer, taking it off line, removing the paper, and so on. As you do so, the program should display different messages on your screen, as shown here:

```
#include <bios.h>
#include <conio.h>
#include <stdio.h>

void main(void)
  {
    int status = 0;
    int old_status = 0;

    do
     {
       status = biosprint(2, 0, 0);  // Read LPT1
       if (status != old_status)
         {
           if (status & 1)
             printf ("Time-out\t");
           if (status & 8)
             printf ("Output Error\t");
           if (status & 16)
             printf ("Printer Selected\t");
           if (status & 32)
             printf ("Out of Paper\t");
           if (status & 64)
             printf ("Acknowledge\t");
           if (status & 128)
             printf ("Printer Not Busy");
           printf ("\n");
           old_status = status;
         }
     }
    while (! kbhit());
  }
```

***Note:** Many compilers provide a function named **_bios_printer** that is similar to **biosprint**. To access a printer connected to a serial port, you must use the **_bios_serialcom** function.*

***Note:** The companion CD-ROM that accompanies this book includes the program **win_print.cpp**, which uses the Windows API to send information to a printer.*

555 CONTROL+BREAK INFORMATION

When you work within the DOS environment, the DOS BREAK command lets you enable and disable extended CTRL+BREAK checking. When you enable extended checking, DOS increases the number of operations after which it checks for a user-entered CTRL+C or CTRL+BREAK. When you disable extended CTRL+BREAK checking, DOS only checks for a CTRL+BREAK after performing keyboard, screen, or printer I/O. Many C compilers provide two functions, *getcbrk* and *setcbrk*, which your programs can use to obtain and set the state of CTRL+BREAK checking. You will implement *getcbrk* and *setcbrk*, as shown here:

```
#include <dos.h>

int getcbrk(void);
int setcbrk(int setting);
```

The *getcbrk* function returns the value 0 if you have disabled extended CTRL+BREAK checking, and 1 if it is active. Likewise, the *setcbrk* function uses the values 0 and 1, respectively, to disable and enable extended checking. The *setcbrk* function also returns the value 0 or 1, depending on the state of extended checking you have selected. The following *Turbo C Lite* program, *ctrlbrk.c*, uses the *setcbrk* function to disable extended CTRL+BREAK checking. The program uses the *getcbrk* function's return value to display the previous setting, as shown here:

```
#include <stdio.h>
#include <dos.h>

void main(void)
 {
   printf("Previous extended Ctrl-Break status %s\n",
          (getcbrk()) ? "On": "Off");

   setcbrk(0);  // Turn it off
 }
```

***Note:** The **setcbrk** function sets the state of CTRL+BREAK checking for the system, not just for the current program. When the program ends, the previously selected state remains in effect. Remember, you will handle messages within Windows using different commands, and trapping CTRL+BREAK is generally not helpful within Windows programs.*

556 UNDERSTANDING POSSIBLE DOS SIDE EFFECTS

In Tip 555 you learned how to use the *setcbrk* function to change the state of extended CTRL+BREAK checking. Likewise, in the Disks and Files section, you learned how to change the state of disk verification. Several other Tips have presented ways your programs can change the current drive or directory. When your programs perform such operations, they should save the original settings when the program begins so the program can restore the original settings before the program ends.

Unless it is the program's explicit purpose to change one or more of these settings, the program should not leave the setting changed after it ends. Such changes in settings are called *side effects*, which you should avoid. When a user runs your budget program, for example, the user's default drive and directory should not have changed after the program ends. Likewise, even more subtle changes, such as disabling disk verification or extended CTRL+BREAK checking, should not occur. As you create your program, include the additional statements your system requires to restore original environment settings.

Note: *Remember, you will handle messages within Windows using different commands. Trapping CTRL+BREAK is generally not helpful within Windows programs.*

SUSPENDING A PROGRAM TEMPORARILY 557

In this book's Date and Time section, you will use the *delay* function to pause your program for a specific number of milliseconds. In a similar way, your programs can use the *sleep* function to specify the pausing interval in seconds, as shown here:

```
#include <dos.h>
void sleep(unsigned seconds);
```

Because the *delay* function works with milliseconds, it is more accurate than *sleep*. However, you can use the *sleep* function to increase your program's portability to other operating systems.

Most operating systems provide a *sleep* function, which lets programs enter an inactive state until a time interval has expired or a specific event has occurred. The following *Turbo C Lite* program, *sleep_5.c*, uses the *sleep* function to pause for five seconds:

```
#include <stdio.h>

#include <dos.h>
void main(void)
 {
   printf("About to sleep for 5 seconds\n");
   sleep(5);
   printf("Awake\n");
 }
```

HAVING SOME FUN WITH SOUND 558

Within almost every PC is a small (low-quality) speaker that programs normally use to generate beeps. However, using the *sound* function that many C compilers provide, your programs can generate sounds that emit different frequencies through the speaker. The *sound* function lets your programs turn on the speaker to display a specific frequency sound. The *nosound* function turns off the speaker, as shown here:

```
#include <dos.h>
void sound(unsigned frequency);
void nosound(void);
```

The following *Turbo C Lite* program, *siren.c*, uses the *sound* function to generate a siren-like sound. When you press any key, the program turns off the speaker using the *nosound* function, as shown here:

```
#include <dos.h>
#include <conio.h>
void main(void)
 {
   unsigned frequency;
   do
    {
      for (frequency = 500; frequency <= 1000; frequency += 50)
       {
         sound(frequency);
         delay(50);
       }
```

```
      for (frequency = 1000; frequency >= 500; frequency -= 50)
       {
         sound(frequency);
         delay(50);
       }
    }
   while (! kbhit());
   nosound();
 }
```

559 OBTAINING COUNTRY-SPECIFIC INFORMATION

As you know, countries all over world use the DOS operating system. To support international users, DOS supports different keyboard templates, code pages, and country-specific information. To help your programs determine the current country settings, your programs can use the *country* function, as shown here:

```
#include <dos.h>

struct COUNTRY *country(int code, struct COUNTRY *info);
```

If successful, the function will return a pointer to a structure of type *COUNTRY*, as shown here:

```
struct COUNTRY
 {
   int  co_date;        // Date format
   char co_curr[5];     // Currency symbol
   char co_thsep[2];    // Thousands separator
   char co_desep[2];    // Decimal separator
   char co_dtsep[2];    // Date separator
   char co_tmsep[2];    // Time separator
   char co_currstyle;   // Currency style
   char co_digits;      // Currency significant digits
   char co_time;        // Time format
   long co_case;        // Pointer to case map
   char co_dasep;       // Data separator
   char co_fill[10];    // Filler
 };
```

The *code* value specifies a country code you want to select. If the value of the *info* parameter is –1, the *country* function will set the current country code to the code you specify. If the value of *info* is not –1, the *country* function will assign to the buffer the settings for the current country code. The following *Turbo C Lite* program, *country.c*, displays the current country settings:

```
#include <stdio.h>
#include <dos.h>

void main(void)
 {
   struct COUNTRY info;

   country(0, &info);
   if (info.co_date == 0)
     printf("Date format: mm/dd/yy\n");
   else if (info.co_date == 1)
     printf("Date format: dd/mm/yy\n");
   else if (info.co_date == 2)
     printf("Date format: yy/mm/dd\n");
   printf("Currency symbol %s\n", info.co_curr);
   printf("Decimal separator %s\n", info.co_thsep);
   printf("Date separator %s Time separator %s\n",
     info.co_dtsep, info.co_tmsep);
```

```
    if (info.co_currstyle == 0)
      printf("Currency symbol precedes with no leading spaces\n");
    else if (info.co_currstyle == 1)
      printf("Currency symbol follows with no spaces\n");
    else if (info.co_currstyle == 2)
      printf("Currency symbol precedes with leading space\n");
    if (info.co_currstyle == 4)
      printf("Currency symbol follows with space\n");
    printf("Currency significant digits %d\n", info.co_digits);
    if (info.co_time)
      printf("24 hour time\n");
    else
      printf("12 hour time\n");
    printf("Data separator %s\n", info.co_dasep);
  }
```

Note: *The companion CD-ROM that accompanies this book includes the program **win_country.cpp**, which returns the current country setting within Windows.*

UNDERSTANDING THE DISK TRANSFER ADDRESS 560

Prior to DOS 3.0, programs performed file operations using file control blocks (FCBs). By default, when DOS read or wrote information, DOS did so through an area in memory called the *disk transfer area.* The disk transfer area was 128 bytes by default. The address of the area's first byte is called the *disk transfer address* (DTA). By default, DOS used offset 80H of the program segment prefix as the disk transfer address. As you will learn in the Redirecting I/O and Processing Command Line section of this book, offset 80H of the program segment prefix also contains the program's command line. Because most programs do not use file control block disk operations, many programmers assume they can ignore the disk transfer address. Unfortunately, routines such as *findnext* and *findfirst,* discussed in this book's Files section, place their results in the disk transfer address, overwriting your program's command line. To prevent operations that use the disk transfer address from overwriting your program's command-line parameters, many programmers use a DOS service to set the disk transfer address to point to a different memory buffer. As you will learn in Tip 561, your programs can change and determine the disk transfer address using run-time library functions.

ACCESSING AND CONTROLLING THE DISK TRANSFER AREA 561

In Tip 560 you learned that the disk transfer area is a 128-byte region DOS uses for file control block-based I/O services or *findfirst* and *findnext* operations. To help you control the disk transfer area, most C compilers support the *getdta* and *setdta* functions, as shown here:

```
#include <dos.h>

char *far getdta(void);
void setdta(char far *disk_transfer_address);
```

The *getdta* function returns a *far* (32-bit) pointer to the current disk transfer area. Likewise, the *setdta* function lets you assign the program's disk transfer address to the far address you specify. The following *Turbo C Lite* program, *dta.c,* illustrates the use of the *getdta* and *setdta* functions:

```
#include <stdio.h>
#include <dos.h>
#include <alloc.h>

void main(void)
 {
   char far *dta;
```

```
    dta = getdta();
    printf("Current DTA is %p\n", dta);
    if (MK_FP(_psp, 0x80) == dta)
      printf("DTA is at same location as command line\n");
    dta = farcalloc(128);
    setdta(dta);
    printf("New DTA is %p\n", getdta());
  }
```

Note: *When you program within Windows, you will not need to control the disk transfer address, as Windows uses the virtual memory model—detailed within the section entitled Windows Memory Management—to handle most file I/O.*

562 Using the BIOS Keyboard Services

DOS, the BIOS, and the C run-time library all provide services that let your programs access the keyboard. As a rule, you should first try to use the C run-time library functions. If a C run-time library function will not work, then use the DOS function. If the DOS function does not work, try the BIOS services. Using the C run-time library functions lets your programs remain more portable. To help your programs access the BIOS keyboard services, the C run-time library provides the *_bios_keybrd* function, as shown here:

```
#include <bios.h>

unsigned _bios_keybrd(unsigned command);
```

The *command* parameter specifies the desired operation. Table 562 lists the possible values you can pass for the *command* parameter.

Value	Meaning
_KEYBRD_READ	Directs *_bios_keybrd* to read a character from the keyboard buffer. If the lower byte of the return value is 0, the upper byte contains an extended keyboard code.
_KEYBRD_READY	Directs *_bios_keybrd* to determine if a character is present in the keyboard buffer. If *_bios_keybrd* returns 0, no keystroke is present. If the return value is 0xFFFF, the user has pressed CTRL+C.
_KEYBRD_SHIFTSTATUS	Directs *_bios_keybrd* to return the keyboard's shift state: Bit 7 Set — INS is on Bit 6 Set — CAPSLOCK is on Bit 5 Set — NUMLOCK is on Bit 4 Set — SCROLLLOCK is on Bit 3 Set — ALT key pressed Bit 2 Set — CTRL key pressed Bit 1 Set — Left SHIFT key pressed Bit 0 Set — Right SHIFT key pressed
_NKEYBRD_READ	Directs *_bios_keybrd* to read a character from the keyboard buffer. If the lower byte of the return value is 0, the upper byte contains an extended keyboard code. *_NKEYBRD_READ* directs *_bios_keybrd* to read enhanced keys, such as the cursor arrow keys.

Table 562 *The possible values for the* ***command*** *parameter.*

Value	Meaning
_NKEYBRD_READY	Directs *_bios_keybrd* to determine if a character is present in the keyboard buffer. If *_bios_keybrd* returns 0, no keystroke is present. If the return value is 0xFFFF, the user has pressed CTRL+C. The *_NKEYBRD_READY* value directs *_bios_keybrd* to support enhanced keys, such as the cursor arrow keys.
_NKEYBRD_SHIFTSTATUS	Directs *_bios_keybrd* to return the keyboard's shift state, including enhanced keys: Bit 15 Set SYSREQ key pressed Bit 14 Set CAPSLOCK key pressed Bit 13 Set NUMLOCK key pressed Bit 12 Set SCROLLLOCK key pressed Bit 11 Set Right ALT key pressed Bit 10 Set Right CTRL key pressed Bit 9 Set Left ALT key pressed Bit 8 Set Right ALT key pressed

Table 562 The possible values for the ***command*** *parameter. (Continued from previous page.)*

The following program, *keystate.c*, uses a loop to display changes in the keyboard state until you press any key other than SHIFT, ALT, CTRL, NUMLOCK, and so on. The program only reads the non-enhanced keys, as shown here:

```
#include <stdio.h>
#include <bios.h>

void main(void)
 {
   unsigned int state, old_state = 0;

   do
    {
      state = _bios_keybrd(_KEYBRD_SHIFTSTATUS);
      if (state != old_state)
        {
          old_state = state;
          if (state & 0x80)
            printf("Ins On ");
          if (state & 0x40)
            printf("Caps On ");
          if (state & 0x20)
            printf("Num Lock On ");
          if (state & 0x10)
            printf("Scroll Lock On ");
          if (state & 0x08)
            Printf("Alt pressed ");
          if (state & 0x04)
            Printf("Ctrl pressed ");
          if (state & 0x02)
            printf("Left shift pressed ");
          if (state & 0x01)
            printf("Right shift pressed ");
          printf("\n");
        }
    }
   while (! _bios_keybrd(_KEYBRD_READY));
 }
```

***Note:** Many C compilers provide a function named **bioskey**, which performs processing similar to **_bios_keyboard**. Refer to the documentation that accompanied your compiler to determine which function it supports. The companion CD-ROM provides a version of the program for Turbo C Lite that uses the **bioskey** function.*

***Note:** Windows does not support the **_bios_keyboard** command or its equivalents. Within Windows, you will gain information about the keyboard using different compiler-specific classes. For instance, within some compiler environments, you will determine whether the CAPS LOCK key and other non-keystroke based keys are selected using the **KeyboardManager** class **KeyboardFlags** property.*

563 OBTAINING THE BIOS EQUIPMENT LIST

As your programs increase in complexity, there may be times when your programs must determine specifics about the computer's hardware. In such cases, your programs can use the *_bios_equiplist* function, as shown here:

```
#include <bios.h>

unsigned _bios_equiplist(void);
```

The following function returns an *unsigned int* value, the bits of which have the following meaning:

```
struct Equip
 {
  unsigned floppy_available:1;       // 1 if present
  unsigned coprocessor_available:1;  // 1 if present
   unsigned  system_memory:2;        //  Original  PC
   unsigned  video_mode:2;           // 01 = 40 x 25 mono
                                     // 10 = 80 x 25 color
                                     // 11 = 80 x 25 mono
   unsigned  floppy_disk_count:2;    // Add 1
   unsigned  serial_port_count:2;
   unsigned  game_adapter_available:1// 1 if present
   unsigned  printer_count:2;
 };
```

The following program, *showequip.c*, uses the *_bios_equiplist* to display the system equipment list:

```
#include <stdio.h>
#include <bios.h>

void main(void)
 {
   struct Equip
    {
      unsigned floppy_available:1;
      unsigned coprocessor_available:1;
      unsigned system_memory:2;
      unsigned video_memory:2;
      unsigned floppy_disk_count:2;
      unsigned unused_1:1;
      unsigned serial_port_count:3;
      unsigned game_adapter_available:1;
      unsigned unused_2:1;
      unsigned printer_count:2;
    } ;
   union Equipment
    {
      unsigned list;
      struct Equip list_bits;
    } equip;

   equip.list = _bios_equiplist();
```

```
    if (equip.list_bits.coprocessor_available)
      printf("Math coprocessor available\n");
    else
      printf("No math coprocessor\n");
    printf("System board memory %d\n", (equip.list_bits.system_memory + 1) * 16);
    printf("Number of floppies %d\n", equip.list_bits.floppy_disk_count + 1);
    printf("Number of printers %d\n", equip.list_bits.printer_count);
    printf("Number of serial ports %d\n", equip.list_bits.serial_port_count);
  }
```

Note: *Some C compilers provide a function named* ***biosequip****, which performs similar processing to* ***_bios_equiplist****. Support for the* ***_bios_equiplist*** *and related commands within Windows varies from compiler to compiler. See your compiler's documentation for details. The companion CD-ROM provides a version of this program for Turbo C Lite that uses the* ***biosequip*** *function.*

CONTROLLING SERIAL PORT I/O 564

To help your programs perform I/O operations to a serial port, such as COM1, many DOS-based compilers provide the *bioscom* function, as shown here:

```
#include <bios.h>
unsigned bioscom(int command, int port, char byte);
```

The *command* parameter specifies the operation you want and must be a value listed in Table 564.1.

Value	Description
_COM_INIT	Sets the port's communication settings
_COM_RECEIVE	Receives a byte from the port
_COM_SEND	Sends a byte to the port
_COM_STATUS	Returns the port settings

Table 564.1 *The possible settings for the* ***command*** *parameter.*

The *port* parameter specifies the serial port you want, where 0 corresponds to COM1, 1 to COM2, and so on. The *byte* parameter specifies either a byte to output or the communications settings you want. If the value contains the communication settings you want, the *byte* parameter can contain a combination of the values listed in Table 564.2.

Value	Meaning
_COM_CHR7	7-bit data
_COM_CHR8	8-bit data
_COM_STOP1	1 stop bit
_COM_STOP2	2 stop bits
_COM_NOPARITY	Parity none
_COM_ODDPARITY	Parity odd
_COM_EVENPARITY	Parity even
_COM_110	110 baud
_COM_150	150 baud
_COM_300	300 baud
_COM_600	600 baud

Table 564.2 *The possible values for the* ***byte*** *parameter.*

Value	Meaning
_COM_1200	1,200 baud
_COM_2400	2,400 baud
_COM_4800	4,800 baud
_COM_9600	9,600 baud

Table 564.2 The possible values for the ***byte*** *parameter. (Continued from previous page.)*

Regardless of the command, the most significant byte of the return value has one of the bit meanings listed in Table 564.3.

Bit	Meaning if Set
8	Data ready
9	Overrun error
10	Parity error
11	Framing error
12	Break detect
13	Transmit holding register empty
14	Transmit shift register empty
15	Time out

Table 564.3 The bit meanings of the return value for ***_bios_serialcom.***

For _COM_INIT and _COM_STATUS, *_bios_serialcom* defines the least significant byte of the return value according to the values in Table 564.4.

Bit	Meaning if Set
0	Change in clear to send
1	Change in data set ready
2	Trailing-edge ring detector
3	Change in receive line signal detector
4	Clear to send
5	Data set ready
6	Ring indicator
7	Received line signal detect

Table 564.4 The return values when you use _COM_INIT and _COM_STATUS.

The following program, *setcom1.c*, sets the data communications for COM1 to 9,600 baud, 8 data bits, 1 stop bit, and no parity:

```
#include <stdio.h>
#include <bios.h>

void main(void)
 {
   char i = 0, title[] = "Jamsa\'s C/C++/C# Programmer\'s Bible";
   unsigned status;

   status = _bios_serialcom(_COM_INIT, 0, _COM_9600 |
     _COM_CHR8 | _COM_STOP1 | _COM_NOPARITY);
   if (status & 0x100) // Data ready
     while (title[i])
       {
```

```
            _bios_serialcom(_COM_SEND, 0, title[i]);
            putchar(title[i]);
            i++;
         }
  }
```

***Note:** Some C compilers provide a **bioscom** function, which provides similar processing.*

ACCESSING DOS SERVICES USING BDOS 565

As you have learned, the *intdos* function lets your programs access the DOS services. Some DOS services only use the AX and DX registers. For such services, your programs can use the *bdos* function, as shown here:

```
#include <dos.h>
int bdos(int dos_function, unsigned dx_register, unsigned al_register);
```

The *dos_function* parameter specifies the service you want to invoke. The *dx_register* and *al_register* parameters specify the values the service expects in the DX and AL registers. Upon return, the function returns the value of the AX register upon the service's termination. The following *Turbo C Lite* program, *bdos.c*, uses the *bdos* function to display the current disk drive:

```
#include <stdio.h>
#include <dos.h>

void main(void)
 {
   int drive;

   drive = bdos(0x19, 0, 0);
   printf("Current drive is %c\n", 'A' + drive);
 }
```

***Note:** The **bdos** function passes an unsigned value for the DX register. If you are using a DOS service that requires a pointer, you can use the **bdosptr** function. If you use the small memory model, the second parameter will correspond to DX. In the large memory model, the value will correspond to DS:DX.*

GETTING EXTENDED DOS ERROR INFORMATION 566

When a DOS system service fails, your programs can request additional error information from DOS to determine the source and cause of the error. To help you request extended error information, many C compilers provide the *dosexterr* function, as shown here:

```
#include <dos.h>
int dosexterr(struct DOSERROR *error_info);
```

The *error_info* parameter is a pointer to a structure of type *DOSERROR* that contains the extended error information, as shown here:

```
struct DOSERROR
 {
   int de_exterror;   // Extended error
   int de_class;      // Error class
   int de_action;     // Recommended action
   int de_locus;      // Error locus
 };
```

If the *dosexterr* function returns 0, the previous DOS service call did not experience an error. The extended error value provides a specific error. The error class describes the category of the error, as shown in Table 566.1.

Value	Meaning
01H	Out of resource
02H	Temporary error
03H	Authorization error
04H	System error
05H	Hardware failure
06H	System error not due to current program
07H	Application error
08H	Item not found
09H	Invalid format
0AH	Item locked
0BH	Media error
0CH	Item exists
0DH	Unknown error

*Table 566.1 The error classes **dosexterr** returns within the **de_class** member.*

The *de_action* member (the recommended action) tells your program how to respond to the error, as shown in Table 566.2.

Value	Action
01H	Retry first and then prompt user
02H	Retry after delay and then prompt user
03H	Prompt user for solution
04H	Abort with cleanup
05H	Abort without cleanup
06H	Ignore error
07H	Retry after user intervention

*Table 566.2 The possible return values within the **de_action** member.*

Finally, the *de_locus* member specifies the source of the error, as shown in Table 566.3.

Value	Location
01H	Unknown locus
02H	Block device error
03H	Network error
04H	Serial device error
05H	Memory error

*Table 566.3 The return values within the **de_locus** member.*

When your programs must respond to errors in a careful and considered way, you should use the *dosexterr* structure to obtain additional information.

DETERMINING THE BIOS CONVENTIONAL MEMORY AMOUNT 567

Many older programs do not take advantage of extended and expanded memory. Instead, the programs only use the PC's 640Kb conventional memory. As you examine such programs, you may find calls to the *biosmemory* function, which returns the amount of conventional memory (in Kb) the BIOS reported during your system startup. The memory amount *biosmemory* returns does not include extended, expanded, or upper memory. You will implement *biosmemory* as shown here:

```
#include <bios.h>

int biosmemory(void);
```

In addition to the *biosmemory* function, you may encounter the *_bios_memsize* function, which performs identical processing. You will implement *_bios_memsize* as shown here:

```
#include <bios.h>

int _bios_memsize(void);
```

The following *Turbo C Lite* program, *biosmem.c*, will display the amount of memory the BIOS reports in response to a *biosmemory* and *_bios_memsize* invocation:

```
#include <stdio.h>
#include <bios.h>

void main(void)
 {
   printf("BIOS Memory report %dKb\n", biosmemory());
 }
```

Note: *Because the Windows memory model uses virtual memory, you should use the Windows API calls the Windows Memory Management section of this book details when you write programs for Windows.*

BUILDING A FAR POINTER 568

A *far pointer* consists of a 16-bit segment and a 16-bit offset address. When you work with far pointers, there may be times when you must break the pointers into their segment and offset counterparts. Likewise, there may be times when you must build a far pointer from a segment and offset address. To help you build a far pointer, C provides the *MK_FP* macro, as shown here:

```
#include <dos.h>

void far *MK_FP(unsigned segment, unsigned offset);
```

The following code fragment uses *MK_FP* to build a far pointer out of a near variable's address:

```
long far *fptr;
long variable;
struct SREGS segs;

 // Get the current data segment
segread(&segs);
fptr = MK_FP(segs.ds, &variable);
```

To better understand the *MK_FP* macro, consider the following implementation:

```
#define MK_FP(s, o) ((void far *) (((long) s << 16) | (o)))
```

To create the 32-bit far address, the macro creates a *long* value and shifts the segment address bits into the value's upper 16 bits. Next, the macro uses a *bitwise OR* operation to assign the offset address to the lower 16 bits.

Note: *As you have learned, because far pointers do not apply to the virtual memory model, Windows programs do not use them.*

569 BREAKING A FAR ADDRESS INTO A SEGMENT AND OFFSET

As discussed in Tip 568, a far pointer consists of a 16-bit segment and a 16-bit offset address. When you work with far pointers, there may be times when you must break the address the far pointer references into its segment and offset counterparts. In such cases, your programs can use the *FP_SEG* and *FP_OFF* macros, as shown here:

```
#include <dos.h>

unsigned FP_OFF(void far *pointer);
unsigned FP_SEG(void far *pointer);
```

The following statements illustrate the use of the *FP_SEG* and *FP_OFF* macros:

```
char far *title = "Jamsa\'s C/C++/C# Programmer\'s Bible";
unsigned segment, offset;

segment = FP_SEG(title);
offset = FP_OFF(title);
```

Note: *As you have learned, because far pointers do not apply to the virtual memory model, Windows programs do not use them.*

570 DETERMINING FREE CORE MEMORY

As your programs allocate memory, you can use the *coreleft* function to estimate the amount of conventional memory currently available for allocation. The *coreleft* function does not provide an exact report of the unused memory. Instead, if you are using a small memory model, *coreleft* returns the unused memory between the top of the heap and the stack. If you are using a larger memory model, *coreleft* returns the amount of memory between the top of allocated memory and the end of conventional memory. The *coreleft* function returns the unused memory in bytes. In the case of the small memory model, *coreleft* returns an unsigned value, as shown here:

```
#include <alloc.h>

unsigned coreleft(void);
```

If you are using a larger memory model, *coreleft* returns a value of type *long*, as shown here:

```
#include <alloc.h>

long coreleft(void);
```

The following *Turbo C Lite* program, *coreleft.c*, displays the amount of available memory. The program uses the memory model constants many compilers support to determine the current memory model:

```
#include <stdio.h>
#include <alloc.h>

void main (void)
 {
#if defined(__SMALL__)
   unsigned result;
#else
   long result;
#endif
   result = coreleft();
   printf("The amount of available memory is %dKb\n", result / 1024);
 }
```

Note: *If your compiler does not support the* ***coreleft*** *function, check whether it provides the functions* ***_memavl*** *and* ***_memmax****. The section entitled Windows Memory Management details how to determine available memory within the Windows environment.*

READING THE SEGMENT REGISTER SETTINGS 571

When you work within the DOS environment, the compiler will keep track of your program code, data, and stack using four segment registers. The four segment registers are listed in Table 571.

Name	Description
CS	Code segment register
DS	Data segment register
SS	Stack segment register
ES	Extra segment register

Table 571 *The four segment registers in DOS.*

Depending on your program's memory model, each segment register may point to a unique 64Kb segment, or two or more segment registers may point to the same segment. When your programs use the DOS and BIOS services, there may be times when you must know a segment register's value. For such cases, you can use the *segread* function, as shown here:

```
#include <dos.h>

void segread(struct SREGS *segs);
```

The header file *dos.h* defines the *SREGS* structure, as shown here:

```
struct SREGS
 {
   unsigned int es;
   unsigned int cs;
   unsigned int ss;
   unsigned int ds;
 };
```

The following *Turbo C Lite* program, *showsegs.c*, uses *segread* to display the current segment register contents:

```
#include <stdio.h>
#include <dos.h>

void main(void)
 {
   struct SREGS segs;

   segread(&segs);
   printf("CS %X DS %X SS %X ES %X\n", segs.cs, segs.ds, segs.ss, segs.es);
 }
```

UNDERSTANDING MEMORY TYPES 572

The PC can hold three types of memory: conventional, extended, and expanded. Several of the Tips that follow discuss these memory types in detail. As you program, it is important that you understand the different memory types and their characteristics. The steps you must perform to allocate and use the different memory types will differ. In addition, each memory type has a different access speed, which will affect your program's performance. To determine the amount and types of memory installed within your PC, you can use the DOS 5 (or later) MEM /CLASSIFY command, as shown here:

```
C:\> MEM /CLASSIFY <ENTER>
```

If you are not using DOS 5 or later, you should upgrade your system. DOS 5 provides several memory management capabilities that help you maximize your PC's memory use.

573 UNDERSTANDING CONVENTIONAL MEMORY

When IBM first released the PC in 1981, the computer usually came with between 64Kb and 256Kb of RAM. At the time, that was more than enough memory. This memory became known as the PC's *conventional memory*. Today, the PC's conventional memory is the first 1Mb of RAM. DOS programs typically run within the first 640Kb of conventional memory. The PC uses the 384Kb of memory (called *reserved* or *upper memory*) that resides between 640Kb and 1Mb for your computer's video memory, device drivers, other memory-mapped hardware devices, and the BIOS. For years, however, the operating system did not use large sections of this reserved memory. Beginning with version 5, DOS provides ways that your programs and device drivers can reside in the unused regions as your programs run. Taking advantage of upper memory lets you free up more of the 640Kb conventional memory for DOS to use. For information on how to take advantage of the upper memory region, refer to the CONFIG.SYS DOS=UMB (UMB stands for *upper memory block*) entry in your DOS documentation.

As you have learned, Windows uses the virtual memory model to manage memory, which means that conventional memory issues are not significant within Windows programs. However, conventional memory is important when you run programs within a DOS window under Windows.

574 UNDERSTANDING THE CONVENTIONAL MEMORY LAYOUT

In Tip 573 you learned that *conventional memory* is your computer's first 1Mb of RAM. Your programs and DOS normally reside in the first 640Kb of conventional memory. To help you better understand how DOS uses conventional memory, Figure 574 presents a conventional memory map.

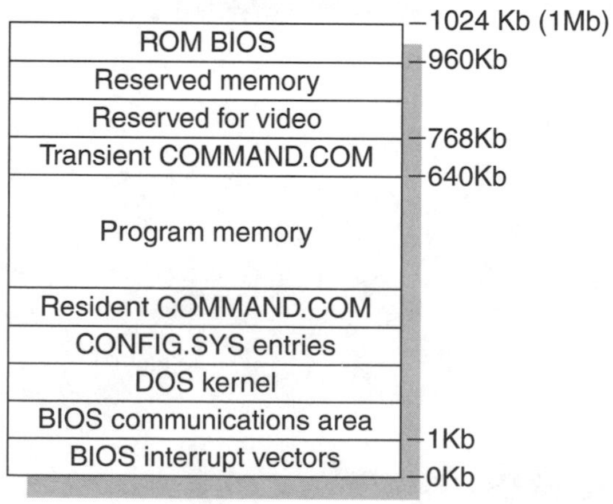

Figure 574 *Map of the PC's conventional memory.*

The DOS and BIOS section of this book explains the BIOS interrupt vectors and BIOS communication area. The *DOS kernel* is the software, *io.sys* and *msdos.sys*, that DOS loads into memory during the system startup. The *config.sys* entries represent the memory region that DOS allocates for device drivers, disk buffers, and so on. The *resident command.com* and the *transient command.com* areas hold the software responsible for displaying the DOS prompt and processing the commands you type. To make more memory available to programs, DOS divides *command.com* into a resident section, which always remains in memory, and a transient section, which each command can overwrite. After the command completes, the resident portion of *command.com* reloads the transient section from disk. The 384Kb of memory between 640Kb and 1Mb is your computer's upper memory, which contains the video memory, upper memory blocks, and the ROM-based BIOS services, as discussed in the DOS and BIOS section of this book.

Note: *Conventional memory considerations are not as important within Windows 95 and Windows NT as they are within DOS. Because Windows uses the virtual memory model, the majority of your program execution will occur outside the first 1Mb of RAM—either in higher RAM locations, or in the virtual memory the computer uses on the hard drive. Later Tips discuss the virtual memory model in detail.*

ACCESSING CONVENTIONAL MEMORY 575

In short, your program's *memory model* defines the program's conventional memory use. Depending on the memory model your program uses, the compiler will allocate one or more 64Kb segments to store your program code and data. When your program must allocate memory dynamically, your program can use C functions, such as *malloc*, to allocate memory from the near heap, or *_fmalloc*, to allocate memory from the far heap. Tips 597 and 598 discuss the near and far heaps. In addition, your programs can use the DOS system services to allocate memory.

Note: *As a rule, your programs should use only one method to allocate and deallocate memory. To improve your program's portability, your programs should try to use C run-time library functions for memory management. Do not mix the C memory allocation functions with those DOS provides. By combining DOS and C memory allocation functions, you increase the possibility of errors and make your program more difficult to understand.*

UNDERSTANDING WHY THE PC AND DOS ARE RESTRICTED TO 1MB 576

Many people often refer to the *640Kb barrier* when they discuss DOS. In short, the 640Kb barrier refers to the conventional memory region within which your programs must run. As you have learned, however, DOS programs actually use the BIOS services and video memory, which reside in the memory range 640K to 1Mb. In addition, beginning with DOS 5, your programs and device drivers can actually reside in the upper memory area, so the DOS memory restriction actually occurs at 1Mb.

The 1Mb memory limit is more a PC limit than a DOS limit. The original PC (which used the 8088 processor) used a 16-bit segment address and a 16-bit offset within the segment. Within the PC's memory, segments occur at 16-byte intervals. The 65,536 unique segment addresses let the PC access 65,536 * 16 bytes (or 1,048,576) unique memory locations. Because DOS must run within this environment, DOS unfairly takes the blame for restricting your program memory.

Note: *Windows uses a special 32-bit type known as a **DWORD** to store segment and offset addresses. 32-bit **DWORD** lets Windows access up to 4Gb of RAM, provided the computer's processor is capable of accessing that much. Most Pentium computers can access up to 128Mb of RAM.*

PRODUCING AN ADDRESS FROM SEGMENTS AND OFFSETS 577

To manage address-specific memory locations, the PC uses a 16-bit *segment* and *offset* address. The segment address normally identifies the start of a 64Kb region. The offset address identifies a specific byte within the region. Segments can begin at 16-byte intervals called *paragraphs*. To address memory, the PC combines the segment and offset address to produce a 20-bit address, which can address 1,048,576 unique memory locations (1Mb). To create the 20-bit address, the PC shifts the 16-bit segment address left by four bit locations and then adds to the result the offset address. For example, assume that the segment address is 1234H. When the PC shifts the address to the left, the result becomes 12340H. Next, if the offset address is 5, the result is 12340H + 5H or 12345H. The following equation better illustrates the processing involved:

```
1234H Segment shifted becomes 12340H
Add the offset of    0005H
Yields 12345H
```

If you examine the operation in binary, the result becomes the following:

```
     0001 0010 0011 0100       Segment
0001 0010 0011 0100 0000       Shifted
                    0101       Offset
========================
0001 0010 0011 0100 0101       Result (20-bit address)
```

578 UNDERSTANDING EXPANDED MEMORY

As you have learned, DOS programs normally run within your computer's 640Kb conventional memory. Many larger programs, however, such as a spreadsheet, require more than 640Kb. The original IBM PC (8088) could not address memory beyond 1Mb. To let the PC access more than 1Mb of memory, the companies Lotus, Intel, and Microsoft created an *expanded memory specification* (EMS), which combines software and a special expanded memory board to trick the PC into accessing large amounts of memory.

To use expanded memory, your computer must contain an expanded memory board. To begin, the expanded memory specification software allocates a 64Kb block within upper memory (the 384Kb region between 640Kb and 1Mb). Next, the software divides the 64Kb region into four 16Kb sections called *pages*. When your program starts, it uses special expanded memory specification functions to allocate and load expanded memory. To do so, your program defines logical (16Kb) pages within the expanded memory region.

For example, if you have a 128Kb spreadsheet, the computer divides the data into eight 16Kb logical pages. When your program must access a specific logical page, the program uses an expanded memory specification function to *map* the logical page into one of the expanded memory specification pages in your computer's upper memory, which your DOS program can then directly access. As your program uses other logical pages, it maps pages in and out of the expanded memory specification area as necessary.

Many DOS programs require expanded memory mapping only because the 8088 processor cannot access memory locations beyond 1Mb. Although the expanded memory provides a way for the 8088 to access large amounts of data, the continual mapping of data introduces considerable overhead, which decreases your system performance. If you are using an 80286 or greater, your computer can access memory beyond 1Mb (called extended memory), which is a much faster process.

Note: *When a program uses expanded memory, the program code remains in the 640Kb conventional memory region. Only the program's data can reside in the expanded memory area.*

Note: *As you learned in Tip 574, conventional memory considerations are not as important within Windows 95 and Windows NT as they are within DOS. Because conventional memory is not as crucial, expanded memory is also not as crucial to newer computers, whether running Windows or DOS. In fact, the newer (x486 and higher) processors no longer support expanded memory, only extended memory, as Tip 580 details.*

USING EXPANDED MEMORY 579

Tip 578 introduced the PC's use of expanded memory. As a rule, to improve performance, your programs should use *extended memory* (see Tip 580) instead of expanded memory. However, if circumstances force you to write a program that must run on an older 8088-based PC, your programs will access the expanded memory specification services using the *int86* function and INT 67H, as the DOS and BIOS section of this book details. There are many different expanded memory specification services that let you allocate, map, deallocate, and manipulate expanded memory. For a complete description and example programs that use these services, refer to Kris Jamsa's book *DOS Programming: The Complete Reference*, Osborne/McGraw-Hill, 1991. The companion CD-ROM that accompanies this book, however, provides the sample program *showems.c*, which illustrates how you can use expanded memory within your programs. The following function, *check_for_ems*, uses memory registers and bitwise operations to determine whether your computer has loaded the expanded memory drivers:

```
int check_for_ems(void)
  {
     union REGS inregs, outregs;
     struct SREGS segs;
     int major, minor;   // DOS version
     struct DeviceHeader {
       struct DeviceHeader far *link;
       unsigned attributes;
       unsigned strategy_offset;
       unsigned interrupt_offset;
       char name_or_number_of_units[8];
     } far *dev;
     int i;
     char driver_name[9];

     // Get the DOS version
     inregs.x.ax = 0x3001;
     intdos (&inregs, &outregs);
     major = outregs.h.al;
     minor = outregs.h.ah;
     if (major < 2)
       return(0);      // Requires DOS 2.0
     else
       {
    // Get the list of lists
    inregs.h.ah = 0x52;
    intdosx (&inregs, &outregs, &segs);
    if (major == 2)
      dev = (struct DeviceHeader far *)
        MK_FP(segs.es + 1, outregs.x.bx + 7);
    else if ((major == 3) && (minor == 0))
      dev = (struct DeviceHeader far *)
        MK_FP(segs.es + 2, outregs.x.bx + 8);
    else
      dev = (struct DeviceHeader far *)
        MK_FP(segs.es + 2, outregs.x.bx + 2);
    while (FP_OFF(dev) != 0xFFFF)
      {
         if (dev->attributes & 0x8000)
          {  // Character device
            for (i = 0; i < 8; i++)
              driver_name[i] = dev->name_or_number_of_units[i];
            driver_name[8] = NULL;
          }
         if (! strcmp(driver_name, "EMMXXXX0"))
```

```
            return(1);   // Found driver
          dev = dev->link;
      }
       }
  return(0);
}
```

580 UNDERSTANDING EXTENDED MEMORY

The original IBM PC (8088) used 20-bit addressing, which restricted its memory access to 1Mb. Beginning with the IBM PC AT (the 80286), the PC gained the ability to use 24-bit addressing, which let it address up to 16Mb. The 386-, 486-, 586-, and 686-based machines increased the addressing to 32-bit, which let the PC address up to 4Gb of memory. When the PC first gained the ability to access memory beyond 1Mb, programmers called the memory beyond 1Mb *extended memory*. Because the original IBM PC cannot access memory beyond 1Mb, it cannot use extended memory.

To access extended memory, you must load an extended memory device driver. In DOS, the driver is usually *himem.sys*. When your DOS-based programs use extended memory, only the program's data can reside in extended memory. The program's code must reside in the 640Kb conventional memory. When your programs use extended memory, however, the system services that provide access to the memory must change your CPU's mode of execution from real mode to protected mode and back again. Changing CPU modes requires some processing time, which introduces overhead. However, the overhead is less than that of expanded memory—making extended memory more desirable.

Note: *As you learned in Tip 574, conventional memory considerations are not as important within Windows 95 and Windows NT as they are within DOS. Computers which run Windows 95 still use so-called "extended memory" to manage the access to the memory beyond the computer's first 1Mb. Computers which run Windows NT manage memory without the transparent use of extended memory that Windows 95 requires. As you will learn in later Tips, your programs should use the virtual memory model to manage all memory from within Windows programs.*

581 UNDERSTANDING REAL AND PROTECTED MODES

DOS is a single-tasking operating system, which means (with the exception of device drivers and memory-resident programs) that there is normally only one program running at a time. Because DOS is a single-tasking operating system, protecting one program from another is not a significant issue. Therefore, DOS lets programs access memory in any way they desire. In other words, a DOS-based program can change any conventional memory location's value. When you run multiple programs at the same time, one program cannot randomly change memory as it can within a single-tasking environment, because the program would likely overwrite the contents of another program in memory. In a multi-program environment, the operating system must protect one program's memory from that of another. To protect programs in memory, the operating system relies on hardware-based memory protection.

Beginning with the 80286 chip, the CPU can run in one of two modes: *real* or *protected*. Real mode exists for compatibility with the original 8088-based IBM PC. DOS uses real mode, which has no memory protection. Other operating systems, such as Unix, OS/2, or Windows, can run in protected mode. In protected mode, one program cannot access another program's memory. In addition, within protected mode the PC changes from its segment and offset addressing scheme to one that lets the CPU use 24-bit addressing within the 80286 and 32-bit addressing within 386-based (and higher) machines. In this way, protected mode lets the PC address extended memory for code and data. When your DOS-based programs use extended memory, the software they use to access the extended memory transparently switches the CPU from real mode (which was running DOS) to protected mode (which can access the extended memory), and then back to real mode again.

ACCESSING EXTENDED MEMORY 582

Before your programs can use extended memory, you must install an extended memory device driver (usually *himem.sys*). Next, using the DOS multiplex interrupt, INT 2FH service 4300H, your programs can obtain an entry point in the computer's memory for the extended memory services. The extended memory driver provides functions that let your programs allocate, deallocate, and manipulate extended memory. To access the services, you assign different parameters to PC registers and then branch to the specified entry point. For a complete description of the extended memory services, refer to Kris Jamsa's book *DOS Programming: The Complete Reference*, Osborne/McGraw-Hill, 1991. To help you better understand the extended memory services, the companion CD-ROM that accompanies this book includes the file *xmsdemo.c*.

UNDERSTANDING THE HIGH MEMORY AREA 583

As you have learned, extended memory is your computer's memory above 1Mb. When DOS programs access extended memory, the CPU changes from real mode to protected mode and back again. If you are using a 386 and DOS 5 or later, you can take advantage of a "glitch" in the 386 processor's design that lets you access the first 64Kb of extended memory from within real mode. As shown in Figure 583, this 64Kb region is called the *high memory area*.

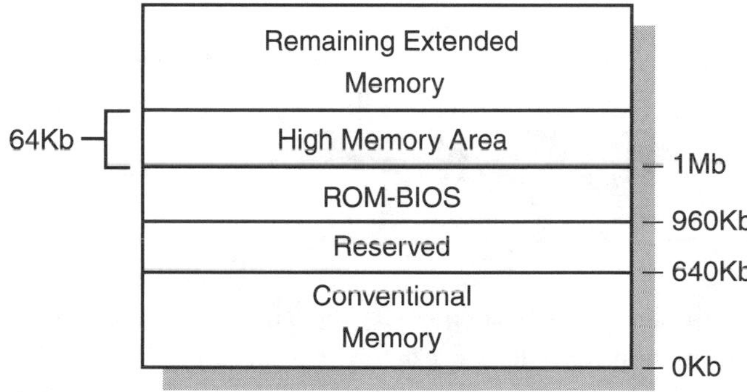

Figure 583 *The high memory area is the first 64Kb of extended memory.*

The best way to use the high memory area is to load the DOS kernel into it, freeing memory within the 640Kb conventional memory. However, if DOS is not using the high memory area, your program can allocate it using an extended memory service. To load DOS into the high memory area, you must install the *himem.sys* driver and then use the DOS=HIGH entry with the *config.sys* file.

Note: *As this Tip indicates, your programs should not use the High Memory Area if they will run on computers more recent than the x386. If your programs run under Windows 95 or Windows NT, your programs should not use the High Memory Area.*

UNDERSTANDING THE STACK 584

The *stack* is a region of memory within which your programs temporarily store data as they execute. For example, when your programs pass parameters to functions, C places the parameters on the stack. When the function completes, C removes the items from the stack. Likewise, when your functions declare local variables, C stores the variables' values on the stack during the function's execution. When the function completes, C discards the variables.

The stack is so named because programs *push* values onto the stack, much as you would stack cafeteria trays one on top of another, and later *pop* the top value off the stack, just as you would remove the top cafeteria tray. Depending on the program's memory model, the amount of stack space the compiler provides will vary. Depending on your program's use of functions and parameters, the amount of stack space your program requires will differ. At a minimum, the compiler will allocate 4Kb of stack space. If your program needs more or less stack space, you can use compiler and linker directives to control the amount of stack space the compiler and linker allocate. The PC uses two registers to locate the stack. The stack segment register (SS) points to the start of the stack, and the stack pointer register (SP) points to the top of the stack.

585 Understanding Different Stack Configurations

In Tip 584 you learned that your program uses the stack to temporarily store information—primarily during function calls. Depending on your program's use of functions and the number and size of parameters your programs pass to those functions, the amount of stack space your program requires will vary from one program to the next. When you use the small memory model, C will allocate the stack space from the top of the data segment, as shown in Figure 585.

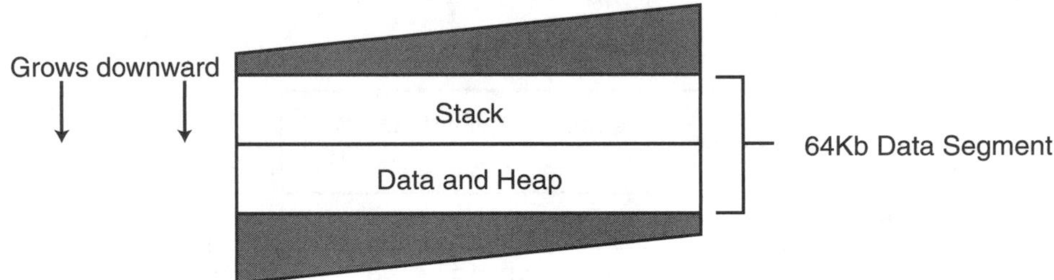

Figure 585 *C's small memory model stack space allocation.*

On the other hand, when you use the large or compact memory model, C will allocate an entire 64Kb segment for the stack. If your program places more information onto the stack than the stack can hold, a *stack-overflow error* will occur. If your program has disabled stack checking, you will be unaware of the error and the data you have placed onto the stack may overwrite your program data. Tip 586 presents ways you can determine your program's current stack size.

Note: *You will learn more about stack checking in this book's Optimization section.*

Windows, however, constructs the stack somewhat differently from a DOS stack. The Windows stack defaults to 1Mb, and its size limit is the limit of virtual memory, which means that a stack could be as large as 250Mb or more. The 250Mb stack size therefore reduces your concern for protecting programs from stack overruns.

586 Determining Your Program's Current Stack Size

Depending on your program's function and parameter use, the amount of stack space your program requires will vary. Using compiler and linker directives, your programs can allocate a specific stack size. As your programs execute, there may be times when you want to know the current stack size. If you are using *Turbo C++ Lite*, you can use the *_stklen* global variable. The following program, *stk_len.c*, uses the *_stklen* global variable to display the current stack size:

```
#include <stdio.h>
#include <dos.h>
```

```
void main(void)
 {
   printf("The current stack size %d bytes\n", _stklen);
 }
```

If you are using Microsoft *Visual C++*, the *stackavail* function returns the amount of available stack space.

CONTROLLING THE STACK SPACE WITH _STKLEN 587

In Tip 586 you learned that the *_stklen* global variable lets your programs determine the current stack size. In addition, your programs can use the *_stklen* variable to control the amount of stack space the compiler allocates. To specify a stack size using *_stklen*, your programs must declare the variable as an external global variable. The following *Turbo C Lite* program, *8kbstack.c*, uses the *_stklen* global variable to allocate an 8Kb stack:

```
#include <stdio.h>
#include <dos.h>

extern unsigned _stklen = 8096;

void main(void)
 {
   printf("The current stack size is %d bytes\n", _stklen);
 }
```

Note: *Within Windows, most compilers will let you set the stack size a certain thread requires within the command that creates the thread. The section of this book entitled Processes and Threads discusses threads in detail.*

ASSIGNING A VALUE TO A MEMORY RANGE 588

When your programs work with arrays and pointers to memory ranges, there may be times when you will want to initialize the memory to a specific value. To do so, your programs can use the *memset* function. You will implement the *memset* function as shown here:

```
#include <mem.h>

void *memset(void *ptr, int character, size_t num_bytes);
```

The *ptr* parameter is a pointer to the first byte in the memory range. The *character* parameter is the byte value you want to assign to the range of memory. Finally, the *num_bytes* parameter specifies the number of bytes in the memory range. The function returns a pointer to the start of the memory range. The following statement uses the *memset* function to initialize a character string array to *NULL*:

```
char string[128];

memset(string, NULL, sizeof(string));
```

COPYING ONE MEMORY RANGE TO ANOTHER 589

When your programs work with character strings, they can use the *strcpy* function to copy one string's contents to another. When you must copy an array of integer or floating-point values, however, your programs can perform similar processing using the *memmove* or *memcpy* functions, as shown here:

```
#include <mem.h>

void *memmove(void *target, const void *source, size_t num_bytes);
void *memcpy(void *target, const void *source, size_t num_bytes);
```

The *target* and *source* parameters are pointers to the array into which the function copies the data (*target*) and from which the function makes the copy (*source*). The *num_bytes* parameter specifies the number of bytes to copy. The primary difference between the two functions is that *memmove* correctly copies data between two byte ranges that may overlap in memory, while *memcpy* might copy the data incorrectly. The following *Turbo C Lite* program, *memmove.c*, uses the *memmove* function to copy the contents of a floating-point array:

```
#include <stdio.h>
#include <mem.h>

void main(void)
 {
   float values[] = { 1.1, 2.2, 3.3, 4.4, 5.5 };
   float empty[5];
   int i;

   memmove(empty, values, sizeof(values));
   for (i = 0; i < 5; i++)
     printf("%3.1f ", empty[i]);
 }
```

590 COPYING A MEMORY RANGE UP TO A SPECIFIC BYTE

When your programs work with arrays, there may be times when you must copy one array's contents to another array. Depending on the array's contents, you might want the copy to move up to *n* bytes, or immediately end if the copy encounters a specific character. To perform such processing, your programs can use the *memccpy* function, as shown here:

```
#include <mem.h>

void *memccpy(void *target, const void *source, int character, size_t num_bytes);
```

The *target* and *source* parameters are pointers to the array into which the function is copying the data (*target*) and from which the function is making the copy (*source*). The *character* parameter specifies the character that, if copied, immediately ends the copy operation. The *num_bytes* parameter specifies the number of bytes to copy. If the function copies *num_bytes*, it returns the value *NULL*. If the function encounters the character you specify, the function returns a pointer to the byte in the target that immediately follows the character. The following *Turbo C Lite* program, *memccpy.c*, uses the *memccpy* function to copy the letters A through K to the array *target*:

```
#include <stdio.h>
#include <mem.h>

void main(void)
 {
   char alphabet[27] = "ABCDEFGHIJKLMNOPQRSTUVWXYZ";
   char target[27];
   char *result;

   result = memccpy(target, alphabet, 'K', sizeof(alphabet));
   if (result)
     *result = NULL;
   printf(target);
 }
```

591 COMPARING TWO ARRAYS OF UNSIGNED CHAR

When your programs work with arrays, there may be times when you want to compare two memory ranges. You will most commonly compare two memory ranges to check two character strings. To compare two memory ranges, your programs can use the *memcmp* or *memicmp* functions, as shown here:

```
#include <mem.h>

int memcmp(const void *block_1, const void *block_2, size_t num_bytes);
int memicmp(const void *block_1, const void *block_2, size_t num_bytes);
```

The difference between the *memcmp* and *memicmp* functions is that *memicmp* ignores character case. The *block_1* and *block_2* parameters are pointers to the start of each memory range. The *num_bytes* parameter specifies the number of bytes to compare. The function returns one of the values listed in Table 591.

Value	Meaning
less than 0	*block_1* is less than *block_2*
0	blocks are the same
greater than 0	*block_1* is greater than *block_2*

Table 591 The return values from ***memcmp*** *and* ***memicmp****.*

The following *Turbo C Lite* program, *memcmp.c,* uses *memcmp* and *memicmp* to compare two character strings:

```
#include <stdio.h>
#include <mem.h>

void main(void)
 {
   char *a = "AAA";
   char *b = "BBB";
   char *c = "aaa";

   printf("Comparing %s and %s with memcmp %d\n",
          a, b, memcmp(a, b, sizeof(a)));
   printf("Comparing %s and %s with memicmp %d\n",
          a, b, memicmp(a, c, sizeof(a)));
 }
```

SWAPPING ADJACENT CHARACTER STRING BYTES 592

When you work with different types of computers, there may be times when you must swap adjacent bytes of memory. To do so, your programs can use the *swab* function, as shown here:

```
#include <stdlib.h>

void swab(char *source, char *target, int num_bytes);
```

The *source* parameter is a pointer to a string whose bytes you want to swap. The *target* parameter is a pointer to a string to which *swab* assigns the bytes it swapped. The *num_bytes* parameter specifies the number of bytes to swap. The following *Turbo C Lite* program, *swab.c,* illustrates the *swab* function:

```
#include <stdio.h>
#include <stdlib.h>
#include <string.h>
#include <mem.h>

void main(void)
 {
   char *source = "aJsm\'a s/C+C/+#CP orrgmaem/'r siBlb e";
   char target[64];

   memset(target, NULL, sizeof(target));
   swab(source, target, strlen(source));
   printf("Source: %s\n Target %s\n", source, target);
 }
```

593 ALLOCATING DYNAMIC MEMORY

When your programs declare an array, the C compiler allocates memory to hold the array. Should your program requirements change and the array size must increase or decrease, you must edit and recompile the program. To reduce the number of changes you must make to your programs for changes in array sizes, your programs can allocate their own memory during run time. When you allocate memory in this way, the C run-time library returns a pointer to the start of the memory range. Your programs can then work with the memory using an array or pointer format, whichever you prefer. When you allocate memory during run time, your program can use the *malloc* run-time library function, as shown here:

```
#include <alloc.h>

void *malloc(size_t number_of_bytes);
```

The *number_of_bytes* parameter specifies the number of bytes you want for the array's size. If the *malloc* function successfully allocates the byte range, *malloc* will return a pointer to the start of the range. If an error occurs, *malloc* will return *NULL*. The following *Turbo C Lite* program, *malloc.c*, uses *malloc* to allocate memory for a character string array, an array of integer values, and a floating-point array:

```
#include <stdio.h>
#include <alloc.h>

void main(void)
 {
   char *string;
   int *int_values;
   float *float_values;

   if ((string = (char *) malloc(50)))
     printf("Successfully allocated a 50 byte string\n");
   else
     printf("Error allocating string\n");
   if ((int_values = (int *) malloc(100 * sizeof(int))) != NULL)
     printf("Successfully allocated int_values[100]\n");
   else
     printf("Error allocating int_values[100]\n");
   if ((float_values = (float *) malloc(25 * sizeof(float))) != NULL)
     printf("Successfully allocated float_values[25]\n");
   else
     printf("Error allocating float_values[25]\n");
 }
```

As you can see, the program invokes *malloc* with the required number of bytes. If *malloc* returns *NULL*, the program will display an error message.

594 REVISITING CASTS

In Tip 593 you learned that you can use the *malloc* run-time library function to instruct your programs to allocate memory during run time. As you learned, the *malloc* function returns a *void* pointer, as shown here:

```
void *malloc(size_t number_of_bytes);
```

When you use *malloc* to allocate memory, your programs should cast *malloc's* result to a pointer of the type you want. For example, the following statement uses *malloc* to allocate a pointer to 100 values of type *int*:

```
int *int_values;

int_values = (int *) malloc(100 * sizeof(int));
```

If you are allocating memory to hold 50 floating-point values, your statements will become the following:

```
float *float_values;
float_values = (float *) malloc(50 * sizeof(float));
```

When you cast *malloc's* return value in this way, you can eliminate compiler warning messages.

RELEASING MEMORY WHEN IT IS NO LONGER NEEDED 595

As you have learned, your programs can use the *malloc* function to allocate memory during execution to hold arrays or other items. When your program no longer needs the memory, your program should release the memory so that your program can reuse it for a different purpose. To release allocated memory, your programs can use the *free* function, as shown here:

```
#include <alloc.h>
void free(void *ptr);
```

The *ptr* parameter is a pointer to the start of the memory range you want to release. The following *Turbo C Lite* program, *free.c*, uses *malloc* to allocate an integer array. The program then uses the array. When the program no longer needs the array, it uses the function *free* to release the memory that corresponds to the array, as shown here:

```
#include <stdio.h>
#include <alloc.h>

void main(void)
 {
   int *int_values;
   int i;

   if ((int_values = (int *) malloc(100 * sizeof(int))) == NULL)
     printf("Error allocating the array\n");
   else
     {
       for (i = 0; i < 100; i++)
         int_values[i] = i;
       for (i = 0; i < 100; i++)
         printf("%d ", int_values[i]);
       free(int_values);
     }
 }
```

***Note:** If your programs do not use **free** to release memory, the program automatically releases the memory when it ends. As a rule, however, your programs should release memory as soon as they no longer need it.*

ALLOCATING MEMORY USING THE CALLOC FUNCTION 596

As you have learned, your programs can use the *malloc* function to allocate memory dynamically during run time. When you use *malloc*, you specify the number of bytes you want to allocate. In addition to using *malloc*, C lets your programs allocate memory using *calloc*. The difference between the two functions is that *malloc* directs you to specify the number of bytes you want, whereas *calloc* directs you to specify the number of elements of a specific size you want, as shown here:

```
#include <alloc.h>
void *calloc(size_t number_of_items, size_t item_size);
```

The *number_of_items* parameter specifies how many elements *calloc* must allocate memory for. The *item_size* parameter specifies each element's size in bytes. If *calloc* successfully allocates the memory, *calloc* will return a pointer to the start of the memory range. If an error occurs, *calloc* will return *NULL*. The following *Turbo C Lite* program, *calloc.c*, uses *calloc* to allocate several different types of arrays:

```
#include <stdio.h>
#include <alloc.h>

void main(void)
 {
   char *string;
   int *int_values;
   float *float_values;

   if ((string = (char *) calloc(50, sizeof(char))))
     printf("Successfully allocated a 50 byte string\n");
   else
     printf("Error allocating string\n");
   if ((int_values = (int *) calloc(100, sizeof(int))) != NULL)
     printf("Successfully allocated int_values[100]\n");
   else
     printf("Error allocating int_values[100]\n");
   if ((float_values = (float *) calloc(25, sizeof(float))) != NULL)
     printf("Successfully allocated float_values[25]\n");
   else
     printf("Error allocating float_values[25]\n");
 }
```

Note: *When your program has finished using the memory* ***calloc*** *allocates, your program should use* ***free*** *to release the memory.*

597 UNDERSTANDING THE HEAP

When your programs allocate memory dynamically, the C run-time library gets the memory from a collection of unused memory called the *heap*. When you compile programs using the small memory model, the heap is the area of memory between the top of your program's data area and the stack, as shown in Figure 597.

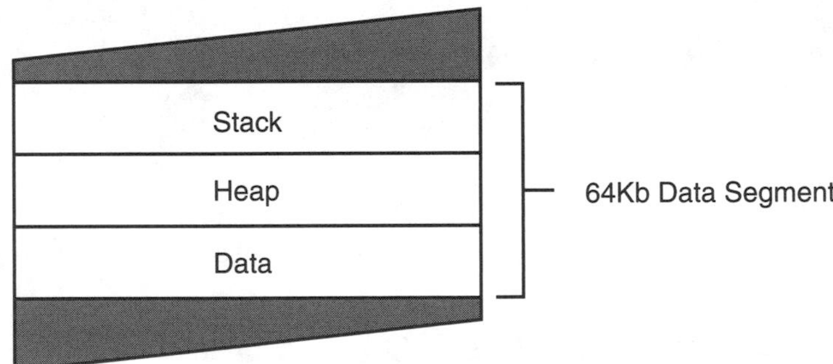

Figure 597 *The heap resides between the program's data area and stack.*

As you can see, the heap resides in your program's data segment. Therefore, the amount of heap space available to your program is fixed for that program, but can vary from one program to the next. When you use *calloc* or *malloc* to allocate memory, the most memory the functions can allocate is 64Kb (assuming the heap contains no data and no stack). The following program, *no_space.c*, tries to allocate three 30Kb arrays. Because the heap does not have 90Kb available, the memory allocation fails, as shown here:

```
#include <stdio.h>
#include <alloc.h>

void main(void)
 {
   char *one, *two, *three;
```

```
  if ((one = (char *) malloc(30000)) == NULL)
    printf("Error allocating array one\n");
  else if ((two = (char *) malloc(30000)) == NULL)
    printf("Error allocating array two\n");
  else if ((three = (char *) malloc(30000)) == NULL)
    printf("Error allocating array three\n");
  else
    printf("All arrays successfully allocated\n");
}
```

In the large memory model, the total heap size is not restricted to 64Kb; however, the largest value you can allocate at any one time is still restricted to a 64Kb segment. To allocate a value larger than 64Kb, you must use the huge model. Try compiling the *no_space.c* program using the large memory model. The program should then be able to satisfy the memory allocation's requirements.

Windows programs use the heap similarly to DOS programs. However, Windows programs have access to two heaps: the *global heap* and the *local heap*. All programs can use the global heap, which Windows uses to handle large memory blocks (256 bytes or more). Windows also gives each program access to its own local heap, which Windows uses to handle small memory blocks (256 bytes or fewer). As a rule, most Windows programs work from the global heap—because it has no effective size limitation. However, your programs may use the local heap for small memory storage purposes.

GETTING AROUND THE 64KB HEAP LIMIT 598

As you have learned, when your DOS-based programs allocate memory from the heap, your programs can allocate, at most, 64Kb of memory. Because the 64Kb limit is a DOS (PC real mode) restriction, many DOS-based compilers provide functions named *farmalloc* and *farcalloc* that let your programs allocate memory from a *far heap*, which resides outside the current data segment, as shown here:

```
#include <alloc.h>

void far *farcalloc(unsigned long number_of_items, unsigned long element_size);
void far *farmalloc(unsigned long number_of_bytes);
```

The parameters your programs pass to *farcalloc* and *farmalloc* are identical in function to those passed to *calloc* and *malloc*. When you allocate memory from the far heap, you will use a far pointer to the data. The following *Turbo C Lite* program, *fmalloc.c*, allocates several arrays from the far heap:

```
#include <stdio.h>
#include <alloc.h>

void main(void)
 {
   char far *string;
   int far *int_values;
   float far *float_values;

   if ((string = (char *) farmalloc(50)))
     printf("Successfully allocated a 50 byte string\n");
   else
     printf("Error allocating string\n");
   if ((int_values = (int *) farmalloc(100 * sizeof(int))) != NULL)
     printf("Successfully allocated int_values[100]\n");
   else
     printf("Error allocating int_values[100]\n");
   if ((float_values = (float *) farmalloc(25 * sizeof(float))) != NULL)
     printf("Successfully allocated float_values[25]\n");
   else
     printf("Error allocating float_values[25]\n");
 }
```

In the large memory model, your program treats all pointers as far pointers. However, far pointer use is transparent within your application.

Note: *When your programs use* ***farcalloc*** *or* ***farmalloc*** *to allocate memory from the far heap, you should use the* ***farfree*** *function to release the memory when your programs no longer need it. If you are using a different compiler, the names of these functions might differ. Refer to the* ***far heap*** *description in your compiler documentation.*

Within Windows, you will get around the heap size restriction by allocating from the global heap, rather than the local heap. In fact, most Windows compilers allocate from the global heap by default and provide a different command you should use to allocate from the local heap.

599 ALLOCATING MEMORY FROM THE STACK

As you have learned, the *malloc* and *calloc* functions let you allocate memory from the heap. When you have finished with the memory, you should release it using *free*. Depending on your program, sometimes you must allocate memory that only exists during a specific function call. To do so, your programs can use *alloca* to allocate the memory from the stack, as shown here:

```
#include <malloc.h>

void *alloca(size_t number_of_bytes);
```

The *number_of_bytes* parameter specifies the size of the memory range your program must allocate. If *alloca* succeeds, it will return a pointer to the start of the memory block. If an error occurs, the function will return *NULL*. Do not use *free* to release memory your program has allocated using *alloca—free* works with the heap, while *alloca* works with the stack. The program releases the memory it allocated automatically when the function containing the allocated memory ends.

Note: *For the program to restore the stack pointer correctly, the function must contain local variables. To ensure a correct stack frame, declare a local variable after you declare the pointer variable to which* ***alloca*** *assigns its result, as shown here:*

```
char *pointer;
char stack_fix[1];

stack_fix[0] = NULL;
pointer = alloca(size)
```

The following program, *alloca.c*, illustrates how to use the *alloca* function:

```
#include <stdio.h>
#include <malloc.h>

void some_function(size_t size)
 {
   int i;
   char *pointer;

   char stack_fix[1];
   stack_fix[0] = NULL;
   if ((pointer = alloca(size)) == NULL)
     printf("Error allocating %u bytes from the stack\n", size);
   else
     {
       for (i = 0; i < size; i++)
         pointer[i] = i;
       printf("Allocated and used a buffer of %u bytes\n", size);
     }
 }

void main(void)
```

```
{
  some_function(1000);
  some_function(32000);
  some_function(65000);
}
```

ALLOCATING HUGE DATA 600

As you have learned, the largest array size you can create is 64Kb. If your applications need a larger array, you can allocate memory for a *huge* array. To help your programs work with huge data structures, many DOS-based C compilers provide the *halloc* and *hfree* functions, as shown here:

```
#include <malloc.h>

void huge *halloc(long number_of_elements, size_t size);
void hfree(void huge *pointer);
```

The *number_of_elements* parameter specifies the number of array elements. The *size* parameter specifies each element's size in bytes. If *halloc* succeeds, it will return a pointer to the start of the memory area. If an error occurs, *halloc* will return *NULL*. The following program, *hugeint.c*, uses *halloc* to allocate an array of 100,000 bytes:

```
#include <stdio.h>
#include <malloc.h>

void main(void)
  {
     long int i;
     int huge *big_array;

     if ((big_array = (int huge *) halloc (100000L,
       sizeof(long int))) == NULL)
       printf ("Error allocating huge array\n");
     else
       {
         printf("Filling the array\n");
         for (i = 0; i < 100000L; i++)
           big_array[i] = i % 32768;
         for (i = 0; i < 100000L; i++)
           printf ("%d ", big_array[i]);
         hfree(big_array);
       }
  }
```

Note: *Again, as you have learned, the limitation on array size does not apply to programs you create within the Windows environment. In fact, some Windows compilers no longer support the* ***huge*** *keyword for array declarations.*

CHANGING THE SIZE OF AN ALLOCATED BLOCK 601

As you have learned, C lets your programs dynamically allocate memory during execution. After you allocate a block of memory, there may be times later when you must change the block's size. In such cases, your programs can use *realloc*, as shown here:

```
#include <stdlib.h>

void *realloc(void *block, size_t desired_bytes);
```

The *block* parameter is a pointer to the previously allocated memory. The *desired_bytes* parameter is the size required for the new block. The *realloc* function can shrink or expand a block. If *realloc* succeeds, it returns a pointer to the block, which may be a different pointer from the original. In other words, *realloc* may move the block to find space (copying the data, as necessary). If an error occurs, *realloc* returns *NULL*. The following *Turbo C Lite* program, *realloc.c*, uses *realloc* to increase the size of a block from 100 bytes to 1,000 bytes:

```
#include <stdio.h>
#include <alloc.h>

void main(void)
 {
   char *string, *new_string;

   if ((string = (char *) malloc(100)))
     {
       printf("Successfully allocated a 100 byte string\n");
       if ((new_string = (char *) realloc(string, 1000)))
         printf("String size increased to 1000\n");
       else
         printf("Error reallocating the string\n");
     }
   else
     printf("Error allocating the 100 byte string\n");
 }
```

602 UNDERSTANDING BRK

As you have learned, the heap starts at the byte location that immediately follows the last byte in the data segment. The *break value* is the address at which the heap starts. The *brk* function lets your programs change the break value, assigning it to a specific address, as shown here:

```
#include <alloc.h>

int brk(void *address);
```

If the *brk* function succeeds, it returns the value 0. If an error occurs, brk returns –1. The following *Turbo C Lite* program, *brk.c*, uses *brk* to set the break value 512 bytes before its current location. The program uses the *coreleft* function to display the amount of available heap before and after the *brk* operation, as shown here:

```
#include <stdio.h>
#include <alloc.h>

void main(void)
 {
   char *ptr;

   printf("Starting heap available %u\n", coreleft());
   ptr = (char *) malloc(1);  // Get pointer to current break value
   if (brk(ptr + 512) == 0)
   printf("Ending heap available %u\n", coreleft());
 }
```

603 VALIDATING THE HEAP

If you are experiencing errors in a program that allocates memory dynamically and you cannot locate the source of the error, then you might consider performing *heap validations*. To help you test the state of the heap, many compilers provide a collection of run-time library routines, such as *heapwalk* and *heapcheck*. Several of the Tips that follow present ways in which your programs can test the heap.

PERFORMING A FAST HEAP CHECK 604

As you have learned, to help you locate errors in your programs that perform dynamic memory allocation, you might want to check the state of the heap. One routine your programs can use to check the heap is *heapcheck,* as shown here:

```
#include <alloc.h>

int heapcheck(void);
```

The *heapcheck* function walks through (moves through one at a time) the heap and examines each of the heap entries. The function returns one of the values listed in Table 604.

Value	Description
_HEAPEMPTY	No heap
_HEAPOK	Heap is verified
_HEAPCORRUPT	One or more corrupted entries

Table 604 The return values for the ***heapcheck*** *function.*

The following *Turbo C Lite* program, *heapchk.c*, uses the *heapcheck* function to test the state of the heap:

```
#include <stdio.h>
#include <alloc.h>

void main(void)
 {
   char *buffer, *second_buffer;
   int i, state;

   buffer = malloc(100);
   second_buffer = malloc(100);
   state = heapcheck();
   if (state == _HEAPOK)
     printf("Heap is ok\n");
   else if (state == _HEAPCORRUPT)
     printf("Heap is corrupt\n");
   for (i = 0; i <= 100; i++)
     buffer[i] = i;
   state = heapcheck();
   if (state == _HEAPOK)
     printf("Heap is ok\n");
   else if (state == _HEAPCORRUPT)
     printf("Heap is corrupt\n");
 }
```

When the program first allocates the memory, *heapcheck* returns a status value stating that the heap is OK. After the program assigns values to *buffer*, however, *heapcheck* returns a status value stating that the heap is corrupted. If you examine the *for* loop closely, you will find that it assigns 101 values to a 100-byte buffer (which corrupts the entry). Using *heapcheck*, you can detect such errors very quickly.

FILLING FREE HEAP SPACE 605

One way to detect memory-use errors in programs that work with dynamic memory is to fill all the free heap space with a specific value. Then, as you perform your memory operations, you can test if the value has changed. To help you fill and test free heap space, many C compilers provide the following functions:

```
#include <alloc.h>

int heapcheckfree(unsigned int value);
int heapfillfree(unsigned int value);
```

The *value* parameter is the value you want to assign to the free heap space. The functions return one of the values listed in Table 605.

Value	Description
_HEAPEMPTY	No heap
_HEAPOK	Heap is verified
_HEAPCORRUPT	One or more corrupted entries
_BADVALUE	A different value was encountered

*Table 605 The return values for **heapcheckfree** and **heapfillfree**.*

The following *Turbo C Lite* program, *fillheap.c*, uses *heapcheckfree* and *heapfillfree* to detect a programming error:

```
#include <stdio.h>
#include <alloc.h>

void main(void)
 {
   char *buffer1, *buffer2, *buffer3;
   int i, state;

   buffer1 = (char *) malloc(100);
   buffer2 = (char *) malloc(200);
   buffer3 = (char *) malloc(300);
   free(buffer2);              // Free space in the middle
   state = heapfillfree('A');
   if (state == _HEAPOK)
     printf("Heap is ok\n");
   else if (state == _HEAPCORRUPT)
     printf("Heap is corrupt\n");
   for (i = 0; i <= 150; i++)
     buffer1[i] = i;
   state = heapcheckfree('A');
   if (state == _HEAPOK)
     printf("Heap is ok\n");
   else if (state == _HEAPCORRUPT)
     printf("Heap is corrupt\n");
   else if (state == _BADVALUE)
     printf("Value has been changed in free space\n");
 }
```

606 Checking a Specific Heap Entry

In Tip 604 you learned how to use the *heapcheck* function to test the status of the entire heap. As you check for errors, you may also want to test the state of individual heap entries. To perform the test, your programs can use the *heapchecknode* function, as shown here:

```
#include <alloc.h>

int heapchecknode(void *block);
```

The *block* parameter is a pointer to a dynamically allocated block of memory. The function will return one of the values shown in Table 606.

Value	Description
_HEAPEMPTY	No heap
_HEAPOK	Heap is verified
_HEAPCORRUPT	One or more corrupted entries
_BADNODE	Block was not found
_FREEENTRY	Block is free
_USEDENTRY	Block is in use

***Table 606** The return values for **heapchecknode**.*

The following *Turbo C Lite* program, *heapnode.c*, illustrates how to use the *heapchecknode* function:

```
#include  <stdio.h>
#include <alloc.h>

void main(void)
 {
   char *buffer, *second_buffer;
   int i, state;

   buffer = (char *) malloc(100);
   second_buffer = malloc(100);
   state = heapchecknode(buffer);
   if (state == _USEDENTRY)
     printf("buffer is ok\n");
   else
     printf("buffer is not ok\n");
   state = heapchecknode(second_buffer);
   if (state == _USEDENTRY)
     printf("second_buffer is ok\n");
   else
     printf("second_buffer is not ok\n");
   for (i = 0; i <= 100; i++)
     buffer[i] = i;
   state = heapchecknode(buffer);
   if (state == _USEDENTRY)
     printf("buffer is ok\n");
   else
     printf("buffer is not ok\n");
   state = heapchecknode(second_buffer);
   if (state == _USEDENTRY)
     printf("second_buffer is ok\n");
   printf("second_buffer is not ok\n");
 }
```

607 WALKING THE HEAP ENTRIES

To help you examine the individual heap entries, many C compilers provide a function named *heapwalk*. *Heapwalk* lets you display the size and state (either in-use or available) of each heap entry, as shown here:

```
#include <alloc.h>

int heapwalk(struct heapinfo *info)
```

The *info* parameter is a pointer to a structure of type *heapinfo*, as shown here:

```
struct heapinfo
 {
   void *pointer;
   unsigned int size;
```

```
    int in_use;
 };
```

Before the first call to *heapwalk*, you must set the *heapinfo* structure's *pointer* member to *NULL*. The *heapwalk* function returns one of the values shown in Table 607.

Value	Description
_HEAPEMPTY	No heap
_HEAPOK	Heap is verified
_HEAPEND	Last heap entry

Table 607 *The return values for* ***heapwalk****.*

The following *Turbo C Lite* program, *heapwalk.c*, walks (travels through) the heap entries using *heapwalk*:

```
#include <stdio.h>
#include <alloc.h>

void main(void)
 {
   char *buffer1, *buffer2, *buffer3;
   struct heapinfo node = { NULL, 0, 0};

   buffer1 = (char *) malloc(100);
   buffer2 = (char *) malloc(200);
   buffer3 = (char *) malloc(300);
   free(buffer2);
   while (heapwalk(&node) == _HEAPOK)
     printf("Size %u bytes State %s\n", node.size,
            (node.in_use) ? "In use": "Free");
 }
```

608 PEEKING INTO A SPECIFIC MEMORY LOCATION

Depending on your program's function, there may be times when you want the program to access specific segment and offset locations in memory. If you are working with *far* pointers, you can combine a segment and offset the address using *MK_FP*. In addition, your programs can use the *peekb* and *peek* functions, as shown here:

```
#include <dos.h>

char peekb(unsigned segment, unsigned offset);
int peek(unsigned segment, unsigned offset);
```

The *segment* and *offset* parameters combine to specify the desired memory location. The following *Turbo C Lite* program, *file_scr.c*, uses the *peekb* function to capture the contents of the current (text mode) screen display and to send the capture to the file *save_scr.dat*. The program peeks at (reads) the character and attribute byte. Therefore, the *file_scr.c* program must *peek* 4,000 characters and 4,000 attributes, as shown here:

```
#include <stdio.h>
#include <dos.h>
#define VIDEO 0xB800    // CGA base

void main(void)
 {
   FILE *fp;
   int offset;

   if ((fp = fopen("SAVE_SCR.DAT", "wb")) == NULL)
     printf("Error opening file\n");
   else
    {
```

```
        for (offset = 0; offset < 8000; offset++)
         fprintf(fp, "%c", peekb(VIDEO, offset));
       fclose(fp);
     }
 }
```

Note: *The file_scr.c program uses the CGA video base address of B800H. If you are using an EGA, VGA, or other video adapter, you may need to change this base address.*

POKING VALUES INTO MEMORY 609

In Tip 608 you learned how to use the *peekb* and *peek* functions to read values from specific segment and offset addresses in memory. In a similar way, most C compilers provide the functions *poke* and *pokeb*, which let your programs place values at specific memory locations, as shown here:

```
#include <dos.h>

void pokeb(unsigned segment, unsigned offset, char value);
void poke(unsigned segment, unsigned offset, int value);
```

The following *Turbo C Lite* program, *scr_poke.c*, uses the *pokeb* function to restore the screen contents that the *file_scr.c* program previously saved:

```
#include <stdio.h>
#include <dos.h>
#define VIDEO 0xB800   // CGA base

void main(void)
 {
   FILE *fp;
   int offset;
   char value;

   if ((fp = fopen("SAVE_SCR.DAT", "rb")) == NULL)
     printf("Error opening file\n");
   else
    {
      for (offset = 0; offset < 8000; offset++)
       {
         fscanf(fp, "%c", &value);
         pokeb(VIDEO, offset, value);
       }
      fclose(fp);
    }
 }
```

UNDERSTANDING PC PORTS 610

The PC uses two techniques to communicate with internal hardware devices. First, the PC can reference memory locations that the device or the PC has previously reserved for the device. The term for input and output operations that occur through such memory locations is *memory-mapped I/O*. The PC uses memory-mapped I/O to perform video output. In addition, the PC can communicate with hardware devices using *ports*. You can best view a port as a register into which the PC or the device can place specific values. Table 610 lists the port addresses different devices use in an EISA system.

Port	Device	Port	Device
00H-1FH	DMA controller	2EFH-2FFH	COM2
20H-3FH	Interrupt controller	300H-31FH	Network cards
40H-5FH	System timer	378H-37FH	LPT1
60H-6FH	Keyboard	380H-38FH	SDLC
70H-7FH	Real-time clock	390H-39FH	Cluster adapter
80H-9FH	DMA page registers	3B0H-3BFH	Monochrome
A0H-BFH	Int controller 2	3C0H-3CFH	EGA
C0H-DFH	DMA controller 2	3D0H-3DFH	CGA
F0H-FFH	Math coprocessor	3FOH-3F7H	Floppy disk
1F0H-1FFH	Hard disk	3F8H-3FFH	COM1
200H-220H	Game adapter	400H-4FFH	DMA
270H-27FH	LPT2	500H-7FFH	Aliases 100H-3FFH
2B0H-2DFH	Alternate EGA	800H-8FFH	CMOS
2E0H-2E7H	COM4	900H-9FFH	Reserved
2E8H-2EFH	COM3	9FFH-FFFFH	Reserved

Table 610 PC port addresses.

The meaning of each port depends on the corresponding device. To obtain port-specific meanings, refer to the technical documentation for your PC or device.

611 ACCESSING PORT VALUES

If your programs perform low-level hardware control, there may be times when they must read or write a port value. To help your programs do so, most DOS-based C compilers provide the following functions:

```
#include <dos.h>

int inport(int port_address);
char inportb(int port_address);
void outport(int port_address, int value);
void outportb(int port_address, unsigned char value);
```

The *port_address* specifies the address of the desired port, as listed in Tip 610. The *value* parameter specifies the word or byte value your program wants to output to the port. Tip 612 illustrates how to use the *inportb* function to read and display the contents of the PC's CMOS memory.

612 UNDERSTANDING THE CMOS

As you know, the PC stores system configuration information in CMOS memory, including your drive types, system date, and so on. The PC does not access the CMOS using standard segment and offset addressing. Instead, the PC uses PC port addresses to communicate with the CMOS. As you learned in Tip 611, most C compilers provide functions such as *inport* and *outport* to help your programs access the PC ports. The following *Turbo C Lite* program, *showcmos.c*, uses the *inportb* function to obtain and display CMOS information:

```
#include <stdio.h>
#include <stdlib.h>
#include <dos.h>

void main(void)
 {
```

```
  struct CMOS {
   unsigned char current_second;
   unsigned char alarm_second;
   unsigned char current_minute;
   unsigned char alarm_minute;
   unsigned char current_hour;
   unsigned char alarm_hour;
   unsigned char current_day_of_week;
   unsigned char current_day;
   unsigned char current_month;
   unsigned char current_year;
   unsigned char status_registers[4];
   unsigned char diagnostic_status;
   unsigned char shutdown_code;
   unsigned char drive_types;
   unsigned char reserved_x;
   unsigned char disk_1_type;
   unsigned char reserved;
   unsigned char equipment;
   unsigned char lo_mem_base;
   unsigned char hi_mem_base;
   unsigned char hi_exp_base;
   unsigned char lo_exp_base;
   unsigned char fdisk_0_type;
   unsigned char fdisk_1_type;
   unsigned char reserved_2[19];
   unsigned char hi_check_sum;
   unsigned char lo_check_sum;
   unsigned char lo_actual_exp;
   unsigned char hi_actual_exp;
   unsigned char century;
   unsigned char information;
   unsigned char reserved3[12];
  } cmos;
  char i, *pointer, byte;
  pointer = (char *) &cmos;
  for (i = 0; i < 0x34; i++)
   {
     outportb(0x70, i);
     byte = inportb(0x71);
     *pointer++ = byte;
   }
  // Extract BCD date and time
  printf("Current date %d%d/%d%d/%0d\n", cmos.current_month >> 4, cmos.current_month
     & 0xF, cmos.current_day >> 4, cmos.current_day & 0xF, cmos.current_year);
  printf("Current time %d%d:%d%d\n", cmos.current_hour >> 4, cmos.current_hour
     & 0xF);
  printf("Hard disk type %d\n", cmos.fdisk_0_type);
 }
```

UNDERSTANDING MEMORY MODELS 613

When you create programs in the PC environment, the compiler uses a *memory model* to determine how much memory the operating system allocates to your program. As you have learned, the PC divides memory into 64Kb blocks called segments. Usually, your program uses one segment for code (the program instructions) and a second segment for data. If your program is very large or uses large amounts of data, there will be times when the compiler must provide multiple code or data segments, or both. A memory model defines the number of segments the compiler can use for each. Memory models are important because, if you use the wrong memory model, your program might not have enough memory to execute.

Usually, the compiler will select a memory model that is large enough for your program. However, as you will learn, the larger the memory model is, the slower your program will execute. Therefore, your goal is always to use the smallest memory model that satisfies your program needs. Most compilers support tiny, small, medium, compact, large, and huge memory models. Several Tips that follow describe these memory models in detail. To select a specific memory model, you usually include a switch within the compiler's command line. Refer to the documentation that accompanied your compiler to determine the memory model switches.

Note: *As you have learned, the different types of memory models that you can use to write C/C++ programs in a DOS environment do not apply to the Windows environment, which only uses the virtual memory model. Although the next seven Tips are useful whether you intend to write DOS programs or not, they are almost purely informational for Windows programmers.*

614 Understanding the Tiny Memory Model

A memory model describes the number of 64Kb memory segments that the compiler allocates for a program. The smallest and fastest memory model is the *tiny* model. Because of its compact nature, the tiny memory model consumes the least amount of memory and loads faster than other models. As Figure 614 shows, the tiny memory model combines your program code and data into one 64Kb segment.

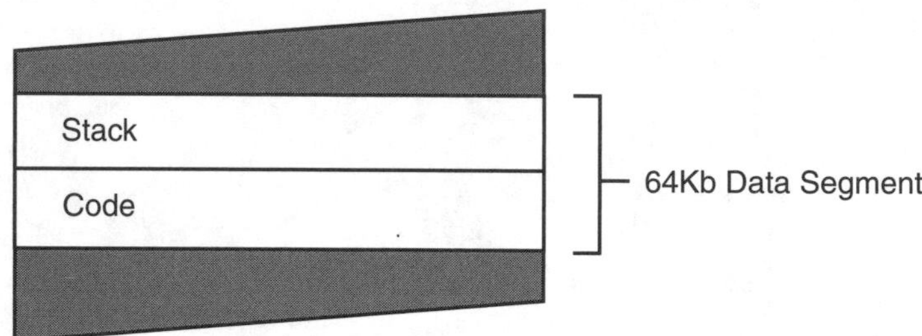

Figure 614 *The tiny memory model places program code and data into one 64Kb segment.*

If you are creating small programs, such as many of the example programs presented throughout this book, you should direct the compiler to use the tiny memory model.

615 Understanding the Small Memory Model

A memory model describes the number of 64Kb memory segments the compiler allocates for a program. The most common memory model is the *small* model. As Figure 615 shows, the small memory model uses one 64Kb segment for your program code and a second for your program data.

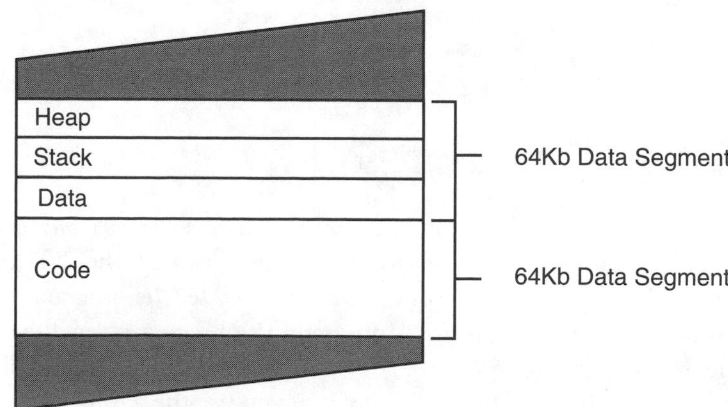

Figure 615 *The small memory model uses one 64Kb segment for program code and another for data.*

The advantage of using the small memory model is that all function calls and all data references use near 16-bit addresses. Such a program will execute faster than those using other, larger memory models.

UNDERSTANDING THE MEDIUM MEMORY MODEL 616

A memory model describes the number of 64Kb memory segments the compiler allocates for a program. If your program requires more than 64Kb of memory for code, but only 64Kb (or less) for data, then your programs can use the *medium* memory model. As Figure 616 shows, the medium memory model allocates multiple code segments and only one data segment.

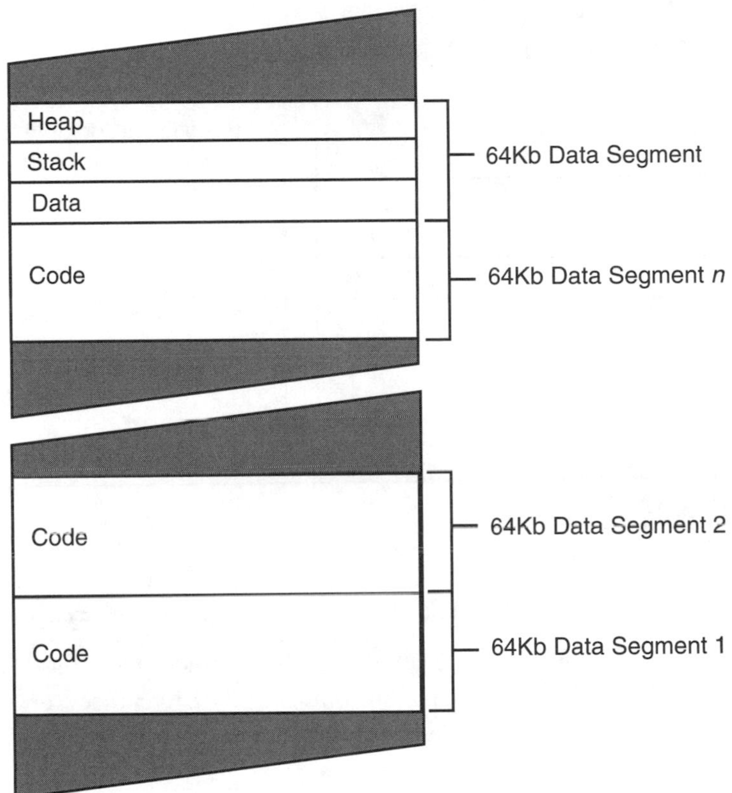

Figure 616 The medium memory model allocates multiple code segments and one data segment.

If your program contains a large number of instructions, then the medium memory model provides fast data access because all data references use near addresses. Because the medium memory model uses multiple code segments, however, all function calls require 32-bit far addresses. The pushing and popping (onto and off of the stack) of the additional segment address for function calls will slightly decrease program performance.

UNDERSTANDING THE COMPACT MEMORY MODEL 617

As you have learned, a memory model describes the number of 64Kb memory segments the compiler allocates for a program. If your program uses a large amount of data but limited instructions, your programs can use the *compact* memory model. As Figure 617 shows, the compact memory model allocates one 64Kb segment for your program code and multiple segments for data.

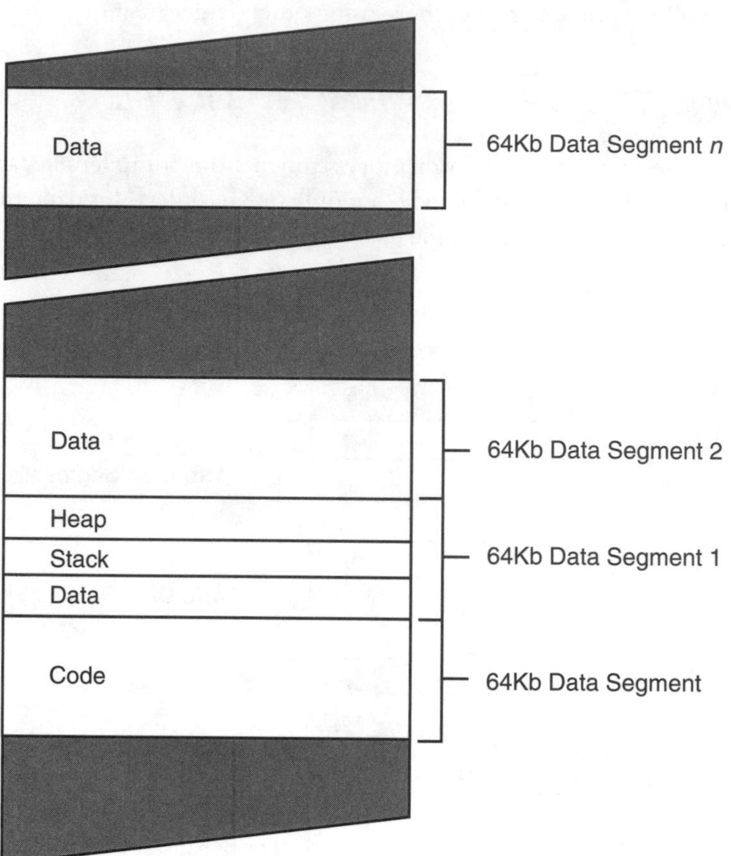

Figure 617 The compact memory model allocates one 64Kb segment for code and multiple segments for data.

Because the compact memory model only uses one code segment, all function calls use 16-bit near addresses. As a result, function calls are faster than in any larger memory model. Data references, on the other hand, require a segment and offset address (a 32-bit far address). The overhead required to work with each data reference's segment and offset address will decrease your program's performance.

618 Understanding the Large Memory Model

A memory model describes the number of 64Kb memory segments the compiler allocates for a program. If your program contains a large amount of code and data, your program can use the *large* memory model. As Figure 618 shows, the large memory model allocates multiple code and data segments.

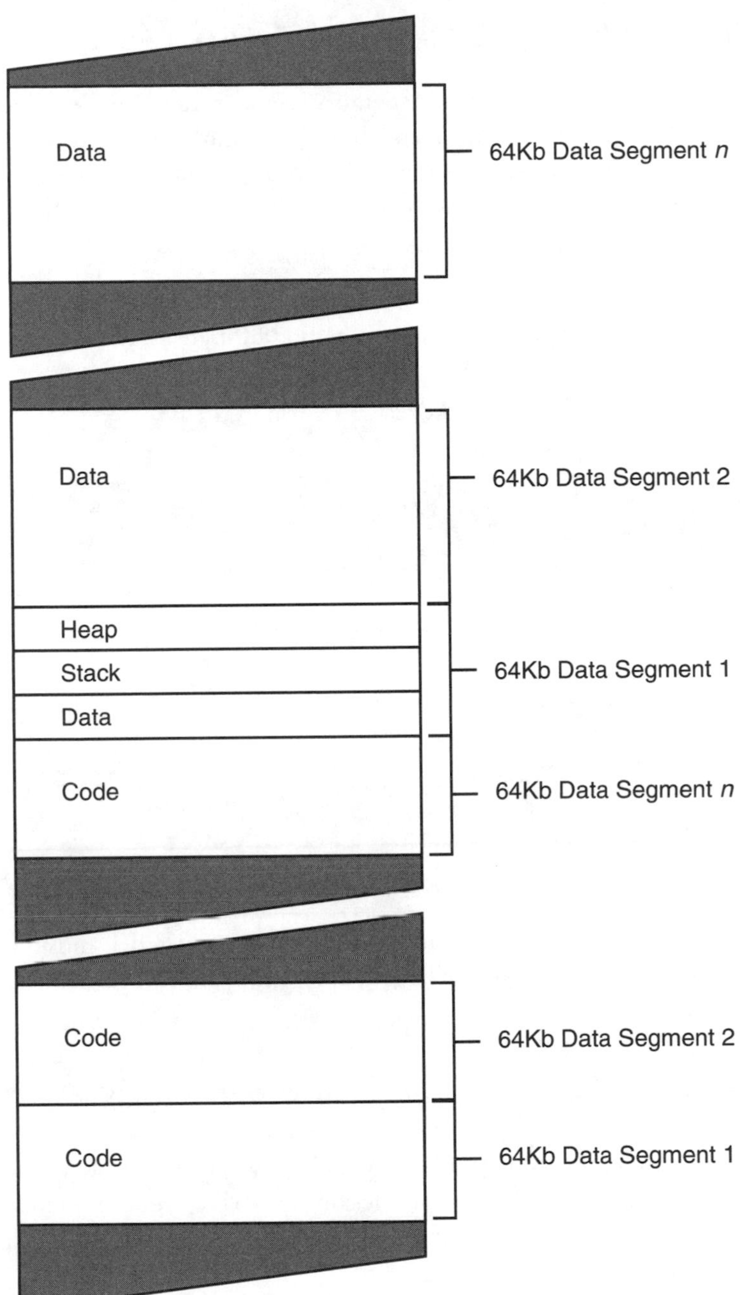

Figure 618 The large memory model allocates multiple code and data segments.

You should use the large memory model only as a last resort. Because the large memory model uses multiple code and data segments, each function call and each data reference requires a 32-bit far address. The overhead associated with constant segment and offset manipulation makes the large memory model the slowest of the models described so far.

619 UNDERSTANDING THE HUGE MEMORY MODEL

A memory model describes the number of 64Kb memory segments the compiler allocates for a program. As you have learned, most PC-based C compilers provide many different memory models to satisfy your program's data and code requirements. A special condition arises, however, when your program uses an array larger than 64Kb. To allocate such an array, your program must use the *huge* keyword to create a pointer, as shown here:

```
int huge *big_array;
```

Next, your program must use the *halloc* function to allocate the memory. The following program, *hugeint.c,* originally presented in Tip 600, uses the *halloc* to allocate a 400,000-byte array:

```
#include <stdio.h>
#include <malloc.h>

void main(void)
  {
    long int i;
    int huge *big_array;

    if ((big_array = (int huge *) halloc (100000L,
      sizeof(long int))) == NULL)
      printf ("Error allocating huge array\n");
    else
      {
        printf("Filling the array\n");
        for (i = 0; i < 100000L; i++)
          big_array[i] = i % 32768;
        for (i = 0; i < 100000L; i++)
          printf ("%d ", big_array[i]);
        hfree(big_array);
      }
  }
```

When you compile and execute a program using the huge memory model, most compilers will use 32-bit far addresses for both code and data (similar to the large memory model). As a result, the program's execution might be slower than you want.

620 DETERMINING THE CURRENT MEMORY MODEL

Depending on your program's processing, there may be times when you must compile the program using a specific memory model. Most C compilers predefine a specific constant to help your programs determine the current memory model. Table 620, for example, lists the constants the *Turbo C++ Lite,* Microsoft C, and Borland C compilers define for different memory models.

Memory Model	Microsoft C Constant	Turbo C++ Lite/Borland C Constant
Small	*M_I86SM*	*__SMALL__*
Medium	*M_I86MM*	*__MEDIUM__*
Compact	*M_I86CM*	*__COMPACT__*
Large	*M_I86LM*	*__LARGE__*

Table 620 *Constants Turbo C++ Lite, Microsoft C and Borland C++ define to indicate the current memory model.*

If your program requires a specific memory model, the program can test for the model, as shown here:

```
#ifndef __MEDIUM__
  printf("Program requires Medium memory model\n");
  exit(1);
#endif
```

OBTAINING CURRENT DATE AND TIME AS SECONDS SINCE 1/1/1970 621

As your programs become more functional, they often must know the current date and time. Most C compilers provide several functions that return the date and time in different formats. One such function is *time*, which returns the current date and time as seconds since 00:00 January 1, 1970. The function returns a value of type *time_t*, as shown here:

```
#include <time.h>

time_t time(time_t *date_time);
```

If you do not want to pass a parameter to *time*, you can invoke the function with *NULL*, as shown here:

```
current_time = time(NULL);
```

The following program, *delay_5.c*, uses the *time* function to implement a five-second delay:

```
#include <stdio.h>
#include <time.h>

void main(void)
 {
    time_t current_time;
    time_t start_time;

    printf("About to delay 5 seconds\n");
    time(&start_time);  // Get starting time in seconds
    do {
      time(&current_time);
    } while ((current_time - start_time) < 5);
    printf("Done\n");
 }
```

CONVERTING A DATE AND TIME FROM SECONDS TO ASCII 622

Tip 621 introduced the *time* function, which returns the current time as seconds since 00:00 January 1, 1970. Using the *ctime* function, your programs can convert the seconds to a character string in the following format:

```
"Fri Oct 31 11:30:00 1997\n"
```

The following program, *ctime.c*, illustrates how to use the *ctime* function:

```
#include <stdio.h>
#include <time.h>

void main(void)
 {
   time_t current_time;

   time(&current_time); // Get the time in seconds;
   printf("The current date and time: %s", ctime(&current_time));
 }
```

623 DAYLIGHT SAVINGS ADJUSTMENT

Several functions this section presents take daylight savings into account. To perform such processing, many C compilers declare a global variable named *daylight*. If daylight savings is in effect, the variable contains the value 1. If daylight savings is not in effect, C sets the variable to 0. The functions *tzset*, *localtime*, and *ftime* control the variable's value. The following code fragment uses the *daylight* variable to determine whether daylight savings or standard time if (daylight) printf("Daylight savings is active\n"); else printf("Daylight savings is not active\n");

Note: *The function **tzset** assigns the value to the **daylight** variable.*

624 DELAYING A SPECIFIC NUMBER OF MILLISECONDS

Depending on your program, there may be times when you need the program to delay for a specific number of milliseconds (1/1000 second). For example, you may want to display a message on your screen for a few seconds, thereby continuing the program's execution without forcing the user to press a key. For such cases, many C compilers provide the *delay* function. The function will delay for the number of milliseconds specified, as shown here:

```
#include <dos.h>

void delay(unsigned milliseconds);
```

Using the *delay* function, your programs can specify a delay period up to 65,535 milliseconds. The following program, *usedelay.c*, uses the *delay* function to delay for five seconds (5,000 milliseconds):

```
#include <stdio.h>
#include <dos.h>

void main(void)
 {
    printf("About to delay 5 seconds\n");
    delay(5000);
    printf("Done\n");
 }
```

625 DETERMINING YOUR PROGRAM'S PROCESSING TIME

When you perform steps to improve your program's performance, you might want to measure the amount of time that different parts of your program consume. You can then determine which sections of your program are the most time consuming. As a rule, you should start optimizing at the section of your program that consumes the most processor time. To help you determine your program's processing time, C provides the *clock* function, which returns the number of clock ticks (which usually occur 18.2 times per second), as shown here:

```
#include <time.h>

clock_t clock(void);
```

The *clock* function returns the program's processing time in clock ticks. To convert the time to seconds, divide the result by the constant *CLK_TCK*, which is defined in the header file *time.h*. The following Turbo C Lite program, *clock.c*, uses the clock function to display the program's processor time in seconds:

```
#include <stdio.h>
#include <time.h>
#include <dos.h>    // Contains the delay prototype

void main(void)
 {
```

```
   clock_t processor_time;

   printf("Processor time consumed %ld\n", clock() / (long) CLK_TCK);
   delay(2000);
   printf("Processor time consumed %ld\n", (long) clock() / (long) CLK_TCK);
   delay(3000);
   printf("Processor time consumed %ld\n", (long) clock() / (long) CLK_TCK);
 }
```

Note: *If your compiler does not provide the **delay** function, you can use one or more **for** loops to implement a delay.*

COMPARING TWO TIMES 626

In Tip 621 you learned how to use the *time* function to obtain the number of seconds since January 1, 1970. When you work with time, your programs often must compare two or more times. To compare times, your programs can use C's *difftime* function, which returns the difference between two times as a floating-point value, as shown here:

```
float difftime(time_t later_time, time_t start_time);
```

The following program, *difftime.c*, uses the *difftime* function to delay until five seconds have passed:

```
#include <stdio.h>
#include <time.h>

void main(void)
 {
   time_t start_time;
   time_t current_time;

   time(&start_time);
   printf("About to delay 5 seconds\n");
   do
    {
       time(&current_time);
    } while (difftime(current_time, start_time) < 5.0);
   printf("Done\n");
 }
```

OBTAINING A DATE STRING 627

In Tip 622 you learned how to use the *ctime* function to create a character string that contains the date and time. To use the *ctime* function, you must first invoke the *time* function to obtain the number of seconds since January 1, 1970. If you only want to obtain the current date, your programs can use the *_strdate* function, as shown here:

```
#include <dos.h>

char *_strdate(char *date_buffer);
```

The character string buffer you pass to the *_strdate* function must be large enough to hold nine characters (the eight-character date and *NULL*). The *_strdate* function places the date in the form *mm/dd/yy*. The following program, *strdate.c*, uses the *_strdate* function to display the current date:

```
#include <stdio.h>
#include <time.h>

void main(void)
 {
   char date[9];

   _strdate(date);
```

```
    printf("The current date is %s\n", date);
  }
```

628 OBTAINING A TIME STRING

In Tip 622 you learned how to use the *ctime* function to create a character string that contains the date and time. To use the *ctime* function, you must first invoke the *time* function to obtain the number of seconds since January 1, 1970. If you only want to obtain the current time, your programs can use *_strtime* function, as shown here:

```
#include <dos.h>

char *_strtime(char *time_buffer);
```

The character string buffer you pass to the *_strtime* function must be large enough to hold nine characters (the eight-character time and *NULL*). The *_strtime* function places the time in the form *hh:mm:ss*. The following program, *strtime.c*, uses the *_strtime* function to display the current time:

```
#include <stdio.h>
#include <time.h>

void main(void)
  {
    char time[9];

    _strtime(time);
    printf("The current time is %s\n", time);
  }
```

629 READING THE BIOS TIMER

The PC BIOS has a built-in timer that ticks 18.2 times per second. The BIOS stores within memory the number of ticks that have occurred since midnight. In the past, many programs used the BIOS timer to delay their programs until a specific number of ticks occurred. As previously discussed, however, your programs can specify a much finer time interval (at the millisecond level) using the *delay* function. The BIOS timer remains useful for generating a seed for a random number generator. Many C compilers provide two functions that let you control the BIOS timer—*biostime* and *_bios_timeofday*. The *biostime* function lets your programs access the number of clock ticks that have occurred since midnight. The format of the *biostime* function follows:

```
#include <bios.h>

long biostime(int operation, long newtime);
```

The *operation* parameter lets you specify whether you want to read or set the BIOS timer, as shown in Table 629.1.

Value	Meaning
0	Read the timer's current value
1	Set the timer's value to the value in *newtime*

Table 629.1 The possible values for the **operation** *parameter.*

The function returns the number of clock ticks. The *_bios_timeofday* function also lets you read or set the timer:

```
#include <bios.h>

long _bios_timeofday(int operation, long *ticks);
```

The *operation* parameter again specifies whether you want to read or set the timer, as shown in Table 629.2.

Value	Meaning
_TIME_GETCLOCK	Read the timer's current value
_TIME_SETCLOCK	Set the timer's value to the value in ticks

Table 629.2 The possible settings for the ***operation*** *parameter.*

The *_bios_timeofday* function returns the value the BIOS timer service stores within the AX register. The following program, *biostime.c*, uses both functions to read the clock ticks (if your compiler does not support one of the functions, simply comment out the function call):

```
#include <stdio.h>
#include <bios.h>

void main(void)
 {
   long ticks;

   ticks = biostime(0, ticks);
   printf("Ticks since midnight %ld\n", ticks);
    _bios_timeofday(_TIME_GETCLOCK, &ticks);
   printf("Seconds since midnight %f\n", ticks / 18.2);
 }
```

WORKING WITH THE LOCAL TIME 630

In Tip 621 you learned that the *time* function returns the current time in seconds since midnight, January 1, 1970. To make the system time easier for your programs to use, the C compiler provides the *localtime* function, which converts the time in seconds to a structure of type *tm*. The *tm* structure is defined in the *time.h* header file and shown here:

```
struct tm
 {
   int tm_sec;          // Seconds 0 thru 59
   int tm_min;          // Minutes 0 thru 59
   int tm_hour;         // Hours 0 thru 24
   int tm_mday;         // Day 1 thru 31
   int tm_mon;          // Month 0 thru 11
   int tm_year;         // Year - 1900
   int tm_wday;         // 0 Sunday thru 6 Saturday
   int tm_yday;         // Day of year 1 thru 365
   int tm_isdst;        // Nonzero if daylight savings time
 };
```

The format of the *localtime* function is as follows:

```
#include <time.h>

struct tm *localtime(const time_t *timer);
```

The *localtime* function uses the global variables *timezone* and *daylight* to adjust the time for your current time zone and to take daylight savings into account. The following program, *localtim.c*, illustrates how to use the *localtime* function:

```
#include <stdio.h>
#include <time.h>

void main(void)
 {
   struct tm *current_date;

   time_t seconds;
   time(&seconds);
   current_date = localtime(&seconds);
   printf("Current date: %d-%d-%d\n", current_date->tm_mon+1,
```

```
         current_date->tm_mday, current_date->tm_year);
   printf("Current time: %02d:%02d\n", current_date->tm_hour,
   current_date->tm_min);
 }
```

631 WORKING WITH GREENWICH MEAN TIME

In Tip 621 you learned that the *time* function returns the current time in seconds since midnight, January 1, 1970. If you work with international users, there may be times when you must work in terms of Greenwich Mean Time. So that you can work with Greenwich Mean Time, the C compiler provides the *gmtime* function, which converts the time in seconds to a structure of type *tm*, as shown in Tip 630. The format of the *gmtime* function is as follows:

```
#include <time.h>

struct tm *gmtime(const time_t *timer);
```

The *gmtime* function uses the global variable *daylight* to take daylight savings into account. The following program, *gmtime.c*, illustrates how to use the *gmtime* function:

```
#include <stdio.h>
#include <time.h>

void main(void)
 {
   struct tm *gm_date;

   time_t seconds;
   time(&seconds);
   gm_date = gmtime(&seconds);
   printf("Current date: %d-%d-%d\n", gm_date->tm_mon+1,
         gm_date->tm_mday, gm_date->tm_year);
   printf("Current time: %02d:%02d\n", gm_date->tm_hour, gm_date->tm_min);
 }
```

632 GETTING THE DOS SYSTEM TIME

If you are using DOS, your programs can use the *gettime* function in order to get the system time from DOS. The *gettime* function assigns the current time to a structure of type *time*, which is defined within the *dos.h* header file and shown here:

```
struct time
 {
    unsigned char ti_min;     // Minutes 0 thru 59
    unsigned char ti_hour;    // Hours 0 thru 59
    unsigned char ti_hund;    // Hundreds of seconds 0 thru 99
    unsigned char ti_sec;     // Seconds 0 thru 59
 };
```

The format of the *gettime* function is as follows:

```
#include  <dos.h>

void gettime(struct time *current_time);
```

The following program, *dostime.c*, uses the *gettime* function to obtain and then display the current system time:

```
#include <stdio.h>
#include <dos.h>

void main(void)
 {
```

```
    struct time curr_time;

   gettime(&curr_time);
   printf("Current time %02d:%02d:%02d.%d\n", curr_time.ti_hour,
          curr_time.ti_min, curr_time.ti_sec, curr_time.ti_hund);
 }
```

Note: *Many DOS-based C compilers also provide the **_dos_gettime** function, which returns a structure of type **dostime_t**, as shown here:*

```
struct dostime_t
 {
   unsigned char hour;        // 0 thru 23
   unsigned char minute;      // 0 thru 59
   unsigned char second;      // 0 thru 59
   unsigned char hsecond;     // 0 thru 99
 };
```

The format of the *_dos_gettime* function is as follows:

```
#include <dos.h>

void _dos_gettime(struct dostime_t *current_time);
```

Note: *The companion CD-ROM that accompanies this book includes the program **wintime.cpp**, which uses the Windows API to obtain the current system time.*

GETTING THE DOS SYSTEM DATE 633

If you are using DOS, your programs can use the *getdate* function to get the system date from DOS. The function assigns the current date to a structure of type *date*, which is defined within the *dos.h* header file and shown here:

```
struct date
 {
  int da_year;   // Current year
  char da_day;   // Current day 1 thru 31
  char da_mon;   // Current month 1 thru 12
 };
```

The format of the *getdate* function is as follows:

```
#include <dos.h>

void getdate(struct date *current_date);
```

The following *Tubo C Lite* program, *dosdate.c*, uses *getdate* to get and then display the current system date:

```
#include <stdio.h>
#include <dos.h>

void main(void)
 {
   struct date curr_date;

   getdate(&curr_date);
   printf("Current date: %d-%d-%d\n", curr_date.da_mon,
          curr_date.da_day, curr_date.da_year);
 }
```

Note: *Many DOS-based C compilers also provide the **_dos_getdate** function, which returns a structure of type **dosdate_t**, as shown here:*

```
struct dosdate_t
 {
  unsigned char day;    // 1 thru 31
```

```
   unsigned char month;// 1 thru 12
  unsigned int year;          // 1980-2099
  unsigned char dayofweek;    // 0 Sunday thru 6 Saturday
 };
```

The format of the *_dos_getdate* function is as follows:

```
#include <dos.h>

void _dos_getdate(struct dosdate_t *current_date);
```

Note: *The companion CD-ROM that accompanies this book includes the program* ***windate.cpp****, which uses the Windows API to obtain the current system date.*

634 SETTING THE DOS SYSTEM TIME

If you are using DOS, your programs can use the *settime* function to set the DOS system time just as if you issued the DOS TIME command. To use *settime*, you assign the time you want to a structure of type *time*, as shown in Tip 632. The format of the *settime* function is as follows:

```
#include <dos.h>

void settime(struct time *current_time);
```

The following program, *settime.c*, uses the *settime* function to set the current system time to 12:30:

```
#include <stdio.h>
#include <dos.h>

void main(void)
 {
   struct time desired_time;

   desired_time.ti_hour = 12;
   desired_time.ti_min = 30;
   settime(&desired_time);
 }
```

Note: *Many DOS-based C compilers also provide the* ***_dos_settime*** *function, which uses a structure of type* ***dostime_t*** *to set the system time, as shown in Tip 632. The format of the* ***_dos_settime*** *function is as follows:*

```
#include <dos.h>

void _dos_settime(struct dostime_t *current_time);
```

Note: *The companion CD-ROM that accompanies this book includes the program* ***wsettim.cpp****, which uses the Windows API to set the current system time.*

635 SETTING THE DOS SYSTEM DATE

If you are using DOS, your programs can use the *setdate* function to get the system date from DOS. Before you invoke the *setdate* function, assign the time you want to a structure of type *date*, as shown in Tip 633. Next, use a pointer to the structure to invoke the function. The format of the *setdate* function is as follows:

```
#include <dos.h>

void setdate(struct date *current_date);
```

The following program, *setdate.c*, uses the *setdate* function to set the current system date to October 31, 1997:

```
#include <stdio.h>
#include <dos.h>
```

```
void main(void)
 {
   struct date desired_date;

   desired_date.da_mon = 10;
   desired_date.da_day = 31;
   desired_date.da_year = 2001;
   setdate(&desired_date);
 }
```

Note: *Many DOS-based compilers also provide the **_dos_setdate** function, which uses a structure of type **dosdate_t** to set the system, as shown in Tip 633. The format of the **_dos_setdate** function is as follows:*

```
#include <dos.h>

unsigned _dos_setdate(struct dosdate_t *date);
```

Note: *The companion CD-ROM that accompanies this book includes the program **winsdat.cpp**, which uses the Windows API to set the current system date.*

CONVERTING A DOS DATE TO UNIX FORMAT 636

In Tip 633 you learned how to use the *getdate* function to get the DOS system date. Likewise, in Tip 632 you learned how to use the *gettime* function to get the DOS system time. If you are working in an environment where you use both DOS and Unix, then there may be times when you must convert a DOS-based date and time format to the date and time format Unix uses. In such cases, your programs can use the *dostounix* function to perform the conversion. The *dostounix* function converts structures of type *date* and type *time* to seconds since midnight, January 1, 1970, as shown here:

```
#include <dos.h>

long dostounix(struct date *DOS_date, struct time *DOS_time);
```

The following program, *dosunix.c*, uses the *dostounix* function to convert the current DOS system date and time to the corresponding Unix format:

```
#include <stdio.h>
#include <dos.h>
#include <time.h>

void main(void)
 {
   struct time dostime;
   struct date dosdate;

   time_t unix_format;
   struct tm *local;
   getdate(&dosdate);
   gettime(&dostime);
   unix_format = dostounix(&dosdate, &dostime);
   local = localtime(&unix_format);
   printf("UNIX time: %s\n", asctime(local));
 }
```

USING TIMEZONE TO COMPUTE THE TIME ZONE DIFFERENCE 637

As you have learned, the C run-time library provides several functions that can convert time values between local and Greenwich Mean Time. To help your programs quickly determine the time difference between the two times, many C compilers provide the *timezone* function, which contains the number of seconds between the two times. The following program, *timezone.c*, uses the *timezone* global variable to display the time difference:

```
#include <stdio.h>
#include <time.h>

void main(void)
 {
   tzset();
   printf("Difference between local and GMT is %d hours\n", timezone / 3600);
 }
```

Note: *The* ***tzset*** *function uses the environment entry TZ to determine the current time zone.*

638 Determining the Current Time Zone

Several Tips in this section presented functions that calculate times based on the current time zone. To help your programs determine the current time zone, many compilers provide the *tzname* global variable. The variable contains two pointers: *tzname[0]* points to the three-character time zone name, and *tzname[1]* points to the three-character daylight savings zone name. The following program, *tzname.c*, uses the *tzname* global variable to display the current time zone names:

```
#include <stdio.h>
#include <time.h>

void main(void)
 {
   tzset();
   printf("Current time zone is %s\n", tzname[0]);
   if (tzname[1])
     printf("Daylight savings zone is %s\n", tzname[1]);
   else
     printf("Daylight savings zone is not defined\n");
 }
```

Note: *The* ***tzset*** *function uses the environment entry TZ to determine the current time zone.*

639 Setting Time Zone Fields with TZSET

Several of the functions and global variables in this section return information about the current time zone. Many functions call the *tzset* function to determine the time zone information, as shown here:

```
#include <time.h>

void tzset(void);
```

The *tzset* function uses the TZ environment entry to determine the current time zone settings. The function then assigns appropriate values to the *timezone*, *daylight*, and *tzname* global variables. The program *tzname.c,* presented in Tip 638, illustrates how to use the *tzset* function.

640 Using the TZ Environment Entry

Many of the Tips presented throughout this section rely on the *tzset* function to provide time zone information. The *tzset* function examines your environment entries for the TZ entry and then assigns the *daylight*, *timezone*, and *tzname* variables, based on the entry's value. You can use either the DOS SET command to assign a value to the TZ set entry or the Windows Date/Time setting within the Control Panel. When you use the DOS SET command, the entry's format is as follows:

```
TZ=SSS[+/-]h[h][DDD]
```

Where *SSS* contains the standard time zone name (for example, EST or PST), the *[+/–]h[h]* specifies the difference in hours between the standard time zone and GMT; and *DDD* specifies the name of the daylight savings time zone (for example, PDT). The following entry sets the time zone for the West coast when daylight savings time is active:

```
C:\> SET TZ=PST8PDT  ENTER
```

Omit the time zone name when daylight savings time is not active, as shown here:

```
C:\> SET TZ=PST8  ENTER
```

Experiment with the TZ environment entry and the time zone programs presented in this section to determine how you want to represent dates and times within your programs. Remember, however, that your programs will need to write the TZ entry to any computer to which you move your applications.

Note: *If you do not specify a TZ entry, the default is EST5EDT.*

SETTING THE TZ ENVIRONMENT ENTRY WITHIN YOUR PROGRAM 641

As you have learned, several of the C run-time library functions use *tzset* to determine the local time zone. As discussed in Tip 640, the *tzset* function uses the TZ environment entry to determine the time zone. In most cases, it is probably unreasonable to expect end users to correctly set the TZ environment entry. If you know the correct setting for a specific user, however, you can use the *putenv* function within the program to create the correct entry for that user, as shown here:

```
putenv("TZ=PST8PDT");
```

The following program, *set_tz.c*, uses the *putenv* function to set the correct time zone. The program then uses the *tzname* global variable to display the time zone settings, as shown here:

```
#include <stdio.h>
#include <stdlib.h>
#include <time.h>

void main(void)
 {
   putenv("TZ=PST8PDT");
   tzset();
   printf("Current time zone is %s\n", tzname[0]);
   if (tzname[1])
     printf("Daylight savings zone is %s\n", tzname[1]);
   else
     printf("Daylight savings zone is not defined\n");
 }
```

GETTING TIME ZONE INFORMATION 642

Several Tips in this section present ways for your programs to determine time zone information. One of the most useful functions your programs can use to obtain time zone information is *ftime,* as shown here:

```
#include <sys\timeb.h>

void ftime(struct timeb *timezone);
```

The *timezone* parameter is a pointer to a structure of type *timeb*, as shown here:

```
struct timeb
 {
   long time;
   short millitm;
   short timezone;
```

```
    short dstflag;
 };
```

The *time* field contains the number of seconds since January 1, 1970 (GMT). The *millitm* field contains the fractional part of the seconds in milliseconds. The *timezone* field contains the difference between the local time zone and GMT in minutes. Finally, the *dstflag* specifies whether daylight savings is active (if the flag's value is 1) or inactive (if the flag's value is 0). The following program, *ftime.c*, uses the *ftime* function to display the current time zone information:

```
#include <stdio.h>
#include <time.h>
#include <sys\timeb.h>

void main(void)
 {
   struct timeb timezone;

   tzset();
   ftime(&timezone);
   printf("Seconds since 1 January 1970 (GMT) %ld\n", timezone.time);
   printf("Fractional seconds %d\n", timezone.millitm);
   printf("Hours difference between GMT and local zone %d\n", timezone.timezone / 60);
   if (timezone.dstflag)
     printf("Daylight savings time active\n");
   else
     printf("Daylight savings time inactive\n");
 }
```

643 Setting the System Time in Seconds Since Midnight 1/1/1970

Several Tips in this section show you ways to set the system time using DOS or the BIOS. In addition to the methods previously discussed, your programs can also use the *stime* function to set the system time using seconds since midnight January 1, 1970, as shown here:

```
#include <time.h>

int stime(time_t *seconds);
```

The *stime* function always returns 0. The following program, *stime.c*, uses the *stime* function to set the date exactly one day ahead of the current date and time:

```
#include <time.h>

void main(void)
 {
   time_t seconds;

   time(&seconds);  // Get current time
   seconds += (time_t) 60 * 60 * 24;
   stime(&seconds);
 }
```

644 Converting a Date to Seconds Since Midnight 1/1/1970

Several of the Tips in this book present run-time library functions that use or return seconds since midnight 1/1/1970. To help you determine the seconds for a specific date, your programs can use the *mktime* function, as shown here:

```
#include <time.h>

time_t mktime(struct tm *time_fields);
```

If the time fields are valid, the function returns the number of seconds for the specified time. If an error occurs, the function returns –1. The *time_fields* parameter is a pointer to a structure of type *tm*, as shown here:

```
struct tm
 {
   int tm_sec;
   int tm_min;
   int tm_hour;
   int tm_mday;
   int tm_mon;
   int tm_year;
   int tm_wday;
   int tm_yday;
   int tm_isdst;
 };
```

The following program, *mktime.c*, uses the *mktime* function to determine the number of seconds between midnight 1/1/1970 and midnight 12/31/2001:

```
#include <stdio.h>
#include <time.h>

void main(void)
 {
   time_t seconds;

   struct tm time_fields;
   time_fields.tm_mday = 31;
   time_fields.tm_mon = 12;
   time_fields.tm_year = 101;
   time_fields.tm_hour = 0;
   time_fields.tm_min = 0;
   time_fields.tm_sec = 0;
   seconds = mktime(&time_fields);
   printf("The number of seconds between 1-1-70 and 12-31-01 is %ld\n", seconds);
 }
```

Note: *When you pass a partial* ***tm*** *structure to the* ***mktime*** *function, the function will fill in the fields that are not correct. The* ***mktime*** *function supports dates in the range of January 1, 1970 through January 19, 2028.*

DETERMINING A DATE'S JULIAN DATE 645

In Tip 644 you used the *mktime* function to determine the number of seconds between a specific date and midnight January 1, 1970. As you learned, the *mktime* function uses a structure of type *tm* to hold the date components. If one or more of the components are not complete, the *mktime* function fills them in. If you examine the *tm* structure, you will see the *tm_yday* member. When you invoke *mktime*, the function will assign to *tm_yday* the Julian date for the day specified. The Julian calendar is identical to the Gregorian calender, except that it commences year one at the Greogrian-equivalent 46 B.C. The computer reports Julian dates in a three-digit format. The following program, *julian.c*, uses the *mktime* function to determine the Julian date for October 31, 2001:

```
#include <stdio.h>
#include <time.h>

void main(void)
 {
   time_t seconds;
   struct tm time_fields;

   time_fields.tm_mday = 31;
   time_fields.tm_mon = 12;
   time_fields.tm_year = 101;
   if (mktime(&time_fields) == -1)
     printf("Error converting fields\n");
```

```
    else
      printf("Julian date for October 31, 2001 is %d\n", time_fields.tm_yday);
  }
```

646 Creating a Formatted Date and Time String

As you have learned, the *_strdate* and *_strtime* functions return the current date and time in a character string format. Many compilers provide you with the following *strftime* function so that you will have better control of a formatted date and time string, as shown here:

```
#include <time.h>

size_t strftime(char *string, size_t max_length,
                const char *format, const struct tm *datetime);
```

The *string* parameter is the character string to which *strftime* writes the formatted date and time string. The *max_length* parameter specifies the maximum number of characters that *strftime* can place into the string. The *format* string uses %*letter* formatting characters similar to *printf* to specify the format you desire. Table 646 lists the valid characters that you can place in the formatted string. Finally, the *datetime* parameter is a pointer to a structure of type *tm* that contains the date and time fields. The *strftime* function returns a count of the number of characters assigned to the *string* parameter, or 1 if the function overflowed the *string* parameter. Table 646 lists the format specifiers for *strftime.*

Format Specifier	Meaning
%%	% character
%a	Abbreviated weekday name
%A	Complete weekday name
%b	Abbreviated month name
%B	Complete month name
%c	Date and time
%d	Two-digit day of month 01 through 31
%H	Two-digit hour 00 through 23
%I	Two-digit hour 01 through 12
%j	Three-digit Julian day
%m	Decimal month 1 through 12
%M	Two-digit minute 00 through 59
%p	AM or PM characters
%S	Two-digit second 00 through 59
%U	Two-digit week number 00 through 53 with Sunday first day of week
%w	Day of week (0 = Sunday 6 = Saturday)
%W	Two-digit week number 00 through 53 with Monday first day of week
%x	Date
%X	Time
%y	Two-digit year 00 through 99
%Y	Four-digit year
%Z	Time zone name

Table 646 Format specifiers for the ***strftime*** *function.*

The following program, *strftime.c*, illustrates the *strftime* function:

```
#include <stdio.h>
#include <time.h>

void main(void)
 {
   char buffer[128];
   struct tm *datetime;
   time_t current_time;

   tzset();
   time(&current_time);
   datetime = localtime(&current_time);
   strftime(buffer, sizeof(buffer), "%x %X", datetime);
   printf("Using %%x %%X: %s\n", buffer);
   strftime(buffer, sizeof(buffer), "%A %B %d, %Y", datetime);
   printf("Using %%A %%B %%d %%Y: %s\n", buffer);
   strftime(buffer, sizeof(buffer), "%I:%M%p", datetime);
   printf("Using %%I:%%M%%p: %s\n", buffer);
 }
```

When you compile and execute the *strftime.c* program, your screen will display output similar to the following (your screen will show output based on the current date and time):

```
Using %x %X: 08/22/97 22:03:13
Using %A %B %m %Y: Friday August 22, 1997
Using %I:%M%p: 10:03PM
C:\>
```

UNDERSTANDING PC CLOCK TYPES 647

Several of the Tips in this section discuss PC dates and times. To understand these functions better, you must know that the PC uses four basic clock types: timers, the CPU clock, the real-time clock, and the CMOS clock, which the following list details:

- The timer clock is a chip inside the PC that generates an interrupt 18.2 times per second. Each time the clock tick occurs, the PC generates *interrupt 8* (a system message). By capturing this interrupt, memory resident programs can activate themselves at specific time intervals.
- The CPU clock controls how fast your programs execute. When users say they are using a 200 MHz system, they are referring to the CPU clock.
- The real-time clock tracks the current date and time. In most cases, the real-time clock contains the same value as the CMOS clock.
- The CMOS clock is maintained by the computer, as opposed to the operating system, which maintains the real-time clock. The CMOS clock generally contains the same entry as the real-time clock.

WAITING FOR A KEYPRESS 648

There are many programs that, when displaying a message, will wait for the user to press any key before removing the message and continuing. To help your programs perform such processing, you can use the following *kbhit* function, which returns true if the user presses a key and false if the user does not press a key:

```
#include <conio.h>

int kbhit(void);
```

The following program, *kbhit.c*, will display a message on the screen that prompts the user to press any key to continue. The program then uses *kbhit* to wait for the keystroke, as shown here:

```
#include <stdio.h>
#include <conio.h>

void main(void)
 {
   printf("Press any key to continue...");
   while (! kbhit());
     ;
   printf("Done\n");
 }
```

649 Prompting the User for a Password

Depending on your programs, there may be times when you must prompt the user for a password. When the user types in the password, the keystrokes the user types should not appear on the screen. Your programs can use the following *getpass* function to perform this activity:

```
#include <conio.h>

char *getpass(const char *prompt);
```

The *getpass* function will display the specified prompt and then wait for the user to type in keystrokes and press ENTER. The *getpass* function then returns a pointer to the password the user typed. The following *Turbo C Lite* program, *getpass.c*, uses the *getpass* function to prompt the user for a password:

```
#include <stdio.h>
#include <conio.h>
#include <string.h>

void main(void)
 {
   char *password;

   password = getpass("Enter Password:");
   if (strcmp(password, "Secret"))
     printf("Password Incorrect\n");
   else
     printf("Password OK\n");
 }
```

***Note:** If your compiler does not provide the **getpass** function, you can use the function **get_password**, shown in Tip 650.*

650 Writing Your Own Password Function

In Tip 649 you learned how to use the *getpass* function to prompt the user for a password. As you learned, *getpass* does not display keystrokes as the user types. Some new users will have difficulty typing a password if the screen does not show any keystrokes, so some programs will display an asterisk (*) each time the user presses a keystroke. To prompt the user for a password and display an asterisk for each entry, you can use the function *get_password*, as shown here within the *Turbo C Lite* program *enter_p.c*:

```
#include <stdio.h>
#include <conio.h>
#include <string.h>

#define BACKSPACE 8
char *get_password(const char *prompt)
 {
   static char buffer[128];
   int i = 0;
   char letter = NULL;
```

```
   printf(prompt);
   while ((i < 127) && (letter != '\r'))
     {
       letter = getch();
       if (letter == BACKSPACE)
         {
           if (i > 0)
            {
              buffer[--i] = NULL;   // Erase previous *
              putchar(BACKSPACE);
              putchar(' ');
              putchar(BACKSPACE);
            }
           else
             putchar(7);   // BELL
         }
       else if (letter != '\r')
         {
           buffer[i++] = letter;
           putchar('*');
         }
     }
   buffer[i] = NULL;
   return (buffer);
 }
void main(void)
 {
   char *password;
   password = get_password("Enter Password: ");
   if (strcmp(password, "Secret"))
     printf("\nPassword Incorrect\n");
   else
     printf("\nPassword OK\n");
 }
```

UNDERSTANDING OUTPUT REDIRECTION 651

Each time you execute a command, the operating system associates the default input device with your keyboard. The operating system refers to the monitor as the standard output device, or *stdout.* Using the output redirection operator (>), you can direct the operating system to route a program's output to a file or to some other device. The following command, for example, directs DOS to redirect the *dir* command's output from the screen display to the printer:

```
C:\> DIR > PRN <     <ENTER>
```

In a similar way, the following command directs DOS to redirect the output of the *chkdsk* command to the file *diskinfo.dat*:

```
C:\> CHKDSK > DISKINFO.DAT  <ENTER>
```

To help you write programs that support output redirection, the header file *stdio.h* defines the constant *stdout*, to which file output operations can direct output. Several of the Tips presented in this book write output to *stdout.*

UNDERSTANDING INPUT REDIRECTION 652

Each time you execute a command, the operating system associates the default output device with your computer screen. The operating system refers to the keyboard as the standard input device, or *stdin.* You can use the input redirection operator (<) to direct the operating system to route a program's input from the keyboard to a file or to some other device. The following command, for example, directs DOS to redirect the input of the *more* command from the keyboard to the file *config.sys*:

```
C:\> MORE < CONFIG.SYS  <ENTER>
```

In a similar way, the following command directs DOS to redirect the input of the *sort* command from the keyboard to the file *autoexec.bat*:

```
C:\> SORT < AUTOEXEC.BAT  <ENTER>
```

To help you write programs that support input redirection, the header file *stdio.h* defines the constant *stdin*, from which file input operations can obtain input. Several of the Tips presented in this book read input from *stdin*.

653 COMBINING INPUT AND OUTPUT REDIRECTION

As discussed in Tips 651 and 652, you can change a program's default input and output source from the keyboard and monitor using the input (<) and output (>) redirection operators. As you create a collection of programs that support input and output redirection, sometimes you will want to redirect a program's input and output sources in the same command. For example, the following command directs DOS to sort the contents of the file *config.sys* and writes the sorted output to the printer:

```
C:\> SORT < CONFIG.SYS > PRN <ENTER>
```

To understand the processing the operating system performs, read the command from left to right. The input redirection operator (<) directs *sort* to obtain its input from the file *config.sys*. Likewise, the output redirection operator (>) directs *sort's* output from the monitor to the printer.

654 USING STDOUT AND STDIN

In Tips 651 and 652, you learned that C defines the file handles *stdin* and *stdout*. The file handles let you write programs that support I/O redirection. The following program, *upper.c*, reads a line of text from the *stdin* file handle and converts the text to uppercase. The program then writes the line of text to *stdout*. The program continues writing text until it detects an end-of-file:

```
#include <stdio.h>
#include <string.h>

void main(void)
 {
   char line[255];                              // Line of text read

   while (fgets(line, sizeof(line), stdin))
     fputs(strupr(line), stdout);
 }
```

Using the *upper* command, you can display the contents of the file *autoexec.bat*, as shown here:

```
C:\> UPPER < AUTOEXEC.BAT  <ENTER>
```

The following command uses the output redirection operator to print the contents of the file *config.sys* in uppercase:

```
C:\> UPPER < CONFIG.SYS > PRN <ENTER>
```

If you invoke the *upper* command without using an I/O redirection operator, as shown in the following example, the *upper* command will read its input from the keyboard and write its output to the screen display:

```
C:\> UPPER   <ENTER>
```

Each time you type a line of text and press ENTER, *upper* will display the corresponding text in uppercase. To end the program, you must press the end-of-file key combination CTRL+Z (under DOS) or CTRL+D (under Unix).

UNDERSTANDING THE PIPE OPERATOR 655

In Tips 651 and 652, you learned how to use the input and output redirection operators to change a program's input source from the keyboard to a file or device. You also learned that you can use input and output redirection operators to route a program's output from the screen display to a file or device. DOS and Unix also provide a third redirection operator, called the *pipe* operator, which lets you redirect one program's output to become another program's input. For example, the following command directs DOS to redirect the output of the *dir* command to become the *sort* command's input:

```
C:\> DIR | SORT  <ENTER>
```

Programs that get their input from another command or file and then change the input in some way are called *filters*. The following command, for example, uses the *find* command to filter the *dir* command's output to only display subdirectory entries:

```
C:\> DIR | FIND "<DIR>"  <ENTER>
```

Just as you can use multiple input and output redirection operators in the same command line, you can also place two or more pipe operators in the same command. For example, the following command uses three pipe operators to display the names of subdirectories in sorted order, one screen at a time:

```
C:\> DIR | FIND "<DIR>" | SORT | MORE <ENTER>
```

UNDERSTANDING GETCHAR AND PUTCHAR 656

Many programs use the *getchar* and *putchar* macros for character input and output. For example, the following program, *lower.c*, will convert each line of user input to lowercase and then display each line of user input on the screen:

```
#include <stdio.h>
#include <ctype.h>  // Contains the tolower prototype

void main(void)
  {
    int letter;

    for (letter = getchar(); ! feof(stdin); letter = getchar())
      putchar(tolower(letter));
  }
```

The following command uses the *lower* program to print the contents of the *autoexec.bat* file in lowercase:

```
C:\> LOWER < AUTOEXEC.BAT > PRN <ENTER>
```

When you use the *getchar* and *putchar* macros, your programs automatically support I/O redirection. To better understand how I/O redirection occurs, examine the header file *stdio.h*. Within *stdio.h*, you will find the *getchar* and *putchar* macros, which define their source of input and output in terms of *stdin* and *stdout*, as shown here:

```
#define getchar()  getc(stdin)
#define putchar(c) putc((c), stdout)
```

657 NUMBERING REDIRECTED INPUT

Depending on a file's contents or a program's output, there may be times when you may want to precede each line of the file's contents or program's output with a line number. The following program, *number.c*, filters its input to precede each line with the line's corresponding line number:

```
#include <stdio.h>

void main(void)
 {
   char line[255];          // Line of input
   long line_number = 0;    // Current line number

    while (fgets(line, sizeof(line), stdin))
    printf("%ld %s", ++line_number, line);
 }
```

For example, the following command prints a copy of the file *number.c*, with a line number preceding each line:

```
C:\> NUMBER < NUMBER.C > PRN <ENTER>
```

658 ENSURING THAT A MESSAGE APPEARS ON THE SCREEN

You can use the *output* and *pipe* redirection operators to redirect a program's output from the screen to a file, to a device, or to the input of another program. Although such output redirection can be a powerful tool, it can also cause users to miss an error message if they do not closely watch their work. To better understand this, consider the following program, *new_type.c*, which will display a file's contents on the screen:

```
#include <stdio.h>
#include <stdlib.h>

void main(int argc, char *argv[])
 {
   char line[255];  // Line read from the file
   FILE *fp;

   if (fp = fopen(argv[1], "r"))
     {
       while (fgets(line, sizeof(line), fp))
         fputs(line, stdout);
       fclose(fp);
       exit(0);  // Successful
     }
   else
     {
       printf("Cannot open %s\n", argv[1]);
       exit (1);
     }
 }
```

If, however, you wanted to send the program's output to a device or file, you could use the output redirection operator to redirect the program's output. The following command, for example, redirects *new_type's* output to print the file *autoexec.bat*:

```
C:\> NEW_TYPE AUTOEXEC.BAT > PRN <ENTER>
```

If *new_type* successfully opens *autoexec.bat*, it will write the file's contents to *stdout*, which, based on output redirection, will make the file print. If the *new_type* program cannot open the file you specify, it uses the *printf* function to display an error message stating that it could not open the file. Unfortunately, due to the output redirection, the message does not appear on the screen; rather, it goes to the printer. Users might mistakenly believe the command was successful unless they immediately check the printout. To prevent your program from inadvertently redirecting your error messages, C defines the *stderr* file handle, which your programs cannot redirect from the screen display. When your program must display an error message on the screen, your program should use *fprintf* to write the message to *stderr*, as shown here:

```
fprintf (stderr, "Cannot open %s\n", argv[1]);
```

659 Writing Your Own more Command

One of the best known filters DOS and Unix provide is the *more* command, which will display its input one screen at a time. Each time *more* displays a screen of output, *more* pauses, waits for the user to press a key, and displays the following message:

```
— More —
```

When the user presses a key, *more* repeats the process and displays the next screen of output. The following *Turbo C Lite* program, *more.c*, implements the *more* command:

```
#include <stdio.h>
#include <dos.h>

void main(void)
  {
     char buffer[256];
     long row_count = 0;
     union REGS inregs, outregs;
     int Ctrl_Key_Pressed, scancode;

     while (fgets (buffer, sizeof(buffer), stdin))
       {
         fputs (buffer, stdout);

         if ((++row_count % 24) == 0)
           {
             printf ("-- More --");
             // get the scancode of the key pressed
             inregs.h.ah = 0;
             int86 (0x16, &inregs, &outregs);
             scancode = outregs.h.ah;
             // get keyboard state in case of Ctrl-C
             Ctrl_Key_Pressed = 0;
             inregs.h.ah = 2;
             int86 (0x16, &inregs, &outregs);
             // Ctrl key flag is bit 2
             Ctrl_Key_Pressed = (outregs.h.al & 4);
             // scancode for C is 0x2E
             if ((Ctrl_Key_Pressed) && (scancode == 0x2E))
             break;      // Ctrl-C pressed
             printf ("\r");
           }
       }
  }
```

Each time *more* pauses for the user to press a key, it invokes the BIOS keyboard interrupt (INT 16H) to get the keystroke. Because DOS defines its input operations in terms of *stdin*, you cannot use *getchar*, *getch*, or *kbhit* to read the keystroke. DOS input functions use the next redirected input, thereby treating the next redirected character as the user keystroke. The BIOS services, however, are not defined in terms of *stdin*, and therefore the redirection operator does not affect the BIOS input services.

660 DISPLAYING A COUNT OF REDIRECTED LINES

Several of the Tips in this section created filter commands that you can use with the *input* and *pipe* redirection operators. The following program, *linecnt.c*, will display a count of the number of redirected input lines:

```
#include <stdio.h>

void main(void)
 {
   char line[256];  // Line of redirected input
   long line_count = 0;

   while (fgets(line, sizeof(line), stdin))
     line_count++;
   printf("The number of redirected lines: %ld\n", line_count);
 }
```

661 DISPLAYING A COUNT OF REDIRECTED CHARACTERS

Several of the Tips in this section created filter commands that you can use with the DOS *input* and *pipe* redirection operators. In a similar way, the following program, *charcnt.c*, will display a count of the number of characters in the redirected input:

```
#include <stdio.h>

void main(void)
 {
   long character_count = 0;

   getchar();
   while (! feof(stdin))
     {
       getchar();
       character_count++;
     }
   printf("The number of redirected characters is %ld\n", character_count);
 }
```

662 CREATING A TIMED MORE COMMAND

Several of the Tips in this section created filter commands that you can use with the DOS *input* and *pipe* redirection operator. In a similar way, the following program, *more15.c*, changes the DOS *more* command to display a screen of redirected input with each keystroke or every 15 seconds, whichever comes first:

```
#include <stdio.h>
#include <time.h>
#include <dos.h>

void main(void)
 {
   char buffer[256];
   char key_pressed = 0;
   long int counter = 1;
   union REGS inregs, outregs;

   time_t start_time, current_time, end_time;
   while (fgets(buffer, sizeof(buffer), stdin))
     {
       fputs (buffer, stdout);
       if ((++counter % 25) == 0)
```

```
          {
            time (&start_time);
            end_time = start_time + 15;
            do
             {
               key_pressed = 0;
               time (&current_time);
               inregs.h.ah = 1;
               int86 (0x16, &inregs, &outregs);
               if ((outregs.x.flags & 64) == 0)
                 {
                  key_pressed = 1;
                  do
                   {
                    inregs.h.ah = 0;
                    int86 (0x16, &inregs, &outregs);
                    inregs.h.ah = 1;
                    int86 (0x16, &inregs, &outregs);
                   } while (! (outregs.x.flags & 64));
                 }
             }
            while ((current_time != end_time) && (! key_pressed));
        }
      }
  }
```

PREVENTING I/O REDIRECTION 663

As you have learned, when you create programs that support I/O redirection, you can build a library of powerful filter commands. However, many programs you will create will not support I/O redirection. Depending on the functions that your program performs, severe errors can result when you let redirection occur. The following program, *no_redir.c*, tests the file handles *stdin* and *stdout* to ensure that they have not been redirected:

```
#include <stdio.h>
#include <dos.h>

void main(void)
  {
     union REGS inregs, outregs;

     // check the stdin handle first
     inregs.x.ax = 0x4400;
     inregs.x.bx = 0;       // stdin is handle 0
     intdos (&inregs, &outregs);
     if ((outregs.x.dx & 1) && (outregs.x.dx & 128))
       fprintf (stderr, "stdin has not been redirected\n");
     else
       fprintf (stderr, "stdin is redirected\n");
     // Now check stdout
     inregs.x.ax = 0x4400;
     inregs.x.bx = 1;       // stdout is handle 1
     intdos (&inregs, &outregs);
     if ((outregs.x.dx & 2) && (outregs.x.dx & 128))
       fprintf (stderr, "stdout has not been redirected\n");
     else
       fprintf (stderr, "stdout is redirected\n");
  }
```

The program uses the DOS service INT 21H function 4400H to examine the file handle. If the handle points to a device, then the service sets bit 7 of the DX registers to 1. If the service sets bit 7 and bit 2, then the handle refers to *stdout*. If the service sets bit 7 and bit 1, then the handle refers to *stdin*. If the service does not set bit 7, then the program has redirected the handle to a file. If the service does not set bit 1 or 2, then the program has redirected the handle to a device other than *stdin* or *stdout*:

Your programs can use INT 21H service 4400H to determine whether the current program, a previously executed program, or the user has redirected the computer's input or output. Depending on the result of the test, the programs can process accordingly.

664 USING THE STDPRN FILE HANDLE

As you have learned, the header file *stdio.h* defines two file handles—*stdin*, which (by default) points to the keyboard, and *stdout*, which points to the screen. If you write input and output operations in terms of *stdin* and *stdout*, your programs will automatically support I/O redirection. In a similar way, *stdio.h* defines the file handle *stdprn*, which points to the standard printer device (PRN or LPT1). Unlike *stdin* and *stdout*, you cannot redirect *stdprn*. The following program, *prt_echo.c*, uses the *stdprn* file to print redirected input as the program displays the output to the screen using *stdout*:

```
#include <stdio.h>
#include <string.h>

void main(void)
 {
   char line[255];  // Line of text read

   while (fgets(line, sizeof(line), stdin))
     {
       fputs(line, stdout);
       strcat(line, "\r");
       fputs(line, stdprn);
     }
 }
```

The following command line uses the *prt_echo* program to print and display a directory listing:

```
C:\> DIR | PRT_ECHO  <ENTER>
```

665 SPLITTING REDIRECTED OUTPUT TO A FILE

When you use the DOS *pipe* operator to redirect one program's output to become another program's input, sometimes you may want to save an intermediate copy of a program's output to a file. The following program, tee.c, saves an intermediate copy of a program's output to a file:

```
#include <stdio.h>
void main(void)
 {
   char buffer[256];
   while (fgets(buffer, sizeof(buffer), stdin))
     {
       fputs(buffer, stdout);
       fputs(buffer, stderr);
     }
 }
```

The *tee* command writes its redirected input to a file you specify, and to *stdout* so your program can redirect its output to another program. The following command, for example, uses *tee* to print an unsorted directory list before the *sort* command displays the sorted directory listing on your screen:

```
C:\> DIR | TEE PRN | SORT  <ENTER>
```

USING THE STDAUX FILE HANDLE 666

As you have learned, the header file *stdio.h* defines three file handles—*stdin*, which (by default) points to the keyboard; *stdout*, which (by default) points to the screen; and *stdprn*, which always points to the printer. If you write your input and output operations in terms of *stdin* and *stdout*, your programs will automatically support I/O redirection. In a similar way, *stdio.h* defines the file handle *stdaux*, which points to standard auxiliary device (AUX or COM1). Unlike *stdin* and *stdout*, you cannot redirect *stdaux*. The following program, *aux_echo.c*, uses the *stdaux* file to send redirected input to COM1 as the program displays output to the screen using *stdout*:

```
#include <stdio.h>
#include <string.h>

void main(void)
 {
   char line[255];  // Line of text read

   while (fgets(line, sizeof(line), stdin))
     {
       fputs(line, stdout);
       strcat(line, "\r");
       fputs(line, stdaux);
     }
 }
```

The following command line uses *aux_echo* print (to a printer attached to COM1) and to display a directory listing:

```
C:\> DIR | AUX_ECHO  <ENTER>
```

FINDING SUBSTRING OCCURRENCES WITHIN REDIRECTED INPUT 667

Several of the Tips presented in this section have created filter commands that you can use with the DOS *input* and *pipe* redirection operators. The following program, *io_find.c*, will display each occurrence of a word or phrase within its redirected input:

```
#include <stdio.h>
#include <string.h>

void main(int argc, char *argv[])
 {
   char string[256];

   while (fgets(string, sizeof(string), stdin))
     if (strstr(string, argv[1]))
       fputs(string, stdout);
 }
```

To display each occurrence of the word *#include* within the file *test.c*, you can invoke *io_find* as follows:

```
C:\> IO_FIND #include < TEST.C <ENTER>
```

To search for two or more words, simply place the words within quotes, as shown here:

```
C:\> IO_FIND "We the people" < CONSTITU.DAT <ENTER>
```

DISPLAYING THE FIRST N LINES OF REDIRECTED INPUT 668

Several of the Tips in this section have created filter commands that you can use with the *input* and *pipe* redirection operators. The following program, *shofirst.c*, displays the number of lines you specify within the command line of redirected input. By default, the program will display the first 10 lines of redirected input, as shown here:

```
#include <stdio.h>

#include <stdlib.h>
void main(int argc, char *argv[])
 {
   char line[255];  // Line read from the file

   int j;

   // Determine the number of lines to display

   if (argc < 2)
      j = 10;

   else
      j = atoi(argv[2]);


   // Get and display the lines

   for (i=0; i < j; i++)
     {
        fgets(line, sizeof(line), stdin);

        fputs(line, stdout);
     }
 }
```

For example, the following command directs *shofirst* to display the first 10 lines of a redirected directory listing:

```
C:\> DIR | SHOFIRST  <ENTER>
```

The following command, on the other hand, directs *shofirst* to display the first 25 lines of a redirected directory listing:

```
C:\> DIR | SHOFIRST 25  <ENTER>
```

669 UNDERSTANDING COMMAND-LINE ARGUMENTS

When you execute commands, the characters you type after the command line and before you press the ENTER key are the program's command line. For example, the following command line invokes a program named *first* by using two arguments, the number of lines to display, and the filename the user desires:

```
C:\> FIRST 15 FILENAME.EXT  <ENTER>
```

Support for command-line arguments increases the number of applications for which you can use your programs. For example, you can use the program *first* to display the contents of an unlimited number of files without having to change your program code. Fortunately, C makes it easy for your programs to support command-line arguments. Each time you invoke a C program, the operating system passes each command-line argument to the program as a parameter to the function *main*. To access the command-line arguments, you must declare *main* as follows:

```
void main(int argc, char *argv[])
  {
    // program statements
  }
```

The first parameter, *argc*, contains a count of the number of distinct command-line entries. Consider the following command line:

```
C:\> FIRST 10 FILENAME.EXT  <ENTER>
```

After this command line invocation, the parameter *argc* will contain the value 3. Because the value C assigns to *argc* includes the command name, *argc* will always contain a value greater than or equal to 1. The second parameter, *argv*, is an array of pointers to character strings that point to each command-line argument. Given the previous command line, the elements of the *argv* array will be pointers to the following:

```
argv[0] contains a pointer to "FIRST.EXE"
argv[1] contains a pointer to "10"
argv[2] contains a pointer to "FILENAME.EXT"
argv[3] contains NULL
```

Many of the programs in this book make extensive use of the command-line arguments.

DISPLAYING A COUNT OF COMMAND-LINE ARGUMENTS 670

Each time you invoke a C program, the operating system passes the number of command-line arguments—as well as pointers to the actual elements—to the function *main*. The following C program, *cmd_cnt.c*, uses the *argc* parameter to display a count of the number of command-line arguments passed to the program:

```
#include <stdio.h>

void main(int argc, char *argv[])
 {
   printf("The number of command line entries is %d\n", argc);
 }
```

Assuming you invoke *cmd_cnt* with no parameters, *cmd_cnt* will display the following:

```
C:\> CMD_CNT   <ENTER>
The number of command line entries is 1
C:\>
```

If you include the command-line arguments *A*, *B*, and *C*, *cmd_cnt* will display the following:

```
C:\> CMD_CNT A B C   <ENTER>
The number of command line entries is 4
C:\>
```

DISPLAYING THE COMMAND LINE 671

As you have learned, each time you invoke a C program, the operating system passes the number of command-line arguments—as well as pointers to the actual elements—to the function *main* as parameters. The following program, *show_cmd.c*, uses the parameter count within a *for* loop to display each of the command-line entries:

```
#include <stdio.h>

void main(int argc, char *argv[])
 {
   int i;

   for (i = 0; i < argc; ++i)
     printf("argv[%d] points to %s\n", i, argv[i]);
 }
```

If you invoke *show_cmd* with no parameters, *show_cmd* will display the following:

```
C:\> SHOW_CMD   <ENTER>
argv[0] points to C:\SHOW_CMD.EXE
C:\>
```

Likewise, if you invoke *show_cmd* with the command-line arguments *A*, *B*, and *C*, *show_cmd* will display the following:

```
C:\> SHOW_CMD A B C   <ENTER>
argv[0] points to C:\SHOW_CMD.EXE
argv[1] points to A
```

```
argv[2] points to B
argv[3] points to C
C:\>
```

672 WORKING WITH QUOTED COMMAND-LINE ARGUMENTS

Each time you invoke a C program, the operating system passes to *main* as parameters the number of command-line entries and an array of pointers to the actual entries. There may be times when your programs must work with parameters that the operating system passes from the command line within quotes. For example, assume that a program named *findtext* searches the user-specified file for specific text, as shown here:

```
C:\> FINDTEXT "We the people" FILENAME.EXT  <ENTER>
```

Most C compilers treat quoted parameters as a single argument. Experiment with the program *show_cmd*, which you wrote in Tip 671, to determine how your compiler treats quoted parameters:

```
C:\> SHOW_CMD "We the people" FILENAME.EXT  <ENTER>
argv[0] points to SHOWCMD.EXE
argv[1] points to We the people
argv[2] points to FILENAME.EXT
C:\>
```

673 DISPLAYING A FILE'S CONTENTS FROM THE COMMAND LINE

Several of the previous Tips have shown you how to use the parameters *argc* and *argv* to access command-line parameters. The following program, *showfile.c*, uses *argv* to display the contents of the file the command line specified:

```
#include <stdio.h>

void main(int argc, char *argv[])
  {
   FILE *fp;        // File pointer

   char line[255];  // Line from file

   if ((fp = fopen(argv[1], "r")) == NULL)
     printf("Error opening %s\n", argv[1]);
   else
     {
       // Read and display each line of the file
       while (fgets(line, sizeof(line), fp))
         fputs(line, stdout);
       fclose (fp);   // Close the file
     }
  }
```

To display the contents of a file, invoke *showfile* with the filename you want, as shown here:

```
C:\> SHOWFILE FILENAME.EXT   <ENTER>
```

Note the *if* statement that opens the file specified within the command line. The *fopen* function call within the *if* statement tries to open the file that *argv[1]* specified. If the file does not exist, *fopen* returns *NULL*, and the program will display a message stating that it could not open the file. If the user does not specify a filename, then *argv[1]* will contain *NULL*, which also causes *fopen* to return *NULL*. If *fopen* successfully opens the file, then the program will use a *while* loop to read and display the file's contents.

TREATING ARGV AS A POINTER 674

Several of the previous Tips have used the array of pointers *argv* to access command-line arguments. Because *argv* is an array, your programs can use a pointer to access its elements. If you use a pointer to access *argv's* elements, *argv* will become a pointer to an array of pointers. The following program, *argv_ptr.c*, will treat *argv* as a pointer to a pointer, then use *argv* to display the command line:

```
#include <stdio.h>

void main(int argc, char **argv)
  {
    while (*argv)
      printf ("%s\n", *argv++);
  }
```

Note the program's declaration of *argv* as a pointer to a pointer to a character string. The program uses a *while* statement to loop through the command-line arguments until the value **argv* points at is *NULL*. As you will recall, C uses *NULL* to indicate the last command-line argument. Within the *while* loop, the *printf* statement will display a string to which *argv* points. The *printf* statement then increments the value in *argv* so that *argv* points to the next command-line argument.

UNDERSTANDING HOW C KNOWS ABOUT THE COMMAND LINE 675

Each time you execute a program, the operating system loads the program into memory. In the case of the DOS operating system, DOS first loads 256 bytes into memory called the *program segment prefix*, which contains such information as the program's file table, environment segment, and command line. Figure 675 illustrates the format of the DOS program segment prefix.

Offset	Contents
0H	Int 20H instruction
2H	Top of memory segment address
4H	Reserved
5H	Far call to DOS dispatcher
AH	Int 22H vector
EH	Int 23H vector
12H	Int 24H vector
16H	Reserved
2CH	Environment copy segment address
2EH	Reserved
5CH	Default FCB 1
6CH	Default FCB 2
7CH	Reserved
80H	Command-line byte length
81H	Command line
FFH	

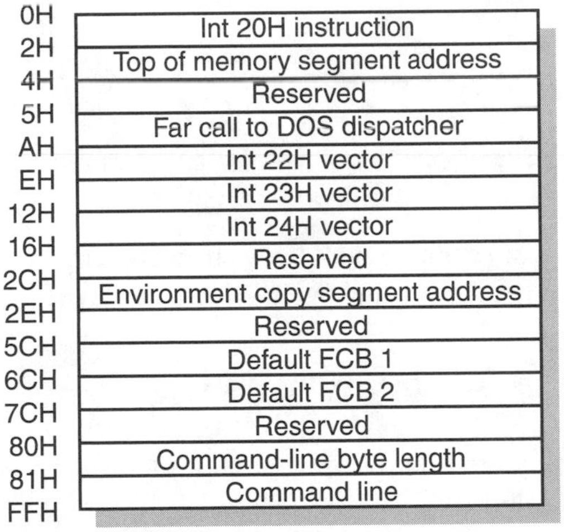

Figure 675 *The DOS program segment prefix.*

As you can see, beginning at offset 80H, DOS stores up to 128 bytes of command-line information. When you compile a C program, the C compiler embeds additional code that parses the command-line information, assigning the code to the *argv* array, which makes the arguments easy for you to access within your C programs.

UNDERSTANDING THE ENVIRONMENT 676

As you know, both DOS and Unix store information in a memory region called the environment. Using the SET command, you can display, add, or change environment entries. Depending on your program's function, there may be times when you must access information the environment contains. For example, many programs use the environment entry TEMP to determine the disk drive and subdirectory within which the programs should create temporary files. C makes it very easy to access the contents of the environment entries. One way to access the environment is to declare *main*, as shown here:

```
void main(int argc, char *argv[], char *env[])
```

Just as C lets you use an array of character string pointers to access a program's command-line arguments, you can access the environment entries in a similar way. The following program, *show_env.c*, uses the *env* array to display the current environment entries:

```
#include <stdio.h>

void main(int argc, char *argv[], char *env[])
 {
   int i;

   for (i = 0; env[i] != NULL; i++)
     printf("env[%d] points to %s\n", i, env[i]);
 }
```

As you can see, the program loops through the *env* array's entries until it finds the *NULL* value, which indicates to the program that it has found the end of the environment.

677 Treating ENV as a Pointer

In Tip 676 you learned that C lets you use the *env* array of pointers to character strings to access the environment contents. Because *env* is an array, you can treat *env* as a pointer. The following program, *env_ptr.c*, will treat *env* as a pointer to a pointer to a character string, then use *env* to display the environment's contents:

```
#include <stdio.h>

void main(int argc, char **argv, char **env)
 {
   while (*env)
     printf("%s\n", *env++);
 }
```

As you can see, the program loops until *env* points to *NULL*, which indicates the end of the environment. Within the loop, the *printf* statement prints the string to which *env* points and then increments *env* to point to the next entry.

678 Use void for main's Parameters

When your program does not use command-line parameters and you do not need to use *argc* and *argv*, you can omit the parameters and declare *main* as follows:

```
void main()
```

However, when a function does not receive parameters, you should use the *void* keyword to make it absolutely clear to the reader that the function receives no parameters, as shown here:

```
void main(void)
```

679 Working with Command-Line Numbers

As you create programs that use command-line arguments, you eventually must work with numbers in the command line. For example, the following command line directs the program *first* to display the first 15 lines of the file *autoexec.bat*:

```
C:\> FIRST 15 AUTOEXEC.BAT    <ENTER>
```

When the command line contains numbers, the *argv* array stores the numbers in ASCII format. To use the number, you must first convert the number from ASCII to an integer or floating-point value. To convert the number, use the functions *atoi*, *atol*, and *atof*, which you learned about in this book's Math Functions section. The following program, *beeps.c*, sounds the computer's built-in speaker the number of times the command line specifies. For example, the following command line directs *beeps* to sound the speaker three times:

```
C:\> BEEPS 3    <ENTER>
```

```
#include <stdio.h>
#include <stdlib.h>

void main(int argc, char *argv[])
 {
   int count;              // The number of times to sound the speaker
   int i;                  // The number of times the program has sounded the speaker

   // Determine the number of times to ring the bell
   count = atoi(argv[1]);
   for (i = 0; i < count; i++)
     putchar(7);           // ASCII 7 sounds the speaker
 }
```

If the user specifies a command-line parameter that is not a valid integer, then the *atoi* function will return the value 0.

UNDERSTANDING EXIT STATUS VALUES 680

Many DOS commands support exit status values that you can test from within your batch files to determine the command's success. For example, the DOS XCOPY command supports the exit status values listed in Table 680.

Exit Value	Meaning
0	Successful file copy operation
1	No files found to copy
2	File copy terminated by user CTRL+C
4	Initialization error
5	Disk write error

***Table 680** The exit status values for XCOPY.*

Using the IF ERRORLEVEL command, your batch files can test a program's exit status to determine if the command succeeded and then continue processing the batch file appropriately. As you create programs, you may want to provide exit status value support. The easiest way to return an exit status value is to use the *exit* function, as shown here:

```
exit(exit_status_value);
```

The following function call, for example, returns the exit status value 1:

```
exit(1);
```

When your program invokes the *exit* function, the program will immediately end and return to the operating system the specified exit status value. The following program, *new_type.c*, will display a file's contents. If the program cannot open the file the command line specifies, the program will return the exit status value 1. If *new_type* successfully displays the file's contents, then it will return the exit status value 0:

```
#include <stdio.h>
#include <stdlib.h>

void main(int argc, char *argv[])
 {
```

```
   char line[255];  // Line read from the file
   FILE *fp;

   if (fp = fopen(argv[1], "r"))
     {
       while (fgets(line, sizeof(line), fp))
         fputs(line, stdout);
       fclose(fp);
       exit(0);  // Successful
     }
   else
     {
       printf("Cannot open %s\n", argv[1]);
       exit (1);
     }
 }
```

Note: *Many C compilers provide a function called **_exit** which, like **exit**, immediately ends a program and returns an exit status value. However, unlike **exit**, which first closes open files and flushes output buffers, the **_exit** function does not close open files possibly resulting in lost data.*

681 USING RETURN FOR EXIT STATUS PROCESSING

Within a C function, the *return* statement ends a function's execution and returns the value specified to the calling function. Within the C function *main*, the *return* statement behaves similarly to its performance within a function, ending your program's execution and returning the value the program specifies to the operating system (the program's caller). The following program, *ret_exit.c*, will display a file's contents on your screen. If the program cannot open the file, *ret_exit* will return the exit status 1. If *ret_exit* successfully displays the file's contents, it will return the exit status value 0:

```
#include <stdio.h>
#include <stdlib.h>

int main(int argc, char *argv[])
 {
   char line[255];  // Line read from the file
   FILE *fp;

   if (fp = fopen(argv[1], "r"))
     {
       while (fgets(line, sizeof(line), fp))
         fputs(line, stdout);
       fclose(fp);
       return(0);  // Successful
     }
   else
     {
       printf("Cannot open %s\n", argv[1]);
       return(1);
     }
 }
```

Note that the program has changed the definition of *main* to indicate that the function will return an integer exit status value.

682 DETERMINING WHETHER TO DECLARE MAIN AS VOID

Several of the programs this book presents define the function *main* as shown in either of the two following implementations:

```
void main(void)
void main(int argc, char *argv[])
```

The keyword *void*, which appears in front of *main*, tells the C compiler (and programmers who are reading your code) that the function *main* does not use a *return* statement to return an exit status value to the operating system. The keyword *void* does not, however, prevent your program from using the exit function to return an exit status value. As a rule, however, if *main* does not use the *return* statement, you must use the keyword *void* to precede the function name. If you do not use the *void* keyword, some compilers may display a warning message similar to the following:

```
Warning Function should return a value in function main
```

SEARCHING THE ENVIRONMENT FOR A SPECIFIC ENTRY 683

In Tip 676 you learned how to use the *env* array of character string pointers to access a program's environment copy, as shown here:

```
void main(int argc, char *argv[], char *env[])
```

When your programs must search the environment for a specific entry, you might find it convenient to use the *getenv* function, which you will implement as shown here:

```
#include <stdib.h>

char *getenv(const char *entry_name);
```

The *getenv* function searches the environment entries for a specific entry, such as "TEMP." The entry name does not include an equal sign. If the specified entry is in the environment, the *getenv* function will return a pointer to the entry's value. If the program does not find the entry, *getenv* will return *NULL*. The following program, *showpath.c*, searches the environment for the PATH entry. If the program finds the entry, the program will display the entry's value:

```
#include <stdio.h>
#include <stdlib.h>

void main(void)
 {
   char *entry;

   entry = getenv("PATH");
   if (*entry)
     printf("PATH=%s\n", entry);
   else
     printf("PATH is not defined\n");
 }
```

HOW DOS TREATS THE ENVIRONMENT 684

When you work with DOS, the environment provides a region in memory in which you can place configuration information, such as your command path or system prompt. The SET command lets you display, add, or change environment entries, as shown here:

```
C:\> SET     <ENTER>
COMSPEC=C:\DOS\COMMAND.COM
PROMPT=$P$G
PATH=C:\DOS;C:\WINDOWS;C:\TCLITE\BIN
C:\>
```

DOS maintains one master copy of the environment that the user can change only with SET. When you invoke a program, DOS makes a copy of the environment's current contents and passes the copy to your programs, as shown in Figure 684.

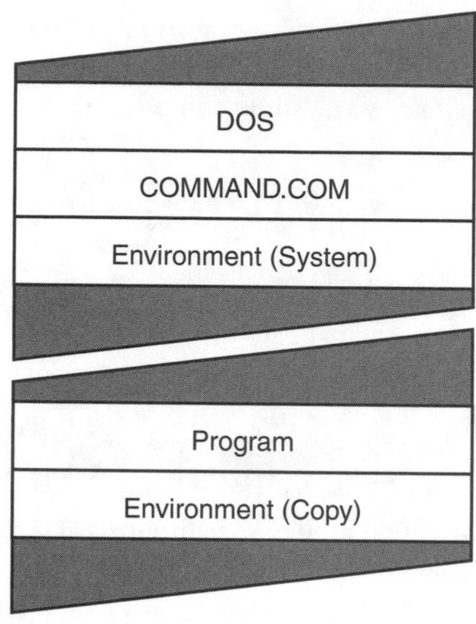

Figure 684 *DOS passes each program its own environment copy.*

Because your programs receive a copy of the environment, changes the programs make to environment entries do not affect the master environment. Several of the Tips in this section present functions that get or set environment entries. In each case, these functions only access the program's environment copy.

Note: *Most Windows programs use the Windows system registry to maintain information about the environment. You will learn more about the registry in later Tips.*

685 USING THE ENVIRON GLOBAL VARIABLE

In Tip 676 you learned that your programs can use the array of character string pointers DOS passes to the program's *main* function to access the program's DOS-provided environment copy, as shown here:

```
void main(int argc, char *argv[], char *env[])
```

In addition to letting you use *env*, C also defines a global variable, named *environ*, that contains your program's environment copy. The following *Turbo C Lite* program, *environ.c*, uses the *environ* global variable to display the environment entries:

```
#include <stdio.h>
#include <dos.h>

void main(void)
 {
   int i;

   for (i = 0; environ[i]; i++)
     printf("%s\n", environ[i]);
 }
```

When your programs use the *putenv* function to add or change an environment entry, you should later access your environment entries by using the *getenv* function or by accessing the *environ* global variable. To place an entry in the program's environment copy, *putenv* may need to move the program's environment copy, which invalidates the *env* pointer DOS passes to *main*.

ADDING AN ENTRY TO THE CURRENT ENVIRONMENT 686

In Tip 684 you learned that DOS keeps one master environment copy. DOS copies the master environment and passes the copy to each program you invoke. As a result, your programs usually cannot change entries in the master environment. Instead, DOS lets any changes your programs make to the environment apply only to the program's environment copy. There maybe times, however, when your programs must nevertheless store an entry in the environment copy.

For example, assume that a program spawns a child process that must know the name of a specific file. The program can first place the filename into its environment copy. When the program spawns the child process, the child will receive a copy of the program's environment and therefore have access to the filename. For such cases, your programs can use the *putenv* function, as shown here:

```
#include <stdlib.h>

int putenv(const char *entry);
```

If *putenv* successfully adds the entry to the program's environment copy, *putenv* will return the value 0. If an error occurs (for example, the environment is full), then *putenv* will return –1. The following *Turbo C Lite* program, *putenv.c*, illustrates how to use the *putenv* function:

```
#include <stdio.h>
#include <stdlib.h>

void main(void)
 {
   if (putenv("BOOK=Jamsa\'s C/C++/C# Programmer\'s Bible"))
     printf("Error writing to environment\n");
   else
     {
       int i;
       for (i = 0; environ[i]; ++i)
         printf("%s\n", environ[i]);
     }
 }
```

Note: *Do not invoke **putenv** with an automatic character string variable or with a string pointer that your program's code may free. Also, if your program uses the array of character string pointers that DOS passes to **main**, be aware that the **putenv** function may move the program's environment copy, thereby invalidating the **env** pointer. To access the environment entries, use **getenv** or the **environ** global variable.*

ADDING ELEMENTS TO THE DOS ENVIRONMENT 687

You have learned that when you run a program, DOS copies the current environment entries and passes the copy to the program. Because the program does not have access to the master copy of the environment entries, DOS cannot change the entries DOS maintains. Because DOS does not provide memory protection that prevents one program from accessing another program's memory, you can write a program that locates and finds the DOS master environment copy. After you know the environment's location, you can change or delete existing entries or add new entries to the environment. Because the environment has a fixed size, there may be times when the entries you want to add will not fit in the environment. In some cases, you can change the environment's size and allocate more memory to avoid running out of space.

The process of locating and sizing the DOS master environment is very complex. If you have an application that must change the entries, however, you can learn how to do so. In the book *DOS Programming: The Complete Reference,* Osborne/McGraw-Hill, 1991, Kris Jamsa devotes an entire chapter to working with the environment. Because the discussion and sample programs would require 20 to 30 pages and are beyond the scope of this work, you should simply recognize that it is possible for your programs to update the master environment copy. If you want to update the master environment, please refer to Jamsa's book on DOS programming.

688 ABORTING THE CURRENT PROGRAM

As your programs become more complex, there may be times when, should a critical error occur, you want the program to immediately terminate and display an error message to *stderr*. In such cases, your programs can use the *abort* function, as shown here:

```
#include <stdlib.h>

void abort(void);
```

When your program invokes the *abort* function, *abort* will display the following message to *stderr* and then invoke the DOS *_exit* function with an exit status value of 3:

```
Abnormal program termination
```

The best way to understand the *abort* function is to consider the following implementation:

```
void abort(void)
 {
   fputs("Abnormal program termination", stderr);
   _exit(3);
 }
```

It is important to note that *abort* invokes *_exit* and not *exit*. As discussed in Tip 680, the *_exit* function does not close any files or flush output.

689 DEFINING FUNCTIONS THAT EXECUTE AT PROGRAM TERMINATION

Depending on your program, sometimes you may want your program to automatically execute one or more functions when the program ends. To do so, your program can use the *atexit* function, which lets your program specify up to 32 functions that your program automatically executes when it ends, as shown here:

```
#include <stdlib.h>

int atexit(void (*function(void));
```

The functions that *atexit* invokes cannot use parameters. If the functions must access specific data, then you must declare the data as global variables. When you define an at-termination function list, your program will execute the functions in order, beginning with the last function you registered and ending with the first function you registered. The following program, *atexit.c*, uses the *atexit* function to register two at-termination functions:

```
#include <stdio.h>
#include <stdlib.h>

void first(void)
 {
   printf("First function registered\n");
 }

void second(void)
 {
   printf("Second function registered\n");
 }

void main(void)
 {
   atexit(first);
```

```
    atexit(second);
  }
```

When you compile and execute the *atexit.c* program, your screen will display the following output:

```
Second function registered
First function registered
C:\>
```

As you can see, the functions in the at-termination list execute in the order opposite the order in which the program registers the functions (in other words, the last function registered is the first function executed).

***Note:** If your program invokes the **exit** function, the program executes the functions in the at-termination list. However, if the program invokes **_exit**, then the program does not execute the functions in the at-termination list.*

UNDERSTANDING LIBRARIES 690

Many of the Tips in this book discuss C run-time library functions. When you use a run-time library function in your program, the linker loads the corresponding code from a library file into your executable program. The obvious advantage of using run-time library functions is that you do not have to write the corresponding code. If you examine the files that accompany your compiler, you will find many files that have the LIB extension. These files contain object libraries. When you compile and link your programs, the linker examines the LIB files to resolve function references.

As you create useful functions, you can build your own libraries. You can then quickly use a function that you created for one program within another program. Most compilers provide a librarian program that lets you build and change libraries. While the *Turbo C++ Lite* program is not capable of creating libraries, Borland's *Turbo C++* library program is called TLI, the Microsoft *C/C++* librarian is called LIB, and both companies' Windows C++ compilers support librarians. Several of the following Tips discuss library operations.

REUSING OBJECT CODE 691

As you create functions, you will often find that a function you wrote for one program will solve a second program's requirement. For example, the program *str_len.c* (located on the CD-ROM that accompanies this book) contains the function *string_length*, as shown here:

```
int string_length(char *str)
 {
   int length = 0;

   while (*str++)
     length++;
   return(length);
 }
```

Compile the file to create the object file *str_len.obj*. After you compile the object file, write the program *find_len.c*, as shown here, which uses the function to display the length of several different strings:

```
#include <stdio.h>

int string_length(char *);

void main(void)
 {
   char *title= "Jamsa\'s C/C++/C# Programmer\'s Bible";
   char *section = "Tools";

   printf("Length of %s is %d\n", title, string_length(title));
```

```
    printf("Length of %s is %d\n", section, string_length(section));
  }
```

Compile the program as follows (replacing your compiler's command for cc) so that it uses the contents of the object file *str_len.obj* (you can also create a project and add both C files to the project, as Tip 268 details):

```
C:\> CC FIND_LEN.C STR_LEN.OBJ <Enter>
```

In this example, you can combine the function's object code with your program to resolve and use the function. As you will learn in Tip 692, however, compiling object files in this way has limited usefulness.

692 PROBLEMS WITH COMPILING C AND OBJ FILES

In Tip 691 you learned that to reuse functions, you can compile the function separately to create an OBJ file and later compile the program's C code and the function's OBJ file to produce an executable program, as shown here:

```
C:\> CC SOMEFILE.C  FUNCTION.OBJ   <Enter>
```

Although this technique lets you resolve and use the function code, it also restricts the number of functions you can resolve to the length of your command line. For example, assume that your program uses 10 functions that reside in separate object files. You will find that remembering which files you must compile—even just fitting all the filenames in the command line—can become very difficult. One solution is to group all your functions into a single object file. A better solution, however, is to create a library file that contains the object code for each function. The library file is preferable to the OBJ file because, as you will learn, most librarian programs let you quickly update the library (by replacing, adding, or deleting object files).

693 CREATING A LIBRARY FILE

Depending on your compiler, the name of your librarian and the command-line switches that the program supports will differ. However, the following list details the operations that most librarians support:

1. Create a library
2. Add one or more object files to the library
3. Replace an object code file with another
4. Delete one or more object files from the library
5. List the routines the library contains

For example, the following command uses a generic LIB librarian to create a library named *my_stuff.lib* and insert into the library the object code for the *string_length* function that *str_len.obj* contains:

```
C:\> LIB MY_STUFF.LIB +STR_LEN.OBJ <Enter>
```

After the library file exists, you can use the library to compile and link the program *find_len.c*, as shown here:

```
C:\> CC FIND_LEN.C MY_STUFF.LIB <Enter>
```

694 UNDERSTANDING COMMON LIBRARY OPERATIONS

Depending on your compiler, the operations your librarian supports may differ. However, most librarians let you add and remove object code files, using the plus symbol (+) to add a file and the minus symbol (–) to remove an object file. Table 694 lists several library operations that use the library file *my_stuff.lib*.

Command	Operation
my_stuff.lib +strcpy.obj	Adds the object file *strcpy.obj* to the library
my_stuff.lib ++strlen.obj+strupr.obj	Adds the object files *strlen.obj* and *strupr.obj* to the library
my_stuff.lib -strlwr.obj	Removes the object file *strlwr.obj* from the library
my_stuff.lib -strlwr.obj+strsiz.obj	Removes the object file *strlwr.obj* from the library while adding the object file *strsiz.obj*
my_stuff.lib +-strlwr.obj	Replaces the object file *strlwr.obj* in the library with the current disk file
*my_stuff.lib *strupr.obj*	Extracts the code for the object file *strupr.obj* from the library to a file with the same name

Table 694 Common library operations.

LISTING THE ROUTINES IN A LIBRARY FILE 695

As you have learned, library files provide convenient storage locations for functions that you may want to use in other programs. Depending on your compiler, the operations your librarian supports may differ. However, most librarians will let you view the routines a library file contains. For example, using LIB librarian, the following command lists the routines contained in the library file *graphics.lib*:

```
C:\> LIB \COMPILER\LIB\GRAPHICS.LIB, CON <Enter>
```

To print the names of the functions the library contains, replace *CON* with *PRN* in the previous command line.

USE LIBRARIES TO REDUCE YOUR COMPILATION TIME 696

As your program increases in size, so too will the program's compilation time. One way you can reduce your program's compilation time is to extract the program's working functions into a library. In this way, when you later compile your program, you will not spend time recompiling the functions. Depending on the number of functions your program contains, removing the functions in this way can significantly speed up your compilation time. In addition, when you remove the function code from your program, your program will become smaller, more manageable, and possibly easier for you to understand.

LEARNING MORE ABOUT YOUR LIBRARIAN'S CAPABILITIES 697

The librarian Tips you just read have provided only an introduction to libraries. The documentation that accompanies your compiler will discuss the librarian's capabilities in detail. For example, the Borland TLIB librarian lets you invoke TLIB without a command line to display the program's switches, as shown here:

```
C:\> TLIB  <Enter>
```

Likewise, Microsoft's LIB librarian lets you use the following command to display the available command-line switches:

```
C:\> LIB /?  <Enter>
```

UNDERSTANDING THE LINKER 698

As you have learned, the compiler converts your C file to machine language. If your programs call functions that library or other object files contain, the linker will load the corresponding code to resolve the calls to the functions.

After resolving all the functions (both internal and external) which the program calls, the compiler will produce the executable file. Assume, for example, that your program contains the following statements:

```
void main(void)
 {
    printf("Jamsa\'s C/C++/C# Programmer\'s Bible");
 }
```

When the compiler compiles the program, it notes that the program code calls the function *printf* but does not define it. The linker, in turn, locates the *printf* function in the run-time library, loads the code into your executable file, and then updates the function call to reference the correct address of the function within your program. If the linker is unable to locate a function, it will display an error message on your screen stating that it encountered an *unresolved external.* The linker's primary function, therefore, is to pull (link) together all the pieces of your program code. However, depending on your linker, you may be able to use the linker to produce a *link map* to describe the layout of your executable file, to specify a stack size, or to control the program's underlying segment use. For specifics on your linker's capabilities, refer to the documentation that accompanied your compiler.

699 Viewing Linker Capabilities

As you learned in Tip 698, the linker's primary role is to pull together all the functions that a program uses into the executable file. Depending on your compiler, your linker may have additional capabilities. If you are using *Turbo C++,* you can list the command-line options for the TLINK linker by invoking TLINK without any parameters, as shown here:

```
C:\> TLINK  <Enter>
```

If you are using Microsoft *C/C++*, you can display the command-line switches for LINK as shown here:

```
C:\> LINK /?  <Enter>
```

700 Using a Link Map

As you have learned, the linker locates external functions within your program, loads the functions into the executable file, and updates the addresses of each function reference. When you debug a program, there may be times when you know the program has an error at a specific memory location. If you use a *link map*, which shows where the linker has loaded each function, you may be able to determine the location of the error. Depending on your linker, the steps you must perform to produce a link map may differ. While *Turbo C++ Lite* does not let you create a link map, you can use the following command line with Borland *Turbo C++* command to create the link map file *find_len.map*:

```
C:\> BCC -lm FILE_LEN.C STR_LEN.OBJ <Enter>
```

The link map is an ASCII file with contents that you can display or print.

701 Using Linker Response Files

As you have learned, while the *Turbo C++ Lite* compiler does not support linker-specific commands, most other compilers do. When you use Borland's TLINK or Microsoft's LINK commands, the general formats of the commands are as follows:

```
TLINK [switches] object_files, exe_file,  map_file, library_files, def_file
LINK [switches] object_files, exe_file, map_file, library_files, def_file
```

Depending on the number of files you are linking, your command lines may become extremely long. Both linkers support *response files* so that you do not have to remember all the filenames, nor the command format. A response file is an ASCII file that contains the filenames that you want the linker to use for each option. The filenames for each file type must appear in the command-line order, with each file type on its own line. For example, consider the following LINK command:

```
C:\> TLINK FIND_STR.OBJ STR_LEN.OBJ, FIND_STR.EXE, FIND_STR.MAP, SOMELIB.LIB
```

Your response file in this case would contain the following:

```
FIND_STR.OBJ STR_LEN.OBJ
FIND_STR.EXE
FIND_STR.MAP
SOMELIB.LIB
```

Assuming you name the response file *find_str.lnk*, you can invoke TLINK as follows:

```
C:\> TLINK @FIND_STR.LNK <Enter>
```

If you are linking multiple object files that cannot fit on one line, your response file can continue on to a second line. To indicate the continuation, place a plus sign at the end of the first line.

SIMPLIFYING APPLICATION BUILDING WITH MAKE 702

As your programs become more complex, they will normally require specific header files, source code modules, object code files, and libraries. When you make a change to your program, there may be times when it will be difficult for you to remember which files the change affects. To simplify your task of rebuilding an executable file after you make changes to your programs, many C compilers provide a MAKE utility.

MAKE is a very powerful programming tool that works with an application-specific file. This file (often called a *make file*) specifies the different files that the compiler will use to build an application and lists the steps that the compiler must perform when you change your program. As you will learn, the files you provide to MAKE are almost like programs themselves (that is, they contain conditions and instructions that MAKE evaluates and possibly executes). There are two common ways to use MAKE. First, you can place the operations you want MAKE to perform in a file named MAKEFILE. Next, you simply invoke MAKE, as shown here, or load the file into *Turbo C++ Lite*:

```
C:\> MAKE  <Enter>
```

When you use a MAKE file, MAKE will read the file's contents and process your programs accordingly. If you are working on several different programs, however, you will probably want to create MAKE files that use the application's name (the second method). For example, you might have a file named *find_len.mak*. When you want to invoke MAKE with a specific file, you must include the *-f* switch, as shown here:

```
C:\> MAKE -f FIND_LEN.MAK <Enter>
```

Several of the following Tips describe MAKE operations in detail.

Note: *When you work with Windows-based C++ development suites, the* ***project*** *file which the suite creates manages the MAKE file for you.*

CREATING A SIMPLE MAKE FILE 703

MAKE is a tool that helps you build executable files or libraries after you have made changes to a file you originally used to build the program or library. You invoke MAKE with a file that contains specifics about an application, such as the files you used to create the application and the files' dependencies. MAKE files follow a specific format. To begin, you will specify a target file and the files you used to create the target. For example, assume that you want to build the program *budget.exe* from the source file *budget.c*. Within your MAKE file, you will specify the dependency as shown here:

```
BUDGET.EXE:      BUDGET.C
```

On the line that immediately follows the dependency, you will specify the command that MAKE must execute to build the target file. In this case, the two lines of the MAKE file will become the following:

```
BUDGET.EXE:      BUDGET.C
    CC BUDGET.C
```

When you execute MAKE with this make file, MAKE first examines the dependency line. If the specified file (in this example, *budget.exe*) does not exist, or if the target file is older than any other file upon which it is dependent (meaning, you changed one of the component files after you last compiled the executable) MAKE will execute the command that follows. In this example, MAKE will invoke the compiler *(bcc budget.c)*, which recompiles the program. To better understand this process, create the following C file, *bible.c*:

```
#include <stdio.h>

void main(void)
 {
   printf("Jamsa\'s C/C++/C# Programmer\'s Bible");
 }
```

Next, create the file *bible.mak,* which contains the following:

```
BIBLE.EXE:BIBLE.C
    TC BIBLE.C
```

Use the *-f* switch to invoke MAKE, as shown here :

```
C:\> MAKE  -f  BIBLE.MAK <Enter>
```

Because the file *bible.exe* does not exist, MAKE will perform the command to build the file. After MAKE ends, invoke MAKE a second time using the same command. Because the *bible.exe* file already exists and because the file is newer than *bible.c*, MAKE will not execute the command. Edit the file *bible.c* and change the *printf* statement in some way (for example, change the output to "This is a test"). Repeat the MAKE command. Because the file *bible.c* is older than *bible.exe* on this execution, MAKE will build a new EXE file.

704 USING MULTIPLE DEPENDENCY FILES WITH MAKE

As you have learned, MAKE is a tool that helps you build applications after one or more of an application's component files change. When you use MAKE, there will be many times when a target file is dependent on several files. For example, assume that the program *budget.exe* is dependent on the C file *budget.c*, the header file *budget.h*, and the library file *budget.lib*. Within MAKE, you can specify the dependencies as shown here:

```
BUDGET.EXE:      BUDGET.C BUDGET.H BUDGET.LIB
    CC BUDGET.C BUDGET.LIB
```

705 COMMENTING YOUR MAKE FILES

All the MAKE files this book has presented so far are small and fairly straightforward. As the complexity of your MAKE files increases, you will want to add comments that explain the processing the files perform. To place a comment within a MAKE file, simply place the pound sign (#) anywhere in your file. MAKE will consider any text that follows (on the current line) as a comment. The following MAKE file illustrates how to use comments:

```
# Build the Budget program BUDGET.EXE
# Make file originally created: 12-31-01 by Kris Jamsa
BUDGET.EXE:       BUDGET.C BUDGET.H BUDGET.LIB
    CC BUDGET.C BUDGET.LIB     # BUDGET.LIB contains General Ledger functions
```

Command Lines and MAKE 706

As you have learned, if a target file is older than a file from which you built the target file, MAKE will execute a specific command. In the examples you have seen so far, MAKE simply invokes the BCC command to compile and link the corresponding files. MAKE can issue any command and, in addition, MAKE fully supports the DOS *input* (<), *output* (>), and *append* (>>) redirection operators. The following command, for example, directs MAKE to compile a specific program, redirecting the compiler's output to the printer:

```
BUDGET.EXE:       BUDGET.C BUDGET.H BUDGET.LIB
    CC BUDGET.C BUDGET.LIB > PRN
```

In addition to supporting the DOS redirection operators, MAKE supports two special operators, << and &&. The double "less-than" signs (<<) direct MAKE to redirect the command's standard input source. Rather than directing the input source to file, however, MAKE uses the text that immediately follows, up to a delimiter you specify, as the redirected input. For example, the following MAKE file directs MAKE to redirect the text "Jamsa's C/C++/C# Programmer's Bible" to the command SHOWMSG:

```
SOMEFILE.EXE:     SOMEFILE.C
    SHOWMSG << ^Jamsa\'s C/C++/C# Programmer\'s Bible
^
```

In this case, the command uses the caret (^) as the input delimiter. MAKE lets you use any character except the pound sign (#) or backslash (\) as the delimiter. The first line that begins with the delimiter you specify marks the end of the text you wish to redirect to the command SHOWMSG. The && operator is similar, but it does not cause a redirection. Instead, the && operator creates a temporary file that contains the text that appears between the specified delimiters. During the command's execution, MAKE replaces the operator itself with the temporary filename replaces. You will use the && operator most commonly to create a linker response file, as shown here:

```
SOMEFILE.EXE:     SOMEFILE.C
    LINK &&^
    SOMEFILE.C
    SOMEFILE.EXE
    SOMEFILE.MAP
    SOMEFILE.LIB
^
```

As in the previous example, the first line that begins with the specified delimiter marks the end of the temporary file.

Note: *You cannot create MAKE files that directly manipulate the linker in* ***Turbo C++ Lite.***

Placing Multiple Dependencies in a MAKE File

If you are building a large system of programs, you can actually have several different executable files. Rather than managing several different MAKE files, you can create one file that includes the related dependencies for the entire program system. For example, the following MAKE file contains the rules you need to build the programs *budget.exe, payroll.exe*, and *taxes.exe*:

```
BUDGET.EXE:       BUDGET.C BUDGET.H
    CC BUDGET.C
```

```
PAYROLL.EXE:     PAYROLL.C PAYROLL.C
    CC PAYROLL.C

TAXES.EXE:TAXES.C TAXES.H
    CC TAXES.C
```

When you execute MAKE with the previously-shown file, MAKE will begin at the first entry in the file. In this example, if MAKE detects that the compiler must rebuild the first target file, MAKE will execute the corresponding command. MAKE will then continue executing the corresponding commands with the second and third files this example specifies, as necessary.

708 Explicit and Implicit MAKE Rules

When you create a MAKE file, the entries you place in the file that tell MAKE the file's dependencies and the corresponding file operations are called *rules*. MAKE supports *explicit* and *implicit* rules. An explicit rule is a rule that defines one or more target names, zero or more dependent files, and zero or more commands. Filenames within explicit rules can be complete DOS pathnames or wildcards. All the example MAKE files this book has presented so far use explicit rules. Implicit rules, on the other hand, are more general. An implicit rule corresponds to all files with a specific extension. MAKE uses an implicit rule when you do not provide an explicit rule for a target file. For example, you might specify that an OBJ file depends on a C file. The following implicit rule directs MAKE to compile all C files whose C source code files are newer than the corresponding OBJ file:

```
.C.OBJ:
    CC $<
```

The syntax for an implicit rule is the dependent file type (C) followed by the target file type (OBJ). The rule uses a special macro ($<) that, as you will learn, directs MAKE to use the corresponding C file's complete name. Placing only an implicit rule within a MAKE file has no effect when MAKE runs—in other words, a MAKE file must include at least one explicit rule. MAKE will only use the implicit rule when MAKE encounters a target file for which you did not provide an explicit rule.

709 Using MAKE Macros

A *MAKE* macro is a symbol that MAKE replaces with a specific value. You can use macros within MAKE for many purposes. For example, the following macro, *MEM_MODEL*, defines the switches the compiler requires to select the small memory model:

```
MEM_MODEL = -ms
```

To use a macro's value within your MAKE file, you place the macro name within parentheses that are preceded by a dollar sign. For example, the following command line uses the *MEM_MODEL* macro:

```
FILENAME.EXE:    FILENAME.C
    CC $(MEM_MODEL) FILENAME.C
```

710 Predefined MAKE Macros

As you have learned, a *MAKE* macro is a symbol that MAKE replaces with a specific value. MAKE provides several predefined macros that you can use within your MAKE files. Depending on whether you are using a macro within an explicit or implicit rule, the value that MAKE substitutes for the symbol will differ. Table 710.1 discusses how to use MAKE's predefined macros in explicit rules. Likewise, Table 710.2 discusses how to use macros in implicit rules.

Macro Name	Value Returned
$*	Dependent basename with path
$&	Dependent basename without path
$.	Dependent fullname without path
$**	Dependent fullname with path
$<	Dependent fullname with path
$?	Dependent fullname with path

Table 710.1 MAKE predefined macro values for explicit rules.

Macro Name	Value Returned
$*	Target basename with path
$&	Target basename without path
$.	Target fullname without path
$**	All dependent filenames
$<	Target fullname with path
$?	All out-of-date dependents

Table 710.2 MAKE predefined macro values for implicit rules.

The following MAKE file, for example, creates an implicit rule that tells MAKE the relationship between files with the OBJ and C extensions:

```
.C.OBJ:
    CC $<
```

In this example, MAKE will use the implicit rule expansion to expand the $< macro into the target filename and path.

PERFORMING CONDITIONAL PROCESSING WITH MAKE 711

In the Macros section of this book, you learned how to use preprocessor directives such as *#if*, *#elif*, *#else*, and *#endf*. In a similar way, MAKE provides conditional processing statements that begin with an exclamation point (!), such as *!if*, *!else*, *!elif*, and *!endif*. You can also use the directives *!ifdef*, *!ifndef*, and *!undef* to test for macros you have defined and to undefine a macro. If a conditional directive evaluates as true, MAKE will perform the rules that follow. If the directive is false, MAKE will not process the corresponding rules. The following statements illustrate several different conditional statements:

```
!ifdef  macro_name      # Test if the macro macro_name is defined
    # statements
!endif

!if  $(Value) > 5# Test if the value of the macro Value is > 5
   # statements
!endif

!if   ! $d(Macro_name) # Test if the macro Macro_name is not defined
   # statements
!endif
```

TESTING FOR A MAKE MACRO 712

As you have learned, MAKE lets you define your own macros. Depending on the processing your MAKE file performs, there may be times when you want to test whether a specific macro is defined. To do so, you can use the *$d(macro)* test. If the macro is defined, the test will return the value 1. If the macro is not defined, the result will be 0. The following statements use MAKE's *!if* conditional operator to determine whether the macro *MEM_MODEL* is defined. If the macro is undefined, the statements will assign it the value of the small memory model, as shown here:

```
!if  ! $d(MEM_MODEL)
MEM_MODEL = -ms
!endif
```

In addition to using the *$d(macro)* test, your macros can perform equivalent processing using the *!ifdef* and *!ifndef* conditional statements. Should you later want to undefine a macro, you can use the *!undef* statement to do so, as shown here:

```
!undef macro_name
```

Note: *As you have learned, this particular assignment would not work in a* ***Turbo C++ Lite*** *MAKE file, because of how* ***Turbo C++ Lite*** *handles memory models.*

713 INCLUDING A SECOND MAKE FILE

If your MAKE files typically take the same form, then you might find it convenient to place your commonly used implicit rules into a MAKE file named *implicit.mak*. At the start of each of your MAKE files, you can include the file using MAKE's !*include* directive, as shown here:

```
!include "IMPLICIT.MAK"
```

714 USING MAKE'S MACRO MODIFIERS

As you learned in Tip 710, MAKE predefines several different macros that your MAKE files can use to obtain the target or dependent file. To give your MAKE files greater control over the filenames these macros return, MAKE lets you use the *B*, *D*, *F*, and *R* modifiers, detailed in Table 714.

Modifier	Purpose
$(macroB)	Returns the basename only
$(macroD)	Returns the drive and directory
$(macroF)	Returns the basename and extension
$(macroR)	Returns the drive, directory, and basename

Table 714 *The modifiers for the* ***MAKE*** *macro.*

The following statement, for example, uses the *D* modifier with the *$<* to copy files from the target file directory to a backup directory:

```
C:\SUBDIR\TIPS.EXE:     TIPS.C
    COPY $(<D)*.C   C:\BACKUP
    CC TIPS.C
```

Note: *This MAKE file executes two commands within the rule. The first command copies files with the C extension to a directory named BACKUP, and the second command compiles the source file.*

715 ENDING A MAKE FILE WITH AN ERROR

Depending on the processing your MAKE file performs, there maybe times when you may want MAKE to end its processing and display an error message to the user. In such cases, you can use the *!error* directive. The following statements, for example, test to see if the macro *MEM_MODEL* is undefined. If the macro is not defined, MAKE will display an error message to the user and end its processing:

```
!ifndef MEM_MODEL
!error Ending program build—define the macro MEM_MODEL
!endif
```

Disabling Command Name Display 716

By default, MAKE will display each command before the command executes. To disable the command display, simply precede the command name with an *at* symbol (@). For example, the following statements use the @ symbol to disable the display of the CC command:

```
TIPS.EXE: TIPS.C
    @CC TIPS.C
```

If you want to disable the display of the command's output as well, simply redirect the output to the NUL device, as shown here:

```
TEST.EXE: TIPS.C
    @CC TIPS.C > NUL
```

Using the File BUILTINS.MAK 717

As you have learned, there may be many implicit rules that you use on a regular basis. One way to ensure that all your MAKE files can use the implicit rules is to place the common rules in a special file named *builtins.mak*. Each time you invoke MAKE, it will look for *builtins.mak*. If the file exists, MAKE will immediately process the information it contains. If the file does not exist, MAKE will continue its processing using MAKEFILE or the file that you specified in the command line. The following *builtins.mak* file contains the implicit rule for converting C files to OBJ files:

```
.C.OBJ:
    CC $<
```

Performing Exit Status Processing in MAKE 718

When MAKE executes a command, there may be times when you want MAKE to evaluate the command's exit status value and then either continue or end. If you precede a command name with a hyphen followed by a value, MAKE will compare the command's exit status value to the value the MAKE file specifies. If the exit status is *greater than* the value, MAKE will abort the current program build. For example, the following statement compares the exit status of the *showfile* command to 3. If the exit status is greater than 3, MAKE will abort the build, as shown here:

```
TEST.EXE: TIPS.C
    -3  SHOWFILE $?
```

If you want MAKE to ignore a command's exit status, precede the command name with a hyphen and no corresponding value:

```
    -  CC  TEST.C
```

Invoking and Changing a Macro at the Same Time 719

As you have learned, MAKE lets you define your own macros. Depending on the processing that your MAKE file performs, there may be times when you will want to change and immediately use a macro. For example, assume you define the macro *INPUT_FILE* as shown here:

```
INPUT_FILE = BUDGET.C
```

You can then use the macro as shown here:

```
BUDGET.EXE:      $(INPUT_FILE)
    CC $(INPUT_FILE)
```

Next, assume that you want to copy the input file to a file with the same name but with the SAV extension. You can change the macro by replacing .C with .SAV and immediately using the new definition. The following command illustrates how to make the copy:

```
COPY $(INPUT_FILE)   $(INPUT_FILE:.C=.SAV)
```

The first part of the COPY command uses the filename *budget.c*. The second part of the command replaces .C with .SAV to create the filename *budget.sav*.

720 Executing a MAKE Command for Multiple Dependent Files

As you have learned, there may be times when a target file is dependent on two or more files. Depending on the processing your MAKE file performs, you might want MAKE to execute a specific command for each file. To do so, simply precede the command name with an ampersand (&). The following rule, for example, directs MAKE to individually compile each out-of-date dependent file:

```
BUDGET.EXE:      BUDGET.C ACCOUNT.C PAYROLL.C
    & CC $?
```

721 Determining Whether a Math Coprocessor is Present

If your programs perform complex mathematical operations, there may be times when you can use the computer's math coprocessor to improve a program's performance. To help your programs take advantage of the math coprocessor, several third-party libraries exist which provide commonly used functions. Before your program uses such functions, however, your program should verify that a math coprocessor is present. For such cases, many C compilers define the global variable *_8087,* which contains the value 1 if a coprocessor is present and 0 if it is not present. The following *Turbo C Lite* program, *chk_math.c,* illustrates how to use the global variable *_8087*:

```
#include <stdio.h>
#include <dos.h>

void main(void)
 {
   if (_8087)
     printf("Math coprocessor found\n");
   else
     printf("No math coprocessor\n");
 }
```

You can use the *87* environment entry to control the value the C compiler assigns to the *_8087* variable. To set the variable's value to 1, assign *87* the value *Yes,* as shown here:

```
C:\> SET 87=Yes  <Enter>
```

Likewise, to set the variable to 0, assign to the environment entry the value *No*.

UNDERSTANDING THE CTYPE.H AND ISTYPE MACROS 722

The Macros section of this book presents several macros that test whether a character is uppercase, lowercase, alphanumeric, and so on. If you examine the header file *ctype.h*, you will find macro definitions similar to the following:

```
#define isalpha(c)   (_ctype[(c) + 1] & (_IS_UPP | _IS_LOW))
#define isascii(c)   ((unsigned)(c) < 128)
#define iscntrl(c)   (_ctype[(c) + 1] & _IS_CTL)
#define isdigit(c)   (_ctype[(c) + 1] & _IS_DIG)
#define isgraph(c)   ((c) >= 0x21 && (c) <= 0x7e)
#define islower(c)   (_ctype[(c) + 1] & _IS_LOW)
```

To decrease the processing time that macro tests require, many C compilers define a global variable named *ctype*, which contains settings that define each ASCII character. Using these settings, the *istype* macros can use fast bitwise operations to perform the necessary testing. The following program, *ctype.c*, displays the settings the compiler uses for each ASCII character:

```
#include <stdio.h>
#include <ctype.h>

void main(void)
 {
   int ascii_char;

   for (ascii_char = 0; ascii_char < 128; ascii_char++)
     if (isprint(ascii_char))
       printf("ASCII value %d setting (hex) %x ASCII %c\n",
              ascii_char, _ctype[ascii_char], ascii_char);
     else
       printf("ASCII value %d setting (hex) %x ASCII %c\n",
              ascii_char, _ctype[ascii_char], ascii_char);
 }
```

CONTROLLING DIRECT VIDEO 723

The header file *conio.h* defines prototypes for functions that perform console I/O, such as *cputs*. To improve these console I/O functions' performance, most PC compilers bypass DOS and the BIOS and write the output directly to your PC's video memory. Although most video operations are standard from one PC to the next, you may encounter a video board that does not support direct video operations. If you experience such errors, you can use the global variable *directvideo* to control whether the PC uses BIOS video routines to perform your program's output, or whether your program performs direct I/O. If you set the variable's value to 1, the routines will perform direct video output. If the value is 0, the routines will perform their output using the BIOS.

DETECTING SYSTEM AND MATH ERRORS 724

Several of C's run-time library functions this book presents assign values to the global variable *errno* when an error occurs. When your programs use these functions, you should test both the function's return value and *errno's* value. Table 724 defines the constants that functions assign to *errno*.

Constant	Meaning
E2BIG	The argument list is too long
EACCES	Permission denied
EBADF	Bad file handle

***Table 724** Constant values that functions assign to **errno**.* *(continued on the following page)*

Constant	Meaning
ECONTRL	Error in memory control blocks
ECURDIR	Attempt to remove current directory
EDOM	An argument violates the domain of supported values
EEXIST	File already exists
EFAULT	Unknown error
EINVACC	Invalid access specifier
EINVAL	Invalid argument value
EINVDAT	Invalid argument data
EINVDRV	Invalid drive specifier
EINVENV	Invalid environment
EINVFMT	Invalid argument format
EINVFNC	Invalid function number
EINVMEM	Invalid memory block specified
ENFILE	Too many open files
ENMFILE	No more files
ENODEV	No such device
ENOENT	Invalid entry (file or directory)
ENOEXEC	Format error in EXEC
ENOFILE	No such file or directory
ENOMEM	Insufficient memory
ENOPATH	Path not found
ENOTSAM	Not same device
ERANGE	Function result is out of range of valid values
EXDEV	Cross-linked device
EZERO	Error zero

Table 724 *Constant values that functions assign to* ***errno****. (continued from previous page)*

Note: *The companion CD-ROM that accompanies this book includes the text file* ***win_math.txt****, which lists the Windows-based math errors.*

725 DISPLAYING PREDEFINED ERROR MESSAGES

In Tip 724 you learned that various math and system functions assign to the global variable *errno* specific status values which your programs can read to gain information about the cause of an error. Depending on your program's processing, you might want to display a predefined message when an error occurs. To help your program process errors, the C compiler provides a global variable named *sys_errlist*, which contains character string error messages for most errors. Additionally, to increase the portability of your programs, the array contains error messages from the Unix environment.

The compiler also assigns to the global variable *sys_nerr* the number of error messages in the array. The following program, *err_msg.c*, uses the *sys_errlist* array to display the predefined error messages:

```
#include <stdio.h>
#include <stdlib.h>

void main(void)
 {
   int error;

   for (error = 0; error < sys_nerr; error++)
```

```
        printf("Error %d %s\n", error, sys_errlist[error]);
  }
```

DETERMINING THE OPERATING SYSTEM VERSION NUMBER 726

If you are developing applications for the DOS environment, there may be times when your programs must know the current operating system version number. In such cases, your programs can use the predefined global variables *_osmajor* and *_osminor*, which contain the operating system's major and minor version numbers. In addition, some compilers provide the variable *_version*. The constant's low byte contains the major version number, and the high byte contains the minor version number. Given DOS 6.0, for example, the variable *_osmajor* will contain the value 6, while the variable *_osminor* will be 0. The following program, *os_ver.c*, uses the global version variables to display the operating system version number:

```
#include <stdio.h>
#include <dos.h>

void main(void)
  {
    printf("Operating system version number %d.%d\n", _osmajor, _osminor);
    printf("Operating system version number %d.%d\n",
           _version & 255, _version >> 8);
  }
```

Note: *If the previous program does not compile on your system, comment out the statements that use* ***_version*** *and include the header file* ***stdlib.h****.*

UNDERSTANDING PORTABILITY 727

Portability is a measure of the ease with which you can move your program from one system to the next. For example, when you write a program using assembly language for the PC, it is very difficult to move that program to a workstation that uses a different assembly language. If you had written the same program in C, however, you might only need to make a few small changes to the program before it would compile and run on the new system. As you program, you should keep portability in mind. You can often use the same code you have written for one program for many other programs, and you can save considerable programming and testing time if you focus on writing portable code. To improve your program's portability, consider the following as you code:

- Avoid operating system services whenever possible. Rely instead on the C run-time library routines.
- Avoid functions and global variables that are specific to your compiler. In most cases, the compiler will precede such function and variable names with an underscore, such as *_8087*.
- Do not make assumptions about the machine's word size. For example, on the PC a variable of type *int* typically holds 16 bits, but on other machines it might hold 32 bits.
- Do not access hardware-specific locations or rely on specific interrupts unless your program's performance absolutely requires it.
- Always try to correct and eliminate compiler warning messages.
- Do not make memory model assumptions that might not exist in a Unix-based environment.
- Restrict hardware- or operating system-dependent code to as few functions as possible.

728 PERFORMING A NONLOCAL GOTO

In the Getting Started with C section of this book you learned that the *goto* statement lets your program's execution branch from one location to another. As you learned, the label to which you want to "goto" must reside in the current function. Depending on your programs, there may be times when you must branch to a label outside the current function (called a *nonlocal goto*). To perform a nonlocal goto, your programs can use the *setjmp* and *longjmp* functions:

```
#include <setjmp.h>

void longjmp(jmp_buf location, int return_value);
void setjmp(jmp_buf location);
```

To begin, your program will use *setjmp* to store the current location (also known as the *task state*) in the buffer *location*. Later, your programs can jump to that location using *longjmp*. The first time your program invokes *setjmp*, the function will return 0. When the program later calls *longjmp*, it will return to the location *setjmp* previously stored and yield the return value specified within the *longjmp* function's *return_value* parameter. The following program, *longjmp.c*, illustrates a nonlocal goto:

```
#include <stdio.h>
#include <setjmp.h>
#include <stdlib.h>

jmp_buf location;    // Global variable

void function(void)
 {
   printf("About to longjmp\n");
   longjmp(location, 1);   // Return 1
 }

void main(void)
 {
   if (setjmp(location) != 0)   // Save the current location
     {
       printf("Returning from longjmp\n");
       exit(1);
     }
   function();
 }
```

729 GETTING THE PROCESS ID (PID)

In a multitasking environment, the operating system assigns each program a unique identifier, called a process ID, or PID. The Unix operating system provides a function named *getpid*, which returns a program's process ID. Many DOS-based compilers provide a similar *getpid* function, as shown here:

```
#include <process.h>

unsigned getpid(void);
```

Within DOS, the process ID is actually the segment address of the program's program segment prefix (PSP). If you have two or more memory-resident programs active, each program will have a unique process ID because each has a unique PSP segment address. The following program, *getpid.c*, displays the program's process ID:

```
#include <stdio.h>
#include <process.h>

void main(void)
 {
```

```
    printf("Process id: %X\n", getpid());
 }
```

Note: *Windows handles processes and their counterparts, threads, slightly differently than the DOS PID. Tips 1376 through 1400 discuss process and thread management in detail.*

INVOKING AN INTERNAL DOS COMMAND 730

Several of the Tips in this section have shown you ways that your programs can invoke executable (EXE and COM) files. Depending on your program, there may be times when you must invoke an internal DOS command or a batch file. In such cases, your programs can use the *system* function, as shown here:

```
#include <stdlib.h>

int system(const char *command);
```

The *command* parameter is a character string that contains the name of the desired internal or external DOS command or batch file. If *system* successfully executes the command, it will return the value 0. If an error occurs, *system* will return the value –1 and assign the global variable *errno* one of the values listed in Table 730.

Value	Meaning
ENOENT	No such file
ENOMEM	Not enough memory
E2BIG	Argument list too long
ENOEXEC	Error in *exec* format

Table 730 *The* ***errno*** *values the* ***system*** *function returns.*

The *system* function spawns a copy of *command.com* to execute the specified command. The function uses the COMSPEC environment entry to locate the command processor. The following program, *system.c*, uses the *system* function and the DOS *dir* command to display a directory listing:

```
#include <stdlib.h>

void main(void)
 {
   if (system("DIR"))
     printf("Error invoking DIR\n");
 }
```

USING THE _PSP GLOBAL VARIABLE 731

Each time you execute a program, DOS loads the program into memory immediately following a 256-byte buffer called the program segment prefix (PSP). The program segment prefix contains information about the command line, a pointer to the program's environment copy, file table information, and so on. Figure 731 illustrates the contents of the program segment prefix.

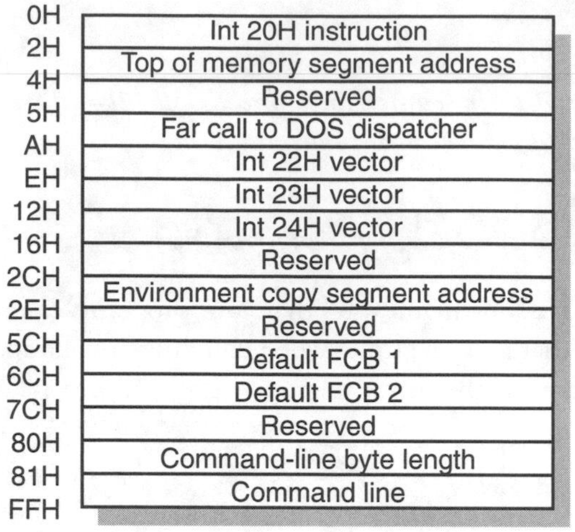

Offset	Contents
0H	Int 20H instruction
2H	Top of memory segment address
4H	Reserved
5H	Far call to DOS dispatcher
AH	Int 22H vector
EH	Int 23H vector
12H	Int 24H vector
16H	Reserved
2CH	Environment copy segment address
2EH	Reserved
5CH	Default FCB 1
6CH	Default FCB 2
7CH	Reserved
80H	Command-line byte length
81H	Command line
FFH	

***Figure** 731 The contents of the program segment prefix.*

As you begin program with DOS internal commands, there may be times when your programs access information that a program's PSP contains. DOS therefore provides a system service that returns the address of the PSP. To simplify such programs, some C compilers define a global variable, *_psp*, which contains the segment address of the program's PSP. The following *Turbo C Lite* program, *psp_addr.c*, uses the *_psp* global variable to display the program's PSP address:

```
#include <stdio.h>
#include <dos.h>

void main(void)
 {
   printf("The Program Segment Prefix begins at %X\n", _psp);
 }
```

Note: *Some C compilers also provide the function* **getpsp**, *which returns the segment address of the program's PSP, as shown here:*

```
#include <dos.h>

unsigned getpsp(void);
```

732 USING THE CONST MODIFIER IN VARIABLE DECLARATIONS

Several function prototypes this book presents use the *const* keyword before parameter names, as shown here:

```
char *strcpy(char *target, const char *source);
```

When you use the *const* keyword before a parameter name, you tell the compiler the program should not change the parameter within the function. If a statement tries to change the parameter, the compiler will generate an error message. C also lets you use the *const* keyword when you declare variables. When you declare a variable as a constant, the C compiler will perform a one-time initialization of the variable. Following the initialization, the compiler will generate an error each time you try to change the constant. The following statements create several different constants:

```
const int number = 1001;
const float price = 39.95;
```

The advantage of using a constant rather than a macro that you created with *#define* is that you can, using the constant, explicitly specify the value's type.

Note: *When you declare a constant, you can still use a pointer alias to change the constant variable's value.*

USING ENUMERATED TYPES 733

As you have learned, using meaningful variable names can significantly improve your program's readability. In addition, replacing constant values (such as 1, 2, and 3) with meaningful names that correspond to the values the variables represent (such as Monday, Tuesday, and Wednesday) can improve your program's readability. To help your programs work with such constants, C supports *enumerated types*. In general, an enumerated type is a list of items, each of which has a unique value. You can use enumerated types to improve your program's readability. For example, the following declaration creates an enumerated type called *weekdays*:

```
enum weekdays { Monday, Tuesday, Wednesday, Thursday, Friday };
```

The enumerated type is similar to a structure definition in that you can declare variables of the type immediately, or you can refer to the type name later, as shown here:

```
enum weekdays { Monday, Tuesday, Wednesday, Thursday, Friday } work_day;

enum weekdays day_off;
```

After you declare an enumerated variable, you can refer to a member name to assign the variable a value, as shown here:

```
day_off = Friday;
work_day = Tuesday;
```

PUTTING AN ENUMERATED TYPE TO USE 734

In Tip 733 you learned that your programs can use enumerated types to improve their readability. The following program, *whichday.c*, illustrates how your programs might use an enumerated type to improve their readability:

```
#include <stdio.h>

void main(void)
 {
   enum { Monday, Tuesday, Wednesday, Thursday, Friday } day;

   for (day = Monday; day <= Friday; day++)
    if (day == Monday)
      printf("No fun--meetings all day Monday\n");
    else if (day == Tuesday)
      printf("No fun--do Monday\'s work today\n");
    else if (day == Wednesday)
      printf("Hump day...");
    else if (day == Thursday)
      printf("Schedule meetings for next Monday\n");
    else
      printf("Meet everyone at happy hour!\n");
 }
```

UNDERSTANDING AN ENUMERATED VALUE 735

As you learned in Tip 733, each member within an enumerated type has a unique value. By default, the C compiler assigns the first member the value 0, the second the value 1, and so on. The following program, *showenum.c*, displays the values that correspond to the enumerated days of the week:

```
#include <stdio.h>

void main(void)
 {
   enum weekdays { Monday, Tuesday, Wednesday, Thursday, Friday };
```

```
   printf("%d %d %d %d %d\n", Monday, Tuesday, Wednesday, Thursday, Friday);
 }
```

When you compile and execute the *showenum.c*, program, your screen will display the values 0 through 4.

736 Assigning a Specific Value to an Enumerated Type

In Tip 735 you learned that the C compiler assigns unique values to each member of an enumerated type. Depending on the function that your program performs, there may be times when you want to specify each member's value. The following declaration, for example, assigns the values 10, 20, 30, 40, and 50 to days of the week:

```
enum weekdays { Monday = 10, Tuesday = 20, Wednesday = 30,
                Thursday = 40, Friday = 50 };
```

The following program, *setenum.c*, assigns these values to the members of the enumerated type and then displays each member's values:

```
#include <stdio.h>

void main(void)
 {
   enum weekdays { Monday = 10, Tuesday = 20, Wednesday = 30,
                   Thursday = 40, Friday = 50 };

   printf("%d %d %d %d %d\n", Monday, Tuesday, Wednesday, Thursday, Friday);
 }
```

In addition to assigning each member a value, you can also assign a value to a specific member. The C compiler will increment each remaining member's value by 1. The following declaration, for example, assigns the values 10, 11, 12, 13, and 14 to the days of the week:

```
enum weekdays { Monday = 10, Tuesday, Wednesday, Thursday, Friday};
```

737 Saving and Restoring Registers

Many compilers let you access register values from within your C programs. Programs that perform such operations often push register values onto the stack before the program changes the register, and later pop the value to restore it to the register. If you have a function that performs such low-level operations, you can direct the C compiler to insert PUSH and POP instructions in the object code. PUSH and POP will automatically save all the registers when the program calls the function and will later restore the registers before the function ends. To direct the compiler to perform such operations, simply include the *_saveregs* modifier in the function header, as shown here:

```
int _saveregs some_function(int parameter);
```

To better understand the processing that the *_saveregs* modifier directs the compiler to perform, create a simple function that uses the *_saveregs* modifier, and then generate an assembly language listing of the source code.

738 Getting Started with Dynamic Lists

In the Structures section of this book you learned how to group related information into a single variable. If your program must work with a fixed number of structure occurrences, your program can create an array of structures. As your programs become more complex, however, there will be many times when you will not know in advance how many structure entries you will need. In such cases, you have two choices. First, your program can allocate memory dynamically for the array of structures. Second, your programs can create a *linked list* of structures, where one entry points to the next entry. Figure 738 illustrates a linked list of filenames.

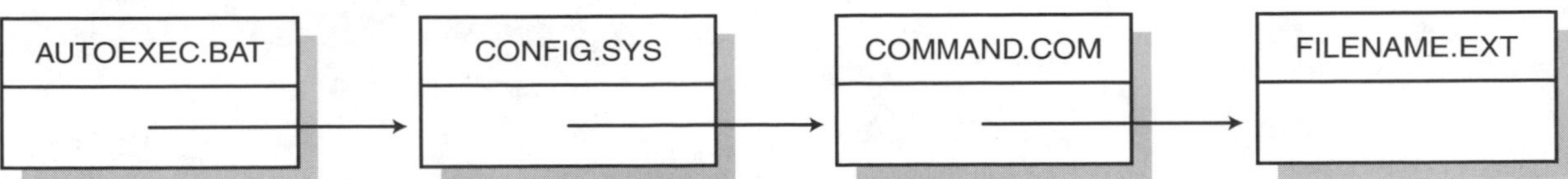

Figure 738 A linked list of file names.

In general, the program maintains a pointer to the start of the list. A pointer to *NULL* indicates the last entry in the list.

DECLARING A LINKED-LIST STRUCTURE 739

To create a linked list, one of the structure members must be a pointer to a structure of the same type. For example, consider the following structure:

```
struct FileList
 {
   char filename[64];
   struct FileList *next;
 };
```

The *filename* member contains a filename. The *next* member is a pointer to the next entry in the list. To create and later traverse a linked list, your programs will usually use at least two variables. The variable *start* is a structure. Its member *next* will contain a pointer to the start of the list or *NULL* if the list is empty. The variable *node* will be a pointer to the current node, as shown here:

```
struct FileList start, *node;
```

BUILDING A LINKED LIST 740

To create a linked list, your programs should perform the following steps:

1. Declare the structure that defines the list entries.
2. Declare the variables *start* and **node.*
3. Assign *start.next* the value *NULL* to signify an empty list.

For each list entry, your programs should perform the following steps:

1. Find the end of the list so that *node->next* is *NULL.*
2. Allocate memory for the new entry and assign the memory's starting location value to the *node->next* pointer member.
3. Assign *node* the value of *node->next.*
4. Assign the member values to *node.*
5. Assign *node->next* the value *NULL* to indicate it is the new end of the list.

A SIMPLE LINKED-LIST EXAMPLE 741

In Tip 740 you learned the steps your programs must perform to create a linked list. The following *Turbo C Lite* program, *1_10list.c*, creates a linked list whose entries contain the numbers 1 through 10:

```
#include <stdio.h>
#include <alloc.h>

void main(void)
 {
   int i;
   struct ListEntry
    {
      int number;
      struct ListEntry *next;
    } start, *node;

   start.next = NULL;  // Empty list
   node = &start;      // Point to the start of the list
   for (i = 1; i <= 10; i++)
    {
       node->next = (struct ListEntry *) malloc(sizeof(struct ListEntry));
       node = node->next;
       node->number = i;
       node->next = NULL;
    }
    // Display the list
    node = start.next;
    while (node)
     {
       printf("%d ", node->number);
       node = node->next;
     }
 }
```

742 Understanding the Linked-List Traversal

In Tip 741 you wrote the program *1_10list.c,* which created a simple linked list with entries that contain the numbers 1 through 10. The program uses the following loop to display the list entries:

```
    // Display the list
    node = start.next;
    while (node)
    {
       printf("%d ", node->number);
       node = node->next;
    }
```

Within this code fragment, the variable *start.next* points to the first entry in the list. As you can see, the code assigns the first entry's address to *node*. Likewise, as you will recall, *NULL* indicates the end of the list. Therefore, the loop simply tests the current value of *node* to see if it is *NULL*. If *node* is not *NULL*, the loop will display the entry's value and assign to *node* the address of the next list entry.

743 Building a More Useful List

In Tip 741 you created a simple linked list containing the numbers 1 though 10. If your compiler supports the directory read functions, the program *filelist.c* creates a linked list of the filenames in the current directory:

```
#include <stdio.h>
#include <dirent.h>
#include <alloc.h>
#include <string.h>

void main(int argc, char *argv[])
```

```
{
  DIR *directory_pointer;
  struct dirent *entry;
  struct FileList
   {
      char filename[64];
      struct FileList *next;
   } start, *node;

  if ((directory_pointer = opendir(argv[1])) == NULL)
    printf("Error opening %s\n", argv[1]);
  else
    {
       start.next = NULL;
       node = &start;
       while (entry = readdir(directory_pointer))
        {
          node->next = (struct FileList *)
          malloc(sizeof(struct FileList));
          node = node->next;
          strcpy(node->filename, entry);
          node->next = NULL;
        }
       closedir(directory_pointer);
       node = start.next;
       while (node)
        {
          printf("%s\n", node->filename);
          node = node->next;
        }
    }
}
```

Filelist.c uses the *readdir* function to read directory entries. *Filelist.c* will then allocate memory to hold the entry and copy the corresponding filename to the list entry. After the program adds all the files to the list, the program will loop through the list and display each entry.

APPENDING A LIST ENTRY 744

Each linked-list program this book has presented so far has built the entire linked list at one time, usually within a *while* or *for* loop. Depending on your program, you will probably at some point want to add entries to the list at different times. The easiest way to add an entry is to append the entry. To append an item to a linked list, you will loop through the list until you find the element that contains the *next* member, which points to *NULL*, as shown here:

```
node = &start;
while (node->next)
  node = node->next;
```

When *node->next* points to *NULL*, you have found the end of the list and you can therefore allocate memory for the new entry, as shown here:

```
node->next = malloc(required_size);
```

Next, assign the entry the value you want (within the *member* item) and assign the new entry's *next* field to point to *NULL*, as shown here:

```
node = node->next;
node->member = some_value;
node->next = NULL;
```

In some cases, you might want your programs to place elements at specific locations in a list. You will learn how to place elements in Tip 745.

745 Inserting a List Entry

In Tip 744 you learned how to append items to a linked list. Depending on your program's function, there will be times when you will want to place items at specific locations in a list. For example, if you want to create a linked list that contains the sorted names of files in the current directory, then your program must place each filename into the list at the correct position. To insert an item at a specific location in a list, your programs will usually track the starting *node*, current *node*, and the previous *node*. When your program must insert a new element, it will perform the following processing:

```
struct ListMember start, *node, *previous, *new;

// Code that performs insert of an entry between
// the elements pointed to by node and previous
new = malloc(sizeof(struct ListMember);
new->next = node;
previous->next = new;
new->member = some_value;
```

746 Displaying a Sorted Directory

In Tip 745 you learned that to insert an element in a singly linked list (where each element contains a pointer to the next element), your programs must track the current and previous nodes (list elements). The following program, *sortlist.c*, inserts elements into a list to create a list that contains the sorted current directory filenames:

```
#include <stdio.h>
#include <dirent.h>
#include <alloc.h>
#include <string.h>
#include <stdlib.h>

void main(int argc, char *argv[])
 {
   DIR *directory_pointer;
   struct dirent *entry;
   struct FileList
    {
       char filename[64];
       struct FileList *next;
    } start, *node, *previous, *new;

   if ((directory_pointer = opendir(argv[1])) == NULL)
     printf("Error opening %s\n", argv[1]);
   else
     {
        start.next = NULL;
        while (entry = readdir(directory_pointer))
          {
            // Find the correct location
            previous = &start;
            node = start.next;
            while ((node) && (strcmp(entry, node->filename) > 0))
             {
               node = node->next;
               previous = previous->next;
             }
```

```
          new = (struct FileList *)
        malloc(sizeof(struct FileList));
        if (new == NULL)
         {
           printf("Insufficient memory to store list\n");
           exit(1);
         }
        new->next = node;
        previous->next = new;
        strcpy(new->filename, entry);
      }
    closedir(directory_pointer);
    node = start.next;
    while (node)
      {
        printf("%s\n", node->filename);
        node = node->next;
      }
  }
}
```

DELETING AN ELEMENT FROM A LIST 747

In Tip 745 you learned how to insert items into a linked list. As your programs manipulate linked lists, there will be times when you must delete an element from a list. Removing an element from a singly linked list is very similar to an insertion operation in that you must track pointers to the current and previous nodes. After your program locates the list element that it wants to delete, it can use code similar to the following to remove the node:

```
previous->next = node->next;
free(node);
```

The following *Turbo C Lite* program, *remove5.c*, creates a linked list of the numbers 1 through 10. The program then searches the list for the element that contains the number five. The program then removes that element:

```
#include <stdio.h>
#include <alloc.h>

void main(void)
 {
   int i;
   struct ListEntry
    {
      int number;
      struct ListEntry *next;
    } start, *node, *previous;

   start.next = NULL;                       // Empty list
   node = &start;                           // Point to the start of the list
   for (i = 1; i <= 10; i++)
     {
       node->next = (struct ListEntry *)
         malloc(sizeof(struct ListEntry));
       node = node->next;
       node->number = i;
       node->next = NULL;
     }
   node = start.next;                       // Remove the number 5
   previous = &start;
   while (node)
      if (node->number == 5)
        {
```

```
          previous->next = node->next;
        free(node);
        break;                          // End the loop
      }
    else
      {
        node = node->next;
        previous = previous->next;
      }
  node = start.next;                    // Display the list
  while (node)
    {
      printf("%d ", node->number);
      node = node->next;
    }
}
```

748 USING A DOUBLY LINKED LIST

A *singly linked list* is so named because each list element contains a pointer to the next element. You have learned that to insert elements in a singly linked list, your programs must maintain pointers to the current and previous elements. To simplify the process of inserting and removing list elements, your programs can use a *doubly linked list*. In a doubly linked list, each element maintains a pointer to the next and previous list elements. Figure 748 illustrates a doubly linked list.

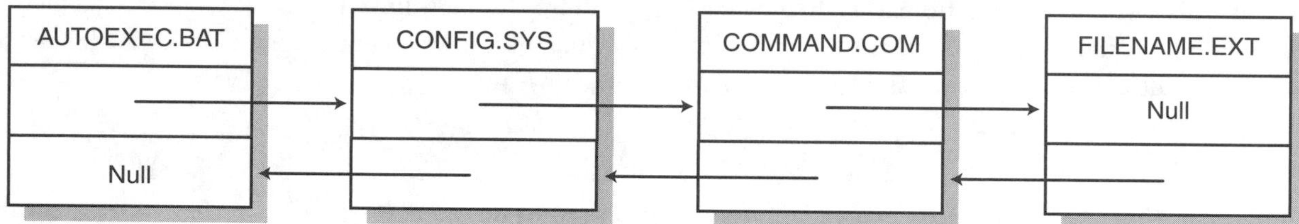

***Figure 748** A doubly linked list maintains two pointers.*

The following structure illustrates a doubly linked list structure:

```
struct FileList
 {
   char filename[64];
   struct FileList *next;
   struct FileList *previous;
 };
```

When your programs use a doubly linked list, the program can traverse the list from left to right or right to left. Therefore, the list must maintain two *NULL* pointers. When your program traverses the list from left to right, it should sense when it reaches the end of the list because *node->next* is *NULL*. Likewise, when your program traverses the list from right to left, it should sense it has reached the end of the list when *node->previous* is *NULL,* which indicates the end of the list.

749 BUILDING A SIMPLE DOUBLY LINKED LIST

In Tip 748 you learned that a doubly linked list simplifies the process of inserting and removing elements. The following *Turbo C Lite* program, *dbl_1_10.c*, uses a doubly linked list to show the numbers 1 through 10 forward and backward:

```
#include <stdio.h>
#include <alloc.h>

void main(void)
```

```
{
  int i;
  struct ListEntry
   {
     int number;
     struct ListEntry *next;
     struct ListEntry *previous;
   } start, *node;
  start.next = NULL;                  // Empty list
  start.previous = NULL;
  node = &start;                      // Point to the start of the list
  for (i = 1; i <= 10; i++)
    {
      node->next = (struct ListEntry *)
      malloc(sizeof(struct ListEntry));
      node->next->previous = node;
      node = node->next;
      node->number = i;
      node->next = NULL;
    }
  node = start.next;                  // Display the list
  do
    {
       printf("%d ", node->number);
       node = node->next;
    } while (node->next);              // Show 10 only one time
  do
    {
       printf("%d ", node->number);
       node = node->previous;
    } while (node->previous);
}
```

UNDERSTANDING NODE->PREVIOUS->NEXT 750

As you have learned, working with doubly linked lists simplifies element insert and delete operations. As you examine programs that work with doubly linked lists, you might encounter statements such as the following:

```
node->previous->next = new_node;
```

As you examine such statements, work from left to right. Figure 750 illustrates how the C compiler resolves the pointer.

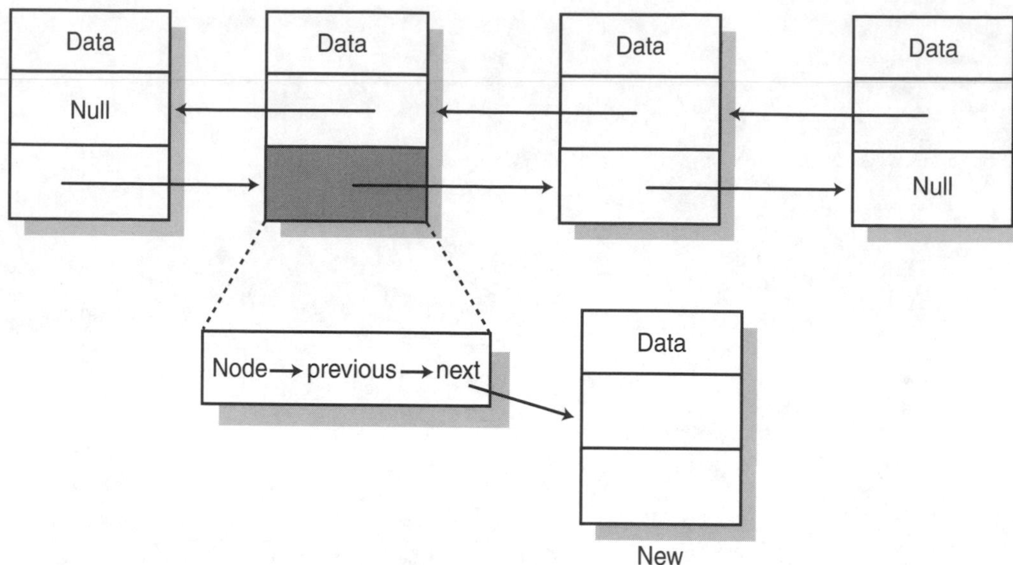

Figure 750 Resolving a complex pointer operation.

751 REMOVING AN ELEMENT FROM A DOUBLY LINKED LIST

A doubly linked list simplifies the process of inserting and removing list elements. The following *Turbo C Lite* program, *remove_7.c*, builds a doubly linked list that contains the numbers 1 through 10. The program then searches the list for the entry containing the number 7 and removes the entry:

```
#include <stdio.h>
#include <alloc.h>

void main(void)
 {
   int i, found;
   struct ListEntry {
     int number;
     struct ListEntry *next;
     struct ListEntry *previous;
   } start, *node;

   start.next = NULL;  // Empty list
   start.previous = NULL;
   node = &start;      // Point to the start of the list
   for (i = 1; i <= 10; i++)
     {
       node->next = (struct ListEntry *)
           malloc(sizeof(struct ListEntry));
       node->next->previous = node;
       node = node->next;
       node->number = i;
       node->next = NULL;
     }
    // Remove the entry
    node = start.next;
    found = 0;
    do {
        if (node->number == 7)
          {
            found = 1;
            node->previous->next = node->next;
```

```
              node->next->previous = node->previous;
           free(node);
         }
       else
         node = node->next;
   } while ((node) && (! found));  // Show 10 only one time
   node = start.next;
   do {
       printf("%d ", node->number);
       node = node->next;
   } while (node);
 }
```

INSERTING AN ELEMENT INTO A DOUBLY LINKED LIST 752

As you have learned, doubly linked lists simplify the insertion and deletion of list elements. The following *Turbo C Lite* program, *bld_1_10.c*, builds a list containing the numbers 1, 3, 5, 7, and 9. The program then inserts the numbers 2, 4, 6, 8, and 10 in the correct locations within the list:

```
#include <stdio.h>
#include <alloc.h>

void main(void)
 {
   int i;
   struct ListEntry {
     int number;
     struct ListEntry *next;
     struct ListEntry *previous;
   } start, *node, *new;

   start.next = NULL;  // Empty list
   start.previous = NULL;
   node = &start;      // Point to the start of the list
   for (i = 1; i < 10; i += 2)
     {
       node->next = (struct ListEntry *)
           malloc(sizeof(struct ListEntry));
       node->next->previous = node;
       node = node->next;
       node->number = i;
       node->next = NULL;
     }
   for (i = 2; i <= 10; i += 2)
     {
       int found = 0;

       new = (struct ListEntry *) malloc(sizeof(struct ListEntry));
       new->number = i;
       node = start.next;
       do {
         if (node->number > new->number)
           {
             new->next = node;
             new->previous = node->previous;
             node->previous->next = new;
             node->previous = new;
             found = 1;
           }
         else
           node = node->next;
       } while ((node->next) && (! found));
```

```
         if (! found)
         if (node->number > new->number)
           {
              new->next = node;
              new->previous = node->previous;
              node->previous->next = new;
              node->previous = new;
           }
         else
           {
              new->next = NULL;
              new->previous = node;
              node->next = new;
           }
     }
    // Display the list
    node = start.next;
    do {
        printf("%d ", node->number);
        node = node->next;
    } while (node);
}
```

753 Understanding Child Processes

When you run a program, that program can run a second program, called a *child process*. The program that runs the second program is called the *parent*. Depending on your needs, the child process can run to completion and the parent can continue, or the child can take the parent's place, overwriting the parent in memory. When the child program runs to completion and the parent continues, the child's execution is called *spawning*. When the child process replaces the parent in memory, the program must *exec* the child. To help your programs perform such processing, the C run-time library provides two types of run-time library functions: *spawn* and *exec*. Tips 754 and 757discuss these run-time library routines in detail.

754 Spawning a Child Process

As you learned in Tip 753, when a program spawns a child task, the parent program suspends its processing while the child process runs, and then later continues. To spawn a child process, your programs can use the *spawnl* function, as shown here:

```
#include <process.h>
#include <stdio.h>

int spawnl(int mode, char *child, char *arg0, ... ,char *argn, NULL);
```

The *child* parameter is a pointer to a character string that specifies the name of the executable file containing the child process. The parameters *arg0* through *argn* specify the child process's command-line arguments. The *mode* parameter specifies how your program runs the child process. Table 754.1 lists the possible mode values.

Value	Mode of Execution
P_NOWAIT	The parent process continues to run in parallel with the child (not available for DOS-based programs)
P_OVERLAY	The child process overwrites the parent in memory
P_WAIT	The parent process resumes after the child ends

Table 754.1 Modes of child process execution.

If the *spawnl* function succeeds, it will return the value 0. If an error occurs, the function will return the value –1 and set the global variable *errno* to one of the values listed in Table 754.2.

Value	Description
E2BIG	Argument list too long
EINVAL	Invalid argument
ENOENT	Child program not found
ENOEXEC	Format error
ENOMEM	Insufficient memory

***Table 754.2** The error values **spawnl** returns.*

To understand child processes better, create the program *child.c*, which will display its command-line arguments and environment entries, as shown here:

```
#include <stdlib.h>
#include <stdio.h>

void main(int argc, char *argv[], char *env[])
 {
   printf("Command line\n");
   while (*argv)
     puts(*argv++);
   printf("Environment entries\n");
   while (*env)
     puts(*env++);
 }
```

Compile the program. Next, create the program *spawnl.c*, that uses the *spawnl* function to execute the child process:

```
#include <process.h>
#include <stdio.h>

void main(void)
 {
   printf("About to call child process\n\n");
   spawnl(P_WAIT, "CHILD.EXE", "CHILD.EXE", "AAA", "BBB", "CCC", NULL);
   printf("\n\nBack from child process\n");
 }
```

When you execute the *spawnl* program, your screen will display a message stating that it is about to call the child process. Next, the child process will run, displaying its command-line arguments and environment entries. After the child process ends, the program will display a message stating that it has returned from the child process.

USING OTHER SPAWNLXX FUNCTIONS

In Tip 754 you learned that the *spawnl* function lets you run a child process. If you examine the C run-time library, you will find several other *spawnlxx* functions, as shown here:

```
#include <stdio.h>
#include <process.h>

int spawnle(int mode, char *child, char *arg0, ..., char *argn, NULL,
            char *environ);
int spawnlp(int mode, char *child, char *arg0, ..., char *argn, NULL);
int spawnlpe(int mode, char *child, char *arg0, ..., char *argn, NULL,
             char *environ);
```

If any *spawnlxx* function succeeds, the function will return the value 0. If an error occurs, each function will return the value –1 and set the global variable *errno* to one of the values listed in Table 755.

Value	Description
E2BIG	Argument list too long
EINVAL	Invalid argument
ENOENT	Child program not found
ENOEXEC	Format error
ENOMEM	Insufficient memory

Table 755 *Error values the* ***spawnlxx*** *functions return.*

The parameters to the *spawnlxx* functions are similar to those *spawnl* uses, as detailed in Tip 754. However, the *spawnlxx* functions also use the *environ* parameter, which contains a pointer to the child's environment entries. The difference between *spawnl* and *spawnlp* is that *spawnlp* and *spawnlpe* will search the command path for the child process. The following program, *spawnlxx.c*, illustrates the use of the *spawnlxx* functions:

```
#include <process.h>
#include <stdio.h>

void main(void)
 {
   char *env[] = { "FILE=SPAWNLXX.C", "LANGUAGE=C", "OS=DOS", NULL};

   spawnle(P_WAIT, "CHILD.EXE", "CHILD.EXE", "Using-spawnle", "BBB", NULL, env);
   spawnlp(P_WAIT, "CHILD.EXE", "CHILD.EXE", "Using-spawnlp", "BBB", NULL);
   spawnlpe(P_WAIT, "CHILD.EXE", "CHILD.EXE", "Using-spawnlpe", "BBB", NULL, env);
 }
```

756 USING THE SPAWNVXX FUNCTIONS

In Tip 754 you learned how to use the *spawnl* function to create a child process. Likewise, in Tip 755 you used the different *spawnlxx* functions, which let you pass an array of environment entries to the child process. The *spawnlxx* functions also let you use the command path to locate the child process. When you use the *spawnl* functions, you pass the command-line arguments as a list of *NULL*-terminated parameters. In addition to the *spawnlxx* functions, C provides a collection of *spawnvxx* functions that let you pass the command-line parameters as an array of character strings, as shown here:

```
#include <stdio.h>
#include <process.h>

int spawnv(int mode, char *child, char *argv[]);
int spawnve(int mode, char *child, char *argv[], char *env[]);
int spawnvp(int mode, char *child, char *argvp[]);
int spawnvpe(int mode, char *child, char *argv[], char *env[]);
```

If any *spawnvxx* function succeeds, the function will return the value 0. If an error occurs, each function will return the value –1 and set the global variable *errno* to one of the values listed in Table 756.

Value	Description
E2BIG	Argument list too long
EINVAL	Invalid argument
ENOENT	Child program not found
ENOEXEC	Format error
ENOMEM	Insufficient memory

Table 756 *The error values* ***spawnvxx*** *returns.*

The parameters to the *spawnvxx* functions are similar to those your program passes to the *spawnlxx* functions, except that the *spawnvxx* functions pass the command-line arguments as an array of character strings. The following program, *spawnvxx.c,* illustrates the *spawnvxx* functions:

```
#include <stdio.h>
#include <process.h>

void main(void)
 {
   char *env[] = { "FILENAME=SPAWNVXX.C", "OS=DOS", "ROUTINES=SPAWNVXX", NULL };
   char *argv[] = { "CHILD.EXE", "AAA", "BBB", NULL };

   spawnv(P_WAIT, "CHILD.EXE", argv);
   spawnve(P_WAIT, "CHILD.EXE", argv, env);
   spawnvp(P_WAIT, "CHILD.EXE", argv);
   spawnvpe(P_WAIT, "CHILD.EXE", argv, env);
 }
```

EXECING A CHILD PROCESS 757

As you learned in Tip 753, when a program execs a child task, the child process overwrites the parent program in memory. Because the child overwrites the parent, the parent process never resumes. To exec a child process, your programs can use the *execl* function, as shown here:

```
#include <process.h>
#include <stdio.h>

int execl(char *child, char *arg0, ...,char *argn, NULL);
```

The *child* parameter is a pointer to a character string that specifies the name of the executable file containing the child process. The parameters *arg0* through *argn* specify the child process's command-line arguments.

If the *execl* function succeeds, it will not return a value. If an error occurs, the function will return the value –1 and set the global variable *errno* to one of the values listed in Table 757.

Value	Description
E2BIG	Argument list too long
EINVAL	Invalid argument
ENOENT	Child program not found
ENOEXEC	Format error
ENOMEM	Insufficient memory

Table 757 The error values **execl** *returns.*

To better understand execing child processes, create the following program, *child.c,* that will display its command-line arguments and environment entries:

```
#include <stdlib.h>
#include <stdio.h>

void main(int argc, char *argv[], char *env[])
 {
   printf("Command line\n");
   while (*argv)
     puts(*argv++);
   printf("Environment entries\n");
   while (*env)
     puts(*env++);
 }
```

Compile the program. Next, create the following program, *execl.c*, which uses the *execl* function to execute the child process:

```
#include <process.h>
#include <stdio.h>

void main(void)
 {
   printf("About to call child process\n\n");
   execl("CHILD.EXE", "CHILD.EXE", "AAA", "BBB", "CCC", NULL);
   printf("\n\nBack from child process—SHOULD NOT APPEAR\n");
 }
```

When you execute the *execl* program, your screen will display a message stating that it is about to call the child process. Next, the child process will run, displaying its command-line arguments and environment entries. The child process overwrites the parent process, so after the child process ends no additional processing occurs.

758 USING OTHER EXECLXX FUNCTIONS

In Tip 757 you learned that the *execl* function lets you run a child process. If you examine the C run-time library, you will find several other *execlxx* functions, as shown here:

```
#include <stdio.h>
#include <process.h>

int execle(char *child, char *arg0, ..., char *argn, NULL, char *environ);
int execlp(char *child, char *arg0, ..., char *argn, NULL);
int execlpe(char *child, char *arg0, ..., char *argn, NULL, char *environ);
```

If any *execlxx* function succeeds, the function will not return a value. If an error occurs, the function will return the value –1 and set the global variable *errno* to one of the values listed in Table 758.

Value	Description
E2BIG	Argument list too long
EINVAL	Invalid argument
ENOENT	Child program not found
ENOEXEC	Format error
ENOMEM	Insufficient memory

Table 758 The error values the ***execlxx*** *functions return.*

The parameters to these functions are similar to those *spawnl* uses, with the exception of the *environ* parameter, which contains a pointer to the child's environment entries. (Tip 754 describes *spawnl* in detail.) The difference between *execl* and *execlp* is that functions containing the letter p will search the command path for the child process. The following program, *execlpe.c*, illustrates the use of the *execlpe* function:

```
#include <process.h>
#include <stdio.h>

void main(void)
 {
   char *env[] = { "FILE=EXECLPE.C", "LANGUAGE=C", "OS=DOS", NULL};

   execlpe("CHILD.EXE", "CHILD.EXE", "Using-execlpe", "BBB", NULL, env);
 }
```

USING THE EXECVXX FUNCTIONS 759

In Tip 757 you learned how to use the *execl* function to create a child process. Likewise, in Tip 758 you used the different *execlxx* functions, which let you pass an array of environment entries to the child process and also let you use the command path to locate the child process. When you use the *execl* functions, you pass the command-line arguments as a list of *NULL*-terminated parameters. In addition to the *execlxx* functions, C provides a collection of *execvxx* functions that let you pass the command-line parameters as an array of character strings, as shown here:

```
#include <stdio.h>
#include <process.h>

int execv(char *child, char *argv[]);
int execve(char *child, char *argv[], char *env[]);
int execvp(char *child, char *argvp[]);
int execvpe(char *child, char *argv[], char *env[]);
```

If any *execvxx* function succeeds, the function will not return a value. If an error occurs, the function will return the value –1 and set the global variable *errno* to one of the values listed in Table 759.

Value	Description
E2BIG	Argument list too long
EINVAL	Invalid argument
ENOENT	Child program not found
ENOEXEC	Format error
ENOMEM	Insufficient memory

***Table 759** The error values the **execvxx** functions return.*

The parameters to the *execvxx* functions are similar to those the program passes to the *execlxx* functions, except that *execvxx* passes the command-line arguments as an array of character strings. The following program, *execvpe.c*, illustrates the *execvpe* function:

```
#include <stdio.h>
#include <process.h>

void main(void)
 {
   char *env[] = { "FILENAME=SPAWNVXX.C", "OS=DOS", "ROUTINE=EXECVPE", NULL };
   char *argv[] = { "CHILD.EXE", "AAA", "BBB", NULL };

   execvpe("CHILD.EXE", argv, env);
 }
```

UNDERSTANDING OVERLAYS 760

As you learned in the Memory section of this book, DOS restricts programs to the first 640Kb of memory. To support larger programs, older programs divided their code into fixed areas called *overlays*. As the program ran, it loaded different overlay sections, as needed. Although overlays let programmers write and compile very large programs, the overlays required the program to keep track of which overlays it currently had loaded, as well as which overlays contained the desired functions. As you can guess, such processing can be difficult because it requires application programs to provide the memory management operations that an operating system often provides.

To help your programs load and execute overlays, DOS provides a system service that loads an overlay file and then transfers control to the start of the file. For specifics on using DOS to load and execute overlays, refer to the book *DOS Programming: The Complete Reference*, Osborne/McGraw-Hill, 1991, by Kris Jamsa, Ph.D. You may find it difficult to use DOS to manage overlays. However, many compilers provide overlay management tools that your programs can use to perform overlay management. For more information on memory managers, refer to the documentation that accompanies your compiler. *Turbo C++ Lite* does not include its own memory manager.

761 Understanding Interrupts

An *interrupt* is an event that causes the computer to temporarily stop the task it is currently performing so that it can work on a second task. When the interrupt's processing concludes, the computer resumes the original task as if the interrupt had never occurred. The PC supports *hardware* and *software* interrupts. The DOS and BIOS section of this book discusses how you can use software interrupts to access the DOS and BIOS interrupts. On the other hand, devices such as the disk drive or the PC's system clock generate hardware interrupts. An *interrupt handler* is software that responds to a specific interrupt. Usually, experienced programmers write interrupt handlers in assembly language. However, newer C compilers let you write handlers within C.

The first 1,024 bytes of the PC's memory contain the segment and offset addresses (called *interrupt vectors*) for the PC's 256 interrupts. When a specific interrupt occurs, the PC pushes onto the stack the current instruction pointer (IP), code segment, and the flags register (machine state). The PC then finds the address of the corresponding interrupt handler using the interrupt vector. The interrupt handler then pushes the PC registers and begins its processing. After the interrupt handler's processing completes, it pops the registers from the stack and then performs an IRET instruction, which pops the flags register, CS and IP. The following assembly language statements, for example, illustrate a typical interrupt handler's layout:

```
; Save the registers on the stack
PUSH AX
PUSH BX
PUSH CX
PUSH DX
PUSH SI
PUSH DI
PUSH DS
PUSH ES

; Perform the interrupt handling instructions

; Pop the registers from the stack
POP ES
POP DS
POP DI
POP SI
POP DX
POP CX
POP BX
POP AX

; Return the previous task
IRET
```

When you define your own interrupt handler with assembly language code, you then update the interrupt vector to point to your own interrupt routine. Before your program ends, you must restore the interrupt vector to its original setting.

762 The PC Interrupts

Within the PC, the first 1,024 bytes of your computer's memory contain the addresses (vectors) of the PC's 256 interrupts. The operating system does not use many interrupts, which leaves them available to your programs for custom purposes. Table 762 lists the PC interrupt vectors and their uses.

Interrupt	Purpose	Interrupt	Purpose
00H	Hardware divide by 0	01H	Hardware single-step trap
02H	Nonmaskable interrupt	03H	Debuggger breakpoint set
04H	Arithmetic overflow	05H	BIOS print screen
08H	IRQ0 clock tick	09H	IRQ1 keyboard
0AH	IRQ2	0BH	IRQ3 COM2
OCH	IRQ4 COM2	0DH	IRQ5 PC/AT LPT1
OEH	IRQ6 Diskette	0FH	IRQ7 LPT1
10H	BIOS video services	11H	BIOS equipment list
12H	BIOS memory size	13H	BIOS disk services
14H	BIOS comm services	15H	BIOS misc services
16H	BIOS keyboard services	17H	BIOS printer services
18H	Invoke ROM-BASIC	19H	System reboot
1AH	BIOS time of day	1BH	Ctrl-Break handler
1CH	Called by 08 handler	1DH	Video parameter table
1EH	Disk parameter table	1FH	Graphics character table
20H	DOS terminate program	21H	DOS system services
22H	Program terminate	23H	DOS CTRL-BREAK
24h	DOS critical error	25H	DOS disk read
26H	DOS disk write	27H	DOS terminate resident
28H	DOS idle	29H	DOS fast putchar
2AH	MS-Net services	2EH	DOS primary loader
2FH	MS-DOS multiplex	33H	Mouse services
40H	Diskette vector	41H	Hard disk parameter table
42H	EGA BIOS redirection	43H	EGA parameter table
44H	EGA character table	4AH	PC/AT Int 70H alarm
5CH	NetBIOS services	67H	EMS services
70H	IRQ8 PC/AT real-time	71H	IRQ9 PC redirect of INT 0AH
75H	IRQ13 PC/AT math coprocessor		

***Table 762** The PC's interrupt vectors and their uses.*

USING THE INTERRUPT KEYWORD 763

As you have learned, DOS lets you create your own interrupt handlers. If you are using *Turbo C++ Lite,* the *interrupt* keyword makes creating an interrupt handler easy, as shown here:

```
void interrupt custom_handler()
 {
      // Handler statements
 }
```

When the compiler encounters the *interrupt* keyword, the compiler inserts statements to push and pop registers, as required, and then later to return from the handler using an IRET (an assembly language statement). To understand the processing the *interrupt* keyword produces, create a program that contains the *custom_handler* function in assembly-language form, if your compiler supports such compilations. If you then examine the assembly language source file, you will see machine instructions similar to those shown in Tip 761.

764 Determining an Interrupt's Vector

As you have learned, an *interrupt vector* is the segment and offset address of the code that handles the interrupt. To help your programs determine an interrupt vector, many DOS-based compilers provide the *_dos_getvect* function, as shown here (some compilers use *getvect*):

```
#include <dos.h>

void interrupt(*_dos_getvect(unsigned interrupt_number))();
```

The *interrupt_number* specifies the interrupt you want (from 0 through 255). The function returns a pointer to an interrupt handler. The following *Turbo C Lite* program, *get_vect.c*, will display the vectors for all the PC's interrupts:

```
#include <stdio.h>
#include <dos.h>

void main(void)
 {
   int i;

   for (i = 0; i <= 255; i++)
     printf("Interrupt: %x Vector: %p\n", i, getvect(i));
 }
```

765 Setting an Interrupt Vector

When your programs create their own interrupt handlers, the programs must assign the interrupt vector to point to the custom interrupt handler. To help your programs assign interrupt vectors most DOS-based compilers provide the *_dos_setvect* function, as shown here:

```
#include <dos.h>

void _dos_setvect(unsigned interrupt_number, void interrupt(*handler)());
```

The *interrupt_number* parameter specifies the interrupt whose vector you want to change. The *handler* parameter is a pointer to the interrupt handler. Tip 766 illustrates the use of the *_dos_setvect* function. When your program changes an interrupt vector, it must save the vector's original settings so it can restore the original vector before your program ends. If a program ends without restoring the interrupt vector, your system can behave erratically (generally, it might stop working).

766 Enabling and Disabling Interrupts

When your programs perform interrupt handling, there will be times when you want your programs to enable and disable interrupts. To help you control interrupts, many DOS-based compilers provide the macros *_disable* and *_enable*, as shown here:

```
#include <dos.h>

void _disable(void);
void _enable(void);
```

To run correctly, the PC must generate key interrupts on a regular basis, so if your programs disable interrupts, your programs should minimize the amount of time that they disable them. Commonly, your programs will use the *_disable* and *_enable* macros when they change an interrupt vector with *_dos_setvect*, as shown here:

```
_disable();
_dos_setvect(interrupt_number, handler);
_enable();
```

CREATING A SIMPLE INTERRUPT HANDLER 767

As you have learned, creating interrupt handlers with the *Turbo C++ Lite interrupt* keyword is much easier than creating interrupt handlers in assembler. The following program, *noprtscr.c*, creates an interrupt handler that replaces the BIOS print-screen interrupt handler, which prints the screen contents when you press the SHIFT+PRTSC keyboard combination. The program uses the *_dos_getvect* function to determine the original vector setting so it can restore the vector before the program ends. Pressing SHIFT+PRTSC while the program is active invokes your interrupt handler, which in turn, will display a screen message stating that you have pressed SHIFT+PRTSC. When you press SHIFT+PRTSC three times, the *noprtscr.c* program will end:

```
#include <stdio.h>
#include <dos.h>
#include <conio.h>

int count = 0;

void interrupt handler(void)
 {
   count++;
 }

void main(void)
 {
   void interrupt (*original_handler)();
   int old_count = 0;

   original_handler = _dos_getvect(5);
   _disable();  // Turn off interrupts during _dos_setvect
   _dos_setvect(5, handler);
   _enable();
   printf("Press SHIFT+PRTSC three times or any key to end\n");
   while (count < 3)
     if (count != old_count)
       {
         printf("SHIFT+PRTSC pressed\n");
         old_count = count;
       }
   _disable();
   _dos_setvect(5, original_handler);
   _enable();
 }
```

CHAINING A SECOND INTERRUPT 768

In Tip 767 you learned how to write an interrupt handler for the BIOS print-screen operation. Depending on the function your program performs, sometimes you may want to execute the original interrupt handler after your handler completes its processing. In such instances, your program can use the *_chain_interrupt* function, as shown here:

```
#include <dos.h>

void _chain_interrupt(void (interrupt far *handler)());
```

The following program, *countdos.c*, for example, keeps a counter of the number of times your program calls specific DOS interrupts (that is, *countdos.c* examines the AH register for INT 21). After the *countdos.c* program ends, it will display a count of the number of services your program called:

```
#include <stdio.h>
#include <dos.h>
#include <dir.h>
```

```
int function[255];    // DOS services
void interrupt far (*original_handler)();

void interrupt far handler(void)
 {
   char i;

   asm { mov i, ah }
   function[i]++;
   _chain_intr(original_handler);
 }

void main(void)
 {
   int i;

   for (i = 0; i < 255; i++)                         // Zero the function counts
     function[i] = 0;
   original_handler = _dos_getvect(0x21);
   _disable();
   _dos_setvect(0x21, handler);
   _enable();
   printf("This is a message\n");
   fprintf(stdout, "This is a second message\n");
   printf("Current disk is %c\n", getdisk() + 'A');
   _disable();
   _dos_setvect(0x21, original_handler);
   _enable();
   for (i = 0; i <= 255; i++)
     if (function[i])
       printf("Function %x called %d times\n", i, function[i]);
 }
```

769 GENERATING AN INTERRUPT

As you learned in the DOS and BIOS section of this book, the C run-time library provides the *intdos* and *int86* functions, which let your programs access the DOS and BIOS services. As your programs handle specific interrupts, there may be times when you want to generate an interrupt to test your handlers, or times when your programs must invoke a specific interrupt. To generate interrupts, your programs can use the function *geninterrupt,* as shown here:

```
#include <dos.h>

void geninterrupt(int interrupt);
```

The *interrupt* parameter specifies the interrupt you want. The following program, *genintr.c,* invokes the infrequently used interrupt 0xFF to notify the program of a specific event:

```
#include <stdio.h>
#include <dos.h>
#include <stdlib.h>

void interrupt far (*original_handler)();

void interrupt far handler(void)
 {
   printf("Some event just happened\n");
   _disable();
   _dos_setvect(0xFF, original_handler);
   _enable();
   exit(0);
 }

void main(void)
 {
```

```
  int i = 0;

  original_handler = _dos_getvect(0xFF);
  _disable();
  _dos_setvect(0xFF, handler);
  _enable();
  while (i++ < 100)
    ;
  geninterrupt(0xFF);
 }
```

***Note:** When you program in Windows, you will catch **messages** and **events**, rather than generating and catching interrupts. You will learn more about messages and events beginning in Tip 1251.*

TRAPPING THE PC TIMER 770

Many devices inside and outside the PC must perform operations at specific intervals. To accomplish these operations, the PC provides a timer chip that generates a signal 18.2 times per second. Each time the signal occurs, the PC generates interrupt 8, which updates the time-of-day clock, and interrupt 1CH, which your programs can trap. The following program, *timer.c*, traps interrupt 1CH each time the interrupt occurs:

```
#include <stdio.h>
#include <dos.h>
#include <conio.h>

int alphanum = 0;
int counter = 0;

void interrupt far handler(void)
 {
   if (++counter == 273)                   // 15 seconds
     {
       alphanum = !alphanum;               // Toggle
       counter = 0;
     }
 }

void main(void)
 {
   int i;
   void interrupt far (*original_handler)();

   original_handler = _dos_getvect(0x1C);
   _disable();
   _dos_setvect(0x1c, handler);
   _enable();
   while (! kbhit())
     if (alphanum)
       for (i = 'A'; i <= 'Z'; i++)
          printf("%c\n", i);
     else
       for (i = 0; i <= 100; i++)
          printf("%d\n", i);
    _disable();
    _dos_setvect(0x1c, original_handler);
    _enable();
 }
```

The interrupt handler counts the number of occurrences and toggles the value of the global variable *alphanum* every 15 seconds. If the value of *alphanum* is 1, the program will repeatedly display the letters of the alphabet. If the value of *alphanum* is 0, the program will display the numbers 1 to 100.

771 UNDERSTANDING CRITICAL ERRORS

As you probably know, when you try to use a floppy drive that does not contain a formatted disk, DOS will display an error message, followed by the familiar:

```
Abort, Retry, Fail?
```

Such errors are called *critical errors* because DOS cannot resolve them without user help. When a critical error occurs, DOS invokes interrupt 24H. When your programs trap interrupt 24H, they can perform critical-error handling, and possibly display a more meaningful or instructional error message to the user. When DOS invokes INT 24H, DOS places considerable information on the stack that describes the cause and source of the error. For a complete discussion of the stack contents and the operations your programs should perform when handling critical errors, refer to Kris Jamsa's *DOS Programming: The Complete Reference*, Osborne/McGraw-Hill, 1991. As you will learn in Tip 772, most DOS-based C compilers provide run-time library functions that simplify critical-error handling.

772 CRITICAL-ERROR HANDLING IN C

As Tip 771 describes, a *critical error* is an error from which DOS cannot continue without user intervention. To help your C programs perform critical-error handling, most C compilers provide the following run-time library functions:

```
#include <dos.h>

void _harderr(int (*handler)());
void _hardresume(int ax_register);
void _hardreturn(int handler_value);
```

The *_harderr* function lets you specify the name of the function that will handle critical errors. The *_hardresume* function lets your programs return a status value to DOS. The *_hardreturn* function, on the other hand, lets you return a value (any value) to your program. The value *_hardresume* returns must be one of those listed in Table 772.

Constant	Meaning
_HARDERR_ABORT	End the current program
_HARDERR_RETRY	Retry service causing the error
_HARDERR_FAIL	Fail the service causing the error
_HARDERR_IGNORE	Ignore the error

*Table 772 Constants **_hardresume** returns.*

The following program, *ersimple.c,* provides a simple critical-error handler that will display a message on the screen and then use the *_hardresume* function to abort the program:

```
#include <stdio.h>
#include <dos.h>
#include <conio.h>

void far handler(unsigned device_error, unsigned error_code,
                 unsigned far *device_header)
 {
   cputs("Critical error ending program\n");
   _hardresume(_HARDERR_ABORT);          // Abort
 }

void main(void)
 {
   FILE *fp;

   _harderr(handler);
```

```
      fp = fopen("A:SOMEFILE.EXT", "r");
    printf("Program message...\n");
    fclose(fp);
  }
```

A More Complete Critical-Error Handler 773

In Tip 772 you created a simple critical-error handler that displayed a message and then ended the program that caused the error. If you take a close look at the critical-error handler, you will find that it supports three parameters, as shown here:

```
void far handler(unsigned device_error, unsigned error_code,
                 unsigned far *device_header)
```

When DOS invokes the critical-error handler, DOS places information about the error onto the stack. The *device_error* parameter contains an error value that DOS would normally pass to a critical-error handler in the AX register. If the failed service sets bit 7 of the *device_error* parameter, the error is a disk error. Table 773.1 lists the values that the failed service may assign to *device_error*.

Bit(s)	Value	Meaning
0	0	Read error
	1	Write error
1–2	00	DOS error
	01	FAT error
	10	Directory error
	11	File error
3	0	Fail operation not allowed
	1	Fail operation allowed
4	0	Retry operation not allowed
	1	Retry operation allowed
5	0	Ignore operation not allowed
	1	Ignore operation allowed
7	0	Disk error
	1	Not disk error

*Table 773.1 Error values the **device_error** variable returns.*

The *error_code* parameter contains the error information DOS would normally pass to the critical-error handler in the DI register. Table 773.2 lists the values DOS passes in DI for disk errors.

Value	Meaning	Value	Meaning
0	Write protected	1	Unknown drive
2	Drive not ready	3	Unknown command
4	CRC data error	5	Invalid request structure
6	Seek error	7	Unknown media type
8	Sector not found	9	Printer out of paper
10	Write fault	11	Read fault
12	General failure	15	Invalid disk change

Table 773.2 Disk error values DOS passes in the DI register.

Finally, the *device_header* parameter is a pointer to the device driver header for the device that generated the error. To help you better understand how your programs can use these values, the companion CD-ROM that accompanies this book contains the following program, *criterr.c*, which will display the values these variables contain.

```
int handler(int errval,int ax,int bp,int si)
{
   static char msg[80];
   unsigned di;
   int drive, errorno;

   di= _DI;                            // if this is not a disk error then it
   if (ax < 0)                         // was another device having trouble
    {
      error_win("Device error");       // report the error
      hardretn(ABORT);                 // and return to the program directly
    }                                  // requesting abort
   drive = ax & 0x00FF;                // otherwise it was a disk error
   errorno = di & 0x00FF;              // report which error it was
   sprintf(msg, "Error: %s on drive %c\r\nA)bort, R)etry, I)gnore: ",
           err_msg[errorno], 'A' + drive);
   hardresume(error_win(msg));         // return to the program via dos interrupt
                                       // interrupt 0x23 with abort, retry, or
                                       // ignore as input by the user.
   return ABORT;
 }
```

774 RESTORING ALTERED INTERRUPTS

When your programs exit, DOS automatically restores the settings of the CTRL+BREAK interrupt handler, the program termination handler, and the critical error handlers to their settings before you ran the program. Depending on your programs, there may be times when you want DOS to restore these settings before your program ends. To restore settings, many compilers provide the *_cexit* function, as shown here:

```
#include <process.h>

void _cexit(void);
```

The *_cexit* function does not terminate your program. Instead, it simply directs DOS to restore the interrupt vectors the previous paragraph discusses. The function will not close files or flush disk buffers. If your compiler does not provide the *_cexit* function, you can write a function that restores the interrupts using the original settings your program saved before it changed the interrupt vectors.

775 CREATING A CTRL+BREAK HANDLER

By default, when the user presses the CTRL+BREAK keyboard combination, your program will end. Often, you will not want the user to be able to press CTRL+BREAK at any time to end the program. As a solution, your programs can define an interrupt handler using the *ctrlbrk* function, as shown here:

```
#include <dos.h>

void ctrlbrk(int (*handler)(void));
```

To create your own CTRL+BREAK interrupt handler, you define a function that you want the program to invoke each time the user presses CTRL+BREAK and then pass the name of the function to the *ctrlbrk* function. The following *Turbo C Lite* program, *ctrlbrk.c*, creates a custom CTRL+BREAK handler:

```
#include <stdio.h>
#include <dos.h>
```

```
int Ctrl_Handler(void)
 {
   printf("\007Press Enter to end the program\n");
   return(1);
 }
void main(void)
 {
   ctrlbrk(Ctrl_Handler);
   printf("Press Enter to end the program\n");
   while (getchar() != '\n');
     ;
 }
```

The program loops until the user presses ENTER. Each time the user presses CTRL+BREAK, the program invokes the function *Ctrl_Handler*. *Ctrl_Handler* beeps and displays a message that directs the user to press ENTER to end the program. In the *ctrlbrk.c* program, the function returns the value 1. If the handler returns any value other than 0, the program will continue. If the handler returns 0, the program will end.

USING DOS SERVICES IN YOUR CRITICAL-ERROR HANDLER 776

When DOS invokes a critical-error handler, you must understand that your system is somewhat unstable—an operating system service has abruptly ended. Within your critical-error handler, you should restrict your use of DOS services to the services listed in Table 776.

Service	Function	Service	Function
01H	Character input	02H	Character output
03H	Aux port input	04H	Aux port output
05H	Printer output	06H	Direct console I/O
07H	Character input	08H	Character input
09H	String output	0AH	Buffered keyboard input
0BH	Test input state	0CH	Flush buffer and input
3300H	Get CTRL+C state	3301H	Set CTRL+C state
3305H	Get startup disk	3306H	Get DOS version
50H	Set PSP	51H	Get PSP
59H	Get extended error	62H	Get PSP

Table 776 *Usable DOS services in a critical-error handler.*

If your programs must perform I/O within a critical-error handler, consider using the I/O functions in the *conio.h* include file.

IMPROVING PERFORMANCE USING INSTRUCTION SET SELECTION 777

By default, most DOS-based compilers generate programs that can run on all Intel-based systems, from the 8088 through the Pentium. If you know in advance that users will only run a program on a specific machine, you can improve the program's performance by using the instruction set of a more advanced machine. For example, the 80386 provides instructions that are not available on an 8088. Using one of these 80386-based instructions might replace several equivalent 8088 instructions. However, when you take advantage of such instructions, your programs will no longer run on the older machines. To generate executable code for a specific machine, refer to your compiler's command-line switches.

778 INLINING INTRINSIC FUNCTIONS

To improve performance, many C compilers let you replace functions with *inline code*. In addition to letting you use the *inline* keyword before the functions you create, many C compilers let you replace intrinsic run-time library functions with an inline counterpart. The intrinsic functions that you can inline will differ from one compiler to the next. Refer to your compiler documentation to determine the available functions. In the case of the Borland C++ compiler, you can place the intrinsic functions listed in Table 778 inline.

Inline-Capable Intrinsic Functions						
alloc	*fabs*	*memchr*	*memcmp*	*memcpy*	*memset*	*rotl*
rotr	*stpcpy*	*strcat*	*strchr*	*strcmp*	*strcpy*	*strlen*
strncat	*strncmp*	*strncpy*	*strnset*	*strrchr*		

***Table** 778 Inline-capable intrinsic functions C supports.*

To direct your compiler to place these functions inline, you can use a command-line switch or the *#pragma intrinsic* function, discussed in Tip 779.

779 ENABLING AND DISABLING INTRINSIC FUNCTIONS

In Tip 778 you learned that many C compilers let you replace specific intrinsic functions with inline code. Using compiler command-line switches, you can direct the compiler to place intrinsic functions inline. In addition, many preprocessors support the *pragma intrinsic*, which lets you enable or disable intrinsic inlining, as shown here:

```
#pragma  intrinsic  function          // Enables inlining
#pragma  intrinsic -function          // Disables inlining
```

The following statement, for example, directs the compiler to generate inline code for the *strlen* function:

```
#pragma  intrinsic  strlen
```

When you use the intrinsic pragma, you must precede the pragma with a function prototype. When the compiler encounters the pragma, the compiler will replace the function name with an equivalent name that begins and ends with an underscore. In the case of the *strlen* function, the compiler will generate the constant *_strlen_*, as shown here:

```
#define strlen _strlen_
```

780 UNDERSTANDING FAST FUNCTION CALLS

When your program invokes a function, C passes parameters to functions on the stack. As discussed in the Functions section of this book, the stack use is responsible for most of the overhead that corresponds to a function call. To try to make function invocations faster, some C compilers provide a *_fastcall* modifier that you can place before a function name, as shown here:

```
int _fastcall some_function(int a, int b);
```

The following program, *fastcall.c*, illustrates the *_fastcall* modifier:

```
#include <stdio.h>
#include <time.h>

int _fastcall add_fast(int a, int b)
 {
   return(a + b);
 }
```

```
int add_slow(int a, int b)
 {
   return(a + b);
 }

void main(void)
 {
   unsigned long int i, result;

   clock_t start_time, stop_time;
   printf("Processing...\n");
   start_time = clock();
   for (i = 0; i < 200000000L; i++)
     result = add_fast(i, -i);
   stop_time = clock();
   printf("Processing time for fast call %d ticks\n", stop_time - start_time);
   start_time = clock();
   for (i = 0; i < 200000000L; i++)
     result = add_slow(i, -i);
   stop_time = clock();
   printf("Processing time for normal call %d ticks\n", stop_time - start_time);
 }
```

RULES FOR _FASTCALL PARAMETER PASSING 781

In Tip 780 you learned that many compilers support the *_fastcall* function modifier, which directs the compiler to pass parameters to the function using registers. Depending on the target machine, the number of registers available for parameters will differ. In the case of Borland's *C++* compiler, your programs can only pass three parameters via registers. Table 781 specifies how the *_fastcall* modifier, when used with Borland's *C++* compiler, passes parameters to functions.

Parameter Type	Registers Used
char (signed and unsigned)	AL, DL, BL
int (signed and unsigned)	AX, DX, BX
long (signed and unsigned)	DX, AX
near pointer	AX, DX, BX
others	Passed on the stack

Table 781 Registers used for parameter passing with the _fastcall modifier.

UNDERSTANDING INVARIANT CODE 782

As you examine compiler directives that affect optimization, you may encounter the term *invariant code*. In general, invariant code refers to statements that appear within a loop whose values do not change. For example, the following *for* loop assigns the result of the multiplication *a * b * c* to each array element:

```
for (i = 0; i < 100; i++)
   array[i] = a * b * c;
```

Because the *a*, *b*, and *c* variables do not change within the loop, the result of the multiplication is *invariant* (it does not change). As you program, you should watch for invariant code. When you find invariant code, you can usually improve your program's performance by changing your program in some way. In the case of the previous *for* loop, you can improve your program's performance by replacing the multiplication with its result, as shown here:

```
result = a * b * c;
for (i = 0; i < 100; i++)
   array[i] = result;
```

To improve program performance, many compilers will test for invariant code, replacing it within the target object code with a non-invariant equivalent. Ideally, you should find and correct invariant code yourself. However, using command-line switches, you may be able to direct your compiler to perform such substitutions for you during compilation.

783 UNDERSTANDING REDUNDANT LOAD SUPPRESSION

As you have learned, to improve performance, the C compiler often loads values into registers. When the compiler performs *redundant load suppression*, the compiler keeps track of the values it is has already placed into registers. The compiler then refers to the registers when it needs a value, rather than loading the value a second time. The compiler uses redundant load suppression to prevent duplicate load operations, which improves your program's performance. The disadvantage of using load suppression is that your programs may take slightly longer to compile. As a rule, however, you should always direct the compiler to perform load suppression.

784 UNDERSTANDING CODE COMPACTION

When you examine your compiler documentation, you might encounter the term *code compaction.* In general, code compaction uses branches to the previous code to eliminate redundant statements. For example, consider the following program, *compact.c*:

```
#include <stdio.h>

void main(void)
 {
   int a = 1, b, c, d;

   switch (a) {
    case 1: a = 5;
            b = 6;
            c = 7;
            d = 8;
            break;
    case 2: b = 6;
            c = 7;
            d = 8;
            break;
    };
 }
```

If you examine the *switch* statement, you will find that the statements the program performs for each case are very similar. Rather than duplicating the assignment statements in both locations, the compiler might place a JMP (jump, which is the assembly language equivalent of *goto*) instruction that branches back to the statement *b = 6*, which occurs in the first case and at the start of the second case.

785 UNDERSTANDING LOOP COMPACTION

If you examine the *for* loops that occur throughout your programs, you may find that most loops manipulate a string or other array. When your program assigns the same value to every element in an array, the C compiler can optimize your program's performance by replacing the loop with one of the 80*x*86 ST*xxx* instructions. For example, the following *for* loop initializes the array *null_string* to *NULL*:

```
for (i = 0; i < sizeof(null_string); i++)
   null_string[i] = NULL;
```

If you examine the assembly language output the compiler produces, you will find that the compiler has eliminated the loop, using the *STOSW* instruction. Such compiler substitutions are called *loop compaction*.

UNDERSTANDING LOOP INDUCTION AND STRENGTH REDUCTION 786

Loop induction and strength reduction are techniques the compiler uses to optimize loops within a program. The compiler usually optimizes loops within a program when the program manipulates arrays within a loop. For example, consider the following loop, which assigns values to array elements:

```
for (i = 0; i < 128; i++)
   array[i] = 0;
```

For each reference to the array, the compiler must perform a multiplication operation to determine the correct element (*base + i * sizeof(array_type)*). Instead of using the array, the compiler may use a pointer, as shown here:

```
end = &array[128];

for (ptr = array; ptr < end; ptr++)
   *ptr = 0;
```

By eliminating the slow multiplication, the compiler improves the program's performance. The process of creating new variables from loop variables is called *loop induction*. Because the inducted variables are usually less complex than the variables they replace, the newly created loop introduces a *strength reduction*.

UNDERSTANDING COMMON SUBEXPRESSION ELIMINATION 787

If your programs work with arrays, sometimes you can improve your program's performance by eliminating common subexpressions. For example, consider the following *if* statement, which tests whether an array element contains an upper or lowercase A:

```
if ((array[i] == 'A') || (array[i] == 'a'))
```

For each test, the compiler must resolve the array element by performing a multiplication (*base + i * sizeof(array_type)*). A faster implementation, however, would replace the array reference with a pointer, as shown here:

```
ptr = &array[i];
if ((*ptr == 'A') || (*ptr == 'a'))
```

Replacing the common subexpression with the faster, alternative code improves the program's performance. In some cases, however, trying to eliminate common subexpressions can make your program more difficult to understand. Many C compilers will detect opportunities for subexpression elimination and perform the operations for you within the resulting (compiled) code. As you work with compound conditions, be aware that you might be able to improve your program performance by eliminating or reducing subexpressions.

UNDERSTANDING STANDARD C CONVERSIONS 788

When you perform arithmetic operation on different value types, the C compiler often *promotes* the lower type value. To help you understand the standard C conversions, consider the following rules, which C applies in order from the first item to the last item in the following list:

- With the exception of *unsigned short*, the C compiler promotes all small integer values to *int*. The C compiler promotes values of type *unsigned short* to *unsigned int*.
- If either operand is *long double*, the C compiler promotes the other to *long double*.
- If either operand is *double*, the C compiler promotes the other to *double*.
- If either operand is *float*, the C compiler promotes the other to *float*.
- If either operand is *unsigned long*, the C compiler promotes the other to *unsigned long*.
- If either operand is *long*, the C compiler promotes the other to *long*.
- If either operand is *unsigned*, the C compiler promotes the other to *unsigned*.
- Otherwise, the C compiler treats both operands as type *int*.

789 Understanding C's Four Basic Types

As you examine complex declarations in C, keep in mind that C supports four basic types: *void*, *scalar*, *function*, and *aggregate*. The type *void* specifies the absence of values. For example, *void* in a parameter list states that a function does not receive any parameters. Likewise, *void* in front of a function name specifies that a function does not return a value. *Scalar* values include arithmetic, enumerated, pointer, and reference values. A *function* type specifies a function that returns a specific type. Finally, an *aggregate* type specifies an array, union, structure, or C++ class. As you examine complex declarations, try to map the declaration to one of the four types C supports.

790 Understanding Fundamental versus Derived Types

As you examine complex declarations in C, understand that C supports *fundamental* and *derived* types. C's fundamental data types include the following: *void*, *char*, *double*, *float*, and *int*. In addition, C lets you apply the modifiers *long*, *short*, *signed*, and *unsigned* to the fundamental types. Derived types, on the other hand, include arrays, classes, functions, pointers, structures, and unions to other types. The types *char*, *int*, *long*, and *short* are integral types. As you examine declarations, the following integral types will be equivalent:

```
char, signed char// Normally a compiler default
int, signed int
unsigned, unsigned int
short, short int, signed short int
unsigned short, unsigned short int
long, long int, signed long int
unsigned long, unsigned long int
```

791 Understanding Initializers

Initializers are the values your programs assign to variables at declaration. When you use initializers, keep the following rules in mind:

- If a program does not explicitly initialize an arithmetic type, most compilers will initialize the variable with 0.
- If a program does not explicitly initialize a pointer type, most compilers will initialize the pointer with *NULL*.
- If the number of initializers exceeds the number of variables to be initialized, the compiler will generate an error.

- All expressions you use to initialize a variable must be constants (C++ does not require this) if the initializers are for a static object, array, structure, or union.
- If the declared variable has block scope and the program has not declared the variable as external, then the declaration cannot have an initializer.
- If the program code provides fewer initializers than the compiler requires to fully initialize the variable, the compiler will initialize the remainder of the values following default initialization technique.

UNDERSTANDING LINKAGE 792

As you have learned, the linker combines code from your programs, object files, and libraries. Depending on the linker you use within a given compilation, there may be times when two or more functions in the linked files have the same name. *Linkage* is the process of determining which function the linker applied to a reference. In C, identifiers have one of three possible linkage attributes: *external*, *internal*, and *none*. An identifier with *external linkage* represents the same object throughout all the files that make up a program. An identifier with *internal linkage* represents the same object within one file. An identifier with *no linkage* is unique throughout the files—meaning that it occurs only once. The linker uses an identifier's linkage to determine with which function C associates the identifier. The linker uses the following linkage rules:

- Identifiers declared as *static* have *internal linkage.*
- If an identifier appears with internal and external linkage, C will use *internal linkage*, and C++ will use *external linkage.*
- If the program declares an identifier with the *extern* keyword, the identifier has the same linkage as any visible declaration with file scope; if no such declaration exists, the identifier has *external linkage.*
- If a function identifier does not have a storage class specifier, the identifier has the same linkage as if the program used *extern* with the identifier.
- Object identifiers declared without a storage class specifier have *external linkage.*
- Identifiers that you declare to be something other than an object or function have no linkage.
- Function parameters have no linkage.
- Identifiers that have block scope declared without a storage class *extern* have no linkage.

UNDERSTANDING TENTATIVE DECLARATIONS 793

A *tentative declaration* is an external data declaration that has no storage class specifier and no initialization. For example, assume the compiler encounters the following declaration:

```
int value;
```

If the compiler later encounters a definition for the variable, the compiler treats the variable as if the *extern* keyword preceded the variable. If the compiler reaches the end of the translation unit without encountering a definition, the compiler allocates memory for the variable. The following program, *tentatve.c*, creates a tentative variable declaration for the variable *value*:

```
#include <stdio.h>
int value;
```

```
void main(void)
 {
   printf("%d\n", value);
 }
int value = 1500;
```

When the compiler encounters the definition of the variable *value* that initializes the variable to 1500, the compiler will convert the first declaration of *value* from a tentative to a full definition.

794 CONTRASTING DECLARATIONS AND DEFINITIONS

Many of the Tips presented throughout this book use the terms *declaration* and *definition*. In general, a *declaration* introduces one or more identifiers within a program. A *definition*, on the other hand, directs the compiler to actually allocate memory for the object. For example, you can consider a function prototype as a declaration and the function header and code as a definition. C classifies declarations as *defining* or *referencing*. A *defining* declaration both declares one or more identifiers and defines the amount of memory the compiler is to allocate to the object. A *referencing* declaration, on the other hand, simply introduces an identifier. Within C, you can declare the objects listed in Table 794.

Objects			
arrays	classes	class members	enumerated
enumerated tags	constants	functions	labels
macros	structures	structure members	types
unions	union members	variables	

Table 794 *Declarable types in C.*

795 UNDERSTANDING LVALUES

When your program works with pointers, you may encounter compiler error messages that state the compiler requires an *lvalue*. An *lvalue* is an expression that the compiler can use to locate an object. You can consider *lvalue*s as expressions that would be valid on the left side of the assignment operator. The following are valid *lvalue*s:

```
variable = value;
*variable = value;
variable[i] = value;
```

It is important to note, however, that each of the *lvalue*s just shown could have been on the right side of the assignment operator as well. An *lvalue* simply provides a value that the compiler can use to locate an object in memory. C supports *modifiable* and *non-modifiable lvalue*s. A *modifiable lvalue* is a pointer value that you can change to point to a different value. Most of the pointers you have used within this book are modifiable *lvalue*s. On the other hand, you cannot change a *non-modifiable lvalue*. A constant pointer, for example, is a non-modifiable *lvalue*.

796 UNDERSTANDING RVALUES

When you compile your programs, you might encounter a compiler error message that states that the compiler encountered an unexpected *rvalue*. An *rvalue* is an expression that appears on the right side of an equal sign. The following expressions are examples of *rvalue*s:

```
result = value;
result = 1001;
```

```
result = value + 1500;
```

The following statements, however, are invalid because they do not specify a memory location to which C can assign a value:

```
1500 = result;
value + 1500 = result;
```

Such declaration errors usually appear when users are trying to create pointers, as shown here:

```
*(value + 1500) = result;
```

USING SEGMENT REGISTER KEYWORDS 797

As you have learned, the PC uses four specific registers to locate your program's code, data, and stack. Your programs can use the *segread* run-time library function to determine the register settings. In addition, many DOS-based compilers provide the keywords listed in Table 797.

Keyword	Meaning
_cs	Creates a pointer to the code segment
_ds	Creates a pointer to the data segment
_es	Creates a pointer to the extra segment
_ss	Creates a pointer to the stack segment

Table 797 Additional register-based keywords many DOS-based compilers provide.

The following declaration creates a pointer to the stack segment:

```
char _ss *my_stack_pointer;
```

Depending on the current memory model, the pointers will contain near or far pointers, as appropriate.

USE FAR POINTERS CAREFULLY 798

In the Pointers section of this book, you learned that a far pointer is a 32-bit pointer that contains a 16-bit segment and a 16-bit offset address. Far pointers let your programs access the PC's 1Mb conventional memory range. When you use far pointers, however, you must understand how the pointer value wraps when the offset value exceeds its 16-bit limit. For example, assume the far *char* pointer *location* contains the following value:

```
location = 0x1000FFFE;    // Segment 0x1000 Offset FFFE
```

If you increment the pointer, the pointer's value will become the following:

```
location = 0x1000FFFF;    // Segment 0x1000 Offset FFFF
```

If you increment the pointer again, an error will possibly result, as shown here:

```
location = 0x10000000;    // Segment 0x1000 Offset 0000
```

Note that the offset value wrapped back to 0, but the segment value did not change. As a result of the incrementing process, the pointer wrapped back to the start of the 64Kb segment address. If you need the pointer to move to the start of the next segment, use a *huge* pointer instead.

799 UNDERSTANDING NORMALIZED POINTERS

As you know, the PC addresses memory locations using segment and offset addresses. The PC supports up to 65,636 segment addresses. Each segment address starts at a 16-byte address: 0, 16, 32, 48, and so on. If you multiply the 16-byte address by the 65,536 segments, the result is a 1Mb address space. Given a segment address, an offset address lets you choose one of 65,536 possible locations within the segment. When you use segment and offset addresses, you can address every location in memory using different segment and offset combinations. For example, assume that you want to address location 48 in memory. To do so, you can use any one of the following segment/offset combinations:

- Segment 0 Offset 48
- Segment 1 Offset 32
- Segment 2 Offset 16
- Segment 3 Offset 0

Because you can reference each memory location differently, it is possible for two far pointers to reference the same memory location but contain different values. Consider the following pointer assignments:

```
char far *ptr1 = 0x00000030;   // Seg 0 offset 48
char far *ptr2 = 0x00030000;   // Segment 3 offset 0
```

Both pointers will reference the same memory location. However, if your program compares the pointer values, the values are not equal. A *normalized* pointer eliminates such disparity by always storing values so that the compiler uses a 16-byte offset. In this way, the compiler always stores segment addresses using the segment nearest to the value. In the previous case, the normalized pointer would contain segment 3 offset 0.

800 MATH COPROCESSOR STATEMENTS

A floating-point (or math) processor is a specialized chip which contains instructions that can perform arithmetic operations quickly—such as division, multiplication, and even square-root calculation—using floating-point values. If you are using an 8088, 80286, or 80386, you must purchase a floating-point coprocessor (an 8087, 80287, or 80387). If you are using a 80486DX or later computer, the floating-point processor is built into the main processor. Because math coprocessors only perform floating-point operations, they can perform the operations quickly. If your computer has a math coprocessor, you should direct your compiler to generate instructions that use the coprocessor. Depending on your compiler, the switches you must use to generate instructions for a floating-point coprocessor will differ. To let programs run on systems that do not have a math coprocessor, by default most compilers do not generate floating-point instructions.

If you are using *Turbo C++ Lite*, the compiler supports several */FPx* switches that let you direct the compiler to always use floating-point instructions. The floating-point switches you select will affect your program's size and speed. Refer to your compiler documentation for more information on your compiler's floating-point switches.

801 UNDERSTANDING CDECL AND PASCAL IN VARIABLES

As you examine programs that use mixed-language modules, such as Pascal and C, you might encounter variables declared with the *pascal* and *cdecl* modifiers. To maintain compatibility with Pascal identifiers, the *pascal* modifier directs the compiler to become case insensitive and not to precede the identifier with a leading underscore. The *cdecl* modifier, on the other hand, directs the compiler to ensure case sensitivity and include leading underscores. The following program, for example, declares an external variable named *number* that a Pascal program defines:

```
#include <stdio.h>
extern int pascal number;
```

```
void main(void)
 {
   printf("The value is %d\n", number);
 }
```

PREVENTING CIRCULAR INCLUDES 802

As your programs make more extensive use of header files (C++ programs often define classes in header files), there may be times when a file that you include will include a second header file which, in turn, will include the first header file. As the preprocessor performs the inclusions, it can end up in a circular operation. To reduce the possibility of circular operations, your header files can declare a macro when your program processes them that will prevent the compiler from processing the files a second time. For example, the following header file uses the macro *MY_STUFF_DEFINED* to determine whether the compiler already processed the contents:

```
#ifndef MY_STUFF_DEFINED
#define MY_STUFF_DEFINED 1

 // Other include statements

#endif
```

In this case, the first time the compiler processes the header file, it defines the macro. If the program includes the header file a second time, the compiler will not process the file's contents, due to the *#ifndef* directive.

INTRODUCING C++ 803

C++ is a programming language Dr. Bjarne Stroustrup developed at AT&T Bell Labs that builds upon the C programming language to add object-oriented capabilities and other enhancements. You can think of C++ as a superset of C because C++ supports the C programming language features that you have learned throughout this book. As you will learn, however, C++ is more than just an "object-oriented C"—C++ actually adds many new features that will improve your program capabilities. If you are using a C++ compiler (such as *Turbo C++ Lite)*, most of the programs presented in the previous 802 Tips should successfully compile and execute with no changes. The following Tips start with the C++ basics and build on your knowledge of C. By the time you reach the end of this book, you should be proficient in C, C++, and C#.

HOW C++ SOURCE FILES DIFFER 804

In general, there are no differences between C and C++ source files. Both languages fully support compiler directives, such as *#include* and *#define*. With respect to naming, many programmers use the CPP extension to differentiate C and C++ source files. Everything you learned in the first section of this book still applies as you begin to create C++ programs, with the exception of some unique structures and constants. That is, in C++, you can still include header files, link object code libraries, and so on.

The CD-ROM that accompanies this book includes the *Turbo C++ Lite* compiler which you can use to compile the C++ applications this book. Borland, however, offers a newer C++ compiler on the Web which you can download for free at the Borland Web site at http://www.borland.com/downloads.

The Borland C++ compiler is fully compliant with the ANSI C++ standard. You should use the compiler as you examine the C++ programming tips that follow.

Many newer C++ compilers support the ANSI C++ Library, a collection of standard routines your programs use to perform specific tasks, such as input and output operations. When you use a C++ compiler that supports the ANSI C++ compiler, you may get warning messages when you compile the following C++ tips programs that state one or more operators has been "deprecated"—in other words, that the compiler considers the operator obsolete. Depending on the C++ compiler you are using, the compiler may still successfully compile the source code. Other compilers, however, may not.

The primary difference between traditional C++ programs and programs written for the Standard C++ Library is that the Standard C++ Library programs do not include the .h file extension for include files. For example, a traditional C++ program will include the files *iostream.h* and *stdlib.h* as using the following *#include* directives:

```
#include <iostream.h>
#include <stdlib.h>
```

As you can see, the #include directives include the .h file extensions. In contrast, a C++ program written to support the ANSI Standard C++ Library will omit the .h extensions as shown here:

```
#include <iostream>
#include <stdlib>
```

Next, within the C++ program written to support the Standard C++ Library, you must include the following *using namespace std* directive, which defines the input and output stream operators you will use throughout the tips that follow:

```
using namespace std;
```

If you omit the *using namespace std* directive, the C++ compiler will generate syntax error, that tell you that the I/O streams such as *cout* and *cin* are not defined.

The following traditional C++ program, *simple.cpp*, uses the *cout* output stream to display a mesage to the screen display:

```
#include <iostream.h>

void main(void)
 {
   cout << "Jamsa's C/C++/C# Programmer's Bible";
 }
```

Note that the *#include* directive includes the *.h* file extension. In a similar way, the following C++ program, *StdSimple.cpp*, written to support the Standard C++ Library, also uses the *cout* output stream to display a mesage to the screen display:

```
#include <iostream>

using namespace std;

void main(void)
 {
   cout << "Jamsa's C/C++/C# Programmer's Bible";
 }
```

Note that this source code does not include *.h* extension and the code provides the *using namespace std* directive.

As you examine C++ programs, you will encounter programs written using the traditional format as well as programs written to support the Standard C++ Library. Because the *Turbo C++ Lite* compiler provided on this CD-ROM supports traditional C++ programs (it does not support programs written for the Standard C++ Library — as does the compiler you can download from Borland for free), most of the C++ tips this book presents use the standard format. However, if you examine the CD-ROM that accompanies this book, you will find that each directory contains two copies of the source code, one for a traditional C++ compiler and one for a compiler that supports the

Standard C++ Library. In each case, the name of the source file for the Standard C++ Compiler has the prefix *new_*, such as *new_simple.cpp*.

As briefly discussed, the *using namespace std* directive defines the common I/O streams to the compiler. By providing the directive, you define all the I/O streams (and other items) to the compiler in one step. As you examine C++ source code, you may periodically encounter code that does not provide the *using namespace std* directive, but instead, precedes stream names and other objects using the *std* namespace identifier, as shown here:

```
std::cout << "Jamsa's C/C++/C# Programmer's Bible" << endl;
```

The following program, *NoUsing.cpp*, uses the *std* identifier to define the *cout* I/O stream to the compiler:

```
#include <iostream>

void main(void)
 {
   std::cout << "Jamsa's C/C++/C# Programmer's Bible";
 }
```

GETTING STARTED WITH A SIMPLE C++ PROGRAM 805

In Tip 2, you created your first C program, which used *printf* to display a message on the screen, as shown here:

```
#include <stdio.h>
void main(void)
 {
    printf("Jamsa\'s C/C++/C# Programmer\'s Bible");
 }
```

The following C++ program, *simple.cpp*, performs identical processing:

```
#include <iostream.h>
void main(void)
 {
   cout << "Jamsa's C/C++/C# Programmer's Bible";
 }
```

The *simple.cpp* program uses the C++ *cout* I/O stream, discussed in Tip 806. You can compile and build the *simple.cpp* program just as you built all your C programs previously. When you execute the *simple.cpp* program, your screen will display the following:

```
Jamsa's C/C++/C# Programmer's Bible
C:\>
```

Note: *As you can see, the C++ code does not use the \' escape character for the apostrophe within the output string. Most C++ compilers do not require the \' escape character, and therefore the remainder of the strings in this book will not use the \' notation.*

UNDERSTANDING THE COUT I/O STREAM 806

In Tip 805, the sample program used the *cout* I/O stream to write a character string to the screen display, as shown here:

```
cout << "Jamsa's C/C++/C# Programmer's Bible"
```

Redirecting output to the *cout* I/O stream is the same as using *printf* to write output to *stdout*. The double less-than (<<) symbol is not the *bitwise left-shift* operator. Instead, the symbol is the output operator that specifies to which stream the program sends the data. The following program, *cout.cpp*, uses the C++ output operator to display several different messages:

```
#include <iostream.h>

void main(void)
 {
   cout << "This is line one.\n";
   cout << "This text is on ";
   cout << "line two.\n";
   cout << "This is the last line.";
 }
```

When you compile and execute the *cout.cpp* program, your screen will display the following:

```
This is line one.
This text is on line two.
This is the last line.
C:\>
```

807 WRITING VALUES AND VARIABLES WITH COUT

As you have learned, the *cout* output stream lets your program display output to the screen display. Previous Tips have used *cout* to display character strings. The following program, *cout_num.cpp*, uses *cout* to display character strings and numbers:

```
#include <iostream.h>

void main(void)
 {
   cout << "cout lets you display strings, ints, and floats\n";
   cout << 1700;
   cout << "\n";
   cout << 1.2345;
 }
```

When you compile and execute the *cout_num.cpp* program, your screen will display the following output:

```
cout lets you display strings, ints, and floats
1700
1.2345
C:\>
```

808 COMBINING DIFFERENT VALUE TYPES WITH COUT

In Tip 807, you learned that the *cout* I/O stream lets your programs display all types of values. The *cout_num.cpp* program presented in Tip 807 used several statements to display its output, as shown here:

```
cout << "cout lets you display strings, ints, and floats\n";
cout << 1700;
cout << "\n";
cout << 1.2345;
```

Fortunately, *cout* lets you place different value types in the output stream in one statement, as shown in the following program, *cout_one.cpp*:

```
#include <iostream.h>

void main(void)
 {
   cout << "cout displays strings " << 1700 << "\n" << 1.2345;
 }
```

DISPLAYING HEXADECIMAL AND OCTAL VALUES 809

As you have learned, the *cout* I/O stream lets your programs display values of type *int* and *float*. When you use *printf* to display integer values, you can use the *%x* and *%o* format specifiers to display the values in hexadecimal and octal. When your programs use *cout* to display output, they can use the *dec*, *oct*, and *hex* modifiers, as shown in the following program, *cout_hex.c*:

```
#include <iostream.h>

void main(void)
 {
   cout << "Decimal value " << dec << 0xFF;
   cout << "\nOctal value " << oct << 10;
   cout << "\nHexadecimal value " << hex << 255;
 }
```

REDIRECTING COUT 810

As you have learned, your programs can use the *cout* I/O stream to display output, just as if they had written the output directly to *stdout*. Therefore, if you have a program that uses *cout*, you can redirect the program's output to a file or to some other device. The following program, *1_to_100.cpp*, uses *cout* to display the numbers 1 to 100:

```
#include <iostream.h>

void main(void)
 {
   int i;

   for (i = 1; i <= 100; i++)
     cout << i  << '\n';
 }
```

Using the DOS output redirection operator, you can redirect the program's output to a file, as shown here:

```
C:\> 1_TO_100 >> FILENAME.EXT <ENTER>
```

IF YOU LIKE PRINTF, USE PRINTF 811

Several previous Tips performed their output using the *cout* I/O stream. If you feel more comfortable using the *printf* function, use *printf*. In later Tips, you will learn how to format output that your program will display better with *cout*. At that time, you might choose to use *cout* for all your output. As a rule, however, to make your programs easier to understand, you should pick one technique and stick with it. The following program, *use_both.cpp*, displays output using *cout* and *printf*:

```
#include <iostream.h>
#include <stdio.h>

void main(void)
 {
   cout << "Jamsa's ";
   printf("C/C++/C# Programmer's ");
   cout << "Bible!";
 }
```

WRITING OUTPUT TO CERR 812

As you know, when your programs write output to the *stderr* file handle, C++ cannot redirect the output from the screen. If you are using C++ I/O streams to perform your input and output, your programs can write output to the *cerr* I/O stream. The following program, *use_cerr.cpp*, uses *cerr* to prevent the user or the program from redirecting the program's output:

```
#include <iostream.h>

void main(void)
 {
   int i;

   for (i = 1; i <= 100; i++)
      cerr << "Can't redirect cerr " << i  << '\n';
 }
```

813 GETTING INPUT WITH CIN

As you have learned, the *cout* I/O stream lets your programs display output to *stdout*. In a similar way, your C++ programs can get input using the *cin* I/O stream. The following program, *use_cin.cpp*, uses *cin* to get input for several different variable types:

```
#include <iostream.h>

void main(void)
 {
   int age;
   float salary;
   char name[128];

   cout << "Enter your first name age salary: ";
   cin >> name >> age >> salary;
   cout << name << ' ' << age << ' ' << salary;
 }
```

814 CIN DOES NOT USE POINTERS

As you saw in Tip 813, the *cin* stream does not use pointers to reference variables. As you have learned, when you input information with a function such as *scanf*, you have to explicitly pass a pointer to the variable that will receive the information (so the function can modify the variable). However, because of the *cin* stream's construction, you should not pass a pointer to a variable to *cin*—if you do, *cin* will return an error. For example, the following program, *cin_err.cpp*, will generate syntax errors when you try to compile the program:

```
#include <iostream.h>

void main(void)
 {
   int age;
   float salary;
   char name[128];

   cout << "Enter your first name age salary: ";
   cin >> &name >> &age >> &salary;
   cout << name << ' ' << age << ' ' << salary;
 }
```

You will learn in later Tips how *cin* places values within a non-pointer variable. For now, simply understand that your programs must pass actual variable names to *cin*.

815 UNDERSTANDING HOW CIN SELECTS DATA FIELDS

In Tip 813, you used the *cin* I/O stream to read the user's name, age, and salary on one line, as shown here:

```
cin >> name >> age >> salary;
```

When your programs use *cin* to read input, you must understand how *cin* parses input. Unless told otherwise, *cin* uses white space (a blank, tab, or *newline*) to delimit input fields. Therefore, if the user typed his full name (such as John Smith) in the previous operation, *cin* would use the first name for the *name* variable and the last name for the *age* variable, and the I/O operation would be in error. In later Tips, you will learn how to perform formatted input using *cin*.

UNDERSTANDING HOW I/O STREAMS KNOW VALUE TYPES 816

As you have learned, your programs can use the *cin*, *cout*, and *cerr* I/O streams to perform I/O operations to the *stdin*, *stdout*, and *stderr* file handles. Using these I/O streams, you can perform I/O operations with character strings, integers, and floating-point values. When your C programs perform I/O operations using *printf* and *scanf*, the functions use the format specifiers to determine the value types (such as *string*, *int*, and so on). When you perform input and output operations with the C++ I/O streams, the compiler provides information about each value's type, so there is no need for a format specifier. As an interesting exercise, you can generate and examine an assembly language listing for a file, such as *use_cin.cpp,* that uses *cin* and *cout,* if your compiler supports that option.

PERFORMING OUTPUT USING CLOG 817

As you have learned, C++ provides the *cin*, *cout*, and *cerr* I/O streams that correspond to *stdin, stdout*, and *stderr*. In addition, C++ provides a fourth I/O stream named *clog*. The *clog* I/O stream is similar to *cerr*, except that it performs buffered output. The following program, *clog.cpp,* uses the *clog* I/O stream to display a message:

```
#include <iostream.h>

void main(void)
  {
    clog << "Some strange processing error";
  }
```

CIN, COUT, CERR AND CLOG ARE CLASS INSTANCES 818

Several of the preceding Tips showed you how to perform I/O operations using *cin*, *cout*, *cerr*, and *clog*. It is important to know that these I/O stream identifiers are not magic operators that are built into C++. Instead, *cin*, *cout*, *cerr*, and *clog* are instances of an *I/O class*. In later Tips, you will learn that a *class* defines a template that contains data and methods (functions or operations that work on the data). The class, therefore, is the C++ fundamental mechanism for object-oriented programming. When you start creating your own classes, you can relax, knowing you have been using several classes since the time you compiled your first program. The double less-than (<<) and double greater-than (>>) symbols are simply class operators. Do not worry if these terms seem confusing, as later Tips will explain them in detail.

FLUSHING OUTPUT WITH FLUSH 819

As you have learned, your programs can use the *cout* I/O stream to output data to *stdout* and the *clog* I/O stream to perform buffered output to *stderr*. When you perform buffered output, the output might not appear on the screen as soon as you would like. Usually, file handles and I/O streams do not flush output until they encounter a carriage return or until an input operation occurs. In such cases, your programs can use *flush* to immediately flush the output from the buffer. The following program, *flush.cpp*, illustrates how to use *flush* to flush data to *stdout* and *stderr*:

```
#include <iostream.h>

void main(void)
```

```
{
  cout << "This immediately appears" << flush;
  clog << "\nSo does this..." << flush;
}
```

820 UNDERSTANDING WHAT IOSTREAM.H CONTAINS

All the C++ programs presented so far include the header file *iostream.h* instead of the file *stdio.h*. As you learned in Tip 818, the I/O streams *cin*, *cout*, *cerr*, and *clog* are actually class instances. The file *iostream.h* defines the corresponding stream class and these four identifiers. Do not look in the *iostream.h* file yet. You will analyze its contents later, after you have the knowledge to better understand its contents. For now, however, just remember that the *iostream.h* file defines the class library for screen and keyboard I/O.

821 C++ REQUIRES FUNCTION PROTOTYPES

In the Functions section of this book, you learned that a *function prototype* specifies the type of parameters a function receives, as well as the type of value the function returns. When you do not specify a function prototype for a function within C, the compiler will generate and display a warning message. However, in C++, you must specify function prototypes because, if you do not, the program will not compile. The following program, *noproto.cpp,* tries to use the *printf* function without providing a function type (contained in the header file *stdio.h*):

```
void main(void)
 {
   printf("This won't compile under C++\n");
 }
```

If you try to compile the *noproto.cpp* program, C++ will generate an error message, and the compilation will end.

822 C++ ADDS NEW KEYWORDS

As you learned in Tip 31, a *keyword* is an identifier that has special meaning to the compiler (such as *for*, *while*, *if*, and so on). In addition to the keywords the C compiler defines, Table 822 shows the new keywords that C++ supports.

C++ Keywords								
asm	*bool*	*catch*	*class*	*delete*	*friend*	*inline*	*mutable*	*namespace*
new	*operator*	*private*	*protected*	*public*	*template*	*this*	*using*	*virtual*

Table 822 *The new keywords C++ supports.*

As is the case with keywords in C, you cannot use C++ keywords for variable, type, or function names.

823 C++ SUPPORTS ANONYMOUS UNIONS

As you have learned, a union is a special data structure for which C maps two or more members to the same memory location. When you declare a union in C, you must declare a variable of type *union*, as shown here:

```
union Values
 {
   unsigned my_data;
   float his_data;
 } solution;
```

When you later want to store data within the union, you must specify the variable name and member, as shown here:

```
solution.my_data = 3;
```

C++, however, lets your programs use anonymous (or unnamed) unions. For example, the following program, *anonym.cpp*, uses a union similar to the one just shown:

```
#include <iostream.h>

void main(void)
 {
   union
    {
      int my_data;
      float his_data;
    };

   my_data = 3;
   cout << "Value of my_data is " << my_data;
   his_data = 1.2345;
   cout << "\nValue of his_data is " << his_data;
 }
```

Using the anonymous union, programs can eliminate the programming overhead of managing a union name and member names. However, the member names of an anonymous union must be different from any other variables within the current scope. You will learn more about anonymous unions in later Tips.

RESOLVING GLOBAL SCOPE 824

As you have learned, a global variable is known from its declaration to the end of your program. When you use global variables, there may be times when a global variable has the same name as a local variable. In such cases, the function will use the local variable. There may be times, however, when you want to refer to the global variable within a function that has a similarly named local variable. For such cases, C++ lets you precede the global variable name with two colons, such as *::variable*. The following program, *global.cpp*, illustrates how to use the C++ *global resolution operator*:

```
#include <iostream.h>

int global_name = 1001;

void main(void)
 {
   int global_name = 1;  // Local variable

   cout << "Local variable value " << global_name << '\n';
   cout << "Global variable value " << ::global_name << '\n';
 }
```

PROVIDING DEFAULT PARAMETER VALUES 825

As you have learned, parameters are the values passed to functions. The primary difference between C and C++ function parameters is that C++ lets your programs provide default values for parameters. Should your program invoke a function without specifying one or more parameters, the program will use the default values. For example, the following program, *default.cpp*, uses the function *show_values* to display three parameters. If the user invokes the function with less than three parameters, the program will use the default values 1, 2, and 3:

```
#include <iostream.h>

void show_values(int one = 1, int two = 2, int three = 3)
 {
   cout << one << ' ' << two << ' ' << three << '\n';
 }
```

```
void main(void)
 {
   show_values(1, 2, 3);
   show_values(100, 200);
   show_values(1000);
   show_values();
 }
```

Note: *When you omit parameters, you cannot skip a parameter. In other words, when you omit a parameter, you must omit all the parameters to the right of the parameter.*

826 CONTROLLING COUT'S OUTPUT WIDTH

Several Tips in this section have used the *cout* I/O stream to display output. When you use *cout*, you can use the *cout width* member to specify the minimum number of characters used to display output. For example, the following program, *setwidth.cpp*, uses the *width* member to select a minimum output width of five characters:

```
#include <iostream.h>

void main(void)
 {
   int i;

   for (i = 0; i < 3; i++)
    {
      cout.width(5);
      cout << i << '\n';
    }
 }
```

When you compile and execute the *setwidth.cpp* program, your screen will display the following:

```
    0
    1
    2
C:\>
```

Note: *When you use the **width** member, you must specify the desired width for each value you output.*

827 USING SETW TO SET COUT WIDTH

In Tip 826, you used the *cout* width member to specify the minimum number of characters used to display a value. In addition, your programs can use the *setw* manipulator to specify the display text with the width you want, as shown here:

```
#include <iomanip.h>

smanip_int _Cdecl _FARFUNC setw(int _desired_width);
```

For now, do not worry about fully understanding the *setw* manipulator's prototype function. The following program, *setw.cpp*, uses *setw* to select different widths:

```
#include <iostream.h>
#include <iomanip.h>

void main(void)
 {
   cout << setw(5) << 1 << '\n' << setw(6) << 2;
   cout << '\n' << setw(7) << 3;
 }
```

When you compile and execute the *setw.cpp* program, your screen will display the following output:

```
      1
        2
           3
C:\>
```

Note: *When you use the* **setw** *manipulator, you must specify the desired width for each value you output.*

SPECIFYING A COUT FILL CHARACTER 828

By default, when you use the *cout* width member or the *setw* manipulator to specify character padding, C++ uses the space character to fill the additional spaces. Using the *cout* fill member, your programs can specify a different fill character. For example, the following program, *coutdots.cpp*, uses a period as a fill character:

```
#include <iostream.h>

void main(void)
 {
   int i;

   for (i = 0; i < 3; i++)
     {
      cout.fill('.');
      cout.width(5 + i);
      cout << i << '\n';
     }
 }
```

When you compile and execute the *coutdots.cpp* program, your screen will display the following output:

```
....0
.....1
......2
C:\>
```

RIGHT- AND LEFT-JUSTIFYING COUT OUTPUT 829

You have learned that using the *setw* manipulator or *cout* width member, your programs can specify the minimum width used to output a specific value. When you specify an output, your programs can select right or left justification using the *setiosflags* manipulator and the *ios* class members, as shown here:

```
#include <iomanip.h>

smanip_long _Cdecl _FARFUNC setiosflags(long flags);
```

To use *setiosflags* to select right justification, place the following manipulator in the *cout* stream:

```
setiosflags(ios::right)
```

Likewise, to select left justification, use the following manipulator in the *cout* stream:

```
setiosflags(ios::left)
```

For now, do not try to understand the format this code uses. Instead, you can use the flags within your programs as shown in the next example. In later Tips, you will learn how to use the *ios* class. The following program, *rghtleft.cpp*, uses the *ios::left* and *ios::right* flags to select right and left justification:

```
#include <iostream.h>
#include <iomanip.h>
```

```
void main(void)
 {
   int i;

   cout << "Right justification\n";
   for (i = 0; i < 3; i++)
     {
       cout.width(5);
       cout << setiosflags(ios::right) << i;
     }
   cout << "\nLeft justification\n";
   for (i = 0; i < 3; i++)
     {
       cout.width(5);
       cout << setiosflags(ios::left) << i;
     }
 }
```

When you compile and execute the *rghtleft.cpp* program, your screen will display the following output:

```
Right justification
    1    2    3
Left justification
1   2     3
C:\>
```

830 CONTROLLING THE NUMBER OF FLOATING-POINT DIGITS COUT DISPLAYS

As you learned in Tip 807, the *cout* I/O stream lets your programs display floating-point values. When you display such values, you can use the *setprecision* manipulator to specify the desired number of digits to the right of the decimal point, as shown here:

```
#include <iomanip.h>

smanip_int _Cdecl _FARFUNC setprecision(int number_of_digits);
```

The following program, *setprec.cpp*, uses the *setprecision* manipulator to change the number of digits displayed to the right of the decimal point:

```
#include <iostream.h>
#include <iomanip.h>

void main(void)
 {
   int i;
   float value = 1.2345;

   for (i = 0; i < 4; i++)
     cout << setprecision(i) << value << '\n';
 }
```

When you compile and execute the *setprec.cpp* program, your screen will display the following output:

```
1.2345
1.2
1.23
1.235
C:\>
```

As you can see, if you specify a precision of 0, *cout* will display all the value's digits.

DISPLAYING VALUES IN FIXED OR SCIENTIFIC FORMAT 831

As you have learned, the *cout* I/O stream lets your programs display floating-point values. When you use *cout*, you can select fixed or scientific (exponential) format to display floating-point values. To control the value's display format, your programs can use the *setiosflags* manipulator's *ios::fixed* and *ios::scientific* flags, as shown here:

```
#include <iomanip.h>

smanip_long _Cdecl _FARFUNC setiosflags(long flags);
```

The following program, *fixed.cpp*, uses the *setiosflags* manipulator to display values in fixed and scientific formats:

```
#include <iostream.h>
#include <iomanip.h>

void main(void)
 {
   float value = 0.000123;

   cout << setiosflags(ios::fixed) << value << '\n';
   cout << setiosflags(ios::scientific) << value << '\n';
 }
```

When you compile and execute the *fixed.cpp* program, your screen will display the following output:

```
0.000123
1.23e-04
C:\>
```

RESTORING COUT TO DEFAULT 832

Several Tips in this section have used the *setiosflags* manipulator to control different *cout* formatting options. To quickly restore *cout's* default settings, you can use the *resetiosflags* manipulator, as shown here:

```
#include <iomanip.h>

smanip_long _Cdecl _FARFUNC resetiosflags(long flag);
```

The *flag* parameter specifies the option you want to set. For example, the following program, *rstio.cpp*, uses the *resetiosflags* modifier to turn off right justification:

```
#include <iostream.h>
#include <iomanip.h>

void main(void)
 {
   cout.width(5);
   cout << setiosflags(ios::left) << 5 << '\n';
   cout.width(5);
   cout << 5 << '\n' << resetiosflags(ios::left);
   cout.width(5);
   cout << 1;
 }
```

When you compile and execute the *rstio.cpp* program, your screen will display the following output:

```
5
5
    1
C:\>
```

833 Setting the I/O Base

You have learned that using the *dec*, *oct*, and *hex* modifiers, you can select decimal, octal, and hexadecimal values. If your program must output several values using a specific base, your programs can use the *setbase* modifier. The following program, *setbase.cpp*, uses the *setbase* modifier to display the value 255 using different bases:

```
#include <iostream.h>
#include <iomanip.h>

void main(void)
 {
   cout << setbase(8) << 255 << '\n';
   cout << setbase(10) << 255 << '\n';
   cout << setbase(16) << 255 << '\n';
 }
```

When you compile and execute the *setbase.cpp* program, your screen will display the following output:

```
377
255
ff
C:\>
```

834 Declaring Variables Where You Need Them

In C, your programs can declare variables following any opening brace. In C++, however, your programs can declare variables at any location in your program. The advantage of such declarations is that your programs can declare variables closer to their use. The following program, *dec_int.cpp*, declares the variable *count* inside a *for* loop:

```
#include <iostream.h>

void main(void)
 {
   cout << "About to start the loop\n";
   for (int count = 0; count < 10; count++)
     cout << count << '\n';
   cout << "count's ending value " << count;
 }
```

When you declare a variable within a block, the variable's scope begins at the declaration and ends with the block. When you compile this program, the compiler will generate a syntax error for the line that tries to display's *count's* value. That's because *count's* scope ends with *for* loop.

835 Placing Default Parameter Values in Function Prototypes

As you have learned, C++ lets you specify default values for function parameters. Usually, such default values appear in the function header, as shown here:

```
void some_function(int a = 1, int b = 2, int c = 3)
  {
    cout << a << b << c;
  }
```

In addition to placing the default parameter values in the function header, however, your programs can specify the defaults within the function prototype, as shown here:

```
void some_function(int a = 1, int b = 2, int c = 3);

void main(void)
 {
    // Statements
 }
```

If the function for which you want to specify default parameters does not reside in the current source file, you can direct the compiler to include the correct defaults by specifying the values in the function prototype, as just shown.

USING BITWISE OPERATORS AND COUT 836

As you have learned, to output data using the *cout* I/O stream, your program uses operators identical to C's *bitwise left-shift* operator (<<). The following program, *bitscout.cpp*, uses both the bitwise operator and *cout*. As you will see, the compiler can tell which operation to perform by the way in which the program uses the operator, as shown here:

```
#include <iostream.h>

void main(void)
 {
   unsigned int value, one = 1;

   value = one << 1;
   cout << "Value: " << value << '\n';
   cout << "Result: " << (one << 1) << '\n';
 }
```

When you compile and execute the *bitscout.cpp* program, your screen will display the following:

```
Value: 2
Result: 2
C:\>
```

By grouping the bitwise operation within parentheses, the compiler associates the operator with the *bitwise left-shift operator*. If you were to remove the surrounding parentheses, the program would produce the following result:

```
Value: 2
Result: 11
C:\>
```

UNDERSTANDING LAZY (OR SHORT-CIRCUIT) EVALUATION 837

Many of the programs presented throughout this book combine conditions within *if* and *while* constructs, as shown here:

```
if ((a > 1) && (b < 3))
while ((letter >= 'A') && (letter <= 'Z'))
```

When your programs combine conditions within *if* and *while* constructs, you must understand that the C++ compiler generates code that performs *lazy* or *short-circuit* evaluation. Short-circuit evaluation means that if the outcome of one of the conditions will make the entire condition true or false, the program will stop performing the remainder of the condition. For example, given the previous *if* statement, the program will not compare the variable *b* to the value 3 if the first part of the *if* condition fails. Performing the comparison would unnecessarily consume processor time. Likewise, given the previous *while* statement, the program will not compare the variable *letter* to 'Z' if the letter is not greater than or equal to 'A.' Performing short-circuit evaluation lets programs save processor time. However, if you are not aware of the fact that your programs perform such processing, your programs can experience errors. For example, consider the following *if* statement:

```
if ((value < 10) && ((letter = getchar()) != 'Q'))
```

In this case, if the variable *value* is not less than 10, the program will not perform the second comparison, which uses the *getchar* macro to assign a value to the variable *letter*. As a result, there are times when the *if* statement will assign a value to *letter* and times when the *if* statement will not.

838 USING THE CONST KEYWORD IN C++

As you have learned, the *const* keyword informs the compiler that the program should not change the variable that follows during the program's execution. When you use the *const* keyword in C++ programs, you can use the corresponding variable in any way that you could normally use a constant expression. For example, the following statements use the constant *array_size* to specify the size of a character string array:

```
const int array_size = 64;
char string[array_size];
```

The advantage of using constants over macros created with *#define* is that constants let you specify type information.

839 USING THE ENUM KEYWORD IN C++

As you have learned, the *enum* keyword lets your programs define enumerated types. The *enum* keyword in C++ is very similar to the one C uses, except that when you declare an enumerated type in C++, your program can later use the type's tag as a type. For example, the following statements declare a variable named *day* in C:

```
enum Days { Monday, Tuesday, Wednesday, Thursday, Friday };
enum Days day;
```

In C++, the declaration becomes the following:

```
enum Days { Monday, Tuesday, Wednesday, Thursday, Friday };
Days day;
```

As you can see, the second declaration does not require the use of the *enum* keyword before the type name *Days*.

840 UNDERSTANDING FREE SPACE

As you learned in the Memory section of this book, your programs can dynamically allocate memory from the heap during run time. As you read C++ documentation, you will find references to the *free space*. The heap and free space are the same. To allocate memory from the free space, C++ programs use *new* and *delete*. It is important to note that unlike *malloc* and *free*, which are functions, *new* and *delete* are operators. Tip 841 shows you how to allocate memory with *new*.

841 ALLOCATING MEMORY WITH NEW

As you have learned, C++ programs allocate dynamic memory from the free space using the *new* operator. To use *new*, a program must specify the desired number of bytes. The following program, *newarray.cpp*, allocates memory for a 256-byte array. The program then fills the array with the letter A and displays the array's contents:

```
#include <iostream.h>

void main(void)
 {
   char *array = new char[256];
   int i;
```

```
   for (i = 0; i < 256; i++)
     array[i] = 'A';
   for (i = 0; i < 256; i++)
     cout << array[i] << ' ';
 }
```

ALLOCATING MULTIPLE ARRAYS 842

In Tip 841, you used the *new* operator to dynamically allocate a 256-byte character string. At that time, the compiler allocated the memory when the program declared the pointer variable, as shown here:

```
char *array = new char[256];
```

When your programs use the *new* operator to allocate memory, however, your programs can do so from any location. The following program, *new_copy.cpp*, uses the *new* operator to allocate three character strings—each at a different location throughout the program:

```
#include <iostream.h>

void main(void)
 {
   char *array = new char[256];
   char *target, *destination;
   int i;

   target = new char[256];
   for (i = 0; i < 256; i++)
    {
     array[i] = 'A';
     target[i] = 'B';
    }
   destination = new char[256];
   for (i = 0; i < 256; i++)
    {
     destination[i] = target[i];
     cout << destination[i] << ' ';
    }
 }
```

TESTING FOR NO FREE SPACE 843

You learned in the Memory section of this book that when the heap cannot satisfy a request, the functions *calloc* and *malloc* return *NULL*. The same is true for the *new* operator and the free space. The following program, *no_free.cpp*, uses the *new* operator to allocate memory until the free space is empty:

```
#include <iostream.h>

void main(void)
 {
   char *pointer;

   do
    {
      pointer = new char[10000];
      if (pointer)
        cout << "Allocated 10,000 bytes\n";
      else
        cout << "Allocation failed\n";
    } while (pointer);
 }
```

844 CONSIDERATIONS ABOUT HEAP SPACE

As you have learned, C++ refers to the heap as the free space. Depending on your memory model, the amount of available heap space will differ. For example, you might compile the program *no_free.cpp* using the large memory model. If you did so, the *no_free.cpp* program compiled with the large memory model will execute longer and will allocate considerably more heap space before failing.

845 USING FAR POINTERS AND THE NEW OPERATOR

As you have learned, the *new* operator lets your programs allocate memory from the free space. If you are using the small memory model, the free space corresponds to the near heap. If your programs must allocate more memory than the near heap provides, your programs can allocate far pointers. The following program, *new_far.cpp*, allocates far pointers from the free space until the far heap runs out of memory:

```
#include <iostream.h>

void main(void)
 {
   char far *pointer;
   do
    {
      pointer = new far char[10000];
      if (pointer)
        cout << "Allocated 10,000 bytes\n";
      else
        cout << "Allocation failed\n";
    } while (pointer);
 }
```

846 RELEASING MEMORY BACK TO THE FREE SPACE

As you have learned, when your programs allocate memory dynamically, your programs should release the memory as soon as they no longer need it. When your C programs allocate memory using *calloc* and *malloc*, your programs release the memory using *free*. When your C++ programs allocate memory using *new*, your programs should later free the memory using *delete*. The following program, *delete.cpp*, uses the *delete* operator to release three dynamically allocated arrays back to the free space:

```
#include <iostream.h>

void main(void)
 {
   char *array = new char[256];
   char *target, *destination;
   int i;

   target = new char[256];
   for (i = 0; i < 256; i++)
    {
     array[i] = 'A';
     target[i] = 'B';
    }
   delete array;
   destination = new char[256];
   for (i = 0; i < 256; i++)
    {
     destination[i] = target[i];
     cout << destination[i] << ' ';
```

```
   }
   delete target;
   delete destination;
}
```

UNDERSTANDING C++ REFERENCES 847

An *alias* is a second name for a variable. In C, your programs can create aliases using pointers. C++ simplifies the creation of aliases using references. To create a reference, you use the reference operator (&), as shown here:

```
int variable;
int& alias = variable;
```

The reference operator is similar to C's address operator. However, note the operator's positioning. The reference operator immediately follows a type (such as *int*, *float*, or *char*). Each alias can correspond to only one variable throughout its lifetime. The following program, *alias.cpp*, creates two aliases and uses them to display the addresses of specific variables:

```
#include <iostream.h>

void main(void)
 {
   int a = 1700;
   int& a_alias = a;
   float price = 54.95;
   float& price_alias = price;

   cout << "The value of a is " << a << " the alias is " << a_alias;
   cout << "\nThe price is " << price << " the alias is " << price_alias;
   a_alias++;
   cout << "\nThe value of a is " << a << " the alias is " << a_alias;
 }
```

The *alias.cpp* program uses the reference variable *a_alias* to increment the value of *a*. When a program refers to a reference, any operations will correspond directly to the aliased variable.

PASSING A REFERENCE TO A FUNCTION 848

As you have learned, to change a variable within a function, your programs must pass a pointer to the variable. When you use C++, you can simplify the changing of a variable within a function by using a *reference*. Using a reference, you eliminate the need for the pointer operator (->). The following program, *functref.cpp*, passes a reference to the variable *value* to the function *change_value*, which assigns the value 1700 to the variable:

```
#include <iostream.h>

void change_value(int& reference_value)
 {
   reference_value = 1700;
 }

void main(void)
 {
   int value = 10;
   int& alias = value;

   cout << "Value before function: " << value << '\n';
   change_value(alias);
   cout << "Value after function: " << value << '\n';
 }
```

As you can see, using a reference simplifies the process of changing a value within a function.

849 WATCHING OUT FOR HIDDEN OBJECTS

As you have learned, a reference creates a second name for a variable—an alias. When you create references, you must ensure that the reference type is identical to the type it will reference. For example, the following statement creates a reference to a variable of type *int*:

```
int value;
int& alias = value;
```

If the reference type and variable type differ, C++ will create a *hidden object* that does not alias the specified value, but instead holds the value for an unnamed variable of the reference type. For example, the following statement creates a hidden object of type *float*:

```
int value;
float& alias = value;
```

As you can see, the reference and variable types differ. Therefore, the compiler will not alias the variable *value*, but rather will allocate memory for a floating-point value, aliasing the memory with the given reference. The reason you must be aware of hidden objects is that they can lead to errors that are very difficult to detect. If you change the type of a variable, make sure you also change the type of the corresponding reference, if a reference exists.

850 USING THREE WAYS TO PASS PARAMETERS

In C, your programs can pass parameters to functions using *call by value* or *call by pointer reference*. As you have learned, in C++ your programs can use a third technique, *call by reference*. The following program, *call_3.cpp*, illustrates how to use all three calling techniques:

```
#include <iostream.h>
#include <iomanip.h>

void call_by_value(int a, int b, int c)
 {
   a = 3; b = 2; c = 1;
 }

void call_by_pointer_reference(int *a, int *b, int *c)
 {
   *a = 3; *b = 2; *c = 1;
 }

void call_by_reference(int& a, int& b, int& c)
 {
   a = 1; b = 2; c = 3;
 }

void main(void)
 {
   int a = 1, b = 2, c = 3;
   int& a_alias = a;
   int& b_alias = b;
   int& c_alias = c;

   call_by_value(a, b, c);
   cout << "By value: " << a << b << c << '\n';
   call_by_pointer_reference(&a, &b, &c);
   cout << "By pointer: " << a << b << c << '\n';
   call_by_reference(a_alias, b_alias, c_alias);
   cout << "By reference: " << a << b << c << '\n';
 }
```

RULES FOR WORKING WITH REFERENCES 851

In C++, a reference lets you create an alias for a variable. When you use references, keep the following rules in mind:

1. After initialization, your program cannot change a reference value.
2. The reference type and variable type must be the same.
3. You cannot create a pointer to a reference.
4. You cannot compare the value of two references—the comparisons would compare the values of the referenced variables.
5. You cannot increment, decrement, or change a reference value—the operations will apply to the value of the referenced variable.
6. You can distinguish the reference operator from the address operator because the reference operator always follows the type (for example, *int&*).

FUNCTIONS CAN RETURN REFERENCES 852

In C++, a reference is an alias for a variable. As you have learned, references can simplify parameter passing by eliminating the need to perform pointer operations. All the Tips presented so far have initialized reference variables at declaration, near the top of *main*. However, C++ also lets functions return references. Because C++ lets your programs declare variables at any location, your programs can therefore create and initialize a reference at any location in your program by returning a reference from a function. The following program, *rtn_ref.cpp*, invokes the function *get_book*, which returns a reference to a variable of type *book*:

```
#include <iostream.h>

struct book
 {
   char author[64];
   char title[64];
   float price;
 };

book library[3] = {
  {"Jamsa", "Jamsa's C/C++/C# Programmer's Bible", 54.95},
  {"Klander", "Hacker Proof", 54.95},
  {"Jamsa and Klander", "1001 Visual Basic Programmer's Tips", 54.95}};

book& get_book(int i)
 {
   if ((i >= 0) && (i < 3))
     return(library[i]);
   else
     return(library[0]);
 }

void main(void)
 {
   cout << "About to get book 0\n";
   book& this_book = get_book(0);
   cout << this_book.author << ' ' << this_book.title;
   cout << ' ' << this_book.price;
 }
```

853 USING THE C++ INLINE KEYWORD

As you have learned, programs pass parameters to functions using the stack. Each time your program invokes a function, the computer must push the function's parameters (and the program's return address) onto the stack, and later pop those same values off the stack. These push and pop operations lead to overhead that make using functions slightly slower than using inline code. If your programs have one or two critical functions that must execute quickly, you should use the *inline* keyword to direct the compiler to place the corresponding code inline in the program at each function call, rather than creating separate function code. If your programs call the inline function from five different locations, the compiler will insert the corresponding function in the program five times. If your programs call the inline function from 50 different locations, the compiler inserts the code 50 times. Therefore, inline code has a tradeoff between time and space. Using inline code creates a faster program, but also makes the program code larger (which, in theory, can slow down the program). The following program, *inline.cpp*, uses two similar functions, placing one function inline and calling the second function. The program displays the amount of time required to call each function 3,000,000 times:

```
#include <iostream.h>
#include <time.h>

inline void swap_inline(int *a, int *b, int *c, int *d)
 {
   int temp;

   temp = *a;
   *a = *b;
   *b = temp;
   temp = *c;
   *c = *d;
   *d = temp;
 }

void swap_call(int *a, int *b, int *c, int *d)
 {
   int temp;

   temp = *a;
   *a = *b;
   *b = temp;
   temp = *c;
   *c = *d;
   *d = temp;
 }
void main(void)
 {
   clock_t start, stop;
   long int i;
   int a = 1, b = 2, c = 3, d = 4;

   start = clock();
   for (i = 0; i < 30000000L; i++)
     swap_inline(&a, &b, &c, &d);
   stop = clock();
   cout << "Time for inline: " << stop - start;
   start = clock();
   for (i = 0; i < 30000000L; i++)
     swap_call(&a, &b, &c, &d);
   stop = clock();
   cout << "\nTime for called function: " << stop - start;
 }
```

USING THE C++ ASM KEYWORD 854

As you have learned, depending on your program's purpose, there may be times when you must perform low-level assembly language programming. In such cases, you can create an assembly language function and link the function to your program, or you can use the *asm* keyword to insert assembly language statements into your C++ code. The following program, *asm_demo.cpp*, uses the *asm* keyword to include the assembly language statements required to sound the computer's built-in speaker:

```
#include <iostream.h>

void main(void)
 {
   cout << "About to sound the speaker...\n";
   asm
    {
      MOV AX,0x0200
      MOV DL,7
      INT 0x21
    };
   cout << "Done...\n";
 }
```

READING A CHARACTER USING CIN 855

Several preceding Tips have used the *cin* I/O stream to read input from the keyboard. To improve your control over keyboard or redirected input, your programs can use *cin.get* to read input a character at a time, as shown here:

```
character = cin.get();
```

The following program, *cin_get.cpp*, uses *cin.get* to assign characters up to, but not including, the newline character to the character string *str*:

```
#include <iostream.h>
#include <stdio.h>

void main(void)
 {
   char str[256];
   int i = 0;

   while ((str[i] = cin.get()) != '\n')
     i++;
   str[i] = NULL;
   cout << "The string was: " <<  str;
 }
```

WRITING A CHARACTER WITH COUT 856

In Tip 855 you learned that your programs can input characters one at a time using *cin.get*. In a similar way, your programs can use *cout.put* to write one character, as shown here:

```
cout.put(character);
```

The following program, *cout_put.cpp*, uses *cout* to output a character string one character at a time:

```
#include <iostream.h>

void main(void)
 {
```

```
   char *title = "Jamsa's C/C++/C# Programmer's Bible";

   while (*title)
     cout.put(*title++);
 }
```

857 WRITING A SIMPLE FILTER PROGRAM

As you have learned, *cout.put* and *cin.get* let your programs perform character I/O. The following program, *to_upper.cpp*, converts redirected input to uppercase. To perform the conversion, the program simply loops until *cin.get* returns –1, indicating the end of the file:

```
#include <iostream.h>
#include <ctype.h>

void main(void)
 {
   char letter;

   while ((letter = cin.get()) != -1)
     cout.put((char) toupper(letter));
 }
```

858 WRITING A SIMPLE TEE COMMAND

As you have learned, C++ lets you redirect the output of the *cout* I/O stream. The following program, *tee.cpp*, writes its redirected input to the *cout* and *cerr* I/O streams. Because the program uses two I/O streams, you can view the program's input on the screen and still redirect the output to a different source:

```
#include <iostream.h>

void main(void)
 {
   char letter;

   while ((letter = cin.get()) != -1)
    {
       cout.put(letter);
       cerr.put(letter);
    }
 }
```

859 WRITING A SIMPLE FIRST COMMAND

As you have learned, the *cin* and *cout* I/O streams support I/O redirection. The following program, *simp1st.cpp*, uses these input streams to write the first ten lines of redirected input to the screen:

```
#include <iostream.h>

void main(void)
 {

   char letter;
   int count = 0;

   while ((letter = cin.get()) != -1)
    {
       cout.put(letter);
       if ((letter == '\n') && (++count == 10))
         break;
    }
 }
```

WRITING A BETTER FIRST COMMAND 860

In Tip 859, you created the *simp1st.cpp* program, which displayed the first ten lines of redirected input. A more flexible command would let the user specify, as a command-line argument, the number of lines the user wants to display. The following program, *first.cpp,* lets the user do just that:

```
#include <iostream.h>
#include <stdlib.h>

void main(int argc, char **argv)
 {
   char letter;
   int count = 0;
   int line_limit;

   line_limit = atoi(argv[1]);
   while ((letter = cin.get()) != -1)
    {
       cout.put(letter);
       if ((letter == '\n') && (++count == line_limit))
         break;
    }
 }
```

If the user does not specify the number of lines for the program to output, or if the user specifies an invalid line count, the program will display all the redirected input.

TESTING FOR END OF FILE 861

Several previous Tips have used *cin.get* to determine the end of the redirected input, as shown here:

```
while ((letter = cin.get()) != -1)
```

In addition to testing the *cin.get* method's return value, your programs can test *cin.eof* as follows:

```
while (! cin.eof())
```

The following program, *firsteof.cpp*, changes the *first.cpp* program to test for an end of file with the *cin.eof* method:

```
#include <iostream.h>
#include <stdlib.h>

void main(int argc, char **argv)
 {
   char letter;
   int count = 0;
   int line_limit;

   line_limit = atoi(argv[1]);
   while (! cin.eof())
    {
       letter = cin.get();
       cout.put(letter);
       if ((letter == '\n') && (++count == line_limit))
         break;
    }
 }
```

GENERATING A NEWLINE WITH ENDL 862

Many preceding Tips have placed the newline character (\n) in the *cout* output stream to generate a carriage return and linefeed. In addition to using the newline character, your programs can use *endl,* as shown here:

```
cout << "Hello, world!" << endl;
```

The following program, *endl.cpp*, uses *endl* several times to generate a carriage return and linefeed:

```
#include <iostream.h>

void main(void)
 {
   cout << "This is line one" << endl;
   cout << "This is line two" << endl;
   cout << "This is line three--";
   cout << "It is the last line" << endl;
 }
```

When you compile and execute the *endl.cpp* program, your screen will display the following output:

```
This is line one
This is line two
This is line three--It is the last line
C:\>
```

863 UNDERSTANDING LINKAGE SPECIFICATIONS

As you have learned, C++ requires function prototypes for every function your program uses. The C++ compiler uses the prototypes to verify parameter and return value types. During compilation, the C++ compiler changes the names of functions and their parameters in the resulting object code. The linker, in turn, uses these new names to resolve external references. Unfortunately, if you are linking to code which a C compiler has previously compiled, the function names in the object code will not be in the same "C++ function name format." To prevent the C++ compiler from changing the names of C functions, you can use a *linkage specifier.* In short, the linkage specifier tells the C++ compiler the correct format it should use for naming functions in the object file. Assume, for example, that you have a function named *calculate_payroll* that you (or another programmer) previously wrote in C. To direct the C++ compiler not to change the function name format, you would use the following linkage specifier:

```
extern "C"
 {
   float calculate_payroll(int employee_count, char *employee_file);
 };
```

Note: *If you examine the header files your compiler provides, you will find several linkage specifiers in the files that are similar to the one shown in this Tip.*

864 UNDERSTANDING OVERLOADING

Overloading is the process of assigning more than one operation to an operator or providing two or more functions with the same name. For example, C and C++ use the plus symbol (+) as the *addition* operator. You can use overloading to direct C++ to also use the plus symbol to concatenate strings, as shown here:

```
pathname = directory_name + filename;
```

Depending on how your program uses the plus symbol, the C++ compiler will determine if the program statement performs addition or string concatenation. You have also learned that when you use C, sometimes you have to create differently named functions that work with values of different types. For example, if you have created a function that returns the sum of values in an integer array, you must create a function with a different name if you want to sum values in an array of type *float.* As you will learn, C++ lets you overload functions and operators, which simplifies many operations.

OVERLOADING FUNCTIONS 865

As you have learned, C++ lets you have multiple functions with the same name. During compilation, the C++ compiler determines which function to call, based on the number and types of parameters the calling statement passes to the function. For example, the following program, *overload.cpp*, creates two functions named *sum* that return the sum of the number of elements in an array. The first function supports arrays of type *float*, while the second supports arrays of type *int*:

```
#include <iostream.h>

int sum(int *array, int element_count)
 {
   int result = 0;
   int count;

   for (count = 0; count < element_count; count++)
     result += array[count];
   return(result);
 }

float sum(float *array, int element_count)
 {
   float result = 0;
   int count;

   for (count = 0; count < element_count; count++)
     result += array[count];
   return(result);
 }

void main(void)
 {
   int a[5] = { 1, 2, 3, 4, 5 };
   float b[4] = { 1.11, 2.22, 3.33, 4.44 };

   cout << "Sum of int values: " << sum(a, 5) << '\n';
   cout << "Sum of float values: " << sum(b, 4) << '\n';
 }
```

OVERLOADING FUNCTIONS: A SECOND EXAMPLE 866

As you have learned, C++ lets you overload functions, creating two or more functions in your programs that have the same name. The following program overloads the function *swap*. The first function swaps two values while the second function swaps four. During compilation, the compiler uses the number of parameters to determine which function to call. The *useparam.cpp* program is your first overloaded function program, as shown here:

```
#include <iostream.h>

void swap(int *a, int *b)
 {
   int temp = *a;
   *a = *b;
   *b = temp;
 }

void swap(int *a, int *b, int *c, int *d)
 {
   int temp = *a;
   *a = *b;
   *b = temp;
   temp = *c;
   *c = *d;
```

```
    *d = temp;
  }

void main(void)
  {
    int a = 1, b = 2 , c = 3, d = 4;

    swap(&a, &b);
    cout << "Just swapped a and b " << a << b << '\n';
    swap(&a, &b, &c, &d);
    cout << "Just swapped four " << a << b << c << d << '\n';
  }
```

867 AVOIDING OVERLOAD AMBIGUITY

When you create overloaded functions, it is possible to create a situation in which the compiler is unable to distinguish between two (or more) overloaded functions. When you create two or more overloaded functions between which the compiler is unable to distinguish, the compiler considers the functions *ambiguous*. Ambiguous function calls are errors, and the compiler will not compile your program.

By far, the most common cause of ambiguity involves the C++ automatic type conversions. As you have learned, C++ automatically tries to convert the arguments the program uses to call a function into the type of arguments the function expects. For example, consider the following code fragment:

```
int myfunc(double d);
//
// Program code here
//
cout << myfunc('c');           // C++ converts to integer
```

As the comment within the code fragment indicates, the function call in the example does not cause an error because C++ automatically converts the character to its double equivalent. C++ disallows few type conversions of the sort shown in the previous example. While type conversions are convenient, they can cause serious problems when you overload functions. The following program, *over_err.cpp*, overloads the *sample_function* function with two different parameter types, *float* and *double*. The program calls the function twice—once with the value 1700.1, which is a *double* parameter type and therefore causes no ambiguity; and once with the value 1700, which causes ambiguity because the compiler does not know whether it should convert the value to a *float* or a *double*. When you compile the following program, you will receive a compiler error message:

```
#include <iostream.h>

float sample_function(float i);
double sample_function(double i);

void main(void)
  {
    cout << sample_function(1700.1) << " "; // unambiguous, calls
                                            //  sample_function(double)
    cout << sample_function(1700);          // ambiguous
  }

float sample_function(float i)
  {
    return i;
  }

double sample_function(double i)
  {
    return -i;
  }
```

***Note:** In Tip 814, you learned not to pass a reference to **cin**. As you have learned within this Tip, passing a reference to an overloaded function will confuse the compiler. Because it can process multiple value types, **cin** must also be an overloaded function—therefore explaining why you cannot use references with it.*

READING A LINE AT A TIME WITH CIN 868

As you have learned, your programs can read keyboard input using *cin*. When your programs want to read input a character at a time, your programs can use *cin.get*. In some cases, your programs might need to perform input operations a line at a time. For such cases, your program can use *cin.getline*, as shown here:

```
char string[256];
cin.getline(string, sizeof(string), '\n');
```

The *string* parameter is a pointer to the string you want *cin.getline* to read. The *sizeof* operator specifies the number of bytes the string can hold. Finally, the newline character specifies the character that will terminate the read. The following program, *getline.cpp*, uses *cin.getline* to read a line of input:

```
#include <iostream.h>

void main(void)
 {
   char string[256];

   cout << "Type in your full name and press Enter\n";
   cin.getline(string, sizeof(string), '\n');
   cout << string;
 }
```

To experiment with *cin.getline*, change the termination character to (for example) a letter of the alphabet, and then note your program's results.

USING CIN.GETLINE IN A LOOP 869

In Tip 857, your programs used *cin.get* and *cout.put* to display redirected input as uppercase. The following program, *allupper.cpp*, uses *cin.getline* and *cout* to perform similar processing:

```
#include <iostream.h>
#include <string.h>

void main(void)
 {
   char string[256];

   while (cin.getline(string, sizeof(string), '\n'))
     cout << strupr(string) << '\n';
 }
```

As you can see, the program loops until *cin.getline* returns 0, which indicates the end of the redirected input. Because the *cin.getline* function does not place the newline character into the string, *cout* must write the newline for each line.

CHANGING THE NEW OPERATOR'S DEFAULT HANDLING 870

As you have learned, when the *new* operator cannot allocate enough memory to satisfy a memory request, *new* returns *NULL*. Depending on your program's function, there may be times when you want *new* to perform other processing when it cannot allocate memory. As it turns out, when *new* cannot allocate memory it can call the function pointed to by a global pointer to a function named *_new_handler*. By assigning the *_new_handler* variable to point to a custom function, you can direct *new* to call your own function when *new* cannot allocate memory. The following program, *new_hand.cpp*, uses the *_new_handler* function pointer to direct *new* to call the function *no_memory* when *new* cannot satisfy a memory allocation request:

```
#include <iostream.h>
#include <stdlib.h>

extern void (*_new_handler)();

void no_memory(void)
 {
   cerr << "There is no more memory to allocate...\n";
   exit(0);
 }

void main(void)
 {
   _new_handler = no_memory;
   char *ptr;

   do
    {
      ptr = new char[10000];
      if (ptr)
        cout << "Just allocated 10,000 bytes\n";
    } while (ptr);
 }
```

Note: *If your handler cannot allocate the memory for the program, the handler must end the program, or an infinite loop will occur.*

871 SETTING A NEW HANDLER WITH SET_NEW_HANDLER

In Tip 870, you learned that C++ lets you define your own handler that your programs will call when the *new* operator cannot satisfy a memory request. To assign a new handler, your program assigns your handler function's address to a global variable named *_new_handler.* To simplify the new handler assignment process, many C++ compilers provide a function named *set_new_handler,* as shown here:

```
#include <new.h>

void (* set_new_handler(void (* custom_handler)()))();
```

The following program, *set_newh.cpp*, uses the *set_new_handler* function to install a custom handler:

```
#include <iostream.h>
#include <stdlib.h>
#include <new.h>

void no_memory(void)
 {
   cerr << "There is no more memory to allocate...\n";
   exit(0);
 }

void main(void)
 {
   char *ptr;

   set_new_handler(no_memory);
   do
    {
      ptr = new char[10000];
      if (ptr)
        cout << "Just allocated 10,000 bytes\n";
    } while (ptr);
 }
```

DETERMINING A C++ COMPILATION 872

Many C++ compilers let you compile standard C programs. Depending on your program's statements, there may be times when you want to terminate a standard C compilation. When many C++ compilers compile a C++ program, they might define a constant that you can test within your programs. For example, the *Turbo C++ Lite* compiler sets the constant *__cplusplus* when it is compiling a C++ program. The following program, *testcpp.cpp*, uses the *__cplusplus* constant to determine if the compiler is performing a C or C++ compilation:

```
#ifdef __cplusplus
#include <iostream.h>
#else
#include <stdio.h>
#endif

void main(void)
 {
#ifdef __cplusplus
  cout << "C++ compilation";
#else
  printf("C compilation\n");
#endif
 }
```

Compile the *testcpp.cpp* program as a file with the CPP extension. Next, copy the contents to a file with the C extension and compile the file. Note the processing the compiler performs.

UNDERSTANDING STRUCTURES IN C++ 873

As you have learned, a structure lets your programs group related information of different types. When you declare a structure in C, you can specify a *tag* (or name) with which you can later declare variables of the structure type, as shown here:

```
struct tag
 {
   int member_a;
   float member_b;
   char member_c[256];
 };

struct tag variable_one, variable_two;
```

When you declare a structure in C++, however, the structure tag becomes a type with which your programs can later declare variables without specifying the *struct* keyword, as shown here:

```
struct tag
 {
   int member_a;
   float member_b;
   char member_c[256];
 };

tag variable_one, variable_two;
```

As you can see, the C++ structure does not require the keyword *struct* before the tag in variable declaration.

INTRODUCING FUNCTIONS AS STRUCTURE MEMBERS 874

When you create C programs, the C compiler lets you use pointers to functions as structure members, as shown here:

```
struct tag
 {
   int member_a;
   int (*member_b)();  // Pointer to function that returns int
 };
```

C++ lets you take the previous concept one step further, letting you place actual functions within a structure, as shown here:

```
struct tag
 {
   int member_a;
   int member_b();
 };
```

When you declare the corresponding function, you have two choices. As shown in Tip 875, you can define the function code immediately within the structure, or you can define the function code outside the structure, as shown in Tip 876. To invoke the function, your program simply refers to the structure member, as shown here:

```
variable.member_b(parameters);
```

875 Defining a Member Function Within a Structure

As you have learned, C++ lets your programs place functions as structure members. When your structure contains a member that is a function, you can define the corresponding function code within the structure. The following program, *func_mbr.cpp*, defines the function that corresponds to the member *show_msg*:

```
#include <iostream.h>

struct Msg
 {
   char message[256];
   void show_message(void) { cout << message; }
 };

void main(void)
 {
   struct Msg book = { "Jamsa's C/C++/C# Programmer's Bible" };

   book.show_message();
 }
```

The *func_mbr.cpp* program invokes the member function *show_message*, which in turn displays the *message* member. As the previous code shows, *func_mbr.cpp* defines *show_message* within the structure. In Tip 876, you will learn how to define the function outside the structure.

876 Declaring a Member Function Outside a Structure

As you have learned, C++ lets you place functions as members in a structure. In Tip 875, you defined the member function *show_message* within the structure itself. The following program, *func_two.cpp*, defines the function outside the structure. To correspond the function with the *Msg* structure, the program precedes the function name with the structure name, followed by two colons:

```
#include <iostream.h>

struct Msg
 {
   char message[256];
   void show_message(char *message);
```

```
 };

void Msg::show_message(char *message)
 {
   cout << message;
 }

void main(void)
 {
   struct Msg book = { "Jamsa's C/C++/C# Programmer's Bible" };

   book.show_message(book.message);
 }
```

PASSING PARAMETERS TO A MEMBER FUNCTION 877

As you have learned, C++ lets you place functions as a member of a structure. Tips 875 and 876 used the member function *show_message* within a *Msg* structure. When you place a function as a structure member, you can treat the function just as you would any other C++ function. In other words, you can pass parameters to the function and declare local variables within the function. The following program, *func_thr.cpp*, passes the value 1700 to the function *show_title*:

```
#include <iostream.h>

struct Msg
 {
   char first[256];
   void show_title(int value)
    {
      cout << first << value << " C/C++/C# Tips";
    }
 };

void main(void)
 {
   struct Msg book = { "This book has " };

   book.show_title(1700);
 }
```

MULTIPLE VARIABLES OF THE SAME STRUCTURE 878

In Tip 875, you defined a structure that contains a function as a member. The *func_mbr.cpp* program then declared a variable of the structure type. The following program, *multstru.cpp*, declares several variables of the type *Msg*, assigns each a unique character string, and then displays the message using the *show_message* function:

```
#include <iostream.h>

struct Msg
 {
   char message[256];
   void show_message(void) { cout << message; }
 };

void main(void)
 {
   struct Msg book = { "Jamsa's C/C++ Programmer's Bible\n" };
   struct Msg section = { "Getting Started with C++" };

   book.show_message();
   section.show_message();
 }
```

879 DIFFERENT STRUCTURES WITH SAME FUNCTION MEMBER NAMES

As you have learned, C++ lets your programs place functions as members within a structure. When your programs use different structures, there may be times when two structures have the same member names. The following program, *samename.cpp*, creates two different structures, *Msg* and *UpperMsg*. Both structures use the member function *show_message*. The C++ compiler differentiates between the function names of the two structures the same way it does with non-function structure members, as shown here:

```
#include <iostream.h>
#include <string.h>

struct Msg
 {
   char message[256];
   void show_message(void) { cout << message; }
 };

struct UpperMsg
 {
   char message[256];
   void show_message(void) { cout << strupr(message); }
 };

void main(void)
 {
   Msg book = { "Jamsa's C/C++/C# Programmer's Bible\n" };
   UpperMsg book_upr = { "C/C++/C# PROGRAMMER'S BIBLE\n" };

   book.show_message();
   book_upr.show_message();
 }
```

The structures in the *samename.cpp* program define the functions within the structure itself. In Tip 880, however, you will learn how to differentiate between the functions that your programs define outside the structures.

880 DIFFERENT FUNCTIONS WITH SAME MEMBER NAMES

In Tip 879, you learned that the C++ compiler will distinguish between function members of different structure types. The following program, *diffname.cpp*, defines member functions outside their corresponding structures. To differentiate between the member functions, the program precedes each function definition with the appropriate structure name, followed by two colons:

```
#include <iostream.h>
#include <string.h>

struct Msg
 {
   char message[256];
   void show_message(void);
 };

struct UpperMsg
 {
   char message[256];
   void show_message(void);
 };

void Msg::show_message(void)
 {
   cout << message;
 }
```

```
void UpperMsg::show_message(void)
  {
    cout << strupr(message);
  }

void main(void)
  {
    Msg book = { "Jamsa's C/C++/C# Programmer's Bible\n" };
    UpperMsg book_upr = { "C/C++/C# PROGRAMMER'S BIBLE\n" };

    book.show_message();
    book_upr.show_message();
  }
```

UNDERSTANDING OBJECTS 881

In the simplest sense, an *object* is a thing or a real-world entity. When programmers create programs, they write instructions that work with different things, such as variables or files. Different objects have different *operations* that your programs perform on the objects. For example, given a *file* object, your program might perform such operations as reading, writing, or printing the file. As you will learn, C++ programs define objects in terms of a *class*. An *object class* (or simply a "class") defines the data the object will store and the functions that will operate on the data. C++ programs often refer to the functions that manipulate the class data as *methods*. For example, most of your C++ programs have already used the *cin* and *cout* objects. In the case of *cin* and *cout*, the I/O stream was the object, and functions such as *cin.get* and *cout.put* were the *operations* on the object.

UNDERSTANDING OBJECT-ORIENTED PROGRAMMING 882

To programmers, an *object* is a collection of data and a set of operations, called *methods*, that manipulate the data. *Object-oriented programming* is a way of looking at programs in terms of the objects (things) that make up a system. After you have identified the objects, you can determine the operations the system commonly performs on the object. If you have a *document* object, for example, common operations might include printing, spell-checking, faxing, or even discarding. Object-oriented programming does not require a special programming language, such as C++. You can write object-oriented programs in such languages as COBOL or FORTRAN. However, as you will learn, languages programmers describe as "object-oriented" usually provide class data structures that let your programs group data and methods into one variable.

As you will learn, object-oriented programming has many advantages, two of which are object reuse and ease of understanding. As you write more programs, you will find that you can often use the objects you write for more than one program. Rather than building a collection of function libraries, object-oriented programmers build *class libraries*. Likewise, when you create your programs around groups of objects and their data and methods, you (and others who read your programs) will more readily understand object-oriented programs than nonobject-based programs (at least, after you learn the syntax of the programming language you use). Programmers and computer professionals often call C++ an object-oriented extension of C. Many of the Tips that follow examine the C++ object-based capabilities.

UNDERSTANDING WHY YOU SHOULD USE OBJECTS 883

As you begin working with C++, you must understand why basing your programming around objects is important. There are several software engineering terms that programmers often use in object-based programming. Although software engineers are far from agreeing on the best use of objects, most of them will agree that object use offers the following advantages:

- **Ease of design and code reuse**—After the code works properly, the use of objects increases your ability to reuse a design or code you created for one application within a second application.
- **Increased reliability**—After you have properly tested object libraries, your use of existing (working) code will improve your program's reliability.
- **Ease of understanding**—Object use helps programmers focus on and understand key system components. The use of objects lets designers and programmers focus on the smaller pieces of a system and provides a framework within which designers can focus more on the operations programs perform on the objects, the information objects must store, and other key system components.
- **Increased abstraction**—Abstraction lets designers and programmers "look at the big picture"—temporarily ignoring underlying details so they can work with system elements they can more easily understand. For example, by focusing only on the word processor objects discussed in the next Tip, the implementation of a word processor can become much less intimidating.
- **Increased encapsulation**—*Encapsulation* (discussed in Tip 887) groups all the pieces of an object into one neat package. For example, the *Book* class previously defined combines the functions and data fields a program must have to work with a book. The programmers who are working with the *Book* class do not need to know each piece of the class, only that they need to use the class within their program. The class, in turn, will bring with it all the necessary pieces.
- **Increased information hiding**—*Information hiding* is the ability for your program to treat a function, procedure, or an object as a "black box," using the item to perform a specific operation without having to know what goes on inside. In Chapter 1, for example, your programs used I/O stream objects for input and output without having to understand how the streams work.

As you examine different C++ programming concepts throughout this book, you will learn how the concepts relate to these definitions.

884 Breaking Programs into Objects

In the simplest sense, an object is a thing. Dogs, books, and computers are all objects. In the past, programmers viewed programs as long lists of instructions that performed a specific task. When you create object-oriented programs, you look instead at the objects that make up your program. For example, assume that you are writing a program that implements a simple word processor. If you think of all the functions a word processor performs, you can quickly become overwhelmed. However, if you view the word processor as a collection of distinct objects, the program becomes less intimidating. For example, Figure 884.1 illustrates the major objects in the word processing system.

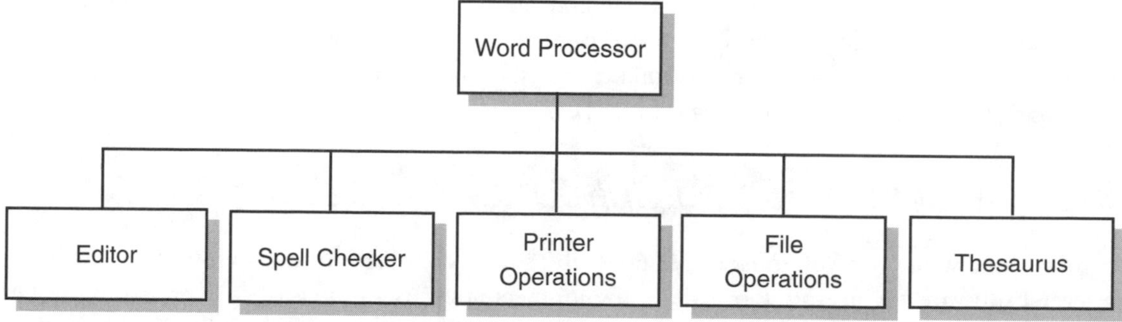

Figure 884.1 Viewing a word processor as a collection of objects.

As you examine each new object, you may find that it, too, is composed of other objects, as shown in Figure 884.2.

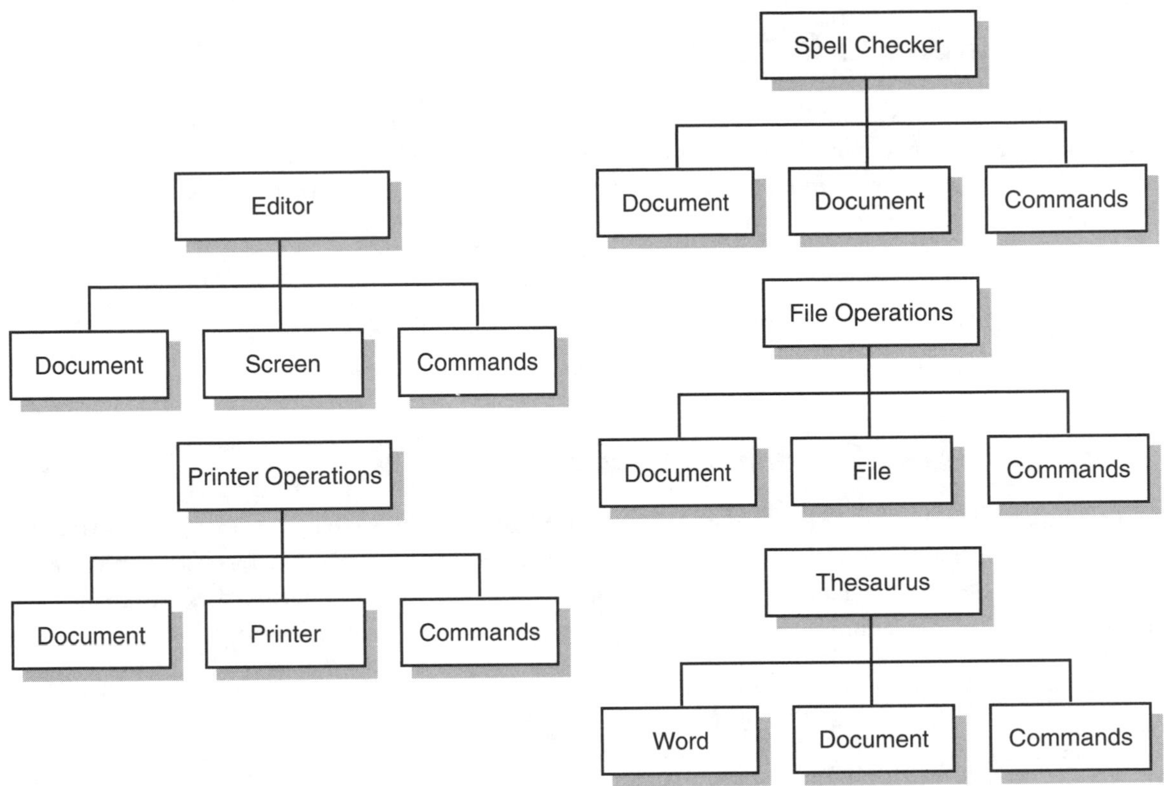

Figure 884.2 Identifying additional objects within your word processor.

As you begin to identify the objects your system uses, you will find that many different parts of your program use the same types of objects. As a result, when you write your programs in terms of objects, you can easily (and quickly) *reuse* the code you write for one section in a different section of your program or maybe even in a different program. The reusability of code is one of the most powerful C++ features.

After you identify the objects, you must determine each object's purpose. To do so, think of the operations an object performs or the operations that the program performs on the object. For example, given a *file* object, a program can copy, delete, or rename the file. It is important to note that, in general, these operations apply to each file on your disk, regardless of the file's contents. These operations will become the object's *member functions*, for which you will later write C++ functions within your program. Next, identify the information you must know about the object. In the case of the file object, you must know the filename, size, protection attribute, and possibly the date and time the program created or last changed the file. These data items will become the file object's *member variables*. Conceptually, you can now view your file object, as shown in Figure 884.3.

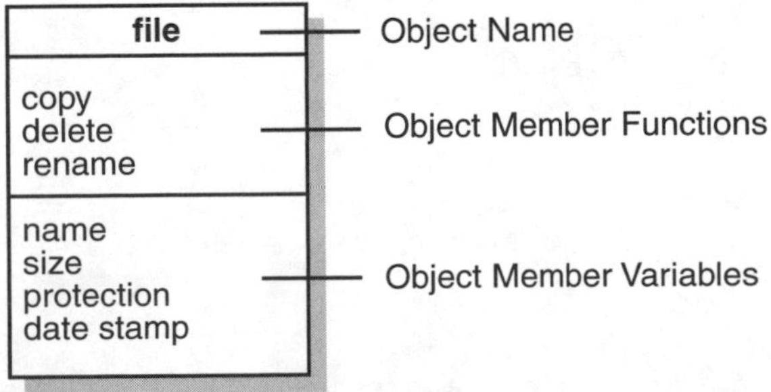

Figure 884.3 A file object's member functions and variables.

885 UNDERSTANDING OBJECTS AND CLASSES

As you read articles and books about C++ and object-oriented programming, you will encounter the terms *class* and *object*. A *class* provides a template, which defines the member functions and data members that the class type requires. An *object*, on the other hand, is an *instance*, or specific example of a class—essentially an *object* variable. You must define the class prior to the object declaration.

To declare an object variable, you simply specify the class type, followed by the object's variable name, as shown here:

```
class_name   object_name;
```

Programmers often refer to the process of creating an object as *instantiating an object* or *creating an object instance.*

886 UNDERSTANDING C++ CLASSES

Throughout this book, programs have used structures to group related data. As you have learned, C++ lets your programs use functions as structure members. You can best view a C++ *class* as a structure extension. A class, like a structure, describes a template for future variable declarations—it does not allocate memory for a variable. A class has a name (tag) and member fields. The following definition, for example, illustrates a simple class named *Book*:

```
class Book
 {
   public:
     char title[256];
     char author[64];
     float price;
     void show_title(void) { cout << title << '\n'; };
     float get_price(void) { return(price); };
 };
```

As you can see, the class definition is very similar to a structure. The only new item is the *public* label. Tip 896 discusses the public label's purpose. The following program, *1stclass.cpp*, uses the *Book* class to display information about a book:

```
#include <iostream.h>
#include <iomanip.h>
#include <string.h>

class Book
 {
   public:
     char title[256];
     char author[64];
     float price;
     void show_title(void) { cout << title << '\n'; };
     float get_price(void) { return(price); };
 };

void main(void)
 {
   Book tips;

   strcpy(tips.title, "Jamsa's C/C++/C# Programmer's Bible");
   strcpy(tips.author, "Jamsa");
   tips.price = 54.95;
   tips.show_title();
   cout << "The book's price is " << setprecision(2) << tips.get_price();
 }
```

UNDERSTANDING ENCAPSULATION 887

As you read articles and books about object-oriented programming and C++, you might encounter the term *encapsulation*. In the simplest sense, encapsulation is the combination of data and methods into a single data structure. Encapsulation groups together all the components of an object. In the "object-oriented" sense, encapsulation also defines how both the object itself and the rest of the program can reference an object's data. As you have learned, C++ classes let you divide your data into public and private sections. Programs can only access an object's private data using defined public methods. Grouping together an object's data and dividing your data into public and private sections protects the data from program misuses. In C++, the class is the fundamental tool for encapsulation.

UNDERSTANDING POLYMORPHISM 888

As you read other books and articles on C++, you will often encounter the term *polymorphism*. Polymorphism lets programs apply the same operation to objects of different types. Because polymorphism lets programmers apply the same operation to multiple types, polymorphism lets programmers use the same interface to access different objects. In C++, *virtual functions* provide access to polymorphism. In the simplest sense, a virtual function is a pointer to a function that the compiler resolves at run time. Depending on the function to which a virtual function points, the operation the program performs will differ. As a result, a single interface (the virtual function) can provide access to different operations. Tip 1090 discusses virtual functions in detail.

UNDERSTANDING INHERITANCE 889

As you derive classes using C++ inheritance support, drawing pictures might help you to understand the relationships between classes. You will find that one class you derive from one or more base classes might well become the base class for other classes. As you begin to define your classes, start with general characteristics and work toward specifics as you derive new classes. For example, if you are deriving classes for types of dogs, your first base class might simply be *Dogs*. *Dogs* would contain characteristics common to all dog breeds, such as name, origin, height, weight, and color. Your next level might become more refined when you create the classes. The second level class types, *DogsWithSpots* and *SpotlessDogs,* for example, would inherit the common characteristics that you defined in the *Dogs* base class. As you further refine pedigrees (for example, between Dalmatians and Labradors), however, you can use these second level classes as base classes for other class definitions. Your base class levels will grow, conceptually similar to a family tree's growth, as shown in Figure 889.

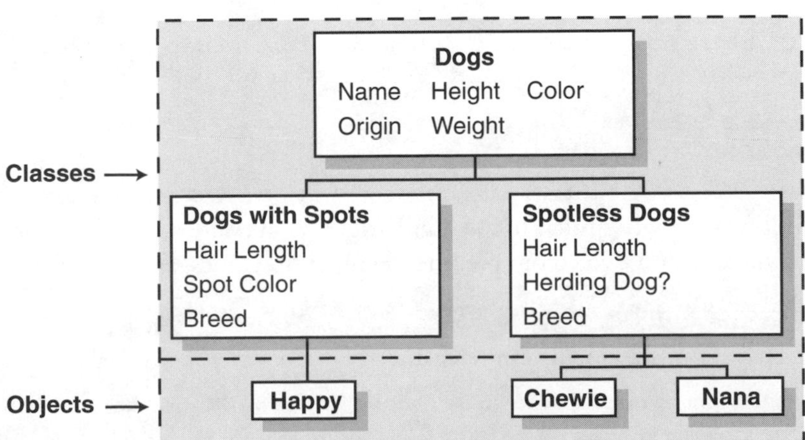

*Figure 889 The **Dogs** inheritance tree.*

890 DECIDING BETWEEN CLASSES AND STRUCTURES

Earlier in this book, you learned about C structures. As you are learning about classes, you should recognize that the syntax for working with classes is very similar to syntax you use with C structures. You may be wondering when you should use classes, as opposed to structures, or even unions. As you know, classes, structures, and unions each let your programs store related data. Your programs should use classes whenever the programs perform specific operations on the data. Consider, for example, that if you simply need to store a date, you can use a structure or a union. However, if you want the program to format and display the date, file the date, or compare two dates, you should use a class. Likewise, if you must choose between using a structure or a union, you should base your decision on the number of values the data structure must store at any given time. Finally, keep in mind that by default, class members are private, and structure and union members are public.

If you experiment with C++ structures, you will find that they support many of the same features as C++ classes, such as public and private data, member functions, and so on. As a rule, if you are creating objects, use a class.

891 CREATING A SIMPLE CLASS MODEL

The best way to understand C++ classes and objects is to create a simple program. In the next Tip, the sample program, *movies.cpp,* creates a class named *movie.* It then creates two objects of type *movie*, named *fugitive* and *sleepless*. The program defines the *movie* class, as shown here:

```
class movie
 {
   public:
    char name[64];
    char first_star[64];
    char second_star[64];
    void show_movie(void);
    void initialize(char *name, char *first, char *second);
 };
```

As you can see, the *movie* class uses three variable members and two member functions. Following the class definition, the program must define the *show_movie* and *initialize* member functions, as shown here:

```
void movie::show_movie(void)
 {
   cout << "Movie name: " << name << endl;
   cout << "Starring: " << first_star << " and " << second_star << endl << endl;
 }

void movie::initialize(char *movie_name, char *first, char *second)
 {
   strcpy(name, movie_name);
   strcpy(first_star, first);
   strcpy(second_star, second);
 }
```

The class function definitions are very similar to the standard function definitions. However, there are two primary differences. First, the class name and two colons precede the function names, as shown here:

```
void movie::initialize(char *movie_name, char *first, char *second)
```

Second, within the class function, statements can refer directly to class member variables, as shown here:

```
void movie::initialize(char *movie_name, char *first, char *second)
 {
   strcpy(name, movie_name);
```

```
    strcpy(first_star, first);
    strcpy(second_star, second);
  }
```

IMPLEMENTING THE SIMPLE CLASS PROGRAM 892

In Tip 891, you learned about the components of the simple *movie* class. Now that you understand how to create a class, you must implement the class to better understand how you will work with the class within your programs. The *movies.cpp* program implements the *movies* class, as shown here:

```
#include <iostream.h>
#include <string.h>

class movie
 {
   public:
     char name[64];
     char first_star[64];
     char second_star[64];
     void show_movie(void);
     void initialize(char *name, char *first, char *second);
 };

void movie::show_movie(void)
 {
   cout << "Movie name: " << name << endl;
   cout << "Starring: " << first_star << " and " << second_star << endl << endl;
 }

void movie::initialize(char *movie_name, char *first, char *second)
 {
   strcpy(name, movie_name);
   strcpy(first_star, first);
   strcpy(second_star, second);
 }

void main(void)
 {
   movie fugitive, sleepless;

   fugitive.initialize("The Fugitive", "Harrison Ford", "Tommy Lee Jones");
   sleepless.initialize("Sleepless in Seattle", "Tom Hanks", "Meg Ryan");
   fugitive.show_movie();
   sleepless.show_movie();
 }
```

As you can see, the program creates two objects of type *movie*:

```
movie fugitive, sleepless;
```

In the *movies.cpp* program, the program uses the *initialize* member function to initialize class member variables. In later Tips, you will learn how to use *constructor* functions to initialize member variables in a more natural way.

DEFINING THE COMPONENTS OF A CLASS 893

As you have learned, a class consists of one or more distinct components, which may be variables, functions, or both. A *class declaration* defines a new class type that links code and data. Your programs will use the new type to declare objects of that class. Therefore, a class is a logical abstraction, but an object has physical existence. In other words, an object is an *instance* of a class.

As you have seen in previous Tips, a class declaration is similar in syntax to a structure declaration. The following code shows the general form of a class:

```
class class-name
 {
  private data and functions
access-specifier:
  data and functions
access-specifier:
  data and functions
.
.
.
access-specifier:
  data and functions
} object-list;
```

The *object-list* is optional. If *object-list* is present, it declares objects of the class. The *access-specifier* is one of the three C++ class definition keywords you have previously learned: *public*, *private*, or *protected*.

The data and public functions that comprise the class are commonly referred to as *properties* and *methods*. As you have learned, methods are also known as *interface functions*.

894 UNDERSTANDING THE SCOPE RESOLUTION OPERATOR

As you have learned, your programs will use the *scope resolution operator* (the :: operator) to link a class name with a member name to tell the compiler to what class the member belongs. The scope resolution operator can also let your programs access a name in an enclosing scope that a local declaration of the same name hides. For example, consider the following code fragment:

```
// Program statements
//
int i;                     //global

void f();
  int i;                   // local

  i = 10;                  // refers to local i.
//
// Program statements
```

However, if function *f* requires a reference to global *i*, rather than local *i*, you could rewrite the fragment as shown here:

```
// Program statements
//
int i;                     //global

void f();
  int i;                   // local

  ::i = 10;                // refers to global i, not local i.
//
// Program statements
```

895 USING OR OMITTING THE CLASS NAME IN DECLARATIONS

As you have learned, a class defines a template for future variable declarations. After you define a class, your program can declare a class in one of two ways—your program can use the class itself, or your program can simply specify the class name (tag), as shown here:

```
class Book
 {
   public:
     char title[256];
     char author[64];
     float price;
     void show_title(void) { cout << title << '\n'; };
     float get_price(void) { return(price); };
 };

// Declare variables of the class type
class Book tips;
Book diary;
```

As you can see, C++ uses class tags in the same manner as it uses structure tags: to create a type with which you can later declare other variables.

UNDERSTANDING THE PUBLIC: LABEL 896

In Tip 895 you created a simple class, named *Book*, that contains the *public:* label, as shown here:

```
class Book
 {
   public:
     char title[256];
     char author[64];
     float price;
     void show_title(void) { cout << title << '\n'; };
     float get_price(void) { return(price); };
 };
```

Unlike a structure, whose members are all accessible to a program, a class can have members the program can directly access using the *dot* operator, and other members (called *private members*) that the program cannot access directly. The *public:* label identifies the class members the program can use the *dot* operator to access. If you want the program to access a specific member directly, you must declare the member within the class's public members.

UNDERSTANDING INFORMATION HIDING 897

Information hiding is the process of hiding underlying implementation details of a function, program, or a class. Information hiding lets programmers treat functions and classes as *black boxes*. In other words, if a programmer passes a value to a function, the programmer knows a specific result will occur. The programmer does not need to know how the function calculates the result, but simply that the function works. For example, most programmers do not know the mathematics behind the *tanh* function, which returns an angle's hyperbolic tangent. However, the programmers know that if they pass a specific value to the function, a known result will occur. To use the function, the programmer only must know the input parameters and the values the function will return.

In object-oriented programming, an object may have underlying implementation details. For example, Microsoft *Word*®, *Excel*®, or some other program may store data in a *document*. To use the *document* object, however, the program should not need to know the format. Instead, the program should perform read, write, print, and fax operations without knowing the object details. To help programmers hide an object's underlying details, C++ lets you divide a class definition into private and public parts. The program can directly access public data and methods, but it cannot directly access private data and methods.

898 UNDERSTANDING THE PRIVATE: LABEL

As you have learned, C++ lets you divide a class definition into public and private sections. The program can use the *dot* operator to access public data and methods. However, the program cannot use the *dot* operator to directly access the private data and methods. The following class definition expands the *Book* class to include public and private data and methods:

```
class Book
 {
   public:
     char title[256];
     char author[64];
     float price;
     void show_title(void) { cout << title << '\n'; };
     float get_price(void) { return(price); };
     void show_book(void)
       {
        show_title();
        show_publisher();
       };
     void assign_publisher(char *name) { strcpy(publisher, name); };
   private:
     char publisher[256];
     void show_publisher(void) { cout << publisher << '\n'; };
 };
```

The program can use the *dot* operator to directly access the data and methods that reside in the public section. The only way to access the private data and methods, however, is through the public methods. Tip 900 presents a program that manipulates both the public and private data.

899 USING THE PROTECTED: LABEL

As you have learned, C++ lets you categorize class members as public or private. A member's public or private categorization controls how your program can access the member. When you use inheritance to derive one class from another, C++ adds a third member category: *protected*. A *protected member* is essentially in the middle of a private and public member. For a base class, derived objects can access the protected members just as though the members were public. Outside of the derived objects, however, only public interface routines can access the protected members. The following code adds two protected members to the *Book* class definition:

```
class Book
 {
   public:
     Book(char *title) { strcpy(Book::title, title); };
     void show_title(void) { cout << title << endl; };
   protected:
     float cost;
     void show_cost(void) { cout << cost  << endl};
   private:
     char title[64];
 };
```

In the previous example, objects derived from the *Book* class can access the members *cost* and *show_cost* just as though the members were public. Outside of the derived classes, however, the program must treat the members as though they were private.

USING PUBLIC AND PRIVATE DATA 900

As you have learned, C++ lets you divide a class definition into public and private data and methods. Programs can access the public data and methods using the *dot* operator. To access the private data and methods, however, the program must call the public methods. The program cannot directly manipulate or invoke private data and methods. The following program, *pub_priv.cpp*, illustrates the use of public and private data:

```
#include <iostream.h>
#include <iomanip.h>
#include <string.h>

class Book
 {
   public:
     char title[256];
     char author[64];
     float price;
     void show_title(void) { cout << title << '\n'; };
     float get_price(void) { return(price); };
     void show_book(void)
       {
        show_title();
        show_publisher();
       };
     void assign_publisher(char *name) { strcpy(publisher, name); };
   private:
     char publisher[256];
     void show_publisher(void) { cout << publisher << '\n'; };
 };

void main(void)
 {
   Book bible;

   strcpy(bible.title, "Jamsa's C/C++ Programmer's Bible");
   strcpy(bible.author, "Jamsa");
   bible.price = 54.95;
   bible.assign_publisher("Delmar");
   bible.show_book();
 }
```

As you can see, the public method *assign_publisher* initializes the private member *publisher*. If the program had tried to access the *publisher* member directly, the compiler would have generated an error. In a similar way, the program uses the public method *show_book*, which in turn invokes the private method *show_publisher*. Again, the program cannot directly access the private method.

DETERMINING WHAT TO HIDE AND WHAT TO MAKE PUBLIC 901

As you have learned, C++ lets you divide class definitions into private and public sections. One of the most difficult tasks programmers who are new to object-oriented programming face is to determine which members of each class they should hide and which members they should make public. As a general rule, the less a program knows about a class, the better. Therefore, you should try to use private data and methods as often as possible. When you use private data and methods, programs which use the object have to use the object's public methods to access the object data. As you will learn in Tip 902, forcing programs to manipulate object data using only public methods can decrease programming errors. In other words, you usually do not want a program to manipulate an object's data directly using only the dot operator. Making use of private data in this way improves information hiding.

902 PUBLIC METHODS ARE OFTEN CALLED INTERFACE FUNCTIONS

As you learned in Tip 901, your programs should try to place most of an object's data in the private section of a class definition. When your programs place object data in private sections, other programs can access the data only by calling the class public methods. In this way, the public methods provide your program's *interface* to the object data. Using such interface functions, your programs can verify that the value the program wants to assign to a member is valid. For example, assume the member *melt_down* in the *NuclearReactor* class should only contain the values 1 through 5. If the member is public, a program can assign an invalid value using the dot operator, as shown here:

```
nuke.melt_down = 99;
```

By restricting access to the *melt_down* member to the public method *set_melt_down*, the object can verify the value, as shown here:

```
int set_melt_down(int value)
 {
   if ((value >= 1) && (value <= 5))
    {
      nuke.melt_down = value;
      return(0);
    }
   else
    return(-1);    // Invalid value
 }
```

By restricting access to object data to the public methods, the only operations a program can perform on the data within the object are those operations the object itself defines.

903 DEFINING CLASS FUNCTIONS OUTSIDE OF THE CLASS

Several preceding Tips created simple classes that defined function members within the classes themselves. As the size of your class functions increases, you will eventually define the functions outside of the class. The following program, *bookfunc.cpp*, defines the functions for the *Book* object outside of the class itself. As you will see, the program identifies the class functions by preceding each function name with the class name and double colons, as shown here:

```
#include <iostream.h>
#include <iomanip.h>
#include <string.h>

class Book
 {
  public:
    char title[256];
    char author[64];
    float price;
    void show_title(void);
    float get_price(void);
    void show_book(void);
    void assign_publisher(char *name);
  private:
    char publisher[256];
    void show_publisher(void);
 };

void Book::show_title(void)
  { cout << title << '\n'; };

float Book::get_price(void)
  { return(price); };
```

```
void Book::show_book(void)
 {
   show_title();
   show_publisher();
 };

void Book::assign_publisher(char *name)
  { strcpy(publisher, name); };

void Book::show_publisher(void)
  { cout << publisher << '\n'; };

void main(void)
 {
   Book tips;

   strcpy(tips.title, "Jamsa's C/C++/C# Programmer's Bible");
   strcpy(tips.author, "Jamsa");
   tips.price = 54.95;
   tips.assign_publisher("Delmar");
   tips.show_book();
 }
```

DEFINING METHODS INSIDE AND OUTSIDE OF CLASSES 904

As you have learned, C++ lets you define methods inside and outside of the class declaration. The decision you make about where to define the method function affects the code the compiler creates for the program. When you define a method within the class, the compiler will treat each method reference as an *inline* function call, placing the function's corresponding instructions into the object code at each method reference. As you have learned, using inline code can improve your program's performance, but it also can also increase your program size. When you define a function outside of the class, on the other hand, the compiler does not use inline code. Instead, the compiler will generate code for a function that the program will call at each method reference. Therefore, if your class has a common operation that is small, you might want the compiler to generate inline code for the method. If the method is larger, do not direct the compiler to generate inline code.

UNDERSTANDING OBJECT INSTANCES 905

Many C++ books and articles refer to *object instances.* In short, an object instance is an object variable. As you have learned, a class defines a template for future variable declarations. When you later declare an object, you create an object instance. In other words, when the compiler allocates memory to a variable, your program creates an object instance. All instances of the same class have the same characteristics. For the purposes of this book, an instance is a variable of a specific class.

OBJECT INSTANCES SHOULD SHARE CODE 906

As you have learned, C++ lets you define class methods within the class or outside of the class. When you declare class methods outside of the class, instances share the same copy of the methods. If, for example, you have a class with three methods, and you create 100 instances of that class, your program will contain only the three methods. If you include code inline, however, instances will not share the code. Therefore, you should reserve inline code for small, commonly performed operations, where operation performance is more important than program size. For example, the following program, *shrbooks.cpp*, creates two instances of the *Book* class:

```
#include <iostream.h>
#include <iomanip.h>
#include <string.h>

class Book
 {
  public:
    char title[256];
    char author[64];
    float price;
    void show_title(void) { cout << title << '\n'; };
    float get_price(void) { return(price); };
    void show_book(void);
    void assign_publisher(char *name) { strcpy(publisher, name); };
  private:
    char publisher[256];
    void show_publisher(void) { cout << publisher << '\n'; };
 };

void Book::show_book(void)
 {
   show_title();
   show_publisher();
 };

void main(void)
 {
    Book tips, diary;

    strcpy(tips.title, "Jamsa's C/C++/C# Programmer's Bible");
    strcpy(tips.author, "Jamsa");
    tips.price = 54.95;
    tips.assign_publisher("Delmar");

    strcpy(diary.title, "All My Secrets...");
    strcpy(diary.author, "Kris Jamsa");
    diary.price = 9.95;
    diary.assign_publisher("None");

    tips.show_book();
    diary.show_book();
 }
```

If you compile the *shrbooks.cpp* program using Borland's *Turbo C++ Lite* to produce an assembly language listing, you will find that the instances do not share code for inline methods, but do share the code defined outside of the class.

907 Accessing Class Members

In previous Tips, you have used the dot operator to invoke class member functions. When your programs place object members after a *public* label, your programs can access the members using the dot operators. For example, in Tip 892 you created a simple *movie* class and accessed its *show_movie* member function. The following program, *public.cpp*, uses the *initialize* function to assign values to the members of the objects *fugitive* and *sleepless*. The program then displays the different member values by referencing the member using the dot operator, as shown here:

```
#include <iostream.h>
#include <string.h>

class movie
 {
    public:
       char name[64];
       char first_star[64];
```

```
    char second_star[64];
    void show_movie(void);
    void initialize(char *name, char *first, char *second);
 };

void movie::show_movie(void)
 {
   cout << "Movie name: " << name << endl;
   cout << "Starring: " << first_star << " and " << second_star << endl << endl;
 }

void movie::initialize(char *movie_name, char *first, char *second)
 {
   strcpy(name, movie_name);
   strcpy(first_star, first);
   strcpy(second_star, second);
 }

void main(void)
 {
   movie fugitive, sleepless;

   fugitive.initialize("The Fugitive", "Harrison Ford", "Tommy Lee Jones");
   sleepless.initialize("Sleepless in Seattle", "Tom Hanks", "Meg Ryan");
   cout << "The last two movies I've watched are: " << fugitive.name << " and "
        << sleepless.name << endl;
   cout << "I thought " << fugitive.first_star << " was great!" << endl;
 }
```

Because the class members are *public*, programs can access the members directly. When you compile and execute the *public.cpp* program, your screen will display the following:

```
The last two movies I've watched are: The Fugitive and Sleepless in Seattle
I thought Harrison Ford was great!
C:\>
```

When a class defines member variables as public, your programs can access the member variables using the dot operators. However, as you will learn in later Tips, such direct variable member access is not always preferable.

908 Reviewing the Global Resolution Operator

As you have learned, C++ lets you repeat the names of functions and variables throughout multiple classes. In other words, the existence of the *next_day* function in the *week* class does not preclude its existence within the *month* class. Within your programs, you can use the *global resolution operator* (::) to protect against data and function name confusion. When you must refer to a class member (either data or function), you simply precede the member name with the class name and two colons, as shown here:

```
void movie::initialize(char *name, char *first_star, char *second_star)
 {
   strcpy(movie::name, name);
   strcpy(movie::first_star, first_star);
   strcpy(movie::second_star, second_star);
 }
```

909 Initializing Class Values

As you have seen in the previous Tips, it is important to initialize values within your classes each time you instantiate a new object (that is, create a new instance) from the class. C++ provides several ways to initialize values within your classes. In later Tips you will learn about constructor functions, which C++ programmers commonly use to initialize new class instances. However, you can also create a member function to initialize a class, as shown here:

```
void movie::initialize(char *name, char *first_star, char *second_star)
 {
   strcpy(movie::name, name);
   strcpy(movie::first_star, first_star);
   strcpy(movie::second_star, second_star);
 }
```

910 USING ANOTHER METHOD TO INITIALIZE CLASS VALUES

As you have learned, you can initialize class values within an initialization function. As you will learn in later Tips, you can also initialize your classes within constructor functions. However, as you examine C++ programs, you might encounter a unique member initialization technique. Assume, for example, that you want the *counter* constructor to initialize the variable *count* to 0, as shown here:

```
counter::counter(void)
 {
    count = 0;
    // Other statements
 }
```

Rather than forcing you to create a separate initialization function for each class, C++ lets you initialize class member variables by placing the variable name and the desired value after a colon and before the function statements, as shown here:

```
counter::counter(void) : count(0)
 {
    // Other statements
 }
```

The following program, *con_init.cpp*, uses the constructor initialization format to initialize three member variables to the values 1, 2, and 3:

```
#include <iostream.h>

class object
 {
   public:
     object::object(void);
     void show_object(void);
   private:
     int a;
     int b;
     int c;
 };

object::object(void) : a(1), b(2), c(3) { };

void object::show_object(void)
 {
   cout << "a contains: " << a << endl;
   cout << "b contains: " << b << endl;
   cout << "c contains: " << c << endl;
 }

void main(void)
 {
   object numbers;

   numbers.show_object();
 }
```

When you compile and execute the *con_init.cpp* program, your screen will display the following output:

```
a contains 1
b contains 2
c contains 3
C:\>
```

UNDERSTANDING STATIC CLASS MEMBERS 911

Within your C++ classes, you can define both data members and functions as *static*. As you will learn, declaring a data member or a function as static has important implications for your C++ classes. C++ governs static data members and functions by different rules than it does normal members and functions. For example, a static function can only access other static members within the same class (as well as global functions and data members). Before you use static members within your classes, it is important to understand the implications. Tips 912 and 913 discuss static data members and functions in detail.

USING STATIC DATA MEMBERS 912

As you learned in Tip 911, the C++ compiler treats data members that you precede with the *static* keyword differently than it does normal data members. In fact, when you precede a member variable's declaration with the *static* keyword, you are telling the compiler that only one copy of that variable will exist and that all objects of the class will share that variable. Unlike regular data members, the program does not create individual copies of a static member for each object. No matter how many objects of a class the program creates, only one copy of each static member variable exists. Therefore, all objects of that class use the same variable. The compiler initializes all static variables to zero when the program creates the first object instance.

When you declare a static data member within a class, you do not *define* that member. In other words, you do not allocate memory storage for the member. Instead, you must provide a global definition for the static data member elsewhere, outside the class. To provide the global definition, you will redeclare the static data member using the scope resolution operator. Doing so instructs the compiler to allocate storage for the static member. To better understand the usage and effect of a static data member, consider the following program, *stat_mem.cpp*:

```
#include <iostream.h>

class shared
 {
    static int a;
     int b;
   public:
     void set(int i, int j)
       {
         a=i;
         b=j;
       }
     void show();
 };

int shared::a;          //     Define the a global variable

void shared::show()
 {
   cout << "This is static a: " << a << endl;
   cout << "This is non-static b: " << b << endl;
 }

void main(void)
 {
   shared x, y;
```

```
  x.set(1,1);
  x.show();
  y.set(2,2);
  y.show();
  x.show();
 }
```

When you compile and execute the *stat_mem.cpp* program, it will generate the following output:

```
This is static a: 1
This is non-static b: 1
This is static a: 2
This is non-static b: 2
This is static a: 2
This is non-static b: 1
```

913 USING STATIC MEMBER FUNCTIONS

As you have learned, you can declare data members within your classes as static. You can also declare member functions within your classes as static. The C++ compiler restricts functions that you declare as static in several ways:

1. Static functions may access only other static members of the class.
2. Static member functions do not have a *this* pointer.
3. You cannot overload a static function with a non-static function, or vice versa.

The following program, *stat_fun.cpp*, is a reworked version of the *stat_mem.cpp* program which appears in Tip 912. The *stat_fun.cpp* program declares *show* as static, so that the program can access *show* either by itself, using only the class resolution operator, or in connection with a single object, as shown here:

```
#include <iostream.h>

class shared
 {
    static int a;
    int b;
  public:
    void set(int i, int j)
     {
       a=i;
       b=j;
     }
    static void show();
 };

int shared::a;          //    Define the a global variable

void shared::show()
 {
  cout << "This is static a: " << a << endl;
 }

void main(void)
 {
  shared x, y;

  x.set(1,1);
  y.set(2,2);
  shared::show();
  y.show();
  x.show();
 }
```

UNDERSTANDING MEMBER FUNCTION DECLARATIONS 914

As you have learned, a class contains member variables and member functions. When you define class functions, you can define the functions outside of the class definition, as shown here:

```
class movie
 {
   public:
     char name[64];
     char first_star[64];
     char second_star[64];
     void show_movie(void);
     void initialize(char *name, char *first, char *second);
 };
void movie::show_movie(void)
 {
   cout << "Movie name: " << name << endl;
   cout << "Starring: " << first_star << " and " << second_star << endl << endl;
 }
void movie::initialize(char *movie_name, char *first, char *second)
 {
   strcpy(name, movie_name);
   strcpy(first_star, first);
   strcpy(second_star, second);
 }
```

In this case, the class definition must contain prototypes that describe each class member function. Likewise, the function definitions must specify the class name prior to the function name.

USING INLINE FUNCTION DECLARATIONS 915

In Tip 914, you learned how to define class functions outside of a class definition. You can also define your class member functions within the class, actually placing the function statements within the class declaration. For example, the following program, *inline.cpp*, defines the class member functions inline, within the class declaration:

```
#include <iostream.h>
#include <string.h>
class movie
 {
   public:
     char name[64];
     char first_star[64];
     char second_star[64];
     void show_movie(void)
      {
        cout << "Movie name: " << name << endl;
       cout << "Starring: " << first_star << " and " << second_star << endl << endl;
      }
   void initialize(char *movie_name, char *first, char *second)
    {
      strcpy(name, movie_name);
      strcpy(first_star, first);
      strcpy(second_star, second);
    }
 };
void main(void)
```

```
{
  movie fugitive, sleepless;

  fugitive.initialize("The Fugitive", "Harrison Ford", "Tommy Lee Jones");
  sleepless.initialize("Sleepless in Seattle", "Tom Hanks", "Meg Ryan");
  cout << "The last two movies I've watched are: " << fugitive.name <<
         " and " << sleepless.name << endl;
  cout << "I thought " << fugitive.first_star << " was great!" << endl;
}
```

916 Determining When to Use Inline and Out-of-Line Functions

As you learned in Tip 915, when you declare a member function inline, the function statements reside within the class itself. An advantage of declaring member functions inline is that inline functions help consolidate the entire class at one location within your program code. Unfortunately, using inline functions in this way also increases the size and complexity of your class definitions. Simply put, the larger your class definitions become, the more difficult your definitions are to understand. In addition, similar object types do not share the code for inline functions.

On the other hand, when you define member functions outside of a class, the C++ compiler creates one copy of each function's instructions. Every object your programs later create from that class use the single copy of the function. In other words, if you create 1,000 objects, each object shares the single copy of the function code. Such function sharing is desirable because it significantly reduces your program's memory overhead.

917 Understanding Classes and Unions

As you have learned, C++ structures are essentially C++ classes. In the same fashion, you can also use C++ unions to declare a class. Unions may also include constructor and destructor functions. A union in C++ retains all its C-like features (which you learned about in Tips 481 through 487), including the characteristic that forces all data elements into the same location in memory. Like a structure, and unlike a class, union members are public by default. As you have learned, one of the best uses for unions is to manipulate numbers using bitwise operations. The following program, *un_class.cpp*, uses the *swap_byte* union to manipulate numbers using bitwise operations:

```
#include <iostream.h>

union swap_byte
 {
   void swap();
   void set_byte(unsigned i);
   void show_word();

   unsigned u;
   unsigned char c[2];
 };

void swap_byte::swap()
 {
   unsigned char t;

   t = c[0];
   c[0] = c[1];
   c[1] = t;
 }

void swap_byte::show_word()
 {
   cout << u;
 }

void swap_byte::set_byte(unsigned i)
 {
```

```
    u = i;
  }

void main(void)
  {
    swap_byte b;

    b.set_byte(49034);
    b.swap();
    b.show_word();
  }
```

INTRODUCING ANONYMOUS UNIONS 918

C++ supports a special type of union called an *anonymous union*. An anonymous union does not contain a type name, and your program cannot declare variables from an anonymous union. Instead, an anonymous union tells the compiler that the member variables of the union are to share the same location. However, the program will refer to the variables directly, without the normal dot operator syntax. To understand anonymous unions better, consider the *anon_un.cpp* program, as shown here:

```
#include <iostream.h>
#include <string.h>

void main(void)
  {
   // define anonymous union
   union
     {
      long l;
      double d;
      char s[4];
     };

   // now, your program can reference elements directly

   l = 100000;
   cout << l << " ";
   d = 123.2342;
   cout << d << " ";
   strcpy(s, "hi");
   cout << s;
  }
```

INTRODUCING FRIEND FUNCTIONS 919

It is possible to grant a non-member function access to the private members of a class. To do this within your programs, you can declare a *friend* function. Friend functions have access to all private and protected members of the class for which they are a friend. To declare a friend function, include its prototype within the class, preceding it with the keyword *friend*, as shown in the *frd_fun.cpp* program:

```
#include <iostream.h>

class sample
  {
   int a, b;
   public:
     friend int sum(sample object);
     void set_ab(int i, int j);
  }

void sample::set_ab(int i, int j)
  {
```

```
   a = i;
   b = j;
 }
int sum(sample object)
 {
   /* Because sum is a friend of sample, it can access a and b directly */
   return object.a + object.b;
 }
void main(void)
 {
   sample integer;
   cout << "Adding 3 and 4:" << endl;
   integer.set_ab(3,4);
   cout << "Result = " << sum(integer);
 }
```

When you compile and execute the *frd_fun.cpp* program, the program will display the following output:

```
Adding 3 and 4:
Result = 7
C:\>
```

Although there is no particular reason to make *sum* a friend function and not a member function in the previous case, there are generally three good reasons for using a friend function within your classes:

1. Friends can be useful when your program overloads certain operators because of the added control friend functions give you over the overloaded operator's actions.
2. Friend functions make the creation of some I/O functions easier.
3. Friend functions may also be useful in cases where two or more classes contain members interrelated relative to other program parts, as they can help you avoid declaring multiple functions with identical program code.

In later Tips, you will learn more about how to use friend functions in each of these cases.

920 INTRODUCING FRIEND CLASSES

As you have learned, a class can contain public and private data and methods. Usually, the only way to access private members is through the public or interface methods. As your programs begin to work with more than one type of object, there may be times when one object calls another object or uses another object's data members. In previous Tips, for example, the *Reader* object used the *Book* object method *show_book* to display the title of a book. The only way the *Reader* object could access the *Book* object's private data was through the *show_book* method. Depending on your program, there may be times when you want one object to have access to another object's public and private data. In such cases, you can specify a *friend object*. Given the previous *Reader* and *Book* program, the *Book* object could declare the *Reader* object as a friend. The *Reader* object could then directly access the *Book* object's private data, displaying the book's title without having to call the *show_title* method. The remainder of the program code could not directly access the *Book* object's private data. The only object that could access the private data would be the *Book* object's friend, the *Reader* object. Before you can specify a friend, however, you should tell the compiler about the friend's class, as later Tips detail.

921 UNDERSTANDING CONSTRUCTOR FUNCTIONS

When your program creates an object instance, the program will then usually assign initial values to the object data members. To simplify the process of initializing object members, C++ supports a special function, called a constructor, that automatically executes each time your program creates a class instance. The constructor function is a public method that uses the same name as the class. For example, using the *Book* class, the constructor function would have the name *Book*, as shown here:

```
class Book
 {
   public:
     Book(char *title, char *author, char *publisher, float price); // Constructor
     char title[256];
     char author[64];
     float price;
     void show_title(void) { cout << title << '\n'; };
     float get_price(void) { return(price); };
     void show_book(void)
      {
        show_title();
        show_publisher();
      };
     void assign_publisher(char *name) { strcpy(publisher, name); };
   private:
     char publisher[256];
     void show_publisher(void) { cout << publisher << '\n'; };
 };
```

Your programs can define the constructor function within the class itself or outside of the class. When your program later declares an object, the program can pass parameters to the constructor function. Next, the constructor function will execute automatically. You can pass parameters to the constructor as shown here:

```
Book tips("Jamsa's C/C++/C# Programmer's Bible", "Jamsa", "Delmar",
          54.95);
```

Tip 922 presents a program that uses a constructor function to initialize instances of the *Book* class.

USING CONSTRUCTOR FUNCTIONS WITH PARAMETERS 922

In Tip 921, you learned that a program can pass parameters to the constructor function. It is also possible to pass arguments to constructor functions. Typically, your programs will use these arguments to help initialize an object which the program creates. To create a *parameterized constructor*, you simply add parameters to the constructor's declaration as you would to any other function. When you define the constructor's body, use the parameters to initialize the object. For example, the following class declaration initializes the *Book* class within the class constructor:

```
class Book
 {
   public:
     char title[256];
     char author[64];
     float price;
     void Book(char *title, char *author, char *publisher, float price);
     void show_title(void) { cout << title << '\n'; };
     float get_price(void) { return(price); };
     void show_book(void)
      {
        show_title();
        show_publisher();
      };
     void assign_publisher(char *name) { strcpy(publisher, name); };
   private:
     char publisher[256];
     void show_publisher(void) { cout << publisher << '\n'; };
 };
```

If you do not use a parameterized constructor function within a class, you can always initialize the values within the class after the program constructs the object.

923 Using a Constructor Function

As you have learned, a constructor function is a special class function that automatically executes when you create an instance of a class. Programs normally use constructor functions to initialize member values. The following program, *construc.cpp*, uses the constructor function *Book* to initialize members of instances of the *Book* class:

```
#include <iostream.h>
#include <iomanip.h>
#include <string.h>

class Book {
  public:
    char title[256];
    char author[64];
    float price;
    Book(char *btitle, char *bauthor, char *bpublisher, float bprice);
    void show_title(void) { cout << title << '\n'; };
    float get_price(void) { return(price); };
    void show_book(void)
     {
       show_title();
       show_publisher();
     };
    void assign_publisher(char *name) { strcpy(publisher, name); };
  private:
    char publisher[256];
    void show_publisher(void) { cout << publisher << '\n'; };
};

Book::Book(char *btitle, char *bauthor, char *bpublisher, float bprice)
 {
   strcpy(title, btitle);
   strcpy(author, bauthor);
   strcpy(publisher, bpublisher);
   price = bprice;
 }

void main(void)
 {
   Book tips("Jamsa's C/C++/C# Programmer's Bible", "Jamsa",
             "Delmar", 54.95);
   Book diary("All My Secrets...", "Kris Jamsa", "None", 9.95);
   tips.show_book();
   diary.show_book();
 }
```

In the *construc.cpp* program, the constructor function *Book* precedes each of its parameter names with the letter *b* to distinguish the names from the class members. As you have learned, however, your programs can precede variable names with the class name to resolve name conflicts.

924 Understanding When a Program Executes a Constructor

As a general rule, the compiler will call an object's constructor when the program code declares the object. However, the actual time when the compiler calls the constructor code may vary depending upon the class type and location within the project. For example, a local object's constructor function executes when the program counter encounters the object's declaration statement. Further, if your program creates two or more objects within the same statement, the program will execute the constructor functions for each object in declaration order, from left to right.

For global objects, the program executes the constructor function before *main* begins its execution. Just as with local objects, global constructors execute in order, left to right and top to bottom. It is impossible to know the order of execution for a series of global constructors spread across multiple source code files. The following program, *show_con.cpp*, shows the execution of constructor functions:

```
#include <iostream.h>

class sample
 {
 public:
   int who;
   sample(int id);
 } global_obj1(1), global_obj2(2);

sample::sample(int id)
 {
   cout << "Initializing " << id << "\n";
   who = id;
 }

void main(void)
 {
   sample local_obj(3);
   cout << "This is NOT the first line displayed.\n";
   sample local_obj2(4);
 }
```

Note: *A constructor function is a special class function that automatically executes when the program creates an instance. Constructor functions do not return a value. However, you do not define constructor functions as returning type* ***void****. Instead, the C++ compiler can determine that the function is a constructor by how you use the function. By definition, you cannot return a value from a constructor.*

USING CONSTRUCTOR FUNCTIONS WITH PARAMETERS 925

As you learned in Tip 921, it is possible to pass arguments to constructor functions. Typically, your programs will use these arguments to help initialize an object when the program creates it. To create a parameterized constructor, you add parameters to the constructor's declaration as you would to any other function. When you define the constructor's body, use the parameters to initialize the object. For example, the following class declaration initializes objects of the *Book* class:

```
Book::Book(char *btitle, char *bauthor, char *bpublisher, float bprice)
 {
   strcpy(title, btitle);
   strcpy(author, bauthor);
   strcpy(publisher, bpublisher);
   price = bprice;
 }
```

If the constructor function within the previous example is the only constructor function you create for the *book* object, your programs must declare each instance with values within the declaration which correspond to the *btitle, bauthor, bpublisher,* and *bprice* parameters. If you fail to do so, the compiler will return an error.

926 Resolving Name Conflicts in a Constructor Function

In previous Tips, you created the *Book* constructor function, then modified it to initialize members for instances of the *Book* class. To differentiate between parameter and class member names, the program preceded each parameter name with the letter b, as shown here:

```
Book::Book(char *btitle, char *bauthor, char *bpublisher, float bprice)
 {
   strcpy(title, btitle);
   strcpy(author, bauthor);
   strcpy(publisher, bpublisher);
   price = bprice;
 }
```

In the previous case, the parameter names *title*, *author*, *publisher*, and *price* are more meaningful and preferable than *btitle*, *bauthor*, *bpublisher*, and *bprice*. However, because the parameter names without the initial *b* conflict with member names, the function must resolve them using the class name and double colons, as shown here:

```
Book::Book(char *title, char *author, char *publisher, float price)
 {
   strcpy(Book::title, title);
   strcpy(Book::author, author);
   strcpy(Book::publisher, publisher);
   Book::price = price;
 }
```

This book's companion CD-ROM contains the program *constr2.cpp*, which presents the complete program source code you use to access objects of the type *Book* when you use the constructor function for the *Book* class, as just shown.

927 Using a Constructor to Allocate Memory

As you have learned, constructor functions let your programs initialize member variables. If the member variable uses arrays, the constructor function can allocate the memory amount you want. For example, the following program, *cons_new.cpp*, uses the *new* operator within the *Book* constructor function to allocate memory for the character string arrays:

```
#include <iostream.h>
#include <iomanip.h>
#include <string.h>
#include <stdlib.h>

class Book {
  public:
    char *title;
    char *author;
    float price;
    Book(char *title, char *author, char *publisher, float price);
    void show_title(void) { cout << title << '\n'; };
    float get_price(void) { return(price); };
    void show_book(void)
     {
       show_title();
       show_publisher();
     };
    void assign_publisher(char *name) { strcpy(publisher, name); };
  private:
    char *publisher;
    void show_publisher(void) { cout << publisher << '\n'; };
};
```

```
Book::Book(char *title, char *author, char *publisher, float price)
 {
   if ((Book::title = new char[256]) == 0)
     {
       cerr << "Error allocating memory\n";
       exit(0);
     }
   if ((Book::author = new char[64]) == 0)
     {
        cerr << "Error allocating memory\n";
       exit(0);
     }
   if ((Book::publisher = new char[128]) == 0)
     {
       cerr << "Error allocating memory\n";
       exit(0);
     }
   strcpy(Book::title, title);
   strcpy(Book::author, author);
   strcpy(Book::publisher, publisher);
   Book::price = price;
 }

void main(void)
 {
    Book tips("Jamsa's C/C++/C# Programmer's Bible", "Jamsa",Delmar",
      54.95);
   Book diary("All My Secrets...", "Kris Jamsa", "None", 9.95);

   tips.show_book();
   diary.show_book();
 }
```

HANDLING MEMORY ALLOCATION IN A CLEANER WAY 928

In Tip 927, the *cons_new.cpp* program used the *new* operator within a constructor function to allocate memory for character string members. The code for each memory allocation was similar, as shown here:

```
   if ((Book::title = new char[256]) == 0)
     {
       cerr << "Error allocating memory\n";
       exit(0);
     }
   if ((Book::author = new char[64]) == 0)
     {
       cerr << "Error allocating memory\n";
       exit(0);
     }
   if ((Book::publisher = new char[128]) == 0)
     {
       cerr << "Error allocating memory\n";
       exit(0);
     }
```

One way to reduce the amount of duplicate code is to try to allocate memory for each variable and then test after the last allocation to see whether the allocations were successful, as shown here:

```
Book::title = new char[256];
Book::author = new char[64];
Book::publisher = new char[128];

if ((Book::title && Book::author && Book::publisher) == 0)
  {
     cout << "Error allocating memory\n";
     exit(1);
  }
```

A second way to reduce the code is first to assign a custom handler that will display the error message and exit when the free space can satisfy the request. You learned how to assign a custom handler in Tip 871, which discussed the *set_new_handler* function in detail.

929 DEFAULT PARAMETER VALUES FOR CONSTRUCTORS

A constructor function is a special class method that automatically executes when your program creates an instance of an object. As you have learned, C++ lets your programs provide default values for function parameters. Constructor functions are no exception. The following program, *def_cons.cpp*, uses the default values 1, 2, and 3 for members of the class *MagicNumbers*:

```
#include <iostream.h>
#include <iomanip.h>

class MagicNumbers {
 public:
   MagicNumbers(int a = 1, int b = 2, int c = 3)
     {
        MagicNumbers::a = a;
        MagicNumbers::b = b;
        MagicNumbers::c = c;
     };
   void show_numbers(void)
     {
       cout << a << ' ' << b << ' '<< c << '\n';
     };
 private:
   int a, b, c;
};

void main(void)
 {
   MagicNumbers one(1, 1, 1);
   MagicNumbers defaults;
   MagicNumbers happy(101, 101, 101);

   one.show_numbers();
   defaults.show_numbers();
   happy.show_numbers();
 }
```

As you can see, the instances *one* and *happy* specify their own member values. The instance named *defaults*, however, uses the default values 1, 2, and 3. By providing default values to your constructor function in this way, you can ensure that your program always initializes class members to meaningful values.

930 OVERLOADING CONSTRUCTOR FUNCTIONS

As you have learned, a constructor function is a special class method that automatically executes when your program creates an instance of an object. As you have learned, C++ lets your programs overload functions so that the C compiler will decide which function to invoke, depending on the parameters passed. Constructor functions are no exception. The following program, *needvals.cpp*, provides two constructor functions for the class *Book*. The first constructor function assigns the values passed as parameters. The second constructor first displays a message stating that the program must provide initial values for each parameter, and then exits. In the *needvals.cpp* program, the second constructor executes only if the program tries to execute a function without specifying initial values, as shown here:

```
#include <iostream.h>
#include <iomanip.h>
#include <string.h>
#include <stdlib.h>

class Book {
  public:
    Book(char *title, char *author, char *publisher, float price);
    Book(void);
    void show_title(void) { cout << title << '\n'; };
    float get_price(void) { return(price); };
    void show_book(void)
     {
       show_title();
       show_publisher();
     };
  private:
    char title[256];
    char author[64];
    float price;
    char publisher[256];
    void show_publisher(void) { cout << publisher << '\n'; };
};

Book::Book(char *title, char *author, char *publisher, float price)
 {
   strcpy(Book::title, title);
   strcpy(Book::author, author);
   strcpy(Book::publisher, publisher);
   Book::price = price;
 }

Book::Book(void)
 {
   cerr << "You must specify initial values for Book instance\n";
   exit(1);
 }
void main(void)
 {
   Book tips("Jamsa's C/C++ Programmer's Bible", "Jamsa",
             "Delmar", 54.95);
   Book diary;

   tips.show_book();
   diary.show_book();
 }
```

FINDING THE ADDRESS OF AN OVERLOADED FUNCTION 931

As you have learned, you can assign the address of a function to a pointer and then invoke that pointer to call the function. However, when you overload functions within your programs, getting a function's address is more complex. To understand better why getting a function's address is more difficult with overloaded functions, consider the following statement, which assigns the address of the function *somef* to a pointer named *p:*

```
p = somef;
```

If the *somef* function has no overloaded functions, the previous statement alone is sufficient for the address assignment. If, on the other hand, there are multiple *somef* functions, the compiler will refuse to compile because it cannot resolve the reference. The answer to this problem is in the declaration of the variable itself, as shown in the following program, *over_pt.cpp*:

```
#include <iostream.h>

int somef(int a);
int somef(int a, int b);

void main(void)
 {
   int (*fp)(int a);    // pointer to int xxx(int)

   fp = somef;          // points to somef(int)
   cout << fp(5);
 }

int somef(int a)
 {
   return a;
 }

int somef(int a, int b)
 {
   return a*b;
 }
```

Because the program declares *fp* as a pointer to a function that returns an *int* value and receives a single *int* parameter, the compiler can resolve the pointer when another program statement references the pointer within the program. If, on the other hand, the *fp* declaration were as follows, the pointer would point to the second function:

```
int (*fp)(int a, int b);
```

932 USING CONSTRUCTOR FUNCTIONS WITH A SINGLE PARAMETER

In previous Tips, you have learned how to parameterize constructor functions to initialize the data members of a class each time the program creates an instance of that class. However, your programs can also manipulate constructor functions with only a single parameter, as if the class declaration were a normal assignment to a type. To better understand how your programs will implement constructor functions with a single parameter, consider the *sing_par.cpp* program, shown here:

```
#include <iostream.h>

class simple
 {
   int a;
   public:
     simple(int j) {a = j;}
     int geta() {return a;}
 };

void main(void)
 {
   simple ob = 99;                  // passes 99 to j

   cout << ob.geta();
 }
```

933 UNDERSTANDING DESTRUCTOR FUNCTIONS

As you have learned, each time you create an object instance, your program can automatically execute a constructor function that you can use to initialize the instance members. In a similar way, C++ lets you define a destructor function that automatically runs when the program destroys the instance. Destructor functions typically run at one of two times: either when your program ends or when you use the *delete* operator to free memory previously allocated to hold an instance. Destructor functions have the same name as the class. You will differentiate destructors from constructors with the tilde character (~), which must precede every destructor function's name. For example, the following code fragment shows declarations for a function's constructor and destructor:

```
Book(char *title, char *author, char *publisher, float price);
~Book(void);
```

As you can see, a destructor function accepts no parameters and, similar to constructor functions, you will declare the destructor function with no return value. You will learn more about destructor functions in the following Tips.

USING A DESTRUCTOR FUNCTION 934

In Tip 933, you learned about the constructor function's counterpart, the destructor function. C++ automatically calls the destructor function each time a program destroys a class instance. To understand this processing better, consider the following program, *destruct.cpp*, which creates a simple destructor function that displays a message stating the program is destroying an instance. The program automatically invokes the destructor function for each instance as the program ends, as shown here:

```
#include <iostream.h>
#include <iomanip.h>
#include <string.h>

class Book {
  public:
    char title[256];
    char author[64];
    float price;
    Book(char *title, char *author, char *publisher, float price);
    ~Book(void);
    void show_title(void) { cout << title << '\n'; };
    float get_price(void) { return(price); };
    void show_book(void)
     {
       show_title();
       show_publisher();
     };
    void assign_publisher(char *name) { strcpy(publisher, name); };
  private:
    char publisher[256];
    void show_publisher(void) { cout << publisher << '\n'; };
};

Book::Book(char *title, char *author, char *publisher, float price)
 {
   strcpy(Book::title, title);
   strcpy(Book::author, author);
   strcpy(Book::publisher, publisher);
   Book::price = price;
 }

Book::~Book(void)
 {
   cout << "Destructing the instance " << title << '\n';
 }

void main(void)
 {
    Book tips("Jamsa's C/C++/C# Programmer's Bible", "Jamsa",
          "Delmar", 54.95);
```

```
   Book diary("All My Secrets...", "Kris Jamsa", "None", 9.95);
   tips.show_book();
   diary.show_book();
}
```

935 Understanding Why You Should Use Destructor Functions

In Tips 933 and 934, you learned about destructor functions. As you have learned, C++ automatically calls the destructor function each time your programs discard a class instance, just as it automatically calls the constructor function each time your programs create a class instance. In many cases, you will find the destructor function performs no special processing. However, as your programs become more complex, you will find two general situations where a class must have a destructor function.

As a general rule, destructor functions are most important to your program structure when your classes allocate memory dynamically. If your class creates all data members, arrays, structures, and so on at the creation of each instance, C++ will automatically handle much of the destruction processing. If, on the other hand, your objects allocate memory as they go (for example, a linked-list object), you should make sure your program releases that memory (a process which programmers often refer to as *garbage collection).*

Additionally, if your program uses a series of linked objects, the destructor function will help you maintain the list after the destruction of an object within the list. As you learned, in a singly linked list, each object maintains the address of the object following it within the list. In a doubly linked list, each object maintains the address of both the object following it and the object preceding it within the list. Each time you delete an element (or node) from within a linked list, your program must update the links within the list to avoid a broken list. As you will learn in later Tips, you can perform much of the housekeeping for a linked list within a destructor function.

936 Understanding When a Program Invokes a Destructor Function

When you create destructor functions for your programs, you should understand when your programs will invoke the destructor functions that you create for your classes. Simply put, C++ invokes the object's destructor function just before it discards the object. To understand better how destructor functions work, consider Figure 936, a simple logical sketch of an object's life cycle.

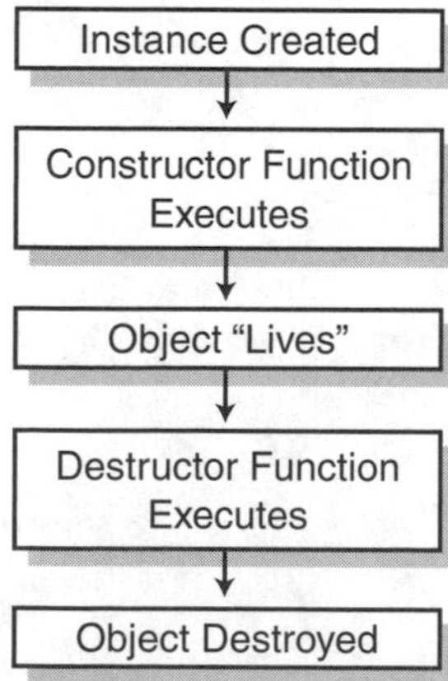

Figure 936 *The logical model of an object's lifetime.*

Therefore, within your programs, you should ensure both that your destructor functions perform only appropriate activities on the about-to-be-destroyed object, and that your programs do not plan on the destructor function executing before the end of a given object's lifetime. To understand the concept of object life cycles better, consider the following program, *stack_cd.cpp*, which constructs and then destroys some objects of type *stack*:

```
#include <iostream.h>
#define ARR_SIZE 100

class stack {
   int stck[ARR_SIZE];
   int stack_top;
 public:
   stack();
   ~stack();
   void push(int i);
   int pop();
 };

stack::stack(void)
 {
   stack_top = 0;
   cout << "Stack Initialized" << endl;
 }

stack::~stack(void)
 {
   cout << "Stack Destroyed" << endl;
 }

void stack::push(int i)
 {
   if (stack_top==ARR_SIZE)
     {
       cout << "Stack is full." << endl;
       return;
     }
   stck[stack_top] = i;
   stack_top++;
 }

int stack::pop(void)
 {
   if (stack_top==0)
     {
       cout << "Stack underflow." << endl;
       return 0;
     }
   stack_top--;
   return stck[stack_top];
 }

void main(void)
 {
   stack obj1, obj2;

   obj1.push(1);
   obj2.push(2);
   obj1.push(3);
   obj2.push(4);
   cout << obj1.pop() << endl;
   cout << obj1.pop() << endl;
   cout << obj2.pop() << endl;
   cout << obj2.pop() << endl;
 }
```

When you compile and execute the *stack_cd.cpp* program, you will notice that the program automatically calls the destructor function for the two stack objects just before the program finishes execution. When you work with arrays of class objects, remember that your program will fire the destructor function for each element in the class array which the program destroys. In other words, if you have an array of 100 elements of the class, when your program destroys the array, it will call the destructor function 100 times—one time for each element. You will learn more about manipulating class arrays in later Tips.

937 USING A COPY CONSTRUCTOR

By default, when C++ copies an object, it performs a *bitwise copy*, which means that the new object is an exact copy of the original object. In some cases, however, a bitwise copy may cause more problems than it solves. For example, if a function receives an object instance by value, then makes a local copy of the instance within the function, when the program exits the function, it will delete the local copy of the object, as you would expect. However, when C++ deletes the local copy, it will also delete the memory the outside copy uses. You can prevent problems of this nature by writing a *copy constructor* function. The general form of the copy constructor function is as shown here:

```
classname (const classname &object)
 {
   // body of constructor
 }
```

In this case, the *object* parameter is the object instance you are copying. You can also include initialization parameters within a copy constructor, although you will have to supply default values for each parameter. The following program, *copy_con.cpp*, uses a copy constructor with an array class:

```
#include <iostream.h>
#include <stdlib.h>

class array {
   int *p;
   int size;
 public:
   array(int sz) {              // simple constructor
      p = new int[sz];
      if(!p) exit(1);
      size = sz;
    }
   ~array() {delete [] p;}     // destructor
   array(const array &object);// copy constructor
   void put(int i, int j){
      if(i>=0 && i<size)
        p[i] = j;
    }
   int get(int i) {return p[i];}
 };

array::array(const array &object)
 {
   int lcl_i;

   p = new int[object.size];
   if (!p)
      exit(1);
   for(lcl_i=0; lcl_i < object.size; lcl_i++)
      p[lcl_i] = object.p[lcl_i];
 }

void main(void)
 {
   array num(10);
   int lcl_i;

   for (lcl_i=0; lcl_i<10; lcl_i++)
      num.put(lcl_i, lcl_i);
   for (lcl_i=9; lcl_i>=0; lcl_i--)
      cout << num.get(lcl_i);
```

```
   cout << endl;

   //  Create another array using the copy constructor
   array x=num;
   for (lcl_i=0; lcl_i<10; lcl_i++)
      cout << x.get(lcl_i);
}
```

When the *copy_con.cpp* program executes, it first creates the object *num*, which it initializes and displays. Next, the program uses *num* to initialize *x*, calling the copy constructor in the process. The copy constructor copies all the data within the *num* object to the *x* object, but in so doing, it creates the *x* copy's own memory space, independent of the *num* object.

Note: *Your programs can only call a copy constructor during initializations. If your program has created an object and then tries to copy another object to that object, the copy constructor will not intervene.*

USING EXPLICIT CONSTRUCTORS 938

You can use *explicit constructors* within your programs to force all declarations into the form your constructor explicitly states. Usually, when you create a constructor, as the following code fragment shows, multiple initialization styles are acceptable to the compiler:

```
sampleclass(int j) {I=j;}
//
// Program statements here

sampleclass obj1(10);
sampleclass obj2 = 10;
//
// More program statements
```

However, if you declare the class as explicit, the compiler will let the program use constructors only of the stated type and format. Using explicit classes is probably most helpful in class libraries and other semi-fixed class locations. You will use the *explicit* keyword to define a class, as shown here:

```
explicit sampleclass(int j) {I=j;}
```

UNDERSTANDING CLASS SCOPE 939

As you have learned, an identifier's *scope* defines the locations within the program for which the identifier is known. C++ classes, like types and variables, have a scope that begins at their definition within the program file and exists to the end of the block within which the classes were defined. To increase the scope of a class, you can define the class outside of all program blocks. In addition, if you define a class as *extern*, the class is known throughout the entire program. If you define a class as *static*, the class scope remains the same as if the class were automatic, but the class existence remains for the duration of the program.

UNDERSTANDING NESTED CLASSES 940

As you learned in earlier sections of this book, you can define one structure within another. Likewise, it is also possible to define one class within another. Defining a class within another class creates a *nested class*. Because a class declaration does, in fact, define a scope, a nested class is valid only within the scope of the enclosing class. For this reason, you should rarely use nested classes within your programs. Because you can use the built-in flexibility of C++, especially its inheritance mechanisms (which you will learn about in later Tips), there is really no need for nested classes.

941 UNDERSTANDING LOCAL CLASSES

Just as you can define local variables within a function, so too is it possible to define a class within a function. When you declare a class within a function, the class is known only within that function. The following program, *local_cl.cpp*, defines a valid local class:

```
#include <iostream.h>

void f(void);
void main(void)
 {
   f();
 }

void f(void)
 {
   class mylocalclass
    {
       int i;
     public:
      void put_i(int n) {i=n;}
      int get_i(void) {return i;}
    } ob;

   ob.put_i(10);
   cout << ob.get_i();
 }
```

C++ applies several restrictions to local classes which make local classes uncommon within C++ programs, including the following:

1. You must define all member functions within the class declaration (in other words, all member functions must be *inline)*.
2. The local class may not use or access local variables of the function in which you declare it.
3. You may not declare any *static* variable inside a local class.

942 RESOLVING MEMBER AND PARAMETER NAME CONFLICTS

Within your member functions, there may be times when a class member name conflicts with the name of a parameter passed to the function. By default, C++ resolves such name conflicts by using the parameter (the local variable) and hiding the existence of the class member. To prevent such name conflicts, precede the class member references with the class name and double colons, as shown here:

```
void dogs::assign_dogs(char *breed, int height, int weight)
 {
   strcpy(dogs::breed, breed);
   dogs::height = height;
   dogs::weight = weight;
 }
```

In this case, the names that *dogs::* precedes correspond to class member names. The other names correspond to the local variables.

943 CREATING AN ARRAY OF CLASS VARIABLES

Several Tips presented in this book created arrays of structures. In a similar way, your programs can create an array of class instances. The following program, *library.cpp*, creates an array that contains the specifics about four books:

```
#include <iostream.h>
#include <iomanip.h>
#include <string.h>

class Book
 {
   public:
     void show_title(void) { cout << title << '\n'; };
     void show_book(void)
      {
        show_title();
        show_publisher();
      };
     void assign_members(char *, char *, char *, float);
   private:
     char title[256];
     char author[64];
     float price;
     char publisher[256];
     void show_publisher(void) { cout << publisher << '\n'; };
 };

void Book::assign_members(char *title, char *author, char *publisher, float price)
 {
   strcpy(Book::title, title);
   strcpy(Book::author, author);
   strcpy(Book::publisher, publisher);
   Book::price = price;
 }

void main(void)
 {
   Book Library[4];

   Library[0].assign_members("Jamsa's C/C++/C# Programmer's Bible",
                             "Jamsa", "Delmar", 49.95);
   Library[1].assign_members("Hacker Proof", "Klander", "Jamsa Press", 54.95);
   Library[2].assign_members("ActiveX Programmer's Library",
                             "Lalani and Chandak", "Jamsa Press", 49.95);
   Library[3].assign_members("Rescued by C++, Third Edition", "Jamsa",
                            "Jamsa Press", 29.95);

   for (int i = 0; i < 4; i++)
     Library[i].show_book();
 }
```

CONSTRUCTORS AND CLASS ARRAYS 944

As you have learned, C++ lets your programs declare arrays of a specific class type. When you declare an array, C++ automatically invokes the constructor function for each array entry. For example, the following program, *arrclass.cpp*, creates an array of the class type *Employee*:

```
#include <iostream.h>

class Employee {
  public:
    Employee(void) { cout << "Constructing an instance\n"; };
    void show_employee(void) { cout << name; };

  private:
    char name[256];
    long id;
 };
```

```
void main(void)
 {
   Employee workers[5];

   // Other statements here
 }
```

When you compile and execute the *arrclass.cpp* program, you will see that the program automatically calls the constructor function five times, one time for each array element.

945 OVERLOADING AN OPERATOR

As you have learned, when you overload a function, the C++ compiler determines which function to invoke, based on the parameter number and types. When you create a class, C++ also lets you overload operators. When you overload an operator, you must continue to use the operator in its standard format. For example, if you overload the plus operator (+), the overload must still use the operator in the form *operand + operand*. In addition, you can only overload existing operators. C++ will not let you define your own operators. The overload you create applies only to instances of the specific class. For example, assume that you create a *String* class and overload the plus operator so that the operator concatenates two strings, as shown here:

```
new_string = string + target;
```

If you use the overloaded plus operator with two integer or floating-point values, the overload will not apply. In addition, C++ will not let you overload the operators listed in Table 945.

Operator	Function
.	Class member operator
.*	Pointer-to-member operator
::	Scope-resolution operator
?:	Conditional-expression operator

Table 945 Operators C++ will not let you overload.

946 CREATING A MEMBER OPERATOR FUNCTION

When you create member *operator* functions to overload an operator's functionality, your member operator declarations will take the general form shown here:

```
return-type class-name::operator #(argument-list)
 {
   // Operations
 }
```

Often, *operator* functions return an object of the class they operate upon. However, C++ lets you define *return-type* as any valid type. The # symbol represents a placeholder for the operator you want to overload. For example, in Tip 947, you will overload the plus operator using a function declaration similar to that shown here:

```
char *operator +(char *append_str)
```

In this function declaration, the member operator is the plus sign. When you are overloading a unary operator (that is, an operator which acts only on a single value), the *argument-list* must be empty. When you are overloading a binary operator, the *argument-list* must contain only a single parameter.

The reason for this seemingly odd construction is that C++ automatically passes the value on the left-hand side of the operator to the overloaded function. Therefore, when you invoke a unary operator, C++ automatically passes the value the function is operating on to the overloaded function. As you will learn later, understanding how C++ passes values to overloaded functions is especially important when you manipulate the prefix and postfix increment and decrement operators.

OVERLOADING THE PLUS OPERATOR 947

You have learned that to overload an operator, you must create a class to which you want the overload to apply. After you have created the class, you must place within the public methods for the class a header line that defines the operator. For example, the following program, *over_pls.cpp* creates a *String* class and overloads the plus operator (+) so that it concatenates strings:

```
#include <iostream.h>
#include <iomanip.h>
#include <string.h>

class String
 {
   public:
     String operator+(char *append_str);
     String(char *in_str)
      { strcpy(buffer, in_str);
        length = strlen(buffer); }
     String(void) {length = 0;};
     void show_string() { cout << buffer; };
   private:
     char buffer[256];
     int length;
 };

String String::operator+(char *append_str)
 {
   String temp;
   int templen;

   templen = strlen(buffer) + strlen(append_str) + 1;
   if(templen>256)
    {
      cout << "String too large!" << endl;
      strcpy(temp.buffer, buffer);
      return temp;
    }
   length = templen;
   strcpy(temp.buffer, buffer);
   strcat(temp.buffer, append_str);
   return temp;
 }

void main(void)
 {
   String title("Jamsa's C/C++/C# ");

   title = title + "Programmer's Bible";
   title.show_string();
 }
```

When you run the *over_pls.cpp* program, it will begin by assigning to the member *buffer* the string "Jamsa's C/C++ ". The program then uses the overloaded plus operator to concatenate the characters "Programmer's Bible". Note that the operator overload is simply a function that receives a parameter. The function receives only one parameter. The parameter is the second operand. The operation itself implies the instance operand.

The overloaded plus operator uses the *strcpy* and *strcat* functions to copy the quoted string onto the *title* object. Note that code within the plus operator overload function refers to the *title* object's data members implicitly, with commands such as the following, which places title's current value into the *temp* object:

```
strcpy(temp.buffer, buffer);
```

The program could just as easily refer to the object explicitly, using the *this* pointer, as shown here:

```
strcpy(temp.buffer, this.buffer);
```

948 OVERLOADING THE MINUS-SIGN OPERATOR

In Tip 947, you created a *String* class and overloaded the plus operator. The following program, *strminus.cpp*, overloads the minus operator (–), then uses the overloaded operator to remove all occurrences of a specified character from the class *buffer* member:

```
#include <iostream.h>
#include <iomanip.h>
#include <string.h>

class String
 {
   public:
     String operator+(char *append_str);
     String operator-(char *substr);
     String(char *in_str)
      { strcpy(buffer, in_str);
        length = strlen(buffer); }
     String() {length = 0;};
     void show_string() { cout << buffer; };
   private:
     char buffer[256];
     int length;
 };

String String::operator+(char *append_str)
 {
   String temp;
   int templen;

   templen = strlen(buffer) + strlen(append_str) + 1;
   if(templen>256)
    {
      cout << "String too large!" << endl;
      strcpy(temp.buffer, buffer);
      length = strlen(buffer);
      return temp;
    }
   length = templen;
   strcpy(temp.buffer, buffer);
   strcat(temp.buffer, append_str);
   return temp;
 }

String String::operator-(char *substr)
 {
   String temp;
   char *s1;
   int i,j;

   s1 = buffer;
   for(i=0; *s1; i++)
    {
```

```
        if(*s1!=*substr)
         {
           temp.buffer[i] = *s1;
           s1++;
         }
        else
         {
           for(j=0; substr[j]==s1[j] && substr[j]; j++)
              ;
           if(!substr[j])
            {
              s1 += j;
              i--;
            }
           else
            {
              temp.buffer[i] = *s1;
              s1++;
            }
         }
     }
     temp.buffer[i] = '\0';
     temp.length = strlen(temp.buffer);
     return temp;
  }
void main(void)
  {
    String title("Jamsa's C/C++/C# ");
    title = title + "Programmer's Bible";
    title.show_string();
    title = title - "'s";
    title.show_string();
  }
```

When you run the *strminus.cpp* program, it will begin by assigning to the member *buffer* the string "Jamsa's C/C++/C# ". The program then uses the overloaded plus operator to concatenate the characters "Programmer's Bible". Note that the operator overload is simply a function that receives a parameter. The function receives only one parameter. The parameter is the second operand. The operation itself implies the instance operand.

The overloaded minus operator uses a simple *for* loop to move through the base char array one element at a time. If the element does not match the first letter of the substring, it copies the element to the *temp* object and moves to the next element. If the element does match the first letter of the substring, it enters a second loop, which checks the elements against the substring elements. If they match, the copy process skips the entire substring. If they do not match, the copy process reverts to the starting element and processes from that position in the *char* array. Note that code within the minus operator overload function refers to the *title* object's data members implicitly, with commands such as the following, which places title's current value into the *temp* object:

```
s1 = buffer;
```

The program could just as easily refer to the object explicitly, using the *this* pointer, as shown here:

```
s1 = this.buffer;
```

OVERLOADING THE PREFIX AND POSTFIX INCREMENT OPERATORS 949

As you have learned, it is possible to overload operators and functions in C++. One of the most commonly used sets of operators in C++ is the prefix and postfix increment operators. As you have learned, if you place the increment operator (++) before a variable, C++ increments the variable before interpreting it; if you place the operator after the variable, C++ interprets the variable before incrementing it.

Early versions of C++ did not provide the programmer with a means of overloading the prefix and postfix increment operators separately. However, modern versions of C++ provide you with a means of determining whether an increment prefixes or postfixes its operand. To overload prefix and postfix incrementation, you will define two versions of the *operator++* function, as shown here:

```
class-name operator++();
class-name operator++(int x);
```

If the increment operator precedes its operand, the compiler will call the *operator++()* function. If, on the other hand, the increment operator follows its operand, the compiler will call the *operator++(int x)* function. The following program, *over_inc.cpp*, overloads the increment operator for the *string* class:

```
#include <iostream.h>
#include <iomanip.h>
#include <string.h>

class String
 {
   public:
     String String::operator++()
      { strcat(buffer, "X");
        return *this; };

     String String::operator++(int x)
      { strcat(buffer, "X");
        return *this; };

     String(char *string)
      { strcpy(buffer, string);
        length = strlen(buffer); }

     void show_string(void) { cout << buffer << endl; };
   private:
     char buffer[256];
     int length;
 };

void main(void)
 {
   String title("Jamsa's C/C++/C# Programmer's Bible");

   title++;
   title.show_string();
   ++title;
   title.show_string();
 }
```

950 OVERLOADING THE PREFIX AND POSTFIX DECREMENT OPERATORS

As you learned in Tip 949, modern versions of C++ provide you with a way to overload both the prefix and postfix types of a given operator. Just as you will declare two operator functions to overload the increment operator, so too will you create two operator functions to overload the decrement operator, as shown here:

```
class-name operator--();
class-name operator--(int x);
```

If the decrement operator precedes its operand, the compiler will call the *operator--()* function. If, on the other hand, the decrement operator follows its operand, the compiler will call the *operator--(int x)* function. The following program, *over_dec.cpp*, overloads the decrement operators:

```
#include <iostream.h>
#include <iomanip.h>
#include <string.h>
```

```
class String
 {
   public:
     String String::operator--()
      { buffer[length-1] = NULL;
        length--;
        return *this; };

     String String::operator--(int x)
      { buffer[length-1] = NULL;
        length--;
        return *this; };

     String(char *string)
      { strcpy(buffer, string);
        length = strlen(buffer); }

     void show_string(void) { cout << buffer << endl; };
   private:
     char buffer[256];
     int length;
 };

void main(void)
 {
   String title("Jamsa's C/C++/C# Programmer's Bible");

   title--;
   title.show_string();
   --title;
   title.show_string();
 }
```

REVISITING THE RESTRICTIONS ON OPERATOR OVERLOADING 951

As you have learned previously, C++ limits which operators your programs can overload. As you also learned, you cannot overload the *dot* operator, the *scope resolution* operator, the *condition* operator, or the *pointer indirection* operator. Beyond these operators, however, you are free to overload any operator you want.

For example, if you want to overload the plus operator in such a way that it writes "Happy is a Dalmatian" to the screen ten times, you can do so. However, you should generally not overload an operator in a way so fundamentally different from its normal use. When another programmer reads your code and sees *a + b*, that programmer should reasonably be able to expect that the overloaded plus operator performs some type of addition activity—not a series of screen outputs.

Except for the assignment operator, derived classes will inherit all overloaded operators from the base class. However, your derived classes remain free to overload any operator themselves (even operators that the base class overloads).

USING A FRIEND FUNCTION TO OVERLOAD OPERATORS 952

As previous Tips have indicated, you can use *friend* functions to overload operators within your classes. It is important, however, to understand that there are some differences between normal operator overloading and *friend* function operator overloading. The most important difference is that *friend* functions have no access to the *this* pointer for the class. Therefore, your program must explicitly pass the operands to the overloaded *friend* operator function. In other words, a *friend* that overloads a *unary* operator receives one parameter, and a friend that overloads a *binary* operator receives two parameters. As you have learned, an in-class overloaded operator receives one less parameter than the operator expects because it uses the *this* pointer. When your programs overload a *binary* operator using a *friend* function, your programs must pass the left operand in the first parameter and the right operand in the second parameter. The following program, *frn_plus.cpp*, uses a *friend* function to overload the + operator:

```
#include <iostream.h>

class loc {
   int longitude, latitude;
 public:
   loc(void) {}                    //  Used to construct temporaries
   loc (int lg, int lt)
    {
      longitude = lg;
      latitude = lt;
    }

   void show(void)
    {
      cout << longitude << " ";
      cout << latitude << endl;
    }

   friend loc operator+(loc op1, loc op2);    //  Friend overloader
   loc operator=(loc op2);
 };

loc operator+(loc op1, loc op2)
 {
   loc temp;

   temp.longitude = op1.longitude + op2.longitude;
   temp.latitude = op1.latitude + op2.latitude;
   return temp;
 }

loc loc::operator=(loc op2)
 {
   longitude = op2.longitude;
   latitude = op2.latitude;
   return *this;
 }

void main(void)
 {
   loc ob1(10,20), ob2(5,30);

   ob1 = ob1+ob2;
   ob1.show();
 }
```

953 Restrictions on Friend Function Operator Overloading

Just as there are restrictions on overloading operators within your classes, C++ imposes two restrictions on overloading operators with *friend* functions. First, you must use a reference parameter to a class when you overload the *increment* or *decrement* operator with a *friend* function. Tip 954 explains how to use a *friend* function to overload the *increment* or *decrement* operator in detail. Second, you cannot use a *friend* function to overload the operators listed in Table 953.

Restricted Operators	
=	()
[]	->

***Table 953** Operators that you cannot overload using a **friend** function.*

USING A FRIEND FUNCTION TO OVERLOAD THE ++ OR -- OPERATORS 954

If you want to use a *friend* function to overload the *increment* or *decrement* operators, you must pass the operand as a reference parameter. You must pass the reference parameter because, as you have learned previously, *friend* functions cannot access the *this* pointer. In addition, you must be sure to pass the operand as a reference parameter—otherwise, C++ will treat the operand as a *by value* parameter and will not perform the operations on the parameter that you want. Instead, your overloaded operator function must modify the *by reference* parameter before it exits. To understand this processing better, consider the following program, *frn_inc.cpp*, which overloads the *increment* and *decrement* operators for the *loc* class:

```
#include <iostream.h>

class loc
{
   int longitude, latitude;
 public:
   loc(void) {}                          //  Used to construct temporaries
   loc (int lg, int lt)
    {
      longitude = lg;
      latitude = lt;
    }

   void show(void)
    {
      cout << longitude << " ";
      cout << latitude << endl;
    }

   loc operator=(loc op2);
   friend loc operator++(loc &op1);      //  Friend overloaded
   friend loc operator--(loc &op1);      //  Friend overloaded
 };

loc loc::operator=(loc op2)
 {
   longitude = op2.longitude;
   latitude = op2.latitude;
   return *this;
 }

loc operator++(loc &op)
 {
   op.longitude++;
   op.latitude++;
   return op;
 }

loc operator--(loc &op)
 {
   op.longitude--;
   op.latitude--;
   return op;
 }

void main(void)
 {
   loc ob1(10,20), ob2;

   ob1.show();
   ++ob1;
   ob1.show();                          // Displays 11 and 21
   ob2 = ++ob1;
   ob2.show();                          // Displays 12 and 22
```

```
   --ob2;
   ob2.show();                            // Displays 11 and 21 again
 }
```

955 REASONS TO OVERLOAD OPERATORS WITH FRIEND FUNCTIONS

In many cases, whether you use a *friend* function or a *member* function to overload an operator makes no functional difference in your program. While overloading with a *friend* function is not materially different from using a *member* function to overload, you should use the *member* function to achieve the greatest encapsulation. However, there are some situations, and one situation in particular, as you will learn in this Tip, in which a *friend* function is extremely helpful.

As you have learned, when you use a *member* function to overload a *binary* operator, the object on the operator's left side generates the call to the overloaded operator function. In addition, C++ passes a pointer to the object on the left side within the *this* pointer. Therefore, if you create a class named *Dogs* and overload the plus operator, the following statement is valid, assuming you have created an object instance named *happy*:

```
happy + 100
```

In the previous example, *happy* generates the call to the overloaded plus function, which performs the addition and returns the value in the *this* pointer to *happy*. However, if you write the expression as shown in the following example, the compiler will return an error:

```
100 + happy
```

Because the overloaded plus function expects to receive a class object that it can reference with the *this* pointer, the constant that it receives in the previous example causes the compiler to return an error. If, on the other hand, you overload the plus operator with a pair of *friend* functions, you can accomplish the same goal without causing an error. The following program, *two_frn.cpp*, uses the *loc* class to show you how to use two *friend* functions:

```
#include <iostream.h>

class loc {
   int longitude, latitude;
 public:
   loc(void) {}                   //  Used to construct temporaries
   loc (int lg, int lt)
    {
      longitude = lg;
      latitude = lt;
    }

   void show(void)
    {
      cout << longitude << " ";
      cout << latitude << endl;
    }

   loc operator=(loc op2);
   friend loc operator+(loc op1, int op2);    //  Friend overloaded
   friend loc operator+(int op1, loc op2);    //  Friend overloaded
 };

loc loc::operator=(loc op2)
 {
   longitude = op2.longitude;
   latitude = op2.latitude;
   return *this;
 }
```

```
loc operator+(loc op1, int op2)
 {
   loc temp;

   temp.longitude = op1.longitude + op2;
   temp.latitude = op1.latitude + op2;
   return temp;
 }

loc operator+(int op1, loc op2)
 {
   loc temp;

   temp.longitude = op1 + op2.longitude;
   temp.latitude = op1 + op2.latitude;
   return temp;
 }

void main(void)
 {
   loc ob1(10,20), ob2( 5,30), ob3( 7,14);

   ob1.show();
   ob2.show();
   ob3.show();
   ob1 = ob2 + 10;
   ob3 = 10 + ob3;
   ob1.show();
   ob3.show();
 }
```

OVERLOADING THE NEW OPERATOR 956

As you have learned, your programs can overload almost any function or operator. In fact, your programs can overload both the *new* and the *delete* operators. You might choose to overload either one of these operators if you want your programs to use some special allocation method. For example, you might want to write an allocation routine that uses the hard disk for virtual memory if the program expends all the memory available on the heap. For whatever reason you want to overload the *new* function, the process is relatively easy, as shown here:

```
#include <stdlib.h>

void *operator new(size_t size)
 {
   //perform allocation
   return pointer_to_memory;
 }
```

The type *size_t* must be a type capable of containing the largest single piece of memory that the overloaded *new* function can allocate. The *stdlib.h* include file defines the *size_t* type. The parameter *size* should contain the number of bytes *new* will require to hold the newly allocated object. Finally, the *new* function must return a pointer to the memory it allocates or return *NULL* if it fails.

OVERLOADING THE DELETE OPERATOR 957

Just as you can overload the *new* operator to handle specific memory allocation needs, you can overload the *delete* operator to release the memory an overloaded *new* operator allocates. The *delete* operator must receive a pointer to the memory that the *new* operator previously allocated for the object. You can overload both the *new* and *delete* operators, either globally or relative to one or more classes, or both globally and in relation to one or more classes. The following program, *cus_newd.cpp*, uses overloaded *new* and *delete* functions relative to the *loc* class:

```
#include <iostream.h>
#include <stdlib.h>

class loc {
   int longitude, latitude;
  public:
    loc(void) {}                    //  Used to construct temporaries
    loc (int lg, int lt)
      {
        longitude = lg;
        latitude = lt;
      }
    void show(void)
      {
      cout << longitude << " ";
      cout << latitude << endl;
      }
    void *operator new(size_t size);
    void operator delete(void *p);
 };

void *loc::operator new(size_t size)
 {
   cout << "In the custom new function." << endl;
   return malloc(size);
 }

void loc::operator delete(void *p)
 {
   cout << "In the custom delete function." << endl;
   free(p);
 }
void main(void)
  {
  loc *p1, *p2;

  p1 = new loc(10,20);
  if (!p1)
   {
     cout << "Allocation error\n";
     exit(1);
   }

  p2 = new loc(-10,-20);
  if (!p2)
   {
      cout << "Allocation error\n";
      exit(1);
   }
  p1->show();
  p2->show();
  delete p1;
  delete p2;
  exit(0);
 }
```

When you compile and execute the *cus_newd.cpp* program, your screen will display the following output:

```
In the custom new function.
In the custom new function.
10 20
-10 -20
In the custom delete function.
In the custom delete function.
C:\>
```

OVERLOADING NEW OR DELETE FOR ARRAYS

958

In the previous Tips, you learned how to overload the *new* and *delete* operators to perform custom memory allocation within your programs. However, if you want to allocate arrays of objects, you must overload the *new* and *delete* functions again, using a special operator form which instructs the compiler that the overload is for arrays. The prototype for the overloaded *new* function that you can use to allocate arrays within your programs is shown here:

```
#include <stdlib.h>

void *operator new[](size_t size)
 {
   // perform allocation
  return pointer_to_memory;
 }
```

When you allocate arrays, C++ will automatically call the class constructor function for each object in the array. When you free an array, C++ will automatically call the object's destructor function. The following program, *cus_arr.cpp*, uses overloaded *new* and *delete* functions to allocate and free array space:

```
#include <iostream.h>
#include <stdlib.h>

class loc {
  int longitude, latitude;
  public:
    loc(void) {}                   //  Used to construct temporaries
    loc (int lg, int lt)
     {
       longitude = lg;
       latitude = lt;
     }

    void show(void)
     {
       cout << longitude << " ";
       cout << latitude << endl;
     }

    void *operator new(size_t size);
    void operator delete(void *p);
    void *operator new[](size_t size);
    void operator delete[](void *p);
 };

void *loc::operator new(size_t size)
 {
   cout << "In the custom new function." << endl;
   return malloc(size);
 }

void loc::operator delete(void *p)
 {
   cout << "In the custom delete function." << endl;
   free(p);
 }

void *loc::operator new[](size_t size)
 {
   cout << "In the custom new ARRAY allocator function." << endl;
   return malloc(size);
 }

void loc::operator delete[](void *p)
 {
```

```
   cout << "Freeing the ARRAY in the custom delete function." << endl;
   free(p);
 }

void main(void)
 {
   loc *p1, *p2;
   int i;

   p1 = new loc(10,20);
   if (!p1)
    {
      cout << "Allocation error\n";
      exit(1);
    }

   p2 = new loc[10];
   if (!p2)
    {
      cout << "Allocation error\n";
      exit(1);
    }
   p1->show();
   for(i=0; i<10; i++)
     p2[i].show();
   delete p1;
   delete [] p2;
 }
```

The *cus_arr.cpp* program overloads the *new* and *delete* operators for both arrays and individual objects. When the program creates an array instance of an object, it invokes the custom *new* operator for the array and the custom *new* operator for each element in the array. The program performs similar processing when it deletes an individual instance or an array. When you compile and execute the *cus_arr.cpp* program, your screen will display the following output:

```
In the custom new function.
In the custom new ARRAY allocator function.
10 20
4258096 4258096

0 0
0 0
0 0
0 0
0 0
0 0
0 0
0 0
0 0
In the custom delete function.
Freeing the ARRAY in the custom delete function.
C:\>
```

959 OVERLOADING THE [ ] ARRAY OPERATOR

As your programs become more complex, there may be times when you must overload the [ ] *array* operator. C++ considers the *array* operator as a *binary* operator for overloading purposes. Therefore, the general form of a *member* operator function overload for the [ ] *array* operator is as shown here:

```
type class-name::operator[](int i)
 {
   // . . .
 }
```

Technically, the *i* parameter in the previous example does not have to be of type *int*, but because you will typically define arrays with an integer parameter, you should avoid using a parameter of type *float* or some other type. When you call the overloaded operator function, C++ will assign the *this* pointer to the object and use the parameter to control size. To better understand the processing the overloaded [ ] *array* function performs, consider the following program, *arr_over.cpp*:

```
#include <iostream.h>

class sometype {
   int a[3];
  public:
   sometype(int i, int j, int k)
    {
      a[0] = i;
      a[1] = j;
      a[2] = k;
    }
   int operator[](int i) {return a[i];}
  };

void main(void)
 {
   sometype ob(1, 2, 3);

   cout << ob[1];
 }
```

Overloading the [ ] *array* operator gives you some important control over the creation of arrays with classes. In addition to letting you assign individual values to individual members, you can use the overloaded function to create a program that performs safe array indexing. Safe array indexing helps you prevent your programs from overrunning or underrunning array boundaries at run time. The following program, *safe_arr.cpp*, extends the *arr_over.cpp* program to include safe array indexing:

```
#include <iostream.h>
#include <stdlib.h>

class sometype {
   int a[3];
  public:
   sometype(int i, int j, int k)
    {
      a[0] = i;
      a[1] = j;
      a[2] = k;
    }
   int &operator[](int i);
  };

int &sometype::operator[](int i)
 {
   if (i<0 || i>2)
    {
      cout << "Boundary error.\n";
      exit(1);
    }
   return a[i];
 }

void main(void)
 {
   sometype ob(1, 2, 3);

   cout << ob[1];
   cout << endl;
```

```
    ob[1] = 25;
   cout << endl;
   cout << ob[1];
   ob[3] = 44;
 }
```

When you try to access an object outside the array's boundaries, it will return an error. In the case of the *safe_arr.cpp* program, trying to access the element at index 3 is outside the boundaries, and so the program returns an error. When you execute the *safe_arr.cpp* program, it will generate the following output:

```
2
25
Boundary error.
C:\>
```

960 OVERLOADING THE ( ) FUNCTION CALL OPERATOR

As you have learned, C++ lets you overload many of its operators within your programs. When you overload the ( ) *function call* operator, you are not creating a new way to call a function. Rather, you create an *operator* function to which your programs can pass an arbitrary number of parameters. In general, when you overload the ( ) *function call* operator, you define the parameters that you want your programs to pass to the overloaded function. To better understand how C++ overloads the ( ) *function call* operator, consider the following program, *fun_over.cpp*, which uses the overloaded ( ) *function call* operator with the *loc* class:

```
#include <iostream.h>

class loc {
  int longitude, latitude;
  public:
    loc(void) {}                  //  Used to construct temporaries
    loc (int lg, int lt)
     {
       longitude = lg;
       latitude = lt;
     }
    void show(void)
     {
       cout << longitude << " ";
       cout << latitude << endl;
     }
    loc operator+(loc op2);
    loc operator()(int i, int j);
  };

loc loc::operator()(int i, int j)
 {
   longitude = i;
   latitude = j;
   return *this;
 }

loc loc::operator+(loc op2)
 {
   loc temp;

   temp.longitude = op2.longitude + longitude;
   temp.latitude = op2.latitude + latitude;
   return temp;
 }
```

```
void main(void)
 {
   loc ob1(10,20), ob2( 1,1);

   ob1.show();
   ob1(7,8);
   ob1.show();
   ob1 = ob2 + ob1(10,10);
   ob1.show();
 }
```

In the *fun_over.cpp* program, the overloaded ( ) *function call* operator for the *loc* class lets you assign new values to an object which you follow with the ( ) *function call* operator. In the program, the next to last statement, *ob1 = ob2 + ob1(10,10)* takes advantage of the overloaded ( ) *function call* to assign a value to *ob1* dynamically. In this particular example, the program does not maintain the newly assigned value. However, if you used the ( ) *function call* operator with a different object, perhaps *ob3*, the program would maintain the newly assigned value within that object.

When you compile and execute the *fun_over.cpp* program, your screen will display the following output:

```
10 20
7 8
11 11
C:\>
```

OVERLOADING THE -> POINTER OPERATOR 961

As you have learned, C++ lets you overload many of its operators. As your programs become more complex, there may be times when you must overload the *pointer* operator. When you do so, you must first understand that C++ treats the *pointer* operator as a unary operator (that is, an operator with only a single operand) when you overload it. When you overload the *pointer* function, it must return a pointer to an object of the invoking class. When you overload the -> *pointer* operator, its return value is the same as what your program would receive if it invoked the dot operator against the object. In other words, the following statements are equivalent:

```
obj->i = 10;
obj.i = 10;
```

To better understand how C++ processes overloaded -> *pointer* operators, consider the program, *over_ptr.cpp*, as shown here:

```
#include <iostream.h>

class sample {
 public:
   int i;
   sample *operator->(void) {return this;}
 };

void main(void)
 {
   sample obj;

   obj->i = 10;                        //    same as obj.i
   cout << obj.i << " " << obj->i;
 }
```

Note: *There does not seem to be any truly useful purpose for overloading the -> operator. However, if you should discover that you must do so, you will use the form shown within this Tip.*

962 OVERLOADING THE , COMMA OPERATOR

As you have learned, your programs can overload many of the C++ operators. As your programs and classes become more complex, you may find that your programs must overload the *comma* operator. The *comma* operator in C++ evaluates each operand in a comma-separated list and returns only the rightmost operand in the list.

In other words, if your code has a list, *E1, E2* the program will evaluate the left operand *E1* as a void expression, and return its evaluation of E2 as the result and type of the comma expression. By recursion, the comma operator results in your program evaluating the expression *E1, E2, ..., En* left-to-right. The comma operator evaluates each *Ei* in turn, and returns the value and type of *En* as the result of the whole expression. To avoid ambiguity between the comma operator and the comma delimiter in function argument and initializer lists, use parentheses, as shown here:

```
func(i, (j = 1, j + 4), k);
```

The previous code fragment calls *func* with three arguments (*i*, 5, *k*), not four. While your programs can overload the *comma* operator in any fashion you want, you should try to maintain consistency with the default C++ operation for the comma. To understand this better, consider the following program, *com_over.cpp*, which overloads the *comma* operator but maintains its normal operation:

```
#include <iostream.h>

class loc {
  int longitude, latitude;
  public:
    loc(void) {}              //  Used to construct temporaries
    loc (int lg, int lt)
     {
       longitude = lg;
       latitude = lt;
     }
    void show(void)
     {
       cout << longitude << " ";
       cout << latitude << endl;
     }
    loc operator+(loc op2);
    loc operator,(loc op2);
  };

loc loc::operator,(loc op2)
 {
   loc temp;

   temp.longitude = op2.longitude;
   temp.latitude = op2.latitude;
   cout << op2.longitude << "  " << op2.latitude << endl;
   return temp;
 }

loc loc::operator+(loc op2)
 {
   loc temp;

   temp.longitude = op2.longitude + longitude;
   temp.latitude = op2.latitude + latitude;
   return temp;
 }

void main(void)
 {
   loc obj1(10,20), obj2( 5,30), obj3(1,1);
```

```
    obj1.show();
   obj2.show();
   obj3.show();
   cout << endl;
   obj1 = (obj1, obj2 + obj2, obj2 + obj3);
   obj1.show();                            // Will display 6,31, obj2 + obj3's value
 }
```

The assignment of the comma-operated list in the *com_over.cpp* program's next to last line results in the assignment statement performing the assignment as shown here:

```
obj1 = (obj1, obj2 + obj2, obj2 + obj3);
obj1 = obj1;
obj1 = obj2 + obj2;
obj1 = obj2 + obj3;
obj1 = (6, 31)
```

UNDERSTANDING ABSTRACTION 963

Abstraction is the process of looking at an object in terms of its methods (operations), while temporarily ignoring the underlying details of the object's implementation. Programmers use abstraction to simplify the design and implementation of complex programs. For example, if you want to write a word processor program, the task might at first seem very difficult. However, using abstraction, you will begin to realize that a word processor actually consists of objects, such as a document object that you will create, save, spell check, and print. By viewing programs in abstract terms, you can better understand the required programming. In C++, the primary tool for supporting abstraction is the class.

ALLOCATING A POINTER TO A CLASS 964

As you work with class variables, there may be times when you want to allocate dynamic arrays or dynamic lists of the class type. As you have learned, you can use dynamic arrays and dynamic lists within your programs when you do not know at compile time how many elements an array or list will require. Declaring a dynamic array of class objects is fundamentally identical to declaring a dynamic array of any of the C or C++ basic types. The following program, *dynclass.cpp*, for example, creates an array of pointers to variables of the class type *Book*:

```
#include <iostream.h>
#include <iomanip.h>
#include <string.h>

class Book
 {
   public:
    void show_title(void) { cout << title << '\n'; };
    void show_book(void)
     {
       show_title();
       show_publisher();
     };
    Book(char *title, char *author, char *publisher, float price);
  private:
    char title[256];
    char author[64];
    float price;
    char publisher[256];
    void show_publisher(void) { cout << publisher << '\n'; };
};

Book::Book(char *title, char *author, char *publisher, float price)
```

```
{
  strcpy(Book::title, title);
  strcpy(Book::author, author);
  strcpy(Book::publisher, publisher);
  Book::price = price;
  cout << "In constructor." << endl;
}
void main(void)
{
  Book *Library[4];
  int i;
  Library[0] = new Book("Jamsa's C/C++/C# Programmer's Bible", "Jamsa",
                         "Delmar", 54.95);
  Library[1] = new Book("Hacker Proof", "Klander", "Jamsa Press", 54.95);
  Library[2] = new Book("ActiveX Programmer's Library", "Lalani and Chandak",
                         "Jamsa Press", 49.95);
  Library[3] = new Book("Rescued by C++, Third Edition", "Jamsa",
                         "Jamsa Press", 29.95);
  for (i = 0; i < 4; i++)
    Library[i]->show_book();
}
```

When you compile and execute the *dynclass.cpp* program, your screen will display the following output:

```
In constructor.
In constructor.
In constructor.
In constructor.
Jamsa's C/C++/C# Programmer's Bible
Delmar

Hacker Proof
Jamsa Press

ActiveX Programmer's Library
Jamsa Press

Rescued by C++, Third Edition
Jamsa Press
C:\>
```

As you can see, each time you create an instance using *new*, C++ invokes the class constructor function.

965 DISCARDING A POINTER TO A CLASS

In Tip 964 you created an array of pointers to objects of type *Book*. Each time the program created an instance, C++ automatically invoked the *Book* constructor function. In a similar way, if the class has a destructor, C++ will automatically invoke the function each time the program destroys an instance. The following program, *dyndestr.cpp*, adds a destructor function to the *Book* class. The program also uses the *delete* operator to discard the pointer to each instance, as shown here:

```
#include <iostream.h>
#include <iomanip.h>
#include <string.h>
class Book
 {
   public:
    void show_title(void) { cout << title << '\n'; };
    void show_book(void)
```

```
      {
        show_title();
        show_publisher();
      };
    Book(char *title, char *author, char *publisher, float price);

    ~Book(void) { cout << "Destroying the entry for " << title << '\n'; };
  private:
    char title[256];
    char author[64];
    float price;
    char publisher[256];
    void show_publisher(void) { cout << publisher << '\n'; };
};

Book::Book(char *title, char *author, char *publisher, float price)
 {
   strcpy(Book::title, title);
   strcpy(Book::author, author);
   strcpy(Book::publisher, publisher);
   Book::price = price;
 }

void main(void)
 {
   Book *Library[4];
   int i = 0;

   Library[0] = new Book("Jamsa's C/C++/C# Programmer's Bible", "Jamsa",
     "Delmar", 54.95);
   Library[1] = new Book("Hacker Proof", "Klander", "Jamsa Press", 54.95);
   Library[2] = new Book("ActiveX Programmer's Library", "Lalani and Chandak",
     "Jamsa Press", 49.95);
   Library[3] = new Book("Rescued by C++, Third Edition", "Jamsa",
     "Jamsa Press", 29.95);
   for (i = 0; i < 4; i++)
     Library[i]->show_book();
   for (i = 0; i < 4; i++)
     delete Library[i];
 }
```

When you compile and execute the *dyndestr.cpp*, your screen will display the following output:

```
In constructor.
In constructor.
In constructor.
In constructor.
Jamsa's C/C++ Programmer's Bible
Jamsa Press

Hacker Proof
Jamsa Press

ActiveX Programmer's Library
Jamsa Press

Rescued by C++, Third Edition
Jamsa Press

Destroying the entry for Jamsa's C/C++ Programmer's Bible
Destroying the entry for Hacker Proof
Destroying the entry for ActiveX Programmer's Library
Destroying the entry for Rescued by C++, Third Edition
C:\>
```

966 DISCARDING LEADING WHITESPACE ON INPUT

As you have learned, the *cin* I/O stream uses whitespace as a delimiter for input data. When you use *cin*, there may be times when you will want *cin* to ignore leading whitespace (in other words, whitespace which precedes text). In such cases, your programs can use the *ws* manipulator, as shown here:

```
cin >> ws >> buffer;
```

The following program, *ws.cpp*, uses the *ws* manipulator to remove leading whitespace:

```
#include <iostream.h>

void main(void)
 {
   char buffer[256];

   cout << "Enter a word with leading blanks\n";
   cin >> ws >> buffer;
   cout << "==" << buffer << "==";
 }
```

To experiment with the *ws.cpp* program, remove the *ws* manipulator and change the leading whitespace input. For example, when you compile and execute the *ws.cpp* and enter the word "   Jamsa", your output is the same is it would be if you entered "Jamsa," as shown here:

```
Enter a word with leading blanks
       Jamsa
==Jamsa==
C:\>ws
Enter a word with leading blanks
Jamsa
==Jamsa==
C:\>
```

967 UNDERSTANDING CLASS LIBRARIES

As you have learned, object libraries make reusing functions very easy for your programs. A *class library* is similar to an object library in that it contains code to which your programs can link. Unlike an object code library, which contains a collection of callable functions, a class library contains class methods. To use the methods, your programs must use the corresponding class structures. In other words, your programs cannot simply call class library functions without using a class. Throughout this book, your programs have made extensive use of the C++ *iostream* class library to perform I/O operations using *cin* and *cout*. Just as you may eventually create object code libraries that contain the functions you create, you can also create class libraries. By creating your own class libraries, you make using existing objects easy for future programs.

968 PLACE YOUR CLASS DEFINITIONS IN A HEADER FILE

When you create a class that other programs might use, you should place the class declaration in a header file that you base on the class name. For example, the header file, *iostream.h*, contains the class declaration for the *iostream* class. Do not place your class methods in the header file. Instead, compile the class methods and place them into a class library, as you have learned previously. By placing the class declaration in a header file, you make it much easier for a program to use a class. Rather than requiring knowledge of the complete class structure, the program must simply include the class header file and then it only has to use those members it requires.

USING THE INLINE KEYWORD WITH CLASS MEMBER FUNCTIONS 969

As you know, the *inline* keyword instructs the compiler to place a function's code inline at each reference. Using the *inline* keyword lets you trade increased program size for improved performance. When you define class member functions, C++ lets you place the functions within the class itself or outside of the class. When you place a function definition inside the class, C++ generates inline code each time it later encounters an invocation of the method. When you define the function outside of the class, however, C++ does not use inline code. If you have a method that you have defined outside of a class for which you want the compiler to generate inline code, simply precede the function name with the *inline* keyword. For example, the following function definition directs the compiler to generate inline code for each invocation of the method *show_book*:

```
inline void Book::show_book(void) { show_title(); show_publisher(); };
```

INITIALIZING A CLASS ARRAY 970

As you have learned, C++ lets you declare a class array. When you declare a class array, the program will automatically invoke the constructor function for each element. When you declare an array of structures, C++ lets you initialize the array member, as shown here:

```
struct Employee
 {
   char name[64];
   long id;
 } workers [2] = {{ "Kris", 1} , {"Happy", 2}};
```

When you declare a class array, C++ does not let you provide initial values. Therefore, you might find assigning the member values to be challenging. The following program, *arrayasn.cpp*, uses the constructor function to assign each array element:

```
#include <iostream.h>
#include <string.h>

class Employee
 {
  public:
    Employee(void);
    void show_employee(void) { cout << name << endl; };
  private:
    char name[256];
    long id;
 };

Employee::Employee(void)
 {
   static int index = 0;

   switch (index++) {
     case 0: strcpy(Employee::name, "Kris");
             Employee::id = 1;
             break;
     case 1: strcpy(Employee::name, "Happy");
             Employee::id = 2;
             break;
     };
 }

void main(void)
 {
   Employee workers[2];
```

```
    workers[0].show_employee();
   workers[1].show_employee();
 }
```

The constructor function uses the static variable *index* to determine which element it is initializing. As you can guess, the constructor could become quite messy, depending on the number of elements and class members.

971 DESTROYING A CLASS ARRAY

As you have learned, your programs can overload the *delete* operator to release a class array's memory back to the heap. You have also learned that your programs will most commonly use destructor functions to release memory or save information about a class to disk. When you use an overloaded *delete* operator to release arrays, the *delete* operator will call the destructor function for each element within the array. When you write code within your destructor function, carefully optimize the code as much as possible to avoid program slowdowns during array destruction sequences. To better understand the special destructor code you must use with class arrays, consider the *arr_des.cpp* program, which adds a destructor function to the example program you wrote in Tip 958, as shown here:

```
#include <iostream.h>
#include <stdlib.h>

class loc
 {
    int longitude, latitude;
  public:
    loc(void) {}                     //  Used to construct temporaries
    ~loc(void);
    loc (int lg, int lt)
     {
       longitude = lg;
       latitude = lt;
     }

    void show(void)
     {
       cout << longitude << " ";
       cout << latitude << endl;
     }

    void *operator new(size_t size);
    void operator delete(void *p);
    void *operator new[](size_t size);
    void operator delete[](void *p);
 };

loc::~loc(void)
 {
   cout << "In the destructor function" << endl;
 }

void *loc::operator new(size_t size)
 {
   cout << "In the custom new operator function." << endl;
   return malloc(size);
 }

void loc::operator delete(void *p)
 {
   cout << "In the custom delete operator function." << endl;
   free(p);
 }

void *loc::operator new[](size_t size)
 {
```

```
    cout << "In the custom new ARRAY allocator function." << endl;
    return malloc(size);
 }
void loc::operator delete[](void *p)
 {
    cout << "Freeing the ARRAY in the custom delete function." << endl;
    free(p);
 }
void main(void)
 {
    loc *p1, *p2;
    int i;
    p1 = new loc(10,20);
    if (!p1)
     {
       cout << "Allocation error\n";
       exit(1);
     }
    p2 = new loc[10];
    if (!p2)
     {
       cout << "Allocation error\n";

       exit(1);
     }
    p1->show();
    for(i=0; i<10; i++)
      p2[i].show();
    delete p1;
    delete [] p2;
 }
```

When you execute the *arr_des.cpp* program, the program will invoke the destructor for each element in the class array, and then call the custom array destructor, and your screen will display the following output:

```
In the destructor function
In the custom delete operator function.
In the destructor function
In the destructor function
In the destructor function
In the destructor function
In the destructor function
In the destructor function
In the destructor function
In the destructor function
In the destructor function
In the destructor function
Freeing the ARRAY in the custom delete function.
```

When the program calls the custom delete function, it immediately fires the destructor function for each object in the array before it deletes the pointer to the array. Therefore, you can use destructor functions with arrays to store values, generate other information, or remove the value from the object.

CREATING INITIALIZED CLASS ARRAYS 972

As you have learned, C++ lets you create arrays of objects. When you create an array of objects, you can initialize the objects in two ways: either after creating the objects (by using a *for* loop to step through the array and initialize each element within the array) or in the object's constructor function. For example, the following program, *simp_ini.cpp*, uses a *for* loop to initialize an array of object variables of type *sample*:

```
#include <iostream.h>

class sample
 {
   int value;
 public:
   void set_value(int j) {value = j;}
   int get_value(void) {return value;}
 };

void main(void)
 {
   sample obj[3];
   int loop;

   for(loop = 0; loop < 3; loop++)
      obj[loop].set_value(loop+1);
   for(loop = 0; loop < 3; loop++)
      cout << obj[loop].get_value() << endl;
 }
```

When you compile and execute the *simp_ini.cpp* program, your screen will display the following output:

```
1
2
3
C:\>
```

As you can see, the program loops through the array, initializing each element. Then, the program loops through the array again, displaying each element. While *for* loops are useful when your arrays are small, it is helpful to initialize your array values within the array's declaration, thereby simplifying your program. The following program, *onep_ini.cpp*, uses the same array as the *simp_ini.cpp* program previously shown, but *onep_ini.cpp* initializes the array within the constructor:

```
#include <iostream.h>

class sample
 {
   int value;
  public:
   sample(int j) {value = j;}                // constructor
   int get_value(void) {return value;}
 };

void main(void)
 {
   int loop;
   sample obj[3] = {1 , 2, 3};               // initializers

   for(loop = 0; loop < 3; loop++)
      cout << obj[loop].get_value() << endl;
 }
```

When you compile and execute the *onep_ini.cpp* program, it will display the same output as the *simp_ini.cpp* program, as shown here:

```
1
2
3
C:\>
```

As you can see, the *onep_ini.cpp* program is clearer and shorter than the *simp_ini.cpp* program. As a rule, you should initialize your arrays within constructors if you know what the array's initial values will be. In Tip 974, you will learn how to write a class that supports both initialized and uninitialized arrays.

INITIALIZING AN ARRAY WITH A MULTI-ARGUMENT CONSTRUCTOR 973

In Tip 972, you learned how to initialize an array of objects from within the object's constructor function. The simple example shown in Tip 972 initialized only a single value. However, most classes you create will contain multiple data members that you may want your programs to initialize automatically. The following program, *twop_ini.cpp*, initializes an array that contains two values within the class:

```
#include <iostream.h>

class sample
 {
    int value1;
    int value2;
  public:
    sample(int j, int k)                         // constructor
     {
       value1 = j;
       value2 = k;
     }
    int get_value2(void) {return value2;}
    int get_value1(void) {return value1;}
 };

void main(void)
 {
   sample obj[3] = {   sample(1,2),
                       sample(3,4),
                       sample(5,6)   };          // initializers
   int loop;

   for(loop = 0; loop < 3; loop++)
    {
      cout << "Value 1, Value 2: ";
      cout << obj[loop].get_value1();
      cout << ", ";
      cout << obj[loop].get_value2() << endl;
    }
 }
```

When you compile and execute the *twop_ini.cpp* program, your screen will display the following output:

```
Value 1, Value 2: 1, 2
Value 1, Value 2: 3, 4
Value 1, Value 2: 5, 6
```

CREATING INITIALIZED VERSUS UNINITIALIZED ARRAYS 974

As your programs get more complex, you will often find that your programs must initialize some arrays of an object, but not necessarily other arrays. For example, you may initialize two arrays which your program will use and leave uninitialized a third which your program will use only for temporary storage. You can easily overload your class constructor functions so that your classes support both initialized and uninitialized declarations. The following program, *init_uni.cpp*, changes the design of the *sample* class to support both declaration types:

```
#include <iostream.h>

class sample
 {
     int value;
  public:
     sample(void) {value = 0;};                                // uninitialized constructor
     sample(int j) {value = j;}                                // initialized constructor
     int get_value(void) {return value;}
 };

void main(void)
 {
   int loop;
   sample obj1[3] = {1, 2, 3};
   sample obj2[32];

   cout << "Entering first loop: " << endl;
   for(loop = 0; loop < 3; loop++)
     cout << obj1[loop].get_value() << endl;

   cout << "Entering second loop: " << endl;
   for(loop = 0; loop < 32; loop++)
     cout << obj2[loop].get_value() << ", ";
   cout << endl;
 }
```

As you can see, the *init_uni.cpp* program defines both an uninitialized (default) constructor and an initialized constructor. The definitions in *main* create an array of three initialized objects and an array of 32 uninitialized objects. When the program loops through the arrays, the first array outputs the values to which the program originally initialized the elements. The second array, however, outputs 0 for each element, because the default constructor initializes each element to 0. When you compile and execute the *init_uni.cpp* program, your screen will display the following output:

```
Entering first loop:
1
2
3
Entering second loop:
0, 0, 0, 0, 0, 0, 0, 0, 0, 0, 0, 0, 0, 0, 0, 0, 0, 0, 0, 0, 0, 0, 0, 0, 0, 0, 0, 0, 0, 0,
0, 0,
```

975 WORKING WITH CLASS ARRAYS

When you work with class arrays, you will treat the class arrays much as you treated structure arrays when you worked with them in C. As you know, arrays and classes let you group related information. As you have learned, C++ lets you create arrays of objects or use arrays as class members. In general, C++ does not place a limit on the depth to which your programs can go with respect to nested data structures. For example, the following declaration creates an array of 100 employee objects. Within each object is an array of *Date* structures that correspond to the employee's hire date, first review, and last review:

```
class employee
 {
  public:
    char name[64];
    int age;
    char ssan[11];    // Social security number
    int pay_grade;
```

```
    float salary;
    unsigned employee_number;
    struct Date
     {
       int month;
       int day;
       int year;
     } emp_dates[3];
    int review_time(date current);
 };
// program statements
void main(void);
 {
   employee staff_members[100];
   // program code
 }
```

To access members and array elements, you simply work from left to right, starting from the outside and working inward. For example, the following statements assign an employee's hire date:

```
staff_members[10].emp_dates[0].month = 10;
staff_members[10].emp_dates[0].day = 31;
staff_members[10].emp_dates[0].year = 97;
```

Although nesting objects and arrays as this Tip describes is often convenient, keep in mind that the deeper your programs nest such data structures, the more difficult your data will become for other programmers to understand.

Note: *In addition to concerns about the clarity of nested classes, the class definition for the* ***employee*** *class this Tip defines does not encapsulate the class well. As you have learned, your program should generally use interface functions to manipulate and return the information within the class. In fact, effective data encapsulation makes nested classes more difficult to use.*

UNDERSTANDING HOW CLASS ARRAYS USE MEMORY 976

As you have learned, despite the sheer quantity of memory available to your programs when you use a newer computer, such as a Pentium, there are still limitations on how much memory your programs can access. Additionally, the larger an object is, the longer processing will require on each function which manipulates that object. Therefore, it is important to understand how much memory your classes will consume. You will compute the amount of memory a single class object requires similarly to the technique you used to determine how much memory a single structure instance requires. Computing the amount of memory a class array will consume is relatively simple. First, you must determine the maximum amount of space that each class instance will require. To better understand a class instance's memory requirements, consider the following declaration:

```
class sample
 {
   public:
     int i, j, k;
     float a, b;
     char c[64];
 }
```

The *sample* class will use 78 bytes of memory for each instance: 6 bytes for the integers, 8 bytes for the floating-point variables, and 64 bytes for the character array. An array of *sample* that contains 10 elements will therefore use 780 bytes of memory, as Figure 976 shows.

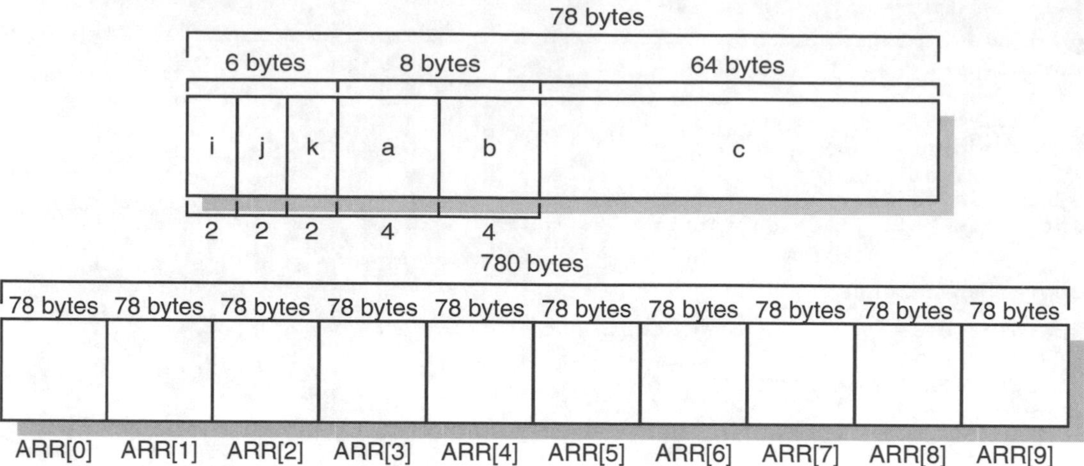

Figure 976 The logical model of the class array's memory consumption.

977 Inline Class Code Allows Changes

As you have learned, C++ lets you place method functions inline within the class or outside of the class. When you are determining which functions to place inline or if any functions should be inline, keep in mind that placing method code inline exposes the code to changes. For example, the following class uses several inline functions:

```
class Book
 {
   public:
    char title[256];
    char author[64];
    float price;
    void show_title(void) { cout << title << '\n'; };
    float get_price(void) { return(price); };
    void show_book(void);
    void assign_publisher(char *name) { strcpy(publisher, name); };
  private:
    char publisher[256];
    void show_publisher(void) { cout << publisher << '\n'; };
 };
```

When another programmer uses the previous class, the programmer can easily change the class methods because code is contained within the class itself. If, instead, you place the class methods within a class library, the programmer must have access to the library source code in order to change the class methods. By using the class library, you can insulate the class from "spur-of-the-moment" changes.

978 Understanding the Static Store

As you have learned, C++ documentation refers to the heap space as the *free store*. As you read through articles and books on C++, you might encounter the term *static store*. In the simplest sense, the static store is a global memory region from which the compiler can allocate data. When you create global or static variables, the compiler can allocate memory for the variables from the static store. In most cases, the scope of objects the compiler allocates from the static store is the entire program. In other words, the objects are global.

SYNCHRONIZING I/O STREAM OPERATIONS WITH STDIO 979

As you have learned, C++ programs can use the standard output functions, such as *printf* and *scanf*, that are defined in *stdio.h*, or they can use the extractor and inserter operators with the I/O streams *cout* and *cin*. To improve your program's readability, usually you should choose one technique or the other. However, there may be times when you cannot avoid using both. For such cases, you can synchronize operations between *cout* and *cin* using the *sync_with_stdio* function. This function directs the two I/O techniques to use the same input buffer and the same output buffer so that the same data is accessible to both. The following program, *syncio.cpp*, illustrates how to use the *sync_with_stdio* function:

```
#include <iostream.h>
#include <stdio.h>

void main(void)
 {
   ios::sync_with_stdio();

   printf("This book is ");
   cout << "Jamsa's C/C++/C# Programmer's Bible\n";
 }
```

UNDERSTANDING C++ I/O STREAMS 980

Almost all the C++ Tips presented throughout this book have made extensive use of the *cin*, *cout*, and *cerr* I/O streams. In the Advanced C++ section of this book, you will learn about *inheritance*, which lets objects of one class inherit the characteristics of another class. Fortunately, C++ provides the base class *ios* (input–output stream), which defines the fundamental I/O operations. Using the *ios* stream, C++ derives an output stream class and an input stream class. When you closely examine the header file *iostream.h* in the Advanced C++ section of this book, you will find a class definition for *ios*

iostream.h, *fstream.h*, and *strstream.h* include files.

UNDERSTANDING THE C++ OUTPUT STREAMS 981

Throughout this book, your programs have made extensive use of the *cout* output stream. In the simplest sense, an *output stream* is a destination for bytes. For most of the previous discussions, you may have assumed that C++ provides an output stream that *cout*, *cerr*, and *clog* use, and an input stream that *cin* uses. Actually, the stream-based header files define three different output streams. Table 981 briefly describes each output stream's use.

Output Stream	Function
ostream	Used for output to *cout*, *cerr*, and *clog*
ofstream	Used for file output to disk
ostrstream	Used to perform buffered output to a string

Table 981 Output streams defined in ***iostream.h, fstream.h,*** *and* ***strstream.h.***

Several Tips presented in this section discuss ways your programs can use these streams.

UNDERSTANDING THE C++ INPUT STREAMS 982

As you learned in Tip 981, the header file *iostream.h* actually defines three different output streams, one for screen output, one for file output, and one for use with strings. As you might guess, the stream header files also define three input streams. Again, in the simplest sense, an *input stream* is a byte source. Table 982 briefly describes each input stream function.

Input Stream	Function
istream	Used for input from *cin*
ifstream	Used for file input from disk
istrstream	Used to read buffered input from a string

Table 982 Input streams defined in ***iostream.h, fstream.h,*** *and* ***strstream.h.***

Several Tips that follow discuss ways your programs can use the streams this Tip and the previous Tip detail.

983 USING THE IOS MEMBERS TO FORMAT INPUT AND OUTPUT

As you have learned, you can use the *ios* members to format input and output. You can also use the *ios formatting flags'* enumerated types. The *iostream.h* include file includes the definitions for the anonymous flags' enumerated types. The possible values for the flags are shown in Table 983.

Flag	Value	Meaning
skipws	0x0001	skip whitespace on input
left	0x0002	left-adjust output
right	0x0004	right-adjust output
internal	0x0008	padding after sign or base indicator
dec	0x0010	decimal conversion
oct	0x0020	octal conversion
hex	0x0040	hexadecimal conversion
showbase	0x0080	use base indicator on output
showpoint	0x0100	force decimal point (floating output)
uppercase	0x0200	uppercase hex output
showpos	0x0400	add '+' to positive integers
scientific	0x0800	// use 1.2345E2 floating notation
fixed	0x1000	use 123.45 floating notation
unitbuf	0x2000	flush all streams after insertion
stdio	0x4000	flush *stdout*, *stderr* after insertion
boolalpha	0x8000	insert/extract bools as text or numeric

Table 983 The enumerated ***iosflags*** *types.*

For example, if you set the *skipws* flag, output will skip leading whitespace characters when your program processes it against a stream. If you set the *hex* flag, you can output information as its hexadecimal representation.

984 SETTING THE FORMAT FLAGS

There are several ways to use format flags with a stream. However, the simplest, and by far the most common way is the *setf* member function, which your programs will implement as shown here:

```
#include <iostream.h>
long setf(long flags);
```

If you invoke the function with an empty parameter, it returns the previous settings of the format flags. If you invoke the function with an enumerated parameter, it turns on the flags that you specify within the parameter. For example, to turn on the *showpos* flag, you will use the following function call:

```
stream.setf(ios::showpos);
```

The *stream* corresponds to the stream to which you want to output—for example, *cout*, *cerr*, and so on. To understand this processing better, consider the following program, *show_hex.cpp*, which displays a value in hexadecimal format:

```
#include <iostream.h>

void main(void)
 {
   cout.setf(ios::hex);
   cout.setf(ios::showbase);
   cout << 100;
 }
```

CLEARING THE FORMAT FLAGS 985

As you have learned, you can use the *setf* function to set the *ios* format flags. The *unsetf* function clears flags that you have previously set with *setf*. Like *setf*, *unsetf* is easy to implement, as shown here:

```
#include <iostream.h>

long unsetf(long flags);
```

For example, you might want to set a flag within a program that forces the computer to output everything in scientific notation. However, certain functions in your program might want to disable that flag before they generate their output. The following program, *unsetf.cpp*, sets and then clears the *uppercase* flag:

```
#include <iostream.h>

void main(void)
 {
   cout.setf(ios::showpos | ios::showprint);
   cout << 100.12;
   cout.unsetf(ios::uppercase);
   cout << endl << 100.12;
 }
```

USING THE OVERLOADED SETF FUNCTION 986

In previous Tips, you have used the *setf* function to control screen output. In addition to the single-parameter implementation of *setf*, *iostream.h* provides an overloaded version of the function, which you will implement within your programs as shown here:

```
#include <iostream.h>

long setf(long flags1, long flags2);
```

When you use the overloaded version of the *setf* function, your program will reset the flags you specify in the *flags2* parameter, and then set those flags equal to the *flags1* parameter. For example, the following program, *setf_ovr.cpp*, clears the *showpos* and *showpoint* flags and sets the *showpoint* flag again:

```
#include <iostream.h>

void main(void)
 {
   cout.setf(ios::showpos | ios::showpoint);
   cout << 100 << endl;
   cout.setf(ios::showpoint, ios::showpos | ios::showpoint);
   cout << 100;
 }
```

987 EXAMINING THE CURRENT FORMATTING FLAGS

As you continue to work with formatting flags within your programs, there may be times when you only want to know what the current format settings are, without altering any settings. To help you determine the current format settings, the *ios* class provides the member function *flags*, which returns the current setting of each format flag encoded into a long integer. The following program, *showflag.cpp*, uses a custom function named *showflags*, together with the *ios flags* function, to generate information about the current system settings:

```
#include <iostream.h>
void showflags(void);

void main(void)
 {
   showflags();
   cout.setf(ios::right | ios::showpoint | ios::fixed);
   showflags();
 }

void showflags(void)
 {
   long flag_set, i;
   int j;
   char flags[15][12] = {
      "skipws", "left", "right", "internal", "dec",
      "oct", "hex", "showbase", "showpoint", "uppercase",
      "showpos", "scientific", "fixed", "unitbuf",
   };

   flag_set = cout.flags();
   for (i=1, j=0; i<0x2000; i = i<<1, j++)
      if (i & flag_set)
         cout << flags[j] << " is on." << endl;
      else
         cout << flags[j] << " is off." << endl;
   cout << endl;
 }
```

When you compile and execute the *showflag.cpp* program, your screen will display the following output:

```
skipws is on.
left is off.
right is off.
internal is off.
dec is off.
oct is off.
hex is off.
showbase is off.
showpoint is off.
uppercase is off.
showpos is off.
scientific is off.
fixed is off.

skipws is on.
left is off.
right is on.
internal is off.
dec is off.
oct is off.
hex is off.
showbase is off.
showpoint is on.
```

```
uppercase is off.
showpos is off.
scientific is off.
fixed is on.
C:/>
```

SETTING ALL FLAGS 988

Just as the *ios* class overloads the *setf* function, it also overloads the *flags* function. The overloaded *flags* function lets you set all format flags associated with a given stream. In other words, your program can create a flag mask, call the *flags* function, and set all the flags detailed within the mask. For example, the following program turns on the *showpos, showbase, oct,* and *right* flags using the *flags* function. It also displays the flags before and after it changes the settings. The *showflags* function is the same as the function called in the *showflag.cpp* program in Tip 987 (which is not reprinted here in the interest of space, but is included within the *set_allf.cpp* program file on this book's companion CD-ROM). However, the *set_allf.cpp* program's *main* function is as shown here:

```
void main(void)
 {
   showflags();
   long f = ios::showpos | ios::showbase | ios::oct | ios::right;
   cout.flags(f);
   showflags();
 }
```

USING THE PRECISION FUNCTION 989

In Tip 830, you used the *setprecision* function to control the precision with which your programs output floating-point data. You can also control the precision with which *cout* displays your floating-point data with the *ios precision* member function. You will implement the *precision* member function as shown here:

```
#include <iostream.h>

int precision(int p);
```

When your program calls the *precision* function, it will set the new precision to the number of decimal places *p* specifies, and it will return the previous precision setting to the calling function. The default precision setting *cout* uses is six places. Depending on your compiler version, you may have to reset the precision before each call to *cout*, or *cout* will use the default precision settings. The following program, *ios_prec.cpp*, uses the *precision* function to format output:

```
#include <iostream.h>

void main(void)
 {
   int i;
   float value = 1.2345;

   for (i = 0; i < 4; i++)
    {
      cout.precision(i);
      cout << value << endl;
    }
 }
```

990 USING THE FILL FUNCTION

As you learned in Tip 828, your programs can use the *fill* member function to change the character that your output streams use to fill whitespace (the default is the space character). As your programs become more complex, using the *fill* member function can provide you with significant assistance in making your output more readable and useful to the user. For example, the following program, *nice_tex.cpp,* uses the *width, precision*, and *fill* member functions to generate formatted output:

```
#include <iostream.h>

void main(void)
 {
   cout.precision(4);
   cout.width(10);
   cout << 10.12345 << endl;
   cout.width(10);
   cout.fill('-');
   cout << 10.12345 << endl;
   cout.width(10);
   cout << "Hi!" << endl;
   cout.width(10);
   cout.setf(ios::left);
   cout << 10.12345;
 }
```

When you compile and execute the *nice_tex.cpp* program, it will generate the following output:

```
     10.12
----10.12
------Hi!
10.12----
C:\>
```

991 UNDERSTANDING MANIPULATORS

In previous Tips, you learned that you can use commands within the *cout* output stream to control text output. These commands are known as *manipulators*. Manipulators are useful because they let you format text while minimizing the number of commands you use. For example, to set the output width to 10 and the fill character to '*', your programs can execute the following commands:

```
cout.width(10);
cout.fill('*');
cout << "Sample" << endl;
```

Alternatively, your programs can use manipulators to achieve the same goal, with fewer programming lines, as shown here:

```
cout << setw(10) << setfill('*') << "Sample" << endl;
```

992 USING MANIPULATORS TO FORMAT I/O

As you have learned, your programs can use manipulators, rather than the stream member functions, to format I/O. The *iostream.h* and *iomanip.h* header files define the manipulators. The *iomanip.h* header file defines only manipulators that receive parameters (such as *setw*). Table 992 lists the manipulators you can use within your programs, their purpose, and whether you can use them for input, output, or both.

Manipulator	Input/Output	Purpose
dec	Input/Output	Instructs the stream to input/output data in decimal format
endl	Output	Instructs the stream to output a newline character and flush itself
ends	Output	Instructs the stream to output a *NULL*
flush	Output	Instructs the stream to flush itself
hex	Input/Output	Instructs the stream to input/output data in hexadecimal format
oct	Input/Output	Instructs the stream to input/output data in octal format
resetiosflags(long f)	Input/Output	Turns off the flags *f* specifies
setbase(int base)	Output	Sets the number base to *base*
setfill(int ch)	Output	Sets the fill character to *ch*
setiosflags(long f)	Input/Output	Turns on the flags *f* specifies
setprecision(int p)	Output	Sets the number of digits of precision
setw(int w)	Output	Sets the field width to *w*
ws	Input	Skips leading whitespace

Table 992 *The stream manipulators you can use to format I/O.*

COMPARING MANIPULATORS AND MEMBER FUNCTIONS

993

As you have seen, your programs can use manipulators, member functions, or both within their code. For example, the following program, *use_man.cpp*, uses a manipulator to control stream output:

```
#include <iostream.h>
#include <iomanip.h>

void main(void)
 {
   cout << hex << 100 << endl;
   cout << setfill('?') << setw(10) << 2343.10 << endl;
 }
```

While the *use_man.cpp* program may take slightly longer to compile (because it includes *iomanip.h)*, its code is more compact than the program *use_mem.cpp*, as shown here:

```
#include <iostream.h>

void main(void)
 {
   cout.setf(ios::hex);
   cout << 100 << endl;
   cout.unsetf(ios::hex);
   cout.fill('?');
   cout.width(10);
   cout << 2343.10 << endl;
 }
```

When you compile the *use_man.cpp* and *use_mem.cpp* programs, both should be identical in size. Both programs will take an identical amount of time to execute. In terms of how the methods will affect your programs, both methods are equivalent. You should use whichever method you find clearer, and make your choice your standard.

994 CREATING YOUR OWN INSERTER FUNCTIONS

In many previous Tips you used the inserter function (the << operator) to place output within a stream. As your classes become more complex, you may want to override the inserter function. When you override the inserter function, the insertion operator must return a reference to an I/O stream. All overloaded inserter functions that you create will be in the following general format:

```
ostream& operator<<(ostream &stream, class_type object)
 {
   // body of inserter
   return stream;
 }
```

The function returns a reference of type *ostream*—a class C++ derives from the *ios* class that supports output. The first parameter to the operator is the stream to insert to, and the second parameter is the object to output to the stream.

Generally, your inserter functions will perform much the same steps as they usually do, except they will do it with the stream object, rather than *cout, cerr,* or *clog* directly. Typically, you will format and structure class data member output within an overloaded insertion function, as you will see in Tip 996.

995 OVERLOADING THE EXTRACTION OPERATOR

In previous Tips, you have overloaded several operators in relation to specific classes. When your classes contain several members whose value you want to output in a specific way, using *cout* and the *extraction* operator can lead to considerable amounts of code. For example, assume you are working with a class whose members include a name, sex (M or F), age, and phone number that you want to output as follows:

```
Name: John Doe          Sex: M  Age: 43     Phone: 555-1212
```

Using *cout*, your programs must include the following statement each time they want to display the output:

```
cout << "Name: " << name << "\tSex: " << sex << "\tAge: " << age <<  "\tPhone: "
     << phone << endl;
```

A better alternative to the form in the previous example would be to overload the extraction operator, as shown in the following program, *outover.cpp*:

```
#include <iostream.h>
#include <string.h>

class Employee
 {
  public:
    Employee(char *name, char sex, int age, char *phone)
     {
       strcpy(Employee::name, name);
       Employee::sex = sex;
       Employee::age = age;
       strcpy(Employee::phone, phone);
```

```
    };
    friend ostream& operator<< (ostream& cout, Employee emp);
  private:
    char name[256];
    char phone[64];
    int age;
    char sex;
 };

ostream& operator<< (ostream& cout, Employee emp)
 {
   cout << "Name: " << emp.name << "\tSex: " << emp.sex;
   cout << "\tAge: " << emp.age << "\tPhone: " << emp.phone << endl;
   return cout;
 }

void main(void)
 {
   Employee worker("Happy", 'M', 4, "555-1212");

   cout << worker ;
 }
```

The *outover.cpp* program overloads the extraction operator within the *ostream* class. The program only uses the overloaded extraction operator when the program invokes the operator with the *Employee* class. Therefore, C++ does not call the overloaded function when the overload performs output operations within the overload itself, but works instead with the regular output stream extractor. Because the extraction operator must access the data members of the *Employee* class, the program declares the operator as a friend of the class.

ANOTHER WAY TO OVERLOAD COUT'S INSERTION OPERATOR 996

Most of the programs presented throughout the last two sections of this book have made extensive use of the *cout* *iostream* to display output to the screen. The following program, *cout_upr.cpp*, overloads the *cout's* insertion operator for character strings, directing *cout* to always display character strings in uppercase:

```
#include <iostream.h>
#include <string.h>

ostream& operator<<(ostream& cout, char *string)
 {
   char *str = strupr(string);

   while (*str)
     cout.put(*str++);

   return(cout);
 }

void main(void)
 {
   cout << "This is a test";
   cout << "\nJamsa's C/C++/C# Programmer's Bible";
 }
```

Within the overload function itself, the statements use the *strupr* function to convert the string to uppercase and then use *cout.put* to display characters one at a time. The overload function could not use *cout's* insertion operator to display the string, or an endless series of recursive calls would result.

997 CREATING YOUR OWN EXTRACTOR FUNCTIONS

Just as you can create your own inserter functions, so too can you overload the extraction (>>) operator for your classes. Generally, you will overload the extraction operator to obtain better control over user input into the data members that comprise a class. You will overload the extractor function using the following general form:

```
istream &operator>>(istream &stream, class_type &object)
 {
   // body of extractor
   return stream;
 }
```

Notice that unlike the overloaded inserter functions, you must pass a reference to the object within the overloaded extractor function. Extractor functions return a reference to a stream of type *istream*, which is the *ios*-derived input stream class. The first parameter is a reference to a stream (generally *cin*).

998 CREATING AN EXTRACTOR EXAMPLE

In Tip 997, you learned the basic format for creating an overloaded extractor function. While you can create globally-overloaded extractor functions, you will typically create an overloaded extractor function specific to each class. For example, the following program, *inover.cpp*, uses the *Employee* class from Tip 995 and receives input tailored specifically for that class:

```
#include <iostream.h>
#include <string.h>

class Employee
 {
  public:
    Employee(void) {};
    Employee(char *name, char sex, int age, char *phone)
     {
       strcpy(Employee::name, name);
       Employee::sex = sex;
       Employee::age = age;
       strcpy(Employee::phone, phone);
     };
    friend ostream &operator<<(ostream &cout, Employee emp);
    friend istream &operator>>(istream &stream, Employee &emp);
  private:
    char name[256];
    char phone[64];
    int age;
    char sex;
 };

ostream &operator<<(ostream &cout, Employee emp)
 {
   cout << "Name: " << emp.name << "\tSex: " << emp.sex;
   cout << "\tAge: " << emp.age << "\tPhone: " << emp.phone << endl;
   return cout;
 }

istream &operator>>(istream &stream, Employee &emp)
 {
   cout << "Enter Name: ";
   stream >> emp.name;
   cout << "Enter Sex: ";
   stream >> emp.sex;
```

```
    cout << "Enter Age: ";
    stream >> emp.age;
    cout << "Enter Phone: ";
    stream >> emp.phone;
    return stream;
  }

void main(void)
  {
    Employee worker;

    cin >> worker;
    cout << worker;
  }
```

CREATING YOUR OWN MANIPULATOR FUNCTION 999

In addition to overloading the insertion and extraction operators, your programs can create their own manipulator functions. Creating custom manipulators may be helpful to you for two reasons. First, you can consolidate a sequence of several separate I/O functions into a single member-function call, enhancing your code's ease of use and readability. Second, you can create custom manipulators to help you manage I/O to non-standard devices. In general, you will construct your custom manipulators in the following manner:

```
stream-name &manipulator-name(stream-name &stream [, parameters])
  {
    // specific manipulator code
   return stream;
  }
```

In the next two Tips, you will create some custom output manipulator functions. However, you can apply the techniques you will use in those Tips to all custom manipulator functions.

CREATING PARAMETERLESS MANIPULATORS 1000

As you learned in Tip 999, you can create custom manipulators within your programs. You will either create parameterless or parameterized manipulators. Tip 1001 discusses parameterized manipulators. However, if your manipulator is performing a standard activity which does not require further input from the calling statement, you will use *parameterless* manipulators. The following program, *sethex.cpp*, creates and uses a parameterless manipulator:

```
#include <iostream.h>
#include <iomanip.h>

ostream &sethex(ostream &stream)
  {
    stream.setf(ios::showbase);
    stream.setf(ios::hex);
    return stream;
  }

void main(void)
  {
    cout << 256 << "    " << sethex << 256;
  }
```

USING PARAMETERS WITH MANIPULATORS 1001

In Tip 1000, you learned how to create a parameterless manipulator function. As you learned, creating a parameterless manipulator is relatively simple. Unfortunately, creating a manipulator that accepts one or more arguments, such as *setw(5)*, is not as simple. To create a manipulator that accepts one or more parameters, you must create the manipulator using a *generic class*. As you will learn in later Tips, generic classes let you write classes and functions that accept multiple types without overloading the class or function for each type. You will learn more about generic classes in later Tips, and you will learn how to write a parameterized manipulator function in Tip 1124. For now, however, consider the following construction of a typical overloaded parameterized manipulator:

```
ostream &indent(ostream &stream, int length)
 {
   register int i;
   for(i = 0; i < length; i++)
      cout << " ";
   return stream;
 }
```

When you call the parameterized *indent* manipulator, the value you pass into the manipulator sets the number of spaces for the program to indent within the stream.

1002 UNDERSTANDING THE OLD STREAM CLASS LIBRARY

As you have learned, many C++ programs will use the *iostream.h* include file, which includes the definitions for many of the C++ style output controls. When Bjarne Stroustrup originally invented C++, the language used a smaller and somewhat different I/O class library, called *stream.h*. As C++ continued to evolve, it replaced the *stream.h* library with the *iostream.h* library this book describes. To maintain backward compatibility, most C++ compilers continue to support the *stream.h* library. However, you should always use the newer, more powerful *iostream.h* library within programs you write in C++.

1003 OPENING A FILE STREAM

As you have learned, C++ provides the *ifstream* and *ofstream* (input and output) file streams. Within your C++ programs, you can perform I/O using these two stream classes, or you can use standard C file I/O operations using *fopen*, *fgets*, *fputs*, and so on. To open a file stream, you must declare a corresponding class variable (*ifstream* or *ofstream*). Assuming you want to perform input and output operations, your declarations will appear as follows:

```
ifstream input;
ofstream output;
```

Then, to open a file stream, you might use the *open* member, as shown here:

```
input.open("FILENAME.EXT", ios::in);
output.open("FILENAME.OUT", ios::out);
```

The generalized form of the *open* statement is shown here:

```
ifstream.open(const char *FILENAME, int nMode=ios::in,
              int nProt=filebuf::openprot);
ofstream.open(const char *FILENAME, int nMode=ios::out,
              int nProt=filebuf::openprot);
```

As you can see, the default value for the *nMode* parameter is *ios::in* or *ios::out*, depending on the stream you are manipulating, so the following constructions are just as valid as those shown previously:

```
input.open("FILENAME.EXT");
output.open("FILENAME.OUT");
```

The *nProt* parameter lets you control shared access to the file. In addition to using the *open* member, you can use the stream object's constructor function when you declare the stream variable, as shown here:

```
ifstream input("FILENAME.EXT", ios::in);
ofstream output("FILENAME.OUT", ios::out);
```

The *ios::in* or *ios::out* parameters select input or output. In addition to these two values, your programs can use combinations of the values listed in Table 1003.

Value	Meaning
ios::app	Opens the stream in append mode
ios::ate	Opens a file for either input or output, moving the file pointer to the end of the file
ios::in	Opens a file for input
ios::out	Opens a file for output
ios::nocreate	Opens a file only if it already exists
ios::noreplace	Opens a file only if it does not already exist
ios::trunc	Truncates an existing file
ios::binary	Opens a file in binary mode

*Table 1003 Open mode values for **ifstream** and **ofstream**.*

Closing a File Stream 1004

As you learned in Tip 1003, your programs can open an *iostream* using the *open* member or using a constructor function when you declare a stream variable. When you finish using the file stream, your programs should close the stream using the *close* member function, as shown here:

```
input.close();
output.close();
```

By default, when your program ends or if you destroy the class variable, C++ closes the file stream. If you want to associate the file stream with a different file, however, you must first use the *close* member, then use the new filename to reopen the stream.

Reading and Writing File Stream Data 1005

As you have learned, C++ lets your programs open file streams of the class *istream* for input operations. To read data from an input file stream, your programs use the *read* member, as shown here:

```
input.read(buffer, number_of_bytes);
```

As you can see, the program must provide a buffer into which the *read* member function can store the data it reads, and the program must inform the *read* member function of the number of bytes the program wants it to read. In a similar way, if you have opened an output stream, your programs can use the *write* member to write data to the output stream, as shown here:

```
output.write(buffer, number_of_bytes);
```

Note that the *read* and *write* members *do not* return the number of bytes the member function successfully read or wrote. Instead, the functions return references to the stream. To determine an operation's success or failure, you must check the status members that Tip 1006 details. Operations may fail for any number of reasons, including full disks, nonexistent files, hard file errors, and more. You should be sure that your programs regularly check the success or failure of file I/O activity to protect against unknown errors.

1006 Checking the Status of a File Operation

When you perform file I/O operations using the *ifstream* and *ofstream* class streams, your programs can use the members listed in Table 1006 to determine whether an open, read, or write operation was successful.

Member	Example	Function
bad	*stream.bad()*	Returns true if the I/O operation encounters an unrecoverable error
fail	*stream.fail()*	Returns true if the I/O operation encounters a recoverable or expected error, such as file not found
good	*stream.good()*	Returns true if the I/O operation is successful
eof	*stream.eof()*	Returns true if the I/O operation encounters an end of file
clear	*stream.clear()*	Clears the status flags
rdstate	*stream.rdstate()*	Returns the current error state

Table 1006 *Member functions that return I/O success or failure information.*

1007 Putting File Stream Operations Together

Several preceding Tips discussed C++ file streams and the various member functions your programs can use to perform file I/O. The following program, *filecopy.cpp*, uses several of these member functions to create a simple file copy program, which copies text files a character at a time:

```
#include <iostream.h>
#include <stdlib.h>
#include <fstream.h>

void main(int argc, char **argv)
 {
   char buffer[1];
   ifstream input(argv[1], ios::in);

   if (input.fail())
    {
      cout << "Error opening the file " << argv[1];
      exit(1);
    }

   ofstream output(argv[2], ios::out);
   if (output.fail())
    {
      cout << "Error opening the file " << argv[2];
      exit(1);
    }
   do {
     input.read(buffer, sizeof(buffer));
     if (input.good())
       output.write(buffer, sizeof(buffer));
   } while (! input.eof());
   input.close();
   output.close();
 }
```

The *filecopy.cpp* program first opens the file which the command-line specifies as the file to copy. If the file open action fails, the program exits with a failure status value. If the program successfully opens the first file, the program next opens the target file that your command-line specified as its second argument. If the program cannot open the second file, the program again exits with a failure status value. If the program successfully opens both files, the program next copies the first file, one character at a time, to the second file. After the loop which copies characters reaches the end of the first file, and thus terminates, the program cleans up after itself (in other words, it closes the open file streams) and ends normally.

To use the *filecopy.cpp* program, you will invoke the *filecopy* program as shown in the following listing to copy the *filecopy.cpp* file to *filecopy.sav*,:

```
C:\> FILECOPY  FILECOPY.CPP  FILECOPY.SAVENTER>
```

PERFORMING A BINARY COPY OPERATION 1008

In Tip 1007, you created the *filecopy.cpp* program, which copies the first text file specified in the command line to a file with the name the second command-line argument specifies. If you want the program to copy binary files (such as an EXE file that contains a program), you must change the open operations to use the *ios::binary* flag, as shown here:

```
ifstream input(argv[1], ios::in | ios:binary);
if (input.fail())
 {
   cout << "Error opening the file " << argv[1];
   exit(1);
 }
ofstream output(argv[2], ios::out | ios::binary);
if (output.fail())
 {
   cout << "Error opening the file " << argv[2];
   exit(1);
 }
```

This book's companion CD-ROM contains the program *bin_copy.cpp*, which performs a binary file copy operation.

UNDERSTANDING THE STREAMBUF CLASS 1009

In previous Tips, you have worked with different file and other I/O streams to manipulate data. When you *buffer* data within your programs, you place it into an interim location before or after a read or write operation. C++ derives the classes that buffer I/O data (buffering I/O streams) from the *streambuf* base class. A buffering I/O stream provides a buffer interface between your data and storage areas, such as memory or physical devices. The buffers *streambuf* objects create are known as *get, put*, and *reserve* areas. Your programs access and manipulate the contents of the *streambuf* object-created areas with pointers that point to the characters within those *get, put* and *reserve* areas.

The buffering actions that *streambuf* objects perform are rather primitive. Because the buffering *streambuf* objects perform is not as useful as the buffering the higher-level stream classes C++ derives from *streambuf* can perform, your applications will normally gain access to buffers and buffering functions through a pointer to *streambuf*. Your programs will use that pointer indirectly, within the definition of an *ios*-based object. The *ios* class provides a pointer to *streambuf* that provides transparent access to buffer services for high-level classes—in other words, your programs should use the *ios* class and let *streambuf* buffer I/O "behind the scenes," on its own. The high-level classes provide I/O formatting. In other words, your programs should generally use high-level stream classes to control your I/O manipulations. However, there may be times when your programs must perform low-level access to a *streambuf* object.

Tip 1010 shows how to access *streambuf* objects from within your programs. Table 1009 lists some of the public member functions for the *streambuf* class and their descriptions.

Function	Description
in_avail	The *in_avail* member function returns the number of characters remaining in the internal input buffer.
out_waiting	The *out_waiting* member function returns the number of characters remaining in the internal output buffer.
pbump(n)	The *pbump* member function increments the put pointer (*pptr*) by *n*, which may be a positive or negative value.
sbumpc	The *sbumpc* member function returns the current character from the internal input buffer, then advances the buffer's internal pointer to the next character.
sgetc	The *sgetc* member function peeks at the next character in the internal input buffer.
snextc	The *snextc* member function advances the input buffer's internal pointer to the next character and returns that character.
sputbackc	The *sputbackc* member function returns a character to the internal input buffer.
sputc	The *sputc* member function puts one character into the internal output buffer.
stossc	The *stossc* member function advances the input buffer's internal pointer to the next character in the input buffer.

Table 1009 *Member functions of the* ***streambuf*** *class.*

If you observe the member function descriptions carefully, you will see that the majority of the member functions for the *streambuf* class perform activities similar to the higher-level tasks that you use *get*, *put*, *read*, and *write* to perform. In fact, because all stream-based classes derive from *streambuf*, all the higher-level functions you use to access streams derive from the *streambuf* member functions.

1010 WRITING A SIMPLE STREAMBUF EXAMPLE

As you learned in Tip 1009, C++ provides the *streambuf* base class to help you manage stream input and output. While your programs should generally use the higher-level classes and their member functions to control streams, you can use an instance of the *streambuf* class within your programs much as you would use an instance of a higher-level class. The following program, *use_sbuf.cpp*, uses a *streambuf* object to output text entered from the keyboard to a disk file:

```
#include <iostream.h>
#include <fstream.h>
#include <stdlib.h>

void main(void)
 {
   int c; const char *filename = "_junk_.$$$";
   ofstream outfile;
   streambuf *out, *input = cin.rdbuf();

   // Position at the end of file. Append all text.
   outfile.open( filename, ios::ate | ios::app);
   if (!outfile)
    {
      cerr << "Could not open " << filename;
      exit(-1);
    }

   out = outfile.rdbuf();  // Connect ofstream and streambuf.

   clog << "Input some text. Use Control-Z to end." << endl;
   while ( (c = input -> sbumpc() ) != EOF)
```

```
      {
        cout << char(c);                              // Echo to screen.
        if (out -> sputc(c) == EOF)
           cerr << "Output error";
      }
   }
```

When you compile and execute the *use_sbuf.cpp* program, it will open the *_junk_.$$$* file in the current directory. The program then attaches the *out streambuf* object to the buffer for the *junk* file. The program checks each letter that the user enters at the keyboard for a CTRL+Z (*EOF*) entry. The program uses the *streambuf* member to output each entry that is not *EOF* to the screen and to the file. As you can see, this program is not fundamentally different from programs that you have written previously, which used the *get* and *put* member functions against the other *stream*-based classes.

READING BINARY DATA USING READ 1011

As you learned in Tip 1009, your programs can use the *istream* class *get* method to read information a byte at a time from a binary file. Your programs can also use the *read* member function to read binary data from a file a block (that is, some number of bytes that you determine) at a time. You will use the *read* function within your programs as shown here:

```
#include <iostream.h>
#include <fstream.h>

istream &read(unsigned char *buffer, int num);
```

The *read* member function reads *num* bytes from the input stream. The *read* member function then places those *num* bytes into the memory buffer that begins at the address to which the *buffer* parameter points. The *num* parameter determines the size of the block *read* returns from the file. Generally, your blocks should be relatively small, both to avoid memory overruns and keep the file manageable in the event a single *read* fails. Clearly, the *read* method is significantly more powerful and more useful than *get* (because you can read data sequences, rather than single characters), provided you have some advance knowledge of the construction of the file that your program will read.

WRITING BINARY DATA USING WRITE 1012

As you learned in Tip 1010, your programs can use the *ostream* class *put* member function to write binary data to a file. As you learned in Tip 1011, your programs can use the *istream* class *read* method to read a predetermined number of bytes from a file. Generally, you will use the *read* method in conjunction with a file that you wrote using the *ostream* class *write* member function. You will use the *write* member function within your programs as shown here:

```
#include <iostream.h>
#include <fstream.h>

ostream& write(const unsigned char *buffer, int num);
```

The *write* function writes the number of bytes the *num* parameter specifies—beginning at the memory address to which the *buffer* points—to the stream with which you invoke the function. To better understand both the *read* and *write* methods, consider the following program, *rw_struc.cpp*, which writes a structure to the disk, and then reads the structure back in and displays it:

```
#include <iostream.h>
#include <fstream.h>
#include <string.h>
#include <stdlib.h>
```

```
struct  status
 {
   char name[80];
   float balance;
   unsigned long account_number;
 };

void main(void)
 {
   struct status acc;

   strcpy(acc.name, "Kris Jamsa");
   acc.balance = 1234.56;
   acc.account_number = 98765432;
   ofstream outbal("balance.asc", ios::out | ios::binary);
   if(!outbal)
    {
      cout << "Cannot open output file." << endl;
      exit (1);
    }
   outbal.write((unsigned char *) &acc, sizeof(struct status));
   outbal.close();
   ifstream inbal("balance.asc", ios::in | ios::binary);
   if(!inbal)
    {
      cout << "Cannot open file." << endl;
      exit (1);
    }
   inbal.read((unsigned char*) &acc, sizeof(struct status));
   cout << acc.name << endl;
   cout << "Account number: " << acc.account_number << endl;
   cout.precision(2);
   cout.setf(ios::fixed);
   cout << "Balance: $" << acc.balance << endl;
   inbal.close();
 }
```

The program writes a single record of type *status* to the file *balance.asc*. As you can see, the *balance.asc* file contains the name on an account, the account's current balance, and the account number. The program writes the values the object of type *status* contains to the file *balance.asc* as a series of binary values. The program then closes the output stream and opens an input stream. The input stream reads the saved binary values into a different instance of the *status* structure. Finally, the program outputs the contents read back in from the file to the screen for the user to review.

1013 USING THE GCOUNT MEMBER FUNCTION

As you learned in Tip 1011, your programs can use the *istream* class *read* method to read a certain number of bytes from a file. There may be times, however, when a *read* operation fails and you must determine how many characters *read* obtained from the stream before it stopped. To do so, your programs can use the *gcount* member function, which you will use within your programs as shown here:

```
#include <iostream.h>
#include <fstream.h>

int gcount(void);
```

When you invoke the *gcount* member function with an input stream, *gcount* returns the number of characters the last binary input operation read. To better understand this processing, consider the following program, *cnt_read.cpp*, which illustrates how you might use *gcount*:

```
#include  <iostream.h>
#include <fstream.h>
#include <sdtlib.h>

void main(void)
 {
   float fnum[4] = {99.75, -34.4, 1776.0, 200.1};
   int i;

   ofstream out("numbers.asc", ios::out | ios::binary);
   if(!out)
    {
      cout << "Cannot open file.";
      exit (1);
    }
   out.write((unsigned char *) &fnum, sizeof(fnum));
   out.close();
   for (i = 0; i < 4; i++)
      fnum[i] = 0.0;
   ifstream in("numbers.asc", ios::in | ios::binary);
   if(!in)
    {
      cout << "Cannot open file.";

       exit (1);
    }
   in.read((unsigned char *) &fnum, sizeof(fnum));
   cout << in.gcount() << " bytes read." << endl;
   for (i = 0; i < 4; i++)
      cout << fnum[i] << " ";
   in.close();
 }
```

The *cnt_read.cpp* program defines an array of four floating-point values. Next, the program writes the four values to an ASCII file. The program then closes the output stream it used to write the ASCII file. After closing the stream, the program clears the values in the *fnum* array and opens an input stream to read the values back from the file. The program alerts you of how many bytes it read, then outputs the floating-point values it read from the disk. When you compile and execute the *cnt_rad.cpp* program, your screen will display the following output:

```
16 bytes read.
99.75 -34.4 1776 200.1
C:\>
```

USING THE OVERLOADED GET FUNCTIONS 1014

As you have learned, your programs can use the *istream* class *get* member function to read a single character at a time from a text file. However, most C++ compilers also include two overloaded versions of the *get* method. The first overloaded version reads a series of characters; the second reads a single *int* at a time. You will use the overloaded versions of the *get* method within your programs as shown here:

```
#include <iostream.h>
#include <fstream.h>

ostream &get(char *buffer, int num, char delimiter ='\n');
int get(void);
```

The first overloaded function reads characters into the array to which the *buffer* parameter points. The function reads characters until one of two events occurs: either the function reads the number of characters the *num* parameter specifies, or it encounters the character the *delimiter* parameter specifies. (The *delimiter* parameter is the newline character by default.) If the overloaded *get* function encounters the delimiter, it will stop reading immediately, without removing the delimiter from the input stream.

The second overloaded function returns the next character from the stream as an integer value. It will return the *EOF* constant if it encounters the end of file marker. The second overloaded function should look very familiar to you, as it is similar to C's *getc* function.

1015 USING THE GETLINE METHOD

In previous Tips, you learned about several methods your programs can use to read information in from an input stream. C++ also provides the *getline* method, which lets your programs read in data from a file one line at a time. You will use the *getline* method within your programs as shown here:

```
#include <iostream.h>
#include <fstream.h>

ostream &getline(char *buffer, int num, char delimiter ='\n');
```

As you can see, the *getline* method is almost identical to the first overloaded *get* method. The *getline* method reads data either until it reaches the *delimiter* or until it reads the number of characters *num* specifies. The *getline* method places the characters within the buffer that begins at the *buffer* pointer. For example, the following program, *getlines.cpp*, reads the contents of a text file one line at a time and outputs it to the screen:

```
#include <iostream.h>
#include <fstream.h>
#include <stdlib.h>

void main(int argc, char *argv[])
 {
   if(argc!=2)
    {
      cout << "Usage: getlines <filename>" << endl;
      exit (1);
    }
   ifstream in(argv[1]);
   if(!in)
    {
      cout << "Cannot open file.";
      exit (1);
    }
   char str[255];
   while(in)
    {
      in.getline(str, 255);       // Delimiter defaults to newline
      cout << str << endl;
    }
   in.close();
 }
```

1016 DETECTING THE END OF FILE

As you have learned, when you work with streams, your programs can determine when the file pointer reaches the end-of-file marker by comparing the value that *get* or a counterpart function returns to the *EOF* constant. Your programs can also check for the end-of-file marker by checking the value of the *eof* method, which you will use within your programs as shown here:

```
#include  <iostream.h>
#include <fstream.h>

int eof(void);
```

The *eof* function returns 0 (false) normally, and returns a non-zero (true) value if the pointer is at the end of file. The following program, *chk_eof.cpp*, uses *eof* to read a file in from a stream:

```
#include <iostream.h>
#include <fstream.h>
#include <ctype.h>
#include <iomanip.h>
#include <stdio.h>
#include <stdlib.h>

void main(int argc, char *argv[])
 {
   if(argc!=2)
    {
      cout << "Usage: chk_eof <filename>" << endl;
      exit (1);
    }
   ifstream in(argv[1], ios::in | ios::binary);
   if(!in)
    {
      cout << "Cannot open input file." << endl;
      exit (1);
    }

   register int i, j; int count = 0;
   char c[16];

   cout.setf(ios::uppercase);
   while(!in.eof())
    {
      for(i = 0; i < 16 && !in.eof(); i++)
         in.get(c[i]);
      if(i<16)
         i--;                                      // Don't print EOF
      for(j=0; j < i; j++)
         cout << setw(3) << hex << (int) c[j];
      for(; j < 16; j++)
         cout << "   ";
      cout << "\t";
      for(j = 0; j < i; j++)
         if(isprint(c[j]))
            cout << c[j];
         else
            cout << ".";
      cout << endl;
      count ++;
      if(count==16)
       {
         count = 0;
         cout << "Press ENTER to continue: ";
         cin.get();
         cout << endl;
       }
    }
   in.close();
 }
```

The *check_eof.cpp* program reads an input file 16 bytes at a time. Each time the program reads 16 bytes, it outputs those 16 bytes to the screen as a sequence of hexadecimal numbers. Next, the program outputs the same 16 bytes to the screen as a sequence of letters. After the program performs the output process 16 times, it stops and waits for the user to press a key. Then it reads 16 bytes 16 more times (for a total of 256 bytes each cycle) until it reads the end of file marker. When you compile and execute the *chk_eof.cpp* program against the *chk_eof.cpp* file, the program will display the following output, as shown here in part:

```
23 69 6E 63 6C 75 64 65 20 3C 69 6F 73 74 72 #include <iostre
61 6D 2E 68 3E  D  A 23 69 6E 63 6C 75 64 65 20 am.h>..#include
3C 66 73 74 72 65 61 6D 2E 68 3E  D  A 23 69 6E <fstream.h>..#in
63 6C 75 64 65 20 3C 63 74 79 70 65 2E 68 3E  D clude <ctype.h>.
 A 23 69 6E 63 6C 75 64 65 20 3C 69 6F 6D 61 6E .#include <ioman
69 70 2E 68 3E  D  A 23 69 6E 63 6C 75 64 65 20 ip.h>..#include
3C 73 74 64 69 6F 2E 68 3E  D  A  D  A 69 6E 74 <stdio.h>....int
20 6D 61 69 6E 28 69 6E 74 20 61 72 67 63 2C 20  main(int argc,
63 68 61 72 20 2A 61 72 67 76 5B 5D 29  D  A 20 char *argv[])..
7B  D  A 20 20 20 69 66 28 61 72 67 63 21 3D 32 {..   if(argc!=2
29  D  A 20 20 20 20 7B  D  A 20 20 20 20 20 20 )..    {..
63 6F 75 74 20 3C 3C 20 22 55 73 61 67 65 3A 20 cout << "Usage:
63 68 6B 5F 65 6F 66 20 3C 66 69 6C 65 6E 61 6D chk_eof <filenam
65 3E 22 20 3C 3C 20 65 6E 64 6C 3B  D  A 20 20 e>" << endl;..
20 20 20 20 72 65 74 75 72 6E 20 31 3B  D  A 20     return (1);..
20 20 20 7D  D  A 20 20 20 69 66 73 74 72 65 61    }..   ifstrea
Press ENTER to continue:
```

1017 USING THE IGNORE FUNCTION

As you have learned, C++ provides you with many tools that you can use to read from and write to I/O streams. However, none of the functions that you have learned about so far let you read and discard characters from within the input stream. To perform read-and-discard processing, your programs can use the *ignore* function, which you will use within your programs as shown here:

```
#include <iostream.h>
#include <fstream.h>

istream &ignore(int num=1, int delimiter=EOF);
```

The *ignore* function reads and discards characters until it has either removed *num* characters from the input stream or it encounters the *delimiter* parameter, which defaults to the *EOF* pointer. If *ignore* encounters the *delimiter*, it will not remove the *delimiter* from the input stream. Instead, *ignore* will stop its own processing immediately. The following program, *ignorech.cpp*, reads itself from the disk. The *ignorech.cpp* program ignores characters until it encounters a space or until it has read 10 characters, and then the program displays the rest of the file:

```
#include <iostream.h>
#include <fstream.h>

void main(void)
 {
   ifstream in("ignorech.cpp");

   if(!in)
    {
      cout << "Cannot open file." << endl;
      exit(1);
    }
```

```
    in.ignore(10, ' ');
   char c;
   while(in)
    {
      in.get(c);
      cout << c;
    }
   in.close();
 }
```

The *ignorech.cpp* program code uses the *ignore* member function to skip the first 10 characters, or up to the first space, whichever comes first, within the named file. The program then uses the *get* function to read the remaining characters within the file. The program outputs each character it reads to the screen.

USING THE PEEK FUNCTION 1018

As you have learned, C++ provides you with many functions you can use to access input streams. As your programs work with input streams, there may be times when you must "see" the next character within a stream without removing that character from the stream. To perform such processing, your programs can use the *peek* method, which you will use within your program as shown here:

```
#include <iostream.h>
#include <fstream.h>
int peek(void);
```

The *peek* function returns the next character in the stream or the constant *EOF* if the file pointer is at the end of file.

USING THE PUTBACK FUNCTION 1019

As you work with streams within your programs, there may be times when your programs must return a character they have read to an input stream. You can use the *putback* function within your programs to return the last character your program has read from a given input stream to that input stream. You will use the *putback* function within your programs as shown here:

```
#include <iostream.h>
#include <fstream.h>
istream &putback(char c);
```

The *putback* function places the character *c* back into the stream your code called it with and returns a reference to that stream. Note that *putback* only returns a character to the stream from which the program retrieved it; you cannot use *putback* to place characters in streams other than those from which the program originally retrieved the characters.

FINDING THE CURRENT POSITION IN THE FILE STREAM 1020

As you have learned, there are many cases when your programs may want to know the file pointer's current position. In Tip 1016, you used the *eof* member function to determine whether the file pointer was at the end-of-file marker. However, your programs often must know where they are within a file. When your programs must determine the current file pointer position, they can use the *tellg* and *tellp* members, which you will use within your programs as shown here:

```
#include <iostream.h>
#include <fstream.h>
```

```
long  output.tellp(void);
long input.tellg(void);
```

It is important to note that *tellp* and *tellg* perform identical processing. However, your programs must always use *tellp* with output streams and *tellg* with input streams.

1021 CONTROLLING THE FILE STREAM POINTER

As you have learned, the *ifstream* and *ofstream* file streams provide member functions that your programs can use to perform file I/O. When your programs perform file I/O operations, there may be times when the programs must set the position of the file-stream pointer. To position the file pointer, your programs can use the *seekg* (for input) and *seekp* (for output) methods, as shown here:

```
#include <iostream.h>
#include <fstream.h>

istream &seekg(streamoff byte_offset [, seek_dir origin]);
ostream &seekp(streamoff byte_offset [, seek_dir origin]);
```

The byte offset is a *long* offset value (enumerated within the *iostream.h* file) that, unless you specify otherwise within the optional *origin* parameter, C++ applies from the start of the file. To apply the byte offset from a location other than the start of the file, use one of the following enumerated values for the *origin* parameter:

```
enum seek_dir { beg=0, cur=1, end=2 };
```

1022 USING SEEKG AND SEEKP FOR RANDOM ACCESS

In Tip 1021, you learned about the *seekg* and *seekp* methods, which your programs can use to manipulate the file pointer's location within a random access file. To better understand how you will use the *seekg* and *seekp* methods within your programs, consider the following program, *change.cpp*, which uses *seekp* to write over a specific character within a file:

```
#include <iostream.h>
#include <fstream.h>
#include <stdlib.h>

void main(int argc, char *argv[])
 {
   if(argc!=4)
    {
      cout << "Usage: change <filename> <byte> <character>" << endl;
      exit (1);
    }
   fstream out(argv[1], ios::in | ios::out | ios::binary);
   if(!out)
    {
      cout << "Cannot open file!";
      exit (1);
    }
   out.seekp(atoi(argv[2]), ios::beg);
   out.put(*argv[3]);
   out.close();
 }
```

When you compile and execute the *change.cpp* program, it searches for the location you specify within the *byte* command-line parameter. It then replaces that byte with the character you specify in the *character* command-line parameter.

***Note:** If you specify a value for **byte** which is beyond the **EOF** pointer, the program will write a character to that byte and move the **EOF** pointer to that location. If you intend to use a program which uses processing similar to **change.cpp** for some greater purpose, be sure to put a test into the program which ensures that the **byte** value is not larger than the file's length.*

MANIPULATING THE FILE POINTER'S POSITION WITHIN A FILE 1023

As you have learned, the C++ *iostream* classes provide you with significant control and information over the file pointer's placement and activities within a file. When you combine the random access information with the information which the streams can return to your program, you have a powerful tool for managing random access files.

The following program, *seek_tel.cpp*, moves to several locations within a file, tells you its current position within the file, and outputs the manipulated file when it completes its activities. While the *seek_tel.cpp* program does not particularly accomplish much, it does make more clear the relationship between the file pointer and the file data, as shown here:

```
#include <iostream.h>
#include <fstream.h>
#include <stdlib.h>

void main(int argc, char *argv[])
 {
   if(argc!=3)
    {
      cout << "Usage: seek_tel <filename> <byte> " << endl;
      exit (1);
    }

   fstream out(argv[1], ios::in | ios::out | ios::binary);

   if(!out)
    {
      cout << "Cannot open file!";
      exit (1);
    }

   out.seekp(atoi(argv[2]), ios::beg);

   cout << "Current position is: " << out.tellp() << endl;
   out.close();
 }
```

The *seek_tel.cpp* program receives three command-line parameters from the user: the program name, the file it is to operate on, the number of bytes it is to offset from the file's beginning.

The program then moves to the appropriate location within the file, outputs the character, and lets you know that it has done so and what its current position is. You will run the *seek_tel.cpp* program with command-line instructions similar to the following:

```
C:\> seek_tel TEXT.TXT 10
```

DETERMINING THE CURRENT STATUS OF AN I/O STREAM 1024

The C++ I/O system maintains status information about the outcome of each I/O operation your program performs. C++ holds the current state of the I/O system within an integer whose value your programs can access by calling the *rdstate* member function. The integer includes the bitwise flags listed in Table 1024.

Flag	Meaning
eofbit	1 when the file pointer reaches the end of file marker 0 otherwise
failbit	1 when a (possibly) non-fatal error occurs 0 otherwise
badbit	1 when a fatal I/O error occurs 0 otherwise

*Table 1024 The flag bits **ios** enumerates.*

Within your programs, you can use the *rdstate* member function to test the current C++ I/O status. You will implement the *rdstate* member function as shown here:

```
#include <iostream.h>
#include <fstream.h>

int rdstate(void);
```

To check the current status of an I/O stream, you will use code similar to the following fragment:

```
status = in.rdstate();
if(status & ios::eofbit)
  cout << "EOF encountered." << endl;
if(status & ios::failbit)
  cout << "Non-Fatal I/O error." << endl;
if(status & ios::badbit)
  cout << "Fatal I/O error." << endl;
```

However, just as with exception handling, your programs will probably pursue more useful processing than simple output in the event of an error of this nature. You may construct a *while* loop which does not let the user exit until the user resolves the error or selects cancel from an options list. Just as exception handling let your programs perform more complex processing with less risk of a fatal systems crash, so too does constant checking of the value *rdstate* returns help protect your program from serious I/O which can lead to a program failure.

1025 UNDERSTANDING THE ARRAY I/O CLASSES

C++ supports three array-based I/O classes, the counterparts to the file-based I/O classes. The three array-based classes are *istrstream, ostrstream,* and *strstream.* C++ derives all three classes from *strstreambuf,* among other base classes. The *strstreambuf* base class defines several low-level details that the derived classes use. In addition to *strstreambuf,* C++ also derives *istrstream* from *istream*, *ostream*, and *strstream*.

Because of their place within the inheritance structure, all three array-based stream classes have access to the same member functions as file-based stream classes. You can therefore use stream arrays within your programs to perform many of the same operations that you would perform within a file stream.

Note: *Array-based I/O classes let your C++ programs perform activities similar to those your C programs used **sprintf** to accomplish. Array-based I/O classes let your programs buffer and format output before they display the output.*

1026 UNDERSTANDING CHARACTER STRING STREAMS

In the C section of this book, you learned that your programs can use the *sprintf* and *sscanf* functions to output and input data to and from a character string. To help you perform similar operations, the header file *ostrstream.h* defines the *ostrstream* class. When your programs create a string output stream, you essentially bind the stream to a specific character string. The following program, *fill_str.cpp*, creates a variable of type *ostrstream* and fills it with the characters "Jamsa's C/C++/C# Programmer's Bible":

```
#include  <iostream.h>
#include <strstrea.h>

void main(void)
 {
   char string[256];

   ostrstream str(string, 256);                         // Bind the string
   str << "Jamsa's C/C++/C# Programmer's Bible" << ends;
   cout << string;
 }
```

USING ISTRSTREAM TO WRITE A CHARACTER STRING 1027

In Tip 1026, you learned how to declare a variable of type *ostrstream* and to use that string stream to output information from within your programs. The C++ libraries also define the *istrstream* class, which your programs can use to bind an input stream to an array. Within your programs, you will declare the *istrstream* type as shown here:

```
#include <iostream.h>
#include <strstrea.h>

istrstream istr(char *buffer);
```

In the *istrstream* constructor, the *buffer* parameter is a pointer to an array that the stream will use as its source for characters. To help you better understand the processing that the *istrstream* class performs, consider the following program, *first_in.cpp:*

```
#include <iostream.h>
#include <strstrea.h>

void main(void)
 {
   char in_string[] = "10 Hello 0x88 12.23 done";
   istrstream ins(in_string);
   int i;
   char str[80];
   float f;

   ins >> i;
   ins >> str;
   cout << i << " " << str << endl;
   ins >> i;
   ins >> f;
   ins >> str;
   cout << hex << i << " " << f << " " << str;
 }
```

Because C++ input streams stop accepting data when they encounter whitespace, you can use the insertion operator with the character string *in_string* to fill the other variables the program defines later while processing. When the program executes, the program fills the *i* variable with the value 10 and the *str* variable with the value "Hello." The program then outputs those values on the first line. After the program generates the first line of output, the program again fills *i*, this time with *0x88*. Next, the program fills *f* with 12.23, and *str* with "done." When you compile and execute the *first_in.cpp* program, your screen will display the following output:

```
10 Hello
88 12.23 done
C:\>
```

1028 BETTER UNDERSTANDING OSTRSTREAM

As you learned in Tip 1026, your programs can use *ostrstream* output arrays to format and design output before you actually display the output. When you declare an instance of *ostrstream* within your programs, you will use the following general form:

```
#include <iostream.h>
#include <strstrea.h>

ostrstream ostr(char *buffer, int size, int mode=ios::out);
```

The *buffer* parameter contains the starting address of the array to which *ostr* writes the output string. The *size* parameter contains the size (in bytes) of the buffer. Finally, the optional *mode* parameter lets you control how C++ opens the stream. By default, C++ opens the stream for normal output, but you can, for example, change the stream so that C++ automatically appends all output to the stream.

When you declare a stream as array-based, your programs will write all information directed to the stream into the array. However, you should be careful that your streams do not override the size of the buffer, or your programs may return an error, or even cause a system crash.

1029 USING THE OVERLOADED ISTRSTREAM FORMS

As you learned in Tip 1027, usually you will declare an array-based input stream with only a single parameter, a pointer to the array that holds the stream's input characters. However, C++ also supports an overloaded version of the *istrstream* constructor, the general form of which is shown here:

```
#include <iostream.h>
#include <strstrea.h>

istrstream istr(char *buffer, int size);
```

You can use the overloaded constructor to limit the access of the stream to only the first *size* elements within the array that the *buffer* parameter points to. You might use the overloaded constructor when you know information about the array in advance, or when you only want the first *size* elements and do not care about discarding the rest. For example, the following program, *stre_in.cpp* uses the overloaded version of the *istrstream* constructor:

```
#include <iostream.h>
#include <strstrea.h>

void main(void)
 {
   char in_string[] = "10 Hello 0x88 12.23 done";
   istrstream ins(in_string, 8);
   int i;
   char str[80];
   float f;

   ins >> i;
   ins >> str;
   cout << i << " " << str << endl;
   ins >> i;
   ins >> f;
   ins >> str;
   cout << hex << i << " " << f << " " << str;
 }
```

The *stre_in.cpp* program limits the size of the array which the stream can access to the first 8 bytes. The first read performs exactly as it did in the *first_in.cpp* program presented in Tip 1027, yielding 10 and "Hello." However, because the stream can no longer access the array (because it has reached the limit of the array's attachment) after the first 8 bytes, the stream begins to read in garbage data from some unknown point in memory into the stream. When you compile and execute the *str_in.cpp*, your screen will display the following output, which is clearly not consistent with the intended output:

```
10 Hello
a 5.9801e-39
C:\>
```

USING PCOUNT WITH OUTPUT ARRAYS 1030

When your programs work with output arrays, there may be times when your programs must know how many characters are in the output array. C++ lets you use the *pcount* member function to determine the number of characters an output array contains. You will implement the *pcount* member function as shown here:

```
#include <iostream.h>
#include <strstrea.h>

int pcount(void);
```

If the output array includes a *NULL* terminator, *pcount* will include the terminator within its returned value (in other words, if the string is 15 characters long plus the *NULL* terminator, *pcount* will return 16). You can use *pcount* with output arrays to better control program processing. The *pcounter.cpp* program uses *pcount* to output the number of characters within an output stream, as shown here:

```
#include <iostream.h>
#include <strstrea.h>

void main(void)
 {
   char str[80];
   ostrstream outs(str, sizeof(str));

   outs << "Hello ";
   outs << 34 << "   " << 9876.98;
   outs << ends;
   cout << "String length: " << outs.pcount() << endl;
   cout << str;
 }
```

The *pcounter.cpp* program adds several characters and values to the *outs* stream. After adding the information to the stream, *pcounter.cpp* outputs the stream's length and then the stream itself. When you compile and execute the *pcounter.cpp* program, your screen will display the following output:

```
String length: 18
Hello 34   9876.98
C:\>
```

MANIPULATING STREAM ARRAYS WITH THE IOS MEMBER FUNCTIONS 1031

As your programs work with array-based streams, you will find that you can use the standard *ios* member functions, such as *get, put, rdstate, eof,* and so on, with your array-based streams. For example, the following program, *arr_get.cpp*, reads the contents of an array with the *get* method:

```
#include <iostream.h>
#include <strstrea.h>

void main(void)
 {
```

```
    char str[] ="abcdefghijklmnop";
   istrstream ins(str);
   char ch;

   while(!ins.eof())
    {
      ins.get(ch);
      cout << ch << " ";
    }
 }
```

The *arr_get.cpp* program attaches the *ins* stream to the *str* array, then uses a *while* loop to read through the *str* array one character at a time and output that character to the screen. When you compile and execute the *arr_get.cpp* program, your screen will display the following output:

```
a b c d e f g h i j k l m n o p
C:\>
```

1032 USING STRSTREAM

In previous Tips you have learned how to create both input and output array-based streams. However, there may be times when your programs must create a single stream to handle both input and output. Your programs can declare a single stream that handles full I/O using the *strstream* constructor. Within your programs, you will declare objects of type *strstream* as shown here:

```
#include <iostream.h>
#include <strstrea.h>

strstream iostr(char *buffer, int size, int mode);
```

Just as with the output and input streams, *buffer* is a pointer to the array's beginning, while *size* represents the number of bytes within *buffer*. You can set the mode using standard *ios*-enumerated values. However, you will most commonly set the mode to *ios::in* | *ios::out*. Additionally, you must *NULL*-terminate your arrays. To better understand how you will implement array-based I/O streams, consider the following program, *array_io.cpp*, which uses a single stream to perform input and output to an array:

```
#include <iostream>
#include <strstrea>
using namespace std;

void main(void)
 {
   char iostr[80], str[80];
   strstream ios(iostr, sizeof(iostr));
   int a, b;

   ios << "10 20 testingtesting";
   ios >> a >> b >> str;
   cout << a << "  " << b << "  " << str << endl;
 }
```

The *array_io.cpp* program declares an array of length 80, then attaches the new *ios* stream to the array. Using the *ios* variable, the program first reads values into the stream, then outputs the values from the string into variables, then outputs the variables to the screen. When you compile and execute the *array_io.cpp* program, your screen will display the following output:

```
10 20 testingtesting
C:\>
```

PERFORMING RANDOM ACCESS WITHIN A STREAM ARRAY 1033

As you have learned, you can use all standard I/O operations (such as formatting and string generation) against a stream array. Therefore, you can use the *seekg* and *seekp* methods to manipulate stream arrays just as you do with file-based streams. The following program, *arr_rand.cpp*, uses the *seekg* to move the stream pointer within a stream array:

```
#include <iostream>
#include <strstrea>
using namespace std;

void main(void)
 {
   char name[]="Jamsa's C/C++/C# Programmer's Bible";
   char iostr[80], ch;
   strstream ios(iostr, sizeof(iostr));

   ios << name;
   ios.seekg(7, ios::beg);
   ios >> ch;
   cout << "Name: " << name << endl;
   cout << "Character at position 7: " << ch;
 }
```

When you compile and execute the *arr_rand.cpp* program, your screen will display the following output:

```
Name: Jamsa's C/C++/C# Programmer's Bible
Character at position 7: C
C:\>
```

USING MANIPULATORS WITH STREAM ARRAYS 1034

As you have learned previously, C++ lets you treat stream arrays just as you would a normal file-based I/O stream. You will use manipulators with stream arrays in the same fashion you have previously used them with file streams. The following program, *arr_arrw.cpp*, uses a pair of custom manipulators (*la* for left-arrow and *ra* for right-arrow) on a stream array:

```
#include <iostream.h>
#include <strstrea.h>

ostream &ra(ostream &stream)
 {
   stream << "-------->";
   return stream;
 }

ostream &la(ostream &stream)
 {
   stream << "<--------";
   return stream;
 }

void main(void)
 {
   char str[80];
   ostrstream outs(str, sizeof(str));

   outs << ra << "Look at this number: ";
   outs << 100000 << la << ends;
   cout << "  " << str << endl;
 }
```

When you compile and execute the *arr_arrw.cpp* program, it will display the following output:

```
-------->Look at this number: 100000<--------
C:\>
```

1035 USING A CUSTOM INSERTION OPERATOR WITH STRING ARRAYS

As you have learned, C++ lets you treat array-based streams just as you would treat a file-based stream. Because stream arrays are identical in most ways to file-based streams, you can create your own extraction and insertion operators for stream arrays in the same way you would for file streams. For example, the following program, *plot_arr.cpp*, creates a class called *plot* and uses custom insertion operators to display the class output:

```
#include <iostream.h>
#include <strstrea.h>

const int size = 5;

class plot
 {
    int x, y;
  public:
    plot(int i, int j)
     {
       if(i > size)
          i = size;
       if(i < 0)
          i = 0;
       if(j > size)
          j = size;
       if(j < 0)
          j = 0;
       x = i;
       y = j;
     }

   friend ostream &operator<<(ostream &stream, plot obj);
};

ostream &operator<<(ostream & stream, plot obj)
 {
   register int i, j;

   for(j = size; j >= 0; j--)
    {
      stream << j;
      if(j == obj.y)
       {
         for(i = 0; i < obj.x; i++)
           stream << "  ";
         stream << '*';
       }
      stream << endl;
    }
   for(i = 0; i <= size; i++)
      stream << " " << i;
   stream << endl;
   return stream;
 }

void main(void)
 {
   plot a(2,3), b(1,1);
   char str[200];
```

```
   cout << "Output using cout:" << endl;
   cout << a << endl << b << endl << endl;

   ostrstream outs(str, sizeof(str));
   outs << a << b << ends;
   cout << "output using in-RAM formatting:" << endl;
   cout << str;
 }
```

In the *plot_arr.cpp* program, the custom insertion operator creates a simple grid. It then uses the information contained within the *plot* object to plot the single point in the grid. The program performs both standard output and stream output to show that the same insertion operator works equally well with both streams. When you compile and execute the *plot_arr.cpp* program, your screen will display the following output:

```
Output using cout:
5
4
3     *
2
1
0
 0 1 2 3 4 5
5
4
3
2
1  *
0
 0 1 2 3 4 5
output using in-RAM formatting:
5
4
3     *
2
1
0
 0 1 2 3 4 5
5
4
3
2
1  *
0
 0 1 2 3 4 5
C:\>
```

USING CUSTOM EXTRACTION OPERATORS WITH STREAM ARRAYS 1036

As you learned in Tip 1035, you will create custom insertion operators for your stream arrays in the exact same manner as you would create custom insertion operators for a file-based stream. Creating a stream array extraction operator is similarly easy. You can create extraction operators for your stream arrays as easily as you can create extraction operators for your file-based streams. The following program, *plt_arr2.cpp*, adds a custom extraction operator to the *plot_arr.cpp* program that Tip 1035 details:

```
#include <iostream.h>
#include <strstrea.h>
```

```
const int size = 5;

class plot
 {
    int x, y;
  public:
    plot(void);
    plot(int i, int j)
     {
       if(i > size)
          i = size;
       if(i < 0)
          i = 0;
       if(j > size)
          j = size;
       if(j < 0)
          j=0;
       x = i;
       y = j;
     }
    friend ostream &operator<<(ostream &stream, plot obj);
 };

ostream &operator<<(ostream &stream, plot obj)
 {
   register int i, j;

   for(j = size; j >= 0; j--)
    {
      stream << j;
      if(j == obj.y)
       {
         for(i = 0; i < obj.x; i++)
           stream << "  ";
         stream << '*';
       }
      stream << endl;
    }
   for(i = 0; i <= size; i++)
     stream << " " << i;
   stream << endl;
   return stream;
 }

plot::plot(void)
 {
   cout << "Enter x value: ";
   cin >> this->x;
   cout << "\nEnter y value: ";
   cin >> this->y;
 }

void main(void)
 {
   plot a(2,3), b(1,1), c;
   char str[200];
   ostrstream outs(str, sizeof(str));

   cout << "Output using cout:" << endl;
   cout << a << endl << b << endl << c << endl << endl;
   outs << a << b << c << ends;
   cout << "output using in-RAM formatting:" << endl;
   cout << str;
   istrstream ins(str);
 }
```

The addition of the custom extraction operator lets you input into the class the values you want your programs to plot within the grid. When you compile and execute the program, it will display both the grids that Tip 1035 displays, and a third grid which contains your custom-entered value.

USING DYNAMIC ARRAYS WITH I/O STREAMS 1037

In previous Tips you learned how to use the *ostrstream* constructor to create a stream array. Each time you used the *ostrstream* constructor, you declared the array's starting point and size. However, as your programs become more complex and you become more comfortable with C++, you will often create dynamic arrays, rather than arrays that are preset in size. To create an output stream that uses a dynamic array, you must use the *ostrstream* constructor slightly differently within your program, as shown here:

```
#include <iostream.h>
#include <strstrea.h>

ostrstream(void);
```

When you use the parameterless *ostrstream* constructor within your programs, *ostrstream* will create and maintain a dynamically allocated array. Notice that the *ostrstream* constructor does not return a pointer to the array. Rather, you must use a second function, *str*, which "freezes" the array and returns a pointer to it. After you freeze a dynamic array, you cannot use it for output again—instead, you must create a new array. The following program, *dyn_out.cpp*, uses a dynamically allocated output stream array:

```
#include <iostream.h>
#include <strstrea.h>

void main(void)
 {
   char *p;
   ostrstream outs;

   outs << "Jamsa's C/C++/C# ";
   outs << "Programmer's Bible ";
   outs << "has cool programs." << ends;
   p = outs.str();
   cout << p;
   delete p;
 }
```

When you compile and execute the program, it places the information the three extraction operators contain into the *outs* stream. Next, the program gets a pointer to the stream's beginning and outputs the stream using the pointer, rather than the actual stream variable itself. When you execute the *dyn_out.cpp* program, your screen will display the following output:

```
Jamsa's C/C++/C# Programmer's Bible has cool programs.
C:\>
```

UNDERSTANDING THE USES FOR STREAM ARRAY FORMATTING 1038

As you have learned, C++ has the ability to overload extraction operators, manipulators, and almost all other formatting tools against a "live" (in other words, file-based) stream, which eliminates much of the need for RAM-based formatting (in other words, stream arrays). However, there are several good reasons why you may want to use stream array formatting within your programs, which this Tip details.

One of the most common uses of array-based formatting is when your programs must construct a string that the program will later use within either a standard library or a third-party function. For example, you might need to create a string that the *strtok* standard library function will later parse. Another way your programs might use array-based I/O is when you create a text editor that performs complex formatting operations—formatting operations which would take significantly longer if performed against a file rather than an array.

Finally, because Windows does not contain standard functions that you can use to format output within a window, you must format all output before you send it to the window. The use of stream arrays in Windows is often very helpful in the creation of attractive, useful Windows applications.

1039 UNDERSTANDING THE ENDS MANIPULATOR

In previous Tips you created a string output stream and used the insertion operator to output the *ends* manipulator to the stream. The *ends* manipulator places a *NULL* character into the stream, much like the *endl* manipulator inserts the newline character. When you use the insertion operator to insert text into a string buffer, you will use the *ends* manipulator on a regular basis. The following program, *ends.cpp*, uses the *ends* manipulator with several string output streams:

```
#include <iostream.h>
#include <strstrea.h>

void main(void)
 {
   char title[64], publisher[64], author[64];

   ostrstream title_str(title, sizeof(title));
   ostrstream pub_str(publisher, sizeof(publisher));
   ostrstream author_str(author, sizeof(author));

   title_str << "Jamsa's C/C++/C# Programmer's Bible" << ends;
   pub_str << "Delmar" << ends;
   author_str << "Jamsa" << ends;
   cout << "Book: " << title << " Publisher: " << publisher << " Author: " <<
           author << endl;
 }
```

Although the character strings would normally terminate with a *NULL* value, when you work with a stream array you should explicitly provide the *ends* manipulator to instruct the stream to end the stream. Otherwise, the stream will remain open until your program takes a "snapshot" of the stream. The *ends* manipulator is not absolutely necessary in simpler situations, such as that shown in the *ends.cpp* program, but can be useful in closely managing stream arrays for more complex output operations.

1040 INVOKING ONE OBJECT FROM ANOTHER

As your C++ programs become more complex, they will begin to use more than one type of object, and there will be times when one object will use another. For example, the following program, *two_objs.cpp*, creates two different object types: an object that contains information about a reader and an object that contains information about a book. The *Reader* object invokes the *Book* object to display information about a reader's favorite book, as shown here:

```
#include <iostream.h>
#include <string.h>

class Book {
  public:
    Book(char *title) { strcpy(Book::title, title); } ;
    void show_book(void) { cout << title; };
  private:
```

```
      char title[64];
};

class Reader {
  public:
    Reader(char *name) { strcpy(Reader::name, name); };
    void show_reader(class Book book)
     {
       cout << "Reader: " << name << endl << "Book: ";
       book.show_book();
     };
  private:
    char name[64];
};

void main(void)
 {
    Reader reader("Kris Jamsa");
    Book favorite_book("Compiler Internals");

    reader.show_reader(favorite_book);
 }
```

The *two_objs.cpp* program first creates the *reader* instance of the *Reader* object and attaches Kris Jamsa's name to the private *name* member. Next, the program creates an instance of the *Book* object named *favorite_book*. When the program invokes the *reader.show_reader* member function with the *favorite_book* object as the member function's parameter, the program calls the *Book.Show_Book* member function for the *favorite_book* object. When you compile and execute the *two_objs.cpp* program, your screen will display the following output:

```
Reader: Kris Jamsa
Book: Compiler Internals
C:\>
```

TELLING THE COMPILER ABOUT A CLASS 1041

When one class references the identifier for a second class that you have not yet declared within your program, you must tell the compiler that the identifier corresponds to a class that you will declare later in the program. To do so, you can place a statement in your program that contains the *class* keyword and the class name, as shown here:

```
class class_name;
```

For example, assume that the *Book* class wants to tell the compiler that the *Reader* class is its friend. If you have not yet declared the *Reader* class, you can place the following statement in your program that tells the compiler that you will define the class later in the source code:

```
class Reader;
```

Tip 1043 uses this technique to inform the compiler about the *Reader* class before referencing the *Reader* class within the *Book* class.

REVISITING FRIENDS 1042

As you have learned, you can declare *friend* functions within your programs. Often, you will use *friend* functions to overload member functions and operators within a given class. However, as you have learned, C++ also lets you specify one class as the friend of another, which lets the friend access the private data and methods of the class. While *friend* functions (which also have access to the private data and methods of the class) are usually sufficient for most processing, there may be times when a class within your programs must have complete access to objects of another class while not deriving from that class. In such cases, you can use a *friend* class. Tips 1043 and 1044 detail some sample *friend* classes.

1043 DECLARING THE READER CLASS AS A FRIEND

In previous Tips you have created and used the *Book* class, which maintains information about a book's title, author, publisher, and price. The following program uses the *friend* keyword to specify that the *Reader* class is a friend of the *Book* class. Therefore, objects of type *Reader* can access the private members of a *Book* object. In the *readbook.cpp* program, the *Reader* class accesses the *Book* class private data member *title*, as shown here:

```
#include <iostream.h>
#include <string.h>

class Reader;                                  // Forward declaration

class Book
 {
  public:
    Book(char *title) { strcpy(Book::title, title); } ;
    void show_book(void) { cout << title; };
    friend Reader;
  private:
    char title[64];
};

class Reader {
  public:
    Reader(char *name) { strcpy(Reader::name, name); };
    void show_reader(class Book book)
     {
         cout << "Reader: " << name << ' ' << "Book: " << book.title;
     };
  private:
    char name[64];
};

void main(void)
 {
    Reader reader("Kris Jamsa");
    Book favorite_book("Compiler Internals");

    reader.show_reader(favorite_book);
 }
```

As you can see, within the *Book* object, the following statement tells the compiler that the *Reader* object is a friend, which lets the *Reader* object access the *Book* object's private members:

```
friend Reader;
```

1044 ANOTHER FRIEND CLASS EXAMPLE

As you learned in previous Tips, you can declare an entire class to be the *friend* of another class. Granting the second class *friend* status lets that class access all members of the first class, including such things as type names and enumerated constants. The following program, *make_chng.cpp*, uses a *friend* class to access a private enumeration:

```
#include <iostream.h>

class amount;

class coins
 {
    enum units {penny, nickel, dime, quarter, half_dollar};
    friend amount;
 };
```

```
class  amount
 {
    coins::units money;
  public:
    void setm(void);
    int getm(void);
 } object;

void amount::setm(void)
 {
   money = coins::dime;
 }

int amount::getm(void)
 {
   return money;
 }

void main(void)
 {
   object.setm();
   cout << object.getm();
 }
```

In the *make_chng.cpp* program, the program declares the enumerated unit types within the first class (*coins*), then declares the second class (*amount*) as maintaining a certain number or type of *units*. Because the program declares the *amount* class as a friend of the *coins* class, the *amount* class has access to the private *units* member.

ELIMINATING THE NEED FOR THE CLASS CLASS_NAME STATEMENT 1045

As you have learned, when one class references an identifier for a class that your program has not yet declared, you can inform the compiler that the identifier corresponds to a class using a statement similar to the following:

```
class class_name;
```

By including such a statement, the class can refer to a second class using only the class identifier, as shown here:

```
friend class_name;
```

If you place *class* between *friend* and *class_name*, you eliminate the need for the "forward" declaration:

```
friend class class_name;
```

The following declaration of the *Book* class uses the last technique to inform the compiler that the *Reader* class is a friend:

```
class Book
 {
   public:
     Book(char *title) { strcpy(Book::title, title); };
     void show_book(void) { cout << title; };
     friend class Reader;
   private:
     char title[64];
 };
```

RESTRICTING A FRIEND'S ACCESS 1046

As you have learned, C++ lets you specify that one object is a *friend* of another, which lets the *friend* have access to the object's private members. To better control the *friend's* access to the object's private members, C++ lets you specify the specific methods within the *friend* that can access the private members. The *friend's* other methods have no access to the members. For example, assume that only the *show_book* member of the *Reader* object requires access to the *Book* object's private members. Within the *Book* class, you can place the following statement:

```
friend class Reader Reader::show_book(void);
```

The following program, *friend2.cpp*, uses the restricted *friend* format to restrict the access of the *Reader* class to the *Book* object so that only the *show_reader* function may access the *Book* object:

```
#include <iostream.h>
#include <string.h>

class Book
{
  public:
    Book(char *title) { strcpy(Book::title, title); } ;
    void show_book(void) { cout << title; };
    friend class Reader show_reader(Book book);
  private:
    char title[64];
};

class Reader
{
  public:
    Reader(char *name) { strcpy(Reader::name, name); };
    void show_reader(class Book book)
      { cout << "Reader: " << name;
        book.show_book(); }
  private:
    char name[64];
};

void main(void)
 {
   Reader reader("Kris Jamsa");
   Book favorite_book("Compiler Internals");
   reader.show_reader(favorite_book);
 }
```

Although the *friend2.cpp* program yields the same output as the *two_objs.cpp* program Tip 1040 described, it is a more careful program, because it gives the *Reader* class only limited access to the *Book* class. As you have learned, one of the most significant dangers of *friend* classes is their unlimited access to the internal members of the befriended class. Limiting the access of the *friend* class is an important step in protecting against unlimited access.

1047 NAME CONFLICTS AND FRIENDS

When your classes use *friends* to access the members of another class, there may be times when member names conflict between the two classes. When such conflicts occur, the program uses the current class member. The following program, *memconfl.cpp*, illustrates a member name conflict between *friends*:

```
#include <iostream.h>
#include <string.h>

class Book
 {
  public:
    Book(char *title) { strcpy(Book::title, title); } ;
    void show_book(void) { cout << title; };
    friend class Reader;
  private:
```

```
      char title[64];
  };

class Reader
  {
   public:
     Reader(char *name) { strcpy(Reader::name, name); };
     void show_reader(class Book book) {
           cout << "Reader: " << name << ' ' << "Book: " << book.title; };
     void show_book(void) { cout << "The book's reader is " << name << endl; } ;

    private:
     char name[64];
  };

void main(void)
  {
     Reader reader("Kris Jamsa");
     Book favorite_book("Compiler Internals");
     reader.show_book();
     reader.show_reader(favorite_book);
  }
```

As you can see, both classes use the member name *show_book*. When you compile and execute the *memconfl.cpp* program, it uses the *Reader* class member, as shown here:

```
The book's reader is Kris Jamsa
Reader: Kris Jamsa Book: Compiler Internals
C:\>
```

INHERITANCE IN C++ 1048

As you learned briefly in previous Tips, C++ supports *inheritance,* which lets you derive a new class from an existing or base class. When you derive one class from another in C++, you will use the following format:

```
class derived_class: public base_class {
     public:
        // Derived class public members
     private:
        // Derived class private members
};
```

As an example of derived classes, the following program, *bld_lib.cpp*, creates a base class named *Book* and then derives a class named *LibraryCard* from the base class *Book*:

```
#include <iostream.h>
#include <string.h>

class Book
  {
   public:
     Book(char *title) { strcpy(Book::title, title); };
     void show_title(void) { cout << title << endl; };
   private:
     char title[64];
  };

class LibraryCard : public Book
  {
   public:
     LibraryCard(char *title, char *author, char *publisher) : Book(title)
       {
         strcpy(LibraryCard::author, author);
```

```
        strcpy(LibraryCard::publisher, publisher);
      };
     void show_library(void)
      {
        show_title();
        cout << author << ' ' << publisher;
      };
   private:
     char author[64];
     char publisher[64];
};

void main(void)
 {
   LibraryCard card("Jamsa's C/C++/C# Programmer's Bible", "Jamsa",
                    "Delmar");
   card.show_library();
 }
```

Inheritance is critically important to object-oriented programming. The ability to create a class hierarchy, from most general to most specific, lets you better control your programs. Moreover, it makes your programs easier for other readers to understand and makes your programs more expandable. While classes are a valuable programming tool in and of themselves, inheritance and its associated capabilities are what really unleash the power of C++ programming. You will learn extensively about inheritance in the following Tips.

1049 Understanding Base and Derived Classes

One foundation of the inheritance concept is the relationship between base classes and derived classes. When you build one class using an existing class, the new class inherits the characteristics of the existing class. The existing class characteristics include data and methods, as well as their access (public and private). As you read through magazines and books on object-oriented programming, you will encounter the terms *base class* and *derived class*. The base class is the original class whose characteristics other classes inherit. The derived class is the class your program creates from the base class. Many different classes can use a base class. Conversely, you can build a derived class from several different base classes. Figure 1049 shows a simple derivation from a base class.

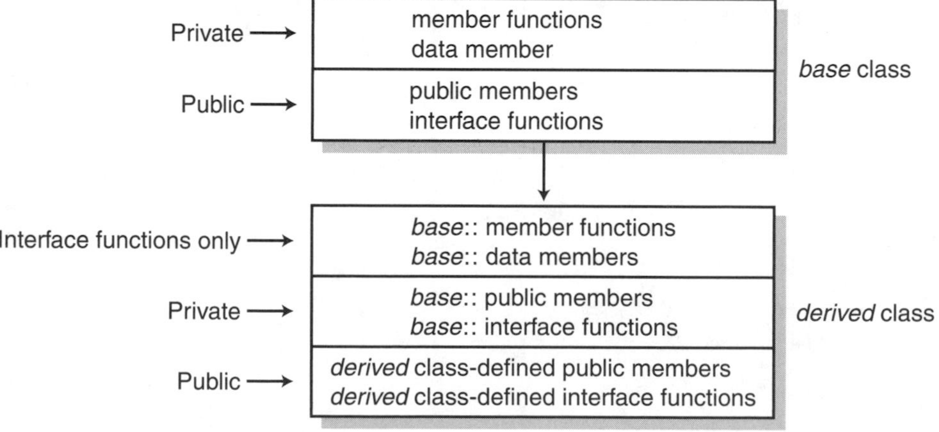

***Figure 1049** A simple inheritance example.*

DERIVING A CLASS 1050

In Tip 1048, the *bld_lib.cpp* program derived the class *LibraryCard* using the base class *Book*. The first line of the *LibraryCard* class declaration informs the compiler that *LibraryCard* is a derived class that uses the base class *Book*, as shown here:

```
class LibraryCard : public Book {
```

The remainder of the class declaration is very similar to those with which you have worked, with the exception of the constructor function. As you can see, immediately following the *LibraryCard* constructor declaration is an invocation of the *Book* constructor:

```
LibraryCard(char *title, char *author, char *publisher) : Book(title)
 {
   strcpy(LibraryCard::author, author);
   strcpy(LibraryCard::publisher, publisher);
 };
```

When the program creates a new instance of the *LibraryCard* class (and therefore calls the *LibraryCard* constructor), the *LibraryCard* constructor will first call the constructor for the base class (for *Book*). If you do not specify the base class constructor after the *LibraryCard* constructor as is shown in the previous code fragment, the compiler will generate a syntax error. As you will learn in later Tips, the compiler will generate the syntax error because the *Book* class constructor requires a single parameter. In the event the *Book* class constructor was parameterless, the compiler would not require that you specify the base class as shown in the previous code fragment.

UNDERSTANDING BASE AND DERIVED CONSTRUCTORS 1051

When you derive a class from a base class that has a constructor function, you must invoke the base class constructor from within the constructor of the derived class. The following program, *basederi.cpp*, derives a class named *Derived* from the base class *Base*. Within the *Derived* class constructor function, the code invokes the *Base* class constructor function, as shown here:

```
#include <iostream.h>

class Base
 {
  public:
    Base(void) { cout << "Base class constructor\n"; };
 };

class Derived:Base
 {
  public:
    Derived(void): Base()
      { cout << "Derived class constructor\n"; };
 };

void main(void)
 {
   Derived object;
 }
```

Each time you create an instance of a derived class, the program will execute both the constructor for that class and the constructor for its base class. When you compile and execute the *basederi.cpp* program, your screen will display the following:

```
Base class constructor
Derived class constructor
C:\>
```

As you can see, the base class constructor executes before the derived class constructor.

1052 PUTTING PROTECTED MEMBERS TO USE

As you have learned, C++ lets your programs declare *protected* class members, which are fully accessible to classes your programs derive from a base class. The following program, *protect.cpp*, illustrates how to use *protected* members. In the *protect.cpp* program, the *Book* class defines several protected members. Next, the program derives the *LibraryCard* class from the *Book* class. Remember, the derived *LibraryCard* class can access protected members within the *Book* class, which lets the *LibraryCard* invocation of the *show_cost* function and modification of the *cost* member execute successfully. The following code implements the *protect.cpp* program:

```
#include <iostream.h>
#include <string.h>

class Book
 {
  public:
    Book(char *title) { strcpy(Book::title, title); };
    void show_title(void) { cout << title << endl; };
  protected:
    float cost;
    void show_cost(void) { cout << cost << endl; };
  private:
    char title[64];
 };

class LibraryCard : public Book
 {
  public:
    LibraryCard(char *title, char *author, char *publisher) : Book(title)
     {
       strcpy(LibraryCard::author, author);
       strcpy(LibraryCard::publisher, publisher);
       cost = 54.95;
     };
    void show_library(void)
     {
       show_title();
       show_cost();
       cout << author << ' ' << publisher;
     };
  private:
    char author[64];
    char publisher[64];
 };

void main(void)
 {
   LibraryCard card("Jamsa's C/C++/C# Programmer's Bible", "Jamsa",
                    "Delmar");
   card.show_library();
 }
```

When you compile and execute the *protect.cpp* program, your screen will display the following output:

```
Jamsa's C/C++/C# Programmer's Bible
54.95
Jamsa Delmar
```

As you can see, the class statements within the derived class *LibraryCard* have full access to the protected members within the base class *Book*.

UNDERSTANDING WHEN TO USE PROTECTED MEMBERS 1053

As you have learned, derived classes can access protected class members in their base class. As you create your classes, you must decide which members to make public, private, or protected. As a rule, you should create each class with the intention that a derived class will later use it. If you never use the class as a base class, the protected members will be essentially private. Should you later decide to use the class as a base class, predetermining the protected members will save you programming time.

REVIEWING PUBLIC AND PRIVATE BASE CLASS INHERITANCE 1054

As you have learned, your programs can derive a second class from a first class. Thus far, you have derived the second class from the first class as a *public* derivation, which means that objects of the derived class can access members of the base class. To understand this better, consider the *baseder2.cpp* program, as shown here:

```
#include <iostream.h>

class Base
 {
  public:
    Base(void) { cout << "Base class constructor\n"; };
    int data;
 };

class Derived: public Base
 {
  public:
    Derived(void): Base() { cout << "Derived class constructor\n"; };
 };

void main(void)
 {
   Derived object;
   object.data = 5;
   cout << object.data << endl;
 }
```

When you compile and execute the *baseder2.cpp* program, the *object* which you created from the *Derived* class can directly access the *data* member of the *Base* class. As you know, such direct access violates encapsulation rules, and therefore should be avoided. Alternately, your program could inherit the *Base* class as private, which has the effect of making all the members of the *Base* class private, as shown in the *baseder3.cpp* program:

```
#include <iostream.h>

class Base
 {
  public:
    Base(void) { cout << "Base class constructor\n"; };
    int data;
};

class Derived: private Base
 {
  public:
    Derived(void): Base() { cout << "Derived class constructor\n"; };
 };

void main(void)
 {
   Derived object;
   object.data = 5;
   cout << object.data << endl;
 }
```

When the *baseder3.cpp* program tries to access the *data* member of the *Base* class, the compiler fails and returns an error. However, the *Derived* class can still access the base class constructor from within the object. Additionally, an interface function would let your program access the data member as well. In Tip 1055, you will learn about the third method of base class inheritance—*protected* base class inheritance.

1055 Understanding Protected Base Class Inheritance

As you have learned, when you create classes within your programs, you can use the *protected* keyword to prevent other parts of the program from accessing certain members of the class, but still make the members of the class available to inherited classes. You can also inherit an entire base class as *protected.* When you inherit an entire base class as protected, all public and protected members of the base class become protected members of the derived class. To better understand how derived classes inherit *public* and *protected* members, consider the following program, *prot_bas.cpp*:

```
#include <iostream.h>

class base
 {
  protected:
    int i, j;
  public:
    void setij(int a, int b)
     {
       i = a;
       j = b;
     }
    void showij(void) {cout << i << " " << j << endl;}
 };

class derived : protected base
 {
  private:
    int k;
  public:
    void setk(void)
     {
       setij(10,12);
       k = i * j;
     }
    void showall(void)
     {
       cout << k << " ";
       showij();
     }
 };

void main(void)
 {
   derived object;

   // object.setij(2,3);      This is an illegal function call, because setij is
   //                         a protected member of the derived class.
   // object.showij();        This is also an illegal command.

   object.setk();
   object.showall();
 }
```

As you can see, the *derived* class uses the *protected* keyword to derive the *base* class. Doing so results in all the members of the *base* class being *private* within the *derived* class. If the program then tries to access a *private* inherited member (such as *setij* or *showij*) the compiler will return an error because that member is encapsulated. The program code notes that both the *object.setij(2,3)* and *object.showij()* functions are illegal function calls, because the members are protected within the derived class—meaning that the only objects that can access the functions *derived* inherits from *base* are interface functions or classes which your program derives from the *derived* class. When you compile and execute the *prot_bas.cpp* program, your screen will display the following output:

```
120 10 12
C:\>
```

UNDERSTANDING MULTIPLE INHERITANCE 1056

As you have learned, inheritance is the ability of one class to inherit the characteristics of another class. Multiple inheritance is the ability of one class to inherit the characteristics of more than one base class. C++ supports multiple inheritance. When a derived class inherits characteristics from more than one base class, you simply separate the base class names using commas, as shown here:

```
class derived_class: public base_class_1, public base_class_2
 {
    public:
       // Derived class public members
     private:
       // Derived class private members
 };
```

Likewise, when you later declare the constructor function for the derived class, you must call the constructor functions for each base class. Tip 1057 illustrates a simple use of multiple inheritance.

A SIMPLE MULTIPLE INHERITANCE 1057

As you have learned, multiple inheritance is the ability of a derived class to inherit the characteristics of two or more base classes. For many programmers, understanding how a single class can inherit the characteristics of two other classes may be confusing. The following program, *simpmult.cpp*, illustrates how multiple inheritance creates a class named *Book*, which inherits the base classes *Page* and *Cover*:

```
#include <iostream.h>
#include <string.h>

class Cover
 {
  public:
    Cover(char *title) { strcpy(Cover::title, title); };
  protected:
    char title[256];
 };

class Page
 {
  public:
    Page(int lines = 55) { Page::lines = lines; };
  protected:
    int lines;
    char *text;
 };

class Book: public Cover, public Page {
  public:
    Book(char *author, char *title, float cost): Cover(title), Page(60)
     {
```

```
        strcpy(Book::author, author);
      strcpy(Book::title, title);
      Book::cost = cost;
    };
   void show_book(void)
    {
      cout << title << endl;
      cout << author << '\t' << cost;
    };
  private:
   char author[256];
   float cost;
 };

void main(void)
 {
   Book text("Jamsa", "Jamsa's C/C++/C# Programmer's Bible", 54.95);

   text.show_book();
 }
```

The *simpmult.cpp* program defines two classes, *Cover* and *Page*. Next, the program derives a third class, *Book*, from both the original classes. The *Book* constructor passes parameters to the constructors for the two base classes. After the program creates an instance of the *Book* object, it generates output with information about the object. The *show_book* function displays values, one each from the *Cover*, *Page*, and *Book* classes, as shown in the following output:

```
Jamsa's C/C++/C# Programmer's Bible
Jamsa     54.95
C:\>
```

1058 Understanding Constructor Order and Base Classes

Multiple inheritance is the ability of a derived class to inherit the characteristics of more than one base class. When you use multiple inheritance to derive a class, the derived class must invoke the constructor functions for each base class. The constructor invocation order depends on the order in which the derived class specifies the base classes. In other words, if the derived class specifies its base classes as *One* and *Two*, the program will invoke the constructors in the following order: *One,* then *Two,* then *derived*, which means that your code would be similar to the following:

```
derived(void) : One(),Two(int i);
```

To better understand the order in which your programs should state constructor parameters, consider the following program, *multinv.cpp*, which illustrates the order of constructor invocation when you derive a class from three base classes:

```
#include <iostream.h>

class One
 {
  public:
    One(void) { cout << "Constructor for One\n"; };
 };

class Two
 {
  public:
    Two(void) { cout << "Constructor for Two\n"; };
 };

class Three
 {
  public:
```

```
      Three(void) { cout << "Constructor for Three\n"; };
  };

class Derived: public One, public Three, public Two
  {
   public:
     Derived(void) : One(), Two(), Three()
       { cout << "Derived constructor called\n"; };
  };

void main(void)
  {
    Derived my_class;
  }
```

When you compile and execute the *multinv.cpp* program, your screen will display the following output:

```
Constructor for One
Constructor for Three
Constructor for Two
Derived constructor called
C:\>
```

As you can see, the program invokes the constructor functions in the same order as the base class names appear in the class header:

```
class Derived: public One, public Three, public Two
```

DECLARING A BASE CLASS AS PRIVATE 1059

As you have learned, your programs can inherit a single class using the *public*, *private*, or *protected* keywords. When you derive a class, you can precede the base class name with *private* or *public*. When your program uses the *public* keyword to derive a base class, your program can use the derived class to directly access public members within the base class. When you use the *private* keyword, however, your program can only access the base class members through the derived class members. Finally, when your programs use the *protected* keyword, the deriving class inherits each of the *public* members within the inherited class as *protected* members, which lets the derived class access base class members, but also lets other classes derive the *protected* members. However, each of the previous Tips that illustrated multiple inheritance used the keyword *public* in front of the base class names, as shown here:

```
class Derived: public One, public Three, public Two {
```

Your programs can use the *public, private*, and *protected* keywords when your programs perform multiple inheritance, just as they can when they perform single inheritance. The following program, *privmult.cpp*, derives a class from two base class names preceded by the keyword *private* and one base class name preceded by the keyword *public*:

```
#include <iostream.h>

class One
  {
   public:
     One(void)
      {
        cout << "Constructor for One\n";
        one = 1;
      };
     int one;
  };
```

```
class Two
 {
  public:
    Two(void)
     {
        cout << "Constructor for Two\n";
        two = 2;
     };
    int two;
 };
class Three
 {
  public:
    Three(void)
     {
       cout << "Constructor for Three\n";
       three = 3;
     };
    int three;
 };
class Derived: private One, private Three, public Two
 {
  public:
     Derived(void) : One(), Two(), Three()
      { cout << "Derived constructor called\n"; };
     void show_value(void) { cout << one << two << three << endl; };
 };
void main(void)
 {
   Derived my_class;
   my_class.show_value();
   cout << my_class.two;
 }
```

Because the derived class declares the base class *Two* as public, the *privmult.cpp* program can directly access the member *two* without having to use interface functions. However, the program cannot directly access the values *one* or *three,* because the derived class declares the bases classes *One* and *Three* as private. Instead, the program must use a member function, such as the *show_value* function, to display the *one* or *three* values.

1060 DESTRUCTOR FUNCTIONS AND MULTIPLE INHERITANCE

As you have learned, when you derive a class from a base class, C++ calls the base class constructor function before it calls the constructor function of the derived class. In the case of destructor functions, however, the opposite is true: C++ calls the derived class destructor and then calls the destructors for each base class. The following program, *destmult.cpp*, illustrates the calling sequence of base and derived class destructor functions:

```
#include <iostream.h>
class One
 {
  public:
    One(void) { cout << "Constructor for One\n"; };
    ~One(void) { cout << "Destructor for One\n"; };
 };
class Two
 {
  public:
```

```
    Two(void) { cout << "Constructor for Two\n"; };
    ~Two(void) { cout << "Destructor for Two\n"; };
 };

class Three
 {
  public:
    Three(void) { cout << "Constructor for Three\n"; };
    ~Three(void) { cout << "Destructor for Three\n"; };
 };

class Derived: public One, public Two, public Three
 {
  public:
    Derived(void) : One(), Two(), Three()
     { cout << "Derived constructor called\n"; };
    ~Derived(void)
     { cout << "Derived destructor called\n"; };
 };

void main(void)
 {
   Derived my_class;
 }
```

When you compile and execute the *destmult.cpp* program, your screen will display the following output:

```
Constructor  for  One
Constructor for Two
Constructor for Three
Derived constructor called
Derived destructor called
Destructor for Three
Destructor for Two
Destructor for One
C:\>
```

As you can see, C++ invokes the destructor functions in the opposite order from the constructor functions.

NAME CONFLICTS BETWEEN BASE AND DERIVED CLASSES 1061

When you derive a new class using one or more base classes, it is possible that a member name in the derived class will be the same as a member name in one or more of the base classes. When such conflicts occur, C++ uses the derived class member name. The following program, *conflict.cpp*, illustrates a member name conflict between a base class member name and a derived class member name:

```
#include <iostream.h>

class Base
 {
  public:
    void display(void) { cout << "This is the base class" << endl; };
 };

class Derived: public Base
 {
  public:
    void display(void) { cout << "This is the derived class" << endl; };
 };

void main(void)
 {
```

```
    Derived my_class;

   my_class.display();
 }
```

When you compile and execute the *conflict.cpp* program, your screen will display the following output:

```
This is the derived class
C:\>
```

Because the *display* member function is part of both the *Base* class and the *Derived* class, a potential conflict exists. However, because the *my_class* variable is an object of the *Derived* class, the compiler will use the *Derived* version of the *display* function.

1062 RESOLVING CLASS AND BASE NAME CONFLICTS

As you learned in Tip 1061, when a derived class member name conflicts with a base class member name, C++ uses the derived class member name. However, there may be times when your program must access the base class member. To do so, your program can use the global resolution operator (::). The following program, *resname.cpp*, uses the global resolution operator to access the base class *display* member:

```
#include <iostream.h>

class Base
 {
  public:
    void display(void) { cout << "This is the base class" << endl; };
 };

class Derived: public Base
 {
  public:
    void display(void) { cout << "This is the derived class" << endl; };
 };

void main(void)
 {
   Derived my_class;

   my_class.display();
   my_class.Base::display();
 }
```

When you compile and execute the *resname.cpp* program, your screen will display the following output:

```
This is the derived class
This is the base class
C:\>
```

As you saw in Tip 1061, the reference simply to the *display* function invoked the derived class function. However, because the second statement precedes the function name with the global resolution operator and the *Base* class name, the second statement calls the *display* member function within the *Base* class.

1063 UNDERSTANDING WHEN INHERITED CLASSES EXECUTE CONSTRUCTORS

As you have learned, the derived class constructor function will always invoke the constructor function of the class from which it derives. However, derived classes can also perform their own processing within the derived class constructor function. In fact, when your programs derive classes, each instance of the class will invoke the constructor function of every class in the hierarchical tree above the derived class before executing the constructor function for the derived class. For example, suppose your *derived5* class is derived from four other derived classes and a *base* class. If each class constructor displays a message, when you create any instance of the *derived5* class, your constructors will output the following messages:

```
Constructing base class.
Constructing derived1 class.
Constructing derived2 class.
Constructing derived3 class.
Constructing derived4 class.
Constructing derived5 class.
```

As you have learned, the program executes the destructor functions in the opposite sequence of the constructor functions, so that they will display a message series similar to the following:

```
Destructing derived5 class.
Destructing derived4 class.
Destructing derived3 class.
Destructing derived2 class.
Destructing derived1 class.
Destructing base class.
```

AN INHERITED CLASS CONSTRUCTOR EXAMPLE 1064

As you have learned, C++ programs invoke the class constructor and destructor for each class above a derived class in the hierarchical tree. To better understand how C++ invokes constructor functions and destructor functions for derived classes, consider the program *cons_des.cpp*. The *cons_des.cpp* program constructs and destroys a single object of type *derived2* and invokes the constructors and destructors from the classes above *derived2* in the tree, as shown here:

```
#include <iostream.h>

class base
 {
  public:
    base(void) {cout << "Constructing base.\n";}
    ~base(void) {cout << "Destructing base.\n";}
 };

class derived1 : public base
 {
  public:
    derived1(void) {cout << "Constructing derived1.\n";}
    ~derived1(void) {cout << "Destructing derived1.\n";}
 };

class derived2 : public derived1
 {
  public:
    derived2(void) {cout << "Constructing derived2.\n";}
    ~derived2(void) {cout << "Destructing derived2.\n";}
 };

void main(void)
 {
   derived2 object;
 }
```

When you compile and execute the *cons_des.cpp* program, your screen will display the following output:

```
Constructing base.
Constructing derived1.
Constructing derived2.
```

```
Destructing  derived2.
Destructing derived1.
Destructing base.
C:\>
```

1065 How to Pass Parameters to Base Class Constructors

As you know, each time you create an instance of a derived class, your program will invoke the constructor for each class from which the derived class derives, in addition to the derived class constructor. In many cases, the base classes from which the class derives will include constructor functions that expect parameters. If your derived class does not pass parameters to the constructor functions of the classes higher up in the hierarchical tree (that is, the base class from which the class derives), an error will occur and the compiler will not compile the program. Therefore, C++ both lets and expects you to pass parameters to constructor functions above an object in the tree. To support the declarations for the base class or classes above the derived class in the tree, you can use the derived class constructor declaration's expanded general form, as shown here:

```
derived constructor(argument-list):         base1(argument-list),
                                            base2(argument-list),
                                            base3(argument-list),
                                            .
                                            .
                                            .
                                            baseN(argument-list)
 {
   // body of derived constructor
 }
```

In the preceding declaration, *base1* through *baseN* represent classes above the derived class within the hierarchical tree. To better understand how you will invoke the base class constructors for a derived class, consider the following program, *cls_parm.cpp:*

```
#include <iostream.h>

class base
 {
  protected:
    int i;
  public:
    base(int x)
    {
      i=x;
      cout << "Constructing base.\n";
    }
    ~base(void) {cout << "Destructing base.\n";}
 };

class derived : public base
 {
    int j;
  public:
    // derived uses x; base uses y.
    derived(int x, int y): base(y)
     {
        j = x;
        cout << "Constructing derived.\n";
     }
    ~derived(void) {cout << "Destructing derived.\n";}
    void show(void) {cout << i << ", " << j << endl;}
 };
```

```
void main(void)
 {
   derived object(3,4);
   object.show();                    // Displays 4, 3
 }
```

The *cls_parm.cpp* program derives the *derived* class from the *base* class. When the program begins execution, it creates an instance of the *derived* class and passes the constructor the values 3 and 4. The *derived* constructor passes the second value (in this case, 4) to the *base* constructor. After the base constructor completes its processing, the *derived* constructor uses the first value (in this case, 3) as its parameter. When you compile and execute the *cls_parm.cpp* program, your screen will display the following output:

```
Constructing base.
Constructing derived.
4, 3
Destructing derived.
Destructing base.
C:\>
```

UNDERSTANDING ACCESS DECLARATIONS WITH DERIVED CLASSES 1066

As you learned in Tip 1054, when a derived class inherits a base class as private, all public and protected members of that class become private members of the derived class, meaning that the members are encapsulated within the class and your programs can only access the members through the derived class interface functions. However, in certain circumstances, you may want to restore one or more inherited members to their original access specifications. For example, you might want to grant certain base class *public* members public status in the derived class even though the derived class inherited the base class as *private*. To do so, you must use an *access declaration* within the derived class. Access declarations within your programs will take the following general form:

```
base-class::member;
```

To better understand how an access declaration works, consider the following code fragment:

```
class base
 {
  public:
    int j;
    int k;
 };
class derived : private base
 {
  public:
    base::j;
    // More declarations here
 };
```

In the previous example, the variable *j* would usually be private within the *derived* class (because the *derived* class used the *private* keyword to inherit the *base* class). However, the *base::j* statement, which redeclares *j* as *public* within the *derived* class, returns *j* to public status, without impacting the remainder of the class. On the other hand, the *k* variable will remain private within the *derived* class.

USING ACCESS DECLARATIONS WITH DERIVED CLASSES 1067

In Tip 1066, you learned about access declarations within derived classes. In the brief code fragment within that Tip, you learned the basics of using access declarations. The following program, *acc_decl.cpp*, uses several additional declarations to take the concept of access declarations further. In the *acc_decl.cpp* program, the code converts three *private* functions within the *derived* class back to *public* access. However, as the last declaration (within a comment) shows, the program cannot declare the *i* member back to *public* because it is *private* within the *base* class. Making *i public* within the *derived* class would violate encapsulation rules, and therefore the compiler will return an error at the assignment. The *acc_decl.cpp* program provides worthwhile insight into how access declarations work:

```
#include <iostream.h>

class base
 {
    int i;          // private in base class
  public:
    int j, k;
    void seti(int x) {i = x;}
    int geti(void) {return i;}
 };

class derived : private base
 {
  public:
                          // The next statements override the private inheritance.
    base::j;              // makes j public again
    base::seti;           // makes seti() public
    base::geti;           // makes geti() public
   // base::i; is an illegal statement, you cannot promote access.
    int a;
 };

void main(void)
 {
   derived object;

   //object.i = 10;  Illegal statement; i is private to base.
   object.j = 20;         // legal because j is public

   //object.k = 30; Illegal because k is private to derived
   object.a = 40;
   object.seti(10);
   cout << object.geti() << ", " << object.j << ", " << object.a;
 }
```

As you can see, the *acc_decl.cpp* program inherits the *derived* class using the *private* keyword, and then makes much of the *derived* class public again. As a general rule, when you use access declarations within your derived classes, you should keep the number of access declarations to a minimum, or reconsider how you have written your code. Keeping access declarations to a minimum avoids confusion (both your own, and that of another programmer who reads your code) and makes your programs more consistent with encapsulation rules.

1068 AVOIDING AMBIGUITY WITH VIRTUAL BASE CLASSES

When your program derives a class which inherits multiple classes previously derived from a single base class, it is possible that class may include members which, while unique to each of the parent classes, share the same name within the derived class. When such a situation occurs, the class member ambiguity will cause the compiler to fail. For example, if you write a program that uses the *base* class, and derives from it two classes, *derived1* and *derived2*, your program has no ambiguity. If, however, your program later derives the *derived3* class from both *derived1* and *derived2*, each object of the *derived3* class will actually contain two objects of the *base* class, as shown in Figure 1068.

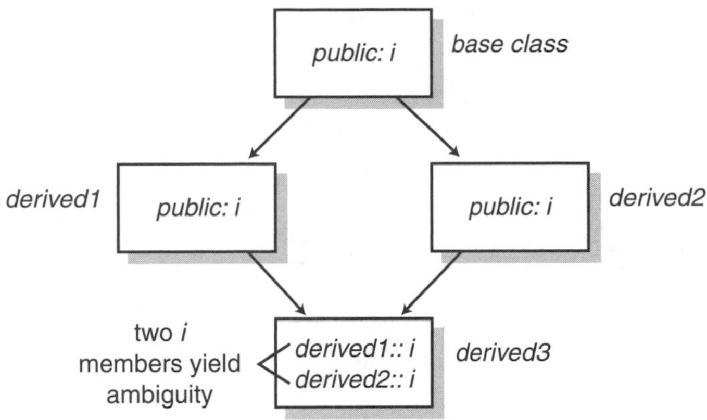

Figure 1068 Deriving from multiple classes can create ambiguity.

Not only is it unnecessary for each object of the *derived3* class to contain two objects of the *base* class, such ambiguity is likely to confuse both you and your compiler. For example, the following program, *ambig_cl.cpp*, creates ambiguity in the *derived3* class. Because of the ambiguity, the compiler will not complete its execution and will instead return an error, which will force the compiler to halt with an error:

```
// This program contains three errors and will not compile.
#include <iostream.h>

class base
 {
  public:
    int i;
 };

class derived1 : public base
 {
  public:
    int j;
 };

class derived2 : public base
 {
  public:
    int k;
 };

class derived3 : public derived1, public derived2
 {
  public:
    int sum;
 };

void main(void)
 {
   derived3 object;

   object.i = 10;        // This causes the compiler to halt because
                              // it doesn't know which i you are referring to.
   object.j = 20;
   object.k = 30;
   object.sum = object.i + object.j + object.k;
   cout << object.i << " ";
   cout << object.j << " " << object.k << " ";
   cout << object.sum << endl;
 }
```

Because the *Base* class includes the public member *i* and both the *derived1* and *derived2* classes derive from the *Base* class, the *derived3* class actually includes two instances of the public member *i*. Because the compiler cannot resolve the ambiguity of the *i* member, the *ambig_cl.cpp* program should return three compiler errors, all from the ambiguous use of the *i* member within the *derived3 object* variable. As you will learn in Tip 1069, your programs can use virtual base classes to avoid ambiguity errors within derived classes.

1069 UNDERSTANDING VIRTUAL BASE CLASSES

As you learned in Tip 1068, when your programs derive two or more objects from a common base class, your programs may encounter ambiguity errors. To avoid ambiguity, you must avoid having multiple copies of the base class within a given object. To prevent your programs from inheriting multiple copies of a given base class, you can declare that base class using the *virtual* keyword, as shown in the following program, *virclass.cpp:*

```
#include <iostream.h>

class base
 {
  public:
    int i;
 };

class derived1 : virtual public base
 {
  public:
    int j;
 };
class derived2 : virtual public base
 {
  public:
    int k;
 };

class derived3 : public derived1, public derived2
 {
  public:
    int sum;
 };

void main(void)
 {
   derived3 object;

   object.i = 10;             // now i is unambiguous
   object.j = 20;
   object.k = 30;
   object.sum = object.i + object.j + object.k;
   cout << object.i << " ";
   cout << object.j << " " << object.k << " ";
   cout << object.sum << endl;
 }
```

In the case of the *virclass.cpp* program, the fact that each of the two intermediate derived classes inherits the base class virtually lets the final derived class (*derived3*) inherit both classes without concern for ambiguity. As you saw in Tip 1068, without the *virtual* keyword, the *derived3* class effectively has two *i* members. As a rule, if your programs derive classes from multiple classes sharing a single parent class, your programs should use the *virtual* keyword in the class definition to avoid ambiguity.

MUTUAL FRIENDS 1070

As you have learned, C++ lets you specify other classes as friends of a class, which lets functions in the *friend* class access a class's private methods. As you examine C++ programs, you might encounter cases in which two classes are *mutual friends*. In other words, the functions in one class can access the private data of another class, and the functions in the other class can access the private data of that class. The following program, *mutual.cpp*, illustrates two classes that are mutual friends:

```
#include <iostream.h>
#include <string.h>

class Curly
 {
  public:
    Curly(char *msg) { strcpy(message, msg); };
    void show_message(void) { cout << message << endl; };
    friend class Moe;
    void show_moe(class Moe moe);
  private:
    char message[256];
 };

class Moe
 {
  public:
    Moe(char *msg) { strcpy(message, msg); };
    void show_message(void) { cout << message << endl; };
    friend class Curly;
    void show_curly(class Curly curly);
  private:
    char message[256];
 };

void Curly::show_moe(class Moe moe) { cout << moe.message << endl; };

void Moe::show_curly(class Curly curly) { cout << curly.message << endl; };

void main(void)
 {
   class Moe moe("Nuck, nuck, nuck...");
   class Curly curly("Whoop, whoop, whoop...");

   moe.show_message();
   moe.show_curly(curly);
   curly.show_message();
   curly.show_moe(moe);
 }
```

Because the *mutual.cpp* program uses the *friend* qualifier in declaring both classes, the mutual friends *Moe* and *Curly* can access each other's private data. When you compile and execute the *mutual.cpp* program, your screen will display the following output:

```
Nuck, nuck, nuck...
Whoop, whoop, whoop...
Whoop, whoop, whoop...
Nuck, nuck, nuck...
C:\>
```

HOW A DERIVED CLASS CAN BECOME A BASE CLASS 1071

As you have learned, C++ lets you create an inheritance hierarchy, which lets one class inherit the characteristics of a base class that itself may have inherited characteristics from a third base class. The following program, *threelvl.cpp*, derives three levels of classes. As you will see, each successive class inherits the characteristics of each class that has gone before it:

```
#include <iostream.h>

class Base
 {
  public:
    void show_base(void) { cout << "Base class message\n"; };
 };

class Level1 : public Base
 {
  public:
    void show_level1(void)
     {
       show_base();
       cout << "Level 1 message\n";
     };
 };

class Level2 : public Level1
 {
  public:
    void show_level2(void)
     {
       show_level1();
       cout << "Level 2 message\n";
     };
 };

class Level3 : public Level2
 {
  public:
    void show_level3(void)
     {
       show_level2();
       cout << "Level 3 message\n";
     };
 };

void main(void)
 {
   Level3 my_data;

   my_data.show_level3();
 }
```

As you can see from the code for the *threelvl.cpp* program, the *Level3* class derives from three previous classes. To understand how *Level3* derives from all three previously-defined classes, consider each derivation in order. The *Level3* class derives directly from the *Level2* class. In turn, the *Level2* class derives directly from the *Level1* class. Finally, the *Level1* class derives from the *Base* class. When you compile and execute the *threelvl.cpp* program, your screen will display the following output:

```
Base class message
Level 1 message
Level 2 message
Level 3 message
C:\>
```

1072 USING PROTECTED MEMBERS IN DERIVED CLASSES

Each time you create a new class, you should assume that the class might eventually become the base class for other class derivations. Therefore, you should take advantage of protected members to limit program access to class members, both for instances of the current class and instances of any future derived classes. As you have learned, protected members let the deriving class access the members of a base class as if those members were *public*, but does not let derivations beyond the first access the base class members as *public* members. The following program, *protderi.cpp*, illustrates how to use protected data in a derived class, which in turn has become a base class:

```
#include <iostream.h>
#include <string.h>

class Base
 {
  public:
    Base(char *str) { strcpy(message, str); };
    void show_base(void) { cout << message << endl; };
  protected:
    char message[256];
 };

class Level1 : public Base
 {
  public:
    Level1(char *str, char *base) : Base(base) { strcpy(message, str); };
    void show_level1(void) { cout << message << endl; } ;
  protected:
    char message[256];
 };

class Lowest : public Level1
 {
  public:
    Lowest(char *str, char *level1, char *base) : Level1(level1, base)
      { strcpy(message, str); };
    void show_lowest(void)
      {
        show_base();
        show_level1();
        cout << message << endl;
      };
  protected:
    char message[256];
 };

void main(void)
 {
   Lowest bottom("Lowest message", "Level1 message", "Base message");

   bottom.show_lowest();
 }
```

As you can see from the code, each class defines the *message* member as protected—which lets the derived classes derive that member, but prevents outside code from modifying the *message* member directly. Because each derived class defines the *message* member as *protected*, the derived classes can directly access the data if they want. In the *protderi.cpp* program, however, the derived classes still use the interface functions. As a rule, your programs should enforce encapsulation principles, even when working with *protected* members from a derived class. When you compile and execute the *protderi.cpp* program, your screen will display the following output:

```
Base class message
Level 1 message
Lowest message
C:\>
```

1073 DEFINING STATIC CLASS DATA

As you have learned, when you declare class members, C++ lets you precede a definition with the *static* qualifier. For example, the following class definition uses *static* and non-static data members:

```
class SomeClass
 {
  public:
    static int count;
    SomeClass(int value)
     {
       count++;
       my_data = value;
     };
    ~SomeClass(void) { count--; };
    int my_data;
};
```

Normally, each object instance receives its own data members. However, if you precede a member definition with the *static* keyword, all object instances will share the member. If one instance changes the data, all instances immediately recognize the changed member's new value. A static member definition does not allocate memory for the member. Instead, you must declare the *static* variable outside of the class, as shown here:

```
int SomeClass::count;
```

The following program, *share_it.cpp*, uses the *static* keyword to share the member variable *count* (which tracks the number of object instances):

```
#include <iostream.h>

class SomeClass
 {
  public:
    static int count;
    SomeClass(int value)
     {
       count++;
       my_data = value;
     };
    ~SomeClass(void) { count--; };
    int my_data;
 };

int SomeClass::count;

void main(void)
 {
   SomeClass One(1);
   cout << "One: " << One.my_data << ' ' << One.count << endl ;

   // Declare another instance
   SomeClass Two(2);
   cout << "Two: " << Two.my_data << ' ' << Two.count << endl ;

   // Declare another instance
   SomeClass Three(3);
   cout << "Three: " << Three.my_data << ' ' << Three.count << endl ;
 }
```

Each time the program creates a new object instance, the class constructor increments the *static* variable *count*. After three instances, *count* contains the value 3. As you can see, all three instances share the *static* member. When you compile and execute the *share_it.cpp* program, your screen will display the following output:

```
One: 1 1
Two: 2 2
Three: 3 3
C:\>
```

INITIALIZING A STATIC DATA MEMBER 1074

In Tip 1073, you learned that C++ lets you declare *static* data members that are accessible to all instances of the class. When you use *static* data members, you must determine the best way to initialize the members. One way is to let the first instance pass the desired value to the constructor function. To do this, you overload the constructor function to support one or two parameters. If the object's initializer passes two parameters, the constructor function assigns the second parameter to the *static* variable. The following program, *stat_ini.cpp*, overloads the constructor function in this way to initialize the *static* member to 999:

```
#include <iostream.h>

class SomeClass
 {
  public:
    static int count;
    SomeClass(int value)
     {
       count++;
       my_data = value;
     };
    SomeClass(int value, int static_value)
     {
       count = static_value;
       my_data = value;
     };
    ~SomeClass(void) { count--; };
    int my_data;
};

int SomeClass::count;

void main(void)
 {
   SomeClass One(1, 999);
   cout << "One: " << One.my_data << ' ' << One.count << endl ;

   // Declare another instance
   SomeClass Two(2);
   cout << "Two: " << Two.my_data << ' ' << Two.count << endl ;

   // Declare another instance
   SomeClass Three(3);
   cout << "Three: " << Three.my_data << ' ' << Three.count << endl ;
 }
```

As you will learn in Tip 1075, your programs can also directly access a *public static* member to assign or reference the member's value. When you compile and execute the *stat_ini.cpp* program, your screen will display the following output:

```
One: 1 999
Two: 2 1000
Three: 3 1001
C:\>
```

DIRECT ACCESS OF A STATIC DATA MEMBER 1075

In Tip 1074 you overloaded a constructor function to help your programs initialize a *static* data member. When a *static* data member is public, your programs can directly access the member's value. Therefore, the *stat_ini.cpp* program could have initialized the member using two different techniques. First, the program could have assigned the value when it declared the member outside of the class, as shown here:

```
int SomeClass::count = 999;
```

Second, within *main* itself, the program could have accessed the *static* member directly, as shown here:

```
void main(void)
 {
   SomeClass::count = 999;
   SomeClass One(1);
   // Other statements
 }
```

When you declare a *public static* member, the program can directly access the member's value *even if* no instances of the class exist. To better protect *static* members, use *static private* data members, as Tip 1076 discusses.

1076 UNDERSTANDING STATIC PRIVATE DATA MEMBERS

As you have learned, C++ lets you declare *static* class members that all instances of a class can access. If the member is also *public*, the program itself can access the member, bypassing the class instances. To better protect the *static* member, you can declare it as *private*. When the *static* member is *private*, only class member functions can access it. The following program, *privstat.cpp*, illustrates how to use a *private static* member:

```
#include <iostream.h>

class SomeClass {
  public:
    SomeClass(int value)
     {
       count++;
       my_data = value;
     };
    SomeClass(int value, int static_value)
     {
       count = static_value;
       my_data = value;
     };
    ~SomeClass(void) { count--; };
    void show_values(void) { cout << my_data << ' ' << count << endl; };
  private:
    static int count;
    int my_data;
};

int SomeClass::count;

void main(void)
 {
   SomeClass One(1, 999);
   One.show_values();

   // Declare another instance
   SomeClass Two(2, 1000);
   Two.show_values();
   // Declare another instance
   SomeClass Three(3);
   Three.show_values();
 }
```

When you declare a *static private* member as shown in the *privstat.cpp* program, your program can use a constructor to initialize the member, or your program can assign a value at the initialization that appears outside of the class definition.

Understanding static Member Functions 1077

As you have learned, C++ lets you use *static* class data members whose values each instance shares. In addition to supporting *static* data members, C++ also supports *static* function members. However, C++ programmers do not often use *static* function members. In general, the only use for *static* function members is to manipulate *static* data members. Unlike other function members, which can use the *this* pointer to access instance data, *static* function members cannot access the *this* pointer or instance data. Therefore, the only time you will use a *static* function member is when you have a function that does not manipulate instance data. The following program, *static.cpp*, illustrates how to use a *static* function member:

```
#include <iostream.h>

class SomeClass
 {
  public:
    SomeClass(int value) { some_value = value; };
    void show_data(void) { cout << data << ' ' << some_value << endl; };
    static void set_data(int value) { data = value; };
  private:
    static int data;
    int some_value;
};

int SomeClass::data;

void main(void)
 {
   SomeClass my_class(1001);
   my_class.set_data(5005);
   my_class.show_data();
 }
```

In the *static.cpp* program, both the *data* member and the *set_data* member are static members. Although the *static.cpp* program creates only a single instance of a *Some_Class* object, it could easily create more—and all would share the same static value, 5005.

Direct Access of a public static Function 1078

As you learned in Tip 1077, C++ lets you define *public static* functions within a class. When you declare such functions as *public*, the functions are fully accessible throughout the program, even if the program has not yet created an instance of the class. To access a *public static* member function, your program will use the global resolution operator (::), as shown here:

```
class_name::member_name(parameters);
```

The following program, *globstat.cpp*, illustrates the direct access of a *public static* member function. Note that the program uses the *message* function, even though no instances of a *SomeClass* object exist:

```
#include <iostream.h>

class SomeClass
 {
   public:
       static void message(void) { cout << "Hello, world!\n"; };
 };
```

```
void main(void)
 {
   SomeClass::message();
 }
```

1079 USING ENHANCED TYPES AS CLASS MEMBERS

For simplicity, most of the examples presented throughout this section have used class members that were *int*, *float*, or *char* values. As your class definitions become more complex, however, your class members might be pointers, references, enumerated types, and even nested classes. The following program, *neat_mbr.cpp*, illustrates how to use more complex class members:

```
#include <iostream.h>

enum Days { Monday, Tuesday, Wednesday, Thursday, Friday };

class NeatClass
 {
  public:
    int *lucky_number;
    enum Days lucky_day;
 };

void main(void)
 {
   NeatClass wow;
   int lucky = 1700;

   wow.lucky_day = Monday;
   wow.lucky_number = &lucky;
   cout << "My lucky number is " << *(wow.lucky_number) << endl;
   switch (wow.lucky_day)
    {
      case Monday: cout << "My lucky day is Monday\n";
                   break;
           default: cout << "My lucky day ain't any day but Monday\n";
    };
 }
```

The *NeatClass* class defines a single public member of type *int* pointer, and a single member of the enumerated *Days* type defined just before the *NeatClass* definition within the program file. When the program executes, it declares an instance of the *NeatClass* called *wow* and an *int* variable called *lucky*. The program assigns the *lucky_day* member the enumerated value *Monday* and assigns the *lucky_number* member a reference to the *lucky* variable. Then, the program generates output based on the values it assigns to the *NeatClass* members. When you compile and execute the *neat_mbr.cpp* program, your screen will display the following output:

```
My lucky number is 1500
My lucky day is Monday
C:\>
```

1080 NESTING A CLASS

As you learned in Tip 1079, C++ lets your class members be any type, including other classes. The following program, *nestclas.cpp*, illustrates how to use a *nested* class:

```
#include <iostream.h>

class Outer
 {
```

```
  public:
    Outer(void)
     {
        cout << "Just instantiated an outer\n";

          outer_data = 2002;
     };
    class Inner
     {
      public:
        Inner(void)
         {
           cout << "Just instantiated an inner\n";
           inner_data = 1001;
         };
        void show_data(void) { cout << "Inner: " << inner_data << endl; };
      private:
        int inner_data;
     } inside_stuff;
    void show_all_data(void)
     {
       inside_stuff.show_data();
       cout << "Outer: " << outer_data << endl;
     };
  private:
    int outer_data;
 };

void main(void)
 {
   Outer my_data;

   my_data.show_all_data();
 }
```

When you compile and execute the *nestclas.cpp* program, your screen will display the following:

```
Just instantiated an inner
Just instantiated an outer
Inner: 1001
Outer: 2002
C:\>
```

As a rule, nested classes can become very difficult to understand, and they can decrease your program's reuse. A better solution would be to implement two distinct classes, using the class *Inner* as a base class from which you can derive the *Outer* class.

UNDERSTANDING SUBCLASSES AND SUPERCLASSES 1081

As you read other articles and books on C++ or Java, another object-oriented programming language, you might encounter the terms *subclass* and *superclass*. The terms relate to class inheritance. In general, the terms *subclass* and *base class* are interchangeable. Likewise, the terms *superclass* and *derived class* are also interchangeable. For your discussions on C++, stick with the terms *base* and *derived* class.

INLINE ASSEMBLY LANGUAGE STATEMENTS IN A METHOD FUNCTION 1082

As you have learned, many C and C++ compilers let your programs place inline assembly language statements within the program code. As you will learn in this Tip, your programs can also place inline assembly language statements within class methods. The following program, *clasbeep.cpp*, creates two members, *beep* and *beepbeep*, that use inline assembly language statements to sound the computer's built-in speaker:

```
#include  <iostream.h>

class Beepers
 {
  public:
    void beep(void);
    void beepbeep(void);
 };

void Beepers::beep(void)
 {
   asm
    {
      mov ah,2;
      mov dl,7;
      int 0x21;
    }
 }

void Beepers::beepbeep(void)
 {
   asm
    {
      mov ah,2;
      mov dl,7;
      int 0x21;
      mov ah,2;
      mov dl,7;
      int 0x21;
    }
 }

void main(void)
 {
   Beepers noise;

   noise.beep();
   noise.beepbeep();
 }
```

When you use inline assembly language, most C++ compilers will require you to declare the corresponding member functions outside of the class.

1083 CLASS MEMBERS CAN BE RECURSIVE

As you learned in the Functions section of this book, a recursive function calls itself to perform a task until a specific ending condition is met. When you define class functions, the functions can be recursive. The following program, *strclass.cpp*, creates a string class with two recursive functions, *str_reverse* and *str_length*:

```
#include <iostream.h>
#include <string.h>

class StringClass {
 public:
   void str_reverse(char *string)
    {
      if (*string)
       {
         str_reverse(string+1);
         cout.put(*string);
       }
    };
   int str_length(char *string)
    {
```

```
        if (*string)
          return (1 + str_length(++string));
      else
          return(0);
    };
   StringClass(char *string) { strcpy(StringClass::string, string); };
   char string[256];
};
void main(void)
 {
   StringClass title("Jamsa's C/C++/C# Programmer's Bible");
   title.str_reverse(title.string);
   cout << endl << "The title is " << title.str_length(title.string)
        << " bytes long.";
 }
```

The *strclass.cpp* program first creates an instance of the *StringClass* class called *title.* The program then reverses the title and outputs it, and then outputs a string with the title's length. When you compile and execute the *strclass.cpp* program, your screen will display the following output:

```
elbiB s'remmargorP ++C/C s'asmaJ
The title is 32 bytes long.
C:\>
```

UNDERSTANDING THE THIS POINTER 1084

Each time your program creates a class instance, C++ creates a special pointer called *this*, which contains the address of the current object instance. C++ recognizes the *this* pointer only when a non-static member of the object instance is executing. The instances, in turn, use the *this* pointer to access the different methods. Normally, however, the use of *this* is transparent (in other words, it is not necessary). The compiler assigns *this* and performs the necessary redirections automatically. Your programs usually do not need to use the *this* pointer, but they can, and many programmers do so for clarity. The following program, *showthis.cpp*, uses the *this* pointer to display the values of several instance members. The program also displays the values without using the *this* pointer to illustrate that the compiler automatically inserts instructions to perform the correct indirection for you:

```
#include <iostream.h>
#include <string.h>
class SomeClass {
 public:
   void show_with_this(void)
     {
       cout << "Book: " << this->title << endl;
       cout << "Author: " << this->author << endl;
     };
   void show_without_this(void)
     {
       cout << "Book: " << title << endl;
       cout << "Author: " << author << endl;
     };
   SomeClass(char *title, char *author)
     {
       strcpy(SomeClass::title, title);
       strcpy(SomeClass::author, author);
     };
```

```
private:
    char title[256];
    char author[256];
 };

void main(void)
 {
   SomeClass book("Jamsa's C/C++/C# Programmer's Bible", "Jamsa");

   book.show_with_this();
   book.show_without_this();
 }
```

When you compile and execute the *showthis.cpp* program, your screen will display the following output:

```
Book: Jamsa's C/C++/C# Programmer's Bible
Author: Jamsa
Book: Jamsa's C/C++/C# Programmer's Bible
Author: Jamsa
C:\>
```

1085 How the this Pointer Differs from Other Pointers

In Tip 1084 you learned that each time your program invokes an instance method, the compiler preassigns a special pointer named *this* to point to the object instance. The *this* pointer is unlike other pointers because its value changes with different instance invocations, so your programs must use the *this* pointer with care. As you examine C++ programs, you might encounter statements that return the value to which *this* points, as shown here:

```
return(*this);
```

In many cases, the compiler will convert the pointer's value to a reference, letting a method return a reference to an instance. You must check the member's return value closely to determine if the method is returning a pointer or a reference value.

1086 Understanding Early and Late Binding

When you read articles and books about resolving function calls, you might encounter terms such as *early* (compile time) and *late* (run-time) *binding*. The terms describe when the address for each function your programs will call is *resolved* (made known to the program). Up to this point, the compiler has resolved all the class member function addresses you have used at either compile or link time. Address resolution at this time is called early (sometimes static) binding. C++ also supports *dynamic binding* through the use of *virtual* functions. Late binding (often called *dynamic binding*) occurs at run time and gives programs that use multiple inheritance great flexibility. Several of the Tips that follow discuss virtual functions in detail. Virtual functions enable C++ to support polymorphism, which was introduced in the Objects section.

1087 Pointers to Classes

As your programs become more complex, you might eventually work with pointers to objects. For example, the following program, *ptr_obj.cpp*, creates a simple base class and a simple derived class. The program uses the *new* operator to dynamically allocate instances of each class type and uses pointer indirection to invoke each instance's methods:

```
#include <iostream.h>

class Base
 {
```

```
   public:
     void base_message(void) { cout << "This is the base class\n"; };
 };
class Derived: public Base
 {
  public:
     void derived_message(void) { cout << "This is the derived class\n" ; };
 };
void main(void)
 {
   Base *base_pointer = new Base;
   Derived *derived_pointer = new Derived;
   base_pointer->base_message();
   derived_pointer->derived_message();
 }
```

As you can see, accessing the members of each class using a pointer and the member operator is fundamentally identical to accessing the members of each class using an object instance and the dot operator. However, as you will learn in later Tips, if you use a pointer to access the base class, you can later use the same pointer to access instances of the derived class. Alternatively, your programs can use a pointer to a derived class to point to a base class. Tip 1088 discusses the use of a single pointer with different classes in detail.

USING THE SAME POINTER TO DIFFERENT CLASSES 1088

In Tip 1087, you dynamically created instances of the *Base* and *Derived* classes. To do so, the *ptr_obj.cpp* program used two different pointer variables—one it declared as a pointer to type *Base* and one it declared as a pointer to type *Derived*. Fortunately, when your programs use inheritance, C++ lets you use a pointer to the base class to point to a derived class. However, when you use the base class pointer, you can only access members of the original base class; you cannot access the derived class members. The following program, *baseptr.cpp*, assigns the base class pointer to point to the derived class. The program then uses the pointer to access the base class member *base_message*:

```
#include <iostream.h>
class Base
 {
  public:
     void base_message(void) { cout << "This is the base class\n"; };
 };
class Derived: public Base
 {
  public:
     void derived_message(void) { cout << "This is the derived class\n"; };
 };
void main(void)
 {
   Base *base_pointer = new Base;
   base_pointer->base_message();
   base_pointer = new Derived;
   base_pointer->base_message();
 }
```

As you will learn in Tip 1089, when a derived class and a base class have the same member names and you use a pointer to the base class to point to a derived class, you might not get the results you expect.

1089 BASE AND DERIVED NAME CONFLICTS WITH POINTERS

As you learned in Tip 1088, C++ lets you point to a derived class using a pointer you declared as a pointer to the base class. The following program, *basename.cpp*, uses a base class pointer to point to a derived class. The base and derived classes both have the member name *show_message*:

```
#include <iostream.h>

class Base
 {
  public:
    void show_message(void) { cout << "This is the base class\n"; };
 };

class Derived: public Base
 {
  public:
    void show_message(void) { cout << "This is the derived class\n"; };
 };

void main(void)
 {
   Base *base_pointer = new Base;
   base_pointer->show_message();

   base_pointer = new Derived;
   base_pointer->show_message();
 }
```

When you compile and execute the *basename.cpp* program, your screen will display the following output:

```
This is the base class
This is the base class
C:\>
```

By default, when the base and derived classes use the same function names and you use a pointer to the base class, the C++ compiler will resolve the pointer to the base class function. However, there may be times when you will want the compiler to invoke a derived class member instead. As you will learn in Tip 1090, you must use virtual functions to do so.

1090 UNDERSTANDING VIRTUAL FUNCTIONS

As you have learned, when one class inherits the methods of another class, there may be times when the names of class members will conflict. If you are using a base class pointer to access a derived class and you invoke one of the members with the same name as a base class member, the base class member will execute. If you want C++ to invoke the derived class member, however, you must define a *virtual function* for that base class member. Using virtual functions is not much different from using the operations you have already performed. To create a virtual function, you simply precede the function name with the *virtual* keyword. The function return type and parameter list must be identical for each virtual function. The following program, *virt_one.cpp*, defines the *show_message* function as a virtual function within the *base* and *derived* classes:

```
#include <iostream.h>

class Base
 {
  public:
    virtual void show_message(void) { cout << "This is the base class\n"; };
 };

class Derived: public Base
 {
```

```
   public:
      virtual void show_message(void) { cout << "This is the derived class\n"; };
 };
void main(void)
 {
   Base *base_pointer = new Base;
   base_pointer->show_message();
   base_pointer = new Derived;
   base_pointer->show_message();
 }
```

When you compile and execute the *virt_one.cpp* program, your screen will display the following output:

```
This is the base class
This is the derived class
C:\>
```

As you can see, because both classes use virtual functions, the pointer can correctly invoke the *Base* and *Derived* class methods.

INHERITING THE VIRTUAL ATTRIBUTE 1091

When classes within your programs inherit a virtual function, the inherited function maintains the virtual nature of the base function. In other words, no matter how many times your program inherits a virtual function, it will remain virtual. For example, when a derived class inherits a virtual function and then your program uses the derived class as a base class, the twice-removed derived class can still override the virtual function. To better understand how the member within the derived class can override the virtual function, consider the following program, *vfunc_in.cpp:*

```
#include <iostream.h>
class base
 {
  public:
    virtual void vfunc(void) { cout << "This is base's vfunc()." << endl; }
 };
class derived1 : public base
 {
  public:
    void vfunc(void) { cout << "This is derived1's vfunc()." << endl; }
 };
class derived2 : public derived1
 {
  public:
    void vfunc(void) { cout << "This is derived2's vfunc()." << endl; }
 };
void main(void)
 {
   base *p, b;
   derived1 d1;
   derived2 d2;
   p = &b;         // Point to base class
   p->vfunc();
   p = &d1;        // Point to first derived class
   p->vfunc();
   p = &d2;        // Point to second derived class
   p->vfunc();
 }
```

In both cases (*derived1* and *derived2*), the in-class definitions of *vfunc* override the virtual definition within *base*. Therefore, when the *vfunc_in.cpp* program changes the pointer to point at the derived classes in sequence, the access of the *vfunc* member executes the local *vfuncs* each class defines. When you compile and execute the *vfunc_in.cpp* program, your screen will display the following output:

```
This is base's vfunc().
This is derived1's vfunc().
This is derived2's vfunc().
C:\>
```

1092 VIRTUAL FUNCTIONS ARE HIERARCHICAL

As you have learned in previous Tips, inheritance is hierarchical. Because inheritance is hierarchical, virtual functions must also be hierarchical. Because the virtual functions are hierarchical, if a derived class does not override the virtual function, the compiler will use the nearest derived version above that class in the tree. In the following program, for example, *derived2* is derived from *derived1*, which in turn, is derived from *base*. However, *derived2* does not override the *vfunc* function, so the compiler will instead use the overridden *vfunc* within the *derived1* class definition. As you work with virtual functions, you may want to test your inheritance hierarchy using a program like *virt_hie.cpp*, as shown here:

```
#include <iostream.h>

class base
 {
  public:
    virtual void vfunc(void) { cout << "This is base's vfunc()." << endl; }
 };

class derived1 : public base
 {
  public:
    void vfunc(void) { cout << "This is derived1's vfunc()." << endl; }
 };

class derived2 : public derived1 { };

void main(void)
 {
   base *p, b;
   derived1 d1;
   derived2 d2;

   p = &b;         // Point to base class
   p->vfunc();
   p = &d1;        // Point to first derived class
   p->vfunc();
   p = &d2;        // Point to second derived class
   p->vfunc();     // However, uses derived1's vfunc function
 }
```

In the case of the *virt_hie.cpp* program, the changing of the pointer to the *derived2* class does not impact which function the member operator calls, because the *derived2* class does not define its own implementation of the *vfunc* member. Instead, the last access invokes the *vfunc* member for the *derived1* class. When you compile and execute the *virt_hie.cpp* program, your screen will display the following output:

```
This is base's vfunc().
This is derived1's vfunc().
This is derived1's vfunc().
C:\>
```

IMPLEMENTING POLYMORPHISM 1093

As you read in the Objects section, polymorphism is the ability of the same object to take on different forms. C++ supports polymorphism using virtual functions. Using virtual functions, the same pointer can point to different classes to perform different operations. The following program, *polymorp.cpp*, creates a base class and two derived classes. The *polymorp.cpp* program then uses the pointer *poly* to invoke different methods:

```
#include <iostream.h>
#include <stdlib.h>

class Base
 {
  public:
    virtual int add(int a, int b) { return(a + b); };
    virtual int sub(int a, int b) { return(a - b); };
    virtual int mult(int a, int b) { return(a * b); };
  };

class ShowMath : public Base
 {
    virtual int mult(int a, int b)
     {
       cout << a * b << endl;
       return(a * b);
     };
 };

class PositiveSubt : public Base
 {
   virtual int sub(int a, int b) { return(abs(a - b)); };
 };

void main(void)
 {
   Base *poly = new ShowMath;

   cout << poly->add(562, 531) << ' ' << poly->sub(1500, 407) << endl;
   poly->mult(1093, 1);

   poly = new PositiveSubt;
   cout << poly->add(892, 201) << ' ' << poly->sub(0, 1093) << endl;
   cout << poly->mult(1, 1093);
 }
```

The *polymorp.cpp* program shows clearly how polymorphism and virtual functions expand the power of inheritance to let your programs achieve interesting results. Note that, depending on the instance pointed to by the *poly* pointer, the operations *poly* performs might differ. However, when you compile and execute the *polymorp.cpp* program, your screen will display the following output:

```
1093 1093
1093
1093 1093
1093
C:\
```

UNDERSTANDING PURE-VIRTUAL FUNCTIONS 1094

You have learned that when you derive one class from another, C++ lets you use virtual functions to control which class's functions the program invokes when you use a base class pointer to point to a derived class. When you read more about virtual functions, you might encounter the term *pure-virtual function.* A pure-virtual function is similar to a prototype that you declare in the base class for which the base class requires the derived class to provide an implementation. Within the base class, a pure-virtual function appears as shown here:

```
virtual type function_name(parameters) = 0;
```

The equals (=) symbol and zero value (0) that follow the prototype indicate that the function is a pure-virtual function for which the program must provide an implementation. The following program, *purevirt.cpp*, illustrates a pure-virtual function:

```
#include <iostream.h>
#include <string.h>

class Base
 {
  public:
    virtual void show_message(void) { cout << "Base class message" << endl; };
    virtual void show_reverse(void) = 0;
 };

class Derived : public Base
 {
  public:
    virtual void show_message(void) { cout << "Derived class message" << endl; };
    virtual void show_reverse(void)
      { cout << strrev("Derived class message") << endl; };
 };

void main(void)
 {
   Base *poly = new Derived;
   poly->show_message();
   poly->show_reverse();
 }
```

It is important you note that a pure-virtual function *requires* that each derived class define its own implementation of the function, while a virtual function lets the derived classes use the function the base class derives. In the *purevirt.cpp* program, for example, if *poly* pointed to *base*, your program could not access the *show_reverse* member function. The *poly* pointer can only access *show_reverse* in the *derived* class.

1095 Understanding Abstract Classes

In Tip 1094 you learned that a pure-virtual function is a function prototype for which the base class requires a derived class to provide an implementation. When a class contains only pure-virtual functions, C++ refers to that class as an *abstract class*. In general, an abstract class provides a template from which your programs can later derive other classes. C++ will not let you create a variable of an abstract class type. If you try to do so, the compiler will generate a syntax error.

Abstract classes are also known as *class factories*, because your programs use them as central locations from which to derive (or manufacture, to keep the analogy) other classes within your program.

1096 Using Virtual Functions

As you have learned, one central aspect of object-oriented programming is "one interface, multiple methods." In other words, you can create base classes in C++ that your programs will use to define the nature of an interface to a general class. Each class you then derive from a base class implements the specific operations as they relate to the type of data the derived type uses.

One of the most powerful ways your programs can meet the goal of "one interface, multiple methods" is to use virtual functions, abstract classes, and run-time polymorphism. Programs that use all these features let you create a class hierarchy which moves from general to complex (base to derived). You will create all common features and interfaces within a base class. In cases where you can implement certain actions only within a derived class, the base class should define a virtual function to create the interface the derived classes will use. To better understand the "one interface, multiple methods" concept, consider the following program, *vfunc_ex.cpp*, which applies the "one interface, multiple methods" concept to some simple classes:

```
#include <iostream.h>

class convert
 {
  protected:
    double val1;
    double val2;
  public:
    convert(double i) { val1 = i; }
    double getconv(void) {return val2;}
    double getinit(void) {return val1;}
    virtual void compute(void) = 0;
 };

// liters to gallons
class l_to_g : public convert
 {
  public:
    l_to_g(double i) : convert(i) { }
    void compute(void) { val2 = val1 / 3.7854; }
 };

// Fahrenheit to Celsius
class f_to_c : public convert
 {
  public:
    f_to_c(double i) : convert(i) { }
    void compute(void) { val2 = (val1 - 32) / 1.8; }
 };

void main(void)
 {
   convert *p;                                    // pointer to base class
   l_to_g lgob(4);
   f_to_c fcob(70);

   p = &lgob;                                     // convert liters to gallons
   cout << p->getinit() << " liters is ";
   p->compute();
   cout << p->getconv() << " gallons." << endl;
   p = &fcob;                                     // convert fahrenheit to celsius
   cout << p->getinit() << " in Fahrenheit is ";
   p->compute();
   cout << p->getconv() << " Celsius." << endl;
 }
```

The *vfunc_ex.cpp* program derives the *l_to_g* (liters to gallons) and *f_to_c* (Fahrenheit to Celsius) classes from the base *Convert* class. Both derived classes initialize the member *i* within the *Convert* class. Both functions also override the pure-virtual function *compute*. However, both derived classes use the *getconv* and *getinit* function, which the base class defines. And both derived classes use the *val1* and *val2* members of the base class. In fact, the only function that differs between the two derived classes is the *compute* member, which uses different computations for each derived class. When you compile and execute the *vfunc_ex.cpp* program, your screen will display the following output:

```
4 liters is 1.05669 gallons.
70 in Fahrenheit is 21.1111 Celsius.
C:\>
```

1097 MORE ON EARLY AND LATE BINDING

As you have learned, when you discuss C++ programs with other programmers, you will often hear the phrases *early binding* and *late binding*. Early binding refers to events that happen at compile time. In other words, the compiler knows all the information it must know to compile the program during compilation. Typical examples of early binding include normal function calls (such as standard library calls), overloading function calls, and overloaded operators.

On the other hand, late binding refers to events that your program does not resolve until run time. For your programs to use late binding, classes within those programs must declare virtual functions. As you know, when you access a virtual function through a base pointer, the program determines which virtual function to call based on the actual object the base pointer references. Because the compiler has no way of knowing what object the pointer references at compile time, the program must respond to the pointer references at run time.

1098 DECIDING BETWEEN EARLY AND LATE BINDING

As you learned in Tip 1097, your programs will generally include early binding, late binding, or both. As you write more complex programs, you will find that you use late binding more frequently with your more complex programs than you do with simpler programs. As you add more virtual functions to your programs, you will begin to decide between early and late binding for your objects more frequently.

The main advantage of early binding is efficiency. Because the compiler can resolve the entire function well before program execution, functions bound early are very efficient. Also, early binding is less subject to errors at run time because the compiler can detect many problems well in advance. On the other hand, the main advantage to late binding is flexibility. Unlike early binding, late binding lets your programs respond to events that occur during program execution without creating a large amount of "contingency code."

1099 AN EARLY AND LATE BINDING EXAMPLE

In previous Tips, you have created classes that use virtual functions and classes that do not. As you learned in Tips 1097 and 1098, programmers describe programs that use virtual functions as late-binding programs. Programs that use no virtual functions are early-binding programs. The following code fragment, for example, shows how you might declare the *convert* class defined in Tip 1096 with a virtual function:

```
class convert
 {
  protected:
    double val1;
    double val2;
  public:
    convert(double i) { val1 = i; }
    double getconv(void) {return val2;}
    double getinit(void) {return val1;}
    virtual void compute(void) = 0;
 };
// liters to gallons
class l_to_g : public convert
 {
  public:
    l_to_g(double i) : convert(i) { }
    void compute(void) { val2 = val1 / 3.7854; }
 };
```

Alternatively, your program might declare the *l_to_g* class with early binding, as shown here:

```
class l_to_g
 {
  protected:
    double val1;
    double val2;
  public:
    l_to_g(double i) { val1 = i; }
    double getconv(void) {return val2;}
    double getinit(void) {return val1;}
    void compute(void) { val2 = val1 / 3.7854; }
 };
```

As you can see, in the preceding code fragment, late binding is not particularly helpful. However, if you were to add other conversions to the program, you would eventually create many repetitive classes that share repeated processing. Just as you would probably use inheritance to solve the repetitive class problem, you should use late binding to make your programs more efficient, and clearer to the reader.

DEFINING AN OUTPUT STREAM MANIPULATOR 1100

Several of the Tips in the Getting Started with C++ section of this book used output stream manipulators such as *hex* and *endl*. As you learned previously, you can create your own output stream manipulators. For example, you might want to write a program that creates a new output stream manipulator called *attention*, which sounds the computer's built-in speaker to get the user's attention. You could then use the *attention* manipulator within the *cout* stream, as shown here:

```
cout << attention << "I think your disk is bad!";
```

The following code implements *manipul.cpp*, which creates the *attention* manipulator:

```
#include <iostream.h>

ostream& attention(ostream& cout) { return(cout << '\a'); };

void main(void)
 {
   cout << attention << "The boss is coming to your office...\n";
 }
```

IT IS TIME TO TAKE A LOOK AT IOSTREAM.H 1101

In the Getting Started with C++ section of this book, you were warned not to take a look inside the header file *iostream.h*. Now that you have mastered classes, overloading, virtual functions, and the basics of object-oriented programming, you should not only take a look in the file, you should start analyzing each line. C++ compiler writers used several interesting programming techniques when they wrote *iostream.h*—techniques you can use within your programs, too. First, print a hard copy of the file to which you can later refer as you create your own classes. Second, make a copy of the file and name it *iostream.nts*. Next, read part of the file every day and add comments that explain the file's processing. If you cover two pages of the file daily, you will get through it in less than a week. Not only will you be an expert in C++ I/O operations, you will learn a great deal more about the ins and outs of complex class definitions.

USING SIZEOF WITH A CLASS 1102

As you know, the *sizeof* operator returns the number of bytes required to store an object. As your programs work with objects, there may be times when you must know an object's size. For example, assume that you are reading a file of objects. You can use the *sizeof* operator to determine the size of each object within the file. The *sizeof* operator returns only the size of the class data members. The following program, *sizeof.cpp*, uses the *sizeof* operator to determine the size of two classes. The first class is a base class and the second class is a derived class:

```
#include  <iostream.h>
#include <string.h>

class Base {
 public:
   Base(char *message) { strcpy(Base::message, message); };
   void show_base(void) { cout << message << endl; };
 private:
   char message[256];
};

class Derived: public Base {
 public:
   Derived(char *dmsg, char *bmsg) : Base(bmsg) {
     strcpy(message, dmsg); };
   void show_derived(void)
    {
      cout << message << endl;
      show_base();
    };
 private:
   char message[256];
};

void main(void)
 {
   Base some_base("This is a base");
   Derived some_derived("Derived message", "Base message");

   cout << "The size of the base class is " << sizeof(some_base) << " bytes" << endl;
   cout << "The size of the derived class is " << sizeof(some_derived) <<
           " bytes" << endl;
 }
```

When you compile and execute the *sizeof.cpp* program, your screen will display the following output:

```
The size of the base class is 256 bytes
The size of the derived class is 512 bytes
C:\>
```

1103 PRIVATE, PUBLIC, AND PROTECTED CAN APPLY TO STRUCTURES TOO

Several of the Tips in this section have used private, public, and protected class members. When your C++ programs use structures, you can also have private, public, and protected members. By default, all structure members are public. However, you can use the private and protected labels to identify members for which you want to control access. The following program, *privstru.cpp*, illustrates the use of private members within a structure. As you will see, most of the capabilities C++ provides for classes also exist for structures:

```
#include <iostream.h>
#include <string.h>

struct MyBook {
  char title[64];  // Public by default
  void show_book(void)
    {
      cout << "Book: " << title << " Price: $" << price ;
    };

   void set_price(float amount) { price = amount; };
   void assign_title(char *name) { strcpy(title, name); };
```

```
   private:
      float price;
};

void main(void)
 {
   MyBook book;

   book.assign_title("Jamsa's C/C++/C# Programmer's Bible");
   book.set_price(54.95);
   book.show_book();
 }
```

When you compile and execute the *privstru.cpp* program, it will construct the *book* object, assign it a title and price amount, and then display the *book* object's title and price information on your screen, as shown here:

```
Book: Jamsa's C/C++/C# Programmer's Bible Price: $54.95
C:\>
```

In the *privstru.cpp* program, the *MyBook* structure specifies the *price* member as private. Therefore, the only way to access the member is to use one of the structure's public methods. If you find that your structures require this type of member protection, you should start working with classes instead.

UNDERSTANDING CLASS CONVERSIONS 1104

As you know, when you pass a value of type *int* to a function that requires a *long* value, C++ promotes the *int* value to the correct type. Likewise, if you pass a parameter of type *float* to a function that requires a value of type *double*, C++ makes a similar conversion. When you work with C++ classes, you can also specify the conversions that C++ should perform to convert class values to a standard data type (such as *int* or *long*) or even to a different class. Conversions are common when you pass a parameter of one type to a constructor function for a class of a different type. For example, assume that the class *BookStuff* normally receives three character strings as parameters to its constructor:

```
BookStuff (char *title, char *author, char *publisher)
  {
      // Statements
  }
```

Assume, however, that the program periodically invokes the constructor with a structure of type *BookInfo*, as shown here:

```
struct BookInfo
 {
   char title[64];
   char author[64];
   char publisher[64];
   float price;
   int pages;
 };
```

The class can create a second constructor that converts the data accordingly. Tip 1105 illustrates a program that uses a structure and a second constructor.

CONVERTING DATA IN A CONSTRUCTOR 1105

As you have learned, there may be times when you must convert data from one format into a format that the class expects. An easy way to perform such a conversion is to use different constructor functions. The following program, *convert.cpp*, uses such a constructor function to convert information a structure of type *BookInfo* contains:

```
#include  <iostream.h>
#include <string.h>

struct BookInfo {
  char title[64];
  char publisher[64];
  char author[64];
  float price;
  int pages;
};

class BookStuff {
  public:
    BookStuff(char *title, char *publisher, char *author);
    BookStuff(struct BookInfo);
    void show_book(void)
     {
       cout << "Book: " << title << " by " << author << endl << " Publisher: " <<
               publisher << endl;
     };
  private:
    char title[64];
    char author[64];
    char publisher[64];
};

BookStuff::BookStuff(char *title, char *publisher, char *author)
 {
   strcpy(BookStuff::title, title);
   strcpy(BookStuff::publisher, publisher);
   strcpy(BookStuff::author, author);
 }

BookStuff::BookStuff(BookInfo book)
 {
   strcpy(BookStuff::title, book.title);
   strcpy(BookStuff::publisher, book.publisher);
   strcpy(BookStuff::author, book.author);
 }

void main(void)
 {
   BookInfo book = {"Rescued by C++, Third Edition", "Jamsa Press", "Jamsa",
                    29.95, 256};
   BookStuff big_book("Jamsa's C/C++/C# Programmer's Bible", "Delmar",
                      "Jamsa");
   BookStuff little_book(book);

   big_book.show_book();
   little_book.show_book();
 }
```

The *convert.exe* program begins its processing by instantiating the *BookInfo* class in the *book* object. The declaration calls the *BookInfo* constructor, which assigns the parameters to the *book* object's member variables. The program then performs similar processing with the *big_book* object. The third declaration, *little_book(book),* invokes the overloaded constructor for the *BookStuff* class, which then pulls its needed values from within the *book* object and assigns the values to the new *little_book* object. When you compile and execute the *convert.exe* progam, your screen will display the following output:

```
Book: Jamsa's C/C++/C# Programmer's Bible by Jamsa
Publisher: Delmar
Book: Rescued by C++, Third Edition by Jamsa
Publisher: Jamsa Press
C:\>
```

Using the constructor function to perform the conversion as shown in this Tip is really no different from the overload operations you have been performing throughout this book.

ASSIGNING ONE CLASS TO ANOTHER 1106

In Tip 1105, you created a constructor function that converted the data in a structure of type *BookInfo* into the data fields for an instance of the *BookStuff* class. As you examine C++ programs, you might encounter statements where the program assigns one class type to another, as shown here:

```
class BookStuff big_book;
char title[256];
title = big_book;
```

In this case, the program assigns the class *big_book* to a character string variable. To support such operations, your program must tell the compiler the correct conversion to apply. To do so, you must create a member function within the *BookStuff* class that performs the conversion. In this case, the function will assign the book title to the string. There are two rules the conversion function must follow. First, within the class, the program must define the function as an operator overload function, as shown here:

```
operator char *(void);
```

Second, the corresponding function code must return a value of the converted type which, in this case, is a pointer to a character string. The following program, *classasn.cpp*, uses a member conversion function to assign the class to a character string. In this case, the conversion function assigns the book's title to the string:

```
#include <iostream.h>
#include <string.h>
class BookStuff {
  public:
    BookStuff(char *title, char *publisher, char *author);
    void show_book(void)
     {
       cout << "Book: " << title << " by " << author << " Publisher: " <<
               publisher << endl;
     };
    operator char *(void);
  private:
    char title[64];
    char author[64];
    char publisher[64];
};
BookStuff::BookStuff(char *title, char *publisher, char *author)
 {
   strcpy(BookStuff::title, title);
   strcpy(BookStuff::publisher, publisher);
   strcpy(BookStuff::author, author);
 }
BookStuff::operator char *(void)
 {
   char *ptr = new char[256];
   return(strcpy(ptr, title));
 }
void main(void)
 {
   BookStuff big_book("Jamsa's C/C++/C# Programmer's Bible", "Delmar",
```

```
                           "Jamsa");
   char *title;

   title = big_book;
   cout << "The book's title is " << title << endl;
 }
```

Using conversion functions such as *classasn.cpp* uses, your programs can convert from one class to another, as needed. When you compile and execute the *classasn.cpp* file, your screen will display the following output:

```
The book's title is Jamsa's C/C++/C# Programmer's Bible.
C:\>
```

1107 USE FRIENDS FOR CONVERSION

When you perform conversions between one class and another, there may be times when you must access the private members of another class. As you have learned, you can specify a *friend* class or *friend* function to let another class access the private data without making the data visible to other parts of your program. If a class must access the private data of another class to perform a conversion, specify the conversion function as a *friend* of the class, as shown here:

```
friend void Convert(BookStuff &new, BookInfo book);
// Class definitions
void Convert(BookStuff &new, BookInfo book)
 {
   strcpy(new.title, book.title);
   strcpy(new.publisher, book.publisher);
   strcpy(new.author, book.author);
 }
```

1108 DETERMINING WHEN OPERATORS IMPROVE OR REDUCE READABILITY

As you have learned, when you define a class, C++ lets you overload one or more operators. Before you overload an operator, however, you must determine if the overload will make the program easier or more difficult to understand. For example, the following program, *str_plus.cpp*, overloads the plus operator for the *String* class. The program then uses the operator, as well as the function *strapd*, to append one string's contents to another. Examine the program and determine which technique is most understandable:

```
#include <iostream.h>
#include <iomanip.h>
#include <string.h>

class String {
  public:
    char *operator +(char *append_str)
      { return(strcat(buffer, append_str)); };

    String(char *string)
     {
       strcpy(buffer, string);
       length = strlen(buffer);
     }

    void show_string(void) { cout << buffer; };
    void strapd(char *source) { strcat(buffer, source); };

  private:
    char buffer[256];
     int length;
};
```

```
void main(void)
 {
   String title("Jamsa's C/C++ ");
   title = title + "Programmer's Bible\n";
   title.show_string();

   String book2("Rescued by C++");
   book2.strapd(", Third Edition");
   book2.show_string();
 }
```

Many new C++ programmers commonly overload more operators than are necessary. The decision you must make when you write your programs is whether or not the overload improves your program's readability.

UNDERSTANDING TEMPLATES 1109

As you have learned, there will probably be many times when your programs must duplicate a function so that it supports parameters of a different type. For example, the following function, *compare_values*, compares two values of type *int*, and returns the larger value:

```
int compare_values(int a, int b)
 {
    return ((a > b) ? a: b);
 }
```

If your program must later compare two floating-point values, you must create a second function. The second function is identical in processing but supports different types, as shown here:

```
float compare_values(float a, float b)
 {
    return ((a > b) ? a: b);
 }
```

If the two functions appear in the same program, you must either overload the original function (creating the possibility for confusion) or select a unique name for each function. To help you reduce such duplicate function definitions and make your programs easier to understand, C++ supports the definition of *templates*. A template provides the formats of functions and type placeholders. The following code shows the general format of a function template, where *T* is a type that the compiler will later replace:

```
template<class T>  T function_name(T param_a, T param_b)
  {
     // Statements
  }
```

For example, consider the following function template for *compare_values*:

```
template<class T> T compare_values(T a, T b)
 {
    return((a > b) ? a: b);
 }
```

As you can see, the compiler can replace the letter *T* with either the type *float* or *int* to create the functions previously shown. Several of the Tips that follow discuss templates in detail.

***Note:** The **Turbo C++ Lite** compiler contained on the companion CD-ROM that accompanies this book does not support generic definitions. To write programs that use generic definitions, you must use another compiler, such as Microsoft **Visual** C++.*

1110 PUTTING A SIMPLE TEMPLATE TO USE

As you learned in Tip 1109, C++ supports the use of template functions. The following program, *compare.cpp*, uses the *compare_values* function template to compare values of different types:

```
#include <iostream.h>

template<class T> T compare_values(T a, T b)
 {
   return((a > b) ? a: b);
 }

//float compare_values(float a, float b);  // Some compilers require explicit
//int compare_values(int a, int b);        // function prototypes
//long compare_values(long a, long b);

void main(void)
 {
   float a = 1.2345, b = 2.34567;
   cout << "Comparing " << a << ' ' << b  << ' ' << compare_values(a, b) << endl;

   int c = 1, d = 1001;
   cout << "Comparing " << c << ' ' << d  << ' ' << compare_values(c, d) << endl;

   long e = 1010101L, f = 2020202L;
   cout << "Comparing " << e << ' ' << f  << ' ' << compare_values(e, f) << endl;
 }
```

In the previous example, the template at the start of the *compare.cpp* program specifies the function statements and the type placeholders. When C++ encounters the prototypes that appear before *main*, C++ creates the necessary functions. Later, the compiler determines which function or functions to use based on the type of the parameters the program passes to the generic template (*float*, *int*, or *long*).

1111 BETTER UNDERSTANDING GENERIC FUNCTIONS

As you have learned, C++ supports generic functions. A generic function defines a general set of operations that the function will apply to various data types. A generic function receives the type of data on which the function will operate as a parameter. Because the generic function is, by nature, not explicitly typed, you can use the same general procedure within the function on a wide variety of data types. Creating generic functions within your programs can be useful because many algorithms are fundamentally the same in their processing yet independent of the data type on which the algorithm operates. In later Tips you will learn about the Standard Template Library (STL), which applies the concept of generic functions to a variety of generalized algorithms. For example, as you learned in earlier Tips, the quicksort algorithm is always the same, no matter what type of data your program is sorting. Because the header file created the quicksort algorithm as a generic function, you can eliminate the need for creating multiple functions that perform essentially the same processing.

Within your programs, you will use the keyword *template* to create generic functions. The meaning of the word "template" is accurate for the purpose of generic functions—in other words, you will use generic functions to create a processing template that your programs will use with specific information. The general form of the generic function declaration is shown here:

```
template <class Ttype> return-type function-name([parameter list])
 {
   // function body
 }
```

Ttype is a placeholder name for the data type the function will use. You may also use the *Ttype* name within the function definition. In other words, *Ttype* represents a placeholder that the compiler will automatically replace with the correct data type each time the compiler creates a specific version of the function. In the *compare.cpp* program detailed in Tip 1110, for example, the compiler will replace the generic function with three specific functions: one that returns a *float* value, one that returns an *int* value, and one that returns a *long* value. In Tip 1112, you will learn about templates that support multiple types.

TEMPLATES THAT SUPPORT MULTIPLE TYPES

When you work with C++ templates, there may be times when the template will require more than one data type. For example, consider the following function, *add_values*, which adds a value of type *long* and *int* and returns a *long* result:

```
long add_values(long a, int b)
 {
    return(a + b);
 }
```

To create a template for the *add_values* function, you must specify two types (such as *T* and *T1*), as shown here:

```
template<class T, class T1> T add_values(T a, T1 b)
 {
    return(a + b);
 }
```

In the previous example, the compiler will substitute each occurrence of class *T* with the type you specify. Likewise, the compiler will substitute each occurrence of class *T1* with the corresponding class your invocation specifies.

In the following example, the compiler will substitute each occurrence of class *T* with the type you specify. Likewise, the compiler will substitute each occurrence of class *T1* with the corresponding class your invocation specifies. The following program, *template.cpp*, uses the *add_values* template:

```
#include <iostream.h>

template<class T, class T1> T add_values(T a, T1 b)
  {
    return(a + b);
  }

//long add_values(long a, int b);          // Some compilers require prototypes
//double add_values(double a, float b);

void main(void)
 {
   long a = 320000L;
   int b = 31000;

   double c = 22.0 / 7.0;
   float d = 3.145;

   cout << "Adding " << a << ' ' << b << ' ' << add_values(a, b) << endl;
   cout << "Adding " << c << ' ' << d << ' ' << add_values(c, d) << endl;
 }
```

When you compile and execute the *template.cpp* program, your screen will display the following output:

```
Adding 320000 31000 351000
Adding 3.14286 3.145 6.28786
C:\>
```

MORE ON TEMPLATES WITH MULTIPLE GENERIC TYPES

As you have learned, you can declare templates that support multiple types. In Tip 1112, for example, you created a function that supported two types and returned the result as the first type. However, your generic functions can support an unlimited number of generic types. The generic declaration form for generic functions that support multiple types is shown here:

```
template <class Ttype1, class Ttype2, … class TtypeN>
  return-type function-name ([parameter list])
 {
   // function body
 }
```

When you declare a generic function that supports multiple types, you should avoid declaring too many generic types within the function, as declaring too many types may create more confusion than solutions. You must also be sure that your function can always derive its *return-type* from the generic types the function receives as parameters.

1114 EXPLICITLY OVERLOADING A GENERIC FUNCTION

In previous Tips, you have learned how to define generic functions. You have also learned that a generic function is essentially "self-overloading"—in other words, it creates as many versions for itself as are necessary. However, there may be times when you will want to explicitly overload a template. For example, you may have a generalized *sum* function, which acts differently when it sums *long* values. If you overload a generic function, the overloaded function hides the generic function for instances of the specified values. The following program, *over_tmp.cpp*, overloads a template function explicitly:

```
#include <iostream.h>

template <class X> void swap(X &a, X &b);
void swap(int &a, int &b);

void main(void)
 {
   int i = 10, j = 20;
   float x = 10.1, y = 23.3;
   char a = 'x', b = 'z';

   cout << "Original i, j: " << i << " " << j << endl;
   cout << "Original x, y: " << x << " " << y << endl;
   cout << "Original a, b: " << a << " " << b << endl;
   swap(i,j);                                  // explicitly overloaded swap
   swap(a,b);
   swap(x,y);
   cout << "Swapped i, j: " << i << " " << j << endl;
   cout << "Swapped x, y: " << x << " " << y << endl;
   cout << "Swapped a, b: " << a << " " << b << endl;
 }

template <class X> void swap(X &a, X &b)
 {
   X temp;

   temp = a;
   a = b;
   b = temp;
 }

void swap(int &a, int &b)
 {
   int temp;

   temp = a;
   a = b;
   b = temp;
   cout << "Inside overloaded swap function." << endl;
 }
```

When you compile and execute the *over_tmp.cpp* program, your screen will display the following output:

```
Original i, j: 10 20
Original x, y: 10.1 23.3
Original a, b: x z
Inside overloaded swap function.
Swapped i, j: 20 10
Swapped x, y: 23.3 10.1
Swapped a, b: z x
C:\>
```

UNDERSTANDING THE RESTRICTIONS ON GENERIC FUNCTIONS 1115

As you have learned, generic functions are similar to overloaded functions. However, together with the power of generic functions, C++ also applies more restrictions to generic functions than it does to overloaded functions. When you overload functions, your overloaded functions may perform different actions depending on the overload criteria. When you write a generic function, however, your function must perform the same processing on all data, independent of type. For example, in the following program, *no_temp.cpp*, you cannot replace the overloaded functions with a generic function because the activities of the overloaded functions differ:

```
#include <iostream.h>
#include <math.h>

void non_template(int i)
 {
   cout << "value is: " << i << endl;
 }

void non_template(double d)
 {
   double intpart;
   double fracpart;

   fracpart = modf(d, &intpart);
   cout << "Fractional part: " << fracpart << endl;
   cout << "Integer part: " << intpart << endl;
 }

void main(void)
 {
   non_template(1);
   non_template(12.2);
 }
```

USING A GENERIC FUNCTION 1116

As you can see, generic functions are one of C++'s most useful features. You can apply generic function to all types of situations. Any time you have a function that defines a generalizable algorithm, you can make the function a template function. After you have created the function, you can use the generic function with any type of data without having to rewrite the function. In later Tips you will learn how to use templates to define classes. In Tip 1117 you will create a generic bubble sort function. For now, be sure you clearly understand the generic function's processing before you move on to later Tips. To better understand generic function processing, consider the following simple program, *show_txt.cpp*, which writes a generic output function:

```
#include <iostream.h>

template <class T1, class T2> void sample(T1 x, T2 y);

void main(void)
 {
    sample(10, "hi");
```

```
    sample(0.23, 10L);
    sample("Jamsa's", "C/C++/C#");
 }
template <class T1, class T2> void sample(T1 x, T2 y)
 {
   cout << x << " " << y << endl;
 }
```

When you compile and execute the *show_txt.cpp* program, the compiler will create three versions of the *sample* function: one that accepts an integer and a string, one that accepts a *float* and a *long*, and one that accepts two strings. When you execute the *show_txt.cpp* program, however, the fact that the compiler created three versions of the function (rather than the program itself including three versions of the function) is invisible to the user, who will see only the following output:

```
10 hi
0.23 10
Jamsa's C/C++/C#
C:\>
```

To perform the same steps without a generic function, however, you would have to write a program such as *show_not.cpp*, shown here:

```
#include <iostream.h>

void sample(int x, char *y);
void sample(float x, long y);
void sample(char *x, char *y);

void main(void)
 {
   sample(10, "hi");
   sample(0.23, 10L);
   sample("Jamsa's", "C/C++/C#");
 }
void sample(int x, char *y)
 {
   cout << x << " " << y << endl;
 }
void sample(float x, long y)
 {
   cout << x << " " << y << endl;
 }
void sample(char *x, char *y)
 {
   cout << x << " " << y << endl;
 }
```

1117 USING A GENERIC BUBBLE SORT FUNCTION

As you have learned, generic functions can be helpful in making your programs easier to understand and can save you significant programming time. To better understand the usefulness of the generic function, consider a program you have previously written without generics and determine how a generic might make the program more useful. For example, in Tips 492 and 493, you created a simple bubble sort function that worked only with integer values. That bubble sort function would be significantly more useful if it were to use a generic template, rather than a specific function. You wrote the original function to accept an array of integers for sorting, as shown here:

```
void bubble_sort(int array[], int size)
 {
```

```
    int temp, i, j;

    for (i = 0; i < size; i++)
     for (j = 0; j < size; j++)
      if (array[i] < array[j])
        {
          temp = array[i];
          array[i] = array[j];
          array[j] = temp;
        }
  }
```

Using your knowledge of generic functions, you can now write the code for the bubble sort so that it handles all value types, not just *int* arrays, as shown here within the *bubble.cpp* program:

```
#include <iostream.h>

template <class X> void bubble_sort(X *items, int size);
template <class X> void show_items(X *items, int size);

void main(void)
 {
   int iarray[7] = {7, 5, 4, 3, 9, 8, 6};
   double darray[5] = {4.2, 2.5, -0.9, 100.2, 3.0};

   cout << "Here is unsorted integer array: " << endl;
   show_items(iarray, 7);
   cout << "Here is unsorted double array: " << endl;
   show_items(darray, 5);
   bubble_sort(iarray, 7);
   bubble_sort(darray, 5);
   cout << "Here is sorted integer array: " << endl;
   show_items(iarray, 7);
   cout << "Here is sorted double array: " << endl;
   show_items(darray, 5);
 }

template <class X> void bubble_sort(X *items, int size)
 {
   register int i, j;
   X temp;

   for (i = 1; i < size; i++)
    for (j = size-1; j >= i; j--)
      if (items[j-1] > items[j])
        {
          temp = items[j-1];
          items[j-1] = items[j];
          items[j] = temp;
        }
  }

template <class X> void show_items(X *items, int size)
 {
   int i;

   for(i=0; i < size; i++)
      cout << items[i] << ", ";
   cout << endl;
 }
```

When you compile and execute the *bubble.cpp* program, your screen will display the following output:

```
Here is unsorted integer array:
7, 5, 4, 3, 9, 8, 6,
Here is unsorted double array:
4.2, 2.5, -0.9, 100.2, 3,
```

```
Here is sorted integer array:
3, 4, 5, 6, 7, 8, 9,
Here is sorted double array:
-0.9, 2.5, 3, 4.2, 100.2,
C:\>
```

1118 USING GENERIC FUNCTIONS TO COMPACT AN ARRAY

As your programs work with arrays, sometimes you will want to remove elements from the middle of an array and move the remaining elements in the array "down" to fill the newly created space—a process called *compacting*. One way to avoid the need for array compacting is to use a linked list, which you have learned about previously. You may often find it more convenient, however, to work with an array. The following program, *compact.cpp*, uses a generic function to compact arrays of multiple types:

```
#include <iostream.h>

template <class X> void compact(X *items, int count, int start, int end);
template <class X> void show_items(X *items, int size);

void main(void)
 {
   int nums[7] =  {0, 1, 2, 3, 4, 5, 6};
   char str[18] = "Generic Functions";

   cout << "Here is uncompacted integer array: ";
   show_items(nums, 7);
   cout << "Here is the uncompacted char array: ";
   show_items(str, 18);
   compact(nums, 7, 2, 4);
   compact(str, 18, 6, 10);
   cout << "Here is compacted integer array: ";
   show_items(nums, 7);
   cout << "Here is the compacted char array: ";
   show_items(str, 18);
 }

template <class X> void compact(X *items, int count, int start, int end)
 {
   register int i;

   for(i = end+1; i < count; i++, start++)
      items[start] = items[i];

   for( ; start < count; start++)
      items[start] = (X) 0;
 }

template <class X> void show_items(X *items, int size)
 {
   int i;

   for(i = 0; i < size; i++)
      cout << items[i];
   cout << endl;
 }
```

In the previous example, the *compact* function fills the remaining elements within the array with zero. When you compile and execute the *compact.cpp* program, your screen will display the following output:

```
Here is uncompacted integer array: 0123456
Here is the uncompacted char array: Generic Functions
Here is compacted integer array: 0100000
Here is the compacted char array: Generinctions
C:\>
```

WHERE TO PLACE TEMPLATES 1119

As you have learned, C++ templates let you reduce your programming for functions that differ only in parameter and return types. As you create templates, you should place them in meaningfully named header files so you can easily reuse them in other programs. Because you will probably create only a few templates initially, you may want to use the header file *template.h*. As you create more templates, you can assign them to other header files, based on their functions.

TEMPLATES ALSO ELIMINATE DUPLICATE CLASSES 1120

As you have learned, C++ templates can reduce your programming for functions that differ only in parameter and return types. Your programs can also use templates to eliminate similar classes. For example, consider the following classes:

```
class ShortDistance {
   public:
        ShortDistance(int distance) { ShortDistance::distance = distance; };
       ShowDistance(void)
        { cout << "The distance is " << distance << " miles" << endl; };
   private:
       int distance;
 };
class LongDistance {
   public:
        LongDistance(long distance) { LongDistance::distance = distance; };
       ShowDistance(void)
        { cout << "The distance is " << distance << " miles" << endl; };
   private:
        long distance;
 };
```

Both classes perform similar processing, the only difference is that the *LongDistance* class maintains values of type *long int*, and the *ShortDistance* class only maintains values of type *int*. Because the classes are fundamentally identical and perform the same processing, but differ only in the value the classes maintain, the classes are good candidates for a single generic class.

The following program, *classtmp.cpp*, combines the two class types, *LongDistance* and *ShortDistance,* into a generic *Distance* class:

```
#include <iostream.h>

template<class T> class Distance {
  public:
     Distance(T distance);
     void show_distance(void)
      { cout << "The distance is " << distance << " miles\n"; };
  private:
    T distance;
};

template<class T>
Distance<T>::Distance(T distance) { Distance::distance = distance; };

void main(void)
 {
     Distance<int> short_distance(100);
     Distance<long> long_distance(2000000L);

     short_distance.show_distance();
     long_distance.show_distance();
 }
```

When you compile the *classtmp.cpp* program, the C++ compiler will create classes using the correct types. As you will learn in Tip 1121, class templates such as *Distance* are often called a *generic class* or *class generator*. When you execute the *classtmp.cpp* program, your screen will display the following output:

```
The distance is 100 miles
The distance is 2000000 miles
C:\>
```

1121 UNDERSTANDING GENERIC CLASSES

As you learned in Tip 1120, C++ lets you use templates to define generic classes. When you specify a class template, you must always specify the corresponding name, followed by angle brackets and a type (either the placeholder type or the actual type), as shown here:

```
Distance<T>
Distance<int>
```

As you examine C++ programs that use class templates, you might find some programs that define templates that accept parameters, as shown here:

```
template<class T, int array_size = 64> class SomeClass {
  // Statements
};
```

In this case, the template not only specifies a type placeholder, it also specifies a parameter that the program can use within the template. When the program later uses the template, it can pass a parameter value to the template, as shown here:

```
SomeClass<int, 1024> this_instance;
```

1122 USING GENERIC CLASSES

In previous Tips you have created and used several generic functions. However, as you learned in Tip 1121, C++ also supports generic classes. Creating a generic class, while often more difficult than creating a typed class, also increases your programming power. You can use generic classes to let your programs access a greatly increased set of value types without significant additional coding. As you will learn in later Tips, the Standard Template Library (STL) is built around generic classes. To better understand generic classes, consider the following program, *gen_stck.cpp*:

```
#include <iostream.h>

const int SIZE = 100;

template <class SType> class stack {
   SType stck[SIZE];
   int tos;
 public:
   stack(void);
   ~stack(void);
   void push(SType i);
   SType pop(void);
 };

template <class SType> stack<SType>::stack(void)
 {
   tos = 0;
   cout << "Stack Initialized." << endl;
 }

template <class SType> stack<SType>::~stack(void)
 {
```

```
   cout << "Stack Destroyed." << endl;
 }
template <class SType> void stack<SType>::push(SType i)
 {
   if(tos == SIZE)
    {
      cout << "Stack is full." << endl;
      return;
    }
   stck[tos++] = i;
 }
template <class SType> SType stack<SType>::pop(void)
 {
   if(tos == 0)
    {
      cout << "Stack underflow." << endl;
      return 0;
    }
   return stck[--tos];
 }
void main(void)
 {
   stack<int> a;
   stack<double> b;
   stack<char> c;
   int i;

   a.push(1);
   a.push(2);
   b.push(99.3);
   b.push(-12.23);

   cout << a.pop() << " ";
   cout << a.pop() << " ";
   cout << b.pop() << " ";
   cout << b.pop() << endl;

   for(i = 0; i < 10; i++)
      c.push((char) 'A' + i);

   for(i = 0; i < 10; i++)
      cout << c.pop();
   cout << endl;
 }
```

The *gen_stck.cpp* program defines the generic class *stack*, which maintains an array of 100 elements of its defined type and an integer value that you can use to access the elements within the stack. As you saw within your previous definitions of the *stack* class, it also defines *push* and *pop* member functions that you can use to place data into and retrieve data from a *stack* object. When the program begins its execution, it creates three instances of the generic *stack* class: an *int* stack, a *double* stack, and a *char* stack. The remainder of the program's processing manipulates each stack in turn. When you compile and execute the *gen_stck.cpp* program, your screen will display the following output:

```
Stack Initialized.
Stack Initialized.
Stack Initialized.
2 1 -12.23 99.3
JIHGFEDCBA
Stack Destroyed.
Stack Destroyed.
Stack Destroyed.
C:\>
```

1123 CREATING A GENERIC CLASS WITH TWO GENERIC DATA TYPES

As you have learned, your programs can declare generic functions that support multiple generic data types. Similarly, your programs can declare generic classes that support multiple generic data types. To declare a generic class that supports multiple generic data types, you will use an expanded form of the standard generic class declaration, as shown here:

```
template <class T1, class T2, … class TN> class class-name {
   // member definitions
 }
```

The *T1, T2,* and *TN* placeholders represent the generic types your class will accept. A class can, in theory, accept an infinite number of generic types within its definition. However, just as with generic functions that support multiple types, you should be careful not to define too many generic types in your classes, as doing so will create confusion within your programs. To better understand how you will declare generic classes that support two or more generic data types, consider the program *two_gen.cpp*, as shown here:

```
#include <iostream.h>

template <class T1, class T2> class two_gen {
   T1 i;
   T2 j;
 public:
   two_gen(T1 a, T2 b)
      { i=a; j=b;}
   void show(void)
      { cout << i << " " << j << endl; }
 };

void main(void)
 {
   two_gen<int, double> obj1(10, 0.23);
   two_gen<char, char *> obj2('X', "This is a test.");

   obj1.show();
   obj2.show();
 }
```

1124 CREATING A PARAMETERIZED MANIPULATOR

As you learned in Tip 1001, you will use generic classes to create parameterized manipulators. When you create a parameterized manipulator, you must include the *iomanip.h* file within your program. The *iomanip.h* file defines three generic classes: *omanip, imanip,* and *smanip.* In general, when you must create a manipulator that takes an argument, your program must create two overloaded manipulator functions. The first must define two parameters: a reference to the stream and a parameter that the first function passes to the second function (the generic function). The generic function accepts a single parameter and generates a call to the first function. To better understand the relationship between the two functions, consider the generalized form of the parameterized output manipulator, as shown here:

```
ostream &manip-name(ostream &stream, type param)
 {
   // your code here
   return stream;
 }

// Overloaded generic function
omanip<type> manip-name(type param)
 {
   return omanip<type> (manip-name, param);
 }
```

In the generalized form, *manip-name* is the manipulator's name, and *type* specifies the parameter type the manipulator uses. Because *omanip* is also a generic class, *type* is also the type of data that the *omanip* object (which the second overloaded function returns) acts upon. (Some compilers provide an OMANIP macro you can use to simplify the process of defining an operator.) Writing parameterized manipulators is a difficult and often confusing task at best. To better understand how you will use parameterized manipulators, consider the *parm_man.cpp* program, shown here:

```
#include <iostream.h>
#include <iomanip.h>

ostream &indent(ostream &stream, int length)
 {
   register int i;

   for(i = 0; i < length; i++)
      cout << " ";
   return stream;
 }

omanip<int> indent(int length)
 {
   return omanip<int>(indent, length);
 }

void main(void)
 {
   cout << indent(10) << "This is a test." << endl;
   cout << indent(20) << "of the new indent manipulator." << endl;
   cout << indent(5) << "It works!" << endl;
 }
```

When you compile and execute the *parm_man.cpp* program, it will display the output shown here:

```
          This is a test.
                    of the new indent manipulator.
     It works!
C:\>
```

1125 CREATING A GENERIC ARRAY CLASS

In previous Tips you have learned about ways you can create generic classes and functions and activities that the templates let, or help you, perform. As you have seen in many previous Tips in this book, you will generally create an array of class types within most programs. Therefore, you must understand how you manipulate a generic class to initialize an array of that class.

Creating a generic array is as easy as creating an array of a normal type. For example, if all your arrays are the same size, you might use a generic class without a parameter to declare an integer array, as shown here:

```
atype<int> int_array;
```

Using a generic class to create arrays lets you operate on arrays differently. As you learned earlier, your programs can overload the [ ] operator to create "safe arrays"—arrays whose boundaries your programs cannot exceed, among other criteria for safe arrays. If you create a generic class and overload the [ ] operator relative to that class, you can force all arrays within your programs into safe arrays. The following program, *gen_safe.cpp*, uses a generic class and an overloaded [ ] operator to create safe arrays:

```
#include <iostream.h>
#include "stdlib.h"

const int SIZE = 10;

template <class AType> class atype {
   AType a[SIZE];
```

```
   public:
     atype(void)
      {
        int i;

        for(i = 0; i < SIZE; i++)
           a[i] = i;
      }
     AType &operator[](int i);
  };

template <class AType> AType &atype<AType>::operator[](int i)
 {
   if(i < 0 || i > SIZE-1)
    {
      cout << endl << "Index value of ";
      cout << i << " is out of bounds." << endl;
    }
   return a[i];
 }

void main(void)
 {
   atype<int> int_array;
   atype<double> double_array;
   int i;

   cout << "Integer array: ";
   for(i = 0; i < SIZE; i++)
      int_array[i] = i;

   for(i = 0; i < SIZE; i++)
      cout << int_array[i] << " ";
   cout << endl;

   cout << "Double array: ";
   cout.precision(2);
   for(i = 0; i < SIZE; i++)
      double_array[i] = (double)i/3;

   for(i = 0; i < SIZE; i++)
      cout << double_array[i] << " ";
   cout << endl;

   int_array[12] = 100;                  // Calls overloaded array operator
 }
```

The *gen_safe.cpp* program creates the generic class *atype* and overloads the [ ] operator within the class. The overloaded function checks to make sure the user does not try to access or initialize any element of the array that is outside the array's previously-defined boundaries. When the *gen_safe.cpp* program executes, it creates two arrays (one *int* and one *double*), fills the arrays, and outputs their values. Finally, the last statement tries to initialize a value outside the array's boundaries. When you compile and execute the *gen_safe.cpp* program, your screen will display the following output:

```
Integer array: 0 1 2 3 4 5 6 7 8 9
Double array: 0 0.33 0.67 1 1.3 1.7 2 2.3 2.7 3

Index value of 12 is out of bounds.
C:\>
```

1126 UNDERSTANDING EXCEPTION HANDLING

As you create class libraries, sometimes you can anticipate in advance the types of run-time errors a program will encounter while working with the class (such as overwriting array bounds or passing too large a value). Unfortunately, there may be many times when you cannot write code to trap such errors when the errors occur within another program. For such cases, many C++ compilers support *exception handlers*. In general, an exception handler is software that executes when such an error occurs. Within the class library code, you would test for the possible error. If the error has occurred, you would raise (C++ calls it *throw*) an exception. The user program that experienced the error is responsible for *catching* and handling the exception—which means the program must provide exception-handling software. To support exception handling, the C++ language provides the keywords shown in Table 1126.

Keyword	Meaning
catch	Catches the thrown exception
throw	Initiates an exception handler
try	Attempts an operation to test for a possible exception

Table 1126 C++ exception-handling keywords.

UNDERSTANDING THE BASIC EXCEPTION-HANDLING FORM 1127

As you learned in Tip 1126, C++ provides you with several statements that your programs can use together to catch exceptions within your program code. The three basic components of every exception handler are the *try, catch,* and *throw* statements. When your programs perform exception processing, you must include the statements that you want to monitor for exceptions (that is, errors) within the *try* block. If a statement processes incorrectly, you must *throw* an error appropriate to the function's action. The user program catches the *throw* statement within the *catch* block, which performs the actual exception handling itself. The generalized form of the exception-handling block is shown here:

```
try {
   // try block
   // if(error) throw exception-value;
 }
catch(exception-type variable-name) {
   // exception-handling processing
 }
```

Within the generalized form of the exception handler, the *thrown exception-value* must match the caught *exception-type*, as you will learn in later Tips. In Tip 1128, you will write a simple exception handler based on the generalized form shown in this Tip.

WRITING A SIMPLE EXCEPTION HANDLER 1128

As you learned in Tip 1127, each exception handler you write within your programs will include a *try* statement, one or more *throw* statements, and one or more *catch* statements. To better understand the processing that an exception handler performs, consider the following program, *simple_e.cpp*, which uses an exception handler to throw and process an error within *main*:

```
#include <iostream.h>

void main(void)
 {
   cout << "Start" << endl;
   try {
      cout << "Inside try block." << endl;
      throw 100;
      cout << "This will not execute.";
    }
```

```
    catch(int i) {
      cout << "Caught an exception -- value is: ";
      cout << i << endl;
    }
   cout << "End";
 }
```

The *simple_e.cpp* program implements a simple *try-catch* block. Rather than waiting for the program to commit an error, the program uses the *throw* statement (which you will learn about in Tip 1129) to cause an error. After the *try* block throws the error, the *catch* block catches it and processes the value the *throw* statement passes. When you compile and execute the *simple_e.cpp* program, your screen will display the following output:

```
Start
Inside try block.
Caught an exception -- value is: 100
End
C:\>
```

1129 UNDERSTANDING THE THROW STATEMENT

As you have learned, the third component of an exception handler is the *throw* statement, which invokes the *catch* statement. You must include at least one *throw* statement within a *try* block for the exception handler to perform any valuable processing. As you saw in previous Tips, the generalized format of the *throw* statement is shown here:

```
throw exception;
```

As you have learned, the *exception* must match the *exception-type* that the *catch* block specifies. As your programs become more complex, and especially as you do more Windows programming, you will find that you can use *try-catch* blocks (often simply called *try* blocks) with many functions to provide more stable and secure programs.

1130 EXCEPTIONS ARE TYPE-SPECIFIC

As you have learned, your programs can perform exception-handling activities using a *try* block. The first *try* block that you saw (in Tip 1127) caught an integer exception. However, your *catch* statements can catch any exception type. As a result, you must be careful that the exception your program throws within the *try* block matches the type the *catch* block specifies. For example, the following program, *catch_d.cpp*, catches a *double* exception, while the *try* block throws an integer exception:

```
#include <iostream.h>

void main(void)
 {
   cout << "Start" << endl;
   try {
      cout << "Inside try block." << endl;
      throw 100;
      cout << "This will not execute.";
    }
   catch(double d) {
      cout << "Caught a double exception -- value is: ";
      cout << d << endl;
    }

   cout << "End";
 }
```

Depending on your compiler, when you compile and execute the *catch_d.cpp* program, you may receive a compiler warning that the *catch* statement is unreachable code. If you receive such a message when you compile your programs, you should check your code carefully for logic errors such as the one in the *catch_d.cpp* program. If you receive such a compiler error message and run the program nevertheless, your program will generate output similar to the *catch_d.cpp* program, as shown here:

```
Start
Inside try block.
Abnormal program termination.
C:\>
```

Throwing Exceptions from a Function Within a try Block 1131

As you have learned, your programs can perform extensive exception-handling activities using a *try* block. However, your programs may often call functions from within a *try* block. When your programs call functions from within a *try* block, C++ will pass the exception to the *try* block outside the function (provided there is not a second *try* block within the function, as you will learn in later Tips). The following program, *out_func.cpp*, uses a *try* block inside *main* to call the *XHandler* function:

```
#include <iostream.h>

void XHandler(int test)
 {
   cout << "Inside XHandler, test is:" << test << endl;
   if(test)
     throw test;
 }

void main(void)
 {
   cout << "Start: " << endl;
   try {
      cout << "Inside try block." << endl;
      XHandler(1);
      XHandler(2);
      XHandler(0);
    }
   catch(int i) {
      cout << "Caught an exception. Value is: ";
      cout << i << endl;
    }
   cout << "End";
 }
```

When you compile and execute the *out_func.cpp* program, your screen will display the following output:

```
Start:
Inside try block.
Inside XHandler, test is:1
Caught an exception. Value is: 1
End
C:\>
```

Localizing a try Block to a Function 1132

As you have learned, your programs can use *try* blocks to catch exceptions during processing. In previous Tips, you created a simple *try* block within *main*, and you created a *try* block within *main* that called a function from within itself. However, you can also use *try* blocks locally within functions. When you localize a *try* block to a function, C++

resets the block each time you enter the function. The following program, *fun_catc.cpp*, shows how you can localize *try* blocks within functions:

```
#include <iostream.h>

void XHandler(int test)
 {
   try
    {
      if(test)
        throw test;
    }
   catch(int i)
    {
      cout << "Caught exception #: " << i << endl;
    }
 }

void main(void)
 {
   cout << "Start: " << endl;
   XHandler(1);
   XHandler(2);
   XHandler(0);
   XHandler(3);
   cout << "End";
 }
```

When you compile and execute the *fun_catc.cpp* program, your screen will display the following output (notice that the function only throws three exceptions because the third call, with its zero value, evaluates false):

```
Start:
Caught exception #: 1
Caught exception #: 2
Caught exception #: 3
End
C:\>
```

1133 UNDERSTANDING WHEN THE PROGRAM EXECUTES CATCH

As you have learned, your programs can use a *try-catch* sequence to control exception handling and protect against abnormal program termination. In previous Tips, you learned how to *catch* exceptions within a *catch* block. However, you may be concerned that your programs will execute statements within a *catch* block even if the program does not throw an exception. As it turns out, your programs will only execute the statements within *catch* blocks if the program throws an exception within the immediately preceding *try* block. The following program, *no_catch.cpp*, illustrates how your programs will skip over the statements within a *catch* block unless the program throws an exception:

```
#include <iostream.h>

void main(void)
 {
   cout << "Start" << endl;
   try
    {
      cout << "Inside try block." << endl;
      cout << "Still inside try block." << endl;
    }
   catch(int i)
    {
      cout << "Caught an exception--value is: " << endl;
      cout << i << endl;
```

```
   }
  cout << "End";
}
```

When you compile and execute the *no_catch.cpp* progam, the program will skip the statements the *catch* block contains because the program did not throw an exception in the *try* block. The *no_catch.cpp* program will display the following output on your screen:

```
Start
Inside try block.
Still inside try block.
End
C:\>
```

USING MULTIPLE CATCH STATEMENTS WITH A SINGLE TRY BLOCK 1134

As your exception handling becomes more complex, there may be times when a single *try* block throws exceptions of multiple types. Within your programs, you will want to construct your exception handler so that it supports the catching of multiple exceptions. When you must catch multiple exceptions, you will use the following generalized format:

```
try
 {
   // statements
 }
catch(type1)
 {
   // exception processing
 }
catch(type2)
 {
   // exception processing
 }
 .
 .
 .
catch(typeN)
 {
   // exception processing
 }
```

As you will learn in later Tips, your *catch* statements can recognize any returned type, not just the basic types that C++ supports. In fact, your *catch* statements can even catch thrown user-defined types. The following program, *catch_3.cpp*, uses multiple *catch* statements to catch several exceptions of different types:

```
#include <iostream.h>

void XHandler(int test)
 {
   try
    {
      if(test==0)
        throw test;
      if(test==1)
        throw "String";
      if(test==2)
        throw 123.23;
    }
   catch(int i)
    {
      cout << "Caught exception #: " << i << endl;
    }
```

```
      catch(char *str)
       {
         cout << "Caught string exception: " << str << endl;
       }
     catch(double d)
       {
         cout << "Caught exception #: " << d << endl;
       }
  }
void main(void)
  {
    cout << "Start: " << endl;
    XHandler(0);
    XHandler(1);
    XHandler(2);
    cout << "End";
  }
```

The *catch_3.cpp* program uses a series of *if* statements within the *try* block to throw three different exceptions in three different function calls. When you compile and execute the *catch_3.cpp* program, your screen will display the following output:

```
Start:
Caught exception #: 0
Caught string exception: String
Caught exception #: 123.23
End
C:\>
```

1135 USING THE (...) ELLIPSIS OPERATOR WITH EXCEPTIONS

As you have learned in previous Tips, your programs can catch exceptions within multiple *try* blocks or use multiple *catch* statements within a single *try* block. However, your programs can also use the *ellipsis* operator to generically catch all errors that occur within a single *try* block. To catch all errors within a single *try* block, you will construct the *try* block in the following generalized format:

```
try
 {
   // statements
 }
catch(...)
 {
   // exception processing
 }
```

1136 CATCHING ALL EXCEPTIONS WITHIN A SINGLE TRY BLOCK

As you learned in Tip 1135, your programs can use the *ellipsis* operator to catch multiple exceptions. As you will learn in Tip 1137, you can use the *ellipsis* operator together with a standard *try-catch* block. However, you can also use the *ellipsis* operator by itself to catch multiple exceptions of unknown or differing types. The following program, *catch_ml.cpp*, uses the *ellipsis* operator to catch three exceptions of three different types:

```
#include <iostream.h>

void XHandler(int test)
 {
   try
    {
```

```
        if(test==0)
          throw test;
       if(test==1)
          throw 'a';
       if(test==2)
          throw 123.23;
     }
   catch(...)
     {
       cout << "Caught one." << endl;
     }
 }
void main(void)
 {
   cout << "Start: " << endl;
   XHandler(0);
   XHandler(1);
   XHandler(2);
   cout << "End";
 }
```

Notice that, unlike the *catch_3.cpp* program in Tip 1134, the *catch_ml.cpp* program uses only a single catch statement to catch all three errors. When you compile and execute the *catch_ml.cpp* program, your screen will display the following output:

```
Start:
Caught one.
Caught one.
Caught one.
End
C:\>
```

You should note that, in the *catch_ml.cpp* program, your *catch* block that uses the *ellipsis* operator cannot discern the type of exception the program throws—meaning that your processing within the *catch* block must be error-independent. Tip 1137 explains how to include error-dependent and error-independent exception handling within a single block.

CATCHING EXPLICIT AND GENERIC EXCEPTIONS IN A SINGLE TRY BLOCK 1137

As you have learned, you can use the *ellipsis* operator to catch multiple exceptions of unknown types. Most often, however, you will want your programs to catch explicit exceptions and respond to the exceptions in specific ways. If you are processing specific, explicit exceptions within a given *try* block, you may additionally want to use the *ellipsis* operator to catch all exceptions that are not of the expected type or types. Because C++ lets you catch multiple exception types within a given *try* block, you can easily create exception sequences that let you catch both explicit and generic exceptions and handle them differently. The following program, *exp_unex.cpp*, catches both integer and unknown exceptions within a single *try* block:

```
#include <iostream.h>
void XHandler(int test)
 {
   try
    {
      if(test==0)
         throw test;
      if(test==1)
          throw 'a';
        if(test==2)
         throw 123.23;
```

```
    }
   catch(int i)
    {
      cout << "Caught an integer." << endl;
    }
   catch(...)
    {
      cout << "Caught one." << endl;
    }
 }
void main(void)
 {
   cout << "Start: " << endl;
   XHandler(0);
   XHandler(1);
   XHandler(2);
   cout << "End";
 }
```

The *exp_unex.cpp* program will catch the *int* throw within the explicit block, and the *double* and *char* throws within the generic block. When you compile and execute the *exp_unex.cpp* program, your screen will display the following output:

```
Start:
Caught an integer.
Caught one.
Caught one.
End
C:\>
```

1138 RESTRICTING EXCEPTIONS

As your programs become more complex, they will frequently call functions from within a *try* block. When your programs call functions from within a *try* block, you can restrict what type of exceptions the called function can throw. You can also prevent that function from throwing any exceptions at all. To restrict the exceptions your functions can throw, you must add a *throw* clause to the function's definition. The general form of the restricted-throwing function is shown here:

```
return-type function-name(argument-list) throw(type-list)
 {
   // Function code here
 }
```

When you declare a function with the *throw* clause, it can only throw those types you detail within the *type-list*. If the function throws any other exception type, the program will terminate abnormally (as you saw in Tip 1130 with the *catch_d.cpp* program). If you do not want a function to throw any exceptions, use an empty *type-list*. To better understand how your programs can limit a function's thrown exceptions, consider the following program, *fun_thrw.cpp*, which limits the *XHandler* function to throws of type *int, char*, and *double*:

```
#include <iostream.h>

void XHandler(int test) throw(int, char, double)
 {
   if(test==0)
      throw test;
   if(test==1)
      throw 'a';
   if(test==2)
      throw 123.23;
 }
```

```
void main(void)
 {
   cout << "Start: " << endl;
   try {
      XHandler(0);                    // try passing 1 and 2 for different responses
    }
   catch(int i) {
      cout << "Caught an integer." << endl;
    }
   catch(char c) {
      cout << "Caught a character." << endl;
    }
   catch(double d) {
      cout << "Caught a double." << endl;
    }
   cout << "End";
 }
```

It is important for you to understand that using the *throw* clause in a function declaration only limits the exceptions the function can throw back to the *try* block from within which the program called the function. Within the function, you can still use a *try* block to catch any exception, even exceptions not listed within the *type-list*. When you compile and execute the *fun_thrw.cpp* program, your screen will display the following output:

```
Start:
Caught an integer.
End
C:\>
```

The following *XHandler* function declaration will prevent the function from throwing any exceptions to the calling block:

```
void XHandler(int test) throw()
 {
   if(test==0)
      throw test;
   if(test==1)
      throw 'a';
   if(test==2)
      throw 123.23;
 }
```

Because the *throw* statement in the function declaration includes no values within the *type-list*, the function will throw no exceptions. If you compile and execute the *fun_nthr.cpp* file, it will display the following output to your screen:

```
Start:
Abnormal program termination
C:\>
```

RE-THROWING AN EXCEPTION 1139

As your exception handlers become more complex, there may be times when your programs must re-throw an exception from inside an exception handler. If you re-throw an exception, C++ will pass the exception to an outer *try* block. The most likely reason that you will perform such processing within your programs will be that you want to handle an exception within two separate handlers. For example, the handler inside a function might handle one aspect of the exception, and the handler outside the function might handle the other. It is important to understand that re-throwing an exception will *immediately* pass the exception to the outside handler—the inner handler will not process the re-thrown exception all. To better understand the processing C++ performs when you re-throw an exception, consider the following program, *two_excp.cpp*, which throws an exception inside a function, re-throws the exception, and catches the exception again outside the function:

```
#include  <iostream.h>
void XHandler(void)
 {
   try {
      throw "hello";
   }
   catch(char *) {
      cout << "Caught char * inside XHandler." << endl;
      throw;
   }
 }
void main(void)
 {
   cout << "Start: " << endl;
   try
    {
      XHandler();
    }
   catch(char *)
    {
      cout << "Caught char * inside main." << endl;
    }
   cout << "End";
 }
```

When you compile and execute the *two_excp.cpp* program, it will call the *XHandler* function. The internal *try* block will catch the *throw* statement inside *Xhandler* and generate a message. The internal *try* block will re-throw the exception, and the *try* block in *main* will then re-catch the exception. When you execute the *two_excp.cpp* program, your screen will display the following output:

```
Start:
Caught char * inside XHandler.
Caught char * inside main.
End
C:\>
```

1140 APPLYING EXCEPTION HANDLING

As you have learned, C++ provides exception handling capabilities to provide you with a structure that you can use to make your programs respond to abnormal or unexpected events. Because you will use *try* to help you handle errors, you will typically process those errors within your programs in a way that is useful to your programs (as opposed to the simple error messages you have generated in the *catch* block so far). To better understand exception handling using *try*, consider the following program, *catch_dz.cpp*, which generates and captures a divide-by-zero error:

```
#include <iostream.h>
void divide(double a, double b)
 {
   try
    {
      if( !b )                                     // checks for divide-by-zero
        throw b;
      cout << "Result: " << a/b << endl;
    }
   catch(double b)
    {
      cout << "Can't divide by zero." << endl;
    }
 }
```

```
void main(void)
 {
   double i, j;
   do {
      cout << "Enter numerator (0 to stop):" << endl;
      cin >> i;
      cout << "Enter denominator: " << endl;
      cin >> j;
      divide (i,j);
    }
   while (i != 0);
 }
```

Compare the *try* block in this Tip to the C math error handler you designed in Tip 351. The *try* block is easier to implement, simpler to understand, and more efficient than the math error handler you designed previously. Because exception handling provides more power and is easier to use, as your programs become more complex, you will find that you use it frequently to protect against unexpected events, particularly user input.

USING DEFAULT FUNCTION ARGUMENTS 1141

As you have learned, C++ lets you create functions with default values for the function parameters. Your functions will only use the default values for the parameters when the calling statement does not explicitly include values for its parameters, or when the program has not yet initialized the variables the calling statement uses for the parameters. Default values are often helpful to your programs, as they can reduce or eliminate the need for certain exception handlers.

However, you will generally use default parameter values within your functions when your program repeatedly calls the function with the same value and only occasionally calls it with a different value. For example, the following program, *clr_wnew.cpp*, uses the newline character to clear the screen (as you have learned, using the newline character is not the most efficient way to clear the screen, but it is a useful technique for this example). Because many screens are twenty-five lines tall, the *clrscr* function defaults to twenty-five lines:

```
#include <iostream.h>

void clrscr(int size = 25)
 {
   while(size > 0)
    {
      cout << endl;
      size--;
    }
 }

void main(void)
 {
   int i;

   for(i = 0; i < 30; i++)
      cout << i << endl;
   cin.get();
   clrscr();                        // clears 25 lines
   for(i = 0; i < 30; i++)
      cout << i << endl;
   cin.get();
   clrscr(10);                      // clears 10 lines
 }
```

1142 AVOIDING ERRORS WITH DEFAULT FUNCTION ARGUMENTS

As you have learned, you can frequently use default arguments within your programs. Many programmers even use default arguments within constructor functions, as shown here:

```
class cube {
   int x, y, x;
 public:
   cube(int i = 0, int j = 0, int k = 0)
    {
      x = i;
      y = j;
      z = k;
    }
   // remainder of definition
 }
```

Using default values within constructor functions can help you avoid overloading constructor functions. However, as with many of the other simplifying technologies that you have learned (such as generic functions), you should be careful to limit the number of parameters to which you set default values. As a general rule, you should not declare a default value for a parameter unless the function uses the default value 75 percent of the time or more. If your function rarely uses the default value, specifying the default value is not only not helpful, it is also often destructive to your programs and misleading to other programmers who read your code. Use default values as necessary within your programs, but be careful not to overuse them.

1143 DEFAULT ARGUMENTS VERSUS FUNCTION OVERLOADING

As you have learned, your programs can use default arguments to simplify the process of handling function parameters. As the previous Tip implies, you can even use default arguments to avoid overloading functions such as constructor functions. You can also use default arguments as an effective shorthand method of providing for overloaded functions. Imagine that you want to create two versions of your own *concatstr* function, which concatenates two strings. The first version performs its processing in the exact same manner as *strcat*, concatenating the entire contents of one string to the end of another. The second version accepts a third parameter that specifies the number of characters the function should concatenate (similar to *strncat*). To overload the function, you would declare the headers for your *concatstr* function, as shown here:

```
void concatstr(char *s1, char*s2);
void concatstr(char *s1, char *s2, int len);
```

Alternately, you could use a default value for the *len* parameter, which the function would check before it began its processing, as shown here:

```
void concatstr(char *s1, char *s2, int len = 0);
 {
   if(len = 0)
    {
      // processing
    }
   else
    {
      // other processing
    }
 }
```

In the previous example, using a default parameter may make your program easier to understand. As a rule, however, try to use whichever function form that makes the most sense and will make its processing most obvious—to you, and to another programmer reading your code.

CREATING CONVERSION FUNCTIONS 1144

As your programs become more complex, there may be times when you want to use an object of a class in an expression that involves other types of data. As you have learned, you can often use overloaded operator functions to help you with such processing. However, in other cases, you may simply want a type conversion from the class type to the target type. To help you with type conversions, C++ lets you create custom *conversion functions*. A conversion function converts your class into a type compatible with the rest of an expression. The generalized format of a type conversion function is shown here:

```
operator type(void) { return value; }
```

In the generalized format, *type* corresponds to the target type, and *value* is the object's value after conversion. Conversion functions accept no parameters, and only return a value of type *type*. To better understand how conversion functions work, consider the following example, which converts values of type *stack* to values of type *int* by returning the *tos* (top of stack) value to the expression. The *conv_stk.cpp* program uses the stack class that you created in previous Tips:

```
#include <iostream.h>

const int SIZE=100;

class stack {
   int stck[SIZE];
   int tos;
 public:
   stack(void) { tos=0; }
   void push(int i);
   int pop(void);
   operator int(void) { return tos; }        // converts stack to int
 };

void stack::push(int i)
 {
   if (tos==SIZE)
    {
      cout << "Stack is full." << endl;
      return;
    }
   stck[tos++] = i;
 }

int stack::pop(void)
 {
   if (tos==0)
    {
      cout << "Stack underrun." << endl;
      return 0;
    }
    return stck[--tos];
 }

void main(void)
 {
   stack stck;
   int i, j;
    for(i = 0; i < 20; i++)
```

```
      stck.push(i);
   j = stck;                                       // convert to int
   cout << j << " items on stack." << endl;
   cout << (SIZE - stck) << " spaces open." << endl;
 }
```

1145 USING CONVERSION FUNCTIONS TO IMPROVE TYPE PORTABILITY

As you learned in Tip 1144, your programs can use conversion functions to convert objects of a class to another type. One of the best uses of conversion functions is to make your object types more portable, and therefore more useful. For example, the following program, *pow_doub.cpp*, uses the *pwr* class that you created previously. However, the *pow_doub.cpp* program also converts *pwr* to a *double* within expressions, which lets you use the result of a *pwr* operation within other mathematical equations, as shown here:

```
#include <iostream.h>

class pwr {
   double b;
   int e;
   double val;
 public:
   pwr(double base, int exp);
   pwr operator+(pwr obj)
    {
      double base;
      int exp;
      base = b + obj.b;
      exp = e + obj.e;
      pwr temp(base, exp);
      return temp;
    }
    operator double(void) {return val;}  // converts to double
 };

pwr::pwr(double base, int exp)
 {
   b = base;
   e = exp;
   val = 1;
   if (exp != 0)
      while(exp-- > 0)
         val *= b;
 }

void main(void)
{
   pwr pwr1(4.0, 2);
   double doub1;

   doub1 = pwr1;                         // convert x to double
   cout << (doub1 + 100.2) << endl;
   pwr pwr2(3.3, 3), pwr3(0,0);
   pwr3 = pwr1 + pwr2;                   // no conversion
   doub1 = pwr3;                         // convert to double
   cout << doub1;
 }
```

The *pow_doub.cpp* program declares the *pwr1* variable (of type *pwr*), and the *doub1* variable, of type *double*. It then converts *pwr1* to a *double* value, and displays the converted value. Next, the program declares two more variables of type *pwr*, *pwr2* and *pwr3*, on which it performs processing and then displays. When you compile and execute the *pow_doub.cpp* program, your screen will display the following output:

```
116.2
20730.7
C:\>
```

CONVERSION FUNCTIONS VERSUS OVERLOADED OPERATORS 1146

As you have learned, you could perform essentially the same processing as you performed in Tips 1144 and 1145 by overloading operators relative to the *stck* and *pwr* classes, respectively. Unfortunately, as you also learned, you must overload the operators differently for your program to perform each of the two following assignments:

```
x = pwr + 102.65;
x = 102.65 + pwr;
```

However, instead of using overloaded operators, you can create a conversion function, as Tip 1145 details. A conversion function would help you avoid overloading multiple operators, using *friend* functions, or performing other complex, repetitive processing to convert an object to another value. On the other hand, if you were working with a class and wanted to add *102.65* to each member of that class, the conversion function would not help you. Just as you have seen with default parameters and overloaded functions, the decision to use a conversion function versus an overloaded operator will vary from program to program and class to class. As with the previous examples, you should make your decision based on the specific application of your class within your program.

UNDERSTANDING THE NEW C++ CASTING OPERATORS 1147

As you learned in previous Tips, C supports a casting operator, which your programs can use to cast a value to a target type. C++ defines four new casting operators, listed in Table 1147.

Name	General Form
const_cast	const_cast<type>(object)
dynamic_cast	dynamic_cast<type>(object)
reinterpret_cast	reinterpret_cast<type>(object)
static_cast	static_cast<type>(object)

***Table 1147** The new casting operators C++ supports.*

Your programs can continue to use the C casting operators and should also use the new C++ casting operators, as appropriate to the program's processing. You will learn more about each new casting operator in the following Tips.

USING THE CONST_CAST OPERATOR 1148

As you have learned, C++ defines several new casting operators that you can use within your programs. Within your programs, you will use the *const_cast* operator to explicitly override a previous *const* or *volatile* declaration. The target type of the cast must match the source type, except for the alteration of its *const* or *volatile* attribute. You will most often use *const_cast* to remove a value's constant attribute—in other words, to make changeable a value which you previously defined as constant.

Your program can use the *const_cast* operator to explicitly convert a pointer to any object type or a pointer to a data member to a type that is identical except for the *const*, *volatile*, and *__unaligned* qualifiers. For pointers and references, the result will refer to the original object. For pointers to data members, the result will refer to the same member as the original (uncast) pointer to a data member. For example, the following program, *const.cpp*, will not compile because you cannot assign a constant pointer value to a normal pointer value:

```
#include <stdio.h>

class c {
 public:
  int j;
  c (void) {j = 10;}
 };

void PrintInt( const c* Object)
{
   c* New = Object;
   New->j += 5;
   printf( "%d\n", New->j);
}

void main(void)
{
  const c Sample;

  PrintInt(&Sample);
}
```

If you change the *PrintInt* function, however, to use the *const_cast* operator, the program will compile and execute correctly, as shown here in a code fragment from the *const_cast.cpp* program:

```
void PrintInt( const c* Object)
{
   c* New = const_cast<c*> (Object);
   New->j += 5;
   printf( "%d\n", New->j);
}
```

The cast lets you create the *New* pointer as a modifiable pointer. However, you should be careful with *const_cast*. Depending on the type of the referenced object, a write operation through the resulting pointer, reference, or pointer to a data member might produce unknown or unexpected behavior. The following code fragment will result in an error:

```
class X {};
class Y : public X {};

const X x;
Y y = const_cast<Y> (x); // error
```

The code results in an error because you cannot use the *const_cast* operator to convert from an X to a Y. You can only use the *const_cast* operator to remove the *const*, *volatile*, and *_unaligned* modifiers. If you wanted to convert the constant object *x* to a non-constant object *y*, you must construct the casting operation as shown here:

```
Y y = (const_cast<Y>)(static_cast<const Y>(x));
```

Note: *The **const_cast** operator converts a NULL pointer value to the NULL pointer value of the destination type.*

1149 USING THE DYNAMIC_CAST OPERATOR

As you have learned, C++ provides several new casting operators that you can use within your programs. The *dymanic_cast* operation performs a run-time cast and verifies the validity of the cast. If the program cannot make the cast, the cast will fail and the expression will evaluate to *NULL*. Generally, you will use the *dynamic_cast* operator to perform casts on polymorphic object types. For example, *dynamic_cast* can return a pointer to a derived object given a pointer to the polymorphic base class. To better understand the processing that the *dynamic_cast* operator performs, consider the following example program, *dyn_cast.cpp*, which generates a class from two base classes and tries to cast pointers from the base classes to the derived class:

```
#include <iostream.h>
#include <typeinfo.h>

class Base1
 {
   virtual void f(void) { /* A virtual function makes the class polymorphic */ }
 };

class Base2 { };
class Derived : public Base1, public Base2 { };

void main(void)
 {
   try
    {
      Derived d, *pd;
      Base1 *b1 = &d;

      // Perform a downcast from a Base1 to a Derived.
      if ((pd = dynamic_cast<Derived *>(b1)) != 0)
       {
         cout << "The resulting pointer is of type " << typeid(pd).name() << endl;
       }
      else
         throw Bad_cast();

      // Cast from the first base to the most derived class and then back
      // to another accessible base.
      Base2 *b2;
      if ((b2 = dynamic_cast<Base2 *>(b1)) != 0)
       {
         cout << "The resulting pointer is of type " << typeid(b2).name() << endl;
       }
      else
         throw Bad_cast();
    }
   catch (Bad cast)
    {
      cout << "dynamic_cast failed" << endl;
      exit(1);
    }
   catch (...)
    {
      cout << "Exception handling error." << endl;
      exit(1);
    }
 }
```

When you execute the *dyn_cast.cpp* program, it will perform two cast tries. First, it will try to downcast from a *Base1* pointer to a *Derived* pointer. Next, it will cast from the first base class to the most derived class, and then try to cast back up the tree to another base class. When you compile and execute the *dyn_cast.cpp* program, your screen will display the following output:

```
The resulting pointer is of type Derived *
The resulting pointer is of type Base2 *
C:\>
```

***Note**: You must compile the **dyn_cast.cpp** program with your compiler's Generate RTTI option enabled, or the program will be unable to perform run-time type identification.*

1150 USING THE REINTERPRET_CAST OPERATOR

As you have learned, C++ provides several new casting operators that you can use within your programs. The *reinterpret_cast* operator casts one type to a fundamentally different, incompatible type. For example, the *reinterpret_cast* operator could cast a pointer of type *pwr* to a pointer of type *int*. Clearly, *reinterpret_cast* presents the opportunity for a great deal of confusion within your programs, and you should avoid it unless its use is absolutely necessary. The following program, *ren_cast.cpp*, uses the *reinterpret_cast* operator to convert a *char* pointer to an *int* pointer:

```
#include <iostream.h>

void main(void)
 {
   int i;
   char *p = "This is a string.";
   i = reinterpret_cast<int> (p);          // cast char pointer to integer
   cout << i;
 }
```

When you compile and execute the *ren_cast.cpp* program, it will display meaningless output because the cast from the *char* pointer to an *int* pointer causes the *int* pointer to return odd results. If you run the program on a 16-bit system, *int* will return *char('T')*. If, on the other hand, you run the program on a 32-bit system with 32-bit integer support, the *i* pointer returns the bitwise equivalent of the 'Th' constant string. On a 32-bit system, the output looks similar to the following:

```
4247824
C:\>
```

1151 USING THE STATIC_CAST OPERATOR

As you have learned, C++ provides several new casting operators that you can use within your programs. Your programs will use the *static_cast* operator to perform a non-polymorphic cast. In other words, you can use the *static_cast* operator to cast a base class pointer to a derived class pointer. You will use the *static_cast* operator within your programs as shown in the following prototype:

```
static_cast< T > (argument)
```

In the generalized form shown here, *T* is a pointer, reference, arithmetic type, or enumerated type. The *argument* type must match the *T* type. The compiler must know both *T* and *argument* at compile time. If your program can convert a complete type to another type by some conversion method the language already provides, making such a conversion using the *static_cast* operator instead will achieve the same conversion. Additionally, your programs can use *static_cast* to convert integral types to enumerated types. A request to convert *argument* to a value that is not an element of *enum* returns a value of *undefined*. The *NULL* pointer converts itself to itself (in other words, a *static_cast* of *NULL* still yields *NULL*), and therefore your programs do not have to use *static_cast* with the *NULL* pointer.

Your program can convert a pointer to one object type to a pointer to another object type. Note that merely pointing to similar types can cause access problems if you do not align the similar types correspondingly. You can explicitly convert a pointer to a class *base* to a pointer to some other class *derived*, if *base* is a base class for *derived*. Your program can make a static conversion only under the following conditions:

- If an unambiguous conversion exists from *derived* to *base*
- If *base* is not a virtual base class

The *static_cast* operator can explicitly convert an object to the reference type *base&* if the compiler can explicitly convert a pointer to that object to the pointer type *base**. The result of the *static_cast* conversion is an *lvalue*. Your program does not call constructors or conversion functions as the result of a cast to a reference. Rather, your program can convert an object or a value to a class object only if you have declared an appropriate constructor or conversion operator for that specific cast. You can explicitly convert a pointer to a member into a different pointer-to-member type only if both types are pointers to members of the same class or pointers to members of two classes. If both the argument and the case result are pointers to members of two classes, your program must derive the class one pointer references unambiguously from the other class. When *T* is a reference, the result of a *static_cast* is an *lvalue*. The result of a pointer or reference cast refers to the original expression.

Understanding Namespaces 1152

As your programs become more complex, your applications may consist of more than one source file. Further, more than one developer may author and maintain source files. Eventually, you will organize and link the separate files to produce the final application. Traditionally, the file organization requires that all names that a source file does not encapsulate within a defined *namespace* (in other words, that do not have a limited scope, such as a function body, class body, or translation unit) must share the same global namespace. Therefore, multiple definitions of names the compiler discovers while linking separate modules require some way to distinguish each name. The C++ *namespace* keyword provides the solution to such "name clashes" in the global scope. Throughout this book's C++ tips, you have seen the *std* namespace, which defines I/O streams such as *cout* and *cin*. As your C++ applications increase in complexity, you can create your own namespaces. The *namespace* keyword lets you partition an application into multiple subsystems. Each subsystem can define and operate within its own scope. Each developer can introduce convenient identifiers in a subsystem without worrying about whether someone else is also using such identifiers within their own subsystem. Each namespace uses a unique identifier. When your programs define namespaces, the compiler knows the subsystem scope throughout the application by each namespace's unique identifier. Using C++ namespaces requires only two steps. The first step is to use the keyword *namespace* to uniquely identify a namespace. The second step is to invoke the *using* keyword to access the elements of a previously identified namespace.

Using Namespaces 1153

As you learned in Tip 1152, issues of scope and naming become more important as your programs become more complex. To help you prevent variable and other name conflicts, C++ provides the *namespace* keyword. You will use the *namespace* keyword within your programs much as you would use a structure, enumerated type, union, or class definition, as shown here in the *namespace* implementation's generalized form:

```
namespace name {
   // object declarations
 }
```

Within your own programs, you can include variables of any type the compiler already knows (either simple C and C++ types or previously defined classes, structures, and unions) within the *namespace* definition. You can also declare functions inline within a namespace. For example, the following code fragment declares two variables and a function within the namespace *limited*:

```
namespace limited {
   int i, k;
   void sample(int j) { cout << j << endl; }
 }
```

In the previous code fragment, *i*, *k*, and the *sample* function are part of the *limited* namespace. Because a namespace defines a scope, you must use the scope resolution operator to refer to objects you define within a namespace. For example, to assign the value 10 to the *k* variable, you must use a statement similar to the following:

```
limited::k = 10;
```

1154 USING THE USING STATEMENT WITH NAMESPACE

As you learned in Tip 1153, when your programs use namespaces, you must resolve references to objects within a namespace with the scope resolution operator. However, if your program will frequently use the members of a namespace, you can use a *using* statement to simplify program access to those members. The *using* statement has two general forms, as shown here:

```
using namespace name;
using name::member;
```

The first form lets you access the entire namespace. In the second form, you define only specific members of the namespace you want to access. Essentially, the first form makes the entire namespace public and the second form lets you encapsulate certain members within the namespace. To better understand the *using* statement's two forms, consider the following code fragment, which uses both forms:

```
using limited::k;                    // only makes k visible
k = 10;

using namespace limited;             // makes entire limited namespace visible
k = 10;
```

1155 UNDERSTANDING RUN-TIME TYPE IDENTIFICATION

An important addition to newer C++ compilers is run-time type identification (RTTI). Run-time type identification lets you write portable code that can determine the actual type of a data object at run time even when the code can access only a pointer or a reference to that object. Run-time type identification makes it possible, for example, to convert a pointer to a virtual base class into a pointer to the derived type of the actual object. As you learned in Tip 1149, you can use the *dynamic_cast* operator together with run-time type identification to make run-time casts.

1156 USING TYPEID FOR RUN-TIME TYPE IDENTIFICATION

As you learned in Tip 1155, run-time type identification lets your programs manipulate pointers and references in entirely new ways. The run-time type identification mechanism also lets you check whether an object is of some particular type and whether two objects are of the same type. You can check objects with the *typeid* operator. The *typeid* operator determines the actual type of its argument and returns a reference to an object of type *const typeinfo*, which describes that type.

You can also use a type name as the argument to *typeid*, and *typeid* will return a reference to a *const typeinfo* object for that type. The class *typeinfo* provides an == operator and an != operator that you can use to determine whether two objects are of the same type. Class *typeinfo* also provides a member function, *name*, that returns a pointer to a character string that holds the name of the type. You must include the header file *typeinfo.h* within your programs to access the *typeid* function. The general form of the *typeid* function is shown here:

```
#include <typeinfo.h>

const typeinfo typeid(object);
```

Within the generalized form, *object* corresponds to the object whose type you want *typeid* to return. When you apply *typeid* to a base class pointer of a polymorphic class, *typeid* will automatically return the type of the object *the pointer points to*, including any classes derived from the base class. To better understand the processing that *typeid* performs, consider the following program, *typeid_1.cpp*:

```
#include <iostream.h>
#include <typeinfo.h>

class A { };
```

```
class B : A { };

void main(void)
 {
   char C;
   float X;

    // USE THE typeinfo::operator==()TO MAKE COMPARISON
   if (typeid( C ) == typeid( X ))
      cout << "C and X are the same type." << endl;
   else
      cout << "C and X are NOT the same type." << endl;

    // USE true AND false LITERALS TO MAKE COMPARISON
   cout << typeid(int).name();
   cout << " before " << typeid(double).name() << ": " <<
        (typeid(int).before(typeid(double)) ? true : false) << endl;
   cout << typeid(double).name();
   cout << " before " << typeid(int).name() << ": " <<
        (typeid(double).before(typeid(int)) ? true : false) << endl;
   cout << typeid(A).name();
   cout << " before " << typeid(B).name() << ": " <<
        (typeid(A).before(typeid(B)) ? true : false) << endl;
 }
```

The *typeid_1.cpp* program declares two classes, a base class (A) and a derived class (B). When *typeid_1.cpp* begins its execution, it defines two variables, one of type *char* and one of type *float*. The program next tests the types of both *X* and *C*. If they are the same type, the program will display a message saying so on the screen; if not, the program will display a message that they are not the same type. Clearly, because *char* and *float* are not the same type, the program will display the "C and X are NOT the same type." message. The program next uses the *before* member to compare some basic types with a lexical comparison. A lexical comparison is a comparison based on the alphabet: a is before b, c is before d and after b, and so on. Therefore, when the program compares *int* and *double*, it returns false when it tests whether *double* is after *int*, but true when it tests whether *double* is before *int*. Finally, the program tests whether A is before B. When you compile and execute the *typeid_1.cpp* program, it will display the following output on your screen:

```
C and X are NOT the same type.
int before double: 0
double before int: 1
A before B: 1
C:\>
```

UNDERSTANDING THE TYPE_INFO CLASS 1157

As you learned in Tip 1156, the *typeid* function returns a value of type *const typeinfo*. The *const typeinfo* value is a string that represents type information about the class. Table 1157 lists the possible return values for the *typeid* function.

Value	Meaning
[ARRAY]	The value is an array. The *typeid* function always returns [ARRAY] together with another keyword.
classname	The name of the user-defined class for the object.
[INTEGER]	The object is an integer or long integer.
[NULL]	The object is a *NULL* value (generally a *NULL* pointer).
[REAL]	The object represents a *float, double*, or *long double* object.
[STRING]	The object is a character string.
[UNINITIALIZED]	The object is uninitialized (generally a pointer to a polymorphic class).

***Table 1157** The possible values that **typeid** returns.*

The *typeinfo* class also defines four public members, in addition to the *typeinfo* function. You will implement those member functions within your programs, as shown here:

```
bool operator==(const type_info &obj) const;
bool operator!=(const type_info &obj) const;
bool before(const type_info &obj) const;
const char *name(void) const;
```

As you learned in Tip 1156, the overloaded == and != operators let you compare types that *typeid* returns. The compiler primarily uses the *before* function internally. It returns *true* if the invoking object is before the *obj* object in lexical order (a compiler-derived list). It returns no information regarding class hierarchies or other useful type information. The *name* function returns a pointer to the type's name. The following program, *typeid_2.cpp*, uses the *name* function to return further information from a *typeid* action:

```
#include <iostream.h>
#include <typeinfo.h>

class Base {
   int a, b;
   virtual void func(void) {};
 };

class Derived1: public Base {
   int i, j;
 };

class Derived2: public Base {
   int k;
 };

void main(void)
 {
   int i;
   Base *p, baseobj;
   Derived1 obj1;
   Derived2 obj2;

   cout << "Typeid of i is: ";
   cout << typeid(i).name() << endl;
   p = &baseobj;
   cout << "p is currently pointing to an object of type: ";
   cout << typeid(*p).name() << endl;
   p = & obj1;
   cout << "p is now pointing to an object of type: ";
   cout << typeid(*p).name() << endl;
   p = & obj2;
   cout << "p is finally pointing to an object of type: ";
   cout << typeid(*p).name() << endl;
 }
```

The *typeid_2.cpp* program defines a single base class and two derived classes. When the program begins its execution, it declares two variables of the derived types, a variable of the base type, a pointer to the base type, and a simple integer variable.

Next, the program checks the *typeid* of the integer variable and returns its name in text. After returning information about the simple *int* type variable, the program manipulates *p* to point to each of the three custom classes. Each time the type *p* points to changes, the program outputs information about the new type name. When you compile and execute the *typeid_2.cpp* program, it will display the following output:

```
Typeid of i is: int
p is currently pointing to an object of type: Base
p is now pointing to an object of type: Derived1
```

```
p is finally pointing to an object of type: Derived2
C:\>
```

UNDERSTANDING THE MUTABLE KEYWORD 1158

As you have learned, C++ adds new specifiers to variable declarations, such as the *long double* specifier, which you can use within your programs. In addition to the new simple data types, your programs can use the *mutable* keyword with a variable of any type to make the variable modifiable although it is in a *const*-qualified expression. For example, the following declaration makes *j* a *mutable int* variable:

```
class ShowExample {
   mutable int j;
 }
```

Your programs can declare only class data members as *mutable*. You cannot use the *mutable* keyword on *static* or *const* names. The purpose of *mutable* is to specify which data members *const* member functions can modify because a *const* member function cannot normally modify data members.

USING THE MUTABLE KEYWORD WITHIN A CLASS 1159

As you learned in Tip 1158, your programs can use the *mutable* keyword within a class definition to make a member variable modifiable although it is in a *const*-qualified expression, even when the object of which it is a member is *const*. You can only use *mutable* on members within a *const* class. To better understand the *mutable* keyword's use, consider the following program, *mutable.cpp*, which declares two *mutable* members within a class:

```
#include <iostream.h>
class Alpha {
   mutable int count;
   mutable const int* iptr;
public:
   Alpha(void) {count = 0;}
   int func1(int i = 0) const { // Promises not to change const arguments.
      count = i++;  // But count can be changed.
      iptr = &i;
      cout << "i is: " << *iptr << endl;
      return count;
      }
   void show_count(void) { cout << "Count is: " << count << endl;}
};
void main(void)
 {
   Alpha a;
   a.show_count();
   a.func1(10);
   a.show_count();
 }
```

Rather than leaving the *count* member unmodified, as the *func1* declaration's *const* modifier promises, the *mutable* keyword lets the member function modify the *count* member. Instead of maintaining its initial value of 0, *count* has a value of 10 at the program's end because *count* is a *mutable* value. When you compile and execute the *mutable.cpp* program, your screen will display the following output:

```
Count is: 0
i is: 11
Count is: 10
C:\>
```

1160 CONSIDERATIONS ABOUT THE MUTABLE KEYWORD

As you have learned, your programs can use the *mutable* keyword to override *constness* as it applies to a member of a class. While the *mutable* keyword provides you with a means to better control what parts of a class should remain unmodified and what parts should be modifiable, the *mutable* keyword can also introduce significant, hard-to-track errors. For example, in the *mutable.cpp* program presented in Tip 1159, the *Alpha* class declared the variable *count* as mutable. Later, the class defined a *const* function, which should not have changed the values of any variables the function used. Instead, because the class declared the *count* variable with the *mutable* keyword, the *func1* function changed the *a.count* member's value to 10.

You should be very careful about how and when you use the *mutable* keyword within your programs, as doing so may create hard-to-track errors. As with many features C++ adds, your decision to use or not to use *mutable* will generally come down to clarity. The *mutable.cpp* program in the previous Tip illustrates the importance of clarity in that it is much less clear with the *mutable* keywords than it would have been otherwise.

1161 INTRODUCING THE BOOL DATA TYPE

Throughout the previous Tips, you have used the *int* data type within your logical comparisons. Newer C++ compilers support the *bool* data type for processing Boolean data. The *bool* data type accepts values of *true* and *false*. You will declare and use variables of the *bool* data type as shown here:

```
bool logical_var;
logical_var = true;
```

Because the *bool* data type supports the *true* and *false* keywords as *rvalues*, you should not define variables named *true* or *false* within your programs.

1162 USING THE BOOL DATA TYPE

As you learned in Tip 1161, C++ supports the new *bool* data type, which your programs can use to maintain logical information. The new *bool* data type uses two new C++ keywords, *true* and *false*. The *true* and *false* keywords correspond to the true (1) and false (0) values for integers. You should use the *bool* data type within your programs to make your program code more clear. For example, you can assign the result of a logical test to a *bool* data type, as shown here:

```
bool result;
result = (A && B)
```

The following program, *bool_fun.cpp*, uses a function that returns a value of type *bool*:

```
#include <iostream.h>

bool func(void)
 {                          // Function returns a bool type
   return false;
   //  return NULL;         // NULL is converted to Boolean false
 }

void main(void)
 {
   bool val = false;        // Boolean variable
   int i = 1;               // i is neither Boolean-true nor Boolean-false
   int g = 3;
   int *iptr = 0;       // null pointer
```

```
    float j = 1.01;      // j is neither Boolean-true nor Boolean-false

   if (i == true)
      cout << "True: value is 1" << endl;
   if (i == false)
      cout << "False: value is 0" << endl;
   if (g)
      cout << "g is true.";
   else
      cout << "g is false.";
   // Test on pointer
   if ((bool) iptr == false)
      cout << "Invalid pointer." << endl;
   if ((bool) iptr == true)
      cout << "Valid pointer." << endl;
   // To test j's truth value, cast it to bool type.
   if (bool(j) == true)
      cout << "Boolean j is true." << endl;
   // Test Boolean function return value
   val = func();
   if (val == false)
      cout << "func() returned false.";
   if (val == true)
      cout << "func() returned true.";
}
```

In addition to using a *bool* function, the program performs a series of comparisons using *bool* values rather than integers, and converts an *int* value to a *bool* result. When you compile and execute the *bool_fun.cpp* program, your screen will display the following output:

```
True: value is 1
Unknown truth value for g.
Invalid pointer.
Boolean j is true.
func() returned false.
C:\>
```

CREATING A STRING TYPE 1163

As you have learned, C++ implements strings as *NULL*-terminated character arrays and not as a separate data type. Most current versions of C++ implement the C++ standard library *<string.h>*, which creates the separate data type for strings. To better understand how you implement and manipulate entire classes within your program, you will learn to create a custom string type. In the next several Tips, you will use what you have learned from previous Tips to create a full-fledged string data type.

The first step in creating any new type is determining what you want to accomplish. In short, you must *define* the type. In Tip 1164, you will define your custom string type's characteristics—the operators your programs can use on the type, the member functions the type supports, and so on.

DEFINING THE STRING TYPE'S CHARACTERISTICS 1164

As you know, when you work with a *char* array in C++, you are actually working with several *char* elements that C++ stores in sequence within the computer's memory. For example, the definition shown here actually creates 32 separate *char* elements within the *sample* array:

```
        cout << "Allocation error!" << endl;
        exit(1);
     }
     *p = '\0';
  }
Strings::Strings(char *str)
  {
     size = strlen(str) + 1;
     p = new char[size];
     if(!p)
      {
        cout << "Allocation error!" << endl;
        exit(1);
      }
     strcpy(p, str);
  }
Strings::Strings(const Strings &obj)
  {
     size = obj.size;
     p = new char[size];
     if(!p)
      {
        cout << "Allocation error!" << endl;
        exit(1);
      }
     strcpy(p, obj.p);
  }
```

As you can see, each constructor allocates memory, creates a *char* array of the requested size, and initializes *p* to point to the array's beginning. When the program destroys the *Strings* object, the inline destructor function simply releases the memory to which *p* points.

1167 PERFORMING I/O WITH THE STRINGS CLASS

In Tip 1165, you created the basic definition for the *Strings* class. The definition included references for two *friend* functions that overload the insertion and extraction operators, as shown here:

```
friend ostream &operator<<(ostream &stream, Strings &obj);
friend istream &operator>>(istream &stream, Strings &obj);
```

Because input and output are the most common operations your programs will perform on strings, you must implement an overloaded version within the class. The output stream handles the *Strings* object easily, sending the information directly to the stream. Note that the extractor receives the *Strings* object by reference. Because the *Strings* objects can be conceptually quite large, it will improve your program's performance to pass the *Strings* object as a pointer, rather than by value, as shown here:

```
ostream &operator<<(ostream &stream, Strings &obj)
  {
     stream << obj.p;
     return stream;
  }
```

1168 WRITING THE ASSIGNMENT FUNCTIONS FOR THE STRINGS CLASS

As you saw in Tip 1165, writing the assignment functions for the *Strings* class is a repetitive task, similar to writing the *Strings* constructor functions. The two overloaded operators must assign the *Strings* object either the value of another *Strings* object or the value of a quoted string and must handle the two situations slightly differently. As you can see, in both cases the overloaded assignment operator clears the memory currently allocated to *p*, creates new memory for *p*, and then assigns the new value to the new memory. The only significant difference between the two functions is that the first receives an operand of type *Strings*, and the second receives an operand of type *char* *, as shown here:

```
Strings Strings::operator=(Strings &obj)
  {
   Strings temp(obj.p);

   if(obj.size > size)
    {
      delete p;
      p = new char[obj.size];
      size = obj.size;
      if(!p)
       {
         cout << "Allocation error!" << endl;
         exit(1);
       }
    }
   strcpy(p, obj.p);
   strcpy(temp.p, obj.p);
   return temp;
 }
Strings Strings::operator=(char *s)
  {
   int len = strlen(s) + 1;
   if(size < len)
    {
      delete p;
      p = new char[len];
      size = len;
      if(!p)
       {
         cout << "Allocation error!" << endl;
         exit(1);
       }
    }
   strcpy(p, s);
   return *this;
 }
```

Both overloaded versions of the = operator use the *strcpy* function to place the value of the *rvalue* string into the *lvalue* string. Because the assignment directly modifies the *lvalue* string, both functions use the *this* pointer to assign a new value to the left-hand *Strings* object.

OVERLOADING THE + OPERATOR TO CONCATENATE STRINGS OBJECTS 1169

As you determined in Tip 1164, your *Strings* class should let you use the + operator to concatenate two *Strings* objects together. Just as you overloaded the assignment operator to let the *Strings* class handle both the other *Strings* objects and quoted character strings, you must also overload the + operator to handle both situations. However, because the + operator is a binary operator and not a unary operator, you must create three overloaded versions to handle all concatenation activities: one that handles two *Strings* objects, one that handles a *Strings* object and a quoted string to its right, and one that handles a quoted string and a *Strings* object to its right. As you have learned, you can use a template *friend* function to avoid rewriting the same algorithm three times, but you will not do so here because it is useful to understand each situation your code must handle. Just as with previous Tips, the companion CD-ROM that accompanies this book contains within the *strings.cpp* file the code for the implementation of the overloaded + operator functions. You will implement these functions, as shown here:

```
Strings Strings::operator+(Strings &obj)
 {
   int len;
   Strings temp;

   delete temp.p;
   len = strlen(obj.p) + strlen(p) + 1;
   temp.p = new char[len];
   temp.size = len;
   if(!temp.p)
    {
      cout << "Allocation error!" << endl;
      exit(1);
    }
   strcpy(temp.p, this.p);
   strcat(temp.p, obj.p);
   return temp;
 }
Strings Strings::operator+(char *s)
 {
   int len;
   Strings temp;

   delete temp.p;
   len = strlen(s) + strlen(p) + 1;
   temp.p = new char[len];
   temp.size = len;
   if(!temp.p)
    {
      cout << "Allocation error!" << endl;
      exit(1);
    }
   strcpy(temp.p, this.p);
   strcat(temp.p, s);
   return temp;
 }
Strings operator+(char *s, Strings &obj)
 {
   int len;
   Strings temp;

   delete temp.p;
   len = strlen(s) + strlen(obj.p) + 1;
   temp.p = new char[len];
   temp.size = len;
   if(!temp.p)
    {
      cout << "Allocation error!" << endl;
      exit(1);
    }
   strcpy(temp.p, s);
   strcat(temp.p, obj.p);
   return temp;
 }
```

Note that the third overloaded function is a *friend* function rather than a member operator overload. As you have learned, when you use a member operator, it explicitly passes the object on the left—which will cause an error if the object on the left is not of the correct class type. When you add the *Strings* object to a quoted string, you must instead pass the string explicitly and append the object to the quoted string's value.

REMOVING A STRING FROM WITHIN A STRINGS OBJECT 1170

A useful string function that you will want to add to your *Strings* object is substring subtraction. When the *Strings* object implements the substring subtraction operation, it removes all occurrences of the substring from the *Strings* object. To better understand how *Strings* implements the substring subtraction operation, consider the following code fragment:

```
sample = "Jamsa's C/C++/C# Programmer's Bible";
sample = sample - "C";
// sample now equals "Jamsa's /++/# Programmer's Bible"
```

Unlike the three overload functions you implemented to overload the + operator, you will use only two overload functions to overload the - operator (because you will not subtract a *Strings* object from a quoted string). The companion CD-ROM that accompanies this book includes the implementation code within the *strings.cpp* file, as shown here:

```
Strings Strings::operator-(Strings &substr)
 {
   Strings temp(p);
   char *s1;
   int i,j;

   s1 = p;
   for(i=0; *s1; i++)
    {
      if(*s1!=*substr.p)
       {
         temp.p[i] = *s1;
         s1++;
       }
      else
       {
         for(j=0; substr.p[j]==s1[j] && substr.p[j]; j++)
           ;
         if(!substr.p[j])
          {
            s1 += j;
            i--;
          }
         else
          {
            temp.p[i] = *s1;
            s1++;
          }
       }
    }
    temp.p[i] = '\0';
    return temp;
 }
Strings Strings::operator-(char *substr)
 {
   Strings temp(p);
   char *s1;
   int i,j;

   s1 = p;
   for(i=0; *s1; i++)
    {
      if(*s1!=*substr)
       {
         temp.p[i] = *s1;
         s1++;
       }
```

```
        else
         {
           for(j=0; substr[j]==s1[j] && substr[j]; j++)
              ;
           if(!substr[j])
            {
              s1 += j;
              i--;
            }
           else
            {
              temp.p[i] = *s1;
              s1++;
            }
         }
     }
    temp.p[i] = '\0';
    return temp;
  }
```

Both overload functions copy the left-hand operand's contents into the *temp* variable. As each function copies the left-hand operand, it removes any occurrences of the substring the right-hand operand specifies during the process. The overloaded operator functions then return the resulting *Strings* object. Because of how you have defined the *Strings* class so far, all the following instructions are valid for use with the *Strings* subtraction operator:

```
Strings x("ABCABCD"), y("A");
Strings z;
z = x - y;// z = "BCBCD"
```

1171 Overloading the Relational Operators

In previous Tips you have overloaded many operators that your programs could effectively use to manipulate strings. Another important aspect of the power of the *Strings* class is its ability to compare two strings easily and return a result more meaningful than -1, 0, or 1 (the return values of the *strcmp* function). However, the relational operators for the *Strings* class, which return such values, are simple. They use the *strcmp* function against the *p* member on the operator's left-hand side and either another object or a *char* pointer to evaluate the operator's right-hand side. Because of the simplicity of the activities the relational operators perform, the *Strings* class definition defines all the relational operators inline within the class definition. The relational operator definitions are within the *strings.cpp* class and are as shown here:

```
  /* relational operators between Strings objects. Note that the operators could
     just as easily return bool, rather than int */
    int operator==(Strings &obj) {return !strcmp(p, obj.p);}
    int operator!=(Strings &obj) {return strcmp(p, obj.p);}
    int operator<(Strings &obj) {return strcmp(p, obj.p) < 0;}
    int operator>(Strings &obj) {return strcmp(p, obj.p) > 0;}
    int operator<=(Strings &obj) {return strcmp(p, obj.p) <= 0;}
    int operator>=(Strings &obj) {return strcmp(p, obj.p) >= 0;}
  /* relational operators between Strings object and a quoted character string.
     Note that the operators could just as easily return bool, rather than int */
    int operator==(char *s) {return !strcmp(p, s);}
    int operator!=(char *s) {return strcmp(p, s);}
    int operator<(char *s) {return strcmp(p, s) < 0;}
    int operator>(char *s) {return strcmp(p, s) > 0;}
    int operator<=(char *s) {return strcmp(p, s) <= 0;}
    int operator>=(char *s) {return strcmp(p, s) >= 0;}
```

All the definitions for the relational operators within the *Strings* class assume that you will compare a *Strings* object with either another *Strings* object or a quoted string. If you want your *Strings* class to let your programs compare quoted strings with a *Strings* object, you must define another set of *friend* functions to overload the relational operators again. For example, you would declare the == operator's third overload, as shown here:

```
friend Strings operator==(char *s, Strings &obj);
```

Outside the class definition, you must create an overloaded operator function, as shown here:

```
Strings operator==(char * s, Strings &obj)
 {
 return !strcmp(s, obj.p);
 }
```

Determining a Strings Object's Size 1172

One of the most common activities you will perform with strings is determining their current size. Often, you will use a string's current size to perform additional processing on the string, to format output, and so on. To help you determine the length of a *Strings* object, the *Strings* class provides the following member function:

```
int strsize(void) {return strlen(p);}      // return string size
```

When you invoke the *strsize* member function against a *Strings* object, the member function returns an integer value that represents the length of the *Strings* object. You will invoke the member function against a *Strings* object as shown here:

```
Strings sample;
int x:
x = sample.strsize();
```

Converting a Strings Object to a Character Array 1173

The *Strings* class definition provides two member functions your programs can use to manipulate the *char* array within a *Strings* object without manipulating the object itself. The first function, *makestr*, copies the *char* array within a *Strings* object to a standard *char* array. You will most often use the *makestr* function to obtain a *NULL*-terminated string from a *Strings* object. Your programs will use the *makestr* member function against a *Strings* object, as shown here:

```
Strings sample = "Jamsa's C/C++ Programmer's Bible"
char array[256];
sample.makestr(array);
```

With the *makestr* member function, you can choose between *char* arrays and *Strings*, and easily convert one to the other.

Using a Strings Object as a Character Array 1174

In Tip 1173, you learned that the *Strings* class supports two member functions that help you use your *Strings* objects as *char* arrays within your program. The first member function, *makestr*, copies the *Strings* object's component string to a *char* array. The second member function, an overload of the ( ) operator, lets you directly use a *Strings* object within any function that expects a normal, *NULL*-terminated character array. For example, the following code will compile and work correctly:

```
Strings x("hello");
puts(x);
```

Because you want to avoid overloading every function that uses or expects to use a *char* array or pointer, overloading the ( ) operator so that it returns a *char* array, rather than returning specific information about the *Strings* object, keeps your code cleaner and makes your *Strings* object more useful.

1175 DEMONSTRATING THE STRINGS OBJECT

Using the *Strings* class that you have created in previous Tips is surprisingly easy. After you have completed the definition, using the many capabilities of the *Strings* objects is a simple process. The following code fragment uses your *Strings* class definition and demonstrates some of the class's many capabilities:

```
void main(void)
  {
   Strings s1("A sample program which uses string objects.\n");
   Strings s2(s1);
   Strings s3;
   char s[80];

   cout << s1 << s2;
   s3 = s1;
   cout << s3;
   s3.makestr(s);
   cout << "Converted to a string: " << s;

   s2 = "This is a new string.";
   cout << s2 << endl;

   Strings s4("This is a new string, too.");
   s1 = s2 + s4;
   cout << s1 << endl;

   if(s2 == s3)
      cout << "Strings are equal." << endl;
   if(s2 != s3)
      cout << "Strings are not equal." << endl;
   if(s1 < s4)
      cout << "s1 is less than s4." << endl;
   if(s1 > s4)
      cout << "s1 is greater than s4." << endl;
   if(s1 <= s4)
      cout << "s1 is less than or equal to s4." << endl;
   if(s1 > s4)
      cout << "s1 is greater than or equal to s4." << endl;

   if(s2 > "ABC")
      cout << "s2 is greater than 'ABC'" << endl << endl;

   s1 = "one two three one two three\n";
   s2 = "two";
   cout << "Initial string: " << s1;
   cout << "String after subtracting two: ";
   s3 = s1 - s2;
   cout << s3;

   cout << endl;
   s4 = "Jamsa's C/C++/C# ";
   s3 = s4 + "Programmer's Bible\n";
   cout << s3;
   s3 = s3 - "C/C++";
   s3 = "This is " + s3;
   cout << s3;

   cout << "Enter a string: ";
   cin >> s1;
```

```
   cout << s1 << endl;
   cout << "s1 is " << s1.strsize() << " characters long." << endl;
   puts(s1);
   s1 = s2 = s3;
   cout << s1 << s2 << s3;
   s1 = s2 = s3 = "Program finished.\n";
   cout << s1 << s2 << s3;
 }
```

The *main* function printed within the previous code fragment appears inside the program *use_str.cpp* on the companion CD-ROM that accompanies this book.

Creating a Header for the Strings Class 1176

In previous Tips you have created an extensive and useful *Strings* class. In case your compiler does not include a *string* class definition, or if you find that you like your *Strings* class better, you may want to move the class definition into a header file so that it does not consume too much space within your programs. To convert the *Strings* class to a header file, delete *main* and all its components from *use_str.cpp*. Next, save the remaining code into a header file called *strings.h*. Save the header file within your compiler's *include* directory. Later, when you want to use the *Strings* class within your programs, you can simply reference the header file from within your program's code, as shown here:

```
#include <strings.h>
```

Another Strings Example 1177

In Tip 1175, you wrote a program that implemented your *Strings* class. In Tip 1176, you learned how to create a header file for your *Strings* class. To help ensure that you understand the power that C++ support for custom creation of header files grants you, it is valuable to write another program that uses the new *strings.h* header file. The program in this Tip, *chk_file.cpp*, accepts a single command-line parameter that corresponds to the name of an executable program you want to find. The program will then try to open a program file on the local drive that has the same name. If the program is successful, the program lets you know that the file on the local drive exists. If the program is unsuccessful, the program lets you know that the file does not exist. You will implement the *chk_file.cpp* program as shown here:

```
#include "strings.h"
#include <iostream.h>
#include <fstream.h>

// executable file extensions

char ext[3][4] = { "EXE", "COM", "BAT" };

void main(int argc, char *argv[])
 {
   Strings fname;
   int i;

   if(argc!=2)
    {
      cout << "Usage: fname name" << endl;
      exit(1);
    }
   fname=argv[1];
   fname = fname + ".";
   for(i = 0; i < 3; i++)
    {
```

```
        fname = fname + ext[i];
        cout << "Trying " << fname << " ";
        ifstream f.open(fname);
        if(f)
         {
           cout << " - Exists" << endl;
           f.close();
         }
        else
           cout << " - Not found" << endl;
        fname = fname - ext[i];
      }
   }
```

It is especially important to notice that the compiler accepts the *f.open* call even though it receives a *Strings* object, not a *char* pointer. The compiler accepts the *f.open* call because your overload of the () operator within the *Strings* class lets the compiler interpret the information as *char* *, rather than as a *Strings* object. When you compile and execute the *chk_file.cpp* program, your screen will display output similar to the following:

```
C:\> chk_file test
test.EXE - Not Found
test.OBJ - Not Found
test.COM - Not Found

C:\> chk_file chk_file
chk_file.EXE - Found
chk_file.OBJ - Found
chk_file.COM - Found
C:\>
```

1178 Using a C++ Class to Create a Doubly linked List

In recent Tips you have expanded your knowledge of how you should create and use your own classes. In the *Strings* example series, you learned how to derive your own class and use it throughout your program. Another useful feature of classes is their ability to handle several different roles to help you process your data. For example, if you recall Tip 749, you created the *ListEntry* structure to help you process a doubly linked list, as shown here:

```
struct ListEntry {
     int number;
     struct ListEntry *next;
     struct ListEntry *previous;
   } start, *node;
```

However, to perform operations on any node within the list (such as moving through the list), you needed to access a series of exterior functions. In the following Tips, you will learn how to build the *list_object* class, which you can use within your programs to maintain information about doubly linked lists.

1179 Understanding the dbllinkob Class Members

In Tip 1178, you briefly revisited the *ListEntry* structure that you previously created. When you work with the *list_object* class, the members will be slightly different. In fact, you will not add any new data members at all; the remaining members you will add to the *list_object* class are the headers for a series of functions your programs will use to better navigate and manipulate the linked list. The companion CD-ROM that accompanies this book includes the *list_object* class members within the *dblinkcl.cpp* file, as shown here:

```
#include <iostream.h>
#include <string.h>
#include <stdlib.h>

class list_object {
 public:
   char info;
   list_object *next;
   list_object *previous;
   list_object(void) {
      info = 0;
      next = NULL;
      previous = NULL;
    }
   list_object *getnext(void) {return next;}
   list_object *getprevious(void) {return previous;}
   void getinfo(char &c) { c = info;}
   void change(char c) {info = c;}
   friend ostream &operator<<(ostream &stream, list_object o)
    {
      stream << o.info << endl;
      return stream;
    }
   friend ostream &operator<<(ostream &stream, list_object *o)
    {
      stream << o->info << endl;
      return stream;
    }
   friend istream &operator>>(istream &stream, list_object &o)
    {
      cout << "Enter information: " << endl;
      stream >> o.info;
      return stream;
    }
 };
```

One important thing to notice about the *list_object* definition is how it defines its *friend* functions (which it uses to manipulate output to and input from streams) inline within the class definition, a slightly different construction than those you have used before. You could just as easily declare the *friend* functions as you normally would. For example, you could declare the overloaded insertion operator, as shown here:

```
friend istream &operator>>(istream &stream, list_object &o);

// Class code

istream &operator>>(istream &stream, list_object &o);
    {
      cout << "Enter information: " << endl;
      stream >> o.info;
      return stream;
    }
```

UNDERSTANDING THE GETNEXT AND GETPREVIOUS FUNCTIONS 1180

In the class definition you saw in Tip 1179, there were two inline function definitions for the *getnext* and *getprevious* functions. Just as with the doubly linked list you built previously using the *ListEntry* structure, each instance of the *list_object* class holds two pointers: one to the item before the object in the list and one to the item following the object in the list. Rather than calling outside functions as you did with the structure-based list, the class-based list lets you build in the movement functions within the class itself, as shown here:

```
list_object *getnext(void) {return next;}
list_object *getprevious(void) {return previous;}
```

Therefore, when your programs navigate the list, they will use the *getnext* and *getprevious* methods to move from one node in the list to the next, as shown here:

```
node = node.getnext();
node = node.getprevious();
```

1181 Understanding the Operator Overload Functions

Unlike the *Strings* class, which required you to overload most of the relational and computational operators, the doubly linked list requires only that you overload the input and output operators. In the case of the particular linked list you are building, the stream operators must be able to handle three situations: a *list_object*, a pointer to a *list_object*, and input to a *list_object*. The first and last situations are relatively intuitive and do not require explanation beyond that they both manipulate the *info* data member (a *char* member in this situation). However, you may wonder why the class definition requires the second overloaded extractor function. As you may remember from your earlier work with a linked list, you will often manipulate pointers to other objects in the list—so, creating an extractor function capable of handling a pointer to a *list_object* object is useful. You will implement the three operator overload functions within your programs, as shown here:

```
friend ostream &operator<<(ostream &stream, list_object o)
 {
   stream << o.info << endl;
   return stream;
 }
friend ostream &operator<<(ostream &stream, list_object *o)
 {
   stream << o->info << endl;
   return stream;
 }
friend istream &operator>>(istream &stream, list_object &o)
 {
   cout << "Enter information: " << endl;
   stream >> o.info;
   return stream;
 }
```

The first two overloaded functions output what the program has previously stored within the node's *info* member to the display. The third function lets you enter information into the node's *info* member. You should pay special attention to how the *list_object* definition overloads the stream operators—*list_object* defines its *friend* functions (which it uses to manipulate output to and input from streams) inline within the class definition. The *list_object* definition uses a slightly different construction than you have used before, but you could just as easily declare the *friend* functions as you normally would. For example, you could declare the overloaded insertion operator as shown here:

```
   friend istream &operator>>(istream &stream, list_object &o);
// Class code
   istream &operator>>(istream &stream, list_object &o);
    {
      cout << "Enter information: " << endl;
      stream >> o.info;
      return stream;
    }
 };
```

INHERITING THE LIST_OBJECT CLASS 1182

In previous Tips you have defined and studied the *list_object* class from which you will create objects within a doubly linked list. However, you must understand that the *list_object* class only defines information about each object within the list—the *list_object* class does not provide your programs with information about the list itself. In the next Tip you will derive the *linked_list* class from the *list_object* class to maintain information about the list itself. However, before you derive the *linked_list* class, be sure you understand the relationship between each object and the list itself, as Figure 1182 illustrates.

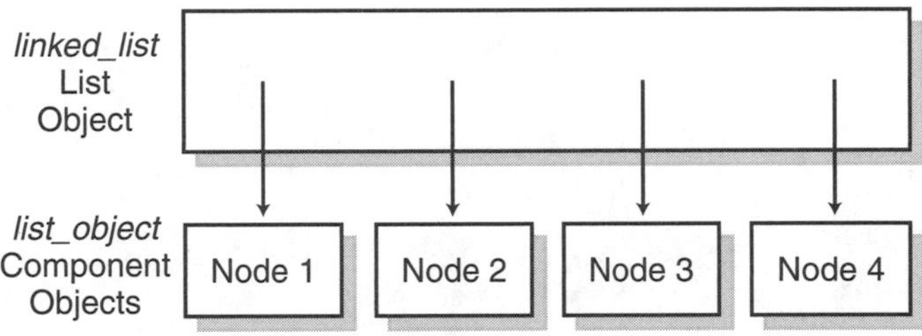

***Figure 1182** The relationship between the **list_object** and **linked_list** classes.*

UNDERSTANDING THE LINKED_LIST CLASS 1183

As you learned in Tip 1182, the *list_object* class does not provide the program with information about the list itself, only about each item in the list. To maintain information about a given list of items, you must derive a second class, the *linked_list* class, from the *list_object* class. The *linked_list* class must maintain two pointers, one to the start of the list and one to the list's ending item. Both the start and end pointers are pointers to *list_object* objects. The *linked_list* constructor initializes both pointers to *NULL* each time your program creates a new list. You will implement the entire *linked_list* class, as shown here:

```
class linked_list : public list_object {
   list_object *start, *end;
 public:
   linked_list(void) {start = end = NULL;}
   void store(char c);
   void remove(list_object *ob);
   void frwdlist(void);
   void bkwdlist(void);
   list_object *find(char c);
   list_object *getstart(void) {return start;}
   list_object *getend(void) {return end;}
 };
```

In addition to the *start* and *end* pointers, the *linked_list* class implements several functions your programs can use to negotiate and manipulate the list. The additional member functions let your programs perform the following actions:

- Put an item in the list
- Remove an item from the list
- Display the list in forward or backward order
- Find a specific element in the list
- Obtain pointers to the start and end of the list

The following Tips examine the implementation of each action in the previous list in detail.

1184 UNDERSTANDING THE LINKED_LIST STORE FUNCTION

As you have learned, you can use classes to more easily implement a doubly linked list similar to the one you created in earlier Tips. As you know, one of the most important actions your programs must perform with any linked list is the insertion of objects into the list. In the *linked_list* class, the *store* member function handles the insertion of objects into the list. You will implement the *store* member function within your *linked_list* class, as shown here:

```
void linked_list::store(char c)
 {
   list_object *p;

   p = new list_object;
   if(!p)
    {
      cout << "Allocation error." << endl;
      exit(1);
    }
   p->info = c;
   if(start == NULL)
    {
     end = start = p;
    }
   else
    {
     p->previous = end;
     end->next = p;
     end = p;
    }
 }
```

Before your program can insert a new item (in this case, a *char*) into the list, the function must create a new *list_object* to hold the item. The *store* function will try to create a new *list_object* object, and will exit the program if the function fails to do so. If it succeeds, the *store* function will store the value in its *c* parameter within the newly created *list_object* object. The *store* function will then add the new *list_object* to the end of the list. The function will also update the *linked_list* object's *start* and *end* pointers, as appropriate. As implemented within your *linked_list* class, *store* will always add new members to the end of the list. However, you can easily modify the function or the class so *store* inserts new objects in the correct location within the list to create a sorted list, as you did with other linked lists in earlier Tips.

As the *store* function clearly shows, *linked_list* manages a set of objects of type *list_object*. The type of data the list stores is immaterial to the *linked_list* class. In other words, you must modify only the *list_object* class to support storage of more useful data. As you will learn in Tip 1191, the *linked_list* class lends itself well to a generic construction, which you will use to create multiple linked lists of multiple types in a single program.

1185 UNDERSTANDING THE LINKED_LIST REMOVE FUNCTION

In Tip 1184, you added the *linked_list store* function to the *linked_list* class. The *store* function lets you add new *list_object* objects to your list. Another important task your *linked_list* list manager must perform is removing *list_object* objects from within the list. The *linked_list* class performs object removals within the *remove* member function, which you will implement as shown here:

```
void linked_list::remove(list_object *ob)
 {
   if(ob->previous)
    {
      ob->previous->next = ob->next;
      if(ob->next)
```

```
          ob->next->previous = ob->previous;
       else
          end = ob->previous;
    }
  else
    {
       if(ob->next)
        {
          ob->next->previous = NULL;
          start = ob->next;
        }
       else
          start = end = NULL;
    }
 }
```

The *remove* function deletes the object to which the *ob* pointer points within the list. As you learned in the previous list example, an object that you must delete from the list resides in one of three places within the list: it is either the first item, the last item, or an item in the middle, as shown in Figure 1185.

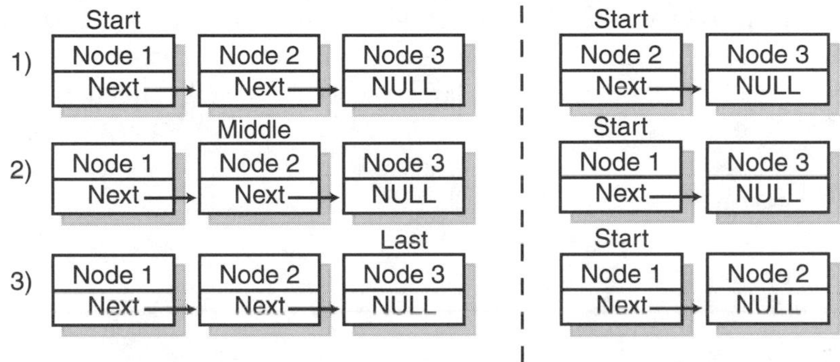

Figure 1185 *The three possible cases that **remove** handles.*

The *remove* member function handles all possibilities and condenses the list after it removes the object you no longer want.

UNDERSTANDING THE GETSTART AND GETEND FUNCTIONS 1186

The only two inline functions that the *linked_list* class defines are the *getstart* and *getend* functions, shown here:

```
list_object  *getstart(void)  {return  start;}
list_object *getend(void) {return end;}
```

Both *getstart* and *getend* are purely interface functions. You can call *getstart* and *getend* from anywhere in the program to move the list pointer to the list's beginning or end. Other member functions within the class can also call *getstart* and *getend* to help the functions traverse the list, as you will learn in the next Tip. Both the *frwdlist* and *bkwdlist* functions use *getstart* and *getend* to position themselves within the list before *frwdlist* and *bkwdlist* generate output.

DISPLAYING THE LINKED_LIST IN FORWARD ORDER 1187

A common activity your programs will perform with any linked list is to display the list. In Tip 1188, you will learn how to display a list from the back forward. However, your programs will frequently display a list in the order that the program created the list. To support forward output of a list, the *linked_list* class defines the *frwdlist* function, as shown here:

```
void linked_list::frwdlist(void)
 {
   list_object *temp;

   temp = getstart();
   do {
      cout << temp->info << " ";
      temp = temp->getnext();
    } while(temp);
   cout << endl;
 }
```

The *frwdlist* function accepts no parameters because it always traverses the entire list. When you invoke *frwdlist* against a *linked_list* object, *frwdlist* first uses the *getstart* function to obtain the pointer to the first object in the list. Next, the *frwdlist* function uses a *do* loop and the *getnext* function to step through the list one item at a time and output the data each item contains, in sequence.

1188 DISPLAYING THE LINKED_LIST IN REVERSE ORDER

In Tip 1187, you created the *frwdlist* member function, which your programs can use to display a *linked_list* from first to last. As you know, your programs must often output a doubly linked list in reverse order. To help your programs output lists in reverse order, the *linked_list* class provides the *bkwdlist* function, as shown here:

```
void linked_list::bkwdlist(void)
 {
   list_object *temp;

   temp = getend();
   do {
      cout << temp->info << " ";
      temp = temp->getprevious();
   } while(temp);
   cout << endl;
 }
```

Just as the *frwdlist* function uses the *getstart* function to get a pointer to the first member in the list and then moves through the list forward, the *bkwdlist* function uses the *getend* function to get a pointer to the last member in the list. Next, the *bkwdlist* function uses a *do* loop and the *getprevious* function (which returns a pointer to the previous item in the list) to move through the list backwards.

1189 SEARCHING THE LIST

As you learned in previous Tips, the *linked_list* object adds new items to the end of the list. Therefore, when you must find an item within the list, your program must be able to search the list for the item, as you will otherwise have no permanent record of where the list stores the object. To help you search a list for an item, the *linked_list* function implements the *find* function to help your programs search for an element within the list, as shown here:

```
list_object *linked_list::find(char c)
 {
   list_object *temp;

   temp = getstart();
   while(temp) {
     if(c == temp->info)
       return temp;
     temp = temp->getnext();
    }
   return NULL;
 }
```

The *find* function begins its processing by invoking the *getstart* function. After *find* obtains a pointer to the first object in the list, it loops through the list one object at a time, trying to match the parameter it receives against the item. If *find* matches the item, it returns a pointer to the item; if *find* does not match the item, it returns a *NULL* pointer.

IMPLEMENTING A SIMPLE LINKED_LIST PROGRAM 1190

In the previous Tips, you have created the simple *list_object* and *linked_list* classes. To implement the simple classes, your programs must create an object of type *linked_list* and a pointer to an object of type *list_object*. The companion CD-ROM that accompanies this book includes the program *use_link.cpp*, which implements both classes. The following code list shows the *main* function from the *use_link.cpp* program. The *use_link.cpp* program creates a simple, three item list, then moves through the list forward and backward, adds and deletes items, and moves through the list "manually" (using *getstart* and *getnext*):

```
void main(void)
 {
   linked_list list;
   char c;
   list_object *p;

   list.store('1');
   list.store('2');
   list.store('3');

   cout << "Here is list backwards, then forwards." << endl;
   list.bkwdlist();
   list.frwdlist();
   cout << endl;

   cout << "'Manually' walk through the list." << endl;
   p = list.getstart();
   while(p) {
      p->getinfo(c);
      cout << c << " ";
      p = p->getnext();
    }
   cout << endl << endl;

   cout << "Looking for item 2." << endl;
   p = list.find('2');
   if(p)
    {
      p->getinfo(c);
      cout << "Found: " << c << endl;
    }
   cout << endl;

   p->getinfo(c);
   cout << "Removing item: " << c << endl;
   list.remove(p);
   cout << "Here is new list forwards." << endl;
   list.frwdlist();
   cout << endl;

   cout << "Adding an item." << endl;
   list.store('4');
   cout << "Here is list forwards." << endl;
   list.frwdlist();
   cout << endl;
   p = list.find('1');

   if(!p)
    {
```

```
      cout << "Error, item not found." << endl;
     exit (1);
   }
   p->getinfo(c);
   cout << "Changing " << c << " to 5." << endl;
   p->change('5');
   cout << "Here is list forwards, then backwards." << endl;
  list.frwdlist();
  list.bkwdlist();
   cout << endl;
   cin >> *p;
   cout << p;
   cout << "Here is list forwards again." << endl;
  list.frwdlist();
   cout << endl;
   cout << "Here is list after removing head of list." << endl;
   p = list.getstart();
   list.remove(p);
   list.frwdlist();
   cout << endl;
   cout << "Here is list after removing end of list." << endl;
   p = list.getend();
   list.remove(p);
  list.frwdlist();
}
```

When you compile and execute the *use_link.cpp* program, it will display the following output on your screen:

```
Here is list backwards, then forwards.
3 2 1
1 2 3
'Manually' walk through the list.
1 2 3
Looking for item 2.
Found: 2
Removing item: 2
Here is new list forwards.
1 3
Adding an item.
Here is list forwards.
1 3 4
Changing 1 to 5.
Here is list forwards, then backwards.
5 3 4
4 3 5
Enter information:
1
Here is list forwards again.
1 3 4
Here is list after removing head of list.
3 4
Here is list after removing end of list.
3
C:\>
```

CREATING A GENERIC DOUBLY LINKED LIST CLASS 1191

In previous Tips you created the *list_object* and *linked_list* classes, which accepted a single *char* and maintained it within a list. However, as you know, a linked-list class is much more useful if it accepts information of various types and stores the information within a list. For example, a program that uses three different lists might maintain one list with *ints*, one list with *floats,* and one list that holds a custom type (such as the *Book* class you designed in previous Tips). Rather than creating a separate *list_object* and *linked_list* for each type, you should create a generic *list_object* and *linked_list* class. As you have learned, when you create a generic class, you can use the class with any C++ or custom type. Your generic list class, or *class template*, can handle any data type.

One advantage of making the class a generic class is that it separates the *mechanism* of the list (that is, the different functions that *linked_list* and *list_object* implement) from the data the list actually stores. Using a generic class to separate the mechanism from the data lets you create the mechanism once and use it over and over again.

UNDERSTANDING THE GENERIC LIST_OBJECT CLASS MEMBERS 1192

As you learned in Tip 1191, a better implementation of your two doubly linked list classes is one that uses generic definitions. However, because your *linked_list* class derives directly from your *list_object* class, you must first make the *list_object* class generic before you can make the *linked_list* class generic (as you will do in Tip 1193). The implementation of the generic *list_object* class is shown here:

```
#include <iostream.h>
#include <string.h>
#include <stdlib.h>

template <class DataT> class list_object {
 public:
   DataT info;
    list_object<DataT> *next;
    list_object<DataT> *previous;
   list_object(void)
   {
     info = 0;
      next = NULL;
      previous = NULL;
   }
    list_object(DataT c)
   {
     info = c;
      next = NULL;
      previous = NULL;
   }
    list_object<DataT> *getnext(void) {return next;}
    list_object<DataT> *getprevious(void) {return previous;}
   void getinfo(DataT &c) {c = info;}
   void change(DataT c) {info = c;}
    friend ostream &operator<<(ostream &stream, list_object<DataT> o)
   {
      stream << o.info << endl;
     return stream;
   }
    friend ostream &operator<<(ostream &stream, list_object<DataT> *o)
   {
      stream << o->info << endl;
     return stream;
   }
    friend istream &operator>>(istream &stream, list_object<DataT> &o)
```

```
     {
       cout << "Enter information: " << endl;
       stream >> o.info;
       return stream;
     }
  };
```

As you learned in Tip 1120, the following declaration creates the generic class *list_object*:

```
template <class DataT> class list_object {
```

The *template* keyword alerts the compiler that a generic class description follows. The *<class DataT>* operator lets the compiler know that the *list_object* class supports a single generic type. When you declare objects of type *list_object*, you must tell the compiler which data type that instance will use, as shown here:

```
list_object<char> char_list;
list_object<float> float_list;
list_object<custom> custom_list;
```

The commands in the previous code fragment create three *list_object* objects: one of type *char*, one of type *float*, and one of type *custom*. In fact, given the following definitions, your *list_object* object can maintain list members of type *BookStuff*:

```
class BookStuff {
  public:
     BookStuff(char *title, char *publisher, char *author);
      void show_book(void)
     {
         cout << "Book: " << title << " by " << author << " Publisher: " <<
              publisher << endl;
     };
     operator char *(void);

   private:
    char title[64];
    char author[64];
    char publisher[64];
};
// Additional program code here

void main(void)
 {
    list_object<BookStuff> object1;
```

1193 UNDERSTANDING THE GENERIC LINKED_LIST CLASS

As you learned in Tip 1192, you must precede generic class definitions with the *template* keyword and an operator that corresponds to the type that the compiler should implement against the generic type. When you implement the *linked_list* class, which derives from the *list_object* class, you must also provide the generic type information to the compiler. However, when you initialize a *linked_list* object with a specific type, it will automatically initialize all *list_object* objects the *linked_list* object uses as the same specific type. You will implement the generic *linked_list* class definition, as shown here:

```
template <class DataT> class linked_list : public list_object<DataT> {
    list_object<DataT> *start, *end;
   public:
    linked_list(void) {start = end = NULL;}
    void store(DataT c);
    void remove(list_object<DataT> *ob);
    void frwdlist(void);
    void bkwdlist(void);
```

```
    list_object<DataT> *find(DataT c);
    list_object<DataT> *getstart(void) {return start;}
    list_object<DataT> *getend(void) {return end;}
};
```

As you can see, both within the *linked_list* class this Tip defines and within the *list_object* class Tip 1192 defines, the class defines each member function and data member in terms of the *DataT* generic object. Additionally, member functions that receive an object of either type as a parameter (such as the *linked_list remove* member function) use the *DataT* generic object together with the parameter definition.

USING THE GENERIC CLASSES WITH A CHAR LIST 1194

In Tip 1190, you created the *use_link.cpp* program, which used the original *list_object* and *linked_list* classes to maintain a short list of character variables. The companion CD-ROM that accompanies this book includes the *use_glink.cpp* program, which uses the generic list classes to create a *char* list identical to the list the *use_link.cpp* program creates.

As you will see later, the only significant difference between the *use_link.cpp* and *use_glink.cpp* programs is the following declaration:

```
linked_list<char> list;
char c;
list_object<char> *p;
```

As you know, the *linked_list<char> list* declaration creates the *list* object. The *list* object supports *info* of type *char*. The remainder of the code within the *main* function is fundamentally the same because the *use_glink.cpp* program knows that it is dealing with a *char* list, just as *use_link.cpp* was in Tip 1190. In Tip 1195, however, you will create a list of type *double* to prove that your generic classes work with multiple types.

USING THE GENERIC CLASSES WITH A DOUBLE LIST 1195

In Tip 1194, you used the generic list classes to create a *char*-based list. However, as you know, one significant benefit of generic classes is that they let your programs use the same basic definition to create multiple types of classes. The companion CD-ROM that accompanies this book includes the program *dbl_link.cpp*, which uses the generic classes to store *double* values.

The *dbl_link.cpp* program performs processing similar to the *use_glink.cpp* program that Tip 1194 implemented. However, the declarations differ slightly, which impacts all the actions the program takes against the list, as shown here:

```
linked_list<double> list;
double c;
list_object<double> *p;
```

When the commands within the *main* function manipulate the *list*, each command handles *double* values, rather than the *char* values Tip 1195 uses.

USING THE GENERIC CLASSES WITH A STRUCTURE 1196

In previous Tips, you have created simple doubly linked lists with your generic classes and simple data types. As you may discover, the generic template requires extra work before you can use the template with more complex struc-

Unfortunately, the compiler does not recognize the constructor function's existence and will return an error if you try to create new list elements in the manner shown in the previous code fragment. In addition, the code is more confusing than it is helpful. The following code, on the other hand, is clearer—especially if your program processes the *Book* information before it tries to store the information within a list:

```
Book cbib("Jamsa's C/C++/C# Programmer's Bible", "Jamsa",
        "Delmar", 54.95);
// Program code here
list.store(cbib);
```

In the second code fragment, you first create the object, then pass the object to the store function. Because of how you designed the generic structure, it processes the *cbib* object perfectly and adds it to the list. The companion CD-ROM that accompanies this book includes the *bk_list.cpp* program, which adds three objects to the list in the manner the second code fragment in this Tip uses. Because of the need to rework the way the class handles output, the program does not perform all the output actions that the list programs in previous Tips did. However, the *bk_list.cpp* program does step through the list one item at a time and generates output. When you compile and execute the *bk_list.cpp* program, your screen will display the following output:

```
Here are some items.
'Manually' walk through the list.
Title: Jamsa's C/C++/C# Programmer's Bible
Publisher: Delmar
Title: 1001 Visual Basic Programmer's Tips
Publisher: Jamsa Press
Title: Hacker Proof
Publisher: Jamsa Press
C:\>
```

1200 Writing a Function to Determine the List's Length

In Tip 1198, you learned about improvements your programs can make to the generic list classes. An improvement Tip 1198 discusses is the addition of a function that traverses the list and returns a count of the number of items within the list. The following code fragment provides a sample implementation of the *listlength* member function, which counts items:

```
template <class DataT> int linked_list<DataT>::listlength(void)
 {
    list_object<DataT> *temp;
   int count = 0;

   temp = start;
   do {
     temp = temp->getnext();
     count = count + 1;
     } while(temp);
   cout << "Number of list items: " << count << endl;
   return count;
 }
```

As you can see, the *listlength* member function performs essentially the same processing as the *frwdlist* function does, except it maintains a count as it traverses the list. The *listlength* member function then outputs the count at the end of its process and returns the count to the *lvalue* within the calling statement. The companion CD-ROM that accompanies this book includes the program *cnt_lst.cpp*, which performs the same processing as the *use_glink.cpp* program presented in Tip 1194. However, the *cnt_lst.cpp* program also invokes *listlength* at different points within the program's execution.

INTRODUCING THE STANDARD TEMPLATE LIBRARY 1201

The Standard Template Library, or STL, is a C++ library of container classes (such as linked lists), algorithms, and iterators; it provides many of computer science's basic algorithms and data structures (such as sorting, mapping, and mathematical functions). The Standard Template Library is a generic library, meaning that its designer heavily parameterized the library's components: almost every component in the Standard Template Library is a template. You should make sure that you understand how templates work in C++ before you use the Standard Template Library. You can use the Standard Template Library to quickly and easily:

- Create sorted lists of objects
- Create sorted lists of related objects sharing a single key
- Manipulate complex data structures in a simple and straightforward manner
- Perform complex manipulations of the information stored within a container using pre-defined algorithms

In the next fifty Tips, you will use the Standard Template Library components to write several programs. Before you begin, you should note the following important points:

1. Because the *Turbo C++ Lite* compiler does not support generic definitions, you cannot use the Standard Template Library with the *Turbo C++ Lite* compiler.
2. Both *Visual C++* and Borland's *C++ for Windows* include the header files for most of the Standard Template Library, meaning you can use many features of the Standard Template Library with either compiler without further modification.
3. If your compiler does not include the Standard Template Library, you can download the necessary header files. To download the Standard Template Library files, visit the Silicon Graphics Web site at *http://216.32.174.40/tech/stl*, as shown in Figure 1201.

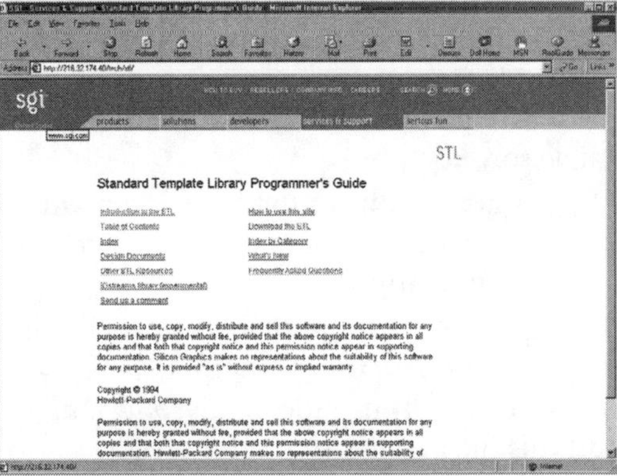

Figure 1201 *The Silicon Graphics STL home page.*

UNDERSTANDING THE STANDARD TEMPLATE LIBRARY HEADER FILES 1202

As you learned in the previous Tip, the Standard Template Library provides generic classes and functions that your programs can use to create additional C++ functionality. Because of the number of classes the Standard Template Library provides, the Library's designers divided the library into multiple header files to reduce compilation time. Table 1202 details the header files.

File Name	Description
algo	Includes all the Standard Template Library algorithms. Later Tips detail the Standard Template Library algorithms (also named *algorith.h* in some STL implementations).
bool	Defines the *bool* data type.
bvector	Defines *bit_vector* objects your programs can use to maintain array-like structures of bit data.
deque	Defines *deque* objects your programs can use to create array-like structures that you can manipulate at both beginning and end.
function	Includes the operators, function objects, and function adapters that manage Standard Template Library classes.
iterator	Defines the iterator tags, stream iterators, and iterator adapters for the Standard Template Library classes.
list	Defines a generic doubly linked list object.
map	Defines the *map* class, a doubly linked, sorted list with a key value and a data value.
multimap	Defines the *multimap* class, a doubly linked, sorted list with one or more key values and a data value. The *multimap* objects support one or more sorted fields within the list.
multiset	Defines the *multiset* class, which lets your programs represent sorted linked lists in a way that permits lookup, insertion, and removal of an arbitrary element with a number of operations proportional to the logarithm of the number of elements in the sequence (a value known as *logarithmic time*). Unlike the *set* class, *multisets* can use multiple sort keys.
pair	Defines the *pair* class, which lets your programs store two values (of the same or different types) within a single object.
random.c	Defines a random-number generator. You should include *random.c* if your programs use the *random_shuffle* algorithm.
set	Defines the *set* class, which lets your programs represent sorted linked lists in a way that permits lookup, insertion, and removal of an arbitrary element in logarithmic time. Unlike the *multiset* class, *sets* must use only a single sort key.
stack	Defines a stack object that your programs can use to control a varying length sequence of elements. The object allocates and frees storage for the sequence it controls.
tempbuf.c	A program file to support an auxiliary buffer for the *get_temporary_buffer stable_partition, inplace_merge,* and *stable_sort* algorithms. The *algo.h* header file automatically includes both *tempbuf.c* and *tempbuf.h*. You should never add this file to your programs directly.
tempbuf	Contains function prototypes and class definitions for the *tempbuf.c* program file.
vector	Defines the *vector* template class, an object that controls a varying-length sequence of elements. Unlike the *list, map*, and *set* classes, the *vector* class is a singly linked list that the compiler treats as an array.

***Table 1202** The include files for the Standard Template Library.*

***Note:** There are some other Standard Template Library include files besides these but they are mainly for use with specific DOS/Windows C++ compilers that do not automatically support multiple memory models. You will not require those other include files if you use **Visual C++** or Borland's **C++ for Windows**. However, if you use a different compiler, check your compiler's documentation to determine whether it requires the additional include files.*

UNDERSTANDING CONTAINERS 1203

One of the fundamental building blocks of the Standard Template Library is the *container*. A container is an object that stores collections of other (similarly typed) objects. For example, the *linked_list* class that you created in Tip 1190 is a container for objects of the *list_object* class. Figure 1203.1 depicts a model of a container and the objects the container stores inside itself.

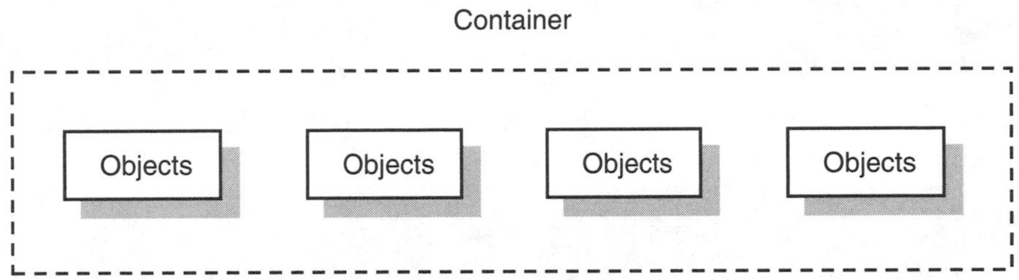

***Figure 1203.1** The logical model for a container and the objects it contains.*

As you continue to write more complex programs, you will find that using containers (whether from the Standard Template Library, other libraries, or of your own design) becomes more important to your programs. Rather than trying to manipulate large numbers of individual objects, which requires a similar number of variables, loops, tests, and so on, you can instead include many objects within a single container, which makes managing the objects simpler. For example, it is much easier to manage an array of 10 integers than it is to manage 10 integer variables. In fact, many of the programs you have written so far have used the simplest type of container: an array. Figure 1203.2 depicts an array as a container.

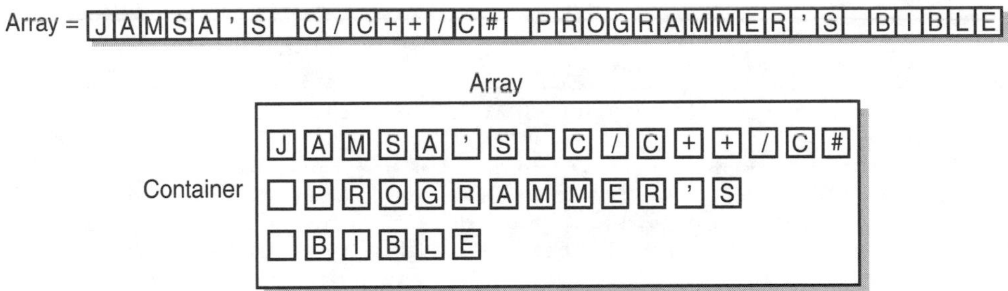

***Figure 1203.2** The array as a container.*

In later Tips, you will learn more about and implement the Standard Template Library containers. In Tip 1204, however, you will use the *linked_list* and *list_object* classes to review the concept of containers.

USING A CONTAINER EXAMPLE 1204

In Tip 1203 you learned about containers and how your programs will use container classes to maintain information about groups of objects. As you learned in Tip 1201, the Standard Template Library includes several different container types (including *lists, sets, maps*, and *deques*). In Tip 1205, you will learn the basics about the Standard Template Library types.

However, before you continue to learn about the Standard Template Library containers, it is valuable to analyze a generic class you previously created to determine whether it qualifies as a container. In previous Tips you designed and implemented the *linked_list* class, which you defined generically as shown here:

```
template <class DataT> class linked_list : public list_object<DataT>
```

```
{
    list_object<DataT> *start, *end;
  public:
    linked_list(void) {start = end = NULL;}
    void store(DataT c);
    void remove(list_object<DataT> *ob);
    void frwdlist(void);
    void bkwdlist(void);
    list_object<DataT> *find(DataT c);
    list_object<DataT> *getstart(void) {return start;}
    list_object<DataT> *getend(void) {return end;}
};
```

The *linked_list* class is, of course, derived from the *list_object* class, which you defined as shown here (for clarity, this reprint relocates the inline friend functions outside the class definition):

```
template <class DataT> class list_object
 {
  public:
    DataT info;
    list_object<DataT> *next;
    list_object<DataT> *previous;
    list_object(void)
   {
     info = 0;
      next = NULL;
       previous = NULL;
   }
    list_object(DataT c)
   {
      info = c;
      next = NULL;
      previous = NULL;
   }
    list_object<DataT> *getnext(void) {return next;}
    list_object<DataT> *getprevious(void) {return previous;}
    void getinfo(DataT &c) { c = info;}
    void change(DataT c) {info = c;}
    friend ostream &operator<<(ostream &stream, list_object<DataT> o)
    friend ostream &operator<<(ostream &stream, list_object<DataT> *o)
    friend istream &operator>>(istream &stream, list_object<DataT> &o)
};
```

Because the *list_object* class is generic, it will accept data of any type. Each of the following declarations is valid:

```
list_object<int> int_obj;
list_object<float> float_obj;
list_object<char> char_obj;
list_object<Book> book_obj;
```

The generic definition of the class means that its type is unimportant (in other words, your programs can just as easily define *int lists* as they can *float lists*, and the *list* class member functions will work equally well on either). As you know, a container manages individual objects of known or unknown types without concern for the object's type. For example, each of the following array declarations is valid:

```
int intarray[10];
float floatarray[10];
char chararray[10];
Book bookarray[10];
```

While the information each element within the arrays stores is different, the arrays themselves are identical. That is, each array contains ten elements of some type which the user can step through the array's index to access.

The *linked_list* class is similar to an array. Because it derives from the *list_object* type, the compiler explicitly types each instance of the list (in other words, if the *list_objects* are all *doubles*, the *linked_list* object must be of type *double* as well). However, the *linked_list* class itself is unconcerned about the nature of the objects it stores; it simply stores those objects and provides you with ways to navigate the objects within the *linked_list*. In other words, the *linked_list* class is clearly a container.

INTRODUCING THE STANDARD TEMPLATE LIBRARY CONTAINERS 1205

As you have learned, the container is one of the fundamental building blocks of the Standard Template Library. In Tip 1204, you analyzed how you had previously designed and implemented many containers within your programs, even when you did not realize they were containers. As you will learn, the Standard Template Library defines similar or related implementations to many of the containers that you have previously used, as well as some containers you may not have considered. The Standard Template Library supports two basic types of containers: *Sequence Containers* and *Associative Containers*. Sequence Containers are objects that store collections of other objects in a strictly linear arrangement. The Standard Template Library supports the following three Sequence Containers:

- *vector<T>*: The *vector<T>* class provides array-like random access to a sequence of objects. As your program executes, the *vector* object's length can vary. Your programs can perform insertions and deletions at the end of the sequence. You will typically use vectors to maintain unsorted information in a series.
- *deque<T>*: The *deque<T>* class provides random access to a sequence of objects. As with *vectors*, as your program executes, the *deque* object's length can vary. Your programs can perform insertions and deletions at both the beginning and the end of the sequence. You will generally use *deques* to maintain unsorted information in a series when you are unsure which end of the series you will add information to.
- *list<T>*: The *list<T>* class provides access to a sequence of objects. As with *vectors* and *deques*, as your program executes, the *list* object's length can vary. Your programs can perform insertions and deletions anywhere in the sequence. You will generally use *lists* to maintain unsorted information in a series when you are unsure where in the series you will insert information.

On the other hand, Associative Containers provide your programs with an easy way to retrieve objects quickly from the object collection the container class contains. Associative Containers use keys to support fast object retrieval. The size of the collection can vary at run time (as can all Standard Template Library containers, making them *dynamic* containers, whereas arrays and other simple containers are *static* containers). An Associative Container maintains the collection in order, based on a comparison function object of type *Compare*. The Standard Template Library supports the following four Associative Containers:

- *set<T, Compare>:* The *set* class supports unique keys (that is, objects of the class contain, at most, one of each key value) and provides for fast retrieval of the keys themselves. You will generally use *sets* to maintain sorted information which uses only a single sort key, such as a simple series of integer numbers.
- *multiset<T, Compare>:* The *multiset* class supports duplicate keys (that is, objects of the class possibly contain multiple copies of the same key value) and provides for fast retrieval of the keys themselves. You will generally use *multisets* to maintain sorted information that uses more than one sort key, which is also the data itself, such as a sorted series of coordinates on a grid.

1208 UNDERSTANDING THE USING NAMESPACE STD STATEMENT

In Tip 1207, you wrote the program *vector1.cpp*, which created a simple vector container and manipulated its component objects. While most of the program is self-explanatory, the following line of code may seem unnecessary or out of place to you:

```
using namespace std;
```

As you know, the *using* statement lets your program access variable names from within a given *namespace*. In this particular case, the *using* statement lets your program access the variables and classes within the *std* namespace, which is the standard namespace for the Standard Template Library. Whenever you write programs that use Standard Template Library components, you must include the *using namespace std* statement or the compiler will not recognize the Standard Template Library class or classes your program uses.

1209 UNDERSTANDING STANDARD TEMPLATE LIBRARY ASSOCIATIVE CONTAINERS

An Associative Container is a variable-sized container that supports efficient element (value) retrieval based on *keys*. A key is a sort value which may or may not be the actual value that the container uses as an index into the container's objects. Associative Containers support element insertion and removal, but differ from Sequence Containers in that Associative Containers do not provide a mechanism for inserting an element at a specific position. As with all containers, the elements in an Associative Container have type *value_type*. Additionally, each element in an Associative Container has a type *key_type* key.

In some Associative Containers, such as *Simple Associative Containers (sets* and *multisets)*, the *value_type* and *key_type* are the same—that is, the elements are their own keys. In others, the key is some specific part of the value. Because Associative Containers store elements according to their keys, it is essential that the key the container associates with each element is immutable (that is, the program cannot change the element). Therefore, in Simple Associative Containers, the elements themselves are immutable. In other Associative Container types, such as Pair Associative Containers, the elements themselves are mutable but the program cannot modify the part of an element that is the element's key.

In Simple Associative Containers, where the elements are the keys, the elements are completely immutable. Therefore, the member types *iterator* and *const_iterator* are the same for Simple Associative Containers. Other Associative Containers types, however, do have mutable elements and do provide iterators through which programs can modify elements.

In some Associative Containers, such as Unique Associative Containers, the Standard Template Library class specification guarantees that no two elements have the same key. In other Associative Containers, such as Multiple Associative Containers, the container will let the program store multiple elements with the same key within the container. To better understand Associative Containers, consider the following program, *firstset.cpp*, which creates and manipulates an object of type *set*:

```
#include <set>
#include <iostream>

using namespace std;
typedef set<int> SET_INT;

void main(void)
 {
   SET_INT s1;
   SET_INT s2;
   SET_INT::iterator i;

   cout << "s1.insert(5)" << endl;
   s1.insert(5);
   cout << "s1.insert(10)" << endl;
```

```
    s1.insert(10);
    cout << "s1.insert(15)" << endl;
    s1.insert(15);
    cout << "s2.insert(2)" << endl;
    s2.insert(2);
    cout << "s2.insert(4)" << endl;
    s2.insert(4);
    cout << "swap(s1,s2)" << endl;
    swap(s1,s2);
    for (i = s1.begin();i != s1.end(); i++)          // Displays: 2,4
      cout << "s1 has " << *i << " in its set." << endl;

    for (i = s2.begin();i != s2.end(); i++)          // Displays: 5,10,15
     cout << "s2 has " << *i << " in its set." << endl;

    cout << "s1.swap(s2)" << endl;
    s1.swap(s2);
    for (i = s1.begin();i != s1.end(); i++)          // Displays: 5,10,15
     cout << "s1 has " << *i << " in its set." << endl;

    for (i = s2.begin();i != s2.end(); i++)          // Displays: 2,4
     cout << "s2 has " << *i << " in its set." << endl;
}
```

The program creates two *sets*, assigning the first *set* the values 5, 10, and 15, and the second *set* the values 2 and 4. The program also defines the iterator *i*, which it later uses to traverse the *sets*. Within the *firstset.cpp* program, the code uses three *set* class member functions. The *swap* function swaps the two controlled sequences. The *begin* function returns a *bi-directional* iterator that points to the first element of the sequence. The *end* function returns a *bi-directional* iterator that points just beyond the sequence's end. (You will learn more about iterators in Tip 1210.) The program inserts elements into the set, then moves through the set swapping elements. When you compile and execute the *firstset.cpp* program, your screen will display the following output:

```
s1.insert(5)
s1.insert(10)
s1.insert(15)
s2.insert(2)
s2.insert(4)
swap(s1,s2)
s1 has 2 in its set.
s1 has 4 in its set.
s2 has 5 in its set.
s2 has 10 in its set.
s2 has 15 in its set.
s1.swap(s2)
s1 has 5 in its set.
s1 has 10 in its set.
s1 has 15 in its set.
s2 has 2 in its set.
s2 has 4 in its set.
C:\>
```

UNDERSTANDING ITERATORS 1210

A key factor in the Standard Template Library's design is the container definitions' consistent use of iterators, which generalize C++ pointers as intermediaries between algorithms and containers. The Standard Template Library defines five iterator categories. The classification is also the main guide for extending the library to include new algorithms that work with Standard Template Library containers, or to include new containers to which you can apply many Standard Template Library generic algorithms.

There are three facts you must take into account when you try to determine which algorithms you can use with which containers and iterators:

- The Standard Template Library classifies iterators into five categories: *forward, input, output, bi-directional,* and *random-access.*
- Each container class description includes the iterator category types the container class provides.
- Each generic algorithm description includes the iterator and container categories with which the generic algorithm works.

Table 1210.1 defines the five iterator type categories.

Iterator Type	Description
forward	Provides for one-directional traversal of a sequence, which your programs will express with the increment operator (++).
input	Similar to *forward* iterators in that your programs can use them to input data into a container. However, *input* iterators may not support all properties of *forward* iterators.
output	Similar to *forward* iterators in that your programs can use them to output data from a container. However, *output* iterators may not support all properties of *forward* iterators.
bi-directional	Provides for traversal in both directions, which your programs will express with ++ (forward) and — (backward). Note that, if the iterator points from the end of the container forward, incrementing the iterator with ++ will actually move the iterator toward the front of the container.
random-access	Provides for bi-directional traversal of a sequence. Additionally, *random-access* iterators provide for bi-directional "long jumps" within a sequence, which you will express as either integer addition, integer subtraction, iterator subtraction, or comparisons, as Table 1210.2 details.

***Table 1210.1** The five Standard Template Library iterator types.*

As Table 1210.1 indicates, the *random-access* iterators let you use several different techniques to move through sequences. Table 1210.2 details the techniques your programs can use with *random-access* iterators to traverse a list.

Technique	Explanation
integer addition	You can perform integer addition and subtraction to a *random-access* iterator using the forms *r* += *n* and *r* -= *n* (where *r* is a *random-access* iterator and *n* is an integer). The operation yields an iterator result.
addition and subtraction	You can add or subtract an integer from an iterator using the forms *r* + *n* and *r* - *n* (where *r* is a *random-access* iterator and *n* is an integer). The operation yields an iterator result.
iterator subtraction	You can subtract an iterator from an iterator using the form *r* - *s* (where *r* is a *random-access* iterator and *s* is another *random-access* iterator). The operation yields an iterator result.
comparisons	You can perform comparisons with *random-access* iterators using the forms *r* < *s*, *r* > *s*, *r* <= *s*, and *r* >= *s*. Iterator comparisons produce *bool* values.

***Table 1210.2** The activities your programs can perform against **random-access** iterators.*

In addition, all five iterator categories also provide for the following actions:

- testing for equality with the == operator, and inequality with the != (not equals) operator
- de-referencing, which means obtaining the data object at the position to which the iterator refers, expressed with * (the pointer de-referencing operator)

The Standard Template Library, however, does not guarantee the following actions will perform as you expect when your programs manipulate iterators:

- The Standard Template Library does not guarantee that you can save an *input* or *output* iterator and use it to start advancing from its current position at a later time.
- The Standard Template Library does not guarantee that you can later assign a value to an object that you previously obtained from a container by applying the pointer redirection operator (*) to an *input* iterator.
- The Standard Template Library does not guarantee that you can read from an object, which you obtained from a container by applying *, to an *output* iterator.
- The Standard Template Library does not guarantee that you can test two *output* iterators for equality or inequality (that is, == and != may not be defined when you apply them against *output* iterators).

USING AN ITERATOR EXAMPLE 1211

As you learned in Tip 1210, your programs will use iterators to traverse or to maintain their position within a container object. The different container types support different iterator types, as Table 1211 details.

Container Type	Iterator Type
vector<T>::iterator	*random-access* iterator
deque<T>::iterator	*random-access* iterator
list<T>::iterator	*bi-directional* iterator
associative containers	All associative containers use *bi-directional* iterators

***Table 1211** The iterator types each container uses.*

As Table 1211 details, the *list* type uses a *bi-directional* iterator. The following program, *use_iter.cpp*, uses a *bi-directional* iterator to traverse an integer list:

```
#include <list>
#include <iostream>

using namespace std;
typedef list<int> LISTINT;

void main(void)
 {
    LISTINT listOne;
    LISTINT::iterator i;

    // Add some data
     listOne.push_front (2);
     listOne.push_front (1);
     listOne.push_back (3);

    for (i = listOne.begin(); i != listOne.end(); ++i) list values: 1 2 3
      cout << *i << " ";
    cout << endl;
    for (i = listOne.end(); i != listOne.begin(); --i) list values: 1 1 1 1
```

```
      cout << *i << " ";
    cout << endl;
  }
```

The *use_iter.cpp* program first creates the *LISTINT* type and then creates that type's *listOne* instance. Next, the program uses the *push_front* and *push_back* methods to add three values to the list (two to the front and one to the back). The program then uses the *i* iterator to traverse the list forward and backward. When you compile and execute the *use_iter.cpp* program, your screen will display the following output (note the first value is a 0 because the *end* member function obtains the location just beyond the vector's end):

```
1 2 3
0 3 2
C:\>
```

1212 Better Understanding the STL Input and Output Iterator Types

As you have learned, the Standard Template Library (STL) supports five iterator types. Two iterator types your programs can use for specific purposes are the *output* iterators (to retrieve data from a container and output the data) and the *input* iterators (to obtain data from another location and store it within the container). As you learned in Tip 1210, neither *output* iterators nor *input* iterators provide the full functionality of the *forward* or *bi-directional* iterators. However, *output* and *input* iterators are useful within your code for performing their specific functions, and making it clear within your code what functions you intend the iterators to perform. The limitations on *output* and *input* iterators are significant. For example, your programs should only use the *output* iterator to return information from a given container, rather than using it to place information into the given container, as shown here:

```
LISTINT::output_iterator out;

for (out = listOne.begin(); out != listOne.end(); ++out)
     cout << *i << " ";

// This code may or may not work correctly
for (out = listOne.begin(); out != listOne.end(); ++out)
     listOne.insert (1);
```

Because the output iterator will not update correctly during an insertion operation, the second *for* loop will have unpredictable effects. Similarly, your programs should use input iterators only when performing insertions into a container and not when retrieving data from a container, as shown here:

```
LISTINT::input_iterator in;

for (in = listOne.begin(); in != listOne.end(); ++in)
    listOne.insert (1);

// This code may or may not work correctly
for (in = listOne.begin(); in != listOne.end(); ++in)
     cout << *i << " ";
```

In addition to yielding untrustworthy effects after you perform an output operation with an *input* iterator and after you perform an *input* operation with an *output* iterator, after you increment either an *input* or *output* iterator you cannot safely compare, de-reference, or increment any other copy of the same iterator.

1213 Understanding the STL's Other Iterator Types

As you learned in Tip 1210 and again in Tip 1212, there are significant limitations on how your programs can use the *input* iterators and *output* iterators. To avoid the *input* and *output* iterators' limitations, your programs should use *forward, bi-directional,* or *random-access* iterators. A *forward* iterator *X* can take the place of an *output* iterator (for writing) or an *input* iterator (for reading). Additionally, you can read (using *V* = **X*) what you just wrote (using **X* = *V*) through a *forward* iterator. Moreover, you can make multiple copies of a *forward* iterator, each of which your program can de-reference and increment independently.

In addition to using a *forward* iterator to traverse and access containers, your programs can use a *bi-directional* iterator. *Bi-directional* iterators perform much the same processing as *forward* iterators. You can, however, also decrement a *bi-directional* iterator, as in *--X, X--*, or *(V = *X--)*.

Finally, you can use a *random-access* iterator *X* rather than a *bi-directional* iterator. *Random-access* iterators perform much the same processing as *bi-directional* iterators—they let you traverse the list in the same manner. In addition to *bi-directional* iteration control, you can also perform the same integer arithmetic on a *random-access* iterator that you can on an object pointer. For *N*, an integer object, you can write *x[N], x + N, x - N*, and *N + X* to navigate a container with non-iterator integers.

Note that an object pointer can take a *random-access* iterator's place or that of any other iterator.

You can easily summarize the hierarchy of iterator categories by visualizing the hierarchy of iterators available to you for any single action. For example, for write-only access to a sequence, you can use any of the following iterators:

```
output iterator ->
   forward iterator ->
     bi-directional iterator ->
          random-access iterator
```

The right arrow means the iterator to the right of and down from the arrow can replace the iterator to the left of and up from the arrow. Therefore, any algorithm that calls for an *output* iterator should work well with a *forward* iterator, for example. However, you cannot conclude that any algorithm that calls for a *forward* iterator should work well with an *output* iterator (because the *forward* iterator is to the right of and down from the *output* iterator).

The hierarchy of iterators is similar for read-only access to a sequence. For read-only access, your programs can use any of the following iterators:

```
input iterator ->
   forward iterator ->
     bi-directional iterator ->
          random-access iterator
```

For read-only actions, an *input* iterator is the weakest of all iterator categories because it can only transverse the list in forward order and is invalidated by any output activities. Again, any algorithm that calls for an *input* iterator should work well with a *forward* iterator or any other iterator below it in the tree. However, you cannot conclude that any algorithm that calls for a *forward* iterator should work well with an *input* iterator.

The hierarchy of iterators is similar for read/write access to a sequence. For read/write access to a sequence, your programs can use any of the following iterators:

```
forward iterator ->
   bi-directional iterator ->
        random-access iterator
```

Remember that an object pointer can always serve as a *random-access* iterator. Therefore, it can serve as any category of iterator, as long as it supports the proper read/write access to the sequence it designates. This iterator "algebra" is fundamental to almost all other processing your programs will perform with Standard Template Library algorithms and containers. It is important to understand each iterator category's capabilities and limitations to see how containers and algorithms in the Standard Template Library use iterators.

```
#include <algorith>
#include <iostream>

using namespace std;

void main(void)
 {
    const int ARRAY_SIZE = 8;
    int IntArray[ARRAY_SIZE] = { 1, 2, 3, 4, 4, 5, 6, 7 };
    int *location;     // stores the position of the first matching element.
    int i;
    int value = 4;

    // print content of IntArray
    cout << "IntArray { ";
    for(i = 0; i < ARRAY_SIZE; i++)
      cout << IntArray[i] << ", ";
    cout << "\b }" << endl;

    // Find the first element in the range [first, last + 1) that matches value.
    location = find(IntArray, IntArray + ARRAY_SIZE, value);

    //print the matching element if any was found
    if (location != IntArray + ARRAY_SIZE)  // matching element found
      cout << "First element that matches " << value << " is at location "
          << location - IntArray << endl;
   else                                     // no matching element found
      cout << "The sequence does not contain any elements with value " << value <<
       << endl;
 }
```

The *arr_find.cpp* program first creates an array of values and then searches for an item within that array. The program uses the *find* algorithm to determine whether the item is within the array. When you compile and execute the *arr_find.cpp* program, your screen will display the following output:

```
IntArray { 1, 2, 3, 4, 4, 5, 6, 7 }
First element that matches 4 is at location 3
C:\>
```

1218 Describing the Algorithms the STL Includes

As you have learned, the Standard Template Library (STL) provides an extended number of algorithms your programs can use when they manipulate both Standard Template Library objects and non-Standard Template Library objects. The Standard Template Library divides its algorithms into four basic algorithm types. The first type is the set of *non-mutating algorithms.* Non-mutating algorithms do not change the containers they operate against and tend to produce linear results. Table 1218.1 lists the Standard Template Library non-mutating algorithms.

Non-Mutating Algorithms			
for_each	*find*	*find_if*	*adjacent_find*
find_first_of	*count*	*count_if*	*mismatch*
equal	*search*	*search_n*	*find_end*

***Table 1218.1** The Standard Template Library non-mutating algorithms.*

The second Standard Template Library algorithm type is the set of *mutating algorithms.* As a rule, mutating algorithms either change the nature of the objects within a container or copy those objects to other containers. Table 1218.2 lists the Standard Template Library mutating algorithms.

Mutating Algorithms			
copy	*copy_n*	*copy_backward*	*swap*
iter_swap	*swap_ranges*	*transform*	*replace*
replace_if	*replace_copy*	*replace_copy_if*	*fill*
fill_n	*generate*	*generate_n*	*remove*
remove_if	*remove_copy*	*remove_copy_if*	*unique*
unique_copy	*reverse*	*reverse_copy*	*rotate*
rotate_copy	*random_shuffle*	*random_sample*	*random_sample_n*
partition	*stable_partition*		

Table 1218.2 *The Standard Template Library mutating algorithms.*

The third important Standard Template Library algorithm type is the set of *sorting algorithms*. Although sorting algorithms are technically a subset of the mutating algorithms, the set of sorting algorithms is extensive enough to stand alone. Table 1218.3 lists the Standard Template Library sorting algorithms.

Sorting Algorithms			
sort	*stable_sort*	*partial_sort*	*partial_sort_copy*
is_sorted	*nth_element*	*lower_bound*	*upper_bound*
equal_range	*binary_search*	*merge*	*inplace_merge*
includes	*set_union*	*set_intersection*	*set_difference*
set_symmetric_difference	*push_heap*	*pop_heap*	*make_heap*
sort_heap	*is_heap*	*min*	*max*
min_element	*max_element*	*lexicographical_compare*	
lexicographical_compare_3way	*next_permutation*	*prev_permutation*	

Table 1218.3 *The Standard Template Library sorting algorithms.*

Finally, the Standard Template Library also supports a set of *generalized numeric algorithms*. Your programs can use the generalized numeric algorithms to perform mathematical activities against an unknown type or types of number (for example, a *float* and a *double* or a *float* and an *int*), depending on the specific algorithm and its purpose. Table 1218.4 lists the Standard Template Library generalized numeric algorithms.

Generalized Numeric Algorithms		
iota	*accumulate*	*inner_product*
partial_sum	*adjacent_difference*	*power*

Table 1218.4 *The Standard Template Library generalized numeric algorithms.*

Clearly, the list of algorithms that the Standard Template Library defines is extensive. Unfortunately, this book will not cover all the algorithms. However, you can check your compiler's documentation or the Standard Template Library documentation for more information on any algorithm listed in this Tip.

STUDYING THE STL FOR_EACH ALGORITHM 1219

As you have learned, the Standard Template Library (STL) defines different types of algorithms. While you have already used the *find* non-mutating algorithm, it is nevertheless worthwhile for you to examine a second sample program, which uses the non-mutating *for_each* algorithm. The *for_each* algorithm calls some function *Func1* for each element in the range [first, last) and returns no value, as shown here:

```
void for_each(first, last, func1);
```

The *for_each* algorithm does not modify any elements in the sequence. For example, the following program, *third_pw.cpp*, uses the *for_each* algorithm to access each element in a vector and output that element's cube:

```
#include <vector>
#include <algorith>
using namespace std;

void PrintCube(int n)
 {
   // prints the cube of integer n
   cout << "The cube of " << n << " is " << n * n * n << endl;
 }

void main(void)
 {
    const int VECTOR_SIZE = 8;
    typedef vector<int> IntVector;                // Define a vector of integers
    typedef IntVector::iterator IntVectorIt;      // Define an iterator type
    IntVector Numbers(VECTOR_SIZE);               // vector containing numbers
    IntVectorIt start, end, it;                   // iterators
    int i;

   for (i = 0; i < VECTOR_SIZE; i++)             // Initialize vector Numbers
     Numbers[i] = i + 1;

   start = Numbers.begin();        // location of first element of Numbers
   end = Numbers.end();            // one past the last element of Numbers

   cout << "Numbers { ";                         // print content of Numbers
   for(it = start; it != end; it++)
     cout << *it << " ";
   cout << " }\n" << endl;

    // for each element in the range [first, last), print the element's cube
    for_each(start, end, PrintCube);
 }
```

The *third_pw.cpp* program initially creates the vector *IntVector*, then assigns a value to each object in the vector. After displaying the vector's values, the program uses the *for_each* algorithm to output each vector element's cube. When you compile and execute the *third_pw.cpp* program, your screen will display the following output:

```
Numbers { 1 2 3 4 5 6 7 8  }

The cube of 1 is 1
The cube of 2 is 8
The cube of 3 is 27
The cube of 4 is 64
The cube of 5 is 125
The cube of 6 is 216
The cube of 7 is 343
The cube of 8 is 512
C:\>
```

1220 Studying the STL generate_n Algorithm

As you learned in Tip 1218, the Standard Template Library (STL) supports an extensive number of algorithms that your programs can use to manipulate containers and other objects. One of the more unique mutating algorithms that the Standard Template Library supports is the *generate_n* algorithm, which fills each object within a range of objects inside a container with the return value from a *generator function*. The generator function returns a value that the algorithm places in that object. To better understand how the generator function returns the value, consider the following program, *gen_fib.cpp*, which uses the *generate_n* algorithm to place numbers from the Fibonacci sequence within a vector:

```
#include  <iostream>
#include <vector>
#include <algorith>

using namespace std;

// return the next Fibonacci number in the Fibonacci series.
int Fibonacci(void)
 {
   static int r;
   static int f1 = 0;
   static int f2 = 1;
   r = f1 + f2;
   f1 = f2;
   f2 = r;
   return f1;
 }

void main(void)
 {
   const int VECTOR_SIZE = 15;

   // Define a template class vector of integers
   typedef vector<int> IntVector;

   //Define an iterator for template class vector of integer
   typedef IntVector::iterator IntVectorIt;

   IntVector Numbers(VECTOR_SIZE);    //vector containing numbers
   IntVectorIt start, end, it;
   int i;

   //Initialize vector Numbers
   for(i = 0; i < VECTOR_SIZE; i++)
     Numbers[i] = i * i;
   start = Numbers.begin();    // location of first element of Numbers
   end = Numbers.end();        // one past the location of last element of Numbers
   cout << "Before calling generate_n" << endl;
   // print content of Numbers
   cout << "Numbers { ";
   for(it = start; it != end; it++)
     cout << *it << " ";
   cout << " }\n" << endl;

   // fill the specified range with a series of
   // Fibonacci numbers using the Fibonacci function
   generate_n(start + 5, Numbers.size() - 5, Fibonacci);

   cout << "After calling generate_n" << endl;

   // print content of Numbers
   cout << "Numbers { ";
   for(it = start; it != end; it++)
      cout << *it << " ";
   cout << "\b }\n" << endl;
}
```

The *gen_fib.cpp* program initializes the *numbers* vector to hold 14 values, each of which contains the square of the value's index within the vector. For example, the value at the 1 index holds 1 (1 * 1) and the value at the 12 index holds 144 (12 * 12). Next, the program replaces numbers within the vector beginning at the sixth number, which it replaces with the first number in the Fibonacci sequence. The program replaces each remaining number within the vector with the next number in the Fibonacci sequence. When you compile and execute the *gen_fib.cpp* program, your screen will display the following output:

```
Before calling generate_n
Numbers { 0 1 4 9 16 25 36 49 64 81 100 121 144 169 196 }
After calling generate_n
Numbers { 0 1 4 9 16 1 1 2 3 5 8 13 21 34 55 }
C:\>
```

1221 Understanding the STL random_shuffle Algorithm

As your programs become more complex, there may be times when you must reshuffle the information within a container in some random order. Generally, you will perform such processing when you write games or other computer programs that simulate intelligent responses from the computer. To simplify such processing, the Standard Template Library (STL) provides the *random_shuffle* algorithm. The *random_shuffle* algorithm shuffles the elements of a sequence (first through last) in a random order. The *random_shuffle* algorithm uses either an internal random-number generator to generate the indices of the elements to swap or seeds the random-number generator with the number of elements the container includes, depending on its invocation. Both the predicate and non-predicate version of *random_shuffle* use the = operator to perform swaps. To better understand the processing the *random_shuffle* algorithm performs, consider the *shuffle.cpp* program, as shown here:

```
#include <iostream>
#include <algorith>
#include <vector>

using namespace std;

void main(void)
 {
   const int VECTOR_SIZE = 8;
   typedef vector<int> IntVector;                // Define a vector of int
   typedef IntVector::iterator IntVectorIt; // Define an iterator for vector
   IntVector Numbers(VECTOR_SIZE);

   IntVectorIt start, end, it;
   Numbers[0] = 4;                               // Initialize vector Numbers
   Numbers[1] = 10;
   Numbers[2] = 70;
   Numbers[3] = 30;
   Numbers[4] = 10;
   Numbers[5] = 69;
   Numbers[6] = 96;
   Numbers[7] = 100;

   start = Numbers.begin();          // location of first element of Numbers
   end = Numbers.end();        // one past the last element of Numbers

   cout << "Before calling random_shuffle\n" << endl;

   cout << "Numbers { ";                    // print content of Numbers
   for(it = start; it != end; it++)
      cout << *it << " ";
   cout << " }\n" << endl;

   random_shuffle(start, end);// shuffle the elements in a random order
   cout << "After calling random_shuffle\n" << endl;
   cout << "Numbers { ";
   for(it = start; it != end; it++)
      cout << *it << " ";
   cout << "\b }\n" << endl;
 }
```

```
Numbers { 4 10 70 30 10 69 96 7  }
Result { 4 7 10 10  }
C:\>
```

UNDERSTANDING THE MERGE ALGORITHM 1223

In Tip 1222, you used the *partial_sort_copy* algorithm to create a partially sorted copy of a vector. In this Tip, you will use the *merge* algorithm to merge two sorted vectors into a third sorted vector that will contain all the values from both original vectors. You will use the *merge* algorithm within your programs as shown in the following prototype:

```
merge(first1, last1, first2, last2, result);
```

The *merge* algorithm merges two sorted sequences: (*first1..last1*) and (*first2..last2*) into a single sorted sequence starting at the *result* iterator. The *merge* algorithm assumes that the ranges *(first1..last1)* and *(first2..last2)* have been previously sorted using the less-than operator. If both ranges contain equal values, *merge* will first store the value from the first range within the resulting range. To better understand the *merge* algorithm's processing, consider the following program, *merge_2v.cpp*:

```
#include <iostream>
#include <algorith>
#include <vector>
#include <list>
#include <deque>

using namespace std;

void main(void)
 {
   const int MAX_ELEMENTS = 8;
   typedef vector<int> IntVector;         // Define a template class vector of int
   typedef IntVector::iterator IntVectorIt;// Define an iterator type
   typedef list<int> IntList;             // Define a template class list of int
   typedef IntList::iterator IntListIt; // Define an iterator type
   typedef deque<int> IntDeque;           // Define a template class deque of int
   typedef IntDeque::iterator IntDequeIt; // Define an iterator type

   IntVector NumbersVector(MAX_ELEMENTS);
   IntVectorIt startv, endv, itv;
   IntList NumbersList;
   IntListIt first, last, itl;
   IntDeque NumbersDeque(2 * MAX_ELEMENTS);
   IntDequeIt itd;

   // Initialize vector NumbersVector
   NumbersVector[0] = 4;
   NumbersVector[1] = 10;
   NumbersVector[2] = 70;
   NumbersVector[3] = 10;
   NumbersVector[4] = 30;
   NumbersVector[5] = 69;
   NumbersVector[6] = 96;
   NumbersVector[7] = 100;

   startv = NumbersVector.begin();   // location of first element of NumbersVector
   endv = NumbersVector.end();// one past the last element of NumbersVector

   // sort NumbersVector, merge requires the sequences to be sorted
   sort(startv, endv);

   // print content of NumbersVector
   cout << "NumbersVector { ";
   for(itv = startv; itv != endv; itv++)
     cout << *itv << " ";
```


First, the *shuffle.cpp* program creates a vector and fills the vector with a series of numbers. Second, the program calls the *random_shuffle* algorithm and instructs it to shuffle all the numbers in the vector. Finally, the program outputs the numbers again after it completes the *random_shuffle*. When you compile and execute the *shuffle.cpp* program, your screen will display output similar to the following:

```
Before calling random_shuffle
Numbers { 4 10 70 30 10 69 96 100 }

After calling random_shuffle
Numbers { 96 4 69 70 10 10 30 100 }
C:\>
```

USING THE PARTIAL_SORT_COPY ALGORITHM

As you learned in Tip 1218, the Standard Template Library supports many algorithms with different functions. In previous Tips, you have written short programs that use both mutating and non-mutating Standard Template Library algorithms. In Tip 1223, you will use the *merge* algorithm to combine the values in two vectors. As you have learned, the *merge* function is one of the Standard Template Library sorting algorithms. In this Tip, you will use another sorting algorithm, the *partial_sort_copy* algorithm.

The *partial_sort_copy* algorithm sorts the smallest *N* elements, where *N* = *min((last1 - first1), (last2 - first2)),* of the sequence [first1, last1) and copies the results to the sequence *[first2, first2 + N]*. In other words, the *partial_sort_copy* algorithm sorts the container to a temporary container in ascending order. The *partial_sort_copy* algorithm then determines the smallest number within the temporary container (the first number). It then copies that number of objects from the temporary sorted container to a third, permanent container. For example, if an array contains the values 15, 25, 7, 8, 10, 12, and 2, the array will initially appear within memory, as shown in Figure 1222.1.

15	25	7	8	10	12	2

Figure 1222.1 The array before sorting or copying.

The *partial_sort_copy* algorithm will then sort the array and store it within a temporary array, as shown in Figure 1222.2.

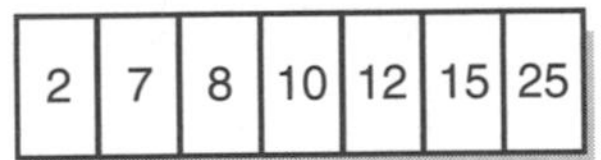

Figure 1222.2 The sorted temporary array.

The *partial_sort_copy* function then determines the smallest number in the sorted result (in this case, 2) and copies the first two values from the temporary array into the permanent array copy, leaving two arrays, as shown in Figure 1222.3.

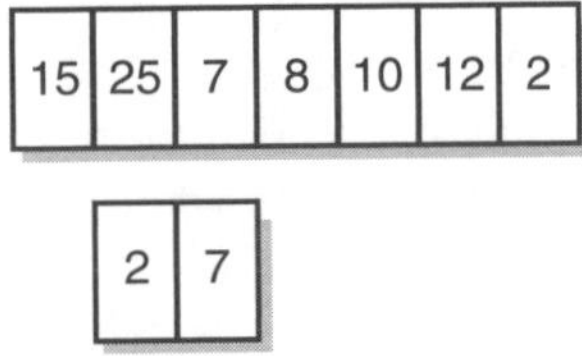

Figure 1222.3 The two arrays that result from the ***partial_sort_copy*** *invocation.*

To better understand the processing that the *partial_sort_copy* algorithm performs, consider the following program, *ps_copy.cpp*, which performs the steps described in the previous figures:

```
#include <iostream>
#include <algorith>
#include <vector>

using namespace std;

void main(void)
 {
   const int VECTOR_SIZE = 8;

   // Define a template class vector of int
   typedef vector<int> IntVector;

   // Define an iterator for template class vector of strings
   typedef IntVector::iterator IntVectorIt;

   IntVector Numbers(VECTOR_SIZE);
   IntVector Result(4);
   IntVectorIt start, end, it;

   // Initialize vector Numbers
   Numbers[0] = 4;
   Numbers[1] = 10;
   Numbers[2] = 70;
   Numbers[3] = 30;
   Numbers[4] = 10;
   Numbers[5] = 69;
   Numbers[6] = 96;
   Numbers[7] = 7;

   start = Numbers.begin();          // location of first element of Numbers
   end = Numbers.end();              // one past the last element of Numbers
   cout << "Before calling partial_sort_copy" << endl;

   cout << "Numbers { ";             // print content of Numbers
   for(it = start; it != end; it++)
      cout << *it << " ";
   cout << " }" << endl;

   // sort the smallest 4 elements in the Numbers vector and copy the results in Result
   partial_sort_copy(start, end, Result.begin(), Result.end());

   cout << "After calling partial_sort_copy\n" << endl;
   cout << "Numbers { ";
   for(it = start; it != end; it++)
      cout << *it << " ";
   cout << " }" << endl;

   cout << "Result { ";
   for(it = Result.begin(); it != Result.end(); it++)
      cout << *it << " ";
   cout << "\b }" << endl;
 }
```

When you compile and execute the *ps_copy.cpp* program, it will create a vector of eight integer elements. It will then assign values to those elements. To show its processing, *ps_copy.cpp* displays the vector elements before it calls *partial_sort_copy*. Within *partial_sort_copy*, the algorithm determines that 4 is the smallest number, and copies the first four values in the sorted array to the output array. When you compile and execute the *ps_copy.cpp* program, your screen will display the following output:

```
Before calling partial_sort_copy
Numbers { 4 10 70 30 10 69 96 7  }

After calling partial_sort_copy
```



```
   cout << "\b }" << endl;

   // Initialize list NumbersList
   for(int i = 0; i < MAX_ELEMENTS; i++)
     NumbersList.push_back(i);
   first = NumbersList.begin();     // location of first element of NumbersList
   last = NumbersList.end();        // one past the last element of NumbersList

   cout << "NumbersList { ";        // print content of NumbersList
   for(itl = first; itl != last; itl++)
     cout << *itl << " ";
   cout << " }" << endl;

   // merge the elements of NumbersVector and NumbersList and place the
   // results in NumbersDeque
   merge(startv, endv, first, last, NumbersDeque.begin());

   // print content of NumbersDeque
   cout << "After calling merge\n" << endl;
   cout << "NumbersDeque { ";
   for(itd = NumbersDeque.begin();itd != NumbersDeque.end(); itd++)
     cout << *itd << " ";
   cout << "\b }" << endl;
 }
```

The *merge_2v.cpp* program first creates and initializes the *NumbersVector* vector. It then calls the *sort* function to place the numbers vector in ascending order. The program then creates the *NumbersList* list, which the code creates fully sorted. After *merge_2v.cpp* creates the *NumbersList* list, it invokes the *merge* function and places the result within the *NumberDeque* container. To show it has completed its processing, *merge_2v.cpp* outputs the numbers the *NumbersDeque* container contains. When you compile and execute the *merge_2v.cpp* program, your screen will display the following output:

```
NumbersVector { 4 10 10 30 69 70 96 100 }
NumbersList { 0 1 2 3 4 5 6 7 }

After calling merge
NumbersDeque { 0 1 2 3 4 4 5 6 7 10 10 30 69 70 96 100 }
C:\>
```

1224 UNDERSTANDING THE INNER_PRODUCT ALGORITHM

In Tip 1223, you used the *merge* algorithm to merge a sorted *vector* and a sorted *list* into a sorted *deque* that contains all the values from both original objects. In fact, in recent Tips you have used Standard Template Library algorithms of three of the four major types: mutating, non-mutating, and sorting. In this Tip, you will use one of the generalized number algorithms, the *inner_product* algorithm. The *inner_product* algorithm multiplies *N* elements within two containers by each other and returns the sum of the multiplications. For example, if you compute the inner product of two arrays beginning at the second element and continuing for three elements, the *inner_product* value will be consistent with that shown in Figure 1224.1.

1	2	3	4	5

5	4	3	2	1

= (2*4)+(3*3)+(2*4)
= 8+9+8
= 25

*Figure 1224.1 How the **inner_product** algorithm computes the "sum of the products" inner product result for two arrays.*

The *inner_product* algorithm also supports a second, overloaded implementation, which is the product of the sums of the inner values. Using the example shown in Figure 1224.1, you will determine the product of the sum's inner product, as shown in Figure 1224.2.

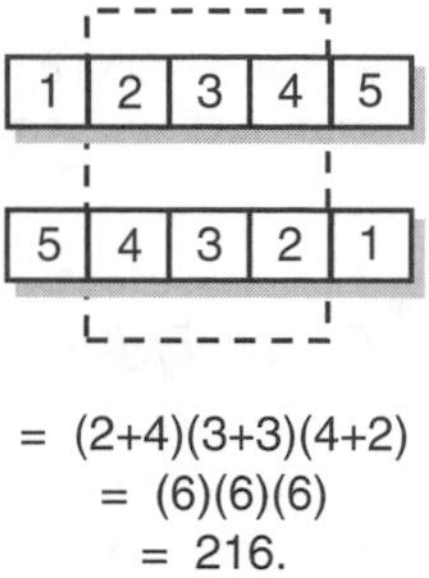

= (2+4)(3+3)(4+2)
= (6)(6)(6)
= 216.

***Figure 1224.2** How the **inner_product** algorithm computes the "product of the sums" inner product result for two arrays.*

To better understand the processing that the *inner_product* algorithm performs, consider the following program, *in_prod.cpp*:

```
#include <iostream>
#include <numeric>
#include <vector>
#include <iterator>

using namespace std;

typedef vector<float> FloatArray;
typedef ostream_iterator<float, char, char_traits<char>>
FloatOstreamIt;

void main(void)
 {
   FloatOstreamIt itOstream(cout," ");

   // Initialize the arrays
   FloatArray rgFl, rgF2;
   for (int i=1; i<=5; i++)
    {
      rgFl.push_back(i);
      rgF2.push_back(i*i);
    };

   // Print the arrays
   cout << "Array 1: ";
   copy(rgFl.begin(),rgFl.end(),itOstream);
   cout << endl;
   cout << "Array 2: ";
   copy(rgF2.begin(),rgF2.end(),itOstream);
   cout << endl;

   // This is the sum of the products (S.O.P.) of the corresponding elements
   float ipl = inner_product(rgFl.begin(),rgFl.end(),rgF2.begin(),0);
   cout << "The inner product (S.O.P) of Arrayl and Array2 is " << ipl << endl;

   // This is the product of the sums (P.O.S.) of the corresponding elements
   float ip2 = inner_product(rgFl.begin(),rgFl.end(),rgF2.begin(),1,
                             multiplies<float>(), plus<float>());
   cout << "The inner product (P.O.S.) of Arrayl and Array2 is " << ip2 << endl;
 }
```

The *in_prod.cpp* program begins by initializing two type *float* vectors and filling the vectors with values—the first, with a straight sequence, and the second with the straight sequence's square. The program outputs the two vectors for your review. Next, the program computes the sum of the product's *inner_product* computation (S.O.P) and completes by computing the product of the sum's *inner_product* computation (P.O.S.). When you compile and execute the *in_prod.cpp* program, your screen will display the following output:

```
Array 1: 1 2 3 4 5
Array 2: 1 4 9 16 25
The inner product (S.O.P) of Array1 and Array2 is 225
The inner product (P.O.S.) of Array1 and Array2 is 86400
C:\>
```

1225 BETTER UNDERSTANDING VECTORS

As you have learned, a *vector* is a Sequence Container that supports random access to elements, constant-time insertion and removal of elements at the end of the container (that is, the amount of time the program uses to insert elements does not vary based on the *vector's* size), and linear-time insertion and removal of elements at the beginning or in the middle of the container (the amount of time the program uses to insert elements varies based on the *vector's* size). The number of elements in a vector may vary dynamically. As you will learn, because the vector can reallocate more memory for itself, its memory management is automatic. The *vector* class is the simplest of the Standard Template Library container classes and, in many cases, the most efficient. As you have learned, a *vector* is similar to an array. The following code fragment, for example, shows how you declare and use a vector:

```
vector<int> V;
V.insert(V.begin(), 3);
```

The first line declares an *int vector V*, which contains no elements. The second line inserts the value 3 at the beginning of the vector. In addition to the *insert* and *begin* member functions, the *vector* class supports many member functions. Table 1225 lists the member functions for the *vector* class and a brief description of each.

Member	Description
value_type	The type of object, *T*, stored in the vector.
pointer	Pointer to *T*.
reference	Reference to *T*.
const_reference	*Const* reference to *T*.
size_type	An unsigned integral type.
difference_type	A signed integral type.
iterator	The definition of the base iterator type your program should use to iterate through a vector.
const_iterator	The definition of the base *Const* iterator type your program should use to iterate through a vector.
reverse_iterator	The definition of the base iterator type your program should use to iterate backward through a vector.
const_reverse_iterator	The definition of the base *Const* iterator type your program should use to iterate backward through a vector.
iterator begin()	Returns an iterator pointing to the beginning of the vector.
iterator end()	Returns an iterator pointing to the end of the vector.
const_iterator begin() const	Returns a *const_iterator* pointing to the beginning of the vector.

***Table 1225** The members of the **vector** class.* *(continued on following page)*

Member	Description
const_iterator end() const	Returns a *const_iterator* pointing to the end of the vector.
reverse_iterator rbegin()	Returns a *reverse_iterator* pointing to the beginning of the reversed vector.
reverse_iterator rend()	Returns a *reverse_iterator* pointing to the end of the reversed vector.
const_reverse_iterator rbegin() const	Returns a *const_reverse_iterator* pointing to the beginning of the reversed vector.
const_reverse_iterator rend() const	Returns a *const_reverse_iterator* pointing to the end of the reversed vector.
size_type size() const	Returns the vector's size in elements.
size_type max_size() const	Returns the largest possible size for the vector in elements.
size_type capacity() const	Number of elements for which the vector has allocated memory. The return value of *capacity* is always greater than or equal to *size*.
bool empty() const	*True* if the vector's size is 0.
reference operator[](size_type n)	Returns the *n*th element in the container.
const_reference operator[](size_type n) const	Returns a constant value representation of the *n*th element in the container.
vector()	Creates an empty vector.
vector(size_type n)	Creates a vector with *n* elements.
vector(size_type n, const T& t)	Creates a vector with *n* copies of the object *t*.
vector(const vector&)	The copy constructor.
vector(InputIterator, InputIterator)	Creates a vector with a copy of a range.
~vector()	The destructor function.
vector& operator=(const vector&)	The assignment operator.
void reserve(size_t n)	If *n* is less than or equal to *capacity*, this call has no effect. Otherwise, it is a request for allocation of additional memory. If the request is successful, then the function sets *capacity* as greater than or equal to *n*; otherwise, *capacity* is unchanged. In either case, the vector's *size* property remains unchanged.
reference front()	Returns the first element in the vector.
const_reference front() const	Returns a constant value representation of the first element in the vector.
reference back()	Returns the last element in the vector.
const_reference back() const	Returns a constant value representation of the last element in the vector.
void push_back(const T&)	Inserts a new element at the end of the vector.
void pop_back()	Removes the last element from the vector.
void swap(vector&)	Swaps the contents of two vectors, provided the vectors are of compatible types. In other words, if a vector Z and Y both have integer values, you can swap the contents of the vectors with the *swap* statement.

***Table 1225** The members of the **vector** class.* *(continued on following page)*

iterator insert(iterator pos, const T& x)	Inserts *x* before *pos* in the vector.
void insert(iterator pos, InputIterator f, InputIterator l)	Inserts the range [first, last) before *pos* in the vector.
void insert(iterator pos, size_type n, const T& x)	Inserts *n* copies of *x* before *pos* into the vector.
iterator erase(iterator pos)	Erases the element at position *pos* from the vector.
iterator erase(iterator first, iterator last)	Erases the range [first, last) from the vector.
bool operator==(const vector&,const vector&)	Tests two vectors for equality. This is a global function, not a member function.
bool operator<(const vector&, const vector&)	Lexicographical comparison. This is a global function, not a member function.

Table 1225 The members of the ***vector*** *class. (continued from previous page)*

As you learned in previous Tips, the Standard Template Library has too many algorithms to discuss them all in the space this book has available. Likewise, the *vector* class has too many members to discuss them all. The subsequent Tips will address some of the most commonly used members of the *vector* class, as well as the other Standard Template Library classes. As you work with other classes in the remaining Tips, you will find that those classes, too, have more members than this book can adequately deal with in the space available.

1226 Using Another Simple vector Program

In Tip 1225, you examined the member functions the *vector* class supports. In Tip 1207, you created a simple program, *vector1.cpp*, which used several *vector* class member functions to manipulate information within a *vector*. Before you learn about and use the *bit_vector* class, it is valuable for you to write another program that uses the *vector* class member functions to manipulate a *vector*. The following program, *vector2.cpp*, uses the *reserve, max_size, resize,* and *capacity* member functions with a *vector* containing only a single integer:

```
#include <iostream>
#include <vector>

using namespace std;
typedef vector<int> INTVECTOR;

void main(void)
 {
   // Dynamically allocated vector begins with 0 elements.
   INTVECTOR theVector;

   // Add one element to the end of the vector, an int with the value 42.
   theVector.push_back(42);

   // Show statistics about vector.
   cout << "theVector's size is: " << theVector.size() << endl;
   cout << "theVector's maximum size is: " << theVector.max_size()<< endl;
   cout << "theVector's capacity is: " << theVector.capacity() << endl;

   // Ensure there's room for at least 1000 elements.
   theVector.reserve(1000);
   cout << endl << "After reserving storage for 1000 elements:" << endl;
   cout << "theVector's size is: " << theVector.size() << endl;
   cout << "theVector's maximum size is: " << theVector.max_size()<< endl;
   cout << "theVector's capacity is: " << theVector.capacity() << endl;

   // Ensure there's room for at least 2000 elements.
   theVector.resize(2000);
   cout << endl << "After resizing storage to 2000 elements:" << endl;
   cout << "theVector's size is: " << theVector.size() << endl;
```

```
    cout << "theVector's maximum size is: " << theVector.max_size()<< endl;
     cout << "theVector's capacity is: " << theVector.capacity() << endl;
  }
```

The *vector2.cpp* program first initializes a *vector*. Second, the program adds a single integer element to the *vector* and shows information about the *vector's* current size, maximum size, and capacity. Third, the program reserves storage for 1,000 elements and displays the same information about the *vector*. Finally, the program resizes the *vector* to 2,000 elements, then displays the information about the *vector*. The program uses the *max_size* member function, which returns the maximum number of elements the *vector* can hold. The program also uses the *capacity* member function to return the number of elements for which the *vector* has actually allocated memory. When you compile and execute the *vector2.cpp* program, your screen will display the following output:

```
theVector's size is: 1
theVector's maximum size is: 1073741823
theVector's capacity is: 1

After reserving storage for 1000 elements:
theVector's size is: 1
theVector's maximum size is: 1073741823
theVector's capacity is: 1000

After resizing storage to 2000 elements:
theVector's size is: 2000
theVector's maximum size is: 1073741823
theVector's capacity is: 2000
C:\>
```

COMPARING VECTORS TO C ARRAYS 1227

As you have seen, a *vector* is similar to a C array. There are, however, several significant differences between C arrays and *vectors*. As a rule, *vectors* are more powerful tools for the following reasons:

- Your programs can dynamically reallocate a *vector's* size at any point during the program's execution, as shown in Tip 1226.
- Because *vectors* use iterators, traversing the *vector* and accessing the elements the *vector* contains is a more powerful process than traversing the array and accessing its elements. The *front*, *back*, *begin*, and *end* member functions, for example, provide you with useful ways of accessing specific items within the *vector* without knowing the exact location of those items within the *vector*.
- While *vectors* allocate capacity, C arrays allocate actual memory. A C array of 1,000 integers allocates 2,000 bytes of memory at its declaration; a *vector* that reserves 2,000 bytes of memory, but stores only a single integer, uses only two bytes of memory, as you saw in Tip 1226 (excluding the object's overhead—the object itself consumes memory).
- If you inadvertently enter too many items into a *vector*, the class will automatically reallocate the *vector's* size to accommodate the additional elements.
- Your programs can use algorithms to manipulate the values within both *vectors* and arrays. However, your programs may be able to more easily transfer values between *vectors* and other Standard Template Library types than they will between arrays and Standard Template Library types.

In short, *vectors* and arrays are similar. As your programs continue to become more complex, the likelihood will increase that you will use *vectors* more often than arrays. However, if you find that you are more comfortable with C arrays than you are with *vectors* and you do not need the *vector's* additional flexibility, do not feel you must use *vectors* rather than arrays.

1228 UNDERSTANDING THE BIT_VECTOR SEQUENCE CONTAINER

In Tips 1225 through 1227, you worked more closely with the *vector* type. The current version of the Standard Template Library also supports the *bit_vector* type. A *bit_vector* is essentially a *vector* of type *bool*. The *bit_vector* Sequence Container has the same interface as a *vector*.

The main difference between a *vector* and a *bit_vector* is that the Standard Template Library's designers optimized the *bit_vector* class for space efficiency. A *vector* always requires at least one byte per element, but a *bit_vector* only requires one bit per element.

It is important to recognize that you must use an iterator of type *bit_vector::iterator* with a *bit_vector*. The reason for the change in iterator types is simple—the *bit_vector* uses only one bit for each item, while a *vector* uses only one byte at a minimum. If you tried to use a *vector* iterator with a *bit_vector*, you would return the next eight items in the *bit_vector* each time you incremented the iterator.

You will implement the *bit_vector* type within your programs as shown here:

```
vector<bool> V(5);

V[0] = true;
V[1] = false;
V[2] = false;
V[3] = true;
V[4] = false;
for (bit_vector::iterator i = V.begin(); i < V.end(); ++i)
  cout << (*i ? '1' : '0');

cout << endl;
```

1229 USING A SIMPLE BVECTOR EXAMPLE

As you learned in Tip 1228, a *bit_vector* is a *vector* you will use specifically to store only a single bit for each element within the *vector*. Because the *bit_vector* uses only a single bit, it is capable of storing only true and false values. You will also use only *bit_vector::iterator* iterators with *bit_vectors*. The following program, *bvectr1.cpp*, changes the *vector1.cpp* program you wrote in Tip 1207 to use a *bit_vector* rather than an integer *vector*:

```
#include <iostream.h>
#include <bvector.h>
using namespace std;
const ARRAY_SIZE = 4;
void main(void)
 {
    // Dynamically allocated vector begins with 0 elements.
    bit_vector theVector(ARRAY_SIZE);
   // Initialize the array to contain the members [100, 200, 300, 400]
    for (int cEachItem = 0; cEachItem < ARRAY_SIZE; cEachItem++)
      if(cEachItem>1)
           theVector.push_back(cEachItem - 2);
     else
           theVector.push_back(cEachItem);
   cout << "First element: " << theVector.front() << endl;
   cout << "Last element: " << theVector.back() << endl;
   cout << "Elements in vector: " << theVector.size() << endl;
```

```
    // Delete the last element of the vector. Remember that the vector
    // is 0-based, so theVector.end() actually points 1 element beyond the end.
    cout << "Deleting last element." << endl;
   theVector.erase(theVector.end() - 1);
   cout << "New last element is: " << theVector.back() << endl;

   // Delete the first element of the vector.
   cout << "Deleting first element." << endl;
   theVector.erase(theVector.begin());
   cout << "New first element is: " << theVector.front() << endl;
   cout << "Elements in vector: " << theVector.size() << endl;
}
```

When you compile and execute the *bvectr1.cpp* program, your screen will display the following output:

```
First element: 0
Last element: 1
Elements in vector: 4

Deleting last element.
New last element is: 0
Deleting first element.
New first element is: 1
Elements in vector: 2
C:\>
```

Note: *Neither* ***Visual*** *C++ nor Borland's* ***C++*** *5.02 include the* ***bit_vector*** *header file* ***bvector.h*** *within their default Standard Template Library implementation. If you want to use the* ***bit_vector*** *type within your programs, you must extract the header file from the Standard Template Library* ***include*** *files and place it within the appropriate* ***include*** *directory of your compiler.*

UNDERSTANDING THE LIST TYPE 1230

As you have learned, the Standard Template Library supports a variety of containers. One of the containers that you will use most often, and that should seem most familiar to you, is the *list* container. A *list* is a doubly linked list (similar to the *linked_list* object you created in earlier Tips). That is, a *list* is a Sequence Container that supports forward and backward traversal and linear-time insertion and removal of elements, at the beginning, at the end, or in the middle. List modification has the following three important properties:

- Insertion does not invalidate iterators to list elements.
- Splicing (inserting an element into the center of the list) does not invalidate iterators to list elements.
- Removing list elements only invalidates the iterators that point to the removed elements.

Your programs may change the ordering of iterators (that is, the *list<T>::iterator* might have a different predecessor or successor after a list operation than it did before), but the class will not invalidate the iterators themselves or make them point to different elements unless the invalidation or mutation is explicit.

Note that singly linked lists, which support only forward traversal, are also sometimes useful. If you do not need backward traversal, the *slist* type may be more efficient than the *list* type. You will learn more about the *slist* type in later Tips.

1231 UNDERSTANDING THE LIST CONTAINER'S GENERIC COMPONENTS

As you learned in Tip 1230, the Standard Template Library supports the *list* container type. You will work with the Standard Template Library *list* type much as you worked with the *linked_list* object you created in previous Tips. Although the *list* class entirely encapsulates the equivalent to the *link_object* class, you can only access members of the *list* class, not the class from which it derives.

The *list* class can be of any type. Your programs can create lists that support simple types and lists that support complex types. For example, the following definition creates a simple integer list, adds several values to the list, and outputs the list's values:

```
list<int> L;

L.push_back(0);
L.push_front(1);
L.insert(++L.begin(), 2);
copy(L.begin(), L.end(), ostream_iterator<int>(cout, " ")); // Outputs 1 2 0
```

The code creates a *list L*, adds three values to the list, and outputs the three values. Note that the last statement uses the Standard Template Library's *copy* algorithm to copy each element in the *list* to *cout* output stream and even inserts a space after each element it copies to the stream. The *list* type supports a single generic type, *T*. You cannot create lists that support multiple types within a single type unless you first encapsulate those types within a class or a structure.

1232 CONSTRUCTING A LIST OBJECT

As you learned in Tip 1230, the Standard Template Library supports the *list* doubly linked list type. Because the *list* type is a class, you will initialize lists using one of several constructors that the Standard Template Library defines for the *list* class. The Standard Template Library defines four different constructors for the *list* type, as shown here:

```
explicit list(void);
explicit list(size_type n, const T& value);
list(const list& FirstList);
list(const_iterator first, const_iterator last);
```

The first constructor specifies an empty initial list. The second constructor specifies a repetition of *n* elements of value *value* (you can assign a default). The third constructor (the copy constructor) instructs the compiler to initialize the new list with a copy of another list *FirstList*. The last constructor specifies the sequence [first, last), which are two iterators from another *list* object. The constructor copies all elements between *last* and *first* from the original *list* object into the newly constructed list object. All constructors store the allocator object *al* or, for the copy constructor, the return value of *FirstList.get_allocator*, in the data member *allocator*. After storing the allocator object, all four constructors initialize the list.

To better understand the different constructors that the Standard Template Library *list* class supports, consider the following program, *list1.cpp*:

```
#include <list>
#include <string>
#include <iostream>

using namespace std;
typedef list<string> LISTSTR;

// Try each of the four constructors
void main(void)
 {
   LISTSTR::iterator i;
   LISTSTR test;                                  // default constructor
```

```
    test.insert(test.end(), "one");
    test.insert(test.end(), "two");

    LISTSTR test2(test);               // construct from another list
    LISTSTR test3(3, "three");                // construct with three elements
                                              // containing the value "three"
    LISTSTR test4(++test3.begin(),test3.end()//; create from part of test3

 // Print them all out
    for (i =  test.begin(); i != test.end(); ++i)
      cout << *i << " ";
    cout << endl;

    for (i =  test2.begin(); i != test2.end(); ++i)
      cout << *i << " ";
    cout << endl;

    for (i =  test3.begin(); i != test3.end(); ++i)
      cout << *i << " ";
    cout << endl;

    for (i =  test4.begin(); i != test4.end(); ++i)
      cout << *i << " ";
    cout << endl;
  }
```

When you compile and execute the *list1.cpp* program, your screen will display the following output:

```
one two
one two
three three three
three three
C:\>
```

Inserting Objects into the List 1233

As you saw in Tip 1232, you can construct the *list* in several ways—one with no initialization and two with initialize. Furthermore, you can assign values to elements of a list in several important ways. As you will learn in Tip 1236, you can use the *push_back* and *push_front* member functions to add values to the *list* object. However, you can also use the *insert* member function to add values to the *list* object. You will use the *insert* member function within your programs as shown in the following prototypes:

```
iterator insert(iterator it, const T& x);
void insert(iterator it, size_type n, const T& x);
void insert(iterator it, const_iterator first, const_iterator last);
```

As you can see, the *list* object provides three overloaded versions of the *insert* member functions. Each member function inserts a sequence the remaining operands specify before the element the *it* iterator points to in the controlled sequence. The first member function inserts a single element with value *x* and returns an iterator that points to the newly inserted element. The second member function inserts a repetition of *n* elements of value *x* (in other words, five elements of value 2, for example). The last member function inserts the sequence [first, last) beginning at the *it* iterator's location.

The CD-ROM that accompanies this book includes the *insert_3.cpp* program, which uses the three *list insert* methods to insert elements into a list.

1234 USING THE ASSIGN MEMBER FUNCTION

In Tip 1233, you learned how your programs can use the *insert* member function to insert elements into a list. Your programs can also use the *assign* member function to assign values to existing elements within a list. Rather than inserting new elements and deleting old elements, you can use *assign* to replace a series of elements within a list with a single statement. You will use the *assign* member function within your programs in one of its two overloaded forms, as shown in the following prototypes:

```
void assign(const_iterator first, const_iterator last);
void assign(size_type n, const T& x = T());
```

The first member function replaces the sequence **this* points to with the sequence [first, last). The second member function replaces the sequence **this* points to with a repetition of *n* elements of value *x*. For example, the following statement replaces the next four items within a list with the value 1:

```
listAnother.assign(4, 1);
```

Depending on where the iterator currently points within the list, the assignment could start at the list's beginning or in the list's middle. If you try to assign a range of values to a series of list elements that exceeds the list's end, the assignment will add additional elements to the list's end.

1235 USING THE REMOVE AND EMPTY MEMBER FUNCTIONS

In Tip 1234 you learned how your programs can use the *assign* member function to replace a series of elements within a list. When you work with lists, you must often remove elements from the list in addition to reassigning values to elements within the list. The *list* class supports two overloaded *erase* member functions that you can use within your programs to remove elements from within a list. You will use the two overloaded *erase* member functions as shown in the following prototypes:

```
iterator erase(iterator it);
iterator erase(iterator first, iterator last);
```

The first member function removes the element of the controlled sequence to which the *it* iterator points. The second member function removes the elements of the controlled sequence in the range [first, last). Both *erase* member functions will return an iterator that designates the first element remaining beyond any elements the functions have removed, or *end* if no such element exists.

Occasionally, you may remove all the elements from within a list. To protect your programs against trying to output or manipulate an empty list, your programs should check after removals to determine whether or not any elements remain within the list. To do so, your programs should use the *empty* member function, whose prototype is shown here:

```
bool empty(void) const;
```

The *empty* member function will return *True* for an empty controlled sequence. To better understand how your programs can use the *erase* and *empty* member functions, as well as the *assign* member function that Tip 1234 discusses, consider the *list_are.cpp* program on the CD-ROM that accompanies this book, which uses all three functions. The program creates two lists, *listOne* and *listAnother*. The program then assigns three values to the *listOne* container, and assigns a single value to the *listAnother* container. Next, the program uses the *assign* method to copy the three elements in *listOne* into *listAnother* and overwrite the element the previous statement assigned. After it displays the *listAnother* container, the program assigns the value one to all the list elements, which replaces the previous element values. The program then displays the *listAnother* container again. The program next erases the first element in the *listAnother* list, and outputs the list again. Finally, the program erases all the elements in the *listAnother* list and generates the "All gone!" message.

TRAVERSING THE LIST OBJECT 1236

In Tip 1235 the *list_are.cpp* program uses the *begin* and *end* member functions with a *list* iterator to traverse the list. Your programs can also use the *front* and *back* member functions to control an iterator's position within a list. The *front* member function returns a reference to the first element of the controlled sequence. The *back* member function returns a reference to the last element of the controlled sequence.

When your programs use the *front* and *back* functions, you may also want to use the *push* and *pop* functions for consistency. Remember that the *push* and *pop* functions place values within the list at the front and back. The *pop_back* member function removes the last element of the controlled sequence. The *pop_front* member function removes the first element of the controlled sequence. All these functions require that the controlled sequence not be empty. The *push_front* member function inserts an element with the value *x* at the beginning of the controlled sequence. The *push_back* member function inserts an element with the value *x* at the end of the controlled sequence. To better understand the processing your programs can perform using *front, back*, and their associated assignment functions, consider the *frntback.cpp* program, shown here:

```
#include <list.h>
#include <string.h>
#include <iostream.h>

using namespace std;
typedef list<string> LISTSTR;
void main(void)
 {
   LISTSTR test;

   test.push_back("back");
   test.push_front("middle");
   test.push_front("front");
   cout << test.front() << endl;     // front
   cout << test.back() << endl;      // back
   test.pop_front();
   test.pop_back();
   cout << test.front() << endl;     // middle
 }
```

When you compile and execute the *frntback.cpp* program, your screen will display the following output:

```
front
back
middle
C:\>
```

UNDERSTANDING THE SLIST TYPE 1237

As you have learned, the Standard Template Library supports the *list* type, a doubly linked list. The Standard Template Library also supports the *slist* type, which is a singly linked list that links each element to the next element in the list, but not to the previous element in the list. That is, an *slist* is a Sequence Container that supports forward, but not backward traversal, and constant-time insertion and removal of elements. Just as with *list* modification, *slist* modification has the following three important properties:

- Insertion does not invalidate iterators to list elements.
- Splicing (insertion within the list, rather than at the end) does not invalidate iterators to list elements.
- Removing list elements invalidates only the iterators that point to the removed elements.

Your programs may change the ordering of iterators (that is, *slist<T>::iterator* might have a different predecessor or successor after a list operation than it did before), but the change will not invalidate the iterators themselves or make them point to different elements unless the invalidation or mutation is explicit.

The main difference between *slist* and *list* is that *list* iterators are *bi-directional* iterators, while *slist* iterators are *forward* iterators. The difference in iterator types means that *slist* is less versatile than *list*. Frequently, however, *bi-directional* iterators are unnecessary. You should use *slist* unless you need the extra functionality of *list*, because singly linked lists are smaller and faster than doubly linked lists.

Note: *Like **bit_vector**, neither **Visual C++** nor Borland's **C++** for Windows include the header file for the **slist** type.*

1238 UNDERSTANDING INSERTIONS INTO AN SLIST SEQUENCE CONTAINER

In Tip 1237 you learned about the *slist* sequence. Like every other Sequence Container, *slist* defines the *insert* and *erase* member functions. Using these member functions carelessly, however, can result in disastrously slow programs. The problem is that *insert's* first argument is a *pos* iterator, and *insert* places the new element(s) before *pos*. In other words, the *insert* function must find the iterator just before the *pos* iterator. Finding the iterator just before *pos* is a constant-time operation for *list* because *list* has *bi-directional* iterators. For *slist*, however, *insert* must traverse the list from the beginning up to *pos* to find that iterator. In other words, *insert* and *erase* are slow operations anywhere but near the beginning of the *slist*.

The *slist* class provides the member functions *insert_after* and *erase_after*, which are constant-time operations; you should use *insert_after* and *erase_after* whenever possible. If you find that *insert_after* and *erase_after* are not adequate for your needs, and that you often must use *insert* and *erase* in the middle of the list, you should probably use a *list* instead of an *slist*.

1239 UNDERSTANDING THE DEQUE CONTAINER

A *deque* is very much like a *vector* because a *deque* is a Sequence Container that lets your programs randomly access elements and perform constant-time insertions, as well as remove elements at the end of the sequence. A *deque* also lets your programs perform linear-time insertion and removal of elements in the middle of the sequence.

The main way in which a *deque* differs from a *vector* is that a *deque* supports constant-time insertion and removal of elements at the beginning of the sequence (in addition to *vector's* support for such activities at the end of the sequence). Additionally, a *deque* does not have any member functions similar to a *vector's capacity* and *reserve* functions and does not provide any of the guarantees of iterator validity that the Standard Template Library associates with *capacity* and *reserve*. You will declare and use *deque* objects within your programs as shown here:

```
deque<int> Q;
Q.push_back(3);                                  // place 3 at end
Q.push_front(1);                                 // place 1 at front
Q.insert(Q.begin() + 1, 2);                      // insert two in the middle
Q[2] = 0;                                        // set the 3rd element to 0
copy(Q.begin(), Q.end(), ostream_iterator<int>(cout, " "));
// The values that are printed are 1 2 0
```

The *push_back, push_front,* and *insert* member functions perform the same tasks with a *deque* that they perform with a *vector*. Additionally, the *copy* algorithm lets your programs copy output directly to the output stream, just as you did previously with the *list* object.

USING THE DEQUE CONTAINER 1240

In Tip 1239 you learned that the *deque* container is similar to the *vector* container and the *list* container. In fact, *deque* uses elements of both container types. As you might expect, the *deque* container includes support for some *vector-style* member functions and some *list-style* member functions. Two of the member functions that the *deque* container supports are the *swap* member function (from the *vector-style* container) and the *assign* member function (from the *list-style* container). You will use the *swap* member function and the *assign* member function within your programs as shown in the following prototypes:

```
void assign(const_iterator first, const_iterator last);
void assign(size_type n, const T& x = T());
void swap(deque& dq);
```

The first *assign* member function replaces the sequence **this* points to with the sequence [first, last). The second *assign* member function replaces the sequence **this* points to with a repetition of *n* elements of value *x*. The *swap* member function swaps the contents between **this* and *dq*. To better understand the processing of these *deque* member functions, consider the following program, *deque1.cpp*:

```
#include <iostream.h>
#include <deque.h>

using namespace std;
typedef deque<char> CHARDEQUE;

void print_contents(CHARDEQUE deque, char*);

void main(void)
 {
   CHARDEQUE  a(3, 'A');       //create a  with  3 As
   CHARDEQUE  b(4, 'B');       //create b with 4 Bs.

   print_contents(a, "a");            //print out the contents
   print_contents(b, "b");
   a.swap(b);                         //swap a and b
   print_contents(a, "a");
   print_contents(b, "b");
   a.swap(b);                         // swap it back
   print_contents(a, "a");
   print_contents(b, "b");
    a.assign(b.begin(),b.end());      //assign  the  contents  of  b  to  a
    print_contents(a,  "a");
   a.assign(b.begin(),b.begin()+2);   //assign the first two items of b to a
   print_contents(a, "a");
   a.assign(3, 'Z');                  //assign 3 'Z's to a
   print_contents(a, "a");
 }
//function to print the contents of deque
void print_contents(CHARDEQUE deque, char *name)
 {
   CHARDEQUE::iterator pdeque;

   cout << "The contents of " << name << " : ";
   for(pdeque = deque.begin(); pdeque != deque.end(); pdeque++)
     cout << *pdeque << " ";
   cout<< endl;
 }
```

When you compile and execute the *deque1.cpp* program, your screen will display the following output:

```
The contents of a : A A A
The contents of b : B B B B
The contents of a : B B B B
```

```
The contents of b : A A A
The contents of a : A A A
The contents of b : B B B B
The contents of a : B B B B
The contents of a : B B
The contents of a : Z Z Z
C:\>
```

1241 USING THE ERASE AND CLEAR MEMBER FUNCTIONS

In Tip 1239 you learned about the *deque* container and its similarity to the *vector* container. In Tip 1240, you learned how to use the *swap* and *assign* member functions with the *deque* container. As you know, one of the most common activities you will perform with an array, *vector*, or other list of elements is to remove elements or remove the value from the elements within the list. The *deque* container provides the *erase* and *clear* member functions to help you manage individual elements and the entire container. The *erase* member functions erase a single element or a range of elements. The *clear* member function erases all the elements within the *deque*. You will use the member functions within your programs as shown in the following prototypes:

```
iterator erase(iterator iter);
iterator erase(iterator first, iterator last);
void clear(void) const;
```

The first *erase* member function removes the element of the container to which *iter* points. The second *erase* member function removes the elements of the container in the range [first, last). Both return an iterator that designates the first element remaining beyond any elements the member functions removed, or *end()* if no such element exists. Removing *N* elements causes *N* destructor calls and an assignment for each element between the insertion point and the nearer end of the sequence. Removing an element at either end invalidates only iterators and references that designate the erased elements. Otherwise, erasing an element invalidates all iterators and references. The *clear* member function calls *erase(begin(), end())*. To better understand the processing these member functions perform, consider the *erase_de.cpp* program on the CD-ROM that accompanies this book, which uses all three.

1242 USING THE [ ] ARRAY OPERATOR WITH A DEQUE

As you have learned, a *deque* is similar to a *vector* and, therefore, similar to an array. Because the *deque* is similar to an array, you might expect that your programs can use the [ ] array operator to access specific elements within a *deque*. However, the programs you have written so far have used iterators with *deque* objects, rather than the *array* operator and an index. As it turns out, you can use an iterator, the *array* operator, or both, to access a *deque* within a given program. Your programs can also use the *at* member function to change from array indexes to iterators.

The member function *operator[ ]* returns a reference to the element of the sequence at position *pos*. If that position is invalid, the function's behavior is unpredictable. The *at* member function returns a reference to the element of the controlled sequence at position *pos*. If that position is invalid, the function will throw an object of class *out_of_range* exception. The *empty* member function returns *True* for an empty controlled sequence. The following program, *deq_arr.cpp*, uses an iterator, the *array* operator, and the *at* member function to access a *deque*:

```
#include <iostream.h>
#include <deque.h>

using namespace std;
typedef deque<char> CHARDEQUE;
void print_contents(CHARDEQUE deque, char*);

void main(void)
 {
```

```
   CHARDEQUE  a;                              // create an empty deque a
    if(a.empty())                             //check whether it is empty
      cout << "a is empty" << endl;
   else
     cout << "a is not empty" << endl;
    a.push_back('A');                         //inset A, B, C, and D  to a
    a.push_back('B');
    a.push_back('C');
    a.push_back('D');
    if(a.empty())                             //check again whether a is empty
        cout << "a is empty" << endl;
   else
        cout << "a is not empty" << endl;
    print_contents(a,"a");                    //print out the contents
   cout << "The first element of a is  " << a[0] << endl;
   cout << "The first element of a is  " << a.at(0) << endl;
   cout << "The last element of a is  " << a[a.size()-1] << endl;
   cout << "The last element of a is  " << a.at(a.size()-1) << endl;
 }
//function to print the contents of deque
void print_contents(CHARDEQUE deque, char *name)
 {
    CHARDEQUE::iterator pdeque;
   cout << "The contents of "<< name << " : ";
    for(pdeque = deque.begin(); pdeque != deque.end(); pdeque++)
     cout << *pdeque <<" ";
   cout << endl;
 }
```

When you compile and execute the *deq_arr.cpp* program, your screen will display the following output:

```
a is empty
a is not empty
The contents of a : A B C D
The first element of a is  A
The first element of a is  A
The last element of a is  D
The last element of a is  D
C:\>
```

USING REVERSE ITERATORS WITH A DEQUE 1243

In previous Tips, you have used both iterators and the *array* operator to manipulate and return the values within a *deque*. However, *deques* also support *reverse* iterators. As you have learned, a *reverse* iterator is an iterator that points backward from the *deque's* end to its front. In other words, when you increment a *reverse* iterator, it moves downward through the *deque*, not upward as a *forward* iterator does. Figure 1243 shows how a *reverse* iterator's movement is different from that of a *forward* iterator.

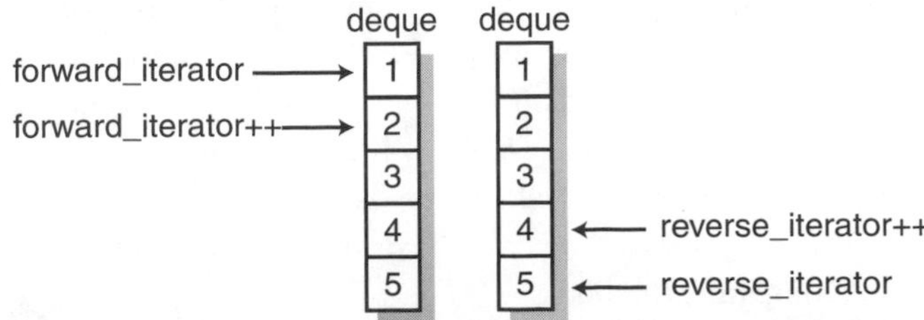

*Figure 1243 The reverse iterator moves downward rather than upward through the **deque**.*

Within your programs, you can use the *rbegin* and *rend* member functions to obtain a *reverse* iterator for a *deque*. The *rbegin* member function returns a *reverse* iterator that points just beyond the end of the controlled sequence. Therefore, *rbegin* designates the beginning of the reverse sequence. The *rend* member function returns a *reverse* iterator that points at the first element of the sequence, or just beyond the end of an empty sequence. Therefore, it designates the end of the reverse sequence. The following program, *rev_iter.cpp*, uses *rbegin* and *rend* to navigate a *deque* in reverse order:

```
#include  <iostream.h>
#include <deque.h>

using namespace std;
typedef deque<int> INTDEQUE;

void main(void)
 {
   // Create A and fill it with elements 1,2,3,4, and 5 using push_back function

   INTDEQUE A;
   A.push_back(1);
   A.push_back(2);
   A.push_back(3);
   A.push_back(4);
   A.push_back(5);

   // Now print the contents in reverse order using reverse_iterator
   // and functions rbegin() and rend()

   INTDEQUE::reverse_iterator rpi;
   for(rpi = A.rbegin(); rpi != A.rend(); rpi++)
      cout << *rpi << " ";
   cout<< endl;
 }
```

The program creates the blank *deque A*, then assigns five values to the *deque*. Next, the program defines a reverse iterator. Finally, the program uses the reverse iterator to output the *deque's* contents in reverse order. When you compile and execute the *rev_iter.cpp* program, your screen will display the following output:

```
5 4 3 2 1
C:\>
```

1244 MANAGING THE DEQUE'S SIZE

In recent Tips, you have used member functions to create and manipulate *deques*. As you learned, the *deque* class has characteristics similar to both the *vector* and *list* classes. One of the most powerful features of *vectors*, which you can also use with *deque* objects, is the management of the object's size. The *deque* type provides three member functions that you can use to manage a *deque's* size, as shown here:

```
size_type size(void) const;
void resize(size_type NewSize, T x = T());
size_type max_size(void) const;
```

The *size* member function returns the length (that is, the current number of elements) of the sequence. The *resize* member function changes the *size* to the number of elements *NewSize* specifies. If *resize* must make the controlled sequence longer, *resize* will append elements with value *x*. If *resize* supplies no value, the default value will depend on the *deque* object's type. For example, if the *deque* is a *deque* of *chars*, the default will be a blank. If the *deque* is a *deque* of integers, the default will be zero. The *max_size* member function returns the length of the longest sequence that the object can control. To better understand the processing the member functions *size, resize,* and *max_size* perform with *deque* objects, consider *deq_size.cpp* program, contained on the CD-ROM that accompanies this book.

The program creates a *deque* object to hold *char* elements. The program then assigns four characters to the *deque* object. After it assigns the four characters to the *deque*, the program prints the *deque*, its maximum size, and its current size. The program then resizes the *deque* to ten characters and fills the fifth through tenth elements with *X*. The program outputs the *deque* again, as well as its current size. The program then resizes the *deque* to five elements and prints those five elements. Finally, the program prints the *deque* object's current size and maximum size to show that the maximum size remains unchanged.

UNDERSTANDING THE MAP OBJECT 1245

A *map* is a Sorted Associative Container that associates objects of type *key* with objects of type *data*. Unlike the Sequence Containers you have worked with previously, a *map* is a Pair Associative Container, meaning that its value type is *pair<const Key, Data>*. In addition to being a Pair Associative Container, a *map* is a Unique Associative Container, meaning that no two elements have the same key.

A *map* lets you insert a new element into a *map* without invalidating iterators that point to existing elements. If you erase an element from a set you do not also invalidate any iterators, except, of course, those iterators that actually point to the element that you are erasing.

Maps differ from the containers you have previously worked with in two ways: one, the *map* sorts elements automatically, and two, the *map* uses a key in addition to the element itself. For example, *maps* are useful for maintaining a sorted numeric representation of a series of strings, such as an employee list organized by social security number. You will work with *maps* in the next several Tips.

A SIMPLE MAP EXAMPLE 1246

As you learned in Tip 1245, your programs can use *maps* to maintain sets of pair objects. To better understand how your programs can use *maps* to manage information, consider the following program, *map1.cpp*:

```
#include <iostream.h>
#include <string.h>
#include <map.h>

using namespace std;
class ltstr
 {
  public:
    bool operator()(const char* s1, const char* s2) const
      { return (strcmp(s1, s2) < 0);}
 };

void main(void)
{
  map<const char*, int, ltstr> months;

  months["January"] = 31;
  months["February"] = 28;
  months["March"] = 31;
  months["April"] = 30;
  months["May"] = 31;
  months["June"] = 30;
  months["July"] = 31;
  months["August"] = 31;
  months["September"] = 30;
  months["October"] = 31;
  months["November"] = 30;
  months["December"] = 31;
```

```
   cout << "june -> " << months["June"] << endl;
   map<const char*, int, ltstr>::iterator cur  = months.find("June");
   map<const char*, int, ltstr>::iterator prev = cur;
   map<const char*, int, ltstr>::iterator next = cur;
   ++next;
   --prev;
   cout << "Previous (in alphabetical order) is " << (*prev).first << endl;
   cout << "Next (in alphabetical order) is " << (*next).first << endl;
}
```

The *map1.cpp* program creates a simple *map* of months. It then displays the numeric value of June within the sorted map (6). Next, the program searches the *map* for the month of June and displays the months whose names immediately precede and follow June (that is, in alphabetical order: July and March) on the screen. When you compile and execute the *map1.cpp* program, your screen will display the following output:

```
June -> 6
Previous (in alphabetical order) is July
Next (in alphabetical order) is March
C:\>
```

1247 USING MEMBER FUNCTIONS TO MANAGE THE MAP

As you learned in Tip 1245, your programs can use *map* containers to maintain sorted lists of objects. In Tip 1246, you used the *find* member function to locate a value within the map. You then assigned two iterators to the *map* and used the iterators to output the mapped values, rather than the key values, within the *map*. In addition to the *find* member function, your programs can use the *end* and *insert* member functions to maintain information within your *maps*. You will use all three functions as shown in the following prototypes:

```
// Key is the data type of template argument #1 for map
iterator map::find(const Key& key);

iterator map::end(void);
pair<iterator, bool> map::insert(const value_type& x);
```

The *end* function returns an iterator that points to one past the end of a sequence. *Find* returns an iterator that designates the first element whose sort key equals *key*. If no such element exists, the iterator equals *end()*. If the key does not already exist, *insert* will add it to the sequence and return *pair<iterator, true>*. If the key already exists, *insert* does not add the key to the sequence and instead returns *pair<iterator, false>*. The following program, *map_ints.cpp*, creates a *map* of integers to strings. In this case, the mapping is from digits to their string equivalents (1 becomes "One", 2 becomes "Two", and so on). The program reads a number from the user, finds the word equivalent for each digit (using the map), and prints the number back as a series of words. For example, if the user enters 25463, the program will respond with: Two Five Four Six Three. The *map_ints.cpp* program, as shown here, uses the *find*, *end*, and *insert* member functions:

```
#include <iostream.h>
#include <string.h>
#include <map.h>

using namespace std;
typedef map<int, string, less<int> > INT2STRING;

void main(void)
 {
   // Create a map of ints to strings
   INT2STRING theMap;
   INT2STRING::iterator theIterator;
   string theString = "";
   int index, continue_loop = 1;

   // Fill theMap with the digits 0 - 9, each mapped to its string counterpart
   // Note: value_type is a pair for maps...
```

```
  theMap.insert(INT2STRING::value_type(0,"Zero"));
  theMap.insert(INT2STRING::value_type(1,"One"));
  theMap.insert(INT2STRING::value_type(2,"Two"));
  theMap.insert(INT2STRING::value_type(3,"Three"));
  theMap.insert(INT2STRING::value_type(4,"Four"));
  theMap.insert(INT2STRING::value_type(5,"Five"));
  theMap.insert(INT2STRING::value_type(6,"Six"));
  theMap.insert(INT2STRING::value_type(7,"Seven"));
  theMap.insert(INT2STRING::value_type(8,"Eight"));
  theMap.insert(INT2STRING::value_type(9,"Nine"));
  // Read a Number from the user and print it back as words
  while (continue_loop)
    {
      cout << "Enter \"q\" to quit, or enter a Number: ";
      cin >> theString;
      if(theString == "q")
        continue_loop = 0;
      // extract each digit from the string, find its corresponding
      // entry in the map (the word equivalent) and print it
      for(index = 0; index < theString.length(); index++)
       {
         theIterator = theMap.find(theString[index] - '0');
         if(theIterator != theMap.end())              // is 0 - 9
           cout << (*theIterator).second << " ";
         else                                        // not 0 - 9
           cout << "[err] ";
       }
      cout << endl;
    }
}
```

The program creates a simple map of numbers to string objects. When it finds a number within the map, the program will display the string equivalent of the number (for example, 21 displays two one). The program creates the map, then enters an infinite loop until the user selects the "q" key, at which point the program exits. When you compile and execute the *map_ints.cpp* program, your screen will display the following output (provided you enter the same values at the prompts):

```
Enter "q" to quit, or enter a Number: 911
Nine One One
Enter "q" to quit, or enter a Number: 1500
One Five Zero Zero
Enter "q" to quit, or enter a Number: 4995
Four Nine Nine Five
Enter "q" to quit, or enter a Number: q
C:\>
```

CONTROLLING THE MAP'S SIZE AND CONTENTS 1248

In Tip 1247, you used the *insert* function to add ten value pairs to a *map*. As your programs become more complex, there may be times when you must erase or replace the elements within a *map*, as well as times when you must determine the *map* object's largest possible size. The Standard Template Library implementation of the *map* type lets you perform all these actions against your *map*. In fact, the four member functions you will use to perform these activities are similar to those you previously used with other Standard Template Library objects, as shown here:

```
size_type max_size(void) const;
void clear(void) const;
bool empty(void) const;
iterator erase(iterator first, iterator last);
```

The *max_size* function lets you set the upper limit on a *map's* size. The *clear* function lets you erase all the elements within a *map*. Similarly, the *empty* function lets your programs determine whether or not the *map* is empty of all elements. Finally, the *erase* function lets you clear a range of elements within a *map*. (Note that using the *clear* statement is identical to invoking *erase* with *[erase(map.begin, map.end)]*—using *clear* simply makes it more obvious what the statement's processing is.) To better understand the processing these member functions perform, consider the *map_mon2.cpp* program, contained on the CD-ROM that accompanies this book, which again creates a *map* of months. However, the *map_mon2.cpp* program then clears the *map* and refills it with a *map* of weekday names. The *map_mon2.cpp* program creates a *map* of strings to integers and fills it first with a *map* of month names to month numbers. The program then empties and refills the map with a *map* of weekday names to corresponding integers.

1249 UNDERSTANDING SETS

A *set* is a Sorted Associative Container that stores objects of type *key*. In addition to its sorted nature, a *set* is a Simple Associative Container, meaning that its value type, as well as its key type, is *key*. Finally, a *set* is also a Unique Associative Container, meaning that no two elements in the container are the same. A *set* is essentially a *map* with only a single type, rather than a *value* type and a *key* type. As you learned, *keys* are constant and unchangeable. When you work with *sets*, each element is unchangeable after you initially assign its value. Your program must instead delete the value and add a replacement element to the *set*.

The *set* and *multiset* types are particularly well-suited to the *set* algorithms the Standard Template Library includes: *set_union, set_intersection, set_difference,* and *set_symmetric_difference*. The reason for this is twofold. First, the *set* algorithms require their arguments to be sorted ranges, and, because *set* and *multiset* are Sorted Associative Containers, each Container always sorts its elements in ascending order. Second, the output range of these algorithms is always sorted, and inserting a sorted range into a *set* or *multiset* is a fast operation: the Unique Sorted Associative Container and Multiple Sorted Associative Container requirements guarantee that inserting a range takes only linear time if the range is already sorted.

A *set* lets you insert a new element into a *set* without invalidating iterators that point to existing elements. A *set* also lets you erase an element from a *set* without invalidating any iterators, except, of course, those iterators that actually point to the element that you are erasing. The following program, *set1.cpp*, creates two simple *sets*, compares their values, and creates a third *set*, as shown here:

```
#include <set.h>
#include <iostream.h>

using namespace std;
class ltstr
 {
  public:
     bool operator()(const char* s1, const char* s2) const
       { return (strcmp(s1, s2) < 0);}
 };

void main(void)
 {
    const int N = 6;
    const char* a[N] = {"aaa", "bbb", "ccc", "ddd", "eee", "fff"};
    const char* b[N] = {"ggg", "hhh", "eee", "iii", "ccc", "aaa"};

    set<const char*, ltstr> A(a, a + N);
    set<const char*, ltstr> B(b, b + N);
    set<const char*, ltstr> C;
    cout << "Set A: ";
    copy(A.begin(), A.end(), ostream_iterator<const char*>(cout, " "));
    cout << endl;
    cout << "Set B: ";
    copy(B.begin(), B.end(), ostream_iterator<const char*>(cout, " "));
```

```
   cout << endl;
   cout << "Union: ";
   set_union(A.begin(), A.end(), B.begin(), B.end(),
           ostream_iterator<const char*>(cout, " "), ltstr());
   cout << endl;
   cout << "Intersection: ";
   set_intersection(A.begin(), A.end(), B.begin(), B.end(),
                   ostream_iterator<const char*>(cout, " "), ltstr());
   cout << endl;
   set_difference(A.begin(), A.end(), B.begin(), B.end(),inserter(C, C.begin()),
                 ltstr());
   cout << "Set C (difference of A and B): ";
   copy(C.begin(), C.end(), ostream_iterator<const char*>(cout, " "));
   cout << endl;
}
```

The *set1.cpp* program creates a union *set*, an intersection *set*, and a difference *set*. The union *set* contains all the values in both *sets* in sorted order. The intersection *set* contains all the values that both *Set A* and *Set B* include, but not the values that either *set* does not include. Finally, the difference *set* holds the values that appear in the first *set*, but not in the second *set*. When you compile and execute the program, your screen will display the following output:

```
Set A: aaa bbb ccc ddd eee fff
Set B: aaa ccc eee ggg hhh iii
Union: aaa bbb ccc ddd eee fff ggg hhh iii
Intersection: aaa ccc eee
Set C (difference of A and B): bbb ddd fff
C:\>
```

A SIMPLE SET EXAMPLE 1250

In Tip 1249, you learned about the *set* object and wrote a simple program, *set1.cpp*, which used the *set* type and generated output to the screen. While there is not sufficient space to explore the *set* type in depth, this Tip does present the *lower_bound*, *upper_bound*, and *equal_range* functions that your programs can use with the *set* type.

The *lower_bound* function returns an iterator to the earliest element in the controlled sequence that has a *key* that does not match the value the program passed to the *lower_bound* function. The *upper_bound* function returns an iterator to the earliest element in the controlled sequence that has a *key* that matches the value the program passed to the *upper_bound* function. If no such element exists, the function returns *end*. In both cases, the program uses the *set::key_comp(key, x)* function to determine if the *keys* match. The *equal_range* function returns a pair value, where *.first* is the result of the *lower_bound* function, and *.second* is the result of the *upper_bound* function. To better understand the processing the *set* member functions perform, consider the following program, *set_rang.cpp*:

```
#include <set.h>
#include <iostream.h>

using namespace std;
typedef set<int, less<int>> SET_INT;

void main(void)
 {
   SET_INT s1;
   SET_INT::iterator i;

   s1.insert(5);
   s1.insert(10);
   s1.insert(15);
   s1.insert(20);
   s1.insert(25);
   cout << "s1 -- starting at s1.lower_bound(12)" << endl;
```

```
// prints: 15,20,25
   for (i = s1.lower_bound(12); i != s1.end(); i++)
      cout << "s1 has " << *i << " in its set." << endl;
   cout << "s1 -- starting at s1.lower_bound(15)" << endl;
// prints: 15,20,25
   for (i = s1.lower_bound(15);i != s1.end(); i++)
      cout << "s1 has " << *i << " in its set." << endl;
   cout << "s1 -- starting at s1.upper_bound(12)" << endl;
// prints: 15,20,25
   for (i = s1.upper_bound(12); i != s1.end(); i++)
      cout << "s1 has " << *i << " in its set." << endl;
   cout << "s1 -- starting at s1.upper_bound(15)" << endl;
// prints: 20,25
   for (i = s1.upper_bound(15); i != s1.end(); i++)
     cout << "s1 has " << *i << " in its set." << endl;
   cout << "s1 -- s1.equal_range(12)" << endl;
// does not print anything
   for (i = s1.equal_range(12).first;i != s1.equal_range(12).second; i++)
      cout << "s1 has " << *i << " in its set." << endl;
   cout << "s1 -- s1.equal_range(15)" << endl;
// prints: 15
   for (i = s1.equal_range(15).first;i != s1.equal_range(15).second; i++)
      cout << "s1 has " << *i << " in its set." << endl;
 }
```

The *set_rang.cpp* program illustrates how to use the *lower_bound* function to get an iterator to the earliest element in the controlled sequence that has a *key* that does not match the value passed to the function. It also illustrates how to use the *upper_bound* function to get an iterator to the earliest element in the controlled sequence that has a *key* that matches the value passed to the function. The last thing it illustrates is how to use the *equal_range* function to get a pair value that contains the *lower_bound* and *upper_bound* results of the *key*. When you compile and execute the *set_rang.cpp* program, your screen will display the following output:

```
s1 -- starting at s1.lower_bound(12)
s1 has 15 in its set.
s1 has 20 in its set.
s1 has 25 in its set.
s1 -- starting at s1.lower_bound(15)
s1 has 15 in its set.
s1 has 20 in its set.
s1 has 25 in its set.
s1 -- starting at s1.upper_bound(12)
s1 has 15 in its set.
s1 has 20 in its set.
s1 has 25 in its set.
s1 -- starting at s1.upper_bound(15)
s1 has 20 in its set.
s1 has 25 in its set.
s1 -- s1.equal_range(12)
s1 -- s1.equal_range(15)
s1 has 15 in its set.
```

1251 INTRODUCING WIN32 PROGRAMMING

In the previous 1,250 Tips, you have learned how to program in both C and C++. While the programs you wrote in the previous Tips will run under Windows, they were designed more for a DOS or Unix environment. In the next 250 Tips, you will learn the fundamentals of Windows programming. Many of the upcoming Tips will revisit earlier Tips with explanations of how to perform similar processing using the Windows Application Program Interface, which this book and most other books about Windows programming refer to as the *Win32 API*. It is important to understand that this book focuses on the Win32 API, not the Win16 API—which means that the function calls this book discusses will work under Windows 95, 98, Windows NT and 2000, but will not work under Windows 3.11 or earlier versions.

Additionally, because of the differences in the two operating systems, there will be cases where a Tip will introduce an API function that either Windows 95/98 or Windows NT/2000, but not both, support. Typically, such Tips will note which operating system does not support the API function, and why it does not. If you have a problem with the program a particular Tip contains, you may want to re-check the Tip to ensure the code is compatible with your operating system. Finally, as you probably know, the output that Windows generates is very different from the output that you have generated throughout this book so far. In fact, a simple DOS program that outputs "Jamsa's C/C++/C# Programmer's Bible" generates output as shown here:

```
Jamsa's C/C++/C# Programmer's Bible
C:\>
```

On the other hand, a simple Windows program which creates a message box that contains similar information would generate the output shown in Figure 1251.

***Figure 1251** A simple Windows message box.*

More Differences between Windows and DOS Programs 1252

As you saw in Tip 1251, there are significant differences in the appearance of output your programs will generate under Windows and the appearance of output your DOS programs generated. However, the differences between DOS and Windows programming are much more significant than just output, although that is one of the most easily understood differences. The following list details some of the other important differences between Windows and DOS programming:

- Because Windows is a multitasking operating system (where two or more programs may run at the same time), your programs will be sharing space in memory with at least one program, and often many other programs. Making sure your programs remain within their allocated memory space and do not modify the memory space of other programs is very important.
- Because Windows processes text-based information differently than DOS, you will no longer generate output using *printf* or *cout*. You may, however, often use stream arrays to format information before you send it to a window.
- Because the Windows graphical interface can receive input from the user at virtually any point in a program's processing, your Windows programs will generally not be as linear as your DOS programs. In other words, you will build your Windows programs around responding to user actions and system messages.
- Your Windows programs will generally consist of multiple classes in multiple class and header files and are often significantly larger than DOS programs because Windows programs are not limited to conventional memory.

- Windows programs that you design in this book are generally more capable of receiving user input than were many of the DOS programs you designed in previous Tips. As a rule, because the graphical nature of Windows requires user interaction for useful processing, most programs you write for Windows will support some type of user interaction.
- Despite all the differences this list discusses, and all the differences that you will learn about in later Tips, it is important to remember that Windows C++ programming is fundamentally the same as the C++ programming that you have done so far—you will just apply many of those concepts in a slightly different way than you did with your DOS programs.

In the next several Tips, you will learn some of the Windows programming basics, and the impact those basics have on all Windows programs. You will then begin to write Windows programs and, by the time you complete this book, you will be able to write Windows programs that manipulate memory, files, graphics, printers, and more.

Note: *The **Turbo C++ Lite** compiler included on the CD-ROM that accompanies this book is not Windows-capable (in other words, you cannot write programs within the compiler that will run in Windows). To write Windows programs, you must have a Windows-capable compiler, such as Microsoft's **Visual C++** or Borland **C++**. Most free compilers you can download across the Web will not create Windows applications.*

1253 INTRODUCING THREADS

As you learned throughout the DOS section of this book, as a general rule, DOS supports the execution of only one program at a time within memory. As you also learned, Windows is not subject to this constraint. In fact, Windows limits the number of programs a computer can run at one time only if there is insufficient memory to open additional programs. Windows manages all these programs in memory using *threads*. Simply put, a thread is a way you can visualize a program's request to use the computer's CPU. When you have multiple programs executing simultaneously, the operating system will manage one or more threads for each of those programs. Windows places each thread in a thread queue (a list). Later, depending on a thread's priority, Windows will take a thread from the queue and assign the thread to the CPU, which, in turn, executes the thread's instruction. A thread's priority determines whether Windows places the thread ahead of or after other threads within the queue. Figure 1253.1 shows a simple graphical depiction of the Windows multitasking thread scheme.

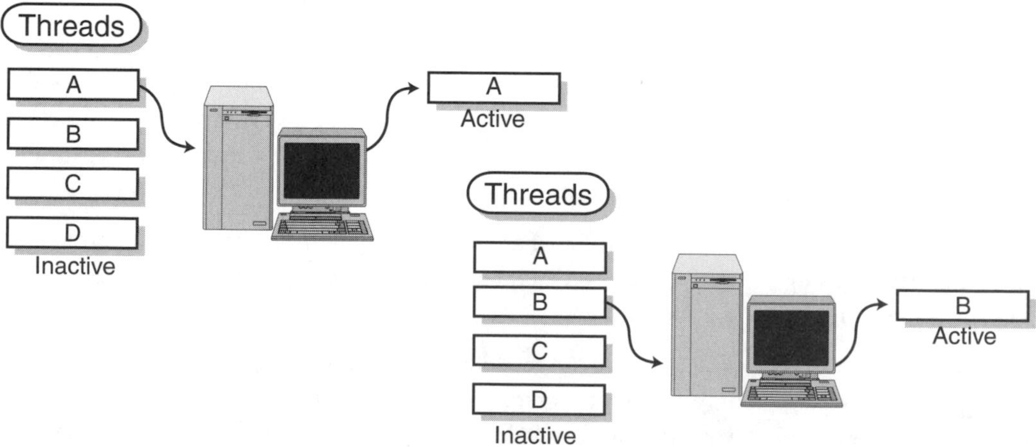

Figure 1253.1 The CPU processes a series of threads.

You will learn about threads in detail in later Tips. However, for now, understand that a thread is the entity to which Windows assigns the CPU to run the program's statements. Windows orders the threads depending on their priority and the computer's CPU executes each thread in turn. Figure 1253.2 shows a simple model of how your program requests processing from the CPU and receives its information when the CPU completes the processing.

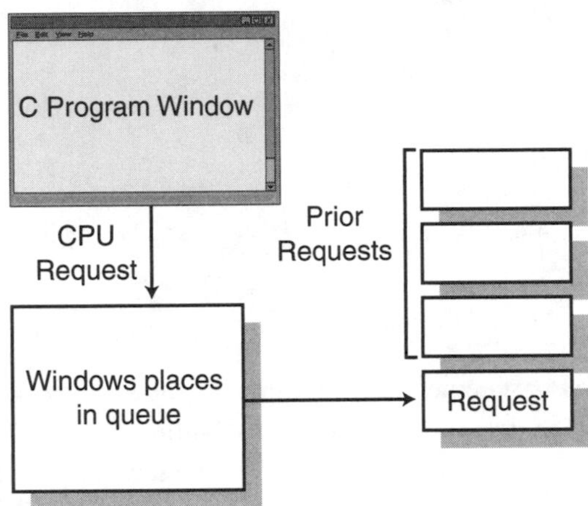

Figure 1253.2 A simple model of the Windows thread-processing scheme.

As you will learn in Tip 1254, Windows returns information to your programs each time it processes a thread for that program in the form of a *message*.

UNDERSTANDING MESSAGES 1254

The fundamental means of communication that Windows and programs written for Windows use is *messages*. Simply put, each time an operation occurs, Windows responds to that operation or action by sending a message to itself or another program. For example, when a user clicks the mouse within your program's window, Windows reads that mouse click and sends a message to your program that the user has clicked the mouse within the program's window at a certain location. Your program, on receiving that message, will begin its own processing as appropriate for the message. If the message is unimportant to the program, your program should simply ignore the message. To better understand the Windows message model, consider Figure 1254, which shows the model in a simplified, linear form.

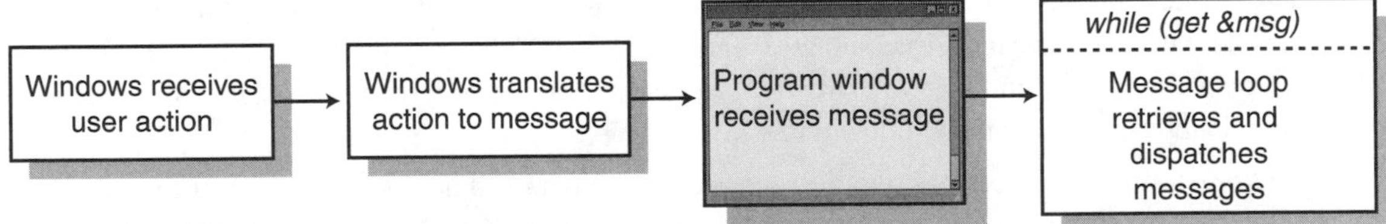

Figure 1254 A simple diagram of the Windows message model.

When you write programs in Windows, the most important routines within your programs will be those that accept and process messages from the operating system. For example, the following code fragment of the *WndProc* function shows a sample message processing and response function:

```
LRESULT CALLBACK WndProc(HWND hWnd, UNIT uMsg, WPARAM wParam, LPARAM lParam)
 {
   switch(uMsg)
    {
```

```
        case WM_COMMAND:
           switch(LOWORD(wParam))
           {
              case IDM_DEMO :
                 break;
              case IDM_EXIT :
                  DestroyWindow(hWnd);
                 break;
           }
           break;
        case WM_DESTROY :
           PostQuitMessage(0);
           break;
        default:
             return  (DefWindowProc(hWnd, UMsg, wParam, lParam));
     }
     return(0L);
  }
```

In general, the code fragment uses a *switch* statement to determine the message size. As you proceed through the following Tips, you will quickly learn and understand the processing performed within the *WndProc* function. In the previous fragment, the function checks for a *WM_COMMAND* or a *WM_DESTROY* message, or passes the function to the default message handler. Based on which message it receives, the function performs appropriate processing.

1255 UNDERSTANDING A WINDOW'S COMPONENTS

As you might have guessed, the building block for all Windows programs is one or more windows or dialog boxes (a special window type). In fact, nearly every object within a window is also a window of some type. In later Tips, you will better understand how the windows your programs will use are all similar, but derived from different locations. For this particular Tip, it is important that you understand the components of a "standard window"—in other words, what your users will consider a window.

Windows generally composes each "standard window" from seven basic pieces. You can break each of those pieces down further, which you will do for most pieces in later Tips. However, it is valuable to understand the bigger picture of the window before you analyze each of the small pieces that make up a single window component. The seven basic components of a window are shown in the following list:

- The *window frame* is the container for everything else within a window. As you will learn, you can make window frames of many different types. The most common type of window frame is a resizable window, similar to that shown in Figure 1255.1. Within your programs, you will manipulate the frame and receive many messages (such as window resizing messages) from the frame. Later Tips discuss the window frame in detail.
- The *title bar* provides information to the user about the program. The title bar extends the width of the window along the top edge inside the border. The title bar identifies what the window shows and lets the user perform many window operations as well. The title bar is the control point for moving the window and the locations of the system menu, *MINIMIZE, MAXIMIZE, RESTORE,* and *CLOSE WINDOW* buttons. As you will learn in later Tips, the title bar's composition will vary greatly depending on the application that creates the window and the purpose of the window within which the title bar falls.
- The *MINIMIZE, MAXIMIZE,* and *CLOSE WINDOW* buttons, while inside the title bar container, are important enough that you should consider them as components of the window itself. The buttons let you control a window's size and close a window when you are finished with it.

- The *client area* is the area within the window that your programs can customize and that your programs should design with the intention of receiving input from the user. In other words, the client area is the section of the window where the majority of the action occurs. For example, if you are working within a Microsoft *Word*® document, the client area is the region where you type in and edit your document within the window.
- The *scroll bars* let the user navigate left and right and up and down within a window. For example, your programs might use the scroll bars to let the user see more of an input form. Your programs also might use the scroll bars to let the user move through a document the user previously saved to disk (as the user would use the scroll bars with Microsoft *Word*). Scroll bars are an important navigational tool within Windows programs. The closest equivalent within a DOS program might be the ARROW keys, depending on how you implemented the keys within a given program.
- The *menu bar* is a component of most parent windows, but is generally not present within most child windows. (Tip 1256 explains parent and child windows in detail.) Windows programs use the menu bar to provide the user with options appropriate to the program. As you will learn, most Windows programs include, at a minimum, a File menu and a Help menu. More complex Windows programs may include 10 or more menus, and each menu may contain 20 or more selections. As your programs become more complex, your menus will also become more complex. Figure 1255.1 shows the Microsoft *Word* menu bar.

Figure 1255.1 The Microsoft ***Word*** *menu bar.*

- The *status bar* is a component of most parent windows, but is generally not present within most child windows. Windows programs use the status bar to provide the user specific information that details the user's current position within the program—be it a location within a document, a line, and so on. Figure 1255.2 shows the Microsoft *Word* status bar, which provides important information about the user's current position within the current document.

Figure 1255.2 The Microsoft ***Word*** *status bar.*

UNDERSTANDING PARENT AND CHILD WINDOWS 1256

In Tip 1255, you learned that most parent windows include the menu bar and the status bar, while most child windows do not. However, you may not yet understand what parent and child windows are. Most Windows programs support the Multiple Document Interface (MDI), which lets a single program maintain and display multiple components within a single window. For example, Figure 1256.1 shows Microsoft *Word* with four document windows open.

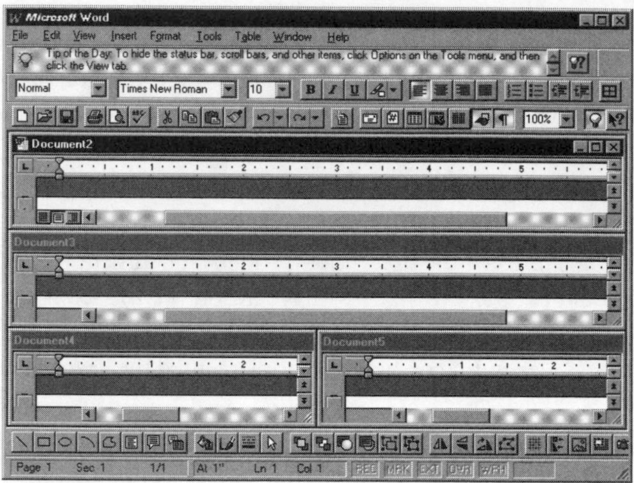

Figure 1256.1 The Microsoft ***Word*** *parent window with four open child windows.*

Much like the class hierarchy that you have learned about and used within several C++ programs, every window in Windows derives from a basic window. Thus, every window has a parent window. In Figure 1256.1, the child windows are the internal document windows, and the parent window is the Microsoft *Word* window. However, you could also consider the Microsoft *Word* window as a child window and the Windows Desktop as its parent window. The Windows Desktop has no parent window.

In the previous figure, the parent window supports the multiple-document interface, which lets the parent window have many child windows. Other Windows programs support the single-document interface, which lets the parent window have only one child window, as shown in Figure 1256.2.

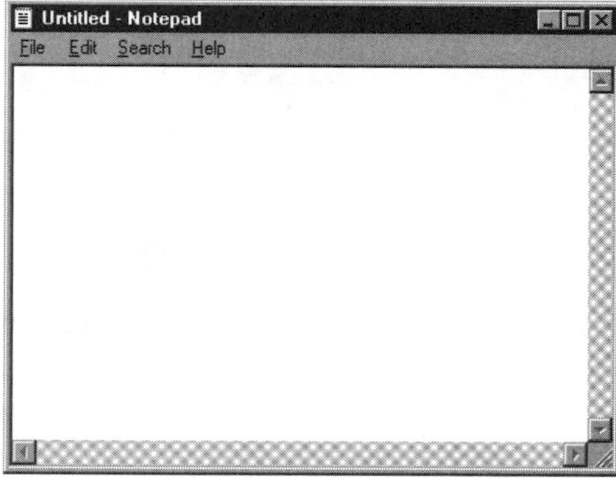

Figure 1256.2 The Windows ***Notepad*** *is a single-document interface program.*

Finally, many newer Windows programs support a special variation on the single-document interface, which is commonly known as the *Explorer-style document interface*. Programmers call the variation the Explorer-style document interface because programmers model the interface after the Windows *Explorer* program. The Explorer-style document interface is similar to the single-document interface in all respects, except the single document window is split down the middle, making it easier for the program to display unique data sets within a single view. Figure 1256.3 shows the Windows *Explorer* and the Explorer-style interface.

*Figure 1256.3 The Windows **Explorer** and the Explorer-style interface.*

Within your programs, you will often manipulate many child windows within a single parent window. As you work with window design in later Tips, you will learn more about the important differences between the various interface styles.

CREATING A GENERIC WINDOWS PROGRAM 1257

As you will discover, almost all Windows programs perform a certain amount of default processing (overhead) that is necessary for the program to run. Simply put, the program must create at least one window, the program must register that window with the operating system, and the program must handle messages sent from the operating system to the window. Additionally, most Windows programs use an additional file called a *resource file*. Resource files tell the compiler about the characteristics of the windows the program creates. Tip 1258 explains resource files in detail. After you create the basic Windows program form, your other programs will generally use that basic form as the stepping stone for the program's creation. To minimize possible confusion with later programs, the basic form program is named *generic.cpp*. The CD-ROM that accompanies this book includes all the dependencies (component files) for *generic.cpp*. However, because subsequent Tips will explain those dependencies in detail, this Tip will only present *generic.cpp*, as shown here:

```
#include <windows.h>
#include "generic.h"

HINSTANCE hInst;          // current instance
LPCTSTR lpszAppName = "Generic";
LPCTSTR lpszTitle =   "Generic Application";
BOOL RegisterWin95(CONST WNDCLASS* lpwc);

int APIENTRY WinMain(HINSTANCE hInstance, HINSTANCE hPrevInstance, LPSTR lpCmdLine,
                     int nCmdShow)
 {
   MSG msg;
   HWND hWnd;
   WNDCLASS wc;

   wc.style            = CS_HREDRAW | CS_VREDRAW;
   wc.lpfnWndProc      = (WNDPROC)WndProc;
   wc.cbClsExtra       = 0;
   wc.cbWndExtra       = 0;
   wc.hInstance        = 0;
   wc.hIcon            = LoadIcon(hInstance, lpszAppName);
   wc.hCursor          = LoadCursor(NULL, IDC_ARROW);
   wc.hbrBackground    = (HBRUSH)(COLOR_WINDOW+1);
   wc.lpszMenuName     = lpszAppName;
   wc.lpszClassName    = lpszAppName;
```

```
    if(!RegisterWin95(&wc))
      return false;
   hInst = hInstance;
   hWnd = CreateWindow (lpszAppName, lpszTitle,
                        WS_OVERLAPPEDWINDOW, CW_USEDEFAULT, 0,
                        CW_USEDEFAULT, 0, NULL, NULL,
                        hInstance, NULL);
   if(!hWnd)
      return false;
   ShowWindow(hWnd, nCmdShow);
   UpdateWindow(hWnd);
   while(GetMessage(&msg, NULL, 0,0))
    {
      TranslateMessage(&msg);
      DispatchMessage(&msg);
    }
   return(msg.wParam);
 }
BOOL RegisterWin95(CONST WNDCLASS* lpwc)
 {
   WNDCLASSEX wcex;
   wcex.style            = lpwc->style;
   wcex.lpfnWndProc      = lpwc->lpfnWndProc;
   wcex.cbClsExtra       = lpwc->cbClsExtra;
   wcex.cbWndExtra       = lpwc->cbWndExtra;
   wcex.hInstance        = lpwc->hInstance;
   wcex.hIcon            = lpwc->hIcon;
   wcex.hCursor          = lpwc->hCursor;
   wcex.hbrBackground    = lpwc->hbrBackground;
   wcex.lpszMenuName     = lpwc->lpszMenuName;
   wcex.lpszClassName    = lpwc->lpszClassName;
   wcex.cbSize           = sizeof(WNDCLASSEX);
   wcex.hIconSm          = LoadIcon(wcex.hInstance, "SMALL");
   return RegisterClassEx(&wcex);
 }
LRESULT CALLBACK WndProc( HWND hWnd, UINT uMsg, WPARAM wParam, LPARAM lParam)
 {
   switch(uMsg)
    {
      case WM_COMMAND:
         switch(LOWORD(wParam))
          {
            case IDM_TEST :
               break;
            case IDM_EXIT :
               DestroyWindow(hWnd);
               break;
          }
         break;
      case WM_DESTROY :
         PostQuitMessage(0);
         break;
      default:
         return (DefWindowProc(hWnd, uMsg, wParam, lParam));
    }
   return(0L);
 }
```

As you can see, the *generic.cpp* program consists of three functions: *WinMain* (the equivalent of *main* for Windows programs), *RegisterWin95*, and *WndProc*. As you saw in Tip 1254, the *WndProc* function handles the messages that the operating system sends to the program. The *RegisterWin95* function handles some extra clean-up (specifically, it adds more information to the *WNDCLASS* object, which Windows 95 requires but Windows NT does not) and registers the new window with the operating system. You will learn more about registering windows in later Tips. When you compile and execute the *generic.cpp* program, your screen will display the output shown in Figure 1257.

***Figure 1257** The **generic.cpp** program's output.*

UNDERSTANDING RESOURCE FILES 1258

As you learned in Tip 1255, even the simplest Windows program has many components. However, many of these components are relatively static—meaning that your program will not often change them. Menus, for example, rarely change within most programs. You will also rarely change icons, title bar information, and so on, within programs, if at all. Additionally, Windows expects your programs to store information about the programs themselves (author, revision number, and so on) together with the actual program. Windows uses this information to provide the user with information about the program. For example, if you select the *generic.exe* program within the Windows *Explorer* and press the ALT+ENTER keys, Windows will display the File Properties dialog box. In addition to displaying basic information about the file, you can click your mouse on the *Version* tab, which results in the File Properties dialog box appearing, as shown in Figure 1258.

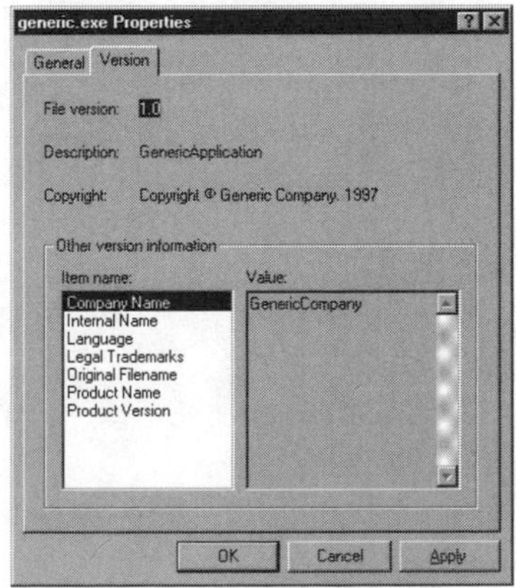

***Figure 1258** The File Properties dialog box for the **generic.exe** program.*

As you can see, Windows stores information about the file within the Properties attribute. In fact, the program's creator assigned those properties to the *generic.exe* program within the *generic.rc* resource file (one of the dependencies mentioned in Tip 1257). The *generic.rc* resource file is a text file that maintains information about many of the static resources the program will use. The *generic.rc* resource file more closely contains information about resources the program will use after execution, including menus, icons, and so on. The *generic.rc* file contains icon listings, menu definitions, and file property information, as shown here:

```
#include "windows.h"
#include "generic.h"
#include "winver.h"

MYAPP ICON  DISCARDABLE "GENERIC.ICO"
SMALL ICON  DISCARDABLE "SMALL.ICO"

GENERIC MENU DISCARDABLE
BEGIN
   POPUP "&File"
   BEGIN
      MENUITEM "E&xit",             IDM_EXIT
   END
   MENUITEM "&Demo!",             IDM_DEMO   POPUP "&Help"
   BEGIN
      MENUITEM "&About Generic...",   IDM_ABOUT
   END
END

VERSIONINFO
 FILEVERSION 3,3,0,0
 PRODUCTVERSION 3,3,0,0
 FILEFLAGSMASK 0X3fl
#ifdef _DEBUG
   FILEFLAGS 0xbl
#else
   FILEFLAGS 0xal
#endif
 FILEOS 0X4L
 FILETYPE 0X1L
 FILESUBTYPE 0X0L
BEGIN
   BLOCK "StringFileInfo"
   BEGIN
      BLOCK "040904B0"
      BEGIN
         VALUE "CompanyName", "GenericCompany\0"
         VALUE "FileDescription", "GenericApplication\0"
         VALUE "FileVersion", "1.0\0"
         VALUE "InternalName", "1.0\0"
         VALUE "LegalCopyright", "Copyright \251 Generic Company. 2001\0"
         VALUE "LegalTrademarks", "Generic Trademark.\0"
         VALUE "OriginalFilename", "\0"
         VALUE "ProductName", "Generic Application.\0"
         VALUE "ProductVersion", "1.0\0"
      END
   END
   BLOCK "VarFileInfo"
   BEGIN
      VALUE "Translation", 0x409, 1200
   END
END
```

The *generic.rc* resource file is a simple resource file. As your programs become more complex, you will find that you design resource files with hundreds or even thousands of entries. The *generic.rc* resource file really handles only two resource types: it stores the information for the File Properties dialog box, as you saw earlier in this Tip, and it stores the information from which to create the menu for the program. When you use the resource compiler to compile the *generic.rc* resource file, the resource compiler will use the information within the resource file to generate the menu and properties information for the executable program file. In the *generic.rc* resource file, the menu consists of two drop-down menus (File and Help), each of which has one or more items.

It is important to understand that programs do not have to use the information the resource file provides. For example, the *generic.cpp* program supports a menu bar because it created the window with a pointer to the *generic* menu bar in the *generic.rc* file. Implementing a resource file is not sufficient; you must implement the resources within your programs in some way.

Note: *Many newer Windows C++ development tools include drag-and-drop capability, which you can use to avoid building text resource files. With drag-and-drop tools, you can add items directly to menus or to a window without manually creating resource file information for those resources. However, the drag-and-drop design is unique to each compiler. Because there is virtually no portability between compilers, the resource file is helpful for ensuring that the programs in this book are consistent from reader to reader.*

UNDERSTANDING WINDOWS HANDLES 1259

As you learned in the Files and I/O section of this book, when your programs work with files and disks, they can do so either at a low-level, with BIOS routines, or at a high-level, with file handles. Within Windows, you will also use handles to maintain information about files. However, you will use a different type of handle, a *window handle*, to obtain or maintain information about a window within your program or somewhere else on the system. A window handle is essentially a *long* value that you maintain within a variable of type *HWND*. When you invoke an API function that requires a window handle, you pass the variable of type *HWND* to the function. Windows, in turn, will check the window handle against its list of valid window handles and will then send the message to, or perform the action against, the appropriate window. Figure 1259 shows a logical model of the Windows window handle evaluation process.

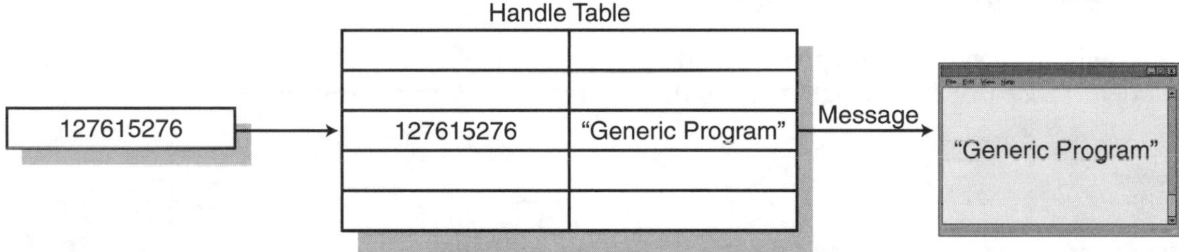

Figure 1259 *The Windows window handle evaluation process.*

Within your programs, you will use handles to windows to help you control a window's appearance, size, and other qualities.

As you will learn in later Tips, Windows C++ defines a great number of additional types, of which *HWND* is only one. For example, the *generic.cpp* program that you wrote in Tip 1257 includes six new types, which your non-Windows programs will not use. Later Tips will examine the new types in detail.

DEFINING THE WINDOWS HANDLE TYPES 1260

In Tip 1259, you learned about the *HWND* type, which you will use within your Windows programs to maintain a *long* value that represents an open window. Windows actually defines twelve handle types. For example, the *generic.cpp* program you created in Tip 1257 included a definition of a variable of type *HWND*, and also included a definition of a variable of type *HINSTANCE*. Because Windows is a multi-tasking operating system, it is possible to have multiple copies, or *instances*, of a program running at one time. Windows maintains a unique number for each instance, which it stores within a handle table of type *HINSTANCE*. Your programs can use a *HINSTANCE* variable to maintain information about which instance is currently running. In some cases, you may use a *HINSTANCE* value together with a Windows API call to determine how many current instances of a program are running. Throughout the remaining Tips in this book, you will encounter different handle types. Later Tips will explain each handle type as you encounter them within your programs. Table 1260 lists the Windows handle types.

Handle Type	Description
HANDLE	Number that uniquely identifies a handle
HBITMAP	Number that uniquely identifies a bitmap
HBRUSH	Number that uniquely identifies a brush
HCURSOR	Number that uniquely identifies a cursor
HFONT	Number that uniquely identifies a font
HGDIOBJ	Number that uniquely identifies a graphical device interface (GDI) object
HICON	Number that uniquely identifies an icon
HINSTANCE	Number that uniquely identifies an instance
HPALETTE	Number that uniquely identifies a palette
HPEN	Number that uniquely identifies a pen
HRGN	Number that uniquely identifies a region
HWND	Number that uniquely identifies a window

Table 1260 *The Windows handle types.*

It is important to recognize that your programs will access most handle types to control the display of information within windows. C++ supports file handles for Windows sufficiently, so Windows needs new handle types only to help it (and you) control the display.

1261 Understanding the Generic Header File

Before you begin to analyze the components of the *generic.cpp* program more closely, it is important to recognize from the original code that the program requires the use of the *generic.h* file, as shown here:

```
#define IDM_EXIT           100
#define IDM_DEMO           200
#define IDM_ABOUT          300
LRESULT CALLBACK WndProc(HWND, UINT, WPARAM, LPARAM);
LRESULT CALLBACK About  (HWND, UINT, WPARAM, LPARAM);
```

The *generic.h* header file defines several constants. You should recognize these constants from the menu declaration within the resource file—the names of the constants correspond to the identifiers within the resource file. Later Tips will use the constants to identify when the user has selected a menu item within the file. The last two lines define the prototypes for the *WndProc* and *About* functions. You created the *WndProc* function in Tip 1257, and you will create the *About* function later. However, you should note that the header file defines the prototypes for both functions as *callback functions*. Tip 1262 discusses callback functions in detail.

1262 Understanding Callback Functions

In the previous Tip and in Tip 1257, you saw several functions declared with the *CALLBACK* keyword. You will refer to functions your programs declare with the *CALLBACK* keyword as *callback functions*. A callback function is a function to which you pass the address of a third function, which that third function then "calls back" with information. You will always define the *WndProc* function as a callback function. Within your programs, you will often use callback functions with specific API functions, such as *EnumFontFamilies* and *EnumWindows*. When you pass the address of a callback function to either of these functions, the function will call the callback function for each item within the list. For example, if you call *EnumWindows*, you will probably pass to *EnumWindows* the address of a callback function that either displays the values or adds them to an array. *EnumWindows*, in turn, will call the callback function for each window within its list of all the windows.

Callback functions are necessary in Windows because your programs will handle many of their important programming actions through the Windows Application Program Interface (API)—which your programs cannot directly modify. Therefore, you must provide the API with a means it can use to call your custom routines when it returns extended information (such as a list). Tip 1263 details the Windows API more closely.

INTRODUCING THE WINDOWS APPLICATION PROGRAMMING INTERFACE 1263

As you have learned, the Windows Application Programming Interface (API) is the foundation of most programs in Windows. In general, any action your program performs in Windows beyond the most simple math operations uses the Windows API, either directly or indirectly (in other words, even defining a menu item within a resource file uses the Windows API, although your program does not call the functions within the API directly). However, you will find that your programs often call functions within the API directly, as shown here:

```
hWnd = CreateWindow (lpszAppName, lpszTitle, WS_OVERLAPPEDWINDOW, CW_USEDEFAULT, 0,
                     CW_USEDEFAULT, 0, NULL, NULL, hInstance, NULL);
```

The *CreateWindow* function is a Windows API function that almost every Windows-based program will use to create the program's windows. As you can see, you will invoke the function (and most other Win32 API functions) within your programs. Your programs can call API functions as though your programs defined the functions for one simple reason: all Windows programs you write will include the *windows.h* header file. The *windows.h* header file, in turn, includes several other Windows header files, notably the *winbase.h* header file, which contains the definitions for the Win32 API functions, as well as several structures and enumerated types (such as *HWND*) that your programs will use.

Unfortunately, the Windows API is too large to list all its functions here—there are currently thousands of base API functions, and about several thousand more operating-system-specific functions add to the API. Microsoft estimates that Windows® 98 operating system includes almost 4,000 API functions.

LOOKING CLOSER AT THE GENERIC.CPP PROGRAM 1264

In Tip 1257, you created the *generic.cpp* program, which implements the basic requirements for a Windows program. Before you begin to learn more about Windows programming and begin to manipulate more complex functions, it is valuable to understand exactly what actions your *generic.cpp* program performed. In the next several Tips, you will closely analyze the *WinMain* function, the window creation process, and more. However, you should also understand the meaning of the global variables the *generic.cpp* program provides, as shown here:

```
#if defined (win32)
   #define IS_WIN32 TRUE
#else
   #define IS_WIN32 FALSE
#endif

HINSTANCE hInst;         // current instance
LPCTSTR lpszAppName = "Generic";
LPCTSTR lpszTitle =   "Generic Application";
BOOL RegisterWin95(CONST WNDCLASS* lpwc);
```

First, the program checks the *win32* compiler constant to determine whether or not the compiler is performing a Win32 compilation. The program checks for the Win32 compilation for many reasons, but primarily because, as you have learned, there are significant differences between the Win32 and Win16 APIs. A program that you design to run on Windows 3.11 but write on a Windows 95 machine must limit itself only to the API calls within the Win16 API, and your programs can use the *IS_WIN32* constant to control what API functions your programs call.

Next, the program defines the *hInst* variable. As you learned in Tip 1260, *HINSTANCE* is a Windows handle that maintains a unique number that corresponds to the currently executing instance of the program and no other instance of this or any other program.

The declarations then define *lpszAppName* and *lpszTitle,* which appear at first glance to be character arrays, or perhaps even string variables. You will use the *LPCTSTR* type (a Windows-defined type) to hold read-only string pointers. When the compiler compiles your program, it will actually convert all *LPCTSTR* declarations to *const char FAR** declarations. However, as you can see, *LPCTSTR* is clearly easier to type and to understand than is *const char FAR**.

Finally, the program makes a prototype declaration for the *RegisterWin95* function. The *RegisterWin95* function accepts a parameter of type *WNDCLASS* and returns a Boolean success value. You will learn more about Windows registration in Tip 1269. However, for now, you should understand that the *WNDCLASS* type contains information that Windows uses each time it registers or creates a new window (such as the caption in the title bar, the frame type, and so on).

1265 Understanding the WinMain Function

In the *generic.cpp* program, which you created in Tip 1257, the program's first function was *WinMain*. As you have learned, the *WinMain* function is the Windows equivalent of the *main* function that all your C and C++ programs used for their primary processing. The *WinMain* function, however, differs in several important ways—not the least of which is in its declaration, as shown here:

```
int WINAPI WinMain(HINSTANCE hInstance, HINSTANCE hPrevInstance,
                   LPSTR lpCmdLine, int nCmdShow)
```

As you can see, the *WinMain* function returns an *int* value, just as many of your C++ programs did. That, however, is where the similarity ends (however, as you will learn, the *WinMain* function header performs processing similar to that of the *main* header). The *WINAPI* keyword indicates the user can only launch (or execute) the program from within Windows. Table 1265 details the parameters your programs must provide with the *WinMain* function.

Parameter Type	Parameter Name	Description
HINSTANCE	*hInstance*	The application's instance handle. Each application instance has a unique instance handle. Your programs will use *hInstance* values as an argument to several Windows functions and can also use the *hInstance* value to distinguish between multiple instances of a given application.
HINSTANCE	*hPrevInstance*	The previous instance of the application's handle. This value is *NULL* if this is the first instance. For Windows 95, this value is always *NULL*.
LPSTR	*lpCmdLine*	A far pointer to a *NULL*-terminated command line. Specify the *lpCmdLine* value when invoking the application from the program manager or from a call to *WinExec*. Note that, under Windows 95, this is a pointer to the entire command line, not an array of pointers to each argument (so your programs must parse the command line before they begin to process it).
int	*nCmdShow*	An integer that specifies the application's window display. Pass this value to *ShowWindow*.

Table 1265 *Parameters* ***WinMain*** *accepts.*

In the *generic.cpp* program, *WinMain's* first action is to define a variable of type *MSG*, a variable of type *HWND*, and a variable of type *WNDCLASS*, as shown here:

```
MSG msg;
HWND hWnd;
WNDCLASS wc;
```

MSG is an enumerated type that you will learn more about in later Tips. *HWND* is a window handle. The *WNDCLASS* type maintains information about the Window class the program uses, as Tips 1267 and 1269 detail. The next several statements within the *generic.cpp* program are all assignments to member values of the *wc* variable. Tip 1269 will explain the assignments in detail. The following Tips will explain the remaining statements within the *WinMain* function.

UNDERSTANDING WINDOW CREATION 1266

As you have learned, the foundation of any Windows program's interaction with the user is through a window that the program creates. Every program that you design in Windows to interact with the user through the Windows interface will create at least one window during its processing. Most programs will register a set of *window classes* and create multiple instances of those classes. You will learn more about registering window classes in Tip 1269.

Creating a window is a relatively simple process: you must first determine the window's components and appearance, and then use the Win32 API *CreateWindow* function to create the window. As you saw in Tip 1257, your programs will implement the *CreateWindow* function with a series of parameters. The generalized form of the *CreateWindow* function is shown here:

```
HWND CreateWindow (LPCTSTR lpszClassName, LPCTSTR lpszWindowName, DWORD dwStyle,
                   int x, int y, int nWidth, int nHeight, HWND hwndParent,
                   HMENU hmenu, HANDLE hinst, LPVOID lpvParam)
```

Clearly, the *CreateWindow* function requires some preparation within your program before the program can invoke the function. Table 1266 details the eleven parameters that you must pass to the *CreateWindow* function.

Parameter Type	Parameter Name	Description
LPCTSTR	*lpszClassName*	A constant pointer to a *NULL*-terminated string that contains a valid window class name. The class name can either be one the program creates with *RegisterClass* or a predefined window type, as Tip 1269 details.
LPCTSTR	*lpszWindowName*	A constant pointer to a *NULL*-terminated string that contains the window's name. Depending on the window style, the window name may display in several different locations.
DWORD	*dwStyle*	A double-WORD value (an unsigned 32-bit integer) that corresponds to the possible styles for a window. Tip 1275 discusses the DWORD type in detail. You will create styles from values that your programs combine together with a binary *OR* operator. For example, an acceptable style would be *WS_CHILD* | *ES_LEFT*. Tips 1279 and 1280 discuss window styles in detail.
int	*x*	The horizontal position of the window's upper left corner. If the position is not important, your program should pass the *CW_USEDEFAULT* value for the *x* parameter.

***Table 1266** The parameters for the **CreateWindow** API function.* *(continued on following page)*

int	*y*	The vertical position of the window's upper left corner. If the position is not important, your program should pass the *CW_USEDEFAULT* value for the *y* parameter.
int	*nWidth*	The horizontal width of the window. If the width is not important, your program should pass the *CW_USEDEFAULT* value for the *nWidth* parameter.
int	*nHeight*	The vertical height of the window. If the height is not important, your program should pass the *CW_USEDEFAULT* value for the *nHeight* parameter.
HWND	*hwndParent*	A handle to the window's parent window. If there is no parent window, pass a *NULL* value for this parameter.
HMENU	*hmenu*	A handle to a window's menu. Pass *NULL* if API should use the menu registered with the window class.
HANDLE	*hinst*	The instance handle of the program that creates the control.
LPVOID	*lpvParam*	A pointer to data that the *CreateWindow* function should pass in the *WM_CREATE* message. For multi-document interface (MDI) child windows, the *lpvParam* value should be a pointer to a *CLIENTCREATESTRUCT* structure. For most non-MDI client windows, pass a *NULL* value.

***Table 1266** The parameters for the **CreateWindow** API function. (continued from previous page)*

While the creation of a window may seem daunting right now, you will discover in Tip 1267 that many of the windows your programs create will share common characteristics—so using *CreateWindow* will become significantly easier.

1267 UNDERSTANDING THE CREATEWINDOW FUNCTION

Tip 1266 introduced you to the basics behind creating windows within your programs. However, with its eleven parameters, understanding the *CreateWindow* function is significantly easier when you consider a real case, rather than the general case Tip 1266 described. Therefore, consider the following code fragment from the *generic.cpp* program:

```
hWnd = CreateWindow(lpszAppName, lpszTitle, WS_OVERLAPPEDWINDOW, CW_USEDEFAULT, 0,
                    CW_USEDEFAULT, 0, NULL, NULL, hInstance, NULL);
if(!hWnd)
   return false;
```

The three statements perform two steps. The first statement tries to create a window. If the statement is successful, it will return a handle to the newly created window. If not successful, *CreateWindow* returns *False*. The second statement checks the window handle to determine whether the window creation was successful. If the window creation was not successful, the third statement ends the program with a false (failed) result.

However, understanding the window's creation is more complex. The *lpszAppName* variable points to a string that contains the application name—an application name that also corresponds with certain information within the resource file, as you learned in previous Tips. The *lpszTitle* pointer corresponds to the "Generic Application" string, which the window will display in its title bar. The *WS_OVERLAPPEDWINDOW* parameter tells the *CreateWindow* function to create an overlapped window—a style you will learn more about in Tip 1279. The next four parameters tell the *CreateWindow* function where and how large to create the window. The first *NULL* parameter lets *CreateWindow* know that this window has no parent, and the second *NULL* parameter tells *CreateWindow* it should use the default menu for this window class. The next-to-last parameter passes in the program's instance, and the final parameter passes a *NULL* value to the *WM_CREATE* message.

Especially in your early programs, many of these parameters will be defaults or will use variables your program initializes at its start-up, to keep the processing simpler. In fact, the only parameter that you will probably change regularly during your early Windows programming efforts is the *dwStyle* (window style) parameter.

Understanding the ShowWindow Function 1268

In Tip 1267, you learned how the *generic.cpp* program creates a window. However, when you program within Windows, you must make the window visible after you create it. To help your programs display windows, the Win32 API provides the *ShowWindow* function. Your programs will use the *ShowWindow* function as shown in the following prototype:

```
BOOL ShowWindow(HWND hWnd, int nCmdShow);
```

The *ShowWindow* function returns a true or false success value that your programs should, as a rule, check the function's success. The *hWnd* parameter is the handle to the window you want to display. In the *generic.cpp* program, *hWnd* refers to the newly created window. The *nCmdShow* parameter controls how your program will display the window. The value you pass in through the *nCmdShow* parameter must correspond to one of the values Table 1268 details.

Value	Meaning
SW_HIDE	Hide the window.
SW_MINIMIZE	Minimize the window and activate the top-level window in the system's window list.
SW_RESTORE	Activate and display the window. If the window is currently minimized or maximized, *ShowWindow* will return the window to its original dimensions and position.
SW_SHOW	Display the window in its current size and position.
SW_SHOWDEFAULT	Display the window in its application default state. *ShowWindow* obtains the application default state from the *STARTUPINFO* structure, which you will learn about in later Tips.
SW_SHOWMAXIMIZED	Display the window as maximized.
SW_SHOWMINIMIZED	Display the window as minimized to an icon.
SW_SHOWMINNOACTIVE	Display the new window minimized and the currently active window will remain active.
SW_SHOWNA	Display the window in its current state, and the active window will remain active.
SW_SHOWNOACTIVE	Display the window in its most recent size and position, and the active window will remain active.
SW_SHOWNORMAL	Display the window in its normal size.

***Table 1268** Valid values for the **nCmdShow** parameter.*

The following modification to the *generic.cpp* program, *tst_max.cpp*, tells Windows to maximize the application window when the user selects the File menu Demo option. Delete the current code within the *WndProc* function and replace it with the following code:

```
LRESULT CALLBACK WndProc( HWND hWnd, UINT uMsg, WPARAM wParam, LPARAM lParam)
 {
   switch(uMsg)
```

```
    {
      case WM_COMMAND:
         switch(LOWORD(wParam))
          {
            case IDM_DEMO :
               ShowWindow(hWnd, SW_SHOWMAXIMIZED);
               break;
            case IDM_EXIT :
               DestroyWindow(hWnd);
               break;
          }
         break;
      case WM_DESTROY :
         PostQuitMessage(0);
         break;
      default:
         return (DefWindowProc(hWnd, uMsg, wParam, lParam));
    }
   return(0L);
  }
```

The only result from the change to the code from the original *generic.cpp* program code is that the window maximizes to fill the screen when the user selects *Demo* from the window's menu. After you compile and execute the modified *generic.cpp* program, test the program's processing by first maximizing the program window, then setting the window back to its normal size. In later Tips, you will perform more extensive manipulations to control a window's appearance, and you will use the settings for *ShowWindow* to better control the window's appearance.

1269 UNDERSTANDING THE REGISTERCLASS FUNCTION

As you learned in Tips 1266 and 1267, when your programs create windows, they can either create a window of a predefined class or they can create their own window styles. When you create your own window styles, you must register the window style with Windows before you can use the style to create windows. You will use the *RegisterClass* API function to register window styles. You will use the *RegisterClass* function within your programs as shown in the following prototype:

```
ATOM RegisterClass(CONST WNDCLASS* lpwc)
```

As you can see, the *RegisterClass* function returns a value of type *ATOM*. An *ATOM* is a *WORD* value that refers to character strings in a case insensitive manner. The fact that *ATOMs* refer to strings in a case insensitive manner means that "happy" is equal to "HAPPY"—an equality which, as you know, is not usually true within C++. Windows stores *ATOMS* in an *ATOM Table*—so the *WORD* value that an *ATOM* maintains is actually very similar to a handle.

In addition to returning an *ATOM* value, the *RegisterClass* function accepts a single parameter—a constant pointer to a structure of type *WNDCLASS*. Windows defines the *WNDCLASS* structure, as shown here:

```
typedef struct _WNDCLASS
 {
   UINT        style;
   WNDPROC     lpfnWndProc;
   int         cbClsExtra;
   int         cbWndExtra;
   HINSTANCE   hInstance;
   HICON       hIcon;
   HCURSOR     hCursor;
   HBRUSH      hbrBackground;
   LPCTSTR     lpszMenuName;
   LPCSTR      lpszClassName;
 } WNDCLASS;
```

As you can see, ten data members compose the *WNDCLASS* structure. Table 1269.1 details the *WNDCLASS* structure data members.

Member Name	Type	Function
style	*UINT*	The style parameter must be one or more of the styles Table 1269.2 details, combined with the binary *OR* operator.
lpfnWndProc	*WNDPROC*	Points to a window callback function that processes messages Windows generates for the window.
cbClsExtra	*int*	The number of extra bytes that *RegisterClass* should allocate at the end of the window class structure for storing information.
cbWndExtra	*int*	The number of extra bytes that *RegisterClass* should allocate following each instance creation for storing information.
hInstance	*HINSTANCE*	A handle to the instance of which the window class is part.
hIcon	*HICON*	A handle to the icon that *CreateWindow* will use for this window class.
hCursor	*HCURSOR*	A handle to the cursor that *CreateWindow* will use for this window class.
hBrush	*HBRUSH*	A handle to a brush that *CreateWindow* will use to create the window's background. You will learn more about brushes in later Tips.
lpszMenuName	*LPCTSTR*	A pointer to a *NULL*-terminated string constant with the default menu name for the class. You should set this value to *NULL* if the window is to have no default menu class.
lpszClassName	*LPCTSTR*	A pointer to a *NULL*-terminated string constant that contains the class name. Your programs will later use the class name in the *CreateWindow* function's *lpszClassName* parameter.

***Table 1269.1** The data members of the **WNDCLASS** structure.*

As Table 1269.1 indicates, the window class *style* data member can correspond to one or more binary constants. Table 1269.2 lists the acceptable values for the *style* data member.

Style Constant	Meaning
CS_BYTEALIGNCLIENT	Aligns a window's client area on the byte boundary horizontally to enhance drawing performance. This affects the window's width and its horizontal position on screen.
CS_BYTEALIGNWINDOW	Aligns a window on a byte boundary horizontally.
CS_CLASSDC	Allocates one device context (DC) that all windows in the class will share. If multiple threads try to access the device context simultaneously, Windows lets only one thread finish successfully. You will learn more about device contexts in later Tips.
CS_DBLCLKS	Notifies a window when the user double-clicks the mouse button.
CS_GLOBALCLASS	Creates a class that is available to all applications while the application that created the class is open. Generally, you will use this setting when you create custom controls for other programs to use.
CS_HREDRAW	Redraws the entire window if the user adjusts the horizontal size.

***Table 1269.2** Valid values for the window class **style** data member.* *(continued on following page)*

Style Constant	Meaning
CS_NOCLOSE	Disables the Close command on the System menu.
CS_OWNDC	Allocates a unique device context for each instance of the window class.
CS_PARENTDC	Each window that your programs create of the class will use the parent window's device context.
CS_SAVEBITS	Saves, as a bitmap, the portion of the screen image a window obscures. Windows uses the bitmap to recreate the screen image when the user removes the window.
CS_VREDRAW	Redraws the entire window if the user adjusts the vertical size.

***Table 1269.2** Valid values for the window class **style** data member. (continued from previous page)*

Now that you understand better how the *RegisterClass* statement works, consider the following assignments from the *generic.cpp* program, which initialize the particular window class the *generic.cpp* program uses:

```
wc.style          = CS_HREDRAW | CS_VREDRAW;
wc.lpfnWndProc    = (WNDPROC)WndProc;
wc.cbClsExtra     = 0;
wc.cbWndExtra     = 0;
wc.hInstance      = hInst;
wc.hIcon          = LoadIcon(hInstance, lpszAppName);
wc.hCursor        = LoadCursor(NULL, IDC_ARROW);
wc.hbrBackground  = (HBRUSH)(COLOR_WINDOW+1);
wc.lpszMenuName   = lpszAppName;
wc.lpszClassName  = lpszAppName;
```

The first assignment tells the operating system to redraw the entire window whenever the user resizes the window in either direction. The second assignment tells the operating system that the callback function is the *WinProc* function. The next two assignment statements tell *RegisterClass* not to allocate extra space and the *wc.hInstance* = statement tells the compiler to use the current program instance.

The next two assignment statements (for *hIcon* and *hCursor*) load the icon and the cursor for the window to use. The assignment statement after those statements (*hbrBackground* =) creates a handle to a color brush, and the final two statements assign the window's default menu name and the window class name.

Again, as you will learn with many of the commands and structures you manipulate for Windows, you often will use the commands and structures the same way, most often with the same values, and will only occasionally change some values when you create a special window.

***Note:** A special case exists when you use **CreateWindow** to create a window using an existing class. The existing class names are **BUTTON, LISTBOX, COMBOBOX, STATIC, EDIT, MDICLIENT,** and **SCROLLBAR**. It is not necessary to register these classes before your program creates a window using one of the classes.*

1270 LEARNING MORE ABOUT MESSAGES

The last piece of the *WinMain* function in the *generic.cpp* program is the *while* loop, which processes messages from the system. As you have learned, you will write your Windows programs to persist, as a rule, until the user specifically instructs the window to close. Each Windows program you write will use a *message loop* to continue processing messages until the user instructs the program to stop. The standard form of the message loop is shown here:

```
while(GetMessage(&msg, NULL, 0,0))
 {
    TranslateMessage(&msg);
    DispatchMessage(&msg);
 }
```

In Tips 1271 and 1272, you will learn more about the *TranslateMessage* and *DispatchMessage* functions. However, it is important that you first understand *GetMessage* and what it returns. You will use the *GetMessage* function within your programs as shown in the following prototype:

```
BOOL GetMessage(LPMSG lpmsg, HWND hwnd, UNIT uMsgFilterMin, UINT uMsgFilterMax)
```

The *GetMessage* function returns a true or false value. *GetMessage* returns true until it retrieves the *WM_QUIT* message. Table 1270 lists the parameters for the *GetMessage* function.

Member Name	Type	Purpose
lpmsg	*MSG*	Returns a pointer to a *MSG* structure. This Tip defines the *MSG* structure after this table.
hwnd	*HWND*	A handle to the window that receives the messages. Usually, you will set this value to *NULL*, which instructs *GetMessage* to catch all messages for the current thread.
uMsgFilterMin	*UINT*	The minimum message value to receive. Usually, you will set this parameter to 0.
uMsgFilterMax	*UINT*	The maximum message value to receive. If you set both *uMsgFilterMin* and *uMsgFilterMax* to 0, *GetMessage* will retrieve all messages.

Table 1270 The parameters for the ***GetMessage*** *function.*

As Table 1270 indicates, the *GetMessage* function receives a parameter of type *MSG*. Windows defines the *MSG* type within the *winuser.h* header file, as shown here:

```
typedef struct tagMSG {
    HWND        hwnd;        // window handle
    UINT        message;     // message ID
    WPARAM      wParam;      // wParam value
    LPARAM      lParam;      // lParam value
    DWORD       time;        // milliseconds since start-up
    POINT       pt;          // screen coordinates of current mouse location
} MSG;
```

While each *MSG* structure component is important, the one that you will most often manipulate is the *message* member, which corresponds to one of many Windows constant definitions for messages. You will learn more about the message constants in later Tips.

USING TRANSLATEMESSAGE TO PROCESS MESSAGES 1271

As you learned in Tip 1270, your program's *WinMain* function will generally end with a *while* loop, which retrieves messages until the user sends the *WM_QUIT* message to the system. Inside the message loop, as you saw in Tip 1270, the program first calls the *TranslateMessage* function. Then *TranslateMessage* takes a virtual-key message (such as *VK_TAB*) that the system generates when the user presses a key, and posts the corresponding *WM_CHAR* code to the application's message queue (*WM_CHAR* stands for *Windows Message, Character*). If the message is not a virtual-key message, the *WM_CHAR* will return false and will not process the message. You will use the *TranslateMessage* function within your programs as shown in the following prototype:

```
BOOL TranslateMessage(CONST MSG* lpMsg);
```

As you have seen, you will generally call *TranslateMessage* immediately after a *GetMessage* call, although you can also use a message that the *PeekMessage* function (explained in later Tips) returns, as shown here:

```
while(GetMessage(&msg, NULL, 0,0))
 {
    TranslateMessage(&msg);
    DispatchMessage(&msg);
 }
```

1272 Using DispatchMessage to Process Messages

Normally, the *WinMain* function within your programs loops, waiting for messages. When a message arrives, *WinMain* dispatches the message to another function that processes the message. As you learned in Tip 1270, your programs will create a message loop to process messages until the program terminates. The last component of the message loop is the *DispatchMessage* function, which sends the message to the processing function after the *TranslateMessage* statement ensures the message is a Windows message. The processing function that *DispatchMessage* calls is the callback function the window class defines in its registration. You will implement the *DispatchMessage* function as shown in the following generalized form:

```
long DispatchMessage(CONST MSG* lpmsg);
```

Although *DispatchMessage* returns a value of type *long*, your programs will generally ignore the result of the *DispatchMessage* call because it does not provide useful information to your program.

Note: *Your message loops **must** include the **DispatchMessage** function or they will not be able to process the messages the system sends.*

1273 Understanding the Components of a Simple Windows Program

Throughout the previous 16 Tips, you have studied the *generic.cpp* program and its components. The *generic.cpp* program includes all the basic components of any Windows program. When you design Windows programs, you must make sure that your Windows programs include all the following components, just as the *generic.cpp* program does:

- A *Resource File*: While you can write Windows programs without using a resource file, it creates additional work for you and is not the standard Windows program construction. Every Windows program you write should include a resource file. However, you must not create more than one resource file per program build. Remember, a resource file may include information about many windows and their contents within a given program, so you should place all information for each program build within a single file.
- The *windows.h* header: All Windows programs *must* include the *windows.h* header which, in turn, includes all the necessary header files for Windows-specific types, functions, and classes.
- The *WinMain* function: Just as all your DOS programs required the *main* function, all your programs for Windows must include the *WinMain* function. Remember, however, that you cannot have a *WinMain* function without its expected four parameters: *HINSTANCE hInstance, HINSTANCE hPrevInstance, LPSTR lpCmdLine,* and *int nCmdShow*.

- A Windows message loop: Every Windows program that requires user interaction (in short, almost all Windows programs) processes messages within a message loop. The message loop retrieves messages from the system message queue and processes them within your programs.
- A message callback function: When you create a window, one of the parameters for that window is where it should send its messages within the program—the message callback function. The message loop, too, sends messages to the callback function. *Every Windows program must have a message callback function.* You will generally name your message callback functions consistently, just as you do *WinMain*, although it is not required. You may also have multiple message callback functions to handle messages within different program windows differently.

Nearly every Windows program will have these five components. Every Windows program you use within this book will have all five components. If you design a program without all five components, you will probably have difficulty making the program work correctly under Windows.

UNDERSTANDING THE LPCTSTR TYPE 1274

As you have learned, your programs will use the Windows *LPCTSTR* type to store a 32-bit pointer to a constant character string. The *LPCTSTR* character type is portable for both Unicode and the double-byte character set (DBCS). You will most often declare strings as *LPCTSTR* if the string is a parameter to a function which the function will not modify. For speed and convenience, if you know a function receives string parameters *by value*, you should declare the string function parameter as type *LPCTSTR*. Your programs should use a *LPTSTR* parameter (a 32-bit pointer to a *NULL*-terminated string) rather than an *LPCTSTR* parameter if the function may return values within the parameter.

For example, the following code fragment declares a class *CName*. The *CName* class includes the two member functions, *SetData* and *GetData*. Because the *SetData* function does not change information within its component strings, *SetData* declares both strings as *LPCTSTR*. *GetData*, on the other hand, modifies the strings, and so declares the strings as *LPSTR*, as shown here:

```
class CName
{
private:
    LPSTR m_firstName;
    char m_middleInit;
    CString m_lastName;
public:
    CName() {}
    void SetData( LPCTSTR fn, const char mi, LPCTSTR ln )
    {
        m_firstName = fn;
        m_middleInit = mi;
        m_lastName = ln;
    }
    void GetData( LPSTR& cfn, char mi, LPSTR& cln )
    {
        cfn = m_firstName;
        mi = m_middleInit;
        cln = m_lastName;
    }
};
```

1275 UNDERSTANDING THE DWORD TYPE

As you have learned in previous Tips, a *DWORD* is a double-*WORD* type. A *WORD* type is a 16-bit unsigned integer (making it the equivalent of the *unsigned long int* in 16-bit C). A double-*WORD* then, is a 32-bit unsigned integer capable of storing values up to 2^{32}-1. Figure 1275.1 shows how your computer allocates memory for a *WORD* and a *DWORD*.

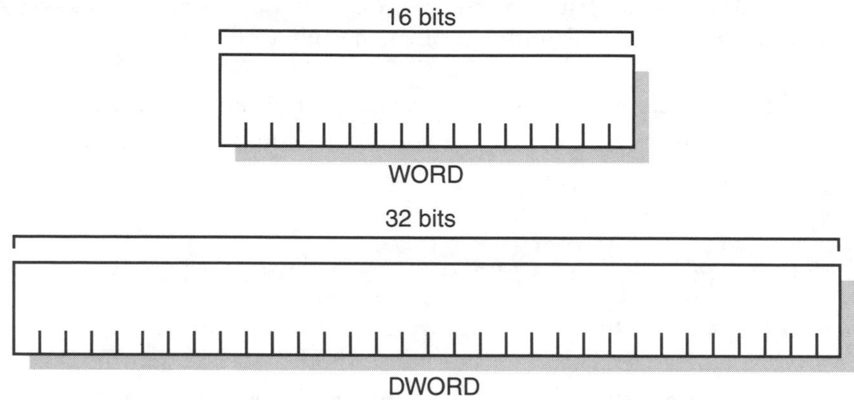

Figure 1275.1 Memory allocation for a ***WORD*** *and a* ***DWORD****.*

While your programs can use a *DWORD* value to store large, unsigned integer numbers, your programs will more often use *DWORD* values to maintain a 32-bit segment and offset address. Because two *WORDs* compose a *DWORD*, most C++ compilers provide several tools for breaking *DWORDs* into a high *WORD* (which contains the two most significant bytes) and a low *WORD* (which contains the two least significant bytes). When you break a *DWORD* into two *WORDs*, the high *WORD* represents the segment address and the low *WORD* represents the offset address, as shown in Figure 1275.2.

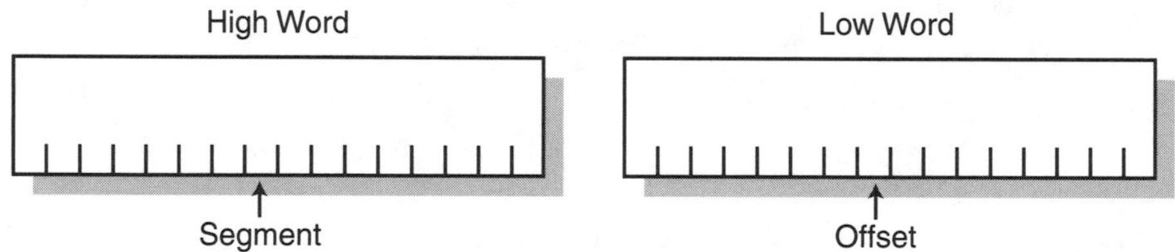

Figure 1275.2 A ***DWORD*** *can represent a segment and offset address.*

1276 UNDERSTANDING WINDOWS PREDEFINED CLASSES

As Tips 1267 and 1269 mention briefly, you can also create windows of several derived types within your programs using the Windows predefined classes. The windows that you will create using the Windows predefined window classes, with the exception of the *MDICLIENT* window (which you will learn about in later Tips), are known as *controls*. Although you will not learn other methods in this book, you can also use ActiveX controls and compiler-specific window classes to create additional windows within your programs. The Windows API supports the predefined window classes listed in Table 1276.

Class	Description
BUTTON	Your programs will use the *BUTTON* control to create buttons within your windows. Buttons may be rectangular push buttons, group boxes, check boxes, radio buttons, or icon windows. Generally, your programs will use a button to start event processing, or the user will use it to tell the system that he or she has completed entry into the form a window displays. Figure 1276.1 shows a sample push button on a window.

***Figure 1276.1** A sample push button in a window.*

LISTBOX	Your programs will use list boxes to maintain lists of information for users. A *LISTBOX* control differs from a *COMBOBOX* control in that it accepts no input other than selections within the list the *LISTBOX* contains. Figure 1276.2 shows a sample *LISTBOX* control.

Figure 1276.2** A sample* ***LISTBOX *control.*

COMBOBOX	A *COMBOBOX* control is a combined *EDIT* control and *LISTBOX* control. Within your programs, you will generally use combo boxes to let your users both select from a list and enter their own new entry into the list. Figure 1276.3 shows a sample *COMBOBOX* control.

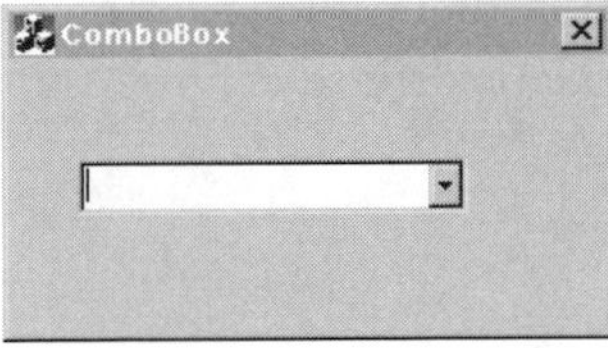

Figure 1276.3** A sample* ***COMBOBOX *control.*

STATIC	Your programs will often use a *STATIC* control, also known as a label, to place information onto your window's client area. Information you place within a *STATIC* control should be information that the user will not edit or otherwise modify during the course of the program's execution. Figure 1276.4 shows a sample *STATIC* control.

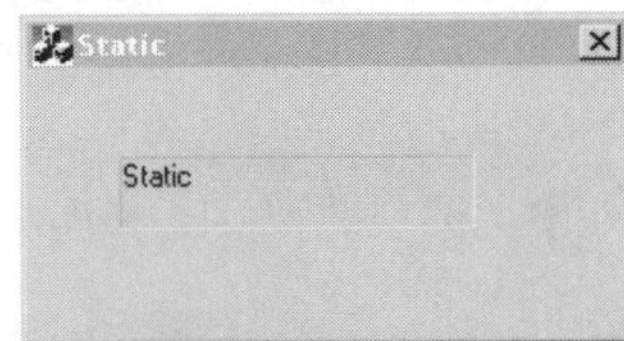

Figure 1276.4** A sample* ***STATIC *control.*

EDIT	You will use *EDIT* controls to let your users enter information into your programs through the Windows interface. Generally, most Windows programs will include one or more *EDIT* controls, especially business application programs. Figure 1276.5 shows a sample *EDIT* control.

***Figure 1276.5** A sample **EDIT** control.*

SCROLLBAR	Your programs will use the *SCROLLBAR* control to place scrollbars at the edges of existing windows, within a window's client area, and to receive user input.

***Table 1276** The preexisting window classes. (continued from previous page)*

In Tip 1277, you will create a simple program that uses several of the predefined window classes. In general, your programs will often use many predefined window classes within a single program, and will often use many different classes even within a single window.

***Note:** Both Microsoft's **Visual C++** and Borland's **C++**, the most commonly used Windows C++ development compilers, include drag-and-drop control support, which lets you add controls to your windows without performing the additional steps necessary to create the controls using the **CreateWindow** command.*

1277 USING PREDEFINED CLASSES TO CREATE A SIMPLE WINDOW

In Tip 1276, you learned about the predefined classes that the *CreateWindow* command supports. As you learned, your programs will generally use predefined classes to create child windows (*controls*) within a program window. Using predefined classes is no more difficult than creating an instance of a window that your program has already registered. Your program will simply invoke the *CreateWindow* function with the first parameter containing the predefined class name, the second parameter containing the text within the predefined control, and the later parameters determining the control's style and placement on the program window.

For example, the CD-ROM that accompanies this book includes the *show_three.cpp* program, which creates the generic opening window you have used previously at the program's beginning. When the user selects the *Test!* option, the program displays a *STATIC* control, an *EDIT* control, and a *BUTTON* control. The *show_three* program uses three files: *show_three.h*, *show_three.cpp*, and *show_three.rc*. In general, the files are the same as for the *generic.cpp* program, with the exception of the *WndProc* function, which calls *CreateWindow* three times when the user selects the *Demo* option.

As will can see from the code listing, the only change between the *WndProc* function in the *show_three.cpp* file and the *WndProc* function in the *generic.cpp* program is how the *show_three.cpp* program handles the user's selection of the *Demo* option. Within the code, if the user selects *Demo*, which sends an *IDM_DEMO* message to *WndProc*, the code will invoke *CreateWindow* three times: once with a *STATIC* control, once with an *EDIT* control, and once with a *BUTTON* control.

1278 WINDOWS SENDS WM_CREATE WHEN IT CREATES A WINDOW

As you have learned, whenever Windows performs a significant activity, it sends a message out to the programs currently executing, informing those programs of the activity Windows completed. Creating windows is no exception. Each time a program invokes the *CreateWindow* function, Windows sends five messages to the *WndProc* function. Typically, your programs will process the *WM_CREATE* message within the *WndProc* function and will let the other four messages pass on to *DefWindowProc*. Your applications will use the *WM_CREATE* message to inform the program to initialize the window. Table 1278 lists the five messages that Windows sends to the program when it successfully creates a window.

Message	Meaning
WM_GETMINMAXINFO	Gets the size and position of the window that Windows is creating.
WM_NCCREATE	Indicates that Windows is about to create the non-client area of the window. The *DefWindowProc* function allocates memory for the window and initializes the scroll bars when it receives the *WM_NCCREATE* message.
WM_NCCALCSIZE	The *DefWindowProc* function calculates the size and position of the window's client area when it receives the *WM_NCCALCSIZE* message.
WM_CREATE	The *WndProc* function should perform initialization of the window when it receives the *WM_CREATE* message, which indicates that Windows is about to create the window.
WM_SHOWWINDOW	Informs the *DefWindowProc* function that Windows is about to show the window.

Table 1278 The messages Windows generates after your program invokes the ***CreateWindow*** *function.*

As a rule, your programs should let the *DefWindowProc* function handle most of the processing when you create windows. (In fact, your programs should let *DefWindowProc* handle all the messages you do not specifically require the programs to handle.) As you have seen, you should make the call to *DefWindowProc* the default for the case statement within your *WndProc* function, as shown here:

```
default:
   return(DefWindowProc(hWnd, uMsg, wParam, lParam));
```

UNDERSTANDING WINDOW AND CONTROL STYLES 1279

As you have learned in previous Tips, your programs will invoke the *CreateWindow* function with a set of parameters that provide the function with information about the new window to create. One parameter that you will pass in every invocation of the *CreateWindow* function is the *DWORD dwStyle* parameter. As you have learned, your programs create the value for the *dwStyle* parameter by using a sequence of *bitwise OR* operators with a sequence of window style constants. As you saw briefly in Tip 1277, your programs will also use different *dwStyle* parameters when they create instances of predefined window classes. Table 1279.1 lists the *dwStyle* parameters for window classes.

Style	Description
WS_BORDER	Creates a window that has a thin-line border.
WS_CAPTION	Creates a window that has a title bar (includes the *WS_BORDER* style).
WS_CHILD	Creates a child window. You cannot use this style with the *WS_POPUP* style.
WS_CHILDWINDOW	Creates the same window as the *WS_CHILD* style.
WS_CLIPCHILDREN	Excludes the area the child windows occupy when drawing occurs within the parent window. You use this style when you create the parent window.

Table 1279.1 The possible values for the ***dwStyle*** *parameter when you create windows within your programs.* *(continued on following page)*

Style	Description
BS_DEFPUSHBUTTON	Creates a push button that behaves like a *BS_PUSHBUTTON* style button, but also has a heavy black border. If the button is in a dialog box, the user can select the button by pressing the ENTER key, even when the button does not have the input focus. This style is useful for letting the user quickly select the most likely (default) option.
BS_GROUPBOX	Creates a rectangle in which your program can then group other controls. Your program displays any text associated with this style in the rectangle's upper left corner. The *GROUPBOX* is also commonly known as a *frame*.
BS_LEFTTEXT	Places text on the left side of the radio button or check box when combined with a radio button or check box style. Same as the *BS_RIGHTBUTTON* style.
BS_OWNERDRAW	Creates an owner-drawn button. The owner window receives a *WM_MEASUREITEM* message when Windows creates the button and a *WM_DRAWITEM* message when a visual aspect of the button has changed. Do not combine the *BS_OWNERDRAW* style with any other button styles.
BS_PUSHBUTTON	Creates a push button that posts a *WM_COMMAND* message to the owner window when the user selects the button.
BS_RADIOBUTTON	Creates a small circle with text. By default, your program displays the text to the right of the circle. To display the text to the left of the circle, combine this flag with the *BS_LEFTTEXT* style (or with the equivalent *BS_RIGHTBUTTON* style). Use radio buttons for groups of related, but mutually exclusive, choices.
BS_USERBUTTON	This style is obsolete, but remains compatible with 16-bit versions of Windows. Win32-based applications should use *BS_OWNERDRAW* instead.
BS_BITMAP	Specifies that the button displays a bitmap.
BS_BOTTOM	Places text at the bottom of the button rectangle.
BS_CENTER	Centers text horizontally in the button rectangle.
BS_ICON	Specifies that the button displays an icon.
BS_LEFT	Left-justifies the text in the button rectangle. However, if the button is a check box or radio button that does not have the *BS_RIGHTBUTTON* style, the text is still left-justified, but on the right side of the check box or radio button.
BS_MULTILINE	Wraps the button text to multiple lines if the text string is too long to fit on a single line in the button rectangle.
BS_NOTIFY	Enables a button to send *BN_DBLCLK*, *BN_KILLFOCUS*, and *BN_SETFOCUS* notification messages to the button's parent window. Note that buttons send the *BN_CLICKED* notification message regardless of whether or not the button has the *BS_NOTIFY* style.
BS_PUSHLIKE	Makes a button (such as a check box, three-state check box, or radio button) look and act like a push button. The button looks raised when it is not in a pushed or checked state, and sunken when it is in a pushed or checked state.

***Table 1279.2** The possible values for the **dwStyle** parameter when you create **BUTTON** windows.* *(continued on following page)*

Style	Description
BS_RIGHT	Right-justifies text in the button rectangle. However, if the button is a check box or radio button that does not have the *BS_RIGHTBUTTON* style, the text is still right-justified, but on the right side of the check box or radio button.
BS_RIGHTBUTTON	Positions a radio button's circle or a check box's square on the right side of the button rectangle. Same as the *BS_LEFTTEXT* style.
BS_TEXT	Specifies that the button displays text.
BS_TOP	Places text at the top of the button rectangle.
BS_VCENTER	Places text in the middle (vertically) of the button rectangle.

***Table 1279.2** The possible values for the **dwStyle** parameter when you create **BUTTON** windows.* *(continued from previous page)*

Just as the *BUTTON* window has its own styles, so too do each of the other predefined window classes. However, this Tip does not list them because the space required to list every one is prohibitive.

***Note:** Most Windows C++ development suites include information on the window styles in their on-line help facility. Check your program's on-line help under **CreateWindow** to find out more.*

CREATING WINDOWS WITH EXTENDED STYLES 1280

In previous Tips, you have learned how to use the *CreateWindow* function to create both custom and predefined windows within your programs. Also, your programs can use the *CreateWindowEx* function to create windows with *extended styles,* such as overlapped, pop-up, or child window styles. You will use the *CreateWindowEx* function within your programs as shown in the following prototype:

```
HWND CreateWindowEx(
   DWORD dwExStyle,          // extended window style
   LPCTSTR lpClassName,      // pointer to registered class name
   LPCTSTR lpWindowName,     // pointer to window name
   DWORD dwStyle,            // window style
   int x,                    // horizontal position of window
   int y,                    // vertical position of window
   int nWidth,               // window width
   int nHeight,              // window height
   HWND hWndParent,          // handle to parent or owner window
   HMENU hMenu,              // handle to menu, or child-window identifier
   HINSTANCE hInstance,      // handle to application instance
   LPVOID lpParam            // pointer to window-creation data
  );
```

As you can see, the *CreateWindowEx* function accepts the same parameters as the *CreateWindow* function, with the exception of the *dwExStyle* parameter, which specifies the window's extended style. Note that your programs can specify an unlimited number of *dwExStyles* for your windows, provided you link the styles with a *bitwise OR* operation. Table 1280 lists the possible values for the *dwExStyle* parameter.

Style	Description
WS_EX_ACCEPTFILES	Specifies that a window the operating system creates with this style accepts drag-drop files.
WS_EX_APPWINDOW	Forces a top-level window onto the taskbar when Windows minimizes the window.

***Table 1280** The possible values for the **dwExStyle** parameter.* *(continued on following page)*

Style	Description
WS_EX_CLIENTEDGE	Specifies that a window has a border with a sunken edge.
WS_EX_CONTEXTHELP	Includes a question mark in the title bar of the window. When the user clicks his or her mouse on the question mark, the cursor will change to a question mark with a pointer. If the user then clicks the mouse on a child window, the child will receive a *WM_HELP* message. The child window should pass the message to the parent window's window procedure, which should call the *WinHelp* function using the *HELP_WM_HELP* command. The help application displays a pop-up window that typically contains help for the child window. Your programs cannot use *WS_EX_CONTEXTHELP* with the *WS_MAXIMIZEBOX* or *WS_MINIMIZEBOX* styles.
WS_EX_CONTROLPARENT	Lets the user navigate among the child windows of the window by using the TAB key.
WS_EX_DLGMODALFRAME	Creates a window that has a double border. You have the option of creating the window with a title bar by specifying the *WS_CAPTION* style in the *dwStyle* parameter.
WS_EX_LEFT	Windows has generic left-aligned properties. This is the default style.
WS_EX_LEFTSCROLLBAR	Places the vertical scroll bar (if present) to the left of the client area if the shell language is Hebrew, Arabic, or another language that supports reading order alignment. For other languages, Windows ignores the style and does not treat it as an error.
WS_EX_LTRREADING	Displays the window text using left to right reading-order properties. This is the default style.
WS_EX_MDICHILD	Creates a Multiple Device Interface (MDI) child window.
WS_EX_NOPARENTNOTIFY	Specifies that a child window created with this style does not send the *WM_PARENTNOTIFY* message to its parent window when Windows creates or destroys it.
WS_EX_OVERLAPPEDWINDOW	Combines the *WS_EX_CLIENTEDGE* and *WS_EX_WINDOWEDGE* styles.
WS_EX_PALETTEWINDOW	Combines the *WS_EX_WINDOWEDGE, WS_EX_TOOLWINDOW*, and *WS_EX_TOPMOST* styles.
WS_EX_RIGHT	Windows has generic right-aligned properties, although the property is dependent upon the window class. This style has an effect only if the shell language is Hebrew, Arabic, or another language that supports reading order alignment; otherwise, Windows ignores the style and does not treat it as an error. Using the *WS_EX_RIGHT* style for static or edit controls has the same effect as using the *SS_RIGHT* or *ES_RIGHT* style, respectively. Using this style with button controls has the same effect as using *BS_RIGHT* and *BS_RIGHTBUTTON* styles.
WS_EX_RIGHTSCROLLBAR	Places the vertical scroll bar (if present) to the right of the client area. This is the default style.

Table 1280 *The possible values for the* ***dwExStyle*** *parameter.* *(continued on following page)*

Style	Description
WS_EX_RTLREADING	Displays the window text using right to left reading-order properties if the shell language is Hebrew, Arabic, or another language that supports reading order alignment. For other languages, Windows ignores the style and does not treat it as an error.
WS_EX_STATICEDGE	Creates a window with a three-dimensional border style intended for Windows to use with items that do not accept user input.
WS_EX_TOOLWINDOW	Creates a tool window, which is a window your program intends for Windows to use as a floating toolbar. A tool window has a title bar that is shorter than a normal title bar, and Windows will use a smaller font to draw the window title. A tool window does not appear in the taskbar or in the dialog box that appears when the user presses ALT+TAB. If a tool window has a system menu, Windows does not display its icon on the title bar. However, you can right-click your mouse or press ALT+SPACE to display the system menu.
WS_EX_TOPMOST	Specifies that Windows should place a window it creates with this style above all non-topmost windows and that the window should stay above them, even when your program deactivates the window. To add or remove this style, use the *SetWindowPos* function.
WS_EX_TRANSPARENT	Specifies that Windows must create the window as transparent. That is, this window does not obscure any windows that are beneath it. A window created with this style receives *WM_PAINT* messages only after your program has updated all the sibling windows beneath it.
WS_EX_WINDOWEDGE	Specifies that a window has a border with a raised edge.

***Table 1280** The possible values for the **dwExStyle** parameter. (continued from previous page)*

As your programs become more complex, and as you include more and more windows within them, you will likely find yourself using *CreateWindowEx* more frequently than *CreateWindow*. For example, the following statement creates a window with a small caption (such as a tool box window display) rather than a full-size caption:

```
hwndMain = CreateWindowEx(WS_EX_SMCAPTION, lpszAppName, lpszTitle,
                          WS_OVERLAPPEDWINDOW | WS_CLIPCHILDREN,
                          CW_USEDEFAULT, 0, CW_USEDEFAULT, 0,
                          NULL, NULL, hInstance, NULL);
```

DESTROYING WINDOWS 1281

Throughout the previous Tips, you have learned how your programs will create custom windows, predefined windows, and windows with extended styles. As you know, each window instance is an object of a *window class*, and as you have learned previously, your programs should always destroy objects after the program completes its processing against that object. However, because you do not use *new, malloc,* or a similar function to allocate memory for the window when your program creates the window, you cannot use *delete* or another memory release function to release that memory and destroy the window. Instead, you must use the *DestroyWindow* API call. *DestroyWindow* deletes the window that you pass to it within its sole parameter. You will use the *DestroyWindow* function within your programs as shown in the following prototype:

```
BOOL DestroyWindow(HWND hWnd);
```

DestroyWindow returns *True* if it is successful, and *False* otherwise. Before it destroys the window, *DestroyWindow* sends *WM_DESTROY* and *WM_NCDESTROY* messages to the window to deactivate it. Your window procedure should respond to the *WM_DESTROY* and *WM_NCDESTROY* messages before the program destroys the window. *DestroyWindow* destroys the window's menu, and flushes the thread message queue.

In addition to using *DestroyWindow* to destroy windows your programs create using *CreateWindow*, your programs can also use *DestroyWindow* to destroy modeless dialogs your programs create using *CreateDialog*. However, you will typically use *DestroyWindow* within a *WndProc* routine to respond to an exit command from within the program. The following code, for example, shows how the *generic.cpp* and the *show_three.cpp* programs use the File menu Exit option to destroy the program's window:

```
case IDM_EXIT:
   DestroyWindow(hWnd);
   break;
```

1282 UNDERSTANDING THE REGISTERCLASSEX *API* FUNCTION

In Tip 1269, you learned that your programs will use the *RegisterClass* API function to register a custom class. The *RegisterClassEx* function differs from *RegisterClass* only in that it lets you register a class with a small icon that Windows will then place in the title bar of all the registered class instances. You will use the *RegisterClassEx* function within your programs as shown in the following prototype:

```
ATOM RegisterClassEx(CONST WNDCLASSEX *lpwcx);
```

The *RegisterClassEx* function returns an *ATOM*, just as the *RegisterClass* function does. However, *RegisterClassEx* receives a single parameter, a pointer to a *WNDCLASSEX* structure (unlike the *WNDCLASS* structure *RegisterClass* receives). You must fill the structure with the appropriate class attributes before passing it to the function. The *WNDCLASSEX* structure is slightly different from the *WNDCLASS* structure, as shown here:

```
typedef struct _WNDCLASSEX
 {
   UINT         style;
   WNDPROC      lpfnWndProc;
   int          cbClsExtra;
   int          cbWndExtra;
   HINSTANCE    hInstance;
   HICON        hIcon;
   HCURSOR      hCursor;
   HBRUSH       hbrBackground;
   LPCTSTR      lpszMenuName;
   LPCSTR       lpszClassName;
   HICON        hIconSm;
 } WNDCLASSEX;
```

The only addition that *WNDCLASSEX* makes to the *WNDCLASS* structure is to add the *HICON* handle as the last item. When you initialize the *WNDCLASSEX* structure, you will generally use the *LoadIcon* command to assign a value to the *hIconSm* member.

Note: *Under Windows 95,* ***RegisterClassEx*** *fails if the* ***cbWndExtra*** *member or the* ***cbClsExtra*** *member of the* ***WNDCLASSEX*** *structure contains more than 40 bytes.*

1283 Attaching Information to a Window with SetProp

As you have learned, creating windows within your programs is a relatively simple process. In addition to creating windows, your programs can attach a list of associated data, known as *property items*, to a window. Generally, you will attach property items to a window to help your programs maintain information without the use of global variables. Any section within the program that knows or can access the handle for a particular window can then obtain the information that you have previously attached to the window within the property item list. Within your programs, you will use the *SetProp* function to add a new entry or change an existing entry in the property list of the specified window. The *SetProp* function adds a new entry to the list if the specified character string does not already exist in the list. The new entry contains the string and the handle. Otherwise, the function replaces the string's current handle with the specified handle. You will invoke the *SetProp* function within your programs as shown here:

```
BOOL SetProp(
   HWND hWnd,                          // handle of window
   LPCTSTR lpString,                   // atom or address of string
   HANDLE hData                        // handle of data
);
```

The *SetProp* function returns *True* if it is successful and *False* otherwise. Table 1283 lists the parameters and their descriptions for the *SetProp* function.

Parameter	Description
hWnd	Identifies the window whose property list receives the new entry.
lpString	Points to a *NULL*-terminated string or contains an *ATOM* that identifies a string. If this parameter is an *ATOM*, it must be a global *ATOM* that a previous call to the *GlobalAddAtom* function created. Your program must place the *ATOM*, a 16-bit value, in the low-order word of *lpString*. The high-order word must be zero.
hData	Identifies data for the *SetProp* function to copy to the property list. The *hData* parameter can identify any value useful to the application.

***Table 1283** The parameters for the **SetProp** function.*

Before destroying a window (that is, before processing the *WM_DESTROY* message), an application must remove all entries it has added to the property list. The application must use the *RemoveProp* function to remove the entries. The CD-ROM that accompanies this book includes the *show_prop.cpp* program, which sets a property for a window at its creation, then displays the property when the user selects the *Demo* option. As in previous programs, the only significant difference between the *generic.cpp* file and the *show_prop.cpp* file is within the *WndProc* function.

The code within the *WndProc* function for the *show_prop* program performs several new activities. First, the *WndProc* function adds a *case* statement to check for the *WM_CREATE* message. When the program receives the *WM_CREATE* message, it adds a property with the *Value for Property* string value to the window's property list. When you later select the *Demo* option, the program will display that single, known property within a *Message Box*. You will learn about Message Boxes in Tip 1286. Tip 1284 details how you can list a window's properties when the program does not know the actual key for the property.

Using EnumProps to List a Window's Properties

In Tip 1283, you learned how to use the *SetProp* function to set a window property and, briefly, how to use the *GetProp* function to return that property elsewhere within your programs. As you saw, the *GetProp* function requires that your program pass either a pointer to a *NULL*-terminated string or an *ATOM* which identifies the property that you want to retrieve. There may be times within your program when it must retrieve all the properties associated with a window without knowing the property names, how many properties there are, and so on. The *EnumProps* function lets you enumerate (list) all entries in the property list of a window by passing each item one at a time to the specified callback function. *EnumProps* continues until it enumerates the last entry or the callback function returns *False*. Within your programs, you will invoke the *EnumProps* function with the following general form:

```
int EnumProps(
   HWND hWnd,                      // handle to window
   PROPENUMPROC lpEnumFunc         // pointer to callback function
);
```

Before the invocation of the *EnumProps* function, you must define the callback function that *EnumProps* will invoke. The general format of the callback function is shown here (note that you can name the function what you want; the *PropEnumProc* header is simply a placeholder):

```
BOOL CALLBACK PropEnumProc(
    HWND hwnd,                     // handle to window
    LPCTSTR lpszString,            // string component of property
    HANDLE hData                   // data handle component of property
);
```

The following restrictions apply to the *PropEnumProc* callback function:

1. The callback function must not yield control or do anything that might yield control to other tasks.
2. The callback function can call the *RemoveProp* function. However, *RemoveProp* can remove only the property passed to the callback function through the callback function's parameters.
3. The callback function should not try to add properties.

When you put the *EnumProps* and *PropEnumProc* functions together, your programs can list each property previously associated with a window in the order in which your program associated the properties with the window. For example, the CD-ROM that accompanies this book includes the program *EnumProps.cpp*, which modifies the *WndProc* program slightly and adds a new function, the *EnumPropsProc* function, to the *generic.cpp* code. To make the *EnumProps* function work correctly, delete the existing *WndProc* function from the *generic.cpp* program and in its place add the following code:

```
BOOL CALLBACK EnumPropsProc(HWND hWnd, LPCTSTR lpszString, HANDLE hData)
 {
   MessageBox( hWnd, (LPCTSTR)hData, lpszString, MB_OK );
   return( TRUE );
 }
LRESULT CALLBACK WndProc(HWND hWnd, UINT uMsg, WPARAM wParam, LPARAM lParam)
 {
   static LPCTSTR szProp1 = "Value for Property 1";
   static LPCTSTR szProp2 = "Value for Property 2";
   static LPCTSTR szProp3 = "Value for Property 3";

   switch( uMsg )
    {
     case WM_CREATE :
               // Add property items that contain string pointers
               // to the main window
               SetProp( hWnd, "Property 1", (HANDLE)szProp1 );
               SetProp( hWnd, "Property 2", (HANDLE)szProp2 );
               SetProp( hWnd, "Property 3", (HANDLE)szProp3 );
               break;

    case WM_COMMAND :
            switch(LOWORD( wParam ))
             {
               case IDM_DEMO :
```

```
                        // Enumerate the properties and display in a message
                        // box the property name and value.
                        EnumProps( hWnd, (PROPENUMPROC)EnumPropsProc );
                        break;

                case IDM_ABOUT :
                        DialogBox(hInst, "AboutBox", hWnd, (DLGPROC)About);
                        break;

                case IDM_EXIT :
                        DestroyWindow(hWnd);
                        break;
               }
               break;

        case WM_DESTROY :
              PostQuitMessage(0);
              break;

        default :
              return(DefWindowProc(hWnd, uMsg, wParam, lParam));
    }
   return(0L);
}
```

When you compile and execute the *EnumProps* program and select the *Demo* option, the program will display three message boxes in sequence, each of which displays the string "Value for Property *n*," where *n* equals 1, 2, or 3, depending on the message box. Note the line that calls the *EnumProps* function, as shown here:

```
EnumProps(hWnd, (PROPENUMPROC)EnumPropsProc);
```

Remember, the *(PROPENUMPROC)* casting operator is an explicit cast to a pointer of the *PROPENUMPROC* type. That pointer then points to the location within memory where the *EnumPropsProc* function begins.

UNDERSTANDING CALLBACK FUNCTIONS 1285

In previous Tips, you used callback functions to achieve various goals. As you have seen, the callback function is critical to windows processing. To understand logically how the callback function fits into a function call, consider the logical diagram shown in Figure 1285.

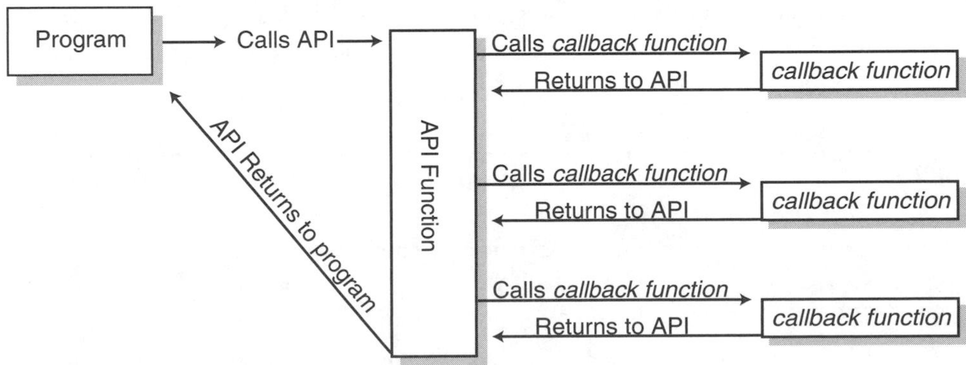

***Figure 1285** The logical model of a callback function's processing.*

Your programs will use callback functions in many ways; however, as previously detailed, you will most frequently use callback functions together with API calls which generate an unknown number of return values. You will often encounter functions that perform callback activities whose names begin with *Enum*, including *EnumFontFamilies*, *EnumWindows*, *EnumProps*, and so on.

1286 UNDERSTANDING THE MESSAGEBOX FUNCTION

In Tips 1283 and 1284, your programs used the *MessageBox* function to display simple dialog boxes that required user interaction before the program would continue its processing. In later Tips you will learn about creating dialog boxes of many types; however, for simple user-input dialog boxes, your programs can use the *MessageBox* function. The *MessageBox* function creates, displays, and operates a message box. The message box contains an application-defined message and title, plus any combination of predefined icons and push buttons. A message box cannot display other windows within itself. Your programs will use the following generalized form of the *MessageBox* function to display message boxes:

```
int MessageBox(
   HWND hWnd,              // handle of owner window
   LPCTSTR lpText,         // address of text in message box
   LPCTSTR lpCaption,      // address of title of message box
   UINT uType              // style of message box
);
```

The *MessageBox* function accepts the parameters Table 1286.1 details.

Flags	Description
hWnd	Identifies the owner window of the message box you want to create. If this parameter is *NULL*, the message box has no owner window.
lpText	Points to a *NULL*-terminated string containing the message the message box must display.
lpCaption	Points to a *NULL*-terminated string used for the dialog box title. If this parameter is *NULL*, your program uses the default title *Error.*
uType	Specifies a set of bit flags that determine the contents and behavior of the dialog box. This parameter can be a combination of flags from the following group of flags. Table 1286.2 details the flags your programs will use to indicate the buttons a message box contains.

***Table 1286.1** The parameters for the **MessageBox** function.*

Aside from the text the message box displays (which you use the *lpText* parameter to set) and the title for the message box (which you use the *lpCaption* parameter to set), your programs will most often manipulate the *uType* parameter to change the buttons and any icons that you want to appear in the message box. Table 1286.2 lists the possible values you can use for the *uType* parameter to control the number of buttons that appear in the message box.

Flag	Meaning
MB_ABORTRETRYIGNORE	The message box contains three push buttons: *Abort*, *Retry*, and *Ignore*.
MB_OK	The message box contains one push button: *OK.* The *MB_OK* flag is the default value for the *uType* parameter.
MB_OKCANCEL	The message box contains two push buttons: *OK* and *Cancel.*
MB_RETRYCANCEL	The message box contains two push buttons: *Retry* and *Cancel.*
MB_YESNO	The message box contains two push buttons: *Yes* and *No.*
MB_YESNOCANCEL	The message box contains three push buttons: *Yes*, *No*, and *Cancel.*

***Table 1286.2** The possible values for the **uType** parameter to control the number of buttons.*

In addition to controlling the number and title of the buttons that appear within the message box, your programs can also control whether or not an icon appears within the message box. The default setting is *MB_NOICON*. Table 1286.3 lists the possible values for the *uType* parameter to control the appearance of the icon within the box.

Flag	Meaning
MB_ICONEXCLAMATION	An exclamation-point icon appears in the message box.
MB_ICONWARNING	An exclamation-point icon appears in the message box.
MB_ICONINFORMATION	An icon consisting of a lowercase letter *i* in a circle appears in the message box.
MB_ICONASTERISK	An icon consisting of a lowercase letter *i* in a circle appears in the message box.
MB_ICONQUESTION	A question-mark icon appears in the message box.
MB_ICONSTOP	A stop-sign icon appears in the message box.
MB_ICONERROR	A stop-sign icon appears in the message box.
MB_ICONHAND	A stop-sign icon appears in the message box.
MB_NOICON	No icon appears in the message box.

***Table 1286.3** The possible values for the **uType** parameter to control the message box's icon.*

The *MessageBox* function supports several dozen additional constants for the *uType* parameter. However, the constants Tables 1286.2 and 1286.3 list are the most commonly used. To find the additional constants, check your compiler's on-line help.

The CD-ROM that accompanies this book includes the *Show_Mess.cpp* program. The *Show_Mess* program generates a beep, which you will learn about in the next Tip, then says "*Hello!*" to the user and waits for a keypress. The program does not perform useful processing of the selection the user makes within the message box, but could do so quite easily.

1287 UNDERSTANDING THE MESSAGEBEEP FUNCTION

In Tip 1286, you used the *MessageBox* function to generate a simple dialog box. The *Show_Mess* program also invoked the *MessageBeep* function when it created the message box. The *MessageBeep* function plays a waveform sound. An entry in the *[sounds]* section of the system registry identifies the waveform sound for each sound type. Your programs will use the *MessageBeep* function in accordance with the following general form:

```
BOOL MessageBeep(UINT uType);
```

The *uType* parameter specifies the sound type, as identified by an entry in the *[sounds]* section of the registry. You can play the sounds from within the Windows control panel if you are unsure about their waveforms. The *uType* sound parameter can be one of the values listed in Table 1287.

Value	Sound
0xFFFFFFFF	Standard beep using the computer speaker
MB_ICONASTERISK	SystemAsterisk
MB_ICONEXCLAMATION	SystemExclamation
MB_ICONHAND	SystemHand
MB_ICONQUESTION	SystemQuestion
MB_OK	SystemDefault

***Table 1287** The possible values for the **uType** sound parameter.*

Your programs should use the *MessageBeep* function to create simple sounds, which you will typically use to alert the user of some event.

1288 REVISITING MESSAGES

As you have learned, message handling is at the "heart" of what makes Windows applications work. Both the operating system and the applications that run over the operating system generate messages for every event that occurs in Windows. Messages are fundamentally important to the value of Windows as a multitasking operating system. As you will learn, each task (or program) uses one or more threads within the operating system. The 32-bit Windows platforms (Windows NT and Windows 95) maintain a separate set of messages (a *message queue)* for each thread executing on the operating system.

Windows generates messages for every hardware event that occurs, such as a key press on the keyboard or a user's mouse click. Windows then passes each message to the appropriate message queue. In other words, if the user clicks the mouse, but not within your application, your application will not know that the user clicked the mouse. Occasionally, the system will generate several copies of a message that it simultaneously places in multiple message queues. Figure 1288 shows a simple diagram of how Windows processes messages throughout several message queues.

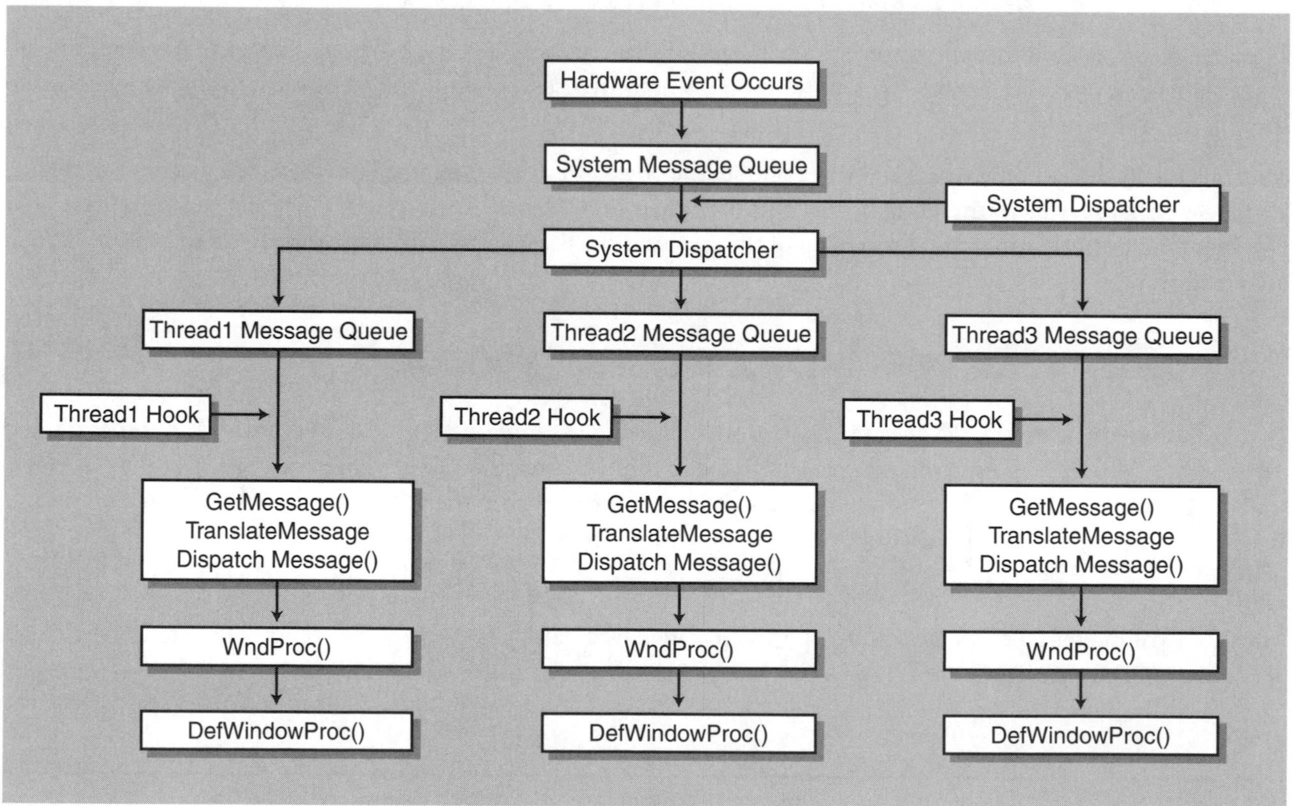

***Figure 1288** Windows processes messages throughout several message queues.*

1289 UNDERSTANDING THE FLOW OF MESSAGES

As you have learned, the Windows message structure is fundamental to the way that Windows manages multiple tasks in close sequence. You should understand the flow of messages, not just to the program's message queue, but also from the program's message queue into your program's message loop. When Windows accepts a message from the computer's hardware, it determines internally to which message queues it will pass that message. After Windows passes the message into the program's message queue, the program processes each message in turn. For example, sometimes when you are typing on a word processor you will type faster than the screen can display your keystrokes. However, the program is able to maintain your typing, even while the screen tries to catch up, because Windows is storing each keystroke you make into the program's message queue, as shown in Figure 1289.1.

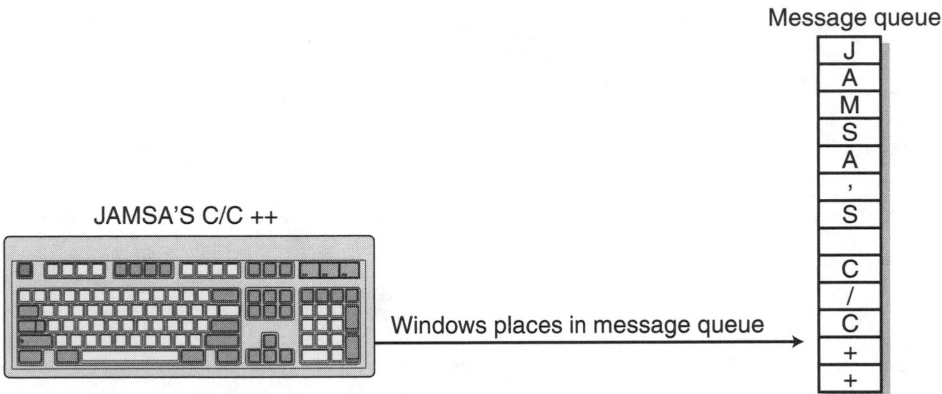

Figure 1289.1 Windows processes keystrokes and places them into the message queue.

After Windows places the keystrokes within the message queue, the program pulls the messages one message at a time from the message queue, retrieving the earliest message first and continuing in order until it retrieves the last message in the queue. After it retrieves each message, the program then uses the message loop to call the program's message callback function, which processes each of the entries, as shown in Figure 1289.2.

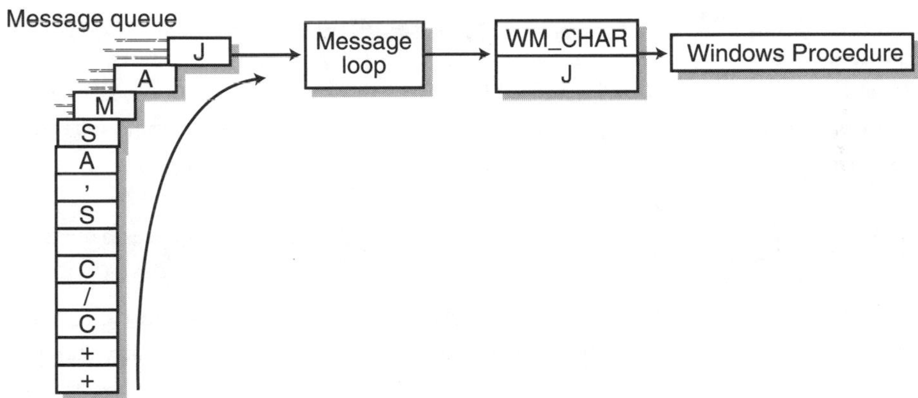

Figure 1289.2 The program's message loop calls the program's message callback function.

As you have learned, your programs will then check the values within the messages to determine how to respond to them. If, for example, the command is a keystroke which the user wants to place within the word processor's actual document, the message loop will then dispatch the character to the current window and add the character to the word processing document at its current location, as shown in Figure 1289.3.

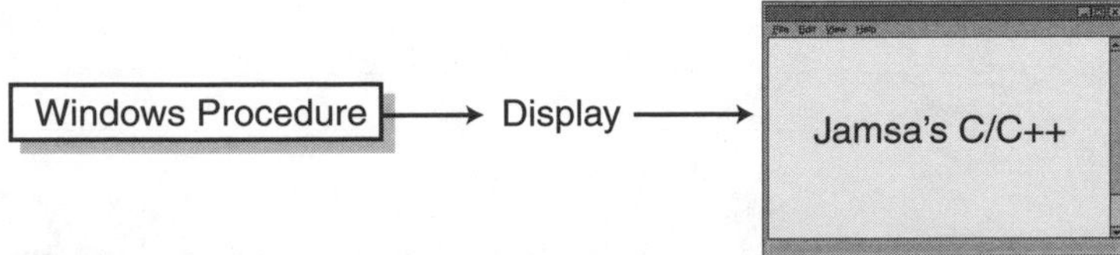

Figure 1289.3 The message callback function inserts the character into the word processing document.

1290 BETTER UNDERSTANDING THE MSG STRUCTURE'S COMPONENTS

As you have learned, the Windows message is one of the fundamental building blocks on which you will create your Windows programs. You have also learned that Windows defines the MSG structure, as shown here:

```
typedef struct tagMSG {
    HWND        hwnd;               // window handle
    UINT        message;            // message ID
    WPARAM      wParam;             // wParam value
    LPARAM      lParam;             // lParam value
    DWORD       time;               // milliseconds since start-up
    POINT       pt;                 // screen coordinates of current mouse location
} MSG;
```

While each of the components of the message structure are important for different reasons, the two components you will most commonly use within this book are the *message* (which *WndProc* receives as *uMsg*) and *wParam* values. When you call the *DispatchMessage* function, as your messaging loops always will, the *DispatchMessage* function sends the message to the *WndProc* function, whose header you will define as shown here:

```
LRESULT CALLBACK WndProc(HWND hWnd, UINT uMsg, WPARAM wParam, LPARAM lParam);
```

The *DispatchMessage* function transmits only the *hWnd, message, wParam,* and *lParam* members of the *MSG* structure to the callback function. As you have learned, the callback function then uses the *uMsg* parameter to make its initial determination of the message type. If the message type is a Windows command, without parameters, the callback function performs its processing appropriately (for example, *WM_DESTROY*). On the other hand, if the message is a *WM_COMMAND* message, the callback function must then check the low word (the least significant 16 bits) of the *wParam* parameter to determine which specific command the system received.

As you have learned, you will define constant identifiers for command-based options within your programs so that your callback function can correctly process the commands when it receives them. For example, most of the programs you have created so far used the *IDM_DEMO* identifier to determine when a user selects the *Demo* menu option. As you continue to expand your programs, you will find that you use identifiers with nearly all a window's component members to ensure that your callback function can process their selections correctly.

1291 UNDERSTANDING THE PEEKMESSAGE FUNCTION

As you have learned, your programs can retrieve messages from the message queue using either the *GetMessage* function or the *PeekMessage* function. The *PeekMessage* function checks a thread message queue for a message and places the message (if any) in the specified structure, as shown here:

```
BOOL PeekMessage (
   LPMSG lpMsg,              // pointer to structure for message
   HWND hWnd,                // handle to window
   UINT wMsgFilterMin,       // first message
   UINT wMsgFilterMax,       // last message
   UINT wRemoveMsg           // removal flags
 );
```

The *PeekMessage* function accepts the parameters Table 1291.1 details.

Parameters	Description
lpMsg	Points to an *MSG* structure that contains message information from the Windows-based application queue.
hWnd	Identifies the window with messages your program must examine.
wMsgFilterMin	Specifies the value of the first message in the range of messages your program must examine.
wMsgFilterMax	Specifies the value of the last message in the range of messages your program must examine.
wRemoveMsg	Specifies how your program handles messages. This parameter can be one of the values Table 1291.2 specifies.

***Table 1291.1** The parameters for the **PeekMessage** function.*

Table 1291.2 details the possible values for the *wRemoveMsg* parameter within the *PeekMessage* function.

Value	Description
PM_NOREMOVE	Does not remove messages from the queue after *PeekMessage* completes its processing.
PM_REMOVE	Removes messages from the queue after *PeekMessage* completes its processing.

***Table 1291.2** The possible values for the **wRemoveMsg** parameter.*

You have the option to combine the value *PM_NOYIELD* with either *PM_NOREMOVE* or *PM_REMOVE*. However, *PM_NOYIELD* has no effect on 32-bit Windows applications. It is defined in Win32 solely to provide compatibility with applications written for previous versions of Windows, wherein Windows used it to prevent the current task from halting and yielding system resources to another task. The 32-bit Windows applications always run simultaneously.

Unlike the *GetMessage* function, the *PeekMessage* function does not wait for Windows to place a message in the queue before returning to the calling location. *PeekMessage* only retrieves messages associated with the window the *hWnd* parameter, or any of its children, identifies, as the *IsChild* function specifies, and within the range of message values the *wMsgFilterMin* and *wMsgFilterMax* parameters specify. If *hWnd* is *NULL*, *PeekMessage* retrieves messages for any window that belongs to the current thread making the call. (*PeekMessage* does not retrieve messages for windows that belong to other threads.) If *hWnd* is -1, *PeekMessage* will only return messages with an *hWnd* value of *NULL*. If *wMsgFilterMin* and *wMsgFilterMax* are both zero, *PeekMessage* will return all available messages (that is, it does not perform any range filtering).

You can use the *WM_KEYFIRST* and *WM_KEYLAST* constants as filter values to retrieve all keyboard messages; you can use the *WM_MOUSEFIRST* and *WM_MOUSELAST* constants to retrieve all mouse messages.

The *PeekMessage* function usually does not remove *WM_PAINT* messages from the queue. *WM_PAINT* messages remain in the queue until the program processes them. However, if a *WM_PAINT* message has a *NULL* update region, *PeekMessage* will remove it from the queue.

UNDERSTANDING THE POSTMESSAGE FUNCTION 1292

In previous Tips, you have learned various ways your programs can retrieve messages from the message queue. Sometimes, you may want to place messages within the program's message queue. The *PostMessage* function places (posts) a message in the message queue associated with the thread that created the specified window and then returns without waiting for the thread to process the message. Calls to the *GetMessage* or *PeekMessage* function retrieve messages from the message queue you previously posted with *PostMessage*. You will use the *PostMessage* function within your programs as shown in the following prototype:

```
BOOL PostMessage(
    HWND hWnd,          // handle of destination window
    UINT Msg,           // message to post
    WPARAM wParam,      // first message parameter
    LPARAM lParam       // second message parameter
);
```

The *PostMessage* function accepts the parameters Table 1292.1 details.

Parameter	Description
hWnd	Identifies the window whose window procedure is to receive the message. Two values have special meanings, as Table 1292.2 details.
Msg	Specifies the message for *PostMessage* to place within the queue.
wParam	Specifies additional message-specific information.
lParam	Specifies additional message-specific information.

***Table 1292.1** The parameters for the **PostMessage** function.*

Two values for the *hWnd* parameter have special meanings to the *PostMessage* function, as Table 1292.2 details.

Value	Meaning
HWND_BROADCAST	The *PostMessage* function posts the message to all top-level windows in the system, including disabled or invisible un-owned windows, overlapped windows, and pop-up windows. The *PostMessage* function does not post the message to child windows.
NULL	The function behaves like a call to *PostThreadMessage* with the *dwThreadId* parameter set to the identifier of the current thread.

***Table 1292.2** The special values for the **PostMessage** function.*

Applications that must use *HWND_BROADCAST* to communicate should use the *RegisterWindowMessage* function to obtain a unique message for inter-application communication.

If you send a message in the range below *WM_USER* to the asynchronous message functions (*PostMessage, SendNotifyMessage,* and *SendMessageCallback*), make sure that the message parameters do not include pointers. Otherwise, the functions will return before the receiving thread has had a chance to process the message and the sender will free the memory before the program uses it.

1293 UNDERSTANDING THE SENDMESSAGE FUNCTION

In Tip 1292, you learned how to use the *PostMessage* function to place a message within a thread's message queue. As you learned, *PostMessage* returns processing to the calling program without waiting for a response from the thread that receives the message. Sometimes, your programs may require a response from the receiving thread before processing can continue within the calling thread. You can use the *SendMessage* function to control when application processing returns to the calling application. The *SendMessage* function sends a message you specify to a window or windows. The *SendMessage* function calls the window procedure for the specified window and does not return processing to the calling program until the window procedure processes the message. The *PostMessage* function, in contrast, posts a message to a thread's message queue and returns immediately. Your programs will use the *SendMessage* function as shown in the following prototype:

```
LRESULT SendMessage(
    HWND hWnd,              // handle of destination window
    UINT Msg,               // message to send
    WPARAM wParam,          // first message parameter
    LPARAM lParam           // second message parameter
);
```

The parameters for the *SendMessage* function are identical to the parameters for the *PostMessage* function. However, if you set the *hWnd* parameter to *HWND_BROADCAST*, your program will send the message to all top-level windows in the system, including disabled or invisible un-owned windows, overlapped windows, and pop-up windows. However, your program will not send the message to child windows.

Applications that must use *HWND_BROADCAST* to communicate should use the *RegisterWindowMessage* function to obtain a unique message for interapplication communication.

If the calling thread created the specified window, Windows calls the window procedure immediately, as a subroutine. If a different thread created the specified window, Windows switches to that thread and calls the appropriate window procedure. Windows processes messages sent between threads only when the receiving thread executes message retrieval code. Windows blocks the sending thread until the receiving thread processes the message.

USING THE REPLYMESSAGE FUNCTION 1294

In previous Tips you learned how to use the *PostMessage* and *SendMessage* functions within your programs. Your programs should use the *ReplyMessage* function to reply to a message sent through the *SendMessage* function without returning control to the function that called *SendMessage*, as shown here:

```
BOOL ReplyMessage(
   LRESULT lResult // message-specific reply
);
```

By calling the *ReplyMessage* function, the window procedure that receives the message lets the thread that called *SendMessage* continue to run as though the thread receiving the message had returned control. The thread that calls the *ReplyMessage* function also continues to run.

If the program did not send the message through *SendMessage,* or if the same thread sent the message, *ReplyMessage* has no effect.

HOOKING MESSAGES 1295

As you have learned, messages are at the heart of how the Windows operating system manages programs. As you have also learned, you can use different API functions to pass messages between two threads. Sometimes, however, you may want a certain program to intercept all messages that Windows sends to another program. *Message hooks* let you intercept messages going to one program within a different (hooking) program. The hooking program can act upon, modify, or even stop the messages it intercepts. You will often hook messages within a data-link library (DLL), a special type of support file, to provide additional message processing services without slowing down the main program. The scope of a hook depends on the hook type. You can set some hooks only with system scope, while others you can set for only a specific thread, as shown in the following list. Table 1295 lists the hook types in the order in which the operating system catches the hooks.

Hook	Scope
WH_CALLWNDPROC	Thread or system
WH_CALLWNDPROCRET	Thread or system
WH_CBT	Thread or system
WH_DEBUG	Thread or system
WH_GETMESSAGE	Thread or system
WH_JOURNALPLAYBACK	System only
WH_JOURNALRECORD	System only
WH_KEYBOARD	Thread or system
WH_MOUSE	Thread or system
WH_MSGFILTER	Thread or system
WH_SHELL	Thread or system
WH_SYSMSGFILTER	System only

Table 1295 *The scope of system hooks.*

For a specified hook type, Windows calls thread hooks first, then system hooks.

1296 *Using the* SetWindowsHookEx *Function*

In Tip 1295, you learned about message hooks, which your programs can use to intercept hooks intended for other applications. The *SetWindowsHookEx* function installs an application-defined *hook procedure* into a *hook chain*. Your programs can use a hook procedure to monitor the system for certain types of events. When you install a hook procedure, Windows associates the events that hook procedure monitors either with a specific thread or with all the threads in the system. You will use the *SetWindowsHookEx* function within your programs as shown in the following prototype:

```
HHOOK SetWindowsHookEx(
    int idHook,          // type of hook to install
    HOOKPROC lpfn,            // address of hook procedure
    HINSTANCE hMod,      // handle of application instance
   DWORD dwThreadId      // identity of thread to install hook for
);
```

The *idHook* parameter specifies the type of hook procedure for the program to install. This parameter can be one of the values listed in Table 1296. The *lpfn* parameter points to the hook procedure. If the *dwThreadId* parameter is zero or specifies the identifier of a thread a different process created, the *lpfn* parameter must point to a hook procedure in a dynamic-link library (DLL). Otherwise, *lpfn* can point to a hook procedure in the code associated with the current process. The *hMod* parameter identifies the dynamic-link library containing the hook procedure the *lpfn* parameter points to. You must set the *hMod* parameter to *NULL* if the *dwThreadId* parameter specifies a thread the current process created and if the hook procedure is within the code associated with the current process. The *dwThreadId* parameter specifies the identifier of the thread with which your program associates the hook procedure. If this parameter is zero, your program associates the hook procedure with all existing threads.

As you might expect, you can set a variety of different hook types. You will determine what hook type the system will set with the value you place within the *idHook* parameter. Table 1296 details the possible values for the *idHook* parameter.

Values	Description
WH_CALLWNDPROC	Installs a hook procedure that monitors messages before the system sends them to the destination window procedure.
WH_CALLWNDPROCRET	Installs a hook procedure that monitors messages after the destination window procedure processes them.
WH_CBT	Installs a hook procedure that receives notifications useful to a computer-based training (CBT) application.
WH_DEBUG	Installs a hook procedure useful for debugging other hook procedures.
WH_GETMESSAGE	Installs a hook procedure that monitors messages posted to a message queue.
WH_JOURNALPLAYBACK	Installs a hook procedure that posts messages a *WH_JOURNALRECORD* hook procedure previously recorded.
WH_JOURNALRECORD	Installs a hook procedure that records input messages posted to the system message queue. This hook is useful for recording macros.
WH_KEYBOARD	Installs a hook procedure that monitors keystroke messages.
WH_MOUSE	Installs a hook procedure that monitors mouse messages.
WH_MSGFILTER	Installs a hook procedure that monitors messages generated as a result of an input event in a dialog box, message box, menu, or scroll bar.
WH_SHELL	Installs a hook procedure that receives notifications useful to shell applications.
WH_SYSMSGFILTER	Installs a hook procedure that monitors messages generated as a result of an input event in a dialog box, message box, menu, or scroll bar. The hook procedure monitors these messages for all applications in the system.

***Table 1296** The possible values for the **idHook** parameter.*

An error may occur if the *hMod* parameter is *NULL* and the *dwThreadId* parameter is zero or specifies the identifier of a thread another process created.

Calling the *CallNextHookEx* function to chain to the next hook procedure is optional. However, as you will learn, if you do not use the *CallNextHookEx* function, other applications that have previously installed hooks will not receive hook notifications and may behave incorrectly as a result. You should call *CallNextHookEx* unless you absolutely must prevent other applications from seeing a notification.

Before terminating, an application must call the *UnhookWindowsHookEx* function to free system resources associated with the hook.

The system hooks are a shared resource, and installing one affects all applications. All system hook functions must be in libraries. You should restrict system hooks to special-purpose applications or use them as a development aid during application debugging. Libraries that no longer need a hook should remove the hook procedure.

UNDERSTANDING THE EXITWINDOWSEX FUNCTION 1297

In previous Tips, you have learned about using messages within your programs. You will find a slightly different way to use messages within the *ExitWindowsEx* function. The *ExitWindowsEx* function either logs off, shuts down, or shuts down and restarts the system. Your programs will use the *ExitWindowsEx* function as shown in the following prototype:

```
BOOL ExitWindowsEx(
   UINT uFlags,                  // shut-down operation
   DWORD dwReserved              // reserved
);
```

Within the *ExitWindowsEx* function, the *uFlags* parameter specifies the type of shut-down. The *uFlags* parameter must be some combination of the values Table 1297 details. Windows currently ignores the *dwReserved* parameter and reserves it for future developments. Table 1297 lists the possible values for the *uFlags* parameter.

Flag	Description
EWX_FORCE	Forces processes to terminate. When the program sets this flag, Windows does not send the messages *WM_QUERYENDSESSION* and *WM_ENDSESSION* to the applications currently running in the system. This can cause the applications to lose data. Therefore, you should only use this flag in an emergency.
EWX_LOGOFF	Shuts down all processes running in the security context of the process that called the *ExitWindowsEx* function. Then it logs the user off.
EWX_POWEROFF	Shuts down the system and turns off the power. The system must support the power-off feature.

***Note**: In Windows NT, the calling process must have the SE_SHUTDOWN_NAME privilege. Windows 9x does not support or require security privileges.*

Flag	Description
EWX_REBOOT	Shuts down the system and then restarts the system.

***Note**: In Windows NT, the calling process must have the SE_SHUTDOWN_NAME privilege.*

Flag	Description
EWX_SHUTDOWN	Shuts down the system to a point at which it is safe to turn off the power. The system flushes all file buffers to disk and stops all running processes.

Table 1297 The possible values for the ***uFlags*** *parameter.*

The *ExitWindowsEx* function returns as soon as it has initiated the shutdown. The shut-down or log-off then proceeds asynchronously (in other words, your programs may perform additional processing during the shut-down process). During a shut-down or log-off operation, the system gives applications that it shuts down a specific amount of time to respond to the shut-down request. If the time expires, Windows will display a dialog box that lets the user forcibly shut down the application, retry the shut-down, or cancel the shut-down request. If you specify the *EWX_FORCE* value, Windows will always force applications to close and will not display the dialog box for the user.

The *ExitWindowsEx* function sends a separate notification message—*CTRL_SHUTDOWN_EVENT* or *CTRL_LOGOFF_EVENT,* as the situation warrants—to *console processes.* A console process routes these messages to its *HandlerRoutine* functions, which calls to the *SetConsoleCtrlHandler* function to add and remove. *ExitWindowsEx* sends these notification messages asynchronously; therefore, an application cannot assume that the *HandlerRoutine* functions have handled the console notification messages when a call to *ExitWindowsEx* returns.

Note:** To shut down or restart a Windows NT system, the calling process must use the* ***AdjustTokenPrivileges *function to enable the* ***SE_SHUTDOWN_NAME*** *privilege. Console processes only run on a Windows NT server. Windows 9x does not support or require security privileges.*

1298 UNDERSTANDING MENU TYPES

As you have seen in previous Tips, one primary component of the Windows architecture is the *menu bar.* The menu bar is, essentially, a container for a menu resource. As you have seen, menu resources let your programs receive textual and mouse-driven input from users into your programs. Typically, you will group menus under specific categories and provide the user with multiple options within those categories.

Windows programs use two basic types of menus: *top-level menus* and *pop-up menus*. The top-level menu (also called the "main" menu of the program) is a set of commands that is visible in the window's menu bar at all times, assuming the program uses a menu. For more complex programs, it may not always be possible or practical to put all the menu options that the program may require onto the menu bar. In such cases, your programs can use pop-up menus (often called "submenus," "pull-down menus," or "drop-down menus"). When you select an option from the top menu bar that spawns a pop-up menu, Windows will automatically extend the pop-up menu as well. Figure 1298 shows a simple top-level menu and a submenu.

Figure 1298 *A simple top-level menu and submenu pair.*

Note: *Do not confuse pop-up menus in the context of the Windows API with* ***context-sensitive menus****, which many books also refer to as pop-up menus. A context-sensitive menu is a special type of menu, which you will learn more about in later Tips.*

Understanding a Menu's Structure 1299

As you have learned, Windows composes a program menu of one or more top-level menu selections, one or more selections under each top-level menu selection (called *options* within this book), and one or more selections under each of the selections (called *sub-options* within this book). For example, Figure 1299 shows a sample structure tree for a set of menus for a simple program. Notice how, in the menu hierarchy, the top-level menus have a variable number of options, and that only some of the options have further sub-options, as shown here.

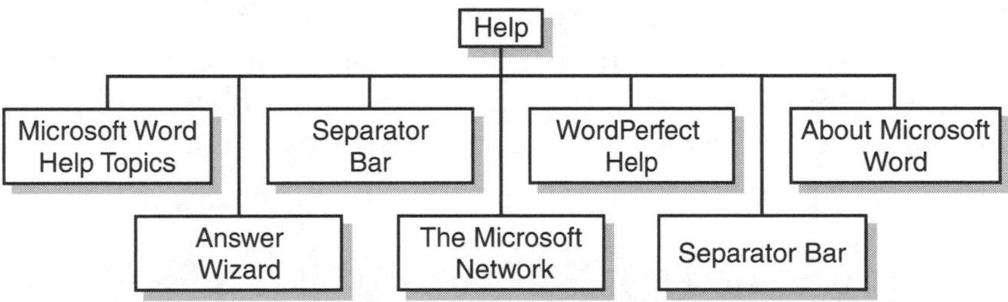

Figure 1299.1 *The hierarchical structure of program menus.*

Additionally, it is important to note that, in a good menu structure, there is only one way to reach any given option or sub-option. Keeping the menu strictly hierarchical protects the user against confusion between sub-options in multiple menus. When you use a strictly hierarchical menu structure, your program menu should look structurally identical to the chart in Figure 1299.1, as shown in Figure 1299.2.

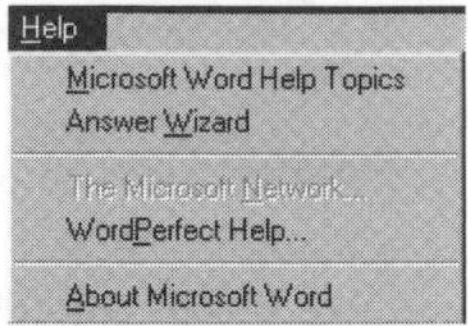

Figure 1299.2 The actual program menu.

1300 Creating a Menu Within a Resource File

As you have learned throughout this book, you will most often create your menus within a resource file. As several previous Tips have mentioned, the most current C++ development suites for Windows will let you create menus using simpler drag-and-drop methods. However, it is important to understand how your programs create menus in case you must create one manually. The following listing, for example, shows the typical menu construction within a resource file:

```
MYAPP MENU DISCARDABLE
BEGIN
    POPUP "&File"
    BEGIN
        MENUITEM "E&xit", IDM_EXIT
    END
    POPUP "&Demo"
    BEGIN
        MENUITEM "Item &1", IDM_ITEM1
        MENUITEM "Item &2", IDM_ITEM2
        MENUITEM "Item &3", IDM_ITEM3
    END
    POPUP "&Help"
    BEGIN
        MENUITEM "&About Generic...", IDM_ABOUT
    END
END
```

In the next several Tips, you will examine in detail the resource file menu definition, as well as how you will use the definition within your programs. Specifically, in Tip 1301, you will examine how your resource files use the *POPUP* and *MENUITEM* descriptors. However, it is important to note the *DISCARDABLE* descriptor within the previous resource declaration. The *DISCARDABLE* descriptor tells the compiler's linker that it should discard the initial resource information about the menu after the program registers the menu within the window class. You will almost always use the *DISCARDABLE* descriptor with your menu definitions to conserve memory and improve your program's processing speed.

1301 Understanding the POPUP and MENUITEM Descriptors

As you saw in Tip 1300, the menu description section of the resource file begins with the *BEGIN* statement and ends with the *END* statement. Within a menu description, you may have any number of *POPUP* and *MENUITEM* items. However, you should be careful not to create too many top-level menu items because they may not all fit onto the window when the window is its normal size.

Within the menu description, the *POPUP* descriptor indicates the menu's top level. For example, the *Test!* item is at the top of a menu in the program in the previous Tip. The use of the ampersand (&) within the string indicates that the user can use an *accelerator key*, rather than the mouse, to access the menu item. Programmers often refer to an accelerator key as a "keyboard shortcut." In the case of the *Test!* menu, the accelerator key is ALT+K.

The resource file uses the *MENUITEM* descriptor to list each item within the menu. For example, the *Test!* menu description that Tip 1300 uses contains three menu items: Item #1, Item #2, and Item #3. Note that each menu item includes a constant reference. In the case of Item #1, the constant reference is *IDM_ITEM1*. Within your *WndProc* routine, your program will first catch the *WM_COMMAND* message. After your program catches the *WM_COMMAND* message, the low word of the *wParam* value that *WndProc* receives will contain the constant identifier for the user-selected menu item. For example, to process the selection of the *IDM_ITEM1* option, your program will use two *switch* statements. The first statement will verify that the message is of type *WM_COMMAND*. The second statement will check the value within the *wParam* value to verify the menu item the user selected was the Item #1 option, as shown here:

```
switch(uMsg )
   {
      case WM_COMMAND :
              switch(LOWORD(wParam ) )
              {
                 case ITEM1 :
                            // some processing
                 case ITEM2 :
                            // more processing
```

ADDING A MENU TO AN APPLICATION'S WINDOW 1302

As you have learned, defining a menu within the resource file (.RC) does not automatically make the menu usable or attach the menu to the application's window. Usually, you will attach the application's menu to the window's class definition within the *WinMain* function before you invoke the *RegisterClass* function, as you have in previous Tips. Remember, you set the *lpszMenuName* element of the *WNDCLASS* and *WNDCLASSEX* structures to perform the early binding attachment. The *RegisterClass* function then associates the menu name with any window the program later creates from the class.

You can also use the *CreateWindow* and *LoadMenu* functions to attach the menu to a window. When you invoke the *CreateWindow* function, you can set the *hMenu* parameter to the value that the *LoadMenu* function returns. Then, you will call *LoadMenu* with the menu name from the resource file as the function's parameter, as shown here:

```
if (LOWORD(wParam ) == IDM_NEW )
  hNewMenu = LoadMenu(hInst, "NEWMENU" );
else
  hNewMenu = LoadMenu(hInst, "OLDMENU" );
```

Finally, you can use the *SetMenu* and *LoadMenu* functions to attach a menu to a window after you have created the window. Again, you will invoke the *SetMenu* function with the *hMenu* handle parameter. *SetMenu* associates the menu the *hMenu* handle points to with the application window that is currently open, as shown here:

```
SetMenu(hWnd, hNewMenu );
```

CHANGING MENUS WITHIN AN APPLICATION 1303

In previous Tips, you have learned how your programs can use a variety of methods to attach menus to an application. However, over the course of your Windows program design, you may develop more complex applications that commonly require changes to the menu while the application is executing. The Win32 API provides an extensive set of functions your programs can use to change a menu during the application's execution. Essentially, the functions let you change every aspect of the menu after you have attached it to the window.

The most common changes your programs will make to menus will be to change the string the menu item displays, place and remove check marks next to menu items, enable and disable menu items, and delete and add menu items. The *ModifyMenu* function lets you perform several of these operations from within a single function call. Alternatively, you can use more specific functions, such as *DeleteMenu* or *CheckMenuItem,* to modify your menu items. The companion CD-ROM that accompanies this book includes the *Delete_Items.cpp* program, which uses menu commands to let the user add and delete items from a menu.

1304 Understanding Menu-Generated Messages

As you have learned, each time a user selects a menu item within your programs, Windows will send the *WM_COMMAND* message to the program's message loop. Within the message loop, your program must then check the low word of the *wParam DWORD* value to determine which menu item the user selected.

As a rule, *WM_COMMAND* is the only message Windows will send to your programs as the result of a user's menu selection. If the user selects a system menu item, for example, Windows will instead send the *WM_SYSCOMMAND* message (which you will then probably handle within the *DefWindowProc* procedure). Your application may, on the other hand, require your message loop to process the *WM_INITMENU* and *WM_INITMENUPOPUP* messages—both of which the system sends to the window right before the system activates a menu (either a Main or Popup menu, depending on the message). Capturing the *WM_INITMENU* and *WM_INITMENUPOPUP* messages lets your application change a menu or menus, should any changes be necessary, immediately before Windows displays the application menu.

The menu will also send the *WM_MENUSELECT* message each time the user selects a menu item. The *WM_MENUSELECT* message is more powerful and versatile than the *WM_COMMAND* message because Windows generates it even if the menu item is currently disabled. However, you will usually only use the *WM_MENUSELECT* message to display context-sensitive menu help.

1305 Understanding the LoadMenu Function

As you learned in Tip 1302, your programs can use the *LoadMenu* function to load a menu that you have previously defined within a resource file. Typically, you will either follow the *LoadMenu* invocation with a call to *SetMenu* or you will use the *LoadMenu* call within a *CreateWindow* invocation. The *LoadMenu* function loads the menu resource the *lpMenuName* parameter specifies from the executable (.EXE) file Windows associates with an application instance. You will use the *LoadMenu* function within your programs as shown in the following prototype:

```
HMENU LoadMenu(
   HINSTANCE hInstance, // handle of application instance
   LPCTSTR lpMenuName          // menu name string or menu-resource identifier
);
```

LoadMenu returns an *HMENU* handle, and accepts as its parameters the handle to the instance of the module containing the menu resource *LoadMenu* is to load, and a pointer to a *NULL*-terminated string containing the name of the menu resource. Instead of using a string pointer to the menu name, you can use a *DWORD* as the second parameter (the pointer to the menu name). In such a case, the *DWORD* can will consist of a *resource identifier*, with the actual identifier in the low-order *WORD* and zero in the high-order *WORD*. To create the *DWORD* value, rather than the string pointer constant, use the *MAKEINTRESOURCE* macro.

To better understand the processing that *LoadMenu* performs, consider the following code fragment from the program *2_menus*, which changes back and forth between two menus, based on the user's selection. Both the resource file and the program file differ from their counterpart *generic* files. The *2_menus.rc* file includes the following menu definitions:

```
OLDMENU MENU DISCARDABLE
BEGIN
    POPUP "&File"
    BEGIN
        MENUITEM "E&xit",       IDM_EXIT
    END
    MENUITEM "&New Menu!"      IDM_NEW
END

NEWMENU MENU DISCARDABLE
BEGIN
    POPUP "&File"
    BEGIN
        MENUITEM "E&xit",       IDM_EXIT
    END
    MENUITEM "&Old Menu!",     IDM_OLD
END
```

The first declaration creates the *OLDMENU* menu with the *New Menu!* option. When the user selects the *New Menu!* option, the program will then change the menu within the program's window from the *OLDMENU* menu to the *NEWMENU* menu. As you might expect, the application processing that changes the application menu from the *OLDMENU* to the *NEWMENU* is within the *WndProc* function of the *2_menus.cpp* file.

As you will see, the program checks the constant value the low word of the *wParam* value contains. Depending on the result, the program loads the other menu (in other words, if the result indicates the program is currently displaying the *OLDMENU* resource, the program will display the *NEWMENU* resource, and vice versa). The program then uses the *SetMenu* function to switch back and forth between the functions. Finally, the program invokes the *DrawMenuBar* function after it completes the switch, which ensures that Windows redraws the newly loaded menu into place.

Note: *Your programs must use the* ***DestroyMenu*** *function before the application closes to destroy the menu and free memory that the loaded menu occupied.*

USING THE MODIFYMENU FUNCTION 1306

As you learned in Tip 1303, your programs can use the *ModifyMenu* function to make a variety of changes to your menus after your programs attach them to a window. The *ModifyMenu* function changes an existing menu item. Your programs can use the *ModifyMenu* function to specify the content, appearance, and behavior of each menu item within a menu. You will use the *ModifyMenu* function within your programs as shown in the following prototype:

```
BOOL ModifyMenu(
   HMENU hMnu,              // handle of menu
   UINT uPosition,          // menu item to modify
   UINT uFlags,             // menu item flags
   UINT uIDNewItem,         // menu item identifier or handle of drop-down menu
   LPCTSTR lpNewItem        // menu item content
);
```

The *ModifyMenu* function accepts the parameters detailed in Table 1306.1.

Parameters	Description
hMnu	Identifies the menu you want to change.
uPosition	Specifies the menu item you want to change in accordance with the *uFlags* parameter values.

Table 1306.1 *The parameters the* ***ModifyMenu*** *function accepts.* *(continued on following page)*

uFlags	Specifies flags that control the interpretation of the *uPosition* parameter and the content, appearance, and behavior of the menu item. This parameter must be a combination of one of the required values Table 1306.2 details and at least one of the values Table 1306.3 details.
uIDNewItem	Specifies either the identifier of the modified menu item or, if the *uFlags* parameter has the *MF_POPUP* flag set, the handle of the drop-down menu or submenu.
lpNewItem	Points to the content of the changed menu item. The interpretation of this parameter depends on whether the *uFlags* parameter includes the *MF_BITMAP, MF_OWNERDRAW,* or *MF_STRING* flag.

***Table 1306.1** The parameters the **ModifyMenu** function accepts. (continued from previous page)*

As Table 1306.1 details, you must pass a value within the required *uFlags* parameter. The *uFlags* value must be a combination of one of the values Table 1306.2 lists and one or more of the values Table 1306.3 lists.

Value	Meaning
MF_BYCOMMAND	Indicates that the *uPosition* parameter gives the identifier of the menu item. The *MF_BYCOMMAND* flag is the default if you specify neither the *MF_BYCOMMAND* nor the *MF_BYPOSITION* flag.
MF_BYPOSITION	Indicates that the *uPosition* parameter gives the zero-based relative position of the menu item.

***Table 1306.2** The possible required values for the **uFlags** parameter.*

In addition to the required value for the *uFlags* parameter, you must also use the *bitwise OR* operator to assign one or more of the values Table 1306.3 details to the menu item.

Value	Meaning
MF_BITMAP	Contains a bitmap handle.
MF_OWNERDRAW	Contains a 32-bit value an application supplies that Windows uses to maintain additional data related to the menu item. The value is in the *itemData* member of the structure pointed to by the *lparam* parameter of the *WM_MEASUREITEM* or *WM_DRAWITEM* messages the program sends when you create the menu item or update its appearance.
MF_CHECKED	Places a check mark next to the item. If your application provides check mark bitmaps (see the *SetMenuItemBitmaps* function), this flag displays a checked bitmap next to the menu item.
MF_DISABLED	Disables the menu item so that the user cannot select it, but this flag does not turn the menu item gray.
MF_ENABLED	Enables the menu item and restores it from its grayed state so that the user can select it.
MF_GRAYED	Disables the menu item and turns it gray so that the user cannot select it.
MF_MENUBARBREAK	Functions the same as the *MF_MENUBREAK* flag for a menu bar. For a drop-down menu, submenu, or shortcut menu, a vertical line separates the new column from the old column.
MF_MENUBREAK	Places the item on a new line (for menu bars) or in a new column (for a drop-down menu, submenu, or shortcut menu) without separating columns.

***Table 1306.3** The possible values for the **uFlags** parameter. (continued on following page)*

MF_OWNERDRAW	Specifies that the item is an owner-drawn item. Before the application displays the menu for the first time, the window that owns the menu receives a *WM_MEASUREITEM* message to retrieve the width and height of the menu item. The program then sends the *WM_DRAWITEM* message to the window procedure of the owner window whenever the program must update the appearance of the menu item.
MF_POPUP	Specifies that the menu item opens a drop-down menu or submenu. The *uIDNewItem* parameter specifies the handle of the drop-down menu or submenu. You will use this flag to add a menu name to a menu bar or a menu item that opens a submenu to a drop-down menu, submenu, or shortcut menu.
MF_SEPARATOR	Draws a horizontal dividing line. You use this flag only when adding items to a drop-down menu, submenu, or shortcut menu. Your programs cannot gray, disable, or highlight the line. Windows ignores the *lpNewItem* and *uIDNewItem* parameters.
MF_UNCHECKED	Does not place a check mark next to the item (the default). If your application supplies check mark bitmaps (see the *SetMenuItemBitmaps* function), this flag displays an unchecked bitmap next to the menu item.
MF_STRING	Contains a pointer to a *NULL*-terminated string (the default).

Table 1306.3 The possible values for the ***uFlags*** *parameter. (continued from previous page)*

If *ModifyMenu* replaces a menu item that opens a drop-down menu or submenu, the function destroys the old drop-down menu or submenu and frees the memory the old menu used. Additionally, the application must call the *DrawMenuBar* function whenever a menu changes, whether or not the menu is in a displayed window. To change the attributes of existing menu items, it is much faster to use the *CheckMenuItem* and *EnableMenuItem* functions.

Note: *The Windows API will not let you use the following groups of flags together when you invoke* ***ModifyMenu:***

- *MF_BYCOMMAND and MF_BYPOSITION*
- *MF_DISABLED, MF_ENABLED, and MF_GRAYED*
- *MF_BITMAP, MF_STRING, MF_OWNERDRAW, and MF_SEPARATOR*
- *MF_MENUBARBREAK and MF_MENUBREAK*
- *MF_CHECKED and MF_UNCHECKED*

To better understand the processing the *ModifyMenu* function performs, consider the *Mod_Menu.cpp* program contained on the CD-ROM that accompanies this book. The *Mod_Menu.cpp* program changes the *Demo* item to the *New Item* menu item when the user selects it, then processes the *New Item* menu item if the user selects the new item. As usual, the *WndProc* function handles the changed processing within the *Mod_Menu.cpp* file.

As you will see, the *Mod_Menu* program uses the *ModifyMenu* function to change the menu item's string value. Additionally, the code checks the menu identifiers to ensure that the function captures the newly created item.

USING ENABLEMENUITEM TO CONTROL MENUS 1307

As you have learned, your programs can use the *ModifyMenu* function to control the appearance of items within the menu. However, as you have also learned, using specific functions to address specific problems often results in quicker program execution. To help you enable, disable, or make gray a menu item you specify, your programs can use the *EnableMenuItem* function rather than the *ModifyMenu* function. You will use the *EnableMenuItem* function within your programs as shown in the following prototype:

MF_POPUP	Specifies that the menu item opens a drop-down menu or submenu. The *uIDNewItem* parameter specifies the handle to the drop-down menu or submenu. You will use the *MF_POPUP* flag to add a menu name to a menu bar, or a menu item that opens a submenu to a drop-down menu, submenu, or shortcut menu.
MF_SEPARATOR	Draws a horizontal dividing line. You will use the *MF_SEPARATOR* flag only in a drop-down menu, submenu, or shortcut menu. Your programs cannot disable, make gray, or highlight the line. The *AppendMenu* function ignores the *lpNewItem* and *uIDNewItem* parameters if you specify the *MF_SEPARATOR* type.
MF_STRING	Specifies that the menu item is a text string; the *lpNewItem* parameter points to the string.
MF_UNCHECKED	Does not place a check mark next to the item (default). If the application supplies check mark bitmaps (see *SetMenuItemBitmaps*), this flag displays the unchecked bitmap next to the menu item.

***Table 1308** The possible values for the **uFlags** parameter. (continued from previous page)*

Note: *Windows will not let you use the following groups of flags together when you invoke the **AppendMenu** function:*

- *MF_BYCOMMAND and MF_BYPOSITION*
- *MF_DISABLED, MF_ENABLED, and MF_GRAYED*
- *MF_BITMAP, MF_STRING, MF_OWNERDRAW, and MF_SEPARATOR*
- *MF_MENUBARBREAK and MF_MENUBREAK*
- *MF_CHECKED and MF_UNCHECKED*

To better understand the processing the *AppendMenu* function performs, consider the *Add_New.cpp* program contained on the CD-ROM that accompanies this book, which adds the *New Item* menu item to the menu on its own line each time the user selects the *Demo* menu item. The *WndProc* function captures the *Demo* selection and uses *AppendMenu* to add a new item, as shown here:

```
case IDM_DEMO :
        // Add new menu option on a new line.
       AppendMenu(GetMenu(hWnd ), MFT_STRING | MFT_MENUBARBREAK, 120, "New Item");
        DrawMenuBar(hWnd);
        break;
```

1309 USING DELETEMENU TO DELETE MENU SELECTIONS

In Tip 1308, you used the *AppendMenu* function to add items to a menu you specify. You can also use the *DeleteMenu* function in your programs to delete an item from the menu you specify. If the menu item opens a menu or submenu, the *DeleteMenu* function destroys the handle to the menu or submenu and frees the memory the menu or submenu was using. You will implement the *DeleteMenu* function within your programs as shown here:

```
BOOL DeleteMenu(
   HMENU hMenu,            // handle to menu
   UINT uPosition,         // menu item identifier or position
   UINT uFlags             // menu item flag
);
```

As always, the application must call the *DrawMenuBar* function whenever it changes a menu, whether or not the menu is in a displayed window. To better understand the processing that the *DeleteMenu* command performs, consider the *Add_Del.cpp* program, contained on the CD-ROM that accompanies this book, which adds three new items to a menu at the window's creation, then deletes those new items in turn as the user selects an item. The program handles the processing within the *WM_CREATE* and *WM_COMMAND* switches of the *WndProc* function, as shown here:

```
case WM_CREATE :
       {
         HMENU hMenu = GetMenu(hWnd);

         AppendMenu(hMenu, MFT_STRING, IDM_ITEM1, "Item&1");
         AppendMenu(hMenu, MFT_STRING, IDM_ITEM2, "Item&2");
         AppendMenu(hMenu, MFT_STRING, IDM_ITEM3, "Item&3");
       }
       break;

case WM_COMMAND :
      switch(LOWORD(wParam))
       {
          case IDM_ITEM1 :
          case IDM_ITEM2 :
          case IDM_ITEM3 :
           {
              HMENU hMenu = GetMenu(hWnd);

              DeleteMenu(hMenu, LOWORD(wParam), MF_BYCOMMAND);
              DrawMenuBar(hWnd);
           }
           break;
```

USING ACCELERATOR KEYS WITH MENU ITEMS 1310

As you have learned, your programs can use the ampersand (&) within a menu definition to define the accelerator key for a menu item. However, your programs can also add accelerators to a menu definition with new menu items as your programs process. To do so, your programs will use the *CreateAcceleratorTable* function, which creates an *accelerator table*. You will implement the *CreateAcceleratorTable* function within your programs in accordance with the following prototype:

```
HACCEL CreateAcceleratorTable(
   LPACCEL lpaccl,      // pointer to structure array with accelerator data
   int cEntries         // number of structures in the array
);
```

The *lpaccl* parameter points to an array of *ACCEL* structures that describes the accelerator table. The *cEntries* parameter specifies the number of *ACCEL* structures in the array. Each *ACCEL* structure element within the array defines, within an *accelerator table*, an accelerator key your program will use. The Win32 API defines the *ACCEL* structure as shown here:

```
typedef struct tagACCEL {
    BYTE   fVirt;
    WORD   key;
    WORD   cmd;
} ACCEL;
```

Table 1310.1 details the members of the *ACCEL* structure.

Members	Description
fVirt	Specifies the accelerator flags. This member can be a combination of the values Table 1310.2 details.
key	Specifies the accelerator key. This member can be either a virtual-key code or an ASCII character code.
cmd	Specifies the accelerator identifier. Windows places the *cmd* value in the low-order word of the *wParam* parameter of the *WM_COMMAND* or *WM_SYSCOMMAND* message when the user presses the accelerator key.

***Table 1310.1** The members of the **ACCEL** structure.*

As Table 1310.1 indicates, the *fVirt* member must be one or more of the values listed in Table 1310.2.

Value	Meaning
FALT	The user must hold down the ALT key while pressing the accelerator key.
FCONTROL	The user must hold down the CTRL key while pressing the accelerator key.
FNOINVERT	Specifies that Windows will not highlight a top-level menu item while the user presses the accelerator key. If the user does not specify the *FNOINVERT* flag, Windows will highlight a top-level menu item, if possible, whenever the user presses the accelerator key.
FSHIFT	The user must hold down the SHIFT key while pressing the accelerator key.
FVIRTKEY	The key member specifies a virtual-key code. If you do not specify this flag, the system will assume the key specifies an ASCII character code.

***Table 1310.2** The possible values for the **fVirt** member.*

To better understand the processing the *CreateAcceleratorTable* function performs, you will create a function in the next Tip that creates a new accelerator table. However, if your programs use accelerator tables, you must also modify the message-handling routine. For example, the following code shows a simple modification to the message loop that now lets the program support accelerator tables:

```
while(GetMessage(&msg, NULL, 0, 0) )
  {
     if (!hAccel || !TranslateAccelerator(hWnd, hAccel, &msg ) )
     {
        TranslateMessage(&msg );
        DispatchMessage(&msg );
     }
  }
```

The *hAccel* variable is a global variable with the initial value of *NULL* that later holds the handle to the accelerator table. The *TranslateAccelerator* function translates any *AcceleratorTable* message to the appropriate *WM_COMMAND* and menu constant combination. Note that the program checks the *TranslateAccelerator* function. If *TranslateAccelerator* handles the message, the loop should not let the normal processing follow, because *TranslateAccelerator* calls the *WndProc* function automatically.

Note: *Before an application closes, it must use the **DestroyAcceleratorTable** function to destroy each accelerator table that it used the **CreateAcceleratorTable** function to create.*

Creating a Sample Accelerator Table 1311

As you learned in Tip 1310, your programs can create accelerator tables during execution to provide additional functionality for menus that you create or dynamically modify. You also learned that the *CreateAccelerator* function uses an array of *ACCEL* structures to create the accelerator table. Within your programs, you will first create the array, and then call the *CreateAccelerator* function. To better understand the process of creating the accelerator table, consider the following *AddNewTestItem* function from the *Create_Accel.cpp* program, contained on the CD-ROM that accompanies this book, which creates a new accelerator table entry each time the user selects the *Demo* menu item:

```
#define IDM_NEWBASE 300

void AddNewTestItem(HWND hWnd)
 {
   static int nNum = 0;

   char    szMenuItem[20];

   HMENU   hMenu       = GetMenu(hWnd);
   ACCEL*  pAccelData  = NULL;
   ACCEL*  pCurAccel   = NULL;
   HANDLE  hAccelData  = NULL;
   int     nNumAccel   = 1;

   if (nNum == 4)                       // Maximum of 4 new items allowed.
      return;

   // If accelerator table exists, get the number of items.
   if (hAccel)
      nNumAccel = CopyAcceleratorTable(hAccel, NULL, 0) + 1;

   // Allocate an array of ACCEL structures.
   hAccelData = GlobalAlloc(GHND, sizeof(ACCEL) * nNumAccel);
   if (hAccelData)
      pAccelData = (ACCEL*)GlobalLock(hAccelData);

   // If an accelerator table exists, copy the items into the newly allocated array.
   if (hAccel && pAccelData)
    {
      CopyAcceleratorTable(hAccel, pAccelData, nNumAccel-1);

      DestroyAcceleratorTable(hAccel);
      hAccel = NULL;
    }

   // Add the new menu option and accelerator key
   if (pAccelData)
    {
      // Get a pointer to the new accelerator key in the array.
      pCurAccel = (ACCEL*)(pAccelData+nNumAccel-1);

      // Create a new menu option on the menu.
      nNum++;
      wsprintf(szMenuItem, "New Item&%d", nNum);
      AppendMenu(hMenu, MFT_STRING, IDM_NEWBASE+nNum, szMenuItem);
      DrawMenuBar(hWnd);

      // Set up a new accelerator of F1,F2,F3,or F4 for the the new menu option.
      pCurAccel->fVirt = FNOINVERT | FVIRTKEY;
      pCurAccel->cmd   = IDM_NEWBASE+nNum;
      pCurAccel->key   = (nNum == 1 ? VK_F1 :
                          nNum == 2 ? VK_F2 :
                          nNum == 3 ? VK_F3 :
                          /*default*/ VK_F4);

      // Create the new accelerator table.
      hAccel = CreateAcceleratorTable(pAccelData, nNumAccel);
```

```
        GlobalUnlock(hAccelData);
      }
    if (hAccelData)
        GlobalFree(hAccelData);
}
```

The *AddNewTestItem* function performs a great deal of processing, most of which the comments within the code explain. However, it is worthwhile to note the creation of the structure element for the new accelerator key. In this particular example, no more than four function keys can serve as shortcuts; however, the limit could just as easily be 10, or the user could determine the shortcut keys.

1312 Better Understanding the Resource File's Structure

As you have learned, virtually every Windows program uses resources. To help keep those resources organized and easily accessible, as well as to keep your code from becoming burdensome, you will generally store information about the resources a program uses within a resource file. Resource files are also efficient because a program usually loads resources into memory only when the program needs the resource.

Typically, a resource compiler (usually named *rc.exe*) compiles the resource file into an *RES* file. The linker then binds the *RES* file to the end of the fully compiled executable program file. All resources that the resource file defines then become available to the program for use during execution.

The resource file can include five single-line resource script types: *BITMAP, CURSOR, ICON, FONT,* and *MESSAGETABLE.* Each of these statements loads a data file of the type its name specifies into the resource table. After you include the resources within the resource file, your program can then use *Load* functions, such as *LoadIcon*, to access those items. You will typically implement a single-line resource type as shown here:

```
MYAPP   ICON    DISCARDABLE     "GENERIC.ICO"
```

In addition to the five single-line resource script types, there are five multiple-line resource script types: *ACCELERATOR, DIALOG, MENU, RCDATA,* and *STRINGTABLE.* You have already learned about *ACCELERATOR* and *MENU* types. You will learn about the *DIALOG* type beginning in Tip 1319. The next six Tips explain how you will use the *STRINGTABLE* and *RCDATA* type.

Multiple-line resource file types are relatively easy to identify. Each multiple-line resource type includes a type statement, a *BEGIN-END* block, and statements within the *BEGIN-END* block that may include additional *BEGIN-END* blocks, as shown here:

```
NEWMENU MENU DISCARDABLE
BEGIN
    POPUP "&File"
    BEGIN
        MENUITEM "E&xit",       IDM_EXIT
    END
    MENUITEM "&Old Menu!",      IDM_OLD
END
```

1313 Introducing String Tables

As you learned in Tip 1312, one of the multiple-line types that resource files support is the *STRINGTABLE* type. Most applications use a series of character strings in messages and character output. Windows provides *string tables* as an alternative to the conventional method of placing strings in the program's static data area. Your programs can therefore define character strings within the resource file and give the string an ID value, as shown here:

```
STRINGTABLE
BEGIN
   IDS_STRING1   "Simple String Sample."
   IDS_STRING2   "Jamsa's C/C++/C# Programmer's Bible"
   IDS_STRING3   "Delmar Publishing"
END
```

In addition to their content definition within the resource, you will typically define the string ID values (such as *IDS_STRING3)* in a separate header file that you will then include within both the resource file and the module (or modules) that will access the strings. When the application must access the data, you will use the *LoadString* function to copy the character data from the resource file into a memory buffer. Strings in a string table can contain control characters (such as tabs and newlines), as well as regular printing characters.

There are a number of significant programming advantages to using string tables. The main programming advantage of using string tables is the reduction in your program's memory usage. Because your program does not load the strings until it needs them, it is not necessary to store the strings within your program's static data area. For this reason, you should avoid copying the string table data into a static memory buffer, because doing so will defeat the purpose of the string table. You should instead copy string table data into a local variable (a stack variable) or into globally allocated memory.

Another significant advantage of using string tables is their support for multiple languages. The Win32 API supports multilingual resources within a single application—meaning that you can distribute the same executable program in several countries without having to change it. Later Tips will use string tables to maintain window information.

UNDERSTANDING CUSTOM RESOURCES 1314

As you learned in Tip 1312, one of the multi-line resource types is the *RCDATA* type. You can use the *RCDATA* type to store other types of static data, specifically raw binary data. For example, the following code listing stores several data pieces of different types within the *DataID* resource:

```
DataID RCDATA
BEGIN

   3
   40
    0x8232
    "String Data (continued)… "
    "More String Data\0"
END
```

While you can include custom resource data directly within the program file, or read it from an outside data file, the best place to store custom resource data is within an external file. The resource compiler can then add the external file's contents to the resource data when it compiles the resource file. For example, the following two statements define the custom resource types *TEXT* and *METAFILE*, assign an identifier to the types, and then add the resources to the resource file:

```
happy              TEXT         "happydog.txt"
picture            METAFILE     "happypic.wmf"
```

When you define custom resource types, you will use the *FindResource* function together with the *LoadResource* function to load custom resources into your programs. Tip 1318 explains the *LoadResource* function.

LOADING STRING TABLES INTO PROGRAMS WITH LOADSTRING 1315

As you have learned, you can create string tables within your resource files. You will assign each string within a string table a specific ID value, which you will usually define within the program's header file, as shown here:

```
#define IDM_EXIT            100
#define IDM_TEST            200
#define IDM_ABOUT           301
```

After you define a string table, your programs will use the *LoadString* function to load a string resource from the executable file associated with a specified module, copy the string into a buffer, and append a terminating *NULL* character to the buffer. You will use the *LoadString* function within your programs as shown in the following prototype:

```
int LoadString(
   HINSTANCE hInstance,        // handle of module containing string resource
   UINT uID,                   // resource identifier
   LPTSTR lpBuffer,            // address of buffer for resource
   int nBufferMax       // size of buffer
);
```

The *hInstance* parameter identifies an instance of the module whose executable file contains the string resource. The *uID* parameter specifies the integer identifier of the string *LoadString* is to load. The *lpBuffer* parameter points to the buffer to receive the string. Finally, the *nBufferMax* parameter specifies the size of the buffer in bytes (ANSI version) or characters (Unicode version). The function truncates and *NULL*-terminates the string if it is longer than the number of characters you specify.

If *LoadString* succeeds, it returns the number of bytes (ANSI version) or characters (Unicode version) it copied into the buffer, not including the *NULL*-terminating character, or zero if the string resource does not exist. To get extended error information, you can call the *GetLastError* function. To better understand the processing *LoadString* performs, consider the following code fragment from the *LoadStrg.cpp* program the CD-ROM that accompanies this book contains:

```
case IDM_DEMO :
        {
          char  szString[40];
          SHORT idx;

          for (idx = IDS_STRINGBASE; idx < IDS_STRINGBASE+3; idx++)
           {
             LoadString(hInst, idx, szString, 40);
             MessageBox(hWnd, szString, "String Loaded", MB_OK | MB_ICONINFORMATION);
           }
        }
        break;
```

The *LoadStrg.cpp* program will load and display three separate strings when the user selects the *Demo* option. The program will display each string within a message box.

1316 Listing a Resource File's Contents

As you have learned, you can store many different types of custom data within a resource file. The *EnumResourceNames* function searches a module for each resource of the type you specify within the *lpszType* parameter and passes the name of each resource it locates to an application-defined callback function. The *EnumResourceNames* function continues to enumerate resource names until the callback function returns *False* or until *EnumResourceNames* has enumerated all resource names. You will use the *EnumResourceNames* function within your programs as shown in the following prototype:

```
BOOL EnumResourceNames(
   HINSTANCE hModule,               // resource-module handling
   LPCTSTR lpszType,                // pointer to resource type
   ENUMRESNAMEPROC lpEnumFunc,// pointer to callback function
   LONG lParam                      // application-defined parameter
);
```

The *EnumResourceNames* function accepts the parameters Table 1316.1 details.

Parameters	Description
hModule	Identifies the module whose executable file contains the resources for which *EnumResourceNames* is to enumerate the names. If this parameter is *NULL*, the function enumerates the resource names in the module used to create the current process.
lpszType	Points to a *NULL*-terminated string that specifies the type name of the resource for which *EnumResourceNames* is enumerating the name. For standard resource types, this parameter can be one of the values Table 1316.2 details.
lpEnumFunc	Points to the callback function *EnumResourceNames* is to call for each enumerated resource name.
lParam	Specifies an application-defined value the *EnumResourceNames* passes to the callback function. Your programs can use this parameter when they are checking errors.

***Table 1316.1** The parameters for the **EnumResourceNames** function.*

As Table 1316.1 notes, the *lpszType* parameter may have one of several values, as Table 1316.2 details.

Value	Meaning
RT_ACCELERATOR	Accelerator table
RT_ANICURSOR	Animated cursor
RT_ANIICON	Animated icon
RT_BITMAP	Bitmap resource
RT_CURSOR	Hardware-dependent cursor resource
RT_DIALOG	Dialog box
RT_FONT	Font resource
RT_FONTDIR	Font directory resource
RT_GROUP_CURSOR	Hardware-independent cursor resource
RT_GROUP_ICON	Hardware-independent icon resource
RT_ICON	Hardware-dependent icon resource
RT_MENU	Menu resource
RT_MESSAGETABLE	Message-table entry
RT_PLUGPLAY	Plug and play resource
RT_RCDATA	Application-defined resource (raw data)
RT_STRING	String-table entry
RT_VERSION	Version resource
RT_VXD	VXD (Virtual Device Driver)

***Table 1316.2** The possible values for the **lpszType**.*

To better understand how the *EnumResourceNames* function works, consider the program *3_pics.cpp*, contained on the companion CD-ROM that accompanies this book. The resource file for the *3_pics.cpp* program includes three bitmaps. The program uses the *EnumResourceNames* function and a callback function named *PaintBitmaps* to enumerate the resources of type *RT_BITMAP* within the *3_pics.rc* resource file. The following code listing shows the *PaintBitmaps* callback function:

```
BOOL CALLBACK PaintBitmaps(HANDLE hModule, LPCTSTR lpszType,
                           LPCTSTR lpszName, LONG lParam)
 {
   HBITMAP hBitmap = (HBITMAP) LoadBitmap((HINSTANCE) hModule, (LPCTSTR) lpszName);
   if (hBitmap)
    {
      BITMAP bm;
      HDC    hMemDC;
      HWND   hWnd = (HWND)lParam;
      HDC    hDC  = GetDC(hWnd );
      HFONT  hOldFont;
      // Get the size of the bitmap.
      GetObject(hBitmap, sizeof(BITMAP), &bm);

      // Create a memory DC to select the bitmap into.
      hMemDC = CreateCompatibleDC(hDC);
      SelectObject(hMemDC, hBitmap);
      // Display the bitmap, stretching it to 50X50 pixels.
      StretchBlt(hDC, gnPos, 0, 50, 50, hMemDC, 0, 0, bm.bmWidth, bm.bmHeight,
                 SRCCOPY );
      // Display the bitmap name.
      hOldFont = SelectObject(hDC, GetStockObject(ANSI_VAR_FONT));
      TextOut(hDC, gnPos, 60, lpszName, strlen(lpszName));
      SelectObject(hDC, hOldFont);
      DeleteDC(hMemDC);
      ReleaseDC(hWnd, hDC);
      DeleteObject(hBitmap);
      gnPos += 100;
    }
   return(TRUE);
 }
```

The *PaintBitmaps* callback function performs significant processing, much of which later Tips will explain in detail. The callback function paints each bitmap that *EnumResourceTypes* enumerates onto the screen.

1317 USING ENUMRESOURCETYPES WITH RESOURCE FILES

In Tip 1316, you learned about the *EnumResourceNames* function, which your programs can use to enumerate the different resources of a given type within a resource file. However, there may be times when your programs will not know in advance the different types the resource file contains. In such cases, your programs can also use the *EnumResourceTypes* function. The *EnumResourceTypes* function searches a module for resources and passes each resource type it finds to an application-defined callback function. The *EnumResourceTypes* function continues to enumerate resource types until the callback function returns *False* or until it has enumerated all resource types. You will implement the *EnumResourceTypes* function within your programs as shown here:

```
BOOL EnumResourceTypes(
   HMODULE hModule,                // resource-module handle
   ENUMRESTYPEPROC lpEnumFunc,// pointer to callback function
   LONG lParam                     // application-defined parameter
);
```

Just as the *EnumResourceNames* function uses a callback function, so too does the *EnumResourceTypes* function. The prototype of the callback function you will use with the *EnumResourceTypes* function is shown here (the *EnumResTypeProc* function name is a placeholder for the application-defined or library-defined function name):

```
BOOL CALLBACK EnumResTypeProc(
   HANDLE hModule,      // resource-module handle
   LPTSTR lpszType,     // pointer to resource type
   LONG lParam          // application-defined parameter
);
```

The *lpszType* parameter points to a *NULL*-terminated string specifying the type name of the resource for which the function is enumerating the type. For standard resource types, the *lpszType* parameter can be one of the values Table 1316.2 detailed.

To better understand the processing the *EnumResourceTypes* function performs, consider the program *EnumResT.cpp*, contained on the CD-ROM that accompanies this book. The *EnumResT.cpp* program fills a list box with all the different types it enumerates within a given resource file. The following code fragment details the *ListResourceTypes* callback function, which the *EnumResourceTypes* function uses as its callback function:

```
BOOL CALLBACK ListResourceTypes(HANDLE hModule, LPTSTR lpszType, LONG lParam )
{
   LPTSTR lpAddString = lpszType;
   HWND   hListBox    = (HWND)lParam;

   // Check to see if the resource type is a predefined type. If it is,
   // set lpAddString to a descriptive string.
   switch(LOWORD(lpszType) )
    {
      case RT_ACCELERATOR   : lpAddString = "Accelerator"; break;
      case RT_BITMAP        : lpAddString = "Bitmap"; break;
      case RT_DIALOG        : lpAddString = "Dialog"; break;
      case RT_FONT          : lpAddString = "Font"; break;
      case RT_FONTDIR       : lpAddString = "FontDir"; break;
      case RT_MENU          : lpAddString = "Menu"; break;
      case RT_RCDATA        : lpAddString = "RC Data"; break;
      case RT_STRING        : lpAddString = "String Table"; break;
      case RT_MESSAGETABLE  : lpAddString = "Message Table"; break;
      case RT_CURSOR        : lpAddString = "Cursor"; break;
      case RT_GROUP_CURSOR  : lpAddString = "Group Cursor"; break;
      case RT_ICON          : lpAddString = "Icon"; break;
      case RT_GROUP_ICON    : lpAddString = "Group Icon"; break;
      case RT_VERSION       : lpAddString = "Version Information"; break;
    }
   SendMessage(hListBox, LB_INSERTSTRING, (WPARAM)-1, (LPARAM)lpAddString );
   return(TRUE );
}
```

The callback function receives the *lpszType* value from the *EnumResourceTypes* function and converts the value to a string value more recognizable to the user, such as "Icon" or "Menu."

LOADING RESOURCES INTO PROGRAMS WITH FINDRESOURCE 1318

As you have learned, your programs can define any number of custom resources within the resource file. In previous Tips, you have used the *EnumResourceNames* and *EnumResourceTypes* functions to list all the resource names of a given type and all the resource types within a given resource file. Alternatively, your programs can use the *FindResource* and *LoadResource* functions to accomplish much the same task. The *FindResource* function determines the location of a resource with the type and name you specify in the module you specify. You will use the *FindResource* function within your programs as shown in the following prototype:

```
HRSRC FindResource(
   HMODULE hModule,     // resource-module handle
   LPCTSTR lpName,      // pointer to resource name
   LPCTSTR lpType       // pointer to resource type—See Table 1316.2
);
```

If the *FindResource* function succeeds, the return value is a handle to the specified resource's information block. To obtain a handle to the resource, pass the handle *FindResource* returns to the *LoadResource* function. If the function fails, the return value is *NULL*.

If the high-order word of the *lpName* or *lpType* parameters is zero, the low-order word will specify the integer identifier of the name or type of the given resource. Otherwise, those parameters will be *long* pointers to *NULL*-terminated strings. If the first character of the string is a pound sign (#), the remaining characters will represent a decimal number that specifies the integer identifier of the resource's name or type. For example, the string "#258" represents the integer identifier 258.

Your applications should reduce the amount of memory the resources require by referring to the resources by integer identifier instead of by name. An application can use the *FindResource* function to find any type of resource, but you should use *FindResource* only if the application must access the binary resource data when it makes subsequent calls to the *LoadLibrary* and *LockResource* functions. To learn more about *LoadLibrary* and *LockResource*, consult your compiler's on-line help documentation.

To use a resource immediately, an application should use one of the resource-specific functions Table 1318 details to find and load the resources in one call.

Function	Action
FormatMessage	Loads and formats a message-table entry
LoadAccelerators	Loads an accelerator table
LoadBitmap	Loads a bitmap resource
LoadCursor	Loads a cursor resource
LoadIcon	Loads an icon resource
LoadMenu	Loads a menu resource
LoadString	Loads a string-table entry

***Table 1318** The resource-specific **Load** functions.*

For example, an application can use the *LoadIcon* function to load an icon for display on the screen. However, the application should use *FindResource* and *LoadResource* if it is loading the icon to copy its data to another application.

To better understand the processing that *FindResource* performs, consider the *FindRes.cpp* program, contained on the companion CD-ROM that accompanies this book. *FindRes.cpp* uses *FindResource* to load raw binary data into a structure. The *FindRes.cpp* program has modifications to the resource file, the header file, and the program file from the original *generic* files. The resource file includes the following additional code, which simply defines a set of hexadecimal values:

```
TestData RCDATA DISCARDABLE
BEGIN
   0x0001,
   0x0002,
   0x0003
END
```

The header file defines a data structure into which the program will read the data within the *TestData* block, as shown here:

```
typedef struct {
   SHORT Value1;
   SHORT Value2;
   SHORT Value3;
 } RESDATA;
```

Finally, the *FindRes.cpp* program file reads the data into the structure within the *Demo* menu item handler in the *WndProc* function, as shown here:

```
case IDM_DEMO :
        {
           HRSRC hres = FindResource(hInst, "DemoData", RT_RCDATA);
           if (hres)
            {
               char szMsg[50];
               DWORD    size = SizeofResource(hInst, hres);
               HGLOBAL  hmem = LoadResource(hInst, hres);
               RESDATA* pmem = (RESDATA*)LockResource(hmem);
               wsprintf(szMsg, "Values loaded: %d, %d, %d\nSize = %d",
                        pmem->Value1, pmem->Value2, pmem->Value3, size);
               MessageBox(hWnd, szMsg, lpszAppName, MB_OK);
            }
        }
        break;
```

If the program finds resource data under the keyword *DemoData*, it will determine the amount of data and save it within the *size* variable. The program will then use the *HGLOBAL* specifier to allocate space off the heap in which to store the data. Finally, it will store the data within an instance of the *RESDATA* structure. When you compile and execute the program and select the *Demo* menu item, the screen will display the value within a message box.

UNDERSTANDING DIALOG BOXES 1319

Throughout previous Tips you have created different windows within your programs. However, most of the windows that you created were top-level windows, whose code you intended to perform processing throughout the program's execution. A top-level window processes messages of many different types to perform the program's processing.

A *dialog box*, on the other hand, is similar to a pop-up window. A dialog box elicits input from the user for some specific task, such as to get a file name or a character string for a search. The main difference between dialog boxes and pop-up windows is that dialog boxes use *templates* that define the controls the dialog box displays. You use the *DIALOG* resource type to define these templates within the resource file. You can also create the templates dynamically in memory during the program's execution.

Dialog boxes also function differently from pop-up windows in that dialog boxes use a special default message-processing function that interprets keystrokes, such as the arrow keys and TAB, which lets the user more easily select controls within the dialog box. As a rule, the additional message-processing function and the dialog box's appearance makes a dialog box most convenient for accepting and processing simple input from the user. In the following Tips, you will learn about the types of dialog boxes and how to use dialog boxes within your programs.

DEFINING DIALOG BOX TYPES 1320

As you learned in Tip 1319, you can use dialog boxes within your programs to provide information to the user and, in turn, to receive information back from the user. There are two basic types of dialog boxes: *modal* dialog boxes and *modeless* (or non-modal) dialog boxes.

When a modal dialog box is on the screen, the user cannot switch to another section of the application, unless he or she first closes the dialog box. By default, a modal dialog box limits access to the other visible windows of the application that called the dialog box. The user, however, can still switch to other applications while a single application displays the modal dialog box. The simplest modal dialog box is the message box, which you have used in previous Tips.

Your programs can also specify *system modal* dialog boxes. A system modal dialog box takes over the entire screen and does not let the user perform additional processing in any program until the user responds to the dialog box. The use of system modal dialog boxes within your programs is appropriate only in the case of a serious problem that the user cannot ignore, such as a system error. To create a system modal dialog box, you can either specify the *WS_SYSMODAL* style in the dialog box template or use the *SetSysModalWindow* function to create the dialog box.

Typically, however, you will use the *DialogBox* function to create modal and system modal dialog boxes. When your programs use the *DialogBox* function to display a dialog box, Windows sends all the calling window's messages to the dialog box.

Less common, but still useful, are *modeless* dialog boxes. Unlike a modal dialog box, a modeless dialog box can get and lose the input focus (that is, it can be the active window and also be an inactive window). A modeless dialog box has an indefinite lifetime. However, because the modeless dialog box can exist for an extended period of time, programs that use modeless dialog boxes must make sure the message loop shares messages with the dialog box. You will typically construct message loops for programs that contain modeless dialog boxes as shown here:

```
while(GetMessage(&msg, NULL, 0, 0))
 {
   if (hDlgModeless || !IsDialogMessage (hDlgModeless, &msg))
    {
      TranslateMessage(&msg);
      DispatchMessage(&msg);
    }
 }
```

In the general construction, the *hDlgModeless* value represents a handle to the modeless dialog box. If the modeless dialog box is not currently open, the *hDlgModeless* value must be *NULL*. The *IsDialogMessage* function determines if a message from Windows is meant for the dialog box. If so, the function will send the message to the dialog box message-handling procedure, and therefore *TranslateMessage* and *DispatchMessage* will not handle the message.

1321 Using the Keyboard with Dialog Boxes

As you have learned, Windows provides built-in logic to handle dialog boxes in the form of the special dialog box message-handling procedure. The built-in logic includes the means to select items within the dialog box with keystrokes, rather than using the mouse. Your programs can provide three sets of keyboard logic: *hot keys* to select items using ALT+letter key combinations, response to the TAB key for movement between controls, and response to the arrow keys for movement between and within controls.

You will provide support for ALT+letter key combinations within your dialog boxes in the same manner you do for menu items. Within the control's text string, precede the letter you want the key combination to use with the ampersand (&) symbol. For example, the following definition for a *DEFPUSHBUTTON* control will use ALT+D as a hot key combination that activates the "Dalmatian" button:

```
CONTROL "&Dalmatian", IDC_DONE, "BUTTON",
    BS_DEFPUSHBUTTON | WS_TABSTOP | WS_CHILD, 45, 66, 48, 12
```

Users may sometimes find keyboard controls more convenient than the mouse. However, as with most other simplified user functions, be careful not to define too many keyboard shortcuts because they may become confusing rather than helpful to users. To support keyboard controls within your dialog boxes, you must set certain elements within the dialog box template, including the *WS_TABSTOP* and *WS_GROUP* styles.

The *WS_TABSTOP* style marks each item that will receive the input focus when the user presses TAB or SHIFT+TAB. The *WS_GROUP* style marks the beginning of a group. All the items you list within the resource script until the next *WS_GROUP* style are part of a single group. The user can use the arrow keys to move among items within a group, but cannot use the arrow keys to move from one group to another.

UNDERSTANDING THE COMPONENTS OF THE DIALOG BOX TEMPLATE 1322

As you have learned, within a resource file, you will precede a series of dialog box control resources with the *DIALOG* statement to define dialog boxes. The general form of the dialog box template is similar to the resource template you have used in previous Tips to design menus and menu items. The general form of the dialog box template i s shown here:

```
DBIdentifier     DIALOG DISCARDABLE  Left, Top, Width, Height
STYLE Style1 | Style2 | . . . | StyleN
CAPTION "Dialog Caption"
FONT FontSize, "FontName"
BEGIN
   ControlType1  "Control Caption", Control_ID, Left, Top, Width, Height
   ControlType2  "Control Caption", Control_ID, Left, Top, Width, Height
. . .
   ControlTypeN  "Control Caption", Control_ID, Left, Top, Width, Height
END
```

A *ControlType* refers to one of several child window types that you can use to create controls on a dialog box. The *ControlType* entry must be one of the values Table 1322 lists.

Possible Control Types			
BUTTON	*CHECKBOX*	*COMBOBOX*	*CONTROL*
CTEXT	*DEFPUSHBUTTON*	*EDITTEXT*	*GROUPBOX*
ICON	*LISTBOX*	*LTEXT*	*PUSHBUTTON*
RADIOBUTTON	*RTEXT*	*SCROLLBAR*	*STATIC*

Table 1322 *The possible values for the* ***ControlType*** *entry within the dialog box definition.*

CREATING A SPECIFIC DIALOG BOX TEMPLATE 1323

In Tip 1322, you learned the generic form of the dialog box declaration, which includes many important components. It may be easier for you to understand the dialog box definition if you analyze a simple dialog box declaration, as shown here:

```
TESTDIALOG DIALOG DISCARDABLE  20, 20, 180, 70
STYLE DS_MODALFRAME | WS_POPUP | WS_VISIBLE | WS_CAPTION | WS_SYSMENU
CAPTION "Test Dialog"
FONT 8, "MS Sans Serif"
BEGIN
    CHECKBOX        "Check box control.",IDC_CHECKBOX,9,7,70,10
    GROUPBOX        "Radio Buttons",-1,7,21,86,39
    RADIOBUTTON     "First", IDC_RADIO1,13,32,37,10,WS_GROUP | WS_TABSTOP
    RADIOBUTTON     "Second",IDC_RADIO2,13,45,39,10
    PUSHBUTTON      "Done",IDCANCEL,116,8,50,14,WS_GROUP
END
```

The previous declaration creates the dialog box shown in Figure 1323.

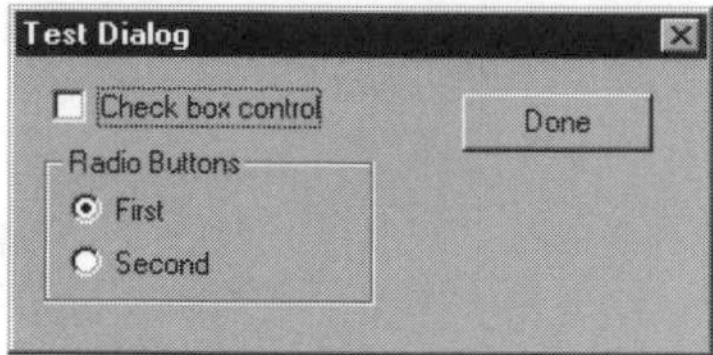

***Figure 1323** The **Test Dialog** dialog box.*

When you look closely at the resulting dialog box, understanding its components within the resource file definition is significantly easier. Each statement within the *BEGIN-END* block creates its own control. The first statement creates the check box at the top left of the form. It also assigns the *IDC_CHECKBOX* identifier to that control so that the message handler can respond appropriately to changes in the control's value (in this case, whether the control is currently marked with a check mark or not).

The second statement creates the Radio Buttons frame, which the user cannot select, so the definition does not assign it an identifier. The third and fourth statements create the *First* and *Second* radio buttons within the frame. These buttons are in a group; the user can select only one button at a time. Notice how the group starts with the *IDC_RADIO1* button, and the resource file then declares another group with the *Done* button.

The last statement within the block uses the identifier *IDCANCEL* to create the *Done* button. Typically, you will assign the *IDCANCEL* identifier to whichever button will close the dialog box without saving changes to some other location.

1324 Understanding the Dialog Box Definition's Components

As you learned in Tip 1323, each dialog box definition performs specific processing within its *BEGIN-END* block to create the controls that the dialog box will display. However, before defining the control definitions, the dialog box defines its own standard attributes. For example, the *TESTDIALOG* dialog box from Tip 1323 begins with the following four-line definition:

```
TESTDIALOG DIALOG DISCARDABLE  20, 20, 180, 70
STYLE DS_MODALFRAME | WS_POPUP | WS_VISIBLE | WS_CAPTION | WS_SYSMENU
CAPTION "Test Dialog"
FONT 8, "MS Sans Serif"
```

The first line identifies the dialog box with the *TESTDIALOG* identifier. The line also declares the dimensions of the dialog box. In the previous example, the dialog box starts at 20 dialog base units (DBU) down and 20 dialog base units (DBU) over from the calling window's inside client edge. The dialog box is also 180 dialog base units (DBU) wide by 70 dialog base units deep.

The second line defines the styles the dialog box will use when you create it. You can use all the window styles Tip 1272 defines that begin with either *WS* or *DS* as the dialog box style. You should always include the *WS_VISIBLE* style to make the dialog box visible. You cannot style dialog boxes as *WS_MINIMIZEBOX* or *WS_MAXIMIZEBOX.*

You will use the *CAPTION* specifier, shown in the third line, with dialog boxes that you declare as having the *WS_CAPTION* style. As a rule, you should caption your dialog boxes for two reasons. First, it reminds the user of the dialog box's purpose. Second, it lets the user move the dialog box around on the screen. To simplify user interaction with your programs, make sure that you place the caption string within quotes.

Finally, the *FONT* specifier determines not only the typeface the dialog box uses, but also the sizing of every control within the dialog box and the dialog box itself. The *FONT* specifier plays such an important role in the dialog box's sizing because Windows computes the dialog base units (DBU) as fractions of the font size. For most dialog boxes, an 8-point font, such as MS Sans Serif, is a good choice. You must make sure the font name within the quotes exactly matches the system-defined font name, or the resource file will not compile correctly.

Defining the Dialog Box's Controls 1325

Windows provides you with two equivalent ways to define the controls within a dialog box. One way you can define the controls is to use an explicit statement, such as the *COMBOBOX* statement, as you saw in Tip 1323. The other way is to use the *CONTROL* statement and include the combo box style within the control as a parameter. For example, the following two statements create the same control:

```
CONTROL "Push if You're Happy", IDC_BUTTON1, "button",
    BS_DEFPUSHBUTTON | WS_TABSTOP | WS_CHILD, 45, 66, 48, 12

DEFPUSHBUTTON "Push if You're Happy", IDC_BUTTON1, 45, 66, 48, 12, WS_TABSTOP
```

Either definition is acceptable; however, as you have learned in many previous examples, you should pick a specific style and use it throughout the resource file.

You will normally define the ID values for each control within a separate header file. For the controls with the optional *style* parameter, the choices include the *WS_TABSTOP* and *WS_GROUP* styles. The *WS_TABSTOP* and *WS_GROUP* styles control the default keyboard interface, as Tip 1321 described. You should use the *bitwise OR* operator (|) to combine all the styles that you assign to a control.

Using the DialogBox Macro to Display a Dialog Box 1326

As you have learned, your programs will define dialog box templates within the resource file. Over the next several Tips, you will learn about different ways your programs can display dialog boxes. Of the methods your programs can use, the simplest one is the *DialogBox* macro. The *DialogBox* macro creates a modal dialog box from a dialog box template resource. *DialogBox* does not return control to the calling program until the specified callback function terminates the modal dialog box by calling the *EndDialog* function. (You will learn more about *EndDialog* in Tip 1334.) The *DialogBox* macro uses the *DialogBoxParam* function, which Tip 1330 discusses in detail. The prototype for the *DialogBox* macro is shown here:

```
int DialogBox(
   HINSTANCE hInstance,        // handle to application instance
   LPCTSTR lpTemplate,         // identifies dialog box template
   HWND hWndParent,            // handle to owner window
   DLGPROC lpDialogFunc        // pointer to dialog box procedure
);
```

The *DialogBox* macro accepts four parameters, as Table 1326 details.

Parameter	Description
hInstance	Identifies the instance of the program whose executable file contains the dialog box template.
lpTemplate	Identifies the dialog box template. This parameter is either the pointer to a *NULL*-terminated character string that specifies the name of the dialog box template or an integer value that specifies the resource identifier of the dialog box template. If the parameter specifies a resource identifier, its high-order word must be zero and its low-order word must contain the identifier. You can use the *MAKEINTRESOURCE* macro to create this value.
hWndParent	Identifies the window that owns the dialog box.
lpDialogFunc	Points to the dialog box procedure. Tip 1327 explains the *DialogProc* callback function in detail.

***Table 1326** The parameters for the **DialogBox** macro.*

If the call to *DialogBox* succeeds, the macro returns the *nResult* parameter. Your program will subsequently use the *nResult* in the call to the *EndDialog* function that closes the dialog box. If the call to *DialogBox* fails, the macro returns the value -1.

The *DialogBox* macro uses the *CreateWindowEx* function to create the dialog box. *DialogBox* then sends a *WM_INITDIALOG* message (and a *WM_SETFONT* message if the template specifies the *DS_SETFONT* style) to the dialog box procedure. The function displays the dialog box (regardless of whether the template specifies the *WS_VISIBLE* style), disables the owner window, and starts its own message loop to retrieve and dispatch messages for the dialog box.

When the dialog box message-processing procedure calls the *EndDialog* function, *DialogBox* destroys the dialog box, ends the message loop, enables the owner window (if previously enabled), and returns to the calling window the *nResult* parameter.

The CD-ROM that accompanies this book includes the *dlgbox.cpp* program, which creates a dialog box into which the user can enter an integer value and a character string. Note that the program defines as global variables the integer and character string values, which your program can change from within the dialog box's processing loop. The following code fragment from the *dlgbox.cpp* program file includes the *WndProc* function that creates the dialog box:

```
case IDM_DEMO :
        DialogBox(hInst, "DemoDialog", hWnd, (DLGPROC)DemoDlgProc);
        break;
```

When you compile and execute the *dlgbox* program, your screen will display the output shown in Figure 1326.

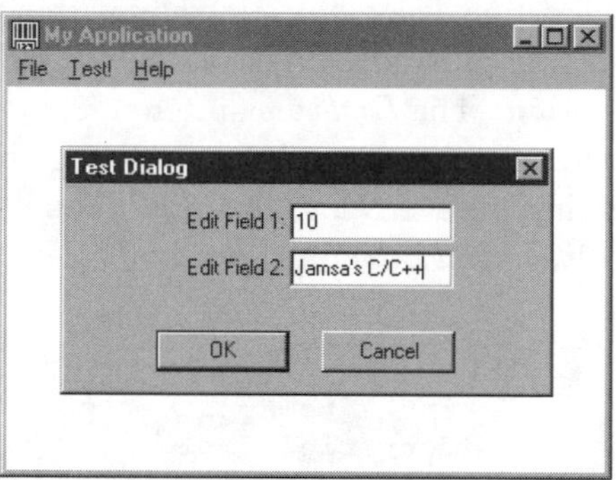

***Figure 1326** The output of the **dlgbox** program.*

1327 UNDERSTANDING THE DIALOG BOX'S MESSAGE LOOP

As you have learned, when your programs create dialog boxes, the dialog boxes will use their own message procedure, rather than the *WndProc* message procedure that your parent windows will use. Your programs may have many different message procedures, even as many as one for each dialog box. You can name message procedures anything you want because you pass the procedure's address each time you create a dialog box. For example, in Tip 1326 the *dlgbox* program creates the dialog box with the following call to the *DialogBox* macro:

```
DialogBox(hInst, "DemoDialog", hWnd, (DLGPROC)DemoDlgProc);
```

The last parameter, the pointer to the message procedure, tells Windows where it should send messages destined for the dialog box. As you can see in the following code listing, the message procedure for the dialog box performs processing similar to that the *WndProc* procedure performs:

```
LRESULT CALLBACK TestDlgProc(HWND hDlg, UINT uMsg, WPARAM wParam, LPARAM lParam)
{
   switch(uMsg)
   {
```

```
        case WM_INITDIALOG :
                SetDlgItemInt  (hDlg, IDC_EDIT1, nEditOne, TRUE);
                SetDlgItemText(hDlg, IDC_EDIT2, szEditTwo);
                break;
        case WM_COMMAND :
                switch(LOWORD(wParam))
                 {
                   case IDOK :
                         {
                           BOOL bTran;
                           nEditOne = GetDlgItemInt(hDlg, IDC_EDIT1, &bTran, TRUE);
                           GetDlgItemText(hDlg, IDC_EDIT2, szEditTwo,
                                           sizeof(szEditTwo)-1);
                           EndDialog(hDlg, IDOK);
                         }
                         break;
                  case IDCANCEL :
                         EndDialog(hDlg, IDCANCEL);
                         break;
                 }
                 break;
       default :
                return(FALSE);
     }
   return(TRUE);
}
```

In the case of the dialog box message procedure the previous code fragment details, the procedure checks for two basic messages: *WM_INITDIALOG* and *WM_COMMAND*. If the procedure receives the *WM_INITDIALOG* message, it will initialize the values the controls on the dialog box will display. On the other hand, if the procedure receives the *WM_COMMAND* message, it will check the low word of the *wParam* value. As you know, the low word of the *wParam* value holds the constant for the command the user selected. If the user clicks the mouse on OK (and *wParam* contains the *IDOK* constant), the procedure will save the values within the edit boxes into two global variables. On the other hand, if the user clicks the mouse on Cancel (and *wParam* contains the *IDCANCEL* constant), the procedure will close the dialog box without saving any changes the user may have made.

MORE ON CONTROL MANIPULATIONS 1328

In Tip 1327, you learned about the dialog box's message procedure, which process message Windows dispatches to the dialog box. In the case of the *dlgbox.cpp* program, the message procedure initializes the controls and, depending on user input, saves each control's value into a global variable. The message procedure uses the *SetDlgItemInt* and *SetDlgItemText* functions to initialize the window's controls, and the *GetDlgItemInt* and *GetDlgItemText* functions to retrieve the values. You will often use these functions within your programs to initialize controls and to retrieve values from controls. The prototypes for these four functions are shown here:

```
BOOL SetDlgItemInt(HWND hDlg, int nIDDlgItem, UINT uValue, BOOL bSigned);
BOOL SetDlgItemText(HWND hDlg, int nIDDlgItem, LPCTSTR lpString);
UINT GetDlgItemInt(HWND hDlg, int nIDDlgItem, BOOL *lpTranslated, BOOL bSigned);
UINT GetDlgItemText(HWND hDlg, int nIDDlgItem, LPTSTR lpString, int nMaxCount);
```

Note that both *Set* functions return a *BOOL* value, which indicates the function's success or failure. Both *Get* functions return a *UINT* (unsigned integer) value. For the *GetDlgItemText* function, the return value indicates the number of characters the function copied to the *lpString* buffer. Table 1328 lists the parameters for the four functions and shows to which functions each parameter applies.

Parameter	Functions	Description
hDlg	*SetDlgItemInt* *SetDlgItemText* *GetDlgItemInt* *GetDlgItemText*	Identifies the dialog box that contains the control.
nIDDlgItem	*SetDlgItemInt* *SetDlgItemText* *GetDlgItemInt* *GetDlgItemText*	Specifies which control to change. Specifies the control from which to obtain the value.
uValue	*SetDlgItemInt*	Specifies the integer value used to generate the item text.
bSigned	*SetDlgItemInt* *GetDlgItemInt*	Specifies whether the *uValue* parameter is signed or unsigned. For *GetDlgItemInt*, specifies whether the return value is signed or unsigned. If this parameter is *True*, *uValue* is signed. If this parameter is *True* and *uValue* is less than zero, *SetDlgItemInt* places a minus sign before the first digit in the string. If this parameter is *False*, *uValue* is unsigned.
lpString	*SetDlgItemText*	For *SetDlgItemText*, *lpString* specifies the string *GetDlgItemText* to place within the control. For *GetDlgItemText*, *lpString* specifies the string buffer into which the function should place the control's return value.
lpTranslated	*GetDlgItemInt*	Points to a Boolean variable that receives a function's success/failure value. *True* indicates success, *False* indicates failure. This parameter is optional because it can be *NULL*. When it is *NULL*, the function returns no information about success or failure.
nMaxCount	*GetDlgItemText*	Specifies the limit on the number of characters the function will read from the control into the string buffer *lpString*.

Table 1328 *The parameters for the* ***SetDlgItemInt, SetDlgItemText, GetDlgItemInt,*** *and* ***GetDlgItemText*** *functions.*

1329 UNDERSTANDING THE CREATEDIALOG MACRO

In Tip 1326, you used the *DialogBox* macro to create a modal dialog box from a dialog box template resource. As you have learned, your programs will often create modeless dialog boxes in addition to modal dialog boxes. You can use the *CreateDialog* macro to create modeless dialog boxes. The *CreateDialog* macro creates a modeless dialog box from a dialog box template resource. The *CreateDialog* macro uses the *CreateDialogParam* function, which Tip 1330 explains in detail. You will implement the *CreateDialog* macro within your programs in accordance with the generalized format shown here:

```
HWND CreateDialog(
   HINSTANCE hInstance,      // handle to application instance
   LPCTSTR lpTemplate,       // identifies dialog box template name
   HWND hWndParent,          // handle to owner window
   DLGPROC lpDialogFunc      // pointer to dialog box procedure
);
```

As you can see, the *CreateDialog* macro accepts four parameters. Table 1326 explains the *CreateDialog* macro's parameters in detail.

The *CreateDialog* macro uses the *CreateWindowEx* function to create the dialog box. *CreateDialog* then sends a *WM_INITDIALOG* message (and a *WM_SETFONT* message if the template specifies the *DS_SETFONT* style) to the dialog box procedure. The function will display the dialog box if the template specifies the *WS_VISIBLE* style.

Finally, *CreateDialog* returns the window handle to the dialog box.

After *CreateDialog* returns from its processing, the application uses the *ShowWindow* function to display the dialog box (if it is not already displayed). The application uses the *DestroyWindow* function to destroy the dialog box. To better understand the processing the *CreateDialog* macro performs, consider the program *Create_Dialog*, contained on the CD-ROM that accompanies this book. When the user selects the *Demo* option, the *WndProc* function creates a simple modeless dialog box, as shown in the following code fragment:

```
case IDM_DEMO :
     if (!hDlgModeless)
      hDlgModeless = CreateDialog(hInst, "TestDialog", hWnd, (DLGPROC)TestDlgProc);
     break;
```

When the user selects the *Demo* option, the *Create_Dialog* program checks to determine whether it is currently displaying the modeless dialog box. If it is, it performs no processing. If it is not, *Create_Dialog* will use the *CreateDialog* macro to display the modeless dialog box.

UNDERSTANDING THE CREATEDIALOGPARAM FUNCTION 1330

In Tip 1320, you learned how to use the *DialogBox* macro to create modal dialog boxes. In Tip 1330, you learned how your programs can use the *CreateDialog* macro to create modeless dialog boxes. When your programs must create a modeless dialog box, they can also use the *CreateDialogParam* function. The *CreateDialogParam* function (which the *CreateDialog* macro invokes as part of its processing) creates a modeless dialog box from a dialog box template resource. The *CreateDialogParam* function also lets your programs pass an application-defined value to the dialog box procedure as the *lParam* parameter of the *WM_INITDIALOG* message. An application can use the *lParam* value to initialize dialog box controls. You will implement the *CreateDialogParam* function within your programs in accordance with the function's generalized form, as shown here:

```
HWND CreateDialogParam(
    HINSTANCE hInstance,          // handle to application instance
    LPCTSTR lpTemplateName,       // identifies dialog box template
    HWND hWndParent,              // handle to owner window
    DLGPROC lpDialogFunc,         // pointer to dialog box procedure
    LPARAM dwInitParam            // initialization value
);
```

The *CreateDialogParam* function expects five parameters, similar to those Table 1326 defines, with the exception of the last parameter. The *dwInitParam* parameter specifies the value to pass to the dialog box procedure in the *lParam* parameter in the *WM_INITDIALOG* message.

The *CreateDialogParam* function uses the *CreateWindowEx* function to create the dialog box. *CreateDialogParam* then sends a *WM_INITDIALOG* message (and a *WM_SETFONT* message if the template specifies the *DS_SETFONT* style) to the dialog box procedure. The function will display the dialog box if the template specifies the *WS_VISIBLE* style. Finally, *CreateDialogParam* will return the window handle of the dialog box if it created the dialog box successfully.

After *CreateDialogParam* returns, your application will use the *ShowWindow* function to display the dialog box (if it is not already displayed). The application will use the *DestroyWindow* function to destroy the dialog box. To better understand the processing that *CreateDialogParam* performs, consider the *CreatePDialog.cpp* program contained on the CD-ROM that accompanies this book. The following code fragment from the *WndProc* function uses the *CreateDialogParam* function:

```
case IDM_DEMO :
        if (!hDlgModeless)
         {
            hDlgModeless = CreateDialogParam(hInst, "DemoDialog", hWnd,
                                             (DLGPROC)DemoDlgProc,
```

```
                                        (LPARAM)lpMem);
        }
        break;
```

The *CreatePDialog.cpp* program defines a structure of type *DLGDATA* to hold the current status of each button on the dialog box, as shown here:

```
typedef struct {
   BOOL bChecked;
   BOOL bRadio1;
 } DLGDATA;
```

When the program initializes (that is, when the *WndProc* procedure receives the *WM_CREATE* message), it allocates a handle to an instance of the structure. When the user selects *Demo*, the program passes the handle as the last parameter in the *CreatePDialog.cpp* program. The *DemoDlgProc* dialog procedure receives the handle as the *lParam* parameter and uses the values to initialize the state of the controls on the dialog.

1331 DEFAULT MESSAGE PROCESSING WITHIN A DIALOG BOX

As you have learned, your programs must create a dialog box message processing function to handle the messages that Windows dispatches to your dialog boxes. As you have also learned, the dialog box message processing function performs steps similar to those of the *WndProc* function, which processes window messages within your program. As you learned with the *WndProc* function, your programs should always define the default Windows message processing function (*DefWindowProc*) as the last case of your *switch* statement, as shown here:

```
default :
            return(DefWindowProc(hWnd, uMsg, wParam, lParam));
   }
```

If you define a separate window class to create the dialog box window, you should also define a default message processing function for your dialog boxes, just as you do within your base window. To define a default message processing function for your dialog boxes, your program will use the *DefDlgProc* function. The *DefDlgProc* function carries out default message processing for a window procedure that belongs to an application-defined dialog box class. You will implement the *DefDlgProc* function within your programs in accordance with the following generalized form:

```
LRESULT DefDlgProc(
    HWND hDlg,          // handle to dialog box
    UINT Msg,           // message
    WPARAM wParam,      // first message parameter
    LPARAM lParam       // second message parameter
);
```

The *DefDlgProc* function sets the default window procedure for the predefined dialog box class. This procedure provides internal processing for the dialog box by forwarding messages to the dialog box procedure and carrying out default processing for any messages that the dialog box procedure returns as *False*. Applications that create custom window procedures for their custom dialog boxes often use the *DefDlgProc* function instead of the *DefWindowProc* function to carry out default message processing.

Applications create custom dialog box classes by filling a *WNDCLASS* structure with appropriate information and registering the class with the *RegisterClass* function. Some applications use the *GetClassInfo* function to fill the structure and specify the name of the predefined dialog box. In such cases, the applications modify at least the *lpszClassName* member before registering the class. In all cases, you must set the *cbWndExtra* member of *WNDCLASS* for a custom dialog box class to at least *DLGWINDOWEXTRA*.

A dialog box procedure must not call the *DefDlgProc* function; doing so will result in recursive execution. To better understand the processing the *DefDlgProc* function performs, consider the program *DefDlgP.cpp*, contained on the CD-ROM that accompanies this book.

When you consider the definition for the *WndProc* function, you should note that the *WM_CREATE* switch initializes the dialog box class. The *SampleDlg* resource definitions make the dialog box a separate class, as shown here:

```
TESTDIALOG DIALOG DISCARDABLE   0, 0, 180, 70
STYLE DS_MODALFRAME | WS_POPUP | WS_VISIBLE | WS_CAPTION | WS_SYSMENU
CAPTION "Demo Dialog"
FONT 8, "MS Sans Serif"
CLASS "SampleDlg"
BEGIN
    CHECKBOX        "Check box control.",IDC_CHECKBOX,9,7,70,10
    GROUPBOX        "Radio Buttons",-1,7,21,86,39
    RADIOBUTTON     "First",IDC_RADIO1,13,32,37,10,WS_GROUP | WS_TABSTOP
    RADIOBUTTON     "Second",IDC_RADIO2,13,45,39,10
    PUSHBUTTON      "Done",IDCANCEL,116,8,50,14,WS_GROUP
END
```

When you execute the *DefDlgP* program and select the *Demo* option, the program will use the *SampleDlg* class to create the dialog box. For any Windows messages the dialog box does not specifically process, *DefDlgP* will call the default dialog processing function (in the case of the *DefDlgP* program, the default dialog processing function is the *TestDlgProc* function).

USING THE DLGDIRLIST FUNCTION TO CREATE A DIALOG LIST BOX 1332

In previous Tips, you have used the *DialogBox* macro, the *CreateDialogParam* function, and the *CreateDialog* function to create simple dialog boxes that contain one or more controls. Often, your programs will use dialog boxes to display information about the files a disk or drive contains. To simplify the process of creating dialog boxes that contain information about files on a disk or drive, the Win32 API provides the *DlgDirList* function. Your programs will use the *DlgDirList* function to fill a specified list box with the names of all files that match the specified path or filename. You will implement the *DlgDirList* function as shown in the following prototype:

```
int DlgDirList(
    HWND hDlg,              // handle to dialog box with list box
    LPTSTR lpPathSpec,      // pointer to path or filename string
    int nIDListBox,         // identifier of list box
    int nIDStaticPath,      // identifier of static control
    UINT uFileType          // file attributes to display
);
```

As you can see, the *DlgDirList* function accepts five parameters. Table 1332.1 explains those parameters in detail.

Parameters	Description
hDlg	Identifies the dialog box that contains the list box.
lpPathSpec	Points to a *NULL*-terminated string that contains the path or filename. *DlgDirList* modifies this string, which should be long enough to contain the modifications.
nIDListBox	Specifies the identifier of a list box. If this parameter is zero, *DlgDirList* will assume that no list box exists and will not try to fill one.
nIDStaticPath	Specifies the identifier of the static control the dialog box uses to display the current drive and directory. If this parameter is zero, *DlgDirList* will assume that no such control is present.
uFileType	Specifies attributes of the filenames the list box is to display. This parameter must be one or more of the values Table 1332.2 details.

***Table 1332.1** The parameters for the **DlgDirList** function.*

As Table 1332.1 details, the *uFileType* parameter accepts several possible values. When you use the possible values for the *uFileType* parameter, you will use the *bitwise OR* operator to combine the values. Table 1332.2 lists the possible values for the *uFileType* parameter.

Value	Description
DDL_ARCHIVE	Includes archived files.
DDL_DIRECTORY	Includes subdirectories. Subdirectory names are enclosed in square brackets ([ ]).
DDL_DRIVES	Includes drives. Drives are listed in the form [-*x*-], where *x* is the drive letter.
DDL_EXCLUSIVE	Includes only files with the specified attributes. By default, read/write files are listed even if you do not specify *DDL_READWRITE*.
DDL_HIDDEN	Includes hidden files.
DDL_READONLY	Includes read-only files.
DDL_READWRITE	Includes read/write files with no additional attribute.
DDL_SYSTEM	Includes system files.
DDL_POSTMSGS	Posts messages to the application's message queue. By default, *DlgDirList* sends messages directly to the dialog box's message procedure.

***Table 1332.2** The possible values for the **uFileType** parameter.*

If the *DlgDirList* function displays a list—even if it is an empty list—the function will return a non-zero value. If the input string does not contain a valid search path (or if an error halts the generation of the list), the return value is zero.

If you specify a zero-length string, or only a directory name with no filename for the *lpPathSpec* parameter, the *DlgDirList* function will change the string to "*.*". The *lpPathSpec* parameter has the following form:

```
[drive:] [[\u]directory[\idirectory]\u] [filename]
```

In this example, *drive* is a drive letter, *directory* is a valid directory name, and *filename* is a valid filename that must contain at least one wildcard (? or *). If *lpPathSpec* includes a drive or directory name, or both, the function will change the current drive and directory to the specified drive and directory before the function fills the list box. The function will also update the static control the *nIDStaticPath* parameter identifies with the new drive or directory name, or both.

After the function fills the list box, *DlgDirList* updates *lpPathSpec* by removing the drive or directory portion, or both, of the path and filename. Then *DlgDirList* sends the *LB_RESETCONTENT* and *LB_DIR* messages to the list box. To better understand the processing the *DlgDirList* function performs, consider the *List_Dir.cpp* program, which the CD-ROM that accompanies this book contains. The *List_Dir.rc* file defines the dialog box with the static text box, the list box, and the command button to close the dialog, as shown here:

```
TESTDIALOG DIALOG DISCARDABLE  20, 20, 150, 110
STYLE DS_MODALFRAME | WS_POPUP | WS_VISIBLE | WS_CAPTION | WS_SYSMENU
CAPTION "Test Dialog"
FONT 8, "MS Sans Serif"
BEGIN
    EDITTEXT        IDC_DIRECTORY,6,5,136,13,ES_AUTOHSCROLL | ES_READONLY |
                                        NOT WS_TABSTOP
    LISTBOX         IDC_LIST,6,20,136,59,LBS_SORT | LBS_NOINTEGRALHEIGHT |
                                    LBS_DISABLENOSCROLL | WS_VSCROLL |
                                    WS_TABSTOP
    PUSHBUTTON      "Done",IDCANCEL,50,87,50,14,WS_GROUP
END
```

The *List_Dir.cpp* file uses the *DlgDirList* function to create the dialog box from within the *WndProc* function. The program creates the dialog box when the user selects the *Demo* function. The dialog box contains a list box that displays the contents of the current directory, and a static text box that displays the directory name. If the user double-clicks the mouse on any file within the list box, the program will change to that directory or drive if the selection is a directory or drive. The program will then display the selection within the static text box, regardless of the selection's type. When you compile and execute the *List_Dir* program, your screen will display the output shown in Figure 1332.

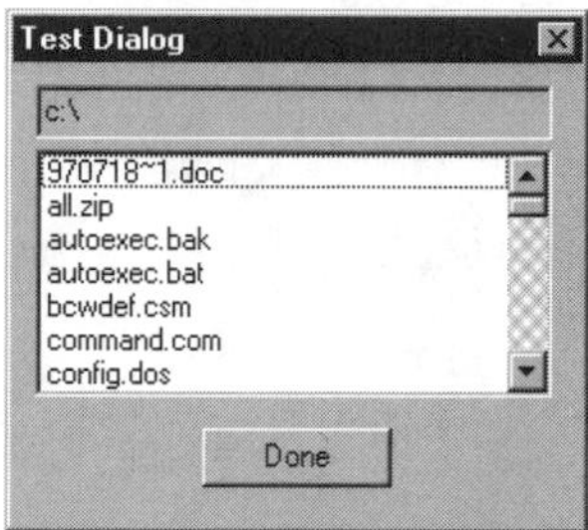

***Figure 1332** The output of the **List_Dir** program.*

RESPONDING TO USER SELECTIONS WITHIN THE LIST BOX 1333

In Tip 1332, you created the *List_Dir* program, which displays the files within the current directory when the program displays a dialog box. As you learned, the program's *DemoDlgProc* function changed the information within the static text box when the user double-clicked the mouse on an item within the list box.

In later Tips, you will learn about capturing user input in detail. However, within the *List_Dir.cpp* program, the processing is relatively simple: the program first checks to determine whether the user double-clicked the mouse button. If the user did, the program checks the *DlgDirSelectEx* function to determine what the user selected—a directory or a file. The *DlgDirSelectEx* function retrieves the current selection from a single-selection list box. The function assumes that the *DlgDirList* function has filled the list box and that the selection is a drive letter, filename, or directory name. Your programs will implement the *DlgDirSelectEx* function as shown here:

```
BOOL DlgDirSelectEx(
    HWND hDlg,          // handle to dialog box with list box
    LPTSTR lpString,    // pointer to buffer for path string
    int nCount,         // number of characters in path string
    int nIDListBox      // identifier of list box
);
```

If the current selection is a directory name, the function returns a non-zero value. If the current selection is not a directory name, the function returns zero. The *DlgDirSelectEx* function copies the selection to the buffer to which the *lpString* parameter points. If the current selection is a directory name or drive letter, *DlgDirSelectEx* will remove the enclosing square brackets (and hyphens, for drive letters), so that the name or letter is ready for the *DlgDirList* function to insert the name or letter into a new path. If there is no selection, *lpString* does not change.

The *DlgDirSelectEx* function sends *LB_GETCURSEL* and *LB_GETTEXT* messages to the list box. The function does not jîÔhthe list box return more than one filename. The list box must not be a multiple-selection list box. If it is a multiple-selection list box, *DlgDirSelectEx* will not return a zero value and *lpString* will remain unchanged. When you invoke the *DlgDirSelectEx* function, your program should test the return value and respond appropriately, as the *TestDlgProc* function from the *List_Dir* program shows:

```
LRESULT CALLBACK TestDlgProc(HWND hDlg, UINT uMsg, WPARAM wParam, LPARAM lParam)
 {
   static char szTmp[255];
   switch(uMsg)
```

```
   {
      case WM_INITDIALOG :
             DlgDirList(hDlg, "*.*", IDC_LIST, IDC_DIRECTORY,
                        DDL_DIRECTORY | DDL_DRIVES);
           break;

    case WM_COMMAND :
           switch(LOWORD(wParam))
           {
              case IDC_LIST :
                      if (HIWORD(wParam) == LBN_DBLCLK)
                      {
                       if (DlgDirSelectEx(hDlg, szTmp, sizeof(szTmp), IDC_LIST))
                         {
                           strcat(szTmp, "*.*");
                           DlgDirList(hDlg, szTmp, IDC_LIST, IDC_DIRECTORY,
                                      DDL_DIRECTORY | DDL_DRIVES);
                         }
                         else
                           MessageBox(hDlg, szTmp, "File Selected",
                                      MB_OK | MB_ICONINFORMATION);
                      }
                      break;

             case IDCANCEL:
                      EndDialog(hDlg, IDCANCEL);
                      break;
           }
           break;

    default :
           return(FALSE);
   }
  return(TRUE);
}
```

1334 CLOSING THE DIALOG BOX

Throughout the previous Tips, you learned how to create and display a variety of dialog boxes. As you saw in the code listings in previous Tips, you must always use the *EndDialog* function to close your modal dialog boxes. The *EndDialog* function destroys a modal dialog box, which causes the system to end any processing for the dialog box. You will use the *EndDialog* function within your programs as shown in the following prototype:

```
BOOL EndDialog(
   HWND hDlg,           // handle to dialog box
   int nResult          // value to return
);
```

The *hDlg* parameter identifies the dialog box that the *EndDialog* function is to destroy. The *nResult* parameter lets your programs specify a return value from the function that created the dialog box to the calling application. You must use the *EndDialog* function to destroy dialog boxes you create with the *DialogBox, DialogBoxParam, DialogBoxIndirect*, and *DialogBoxIndirectParam* functions. An application calls *EndDialog* from within the dialog box procedure. Do not use the *EndDialog* function for any other purpose.

A dialog box procedure can call the *EndDialog* function at any time, even during the processing of the *WM_INITDIALOG* message. If your application calls *EndDialog* while the application is processing the *WM_INITDIALOG* message, Windows will destroy the dialog box before it shows and before Windows sets the input focus to the dialog box.

EndDialog does not destroy the dialog box immediately. Instead, it sets a flag and lets the dialog box procedure return control to the system. The system checks the flag before it tries to retrieve the next message from the application queue. If *EndDialog* has previously set the flag, the system will end the message loop, destroy the dialog box, and use the value in *nResult* as the return value from the function that created the dialog box.

Understanding User Input 1335

In previous Tips, you have created several different types of Windows programs. However, each program has had a single feature in common with each other program you created: each program received user input to perform its processing—even if the user input was simply a *close* command. As you have learned, the vast majority of Windows programs you will write (with the exception of automated server programs) will require user input of some type to perform useful processing. Automated server programs are a notable exception because their input comes entirely from other programs, rather than from users.

In Windows, a program receives user input from any of several devices. However, the two most commonly used devices are the mouse and the keyboard. Other input devices the user might use include a light pen, a touch screen, a joystick, and so on. As you learned previously, each time the user types a key on the keyboard (and each time the user moves or clicks the mouse), Windows generates a message. That message may go to the operating system, it may go to your program window, or it may go to a dialog box. In some cases, that message may go to multiple locations.

In Windows, your programs will generally use predefined controls such as buttons, edit controls, and menus to respond to the keyboard and mouse in the appropriate manner. In the next several Tips, you will learn about additional methods your programs can use to better control their response to input from the keyboard or the mouse.

Responding to Mouse Events 1336

As you have learned, your programs will respond to input from the user in the form of system messages. In previous Tips, your programs have checked the *WM_COMMAND* message to determine whether the user has sent a command to your program and whether to respond to the menu or control selection that generated the command. When your programs work with mouse activities the user performs, your programs must check for the Windows messages Table 1336 lists.

Message	Meaning
WM_CAPTURECHANGED	Windows sends the *WM_CAPTURECHANGED* message to the window that is losing the mouse capture.
WM_LBUTTONDBLCLK	The user has pressed the left mouse button twice.
WM_LBUTTONDOWN	The user has pressed the left mouse button, and it is still pressed.
WM_LBUTTONUP	The user has released the left mouse button.
WM_MBUTTONDBLCLK	The user has pressed the middle mouse button twice (three-button mice only).
WM_MBUTTONDOWN	The user has pressed the middle mouse button, and it is still pressed (three-button mice only).
WM_MBUTTONUP	The user has released the middle mouse button (three-button mice only).
WM_MOUSEACTIVATE	The user has clicked the mouse within a currently inactive window.
WM_MOUSEMOVE	The user moved the mouse within the window.

Table 1336 *Windows mouse messages.* *(continued on following page)*

Message	Meaning
WM_NCLBUTTONDBLCLK	The user has pressed the left mouse button twice within a non-client area of the window.
WM_NCLBUTTONDOWN	The user has pressed the left mouse button within a non-client area of the window, and it is still pressed.
WM_NCLBUTTONUP	The user has released the left mouse button within a non-client area of the window.
WM_NCMBUTTONDBLCLK	The user has pressed the middle mouse button twice within a non-client area of the window (three-button mice only).
WM_NCMBUTTONDOWN	The user has pressed the middle mouse button within a non-client area of the window, and it is still pressed (three-button mice only).
WM_NCMBUTTONUP	The user has released the middle mouse button within a non-client area of the window (three-button mice only).
WM_NCMOUSEMOVE	The user moved the mouse within a non-client area of the window.
WM_NCRBUTTONDBLCLK	The user has pressed the right mouse button twice within a non-client area of the window.
WM_NCRBUTTONDOWN	The user has pressed the right mouse button within a non-client area of the window, and it is still pressed.
WM_NCRBUTTONUP	The user has released the right mouse button within a non-client area of the window.
WM_RBUTTONDBLCLK	The user has pressed the right mouse button twice.
WM_RBUTTONDOWN	The user has pressed the right mouse button, and it is still pressed.
WM_RBUTTONUP	The user has released the right mouse button.

Table 1336 Windows mouse messages. (continued from previous page)

When an application receives a mouse message, the *lParam* value contains the cursor's X and Y positions on the screen. Windows stores the Y position in the high-order word of *lParam*, and the X position in the low-order word. Your program should use the *LOWORD* and *HIWORD* macros to extract the two values. Tip 1337 examines the *WM_MOUSEMOVE* message as an example of a Windows mouse message, and includes a code fragment that shows how you will extract the values from the *lParam* parameter.

1337 USING THE WM_MOUSEMOVE MESSAGE

As you learned in Tip 1336, Windows generates different messages to your applications as the user manipulates the mouse. One of the most common messages that Windows will send to your programs is the *WM_MOUSEMOVE* message. Windows posts the *WM_MOUSEMOVE* message to a window each time the cursor moves. If a window has not previously captured the mouse, Windows posts the *WM_MOUSEMOVE* message to the window that contains the cursor. Otherwise, Windows posts the *WM_MOUSEMOVE* message to the window that has captured the mouse. When your program receives a *WM_MOUSEMOVE* message, it should check the following values before it begins its respondent processing:

```
fwKeys = wParam;          // key flags
xPos = LOWORD(lParam);    // horizontal position of cursor
yPos = HIWORD(lParam);    // vertical position of cursor
```

As you learned in Tip 1336, the *lParam* parameter contains the X and Y positions of the cursor. The *wParam* parameter will contain the *fwKeys* values. The *fwKeys* value indicates whether various virtual keys are down. The *fwKeys* value can be any combination of the values shown in Table 1337.

Value	Description
MK_CONTROL	Set if the CTRL key is down
MK_LBUTTON	Set if the left mouse button is down
MK_MBUTTON	Set if the middle mouse button is down
MK_RBUTTON	Set if the right mouse button is down
MK_SHIFT	Set if the SHIFT key is down

*Table 1337 Possible values for the **fwKeys** value.*

To capture mouse moves from within your message procedure, your programs should use code similar to that in the following fragment:

```
case WM_MOUSEMOVE :
       {
          nXPos = LOWORD(lParam);
          nYPos = HIWORD(lParam);
          // other statements
```

In the preceding fragment, the program assigns the mouse's current position to the *nXPos* and *nYPos* values each time the user moves the mouse.

READING THE MOUSE BUTTONS 1338

In Tip 1337, you learned how to use the *WM_MOUSEMOVE* windows message to have your programs respond to mouse movements on the screen. Similarly, adding code to your programs to make them respond to mouse button activities is also simple. For example, if you want your programs to respond when the user double-clicks the mouse anywhere within the client area of the window, your programs will use code within the *WndProc* function similar to the following:

```
LRESULT CALLBACK WndProc(HWND hWnd, UINT uMsg, WPARAM wParam, LPARAM lParam)
 {
   switch(uMsg)
    {
      case WM_LBUTTONDBLCLK :
              MessageBox(hWnd, "Button Double-Clicked", NULL, MB_OK);
              break;
```

If, on the other hand, you want your program to respond differently if the user holds the CTRL key down while pressing the mouse, you can instead construct the code as shown here:

```
case WM_LBUTTONDBLCLK :
   if ((wParam && MK_CONTROL) == MK_CONTROL)
    {
      MessageBox(hWnd, "Button Double-Clicked with Control Key Down", NULL, MB_OK);
      break;
    }
   else
    {
      MessageBox(hWnd, "Button Double-Clicked w/o Control Key Down", NULL, MB_OK);
      break;
    }
```

As you can see from the code, your programs can check whether the user is currently pressing a given virtual key by performing a *bitwise AND* on the *wParam* value and the value for the virtual key you want to check. As your programs become more complex, you will often test for virtual keys when you process mouse events. While Windows passes only certain virtual keys within the *wParam* parameter in conjunction with a mouse click, there are a great number of virtual keys that Windows supports, as you will learn in Tip 1340.

1339 RESPONDING TO KEYBOARD EVENTS

As you have learned, Windows will dispatch messages to your applications each time the user performs some activity with the mouse. Windows will also dispatch messages to your applications each time the user types a keystroke on the computer's keyboard. The message Windows will dispatch will be one of those listed in Table 1339.

Message	Description
WM_ACTIVATE	The keystroke is within an inactive window
WM_CHAR	The ASCII code for the letter if the user pressed a character on the keyboard
WM_GETHOTKEY	Lets your programs retrieve the previously set "hot key" for a window
WM_HOTKEY	The user has depressed the "hot key"
WM_KEYDOWN	The user has depressed a key
WM_KEYUP	The user has released the depressed key
WM_KILLFOCUS	Sent to a window immediately before the window loses the keyboard focus
WM_SETFOCUS	Sent to a window immediately after the window receives the keyboard focus
WM_SETHOTKEY	Lets your programs set a "hot key" for a window
WM_SYSCHAR	The ASCII code for the letter the user depressed while simultaneously depressing the ALT key
WM_SYSKEYDOWN	The user has depressed a key while simultaneously depressing the ALT key
WM_SYSKEYUP	The user has released the depressed key and the depressed ALT key

***Table 1339** The system messages Windows sends for keyboard processing.*

The *TranslateMessage* function in the message loop of the thread will generate the *WM_CHAR* message if it recognizes the character as an ASCII character. Generally, your applications will use the *WM_KEYDOWN* message to check for function keys, cursor keys, the numeric keypad, and the edit keys, such as PAGEUP and PAGEDOWN. These keys make the best use of the virtual-key codes, which you will learn about in the next Tip.

On the other hand, your programs should use *WM_CHAR* to retrieve letters, numbers, and printable symbols. Using the *WM_CHAR* message to process such keystrokes is best because ASCII assigns different values to lowercase and capital letters. With *WM_KEYDOWN*, your applications must check for the character the user has entered, as well as for the current state of the SHIFT key. The *WM_CHAR* message handles the SHIFT key processing for you.

If the user presses the ALT key while pressing another key, the application will receive the following message sequence: *WM_SYSKEYDOWN, WM_SYSCHAR, WM_SYSKEYUP*. Your programs should use these messages to check for specific ALT key sequences.

1340 UNDERSTANDING VIRTUAL KEYS

As you learned in Tip 1339, if your programs can process a keystroke as an ASCII character, they should use the *WM_CHAR* message to do so. However, many times your programs may require or let the user use other keys beyond normal ASCII keys—for example, the arrow keys, the number keys on the numeric keypad, and so on. When you track such keys, you should use virtual-key values to do so. Virtual keys free you from considering what type of keyboard the user will have because the virtual-key code for the first function key should always be the same, regardless of the keyboard's make or model. The *WM_KEYDOWN, WM_KEYUP, WM_SYSKEYDOWN*, and *WM_SYSKEYUP* messages send the virtual-key codes as the message's *wParam* value.

Within the virtual-key code structure, numeric keypad numbers have their own virtual-key codes. Additionally, the virtual code for the character and number keys is their uppercase ASCII equivalent (in other words, both a and A are

VK_A). Finally, note that both SHIFT keys generate the same virtual-key code. Table 1340 lists the Win32 API-defined virtual-key codes.

Key Code	Value (hex)	Mouse or Keyboard Equivalent
VK_LBUTTON	01	Left mouse button
VK_RBUTTON	02	Right mouse button
VK_CANCEL	03	Control-break processing
VK_MBUTTON	04	Middle mouse button (three-button mouse) or both left and right mouse buttons simultaneously
VK_BACK	08	Backspace
VK_TAB	09	Tab
VK_CLEAR	0C	Clear
VK_RETURN	0D	Enter
VK_SHIFT	10	Shift
VK_CONTROL	11	Ctrl
VK_MENU	12	Alt
VK_PAUSE	13	Pause
VK_CAPITAL	14	Caps lock
VK_ESCAPE	18	Esc
VK_SPACE	20	Space bar
VK_PRIOR	21	PageUp
VK_NEXT	22	PageDown
VK_END	23	End
VK_HOME	24	Home
VK_LEFT	25	Left arrow
VK_UP	26	Up arrow
VK_RIGHT	27	Right arrow
VK_DOWN	28	Down arrow
VK_SELECT	29	Select
VK_EXECUTE	2B	Execute
VK_SNAPSHOT	2C	Print screen
VK_INSERT	2D	Insert
VK_DELETE	2E	Delete
VK_HELP	2F	Help
VK_0-VK_9	30-39	0-9
VK_A-VK_Z	41-5A	A-Z
VK_NUMPAD0	60	Numeric keypad 0
VK_NUMPAD1	61	Numeric keypad 1
VK_NUMPAD2	62	Numeric keypad 2
VK_NUMPAD3	63	Numeric keypad 3
VK_NUMPAD4	64	Numeric keypad 4
VK_NUMPAD5	65	Numeric keypad 5
VK_NUMPAD6	66	Numeric keypad 6
VK_NUMPAD7	67	Numeric keypad 7
VK_NUMPAD8	68	Numeric keypad 8

Table 1340 *The Win32 API-defined virtual-key codes.* *(continued on following page)*

of the keyboard message (such as *WM_KEYDOWN*) that *GetKeyNameText* is to process. The function interprets the portions of *lParam* that Table 1342.2 details.

Bits	Meaning
16-23	Scan code.
24	Extended-key flag. Distinguishes some keys on an enhanced keyboard.
25	"Don't care" bit. The application that calls the *GetKeyNameText* function sets this bit to indicate that the function should not distinguish between left and right CTRL and SHIFT keys, for example.

*Table 1342.2 The bits within **lParam** that **GetKeyNameText** interprets.*

The format of the key-name string depends on the current keyboard layout. The keyboard driver maintains a list of names in the form of character strings for keys with names longer than a single character. The keyboard driver translates the key name according to the layout of the currently installed keyboard. The name of a character key is the character itself. The CD-ROM that accompanies this book includes the program *Show_Keys.cpp*, which displays a key name each time the user presses a key.

As you will see, within the *WM_KEYDOWN* case, the program performs significant processing that relates to device contexts, which you will learn about in later Tips. The important function call for the purpose of this Tip is the call to the *GetKeyNameText* function, as shown here:

```
GetKeyNameText(lParam, szName, 30);
```

The program simply calls *GetKeyNameText*, then outputs the value within the *szName* buffer to the window.

1343 SETTING AND RETURNING THE MOUSE'S DOUBLE-CLICK TIME

As you have learned, your programs will perform significant processing to catch and correctly process user input. One of the most common activities that users will perform within your programs is to *double-click* the mouse on an option. A double-click is a series of two clicks of a mouse button, the second occurring within a specified time after the first. Sometimes, you may want to control the amount of time you will let pass before a second click, yet still count the two clicks as a double-click. Your programs can use the *SetDoubleClickTime* function to control the mouse's double-click speed. The *SetDoubleClickTime* function sets the *double-click time* for the mouse. The double-click time is the maximum number of milliseconds that may occur between the first and second clicks of a double-click. Your programs will use the *SetDoubleClickTime* function as shown in the following prototype:

```
BOOL SetDoubleClickTime(
   UINT uInterval      // double-click interval
);
```

The *uInterval* parameter specifies the number of milliseconds that may occur between the first and second clicks of a double-click. If *uInterval* is set to zero, Windows will use the default double-click time of 500 milliseconds.

Note: *The **SetDoubleClickTime** function alters the double-click time for all windows in the system. If your programs change the double-click time value, they should return it to its original state when the program exits.*

Just as your programs can set the amount of time between two mouse clicks that qualify as a double-click, so too can your programs retrieve the current double-click time for the mouse. The *GetDoubleClickTime* function retrieves the current double-click time for the mouse. As you know, a double-click is a series of two clicks of the mouse button, the second occurring within a specified time after the first. Likewise, the double-click time is the maximum number of milliseconds that may occur between the first and second click of a double-click. Your programs will implement the *GetDoubleClickTime* function in accordance with the following generalized form:

```
UINT GetDoubleClickTime(void)
```

If the *GetDoubleClickTime* function succeeds, the function's return value will specify the current double-click time in milliseconds. To better understand the processing both the *GetDoubleClickTime* and *SetDoubleClickTime* functions perform, consider the *Get_Set_Dbl-Click.cpp* program, contained on the CD-ROM that accompanies this book.

Each time the user double-clicks the mouse, the program will display a message box that the program has received the double-click. However, each time the user selects the *Test!* menu, the program increases the double-click timing by one-tenth of a second. Note that the program returns the double-click time to its original timing when the program exits.

SWAPPING THE MOUSE BUTTONS 1344

As you learned in Tip 1343, there may be times when your programs must change the duration between clicks for a mouse double-click. Similarly, there may be times within your programs when the user must swap the mouse buttons—using the right button to perform left button actions, and using the left button to perform right button actions. The *SwapMouseButton* function reverses or restores the meaning of the left and right mouse buttons. Windows provides button swapping as a convenience to people who use the mouse with their left hands. Usually, only the Control Panel calls the *SwapMouseButton* function, although an application is free to call the function. If your programs must use the *SwapMouseButton* function, you will invoke it as shown in the following prototype:

```
BOOL SwapMouseButton(
   BOOL fSwap    // reverse or restore buttons
);
```

The *fSwap* parameter specifies whether the mouse button meanings are currently reversed or restored. If *fSwap* is *True*, the left button will generate right-button messages and the right button will generate left-button messages. If *fSwap* is *False*, the buttons are restored to their original meanings. The CD-ROM that accompanies this book includes the program *Swap_B.cpp*, which swaps the state of the mouse buttons back and forth each time the user selects the *Test!* menu option.

Note: *The mouse is a shared resource and reversing the meaning of its buttons affects all applications. Your programs should avoid mouse button swapping whenever possible.*

DETERMINING IF THE USER HAS PRESSED A KEY 1345

As you have learned, your programs can perform processing each time the user presses a key or clicks the mouse button. However, some applications may require that the user perform several steps to select an option. For example, a program might respond differently if the user presses the F2 function key and then clicks the mouse on an option than the program will respond if the user simply clicks the mouse on the option. To perform such asynchronous (that is, not at the same time) processing, your programs can use the *GetAsyncKeyState* function. The *GetAsyncKeyState* function determines whether a key is up or down at the time the program calls the function, and whether the user pressed the key after a previous call to *GetAsyncKeyState*. You will implement the *GetAsyncKeyState* function as shown here:

```
SHORT GetAsyncKeyState(
   int vKey       // virtual-key code
);
```

The *vKey* parameter specifies one of 256 possible virtual-key codes, as Table 1340 details. If the function succeeds, the return value will specify whether the user pressed the key since the last call to *GetAsyncKeyState*, and whether the key is currently up or down. If the most significant bit is set, the key is down, and if the least significant bit is set, the user pressed the key after the previous call to *GetAsyncKeyState*. The return value is zero if a window in another thread or process currently has the keyboard focus.

***Note:** Under Windows 95, you can use the virtual-key code constants **VK_SHIFT, VK_CONTROL,** and **VK_MENU** as values for the **vKey** parameter. This gives the state of the SHIFT, CTRL, or ALT keys without distinguishing between left and right.*

***Note:** Under Windows NT, you can use the following virtual-key code constants as values for **vKey** to distinguish between the left and right instances of those keys:*

VK_LSHIFT	VK_RSHIFT
VK_LCONTROL	VK_RCONTROL
VK_LMENU	VK_RMENU

*These left- and right-distinguishing constants are only available when you call the **GetKeyboardState, SetKeyboardState, GetAsyncKeyState, GetKeyState,** and **MapVirtualKey** functions.*

The *GetAsyncKeyState* function works with mouse buttons. However, it checks on the state of the physical mouse buttons, not on the logical mouse buttons to which the physical buttons are mapped. For example, the call *GetAsyncKeyState(VK_LBUTTON)* always returns the state of the left physical mouse button, which the system may have mapped to the left or right logical mouse button. You can determine the system's current mapping of physical mouse buttons to logical mouse buttons by calling *GetSystemMetrics(SM_SWAPBUTTON),* which returns *True* if the operating system has previously swapped the mouse buttons, and *False* otherwise. To better understand the processing that *GetAsyncKeyState* performs, consider the *Get_Async_Keys.cpp* program contained on the CD-ROM that accompanies this book. The *Get_Async_Keys.cpp* program returns the current status of the SHIFT keys when the user selects the *Test!* menu item.

1346 INTRODUCING SCROLL BARS

As you have learned, a window in an application can display many things. From time to time, a window may display a data object, such as a document or a graphic image, that is larger than the window's client area. When the window has a scroll bar, the user can scroll the data object within the client area to view the entire document or graphic image. An application should include a scroll bar on a window whenever the client area's contents extend beyond the size of the window's client area. A standard scroll has three components: the arrows on both sides, the scroll bar's width or height, and the scroll bar's *thumb*, the small box that users drag back and forth along the scroll bar. For example, Figure 1346 shows a program with a scroll bar on the client window's vertical and horizontal edges, as well as the scroll bar's components.

***Figure 1346** A window with two internal scroll bars and their components.*

1347 UNDERSTANDING THE DIFFERENT SCROLL BAR TYPES

As you learned in Tip 1346, every standard scroll bar, regardless of its positioning or direction, shares certain common features with other standard scroll bars. However, as you saw in Figure 1346, there are at least two different scroll bar types. There are actually two categories of scroll bars. The first category of scroll bars is the *client-area scroll bars*. Client-area scroll bars can extend vertically or horizontally, and may appear as a standard scroll bar or as a track bar. The second category of scroll bars is the *non-client area scroll bars.* Windows attaches the non-client-area scroll bars to the window's border as soon as you make the scroll bars visible.

As you know, you can create non-client area scroll bars for your application within the *WNDCLASS* structure that your programs pass to the *CreateWindow* function. When you attach a scroll bar to a window's edge, Windows automatically subtracts the scroll bar's width from the window's client area so that painting onto the client area does not run over the scroll bars. Your programs can also invoke the *ShowScrollBar* function to attach scroll bars to a window's inside edge. You will learn more about *ShowScrollBar* in Tip 1348.

In addition to attaching scroll bars to the inside edge of windows, the operating system will automatically attach scroll bars to list boxes and combo boxes when the list of items exceeds the size of the list window. Your programs can also attach scrollbars to edit boxes. However, single-line edit boxes support only the horizontal scroll bar, while multiple-line edit boxes can support both horizontal and vertical scroll bars.

USING THE SHOWSCROLLBAR FUNCTION 1348

As you learned in Tip 1347, your programs can attach scroll bars to windows both at the window's creation and after the window's creation. To attach or hide a scrollbar on a previously created window, your programs will use the *ShowScrollBar* function. The *ShowScrollBar* function shows or hides the scroll bar you specify. You will use the *ShowScrollBar* function within your programs as shown in the following prototype:

```
BOOL ShowScrollBar(
   HWND hWnd,            // handle of window with scroll bar
   int wBar,             // scroll bar flag
   BOOL bShow            // scroll bar visibility flag
);
```

The *hWnd* parameter identifies a scroll bar control or a window with a standard scroll bar, depending on the value of the *wBar* parameter. The *bShow* parameter specifies whether Windows shows or hides the scroll bar. If *bShow* is *True*, Windows will show the scroll bar. Otherwise, Windows will hide the scroll bar. The *wBar* parameter specifies the scroll bars Windows will show or hide. The *wBar* parameter can be one of the values listed in Table 1348.1.

Value	Meaning
SB_BOTH	Shows or hides a window's standard horizontal and vertical scroll bars.
SB_CTL	Shows or hides a scroll bar control. The *hWnd* parameter must be the handle of the scroll bar control.
SB_HORZ	Shows or hides a window's standard horizontal scroll bars.
SB_VERT	Shows or hides a window's standard vertical scroll bar.

Table 1348.1 The possible values for the ***wBar*** *parameter.*

Note: *You should not call the* ***ShowScrollBar*** *function to hide a scroll bar while processing a scroll bar message.*

After you show the scroll bars, you will often want to control how the user can manipulate the scroll bars. One of the most common controls you will apply to a scroll bar is to enable or disable one or both of the arrows within the scroll bar. To do so, your programs will call the *EnableScrollBar* function. The *EnableScrollBar* function enables or disables one or both scroll bar arrows. You will implement the *EnableScrollBar* function within your programs as shown here:

```
BOOL EnableScrollBar(
   HWND hWnd,              // handle to window or scroll bar
   UINT wSBflags,       // scroll bar type flag
   UINT wArrows            // scroll bar arrow flag
);
```

The *hWnd* parameter identifies a window or a scroll bar control, depending on the value of the *wSBflags* parameter. The *wSBflags* parameter specifies the scroll bar type. This parameter can be one of the values Table 1348.1 details.

The *wArrows* parameter specifies whether the scroll bar arrows are enabled or disabled and indicates which arrows are enabled or disabled. The *wArrows* parameter can be one of the values listed in Table 1348.2.

Value	Meaning
ESB_DISABLE_BOTH	Disables both arrows on a scroll bar
ESB_DISABLE_DOWN	Disables the down arrow on a vertical scroll bar
ESB_DISABLE_LEFT	Disables the left arrow on a horizontal scroll bar
ESB_DISABLE_LTUP	Disables the left arrow on a horizontal scroll bar or the up arrow of a vertical scroll bar
ESB_DISABLE_RIGHT	Disables the right arrow on a horizontal scroll bar
ESB_DISABLE_RTDN	Disables the right arrow on a horizontal scroll bar or the down arrow of a vertical scroll bar
ESB_DISABLE_UP	Disables the up arrow on a vertical scroll bar
ESB_ENABLE_BOTH	Enables both arrows on a scroll bar

Table 1348.2 *The possible values for the* ***wArrows*** *parameter.*

With these two functions alone, your programs can perform most scroll bar manipulations they require. The CD-ROM that accompanies this book includes the *Scroll1.cpp* program, which uses both the *ShowScrollBars* and the *EnableScrollBars* functions to control a scroll bar within your program.

1349 Understanding the Scroll Bar's Position and Range

As you saw in Tip 1348, adding a scroll bar to your windows is a simple process. When you create a scroll bar, the default range of values the two ends of the control represent is 0 to 100. In most cases, an application will change the range to reflect the size of the document or image. The thumb's position is the value within the range where the thumb is located. For example, if the range is 0 to 100 and the thumb position is 50, the thumb will appear halfway between the ends of the control.

You can also set the *page size* for your scroll bar controls. The page size represents the number of increments within the scroll range that the page can display at one time. For example, if the range is 0 to 100 and the page size is set to 50, the page can display half the control range at one time. Your programs can use the *SetScrollInfo* and *GetScrollInfo* functions to set and retrieve a scroll bar's range, thumb position, and page size.

1350 Understanding the Scroll Bar's Messages

In previous Tips, you have learned some of the basics about scroll bars. As the user uses the scroll bar to scroll, the scroll bar will generate messages to your applications, just as the mouse and the keyboard do. When users click the mouse on a scroll bar, Windows sends either a *WM_HSCROLL* or *WM_VSCROLL* message to the application, depending on whether the scroll bar is horizontal or vertical. The low-order word of the *lParam* parameter tells the processing function where the mouse was when the user clicked the mouse button. Depending on where the mouse was located at the time of the click, Windows will send one of ten messages to your programs. Figure 1350 shows the possible positions for the mouse click and the corresponding windows message.

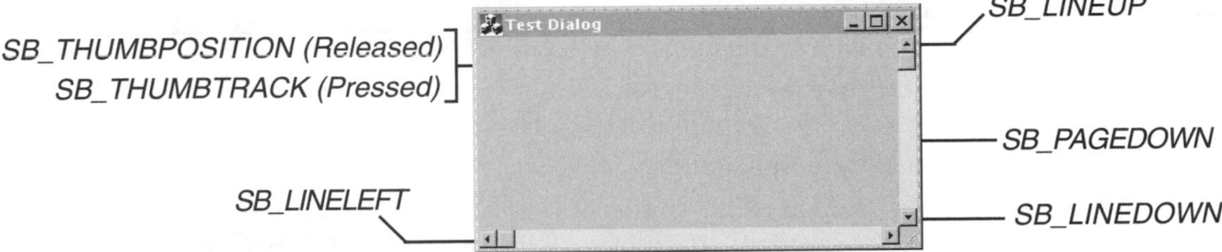

Figure 1350 Some of the messages Windows generates for mouse clicks on scroll bars.

When the user releases the mouse button after clicking anywhere on a scroll bar, Windows will send the *SB_ENDSCROLL* message to the application, unless the user was moving his or her thumb with the mouse. In that case, Windows will generate the *SB_THUMBPOSITION* message when the user releases the mouse button.

OBTAINING THE SCROLL BAR'S CURRENT SETTINGS 1351

As you have learned, scroll bars begin at certain default settings when you add them to windows within your programs. However, as you have also learned, your programs will often change a scroll bar's settings over the course of a program's processing. Your programs must often perform specific processing based on the scroll bar's settings. To help your programs determine a scroll bar's current settings, your programs can use the *GetScrollInfo* function. The *GetScrollInfo* function retrieves the parameters of a scroll bar, including the minimum and maximum scrolling positions, the page size, and the position of the scroll box (thumb). You will use the *GetScrollInfo* function within your programs as shown in the following prototype:

```
BOOL GetScrollInfo(
   HWND hwnd,                // handle of window with scroll bar
   int fnBar,                // scroll bar flag
   LPSCROLLINFO lpsi         // pointer to structure for scroll parameters
);
```

As you can see, the *GetScrollInfo* function accepts three parameters. The *lpsi* parameter, which you will learn more about later in this Tip, returns a value of type *SCROLLINFO*, a system-defined structure with the members shown here:

```
typedef struct tagSCROLLINFO
 {
    UINT cbSize;
    UINT fMask;
    int  nMin;
    int  nMax;
    UINT nPage;
    int  nPos;
    int  nTrackPos;
 } SCROLLINFO;
```

Both *GetScrollInfo* and *SetScrollInfo* use the *SCROLLINFO* structure. Table 1351.1 details the structure's members and their use.

Member	Description
cbSize	Specifies the size, in bytes, of the *SCROLLINFO* structure.
fMask	Specifies the scroll bar parameters to set or retrieve. This member can be a combination of the values Table 1351.2 details.
nMin	Specifies the minimum scrolling position.

***Table 1351.1** The members of the **SCROLLINFO** structure.* *(continued on following page)*

Member	Description
nMax	Specifies the maximum scrolling position.
nPage	Specifies the page size. A scroll bar uses this value to determine the appropriate size of the proportional scroll box.
nPos	Specifies the position of the scroll box.
nTrackPos	Specifies the immediate position of a scroll box that the user is dragging. An application can retrieve this value while processing the *SB_THUMBTRACK* notification message. An application cannot set the immediate scroll position. The *SetScrollInfo* function ignores this member.

Table 1351.1 The members of the **SCROLLINFO** *structure. (continued from previous page)*

As you learned in Table 1351.1, the *fMask* member can have one of several predefined values. Table 1351.2 shows the possible values for the *fMask* member.

Value	Meaning
SIF_ALL	Combination of *SIF_PAGE, SIF_POS,* and *SIF_RANGE.*
SIF_DISABLENOSCROLL	Your programs will use this value only when setting a scroll bar's parameters. If the scroll bar's new parameters make the scroll bar unnecessary, you must disable the scroll bar instead of removing it.
SIF_PAGE	The *nPage* member contains the page size for a proportional scroll bar.
SIF_POS	The *nPos* member contains the scroll box position.
SIF_RANGE	The *nMin* and *nMax* members contain the minimum and maximum values for the scrolling range.

Table 1351.2 The possible predefined values for the **fMask** *member.*

Within the *GetScrollInfo* function, the *hwnd* parameter identifies a scroll bar control or a window with a standard scroll bar, depending on the value of the *fnBar* parameter. The *fnBar* parameter specifies the type of scroll bar for which to retrieve parameters. The *fnBar* parameter can be one of the values Table 1351.3 details.

Value	Meaning
SB_CTL	Retrieves the parameters for a scroll bar control. The *hwnd* parameter must be the handle of the scroll bar control.
SB_HORZ	Retrieves the parameters for the given window's standard horizontal scroll bar.
SB_VERT	Retrieves the parameters for the given window's standard vertical scroll bar.

Table 1351.3 The possible values for the **fnBar** *parameter.*

Finally, the *lpsi* parameter points to a *SCROLLINFO* structure whose *fMask* member, upon entry to the function, specifies the scroll bar parameters to retrieve. Before returning, the function copies the specified parameters to the appropriate members of the structure. The *fMask* member can be a combination of values Table 1351.4 details (note that the *GetScrollInfo* function accepts only certain values for the *fMask* member).

Value	Meaning
SIF_PAGE	Copies the scroll page to the *nPage* member of the *SCROLLINFO* structure to which the *lpsi* parameter points.

Table 1351.4 The possible values for the **SCROLLINFO** *structure's* **fMask** *member.* *(continued on following page)*

Value	Meaning
SIF_POS	Copies the scroll position to the *nPos* member of the *SCROLLINFO* structure to which the *lpsi* parameter points.
SIF_RANGE	Copies the scroll range to the *nMin* and *nMax* members of the *SCROLLINFO* structure to which the *lpsi* parameter points.

Table 1351.4 The possible values for the **SCROLLINFO** *structure's* ***fMask*** *member.*

The *GetScrollInfo* function lets applications use 32-bit scroll positions. Although the messages that indicate scroll-bar position, *WM_HSCROLL* and *WM_VSCROLL*, provide only 16 bits of position data, the functions *SetScrollInfo* and *GetScrollInfo* provide 32 bits of scroll-bar position data. Therefore, an application can call *GetScrollInfo* while processing either the *WM_HSCROLL* or *WM_VSCROLL* messages to obtain 32-bit scroll-bar position data.

The limitation on accessing 32-bit scroll-bar position data applies to real-time scrolling of a window's contents. An application implements real-time scrolling by processing the *WM_HSCROLL* or *WM_VSCROLL* messages that carry the *SB_THUMBTRACK* notification value, thereby tracking the position of the scroll box (thumb) as the user moves it. Unfortunately, there is no function that retrieves the 32-bit scroll-box position as the user moves the scroll box. Because *GetScrollInfo* provides only the static position, an application can only obtain 32-bit position data before or after a scroll operation.

Scrolling the Window Content 1352

As you might imagine, the most important function your scroll bars will perform is to let users scroll the content a window contains. You will control your window's scrolling with the *ScrollWindowEx* function. The *ScrollWindowEx* function scrolls the content of the specified window's client area. You will use the *ScrollWindowEx* function within your programs as shown in the following prototype:

```
int ScrollWindowEx(
   HWND hWnd,                    // handle of window to scroll
   int dx,              // amount of horizontal scrolling
   int dy,              // amount of vertical scrolling
   CONST RECT *prcScroll// address of structure with scroll rectangle
   CONST RECT *prcClip,          // address of structure with clip rectangle
   HRGN hrgnUpdate,              // handle of update region
   LPRECT prcUpdate,             // address of structure for update rectangle
   UINT flags                    // scrolling flags
);
```

As you can see, *ScrollWindowEx* takes eight parameters, most of which are predictable. Table 1352.1 details the parameters for the *ScrollWindowEx* function.

Parameter	Description
hWnd	Identifies the window where the *ScrollWindowEx* function is to scroll the client area.
dx	Specifies the amount, in device units, of horizontal scrolling. This parameter must be a negative value to scroll to the left.
dy	Specifies the amount, in device units, of vertical scrolling. This parameter must be a negative value to scroll up.
prcScroll	Points to the *RECT* structure that specifies the portion of the client area that the *ScrollWindowEx* function is to scroll. If this parameter is *NULL*, the *ScrollWindowEx* function will scroll the entire client area.

Table 1352.1 The parameters for the **ScrollWindowEx** *function.* *(continued on following page)*

Parameter	Description
prcClip	Points to the *RECT* structure that contains the coordinates of the clipping rectangle. The *ScrollWindowEx* function only affects device bits within the clipping rectangle. Windows will paint bits the function scrolls from the outside of the rectangle to the inside; Windows will not paint bits the function scrolls from the inside of the rectangle to the outside.
hrgnUpdate	Identifies the region that the *ScrollWindowEx* function is to modify to hold the region that scrolling invalidates. The *hrgnUpdate* parameter may be *NULL*.
prcUpdate	Points to the *RECT* structure that receives the boundaries of the rectangle that scrolling invalidates. The *prcUpdate* parameter may be *NULL*.
flags	Specifies flags that control scrolling. This parameter has one of the values Table 1352.2 details.

***Table 1352.1** The parameters for the **ScrollWindowEx** function. (continued from previous page)*

As Table 1352.1 indicates, the *flags* parameter may have one of several predefined values. Table 1352.2 lists the possible values for the *flags* parameter.

Value	Meaning
SW_ERASE	Erases the newly invalidated region by sending a *WM_ERASEBKGND* message to the window when you include the *SW_ERASE* flag with the *SW_INVALIDATE* flag.
SW_INVALIDATE	Invalidates the region the *hrgnUpdate* parameter identifies after scrolling.
SW_SCROLLCHILDREN	Scrolls all child windows that intersect the rectangle to which the *prcScroll* parameter points. Windows scrolls the child windows the number of pixels the *dx* and *dy* parameters specify. Windows sends a *WM_MOVE* message to all child windows that intersect the *prcScroll* rectangle, even if they do not move.

***Table 1352.2** The possible values for the **flags** parameter.*

If the function succeeds, the return value is *SIMPLEREGION* (rectangular invalidated region), *COMPLEXREGION* (non-rectangular invalidated region; overlapping rectangles), or *NULLREGION* (no invalidated region). If the function fails, the function returns the value *ERROR*.

If the function call does not specify the *SW_INVALIDATE* and *SW_ERASE* flags, *ScrollWindowEx* does not invalidate the area from which it scrolls. If either of these flags is set, *ScrollWindowEx* will invalidate the scroll-from area. Windows will not update the area until the application calls the *UpdateWindow* function, calls the *RedrawWindow* function (specifying the *RDW_UPDATENOW* or *RDW_ERASENOW* flag), or retrieves the *WM_PAINT* message from the application queue.

If the window has the *WS_CLIPCHILDREN* style, the returned areas that *hrgnUpdate* and *prcUpdate* specify will represent the total area of the scrolled window that Windows must update, including any areas in child windows. If the call to *ScrollWindowEx* specifies the *SW_SCROLLCHILDREN* flag, Windows will not properly update the screen if a user scrolls a section of a child window. Windows does not erase the section of the scrolled child window that lies outside the source rectangle and does not properly redraw the child window in its new destination. To move child windows that do not lie completely within the rectangle *prcScroll* specifies, use the *DeferWindowPos* function. Windows will reposition the cursor if the *SW_SCROLLCHILDREN* flag is set and the caret rectangle intersects the scroll rectangle.

Windows determines all input and output coordinates (for *prcScroll*, *prcClip*, *prcUpdate*, and *hrgnUpdate*) as client coordinates, regardless of whether the window has the *CS_OWNDC* or *CS_CLASSDC* class style. Use the *LPtoDP* and *DPtoLP* functions to convert to and from logical coordinates, if necessary.

To better under the processing the *ScrollWindowEx* function performs, consider the *Scroll_Window.cpp* program contained on the CD-ROM that accompanies this book. The *Scroll_Window.cpp* program lets the user add a line at a time to the program window. After the number of lines exceeds the amount of available space within the window, the scroll bar becomes active. When the user clicks the mouse on the up or down arrow, the thumb moves one line in the appropriate direction.

UNDERSTANDING THE WM_SIZE MESSAGE 1353

In Tip 1352, you learned about the *Scroll_Window* program, which let users select the scroll bar to move up and down through a window. As you learned, the program only activated the scroll bar when the window's contents became too long to display within the currently sized window. If you experiment with the *Scroll_Window* program and make the window larger, you will note that the scroll bar becomes inactive again after you make the window large enough to display the text. The *Scroll_Window* program captures the *WM_SIZE* system message to adjust the scroll bar after you resize the window, as shown here:

```
// Every time the window is sized, recalculate the number of lines
// the client area can display and set the scroll bar accordingly.
Case WM_SIZE :
       {
          RECT rect;
          GetClientRect(hWnd, &rect);
          nDspLines = rect.bottom / 20;
          if (nDspLines < nNumItems)
           {
              SCROLLINFO si;
              si.cbSize = sizeof(SCROLLINFO);
              si.fMask  = SIF_POS | SIF_RANGE | SIF_PAGE;
              si.nMin   = 0;
              si.nMax   = nNumItems-1;
              si.nPage  = nDspLines;
              si.nPos   = nCurPos;
              EnableScrollBar(hWnd, SB_VERT, ESB_ENABLE_BOTH);
              SetScrollInfo(hWnd, SB_VERT, &si, TRUE);
           }
          else
           EnableScrollBar(hWnd, SB_VERT, ESB_DISABLE_BOTH);
       }
       break;
```

In the case of the *Scroll_Window.cpp* program, the *WM_SIZE* case tests the number of lines the window can display against the number of lines it is currently displaying. If the number of possible lines is larger than the number of displayed lines, the program will disable the scroll bars. Otherwise, it will change the thumb size (if necessary) and display the newly sized scroll bar.

Your programs will use the *WM_SIZE* message to check the window's current size and replace any items within the window whose position depends on the window's size. For example, if you have an edit box on the window that displays a title, and the user resizes the window, you may want to resize the edit box concurrently with the window. The action your programs will perform when they receive a *WM_SIZE* message will differ, based on the type of resizing the user performs. The *WM_SIZE* message passes a constant that represents the resize type within the *wParam* parameter. Table 1353 details the possible values for the *wParam* parameter.

Value	Meaning
SIZE_MAXHIDE	Windows sends the message to all pop-up windows when the user maximizes some other window.
SIZE_MAXIMIZED	User has maximized the window.
SIZE_MAXSHOW	Windows sends the message to all pop-up windows when the user restores some other window to its former size.
SIZE_MINIMIZED	User has minimized window.
SIZE_RESTORED	User has resized window, but neither the *SIZE_MINIMIZED* nor *SIZE_MAXIMIZED* value applies.

*Table 1353 The possible values for **wParam** with a **WM_SIZE** message.*

Additionally, the *WM_SIZE* message passes information about the new window within the *lParam* parameter. The low word of *lParam* specifies the new width of the client area, and the high word of *lParam* specifies the new height of the client area.

1354 Understanding the WM_PAINT Message

As you learned in Tip 1353, your applications will receive the *WM_SIZE* message each time the user resizes the application window. Windows will also send your applications the *WM_PAINT* message following the *WM_SIZE* message each time the user resizes the application window. In fact, your application receives the *WM_PAINT* message when Windows or another application makes a request to paint a portion of your application's window. Windows sends the *WM_PAINT* message when your program calls the *UpdateWindow* or *RedrawWindow* function, or it sends the *DispatchMessage* function when your application uses the *GetMessage* or *PeekMessage* function to obtain a *WM_PAINT* message.

The *wParam* parameter for a *WM_PAINT* message contains the *hdc* value. The *hdc* value is a handle to a device context. It identifies the device context in which Windows draws. If the *wParam* parameter is *NULL*, the application should use the default device context (rather than creating a private device context). Some common controls use *wParam* to enable drawing in a device context other than the default device context. Other windows can safely ignore *wParam*. You will learn more about device contexts in later Tips.

The *DefWindowProc* function validates the update region. The function may also send the *WM_NCPAINT* message to the window procedure if Windows must paint the window frame, and additionally send the *WM_ERASEBKGND* message if Windows must erase the window background.

The system sends the *WM_PAINT* message when there are no other messages in the application's message queue. *DispatchMessage* determines where to send the message; *GetMessage* determines which message to dispatch. *GetMessage* returns the *WM_PAINT* message when there are no other messages in the application's message queue, and *DispatchMessage* sends the message to the appropriate window procedure.

A window may receive internal paint messages as a result of calling *RedrawWindow* with the *RDW_INTERNALPAINT* flag set. In this case, the window may not have an update region. An application should call the *GetUpdateRect* function to determine whether the window has an update region. If *GetUpdateRect* returns zero, the application should not call the *BeginPaint* and *EndPaint* functions. If the application does not call the *GetUpdateRect* functio n, sections of the window may not update correctly or at all.

An application must look at its internal data structures for each *WM_PAINT* message to check for any necessary internal painting, because either a non-*NULL* update region or a call to *RedrawWindow* with the *RDW_INTERNALPAINT* flag set may have caused a *WM_PAINT* message.

Windows sends an internal *WM_PAINT* message only once. After *GetMessage* returns an internal *WM_PAINT* message or *UpdateWindow* sends *PeekMessage* to a window, Windows does not post or send further *WM_PAINT* messages until the program invalidates the window or until the program calls *RedrawWindow* again with the *RDW_INTERNALPAINT* flag set.

For some common controls, the default *WM_PAINT* message processing checks the *wParam* parameter. If *wParam* is not *NULL*, the controls will assume that the value is a handle to a device context and will use that device context to paint. To better understand the processing your programs perform when they receive a *WM_PAINT* message, see the *Scroll_Window.cpp* program that Tip 1353 presented.

OTHER SCROLL BAR MESSAGES YOUR PROGRAMS MUST CAPTURE 1355

As you have learned, the scroll bar generates different scroll bar-specific messages, depending on where the user clicks within the scroll bar area. In addition to the *WM_SIZE* and *WM_PAINT* messages that you learned about in Tips 1353 and 1354, your programs should also capture the *WM_VSCROLL* and *WM_HSCROLL* messages.

Windows sends the *WM_VSCROLL* message to a window when a scroll event occurs in the window's standard vertical scroll bar. Windows also sends the *WM_VSCROLL* message to the owner of a vertical scroll bar control when a scroll event occurs in the control. When your program receives a *WM_VSCROLL* message, the low-order word of *wParam* specifies a scroll bar value that indicates the user's scrolling request. The low-order word can be one of the values Table 1355.1 lists.

Value	Meaning
SB_BOTTOM	Scrolls to the lower right.
SB_ENDSCROLL	Ends scroll.
SB_LINEDOWN	Scrolls one line down.
SB_LINEUP	Scrolls one line up.
SB_PAGEDOWN	Scrolls one page down.
SB_PAGEUP	Scrolls one page up.
SB_THUMBPOSITION	Scrolls to the absolute position (an exact numeric offset from the window's beginning, such as "12 lines down from the top"). The *nPos* parameter specifies the current position.
SB_THUMBTRACK	Drags scroll box to the specified position. The *nPos* parameter specifies the current position.
SB_TOP	Scrolls to the upper left.

***Table 1355.1** The possible values for the low-order word of the **wParam** parameter.*

In addition to the value that Windows passes within the low-order word of the *wParam* parameter, your programs should also check the high-order word of *wParam*. The high-order word specifies the current position of the scroll box if the *nScrollCode* parameter is *SB_THUMBPOSITION* or *SB_THUMBTRACK*; otherwise, the high-order word contains an unusable value. Finally, the *lParam* parameter contains the scroll bar control's handle. If a scroll bar does not send the message, *lParam* is *NULL*.

Applications that provide feedback as the user drags the scroll box typically use the *SB_THUMBTRACK* notification message. If an application scrolls the window's content, it must also use the *SetScrollPos* function to reset the position of the scroll box.

Windows sends the *WM_HSCROLL* message to a window when a scroll event occurs in the window's standard horizontal scroll bar. Windows also sends the *WM_HSCROLL* message to the owner of a horizontal scroll bar control when a scroll event occurs in the control. Just as the *WM_VSCROLL* message includes additional information within the *wParam* and *lParam* parameters, so too does the *WM_HSCROLL* message.

With *WM_HSCROLL*, the low-order word of *wParam* specifies a scroll bar value that indicates the user's scrolling request. The low-order word of *wParam* can contain one of the values Table 1355.2 lists.

Value	Meaning
SB_BOTTOM	Scrolls to the lower right.
SB_ENDSCROLL	Ends scroll.
SB_LINELEFT	Scrolls left by one unit.
SB_LINERIGHT	Scrolls right by one unit.
SB_PAGELEFT	Scrolls left by the width of the window.
SB_PAGERIGHT	Scrolls right by the width of the window.
SB_THUMBPOSITION	Scrolls to the absolute position (an exact numeric offset from the window's left side). The *nPos* parameter (the high-order word of *wParam*) specifies the current position.
SB_THUMBTRACK	Drags scroll box to the specified position. The *nPos* parameter (the high-order word of *wParam*) specifies the current position.
SB_TOP	Scrolls to the upper left.

***Table 1355.2** The possible values for the low-order word of the **wParam** parameter.*

Just as with the *WM_VSCROLL* message, the high-order word of *wParam* specifies the current position of the scroll box if the *nScrollCode* parameter is *SB_THUMBPOSITION* or *SB_THUMBTRACK*; otherwise, the high-order word does not contain a usable value. The *lParam* parameter returns the handle of the scroll bar control if a scroll bar sends the message. If a scroll bar does not send the message, *hwndScrollBar* is *NULL*.

Note that both the *WM_HSCROLL* and *WM_VSCROLL* messages carry only 16 bits of scroll box position data. Therefore, applications that rely solely on *WM_HSCROLL* and *WM_VSCROLL* for scroll position data have a practical maximum position value of 65,535. However, because the *SetScrollPos, SetScrollRange, GetScrollPos*, and *GetScrollRange* functions support 32-bit scroll bar position data, there is a way to circumvent the 16-bit barrier of the *WM_HSCROLL* and *WM_VSCROLL* messages within these functions. See your compiler's help file for more information on circumventing the 16-bit barrier.

1356 Enabling and Disabling the Scroll Bars

As you have learned, your programs can administer close and complete control over scroll bars. One of the most common activities your programs will perform with scroll bars is to enable and disable them. As you saw in Tip 1352, for example, your programs may disable the scroll bar if the user resizes the window to a size large enough to display everything the window contains. To enable and disable scroll bars, your programs will use the *EnableScrollBar* function. The *EnableScrollBar* function enables or disables one or both scroll bar arrows. You will use the *EnableScrollBar* function within your programs as shown in the following prototype:

```
BOOL EnableScrollBar(
   HWND hWnd,          // handle to window or scroll bar
   UINT wSBflags,// scroll bar type flag
   UINT wArrows        // scroll bar arrow flag
);
```

As you can see, the *EnableScrollBar* function accepts three parameters. The *hWnd* parameter identifies a window or a scroll bar control, depending on the value of the *wSBflags* parameter. The *wSBflags* parameter specifies the scroll bar type. The *wSBflags* parameter can be one of the values Table 1356.1 shows.

Value	Meaning
SB_BOTH	Enables or disables the arrows on the horizontal and vertical scroll bars associated with the specified window. The *hWnd* parameter must be the handle to the window.
SB_CTL	Identifies the scroll bar as a scroll bar control. The *hWnd* parameter must be the handle to the scroll bar control.
SB_HORZ	Enables or disables the arrows on the horizontal scroll bar associated with the specified window. The *hWnd* parameter must be the handle to the window.
SB_VERT	Enables or disables the arrows on the vertical scroll bar associated with the specified window. The *hWnd* parameter must be the handle to the window.

***Table 1356.1** The possible values for the **wSBflags** parameter.*

Finally, the *wArrows* parameter specifies whether the scroll bar arrows are enabled or disabled and indicates which arrows the *EnableScrollBar* function should enable or disable. The *wArrows* parameter can be one of the values Table 1356.2 shows.

Value	Meaning
ESB_DISABLE_BOTH	Disables both arrows on a scroll bar.
ESB_DISABLE_DOWN	Disables the down arrow on a vertical scroll bar.
ESB_DISABLE_LEFT	Disables the left arrow on a horizontal scroll bar.
ESB_DISABLE_LTUP	Disables the left arrow on a horizontal scroll bar or the up arrow of a vertical scroll bar.
ESB_DISABLE_RIGHT	Disables the right arrow on a horizontal scroll bar.
ESB_DISABLE_RTDN	Disables the right arrow on a horizontal scroll bar or the down arrow of a vertical scroll bar.
ESB_DISABLE_UP	Disables the up arrow on a vertical scroll bar.
ESB_ENABLE_BOTH	Enables both arrows on a scroll bar.

***Table 1356.2** The possible values for the **wArrows** parameter.*

If the function successfully enables or disables the arrows the *wArrows* parameter specifies, the return value is non-zero. If the arrows are already in the requested state or an error occurs, the return value is zero.

USING THE SCROLLDC FUNCTION 1357

Throughout the previous Tips, you have learned about several of the scroll bar functions and the messages they generate. While you have not learned about device contexts yet (you will do so in later Tips), it is important to understand how your programs can scroll windows that display graphics or other non-textual items.

As a rule, when you draw non-textual items on a window, you will use a *device context* to do so. You can also use a device context when you draw text on a window, though you will generally use indirection to access the context. When you scroll a device context within a window, you will use slightly different processing to scroll the device context or a piece of the device context.

The *ScrollDC* function scrolls a rectangle of bits within a device context horizontally and vertically. You will use the *ScrollDC* function within your programs as shown in the following prototype:

paging file as a single block of contiguous memory. To let your programs manage memory in such a way, the virtual memory manager lets your programs work with *maps* to actual memory, rather than working with actual memory itself. Because you are working with the illusion of contiguous memory, rather than a true contiguous memory block, Window's designers named the memory model the *virtual memory model*. Figure 1360 shows a logical model of how Windows uses virtual memory to access physical memory.

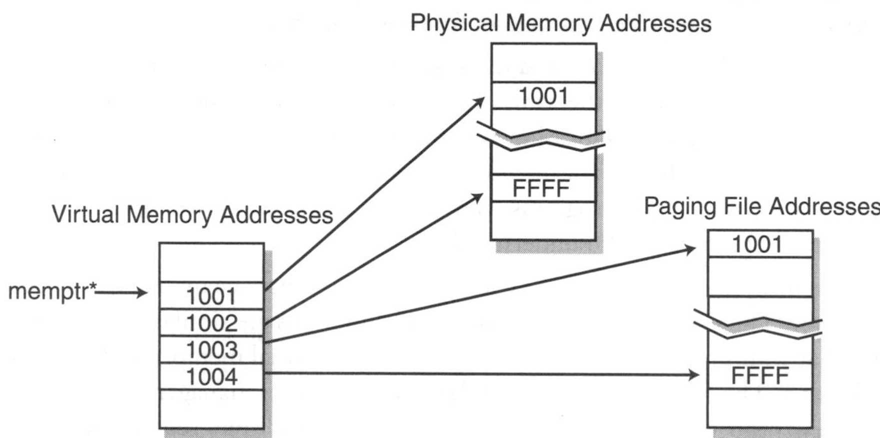

Figure 1360 *Windows uses virtual memory to access physical memory.*

Because of the ways Windows manages virtual memory, the virtual address space for each process is much larger than the total physical address available to all processes. To increase the size of memory, Windows uses the hard drive for additional storage. The total amount of memory available to all processes is the sum of physical memory and the free space available on the disk within the Windows *paging file*. The paging file is a disk file that Windows uses to increase your computer's effective memory. Windows organizes the virtual memory space into *pages*, or units of memory. The page size depends on the host computer. You can invoke the *GetSystemInfo* function to determine the page size for a computer. On x86 computers, the page size is 4Kb.

As you have learned, in the Win32 memory model, the operating system provides each process with its own private address space. When a thread in a process is running, that thread can access only memory that belongs to its process. The memory that belongs to all other processes is hidden and inaccessible to the running thread. Because each process has its own virtual 4Gb of address space, every process views memory as if that memory ran from address 0x00000000 to 0xFFFFFFFF. Two processes running simultaneously may each store memory at address 0x12341234 without interfering with each other. Clearly, such address sharing would not be possible if both processes were using the same address space.

In reality, the virtual memory manager maps the virtual address to an actual, physical address—which may be in the computer's physical memory or may be in the paging file. The important consideration is that, as far as the application is concerned, the address of the memory is 0x12341234–whether the memory is actually at ram location 0x00012345 or at sector 14352 on the computer's hard drive. Therefore, the virtual memory manager lets you execute multiple programs simultaneously without making you ensure that each program does not override the memory of the other executing programs.

In Windows 95, the operating system divides the 4Gb virtual address space for each process into four partitions. Windows 95 uses the partition from *0x00000000* through *0x003FFFFF* (a 4Mb partition at the bottom of the virtual address space) to maintain compatability with MS-DOS and 16-bit Windows. Your Win32 applications should not read from or write to this partition. If your applications try to access this memory, the operating system will return a *NULL* pointer and may cause instability (resulting in program crashes, operating-system lock-ups, and so on).

Windows 95 uses the partition from *0x00400000* to *0x7FFFFFFF* for the process's private, unshared address space. Other Win32 processes cannot read from, write to, or in any way access another process's data stored within this nearly 2Gb partition.For your applications, you should maintain the bulk of your process information within this protected region.

Windows 95 uses the remaining 2Gb of each process's virtual address space to store shared files and operating system files. When you work with memory within your program's address space, you should avoid access of the address space above *0x80000000* unless your program is specifically trying to access a shared or system file.

REVISITING HEAPS 1361

As you learned in Tip 1360, Windows allocates your programs 4Gb of virtual address space for each running process. The Win32 heap functions let a process create a *private heap*, which is a block of one or more pages in the address space of the process. The *HeapCreate* function creates a heap of a given size, and the *HeapAlloc* and *HeapFree* functions allocate and free memory in the heap. When you create heaps within your Windows programs, Windows will create the heap beginning at *0x7FFFFFF* and grow the heap downwards through the 2Gb of reserved process memory.

Heap objects can grow dynamically within the range you specify at their creation with the *HeapCreate* function. The maximum size of the heap determines the number of pages of memory the heap reserves. The initial size determines the number of committed, read/write pages that Windows initially allocates for the heap. Windows automatically allocates additional pages from the reserved space if *HeapAlloc* requirements exceed the current size of committed pages. After Windows commits pages to a heap, it will not release those committed pages until the process terminates or the program destroys the heap with the *HeapDestroy* function. Because memory your programs allocate in the heap with *HeapAlloc* has a fixed location within the process's virtual address space and the system cannot compact the heap, you should always write your applications so that they minimize fragmentation of the heap.

The memory of a private heap is accessible only to the process that created the heap. If a dynamic-link library (DLL) creates a private heap, Windows creates that private heap in the address space of the process that called the dynamic-link library (under Windows 95, above 0x8000000). However, only the process that called the dynamic-link library can access the information in the dynamic-link library-created private heap—meaning that multiple processes which execute the same dynamic-link library may result in multiple dynamic-link library-created private heaps, but each library instance can only access a single private heap.

ALLOCATING A MEMORY BLOCK FROM THE GLOBAL HEAP 1362

As you have learned, your programs can use several different techniques to allocate different memory types within the Windows 95 and Windows NT environments. One of the most common allocations you will perform is allocating memory from the global heap—which is similar to the allocation which you used functions such as *malloc* and *halloc* to perform in DOS programs. You will use the *GlobalAlloc* function to allocate memory from the global heap. The *GlobalAlloc* function allocates the number of bytes you specify from the heap. You will use the *GlobalAlloc* function as shown in the following prototype:

```
HGLOBAL GlobalAlloc(
   UINT uFlags,  // object allocation attributes
   DWORD dwBytes // number of bytes to allocate
);
```

The *uFlags* parameter specifies how to allocate memory. If the *uFlags* parameter specifies zero, the default is the *GMEM_FIXED* flag. Except for the incompatible combinations that the following tables specifically note, your programs can use any combination of the flags the following tables detail when your programs call *GlobalAlloc*. To indicate whether the function allocates fixed or moveable memory, specify one of the flags from Table 1362.1.

Flag	Meaning
GMEM_FIXED	Allocates fixed memory. Your programs cannot combine this flag with the *GMEM_MOVEABLE* or *GMEM_DISCARDABLE* flag. The return value is a pointer to the memory block. To access the memory, the calling process simply casts the return value to a pointer.
GMEM_MOVEABLE	Allocates moveable memory. Your programs cannot combine this flag with the *GMEM_FIXED* flag. The return value is the handle of the memory object. The handle is a 32-bit quantity that is private to the calling process. To translate the handle into a pointer, use the *GlobalLock* function.
GPTR	Combines the *GMEM_FIXED* flag and the *GMEM_ZEROINIT* flag from Table 1362.2.
GHND	Combines the *GMEM_MOVEABLE* flag and the *GMEM_ZEROINIT* flag from Table 1362.2.

***Table 1362.1** Memory allocation types for use with **GlobalAlloc**.*

In addition to the flags specified in Table 1362.1, the *uFlags* parameter can contain any of the values Table 1362.2 details, except where the table notes that a flag specifically conflicts with another flag.

Flag	Meaning
GMEM_DDESHARE	Allocates memory the dynamic-data exchange (DDE) functions will use for a dynamic-data exchange conversation. This flag is available for compatibility purposes with Win16 applications. Some applications may use *GMEM_DDESHARE* to enhance the performance of dynamic-data exchange operations, and your programs should specify the *GMEM_DDESHARE* flag if they will use the memory for dynamic-data exchange. Only processes that use dynamic-data exchange or the clipboard for interprocess communications should specify the *GMEM_DDESHARE* flag.
GMEM_DISCARDABLE	Allocates discardable memory (memory which is not fixed to a specific address within the process's virtual address space). Your programs cannot combine this flag with the *GMEM_FIXED* flag. Some Win32-based applications may ignore this flag.
GMEM_LOWER	Win32 ignores this flag. The Win32 API provides this flag only for compatibility with Windows version 3.*x*.
GMEM_NOCOMPACT	Does not compact or discard memory to satisfy the allocation request.
GMEM_NODISCARD	Does not discard memory to satisfy the allocation request.
GMEM_NOT_BANKED	Win32 ignores this flag. The Win32 API provides this flag only for compatibility with Windows version 3.*x*.
GMEM_NOTIFY	Win32 ignores this flag. The Win32 API provides this flag only for compatibility with Windows version 3.*x*.
GMEM_SHARE	Allocates memory the dynamic-data exchange (DDE) functions will use for a DDE conversation. Same as the *GMEM_DDESHARE* flag.
GMEM_ZEROINIT	Initializes memory contents to zero.

***Table 1362.2** Additional flag values for the **GlobalAlloc** function.*

In addition to specifying the type of memory for *GlobalAlloc* to allocate, your programs must specify the *dwBytes* parameter, which specifies the number of bytes to allocate. If this parameter is zero and the *uFlags* parameter specifies the *GMEM_MOVEABLE* flag, the function will return a handle to a memory object that Windows marks as discarded.

If the function succeeds, the return value will be the handle of the newly allocated memory object. If the function fails, the return value will be *NULL*.

If the heap does not contain sufficient free space to satisfy the request, *GlobalAlloc* will return *NULL*. Because *GlobalAlloc* uses *NULL* to indicate an error, Windows never allocates virtual address zero. It is, therefore, easy to detect the use of a *NULL* pointer. Windows creates all memory with execute access. Windows does not require a special function to execute dynamically generated code. Windows guarantees that memory your programs allocate with the *GlobalAlloc* function will align onto an 8-byte boundary.

Windows limits the *GlobalAlloc* and *LocalAlloc* functions to a combined total of 65,536 handles for *GMEM_MOVEABLE* (for globally alocated memory) and *LMEM_MOVEABLE* (for locally alocated memory) memory per process. This limitation does not apply to *GMEM_FIXED* or *LMEM_FIXED* memory. If the *GlobalAlloc* function succeeds, it will allocate at least the amount of memory the function call requests. If the actual amount that *GlobalAlloc* allocates is greater than the amount the function call requests, the process can nevertheless use the entire amount. To determine the actual number of bytes the *GlobalAlloc* function allocated, use the *GlobalSize* function.

To better understand the processing the *GlobalAlloc* function performs, consider the program *Global_Alloc*, contained on the CD-ROM that accompanies this book. The *Global_Alloc* program allocates memory to store a string. When the application begins, the *WM_CREATE* message handler creates a 27 character buffer (26 bytes and a *NULL* terminator). When the user selects *Test!*, the program displays the buffer and its size on the screen.

Using GlobalReAlloc to Dynamically Change Heap Sizes 1363

As you learned in Tip 1362, your programs can use the *GlobalAlloc* function to allocate memory from the global heap. Often, however, your programs must reallocate a memory block after its initial allocation. You can do so with the *GlobalReAlloc* function. The *GlobalReAlloc* function changes the size or attributes of a specified global memory object. The size can increase or decrease, depending on invocation. You will use the *GlobalReAlloc* function within your programs as shown in the following prototype:

```
HGLOBAL GlobalReAlloc(
    HGLOBAL hMem,       // handle to the global memory object
    DWORD dwBytes,      // new size of the block
    UINT uFlags         // how to reallocate object
);
```

The *hMem* handle identifies the global memory object that *GlobalReAlloc* is to reallocate. Either the *GlobalAlloc* or *GlobalReAlloc* function will have previously returned this handle. The *dwBytes* parameter specifies the new size, in bytes, of the memory block. If the *dwBytes* parameter is zero and the *uFlags* parameter specifies the *GMEM_MOVEABLE* flag, the function will return the handle of a memory object that Windows marks as discarded. If the *dwBytes* parameter is zero and *uFlags* specifies the *GMEM_MODIFY* flag, the API function ignores the *dwBytes* parameter. The *uFlags* parameter specifies how to reallocate the global memory object. If the *uFlags* parameter specifies the *GMEM_MODIFY* flag, the *uFlags* parameter modifies the attributes of the memory object, and the API function ignores the *dwBytes* parameter. Otherwise, the *uFlags* parameter controls the reallocation of the memory object.

When you call the *GlobalReAlloc* function, your program can combine the *GMEM_MODIFY* flag with one or both of the flags Table 1363.1 details.

Flag	Meaning
GMEM_DISCARDABLE	Allocates discardable memory if you also specify the *GMEM_MODIFY* flag. Windows ignores this flag unless you previously allocated the object as moveable or unless you also specify the *GMEM_MOVEABLE* flag.
GMEM_MOVEABLE	*Windows NT only:* Changes a fixed memory object to a moveable memory object if you also specify the *GMEM_MODIFY* flag.

Table 1363.1 Flags compatible with the ***GMEM_MODIFY*** *flag.*

If the *uFlags* parameter does not specify *GMEM_MODIFY*, it can be any combination of the flags Table 1363.2 lists.

Flag	Meaning
GMEM_MOVEABLE	If *dwBytes* is zero, *GMEM_MOVEABLE* will discard a previously moveable and discardable memory block. If the lock count of the object is not zero or if the block is not moveable and discardable, the function will fail. If *dwBytes* is non-zero, *GMEM_MOVEABLE* will enable the system to move the reallocated block to a new location without changing the moveable or fixed attribute of the memory object. If the object is fixed, the handle the function returns may be different from the handle the *hMem* parameter specifies. If the object is moveable, your programs can move the block without invalidating the object's handle, even if a previous call to the *GlobalLock* function is currently locking the object. To get the new address of the memory block, use *GlobalLock*.
GMEM_NOCOMPACT	Prevents Windows from compacting or discarding memory to satisfy the allocation request.
GMEM_ZEROINIT	Causes Windows to initialize the additional memory contents to zero if the memory object is growing in size.

Table 1363.2 Additional flags for the ***uFlags*** *parameter.*

If the function succeeds, the return value will be the handle of the reallocated memory object. If the function fails, the return value will be *NULL*. If *GlobalReAlloc* reallocates a moveable object, the return value will be the handle of the memory object. To convert the handle to a pointer, use the *GlobalLock* function. If *GlobalReAlloc* reallocates a fixed object, the value of the handle returned will be the address of the first byte of the memory block. To access the memory, a process can simply cast the return value to a pointer. If *GlobalReAlloc* fails, it does not free the original memory, and the original handle and pointer will still be valid.

To better understand the processing *GlobalReAlloc* performs, consider the *Global_ReAlloc* program contained on the CD-ROM that accompanies this book. The *Global_ReAlloc* program performs similar processing to the *Global_Alloc* program. However, the *Global_ReAlloc* program also reallocates another 27 bytes of memory to show the lowercase alphabet when the user selects the *Test!* option.

1364 DISCARDING AN ALLOCATED MEMORY BLOCK

In earlier Tips, you used the *free* and *delete* member functions to discard allocated memory within your programs. When you allocate or reallocate memory from a global heap, your programs should use the *GlobalDiscard* function to release that memory after your program has completed its processing on the memory. The *GlobalDiscard* function discards a global memory block that you previously allocated with the *GMEM_DISCARDABLE* flag set. The lock count of the memory object you want to discard must be zero, or the function will fail to discard the memory. You will use the *GlobalDiscard* function as shown in the following prototype:

```
HGLOBAL GlobalDiscard(
   HGLOBAL hglbMem      // handle to the global memory object
);
```

The *hglbMem* parameter identifies the global memory object to discard. If the function succeeds, the function will return the handle of the memory object (that is, the *hglbMem* handle). If the function fails, the function will return the value *NULL*.

The *GlobalDiscard* function discards only global objects that the calling process allocated with the *GMEM_DISCARDABLE* flag. If a process tries to discard a fixed or locked object, the function will fail. Although the *GlobalDiscard* function discards the object's memory block, the object's handle remains valid. A process can subsequently pass the handle to the *GlobalReAlloc* function to allocate another global memory block the same handle identifies.

USING THE GLOBALFREE FUNCTION 1365

In Tip 1364, you learned about the *GlobalDiscard* function, which your programs will use to discard a previously allocated memory block and to maintain the handle to that memory block for possible future use. On the other hand, if you know that your program will not reuse the same memory block, if you want to keep your program from using the same memory block, or if you are unsure whether your program allocated the block with the *GMEM_DISCARDABLE* flag, your programs can use the *GlobalFree* function to free a memory object. The *GlobalFree* function frees the global memory object you specify and invalidates the memory object's handle. You will use the *GlobalFree* function as shown in the following prototype:

```
HGLOBAL GlobalFree(HGLOBAL hMem);
```

The *hMem* parameter identifies the global memory object to release. Unlike the *GlobalDiscard* function, if the *GlobalFree* function succeeds, the return value will be *NULL*. If the *GlobalFree* function fails, the return value will be equal to the global memory object's handle.

Heap corruption or an access violation exception (*EXCEPTION_ACCESS_VIOLATION*) may occur if the process tries to examine or modify the memory after the process has previously freed the memory. If the *hgblMem* parameter is *NULL*, *GlobalFree* will fail and the system will generate an access violation exception. Both *GlobalFree* and *LocalFree* will free a *locked-memory object*. A locked-memory object has a lock count greater than zero. The *GlobalLock* function locks a global memory object (which prevents another function from discarding the memory object) and increments the object's lock count by one. The *GlobalUnlock* function unlocks the memory object and decrements the lock count by one. To get the lock count of a global memory object, use the *GlobalFlags* function.

USING GLOBALLOCK AND GLOBALHANDLE 1366

As you have learned, your programs can use the *GlobalAlloc* and *GlobalReAlloc* functions to allocate memory from the global heap. However, as you saw, both allocation functions return a handle to the allocated memory. On the other hand, you will want most of your programs to use that memory with a pointer. You can use the *GlobalLock* and *GlobalHandle* functions to easily convert allocated memory to a pointer and back to a handle again. The *GlobalLock* function locks a global memory object and returns a pointer to the first byte of the object's memory block. Your programs cannot move or discard the memory block associated with a locked-memory object. For memory objects you allocate with the *GMEM_MOVEABLE* flag, the function will increment the lock count associated with the memory object. You will use the *GlobalLock* function as shown in the following prototype:

```
LPVOID GlobalLock(
   HGLOBAL hMem  // address of the global memory object
);
```

In the *GlobalLock* function, the *hMem* parameter identifies the global memory object. Either the *GlobalAlloc* or *GlobalReAlloc* function will return this handle. If the *GlobalLock* function succeeds, the return value will be a pointer to the first byte of the memory block. If the *GlobalLock* function fails, the return value will be *NULL*.

The internal data structures for each memory object include a lock count that is initially zero. For moveable memory objects, *GlobalLock* increments the count by one, and the *GlobalUnlock* function decrements the count by one. For each call that a process makes to *GlobalLock* for an object, the process must eventually call *GlobalUnlock*. Your program cannot move or discard locked memory, unless the program reallocates the memory object using the *GlobalReAlloc* function. A locked-memory object's memory block remains locked until your program decrements the memory block's lock count to zero, at which time your program can move or discard the memory.

Memory objects previously allocated with the *GMEM_FIXED* flag always have a lock count of zero. For these objects, the value of the returned pointer is equal to the value of the specified handle. If your program has previously discarded the specified memory block or if the memory block has a zero-byte size, the *GlobalLock* function will return *NULL*. Discarded objects always have a lock count of zero. You will generally implement the *GlobalLock* function with code similar to that shown in the following fragment:

```
HGLOBAL hMem = GlobalAlloc(GHND, 27);
LPTSTR  pCur;
if (hMem && (pMem = (LPTSTR)GlobalLock(hMem)) != NULL)
```

The preceding code allocates a handle to 27 bytes of memory (the *hMem* variable), then locks that memory into the *pMem* pointer, which it simultaneously casts as a string type.

As you have learned, your programs will often use *GlobalLock* to convert memory handles to pointers. You will generally use the *GlobalHandle* function to convert a pointer back to a memory handle. The *GlobalHandle* function retrieves the handle associated with the specified pointer to a global memory block. The most common reason for converting a pointer back to a handle is to prepare for a *GlobalFree* command. You will use the *GlobalHandle* function within your programs as in the following prototype:

```
HGLOBAL GlobalHandle(
   LPCVOID pMem  // pointer to the global memory block
);
```

The *pMem* parameter is a pointer to the first byte of the global memory block. The *GlobalLock* function returns this pointer. If the *GlobalHandle* function succeeds, the return value will be the handle of the specified global memory object. If the *GlobalHandle* function fails, the return value will be *NULL*.

When the *GlobalAlloc* function allocates a memory object with the *GMEM_MOVEABLE* flag, it returns the handle of the object. The *GlobalLock* function converts this handle into a pointer to the memory block, and *GlobalHandle* converts the pointer back into a handle. You will generally implement *GlobalHandle* within your code as shown here:

```
HGLOBAL hMem = GlobalHandle(pMem);

GlobalUnlock(hMem);
pMem = NULL;
hMem = GlobalReAlloc(hMem, (26*2)+1, GMEM_MOVEABLE);
```

In this particular case, the program creates the handle, then unlocks the handle (releasing the memory). The program then reallocates the memory, as needed. Later code within your programs will probably convert the handle back to a pointer.

CHECKING THE COMPUTER'S MEMORY 1367

As you have learned, Windows maintains information about both your computer's physical memory and your computer's virtual memory. Often, your programs will require information about the amount of available free memory that your programs can access. Your programs can use the *GlobalMemoryStatus* function to find out about the current state of your computer's memory. The *GlobalMemoryStatus* function returns information about physical and virtual memory. You will use the *GlobalMemoryStatus* function within your programs as shown in the following prototype:

```
VOID GlobalMemoryStatus(
   LPMEMORYSTATUS lpBuffer    // pointer to the memory status structure
);
```

The *lpBuffer* parameter points to a *LPMEMORYSTATUS* structure in which *GlobalMemoryStatus* returns information about current memory availability. Before calling the *GlobalMemoryStatus* function, the calling process should set the *dwLength* member of this structure. The *MEMORYSTATUS* structure contains information about current memory availability. The Windows API defines the *MEMORYSTATUS* structure as shown here:

```
typedef struct _MEMORYSTATUS {
    DWORD dwLength;         // sizeof(MEMORYSTATUS)
    DWORD dwMemoryLoad;     // percent of memory in use
    DWORD dwTotalPhys;      // bytes of physical memory
    DWORD dwAvailPhys;      // free physical memory bytes
    DWORD dwTotalPageFile; // bytes of paging file
    DWORD dwAvailPageFile; // free bytes of paging file
    DWORD dwTotalVirtual;  // user bytes of address space
    DWORD dwAvailVirtual;  // free user bytes
} MEMORYSTATUS;
```

As you can see, the *MEMORYSTATUS* structure stores significant information about the computer's current available memory. Table 1367 explains the members of the *MEMORYSTATUS* structure.

Member	Description
dwLength	Indicates the size of the structure. The calling process should set this member before calling *GlobalMemoryStatus*.
dwMemoryLoad	Specifies a number between 0 and 100 that gives a general idea of current memory utilization, in which 0 indicates no memory use and 100 indicates full memory use.
dwTotalPhys	Indicates the total number of bytes of physical memory.
dwAvailPhys	Indicates the number of bytes of physical memory available.
dwTotalPageFile	Indicates the total number of bytes that all programs can store in the paging file. Note that this number does not represent the actual physical size of the paging file on disk.
dwAvailPageFile	Indicates the number of bytes available in the paging file.
dwTotalVirtual	Indicates the total number of bytes that Windows can describe in the user mode portion of the virtual address space of the calling process.
dwAvailVirtual	Indicates the number of unreserved and uncommitted memory bytes in the user mode portion of the virtual address space of the calling process.

***Table 1367** The members of the **MEMORYSTATUS** structure.*

An application can use the *GlobalMemoryStatus* function to determine how much memory it can allocate without severely impacting other applications. The information the *GlobalMemoryStatus* function returns is volatile, and there is no guarantee that two sequential calls to the *GlobalMemoryStatus* function will return the same information.

To better understand the processing the *GlobalMemoryStatus* function performs, consider the *Global_Mem_Status.cpp* program, contained on the CD-ROM that accompanies this book. The *Global_Mem_Status.cpp* program checks the current state of memory and returns that amount of memory to the program window.

1368 CREATING A HEAP WITHIN A PROCESS

In previous Tips, you have learned how to allocate memory to your programs from the global heap. However, as you have also learned, your programs will also allocate smaller blocks of memory from a local (private) heap. The *HeapCreate* function creates a heap object that the calling process can use. The function reserves a contiguous block in the virtual address space of the process and allocates virtual storage for a specified initial portion of the reserved block. You will use the *HeapCreate* function within your programs as shown in the following prototype:

```
HANDLE HeapCreate(
   DWORD flOptions,          // heap allocation flag
   DWORD dwInitialSize,      // initial heap size
   DWORD dwMaximumSize       // maximum heap size
);
```

The *flOptions* parameter specifies optional attributes (flags) for the new heap. These flags will affect subsequent access to the new heap through calls to the heap functions (*HeapAlloc, HeapFree, HeapReAlloc,* and *HeapSize*). You can specify one or more of the flags Table 1368 lists.

Flag	Description
HEAP_GENERATE_EXCEPTIONS	Specifies that the system will raise an exception to indicate a function failure, such as an out-of-memory condition, instead of returning *NULL*.
HEAP_NO_SERIALIZE	Specifies that the heap will not use mutual exclusion when the heap functions allocate and free memory from the heap. The default, occurring when you do not specify the *HEAP_NO_SERIALIZE* flag, is to serialize access to the heap. Serialization of heap access lets two or more threads simultaneously allocate and free memory from the same heap.

***Table 1368** The possible flags for the **flOptions** parameter.*

The *dwInitialSize* parameter specifies the initial size, in bytes, of the heap. The value of the *dwInitialSize* parameter determines the initial amount of physical storage that *HeapCreate* allocates for the heap. *HeapCreate* rounds the value up to the next page boundary. To determine a page's size on the host computer, use the *GetSystemInfo* function. If the *dwMaximumSize* parameter is a non-zero value, it will specify the maximum size, in bytes, of the heap. The *HeapCreate* function rounds *dwMaximumSize* up to the next page boundary, and then reserves a block of that size in the process's virtual address space for the heap. If the *HeapAlloc* or *HeapReAlloc* functions make allocation requests that exceed the initial amount of physical storage *dwInitialSize* specifies, the system will allocate additional pages of physical storage for the heap, up to the heap's maximum size.

In addition, if *dwMaximumSize* is non-zero, the heap will not be able to grow, and an absolute limitation will arise: the maximum size of a memory block in the heap will be a bit less than *0x0007FFF8* bytes (the size of the process's private address space). Requests to allocate larger blocks will fail, even if the maximum size of the heap is large enough to contain the block. If *dwMaximumSize* is zero, it will specify that the heap is able to grow. Only available memory limits the heap's size. Requests to allocate blocks larger than *0x0007FFF8* bytes do not automatically fail; the system calls *VirtualAlloc* to obtain the memory needed for such large blocks. Applications that must allocate large memory blocks should set *dwMaximumSize* to zero.

If the function succeeds, the return value will be a handle of the newly created heap. If the function fails, the return value will be *NULL*. To get extended error information, you must call *GetLastError*.

The *HeapCreate* function creates a private heap object from which the calling process can use the *HeapAlloc* function to allocate memory blocks. The initial size determines the number of committed pages that *HeapCreate* initially allocates for the heap. The maximum size determines the total number of reserved pages. These committed pages and the reserved pages together create a contiguous block in the process's virtual address space into which the heap can grow. If *HeapAlloc* makes requests that exceed the current size of committed pages, Windows automatically commits additional pages from the reserved pages, assuming that the physical storage is available.

Only the process that created the private heap object can access that memory stored within the private heap. If a dynamic-link library (DLL) creates a private heap, it creates the heap in the address space of the process that called it. Moreover, the heap is accessible only to that process. The system uses memory from the private heap to store heap support structures, so not all the specified heap size is available to the process. For example, if the *HeapAlloc* function requests 64Kb from a heap with a maximum size of 64Kb, the request may fail because of system overhead.

If you do not specify the *HEAP_NO_SERIALIZE* flag (the simple default), the heap will serialize access within the calling process. Serialization ensures mutual exclusion when two or more threads try to simultaneously allocate or free blocks from the same heap. There is a small performance cost to serialization (that is, Windows requires additional processing time), but your programs must use serialization whenever multiple threads allocate and free memory from the same heap.

Setting the *HEAP_NO_SERIALIZE* flag eliminates mutual exclusion on the heap. Without serialization, two or more threads that use the same heap handle might try to allocate or free memory simultaneously, which would likely cause corruption in the heap. Therefore, you can safely use the *HEAP_NO_SERIALIZE* flag only in the following situations:

- The process has only one thread.
- The process has multiple threads, but only one thread calls the heap functions for a specific heap.
- The process has multiple threads, and the application provides its own mechanism for mutual exclusion to a specific heap.

MANAGING PROCESS-SPECIFIC MEMORY 1369

As you have learned, your programs should allocate small amounts of memory from the heap that a specific process requires. Your programs will typically use the *HeapAlloc* function to allocate such memory. The *HeapAlloc* function allocates a block of memory from a heap. The memory you allocate with *HeapAlloc* is not moveable. You will use the *HeapAlloc* function within your programs as shown in the following prototype:

```
LPVOID HeapAlloc(
    HANDLE hHeap,       // handle to the private heap block
    DWORD dwFlags,      // heap allocation control flags
    DWORD dwBytes       // number of bytes to allocate
);
```

The *hHeap* parameter specifies the heap from which *HeapAlloc* will allocate the memory. The *hHeap* parameter is a handle that either the *HeapCreate* or *GetProcessHeap* function returns. The *dwFlags* parameter specifies several controllable aspects of heap allocation. Specifying any of these flags will override the corresponding flag you specified when you created the heap with *HeapCreate*. You can specify one or more of the flags for the *dwFlags* parameter, which Table 1369.1 details.

Flag	Meaning
HEAP_GENERATE_EXCEPTIONS	Specifies that the operating system will raise an exception to indicate a function failure, such as an out-of-memory condition, instead of returning *NULL*.
HEAP_NO_SERIALIZE	Specifies that the heap will not use mutual exclusion while the *HeapCreate* function is accessing the heap.
HEAP_ZERO_MEMORY	Specifies that Windows will initialize the allocated memory to zero.

Table 1369.1 The flags for the ***dwFlags*** *parameter.*

Finally, the *dwBytes* parameter specifies the number of bytes *HeapAlloc* will allocate. If the *hHeap* parameter specifies a "non-growable" heap, *dwBytes* must be less than *0x7FFF8*. You call the *HeapCreate* function with a non-zero value to create a "non-growable" heap. If the function succeeds, the return value will be a pointer to the allocated memory block. If the function fails and you have not specified *HEAP_GENERATE_EXCEPTIONS*, the return value will be *NULL*. If the function fails and you have specified *HEAP_GENERATE_EXCEPTIONS*, the function may generate an exception with the error values Table 1369.2 lists.

Value	Meaning
STATUS_NO_MEMORY	The allocation attempt failed because of a lack of available memory or heap corruption.
STATUS_ACCESS_VIOLATION	The allocation attempt failed because of heap corruption or improper function parameters.

Table 1369.2 The error values from a bad heap allocation.

Note that heap corruption can lead to either exception; it depends on the nature of the heap corruption. If *HeapAlloc* succeeds, it will allocate at least the amount of memory the calling program requests. If the actual amount that *HeapAlloc* allocates is greater than the amount the calling program requests, the process can use the entire amount. You can use the *HeapSize* function to determine the allocated block's actual size.

To free a block of memory *HeapAlloc* has allocated, use the *HeapFree* function. Memory *HeapAlloc* has allocated is not moveable. Because the memory is not moveable, it is possible for the heap to become fragmented. Note that if you do not specify *HEAP_ZERO_MEMORY*, Windows will not initialize the allocated memory to zero.

To better understand the processing *HeapAlloc* performs, consider the *Heap_Strings.cpp* program, contained on the companion CD-ROM that accompanies this book. The *Heap_Strings.cpp* program creates a heap, then uses *HeapAlloc* to allocate memory from that heap, which the program treats as a dynamic array of strings. When the user selects the *Allocate* menu item, the program allocates memory from the heap to store a new string as well as declaring the pointer to the memory that the previous allocation added to the array. If there is no room left in the array, the program will use *HeapReAlloc* to expand the array. When the user selects the *Free* menu item, the program frees the memory for the last allocated string. When the program detects that the user has released sufficient memory to leave a significant amount of unused space, the program reallocates the heap to shrink the array. Moreover, each time the *Heap_Strings.cpp* program reallocates the heap, it uses the *HeapCompact* function to compact the heap.

1370 CHECKING THE SIZE OF MEMORY ALLOCATED FROM A HEAP

As you have learned, your Windows programs will often allocate small amounts of local memory from a private heap. In previous Tips, you have created a heap and allocated memory from that heap. Your programs can also use functions such as *HeapReAlloc* to reallocate space from a heap, and *HeapFree* to free memory you allocate from a heap. Additionally, your programs should always use the *HeapDestroy* function to destroy private heaps they create. Often, however, during your program's execution, you will want to check the size of an allocation that you have made from the heap. Your programs can use the *HeapSize* function to check the size of a heap-allocated memory block. The *HeapSize* function returns the size, in bytes, of a memory block that the *HeapAlloc* or *HeapReAlloc* function has allocated from a heap. You will use the *HeapSize* function within your programs as shown in the following prototype:

```
DWORD HeapSize(
    HANDLE hHeap,        // handle to the heap
    DWORD dwFlags,       // heap size control flags
    LPCVOID lpMem        // pointer to memory to return size for
);
```

The *hHeap* parameter specifies the heap in which the memory block resides. Either the *HeapCreate* or *GetProcessHeap* function will return this handle. The *dwFlags* parameter specifies several controllable aspects of accessing the memory block. Currently, you can only specify the *HEAP_NO_SERIALIZE* flag; however, Windows reserves all other flag values for future use. Specifying the *HEAP_NO_SERIALIZE* flag will override the corresponding flag you specified in the *flOptions* parameter when you used the *HeapCreate* function to create the heap. Finally, the *lpMem* parameter points to the memory block whose size the function will obtain. This is a pointer that the *HeapAlloc* or *HeapReAlloc* function returns.

If the *HeapSize* function succeeds, the return value will be the size, in bytes, of the allocated memory block. If the *HeapSize* function fails, the return value will be *0xFFFFFFFF*.

To better understand the processing that the *HeapSize* function performs, consider the *Heap_Size.cpp* program, contained on the companion CD-ROM that accompanies this book. The *Heap_Size.cpp* program creates a heap and allocates a block of zero memory 20 bytes long. The *Heap_Size.cpp* program then calls the *HeapSize* function to display the size of the allocated block.

ALLOCATING A VIRTUAL MEMORY BLOCK 1371

As you have learned, your Windows programs will generally allocate memory using one of three allocation types—allocation from the global heap, allocation from the private heap, or direct allocation of virtual memory. As you have learned, using virtual memory provides your programs with additional options and control over the allocation process. However, as with other allocation types, you will still perform memory allocation with an *alloc* function. The *VirtualAlloc* function differs from other allocation functions in that it either reserves or commits a region of pages in the virtual address space of the calling process. Windows automatically initializes memory that your programs allocate using the *VirtualAlloc* function to zero. You will use the *VirtualAlloc* function within your programs as shown in the following prototype:

```
LPVOID VirtualAlloc(
   LPVOID lpAddress,                 // address of region to reserve or commit
   DWORD dwSize,                     // size of region
   DWORD flAllocationType,           // type of allocation
   DWORD flProtect                   // type of access protection
)
```

The *VirtualAlloc* function accepts the parameters that Table 1371.1 details.

Parameter	Description
lpAddress	Specifies the starting address of the region you want to allocate. If your program is reserving the memory, Windows rounds down the specified address to the next 64Kb boundary. If your program has already reserved the memory and is now calling *VirtualAlloc* to commit the memory, Windows rounds down to the next page boundary. To determine the size of a page on the host computer, use the *GetSystemInfo* function. If this parameter is *NULL*, the system will determine where to allocate the region.
dwSize	Specifies the size, in bytes, of the region. If the *lpAddress* parameter is *NULL*, Windows rounds up the *dwSize* value to the next page boundary. Otherwise, the allocated pages will include all pages that contain one or more bytes in the range from *lpAddress* to (*lpAddress+dwSize*). This means that a 2-byte range that straddles a page boundary will cause Windows to include both pages in the allocated region.
flAllocationType	Specifies the type of allocation. You can specify any combination of the flags Table 1371.2 details.
flProtect	Specifies the type of access protection. If you are using *VirtualAlloc* to commit the pages, you can specify any one of the flags Table 1371.3 details, along with the *PAGE_GUARD* and *PAGE_NOCACHE* protection modifier flags, as desired.

***Table 1371.1** The parameters for the **VirtualAlloc** function.*

As Table 1371.1 explains, you use the *flAllocationType* parameter to control the allocation that *VirtualAlloc* performs. You can specify any combination of the flags Table 1371.2 details to control virtual memory allocation.

Flag	Meaning
MEM_COMMIT	Allocates physical storage in memory or in the paging file on disk for the specified region of pages—in other words, protects part of the process's virtual address space from other allocation calls within the same process. Trying to commit an already committed page will not cause the function to fail. This means that your programs can commit a range of committed or decommitted pages without having to worry about a failure.
MEM_RESERVE	Reserves a range of the process's virtual address space without allocating any physical storage. Any other allocation operations (such as the *malloc* function, or the *GlobalAlloc* function) cannot use the reserved range until your progam releases the range. Your progam can commit reserved pages in subsequent calls to the *VirtualAlloc* function.
MEM_TOP_DOWN	Allocates memory at the highest possible address.

***Table 1371.2** The possible allocation types for the **VirtualAlloc** function.*

Because of the nature of virtual page allocation, you can control the access to virtual pages that you commit with the *VirtualAlloc* function. As Table 1371.1 notes, you can specify a single type of page security, together with the *PAGE_GUARD* and *PAGE_NOCACHE* modifiers. Table 1371.3 lists the security flags you can use when allocating virtual pages.

Flag	Meaning
PAGE_READONLY	Enables read access only to the committed page region. Trying to write to the read-only pages will result in an access violation. If the system differentiates between read-only access and execute access, trying to execute code in the committed region will result in an access violation.

***Table 1371.3** The possible security flags for virtual page allocations. (continued on following page)*

Flag	Meaning
PAGE_READWRITE	Enables both read and write access to the committed page region.
PAGE_EXECUTE	Enables execute access only to the committed page region. Trying to read or write to the execute-only pages will result in an access violation.
PAGE_EXECUTE_READ	Enables execute and read access only to the committed page region. Trying to write to the read- and execute-only pages will result in an access violation.
PAGE_EXECUTE_READWRITE	Enables execute, read, and write access to the committed page region.
PAGE_GUARD	Pages in the region become guard pages. If a program tries to read from or write to a guard page, the guard page will cause the operating system to raise a *STATUS_GUARD_PAGE* exception and turn off the guard page status. Guard pages therefore act as a one-shot access alarm. The *PAGE_GUARD* flag is a page protection modifier. An application uses it with one of the other page protection flags, with one exception: an application cannot use it with *PAGE_NOACCESS*. After a failed read or write action leads the operating system to turn off guard page status, the underlying page protection takes over. If a guard page exception occurs during a system service, the service will typically return a failure status indicator. Tip 1372 explains guard pages in detail.
PAGE_NOACCESS	Disables all access to the committed page region. Trying to read from, write to, or execute in the disabled-access pages will result in an access violation exception, called a general protection (GP) fault.
PAGE_NOCACHE	Permits no caching of the pages previously allocated with the *MEM_COMMIT* flag. You should specify the hardware attributes for the physical memory as "no cache." Microsoft does not recommend this flag for general usage. It is useful for device drivers (for example, mapping a video frame buffer) which use no caching. This flag is a page protection modifier, only valid when you use it with one of the page protections other than *PAGE_NOACCESS*.

***Table 1371.3** The possible security flags for virtual page allocations. (continued from previous page)*

If the *VirtualAlloc* function succeeds, the return value will be the base address of the allocated region of pages. If the function fails, the return value will be *NULL*.

VirtualAlloc can perform the following operations:

- Commit a region of pages a previous call to the *VirtualAlloc* function has reserved.
- Reserve a region of free pages.
- Reserve and commit a region of free pages.

You can use *VirtualAlloc* to reserve a block of pages and then make additional calls to *VirtualAlloc* to commit individual pages from the reserved block. Reserving a block of pages enables a process to reserve a range of its virtual address space without consuming physical storage until you need it.

Each page in the process's virtual address space is in one of the three states Table 1371.4 details.

State	Meaning
Free	The process has not committed or reserved this page and so it is not accessible to the process. *VirtualAlloc* can reserve, or simultaneously reserve and commit, a free page.
Reserved	Other allocation functions cannot use the range of addresses, but the process cannot access the page and Windows associates no physical storage with the page. *VirtualAlloc* can commit a reserved page, but it cannot reserve it a second time. The *VirtualFree* function can release a reserved page, making it a free page.
Committed	Windows has allocated storage for the page, and a protection code controls access. The system initializes and loads each committed page into physical memory only at the first try to read or write to that page. When the process terminates, the system will release the storage for committed pages. *VirtualAlloc* can commit an already committed page. This means that you can commit a range of pages, regardless of whether you have already committed them, and the function will not fail. *VirtualFree* can decommit a committed page, releasing the page's storage, or it can simultaneously decommit and release a committed page.

***Table** 1371.4 The possible states for virtual memory.*

If the *lpAddress* parameter is not *NULL*, the function will use the *lpAddress* and *dwSize* parameters to compute the region of pages *VirtualAlloc* will allocate. The current state of the entire range of pages must be compatible with the type of allocation the *flAllocationType* parameter specifies. Otherwise, the function will fail and *VirtualAlloc* will allocate none of the pages. This compatibility requirement does not preclude committing an already committed page (see Table 1371.4).

To better understand the processing that *VirtualAlloc* performs, consider the program *Virtual_Allocate*, contained on the CD-ROM that accompanies this book. The *Virtual_Allocate* program reserves 1 megabyte (1Mb) of virtual memory when the program sends the *WM_CREATE* message. When the user selects the *Demo* menu item, the program commits and uses 70Kb of virtual memory. First, the program places values in each Kb block of the allocated memory. Second, the program changes the access of the entire committed memory block to read-only. Third, the program accesses a value in the memory and displays it within a message box. Finally, the program tries to set a value into memory, which causes a protection fault. The *Virtual_Allocate* program uses a *try-catch* block to catch the protection fault.

1372 Understanding Guard Pages

As you learned in Tip 1371, an application sets a memory page's *PAGE_GUARD* page protection modifier flag to establish a *guard page*. You can specify this flag, along with other page protection flags, in the functions *VirtualAlloc, VirtualProtect,* and *VirtualProtectEx*. You can use the *PAGE_GUARD* flag with any other page protection flag, except for the *NO_ACCESS* flag.

If a program tries to access an address within a guard page, the operating system will raise a *STATUS_GUARD_PAGE (0x80000001)* exception. The operating system will also clear the *PAGE_GUARD* flag, removing the memory page's guard page status. The system will not stop the next try to access the memory page with a *STATUS_GUARD_PAGE* exception.

If a guard page exception occurs during a system service, the service will fail and typically return some failure status indicator. Because the system also removes the relevant memory page's guard page status, the next invocation of the same system service will not fail because of a *STATUS_GUARD_PAGE* exception (unless, of course, someone reestablishes the guard page).

A guard page therefore provides a one-shot alarm for memory page access. This can be useful for an application that must monitor the growth of large dynamic data structures. For example, some operating systems use guard pages to implement automatic stack checking.

The CD-ROM that accompanies this book includes the *Guard_Page.cpp* program, which illustrates the one-shot behavior of guard page protection, and how it can cause a system service to fail (the program generates output to a DOS window). When you compile and execute the *Guard_Page.cpp* program, it will display the following output on your screen:

```
Committed 512 bytes at address 003F0000
Cannot lock at 003F0000, error = 0x80000001
2nd Lock Achieved at 003F0000
C:\>
```

Note that the first try to lock the memory block fails, raising a *STATUS_GUARD_PAGE* exception. The second try succeeds because the first try has toggled off the memory block's guard page protection.

BETTER UNDERSTANDING VIRTUAL MEMORY BLOCKS 1373

As you have learned, virtual memory offers you an additional, effective means of allocating large blocks of memory. However, visualizing the programmatical benefits of virtual memory is often difficult. One of the easiest ways to see the benefits of virtual memory is to consider a large array of a complex structure, perhaps a spreadsheet-type structure. If you were to create a two-dimensional array, the definition would look similar to the following:

```
Cell LARGEARRAY[200][256];
```

If the size of the *Cell* structure were 128 bytes, it would require 6,533,600 (200 x 256 x 128) bytes of physical storage. Clearly, that is a significant amount of physical memory to allocate for a spreadsheet—particularly when most spreadsheets will not use anywhere near that many cells.

Alternatively, you could use a linked list to create the spreadsheet-type structure. With the linked-list approach, you only have to create *Cell* structures for the cells in the spreadsheet that actually contain data. Because most cells in a spreadsheet go unused, the linked list method saves a significant amount of storage. However, using a linked-list makes it much more difficult to obtain a cell's contents. For example, if your program required the contents of the cell in row 5, column 10, it must first walk through the linked list to find the cell—thereby slowing down processing.

Virtual memory offers a compromise between declaring the two-dimensional matrix up front and implementing complex, extended linked lists. With virtual memory, you get the fast, easy access of the matrix technique, and the superior storage capability of the linked list. To best realize the benefits of the virtual memory technique, your program would do the following:

1. Reserve a region large enough to contain the entire matrix of *Cell* structures. As you have learned, reserving a region uses no physical memory.
2. Locate the memory address in the reserved region where the *Cell* structure should go when a user enters data into a cell.
3. Commit just enough physical storage to the memory address you located in Step 2 for a *Cell* structure.
4. Set the members of the new *Cell* structure.

After you have mapped physical storage to the proper location, your program can access the storage without creating an access violation. Clearly, the virtual memory technique is a significant improvement over the other techniques because the program commits physical storage only as the user enters data into the spreadsheet's cells. Because most of the cells in the spreadsheet are empty, the program will not use most of the reserved region. You can use the virtual memory allocation to provide your programs with fast access to a great number of members without many of the sacrifices you made with earlier memory models.

1374 FREEING UP VIRTUAL MEMORY

As you have learned, your programs can use the *VirtualAlloc* function to reserve or commit virtual memory pages. After you have reserved or committed virtual memory pages, your programs can use the *VirtualFree* function to release or decommit those pages. The *VirtualFree* function releases or decommits (or both) a region of pages within the virtual address space of the calling process. You will use the *VirtualFree* function as shown in the following prototype:

```
BOOL VirtualFree(
   LPVOID lpAddress,    // address of region of committed pages
   DWORD dwSize,        // size of region
   DWORD dwFreeType     // type of free operation
);
```

The *lpAddress* parameter points to the base address of the region of pages *VirtualFree* is to free. If the *dwFreeType* parameter includes the *MEM_RELEASE* flag, the *lpAddress* parameter must be the base address the *VirtualAlloc* function returned when you reserved the region of pages. The *dwSize* parameter specifies the size, in bytes, of the region *VirtualFree* is to free. If the *dwFreeType* parameter includes the *MEM_RELEASE* flag, the *dwSize* parameter must be zero. Otherwise, the region of affected pages will include all pages that contain one or more bytes in the range from the *lpAddress* parameter to *(lpAddress+dwSize)*. This means that a 2-byte range that straddles a page boundary will cause *VirtualFree* to free both pages. The *dwFreeType* specifies the type of free operation. When you call the *VirtualFree* function, you will use one, but not both, of the flags in Table 1374.

Flag	Meaning
MEM_DECOMMIT	Decommits the specified region of committed pages. Trying to decommit an uncommitted page will not cause the function to fail. This means that your programs can decommit a range of previously committed or uncommitted pages without having to worry about a failure.
MEM_RELEASE	Releases the specified region of reserved pages. If your program specifies this flag, the *dwSize* parameter must be zero or the function fails.

***Table 1374** The flags for the **dwFreeType** parameter.*

Your programs can use the *VirtualFree* function to perform one of the following operations:

- Decommit a region of committed or uncommitted pages
- Release a region of reserved pages
- Decommit and release a region of committed or uncommitted pages

To release a region of pages, the entire range of pages must be in the same state (all reserved or all committed) and the *VirtualFree* function must release the entire originally reserved region at the same time. If you have previously used *VirtualAlloc* to commit only some of the pages in the original reserved region, you must first call *VirtualFree* to decommit the committed pages and then call *VirtualFree* again to release the entire block.

The pages that your programs used *VirtualFree* to release are free pages available for subsequent allocation operations. Trying to read from or write to a free page will result in an access violation exception. *VirtualFree* can decommit an uncommitted page, which means that *VirtualFree* can decommit a range of committed or uncommitted pages without having to worry about a failure. Decommitting a page releases its physical storage, either in memory or in the paging file on disk. If your program decommits a page but does not release the page, that page's state changes to reserved, and a subsequent call to *VirtualAlloc* can commit the page again. Trying to read from or write to a reserved page results in an access violation exception.

The current state of the entire range of pages must be compatible with the type of free operation the *dwFreeType* parameter specifies. Otherwise, the *VirtualFree* function will fail and and it will not release or decommit any pages.

MANAGING VIRTUAL MEMORY PAGES 1375

As you have learned, your programs will use virtual memory pages, both to expand your program's memory access and to help your programs allocate large blocks of memory within the virtual memory space. Often, you will require information about a range of pages within the virtual address space of a process. The *VirtualQuery* function provides information about a range of pages in the virtual address space of the calling process, as shown here:

```
DWORD VirtualQuery(
    LPCVOID lpAddress,                      // address of region
    PMEMORY_BASIC_INFORMATION lpBuffer,     // address of information buffer
    DWORD dwLength                          // size of buffer
);
```

The *VirtualQuery* function accepts the three parameters shown in Table 1375.

Parameter	Description
lpAddress	Points to thc base address of the region of pages *VirtualQuery* is to query. Windows rounds this value down to the next page boundary. To determine the size of a page on the host computer, use the *GetSystemInfo* function.
lpBuffer	Points to a *MEMORY_BASIC_INFORMATION* structure in which *VirtualQuery* returns information about the specified page range.
dwLength	Specifies the size, in bytes, of the buffer to which the *lpBuffer* parameter points.

***Table 1375** The parameters for the **VirtualQuery** function.*

VirtualQuery provides information about a region of consecutive pages beginning at a specified address that share the following attributes:

- The state of all pages is the same with the *MEM_COMMIT, MEM_RESERVE, MEM_FREE, MEM_PRIVATE, MEM_MAPPED*, or *MEM_IMAGE* flag.
- If the initial page is not free, all pages in the region will be part of the same initial allocation of pages that a call to the *VirtualAlloc* function reserves.
- The access of all pages is the same with the *PAGE_READONLY, PAGE_READWRITE, PAGE_NOACCESS, PAGE_WRITECOPY, PAGE_EXECUTE, PAGE_EXECUTE_READ, PAGE_EXECUTE_READWRITE, PAGE_EXECUTE_WRITECOPY, PAGE_GUARD*, or *PAGE_NOCACHE* flags.

The *VirtualAlloc* function determines the attributes of the first page in the region and then scans subsequent pages until it scans the entire range of pages or until it encounters a page with a non-matching set of attributes. *VirtualAlloc* returns the attributes and the size, in bytes, of the region of pages with matching attributes. For example, if there is a 40Mb region of free memory, and your program calls *VirtualQuery* on a page that is 10Mb into the region, the function will obtain a state of *MEM_FREE* and a size of 30Mb.

The *VirtualQuery* function reports on a region of pages in the memory of the calling process, and the *VirtualQueryEx* function reports on a region of pages in the memory of a specified process. The CD-ROM that accompanies this book includes the *Virtual_Query.cpp* program, which first allocates a 70Kb block of virtual memory. The program then invokes the *VirtualQuery* function, which returns the region size the memory actually occupies. Note that the region size is divisible by 4096 bytes (4Kb), which is the page size of all x86 computers.

1376 BETTER UNDERSTANDING PROCESSES

As you have learned, one of the most powerful Windows features is its support of multitasking, or the running of multiple processes in memory simultaneously. Throughout the next thirty Tips, you will learn more about process and thread management, one of the Windows programmer's most important skills.

A *process* is the object that owns all the resources for an application. A Windows process, in turn, can create one or more threads. A *thread* is an independent path of execution within a process that shares its address space, code, and global data.

Each thread has its own set of registers, its own stack, and its own input mechanisms, including a private message queue. Windows 95 and Windows NT allocate CPU time slices on a thread-by-thread basis and perform preemptive multitasking (in other words, move threads in front of each other) based on the priority assigned to each thread.

Additionally, a process will include global memory allocations, virtual pages, and so on. Figure 1376.1 shows a logical model of the relationship between threads and processes.

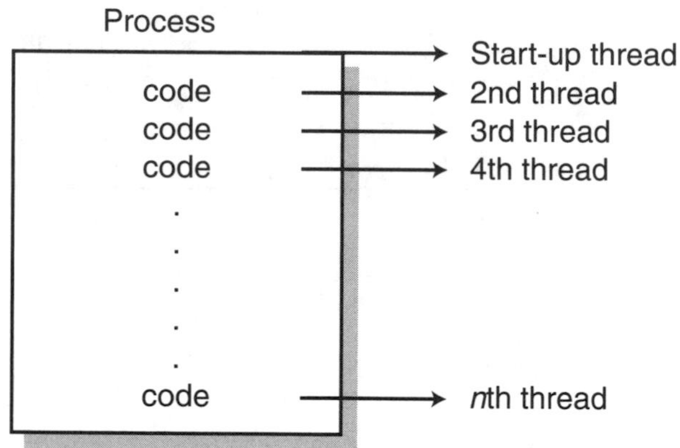

Figure 1376.1 The relationship between threads and processes.

As you have learned, Windows allocates virtual memory to processes as they need it. Therefore, when you consider the memory model for a computer running multiple processes simultaneously, it is important to identify which process is currently active. Determining the currently active process is important because Windows will automatically attach higher priority to most CPU requests that the currently active process performs.

Tip 1400 discusses the operating system's management of thread priority in detail. Additionally, Windows will typically allot additional physical memory to speed the processing of the currently active process. Figure 1376.2 shows a sample memory model for three simultaneous applications, with the *First_App* application currently active.

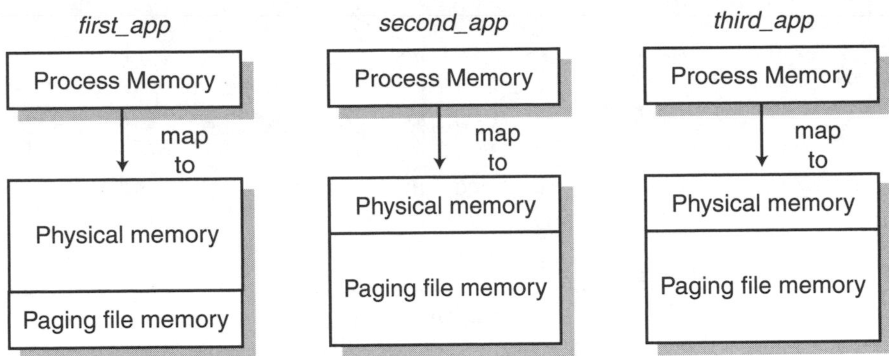

Figure 1376.2 A sample memory model for three simultaneously executing applications.

However, in the event the user or Windows makes the *Second_App* program the active program, Windows will reallocate physical memory in a manner that frees more space for the *Second_App* program's execution, as shown in Figure 1376.3.

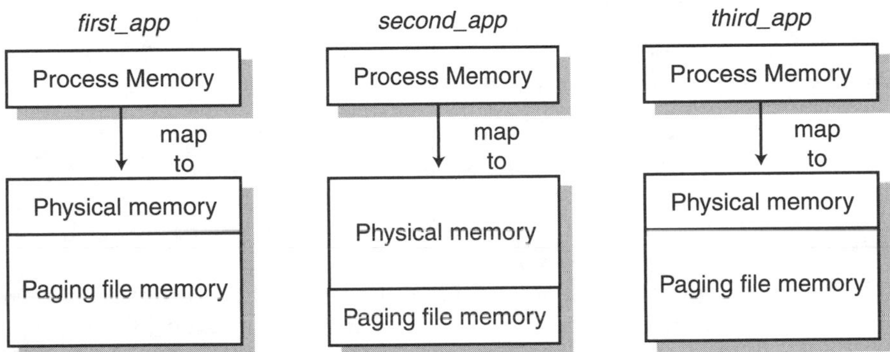

Figure 1376.3 Another sample memory model for three simultaneously executing applications.

In the next five Tips, you will learn the basics of creating and managing processes. In subsequent Tips you will learn the basics of creating and managing threads. Be sure you understand the basics behind processes (which are essentially containers for threads) before you begin to work with threads.

CREATING A PROCESS 1377

As you have learned, your programs generally will run within a single process. That process may contain one or more threads. The *CreateProcess* function creates a new process and its primary thread (often known as a *child process*). The new process executes the executable file you specify. *CreateProcess* lets the parent process (that is, the process that calls the *CreateProcess* function) specify the operating environment of the new process, including its working directory, how it should appear on the screen by default, its environment variables, and its priority. Windows passes the command line and its contents to the child process. You will use the *CreateProcess* function as shown in the following prototype:

```
BOOL CreateProcess(
   LPCTSTR lpApplicationName,                  // name of executable module
   LPTSTR lpCommandLine,                       // command line string
   LPSECURITY_ATTRIBUTES lpProcessAttributes,  // process security attributes
   LPSECURITY_ATTRIBUTES lpThreadAttributes,   // thread security attributes
   BOOL bInheritHandles,                       // handle inheritance flag
```

```
    DWORD dwCreationFlags,                      // creation flags
    LPVOID lpEnvironment,                       // pointer to new environment block
    LPCTSTR lpCurrentDirectory,                 // pointer to current directory name
    LPSTARTUPINFO lpStartupInfo,                // pointer to STARTUPINFO
    LPPROCESS_INFORMATION lpProcessInformation  // pointer to PROCESS_INFORMATION
);
```

As you can see, *CreateProcess* is a heavily parameterized function. Table 1377.1 explains the parameters for the *CreateProcess* function.

Parameters	Description
lpApplicationName	Pointer to a *NULL*-terminated string that specifies the module to execute. The string can specify the full path and filename of the module to execute or the string can specify a partial name. When the string only specifies a partial name, the function uses the current drive and current directory to complete the specification. The *lpApplicationName* parameter can be *NULL*, in which case the module name must be the first whitespace-delimited token in the *lpCommandLine* string. The specified module can be a Win32-based application or it can be some other type of module (for example, MS-DOS or OS/2) if the local computer has the appropriate subsystem available.

Note: *Under Windows NT, if the executable module is a 16-bit application,* ***lpApplicationName*** *should be* ***NULL****, and the string the* ***lpCommandLine*** *points to should specify the executable module.*

Parameters	Description
lpCommandLine	Pointer to a *NULL*-terminated string that specifies the command line to execute. The *lpCommandLine* parameter can be *NULL,* in which case the function uses the string the *lpApplicationName* points to as the command line. If both *lpApplicationName* and *lpCommandLine* are non-*NULL*, **lpApplicationName* specifies the module to execute, and **lpCommandLine* specifies the command line. The new process can use *GetCommandLine* to retrieve the entire command line. Additionally, C run-time processes can use the *argc* and *argv* arguments. If *lpApplicationName* is *NULL*, the first whitespace-delimited token of the command line specifies the module name. If the filename does not contain an extension, Windows assumes *EXE*. If the filename ends in a period (.) with no extension, or the filename contains a path, Windows does not append *EXE*. If the filename does not contain a directory path, Windows searches for the executable file in the following sequence: 1. The directory from which the application loaded. 2. The current directory for the parent process. 3. Windows 95: The Windows system directory. Use the *GetSystemDirectory* function to get the path of this directory. 4. Windows NT: The 32-bit Windows system directory. Use the *GetSystemDirectory* function to get the path of this directory. The directory's name is generally *SYSTEM32*. 5. Windows NT: The 16-bit Windows system directory. There is no Win32 function that obtains the path of this directory, but the function searches it. The directory's name is generally *SYSTEM*.

Table 1377.1 *The parameters for the* ***CreateProcess*** *function.* *(continued on following page)*

Parameters	Description
	6. The Windows directory. Use the *GetWindowsDirectory* function to get the path of this directory. The directory's name is generally *Windows*. 7. The directories that Windows lists in the PATH environment variable. If the process you want *CreateProcess* to create is an MS-DOS-based or Windows-based application, *lpCommandLine* should be a full command line in which the first element is the application name. Because the *CreateProcess* function also works well for Win32-based applications, you should set the *lpCommandLine* parameter to a full command line for Win32 programs as well.
lpProcessAttributes	Pointer to a *SECURITY_ATTRIBUTES* structure that determines whether the child processes can inherit the returned handle. If *lpProcessAttributes* is *NULL*, the child processes cannot inherit the handle.

Note: *Under Windows NT, the **lpSecurityDescriptor** member of the structure specifies a security descriptor for the new process. If **lpProcessAttributes** is **NULL**, the process gets a default security descriptor. Under Windows 95, the **CreateProcess** function ignores the **lpSecurityDescriptor** member of the structure.*

Parameters	Description
lpThreadAttributes	Pointer to a *SECURITY_ATTRIBUTES* structure that determines whether the child processes can inherit the returned handle. If *lpThreadAttributes* is *NULL*, the child processes cannot inherit the handle.

Note: *Under Windows NT, the **lpSecurityDescriptor** member of the structure specifies a security descriptor for the main thread. If **lpThreadAttributes** is **NULL**, the thread gets a default security descriptor. Under Windows 95, the **CreateProcess** function ignores the **lpSecurityDescriptor** member of the structure.*

Parameters	Description
bInheritHandles	Indicates whether the new process inherits handles from the calling process. If *True*, the new process inherits each inheritable open handle in the calling process. Inherited handles have the same value and access privileges as the original handles.
dwCreationFlags	Specifies additional flags that control the priority class and the creation of the process. Your programs can specify the creation flags in any combination of those shown in Table 1377.2, except as noted. The *dwCreationFlags* parameter also controls the new process's priority class, which Windows uses to determine the scheduling priorities of the process's threads. If Windows does not specify any of the priority class flags in Table 1377.2, the priority class defaults to *NORMAL_PRIORITY_CLASS*. However, if the priority class of the creating process is *IDLE_PRIORITY_CLASS*, the default priority class of the child process is also *IDLE_PRIORITY_CLASS*. Your programs can specify any one of the priority flags Table 1377.3 details.
lpEnvironment	Points to an environment block for the new process. If this parameter is *NULL*, the new process uses the environment of the calling process. An environment block consists of a *NULL*-terminated block of *NULL*-terminated strings. Each string is in the form *name=value*. Because the string uses the equal sign as a separator, your program must not use it in the name of an environment variable. If an application provides an environment block, rather than passing *NULL* for this parameter, Windows does not automatically propagate the current directory information of the system drives to the new process.

***Table 1377.1** The parameters for the **CreateProcess** function.* *(continued on following page)*

Flag	Meaning
HIGH_PRIORITY_CLASS	Indicates a process that performs time-critical tasks that Windows must execute immediately for the process to run correctly. The threads of a high-priority class process preempt the threads of normal-priority or idle-priority class processes. An example is the Windows Task List, which must respond quickly when the user calls it, regardless of the load on the operating system. You must take extreme care when using the high-priority class, because a high-priority class CPU-bound application can use nearly all available cycles.
IDLE_PRIORITY_CLASS	Indicates a process whose threads run only when the system is idle and the threads of any process running in a higher-priority class preempt it. An example is a screen saver. Child processes inherit the idle-priority class.
NORMAL_PRIORITY_CLASS	Indicates a normal process with no special scheduling needs.
REALTIME_PRIORITY_CLASS	Indicates a process that has the highest possible priority. The threads of a real-time priority class process preempt the threads of all other processes, including operating system processes performing important tasks. For example, a real-time process that executes for longer than a very brief interval can cause disk caches not to flush or cause the mouse to be unresponsive.

Table 1377.3 *The possible priority flags for the* ***CreateProcess*** *function.*

In addition to creating a process, *CreateProcess* also creates a thread object. The *CreateProcess* function creates the thread with an initial stack whose size it describes in the image header of the specified program's executable file. The thread begins execution at the image's entry point. *CreateProcess* creates the new process and the new thread handles with full access rights. For either handle, if the function does not provide a security descriptor, your program can use the handle in any function that requires an object handle of that type. When the function provides a security descriptor, your program performs an access check on all subsequent uses of the handle before it grants access. If the access check denies access, the requesting process is not able to use the handle to gain access to the thread.

Windows assigns the process a 32-bit process identifier. The identifier is valid until the process terminates. Your program can use it to identify the process or to specify the *OpenProcess* function to open a handle to the process. Windows also assigns a 32-bit thread identifier to the initial thread in the process. The identifier is valid until the thread terminates and your program can use it to uniquely identify the thread within the system. These identifiers are returned in the *PROCESS_INFORMATION* structure.

When you specify an application name in the *lpApplicationName* or *lpCommandLine* strings, it does not matter whether the application name includes the filename extension, with one exception: an MS-DOS-based or Windows-based application whose filename extension is .COM must include the .COM extension. The calling thread can use the *WaitForInputIdle* function to wait until the new process finishes its initialization and is waiting for user input with no input pending. Using *WaitForInputIdle* can be useful for synchronization between parent and child processes because *CreateProcess* returns without waiting for the new process to finish its initialization. For example, the creating process should use *WaitForInputIdle* before trying to find a window associated with the new process.

The Microsoft-recommended way to shut down a process is to use the *ExitProcess* function, because this function notifies all dynamic-link libraries (DLLs) attached to the process of the approaching termination. Other means for you to shut down a process do not notify the attached dynamic-link libraries. Note that when a thread calls *ExitProcess*, the function terminates other process threads without giving them an opportunity to execute any additional code (including the thread termination code of attached dynamic-link libraries).

Windows serializes *ExitProcess, ExitThread, CreateThread, CreateRemoteThread*, and a process that is starting (as the result of a *CreateProcess* call) between each other within a process. These events can only happen one at a time in an address space, which means the following restrictions apply:

- During process start-up and dynamic-link library initialization routines, your program can create new threads, but they do not begin execution until Windows finishes the dynamic-link library initialization for the process.
- Only one thread in a process can be in a dynamic-link library initialization or detach routine at a time.
- The *ExitProcess* function does not return until no threads are in their dynamic-link library initialization or detach routines.

The created process remains in the system until all threads within the process have terminated and calls to the *CloseHandle* function have closed all handles to the process and any of its threads. Calls to *CloseHandle* must close the handles for both the process and the main thread. If your program does not need these handles, it is best to close them immediately after Windows creates the process.

When the last thread in a process terminates, the following events occur:

- Windows implicitly closes all objects the process opened.
- The process's termination status (which *GetExitCodeProcess* returns) changes from its initial value of *STILL_ACTIVE* to the termination status of the last thread that terminates.
- Windows sets the thread object of the main thread to the signaled state, satisfying any threads that were waiting on the object.
- Windows sets the process object to the signaled state, satisfying any threads that were waiting on the object.

If the current directory on drive C is *\CBIBLE\BIBLE*, there is an environment variable called *=C*, whose value is *C:\CBIBLE\BIBLE*. As noted in the previous description of *lpEnvironment*, such current directory information for a system's drives does not automatically propagate to a new process when the *CreateProcess* function's *lpEnvironment* parameter is non-*NULL*. An application must manually pass the current directory information to the new process. To do so, the application must explicitly create the *=X* environment variable strings, alphabetize them (because Windows NT and Windows 95 use a sorted environment), and then put them into the environment block the *lpEnvironment* function specifies. Typically, the *=X* environment variable strings will go at the front of the environment block, due to the previously mentioned environment block sorting.

One way to obtain the current directory variable for a drive X is to call the *GetFullPathName* function, *("X:",. .)*, which lets an application avoid having to scan the environment block. If the full path *GetFullPathName* returns is *X:*, your program does not need to pass that value on as environment data, because the root directory is the default current directory for drive X of a new process. The *CreateProcess* function returns the handle that has *PROCESS_ALL_ACCESS* access to the process object.

The current directory the *lpCurrentDirectory* parameter specifies is the current directory for the child process. The current directory in the *lpCommandLine* parameter is the current directory for the parent process.

Note: *Under Windows NT, when a function creates a process with the* ***CREATE_NEW_PROCESS_GROUP*** *priority flag specified, Windows NT makes an implicit call to* ***SetConsoleCtrlHandler(NULL, True)*** *on behalf of the new process; this means that the new process has the CTRL+C disabled signal. This lets good shells handle CTRL+C themselves and selectively pass that signal on to subprocesses. CTRL+BREAK is not a disabled signal, and Windows NT may use it to interrupt the process and process group.*

1378 Terminating Processes

As you have learned, your programs will execute within a process. In Tip 1377, you learned how to create a process. Just as with most allocations or creations your programs perform, it is your responsibility to exit that process when it completes its functions. You will use the *ExitProcess* function to close a currently executing process. The *ExitProcess* function ends a process and all its threads and returns to the calling location, as shown here:

```
VOID ExitProcess(
   UINT uExitCode// exit code for all threads
);
```

The *uExitCode* parameter specifies the exit code for the process, and for all the threads the process terminates as a result of the *ExitProcess* call. You use the *GetExitCodeProcess* function to retrieve the process's exit value and the *GetExitCodeThread* function to retrieve a thread's exit value.

You should always call *ExitProcess* to end a process. The *ExitProcess* function provides a clean process shut-down, which includes calling the entry-point function of all attached dynamic-link libraries (DLLs) with a value indicating that the process is detaching from the dynamic-link library. If a process calls *TerminateProcess* to terminate, the process does not notify the dynamic-link libraries that Windows may have attached to the process.

After all attached dynamic-link libraries have executed any process termination value, the *ExitProcess* function terminates the current process. Terminating a process has the following effects:

1. The *ExitProcess* function closes all the object handles the process opened.
2. All the threads in the process terminate their execution.
3. The state of the process object becomes signaled, which resumes any threads that were waiting for the process to terminate.
4. The states of all process threads become signaled, which resumes any threads that were waiting for any of the threads comprising the process to terminate.
5. The termination status of the process changes from *STILL_ACTIVE* to the exit value of the process.

Terminating a process does not cause Windows to terminate child processes. Terminating a process does not necessarily remove the process object from the operating system. The system deletes a process object when the program closes the last handle to the process. Windows serializes the *ExitProcess, ExitThread, CreateThread,* and *CreateRemoveThread* functions. Additionally, Windows serializes a process that is starting (as the result of a *CreateProcess* call) with the previously named processes within the calling process. Only one of these events can happen in an address space at a time, which means the process maintains the following restrictions:

- During process start-up and dynamic-link library initialization routines, the program can create new threads, but they do not begin execution until dynamic-link library initialization is done for the process.
- Only one thread in a process can be in a dynamic-link library initialization or detachment routine at a time.
- *ExitProcess* does not return until none of the threads are in their dynamic-link library initialization or detach routines.

1379 Spawning Child Processes

As you have learned, your programs can use the *CreateProcess* command to begin executing another process. When you design more complex applications, you will often encounter situations in which you want another block of code to perform work. Within your programs, you can call a function to perform such work. However, functions are serial in nature—that is, your code cannot continue its processing until after a function completes its processing.

An alternative method for having another block of code perform work within your programs is to create a new thread within your process and have the new thread help with the processing. Using multiple threads lets your program code continue processing while the new thread performs the requested work. Unfortunately, using multiple threads will often cause synchronization problems when your thread must see the results from the new thread. You will learn more about synchronization in later Tips.

The other approach is to spawn off a new process (called a *child process*) to help with the work. Working with processes lets your programs continue processing while the child process works on a specific problem, or lets your programs pause while a child process works on the specific problem. In the next few Tips, you will work with child processes and with threads. Working with both will help you better understand the differences between the two, and how you can use both within your programs.

As your Windows programs become more complex, you will work more and more with unrelated, shared components. These components may be, for example, dynamic-link libraries, which execute within the program's current process (*in-process objects*), or object-linking and embedding (OLE) automation servers, which execute outside the program's current process (*out-of-process objects*). This book will not discuss in-process or out-of-process objects in detail.

WORKING MORE WITH CHILD PROCESSES 1380

As you have learned, your programs can control the way they interact with child processes. However, almost all child processes require access to data contained in the address space of the parent process. In general, when a child process requires access to data in the parent process, you should run the child process in its own address space and simply give the child process access to the relevant data in the parent process's address space. Controlling access to the parent process's address space lets you protect data not relevant to the child process from inadvertent corruption. Win32 gives you several methods for transferring data between different processes: dynamic-data exchange (DDE), object-linking and embedding (OLE), pipes, mailslots, and so on. One of the most convenient (and simplest) ways to share data is to use a *memory-mapped file*.

A memory-mapped file is a special file type that lets you reserve a region of address space and commit physical storage to the region, similar to virtual-memory allocation. Unlike virtual memory, however, the memory-mapped file's physical storage comes from a file that is already on the disk, rather than the system's paging file. After you map the file, you can access the entire file as though your program had loaded it into memory.

You will use memory-mapped files for the following three purposes:

- The system uses memory-mapped files to load and execute executable and dynamic-link library files. Using memory-mapped files greatly conserves both paging file space and the time an application requires to begin executing.
- You can use memory-mapped files to access a data file on the disk. Using a memory-mapped file shelters you from performing file I/O operations on a file and buffering the file's contents.
- You can use memory-mapped files to let multiple processes running on the same machine share data with each other, as you have already learned.

Many Win32 communications objects will use the memory-mapped file structure because it is powerful and easy to use. You will learn more about how to create a memory-mapped file in later Tips.

If you want to create a new process and have it do some work while the parent program waits for the result (for example, write some data to a file that the parent process will later read), the parent process can use code similar to the following:

```
PROCESS_INFORMATION  pi;
DWORD dwExitCode;

BOOL fSuccess = CreateProcess(ProcessName, &pi):
if (fSuccess)
 {
   // Close the thread handle as soon as you no longer need it
   CloseHandle(pi.hThread);
   WaitForSingleObject(pi.hProcess, INFINITE);

   // The process terminated
   GetExitCodeProcess(pi.hProcess, &dwExitCode);

   // Close the process handle
   CloseHandle(pi.hProcess);
 }
```

The code fragment creates the new *ProcessName* process. If successful, the fragment will close the extra thread handle to free CPU time, then wait for the process to complete its processing. After the process completes, the *dwExitCode* variable contains the process's exit code information, and the code fragment closes the process handle. If unsuccessful at launching the process, the code fragment will perform no processing.

1381 Running Detached Child Processes

In Tip 1380, you learned how to create a child process and halt the current thread's execution until the process completed. However, most of the time your programs will start another process as a *detached child process*. In a detached child process, after the parent process creates the child process and the child process begins executing, the parent process either no longer requires communication with the child process or does not require that the child process finish its work before the parent process continues. Running detached child processes lets your programs launch other programs without concern for their administration time or performance. When you execute a program from within the Windows *Explorer,* for example, *Explorer* creates a new process, then ignores the new process and continues its own processing.

When your programs create a detached child process, the program must first create the process, then close its handles to the new process and its primary thread. The following code fragment shows how your programs can create a new detached child process:

```
BOOL fSuccess = CreateProcess(ProcessName, &pi);
if (fSuccess)
 {
   CloseHandle(pi.hThread);
   CloseHandle(pi.hProcess);
 }
```

1382 Better Understanding Threads

In previous Tips you have learned how to create and manage processes. Each Win32 process contains one or more threads. Remember, a thread is a path of execution within a process. Every time Windows initializes a new instance of a process, the operating system creates a new primary thread for that process. The primary thread starts when Windows loads the program. The thread, in turn, will call your *WinMain* function and continue executing until your *WinMain* function ends its processing and the program calls *ExitProcess* to end itself. For many applications, the primary thread the operating system creates is the only thread the application requires. However, processes can create additional threads to help them do their work. The idea behind creating additional threads is to use as much of the CPU's processing time as possible, as efficiently as possible. When you create additional threads, you send additional requests for CPU time to the operating system. As you will learn in later Tips, additional threads will let your programs more efficiently perform background processing, do extended calculations, and perform time- and event-based activities, as well as many other advanced programming tasks. Tips 1383 and 1384 discuss in detail when to create and when not to create a thread.

When you think about what is currently executing on the system, it is helpful to visualize threads as encapsulated within processes. Depending on the thread's *priority*, the system will process some threads more often than others, both among different processes and within a single process. Figure 1382 shows a logical model of several processes, each of which contains multiple threads running on the Windows operating system, as well as a potential order in which the CPU might process those threads.

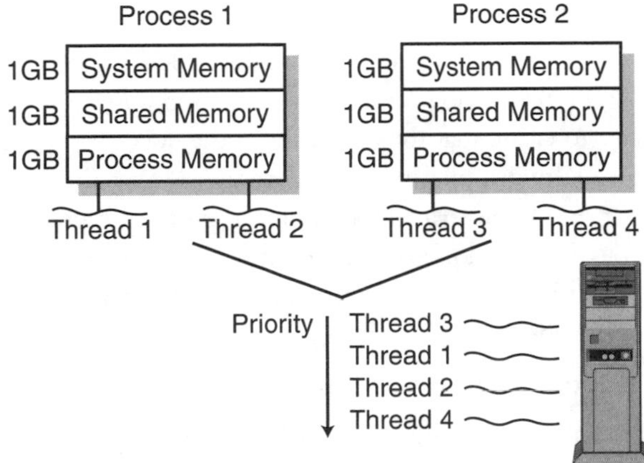

Figure 1382 *The process-thread logical model.*

EVALUATING THE NEED FOR THREADS 1383

As you learned in Tip 1382, your programs will most often use threads to ensure that various sections of a given application get as much access to the computer's CPU as possible. Determining when to use additional threads and when not to use additional threads is one of the most important decisions you will make when you write Windows programs.

For example, consider a spreadsheet program. A spreadsheet program must perform recalculations as the user changes data entries within the cells. Because the recalculations for complex spreadsheets may take several seconds or more to complete, a well-designed application should not recalculate the spreadsheet after each change the user makes. Instead, the application should execute the spreadsheet's recalculation function as a separate thread with a lower priority than that of the primary thread. If you use two threads, the primary thread will always run while the user is typing, so the lower-priority recalculation thread will not access the CPU. When the user stops typing, the lower priority thread will execute while the primary thread waits for the user to begin typing again.

Clearly, you can best apply the usefulness of threads to complex programs that perform multiple task sets. Background activities also make excellent use of additional threads. Other situations that make good use of threads include the following:

- It is useful to create a separate thread to handle printing tasks within an application, which lets the user continue to use the application while the application performs the printing task.
- You can use a separate thread to maintain a modeless dialog box that lets the user interrupt extended tasks, such as copying or printing tasks.
- You can use threads to create applications that simulate real-world events (for example, events that occur at a predetermined time interval).

1384 DETERMINING WHEN NOT TO CREATE A THREAD

The first time many programmers get access to an environment that supports multiple threads, they tend to overuse threads simply because of the new processing power and usability that threads let them exploit. Many programmers start dividing existing applications into smaller pieces, each of which executes its own thread. Despite how useful such thread-based processing might sound at first, it may actually result in a lot of wasted application processing time. Each thread requires a certain amount of overhead to create, and each thread in the queue, even if the operating system does not release the thread to the CPU for processing consumes resources. Additional threads will slow the system's performance because, in addition to the thread overhead, the system must check the priority level of each currently active thread when determining which thread to execute.

As you have learned, threads are incredibly useful and important, and they have a place in all Windows programs. However, it is important to recognize that when you use threads, you can potentially create new problems while trying to solve old problems. For example, if you are developing a word processing program and want to let the printing function run as its own thread, your initial response might be to create that thread and print the document page by page. Unfortunately, the user might change the document while you are printing in the background. Rather than the simple solution you might have expected, you must instead copy the printing file to a temporary file, print the temporary file, and delete the temporary file. Using multiple threads and a temporary file is undoubtedly more efficient for the user, and makes the program more attractive to the user. However, you must be careful to ensure that creating additional threads does not endanger the processing your program performs within its current threads.

There are several rules you should apply when you determine whether or not to create multiple threads:

- All user interface components (controls and windows) should share a common thread, with very rare exceptions.
- Programs should only create threads as they need them, rather than creating additional threads and "holding them in reserve."
- Programs should release threads as the thread's processing completes.
- More complex applications, which use multiple windows, may require extra threads to handle processing within certain windows.
- You should not create threads that might let a user interrupt a critical system process and therefore corrupt memory or other currently executing processes.

As a rule, you should always be sure that you need a thread, or that the thread will significantly improve the program's processing. Using additional threads without making such a determination will lead to wasted processor time and slower programs.

1385 CREATING A SIMPLE THREAD FUNCTION

In previous Tips, you have learned that your programs can use multiple threads to let your programs perform multiple execution activities within a single process. When you create threads, your programs will generally use the *CreateThread* function. The *CreateThread* function creates a thread to execute within the address space of the calling process. You will use the *CreateThread* function within your programs as shown in the following prototype:

```
HANDLE CreateThread(
  LPSECURITY_ATTRIBUTES lpThreadAttributes,  // ptr to thread security attributes
  DWORD dwStackSize,                         // initial thread stack size, in bytes
  LPTHREAD_START_ROUTINE lpStartAddress,     // pointer to thread function
  LPVOID lpParameter,                        // argument for new thread
  DWORD dwCreationFlags,                     // creation flags
  LPDWORD lpThreadId                         // ptr to returned thread identifier
);
```

The *CreateThread* function accepts the parameters Table 1385 details.

Parameter	Description
lpThreadAttributes	Pointer to a *SECURITY ATTRIBUTES* structure that determines whether the child processes can inherit the returned handle. If *lpThreadAttributes* is *NULL*, the child processes cannot inherit the handle. Under Windows NT, the *lpSecurityDescriptor* member of the structure specifies a security descriptor for the new thread. If *lpThreadAttributes* is *NULL*, the thread gets a default security descriptor. Under Windows 95, the *CreateThread* function ignores the *lpSecurityDescriptor* member of the structure.
dwStackSize	Specifies the size, in bytes, of the stack for the new thread. If *dwStackSize* specifies zero, the stack size defaults to the same size as that of the process's primary thread. Windows automatically allocates the process stack in the process memory space and frees the stack when the thread terminates. Note that the stack size grows, if necessary. *CreateThread* tries to commit the number of bytes *dwStackSize* specifies, and it will fail if the size exceeds available memory.
lpStartAddress	The starting address of the new thread. This is typically the address of a function declared with the *LPTHREAD_START_ROUTINE* calling convention that accepts a single 32-bit pointer as an argument and returns a 32-bit exit code. Its prototype is as shown here: *LPTHREAD_START_ROUTINE ThreadFunc(LPVOID);*
lpParameter	Specifies a single 32-bit parameter value passed to the thread.
dwCreationFlags	Specifies additional flags that control the creation of the thread. If *dwCreationFlags* specifies the *CREATE_SUSPENDED* flag, the process creates the thread in a suspended state, and will not run the thread until the program (or another program) calls the *ResumeThread* function. If the parameter is zero, the thread runs immediately after creation. Windows does not currently support other values.
lpThreadId	Points to a 32-bit variable that receives the thread identifier.

Table 1385 *The parameters for the* ***CreateThread*** *function.*

If the *CreateThread* function succeeds, the return value is a handle to the new thread. If the function fails, the return value is *NULL*. Under Windows 95, *CreateThread* succeeds only when the system calls it in the context of a 32-bit program. A 32-bit dynamic-link library cannot create an additional thread when a 16-bit program calls it.

The *CreateThread* function creates the new thread handle with full access to the new thread. If the call to *CreateThread* does not provide a security descriptor, the program can use the returned handle in any function that requires a thread object handle. When the process does provide a security descriptor, the program performs an access check on all subsequent uses of the handle before it grants access. If the access check denies access, the requesting process cannot use the handle to gain access to the thread.

The thread execution begins at the function the *lpStartAddress* parameter specifies. If this function returns, the *CreateThread* function uses the *DWORD* return value to terminate the thread in an implicit call to the *ExitThread* function. Your programs should use the *GetExitCodeThread* function to get the thread's return value after thread termination.

The *CreateThread* function may succeed even if *lpStartAddress* points to data or code, or if it is not accessible. If the start address is invalid when the thread runs, an exception occurs, and the thread terminates. The program handles thread termination due to a invalid start address as an error exit for the thread's process. This behavior is similar to the asynchronous nature of *CreateProcess*, where the function creates the process even if it refers to invalid or missing dynamic-link libraries (DLLs).

The *CreateThread* function creates the thread with a thread priority of *THREAD_PRIORITY_NORMAL*. Use the *GetThreadPriority* and *SetThreadPriority* functions to get and set the priority value of a thread.

The thread object remains in the system until the thread has terminated and the program has closed all handles to it with a call to *CloseHandle*. The program serializes the *ExitProcess, ExitThread, CreateThread, CreateRemoteThread* functions, and a process that is starting (as the result of a *CreateProcess* call) between each other within a process. Only one of these events can happen in an address space at a time, which means the process maintains the following restrictions:

- During process start-up and dynamic-link library initialization routines, the program can create new threads, but they do not begin execution until the process completes its dynamic-link library initialization.
- Only one thread in a process can be in a dynamic-link library initialization or detachment routine at a time.
- *ExitProcess* does not return until no threads are in their dynamic-link library initialization or detach routines.

Note: *A thread that uses functions from the C run-time libraries should use the* ***beginthread*** *and* ***endthread*** *C run-time functions for thread management rather than* ***CreateThread*** *and* ***ExitThread****. Failure to do so results in small memory leaks when the program calls* ***ExitThread****.*

To better understand the processing the *CreateThread* function performs, the following code fragment uses the *CreateThread* routine to create the thread, as shown here:

```
case IDM_DEMO:      // start up a thread.
  {
   DWORD dwChildId;
   CreateThread(NULL,0,(LPTHREAD_START_ROUTINE) ChildThreadProc,hWnd,0, &dwChildId);
  }
break;
```

As you can see, the code passes a pointer to the *ChildThreadProc* function that is the function the new thread will execute. The code casts the function address to a value of type *LPTHREAD_START_ROUTINE.*

1386 VISUALIZING THE THREAD'S START-UP

As you have learned, your programs can create threads to perform additional processing within their own execution space. Each time your create a new thread, Windows performs several basic tasks to initialize the thread and begin its processing.

First, Windows allocates each thread its own stack. Windows allocates the stack to the thread from the owning process's 4Gb address space. When your programs use static and global variables, multiple threads can access those variables simultaneously, potentially corrupting the variable's contents. However, Windows creates local and automatic variables on the thread's stack, therefore making them far less likely to suffer corruption in multi-threaded programs than global variables. As you have learned, you should always try to use local or automatic variables, rather than global variables. This rule applies even more when you work with threads.

Second, Windows allocates each thread its own set of CPU registers, called the thread's *context*. Windows stores the thread's context within the *CONTEXT* structure. Your programs can query the *CONTEXT* structure at any time to determine the state of the thread's CPU registers. When the operating system schedules CPU time for a thread, the system initializes the CPU registers with the thread's context. The CPU registers include both an instruction pointer (that identifies the address of the next CPU instruction for the thread to execute) and a stack pointer (that identifies the address of the thread's stack).

After the thread completes these stack and context initializations, the thread (assuming you did not create it in a suspended state) will begin its execution at the first line of the function you defined within the thread's creation.

STEPS THE OPERATING SYSTEM PERFORMS AT THREAD CREATION 1387

As you learned in Tip 1386, the operating system performs several important steps when it allocates threads, of which your programs will take advantage. However, the operating system actually performs six specific steps whenever it creates a new thread. Those six steps are as follows:

1. Allocates a thread kernel object to identify and manage the newly created thread. The kernel object holds much of the system information to manage the thread. A handle to the thread kernel object is the value *CreateThread* returns.
2. Initializes the thread's exit code (which Windows maintains in the thread kernel object) to *STILL_ACTIVE* and sets the thread's suspend count (which Windows also maintains in the thread kernel object) to 1.
3. Allocates a *CONTEXT* structure for the new thread.
4. Prepares the thread's stack by reserving a region of address space, committing two pages of physical storage to the region, setting the protection of the committed storage to *PAGE_READWRITE*, and setting the *PAGE_GUARD* attribute on the second-to-top page.
5. Places the *lpStartAddr* and *lpvThread* values at the stack's top so the new thread sees them as parameters passed to the *StartOfThread* function (only if your code uses the C run-time library).
6. Initializes the stack pointer register in the thread's *CONTEXT* structure to point to the values Windows placed on the stack in Step 5. Next, the operating system initializes the instruction pointer register to point to the internal function Windows processes before it executes the first instruction in the thread's start-up function.

DETERMINING THE THREAD'S STACK SIZE 1388

As you learned in Tip 1385, your programs can specify the thread's stack size within the *CreateThread* function. It is important to note that after Windows creates the thread, your program cannot safely change the thread's stack size. Instead, Windows will dynamically grow the stack downward as necessary.

If you do not specify a size for the thread's stack, Windows will allocate a stack the same size as the primary thread's size. *CreateThread* will create the stack within the process's memory address space. You can best visualize the stack's creation in a logical model of the allocated stack space, as shown in Figure 1388.

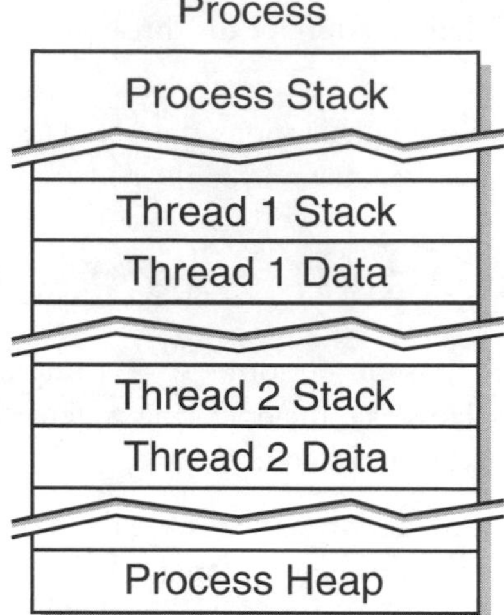

***Figure 1388** The stack space for the process, the primary thread, and a secondary thread.*

When Windows allocates stack space for additional threads, it will generally do so below the process stack space, and a segment away. Windows will allocate the thread's stack space virtually so it can move the thread stack around as necessary. For example, if the process initializes the secondary thread, then creates a very large, multidimensional automatic array (say, [1024,64] of characters), Windows must be able to move the secondary stack around to protect against the primary process accidentally overwriting the secondary stack.

1389 Acquiring a Handle to the Current Thread or Process

As your programs become more complex, there may be times when they must dynamically obtain a handle to the current thread or process. Performing either activity is relatively simple. Your programs can call the *GetCurrentThread* or *GetCurrentProcess* function to obtain the pseudo-handle of the current thread or process (it is called a pseudo-handle because its value is only meaningful within the current thread or process) at any time within your programs. You will implement these functions in accordance with their prototypes, as shown here:

```
HANDLE GetCurrentThread(VOID);
HANDLE GetCurrentProcess(VOID);
```

You should note that the pseudo-handles that both functions return are useless outside the current process. To pass the handle of a process or thread to another process, your programs must use the *DuplicateHandle* function.

To better understand the processing the *GetCurrentThread* and *GetCurrentProcess* functions perform, consider the *Show_Current.cpp* program, contained on the CD-ROM that accompanies this book. The *Show_Current.cpp* program creates threads one at a time, and displays information about the threads and the process.

1390 Handling the Thread's Processing Time

As you might imagine, determining how long your process will take to perform a given activity in a multi-threaded environment is significantly more difficult than it is in a single-threaded environment. Because your process might have a thread that is busy recalculating a complex algorithm while threads in other processes continue to compete for time in the CPU, your process may take markedly longer to execute between any two benchmark situations. Your programs must use a different method to record thread processing time from the simple code to time the program's execution, as you did in Tip 625 with the *clock.c* program.

Instead, your programs must use a function that checks the time of a thread's execution. Within Windows, you will use the *GetThreadTimes* function to perform such processing. The *GetThreadTimes* function obtains timing information about a thread you specify. You will use the *GetThreadTimes* function within your programs as shown in the following prototype:

```
BOOL GetThreadTimes(
    HANDLE hThread,              // specifies the thread of interest
    LPFILETIME lpCreationTime,   // when the thread was created
    LPFILETIME lpExitTime,       // when the thread was destroyed
    LPFILETIME lpKernelTime,     // time the thread has spent in kernel mode
    LPFILETIME lpUserTime        // time the thread has spent in user mode
);
```

The *GetThreadTimes* function accepts the parameters Table 1390 details.

Parameter	Description
hThread	An open handle that specifies the thread whose timing information a program seeks. You must create this handle with *THREAD_QUERY_INFORMATION* access.
lpCreationTime	Points to a *FILETIME* structure that receives the thread's creation time.
lpExitTime	Points to a *FILETIME* structure that receives the thread's exit time. If the thread has not exited, the content of this structure is undefined.
lpKernelTime	Points to a *FILETIME* structure that receives the amount of time that the thread has executed in kernel mode.
lpUserTime	Points to a *FILETIME* structure that receives the amount of time that the thread has executed in user mode.

Table 1390 *The parameters for the* ***GetThreadTimes*** *function.*

If the function succeeds, the return value is non-zero. If the function fails, the return value is zero. The *GetThreadTimes* function uses *FILETIME* data structures to express all times. Such structures contain two 32-bit values that combine to form a 64-bit count of 100-nanosecond time units. Thread creation and exit times are points in time the *FILETIME* structure expresses as the amount of time that has elapsed since midnight on January 1, 1601, in Greenwich, England. The Win32 API provides several functions that an application can use to convert such values to more generally useful forms.

Thread kernel-mode and user-mode times are amounts of time measured in nanoseconds. For example, if a thread has spent one second in kernel mode, the *GetThreadTimes* function will fill the *FILETIME* structure that the *lpKernelTime* parameter specifies with a 64-bit value of ten million, which is the number of 100-nanosecond units in one second.

MANAGING THE PROCESSING TIME OF MULTIPLE THREADS 1391

As you learned in Tip 1390, your programs can use the *GetThreadTimes* function to determine a thread's execution time. Often, however, your programs will require detailed information about the execution time of many threads within a process (to determine if all the threads in a process are slow, or if one or more threads is the cause of a process slowdown). In such cases, your programs can use the *GetProcessTimes* function to obtain timing information about a process you specify. You will use the *GetProcessTimes* function within your programs as shown in the following prototype:

```
BOOL GetProcessTimes(
    HANDLE hProcess,             // specifies the process of interest
    LPFILETIME lpCreationTime,   // when the process was created
    LPFILETIME lpExitTime,       // when the process exited
    LPFILETIME lpKernelTime,     // time the process has spent in kernel mode
    LPFILETIME lpUserTime        // time the process has spent in user mode
);
```

The *GetProcessTimes* function accepts the parameters Table 1390 details.

If the function succeeds, the function returns a non-zero value. If the function fails, the function returns a zero value. The *GetProcessTimes* function uses *FILETIME* data structures to express all times. Such structures contain two 32-bit values that combine to form a 64-bit count of 100-nanosecond time units. Process creation and exit times are points in time the *FILETIME* structure expresses as the amount of time that has elapsed since midnight on January 1, 1601, in Greenwich, England. The Win32 API provides several functions that an application can use to convert such values to more useful forms.

Process kernel mode and user mode times are amounts of time. For example, if a process has spent one second in kernel mode, the *GetProcessTimes* function will fill the *FILETIME* structure that the *lpKernelTime* parameter specifies with a 64-bit value of ten million, the number of 100-nanosecond units in one second.

1392 BETTER UNDERSTANDING THE GETQUEUESTATUS FUNCTION

As you work with threads, you will encounter situations where events occur (such as keystrokes, and so on) that the primary thread will usually process within your program's *WndProc* message function. If you have a suspended child thread, however, the control over messaging information becomes more complicated. However, because the thread is already suspended (for whatever reason), the thread's message function will instead hold those messages for that thread in the thread's message queue. To determine the contents of the message queue after a suspended thread resumes execution, your programs can use the *GetQueueStatus* function. The *GetQueueStatus* function returns flags that indicate the types of messages found in the calling thread's message queue. You will implement the *GetQueueStatus* function within your programs in accordance with the following prototype:

```
DWORD GetQueueStatus(UINT flags);
```

The *flags* parameter specifies queue-status flags giving the types of messages for which the function should check. This parameter can be a combination of the values Table 1392 lists.

Value	Meaning
QS_ALLEVENTS	An input, *WM_TIMER, WM_PAINT, WM_HOTKEY*, or posted message is in the queue.
QS_ALLINPUT	Any message is in the queue.
QS_HOTKEY	A *WM_HOTKEY* message is in the queue.
QS_INPUT	An input message is in the queue.
QS_KEY	A *WM_KEYUP, WM_KEYDOWN, WM_SYSKEYUP*, or *WM_SYSKEYDOWN* message is in the queue.
QS_MOUSE	A *WM_MOUSEMOVE* message or mouse-button message (*WM_LBUTTONUP, WM_RBUTTONDOWN*, and so on) is in the queue.
QS_MOUSEBUTTON	A mouse-button message (*WM_LBUTTONUP, WM_RBUTTONDOWN*, and so on) is in the queue.
QS_MOUSEMOVE	A *WM_MOUSEMOVE* message is in the queue.
QS_PAINT	A *WM_PAINT* message is in the queue.
QS_POSTMESSAGE	A posted message (other than those just listed) is in the queue.
QS_SENDMESSAGE	A message that another thread or application sent is in the queue.
QS_TIMER	A *WM_TIMER* message is in the queue.

***Table 1392** The possible flag values for the **flags** parameter.*

The return value's high-order word indicates the types of messages currently in the queue. The low-order word indicates the types of messages that Windows has added to the queue and that are still in the queue since the last call to the *GetQueueStatus*, *GetMessage*, or *PeekMessage* function.

The presence of a *QS_* flag in the return value does not guarantee that a subsequent call to the *PeekMessage* or *GetMessage* function will return a message. *GetMessage* and *PeekMessage* perform some internal filtering that may cause the program to process the message internally.

For this reason, you should consider the return value from *GetQueueStatus* as only a hint to whether your program should call *GetMessage* or *PeekMessage*.

HANDLING UNHANDLED EXCEPTIONS 1393

The Win32 operating system places a top-level exception handler at the top of each thread and process. The purpose of the top-level exception handler is to ensure that programs gracefully respond to unhandled exceptions (essentially, that programs shut down without negatively impacting other processes). Sometimes, however, you may want to capture all unhandled exceptions within a special routine, perhaps one that saves the user's work to disk in a recovery file. The *SetUnhandledExceptionFilter* function lets an application supersede the top-level exception handler that Win32 places at the top of each thread and process.

After calling the *SetUnhandledExceptionFilter* function, if an exception occurs in a process that Windows is not currently debugging, and the exception makes it to the Win32 unhandled exception filter, that filter will call the exception filter function that the *lpTopLevelExceptionFilter* parameter specifies. You will use the *SetUnhandledExceptionFilter* function as shown with the following prototype:

```
LPTOP_LEVEL_EXCEPTION_FILTER SetUnhandledExceptionFilter(
   LPTOP_LEVEL_EXCEPTION_FILTER lpTopLevelExceptionFilter);
```

The *lpTopLevelExceptionFilter* parameter supplies the address of a top-level exception filter function that Windows will call whenever the *UnhandledExceptionFilter* function gets control, and Windows is not debugging the process. A value of *NULL* for the *lpTopLevelExceptionFilter* parameter specifies default handling within *UnhandledExceptionFilter.*

The filter function has syntax congruent to that of *UnhandledExceptionFilter.* The filter function takes a single parameter of type *LPEXCEPTION_POINTERS* and returns a value of type *LONG*. The filter function should return one of the values Table 1393 lists.

Value	Meaning
EXCEPTION_EXECUTE_HANDLER	Returns from *UnhandledExceptionFilter* and executes the associated exception handler. This value usually results in process termination.
EXCEPTION_CONTINUE_EXECUTION	Returns from *UnhandledExceptionFilter* and continues execution from the point of the exception. Note that the filter function is free to modify the continuation state by modifying the exception information its *LPEXCEPTION_POINTERS* parameter supplies.
EXCEPTION_CONTINUE_SEARCH	Proceeds with normal execution of *UnhandledExceptionFilter,* which means obeying the *SetErrorMod* flags, or invoking the Application Error pop-up message box.

***Table 1393** The possible return values for the filter function.*

The *SetUnhandledExceptionFilter* function returns the address of the previous exception filter established with the function. A *NULL* return value means that there is no current top-level exception filter. Issuing *SetUnhandledExceptionFilter* replaces the existing top-level exception filter for all existing and all future threads in the calling process. The program executes the exception handler the *lpTopLevelExceptionFilter* parameter specifies in the context of the thread that caused the fault. Because the exception handler executes within a thread it may affect the exception handler's ability to recover from certain exceptions, such as an invalid stack.

1394 TERMINATING THREADS

As your programs become more complex, there will be times when a thread will fail to complete its execution normally. As you have learned, your programs must halt or pause a thread for the operating system to stop scheduling that thread for execution. If your programs fail to complete their execution correctly, they will not close. Instead, your programs must instruct the operating system to terminate the thread. To do so, your programs will use the *TerminateThread* function. You will use the *TerminateThread* function as shown in the following prototype:

```
BOOL TerminateThread(
   HANDLE hThread,             // handle to the thread
   DWORD dwExitCode            // exit code for the thread
);
```

The *hThread* parameter identifies the thread to terminate. Under Windows NT, the handle must have *THREAD_TERMINATE* access. The *dwExitCode* parameter specifies the exit code for the thread. Use the *GetExitCodeThread* function to retrieve a thread's exit value. If the function succeeds, the return value is nonzero. If the function fails, the return value is zero.

Your program uses *TerminateThread* to cause a thread to exit. When you use *TerminateThread* to cause a thread to exit, the target thread has no chance to execute any user-mode code and your program does not deallocate its initial stack. Your program does not notify the dynamic-link libraries (DLLs) attached to the thread that the thread is terminating.

TerminateThread is a potentially dangerous function that you should only use in the most extreme cases. You should call *TerminateThread* only if you know exactly what the target thread is doing, and you control all the code that the target thread could possibly be running at the time of the termination. For example, *TerminateThread* can result in the following problems:

- If the target thread owns a critical section, Windows will not release the critical section.
- If the target thread is executing certain kernel calls when the call to *TerminateThread* terminates it, the kernel state for the thread's process could be inconsistent.
- If the target thread is manipulating the global state of a shared dynamic-link library, Windows could destroy the state of the dynamic-link library, affecting other dynamic-link library users.

A thread can only protect itself against *TerminateThread* by controlling access to its handles. The thread handle the *CreateThread* and *CreateProcess* functions return has *THREAD_TERMINATE* access, so any caller holding one of these handles can terminate your thread. If the target thread is the last thread of a process when your program calls the *TerminateThread* function, the program also terminates the thread's process. The state of the thread object becomes signaled, releasing any other threads that were waiting for the thread to terminate. The thread's termination status changes from *STILL_ACTIVE* to the value of the *dwExitCode* parameter.

Terminating a thread does not necessarily remove the thread object from the system. Windows deletes a thread object when your program closes the last thread handle.

To better understand the processing the *TerminateThread* function performs, consider the program *Manip_Threads*, contained on the CD-ROM that accompanies this book. The *Manip_Threads.cpp* program lets the user create a thread, and then suspend that thread (press ALT+S), resume that thread (press ALT+R), or terminate that thread (press

ALT+K). The program avoids the issues surrounding thread termination by having the thread perform no activities and only letting the user access the most recently created thread. The *ThreadProc* and *WndProc* functions of the *Manip_Threads* program contain the code that performs the operative processing.

DETERMINING THE ID OF A THREAD OR PROCESS 1395

As you learned in Tip 1389, your programs will often require a temporary identifier, or pseudo-handle, to the current thread or process. Less frequently, your programs will require a permanent handle or other unique value to represent a thread or process throughout the system. The Win32 API provides two functions, the *GetCurrentThreadId* and the *GetCurrentProcessId* functions, which let your programs obtain a unique *DWORD* value that the operating system uses to represent the thread or process internally. You will use these functions within your programs as shown in the following prototypes:

```
DWORD GetCurrentThreadId(void);
DWORD GetCurrentProcessId(void);
```

The *GetCurrentThreadId* function returns the thread identifier of the calling thread, which is the calling thread's return value. Until the thread terminates, the thread identifier uniquely identifies the thread throughout the system.

Similarly, the *GetCurrentProcessId* function returns the process identifier of the calling process. This function has no parameters. The return value is the process identifier of the calling process. Until the process terminates, the process identifier uniquely identifies the process throughout the system.

The *ShowCurrent* program Tip 1389 presented uses both the *GetCurrentProcessID* and *GetCurrentThreadID* functions.

Note: *Similar to a handle, the* ***DWORD*** *that either* ***GetCurrentProcessID*** *or* ***GetCurrentThreadID*** *returns is a unique value that identifies the thread or process throughout the operating system. Do not confuse the Process ID or Thread ID with the pseudo-handles that* ***GetCurrentProcess*** *and* ***GetCurrentThread*** *return.*

UNDERSTANDING HOW THE OPERATING SYSTEM SCHEDULES THREADS 1396

As you have learned, the Win32 operating system is a multi-threaded operating system. The Win32 operating system is capable of handling a large number of subsequent threads or processes in close order with each other. However, as previous Tips have indicated, it will handle certain threads more quickly, or in a different order than it will handle others. For example, the Win32 operating system will tend to attach greater priority to threads in the current process than it will in processes executing in the background.

In fact, the Win32 operating system schedules every thread requesting CPU processing (that is, all active threads) based on their priority levels. Tip 1397 explains priority levels in detail. When the system assigns a CPU to a thread, it treats all threads of the same priority as equal. In other words, the system assigns the first thread in the queue with priority level 31 to a CPU, and after that thread's time slice expires, the system assigns the next thread with priority level 31 to a CPU. Therefore, it is important to note that if you always have at least one priority 31 thread for each CPU, threads with lower priorities will never execute. Programmers call this priority condition *starvation*. Starvation occurs when some threads use so much of the CPU's time that no other thread can execute.

When the system completes the priority 31 threads, it will begin to assign priority 30 threads. When the system completes the priority 30 threads, it will begin to assign priority 29 threads, and so on. It might seem, then, that low-priority threads will never execute (as a rule) in such a system. As it turns out, however, even the highest-priority threads often do not require CPU time, which frees the CPU to handle lower-priority threads.

Finally, you should understand that if a lower-priority thread is executing—even if it is in the middle of its time slice—and the system determines that a higher-priority thread is waiting to execute, the system will immediately halt the execution of the lower-priority thread and begin the execution of the higher-priority thread. The higher-priority thread will always preempt the lower-priority thread, regardless of what the lower-priority thread is doing or what state it is in its execution.

Note: *Microsoft reserves the right to change the algorithm that Win32-based operating systems use for thread scheduling, and they have—Windows 95, Windows NT 3.51, and Windows NT 4.0 all use different variations of the Win32 scheduling algorithm. While you should understand how thread priority works and should use it carefully, your programs should not be so tied to the method one operating system uses to manage thread priority that it causes significant problems in the event that Microsoft changes the scheduling algorithm in future releases.*

1397 Introducing Priority Levels

As you learned in Tip 1396, the Win32 operating system schedules all active threads based on their current priority levels. Priority levels range from 0 (the lowest priority) to 31 (the highest priority). The operating system assigns the priority level zero to a special system thread known as the *zero page thread*. The zero page thread is responsible for zeroing any free pages in the system when there are no other threads that must perform work in the system. It is not possible for any thread other than the zero page thread to have a priority level of zero.

When you create threads, you do not use numbers to assign them priority levels. Instead, the system uses a two-step process to determine the thread's priority level. The first step is to assign a *priority class* to a process. The process's priority class tells the system the priority the process requires compared to other running processes. Tip 1398 explains priority classes in detail. The second step is to assign a relative priority level to each thread the process owns.

When you first create a thread within a process, the thread's priority level is the same as the process's priority level. In Tip 1400, you will learn how to use the Win API to change a thread's relative priority level.

1398 Understanding the Windows Priority Classes

As Tip 1397 explains, Windows uses a two-step process to determine a thread's priority. The first step is to determine the process's priority class. Win32 supports four different priority classes: idle, normal, high, and real-time. Table 1398 details the priority classes.

Priority	Meaning
HIGH_PRIORITY_CLASS	Indicates a process that performs time-critical tasks that the operating system must execute immediately for the process to run correctly. The threads of a high-priority class process preempt the threads of normal or idle priority class processes. An example is the Windows Task List, which must respond quickly when the user calls, regardless of the load on the operating system. Exercise extreme care when you use the high-priority class, because a high-priority class CPU-bound application can use nearly all available cycles. The priority level of a *HIGH_PRIORITY_CLASS* process is 13.
IDLE_PRIORITY_CLASS	Indicates a process whose threads run only when the system is idle and the threads of any process running in a higher priority class preempt them. An example is a screen saver. Child processes inherit the idle priority class. The priority level of an *IDLE_PRIORITY_CLASS* is 6.

Table 1398 *The possible priority classes.* *(continued on following page)*

Priority	Meaning
NORMAL_PRIORITY_CLASS	Indicates a normal process with no special scheduling needs. The priority level of a *NORMAL_PRIORITY_CLASS* is 8.
REALTIME_PRIORITY_CLASS	Indicates a process that has the highest possible priority. The threads of a real-time priority class process preempt the threads of all other processes, including operating system processes performing important tasks. For example, a real-time process that executes for more than a very brief interval can cause disk caches not to flush or cause the mouse to be unresponsive. The priority level of a *REALTIME_PRIORITY_CLASS* is 24.

***Table 1398** The possible priority classes. (continued from previous page)*

It is very helpful for you to more completely understand the impact of the different class levels. For example, you should only use the *HIGH_PRIORITY_CLASS* setting when absolutely necessary. The most common process to use the *HIGH_PRIORITY_CLASS* setting is the Windows *Explorer*. Even though most of the desktop threads sleep during normal execution, users expect the desktop to be responsive when they want to access it. Therefore, Windows gives the Explorer's threads the highest priority and will preempt almost any other thread when the user selects a desktop option. If you create programs that also use the *HIGH_PRIORITY_CLASS* setting, the desktop may not be as responsive, and Windows may even lock it out.

Generally, your programs will use the *IDLE_PRIORITY_CLASS* for system-monitoring applications. For example, you might write an application that will periodically display the amount of free RAM in the system. Because you will not want the application to interfere with the performance of other, more time-critical tasks, you will set the periodic process's class to *IDLE_PRIORITY_CLASS*.

Windows automatically assigns the *NORMAL_PRIORITY_CLASS* setting to any process to which you do not explicitly assign another setting. Your programs should generally use the *NORMAL_PRIORITY_CLASS* setting. Note that when the user brings a process into the foreground, the operating system increases that process's relative priority to provide for better execution speed. For example, in Windows 95, the operating system adds one to the foreground process's priority count.

In general, your programs should never use the *REALTIME_PRIORITY_CLASS* setting because real-time priority is very high priority—in fact, it is even higher than most operating system management threads. The threads in the system that control the mouse and the keyboard, background disk flushing, even CTRL+ALT+DEL trapping, all execute at lower than real-time priority. Programs that use real-time priority will often have significant adverse effects on the user's system.

ALTERING A PROCESS'S PRIORITY CLASS 1399

As you learned in Tip 1398, the Win32 operating system automatically assigns normal priority to every new process. Often, however, your programs may require changes to a process's current priority class. You can use the *GetPriorityClass* and *SetPriorityClass* functions to manage a process's priority class. The *GetPriorityClass* function returns the priority class for the specified process. You will use the *GetPriorityClass* function within your programs as shown in the following prototype:

```
DWORD GetPriorityClass(HANDLE hProcess);
```

The *hProcess* handle identifies the process. Under Windows NT, the *hProcess* handle must have *PROCESS_QUERY_INFORMATION* access right. If the *GetPriorityClass* function succeeds, the return value is the specified process's priority class. If it fails, the return value is zero. The specified process's priority class is one of the values listed in Table 1398.

Similarly, the *SetPriorityClass* function sets the priority class for the specified process. The priority class, together with the priority value of each process thread determines each thread's base priority level. You will use the *SetPriorityClass* function within your programs in accordance with the following prototype:

```
BOOL SetPriorityClass(
    HANDLE hProcess,           // handle to the process
    DWORD dwPriorityClass      // priority class value
);
```

Just as with the *GetPriorityClass* function, the *hProcess* handle identifies the process. Under Windows NT, the *hProcess* handle must have the *PROCESS_SET_INFORMATION* access right. The *dwPriorityClass* parameter specifies the priority class for the process. The *dwPriorityClass* parameter can be any one of the values listed in Table 1398. If the *SetThreadPriority* function succeeds, the return value is non-zero. If it fails, the return value is zero.

Every thread has a base priority level, which Windows determines based on the thread's priority value and the priority class of the thread's process. The system uses the base priority level of all executable threads to determine which thread gets the next slice of CPU time. The *SetThreadPriority* function lets you set the base priority level of a thread relative to the priority class of its process. Tip 1400 uses the *SetThreadPriorityFunction* to set a thread's priority level.

The CD-ROM that accompanies this book includes the *Get_Set_Priority* program. The *Get_Set_Priority* program lets the user select the process's priority class. After each selection, the process will perform a CPU intensive function and display the results, together with the operating system-returned priority class.

1400 SETTING A THREAD'S RELATIVE PRIORITY

As you have learned, Windows sets a thread's priority level based on the thread's process class and the thread's priority level offset. In Tip 1399, you learned how to use the *SetProcessClass* function to change a process's priority class. To change the priority level of threads within a process, your programs should use the *SetThreadPriority* function. The *SetThreadPriority* function sets the priority value for the specified thread. This value, together with the priority class of the thread's process, determines the thread's base priority level. You will use the *SetThreadPriority* function within your programs as shown in the following prototype:

```
BOOL SetThreadPriority(
   HANDLE hThread,      // handle to the thread
   int nPriority        // thread priority level
);
```

The *hThread* parameter identifies the thread whose priority value the function is to set. Under Windows NT, the handle must have the *THREAD_SET_INFORMATION* access right. The *nPriority* parameter specifies the priority value for the thread. This parameter can be one of the values Table 1400 lists.

Priority	Meaning
THREAD_PRIORITY_ABOVE_NORMAL	Indicates 1 point above normal priority for the priority class.
THREAD_PRIORITY_BELOW_NORMAL	Indicates 1 point below normal priority for the priority class.
THREAD_PRIORITY_HIGHEST	Indicates 2 points above normal priority for the priority class.

***Table 1400** The thread priority level values for the **nPriority** parameter.* *(continued on following page)*

Priority	Meaning
THREAD_PRIORITY_IDLE	Indicates a base priority level of 1 for *IDLE_PRIORITY_CLASS*, *NORMAL_PRIORITY_CLASS*, or *HIGH_PRIORITY_CLASS* processes, and a base priority level of 16 for *REALTIME_PRIORITY_CLASS* processes.
THREAD_PRIORITY_LOWEST	Indicates 2 points below normal priority for the priority class.
THREAD_PRIORITY_NORMAL	Indicates normal priority for the priority class.
THREAD_PRIORITY_TIME_CRITICAL	Indicates a base priority level of 15 for *IDLE_PRIORITY_CLASS*, *NORMAL_PRIORITY_CLASS*, or *HIGH_PRIORITY_CLASS* processes, and a base priority level of 31 for *REALTIME_PRIORITY_CLASS* processes.

***Table 1400** The thread priority level values for the **nPriority** parameter.* *(continued from previous page)*

If the *SetThreadPriority* function succeeds, the return value is non-zero. If it fails, the return value is zero. As you have learned, every thread has a base priority level the thread's priority value and the priority class of its process determine. The system uses the base priority level of all executable threads to determine which thread gets the next slice of CPU time. The system schedules threads in a round-robin fashion at each priority level, and only when there are no executable threads at a higher level will the system schedule threads at a lower level.

The *SetThreadPriority* function lets you set a thread's base priority level relative to its process's priority class. For example, specifying *THREAD_PRIORITY_HIGHEST* in a call to *SetThreadPriority* for a thread of an *IDLE_PRIORITY_CLASS* process sets the thread's base priority level to 6. For *IDLE_PRIORITY_CLASS*, *NORMAL_PRIORITY_CLASS*, and *HIGH_PRIORITY_CLASS* processes, the system dynamically boosts a thread's base priority level when events occur that are important to the thread (such another thread going into an idle state). *REALTIME_PRIORITY_CLASS* processes do not receive dynamic boosts. All threads initially start at *THREAD_PRIORITY_NORMAL*.

Use a process's priority class to differentiate between applications that are time critical and those that have normal or below normal scheduling requirements. Use thread priority values to differentiate between the relative priorities of a process's tasks. For example, a thread that handles input for a window could have a higher priority level than a thread that performs intensive calculations for the CPU.

When manipulating priorities, be very careful to ensure that a high-priority thread does not consume all the available CPU time. A thread with a base priority level above 11 interferes with the normal operation of the operating system. Using *REALTIME_PRIORITY_CLASS* may cause disk caches to not flush, the mouse to be unresponsive, and so on.

OBTAINING A THREAD'S CURRENT PRIORITY LEVEL 1401

As you learned in Tip 1400, your programs can use the *SetThreadPriority* function to change a thread's current priority level. Often, your programs will require information about a thread's current priority level, usually as the step before a call to *SetThreadPriority*. The *GetThreadPriority* function returns the priority value for the specified thread. This value, together with the priority class of the thread's process, determines the thread's base-priority level. You will use the *GetThreadPriority* function within your programs as shown in the following prototype:

```
int GetThreadPriority(
   HANDLE hThread// handle to thread
);
```

As with the *SetThreadPriority* function, the *hThread* parameter identifies the thread. If the function succeeds, the function returns the thread's priority level. If the function fails, the function returns *THREAD_PRIORITY_ERROR_RETURN*. To get extended error information, call *GetLastError*. The thread's priority level is one of the values Table 1400 details.

Remember, every thread has a base-priority level that the operating system determines by the thread's priority value and the priority class of the thread's process. The operating system uses the base-priority level of all executable threads to determine which thread gets the next slice of CPU time. The operating system schedules threads in a round-robin fashion at each priority level, and only when there are no executable threads at a higher level will the operating system schedule threads at a lower level.

Note: *Under Windows NT, the handle must have* ***THREAD_QUERY_INFORMATION*** *access.*

1402 Obtaining a Thread's Context

As you have learned, Windows stores information about each thread within the thread's *CONTEXT* structure. As your program manipulate threads more and more often, your programs may require information about a thread's context. The *GetThreadContext* function retrieves the context of the specified thread. You will use the *GetThreadContext* function as shown in the following prototype:

```
BOOL GetThreadContext(
   HANDLE hThread,                    // handle of thread with context
   LPCONTEXT lpContext                // address of context structure
);
```

The *hThread* parameter identifies an open handle of a thread whose context the function is to retrieve. The *lpContext* parameter points to the address of a *CONTEXT* structure that receives the appropriate context of the specified thread. The value of the *ContextFlags* member of this structure specifies which portions of a thread's context are retrieved. The *CONTEXT* structure is highly computer specific. Currently, there are *CONTEXT* structures defined for Intel, MIPS, Alpha, and PowerPC processors.

You should use the *GetThreadContext* function to retrieve the context of a specified thread. The function lets your programs retrieve a selective context based on the value of the *ContextFlags* member of the *CONTEXT* structure. The thread handle the *hThread* parameter identifies is typically being debugged, but the function can also operate when it is not being debugged. You cannot get a valid context for a running thread. You must use the *SuspendThread* function to suspend the thread before calling *GetThreadContext*.

Note: *Under Windows NT, the handle must have* ***THREAD_GET_CONTEXT*** *access to the thread.*

1403 Pausing and Resuming Threads

In previous Tips, you learned that your programs can create threads in a suspended state (using the *CREATE_SUSPENDED* flag with the *CreateProcess* or *CreateThread* function). When you create a suspended thread, the system creates the kernel object that identifies the thread, creates the thread's stack, and initializes the thread's CPU register members within the *CONTEXT* structure. However, the creating function gives the thread object an initial suspend count of 1, which means the system will never assign CPU time to execute the thread. To let the thread begin executing, another thread must call the *ResumeThread* function and pass to it the handle for the suspended thread. The *ResumeThread* function decrements a thread's suspend count. When *ResumeThread* decrements the suspend count to zero, the program resumes executing the thread. You will use the *ResumeThread* function within your programs as shown in the following prototype:

```
DWORD ResumeThread(HANDLE hThread);
```

The *hThread* parameter specifies a handle to the thread the program is to restart. You can suspend a thread multiple times; however, you must call *ResumeThread* the same number of times that you suspend the thread before the thread will resume execution.

The *ResumeThread* function checks the suspend count of the subject thread. If the suspend count is zero, the thread is not currently suspended. Otherwise, the *ResumeThread* function decrements the subject thread's suspend count. If the resulting value is zero, then the program resumes executing the subject thread. If the return value is zero, you did not suspend the subject thread. If the return value is 1, you did suspend the subject thread, but the program restarted it. If the return value is greater than 1, the subject thread is still suspended.

Note that while reporting debug events, all threads within the reporting process are frozen. The operating system expects debuggers to use the *SuspendThread* and *ResumeThread* functions to limit the set of threads that can execute within a process. It is possible to "single step" a single thread by suspending all threads in a process except for the one reporting a debug event. The other threads are not released by a continue operation if they are suspended.

Note: *Under Windows NT, the **hProcess** handle must have **THREAD_SUSPEND_RESUME** access to the thread.*

UNDERSTANDING THREAD SYNCHRONIZATION 1404

As you have learned in previous Tips, Windows supports multiple thread execution. In an environment where one or more threads may execute concurrently, it will often become important to let your programs synchronize the activities of various threads. The Win32-based operating system provides several synchronization objects that let threads synchronize their actions with other threads. In Tip 1405, you will learn more about the specific synchronization objects.

In general, a thread synchronizes itself with another thread by "putting itself to sleep." When the thread sleeps, the operating system no longer schedules CPU time for the thread, and the thread therefore stops executing. Just before the thread begins sleeping, however, the thread tells the operating system what "special event" (such as a keystroke, mouse click, and algorithm completion) must occur for the thread to start executing again.

The operating system, in turn, remains aware of the thread's request and watches to see whether or when the special event occurs. When the event does occur, the operating system alerts the thread, which then becomes eligible to once again be scheduled to access the CPU. Eventually, the CPU will schedule the thread and continue the thread's execution—meaning the thread is now *synchronized* with the special event's occurrence.

DEFINING THE FIVE MAJOR SYNCHRONIZATION OBJECTS 1405

As you learned in Tip 1404, Windows supports different types of synchronization objects. Of those types, the five most commonly used are *critical sections, mutexes, semaphores, events,* and *waitable timers*. Later Tips will discuss some of these types in detail. However, it is valuable to understand the simple definitions of each type. Table 1405 lists the five major synchronization object types and their actions.

Type	Use and Actions
Critical section	A critical section is a small section of code that requires exclusive access to some shared data before the code can execute. Of all the synchronization objects, critical sections are the simplest to use. However, you can only use critical sections to synchronize threads within a single process.

Table 1405 *The five major thread synchronization types.* *(continued on following page)*

Type	Use and Actions
Mutexes	Mutexes are much like critical sections. However, your programs use mutexes to synchronize data access across multiple processes. In addition, mutexes are *kernel objects*, meaning that your programs will actually create a mutex using an API function, such as *CreateMutex*.
Semaphores	Your programs will use semaphore objects to count resources. A single thread can use a semaphore to count the number of available resources and to allocate resources. For example, if a computer has three serial ports, you can create a semaphore with a resource count of three. Each time a thread accesses a serial port, the semaphore's resource count reduces by one, and each time a thread releases a serial port, the semaphore's resource count increments by one. Threads can, therefore, call the semaphore and wait until it becomes available before they try to access the serial ports. Unlike mutexes and critical sections, *semaphores are not thread-owned.*
Events	Event objects are the most primitive form of synchronization objects and are quite different from mutexes and semaphores. Typically, your programs will use mutexes and semaphores to control access to data or resources. On the other hand, your programs will use events to signal completion of an operation. Your programs will most often use events to start a second thread after the first thread completes some part of its processing.
Waitable timers	A waitable timer is a kernel object that periodically signals itself, either at a specific time or at regular intervals. You can think of a waitable timer as an internal alarm clock for your programs. For example, you might write a scheduling program that alerts the user every hour of new appointments that hour. Rather than looping constantly and waiting for the hour to change, your program could create a waitable timer that signaled the program of the time change each hour. Waitable timers exist only in Windows NT 4 and higher. Windows 95 does not support waitable timers.

Table 1405 *The five major thread synchronization types. (continued from previous page)*

1406 Creating a Critical Section

As you learned in Tip 1405, the simplest type of thread synchronization to use is a critical section. As you also learned, a critical section lets your programs control access to a given piece of data or a function within the program, ensuring that only one thread at a time accesses that data or that all other threads internal to the process have completed their processing before the critical section executes.

Creating a critical section is relatively easy. First, your program must allocate a *CRITICAL_SECTION* data structure within the executing process. Your program must allocate the *CRITICAL_SECTION* data structure globally so that different threads within the current process can access the program's *CRITICAL_SECTION* instance. Usually, that means your *CRITICAL_SECTION* instance will be a global variable.

After your program allocates the *CRITICAL_SECTION* data structure, it must perform two steps to create and enter the critical section. Your program must first call the *InitializeCriticalSection* function to initialize the section, and then your program must call the *EnterCriticalSection* function when it is ready to enter the critical section. The *EnterCriticalSection* function waits for the thread to own the specified critical section object. The function returns when the operating system grants the calling thread ownership. Your programs will use the *EnterCriticalSection* function as shown in the following prototype:

```
void EnterCriticalSection(LPCRITICAL_SECTION lpCriticalSection);
```

The *lpCriticalSection* parameter points to the critical section object. To enable mutually exclusive access to a shared resource, each thread calls the *EnterCriticalSection* or *TryEnterCriticalSection* function to request ownership of the critical section before executing any section of code that accesses the protected resource. The difference is that the *TryEnterCriticalSection* function returns immediately, regardless of whether it obtained ownership of the critical section, while the *EnterCriticalSection* function blocks processing until the thread can take ownership of the critical section. When it has finished executing the protected code, the thread uses the *LeaveCriticalSection* function to relinquish ownership, enabling another thread to become owner and access the protected resource. The controlling thread must call *LeaveCriticalSection* once for each time that it enters the critical section. The thread enters the critical section each time *EnterCriticalSection* or *TryEnterCriticalSection* succeeds.

After a thread owns a critical section, it can make additional calls to the *EnterCriticalSection* function or the *TryEnterCriticalSection* function without blocking its own execution. This prevents a thread from deadlocking itself (that is, stopping its own execution) while waiting for a critical section that it already owns.

Any process thread can use the *DeleteCriticalSection* function to release the system resources that the program allocated when it initialized the critical section object. After the thread has called the *DeleteCriticalSection* function, the program can no longer use the critical section object for synchronization.

Note: *Although the members of the* ***CRITICAL_SECTION*** *data structure are defined within the* ***winbase.h*** *header file, your programs should not try to access the members of the structure because Windows manages that information internally and modifications to members may cause fatal system errors.*

USING A SIMPLE CRITICAL SECTION 1407

As you learned in Tip 1406, creating and using a critical section is, at a minimum, a three-step process. You must create the section, initialize the section, and enter the section. Before you can synchronize threads with a critical section, you must initialize the critical section, passing the address of the *CRITICAL_SECTION* data structure as the sole parameter. When you reach the beginning of the critical section, your program must call either the *EnterCriticalSection* function or the *TryEnterCriticalSection* function, again passing the address of the *CRITICAL_SECTION* data structure as the sole parameter. After the threads have synchronized, the critical section will remain until your program leaves the critical section or until your program deletes the critical section.

To better understand the processing your programs will perform when you manage critical sections, consider the *Crit_Section.cpp* program, contained on the CD-ROM that accompanies this book. The *Crit_Section.cpp* program uses a critical section, which holds a five-second sleep instruction, to let only one thread at a time execute critical code. Every time the user selects the *Test!* option, the program creates another thread which, in turn, waits for its access to the critical section. The *Crit_Section.cpp* program's operative code is within the *ChildThreadProc* and *WndProc* functions.

USING WAITFORSINGLEOBJECT TO SYNCHRONIZE TWO THREADS 1408

As you have learned, many synchronization activities will revolve around waiting for one or more other threads before continuing the current thread's processing. When your current thread is waiting for the CPU to return from another thread, your programs should use the *WaitForSingleObject* function, which returns when one of the following events occurs:

- The specified object is in the signaled state.
- The time-out interval elapses.

Your programs will use the *WaitForSingleObject* function as shown in the following prototype:

```
DWORD WaitForSingleObject(
   HANDLE hHandle,                    // handle of object to wait for
   DWORD dwMilliseconds       // time-out interval in milliseconds
 );
```

The *WaitForSingleObject* function accepts two parameters, *hHandle* and *dwMilliseconds*. The *hHandle* parameter identifies the object for which the function is to wait. The *dwMilliseconds* parameter specifies the time-out interval in milliseconds. The *WaitForSingleObject* function returns if the interval elapses, even if the object's state is non-signaled. If *dwMilliseconds* is zero, the function tests the object's state and returns immediately. If *dwMilliseconds* is *INFINITE*, the function's time-out interval never elapses. Table 1408.1 contains the list of the object types whose handles you can specify (that is, object types that the *WaitForSingleObject* function can wait for).

Object Type	Description
Change notification	The *FindFirstChangeNotification* function returns the handle. A *change notification* object's state is signaled when a specified type of change occurs within a directory or directory tree you specify.
Console input	The *CreateFile* function or the *GetStdHandle* function returns the handle when you specify the *CONIN$* value. The *console input* object's state is *signaled* when there is unread input in the console's input buffer, and it is *non-signaled* when the input buffer is empty.
Event	The *CreateEvent* or *OpenEvent* function returns the handle. Both the *SetEvent* and *PulseEvent* functions explicitly set an *event* object's state to signaled. Your programs must use the *ResetEvent* function to reset each manual-reset *event* object's state to non-signaled. For an auto-reset *event* object, the wait function resets the object's state to non-signaled before returning. You can also use *event* objects in overlapped operations, in which the system sets the state.
Mutex	The *CreateMutex* or *OpenMutex* function returns the handle. A mutex object's state is signaled when no thread owns it. The wait function requests ownership of the mutex for the calling thread, changing the mutex's state to non-signaled when the operating system grants ownership to the mutex.
Process	The *CreateProcess* or *OpenProcess* function returns the handle. A *process* object's state is signaled when the process terminates.
Semaphore	The *CreateSemaphore* or *OpenSemaphore* function returns the handle. A *semaphore* object maintains a count between zero and some maximum value. Its state is signaled when its count is greater than zero, and non-signaled when its count is zero. If the current state is signaled, the wait function decreases the count by one.
Thread	The *CreateProcess*, *CreateThread*, or *CreateRemoteThread* function returns the handle. A *thread* object's state is signaled when the thread terminates.
Timer	The *CreateWaitableTimer* or *OpenWaitableTimer* function returns the handle. Activate the timer by calling the *SetWaitableTimer* function. The state of an active timer is signaled when it reaches its due time. You can deactivate the timer by calling the *CancelWaitableTimer* function.

***Table 1408.1** The objects for which **WaitForSingleObject** can wait.*

Note: *In Windows NT, the handle must have the **SYNCHRONIZE** access right.*

If the *WaitForSingleObject* function fails, the return value will be *WAIT_FAILED*. If the *WaitForSingleObject* function succeeds, the return value will indicate the event that caused the function to return. The return value for a successful function invocation is any one of the values Table 1408.2 details.

Value	Meaning
WAIT_ABANDONED	The specified object is a mutex object that the thread that owned the mutex object did not release before it terminated. The operating system grants ownership of the mutex object to the calling thread, and sets the mutex to non-signaled.
WAIT_OBJECT_0	The state of the specified object is signaled.
WAIT_TIMEOUT	The time-out interval elapsed, and the object's state is non-signaled.

***Table 1408.2** The possible successful return values for the **WaitForSingleObject** function.*

The *WaitForSingleObject* function checks the specified object's current state. If the object's state is non-signaled, the calling thread enters an efficient wait state. The thread consumes very little processor time while waiting for the object state to become signaled or the time-out interval to elapse. Before returning, a wait function modifies the state of some types of synchronization objects. Modification occurs only for the object or objects whose signaled state caused the function to return. For example, the wait function decreases the count of a *semaphore* object by one increment.

You must be careful when you use the wait functions and dynamic-data exchange. If a thread creates any windows, it must process messages. Dynamic-data exchange sends messages to all windows in the system. If you have a thread that uses a wait function with no time-out interval, the system will deadlock. Therefore, if you have a thread that creates windows, use the *MsgWaitForMultipleObjects* or *MsgWaitForMultipleObjectsEx* functions, rather than the *WaitForSingleObject* function.

To better understand the processing that the *WaitForSingleObject* function performs, consider the program *Wait_Events.cpp*, contained on the CD-ROM that accompanies this book. Each time the user selects the *Demo* menu option, the *Wait_Events* program starts a thread that waits for an event, sleeps, and releases the event. The use of *auto-reset events* (events that change their signaling flags automatically) enforces the thread access in serial order because the events automatically reset their values to non-signaled when the first waiting thread acquires the object. The *Wait_Events.cpp* program performs its operative processing within the *ChildThreadProc* and *WndProc* functions.

USING WAITFORMULTIPLEOBJECTS TO SYNCHRONIZE MANY THREADS 1409

In Tip 1408, you learned how your programs can use the *WaitForSingleObject* function to synchronize a thread with a single auto-reset event. More commonly, however, your programs will require that processing continue only when one or more of a certain set of objects occurs. In such cases, your programs can use the *WaitForMultipleObjects* function. The *WaitForMultipleObjects* function returns when one of the following events occurs:

- Any one or all the specified objects are in the signaled state.
- The time-out interval elapses.

You will use the *WaitForMultipleObjects* function within your programs as shown in the following prototype:

```
DWORD WaitForMultipleObjects(
   DWORD nCount,              // number of handles in the object handle array
   CONST HANDLE *lpHandles,   // pointer to the object-handle array
   BOOL bWaitAll,             // wait flag
   DWORD dwMilliseconds       // time-out interval in milliseconds
);
```

The *WaitForMultipleObjects* function accepts the parameters Table 1409.1 details.

Parameter	Description
nCount	Specifies the number of object handles in the array to which *lpHandles* points. The maximum number of object handles is *MAXIMUM_WAIT_OBJECTS* (a system-defined constant which will vary from installation to installation).
lpHandles	Points to an array of object handles. Table 1408.1 lists the object types whose handles you can specify. The *lpHandles* array can contain handles of objects of different types.
Note: *Under Windows NT, the handles must have the SYNCHRONIZE access right.*	
bWaitAll	Specifies the wait type. If *True*, the function returns when the state of all objects in the *lpHandles* array is signaled. If *False*, the function returns when the state of any one of the objects is set to signaled and the return value then indicates the object whose state caused the function to return.
dwMilliseconds	Specifies the time-out interval in milliseconds. The function returns if the interval elapses, even if the conditions the *bWaitAll* parameter specified are unmet. If *dwMilliseconds* is zero, the function tests the states of the specified objects and returns immediately. If *dwMilliseconds* is INFINITE, the function's time-out interval never elapses.

Table 1409.1 *The parameters for the* ***WaitForMultipleObjects*** *function.*

After invocation, *WaitForMultipleObjects* will return upon failure, success, or time-out. If the *WaitForMultipleObjects* function fails, the return value will be *WAIT_FAILED*. If the *WaitForMultipleObjects* function succeeds, the return value will indicate the event that caused the function to return. The return value, upon success, is one of the values Table 1408.2 details.

The *WaitForMultipleObjects* function determines whether one or more of the objects on which the thread is waiting has met the wait criteria. If none of the objects on which the thread is waiting has met the wait criteria, the calling thread enters an efficient wait state, consuming very little processor time while waiting for one or more of the objects on which the thread is waiting to meet the wait criteria. When the *bWaitAll* parameter is *True*, the function's wait operation is completed only when the states of all objects are signaled. The *bWaitAll* parameter does not modify the states of the specified objects until the states of all objects are signaled. For example, a mutex can be signaled, but the thread does not get ownership until the states of the other objects are also signaled. In the meantime, some other thread may get ownership of the mutex, thereby setting its state to non-signaled. Before returning, a wait function modifies the state of some types of synchronization objects. Modification occurs only for the object or objects whose signaled state caused the function to return. For example, some other thread or processes decreases the count of a *semaphore* object by one. The *WaitForMultipleObjects* function can specify in the *lpHandles* array one or more handles of any of the object types listed in Table 1408.1.

Just as with the *WaitForSingleObject* function, you must be careful when using the *WaitForMultipleObjects* function and dynamic-data exchange. If a thread creates any windows, it must process messages. Dynamic-data exchange sends messages to all windows in the system. If you have a thread that uses a wait function with no time-out interval, the system will deadlock. Therefore, if you have a thread that creates windows, use the *MsgWaitForMultipleObjects* or *MsgWaitForMultipleObjectsEx* functions, rather than the *WaitForMultipleObjects* function.

1410 Creating a Mutex

As you have learned, mutexes are similar to critical sections, except that your programs can also use *mutexes* to synchronize data access across multiple objects, rather than within only a single object. To use a mutex, your programs must first create the mutex with the *CreateMutex* function. You will use the *CreateMutex* function within your programs as shown in the following prototype:

```
HANDLE CreateMutex(
    LPSECURITY_ATTRIBUTES  lpMutexAttributes, security  attributes
    BOOL bInitialOwner,                    // flag for initial ownership
    LPCTSTR lpName                         // pointer to mutex-object name
);
```

The *lpMutexAttributes* parameter is a pointer to a *SECURITY_ATTRIBUTES* structure that determines whether child processes can inherit the returned handle. If *lpMutexAttributes* is *NULL*, the child processes cannot inherit the handle. Under Windows NT, the *lpSecurityDescriptor* member of the structure specifies a security descriptor for the new mutex. If *lpMutexAttributes* is *NULL*, the mutex gets a default security descriptor. Under Windows 95, the *CreateMutex* function ignores the *lpSecurityDescriptor* member of the structure. The *bInitialOwner* parameter specifies the initial owner of the mutex object. If *True*, the calling thread requests immediate ownership of the mutex object. Otherwise, no thread owns the mutex object. The *lpName* parameter points to a *NULL*-terminated string specifying the name of the mutex object. Windows limits the name to *MAX_PATH* characters and the name can contain any character except the backslash path-separator (\). Keep in mind that name comparison is case sensitive.

If the *lpName* parameter matches the name of an existing named mutex object, *lpName* requests *MUTEX_ALL_ACCESS* access to the existing object. The *lpName* function ignores the *bInitialOwner* parameter because the creating process has already set it. If the *lpMutexAttributes* parameter is not *NULL*, *CreateMutex* determines whether the handle is inheritable, but ignores the handle's security-descriptor member. If *lpName* is *NULL*, *CreateMutex* creates the mutex object without a name. If the *lpName* function matches the name of an existing *event*, *semaphore*, or *file-mapping* object, the function fails and the *GetLastError* function returns *ERROR_INVALID_HANDLE* access. The function fails because event, mutex, semaphore, and file-mapping objects share the same name space.

The handle *CreateMutex* returns has *MUTEX_ALL_ACCESS* access to the new mutex object, and you can use it in any function that requires a handle to a mutex object. Any calling process thread can specify the mutex-object handle in a call to one of the wait functions (such as *WaitForSingleObject* and *WaitForMultipleObjects*). The single-object wait functions return when the state of the specified object is signaled. You can instruct the multiple-object wait functions to return when any one or when all the specified objects are signaled. When a wait function returns, the operating system releases the waiting thread to continue its execution.

The state of a mutex object is signaled when no thread owns it. The creating thread can use the *bInitialOwner* flag to request immediate ownership of the mutex. Otherwise, a thread must use one of the wait functions to request ownership. When the mutex's state is signaled, the operating system grants one waiting thread ownership, the mutex's state changes to non-signaled, and the wait function returns. Only one thread can own a mutex at any given time. The owning thread uses the *ReleaseMutex* function to release its ownership. The thread that owns a mutex can specify the same mutex in repeated wait function calls without blocking its execution. Typically, a thread will not wait repeatedly for the same mutex, but the repeated-call mechanism prevents a thread from deadlocking itself while waiting for a mutex that it already owns. However, to release its ownership, the thread must call the *ReleaseMutex* function once for each time that the mutex satisfied a wait.

Two or more processes can call *CreateMutex* to create the same named mutex object. The first process creates the mutex, and subsequent processes open a handle to the existing mutex. Multiple processes can use *CreateMutex* to let them get handles of the same mutex object, while relieving the user of the responsibility of ensuring that the creating process started the mutex previously. When using *CreateMutex* in multiple processes, you should set the *bInitialOwner* flag to *False*; otherwise, it can be difficult to know which process has initial ownership. Multiple processes can have handles of the same mutex object, letting your programs use the mutex object for inter-process synchronization. The following object-sharing mechanisms are available to your programs:

- A child process the *CreateProcess* function creates can inherit a handle to a mutex object if the *lpMutexAttributes* parameter of *CreateMutex* enabled inheritance.

- A process can specify the *mutex-object* handle in a call to the *DuplicateHandle* function to create a duplicate handle that another process can use.
- A process can specify a mutex object's name in a call to the *OpenMutex* or *CreateMutex* functions.

Use the *CloseHandle* function to close the handle. The system closes the handle automatically when the process terminates. The operating system destroys the mutex object when the system closes its last handle to the mutex.

1411 Using a Mutex Within a Sample Program

As you learned in Tip 1410, your programs can easily create mutex objects to help with thread synchronization across multiple processes. In Tip 1410, you learned the basic process behind the creation of a mutex object. However, after you create a mutex, your programs must then obtain a handle to that mutex before they can use it within their own code. As you learned, a second process can obtain a handle to a mutex by invoking a *CreateMutex* statement with the same named mutex as the first process. Alternatively, your programs can use the *OpenMutex* function to obtain a handle to a previously created mutex object. You will use the *OpenMutex* function within your programs as shown in the following prototype:

```
HANDLE OpenMutex(
    DWORD dwDesiredAccess,     // access flag
    BOOL bInheritHandle,       // inherit flag
    LPCTSTR lpName             // pointer to mutex-object name
);
```

The *dwDesiredAccess* parameter specifies the requested access to the mutex object. For systems that support object security (such as secure Windows NT installations), the parameter fails if the security descriptor of the specified object does not permit the requested access for the calling process. The *dwDesiredAccess* parameter can be any combination of either the *MUTEX_ALL_ACCESS* value, which specifies all possible access flags for the mutex object; the *SYNCHRONIZE* value, which only Windows NT supports, and which lets a process use the mutex handle in any of the wait functions to acquire ownership of the mutex object, or in the *ReleaseMutex* function to release ownership; or both.

The *bInheritHandle* parameter specifies whether the returned handle is inheritable. If the returned handle is inheritable, a process previously created with the *CreateProcess* function can inherit the handle; otherwise, the handle is not inheritable. The *lpName* parameter points to a *NULL*-terminated string that names the mutex object that *OpenMutex* is to open. Keep in mind that name comparisons are case sensitive.

The *OpenMutex* function lets multiple processes open handles of the same mutex object. The *OpenMutex* function succeeds only if some process has already created the mutex, using the *CreateMutex* function. The calling process can use the returned handle in any functions that require a mutex object handle, such as the wait functions, which are subject to the limitations of the access the *dwDesiredAccess* parameter specifies.

Your programs can use the *DuplicateHandle* function to duplicate a handle and the *CloseHandle* function to close the handle. The system closes the handle automatically when the process terminates. The operating system destroys the mutex object when the system has closed the mutex's last handle.

To better understand the processing your programs will perform with mutex objects, consider the *Simple_Mutex.cpp* program on the CD-ROM that accompanies this book. The *Simple_Mutex.cpp* program creates a simple mutex object. When the user selects the *Test!* menu item, the program starts a thread that waits for access to a mutex, sleeps, and then releases the mutex. If the user chooses the *Test!* option multiple times, the program will demonstrate *mutex contention*. That is, the mutex will force the subsequent threads to wait until each preceding thread completes its execution. While the *Simple_Mutex.cpp* program creates the child thread within the *WndProc* function, the majority of the program's operative processing occurs within the *ChildThreadProc* function.

USING SEMAPHORES 1412

As you have learned, most semaphores use a counter for synchronization. Within your programs, you will generally use semaphores to limit access to an object, some code, or another limited resource. When you create a semaphore, you tell the semaphore how many accesses it should permit, and its initial number of accesses. You will use the *CreateSemaphore* function to create a semaphore. You will use the *CreateSemaphore* function within your programs as shown in the following prototype:

```
HANDLE CreateSemaphore(
    LPSECURITY_ATTRIBUTES lpSemaphoreAttributes,// security attributes
    LONG lInitialCount,                         // initial count
    LONG lMaximumCount,                         // maximum count
    LPCTSTR lpName                              // pointer to semaphore name
);
```

The *CreateSemaphore* function accepts the parameters Table 1412 details.

Parameter	Description
lpSemaphoreAttributes	Pointer to a *SECURITY_ATTRIBUTES* structure that determines whether child processes can inherit the returned handle. If *lpSemaphoreAttributes* is *NULL*, child processes cannot inherit the handle. Under Windows NT, the *lpSecurityDescriptor* member of the structure specifies a security descriptor for the new semaphore. If *lpSemaphoreAttributes* is *NULL*, the semaphore gets a default security descriptor. Under Windows 95, the function ignores the *lpSecurityDescriptor* member of the structure.
lInitialCount	Specifies an initial count for the semaphore object. This value must be greater than or equal to zero and less than or equal to *lMaximumCount*. The state of a semaphore is signaled when its count is greater than zero and non-signaled when it is zero. The wait function decreases the count by one increment whenever it releases a thread that was waiting for the semaphore. The wait function increases the count by a specified amount by calling the *ReleaseSemaphore* function.
lMaximumCount	Specifies the maximum count for the semaphore object. This parameter must be greater than zero.
lpName	Points to a *NULL*-terminated string that specifies the name of the semaphore object. Windows limits the name to *MAX_PATH* characters, and the name can contain any character except the backslash path-separator (\\). Keep in mind that name comparison is case sensitive. If *lpName* matches the name of an existing named semaphore object, *lpName* requests *SEMAPHORE_ALL_ACCESS* access to the existing object. The function ignores the *lInitialCount* and *lMaximumCount* parameters because the creating process has already set them. If the *lpSemaphoreAttributes* parameter is not *NULL*, it determines whether the handle is inheritable, but the function ignores the parameter's security-descriptor member. If *lpName* is *NULL*, the function creates the semaphore object without a name. If *lpName* matches the name of an existing event, mutex, or file-mapping object, the function fails and the *GetLastError* function returns the *ERROR_INVALID_HANDLE* constant. This occurs because event, mutex, semaphore, and file-mapping objects share the same name space.

Table 1412 The parameters for the ***CreateSemaphore*** *function.*

If the *CreateSemaphore* function succeeds, the return value is a handle to the semaphore object. If the named semaphore object existed before the function call, the *GetLastError* function returns the *ERROR_ALREADY_EXISTS* constant. If the function fails, the function returns *NULL*.

The handle the *CreateSemaphore* function returns has *SEMAPHORE_ALL_ACCESS* access to the new semaphore object, and your programs can use it in any function that requires a handle to a semaphore object. Any calling process thread can specify the semaphore-object handle in a call to one of the wait functions. The single-object wait functions return when the state of the specified object is signaled. Your programs can instruct the multiple-object wait functions to return when any one or when all the specified objects are signaled. When a wait function returns, the operating system releases the waiting thread to continue its execution.

The state of a semaphore object is signaled when its count is greater than zero, and non-signaled when its count is equal to zero. The *lInitialCount* parameter specifies the initial count. Each time the operating system releases a waiting thread because of the semaphore's signaled state, the semaphore decreases the count by one. Use the *ReleaseSemaphore* function to increment a semaphore's count by a specified amount. The count can never be less than zero or greater than the value specified in the *lMaximumCount* parameter.

Multiple processes can have handles of the same semaphore object, letting the processes use the object for interprocess synchronization. The following object-sharing mechanisms are available to your programs:

- A child process the *CreateProcess* function created can inherit a handle to a semaphore object if the *lpSemaphoreAttributes* parameter of *CreateSemaphore* enabled inheritance.
- A process can specify the *semaphore-object* handle in a call to the *DuplicateHandle* function to create a duplicate handle that another process can use.
- A process can specify the name of a semaphore object in a call to the *OpenSemaphore* or *CreateSemaphore* functions.

Use the *CloseHandle* function to close the handle. The system closes the handle automatically when the process terminates. The operating system destroys a semaphore object when the system has closed the last handle to the semaphore.

As you learned in Tip 1411, your programs use a mutex by creating the mutex, opening the mutex, and releasing the mutex. You will perform similar steps with a semaphore. To better understand the processing your programs can perform with a semaphore, consider the program *Create_Semaphore.cpp* on the CD-ROM that accompanies this book. The *Create_Semaphore.cpp* program uses semaphores to ensure that only four threads at one time can execute the child thread procedure. Each time the user selects the *Test!* menu option, the program tries to create a new thread. However, the thread procedure makes sure that there are available thread spots (up to four) before it lets the thread perform its processing. If there are no available resources for the semaphore, the thread procedure waits until a process becomes available. The operative processing occurs within the *ChildThreadProc* function of the *Create_Semaphore.cpp* program.

1413 Using a Simple Event Processor

As you have learned, events are the most primitive form of synchronization objects. Your programs will generally use an event to signal one or more threads that an operation has completed. Just as with semaphores and mutexes, your programs will create an event by calling the *CreateEvent* function. You will use the *CreateEvent* function within your programs as shown in the following prototype:

```
HANDLE CreateEvent(
    LPSECURITY_ATTRIBUTES lpEventAttributes,   // security attributes
    BOOL bManualReset,                         // flag for manual-reset event
    BOOL bInitialState,                        // flag for initial state
    LPCTSTR lpName                             // pointer to event-object name
);
```

The *CreateEvent* function accepts the parameters Table 1413 details.

Parameter	Description
lpEventAttributes	Pointer to a *SECURITY_ATTRIBUTES* structure that determines whether child processes can inherit the returned handle. If *lpEventAttributes* is *NULL*, the handle is not inheritable. Under Windows NT, the *lpSecurityDescriptor* member of the structure specifies a security descriptor for the new event. If *lpEventAttributes* is *NULL*, the event gets a default security descriptor. Under Windows 95, *CreateEvent* ignores the *lpSecurityDescriptor* member of the structure.
bManualReset	Specifies whether *CreateEvent* creates a manual-reset or auto-reset event object. If *True*, you must use the *ResetEvent* function to manually reset the state to non-signaled. If *False*, Windows automatically resets the state to non-signaled after the system has released a single waiting thread.
bInitialState	Specifies the initial state of the event object. If *True*, the initial state is signaled; otherwise, it is non-signaled.
lpName	Points to a *NULL*-terminated string that specifies the name of the event object. Windows limits the name to *MAX_PATH* characters and the name can contain any character except the backslash path-separator (\). Keep in mind that name comparison is case sensitive. If *lpName* matches the name of an existing named event object, *lpName* requests *EVENT_ALL_ACCESS* access to the existing object. *CreateEvent* ignores the *bManualReset* and *bInitialState* parameters because the creating process has already set them. If the *lpEventAttributes* parameter is not *NULL*, it determines whether the handle is inheritable, but *CreateEvent* ignores its security-descriptor member. If *lpName* is *NULL*, *CreateEvent* creates the event object without a name. If *lpName* matches the name of an existing semaphore, mutex, or file-mapping object, the function fails and the *GetLastError* function returns the *ERROR_INVALID_HANDLE* constant. The error occurs because event, mutex, semaphore, and file-mapping objects share the same name space.

***Table 1413** The parameters for the **CreateEvent** function.*

If the *CreateEvent* function succeeds, the return value is a handle to the event object. If the named event object existed before the function call, the *GetLastError* function returns the *ERROR_ALREADY_EXISTS* constant. If the function fails, the return value is *NULL*.

The handle *CreateEvent* returns has the *EVENT_ALL_ACCESS* access right to the new event object and any function that requires a handle to an event object can use it. Any calling process thread can specify the event-object handle in a call to one of the wait functions. The single-object wait functions return when the specified object's state is signaled. You can instruct the multiple-object wait functions to return when any one or when all the specified objects are signaled. When a wait function returns, the operating system releases the waiting thread to continue its execution.

The *bInitialState* parameter specifies the event object's initial state. Use the *SetEvent* function to set the state of an event object to signaled. Use the *ResetEvent* function to reset the state of an event object to non-signaled. When the state of a manual-reset event object is signaled, it remains signaled until the *ResetEvent* function explicitly resets it to non-signaled. Your programs can release any number of waiting threads, or threads that subsequently begin wait operations for the specified event object, while the object's state is signaled.

When the state of an auto-reset event object is signaled, it remains signaled until your program or the operating system releases a single waiting thread; the system then automatically resets the state to non-signaled. If no threads are waiting, the event object's state remains signaled. Multiple processes can have handles of the same event object, enabling use of the object for inter-process synchronization. The following object-sharing mechanisms are available to your programs:

- A child process that the *CreateProcess* function created can inherit a handle to an event object if the *lpEventAttributes* parameter of *CreateEvent* enabled inheritance.
- A process can specify the event-object handle in a call to the *DuplicateHandle* function to create a duplicate handle that another process can use.
- A process can specify an event object's name in a call to the *OpenEvent* or *CreateEvent* functions.

Use the *CloseHandle* function to close the handle. The system closes the handle automatically when the process terminates. The operating system destroys the event object when the system has closed its last handle.

To better understand the processing your programs perform when they work with events, consider the *Three_Options.cpp* program, contained on the CD-ROM that accompanies this book. To best see the effects of synchronization, run the *Three_Options* program three times and tile the instances on the desktop. In the first two instances, select the Read menu option and in the third instance, select the Write menu option. While the readers can execute simultaneously, the writer must wait until the readers finish. After the read operations finish, the write operation executes. If you reverse the order and select the write operation first, the read operations must wait until the write operation completes its processing. Each process shows its current state within the window's status bar. The *Three_Options* program performs the operative processing within the *WndProc* function.

1414 UNDERSTANDING THE GRAPHICS DEVICE INTERFACE

The graphics device interface (GDI) is a set of library functions that provides Windows applications with a device-independent interface to the screen and to printers. The graphics device interface is a layer between an application and the different types of hardware. The graphics device interface frees the programmer from having to deal with each type of device directly by letting the graphics device interface resolve differences in hardware instead. A well-designed Windows application will function the same on all types of current hardware and any new hardware manufacturers introduce in the future because of its graphics device interface.

All graphics device interface functions in Win32 use 32-bit values for graphics device interface coordinates; however, in Windows 95 and Win32s, the operating system ignores the high-order word, which results in a 16-bit value for coordinates. Only in Windows NT can your applications use the full 32-bit values.

1415 REASONS TO USE THE GRAPHICS DEVICE INTERFACE

As you might imagine, your programs should use the graphics device interface to generate output rather than using textual or non-graphics output. There are many reasons to use the graphics device interface to manipulate a window's contents, including the following:

- You can apply the same device context to multiple devices.
- You can format output on the device context before you send it to the device.
- You can manipulate graphics and other visual information within the window with a device context.
- You can control the window's appearance (after scrolling, sizing, and so on) more easily with a device context.
- Your programs can easily send output that you previously placed on a window device context to a printer or other device context.

Many of the programs that you have used previously in this book used device contexts to maintain information within the program window. In the following Tips, you will focus more closely on the merits and uses of device contexts.

BETTER UNDERSTANDING DEVICE CONTEXTS 1416

The basic tool that Windows uses to provide device independence for an application is a *device context*, or DC. The device context is an internal structure that Windows uses to maintain information about an output device. Instead of sending output directly to the hardware, an application sends the output to the device context instead. Windows then sends it to the hardware on your program's behalf.

A device context always contains one pen to draw lines, one brush to fill areas, one font to output characters, and a series of other values to control how the device context behaves. If the application requires a different font, the application should select the font onto a device context prior to displaying the text. Selecting a new font does not change existing text on a window's client area.

You can visualize the interface a device context provides as being similar to that shown in Figure 1416.

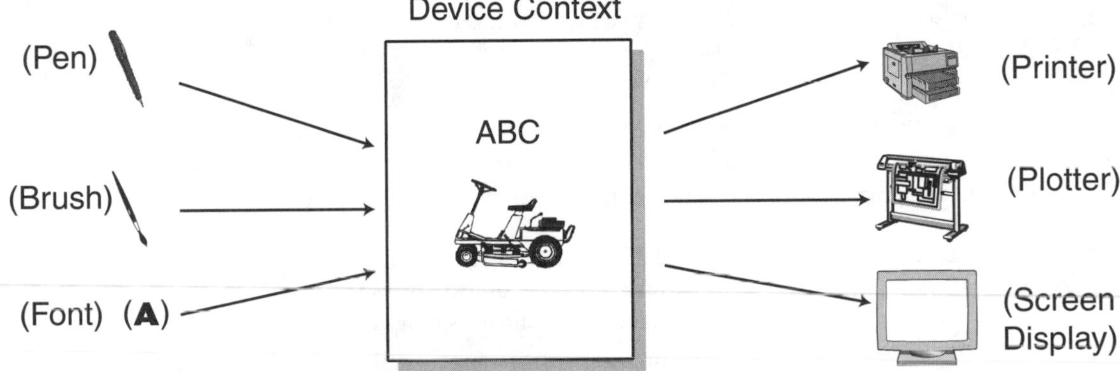

Figure 1416 The logical model for a device context.

USING PRIVATE DEVICE CONTEXTS 1417

Normally, applications share and retrieve device contexts right before use and release them right after use. Handling device contexts for only brief periods of time works best for applications that do not frequently use a device context. An application that requires a device context on an ongoing basis can create a window with its own private device context by specifying the *CS_OWNDC* class style in the class definition for the window. With the *CS_OWNDC* class style, the device context exists throughout a window's life. An application will still use the *GetDC* function to retrieve a handle to the device context, but does not require a call to the *ReleaseDC* function after it completes its device context processing. When you use a private device context with programs that alter the device context settings, you create changes, such as new text colors, pens, and brushes, that remain in effect until the programs alter them again.

UNDERSTANDING ORIGINS AND EXTENTS 1418

A device context has two special mapping modes, *MM_ISOTROPIC* and *MM_ANISOTROPIC*, which a fixed size does not constrain. These mapping modes use two rectangular regions, the *window* and the *viewport*, to derive a scaling factor and an orientation. The window is in logical coordinates, and the viewport is in physical coordinates. Together they determine how the operating system maps logical units to physical units. Both the window and the viewport contain an origin, an x-extent, and a y-extent. The *origin* is a point that describes any one of the four corners. The *viewport origin* is an offset from the *window origin*. The *x-extent* is the horizontal distance from the origin to its opposing corner. The *y-extent* is the vertical distance from the origin to its opposing corner.

Windows defines a horizontal scaling factor by dividing the viewport's x-extent by the window's x-extent. Windows defines a vertical scaling factor by dividing the viewport's y-extent by the window's y-extent. These scaling factors determine the number of logical units that Windows maps to a number of pixels. In addition to determining scaling factors, the window and viewport determine the orientation of an object.

1419 Obtaining a Device Context to a Window

As you have learned, your programs should use device contexts to build graphical and text displays, both on a window device and on a printer device. Your programs can use either the *GetDC* or the *GetDCEx* function to retrieve a device context to a window. The *GetDCEx* function retrieves the handle of a display device context (DC) for the specified window. Your programs can use the display device context in subsequent graphics device interface functions to draw in the client area. The *GetDCEx* function is an extension to the *GetDC* function that gives an application more control over whether and how clipping occurs in the client area. Within your programs, you will use the *GetDCEx* function as shown in the following prototype:

```
HDC GetDCEx(
    HWND hWnd,          // handle of window
    HRGN hrgnClip,      // handle of clip region
    DWORD flags         // device-context creation flags
);
```

The *hWnd* parameter identifies the window where drawing will occur. The *hrgnClip* parameter specifies a clipping region that you may combine with the visible region of the client window. The *flags* parameter specifies how an application will create a device context. The *flags* parameter can be a combination of the values Table 1419 lists.

Value	Meaning
DCX_WINDOW	Returns a device context corresponding to the window rectangle rather than the client rectangle.
DCX_CACHE	Returns a device context from the cache, rather than the *OWNDC* or *CLASSDC* window. Essentially, this value overrides *CS_OWNDC* and *CS_CLASSDC*.
DCX_PARENTCLIP	Uses the visible region of the parent window. This value ignores the parent's *WS_CLIPCHILDREN* and *CS_PARENTDC* style bits. *GetDCEx* sets the device context origin to the upper left corner of the window that *hWnd* identifies.
DCX_CLIPSIBLINGS	Excludes the visible regions of all sibling windows above the window *hWnd* identifies.
DCX_CLIPCHILDREN	Excludes the visible regions of all child windows below the window *hWnd* identifies.
DCX_NORESETATTRS	Does not reset the attributes of this device context to the default attributes when the program releases this device context.
DCX_LOCKWINDOWUPDATE	Lets the program and the user draw onto the device context even if there is a *LockWindowUpdate* call in effect that would otherwise exclude this window. You will use this value to let your programs draw during a tracking operation.
DCX_EXCLUDERGN	Excludes the clipping region *hrgnClip* identifies from the visible region of the device context *GetDCEx* returns.
DCX_INTERSECTRGN	Intersects the clipping region *hrgnClip* identifies with the visible region of the device context *GetDCEx* returns.

*Table 1419 The possible values for the **flags** parameter.* *(continued on following page)*

Value	Meaning
DCX_VALIDATE	When specified with *DCX_INTERSECTUPDATE*, causes the device context *GetDCEx* returns to be completely validated. Using this function with both *DCX_INTERSECTUPDATE* and *DCX_VALIDATE* is identical to using the *BeginPaint* function.

***Table 1419** The possible values for the **flags** parameter. (continued from previous page)*

Unless the display device context belongs to a window class, the application must call the *ReleaseDC* function to release the device context after painting. Because only five common device contexts are available to programs at any given time, failure to release a device context can prevent other programs from accessing it. Both the *GetDC* and *GetDCEx* functions return a device context that belongs to the window's class if the program specified *CS_CLASSDC*, *CS_OWNDC*, or *CS_PARENTDC* as a style in the *WNDCLASS* structure when it registered the class.

To better understand the processing the *GetDCEx* function performs, consider the program *Draw_Hollow.cpp*, contained on the CD-ROM that accompanies this book. The *Draw_Hollow.cpp* program uses the *GetDCEx* function to get a device context for the client area of the window, excluding a region. The program then paints a gray rectangle on the client area, excluding the internal (white) region. The *Draw_Hollow.cpp* program performs the operative processing within the *WndProc* function.

CREATING A DEVICE CONTEXT FOR A PRINTER

As you have learned, your programs should use device contexts to draw output onto a device. While windows have a built-in device context (either private or public), other output devices do not have preexisting contexts. Your programs should use the *CreateDC* function to create a device context for a device using the name you specify. You will use the *CreateDC* function within your programs as shown in the following prototype:

```
HDC CreateDC(
    LPCTSTR lpszDriver,        // pointer to string specifying driver name
    LPCTSTR lpszDevice,        // pointer to string specifying device name
    LPCTSTR lpszOutput,        // do not use; set to NULL
    CONST DEVMODE *lpInitData  // pointer to optional printer data
);
```

The *CreateDC* function accepts the parameters Table 1420.1 details.

Parameter	Description
lpszDriver	Applications written for earlier versions of Windows used this parameter to specify the filename (without extension) of the device driver. In Windows 95 and Win32-based applications, *CreateDC* ignores this parameter, which should be *NULL*, with one exception: you may obtain a display device context by specifying the *NULL*-terminated string *DISPLAY*. If this parameter is *DISPLAY*, all other parameters must be *NULL*. In the Windows NT environment, the *lpszDriver* points to a *NULL*-terminated character string that specifies either *DISPLAY* for a display driver, or the name of a printer driver, which is usually *WINSPOOL*.
lpszDevice	Points to a *NULL*-terminated character string that specifies the name of the specific output device your programs use. The Print Manager will show the output device name, such as "Epson FX-80," which is a Windows-internal name, not necessarily the printer model name. You must use the *lpszDevice* parameter.

***Table 1420.1** The parameters for the **CreateDC** function. (continued on following page)*

Parameter	Description
lpszOutput	Windows ignores this parameter. Do not use it in a Win32 application. Win32-based applications should set this parameter to *NULL* because *lpszOutput* exists to provide compatibility for applications written for earlier versions of Windows.
lpInitData	Points to a *DEVMODE* structure containing device-specific initialization data for the device driver. The *DocumentProperties* function retrieves this structure, and fills the structure as appropriate for the specified device. The *lpInitData* parameter must be *NULL* if the device driver is to use the default initialization (if any) the user specifies.

Table 1420.1 The parameters for the ***CreateDC*** *function. (continued from previous page)*

If the *CreateDC* function succeeds, the return value is the handle to a device context for the specified device. If *CreateDC* fails, the return value is *NULL*.

As you saw in Table 1420.1, the *CreateDC* function expects as its last parameter a pointer to a *DEVMODE* structure that contains device-specific initialization information for the device driver. The Win32 API defines the *DEVMODE* structure as shown here (depending on your operating system, the structure's members may vary):

```
typedef struct _devicemode {
    TCHAR  dmDeviceName[32];
    WORD   dmSpecVersion;
    WORD   dmDriverVersion;
    WORD   dmSize;
    WORD   dmDriverExtra;
    DWORD  dmFields;
    short  dmOrientation;
    short  dmPaperSize;
    short  dmPaperLength;
    short  dmPaperWidth;
    short  dmScale;
    short  dmCopies;
    short  dmDefaultSource;
    short  dmPrintQuality;
    short  dmColor;
    short  dmDuplex;
    short  dmYResolution;
    short  dmTTOption;
    short  dmCollate;
    TCHAR  dmFormName[32];
    WORD   dmUnusedPadding;
    USHORT dmBitsPerPel;
    DWORD  dmPelsWidth;
    DWORD  dmPelsHeight;
    DWORD  dmDisplayFlags;
    DWORD  dmDisplayFrequency;
} DEVMODE;
```

The *DEVMODE* data structure contains information about the device initialization and environment of a printer. Table 1420.2 details the members of the *DEVMODE* data structure.

Members	Description
dmDeviceName	Specifies the name of the device the driver supports (for example, "PCL/HP LaserJet" in the case of *PCL/HP LaserJet®*). This string is unique among device drivers.
dmSpecVersion	Specifies the version number of the initialization data specification on which the device context structure is based.

Table 1420.2 The members of the ***DEVMODE*** *structure. (continued on following page)*

Members	Description
dmDriverVersion	Specifies the printer driver version number the printer driver developer assigns.
dmSize	Specifies the size, in bytes, of the *DEVMODE* structure, except the *dmDriverData* (device-specific) member. If an application manipulates only the driver-independent portion of the data, the application can use this member to determine the length of a structure without having to account for different versions.
dmDriverExtra	Contains the number of bytes of private driver-data that follow the *DEVMODE* structure. If a device driver does not use device-specific information, set this member to zero.
dmFields	Specifies which of the remaining members in the *DEVMODE* structure have been initialized. Bit zero (defined as *DM_ORIENTATION*) corresponds to *dmOrientation*; bit 1 (defined as *DM_PAPERSIZE*), which specifies *dmPaperSize*, and so on. A printer driver supports only those members that are appropriate for the printer's technology.
dmOrientation	Selects the orientation of the paper. This member can be either *DMORIENT_PORTRAIT (1)* or *DMORIENT_LANDSCAPE (2).*
dmPaperSize	Selects the size of the paper to print on. You can set this member to zero if the length and width of the paper are both set by the *dmPaperLength* and *dmPaperWidth* members. Otherwise, you can set the *dmPaperSize* member to one of the predefined values Table 1420.3 lists.
dmPaperLength	Overrides the length of the paper the *dmPaperSize* member specifies, either for custom paper sizes or for devices such as dot-matrix printers which can print on a page of arbitrary length. This value, along with all other values in this structure that specify a physical length, are in tenths of a millimeter.
dmPaperWidth	Overrides the width of the paper the *dmPaperSize* member specifies.
dmScale	Specifies the factor by which the *DEVMODE* data structure is to scale the printed output. This member scales the apparent page size from the physical page size by a factor of *dmScale/100.* For example, a letter-sized page with a *dmScale* value of 50 would contain as much data as a page of 17 by 22 inches because the output text and graphics would be half their original height and width.
dmCopies	Selects the number of copies printed if the device supports multiple-page copies.
dmDefaultSource	Reserved—must be zero.
dmPrintQuality	Specifies the printer resolution. There are four predefined device-independent values: *DMRES_HIGH* *DMRES_MEDIUM* *DMRES_LOW* *DMRES_DRAFT* If this member contains a non-constant, positive value, the value specifies the number of dots per inch (DPI) and is therefore device-dependent.
dmColor	Switches between color and monochrome on color printers. The following are the possible *dmColor* values: *DMCOLOR_COLOR*

***Table 1420.2** The members of the **DEVMODE** structure.* *(continued on following page)*

Members	Description
	DMCOLOR_MONOCHROME
dmDuplex	Selects duplex, or double-sided, printing for printers capable of duplex printing. The following are the possible *dmDuplex* values: *DMDUP_SIMPLEX* *DMDUP_HORIZONTAL* *DMDUP_VERTICAL*
dmYResolution	Specifies the y-resolution, in dots per inch, of the printer. If the printer initializes this member, the *dmPrintQuality* member specifies the x-resolution, in dots per inch, of the printer.
dmTTOption	Specifies how the printer should print *TrueType*® fonts. This member can be one of the following values: *DMTT_BITMAP* — Prints *TrueType* fonts as graphics. This is the default action for dot-matrix printers. *DMTT_DOWNLOAD* — Downloads *TrueType* fonts as soft fonts. This is the default action for Hewlett-Packard printers that use Printer Control Language (PCL). *DMTT_SUBDEV* — Substitutes device fonts for *TrueType* fonts. This is the default action for *PostScript*® printers.
dmCollate	Specifies whether the printer should use collation when printing multiple copies. Using *DMCOLLATE_FALSE* provides faster, more efficient output because it sends the data to a page printer just once, no matter how many copies you require. The program simply tells the printer to print the page again. The *dmCollate* member can be one of the following values: *DMCOLLATE_TRUE* — Collate when printing multiple copies. *DMCOLLATE_FALSE* — Do NOT collate when printing multiple copies.
dmFormName	Specifies the name of the form to use (for example, Letter or Legal). You can retrieve a complete set of names through the Windows *EnumForms* function.
dmUnusedPadding	Aligns the structure to a *DWORD* boundary. You should not use or reference this member. Microsoft reserves its name and usage, and can change in future releases of Windows.
dmBitsPerPel	Specifies in bits per pixel the color resolution of the display device. For example, 4 bits for 16 colors, 8 bits for 256 colors, or 16 bits for 65,536 colors.
dmPelsWidth	Specifies the width, in pixels, of the visible device surface.
dmPelsHeight	Specifies the height, in pixels, of the visible device surface.
dmDisplayFlags	Specifies the device's display mode. The following are valid flags for the *dmDisplayFlags* member: *DM_GRAYSCALE* — Specifies that the display is a non-color device. If you do not set this flag, the program assumes color. *DM_INTERLACED* — Specifies that the display mode is interlaced. If you do not set this flag, the program assumes the display mode is non-interlaced.
dmDisplayFrequency	Specifies the frequency, in hertz (cycles per second), of the display device in a particular mode.

Table 1420.2 *The members of the* **DEVMODE** *structure. (continued from previous page)*

As Table 1420.2 indicates, the *dmPaperSize* parameter accepts predefined constants that correspond to commonly used international paper sizes. Table 1402.3 lists some of the possible paper size constants.

Value	Paper Dimensions
DMPAPER_LETTER	Letter, 8 1/2 by 11 inches
DMPAPER_LEGAL	Legal, 8 1/2 by 14 inches
DMPAPER_A4	A4 Sheet, 210 by 297 millimeters
DMPAPER_LEDGER	Ledger, 17 by 11 inches
DMPAPER_STATEMENT	Statement, 5 1/2 by 8 1/2 inches
DMPAPER_EXECUTIVE	Executive, 7 1/4 by 10 1/2 inches
DMPAPER_FOLIO	Folio, 8 1/2 by 13-inch paper
DMPAPER_QUARTO	Quarto, 215 by 275-millimeter paper
DMPAPER_11X17	11 by 17-inch sheet
DMPAPER_ENV_10	#10 Envelope, 4 1/8 by 9 1/2 inches
DMPAPER_FANFOLD_US	US Std Fanfold, 14 7/8 by 11 inches
DMPA PER_FANFOLD_LGL_GERMAN	German Legal Fanfold, 8 1/2 by 13 inches

***Table 1420.3** Some possible paper size values.*

A device driver's private data will follow the *dmDisplayMode* member. The *dmDriverExtra* member specifies the number of bytes of private data. Applications written for earlier versions of Windows used the *lpszOutput* parameter to specify a port name or to print to a file. Win32-based applications do not need to specify a port name. Win32-based applications can print to a file by calling the *StartDoc* function with a *DOCINFO* structure whose *lpszOutput* parameter specifies the path of the output filename. When you no longer need the device context, call the *DeleteDC* function to delete it.

To better understand the processing *CreateDC* performs, consider the *Print_File.cpp* program contained on the CD-ROM that accompanies this book. The *Print_File.cpp* program prints a single line of text to the printer when the user selects the *Demo* menu item. *CreateDC* creates the device context for the printer. As the CD-ROM code currently reads, the printer name is "HP LaserJet 4 Plus," but you can change it to match the driver for your system, or use the *EnumPrinters* function to determine which printer you have connected to your computer. The *WndProc* function contains the operative processing for the *Print_File.cpp* program.

USING CREATECOMPATIBLEDC TO CREATE A MEMORY DEVICE CONTEXT 1421

In previous Tips, you have learned how your programs can use device contexts to generate output to both the display and to printers. However, your programs cannot generally draw directly onto a device context. The *CreateCompatibleDC* function creates a memory device context (DC) compatible with a device you specify. Before an application can use a memory device context for drawing operations, you must select a bitmap of the correct width and height into the device context. After you have selected a bitmap, you can use the device context to prepare images that a program will copy to the screen or print. Whenever your programs work with bitmaps (which you will learn about in later Tips), your programs will place the bitmap onto a memory device context first, then copy it to specific device context. Your programs will use the *CreateCompatibleDC* function as shown in the following prototype:

```
HDC CreateCompatibleDC(HDC hdc);
```

The *hdc* parameter identifies the device context. If the *hdc* handle is *NULL*, the *CreateCompatibleDC* function creates a memory device context compatible with the application's current screen. If *CreateCompatibleDC* succeeds, the return value is a handle to a memory device context. If *CreateCompatibleDC* fails, the return value is *NULL*.

You can use the *CreateCompatibleDC* function only with devices that support raster operations. An application can determine whether a device supports raster operations by calling the *GetDeviceCaps* function. When you no longer need the memory device context, call the *DeleteDC* function to delete it.

To better understand the processing the *CreateCompatibleDC* function performs, consider the program *Draw_Bitmap.cpp* contained on the CD-ROM that accompanies this book. The *Draw_Bitmap* program loads a bitmap and places it into a memory device context, then copies it onto the client area of the window. As usual, the *WndProc* function contains the *Draw_Bitmap* program's operative processing code.

1422 UNDERSTANDING THE DANGERS OF CREATEDC

As you learned in Tip 1420, your programs will often use the *CreateDC* function to obtain a device context for a printer. However, your programs can also use *CreateDC* to get the device context for the monitor's screen (the hardware screen, not a single window's client area, or even the desktop's client area). When you use *CreateDC* to obtain the device context for the screen, your programs can actually draw anywhere on the screen, not just within a program area's boundaries. In addition to yielding potentially unpredictable results, such processing is not consistent with the Windows standard. When trying to obtain a device context to a window on the screen, your programs should always use the *GetDC* or *BeginPaint* function, not the *CreateDC* function.

1423 USING THE CREATEFONT FUNCTION

As your programs work closely with device contexts and create many interesting screen displays, you may find that your programs must modify the fonts a window displays, or create custom fonts. Your programs can create fonts using either the *CreateFont* or *CreateFontIndirect* function. If you are manipulating multiple fonts, you should generally use the *CreateFontIndirect* function. If you are manipulating only a single font, you should generally use the *CreateFont* function. The *CreateFont* function lets your programs create a logical font (a numeric font definition) with specific characteristics. Subsequently, you can select the logical font as the font for any device. Your programs will use the *CreateFont* function as shown in the following prototype:

```
HFONT CreateFont(
    int nHeight,                // logical height of font
    int nWidth,                 // logical average character width
    int nEscapement,            // angle of escapement
    int nOrientation,           // base-line orientation angle
    int fnWeight,               // font weight
    DWORD fdwItalic,            // italic attribute flag
    DWORD fdwUnderline,         // underline attribute flag
    DWORD fdwStrikeOut,         // strikeout attribute flag
    DWORD fdwCharSet,           // character set identifier
    DWORD fdwOutputPrecision,   // output precision
    DWORD fdwClipPrecision,     // clipping precision
    DWORD fdwQuality,           // output quality
    DWORD fdwPitchAndFamily,    // pitch and family
    LPCTSTR lpszFace            // pointer to typeface name string
);
```

The *CreateFont* function accepts the parameters Table 1423.1 details.

Parameter	Description
nHeight	Specifies the height, in logical units, of the font's character cell or character. The character height value (also known as the em height) is the character cell height value minus the internal-leading value. The font *mapper* interprets the value specified in *nHeight*. If the value for *nHeight* is greater than zero, the font mapper transforms this value into device units and matches it against the cell height of the available fonts.

***Table 1423.1** The parameters for the **CreateFont** function.* *(continued on following page)*

Parameter	Description
	If the value for *nHeight* equals zero, the font mapper uses a default height value when it searches for a match. If the value for *nHeight* is less than zero, the font mapper transforms this value into device units and matches its absolute value against the character height of the available fonts. For all height comparisons, the font mapper looks for the largest font that does not exceed the requested size. This mapping occurs when you use the font for the first time.
nWidth	Specifies the average width, in logical units, of characters in the requested font. If the value is zero, the font mapper chooses a "closest match" value. The *nWidth* parameter determines the "closest match" value by comparing the absolute values of the difference between the current device's aspect ratio and the digitized aspect ratio of available fonts.
nEscapement	Specifies the angle, in tenths of degrees, between the *escapement* vector and the x-axis of the device. The escapement vector is parallel to the base line of a row of text. Under Windows NT, when you set the graphics mode to *GM_ADVANCED*, you can specify the escapement angle of the string, independent of the orientation angle of the string's characters. When you set the graphics mode to *GM_COMPATIBLE*, *nEscapement* specifies both the escapement vector and the orientation angle. In general, you should set *nEscapement* and *nOrientation* to the same value. Under Windows 95, the *nEscapement* parameter specifies both the escapement vector and the orientation angle. However, just as with Windows NT, you should generally set *nEscapement* and *nOrientation* to the same value.
nOrientation	Specifies the angle, in tenths of degrees, between each character's base line and the x-axis of the device.
fnWeight	Specifies the weight of the font in the range 0 through 1000. For example, 400 is normal and 700 is bold. If the value is zero, the operating system uses a default weight. The *fnWeight* parameter can be one of the values Table 1423.2 lists.
fdwItalic	Specifies an italic font if set to *True*.
fdwUnderline	Specifies an underlined font if set to *True*.
fdwStrikeOut	Specifies a strikeout font if set to *True*.
fdwCharSet	Specifies the character set. The values in Table 1423.3 are predefined. The *OEM_CHARSET* value specifies a character set that is operating-system dependent. You can use the *DEFAULT_CHARSET* value to let the name and size of a font fully describe the logical font. If the specified font name does not exist, the operating system can substitute a font from any character set for the specified font; therefore, you should use *DEFAULT_CHARSET* sparingly to avoid unexpected results. Fonts with other character sets may exist in the operating system. If an application uses a font with an unknown character set, it should not try to translate or interpret strings that it renders with that font. The *fdwCharSet* parameter is important in the font mapping process. To ensure consistent results, you must specify a particular character set. If you specify a typeface name in the *lpszFace* parameter, make sure that the *fdwCharSet* value matches the character set of the typeface specified in *lpszFace*.

***Table 1423.1** The parameters for the **CreateFont** function. (continued on following page)*

Parameter	Description
fdwOutputPrecision	Specifies the output precision. The *output precision* defines how closely the output must match the requested font's height, width, character orientation, escapement, pitch, and font type. It can be one of the values Table 1423.4 details. Applications can use the *OUT_DEVICE_PRECIS*, *OUT_RASTER_PRECIS*, and *OUT_TT_PRECIS* values to control how the font mapper chooses a font when the operating system contains more than one font with a specified name. For example, if an operating system contains a font named Symbol in raster and *TrueType* form, specifying *OUT_TT_PRECIS* forces the font mapper to choose the *TrueType* version. Specifying *OUT_TT_ONLY_PRECIS* forces the font mapper to choose a *TrueType* font, even if it must substitute a *TrueType* font of another name.
fdwClipPrecision	Specifies the clipping precision. The *clipping precision* defines how to clip characters that are partially outside the clipping region. It can be one or more of the values Table 1423.5 details.
fdwQuality	Specifies the output quality. The *output quality* defines how carefully the graphics device interface must try to match the logical-font attributes to those of an actual physical font. It can be one of the values Table 1423.6 lists.
fdwPitchAndFamily	Specifies the pitch and family of the font. The two low-order bits specify the pitch of the font and can be one of the following values: *DEFAULT_PITCH* *FIXED_PITCH* *VARIABLE_PITCH* The four high-order bits specify the font family and can be one of the values Table 1423.7 lists. An application can specify a value for the *fdwPitchAndFamily* parameter by using the Boolean *OR* operator to join a pitch constant with a family constant. Font *families* describe the look of a font in a general way. They are intended for specifying fonts when the exact typeface requested is not available.
lpszFace	Points to a *NULL*-terminated string that specifies the typeface name of the font. The length of this string must not exceed 32 characters, including the *NULL* terminator. You can use the *EnumFontFamilies* function to enumerate the typeface names of all currently available fonts. If *lpszFace* is *NULL* or points to an empty string, the graphics device interface uses the first font that matches the other specified attributes.

***Table 1423.1** The parameters for the **CreateFont** function. (continued from previous page)*

As you learned in Table 1423.1, your programs can set a variety of preset font "weights" when creating a logical font. Table 1423.2 details the possible values for the *nWeight* parameter.

Value	Weight	Value	Weight
FW_DONTCARE	0	*FW_SEMIBOLD*	600
FW_THIN	100	*FW_BOLD*	700
FW_EXTRALIGHT	200	*FW_EXTRABOLD*	800
FW_LIGHT	300	*FW_ULTRABOLD*	800

***Table 1423.2** The possible values for the **nWeight** parameter. (continued on following page)*

Value	Weight	Value	Weight
FW_NORMAL	400	*FW_HEAVY*	900
FW_REGULAR	400	*FW_BLACK*	900
FW_MEDIUM	500		

Table 1423.2 The possible values for the ***nWeight*** *parameter.(continued from previous page)*

As you also saw in Table 1423.1, the *CreateFont* function lets you specify a predefined character set for the font. Table 1423.3 lists the possible constant values for the character set.

Predefined Character Sets		
ANSI_CHARSET	*DEFAULT_CHARSET*	*SYMBOL_CHARSET*
SHIFTJIS_CHARSET	*GB2312_CHARSET*	*HANGEUL_CHARSET*
CHINESEBIG5_CHARSET	*OEM_CHARSET*	
Windows 95 only:		
JOHAB_CHARSET	*HEBREW_CHARSET*	*ARABIC_CHARSET*
GREEK_CHARSET	*TURKISH_CHARSET*	*THAI_CHARSET*
EASTEUROPE_CHARSET	*RUSSIAN_CHARSET*	
MAC_CHARSET	*BALTIC_CHARSET*	

Table 1423.3 The predefined character set constants.

As you have learned, the output precision defines how closely the output must match the requested font's height, width, and other characteristics. Your programs can use the constants for the *fdwOutputPrecision* parameter Table 1423.4 lists to control output precision.

Value	Meaning
OUT_CHARACTER_PRECIS	Not used.
OUT_DEFAULT_PRECIS	Specifies the default font mapper behavior.
OUT_DEVICE_PRECIS	Instructs the font mapper to choose a device font when the system contains multiple fonts with the same name.
OUT_OUTLINE_PRECIS	Under Windows NT, this value instructs the font mapper to choose from *TrueType* and other outline-based fonts. Windows 95 does not use this value.
OUT_RASTER_PRECIS	Instructs the font mapper to choose a raster font when the system contains multiple fonts with the same name.
OUT_STRING_PRECIS	The font mapper does not use this value, but the function does return this value when the program enumerates raster fonts.
OUT_STROKE_PRECIS	Under Windows NT, the font mapper does not use this value, but the function does return this value when the program enumerates *TrueType*, other outline-based fonts, and vector fonts. Windows 95 uses this value to map vector fonts, and the function returns the value when the program enumerates *TrueType* or vector fonts.
OUT_TT_ONLY_PRECIS	Instructs the font mapper to choose from only *TrueType* fonts. If there are no *TrueType* fonts installed in the system, the font mapper returns to default behavior.
OUT_TT_PRECIS	Instructs the font mapper to choose a *TrueType* font when the system contains multiple fonts with the same name.

Table 1423.4 The possible values for the ***fdwOutputPrecision*** *parameter.*

Just as your programs can control how precisely Windows should draw fonts onto the device context, so too can your programs control how precisely Windows should clip characters that are partially outside the clipping region. Table 1423.5 details the possible values for the *fdwClipPrecision* parameter.

Value	Meaning
CLIP_DEFAULT_PRECIS	Specifies default clipping behavior.
CLIP_CHARACTER_PRECIS	Not used.
CLIP_STROKE_PRECIS	The font mapper does not use this value, but the function does return this value when the program enumerates raster, vector, or *TrueType* fonts. Under Windows NT, for compatibility, the function always returns this value when enumerating fonts.
CLIP_MASK	Not used.
CLIP_EMBEDDED	You must specify this flag to use an embedded read-only font.
CLIP_LH_ANGLES	When you use this value, the rotation for all fonts depends on whether the orientation of the coordinate system is left-handed or right-handed. If you do not use this value, the device fonts always rotate counterclockwise, but the rotation of other fonts is dependent on the orientation of the coordinate system.
CLIP_TT_ALWAYS	Not used.

***Table 1423.5** The possible values for the **fdwClipPrecision** parameter.*

In addition to the output precision, your programs can control the output quality. There are three possible output qualities the *fdwQuality* parameter uses for your fonts, as Table 1423.6 details.

Value	Meaning
DEFAULT_QUALITY	Appearance of the font does not matter.
DRAFT_QUALITY	Appearance of the font is less important than when you use the *PROOF_QUALITY* value. For graphic device interface raster fonts, the operating system enables scaling, which means that more font sizes are available, but the quality may be lower. The operating system synthesizes bold, italic, underline, and strikeout fonts, if necessary.
PROOF_QUALITY	Character quality of the font is more important than exact matching of the logical-font attributes. For graphics device interface raster fonts, scaling is disabled and the operating system chooses the font closest in size. Although the chosen font size may not be mapped exactly when you use *PROOF_QUALITY*, the quality of the font is high and there is no distortion of appearance. The operating system synthesizes bold, italic, underline, and strikeout fonts, if necessary.

***Table 1423.6** The possible values for the **fdwQuality** parameter.*

Finally, your programs can specify the pitch and family of a font, which helps the *CreateFont* function match the font closely if it cannot match the font exactly. Table 1423.7 details the font families you can use with the *dwPitchAndFamily* parameter.

Value	Description
FF_DECORATIVE	Novelty fonts. Old English is an example.
FF_DONTCARE	Your program does not care or does not know.
FF_MODERN	Fonts with constant stroke width, with or without serifs. Pica, Elite, and Courier New® are examples.
FF_ROMAN	Fonts with variable stroke width and with serifs. MS Serif® is an example.
FF_SCRIPT	Fonts designed to look like handwriting. Script and Cursive are examples.
FF_SWISS	Fonts with variable stroke width and without serifs. MS Sans Serif is an example.

***Table 1423.7** The possible family values for the **fdwPitchAndFamily** parameter.*

If the *CreateFont* function succeeds, the return value is a handle to a logical font. If the *CreateFont* function fails, the return value is *NULL*. When you no longer need the font, call the *DeleteObject* function to delete it.

To help protect the copyrights of vendors who provide fonts for Windows operating systems, applications should always report the exact name of a selected font. Because available fonts can vary from system to system, do not assume that the selected font is always the same as the requested font. For example, if you request a font named "Palatino," but no such font is available on the system, the font mapper will substitute a font that has similar attributes but a different name. Always report the name of the selected font to the user.

To better understand the processing the *CreateFont* function performs, consider the program *Create_BigRoman*, contained on the CD-ROM that accompanies this book. The *Create_BigRoman* program creates a Times New Roman font that is 24 by 16 units in size. The program then uses the newly created font to paint text on the window's client area. The *WndProc* function contains the *Create_BigRoman* program's operative processing.

USING THE ENUMFONTFAMILIES FUNCTION 1424

As you learned in Tip 1423, Windows groups fonts together into families, based on a series of characteristics common to each font within the family. Your programs can use the *EnumFontFamilies* function to enumerate in a specified font family the fonts available on a specified device. Within your programs, you will generally use the *EnumFontFamilies* function to determine what fonts you can use with a specific device and to obtain a pointer to a *LOGFONT* structure that your programs can use together with the *CreateFontIndirect* function to create the font for the specified device. Your programs will use the *EnumFontFamilies* function as shown in the following prototype:

```
int EnumFontFamilies(
   HDC hdc,                          // handle to device context
   LPCTSTR lpszFamily,               // pointer to family-name string
   FONTENUMPROC lpEnumFontFamProc,   // pointer to callback function
   LPARAM lParam                     // address of application-supplied data
);
```

Table 1424.1 details the parameters the *EnumFontFamilies* function accepts.

Parameter	Description
hdc	Identifies the device context.
lpszFamily	Points to a *NULL*-terminated string that specifies the family name of the desired fonts. If *lpszFamily* is *NULL*, the *EnumFontFamilies* function randomly selects and enumerates one font of each available type family.

***Table 1424.1** The **EnumFontFamilies** function's parameters.* *(continued on following page)*

Parameter	Description
lpEnumFontFamProc	Specifies the procedure-instance address of the application-defined callback function. For information about the callback function, see the *EnumFontFamProc* function.
lParam	Points to application-supplied data. The function passes the data to the callback function along with the font information.

*Table 1424.1 The **EnumFontFamilies** function's parameters. (continued from previous page)*

If the *EnumFontFamilies* function succeeds, the return value is the last value the callback function returns. Its meaning is implementation specific. The *EnumFontFamilies* function retrieves the style names associated with a *TrueType* font. With *EnumFontFamilies*, you can retrieve information about unusual font styles (for example, Outline). For each font that has the typeface name the *lpszFamily* parameter specifies, the *EnumFontFamilies* function retrieves information about that font and passes it to the function to which the *lpEnumFontFamProc* parameter points. The application-defined callback function can process the font information as desired. Enumeration continues until there are no more fonts or the callback function returns zero. You will use the callback function as shown here:

```
int CALLBACK EnumFontFamProc(
     ENUMLOGFONT* lpelf,        // pointer to an ENUMLOGFONT structure
     NEWTEXTMETRIC* lpntm,      // pointer to NEWTEXTMETRIC structure
     int nFontType,             // the font type
     LPARAM lParam              // application-defined data
);
```

As you can see, the callback function accepts four parameters. The first is a pointer to an *ENUMLOGFONT* structure. The *ENUMLOGFONT* structure defines the attributes of a font, the complete name of a font, and the style of a font. The Win32 API defines the *ENUMLOGFONT* structure as shown here:

```
typedef struct tagENUMLOGFONT {
   LOGFONT  elfLogFont;
   BCHAR    elfFullName[LF_FULLFACESIZE];
   BCHAR    elfStyle[LF_FACESIZE];
} ENUMLOGFONT;
```

Table 1424.2 defines the members of the *ENUMLOGFONT* structure.

Member	Description
elfLogFont	Specifies a *LOGFONT* structure that defines the attributes of a font.
elfFullName	Specifies a unique name for the font. For example, "ABCD Font Company TrueType Bold Italic Sans Serif."
elfStyle	Specifies the style of the font. For example, "Bold Italic." The *TEXTMETRIC* structure contains basic information about a physical font. All sizes are given in logical units; that is, they depend on the current mapping mode of the display context.

*Table 1424.2 The members of the **ENUMLOGFONT** structure.*

The other three parameters specify *TrueType*-specific information and the font's type (either *DEVICE_FONTTYPE*, *RASTER_FONTTYPE*, or *TRUETYPE_FONTTYPE*, or some combination of the three).

To better understand the processing the *EnumFontFamilies* function performs, consider the *Enum_AllFonts.cpp* program, contained on the CD-ROM that accompanies this book. The program uses the *EnumFontFamilies* function to fill a list box with the available *TrueType* font names. When the user selects the *Test!* menu item, the program displays a sample test string in the currently selected font format. The *Enum_AllFonts.cpp* program's operative processing occurs within the *WndProc*, *EnumFontProc*, and *FindFontProc* functions.

DISPLAYING MULTIPLE FONTS WITH CREATEFONTINDIRECT 1425

As you have learned, your programs can use the *CreateFont* function to create fonts. However, the number of parameters for a single *CreateFont* call is enough to make *CreateFont* burdensome to use. A better alternative is the *CreateFontIndirect* function. The *CreateFontIndirect* function creates a logical font that has the characteristics specified in a particular structure. You can subsequently select the font as the current font for any device context, as shown here:

```
HFONT CreateFontIndirect(CONST LOGFONT *lplf);
```

The *lplf* parameter points to a *LOGFONT* structure that defines the characteristics of the logical font. The *LOGFONT* structure defines the attributes of a font, as shown here:

```
typedef struct tagLOGFONT {
   LONG lfHeight;
   LONG lfWidth;
   LONG lfEscapement;
   LONG lfOrientation;
   LONG lfWeight;
   BYTE lfItalic;
   BYTE lfUnderline;
   BYTE lfStrikeOut;
   BYTE lfCharSet;
   BYTE lfOutPrecision;
   BYTE lfClipPrecision;
   BYTE lfQuality;
   BYTE lfPitchAndFamily;
   TCHAR lfFaceName[LF_FACESIZE];
} LOGFONT;
```

If you look closely at the *LOGFONT* structure, you will see that its members correspond directly to the parameters for the *CreateFont* function in Tip 1423. In fact, the members accept the same values as do the parameters for the *CreateFont* function.

If the *CreateFontIndirect* function succeeds, the return value is a handle to a logical font. If the *CreateFontIndirect* function fails, the return value is *NULL*. The *CreateFontIndirect* function creates a logical font with the characteristics specified in the *LOGFONT* structure. When you select the font with the *SelectObject* function, the graphic device interface's font mapper tries to match the logical font with an existing physical font. If the graphic device interface fails to find an exact match, it provides an alternative whose characteristics match as many of the requested characteristics as possible. When you no longer need the font, call the *DeleteObject* function to delete it.

The *Enum_AllFonts.cpp* program presented in Tip 1424 uses the *CreateFontIndirect* function. Rather than invoking *CreateFont* with fourteen parameters, the program instead invokes *CreateFontIndirect* with only a single parameter, as shown here:

```
hFont = CreateFontIndirect(&lf);
```

RETRIEVING A DEVICE'S CAPABILITIES 1426

As you have learned, your programs will use device contexts to manage text and graphics within your applications, both for on-screen displays and for output to devices, such as printers or plotters. When your program must generate output to a printer or plotter, you will generally require information about that device's capabilities before you send your output to the device. Within your programs, you can use the *GetDeviceCaps* function to retrieve device-specific information about a specified device. You will use the *GetDeviceCaps* function within your programs as shown in the following prototype:

It is important to note that, if the *hdc* parameter identifies the device context of an enhanced metafile, the device technology will be that of the referenced device as previously passed to the *CreateEnhMetaFile* function. To determine whether the context is an enhanced metafile device context, use the *GetObjectType* function.

Finally, if your program must identify the raster characteristics of the device in question, your call to *GetDeviceCaps* will include the *RASTERCAPS* constant. When your program requests the raster characteristics of a device, the function's return value will be one or more of the values in Table 1426.3.

Capability	Meaning
RC_BANDING	Requires banding support
RC_BITBLT	Capable of transferring bitmaps
RC_BITMAP64	Capable of supporting bitmaps larger than 64K
RC_DI_BITMAP	Capable of supporting the *SetDIBits* and *GetDIBits* functions
RC_DIBTODEV	Capable of supporting the *SetDIBitsToDevice* function
RC_FLOODFILL	Capable of performing flood fills
RC_GDI20_OUTPUT	Capable of supporting features of Windows 2.0
RC_PALETTE	Specifies a palette-based device
RC_SCALING	Capable of scaling
RC_STRETCHBLT	Capable of performing the *StretchBlt* function
RC_STRETCHDIB	Capable of performing the *StretchDIBits* function

Table 1426.3 The possible return values for the ***RASTERCAPS*** *request.*

In addition to information about the device's rasterizing capabilities, your programs will often require knowledge of the device's ability to draw or output curved figures. You can check a device's curve-drawing capabilities with the *CURVECAPS* request. The *CURVECAPS* request returns one or more of the values Table 1426.4 lists.

Value	Meaning
CC_NONE	Device does not support curves
CC_CIRCLES	Device can draw circles
CC_PIE	Device can draw pie wedges
CC_CHORD	Device can draw chord arcs
CC_ELLIPSES	Device can draw ellipses
CC_WIDE	Device can draw wide borders
CC_STYLED	Device can draw styled borders
CC_WIDESTYLED	Device can draw borders that are wide and styled
CC_INTERIORS	Device can draw interiors
CC_ROUNDRECT	Device can draw rounded rectangles

Table 1426.4 Return values from the ***CURVECAPS*** *invocation.*

As you have learned, a device may or may not have certain curve-drawing capabilities. Most devices also have certain line-drawing capabilities. Your programs can check a device's line-drawing capabilities with a *LINECAPS* call. Table 1426.5 details the possible return values from the *LINECAPS* call.

Value	Meaning
LC_NONE	Device does not support lines
LC_POLYLINE	Device can draw a polyline
LC_MARKER	Device can draw a marker
LC_POLYMARKER	Device can draw multiple markers
LC_WIDE	Device can draw wide lines
LC_STYLED	Device can draw styled lines
LC_WIDESTYLED	Device can draw lines that are wide and styled
LC_INTERIORS	Device can draw interiors

*Table 1426.5 Return values from the **LINECAPS** invocation.*

In addition to information about a device's ability to draw lines and circles, your programs will often require information about a device's ability to draw polygons, including rectangles, trapezoids, and so on. You can check the device's capabilities with a call to *POLYGONALCAPS*, which will return one or more of the values in Table 1426.6.

Value	Meaning
PC_NONE	Device does not support polygons
PC_POLYGON	Device can draw alternate-fill polygons
PC_RECTANGLE	Device can draw rectangles
PC_WINDPOLYGON	Device can draw winding-fill polygons
PC_SCANLINE	Device can draw a single scanline
PC_WIDE	Device can draw wide borders
PC_STYLED	Device can draw styled borders
PC_WIDESTYLED	Device can draw borders that are wide and styled
PC_INTERIORS	Device can draw interiors

*Table 1426.6 The possible return values from a **POLYGONALCAPS** invocation.*

Finally, your programs will often require information about a device's text display capabilities. Your programs can check a device's text capabilities with a call to *TEXTCAPS*. Calls to *TEXTCAPS* will return one or more of the values Table 1426.7 details.

Bit	Meaning
TC_OP_CHARACTER	Device is capable of character output precision
TC_OP_STROKE	Device is capable of stroke output precision
TC_CP_STROKE	Device is capable of stroke clip precision
TC_CR_90	Device is capable of 90-degree character rotation
TC_CR_ANY	Device is capable of any character rotation
TC_SF_X_YINDEP	Device can scale independently in the x- and y-directions
TC_SA_DOUBLE	Device is capable of doubled character for scaling
TC_SA_INTEGER	Device uses integer multiples only for character scaling
TC_SA_CONTIN	Device uses any multiples for exact character scaling
TC_EA_DOUBLE	Device can draw double-weight characters
TC_IA_ABLE	Device can italicize

*Table 1426.7 The possible return values from a **TEXTCAPS** invocation.* *(continued on following page)*

Bit	Meaning
TC_UA_ABLE	Device can underline
TC_SO_ABLE	Device can draw strikeouts
TC_RA_ABLE	Device can draw raster fonts
TC_VA_ABLE	Device can draw vector fonts
TC_RESERVED	Reserved; must be zero
TC_SCROLLBLT	Device cannot use a bit-block transfer to scroll

Table 1426.7 *The possible return values from a* ***TEXTCAPS*** *invocation. (continued from previous page)*

To better understand the processing the *GetDeviceCaps* function performs, consider the *Get_DevC.cpp* program, contained on the CD-ROM that accompanies this book. The *Get_DevC.cpp* program uses the *CreateIC* function to create an information context for the screen. The *Get_DevC.cpp* program uses the created context to find the number of bits per pixel and the number of color planes for the display. The *Get_DevC.cpp* program uses *GetDeviceCapabilities* to make each determination. The operative processing is, as usual, within the *WndProc* function.

1427 Using the GetSystemMetrics Function to Analyze a Window

As you have learned, your programs can manipulate almost any characteristic of a window, be it a parent window, a child window, or a control window. One of the most common manipulations your programs must perform is to adjust the current display size as appropriate for a computer's display size. You can obtain this information, as well as other information about a computer's system settings, with the *GetSystemMetrics* function. The *GetSystemMetrics* function retrieves various system metrics and system configuration settings. System metrics are the dimensions (widths and heights) of Windows display elements. All dimensions *GetSystemMetrics* retrieves are in pixels. You will use the *GetSystemMetrics* function within your programs as shown in the following prototype:

```
int GetSystemMetrics(int nIndex);
```

The *nIndex* parameter specifies the system metric or configuration setting *GetSystemMetrics* is to retrieve. All *SM_CX** values are widths. All *SM_CY** values are heights. The Windows API defines the values listed in Table 1427.1.

Value	Meaning
SM_ARRANGE	Flags specifying how the system arranged minimized windows.
SM_CLEANBOOT	Value that specifies how the user started the system. The three possible values are 0 Normal boot, 1 Fail-safe boot, and 2 Fail-safe with network boot. A fail-safe boot (also called *SafeBoot*) bypasses the user's start-up files.
SM_CMOUSEBUTTONS	Number of buttons on the mouse, or zero if no mouse is installed.
SM_CXBORDER	The width, in pixels, of a window border. This is equivalent to the *SM_CXEDGE* value for windows with the 3-D look.
SM_CYBORDER	The height, in pixels, of a window border. This is equivalent to the *SM_CYEDGE* value for windows with the 3-D look.
SM_CXCURSOR	Width, in pixels, of a cursor. These are the cursor dimensions the current display driver supports. The system cannot create cursors of other sizes.
SM_CYCURSOR	Height, in pixels, of a cursor. These are the cursor dimensions the current display driver supports. The system cannot create cursors of other sizes.
SM_CXDLGFRAME	Same as *SM_CXFIXEDFRAME.*

Table 1427.1 *The possible system metrics to retrieve. (continued on following page)*

Value	Meaning
SM_CYDLGFRAME	Same as *SM_CYFIXEDFRAME.*
SM_CXDOUBLECLK	Width, in pixels, of the rectangle around the location of a first click in a double-click sequence. The second click must occur within this rectangle for the system to consider the two clicks a double-click. (The two clicks must also occur within a specified time.)
SM_CYDOUBLECLK	Height, in pixels, of the rectangle around the location of a first click in a double-click sequence. The second click must occur within this rectangle for the system to consider the two clicks a double-click. (The two clicks must also occur within a specified time.)
SM_CXDRAG	Width, in pixels, of a rectangle centered on a drag point to allow for limited movement of the mouse pointer before a drag operation begins. This lets the user click and release the mouse button easily without unintentionally starting a drag operation.
SM_CYDRAG	Height, in pixels, of a rectangle centered on a drag point that lets the user move the mouse pointer a limited distance before a drag operation begins. Such a rectangle lets the user click and release the mouse button easily without unintentionally starting a drag operation.
SM_CXEDGE	Width, in pixels, of a 3-D border. *SM_CXEDGE* is a 3-D counterpart of *SM_CXBORDER.*
SM_CYEDGE	Height, in pixels, of a 3-D border. *SM_CYEDGE* is a 3-D counterpart of *SM_CYBORDER.*
SM_CXFIXEDFRAME	Thickness, in pixels, of the frame around the perimeter of a window that has a caption but is not sizable. *SM_CXFIXEDFRAME* is the width of the horizontal border. Same as *SM_CXDLGFRAME.*
SM_CYFIXEDFRAME	Thickness, in pixels, of the frame around the perimeter of a window that has a caption but is not sizable. *SM_CYFIXEDFRAME* is the height of the vertical border. Same as *SM_CYDLGFRAME.*
SM_CXFRAME	Same as *SM_CXSIZEFRAME.*
SM_CYFRAME	Same as *SM_CYSIZEFRAME.*
SM_CXFULLSCREEN	Width of the client area for a full-screen window. To get the coordinates of the portion of the screen the tray does not obscure, call the *SystemParametersInfo* function with the value *GetSystemMetrics* returns.
SM_CYFULLSCREEN	Height of the client area for a full-screen window. To get the coordinates of the portion of the screen the tray does not obscure, call the *SystemParametersInfo* function with the *SPI_GETWORKAREA* value.
SM_CXHSCROLL	Width, in pixels, of the arrow bitmap on a horizontal scroll bar.
SM_CYHSCROLL	Height, in pixels, of a horizontal scroll bar.
SM_CXHTHUMB	Width, in pixels, of the thumb box in a horizontal scroll bar.

***Table 1427.1** The possible system metrics to retrieve.* *(continued on following page)*

Value	Meaning
SM_CXICON	The default width, in pixels, of an icon. This value is typically 32x32, but can vary depending on the installed display hardware. The *LoadIcon* function can only load icons of these dimensions.
SM_CYICON	The default height, in pixels, of an icon. This value is typically 32x32, but can vary depending on the installed display hardware. The *LoadIcon* function can only load icons of these dimensions.
SM_CXICONSPACING	X dimension, in pixels, of a grid cell for items in large icon view. Each item fits into a rectangle of this size when arranged. These values are always greater than or equal to *SM_CXICON* and *SM_CYICON*.
SM_CYICONSPACING	Y dimension, in pixels, of a grid cell for items in large icon view. Each item fits into a rectangle of this size when arranged. These values are always greater than or equal to *SM_CXICON* and *SM_CYICON*.
SM_CXMAXIMIZED	Default X dimension, in pixels, of a maximized top-level window.
SM_CYMAXIMIZED	Default Y dimension, in pixels, of a maximized top-level window.
SM_CXMAXTRACK	Default maximum width, in pixels, of a window that has a caption and sizing borders. The user cannot drag the window frame to a size larger than this width. A window can process the *WM_GETMINMAXINFO* message to override these values.
SM_CYMAXTRACK	Default maximum height, in pixels, of a window that has a caption and sizing borders. The user cannot drag the window frame to a size larger than this height. A window can process the *WM_GETMINMAXINFO* message to override these values.
SM_CXMENUCHECK	Width, in pixels, of the default menu check mark bitmap.
SM_CYMENUCHECK	Height, in pixels, of the default menu check mark bitmap.
SM_CXMENUSIZE	Width, in pixels, of menu bar buttons, such as multiple document (MIDI) child close.
SM_CYMENUSIZE	Height, in pixels, of menu bar buttons, such as multiple document (MIDI) child close.
SM_CXMIN	Minimum width, in pixels, of a window.
SM_CYMIN	Minimum height, in pixels, of a window.
SM_CXMINIMIZED	Width, in pixels, of a normal minimized window.
SM_CYMINIMIZED	Height, in pixels, of a normal minimized window.
SM_CXMINSPACING	Dimensions, in pixels, of a grid cell for minimized windows. Each minimized window fits into a rectangle this size when arranged. These values are always greater than or equal to *SM_CXMINIMIZED* and *SM_CYMINIMIZED*.
SM_CYMINSPACING	Dimensions, in pixels, of a grid cell for minimized windows. Each minimized window fits into a rectangle this size when arranged. These values are always greater than or equal to *SM_CXMINIMIZED* and *SM_CYMINIMIZED*.
SM_CXMINTRACK	Minimum tracking width and height, in pixels, of a window. The user cannot drag the window frame to a size smaller than these dimensions. A window can process the *WM_GETMINMAXINFO* message to override these values.

Table 1427.1 *The possible system metrics to retrieve.* *(continued on following page)*

Value	Meaning
SM_CYMINTRACK	Minimum tracking width and height, in pixels, of a window. The user cannot drag the window frame to a size smaller than these dimensions. A window can process the *WM_GETMINMAXINFO* message to override these values.
SM_CXSCREEN	Width, in pixels, of the screen.
SM_CYSCREEN	Height, in pixels, of the screen.
SM_CXSIZE	Width, in pixels, of a button in a window's caption or title bar.
SM_CYSIZE	Height, in pixels, of a button in a window's caption or title bar.
SM_CXSIZEFRAME	Thickness, in pixels, of the sizing border around the perimeter of a window that the user can resize. *SM_CXSIZEFRAME* is the width of the horizontal border. Same as *SM_CXFRAME*.
SM_CYSIZEFRAME	Thickness, in pixels, of the sizing border around the perimeter of a window that the user can resize. *SM_CYSIZEFRAME* is the height of the vertical border. Same as *SM_CYFRAME*.
SM_CXSMICON	Recommended dimensions, in pixels, of a small icon. Small icons typically appear in window captions and in small icon view.
SM_CYSMICON	Recommended dimensions, in pixels, of a small icon. Small icons typically appear in window captions and in small icon view.
SM_CXSMSIZE	Dimensions, in pixels, of small caption buttons.
SM_CYSMSIZE	Dimensions, in pixels, of small caption buttons.
SM_CXVSCROLL	Width, in pixels, of a vertical scroll bar and height, in pixels, of the arrow bitmap on a vertical scroll bar.
SM_CYVSCROLL	Width, in pixels, of a vertical scroll bar and height, in pixels, of the arrow bitmap on a vertical scroll bar.
SM_CYCAPTION	Height, in pixels, of normal caption area.
SM_CYKANJIWINDOW	For double-byte character set versions of Windows, height, in pixels, of the Kanji window at the bottom of the screen.
SM_CYMENU	Height, in pixels, of single-line menu bar.
SM_CYSMCAPTION	Height, in pixels, of a small caption.
SM_CYVTHUMB	Height, in pixels, of the thumb box in a vertical scroll bar.
SM_DBCSENABLED	*True* or non-zero if the double-byte character set (DBCS) version of *USER.EXE* is installed; *False*, or zero, otherwise.
SM_DEBUG	*True* or non-zero if the debugging version of *USER.EXE* is installed; *False,* or zero, otherwise.
SM_MENUDROPALIGNMENT	*True*, or non-zero if drop-down menus are right-aligned relative to the corresponding menu-bar item; *False*, or zero, if they are left-aligned.
SM_MIDEASTENABLED	*True* if the system is enabled for Hebrew/Arabic languages.
SM_MOUSEPRESENT	*True* or non-zero if a mouse is installed; *False*, or zero, otherwise.
SM_MOUSEWHEELPRESENT	Under Windows NT only, *True* or non-zero if a mouse with a wheel is installed; *False*, or zero, otherwise.
SM_NETWORK	The function returns a value with the least significant bit set if a network is present; otherwise, the function's return value has a cleared least significant bit. Windows reserves the other bits for future use.

Table 1427.1 *The possible system metrics to retrieve.* *(continued on following page)*

Value	Meaning
SM_PENWINDOWS	*True* or non-zero if the Microsoft Windows for Pen computing extensions are installed; *False,* or zero, otherwise.
SM_SECURE	*True* if security is present, *False* otherwise.
SM_SHOWSOUNDS	*True* or non-zero if the user requires an application to present information visually in situations where it would otherwise present the information only in audible form; *False,* or zero, otherwise.
SM_SLOWMACHINE	*True* if the computer has a low-end (slow) processor, *False* otherwise.
SM_SWAPBUTTON	*True* or non-zero if the meanings of the left and right mouse buttons are swapped; *False,* or zero, otherwise.

***Table 1427.1** The possible system metrics to retrieve. (continued from previous page)*

If the function succeeds, the return value will be the requested system metric or configuration setting. If the function fails, the return value will be zero.

System metrics may vary from display to display. The *SM_ARRANGE* setting specifies how the system arranges minimized windows, and consists of a starting position and a direction. The starting position can be one of the values in Table 1427.2.

Value	Meaning
ARW_BOTTOMLEFT	Start at lower left corner of screen (default position).
ARW_BOTTOMRIGHT	Start at lower right corner of screen. Equivalent to *ARW_STARTRIGHT.*
ARW_HIDE	Hide minimized windows by moving them off the visible area of the screen.
ARW_TOPLEFT	Start at upper left corner of screen. Equivalent to *ARV_STARTTOP.*
ARW_TOPRIGHT	Start at upper right corner of screen. Equivalent to *ARW_STARTTOP | SRW_STARTRIGHT.*

***Table 1427.2** The starting position values.*

The direction in which the system arranges minimized windows can be one of the values listed in Table 1427.3.

Value	Meaning
ARW_DOWN	Arrange vertically, top to bottom
ARW_LEFT	Arrange horizontally, left to right
ARW_RIGHT	Arrange horizontally, right to left
ARW_UP	Arrange vertically, bottom to top

***Table 1427.3** The possible arrangement values.*

Tip 1428 details how your programs can use system metrics while processing.

1428 UNDERSTANDING THE USES FOR GETSYSTEMMETRICS

As you learned in Tip 1427, your programs can use the *GetSystemMetrics* function to obtain information about the current Windows metrics on any given computer. As your programs continue to become more complex, and as you build applications that may run on different computers, it becomes important to check system metrics before generating displays because it helps ensure your screen displays look appropriate on both 640x480 screens and 1024x768 screens. The program *Get_Sysm.cpp*, for example, contained on the CD-ROM that accompanies this book, uses *GetSystemMetrics* to enumerate the current status of a window's size, the screen's size, and other mixed information. When the user selects *Demo*, the program will display the information within the menu bar.

GETTING A DEVICE CONTEXT FOR AN ENTIRE WINDOW 1429

As you have learned, your programs will generally work with the device context for a window's client area. However, there may be times when your programs will require a device context throughout the entire window. To get a device context to an entire window, your programs can use the *GetWindowDC* function. The *GetWindowDC* function retrieves the device context (DC) for the entire window, including title bar, menus, and scroll bars. A window device context permits painting anywhere in a window, because the origin of the device context is the upper left corner of the window instead of the client area.

GetWindowDC assigns default attributes to the window device context each time it retrieves the device context. Because *GetWindowDC* assigns default attributes to the window device context, previous attributes are lost. You will use the *GetWindowDC* function within your programs as shown in the following prototype:

```
HDC GetWindowDC(HWND hWnd);
```

The *hWnd* parameter identifies the window with a device context that *GetWindowDC* is to retrieve. If the function succeeds, the return value will be the handle of a device context for the specified window. If the function fails, the return value will be *NULL*, indicating an error or an invalid *hWnd* parameter.

GetWindowDC is intended for special painting effects within a window's non-client area. Painting in non-client areas of any window is not recommended. You can use the *GetSystemMetrics* function to retrieve the dimensions of various parts of the non-client area, such as the title bar, menu, and scroll bars. After painting is complete, the program must call the *ReleaseDC* function to release the device context. Not releasing the window device context has serious effects on painting that applications request subsequently.

To better understand the processing that *GetWindowDC* performs, consider the *GetWinDC.cpp* program on the CD-ROM that accompanies this book. The *GetWinDC.cpp* program uses the *GetWindowDC* function to retrieve a device context for the entire window. It then uses *GetSystemMetrics* to determine the sizes of the borders and the caption. Finally, it uses the device context for the entire window to paint a pattern in the caption bar of the window. The operative processing occurs inside the *WndProc* function.

RELEASING DEVICE CONTEXTS 1430

As you have learned, releasing an object after your program completes its processing on that object is an important part of good C++ programming. Handling device contexts is no exception. The *ReleaseDC* function releases a device context (DC), freeing it so that other applications can use it. The effect of the *ReleaseDC* function depends on the type of device context. *ReleaseDC* frees only common and window device contexts. It has no effect on class or private device contexts. You will use the *ReleaseDC* function within your programs as shown in the following prototype:

```
int ReleaseDC(HWND hWnd, HDC hDC);
```

The *hWnd* parameter identifies the window whose device context *ReleaseDC* is to release. The *hDC* parameter identifies the device context that *ReleaseDC* is to release. The return value specifies whether *ReleaseDC* succesfully released the device context. If *ReleaseDC* succesfully released the device context, the function will return 1. If *ReleaseDC* did not succesfully release the device context, the function will return zero.

The application must call the *ReleaseDC* function for each call to the *GetWindowDC* function and for each call to the *GetDC* function that retrieves a common device context. An application cannot use the *ReleaseDC* function to release a device context that the program called the *CreateDC* function to create. Instead, the application must use the *DeleteDC* function.

1431 GETTING A WINDOW'S HANDLE FROM THE DEVICE CONTEXT

Your programs may often perform generalized processing on any given device context. However, you may want to avoid performing such processing on the device context associated with a given window. With the *WindowFromDC* function, your programs can convert a device context into a window handle. The *WindowFromDC* function returns the handle of the window associated with the given display device context (DC). Output functions that use the device context draw into the window whose handle *WindowFromDC* returns. Your programs will use the *WindowFromDC* function as shown in the following protoype:

```
HWND WindowFromDC(HDC hDC);
```

The *hDC* parameter identifies the device context from which *WindowFromDC* is to retrieve a handle for the associated window. If the function succeeds, the return value will be the handle of the window associated with the given display device context. If the function fails, the return value will be *NULL*.

1432 UNDERSTANDING DEVICE-DEPENDENT BITMAPS

Bitmaps are blocks of data that your programs can output directly to a device, such as a video display. You can think of bitmaps as a way to store the pixel data directly from the screen into a memory buffer. Painting bitmaps onto the screen is much faster than using graphics device interface functions such as *Rectangle* and *LineTo*. The drawbacks to bitmaps are that they take up a lot of memory and disk space, and they do not scale well, particularly if they contain text. When you scale a bitmap, it loses quality and distorts the text.

Windows provides two types of bitmaps: *device-dependent bitmaps* and *device-independent bitmaps*. The device-dependent bitmap is an older format that is, as its name implies, less flexible than the device-independent bitmap. All Win32 applications that you write should use device-independent bitmaps because Windows provides most of the Win32 functions that deal with device-dependent bitmaps purely for legacy purposes with 16-bit Windows applications.

You normally create bitmaps with a paint program such as Microsoft *Paint* or within a bitmap editor component of an Integrated Development Environment (such as Borland *C++ 5.02* or Visual *C++ 5.0*). You then store the bitmap onto a disk with a filename extension of *BMP*. Windows will save the bitmap as a device-independent bitmap and convert it to a device-dependent bitmap when your program calls *LoadBitmap*. You can precede the bitmap filename with the *BITMAP* keyword to add bitmaps to a resource file, as shown here:

```
pen BITMAP pen.bmp
```

To better understand the processing your programs perform to manipulate bitmaps, consider following code fragment that loads a bitmap using *LoadBitmap*. Then, the code displays the bitmap image using a device-independent context:

```
HBITMAP hBitmap;
HDC   hDC;
HDC   hMemDC;

//  Load the bitmap into memory

hBitmap = LoadBitmap(hInst, "pen");

//  Paint the bitmap onto the MemDC and then the screen

hDC = GetDc(hWnd);
hMemDC = CreateCompatibleDC(hDC);
SelectObject(hMemDC, hBitmap);
BitBlt(hDC, 10, 10, 60, 60, hMemDC, 0, 0, SRCCOPY);
```

```
DeleteDC(hMemDC);
ReleaseDC(hWnd, hDC);
DeleteObject(hBitmap);
```

The device-dependent bitmap format works well for copying sections of the screen into memory and pasting those sections back onto other locations of the screen. When the application must save data to a disk file then display it on another device, however, the device-dependent bitmap format falls flat. The device-dependent bitmap format assumes that you will always display the bitmap on a device similar to the one on which it was originally created, and that the colors are the same on the second device as they were on the first. The unfortunate result is that if you display the bitmap on another device, the colors may be different. The device-independent bitmap avoids the problems of the device-dependent bitmap format.

UNDERSTANDING DEVICE-INDEPENDENT BITMAPS 1433

As you learned in Tip 1432, the device-dependent bitmap format has some limitations. The device-independent bitmap format overcomes those limitations. The most significant difference between a device-dependent bitmap and a device-independent bitmap is that a device-independent bitmap includes a table of the colors the bitmap will use. The header of the bitmap is also more complex in the device-independent bitmap format. Unlike a device-dependent bitmap, a device-independent bitmap is not a graphic object; rather, it is a data format. An application cannot select a device-independent bitmap into a device context. The device-independent bitmap format consists of three separate data sections. The first section of the device-independent bitmap-formatted file is the *BITMAPINFOHEADER* structure, which the Windows API defines as shown here:

```
typedef struct tagBITMAPINFOHEADER{
   DWORD  biSize;
   LONG   biWidth;
   LONG   biHeight;
   WORD   biPlanes;
   WORD   biBitCount
   DWORD  biCompression;
   DWORD  biSizeImage;
   LONG   biXPelsPerMeter;
   LONG   biYPelsPerMeter;
   DWORD  biClrUsed;
   DWORD  biClrImportant;
} BITMAPINFOHEADER;
```

The *BITMAPINFOHEADER* structure contains the members Table 1433.1 details.

Member Name	Description
biSize	Specifies the number of bytes the structure requires.
biWidth	Specifies the width of the bitmap, in pixels.
biHeight	Specifies the height of the bitmap, in pixels. If *biHeight* is positive, the bitmap will be a bottom-up device-independent bitmap and its origin will be the lower left corner. If *biHeight* is negative, the bitmap will be a top-down device-independent bitmap and its origin will be the upper left corner.
biPlanes	Specifies the number of planes for the target device. This value must be set to 1.
biBitCount	Specifies the number of bits per pixel. This value must be 1, 4, 8, 16, 24, or 32.

Table 1433.1 The members of the ***BITMAPINFOHEADER*** *structure.* *(continued on following page)*

biCompression	Specifies the type of compression for a compressed bottom-up bitmap (Windows will not compress top-down device-independent bitmaps). It can be one of the values Table 1433.2 details.
biSizeImage	Specifies the size, in bytes, of the image. This may be set to zero for *BI_RGB* bitmaps.
biXPelsPerMeter	Specifies the horizontal resolution, in pixels per meter, of the target device for the bitmap. An application can use this value to select a bitmap from a resource group that best matches the characteristics of the current device.
biYPelsPerMeter	Specifies the vertical resolution, in pixels per meter, of the target device for the bitmap.
biClrUsed	Specifies the number of color indices in the color table that the bitmap actually uses. If this value is zero, the bitmap will use the maximum number of colors corresponding to the value of the *biBitCount* member for the compression mode *biCompression* specifies. If *biClrUsed* is non-zero and the *biBitCount* member is less than 16, the *biClrUsed* member will specify the actual number of colors the graphics engine or device driver will access. If *biBitCount* is 16 or greater, the *biClrUsed* member will specify the size of the color table the device-independent bitmap uses to optimize performance of Windows color palettes. If *biBitCount* equals 16 or 32, the optimal color palette will start immediately following the three double-word masks. If the bitmap is a packed bitmap (a bitmap in which the bitmap array immediately follows the *BITMAPINFO* header and which a single pointer references), the *biClrUsed* member must be either zero or the actual size of the color table.
biClrImportant	Specifies the number of color indices that the device-independent bitmap's definition specifies as important for displaying the bitmap. If this value is zero, all colors are important.

***Table 1433.1** The members of the **BITMAPINFOHEADER** structure. (continued from previous page)*

As you learned in Table 1433.1, you can create compressed bottom-up bitmaps. When your program loads a compressed bitmap, it should check the value of the *biCompressed* member to determine the compression type. Table 1433.2 shows the possible values for the *biCompressed* member.

Value	Description
BI_RGB	An uncompressed format.
BI_RLE8	A run-length encoded (RLE) format for bitmaps with 8 bits per pixel. The compression format is a two-byte format consisting of a count byte followed by a byte that contains a color index.
BI_RLE4	An RLE format for bitmaps with 4 bits per pixel. The compression format is a two-byte format consisting of a count byte followed by two word-length color indices.
BI_BITFIELDS	Specifies that the bitmap is not compressed and that the color table consists of three double-word color masks that specify the red, green, and blue components, respectively, of each pixel. The *BI_BITFIELDS* value is valid for your programs to use with either 16- or 32-bits-per-pixel bitmaps.

***Table 1433.2** The possible values for the **biCompressed** member.*

The *BITMAPINFO* structure combines the *BITMAPINFOHEADER* structure and a color table to provide a complete definition of the dimensions and colors of a device-independent bitmap. An application should use the information the *biSize* member contains to locate the color table in a *BITMAPINFO* structure, as follows:

```
pColor = ((LPSTR)pBitmapInfo + (WORD)(pBitmapInfo-bmiHeader.biSize));
```

Windows supports formats for compressing bitmaps that define their colors with 8 or 4 bits per pixel. Compression reduces the disk and memory storage the bitmap requires. In Table 1433.2, you learned that Windows supports three compression formats.

When the *biCompression* member is *BI_RLE8*, the creating program originally compressed the bitmap with the run-length encoding (RLE) format for an 8-bit bitmap. The creating program can use the run-length encoding format to compress in encoded or absolute modes. Both modes can occur anywhere in the same bitmap.

Encoded mode consists of two bytes. The first byte specifies the number of consecutive pixels Windows is to draw using the color index the second byte contains. In addition, the program that created the bitmap can set the first byte of the pair to zero to indicate an escape that denotes an end of line, end of bitmap, or delta (that is, a change). The interpretation of the escape depends on the value of the second byte of the pair, which can be one of the values in Table 1433.3.

Value	Meaning
0	End of line.
1	End of bitmap.
2	Delta. The two bytes following the escape contain unsigned values that indicate the horizontal and vertical offsets of the next pixel from the current position.

Table 1433.3 The possible values for the second byte of a byte pair.

On the other hand, in absolute mode, the first byte is zero and the second byte is a value in the range 03H through FFH. The second byte represents the number of bytes that follow, each of which contains the color index of a single pixel. When the second byte is 2 or less, the escape has the same meaning as it does in encoded mode. In absolute mode, the creating program must align each run within the device-independent bitmap on a word boundary.

After the *BITMAPINFOHEADER* structure, a device-independent bitmap contains the color table. The color table is a set of *RGBQUAD* data structures that hold the RGB color for each color the bitmap uses. The *RGBQUAD* structure describes a color consisting of relative intensities of red, green, and blue. There will be as many *RGBQUAD* entries as there are color choices in the bitmap. The Windows API constructs the *RGBQUAD* structure as shown here:

```
typedef struct tagRGBQUAD {
    BYTE    rgbBlue;
    BYTE    rgbGreen;
    BYTE    rgbRed;
    BYTE    rgbReserved;
} RGBQUAD;
```

The *RGBQUAD* structure consists of the members Table 1433.4 details.

Member	Description
rgbBlue	Specifies the intensity of blue in the color
rgbGreen	Specifies the intensity of green in the color
rgbRed	Specifies the intensity of red in the color
rgbReserved	Windows reserves this member; must be zero

Table 1433.4 The members of the **RGBQUAD** *structure.*

The remainder of the device-independent bitmap file contains the actual pixel data for the bitmap. You will learn more about creating and displaying both device-dependent bitmap and device-independent bitmap formats in later Tips.

1434 CREATING BITMAPS

As you have learned, one of the three types of graphics files native to the Windows API is the bitmap file. Within your programs, you can create bitmaps to display within a window. To create bitmaps, your programs will use the *CreateBitmap* function. The *CreateBitmap* function creates a bitmap with the width, height, and color format (color planes and bits per pixel) you specify. You will use the *CreateBitmap* function within your programs as shown in the following protoype:

```
HBITMAP CreateBitmap(
    int nWidth,          // bitmap width, in pixels
    int nHeight,         // bitmap height, in pixels
    UINT cPlanes,        // number of color planes used by device
    UINT cBitsPerPel,    // number of bits required to identify a color
    CONST VOID *lpvBits  // pointer to array containing color data
);
```

When your programs call the *CreateBitmap* function, the function will expect the parameters Table 1434.1 details.

Parameter	Description
nWidth	Specifies the bitmap width, in pixels.
nHeight	Specifies the bitmap height, in pixels.
cPlanes	Specifies the number of color planes the device uses.
cBitsPerPel	Specifies the number of bits the device requires to identify the color of a single pixel.
lpvBits	Points to an array of color data the device uses to set the colors in a rectangle of pixels. You must ensure that you align each scan line in the rectangle along a word (you must pad scan lines that do not word align with zeros). If this parameter is *NULL*, the new bitmap will be undefined.

***Table 1434.1** The parameters for the **CreateBitmap** function.*

If the *CreateBitmap* function succeeds, the return value will be a handle to a bitmap. If the function fails, the return value will be *NULL*.

After you create a bitmap, you can call the *SelectObject* function to select the bitmap into a device context. While your programs can use the *CreateBitmap* function to create color bitmaps, for performance reasons applications should use *CreateBitmap* to create monochrome bitmaps and *CreateCompatibleBitmap* to create color bitmaps. When your program selects a color bitmap previously returned from *CreateBitmap* into a device context, Windows must ensure that the bitmap matches the format of the device context you are selecting the bitmap into. Because *CreateCompatibleBitmap* takes a device context, it returns a bitmap that has the same format as the specified device context. Therefore, subsequent calls to *SelectObject* are faster than with a color bitmap *CreateBitmap* returns.

If the bitmap is monochrome, zeros will represent the foreground color and ones will represent the background color for the destination device context. If an application sets the *nWidth* or *nHeight* parameters to zero, *CreateBitmap* will return the handle of a 1-by-1 pixel, monochrome bitmap. When you no longer need the bitmap, call the *DeleteObject* function to delete it.

To better understand the processing the *CreateBitmap* function performs, consider the *Create_Bitmap.cpp* program, contained on the CD-ROM that accompanies this book. The *Create_Bitmap.cpp* program creates a monochrome bitmap when the user selects the *Demo* menu item. The program selects the bitmap into a memory device context and draws a rectangle and an ellipse on the bitmap. The program then displays the resulting bitmap. The program performs the operative processing within the *WndProc* function.

DISPLAYING BITMAPS

1435

As you saw in Tip 1434, your programs must "push" a bitmap onto the display device context to display the bitmap. In the *Create_Bitmap.cpp* program, you first created the bitmap, then added it to the device context to display it. Typically, your programs will use the *BitBlt* function to display a bitmap. The *BitBlt* function performs a bit-block transfer of the color data that corresponds to a rectangle of pixels from the source device context you specify into a destination device context. You must first select the bitmap into a memory device context (which you create with *CreateCompatibleDC*). Next, you will invoke *BitBlt* to copy the bitmap into the actual device context. Because the actual *BitBlt* copy procedure uses raster operations, you should check the *RASTERCAPS* of a device before performing *BitBlt* against the device. You will use *BitBlt* within your programs in accordance with the following prototype:

```
BOOL BitBlt(
   HDC hdcDest,   // handle to destination device context
   int nXDest,    // x-coordinate of destination rectangle's upper left corner
   int nYDest,    // y-coordinate of destination rectangle's upper left corner
   int nWidth,    // width of destination rectangle
   int nHeight,   // height of destination rectangle
   HDC hdcSrc,    // handle to source device context
   int nXSrc,     // x-coordinate of source rectangle's upper left corner
   int nYSrc,     // y-coordinate of source rectangle's upper left corner
   DWORD dwRop    // raster operation code
);
```

The *BitBlt* function accepts the parameters that Table 1435.1 details.

Parameter	Description
hdcDest	Identifies the destination device context.
nXDest	Specifies the logical x-coordinate of the upper left corner of the destination rectangle.
nYDest	Specifies the logical y-coordinate of the upper left corner of the destination rectangle.
nWidth	Specifies the logical width of the source and destination rectangles.
nHeight	Specifies the logical height of the source and the destination rectangles.
hdcSrc	Identifies the source device context.
nXSrc	Specifies the logical x-coordinate of the upper left corner of the source rectangle.
nYSrc	Specifies the logical y-coordinate of the upper left corner of the source rectangle.
dwRop	Specifies a raster-operation code. These codes define how the function is to combine the color data for the source rectangle with the color data for the destination rectangle to achieve the final color. Table 1435.2 details some common raster operation codes.

*Table 1435.1 The parameters for the **BitBlt** function.*

As you have learned, your programs will generally rasterize bitmaps when placing them onto a device context. Table 1435.2 shows some common raster operation codes your programs will use when placing bitmaps onto a device context.

Value	Description
BLACKNESS	Uses the color associated with index zero in the physical palette to fill the destination rectangle. (This color is black for the default physical palette.)
DSTINVERT	Inverts the destination rectangle.

Table 1435.2 Common raster operations your programs will perform with bitmaps. *(continued on following page)*

MERGECOPY	Uses the Boolean *AND* operator to merge the colors of the source rectangle with the specified pattern.
MERGEPAINT	Uses the Boolean *OR* operator to merge the colors of the inverted source rectangle with the colors of the destination rectangle.
NOTSRCCOPY	Copies the inverted source rectangle to the destination.
NOTSRCERASE	Uses the Boolean *OR* operator to combine the colors of the source and destination rectangles and then inverts the resultant color.
PATCOPY	Copies the specified pattern into the destination bitmap.
PATINVERT	Uses the Boolean *XOR* operator to combine the colors of the specified pattern with the colors of the destination rectangle.
PATPAINT	Uses the Boolean *OR* operator to combine the colors of the pattern with the colors of the inverted source rectangle *y*. You can use the Boolean *OR* operator to combine the result of this operation with the colors of the destination rectangle.
SRCAND	Uses the Boolean *AND* operator to combine the colors of the source and destination rectangles.
SRCCOPY	Copies the source rectangle directly to the destination rectangle.
SRCERASE	Uses the Boolean *AND* operator to combine the inverted colors of the destination rectangle with the colors of the source rectangle.
SRCINVERT	Uses the Boolean *XOR* operator to combine the colors of the source and destination rectangles.
SRCPAINT	Uses the Boolean *OR* operator to combine the colors of the source and destination rectangles.
WHITENESS	Uses the color associated with index 1 in the physical palette to fill the destination rectangle. (This color is white for the default physical palette.)

Table 1435.2 Common raster operations your programs will perform with bitmaps. (continued from previous page)

If a rotation or shear transformation (a transformation that alters the apparent length and orientation of vertical or horizontal lines in an object) is in effect in the source device context, *BitBlt* will return an error. If other transformations exist in the source device context (and a matching transformation is not in effect in the destination device context), the *BitBlt* function will stretch, compress, or rotate the rectangle in the destination device context as necessary.

If the color formats of the source and destination device contexts do not match, the *BitBlt* function will convert the source color format to match the destination format. When *BitBlt* is recording an enhanced metafile, an error occurs if the source device context identifies an enhanced-metafile device context. *BitBlt* will return an error if the source and destination device contexts represent different devices (meaning you should always create the source context with the *CreateCompatibleDC* function).

To better understand the processing your programs will perform with the *BitBlt* function, consider the *Happy_Faces.cpp* program on the CD-ROM that accompanies this book. The *Happy_Faces.cpp* program loads a bitmap into memory, then tiles that bitmap onto the client area of the window. The program performs the operative processing within the *WndProc* function.

1436 Creating DIBitmaps

As you have learned, Windows supports device-dependent bitmaps and device-independent bitmaps. In Tip 1435, for example, you created a device-dependent bitmap and displayed the bitmap on the screen. To display device-independent bitmaps, your programs will perform similar processing. You must convert each device-independent bitmap to a device-dependent bitmap before you can display it within a device context. You will use the *CreateDIBitmap* function to create a device-dependent bitmap from a device-independent bitmap and, optionally, set the bitmap bits. You will use the *CreateDIBitmap* function within your programs as shown in the following prototype:

```
HBITMAP CreateDIBitmap(
   HDC hdc,                          // handle to device context
   CONST BITMAPINFOHEADER *lpbmih,   // pointer to bitmap size and format data
   DWORD fdwInit,                    // initialization flag
   CONST VOID *lpbInit,              // pointer to initialization data
   CONST BITMAPINFO *lpbmi,          // pointer to bitmap color-format data
   UINT fuUsage                      // color-data usage
);
```

The *CreateDIBitmap* function accepts the parameters Table 1436.1 details.

Parameter	Description
hdc	Identifies a device context.
lpbmih	Points to a *BITMAPINFOHEADER* structure. If *fdwInit* is the *CBM_INIT* value, the function will use the *BITMAPINFOHEADER* structure to obtain the desired width and height of the bitmap as well as other information. Note that a positive value for the height indicates a bottom-up device-independent bitmap while a negative value for the height indicates a top-down device-independent bitmap. This scenario is compatible with the *CreateDIBitmap* function.
fdwInit	A set of bit flags that specify how the operating system initializes the bitmap's bits. Your programs can specify only one constant for *fdwInit*. The parameter's value must be either *CBM_INIT* or zero. If this flag is set, the operating system will use the data pointed to by the *lpbInit* and *lpbmi* parameters to initialize the bitmap's bits. If this flag is clear, the function will not use the data pointed to by those parameters. If *fdwInit* is zero, the operating system will not initialize the bitmap's bits.
lpbInit	Points to an array of bytes that contain the initial bitmap data. The format of the data depends on the *biBitCount* member of the *BITMAPINFO* structure to which the *lpbmi* parameter points.
lpbmi	Points to a *BITMAPINFO* structure that describes the dimensions and color format of the array pointed to by the *lpbInit* parameter.
fuUsage	Specifies whether the program that created the bitmap initialized the *bmiColors* member of the *BITMAPINFO* structure and, if so, whether *bmiColors* contains explicit red, green, blue (RGB) values or palette indices. The *fuUsage* parameter must contain one of the values Table 1436.2 details.

***Table 1436.1** The parameters for the **CreateDIBitmap** function.*

As you learned, the *fuUsage* parameter accepts one of two constant values. Table 1436.2 lists the possible constant values for the *fuUsage* parameter.

Value	Meaning
DIB_PAL_COLORS	The bitmap file provides a color table and consists of an array of 16-bit indices into the logical palette of the device context into which your program will select the bitmap.
DIB _RGB_COLORS	The bitmap file provides a color table and contains literal RGB values.

***Table 1436.2** The possible values for the **fuUsage** parameter.*

If the function succeeds, the return value will be a handle to the bitmap. If the function fails, the return value will be *NULL*. When you no longer need the bitmap, call the *DeleteObject* function to delete it.

To better understand the processing the *CreateDIBitmap* function performs, consider the *Create_DIB_Bitmap.cpp* program, contained on the CD-ROM that accompanies this book. The *Create_DIB_Bitmap.cpp* program loads a device-independent bitmap, paints the bitmap onto a memory device context, and then transfers the memory device context to a screen device context (the program window). As usual, the program code within the *WndProc* function in the *Create_DIB_Bitmap.cpp* program performs the operative processing.

1437 Filling a Rectangle with a Pattern

As you have learned, your programs can use bitmaps to add graphical information to window displays. Additionally, your programs can use pens and memory device contexts to draw both simple and complex shapes onto a window. For example, your programs can use the *PatBlt* function to draw rectangles on a device context. The *PatBlt* function paints the given rectangle using the brush that is currently selected into the specified device context. The *PatBlt* function uses the given raster operation to combine the brush color and the surface colors. You will use the *PatBlt* function within your programs as shown in the following prototype:

```
BOOL PatBlt(
   HDC hdc,              // handle to device context
   int nXLeft,           // x-coord. of upper left corner of rect. to be filled
   int nYLeft,           // y-coord. of upper left corner of rect. to be filled
   int nWidth,           // width of rectangle to be filled
   int nHeight,          // height of rectangle to be filled
   DWORD dwRop           // raster operation code
);
```

The *PatBlt* function accepts the parameters that Table 1437.1 details.

Parameters	Description
hdc	Identifies the device context.
nXLeft	Specifies the x-coordinate, in logical units, of the upper left corner of the rectangle the function is to fill.
nYLeft	Specifies the y-coordinate, in logical units, of the upper left corner of the rectangle the function is to fill.
nWidth	Specifies the width, in logical units, of the rectangle.
nHeight	Specifies the height, in logical units, of the rectangle.
dwRop	Specifies the raster operation code. This code may be one of the values Table 1437.2 details.

***Table 1437.1** The parameters for the **PatBlt** function.*

As you saw in Table 1437.1, the *dwRop* parameter accepts several different raster operation codes. Table 1437.2 lists the various raster operation codes your programs will accept.

Value	Meaning
PATCOPY	Copies the specified pattern into the destination bitmap.
PATINVERT	Uses the Boolean *OR* operator to combine the colors of the specified pattern with the colors of the destination rectangle.
DSTINVERT	Inverts the destination rectangle.
BLACKNESS	Uses the color associated with index zero in the physical palette to fill the destination rectangle. (This color is black for the default physical palette.)
WHITENESS	Uses the color associated with index 1 in the physical palette to fill the destination rectangle. (This color is white for the default physical palette.)

***Table 1437.2** The possible raster operation codes.*

The values of the *dwRop* parameter for the *PatBlt* function are a limited subset of the full 256 ternary raster-operation codes; in particular, you cannot use an operation code that refers to a source rectangle. For a complete list of the ternary raster-operation codes, refer to your compiler's on-line help. Not all devices support the *PatBlt* function. You should use the *GetDeviceCaps* function to check a device's *RC_BITBLT* capability before you invoke the *PatBlt* function against the device.

To better understand the processing the *PatBlt* function performs, consider the *Paint_Rect.cpp* program, contained on the CD-ROM that accompanies this book. The *Paint_Rect.cpp* program paints a gray box with a three-dimensional border onto the program window. The *Paint_Rect.cpp* program uses three standard brushes to paint the rectangles that compose the box's three-dimensional border. The *Paint_Rect.cpp* program's *WndProc* function performs the operative processing.

USING SETDIBITS 1438

As your programs become more complex, there may be times when you must change a bitmap's color scheme. As you have learned, device-independent bitmaps maintain color information within the bitmap's header. You can use the colors the creating program previously stored within a device-independent bitmap header to change the colors within a given bitmap. The *SetDIBits* function uses the color data it finds in the specified device-independent bitmap to set the pixels in a bitmap. You will use the *SetDIBits* function as shown in the following prototype:

```
int SetDIBits(
   HDC hdc,                    // handle of device context
   HBITMAP hbmp,               // handle of bitmap
   UINT uStartScan,            // starting scan line
   UINT cScanLines,            // number of scan lines
   CONST VOID *lpvBits,        // array of bitmap bits
   CONST BITMAPINFO *lpbmi,    // address of structure with bitmap data
   UINT fuColorUse             // type of color indices to use
 );
```

The *SetDIBits* function accepts the parameters listed in Table 1438.1.

Parameter	Description
hdc	Identifies a device context.
hbmp	Identifies the bitmap that *SetDIBits* is to alter using the color data from the specified device-independent bitmap.
uStartScan	Specifies the starting scan line for the device independent color data in the array pointed to by the *lpvBits* parameter.

***Table 1438.1** The parameters for the **SetDIBits** parameter. (continued on following page)*

cScanLines	Specifies the number of scan lines the function found in the array that contains device-independent color data.
lpvBits	Points to the device-independent bitmap color data, which the function stores as an array of bytes. The format of the bitmap values depends on the *biBitCount* member of the *BITMAPINFO* structure pointed to by the *lpbmi* parameter.
lpbmi	Points to a *BITMAPINFO* data structure that contains information about the device-independent bitmap.
fuColorUse	Specifies whether the bitmap's definition includes the *bmiColors* member of the *BITMAPINFO* structure and, if so, whether *bmiColors* contains explicit red, green, blue (RGB) values or palette indices. The *fuColorUse* parameter must be one of the values shown in Table 1438.2.

Table 1438.1 The parameters for the ***SetDIBits*** *parameter. (continued from previous page)*

As you learned in Table 1438.1, *fuColorUse* accepts several preset constant values. Table 1438.2 details the possible values for the *fuColorUse* parameter.

Value	Meaning
DIB_PAL_COLORS	The color table consists of an array of 16-bit indices into the logical palette of the device context the *hdc* parameter identifies.
DIB_RGB_COLORS	The bitmap provides a color table that contains literal RGB values.

Table 1438.2 The possible values for the ***fuColorUse*** *parameter.*

If the function succeeds, the return value will be the number of scan lines the *SetDIBits* function copied successfully. If the function fails, the return value will be zero. Windows achieves optimal bitmap drawing speed when the bitmap bits are indices into the system palette.

Applications can call thc *GetSystemPaletteEntries* function to retrieve the system palette colors and indices. After the application retrieves the colors and indices, it can create the device-independent bitmap. The function uses the device context the *hdc* parameter identifies only if you set the *DIB_PAL_COLORS* constant for the *fuColorUse* parameter; otherwise, the function ignores the *hdc* parameter. Your program or another program must not have already selected the bitmap the *hbmp* parameter identifies into a device context when the application calls the *GetSystemPaletteEntries* function. The origin for bottom-up device-independent bitmaps is the lower left corner of the bitmap; the origin for top-down device-independent bitmaps is the upper left corner of the bitmap.

To better understand the processing the *SetDIBits* function performs, consider the *Change_Colors.cpp* program, which paints a device-independent bitmap on the window's client area. Initially, the program colors the bitmap black and white. When the user selects the *Test!* menu item, the program alters the bitmap data so that any two black pixels next to each other reflect the blue-black pixel combination. The *WndProc* function contains the program code that performs the program's operative processing.

1439 Using SetDIBitsToDevice to Output a Bitmap to a Given Device

As you learned in Tip 1438, your programs can use the *SetDIBits* function to set the color bits of a bitmap in accordance with the values the header of a given device-independent bitmap contains. Often, your programs must reduce the colors of a device-independent bitmap for a given device. For example, Windows must map a 256-color bitmap to a 20-color bitmap if you try to output that bitmap on a VGA display. The *SetDIBitsToDevice* function draws a device-independent bitmap on the device associated with the given device context. Additionally, the *SetDIBitsToDevice* function uses color data from the original device-independent bitmap to set the pixels in the drawn bitmap. You will use the *SetDIBitsToDevice* function within your programs as shown in the following prototype:

```
int SetDIBitsToDevice(
    HDC hdc,                    // handle of device context
    int XDest,                  // x-coordinate of upper left corner of dest. rect.
    int YDest,                  // y-coordinate of upper left corner of dest. rect.
    DWORD dwWidth,              // source rectangle width
    DWORD dwHeight,             // source rectangle height
    int XSrc,                   // x-coordinate of lower left corner of source rect.
    int YSrc,                   // y-coordinate of lower left corner of source rect.
    UINT uStartScan,            // first scan line in array
    UINT cScanLines,            // number of scan lines
    CONST VOID *lpvBits,        // address of array with DIB bits
    CONST BITMAPINFO *lpbmi,    // address of structure with bitmap info.
    UINT fuColorUse             // RGB or palette indices
);
```

The *SetDIBitsToDevice* function accepts the parameters Table 1439.1details.

Parameter	Description
hdc	Identifies the device context.
XDest	Specifies the x-coordinate, in logical units, of the upper left corner of the destination rectangle.
YDest	Specifies the y-coordinate, in logical units, of the upper left corner of the destination rectangle.
dwWidth	Specifies the width, in logical units, of the device-independent bitmap.
dwHeight	Specifies the height, in logical units, of the device-independent bitmap.
XSrc	Specifies the x-coordinate, in logical units, of the lower left corner of the device-independent bitmap.
YSrc	Specifies the y-coordinate, in logical units, of the lower left corner of the device-independent bitmap.
uStartScan	Specifies the starting scan line in the device-independent bitmap.
cScanLines	Specifies the number of device-independent bitmap scan lines in the array pointed to by the *lpvBits* parameter.
lpvBits	Points to device-independent bitmap color data the bitmap-creating program stored previously as an array of bytes.
lpbmi	Points to a *BITMAPINFO* structure that contains information about the device-independent bitmap.
fuColorUse	Specifies whether the *bmiColors* member of the *BITMAPINFO* structure contains explicit red, green, blue (RGB) values or indices into a palette. The *fuColorUse* parameter must be one of the values detailed in Table 1438.2.

***Table 1439.1** The parameters for the **SetDIBitsToDevice** function.*

As you learned in Table 1439.1, the *fuColorUse* parameter will accept one of two predefined values. Table 1438.2 lists the acceptable values for the *fuColorUse* parameter with the *SetDIBitsToDevice* function.

The function will achieve optimal bitmap drawing speed when the bitmap bits are indices into the system palette. Applications can call the *GetSystemPaletteEntries* function to retrieve the system palette colors and indices. After the application retrieves the colors and indices, it can create the device-independent bitmap.

The origin of a bottom-up device-independent bitmap is the lower left corner of the bitmap; the origin of a top-down device-independent bitmap is the upper left corner. To reduce the amount of memory required to set bits from a large device-independent bitmap on a device surface, an application can repeatedly call *SetDIBitsToDevice* to band the output, which places a different portion of the bitmap into the *lpvBits* array each time. The values of the *uStartScan* and *cScanLines* parameters identify the portion of the bitmap the *lpvBits* array contains. The *SetDIBitsToDevice* function will return an error if a process that is running in the background calls it while a full-screen MS-DOS session runs in the foreground.

To better understand the processing the *SetDIBitsToDevice* function performs, consider the *Draw_2_Boxes.cpp* program, contained on the CD-ROM that accompanies this book. The *Draw_2_Boxes.cpp* program uses a device-independent bitmap and a custom palette. The program paints the device-independent bitmap at its normal size with the *SetDIBitsToDevice* function and then uses the *StretchDIBits* function to display the bitmap at 200 percent its normal size. The operating processing occurs within the *WM_CREATE* and *WM_PAINT* message handlers in the *WndProc* function.

1440 UNDERSTANDING METAFILES

In previous Tips, you have learned about bitmaps. You have also learned that Windows supports three basic graphical types. The second type is known as a *metafile.* Metafiles are coded graphics device interface (GDI) function calls. When a program *plays* a metafile, the result is the same, as though the program had directly used the graphics device interface functions. You can think of a metafile as, effectively, a graphics macro. You can store metafiles in memory (or as a file on a disk), you can reload the metafile, and you can play the metafile within any number of different applications. Additionally, metafiles are more device-independent than bitmaps because the computer (and the device) will interpret the graphics device interface functions based on the output device context.

Windows 95 and Windows NT support the Windows 3.x metafile. However, both Win32 operating systems also support the new *enhanced metafile* type, and your programs should use enhanced metafiles as well. The main differences between a Windows 3.x metafile and an enhanced metafile is that enhanced metafiles are truly device-independent, and enhanced metafiles support the new Win32 GDI API functions.

It is important to understand that the Win32 API fully supports the Win16 metafile type. In fact, the Win32 API contains two sets of metafile functions—one set for the Win16 API and one set for the Win32 API. For example, your programs can call the *PlayMetaFile* function to play a 16-bit metafile. In the same fashion, your programs can call the *PlayEnhMetaFile* function to play a 32-bit metafile. For clarity, Tips in this book will focus on the enhanced metafiles, not the Win16 metafiles.

1441 CREATING AND DISPLAYING METAFILES

As you learned in Tip 1440, a metafile is a series of graphics device interface instructions the creating program previously stored within a saveable structure. You will use graphics device interface functions on a metafile device context to create metafiles. You will use a *reference device context* as the basis for the metafile to maintain picture dimensions across output devices. The reference device corresponds to the device on which the picture first appeared. You will use the *CreateEnhMetaFile* function to create metafiles. The *CreateEnhMetaFile* function creates a device context for an enhanced-format metafile. You can use the created device context to store a device-independent picture. You will use the *CreateEnhMetaFile* function within your programs as shown in the following prototype:

```
HDC CreateEnhMetaFile(
   HDC hdcRef,              // handle to a reference device context
   LPCTSTR lpFilename,      // pointer to a filename string
   CONST RECT *lpRect,      // pointer to a bounding rectangle
   LPCTSTR lpDescription    // pointer to an optional description string
 );
```

The *CreateEnhMetaFile* function accepts the parameters Table 1441 details.

Parameter	Description
hdcRef	Identifies a reference device for the enhanced metafile.
lpFilename	Points to the filename for the enhanced metafile the function is to create. If this parameter is *NULL*, the enhanced metafile will be memory-based and its contents will be lost when you use the *DeleteEnhMetaFile* function to delete it.
lpRect	Points to a *RECT* structure that specifies the dimensions (in .01-millimeter units) of the picture the program will eventually store in the enhanced metafile.
lpDescription	Points to a string that specifies the name of the application that created the picture, as well as the picture's title. The string the *lpDescription* parameter points to must contain a *NULL* character between the application name and the picture name and must terminate with two *NULL* characters—for example, *"XYZ Graphics Editor\0Bald Eagle\0\0"*, where \0 represents the *NULL* character. If *lpDescription* is *NULL*, there is no corresponding entry in the enhanced-metafile header.

***Table 1441** The parameters for the **CreateEnhMetaFile** function.*

Windows uses the reference device the *hdcRef* parameter identifies to record the resolution and units of the device on which a picture originally appeared. If the *hdcRef* parameter is *NULL*, it will use the current display device for reference.

The left and top members of the RECT structure pointed to by the *lpRect* parameter must be less than the right and bottom members, respectively. Points along the edges of the rectangle are included in the picture. If *lpRect* is *NULL*, the graphics device interface (GDI) will compute the dimensions of the smallest rectangle that surrounds the picture the application has drawn. Your programs should provide the *lpRect* parameter whenever possible.

Applications use the device context the *CreateEnhMetaFile* function creates to store a graphics picture in an enhanced metafile. Your programs can pass the handle that identifies this device context to any graphics device interface function.

After an application stores a picture in an enhanced metafile, it can call the *PlayEnhMetaFile* function to display the picture on any output device. When displaying the picture, Windows uses the rectangle pointed to by the *lpRect* parameter and the resolution data from the reference device to position and scale the picture. The device context the *PlayEnhMetaFile* function returns contains the same default attributes associated with any new device context. Applications must use the *GetWinMetaFileBits* function to convert an enhanced metafile to the older Windows metafile format. The filename for an enhanced metafile should use the *EMF* extension.

To better understand the processing the *CreateEnhMetaFile* function performs, consider the *Crosshatch_Box.cpp* program, contained on the CD-ROM that accompanies this book. The *Crosshatch_Box.cpp* program creates an enhanced metafile of a rectangle filled with a cross-hatched brush at start-up. The program then plays the metafile within the *WM_PAINT* message processing routine. The program scales the metafile based on the window's client area—meaning that the rectangle's size will change as you change the window size. The *WM_CREATE* and *WM_PAINT* message handlers perform the program's operative processing.

ENUMERATING THE ENHANCED METAFILES 1442

As you have learned, your programs can use metafiles to reuse stored instructions to a graphics device interface. In fact, you can store multiple sets of instructions within a given metafile. Each set of instructions is known as a record within the metafile. Later, when you access the metafile, you may want to access a specific set of instructions, or even determine what the instructions are within the metafile. To do so, your programs can use the *EnumEnhMetaFile* function to list all records within a metafile. The *EnumEnhMetaFile* function retrieves each record and passes it to the callback function you specify to enumerate the records within an enhanced-format metafile. The application-supplied callback function processes each record as required. The enumeration continues until the application-supplied callback function processes the last record or when that function returns zero. You will use the *EnumEnhMetaFile* function within your programs as shown in the following prototype:

```
BOOL EnumEnhMetaFile(
   HDC hdc,                          // handle to device context
   HENHMETAFILE hemf,                // handle to enhanced metafile
   ENHMFENUMPROC lpEnhMetaFunc,      // pointer to callback function
   LPVOID lpData,                    // pointer to callback-function data
   CONST RECT *lpRect                // pointer to bounding rectangle
 );
```

The *EnumEnhMetaFile* function accepts the parameters that Table 1442.1 details.

Parameter	Description
hdc	Identifies a device context. Your programs must pass this handle to the callback function.
hemf	Identifies an enhanced metafile.
lpEnhMetaFunc	Points to the application-supplied callback function.
lpData	Points to optional callback-function data.
lpRect	Points to a *RECT* structure that specifies the coordinates of the picture's upper left and lower right corners. The *RECT* structure specifies the dimensions of this rectangle in logical units.

***Table 1442.1** The parameters for the **EnumEnhMetaFile** function.*

As you know, an API enumeration function uses a callback function to process the information the enumeration function returns. The *EnumEnhMetaFile* function uses a callback function with the *EnhMetaFileProc* function's generalized format. The *EnhMetaFileProc* function processes enhanced-format metafile records. Your programs will use the *EnhMetaFileProc* function as shown in the following prototype:

```
int CALLBACK EnhMetaFileProc(
   HDC hDC,                          // handle to device context
   HANDLETABLE FAR *lpHTable,        // pointer to metafile handle table
   ENHMETARECORD FAR *lpEMFR,        // pointer to metafile record
   int nObj,                         // count of objects
   LPARAM lpData                     // pointer to optional data
 );
```

The *EnhMetaFileProc* function accepts the parameters Table 1442.2 details.

Parameters	Description
hDC	Identifies the device context your program passed to *EnumEnhMetaFile.*
lpHTable	Points to a table of handles associated with the graphics objects (pens, brushes, and so on) in the metafile. The first entry contains the enhanced-metafile handle.
lpEMFR	Points to one of the records in the metafile. Your progams should not modify this record. (If modification is necessary, your programs should perform it on a copy of the record.)
nObj	Specifies the number of objects with associated handles in the handle table.
lpData	Points to any application-supplied data.

***Table 1442.2** The parameters the **EnhMetaFileProc** function accepts.*

An application must pass its address to the *EnumEnhMetaFile* function to register the callback function. As with other callback functions you have learned about in previous Tips, *EnhMetaFileProc* is a placeholder for the application-supplied function name.

The picture includes points along the edge of the rectangle pointed to by the *lpRect* parameter. If the *hDC* parameter is *NULL*, Windows will ignore *lpRect*. If the callback function calls the *PlayEnhMetaFileRecord* function, *hDC* must identify a valid device context. Windows uses the device context's transformation and mapping mode to transform the picture the *PlayEnhMetaFileRecord* function displays. You can use the *EnumEnhMetaFile* function to embed one enhanced metafile within another.

To better understand the processing the *EnumEnhMetaFile* function performs, consider the *Draw_Shapes.cpp* program, contained on the CD-ROM that accompanies this book. The *Draw_Shapes.cpp* program loads an enhanced metafile and displays it on the client area of the window. When the user selects the *Demo* option, the program again plays the metafile, this time using the *EnumEnhMetaFile* function. The callback function intercepts the *EMR_CREATEBRUSHINDIRECT* metafile record and replaces it with a light gray brush. The operative processing occurs within the *WndProc* and *PaintMetaFile* procedures. When you compile and execute the *Draw_Shapes.cpp* program, it first draws the shapes with a cross-hatch, then redraws the shapes in a solid gray.

USING THE GETWINMETAFILEBITS *FUNCTION* 1443

As you have learned, your Win32 programs should use the enhanced metafile structure. However, as you program across multiple platforms, there may be times when your programs must convert an enhanced metafile to a Windows 3.x-style metafile. The *GetWinMetaFileBits* function converts enhanced-format metafile records into Windows-format metafile records and stores the converted records in the specified buffer. You will use the *GetWinMetaFileBits* function within your program as shown in the following prototype:

```
UINT GetWinMetaFileBits(
   HENHMETAFILE hemf,   // handle to the enhanced metafile
   UINT cbBuffer,       // buffer size
   LPBYTE lpbBuffer,    // pointer to buffer
   INT fnMapMode,       // mapping mode
   HDC hdcRef           // handle of reference device context
 );
```

The *GetWinMetaFileBits* function accepts the parameters shown in Table 1443.

Parameter	Description
hemf	Identifies the enhanced metafile.
cbBuffer	Specifies the size, in bytes, of the buffer into which the *GetWinMetafileBits* function is to copy the converted records.
lpbBuffer	Points to the buffer into which the *GetWinMetafileBits* function is to copy the converted records. If *lpbBuffer* is *NULL*, *GetWinMetaFileBits* will return the number of bytes the function requires to store the converted metafile records.
fnMapMode	Specifies the mapping mode your programs should use with the converted metafile.
hdcRef	Identifies the reference device context.

Table 1443 The parameters for the ***GetWinMetaFileBits*** *function.*

If the *GetWinMetaFileBits* function succeeds and the buffer pointer is *NULL*, the return value will be the number of bytes the function requires to store the converted records. If the function succeeds and the buffer pointer is a valid pointer, the return value will be the size of the metafile data in bytes. If the function fails, the return value will be zero.

The *GetWinMetaFileBits* function converts an enhanced metafile into a Windows-format metafile so that an application that recognizes the older format can display the metafile. Windows uses the reference device context to determine the resolution of the converted metafile. The *GetWinMetaFileBits* function does not invalidate the enhanced metafile handle. An application should call the *DeleteEnhMetaFile* function to release the handle when the application no longer needs the metafile handle.

Due to the limitations of the Windows-format metafile, some information can be lost in the retrieved metafile contents. For example, the function may convert an original call to the *PolyBezier* function in the enhanced metafile into a call to the *Polyline* function in the Windows-format metafile, because there is no equivalent *PolyBezier* function in the Windows format.

Windows 3.x applications define the viewport origin and extent of a picture the Windows-format metafile contains. As a result, the Windows-format records *GetWinMetaFileBits* creates do not contain the *SetViewportOrgEx* and *SetViewportExtEx* functions. However, *GetWinMetaFileBits* does create Windows-format records for the *SetWindowExtEx* and *SetMapMode* functions. To create a scalable Windows-format metafile, specify *MM_ANISOTROPIC* as the *fnMapMode* parameter. Windows always maps the upper left corner of the metafile picture to the origin of the reference device.

To better understand the processing the *GetWinMetaFileBits* function performs, consider the *Convert_EMF_WMF.cpp* program, contained on the CD-ROM that accompanies this book. The *Convert_EMF_WMF.cpp* program converts a given enhanced metafile (in this case, *demo.emf*) to a standard Windows 3.x metafile. The program then saves the converted metafile as *sample.wmf.* The program's operative processing occurs within the *WM_COMMAND* message processing routine.

1444 UNDERSTANDING ICONS

As you have learned, Windows supports three basic graphic file types. As you have also learned, your programs can use the bitmap and the metafile to manage large or small graphics within a window's client area, and even on the window itself. *Icons* are actually a sub-class of bitmaps. However, the set of activities that you will use icons to perform is limited enough that Windows treats icons as a separate type of graphic file.

Icons are small bitmaps that Windows uses as visual representations of objects such as applications, files, and directories. In Windows 95, you will see icons in every aspect of the user interface. In earlier versions of Windows and in Windows NT 3.51, you will see icons primarily in the Program Manager.

A typical Windows 95 or Windows NT application will have at least two icons: a large (32 x 32) icon and a small (16 x 16) icon. Windows will display the small icon in the application window's upper left corner when the application is minimized. Windows uses the large icon for a program's desktop icon and in large icon views.

Typically, you will create icons with the Windows 95 SDK *Image Editor* or with another editor, such as the *Visual C++* icon editor. You will then use the *ICON* statement to add the icons you create to an application's resource script file. A typical example of the use of icons is the registration of the main window class. As you have seen throughout the programs in the previous 187 Tips, programs will register an icon when they register the window class with a call to *RegisterClassEx*, as shown here:

```
int APIENTRY WinMain(HINSTANCE hInstance, HINSTANCE hPrevInstance,
                     LPTSTR lpCmdLine, int nCmdShow)
 {
   MSG msg;
   HWND hWnd;
   WNDCLASS wc;

   // Register the main application window class.
   //.......................................
```

```
    wc.style = CS_HREDRAW | CS_VREDRAW;
    wc.lpfnWndProc = (WNDPROC)WndProc;
    wc.cbClsExtra = 0;
    wc.cbWndExtra = 0;
    wc.hInstance = hInstance;
    wc.hIcon = LoadIcon(hInstance, lpszAppName);
    wc.hCursor = LoadCursor(NULL, IDC_ARROW);
    wc.hbrBackground = (HBRUSH)(COLOR_WINDOW+1);
    wc.lpszMenuName = lpszAppName;
    wc.lpszClassName = lpszAppName;
    if (!RegisterClass(&wc))
       return(FALSE);
 // More code here
```

As you can see, the previous code fragment registers an icon with the name the *lpszAppName* string contains as the main window's icon. In later Tips, you will learn more about how to attach an icon to a program.

CREATING ICONS 1445

Just as with bitmaps and metafiles, Windows lets you create and modify icons at run time. Your programs will generally use the *CreateIcon* function to create an icon at run time. The *CreateIcon* function lets your programs create icons from binary arrays, bitmap data, and device-independent bitmaps. The *CreateIcon* function creates an icon that has the specified size, colors, and bit patterns. You will use *CreateIcon* within your programs as shown in the following prototype:

```
HICON CreateIcon(
   HINSTANCE hInstance,       // handle to application instance
   int nWidth,                // icon width
   int nHeight,               // icon height
   BYTE cPlanes,              // number of planes in XOR bitmask
   BYTE cBitsPixel,           // number of bits per pixel in XOR bitmask
   CONST BYTE *lpbANDbits,    // pointer to AND bitmask array
   CONST BYTE *lpbXORbits     // pointer to XOR bitmask array
);
```

The *CreateIcon* function accepts the parameters Table 1445.1 details.

Parameter	Description
hInstance	Identifies the instance of the module creating the icon.
nWidth	Specifies the width, in pixels, of the icon.
nHeight	Specifies the height, in pixels, of the icon.
cPlanes	Specifies the number of planes in the *XOR* bitmask of the icon.
cBitsPixel	Specifies the number of bits per pixel in the *XOR* bitmask of the icon.
lpbANDbits	Points to an array of bytes that contains the bit values for the *AND* bitmask of the icon. This bitmask describes a monochrome bitmap.
lpbXORbits	Points to an array of bytes that contains the bit values for the *XOR* bitmask of the icon. This bitmask describes a monochrome or device-dependent color bitmap.

*Table 1445.1 The parameters the **CreateIcon** function accepts.*

The *nWidth* and *nHeight* parameters must specify a width and height the current display driver supports, because the system cannot create icons of other sizes. To determine the width and height the display driver supports, use the *GetSystemMetrics* function, specifying the *SM_CXICON* or *SM_CYICON* value.

The *CreateIcon* function creates the icon from two bitmaps (which the function uses as bitmasks), the *AND* bitmask, and the *XOR* bitmask. The *AND* bitmask is always a monochrome bitmap, with one bit per pixel. *CreateIcon* applies the truth table detailed in Table 1445.2 to the *AND* and *XOR* bitmasks.

AND bitmask	*XOR* bitmask	Display
0	0	Black
0	1	White
1	0	Screen
1	1	Reverse screen

Table 1445.2 The truth table ***CreateIcon*** *applies to the* ***AND*** *and* ***XOR*** *bitmasks.*

To better understand the processing *CreateIcon* performs, consider the *Create_Icon.cpp* program, contained on the CD-ROM that accompanies this book. The *Create_Icon.cpp* program directly specifies the bit values of both the *AND* mask and the *XOR* mask of the icon to create a monochrome icon. When the *WndProc* routine receives the *WM_CREATE* message, the program creates the icon; when the *WndProc* routine receives the *WM_PAINT* message, the program paints the icon onto the display.

1446 CREATING ICONS FROM A RESOURCE

As you have learned, your programs can create icons in several different ways. However, your programs will generally not create two bitmaps and two bitmasks in memory, as the *Create_Icon.cpp* program in Tip 1445 did. Just as with string tables and other reusable information, your programs can load an icon's component bits from within a resource file and convert the bits to an actual icon. To perform such processing, your programs will use the *CreateIconFromResource* function. The *CreateIconFromResource* function creates an icon or cursor from resource bits that describe the icon. You will use the *CreateIconFromResource* function as shown in the following prototype:

```
HICON CreateIconFromResource(
   PBYTE presbits,      // pointer to icon or cursor bits
   DWORD dwResSize,     // number of bytes in bit buffer
   BOOL fIcon,          // icon or cursor flag
   DWORD dwVer          // Windows format version
 );
```

The *CreateIconFromResource* function accepts the parameters Table 1446.1 details.

Parameter	Description
presbits	Points to a buffer that contains the icon or cursor resource bits. Calls to the *LookupIconIDFromDirectory* (in Windows 95, you can also call *LookupIconIDFromDirectoryEx*) and *LoadResource* functions typically load these bits.
dwResSize	Specifies the size, in bytes, of the set of bits pointed to by the *presbits* parameter.
fIcon	Specifies whether the function is to create an icon or a cursor. If this parameter is *True*, the function will create an icon. If it is *False*, the function will create a cursor.
dwVer	Specifies the version number of the icon or cursor format for the resource bits pointed to by the *presbits* parameter.

Table 1446.1 The parameters the ***CreateIconFromResource*** *function accepts.*

The *dwVer* parameter can be one of the values shown in Table 1446.2.

Format	dwVer
Windows 2.*x*	0x00020000
Windows 3.*x*	0x00030000

***Table 1446.2** The values for the **dwVer** parameter.*

All Microsoft Win32-based applications use the Windows 3.*x* format for icons and cursors. The *CreateIconFromResource, CreateIconIndirect, GetIconInfo*, and *LookupIconIdFromDirectory* functions (and in Windows 95, the *CreateIconFromResourceEx* and *LookupIconIdFromDirectoryEx* functions) let shell applications and icon browsers examine and use resources from throughout the system.

1447 Using the CreateIconIndirect Function

As you have learned, your programs can create icons from bitmaps or from a resource identifier. Your programs can also create icons from a structure value. Within your programs, you will use the *CreateIconIndirect* function to create icons from components that you do not define within the program or a resource file. The *CreateIconIndirect* function creates an icon or cursor from an *ICONINFO* structure. You will use the *CreateIconIndirect* function within your programs as shown in the following prototype:

```
HICON CreateIconIndirect(PICONINFO piconinfo);
```

The *piconinfo* pointer points to an *ICONINFO* structure the function uses to create the icon or cursor. If the function succeeds, the return value will be the handle to the icon or cursor that the function created. The system copies the bitmaps in the *ICONINFO* structure before creating the icon or cursor. The application must continue to manage the original bitmaps and delete them when they are no longer necessary. The *ICONINFO* structure contains information about an icon or a cursor. The Windows API defines the *ICONINFO* structure as shown here:

```
typedef struct _ICONINFO {
   BOOL    fIcon;
   DWORD   xHotspot;
   DWORD   yHotspot;
   HBITMAP hbmMask;
   HBITMAP hbmColor;
} ICONINFO;
```

Table 1447 details the members of the *ICONINFO* structure.

Member	Description
fIcon	Specifies whether this structure defines an icon or a cursor. A value of *True* specifies an icon; *False* specifies a cursor.
xHotspot	Specifies the x-coordinate of a cursor's hot spot. If this structure defines an icon, the hot spot will always be in the center of the icon, and functions that use the *ICONINFO* structure will ignore the *xHotspot* member.
yHotspot	Specifies the y-coordinate of the cursor's hot spot. If this structure defines an icon, the hot spot will always be in the center of the icon, and functions that use the *ICONINFO* structure will ignore the *yHotspot* member.
hbmMask	Specifies the icon bitmask bitmap. If this structure defines a black and white icon, this bitmask will be formatted so that the upper half is the icon *AND* bitmask and the lower half is the icon *XOR* bitmask. Under this condition, the height should be an even multiple of two. If this structure defines a color icon, this mask will only define the *AND* bitmask of the icon.

***Table 1447** The members of the **ICONINFO** structure.* *(continued on following page)*

Member	Description
hbmColor	Identifies the icon color bitmap. This member can be optional if the structure defines a black and white icon. *CreateIconIndirect* applies the *AND* bitmask of *hbmMask* with the *SRCAND* flag to the destination. Subsequently, *CreateIconIndirect* uses the *SRCINVERT* flag to apply the color bitmap (using *XOR*) to the destination by.

***Table 1447** The members of the **ICONINFO** structure. (continued from previous page)*

In short, the *ICONINFO* structure defines the monochrome bitmap and the color bitmap, and the *CreateIconIndirect* function combines the bitmaps based on the values the *ICONINFO* structure holds. To better understand the processing the *CreateIconIndirect* function performs, consider the *Two_Icons.cpp* program, contained on the CD-ROM that accompanies this book. The *Two_Icons.cpp* program combines two bitmaps into a third icon bitmap. As usual, the *WM_COMMAND* message handler within the *WndProc* function handles the program's primary processing.

1448 USING THE LOADICON FUNCTION

In the previous Tips, you have learned how your programs can use several different methods to create icons at run time. However, as you have seen in other programs, your programs will most commonly use the *LoadIcon* function to load an icon into a program from within the program's resource file. As you have learned, the *LoadIcon* function provides your programs with a simple and efficient way of loading previously created icons into the program. The *LoadIcon* function loads the specified icon resource from the executable (.EXE) file associated with an application instance. You will use the *LoadIcon* function within your programs as shown in the following prototype:

```
HICON LoadIcon(HINSTANCE hInstance, LPCTSTR lpIconName);
```

The *hInstance* parameter identifies an instance of the module whose executable file contains the icon *LoadIcon* is to load. The *hInstance* parameter must be *NULL* when *LoadIcon* is to load a standard icon. The *lpIconName* parameter points to a *NULL*-terminated string that contains the name of the icon resource *LoadIcon* is to load. Alternatively, the *lpIconName* parameter can contain the resource identifier in the low-order word and zero in the high-order word. Use the *MAKEINTRESOURCE* macro to create a resource identifier value. To use one of the Windows predefined icons, set the *hInstance* parameter to *NULL* and the *lpIconName* parameter to one of the values Table 1448 details.

Value	Description
IDI_APPLICATION	Default application icon
IDI_ASTERISK	Asterisk (used in informative messages)
IDI_EXCLAMATION	Exclamation point (used in warning messages)
IDI_HAND	Hand-shaped icon (used in serious warning messages)
IDI_QUESTION	Question mark (used in prompting messages)
IDI_WINLOGO	Windows logo

***Table 1448** The Windows predefined icons.*

LoadIcon loads the icon resource only if the program has not previously loaded the icon resource; otherwise, *LoadIcon* retrieves a handle to the existing resource. The function searches the icon resource for the icon most appropriate for the current display. The icon resource can be a color or monochrome bitmap. *LoadIcon* can only load an icon whose size conforms to the *SM_CXICON* and *SM_CYICON* system metric values. Use the *LoadImage* function to load icons of other sizes.

USING LOADIMAGE TO LOAD MULTIPLE GRAPHICAL TYPES 1449

As you have learned, your programs can use the *LoadIcon* function to load an icon from the program's resource file. Your programs can also use the *LoadBitmap* and *LoadCursor* functions to load bitmaps and cursors, respectively, from the program's resource file. Alternatively, your programs can use the *LoadImage* function, which loads an icon, a cursor, or a bitmap. You will use the *LoadImage* function within your programs as shown in the following prototype:

```
HANDLE LoadImage(
   HINSTANCE hinst,     // handle of the instance that contains the image
   LPCTSTR lpszName,    // name or identifier of image
   UINT uType,          // type of image
   int cxDesired,       // desired width
   int cyDesired,       // desired height
   UINT fuLoad          // load flags
);
```

The *LoadImage* function accepts the parameters Table 1449.1 details.

Parameter	Description
hinst	Identifies an instance of the module that contains the image *LoadImage* is to load. Set this parameter to zero to load an *OEM* image.
lpszName	Identifies the image *LoadImage* is to load. If the *hinst* parameter is non-*NULL* and the *fuLoad* parameter does not include *LR_LOADFROMFILE*, *lpszName* will be a pointer to a *NULL*-terminated string that contains the name of the image resource in the *hinst* module. On the other hand, if *hinst* is *NULL* and you do not specify the *LR_LOADFROMFILE* constant, the low-order word of this parameter must be the identifier of the OEM image *LoadImage* is to load. The *winuser.h* file defines the OEM image identifiers, which have the prefixes listed in Table 1449.2. Under Windows 95, if the *fuLoad* parameter includes the *LR_LOADFROMFILE* value, *lpszName* will be the name of the file that contains the image. Windows NT does not support *LR_LOADFROMFILE.*
uType	Specifies the type of image *LoadImage* is to load. This parameter can be one of the values listed in Table 1449.3.
cxDesired	Specifies the width, in pixels, of the icon or cursor. If this parameter is zero and the *fuLoad* parameter is *LR_DEFAULTSIZE*, the function will use the *SM_CXICON* or *SM_CXCURSOR* system metric value to set the width. If this parameter is zero and the invocation does not use *LR_DEFAULTSIZE*, the function will use the actual resource width.
cyDesired	Specifies the height, in pixels, of the icon or cursor. If this parameter is zero and the *fuLoad* parameter is *LR_DEFAULTSIZE*, the function will use the *SM_CYICON* or *SM_CYCURSOR* system metric value to set the height. If this parameter is zero and the invocation does not use *LR_DEFAULTSIZE*, the function will use the actual resource height.
fuLoad	Determines how the function loads the image. Specifies a combination of the constants listed in Table 1449.4.

***Table 1449.1** The parameters for the **LoadImage** function.*

As Table 1449.1 indicates, your programs can load OEM images. In such cases, your programs can specify one of the image identifiers listed in Table 1449.2.

Prefix	Meaning
*OBM_*	OEM bitmaps
*OIC_*	OEM icons
*OCR_*	OEM cursors

Table 1449.2 The image identifiers.

As you have learned, your programs can load one of several image types from within the *LoadImage* function. The *uType* parameter specifies the image type, and can be one of the values Table 1449.3 details.

Value	Meaning
IMAGE_BITMAP	Loads a bitmap
IMAGE_CURSOR	Loads a cursor
IMAGE_ICON	Loads an icon

Table 1449.3 The possible image types.

As you have learned, you can load the image file with several options for the file's display. The *fuLoad* parameter must be one or more of the values Table 1449.4 lists.

Value	Meaning
LR_DEFAULTCOLOR	The default flag means "not *LR_MONOCHROME*."
LR_CREATEDIBSECTION	When the *uType* parameter specifies *IMAGE_BITMAP*, this parameter causes the function to return a DIB section bitmap rather than a compatible bitmap. This flag is useful for loading a bitmap without mapping it to the colors of the display device.
LR_DEFAULTSIZE	Uses the width or height the system metric values specify for cursors or icons if the invocation sets the values for *cxDesired* or *cyDesired* to zero. If this flag is not specified and the invocation sets the values for *cxDesired* and *cyDesired* to zero, the function will use the actual resource size. If the resource contains multiple images, the function will use the size of the first image.
LR_LOADFROMFILE	Loads the image from the file the *lpszName* parameter specifies. If this flag is not specified, *lpszName* will be the name of the resource.
LR_LOADMAP3DCOLORS	Searches the color table for the image and replaces the shades of gray with the corresponding three-dimensional color, as shown in Table 1449.5.
LR_LOADTRANSPARENT	Retrieves the color value of the first pixel in the image and replaces the corresponding entry in the color table with the default window color (*COLOR_WINDOW*). All pixels in the image that use that entry become the default window color. This value applies only to images that have corresponding color tables. If *fuLoad* includes both the *LR_LOADTRANSPARENT* and *LR_LOADMAP3DCOLORS* values, *LRLOADTRANSPARENT* will take precedence. However, the *fuLoad* function will then replace the color table entry with *COLOR_3DFACE* rather than *COLOR_WINDOW*.
LR_MONOCHROME	Loads the image in black and white.

Table 1449.4 The load options for the image file. *(continued on following page)*

Value	Meaning
LR_SHARED	Shares the image handle if one or more programs load the image multiple times simultaneously. If you do not set *LR_SHARED*, a second call to *LoadImage* for the same resource will load the image again and return a different handle. Do not use *LR_SHARED* for images that have non-standard sizes that may change after loading, or that your program loads from a file.

***Table 1449.4** The load options for the image file. (continued from previous page)*

When you load an image file as a three-dimensional file, the Windows API will map the colors for you. Windows will replace each color in the left column of Table 1449.5 with the color in the right column of the table.

Color	Replaced with
Dk Gray, RGB(128,128,128)	*COLOR_3DSHADOW*
Gray, RGB(192,192,192)	*COLOR_3DFACE*
Lt Gray, RGB(223,223,223)	*COLOR_3DLIGHT*

***Table 1449.5** The 3D map values for **LoadImage**.*

As the text indicates, your programs can use the *LoadImage* function in any situation where they might otherwise use *LoadIcon, LoadBitmap*, or *LoadCursor*.

UNDERSTANDING WINDOWS FILE I/O 1450

Throughout earlier sections of this book, you learned about file I/O. You learned how to perform file I/O in C, and you have learned how to perform file I/O in C++. In the following section, you will learn the fundamentals of file I/O in Windows.

As you know, the traditional concept of a file is a block of data on a storage device. A unique identifier known as a filename identifies the block of data. In the DOS environment, programs will generally save files onto disks and drives. On the other hand, for I/O purposes, the Win32 API treats named pipes, communications resources, disk devices, console input or output, or the traditional disk file as a "file." Each different file type is the same at the base level, but each file type has its own properties and limitations. The Win32 API file functions let your programs access files regardless of the underlying file system or device type. However, as you might expect, the capabilities of a file will vary from file system to file system and device type to device type.

INTRODUCING PIPES, RESOURCES, DEVICES, AND FILES 1451

As you learned in Tip 1450, Windows supports file-style input and output across a wide variety of devices. In subsequent Tips, you will learn not only about file input and output to a standard disk file, but you will also learn some fundamentals of file input and output to other devices. It is useful to know some of the most common devices that you will encounter when you work with Windows programs. Table 1451 lists some of the most common devices.

Device	Most Common Use
File	Persistent storage of data
Directory	Attributes and file compression
Logical Disk Drive	Formatting
Physical Disk Drive	Partition Table Access

***Table 1451** Common devices and their use. (continued on following page)*

Device	Most Common Use
Serial Port	Data transmission over a phone line
Parallel Port	Data transmission to a printer
Mailslot	One-to-many transmission of data, usually over a network to a Windows-based machine
Named pipe	One-to-one transmission of data, usually over a network to a Windows-based machine
Anonymous pipe	One-to-one transmission of data on a single machine (never over a network)
Socket	Datagram or stream transmission of data, usually over a network to any machine supporting sockets
Console	A text window screen buffer

Table 1451 Common devices and their use. (continued from previous page.)

As you will discover, Win32 tries to hide the differences between devices from the user as much as possible. In other words, if you open a mailslot and a file, Windows will generally let you read from or write to either device with similar functions.

You should note, however, that aside from *CreateFile, ReadFile*, and *WriteFile*, Windows provides an extensive collection of device-specific functions that let your programs closely manage specific features of devices. For example, it makes no sense to set a baud rate when you use a named pipe to communicate, although it makes perfect sense to do so when you use a communications port. Therefore, the function *SetCommConfig* will work with a serial port device, but will not work correctly with a named pipe. For that reason, most of the following Tips will focus on generalized use of the *CreateFile, ReadFile,* and *WriteFile* functions, rather than specific application of a function to a given device. Refer to your compiler's and the device's documentation for more information about communicating with a certain specific device.

1452 USING THE CREATEFILE FUNCTION TO OPEN FILES

As you have learned, the Win32 API supports many device types and will normally let your programs manipulate files in all device types. To create a file on any device, your programs will generally use the Win32 API *CreateFile* function. The *CreateFile* function creates or opens the following objects and returns a handle that your programs can then use to access the object:

- Files
- Pipes
- Mailslots
- Communications resources
- Disk devices (Windows NT only)
- Consoles
- Directories (open only)

You will use the *CreateFile* function within your programs as shown in the following prototype:

```
HANDLE CreateFile(
   LPCTSTR lpFileName,                          // pointer to name of the file
   DWORD dwDesiredAccess,                       // access (read-write) mode
   DWORD dwShareMode,                           // share mode
   LPSECURITY_ATTRIBUTES lpSecurityAttributes,  // security attributes
   DWORD dwCreationDistribution,                // how to create
   DWORD dwFlagsAndAttributes,                  // file attributes
   HANDLE hTemplateFile                         // handle to template file
);
```

The *CreateFile* function provides your programs with significant control over the file you create. Table 1452.1 lists the parameters for the *CreateFile* function.

Parameters	Description
lpFileName	Points to a *NULL*-terminated string that specifies the name of the object (file, pipe, mailslot, communications resource, disk device, console, or directory) the *CreateFile* function is to create or open. If **lpFileName* is a path, there is a default string size limit of *MAX_PATH* characters. This limit relates to how the *CreateFile* function parses paths.
dwDesiredAccess	Specifies the type of access to the object. An application can obtain read access, write access, read-write access, or device-query access. This parameter can be any combination of the values Table 1452.2 details.
dwShareMode	Set of bit flags that specifies how programs can share the object. If *dwShareMode* is zero, programs cannot share the object. Subsequent open operations on the object will fail, until the program using the object closes the handle. To share the object, use a combination of one or more of the values Table 1452.3 details.
lpSecurityAttributes	Pointer to a *SECURITY_ATTRIBUTES* structure that determines whether child processes can inherit the returned handle. If *lpSecurityAttributes* is *NULL*, child processes cannot inherit the handle. Windows NT: The *lpSecurityDescriptor* member of the structure specifies a security descriptor for the object. If *lpSecurityAttributes* is *NULL*, the object gets a default security descriptor. The target file system must support security on files and directories for this parameter to have an effect on files. Windows 95: *CreateFile* ignores the *lpSecurityDescriptor* member.
dwCreationDistribution	Specifies which action to take on files that exist, and which action to take when files do not exist. This parameter must be one of the values Table 1452.4 lists.
dwFlagsAndAttributes	Specifies the file attributes and flags for the file. Any combination of the attributes Table 1452.5 lists is acceptable for the *dwFlagsAndAttributes* parameter, except that all other file attributes override *FILE_ATTRIBUTE_NORMAL*. If the *CreateFile* function opens the client side of a named pipe, the *dwFlagsAndAttributes* parameter can also contain Security Quality of Service (SQOS) information. Tip 1453 explains opening named pipes in detail. When the calling application specifies the *SECURITY_SQOS_PRESENT* flag, the *dwFlagsAndAttributes* parameter can contain one or more of the values Table 1453.1 lists.
hTemplateFile	Specifies a handle with *GENERIC_READ* access to a template file. The template file supplies file attributes and extended attributes for the file the *CreateFile* function is creating. Under Windows 95, this value must be *NULL*. If you supply a handle under Windows 95, the call will fail.

***Table 1452.1** The parameters for the **CreateFile** function.*

As you learned in Table 1452.1, your programs can specify the desired access level to files that they open with *CreateFile* within the *dwAccess* parameter. The possible values for the *dwAccess* parameter are shown in Table 1452.2.

Value	Meaning
0	Specifies device-query access to the object. An application can query device attributes without accessing the device.
GENERIC_READ	Specifies read access to the object. Programs can read data from the file and API calls can move the file pointer. Combine with *GENERIC_WRITE* for read-write access.
GENERIC_WRITE	Specifies write access to the object. Programs can write data to the file and the API calls can move the file pointer. Combine with *GENERIC_READ* for read-write access.

*Table 1452.2 The possible values for the **dwAccess** member.*

In addition to access levels, your programs can specify a share mode for the file within the *dwShareMode* parameter. Table 1452.3 lists the possible values for the *dwShareMode* parameter.

Value	Meaning
FILE_SHARE_DELETE	Windows NT only: Subsequent open operations on the object will succeed only if the opening process requests delete access.
FILE_SHARE_READ	Subsequent open operations on the object will succeed only if the opening process requests read access.
FILE_SHARE_WRITE	Subsequent open operations on the object will succeed only if the opening process requests write access.

*Table 1452.3 The possible values for the **dwShareMode** parameter.*

The *dwCreate* parameter determines the action that Windows should take when the file exists or does not exist. The parameter must be one of the values in Table 1452.4.

Value	Meaning
CREATE_NEW	Creates a new file. The function fails if the specified file already exists.
CREATE_ALWAYS	Creates a new file. The function overwrites the file if it exists.
OPEN_EXISTING	Opens the file. The function fails if the file does not exist.
OPEN_ALWAYS	Opens the file, if it exists. If the file does not exist, the function creates the file as though *dwCreationDistribution* were *CREATE_NEW*.
TRUNCATE_EXISTING	Opens the file. Once opened, Windows truncates the file so that its size is zero bytes. The calling process must open the file with at least *GENERIC_WRITE* access. The function fails if the file does not exist.

*Table 1452.4 Possible values for the **dwCreate** parameter.*

When you create any file within the Windows operating system, Windows will attach attributes to that file. You can specify flags and attributes for each new file you create. Table 1452.5 lists the possible flags and attributes.

Attribute	Meaning
FILE_ATTRIBUTE_ARCHIVE	Marks the file for archival. Applications use this attribute to mark files for backup or removal.
FILE_ATTRIBUTE_COMPRESSED	The file or directory is compressed. For a file, this means that all the data in the file is compressed. For a directory, this means that compression is the default for newly created files and subdirectories.

*Table 1452.5 The possible flag and attribute values for the **dwFlagsAndAttributes** parameter.* *(continued on following page)*

Attribute	Meaning
FILE_ATTRIBUTE_HIDDEN	The file is hidden (not to be included in an ordinary directory listing).
FILE_ATTRIBUTE_NORMAL	The file has no other attributes set. This attribute is valid only if your program uses it alone when the program calls *CreateFile*.
FILE_ATTRIBUTE_OFFLINE	The data of the file is not immediately available. Indicates that the file data has been physically moved to offline storage.
FILE_ATTRIBUTE_READONLY	The file is read only. Applications can read the file but cannot write to it or delete it.
FILE_ATTRIBUTE_SYSTEM	The file is part of the operating system, or the operating system exclusively uses it.
FILE_ATTRIBUTE_TEMPORARY	A process is using the file for temporary storage. An application should delete a temporary file as soon as the application no longer needs the file.
FILE_FLAG_WRITE_THROUGH	Instructs the system to write through any intermediate cache and go directly to disk. Windows can still cache write operations, but cannot flush them at its convenience.
FILE_FLAG_OVERLAPPED	Instructs the system to initialize the object so that operations that take a significant amount of time to process return *ERROR_IO_PENDING*. When the operating system finishes the operation, Windows will set the event specified within the *OVERLAPPED* structure to the signaled state. When you specify *FILE_FLAG_OVERLAPPED*, the *ReadFile* and *WriteFile* functions *must* specify an *OVERLAPPED* structure. That is, when you specify *FILE_FLAG_OVERLAPPED*, an application *must* perform overlapped reading and writing. When you specify *FILE_FLAG_OVERLAPPED*, the system does not maintain the file pointer. The calling process must pass the file position as part of the *lpOverlapped* parameter (pointing to an *OVERLAPPED* structure) to the *ReadFile* and *WriteFile* functions. This flag also lets a process perform more than one operation simultaneously with a single handle (a simultaneous read and write operation, for example).
FILE_FLAG_NO_BUFFERING	Instructs the system to open the file with no intermediate buffering or caching. When combined with *FILE_FLAG_OVERLAPPED*, the flag gives maximum asynchronous performance because the I/O does not rely on the synchronous operations of the memory manager. However, some I/O operations will take longer because the operating system is not holding data in the cache.

***Table 1452.5** The possible flag and attribute values for the **dwFlagsAndAttributes** parameter.* *(continues on following page.)*

Attribute	Meaning
	An application must meet certain requirements when working with files you open with *FILE_FLAG_NO_BUFFERING*. File access must begin at byte offsets within the file that are integer multiples of the volume's sector size. File access must be for numbers of bytes that are integer multiples of the volume's sector size. For example, if the sector size is 512 bytes, an application can request reads and writes of 512, 1024, or 2048 bytes, but not of 335, 981, or 7171 bytes. Buffer addresses for read and write operations must be aligned on addresses in memory that are integer multiples of the volume's sector size. One way to align buffers on integer multiples of the volume sector size is to use *VirtualAlloc* to allocate the buffers. It allocates memory that is aligned on addresses that are integer multiples of the operating system's memory page size. Because both memory page and volume sector sizes are powers of 2, this memory is also aligned on addresses that are integer multiples of a volume's sector size. An application can call the *GetDiskFreeSpace* function to determine a volume's sector size.
FILE_FLAG_RANDOM_ACCESS	Indicates that the process will access the file randomly. The system can use this attribute as a hint to optimize file caching for that file.
FILE_FLAG_SEQUENTIAL_SCAN	Indicates that the process will access the file sequentially from beginning to end. The system can use this attribute as a hint to optimize file caching for that file. If an application moves the file pointer for random access, optimum caching may not occur; however, correct operation is still guaranteed. Specifying this flag can increase performance for applications that use sequential access to read large files. Performance gains can be even more noticeable for applications that read large files mostly sequentially, but occasionally skip over small ranges of bytes.
FILE_FLAG_DELETE_ON_CLOSE	Indicates that the operating system is to delete the file immediately after all its handles have been closed, not just the handle for which you specified *FILE_FLAG_DELETE_ON_CLOSE*. Subsequent open requests for the file will fail, unless the calling process uses the *FILE_SHARE_DELETE* flag.
FILE_FLAG_BACKUP_SEMANTICS	Windows NT only: Indicates that the file is being opened or created for a backup or restore operation. The operating system ensures that the calling process overrides file security checks, provided it has the necessary permission to do so. The relevant permissions are *SE_BACKUP_NAME* and *SE_RESTORE_NAME*. You can also set this flag to obtain a handle to a directory. A process can pass a directory handle to some Win32 functions in place of a file handle.

Table 1452.5 *The possible flag and attribute values for the* ***dwFlagsAndAttributes*** *parameter.* *(continues on following page)*

Attribute	Meaning
FILE_FLAG_POSIX_SEMANTICS	Indicates that the process is to access the file according to POSIX rules. This includes allowing multiple files with names, differing only in case, for file systems that support such naming. Be careful when you use this option because applications written for MS-DOS or Windows may not be able to access files you create with this flag.

Table 1452.5 *The possible flag and attribute values for the* ***dwFlagsAndAttributes*** *parameter. (continued from previous page)*

If the function succeeds, the function returns an open handle to the specified file. If the specified file exists before the function call and *dwCreationDistribution* is *CREATE_ALWAYS* or *OPEN_ALWAYS*, a call to *GetLastError* returns *ERROR_ALREADY_EXISTS* (even though the function has succeeded). If the file does not exist before the call, *GetLastError* returns zero. If the function fails, the function returns *INVALID_HANDLE_VALUE*.

As noted previously, specifying zero for *dwDesiredAccess* lets an application query device attributes without actually accessing the device. This type of querying is useful, for example, if an application wants to determine the size of a floppy disk drive and the formats it supports without having a floppy in the drive.

The CD-ROM that accompanies this book includes the *First_File.cpp* program, which creates the file *file.dat* and writes a string to it. When the user selects *Test!*, the program will read the string back in from the file and display the string within a message box.

USING CREATEFILE WITH DIFFERENT DEVICES 1453

As you have learned, your programs can use the *CreateFile* function to create files on a wide variety of devices. However, there are significant differences in how *CreateFile* operates depending on the device your program passes to the function. The following paragraphs describe some of the variations in the *CreateFile* function's processing that depend on device.

When you use the *CreateFile* function to create a new file, the *CreateFile* function performs the following actions:

- Combines the file attributes and flags *dwFlagsAndAttributes* specifies with *FILE_ATTRIBUTE_ARCHIVE.*
- Sets the file length to zero.
- Copies the extended attributes the template file supplies to the new file if you specify the *hTemplateFile* parameter.

When the *CreateFile* function opens an existing file, it performs the following actions:

- Combines the file flags *dwFlagsAndAttributes* specifies with existing file attributes. The *CreateFile* function ignores the file attributes *dwFlagsAndAttributes* specifies.
- Sets the file length according to the value of *dwCreationDistribution.*
- Ignores the *hTemplateFile* parameter.
- Ignores the *lpSecurityDescriptor* member of the *SECURITY_ATTRIBUTES* structure if the *lpSecurityAttributes* parameter is not *NULL. CreateFile* does not use the other structure members. The *bInheritHandle* member is the only way to indicate whether another process can inherit the file handle.

If you try to create a file on a floppy drive that does not have a floppy disk or to open a file on a CD-ROM drive that does not have a CD, the system will display a message box asking the user to insert a disk or a CD, respectively. To prevent the system from displaying this message box, call the *SetErrorMode* function with *SEM_FAILCRITICALERRORS*. For more information about *SetErrorMode*, check your compiler's on-line documentation.

If *CreateFile* opens the client end of a named pipe, the function will use any instance of the named pipe that is in the listening state. The opening process can duplicate the handle as many times as required, but, after *CreateFile* has opened the client end, another client cannot open the named pipe instance. The access you specify when *CreateFile* opens a pipe must be compatible with the access you specify in the *dwOpenMode* parameter of the *CreateNamedPipe* function. When you specify security in connection with opening a named pipe, your program should also specify one or more of the security flags Table 1453.1 details.

Value	Meaning
SECURITY_ANONYMOUS	Specifies that the file will impersonate the client at the Anonymous level.
SECURITY_IDENTIFICATION	Specifies that the file will impersonate the client at the Identification level.
SECURITY_IMPERSONATION	Specifies that the file will impersonate the client at the Impersonation level.
SECURITY_DELEGATION	Specifies that the file will impersonate the client at the Delegation level.
SECURITY_CONTEXT_TRACKING	Specifies that the security tracking mode is dynamic. If you do not specify this flag, Security Tracking Mode is static.
SECURITY_EFFECTIVE_ONLY	Specifies that only the enabled aspects of the client's security context are available to the server. If you do not specify this flag, all aspects of the client's security context are available. This flag lets the client limit the groups and privileges that a server can use while impersonating the client.

***Table 1453.1** Security flags for use when creating a named pipe.*

If the *CreateFile* function opens the client end of a mailslot, the function will return *INVALID_HANDLE_VALUE* if the mailslot client tries to open a local mailslot before the mailslot server has created it using the *CreateMailSlot* function.

The *CreateFile* function can create a handle to a communications resource, such as the serial port COM1. For communications resources, the *dwCreationDistribution* parameter must be *OPEN_EXISTING*, and the *hTemplate* parameter must be *NULL*. You can specify read, write, or read-write access, and you can open the handle for overlapped I/O.

Under Windows NT, you can use the *CreateFile* function to open a disk drive or a partition on a disk drive. The function returns a handle to the disk device. Your programs can later use that handle with the *DeviceIOControl* function. The call must meet the following requirements in order for the call to succeed:

- The caller must have administrative privileges for the operation to succeed on a hard disk drive.
- The *lpFileName* string should be of the form \\.*physicaldrivex* to open the hard disk *x*. Hard disk numbers start at zero. For example, \\.*physicaldrive2* obtains a handle to the third physical drive on the user's computer.
- The *lpFileName* string should be \\.*x:* to open a floppy drive *x* or a partition *x* on a hard disk. For example \\.*A* obtains a handle to drive A on the user's computer, and \\.*C* obtains a handle to drive C on the user's computer.
- Under Windows 95, this technique does not work for opening a logical drive. In Windows 95, specifying a string in this form causes *CreateFile* to return an error.
- The *dwCreationDistribution* parameter must have the *OPEN_EXISTING* value.
- When opening a floppy disk or a partition on a hard disk, you must set the *FILE_SHARE_WRITE* flag in the *dwShareMode* parameter.

The *CreateFile* function can create a handle to console input (*CONIN$*). If the process has an open handle to console input as a result of inheritance or duplication, the process can also create a handle to the active screen buffer (*CONOUT$*). You must attach the calling process to an inherited console or one that the *AllocConsole* API function allocates. For console handles, set the *CreateFile* parameters as Table 1453.2 details.

Parameters	Value
lpFileName	Use the *CONIN$* value to specify console input and the *CONOUT$* value to specify console output. *CONIN$* gets a handle to the console's input buffer, even if the *SetStdHandle* function redirected the standard input handle. To get the standard input handle, use the *GetStdHandle* function. *CONOUT$* gets a handle to the active screen buffer, even if *SetStdHandle* redirected the standard output handle. To get the standard output handle, use *GetStdHandle*.
dwDesiredAccess	Microsoft recommends that you use only *GENERIC_READ* | *GENERIC_WRITE*, but your programs can use either one to limit access.
dwShareMode	If the calling process inherited the console or if a child process should be able to access the console, this parameter must be *FILE_SHARE_READ* | *FILE_SHARE_WRITE*.
lpSecurityAttributes	If you want the child process to inherit the console, the *bInheritHandle* member of the *SECURITY_ATTRIBUTES* structure must be *True*.
dwCreationDistribution	You should specify *OPEN_EXISTING* when you use *CreateFile* to open the console.
dwFlagsAndAttributes	Ignored.
hTemplateFile	Ignored.

Table 1453.2 *Parameter values for* ***CreateFile*** *when you create a console.*

Table 1453.3 shows the effects of various settings of *dwDesiredAccess* and *lpFileName* when *lpFileName* is set to *CON*.

Settings	Result
GENERIC_READ	Opens console for input.
GENERIC_WRITE	Opens console for output.
GENERIC_READ | *GENERIC_WRITE*	Causes *CreateFile* to fail.

Table 1453.3 *The effects of access settings when opening a console.*

An application cannot create a directory with *CreateFile*. Instead, to create a directory, the application must call *CreateDirectory* or *CreateDirectoryEx*. However, under Windows NT, you can set the *FILE_FLAG_BACKUP_SEMANTICS* flag to obtain a handle to a directory. Your programs can pass a directory handle to some Win32 functions in place of a file handle. Some file systems, such as NTFS, support compression for individual files and directories. On volumes formatted for such a file system, a new directory inherits its parent directory's compression attribute.

Using File Handles 1454

Just like DOS and UNIX, Windows assigns a *file handle* for each file that your programs open or create. A file handle is a unique identifier that an application uses within functions that access a file. File handles are valid until your applications close them with the *CloseHandle* function, which closes the file and flushes the file's buffers to disk. When an application first starts, it inherits all open file handles from the process that started the application, provided the parent process opened the files and permitted inheritance. If the parent process opened the files without permitting inheritance, an application will not inherit the open file handles.

As you have learned, an application can open file handles for console input and output. As you learned, instead of a filename, the application passes the *CONIN$* string to the *CreateFile* function as the filename for the console input and the *CONOUT$* as the filename for the console output.

1455 REVISITING FILE POINTERS

As you learned in the Files, Directories, and Disks sections of this book, when your application first opens a file, operating systems generally place a file pointer at the beginning of the file. Windows is no exception. The file pointer marks the current position in the file where the next read or write operation will take place. As your programs read or write each byte from the file, Windows advances the file pointer to the next byte. An application can also move the file-pointer position with the *SetFilePointer* function. You will use the *SetFilePointer* function within your programs as shown in the following prototype:

```
DWORD SetFilePointer(
   HANDLE hFile,                  // handle of file
   LONG lDistanceToMove,          // number of bytes to move file pointer
   PLONG lpDistanceToMoveHigh,    // address of high-order word of distance to move
   DWORD dwMoveMethod             // how to move
);
```

The *SetFilePointer* function uses the parameters Table 1455.1 details.

Parameter	Description
hFile	Identifies the file whose file pointer the function is to move. The file handle must have *GENERIC_READ* or *GENERIC_WRITE* access to the file.
lDistanceToMove	Specifies the number of bytes to move the file pointer. A positive value moves the pointer forward in the file and a negative value moves it backward.
lpDistanceToMoveHigh	Points to the high-order word of the 64-bit distance to move. If the value of this parameter is *NULL*, *SetFilePointer* can operate only on files whose maximum size is $(2^{32} - 2)$. If you specify this parameter, the maximum file size is $(2^{64} - 2)$. This parameter also receives the high-order word of the new value of the file pointer.
dwMoveMethod	Specifies the starting point for the file pointer move. This parameter can be one of the values Table 1455.2 details.

*Table 1455.1 The parameters for the **SetFilePointer** function.*

There are several different methods that you can instruct Windows to use to move the file pointer. Table 1455.2 lists the Window predefined file pointer movement methods.

Value	Meaning
FILE_BEGIN	The starting point is zero or the beginning of the file. If *FILE_BEGIN* is specified, the function interprets *DistanceToMove* as an unsigned location for the new file pointer.
FILE_CURRENT	The current value of the file pointer is the starting point.
FILE_END	The current end-of-file position is the starting point.

*Table 1455.2 The possible movement methods for **SetFilePointer**.*

If the *SetFilePointer* function succeeds, the function returns the low-order double-word of the new file pointer, and if *lpDistanceToMoveHigh* is not *NULL*, the function puts the high-order double-word of the new file pointer into the *long* pointed to by that parameter. If the function fails and *lpDistanceToMoveHigh* is *NULL*, the function returns *0xFFFFFFFF*. If the function fails, and *lpDistanceToMoveHigh* is non-*NULL*, the function returns *0xFFFFFFFF* and *GetLastError* will return a value other than *NO_ERROR*.

You cannot use the *SetFilePointer* function with a handle to a non-seeking device, such as a pipe or a communications device. To determine the file type for *hFile*, use the *GetFileType* function. Be careful when you set the file pointer in a multi-threaded application. An application whose threads share a file handle, update the file pointer, and read from the file must use a critical-section object or mutex object to protect the file-pointer updates. If the *hFile* file handle was opened with the *FILE_FLAG_NO_BUFFERING* flag set, an application can move the file pointer only to *sector-aligned positions*. A sector-aligned position is a position that is a whole number multiple of the volume's sector size. An application can call the *GetDiskFreeSpace* function to obtain a volume's sector size. If an application calls *SetFilePointer* with distance-to-move values that result in a position that is not sector-aligned and a handle that the program originally opened with *FILE_FLAG_NO_BUFFERING*, the function fails, and *GetLastError* returns *ERROR_INVALID_PARAMETER*.

Note: ***SetFilePointer*** *is similar to the* ***lseek*** *function, which Tip 408 explains, and the* ***fseek*** *function, which Tip 450 explains.*

USING WRITEFILE TO WRITE TO THE FILE 1456

As you have learned, your programs can open files with the *CreateFile* function. If you create a file with write access, your programs can then use the *WriteFile* function to write data to that file. The *WriteFile* function writes data to a file and is designed for both synchronous and asynchronous operation. The function starts writing data to the file at the position the file pointer indicates. After *WriteFile* completes the write operation, the function adjusts the file pointer by the number of bytes it actually wrote, except when the file is opened with *FILE_FLAG_OVERLAPPED*. If you created the file handle for overlapped input and output (I/O), the application must adjust the position of the file pointer after the write operation completes. You will use the *WriteFile* function within your programs as shown in the following prototype:

```
BOOL WriteFile(
   HANDLE hFile,                    // handle to file to write to
   LPCVOID lpBuffer,                // pointer to data to write to file
   DWORD nNumberOfBytesToWrite,     // number of bytes to write
   LPDWORD lpNumberOfBytesWritten,  // pointer to number of bytes written
   LPOVERLAPPED lpOverlapped        // pointer to structure needed for overlapped I/O
);
```

The *WriteFile* function accepts the parameters Table 1456 details.

Parameter	Description
hFile	Identifies the file the function is to write to. The file handle must have *GENERIC_WRITE* access to the file. Under Windows NT, for asynchronous write operations, *hFile* can be any handle the *CreateFile* function opens with the *FILE_FLAG_OVERLAPPED* flag, or a socket handle the *socket* or *accept* functions return. Under Windows 95, for asynchronous write operations, *hFile* can be a communications resource, mailslot, or named pipe handle *CreateFile* opened with the *FILE_FLAG_OVERLAPPED* flag, or a socket handle the *socket* or *accept* functions return. Windows 95 does not support asynchronous write operations on disk files.

Table 1456 *The parameters for the* ***WriteFile*** *function.* *(continued on following page)*

Parameter	Description
lpBuffer	Points to the buffer containing the data the function is to write to the file.
nNumberOfBytesToWrite	Specifies the number of bytes to write to the file. Unlike the MS-DOS operating system, Windows NT interprets a value of zero as specifying a *NULL* write operation. A *NULL* write operation does not write any bytes, but does cause the time stamp to change.
lpNumberOfBytesWritten	Points to the number of bytes *WriteCall* writes. *WriteFile* sets this value to zero before doing any work or error checking. If *lpOverlapped* is *NULL*, *lpNumberOfBytesWritten* cannot be *NULL*. If *lpOverlapped* is not *NULL*, *lpNumberOfBytesWritten* can be *NULL*. If this is an overlapped write operation, you can call *GetOverlappedResult* to get the number of bytes written. If *hFile* is associated with an I/O completion port, you can call *GetQueuedCompletionStatus* to get the number of bytes written.
lpOverlapped	Points to an *OVERLAPPED* structure. The function call requires this structure if the process opened *hFile* with the *FILE_FLAG_OVERLAPPED* flag set. If the process opened *hFile* with *FILE_FLAG_OVERLAPPED*, the *lpOverlapped* parameter must not be *NULL*. It must point to a valid *OVERLAPPED* structure. If *hFile* was opened with *FILE_FLAG_OVERLAPPED* and *lpOverlapped* is *NULL*, the function can incorrectly report that the write operation is complete. If the process opened *hFile* with the *FILE_FLAG_OVERLAPPED* flag set and *lpOverlapped* is not *NULL*, the write operation starts at the offset specified in the *OVERLAPPED* structure and *WriteFile* may return before the operating system completes the write operation. In such a case, *WriteFile* returns FALSE and the *GetLastError* function returns *ERROR_IO_PENDING*. Using overlapped file I/O lets the calling process continue processing while the operating system completes the write operation. The operating system sets the event you specify in the *OVERLAPPED* structure to the signaled state upon the write operation's completion. If the process did not open *hFile* with *FILE_FLAG_OVERLAPPED* and *lpOverlapped* is *NULL*, the write operation starts at the current file position and *WriteFile* does not return until the operating system has completed the operation. If the process did not open *hFile* with *FILE_FLAG_OVERLAPPED* and *lpOverlapped* is not *NULL*, the write operation starts at the offset you specified in the *OVERLAPPED* structure and *WriteFile* does not return until the operating system completes the write operation.

Table 1456 *The parameters for the* ***WriteFile*** *function. (continued from previous page)*

If the *WriteFile* function succeeds, the function returns non-zero. If it fails, the function returns zero. If another process locks part of the file and the write operation overlaps the locked portion, this function fails. Applications must not read from nor write to the output buffer that a write operation is using until the write operation completes. Premature access of the output buffer may lead to corruption of the data written from that buffer.

Your programs can use *WriteFile* with a handle to console output to write characters to the screen buffer. The console mode determines the function's exact behavior. The data is written to the current cursor position. The operating system updates the screen cursor position after the write operation. Unlike the MS-DOS operating system, Windows NT interprets zero bytes to write as specifying a *NULL*-write operation and *WriteFile* does not truncate or extend the file. To truncate or extend a file, use the *SetEndOfFile* function.

When an application uses the *WriteFile* function to write to a pipe, the write operation may not finish if the pipe buffer is full. Instead, the write operation will finish when a read operation (using the *ReadFile* function) makes more buffer space available. If a process has closed the anonymous read pipe handle and *WriteFile* tries to use the corresponding anonymous write pipe handle to write, the function returns *False* and *GetLastError* returns *ERROR_BROKEN_PIPE*.

The *WriteFile* function may fail and return *ERROR_INVALID_USER_BUFFER* or *ERROR_NOT_ENOUGH_MEMORY* whenever there are too many outstanding asynchronous I/O requests. To cancel all pending asynchronous I/O operations, use the *CancelIO* function. This function only cancels operations the calling thread for the specified file handle issues. I/O operations that the operating system cancels as a result of a call to *CancelIO* complete with the error *ERROR_OPERATION_ABORTED*.

If you are trying to write to a floppy drive that does not have a floppy disk, the system will display a message box prompting the user to retry the operation. To prevent the system from displaying this message box, call the *SetErrorMode* function with *SEM_NOOPENFILEERRORBOX*. If *hFile* is a handle to a named pipe, the *Offset* and *OffsetHigh* members of the *OVERLAPPED* structure pointed to by *lpOverlapped* must be zero, or the function will fail.

USING READFILE TO READ FROM THE FILE 1457

As you learned in Tip 1456, your programs can use the *WriteFile* function to write to a file from within Windows. Similarly, your programs can use the *ReadFile* function to read from a file. The *ReadFile* function reads data from a file, starting at the position the file pointer indicates. After the *ReadFile* function completes the read operation, the operating system adjusts the file pointer by the number of bytes actually read, unless the process creates the file handle with the overlapped attribute. If the process creates the file handle for overlapped input and output (I/O), the application must adjust the position of the file pointer after the read operation. You will use the *ReadFile* function as shown on the following prototype:

```
BOOL ReadFile(
   HANDLE hFile,                    // handle of file to read
   LPVOID lpBuffer,                 // address of buffer that receives data
   DWORD nNumberOfBytesToRead,      // number of bytes to read
   LPDWORD lpNumberOfBytesRead,     // address of number of bytes read
   LPOVERLAPPED lpOverlapped        // address of structure for data
  );
```

The *ReadFile* function accepts identical parameters to the *WriteFile* function Tip 1456 details, except the *ReadFile* function expects *nNumberOfBytesToRead* parameter instead of the *nNumberOfBytesToWrite* parameter and returns the *lpNumberOfBytesRead* parameter rather than the *lpNumberOfBytesWritten* parameter. The *nNumberOfBytesToRead* parameter specifies the number of bytes that *ReadFile* should read from the file. The *lpNumberOfBytesRead* parameter points to the number of bytes read. *ReadFile* sets this value to zero before doing any work or error checking. If the *lpNumberOfBytesRead* parameter is zero when *ReadFile* returns *True* on a named pipe, the other end of the message-mode pipe calls the *WriteFile* function with *nNumberOfBytesToWrite* set to zero.

If *lpOverlapped* is *NULL*, *lpNumberOfBytesRead* cannot be *NULL*. If *lpOverlapped* is not *NULL*, *lpNumberOfBytesRead* can be *NULL*. If this is an overlapped read operation, you can call *GetOverlappedResult* to get the number of bytes read. If *hFile* is associated with an I/O completion port, you can call *GetQueuedCompletionStatus* to get the number of bytes read.

If *hFile* was opened with *FILE_FLAG_OVERLAPPED* and *lpOverlapped* is not *NULL*, the read operation starts at the offset you specified in the *OVERLAPPED* structure and *ReadFile* may return before the read operation has been completed. In this case, *ReadFile* returns *False* and the *GetLastError* function returns *ERROR_IO_PENDING*. This lets the calling process continue while the read operation finishes. The operating system sets the event you specified in the *OVERLAPPED* structure to the signaled state upon completion of the read operation.

If *hFile* was not opened with *FILE_FLAG_OVERLAPPED* and *lpOverlapped* is *NULL*, the read operation starts at the current file position and *ReadFile* does not return until the operation completes. If the process did not open *hFile* with *FILE_FLAG_OVERLAPPED* and *lpOverlapped* is not *NULL*, the read operation starts at the offset you specified in the *OVERLAPPED* structure. *ReadFile* does not return until the operating system completes the read operation.

If the function succeeds, the function returns non-zero. If the function returns non-zero and the number of bytes read is zero, the file pointer was beyond the current end of the file at the time of the read operation. However, if the process opened the file with *FILE_FLAG_OVERLAPPED* and *lpOverlapped* is not *NULL*, the function's return value is *False* and *GetLastError* returns *ERROR_HANDLE_EOF* when the file pointer goes beyond the current end of file. If the function fails, the function's return value is zero.

ReadFile returns when one of the following is true: a write operation completes on the write end of the pipe, the number of bytes requested has been read, or an error occurs. If another process locks part of the file and the read operation overlaps the locked portion, the *ReadFile* function fails. Applications must not read from nor write to the input buffer that a read operation is using until the read operation completes. A premature access to the input buffer may lead to corruption of the data read into that buffer. Characters can be read from the console input buffer by using *ReadFile* with a handle to console input. The console mode determines the *ReadFile* function's exact behavior.

If the process is reading a named pipe in message mode and the next message is longer than the *nNumberOfBytesToRead* parameter specifies, *ReadFile* returns *False* and *GetLastError* returns *ERROR_MORE_DATA*. A subsequent call to the *ReadFile* or *PeekNamedPipe* function may read the remainder of the message. When you read from a communications device, the current communication timeouts (as set and retrieved using the *SetCommTimeouts* and *GetCommTimeouts* functions) governs the *ReadFile* function's behavior. Unpredictable results can occur if you fail to set the timeout values. If *ReadFile* tries to read from a mailslot whose buffer is too small, the function returns *False* and *GetLastError* returns *ERROR_INSUFFICIENT_BUFFER*.

If a process has closed the anonymous write pipe handle and *ReadFile* tries to use the corresponding anonymous read pipe handle to read, the function returns *False* and *GetLastError* returns *ERROR_BROKEN_PIPE*. The *ReadFile* function may fail and return *ERROR_INVALID_USER_BUFFER* or *ERROR_NOT_ENOUGH_MEMORY* whenever there are too many outstanding asynchronous I/O requests. The *Write_File.cpp* program Tip 1457 presented shows the use of the *ReadFile* function.

Note: *The* ***ReadFile*** *code to check for the end-of-file condition (eof) differs for synchronous and asynchronous read operations. When a synchronous read operation reaches the end of a file,* ***ReadFile*** *returns* ***True*** *and sets* ****lpNumberOfBytesRead*** *to zero. When an asynchronous read operation reaches the end of a file,* ***ReadFile*** *fails and returns* ***ERROR_HANDLE_EOF****.*

1458 Closing the File

As you have learned, you use file handles to work with files from within your Windows programs. As with other handles (such as handles to memory or a handle to a device context), you should always close a file handle after your program completes its processing. The *CloseHandle* function closes an open object handle. You will use the *CloseHandle* function within your programs as shown in the following prototype:

```
BOOL CloseHandle(HANDLE hObject);
```

The *hObject* parameter identifies an open object handle. If the function succeeds, the function returns non-zero. If the function fails, the function returns zero. You can use the *CloseHandle* function to close handles to the following objects:

- Console input or output
- Event file
- File mapping
- Mutex
- Named pipe
- Process
- Semaphore
- Thread
- Token (Windows NT only)

The *CloseHandle* function invalidates the specified object handle, decrements the object's handle count, and performs object retention checks. After the process closes the last handle to an object, the operating system removes the object from the operating system's ongoing processing. The *CloseHandle* function does not close module objects. Closing an invalid handle raises an exception. This includes closing a handle twice, not checking the return value and closing an invalid handle, and trying to use *CloseHandle* on a handle *FindFirstFile* returns.

Note: *Use* ***CloseHandle*** *to close handles the* ***CreateFile*** *function returns. Use* ***FindClose*** *to close handles the* ***FindFirstFile*** *function returns. You will learn more about* ***FindClose*** *in Tip 1476.*

1459 Sharing Data with File Mapping

File mapping is the copying of a file's contents to a process's virtual address space. When you map a file, the copy of the file's contents is known as the *file view*, and the internal structure your program uses to maintain the copy is known as the *file-mapping object*. To share data, another process can use the first process's file-mapping object to create an identical file view in its own virtual address space.

A common example of processes sharing data is dynamic-data exchange (DDE) and its "younger brother," object-linking and embedding (OLE). In Windows 3.x, applications would allocate global memory with the *GMEM_DDESHARE* flag and use the memory handle to share data between processes. In Windows 95 and Windows NT, applications should instead use file mapping. It is not necessary for your applications to actually have a file to map into memory. An application can specify a *0xFFFFFFFF* value for the file handle when it calls the *CreateFileMapping* function, which will then map a view of the operating system's paging file (which you have learned about previously) into memory. Tip 1460 explains the *CreateFileMapping* function in detail.

When you map a file to memory, your programs can essentially access the file's contents as if the file were an array—your programs can even use index values and pointers to access the file's contents.

1460 Mapping a File to Virtual Memory

As you learned in Tip 1459, your programs will often map files into a process's virtual address space. Mapping files will help your programs access the data within the files more quickly and efficiently. When you want to map a file into memory, your programs should use the *CreateFileMapping* function to create a named or unnamed file-mapping object for the file you specify. You will use the *CreateFileMapping* function within your programs as shown in the following prototype:

```
HANDLE CreateFileMapping(
   HANDLE hFile,                                  // handle to file to map
   LPSECURITY_ATTRIBUTES lpFileMappingAttributes, // optional security attributes
   DWORD flProtect,                               // protection for mapping object
```

```
    DWORD dwMaximumSizeHigh,                    // object's size, high-order 32 bits
     DWORD dwMaximumSizeLow,                     // object's size, low-order 32 bits
    LPCTSTR lpName                               // name of file-mapping object
  );
```

The *CreateFileMapping* function accepts the parameters shown in Table 1460.1.

Parameter	Description
hFile	Identifies the file from which to create a mapping object. You must open the file with an access mode compatible with the protection flags the *flProtect* parameter specifies. Microsoft recommends, though Windows does not require, that you open files you intend to map for exclusive access. If *hFile* is *(HANDLE)0xFFFFFFFF*, the calling process must also specify a mapping object size in the *dwMaximumSizeHigh* and *dwMaximumSizeLow* parameters. The function creates a file-mapping object of the size the calling process specifies that the operating-system paging file backs rather than a named file in the file system backs. Processes can share the file-mapping object through duplication, through inheritance, or by name.
lpFileMappingAttributes	Pointer to a *SECURITY_ATTRIBUTES* structure that determines whether child processes can inherit the returned handle. If *lpFileMappingAttributes* is *NULL*, child processes cannot inherit the handle.
flProtect	Specifies the protection you want for the file view, when the file is mapped: *PAGE_READONLY*, which gives read-only access to the committed region of pages. An attempt to write to or execute the committed region results in an access violation. The file the *hFile* parameter specifies must have been created with *GENERIC_READ* access; *PAGE_READWRITE*, which gives read-write access to the committed region of pages. The file *hFile* specifies must have been created with *GENERIC_READ* and *GENERIC_WRITE* access; *PAGE_WRITECOPY*, which gives copy on write access to the committed region of pages. The files the *hFile* parameter specifies must have been created with *GENERIC_READ* and *GENERIC_WRITE* access. In addition, an application can combine (using the *bitwise OR* operator) one or more of the section attribute values listed in Table 1460.2 with one of the preceding three page protection values to specify certain section attributes.
dwMaximumSizeHigh	Specifies the high-order 32 bits of the maximum size of the file-mapping object.
dwMaximumSizeLow	Specifies the low-order 32 bits of the maximum size of the file-mapping object. If this parameter and *dwMaximumSizeHigh* are zero, the maximum size of the file-mapping object is equal to the current size of the file *hFile* identifies.
lpName	Points to a *NULL*-terminated string that specifies the mapping object's name. The name can contain any character except the backslash character (\). If this parameter matches the name of an existing named mapping object, the function requests access to the mapping object with the protection *flProtect* specifies. If this parameter is *NULL*, the function creates the mapping object without a name.

***Table 1460.1** The parameters for the **CreateFileMapping** function.*

As you learned in Table 1460.1, you can combine the page protection values for the file with one or more section attribute values, as shown in Table 1460.2.

Value	Description
SEC_COMMIT	Allocates physical storage in memory or in the paging file on disk for all pages of a section. This is the default setting.
SEC_IMAGE	The file you specified for a section's file mapping is an executable image file. Because the mapping information and file protection are taken from the image file, no other attributes are valid with *SEC_IMAGE.*
SEC_NOCACHE	The function is to set all pages of a section as non-cacheable. On 80x86 and MIPS machines, using the cache for these structures only slows down the performance as the hardware keeps the caches coherent. Some device drivers require non-cached data so that programs can write through to the physical memory. *SEC_NOCACHE* requires either the *SEC_RESERVE* or *SEC_COMMIT* to also be set.
SEC_RESERVE	Reserves all pages of a section without allocating physical storage. No other allocation operations can use the reserved range of pages until it is released. The application can commit reserved pages in subsequent calls to the *VirtualAlloc* function. This attribute is valid only if the *hFile* parameter is *(HANDLE) 0xFFFFFFFF*; that is, a file-mapping object the operating system paging file backs.

Table 1460.2 *The page protection values for the memory-mapped file.*

If the function succeeds, the function returns a handle to the file-mapping object. If the mapping object existed before the function call, the *GetLastError* function returns *ERROR_ALREADY_EXISTS*, and the function returns a valid handle to the existing file-mapping object (with its current size, not the new specified size). If the mapping object did not exist, *GetLastError* returns zero. If the function fails, the function returns *NULL*.

After a process has created a file-mapping object, the size of the file must not exceed the size of the file-mapping object; if it does, not all the file's contents will be available for applications to share. If an application specifies a size for the file-mapping object that is larger than the size of the actual named file on disk, Windows grows the file on disk to match the specified size of the file-mapping object. The handle that *CreateFileMapping* returns has full access to the new file-mapping object. The application can use the handle with any function that requires a handle to a file-mapping object. Processes can share file-mapping objects either through process creation, through handle duplication, or by name.

Note: *Under Windows 95, you must not use file handles that you have used to create file-mapping objects in subsequent calls to file I/O functions, such as* ***ReadFile*** *and* ***WriteFile****. In general, if you have used a file handle in a successful call to the* ***CreateFileMapping*** *function, do not use that handle unless you first close the corresponding file-mapping object.*

Creating a file-mapping object creates the potential for mapping a view of the file but does not map the view. The *MapViewOfFile* and *MapViewOfFileEx* functions map a view of a file into a process's address space. Tip 1461 explains the *MapViewOfFile* function in detail.

With one important exception, file views derived from a single file-mapping object are coherent, or identical, at a given time. If multiple processes have handles of the same file-mapping object, they see a coherent view of the data when they map a view of the file. The exception has to do with remote files. Although *CreateFileMapping* works with remote files, it does not keep them coherent. For example, if two computers map a file as writable, and both change the same page, each computer will see only its own writes to the page. When the data gets updated on the disk, the operating system will not merge the data. Moreover, a mapped file and a file that the input and output (I/O) functions (*ReadFile* and *WriteFile*) access are not necessarily coherent.

To fully close a file-mapping object, an application must call *UnmapViewOfFile* to unmap all mapped views of the file-mapping object, and call *CloseHandle* to close the file-mapping object handle. The order in which an application calls these functions does not matter. The call to *UnmapViewOfFile* is necessary because mapped views of a file-mapping object maintain internal open handles to the object, and a file-mapping object will not close until the process closes all open handles to the file-mapping object.

1461 MAPPING A VIEW OF A FILE INTO THE CURRENT PROCESS

As you learned in Tip 1460, after your programs create a file-mapping object, they must then map the file into the object. To do so, your programs will use the *MapViewOfFile* function. The *MapViewOfFile* function maps a view of a file into the address space of the calling process. You will use the *MapViewOfFile* function within your programs as shown in the following prototype:

```
LPVOID MapViewOfFile(
   HANDLE hFileMappingObject,    // file-mapping object to map into address space
   DWORD dwDesiredAccess,        // access mode
   DWORD dwFileOffsetHigh,       // high-order 32 bits of file offset
   DWORD dwFileOffsetLow,        // low-order 32 bits of file offset
   DWORD dwNumberOfBytesToMap    // number of bytes to map
);
```

The *hFileMappingObject* parameter identifies an open handle of a file-mapping object. The *CreateFileMapping* and *OpenFileMapping* functions return this handle. The *dwDesiredAccess* parameter specifies the type of access to the file view and, therefore, the protection of the pages the file maps. This parameter can be one of the values Table 1461 lists.

Value	Meaning
FILE_MAP_WRITE	Read-write access. The *hFileMappingObject* parameter must have been created with *PAGE_READWRITE* protection. The operating system maps a read-write view of the file.
FILE_MAP_READ	Read-only access. The *hFileMappingObject* parameter must have been created with *PAGE_READWRITE* or *PAGE_READONLY* protection. The operating system maps a read-only view of the file.
FILE_MAP_ALL_ACCESS	Same as *FILE_MAP_WRITE*.
FILE_MAP_COPY	Copy on write access. If you create the map with *PAGE_WRITECOPY* and the view with *FILE_MAP_COPY*, you will receive a view to file. If you write to it, the pages are automatically swappable and the modifications you make will not go to the original data file. Under Windows 95, you must pass *PAGE_WRITECOPY* to *CreateFileMapping*; otherwise, Windows will return an error. Under Windows NT, there is no restriction as to how you must create the *hFileMappingObject* parameter. Copy on write is valid for any type of view. If you use *DuplicateHandle* or *OpenFileMapping* to share the mapping between multiple processes and one process writes to a view, the modification is propagated to the other process. The original file does not change.

***Table 1461** The possible values for the **dwDesiredAccess** parameter.*

The *dwFileOffsetHigh* parameter specifies the high-order 32 bits of the file offset where mapping is to begin. The *dwFileOffsetLow* parameter specifies the low-order 32 bits of the file offset where mapping is to begin. The combination of the high and low offsets must specify an offset within the file that matches the system's memory-allocation granularity, or the function fails.

That is, the offset must be a multiple of the allocation granularity (such as 8 bytes or 16 bytes). The *dwNumberOfBytesToMap* parameter specifies the number of bytes of the file to map. If *dwNumberOfBytesToMap* is zero, the operating system maps the entire file.

If the function succeeds, the function returns the starting address of the mapped view. If the function fails, the function returns *NULL*. Mapping a file makes the specified portion of the file visible in the address space of the calling process.

OPENING A NAMED FILE-MAPPING OBJECT 1462

As you have learned, your programs will use the *CreateFileMapping* function to create a named or unnamed file map and the *MapViewOfFile* function to map the file to memory. In the *Sample_Map.cpp* program, however, a second thread opens the mapped file.

Rather than creating a new file mapping, the second thread uses the *OpenFileMapping* function to open the mapped file for its own use. You will use the *OpenFileMapping* function within your programs as shown in the following prototype:

```
HANDLE OpenFileMapping(DWORD dwDesiredAccess, BOOL bInheritHandle, LPCTSTR lpName);
```

The *dwDesiredAccess* parameter specifies the access to the file-mapping object. Under Windows NT, the operating system checks the access parameter against any security descriptor on the target file-mapping object. This parameter can be one of the values Table 1461 lists.

The *bInheritHandle* parameter specifies whether a new process is to inherit the returned handle during process creation. A value of *True* indicates that the new process inherits the handle. The *lpName* parameter points to a string that names the file-mapping object the process is to open. If there is an open handle to a file-mapping object by this name and the security descriptor on the mapping object does not conflict with the *dwDesiredAccess* parameter, the open operation succeeds.

If the function succeeds, the function returns an open handle to the specified file-mapping object. If the function fails, the function returns *NULL*. You can use the handle *OpenFileMapping* returns with any function that requires a handle to a file-mapping object.

UNDERSTANDING FILE ATTRIBUTES 1463

In the Files, Directories, and Disks sections of this book, you learned about the file attributes that DOS attaches to each file you create. Similarly, Windows associates a set of file attributes with every file. Windows initializes many of the file attributes when you create the file, and later Windows will alter some attributes each time you access the file. Most often, you will not want to change a file's attributes, but simply read them and react accordingly. Similar to DOS, most file attributes have to do with flag settings, file size, and file time stamps. In subsequent Tips, you will manipulate and access some of the most commonly used file attributes.

OBTAINING AND CHANGING A FILE'S ATTRIBUTES 1464

As you learned in Tip 1463, Windows attaches file attributes to each file you create within the operating system. Just as you read file attributes with standard C functions within the Files, Directories, and Disks section of this book, the Windows API provides many functions that you can use to read a file's attributes. The *GetFileAttributes* function returns information from a subset of all possible attributes for a specified file or directory. Later Tips will discuss other functions that return other file attributes, such as the file time or file size. You will use the *GetFileAttributes* function within your programs as shown in the following prototype:

```
DWORD GetFileAttributes(LPCTSTR lpFileName);
```

The *lpFileName* parameter points to a *NULL*-terminated string that specifies the name of a file or directory. Under Windows NT, there is a default string size limit for paths of *MAX_PATH* characters. This limit results from how the *GetFileAttributes* function parses paths. An application can call the wide (W) version of *GetFileAttributes* and insert "\\?\" to the path to transcend this limit and send in paths longer than *MAX_PATH* characters. The "\\?\" tells the function to turn off path parsing; it lets your programs use paths longer than *MAX_PATH* with *GetFileAttributesW.* The *GetFileAttributesW* function also works with UNC names. The operating system ignores the "\\?\"as part of the path. For example, the function sees "\\?\C:\mydocuments\private" as "C:\mydocuments\private", and "\\?\jamsa1\happy\foodstuff" as "\\jamsa1\happy\foodstuff". Under Windows 95, the *lpFileName* string must not exceed *MAX_PATH* characters. Windows 95 does not support the "\\?\" prefix.

If the function succeeds, the return value contains the attributes of the specified file or directory. If the function fails, the function returns *0xFFFFFFFF.* To get extended error information, call *GetLastError.*

The attributes can be one or more of the values Table 1464 lists.

Value	Meaning
FILE_ATTRIBUTE_ARCHIVE	The file or directory is an archive file or directory. Applications use this flag to mark files for backup or removal.
FILE_ATTRIBUTE_COMPRESSED	The file or directory is compressed. For a file, this means that all the data in the file is compressed. For a directory, this means that compression is the default for newly created files and subdirectories.
FILE_ATTRIBUTE_DIRECTORY	The current object is a directory.
FILE_ATTRIBUTE_HIDDEN	The file or directory is hidden (not included in an ordinary directory listing).
FILE_ATTRIBUTE_NORMAL	The file or directory has no other attributes set. This attribute is valid only if used alone.
FILE_ATTRIBUTE_OFFLINE	The file's data is not immediately available. Indicates that the file data has been physically moved to offline storage.
FILE_ATTRIBUTE_READONLY	The file or directory is read-only. Applications can read the file but cannot write to it or delete it. In the case of a directory, applications cannot delete it.
FILE_ATTRIBUTE_SYSTEM	The file or directory is part of the operating system, or the operating system uses the file or directory exclusively.
FILE_ATTRIBUTE_TEMPORARY	A process is using the file for temporary storage. The application should delete a temporary file as soon as the application no longer needs the file.

Table 1464 *The return values from the* ***GetFileAttributes*** *function.*

As you have learned, your programs can use the *GetFileAttributes* function to obtain a file's attributes. Occasionally, however, your programs may also change file attributes. Within your programs, you can use the *SetFileAttributes* function to set a file's attributes, as shown in the following prototype:

```
BOOL SetFileAttributes(
   LPCTSTR lpFileName,        // address of filename
   DWORD dwFileAttributes     // attributes to set
);
```

The *lpFileName* parameter points to a string that specifies the name of the file whose attributes the function is to set. The same limitations apply to the *lpFileName* parameter as applied to the *GetFileAttributes* function's *lpFileName* parameter. The *dwFileAttributes* parameter specifies the file attributes the function is to set. This parameter can be a combination of the values Table 1464 details. However, all other values override *FILE_ATTRIBUTE_NORMAL.*

If the function succeeds, the function returns non-zero. If the function fails, the function returns zero. To get extended error information, call *GetLastError*. You cannot use the *SetFileAttribute* function to set a file's compression state. Setting *FILE_ATTRIBUTE_COMPRESSED* in the *dwFileAttributes* parameter does nothing. Use the *DeviceIoControl* function and the *FSCTL_SET_COMPRESSION* operation to set a file's compression state.

The CD-ROM that accompanies this book includes the *Check_ReadOnly.cpp* program. The program checks the *file.dat* file to determine whether the read-only attribute is set. If it is, the program clears all the attribute settings and then deletes the file.

OBTAINING A FILE'S SIZE 1465

As you have learned, Windows attaches certain attributes to each file it saves to disk. One of the most common requests your programs will make of a file is to request the file's size. The *GetFileSize* function retrieves the size, in bytes, of the specified file. You will use the *GetFileSize* function within your programs as shown in the following prototype:

```
DWORD GetFileSize(
   HANDLE hFile,               // handle of file to get size of
   LPDWORD lpFileSizeHigh      // address of high-order word for file size
);
```

The *hFile* parameter specifies an open handle of the file whose size the function is to return. The process must have created the handle with either *GENERIC_READ* or *GENERIC_WRITE* access to the file. The *lpFileSizeHigh* parameter points to the variable where the function will return the high-order word of the file size. If the application does not require the high-order word, the *lpFileSizeHigh* parameter can be *NULL*.

If the function succeeds, the function returns the low-order double-word of the file size, and if *lpFileSizeHigh* is not *NULL*, the function puts the high-order double-word of the file size into the variable to which the parameter points. If the function fails and *lpFileSizeHigh* is *NULL*, the function returns *0xFFFFFFFF*. To get extended error information, call *GetLastError*. If the function fails and *lpFileSizeHigh* is non-*NULL*, the function returns *0xFFFFFFFF* and *GetLastError* will return a value other than *NO_ERROR*.

You cannot use the *GetFileSize* function with a handle of a non-seeking device such as a pipe or a communications device. To determine the file type for *hFile*, use the *GetFileType* function. The *GetFileSize* function gets the file's uncompressed size. Use the *GetCompressedFileSize* function to obtain a compressed file's size. Note that if the function returns *0xFFFFFFFF* and *lpFileSizeHigh* is not *NULL*, an application must call *GetLastError* to determine whether the function succeeded or failed.

The CD-ROM that accompanies this book includes the *Check_FileSize.cpp* program. The program defines a function, *DisplayFileSize*, that first uses *GetFileType* to check the type of the file handle. If the file is a disk-based file, the *DisplayFileSize* function calls the *GetFileSize* function to obtain the file's size, and display the size within a message box. If the file is not a disk-based file, the function alerts the user that the file is not a disk-based file and exits.

OBTAINING A FILE'S TIME STAMP 1466

In earlier Tips you learned how to check a file's time stamp using C run-time functions. The Windows API also provides a function that your programs can use to obtain a file's timestamp. The *GetFileTime* function retrieves the date and time that a file was created, last accessed, and last modified. You will use the *GetFileTime* function within your programs as shown in the following protoype:

```
BOOL GetFileTime(
   HANDLE hFile,                    // identifies the file
   LPFILETIME lpCreationTime,       // address of creation time
   LPFILETIME lpLastAccessTime,     // address of last access time
```

```
   LPFILETIME lpLastWriteTime        // address of last write time
);
```

The *hFile* parameter identifies the files for which the function is to get dates and times. The process must have created the file handle with *GENERIC_READ* access to the file. The *lpCreationTime* parameter points to a *FILETIME* structure to receive the date and time the file was created. The *lpLastAccessTime* parameter points to a *FILETIME* structure to receive the date and time the file was last accessed. The last access time includes the last time the file was written to, read from, or, in the case of executable files, run. The *lpLastWriteTime* parameter points to a *FILETIME* structure to receive the date and time the file was last written to. If the file does not require information about all three times, the parameters for the times the function does not require information can have a *NULL* value.

If the function succeeds, the function returns non-zero. If the function fails, the function returns the value zero. To get extended error information, call *GetLastError*.

The Windows 95 and Windows NT file systems support the file creation, last access, and last write time values. In Windows 95, the precision for a file time is two seconds (meaning the time Windows records within the attribute will be within two seconds of the actual access time). The time precision for files in other file systems, such as those connected through a network, depends on the remote file system. The remote device may also limit time precision.

The CD-ROM that accompanies this book includes the program *Get_Time.cpp*, which uses *GetFileTime* to obtain a file's current time and then converts that time to local time. The program then converts the local time *FILETIME* structure into a more recognizable DOS date and time format. Finally, the program displays the converted DOS date and time within a message box.

1467 CREATING DIRECTORIES

In Tip 396, you learned how to use a standard C run-time library function to create a directory within DOS. Similarly, your Windows programs can use the Win32 API *CreateDirectory* function to create a new directory. If the underlying file system supports security on files and directories, the function applies a specified security descriptor to the new directory. Remember, each process will have its own current directory, so calls to *CreateDirectory* should not presume the current directory for the process is a constant. You will use the *CreateDirectory* function within your programs as shown in the following prototype:

```
BOOL CreateDirectory(
   LPCTSTR lpPathName,                          // pointer to a directory path string
   LPSECURITY_ATTRIBUTES lpSecurityAttributes  // pointer to a security descriptor
);
```

The *lpPathName* parameter points to a *NULL*-terminated string that specifies the path of the directory to be created. There is a default string size limit for paths of *MAX_PATH* characters. This limit relates to how the *CreateDirectory* function parses paths. You can exceed path limits under Windows NT as Tip 1464 details.

The *lpSecurityAttributes* parameter is a pointer to a *SECURITY_ATTRIBUTES* structure that determines whether child processes can inherit the returned handle. If *lpSecurityAttributes* is *NULL*, child processes cannot inherit it. Under Windows NT, the *lpSecurityDescriptor* member of the structure specifies a security descriptor for the new directory. If *lpSecurityAttributes* is *NULL*, the directory gets a default security descriptor. The target file system must support security on files and directories for this parameter to have an effect. Under Windows 95, the operating system ignores the *lpSecurityDescriptor* member of the structure.

If the function succeeds, the function returns a non-zero value. If the function fails, the function returns the value zero. To get extended error information, call *GetLastError*.

The CD-ROM that accompanies this book includes the *Create_NewDir.cpp* program. The program uses the *CreateDirectory* function to create a new directory named "NewDirectory" on the current drive of the computer on which the program is currently running.

Note: *Some file systems, such as NT File System (NTFS), support compression for individual files and directories. On volumes formatted for such a file system, a new directory inherits the compression attribute of its parent directory. An application can call **CreateFile** with the **FILE_FLAG_BACKUP_SEMANTICS** flag set to obtain a handle to a directory. For a code example, see **CreateFile**.*

Getting and Setting the Current Directory 1468

As you learned in Tip 1467, your programs can create new directories within the file system. However, your programs will likely require information about the current directory more often than they will create new directories. The *GetCurrentDirectory* function retrieves the current directory for the current process. You will use the *GetCurrentDirectory* function within your programs as shown in the following prototype:

```
DWORD GetCurrentDirectory(
   DWORD nBufferLength,        // size, in characters, of directory buffer
   LPTSTR lpBuffer             // address of buffer for current directory
);
```

The *nBufferLength* parameter specifies the length, in characters, of the buffer for the current directory string. The buffer length must include room for a terminating *NULL* character. The *lpBuffer* parameter points to the buffer for the current directory string. This *NULL*-terminated string specifies the absolute path to the current directory. If the function succeeds, the return value specifies the number of characters written to the buffer, not including the terminating *NULL* character.

If the function fails, the function returns zero. To get extended error information, call *GetLastError*. If the buffer pointed to by *lpBuffer* is not large enough, the return value specifies the buffer's required size, including the number of bytes necessary for a terminating *NULL* character.

Similarly, your programs can use the *SetCurrentDirectory* function to change the current directory for the current process to a directory you specify. You will use the *SetCurrentDirectory* function within your programs as shown in the following prototype:

```
BOOL SetCurrentDirectory(LPCTSTR lpPathName);
```

The *lpPathName* parameter points to a *NULL*-terminated string that specifies the path to the new current directory. This parameter may be a relative path or a fully qualified path. In either case, the operating system calculates the fully qualified path of the specified directory and stores that path as the current directory. If the function succeeds, the function returns a non-zero value. If the function fails, the function returns the value zero. To get extended error information, call *GetLastError*.

Each process has a single current directory made up of two parts:

- A disk designator that is either a drive letter followed by a colon, or a server name and share name (for example, *servername**sharename*)
- A directory on the disk designator

1469 Getting the Windows and System Directories

In Tip 1468, you learned how your programs can get the current directory for the current process. As your programs become more complex, you will often require information about the location of the *Windows* directory and the location of the Windows *system* directory. The *GetWindowsDirectory* function retrieves the path of the Windows directory (which may be a directory with a name other than *c:\windows* or *c:\winnt*). The Windows directory contains such files as Windows-based applications, initialization files, and Help files. You will use the *GetWindowsDirectory* function within your programs as shown in the following prototype:

```
UINT GetWindowsDirectory(
   LPTSTR lpBuffer,     // address of buffer for Windows directory
   UINT uSize           // size of directory buffer
);
```

The *lpBuffer* parameter points to the buffer to receive the *NULL*-terminated string that contains the path. This path does not end with a backslash unless the Windows directory is the root directory. For example, if the Windows directory is named *windows* on drive C, the path of the Windows directory this function retrieved is *c:\windows*. If Windows was installed in the root directory of drive C, the path retrieved is C:\. The *uSize* parameter specifies the maximum size, in characters, of the buffer specified by the *lpBuffer* parameter. You should set the *uSize* parameter to at least *MAX_PATH* to allow sufficient room in the buffer for the path.

If the function succeeds, the function returns the length, in characters, of the string the function copied to the buffer, not including the terminating *NULL* character. If the length is greater than the size of the buffer, the function returns the size of the buffer it requires to hold the path. If the function fails, the function returns zero. To get extended error information, call *GetLastError*.

The *Windows* directory is the directory where an application should store initialization and help files. If the user is running a shared version of Windows, the operating system guarantees that the *Windows* directory is private for each user. If an application creates other files that it wants to store on a per-user basis, it should place them in the directory the *HOMEPATH* environment variable specifies. *HOMEPATH* always specifies either the user's home directory, which is guaranteed to be private for each user, or a default directory (for example, *c:\users\default*) where the user will have all access.

Just as your programs will often require information about the location of the Windows directory, so too will they often require information about the system directory. The *GetSystemDirectory* function retrieves the path of the Windows system directory. The system directory contains such files as Windows libraries, drivers, and font files. You will use the *GetSystemDirectory* function within your programs as shown in the following prototype:

```
UINT GetSystemDirectory(
   LPTSTR lpBuffer,     // address of buffer for system directory
   UINT uSize           // size of directory buffer
);
```

The parameters for *GetSystemDirectory* are the same as those for *GetWindowsDirectory*. If the *GetSystemDirectory* function succeeds, the function returns the length, in characters, of the string copied to the buffer, not including the terminating *NULL* character. If the length is greater than the size of the buffer, the function returns the size of the buffer it requires to store the path. If the function fails, the function returns zero. To get extended error information, call *GetLastError*.

The CD-ROM that accompanies this book includes the *Show_Windows.cpp* program. When you compile and execute the *Show_Windows.cpp* program, it will use the *GetWindowsDirectory* and *GetSystemDirectory* functions to obtain the names of the *Windows* and *system* directories. The program will then display the directory names within the window's client area.

***Note:** As a rule, applications should not create files in the **system** directory. If the user is running a shared version of Windows, the application does not have write access to the **system** directory. Applications should create files only in the directory the **GetWindowsDirectory** function returns.*

REMOVING DIRECTORIES 1470

Just as your programs may create temporary directories, or directories that it will use only internally, there may be times when your programs must remove an existing directory. The *RemoveDirectory* function deletes an existing empty directory. You will use the *RemoveDirectory* function as shown in the following prototype:

```
BOOL RemoveDirectory(LPCTSTR lpPathName);
```

The *lpPathName* parameter points to a *NULL*-terminated string that specifies the path of the directory to be removed. The path must specify an empty directory, and the calling process must have delete access to the directory. If the function succeeds, the function returns a non-zero value. If the function fails, the function returns the value zero. To get extended error information, call *GetLastError*.

COPYING FILES 1471

As you have learned, your programs can manipulate files extensively from within Windows. As your programs become more complex, there may be times when your program must copy a file from one location to another location, while maintaining the original file in its original location. The *CopyFile* function copies an existing file to a new file. You will use the *CopyFile* function within your programs as shown here:

```
BOOL CopyFile(
     LPCTSTR lpExistingFileName,      // pointer to name of an existing file
     LPCTSTR lpNewFileName,           // pointer to filename to copy to
     BOOL bFailIfExists               // flag for operation if file exists
);
```

The *lpExistingFileName* parameter points to a *NULL*-terminated string that specifies the name of an existing file. The *lpNewFileName* parameter points to a *NULL*-terminated string that specifies the name of the new file. The *bFailIfExists* parameter specifies how the function is to proceed if a file of the same name as that *lpNewFileName* specifies already exists. If the *bFailIfExists* parameter is *True* and the new file already exists, the function fails. If this parameter is *False* and the new file already exists, the function overwrites the existing file and succeeds.

If the function succeeds, the function returns non-zero. If the function fails, the function returns zero. To get extended error information, call *GetLastError*. Security attributes for the existing file are not copied to the new file.

File attributes *(FILE_ATTRIBUTE_*)* for the existing file are copied to the new file. For example, if an existing file has the *FILE_ATTRIBUTE_READONLY* file attribute, a copy created through a call to *CopyFile* will also have the *FILE_ATTRIBUTE_READONLY* file attribute.

The CD-ROM that accompanies this book includes the *Create_Copy.cpp* program. When you compile and execute the *Create_Copy.cpp* program and select the *Demo* option, the program will copy the *file1.txt* file (which exists on the CD-ROM) to the *file2.txt* file. If the *file2.txt* file already exists, the program will prompt the user to replace the existing file.

MOVING AND RENAMING FILES 1472

As you learned in Tip 1471, your programs can easily make a copy of a file and place it in another location. However your programs will often want to move a file or a directory to another location without keeping the file's original copy. The *MoveFile* function renames an existing file or a directory (including all its children). You will use the *MoveFile* function within your programs as shown in the following prototype:

```
BOOL MoveFile(
   LPCTSTR lpExistingFileName,// address of name of the existing file
   LPCTSTR lpNewFileName      // address of new name for the file
);
```

The *lpExistingFileName* parameter points to a *NULL*-terminated string that names an existing file or directory. The *lpNewFileName* parameter points to a *NULL*-terminated string that specifies the new name of a file or directory. The new name must not already exist. A new file may be on a different file system or drive. A new directory must be on the same drive. If the function succeeds, the function returns a non-zero value. If the function fails, the function returns a zero value. To get extended error information, call *GetLastError*.

The *MoveFile* function will move (rename) either a file or a directory (including all its children) either within the same directory or across directories. The one limitation on the *MoveFile* function is that it will fail on directory moves when the destination is on a different volume.

1473 DELETING FILES

In Tip 1470, you learned how your programs can use the *RemoveDirectory* function to remove a directory from the file system. However, as Tip 1470 noted, the directory must be empty before you invoke *RemoveDirectory* or the function will fail. To remove files within a directory, your programs can use the *DeleteFile* function. You will use the *DeleteFile* function within your programs as shown in the following prototype:

```
BOOL DeleteFile(LPCTSTR lpFileName);
```

The *lpFileName* parameter points to a *NULL*-terminated string that specifies the file the function is to delete. If the function succeeds, the function returns a non-zero value. If the function fails, the function returns a zero value. To get extended error information, call *GetLastError*.

If an application tries to delete a file that does not exist, the *DeleteFile* function fails. Under Windows 95, the *DeleteFile* function deletes a file even if the file is currently open for normal I/O or as a memory-mapped file. To prevent errant results, close files before you try to delete them. Under Windows NT, the *DeleteFile* function fails if an application tries to delete a file that is currently open for normal I/O or as a memory-mapped file.

1474 USING FINDFIRSTFILE TO LOCATE FILES

In Tips 390 and 391, you learned how to use standard C run-time library functions to search the command path and the current directory for a file. Within your Windows programs, you should use the *Find* functions to locate files that match a given criteria. There are two *Find* functions that your programs can use. The *FindFirstFile* function searches a directory for a file whose name matches the filename you specify. *FindFirstFile* examines subdirectory names as well as filenames. You will use the *FindFirstFile* function within your programs as shown in the following prototype:

```
HANDLE FindFirstFile(
   LPCTSTR lpFileName,              // pointer to name of file to search for
   LPWIN32_FIND_DATA lpFindFileData // pointer to returned information
);
```

Under both Windows 95 and Windows NT, the *lpFileName* parameter points to a *NULL*-terminated string that specifies a valid directory or path and filename, which can contain wildcard characters (* and ?). Under Windows 95, however, this string must not exceed *MAX_PATH* characters.

The *lpFindFileData* parameter points to the *WIN32_FIND_DATA* structure that receives information about the found file or subdirectory. Your programs can use the structure in subsequent calls to the *FindNextFile* or *FindClose* functions to refer to the file or subdirectory. The *WIN32_FIND_DATA* structure describes a file the *FindFirstFile*, *FindFirstFileEx*, or *FindNextFile* function found. The Win32 API defines the *WIN32_FIND_DATA* structure as shown here:

```
typedef struct _WIN32_FIND_DATA {
    DWORD dwFileAttributes;
    FILETIME ftCreationTime;
    FILETIME ftLastAccessTime;
    FILETIME ftLastWriteTime;
    DWORD    nFileSizeHigh;
    DWORD    nFileSizeLow;
    DWORD    dwReserved0;
    DWORD    dwReserved1;
    TCHAR    cFileName[ MAX_PATH ];
    TCHAR    cAlternateFileName[ 14 ];
} WIN32_FIND_DATA;
```

Table 1474 explains the *WIN32_FIND_DATA* structure in detail.

Members	Description
dwFileAttributes	Specifies the file attributes of the file found. This member can be one or more of the values Table 1464 lists.
ftCreationTime	Specifies a *FILETIME* structure that contains the time the file was created. *FindFirstFile* and *FindNextFile* report file times in Coordinated Universal Time (UTC) format. These functions set the *FILETIME* members to zero if the file system that contains the file does not support this time member. You can use the *FileTimeToLocalFileTime* function to convert from UTC to local time, and then use the *FileTimeToSystemTime* function to convert the local time to a *SYSTEMTIME* structure that contains individual members for the month, day, year, weekday, hour, minute, second, and millisecond.
ftLastAccessTime	Specifies a *FILETIME* structure that contains the time that the file was last accessed. The time is in UTC format; the *FILETIME* members are zero if the file system does not support this time member.
ftLastWriteTime	Specifies a *FILETIME* structure that contains the time that the file was last written to. The time is in UTC format; the *FILETIME* members are zero if the file system does not support this time member.
nFileSizeHigh	Specifies the high-order *DWORD* value of the file size, in bytes. This value is zero unless the file size is greater than *MAXDWORD*. The size of the file is equal to (*nFileSizeHigh* * *MAXDWORD*) + *nFileSizeLow*.
nFileSizeLow	Specifies the low-order *DWORD* value of the file size, in bytes.
dwReserved0	Reserved for future use.
dwReserved1	Reserved for future use.
cFileName	A *NULL*-terminated string that is the name of the file.
cAlternateFileName	A *NULL*-terminated string that is an alternative name for the file. This name is in the classic 8.3 (*filename.ext*) filename format.

Table 1474 *The components of the* ***WIN32_FIND_DATA*** *structure.*

If a file has a long filename, the complete name appears in the *cFileName* field, and the 8.3 format truncated version of the name appears in the *cAlternateFileName* field. Otherwise, *cAlternateFileName* is empty. As an alternative, you can use the *GetShortPathName* function to find the 8.3 format version of a filename.

If the function succeeds, the function returns a search handle you will use in a subsequent call to *FindNextFile* or *FindClose*. If the function fails, the function returns *INVALID_HANDLE_VALUE*. To get extended error information, call *GetLastError*.

The *FindFirstFile* function opens a search handle and returns information about the first file whose name matches the specified pattern. After the search handle is established, you can use the *FindNextFile* function to search for other files that match the same pattern. When the search handle is no longer needed, use the *FindClose* function to close it. This function searches for files by name only; you cannot use it for attribute-based searches. To learn about attribute-based searches, see Tip 1477.

1475 USING FINDNEXTFILE

In Tip 1474, you learned about the *FindFirstFile* function, which finds the first instance of a file within a directory tree that matches a given filename. However, to continue to search for additional files with the same filename (or matching the wildcard), your programs must use the second *Find* function. The *FindNextFile* function continues a file search from a previous call to the *FindFirstFile* function. You will use the *FindNextFile* function within your programs as shown in the following prototype:

```
BOOL FindNextFile(
   HANDLE hFindFile,                  // handle to search
   LPWIN32_FIND_DATA lpFindFileData  // pointer to structure for found file
);
```

The *hFindFile* parameter identifies a search handle a previous call to the *FindFirstFile* function returns. The *lpFindFileData* parameter points to the *WIN32_FIND_DATA* structure that receives information about the found file or subdirectory. You can use the structure in subsequent calls to *FindNextFile* to refer to the found file or directory.

If the function succeeds, the function returns a non-zero value. If the function fails, the function returns a zero value. To get extended error information, call *GetLastError*. If *GetLastError* finds no matching files, the function returns *ERROR_NO_MORE_FILES*. The *FindNextFile* function searches for files by name only; you cannot use it for attribute-based searches. You will learn about attribute-based searches in Tip 1477.

1476 CLOSING THE SEARCH HANDLE WITH FINDCLOSE

As you have learned, when your programs work with the *FindFirstFile* and *FindNextFile* functions, you will open a different handle (a search handle) for those functions than the operating system would normally return with a *CreateFile* invocation. If you try to close a search handle with *CloseHandle*, your programs will return an error. Instead, your programs must use the *FindClose* function to close the search handle. You will use the *FindClose* function within your programs as shown in the following prototype:

```
BOOL FindClose(HANDLE hFindFile);
```

The *hFindFile* parameter identifies the search handle. Your program must have previously opened the search handle with a call to the *FindFirstFile* function. If the function succeeds, the function returns a non-zero value. If the function fails, the function returns zero. After the program calls the *FindClose* function, the program cannot use the search handle that the *hFindFile* parameter specifies in subsequent calls to either the *FindNextFile* or *FindClose* function. Within the *Walk_Directories.cpp* program detailed in Tip 1475, the program uses the *FindClose* function only when the user finishes walking all the file directories.

SEARCHING BY ATTRIBUTES WITH THE FINDFILE FUNCTIONS 1477

As you learned in Tip 1475, your programs can use a recursive search function (*WalkDirsRecurse* within the *Walk_Directories.cpp* program) to search the entire drive for a file. However, if you are designing an application that only Windows NT 4.0 users will use, you can instead use the *FindFirstFileEx* function (together with the *FindNextFile* function) to search a directory. The *FindFirstFileEx* function searches a directory for a file whose name and attributes match those you specify in the function call. You will use the *FindFirstFileEx* function as shown in the following prototype:

```
HANDLE FindFirstFileEx(
   LPCTSTR lpFileName,                  // pointer to the file's name to search for
   FINDEX_INFO_LEVELS fInfoLevelId,     // information level of the returned data
   LPVOID lpFindFileData,               // pointer to the returned information
   FINDEX_SEARCH_OPS fSearchOp,         // type of filtering to perform
   LPVOID lpSearchFilter,               // pointer to search criteria
   DWORD dwAdditionalFlags              // additional search control flags
);
```

The *lpFileName* parameter points to a *NULL*-terminated string that specifies a valid directory or path and filename, just as with the *FindFirstFile* function, except the *lpFileName* parameter can contain wildcard characters (* and ?). The *fInfoLevelId* parameter specifies a *FINDEX_INFO_LEVELS* enumeration type that gives the returned data's information level. The *lpFindFileData* parameter returns a pointer to the file data. The information level specified in the *fInfoLevelId* parameter determines the pointer type. The *fSearchOp* parameter specifies a *FINDEX_SEARCH_OPS* enumeration type that tells the function the type of filtering to perform beyond wildcard matching. If the enumeration type *fSearchOp* specifies needs structured search information, *lpSearchFilter* points to the search criteria. Currently, none of the supported *fSearchOp* parameter values require extended search information (although future releases of Windows NT may require extended information). Therefore, this pointer must be *NULL*. The *dwAdditionalFlags* parameter specifies additional flags for controlling the search. You can use the *FIND_FIRST_EX_CASE_SENSITIVE* flag for case-sensitive searches. The default search is case insensitive. Windows NT 4.0 defines no other flags, however future releases of Windows NT may define other flags.

If the function succeeds, it returns a search handle that your program can use in a subsequent call to the *FindNextFile* or *FindClose* functions. If the function fails, it returns *INVALID_HANDLE_VALUE*. The *FindFirstFileEx* function lets you open a search handle and return information about the first file whose name matches the pattern and attributes you specify in the function call. If the underlying file system does not support the type of filtering the *fSearchOp* parameter specifies, other than directory filtering, *FindFirstFileEx* will fail and the function will return the error code *ERROR_NOT_SUPPORTED*. The program must then use the *FileExSearchNameMatch* function and perform its own filtering. For more information about the *FileExSearchNameMatch* function, see your compiler's online documentation.

When you establish the search handle, your program can use it in the *FindNextFile* function to search for other files that match the same pattern with functions that perform the same filtering. When your program no longer needs the search handle, it should close the handle using the *FindClose* function.

USING SEARCHPATH INSTEAD OF FIND TO SEARCH 1478

As you have learned, your programs can use the *Find* functions to search a directory for a series of files. Alternatively, your programs can use the *SearchPath* function to search for the file you specify. However, the *SearchPath* function will only search for the file within a certain set of paths, as Table 1478 details. You will use the *SearchPath* function within your programs as shown in the following prototype:

```
DWORD SearchPath(
   LPCTSTR lpPath,           // address of search path
   LPCTSTR lpFileName,       // address of filename
   LPCTSTR lpExtension,      // address of extension
   DWORD nBufferLength,      // size, in characters, of buffer
   LPTSTR lpBuffer,          // address of buffer for found filename
   LPTSTR *lpFilePart        // address of pointer to file component
);
```

The *SearchPath* function accepts the parameters Table 1478 details.

Parameter	Description
lpPath	Points to a *NULL*-terminated string that specifies the path that the function searches for the file. If this parameter is *NULL*, the function searches for a matching file in the following directories in the following sequence: 1. The directory from which the application loaded. 2. The current directory. 3. Under Windows 95, the Windows system directory. Use the *GetSystemDirectory* function to get the path of this directory. Under Windows NT, the 32-bit Windows system directory. Use the *GetSystemDirectory* function to get the path of this directory. The name of the 32-bit Windows system directory is typically *SYSTEM32*. 4. Under Windows NT, the 16-bit Windows system directory. There is no Win32 function that obtains the path of this directory, but the function will search it nevertheless. The name of this directory is typically *SYSTEM*. 5. The Windows directory. Use the *GetWindowsDirectory* function to get the path of this directory. 6. The directories that the Windows *PATH* environment variable lists.
lpFileName	Points to a *NULL*-terminated string that specifies the name of the file for which to search.
lpExtension	Points to a *NULL*-terminated string that specifies an extension the function adds to the filename when searching for the file. The first character of the filename extension must be a period. The function adds the extension only if the filename you specify does not end with an extension. If the program does not require a filename extension or if the filename contains an extension, this parameter can be *NULL*.
nBufferLength	Specifies the length, in characters, of the buffer that receives the valid path and filename.
lpBuffer	Points to the buffer for the valid path and filename of the file found.
lpFilePart	Points to the address (within *lpBuffer*) of the last component of the valid path and filename, which is the address of the character immediately following the final backslash (\) in the path.

Table1478 *The parameters for the* ***SearchPath*** *function.*

If the *SearchPath* function succeeds, it returns the length, in characters, of the string the function copied to the buffer, not including the terminating *NULL* character. If the return value (that is, the string length) is greater than *nBufferLength*, the value the *SearchPath* function returns is the size of the buffer it will require to hold the path. If the function fails, it returns zero. To get extended error information, call *GetLastError*.

The CD-ROM that accompanies this book includes the program *Search_For_Calc.cpp*. When you compile and execute the *Search_For_Calc.cpp* program, it will use the *SearchPath* function to search the drive for the *calc.exe* file. If the program finds the *calc.exe* file, it will display the file path in a message box.

Obtaining a Temporary Path 1479

In Tips 386 and 387 you learned how to create a temporary file within C programs. As you learned, one of the steps you perform when you create a temporary file is to determine what the temporary path is from the *TEMP* or *TMP* environment variable. In Windows, you can use the *GetTempPath* function to retrieve the path of the directory Windows designates for temporary files. You will use the *GetTempPath* function within your programs as shown in the following prototype:

```
DWORD GetTempPath(
   DWORD nBufferLength,        // size, in characters, of the buffer
   LPTSTR lpBuffer             // address of buffer for temp. path
);
```

The *nBufferLength* parameter specifies the size, in characters, of the string buffer the *lpBuffer* identifies. The *lpBuffer* parameter points to a string buffer that receives the *NULL*-terminated string specifying the temporary file path.

If the *GetTempPath* function succeeds, it returns the length, in characters, of the string the program copies to the *lpBuffer* parameter, not including the terminating *NULL* character. If the return value (the path string) is greater than *nBufferLength*, the return value is the size of the buffer the function will require to hold the path. If the function fails, it returns zero.

The *GetTempPath* function gets the temporary file path as follows:

1. The path the *TMP* environment variable specifies.
2. The path the *TEMP* environment variable specifies, if Windows does not define *TMP*.
3. The current directory, if Windows does not define either *TMP* or *TEMP*.

Creating Temporary Files 1480

In Tips 386 and 387 you learned how to create temporary files from within DOS. Creating a temporary file from within a Windows program is similarly easy. The *GetTempFileName* function creates a name for a temporary file. The filename is the concatenation of path and prefix strings you specify, a hexadecimal string formed from an integer you specify, and the *TMP* extension. The integer you specify can be non-zero, in which case, the *GetTempFileName* function creates the filename, but does not create the file. If you specify zero for the integer, the function creates a unique filename and creates the file in the specified directory. You will use the *GetTempFileName* function within your programs as shown in the following prototype:

```
UINT GetTempFileName(
   LPCTSTR lpPathName,         // address of directory name for temporary file
   LPCTSTR lpPrefixString,     // address of filename prefix
   UINT uUnique,               // number used to create temporary filename
   LPTSTR lpTempFileName       // address of buffer that receives the new filename
 );
```

The *lpPathName* parameter points to a *NULL*-terminated string that specifies the directory path for the filename. The *NULL*-terminated string must consist of characters in the ANSI character set. Applications typically specify a period or the result of the *GetTempPath* function for the *lpPathName* parameter. If this parameter is *NULL*, the function fails. The *lpPrefixString* parameter points to a *NULL*-terminated prefix string. The *GetTempPath* function uses the first three characters of the *NULL*-terminated prefix string as the prefix of the filename. This string must consist of characters in the ANSI character set. The *uUnique* parameter specifies an unsigned integer that the function converts to a hexadecimal string for use in creating the temporary filename. If *uUnique* is non-zero, the function appends the hexadecimal string to *lpPrefixString* to form the temporary filename. If *uUnique* is non-zero, the function does not create the specified file, and does not test whether the filename is unique. If *uUnique* is zero, the *GetTempPath* function uses a hexadecimal string it derives from the current system time. If *uUnique* is zero, the function uses different values until it finds a unique filename, and then it creates the file in the *lpPathName* directory. The *lpTempFileName* parameter points to the buffer that receives the temporary filename. This buffer is a *NULL*-terminated string consisting of characters in the ANSI character set. This buffer should be at least the length, in bytes, that the system-defined *MAX_PATH* constant specifies (typically 255) to accommodate the path.

If the *GetTempPath* function succeeds, it returns the unique numeric value used in the temporary filename. If the *uUnique* parameter is non-zero, the function returns that same number. If the function fails, it returns zero.

The *GetTempFileName* function creates a temporary filename of the form *path\preuuuu.TMP*, where *path* represents the path the *lpPathName* parameter specifies, *pre* represents the first three letters of the *lpPrefixString* string, and *uuuu* represents the *uUnique* parameter's hexadecimal value. When Windows shuts down, it does not automatically delete temporary files whose names the *GetTempFileName* function created.

If the *uUnique* parameter is zero, *GetTempFileName* tries to form a unique number based on the current system time. If a file with the resulting filename exists, *GetTempFileName* increases the number by one and repeats the test for the filename's existence. The function continues its testing until it finds a unique filename. The *GetTempFileName* function then creates a file by that unique name and closes it. When *uUnique* is non-zero, the function does not try to create and open the file.

1481 INTRODUCING THE CREATENAMEDPIPE FUNCTION

As you have learned, your programs can use pipes (which you will generally use for communication between two or more computers) in a way similar to file input and output. When you want to use a pipe within your programs, you must first create it. The *CreateNamedPipe* function creates an instance of a named pipe on a pipe server and returns a handle for subsequent pipe operations. A named pipe server process uses the *CreateNamedPipe* function either to create the first instance of a named pipe you specify and establish its basic attributes or to create a new instance of an existing named pipe. Note that you can only use named pipes to communicate across a network. You will use the *CreateNamedPipe* function as shown in the following prototype:

```
HANDLE CreateNamedPipe(
   LPCTSTR lpName,                               // pointer to pipe name
   DWORD dwOpenMode,                             // pipe open mode
   DWORD dwPipeMode,                             // pipe-specific modes
   DWORD nMaxInstances,                          // maximum number of instances
   DWORD nOutBufferSize,                         // output buffer size, in bytes
   DWORD nInBufferSize,                          // input buffer size, in bytes
   DWORD nDefaultTimeOut,                        // time-out time, in milliseconds
   LPSECURITY_ATTRIBUTES lpSecurityAttributes    // pointer to security attributes
);
```

The *CreateNamedPipe* function accepts the parameters Table 1481.1 details.

Parameters	Description
lpName	Points to the *NULL*-terminated string that uniquely identifies the pipe. The string must have the form \\.*pipe\pipename*. The pipename section of the name can include any character other than a backslash, including numbers and special characters. The entire pipe name string can be up to 256 characters long. Pipe names are not case sensitive.
dwOpenMode	Specifies the pipe access mode, the overlapped mode, the write-through mode, and the security access mode of the pipe handle. The *dwOpenMode* parameter must specify one of the pipe access mode flags Table 1481.2 details and it must specify the same mode flag for each instance of the pipe.
dwPipeMode	Specifies the type, read, and wait modes of the pipe handle. The *dwPipeMode* parameter must specify one or more of the type mode flags Table 1481.3 details and you must specify the same type mode or modes for each instance of the pipe. If you specify zero, the parameter defaults to byte-type mode.
nMaxInstances	Specifies the maximum number of instances that the function can create for this pipe. The *nMaxInstances* parameter must specify the same number for all instances. Acceptable values are in the range 1 through *PIPE_UNLIMITED_INSTANCES*. If this parameter is *PIPE_UNLIMITED_INSTANCES*, only the availability of system resources limits the number of pipe instances that *CreateNamedPipe* can create.
nOutBufferSize	Specifies the number of bytes to reserve for the output buffer.
nInBufferSize	Specifies the number of bytes to reserve for the input buffer.
nDefaultTimeOut	Specifies the default time-out value, in milliseconds, if the *WaitNamedPipe* function specifies *NMPWAIT_USE_DEFAULT_WAIT*. Each instance of a named pipe must specify the same value.
lpSecurityAttributes	Pointer to a *SECURITY_ATTRIBUTES* structure that specifies a security descriptor for the new named pipe and determines whether child processes can inherit the returned handle. If *lpSecurityAttributes* is *NULL*, the named pipe gets a default security descriptor and child processes cannot inherit the handle.

Table 1481.1 *The parameters for the* ***CreateNamedPipe*** *function. (continued from previous page)*

As you learned in Table 1481.1, there are several predefined constant values that Windows lets you use for the *dwOpenMode* parameter. Table 1481.2 lists the possible mode values.

Mode	Description
FILE_FLAG_WRITE_THROUGH	Enables write-through mode. This mode affects only write operations on byte-type pipes and, then, only when the client and server processes are on different computers. If *FILE_FLAG_WRITE_THROUGH* enables this mode, functions writing to a named pipe do not return until the system transmits the data written across the network and puts it in the pipe's buffer on the remote computer. If the call to *CreateNamedPipe* does not enable this mode, the system enhances the efficiency of network operations by buffering data until a minimum number of bytes accumulate or until a maximum time elapses.

Table 1481.2 *The possible values for the* ***dwOpenMode*** *parameter. (continued on following page)*

Mode	Description
FILE_FLAG_OVERLAPPED	Enables overlapped mode. If the *CreateNamedPipe* function enables this mode, functions performing read, write, and connect operations that may take significant time to complete can return immediately. For example, in overlapped mode, a thread can handle simultaneous input and output (I/O) operations on multiple instances of a pipe or perform simultaneous read and write operations on the same pipe handle. If the *CreateNamedPipe* function does not enable overlapped mode, functions performing read, write, and connect operations on the pipe handle do not return until the operating system finishes the operation. A program can only use the *ReadFileEx* and *WriteFileEx* functions with a pipe handle in overlapped mode. The *ReadFile, WriteFile, ConnectNamedPipe*, and *TransactNamedPipe* functions can execute either synchronously or as overlapped operations. This parameter can include any combination of the security access mode flags this table details. These mode flags can be different for different instances of the same pipe. You can specify them without concern for what other *dwOpenMode* modes you have already specified.
WRITE_DAC	The calling process will have write access to the named pipe's discretionary access control list (ACL).
WRITE_OWNER	The calling process will have write access to the named pipe's owner.
ACCESS_SYSTEM_SECURITY	The calling process will have write access to the named pipe's system ACL.

Table 1481.2 *The possible values for the* ***dwOpenMode*** *parameter. (continued from previous page)*

The *dwPipeMode* parameter lets you specify how you want to use the current instance of the pipe. You can specify type, read mode, and wait mode within the *dwPipeMode* parameter. Table 1481.3 lists the possible mode values.

Mode	Description
PIPE_TYPE_BYTE	Writes data to the pipe as a stream of bytes. You cannot use this mode with *PIPE_READMODE_MESSAGE.*
PIPE_TYPE_MESSAGE	Writes data to the pipe as a stream of messages. You can use this mode with either *PIPE_READMODE_MESSAGE* or *PIPE_READMODE_BYTE.*
PIPE_READMODE_BYTE	Reads data from the pipe as a stream of bytes. You can use this mode with either *PIPE_TYPE_MESSAGE* or *PIPE_TYPE_BYTE.* You can specify different read modes for different instances of the same pipe. If you specify zero, the parameter defaults to byte-read mode.
PIPE_READMODE_MESSAGE	Reads data from the pipe as a stream of messages. You can only use this mode if you also specify *PIPE_TYPE_MESSAGE.* You can specify different read modes for different instances of the same pipe. If you specify zero, the parameter defaults to byte-read mode.

Table 1481.3 *The possible values for the* ***dwPipeMode*** *parameter.* *(continued on following page)*

Mode	Description
PIPE_WAIT	Enables blocking mode. When you specify the pipe handle in the *ReadFile, WriteFile,* or *ConnectNamedPipe* functions, the operating system does not complete the operations until the function reads data, writes all data, or connects a client, respectively. Use of the *PIPE_WAIT* mode can mean waiting indefinitely in some situations for a client process to perform an action. You can specify different wait modes for different instances of the same pipe. If you specify zero, the parameter defaults to blocking mode.
PIPE_NOWAIT	Enables non-blocking mode. In this mode, *ReadFile, WriteFile,* and *ConnectNamedPipe* always return immediately.
PIPE_ACCESS_DUPLEX	The pipe is bi-directional; both server and client processes can read from and write to the pipe. This mode gives the server the equivalent of *GENERIC_READ* | *GENERIC_WRITE* access to the pipe. The client can specify *GENERIC_READ* or *GENERIC_WRITE*, or both, when it connects to the pipe using the *CreateFile* function. You can specify different access modes for different instances of the same pipe.
PIPE_ACCESS_INBOUND	The flow of data in the pipe goes from client to server only. This mode gives the server the equivalent of *GENERIC_READ* access to the pipe. The client must specify *GENERIC_WRITE* access when it connects to the pipe.
PIPE_ACCESS_OUTBOUND	The flow of data in the pipe goes from server to client only. This mode gives the server the equivalent of *GENERIC_WRITE* access to the pipe. The client must specify *GENERIC_READ* access when it connects to the pipe.

***Table 1481.3** The possible values for the **dwPipeMode** parameter. (continued from previous page)*

If the *CreateNamedPipe* function succeeds, it returns a handle to the server end of a named-pipe instance. If the function fails, it returns *INVALID_HANDLE_VALUE*. To get extended error information, call *GetLastError*. The function returns *ERROR_INVALID_PARAMETER* if the *nMaxInstances* parameter is greater than *PIPE_UNLIMITED_INSTANCES*.

To use *CreateNamedPipe* to create an instance of a named pipe, the user must have *FILE_CREATE_PIPE_INSTANCE* access to the named pipe object. If the *CreateNamedPipe* function is creating a new named pipe, the access control list (ACL) from the security attributes parameter defines the discretionary access control for the named pipe.

All instances of a named pipe must specify the same pipe type (byte-type or message-type), pipe access (duplex, inbound, or outbound), instance count, and time-out value. If the instances use different values, the *CreateNamedPipe* function will fail and *GetLastError* will return *ERROR_ACCESS_DENIED*.

The input and output buffer sizes are advisory. In other words, the actual buffer size the Windows will reserve for each end of the named pipe is either the system default, the system minimum or maximum, or the size you specify rounded up to the next allocation boundary. The pipe server should not perform a blocking read operation until the pipe client has started. Otherwise, a race condition can occur. A race condition typically occurs when initialization code (such as a process start-up routine) must lock and examine inherited handles. A *race condition* is a bug in a multi-threaded process where the code of one thread relies on a second thread to complete some action, but where there is no synchronization between the two threads. The process works if the second thread "wins" the race by completing its action before the first thread needs the second thread, but the process fails if the first thread "wins" the race.

The program always deletes an instance of a named pipe when it closes the last handle to that instance of the named pipe. In Tip 1483, you will use *CallNamedPipe* and *ConnectNamedPipe* together with *CreateNamedPipe* to perform output onto a network server.

1482 CONNECTING A NAMED PIPE

As you learned in Tip 1481, your programs can use the *CreateNamedPipe* function to create or connect to a named pipe. Many times, however, your program will run on the client machine, which forces it to connect to the named pipe to communicate with the server, rather than creating a named pipe (which must occur at the pipe server). In addition, before your program can connect to the server's named pipe, the named pipe must call the *ConnectNamedPipe* function. The *ConnectNamedPipe* function instructs a named pipe server process to wait for a client process to connect to an instance of a named pipe. A client process calls either the *CreateFile* or *CallNamedPipe* function to connect to the instance. You will learn about the *CallNamedPipe* function in Tip 1483. You will use the *ConnectNamedPipe* function within your programs as shown here:

```
BOOL ConnectNamedPipe(
   HANDLE hNamedPipe,                  // handle to named pipe to connect
   LPOVERLAPPED lpOverlapped           // pointer to overlapped structure
);
```

The *hNamedPipe* parameter identifies the server end of a named-pipe instance. The *CreateNamedPipe* function returns this handle. The *lpOverlapped* parameter points to an *OVERLAPPED* structure (which, as you learned in Tip 1451, lets your programs perform asynchronous I/O). If the function succeeds, it returns non-zero. If the function fails, it returns zero. To get extended error information, call *GetLastError*.

A named pipe server process can use *ConnectNamedPipe* with a newly created pipe instance or with an instance that was previously connected to another client process. In this case, the server process must first call the *DisconnectNamedPipe* function to disconnect the handle from the previous client before it can reconnect the handle to a new client. Otherwise, *ConnectNamedPipe* returns *False*, and *GetLastError* returns *ERROR_NO_DATA* if the previous client has closed its handle or *ERROR_PIPE_CONNECTED* if it has not closed its handle.

The behavior of the *ConnectNamedPipe* function depends on two conditions: whether you set the pipe handle's wait mode to blocking or non-blocking and whether you set the function to execute synchronously or in overlapped mode. A server initially specifies a pipe handle's wait mode in the *CreateNamedPipe* function, and you use the *SetNamedPipeHandleState* function to change it.

If the function opens *hNamedPipe* with *FILE_FLAG_OVERLAPPED*, the *lpOverlapped* parameter must not be *NULL*. It must point to a valid *OVERLAPPED* structure. If the function opens *hNamedPipe* with *FILE_FLAG_OVERLAPPED* and *lpOverlapped* is *NULL*, the *ConnectNamedPipe* function can incorrectly report that the connect operation is complete. On the other hand, if the function creates *hNamedPipe* with *FILE_FLAG_OVERLAPPED* and *lpOverlapped* is not *NULL*, the *OVERLAPPED* structure the *lpOverlapped* parameter points to must contain a handle to a manual-reset event object (which the server can use the *CreateEvent* function to create).

If the function does not open *hNamedPipe* with *FILE_FLAG_OVERLAPPED* and *lpOverlapped* is *NULL*, the function does not return until the program connects a client or an error occurs. Successful synchronous operations result in the function returning *True* if a client connects after the program calls the function. If a client connects before the program calls the function, the function returns *False* and *GetLastError* returns *ERROR_PIPE_CONNECTED*. A *False* result can happen if a client connects in the interval between the call to *CreateNamedPipe* and the call to *ConnectNamedPipe*. In this situation, there is a good connection between client and server, even though the function returns *False*.

If the function does not open *hNamedPipe* with *FILE_FLAG_OVERLAPPED* and *lpOverlapped* is not *NULL*, the operation executes asynchronously. The function returns immediately with a *False* return value. If a client process connects before the program calls the function, *GetLastError* returns *ERROR_PIPE_CONNECTED*. Otherwise, *GetLastError* returns *ERROR_IO_PENDING*, which indicates that the operation is executing in the background. When this happens, the program sets the event object in the *OVERLAPPED* structure to the non-signaled state before *ConnectNamedPipe* returns, and the program sets it to the signaled state when a client connects to this instance of the pipe.

The server process can use any of the wait functions or the *SleepEx* function to determine when the state of the event object is signaled. Then, the server can use the *GetOverlappedResult* function to determine the results of the *ConnectNamedPipe* operation. If the pipe handle you specify is in non-blocking mode, *ConnectNamedPipe* always returns immediately. In non-blocking mode, *ConnectNamedPipe* returns *True* the first time the program calls it for a pipe instance that the process has already disconnected from a previous client. This indicates that the pipe is now available for the system to connect it to a new client process. In all other situations, when the pipe handle is in non-blocking mode, *ConnectNamedPipe* returns *False*. In these situations, *GetLastError* returns *ERROR_PIPE_LISTENING* if no client is connected, *ERROR_PIPE_CONNECTED* if a client is connected, and *ERROR_NO_DATA* if a previous client has closed its pipe handle but the server has not disconnected. Note, when a pipe is in non-blocking mode, a good connection between client and server exists only after the program receives the *ERROR_PIPE_CONNECTED* error.

Note: ***ConnectNamedPipe*** *supports non-blocking mode for compatibility with Microsoft LAN Manager 2.0. You should not use it to achieve asynchronous input and output (I/O) with named pipes.*

CALLING A NAMED PIPE 1483

As you have learned, your programs can create a named pipe at the server end of a network connection. The client end must then connect to the named pipe to send information to the pipe. Within your programs, you can use either *CreateFile* or *CallNamedPipe* to connect to a named pipe. The *CallNamedPipe* function connects to a message-type pipe (and waits if an instance of the pipe is not available), writes to and reads from the pipe, and then closes the pipe. You will use the *CallNamedPipe* function as shown in the following prototype:

```
BOOL CallNamedPipe(
   LPCTSTR lpNamedPipeName,    // pointer to pipe name
   LPVOID lpInBuffer,          // pointer to write buffer
   DWORD nInBufferSize,        // size, in bytes, of write buffer
   LPVOID lpOutBuffer,         // pointer to read buffer
   DWORD nOutBufferSize,       // size, in bytes, of read buffer
   LPDWORD lpBytesRead,        // pointer to number of bytes read
   DWORD nTimeOut              // time-out time, in milliseconds
);
```

The *CallNamedPipe* function accepts the parameters Table 1483.1 details.

Parameter	Description
lpNamedPipeName	Pointer to a *NULL*-terminated string specifying the pipe name.
lpInBuffer	Pointer to the buffer containing the data *CallNamedPipe* writes to the pipe.
nInBufferSize	Specifies the size, in bytes, of the write buffer.
lpOutBuffer	Pointer to the buffer that receives the data *CallNamedPipe* reads from the pipe.
nOutBufferSize	Specifies the size, in bytes, of the read buffer.
lpBytesRead	Pointer to a 32-bit variable that receives the number of bytes *CallNamedPipe* reads from the pipe.

Table 1483.1 *The parameters for the* ***CallNamedPipe*** *function.* *(continued on following page)*

Parameter	Description
nTimeOut	Specifies the number of milliseconds to wait for the named pipe to be available. In addition to numeric values, your programs can specify the special values Table 1483.2 lists.

*Table 1483.1 The parameters for the **CallNamedPipe** function. (continued from previous page)*

When you call a named pipe across a network, it is important that you place a limit on the amount of time that your programs wait for the named pipe to respond. A time limit prevents against client computers waiting indefinitely for a named pipe connection that may be busy or even nonexistent. In addition to specifying a fixed number of milliseconds for your programs to wait for a response from the named pipe, you can use the values Table 1483.2 details to provide control over how long your client processes wait for a named pipe connection.

Value	Meaning
NMPWAIT_NOWAIT	Does not wait for the named pipe. If the named pipe is not available, the function returns an error.
NMPWAIT_WAIT_FOREVER	Waits indefinitely.
NMPWAIT_USE_DEFAULT_WAIT	Uses the default time-out specified in a call to the *CreateNamedPipe* function.

*Table 1483.2 The possible wait duration constants for the **CallNamedPipe** function.*

If the function succeeds, it returns non-zero. If the function fails, it returns zero. To get extended error information, call *GetLastError.*

Calling *CallNamedPipe* is equivalent to calling the *CreateFile* (or *WaitNamedPipe,* if *CreateFile* cannot open the pipe immediately), *TransactNamedPipe,* and *CloseHandle* functions. The program calls *CreateFile* with an access flag of *GENERIC_READ | GENERIC_WRITE,* an inherit handle flag of *False*, and a share mode of zero (indicating no sharing of this pipe instance). If the message the server process writes to the pipe is longer than *nOutBufferSize, CallNamedPipe* returns *False*, and *GetLastError* returns *ERROR_MORE_DATA.* The server process discards the remainder of the message, because *CallNamedPipe* closes the handle to the pipe before returning.

Note: *The **CallNamedPipe** function will fail if the pipe the function tries to call is a byte-type pipe.*

1484 Disconnecting a Named Pipe

As you have learned, your programs can use named pipes to communicate between two processes running on different machines (and even on the same machine). After you complete the information sharing between the two processes, however, it is important to close the named pipe, so that it does not slow down the network or the computer. Generally, you will close the handle to a named pipe at the client first, and then disconnect the named pipe at the server. The *DisconnectNamedPipe* function disconnects the server end of a named-pipe instance from a client process. You will use the *DisconnectNamedPipe* function as shown in the following prototype:

```
BOOL DisconnectNamedPipe(HANDLE hNamedPipe);
```

The *hNamedPipe* parameter handle identifies an instance of a named pipe. The *CreateNamedPipe* function must create this handle. If the function succeeds, it returns non-zero. If the function fails, it returns zero. To get extended error information, call *GetLastError*.

If the named pipe's client end is open, the *DisconnectNamedPipe* function forces that end of the named pipe closed. The client receives an error the next time it tries to access the pipe. A client that *DisconnectNamedPipe* forces off a pipe must still use the *CloseHandle* function to close its end of the pipe.

When the server process disconnects a pipe instance, it discards any unread data in the pipe. Before disconnecting, to make sure data is not lost the server can call the *FlushFileBuffers* function, which does not return until the client process has read all the data. The server process must call *DisconnectNamedPipe* to disconnect a pipe handle from its previous client before it can connect the handle to another client by using the *ConnectNamedPipe* function.

BETTER UNDERSTANDING ASYNCHRONOUS PROCESSING 1485

As you have learned, your programs can store and retrieve information to and from files on most devices using both synchronous and asynchronous output. As you have also learned, asynchronous file I/O lets your programs perform non-synchronized file operations. However, it is important to better understand why you may want to perform asynchronous I/O.

Compared to most other operations your computer carries out, device I/O is one of the slowest. The CPU performs arithmetic operations and even paints the screen much faster than it reads data from or writes data to a file or across a network. Asynchronous I/O lets you use multiple threads to tell the operating system to read from or write to a device while the rest of the code within your application continues to execute.

To better understand how asynchronous I/O can improve program performance, suppose that you are developing a simple database application. When the user opens a database, you will have your application read the database's contents into memory, as well as into an index file. After the user selects the database to load, the application must then pause and load all the data into memory (probably displaying an hourglass to the user in the process).

However, if you use asynchronous I/O, a program can start the disk I/O operation, which the disk controller will perform, and let the CPU perform other, unrelated tasks at the same time. Asynchronous file I/O lets you perform multiple I/O activities at the same time—or start a non-critical I/O activity and let it complete in its own time without slowing down your program's execution. As your programs become more complex, and you read more data from and write more data to the hard drive or other storage media, the benefits of asynchronous I/O will increase for you.

USING ASYNCHRONOUS INPUT AND OUTPUT 1486

To access a device asynchronously, you must first call *CreateFile* to open the device and specify the *FILE_FLAG_OVERLAPPED* flag in the *dwFlagsAndAttrs* parameter. The *FILE_FLAG_OVERLAPPED* flag notifies the system that you intend to use the device for asynchronous I/O.

The Win32 operating system lets you use four different techniques to perform asynchronous I/O. The four techniques share a common operation theory. When you perform asynchronous I/O, you must first issue an I/O request to the operating system. The operating system will queue all I/O requests and handle them internally. While the operating system is handling I/O requests, it lets your thread return and continue processing. At some point thereafter, the operating system will complete the I/O task and will notify your application that it has sent and received the data, or that an error has occurred.

As noted previously, there are four different I/O techniques. Table 1486 lists them in order of complexity, from easiest to understand and implement (device handle signaling) to most difficult to understand and implement (I/O completion ports).

Technique	Description
Signaling a device object	Uses a *Wait* function and a device handle to perform asynchronous I/O. Not useful if you are going to perform multiple simultaneous I/O requests against a single device. Lets one thread issue an I/O request and another thread process it.
Signaling an event object	Uses a *WaitForMultipleObjects* function and one or more *event* objects to perform asynchronous I/O. Lets one thread issue an I/O request and another thread process it.
Alertable I/O	Uses a special message queue to process operating system notifications that an asynchronous I/O operation's processing has completed. Provides more flexibility than use of an event kernel object because you can use callback functions and specific message processing to respond to operating system information about the I/O action. The thread that issued the I/O request must also process the response. You can only use alertable I/O on Windows NT systems.
I/O completion ports	Uses a *concurrent threading model* (explained in Tip 1495) to respond to a large number of simultaneous I/O requests. Lets one thread issue an I/O request and another thread process it. A highly scalable technique, completion ports are most often used by professional developers. You can only use I/O completion ports on Windows NT systems.

***Table 1486** The four techniques of asynchronous I/O processing.*

1487 UNDERSTANDING THE OVERLAPPED STRUCTURE

As Table 1486 indicates, the simplest technique for performing asynchronous device I/O is to use device-handle signaling, which you will use in Tip 1488. To issue the I/O request, you use the *ReadFile* and *WriteFile* functions Tips 1456 and 1457 introduced. However, to perform asynchronous device I/O, your programs must pass the address of an initialized *OVERLAPPED* structure as the *lpOverlapped* parameter. The Win32 API defines the *OVERLAPPED* structure as shown here:

```
typedef struct _OVERLAPPED {
    DWORD  Internal;
    DWORD  InternalHigh;
    DWORD  Offset;
    DWORD  OffsetHigh;
    HANDLE hEvent;
} OVERLAPPED;
```

Table 1487 lists the members of the *OVERLAPPED* structure.

Member	Description
Internal	Specifies a system-dependent status. This member is valid when the *GetOverlappedResult* function returns without setting the extended error information to *ERROR_IO_PENDING*. Reserved for operating system use.
InternalHigh	Specifies the length of the data transferred. This member is valid when the *GetOverlappedResult* function returns *True*. Reserved for operating system use.

***Table 1487** The **OVERLAPPED** structure's members. (continued on following page)*

Member	Description
Offset	Specifies a file position at which to start the transfer. The file position is a byte offset from the start of the file. The calling process sets this member before calling the *ReadFile* or *WriteFile* functions. The calling process ignores this member when reading from or writing to named pipes and communications devices.
OffsetHigh	Specifies the high word of the byte offset at which to start the transfer. The calling process ignores this member when reading from or writing to named pipes and communications devices.
hEvent	Identifies an event set to the signaled state when the transfer has been completed. The calling process sets this member before calling the *ReadFile, WriteFile, ConnectNamedPipe,* or *TransactNamedPipe* function.

Table 1487 The ***OVERLAPPED*** *structure's members. (continued from previous page)*

You can use the *HasOverlappedIoCompleted* macro to determine whether an asynchronous I/O operation has completed. You can use the *CancelIo* function to cancel an asynchronous I/O operation.

ASYNCHRONOUS I/O WITH A DEVICE KERNEL OBJECT 1488

As you have learned, your programs can perform asynchronous I/O using four different techniques, the simplest of which is the use of a device kernel object. When you perform asynchronous I/O with a device kernel object, you simply instruct the thread to wait until the I/O has completed.

For example, suppose you read from a file with the *ReadFile* function, perform some interim processing, but the program cannot continue beyond a certain point until the read operation completes. In such a case, you can construct code similar to that shown in the following fragment:

```
ReadFile(hFile, bBuffer, sizeof(bBuffer), &dwNumBytesRead, &Overlapped);

// Processing here
WaitForSingleObject(hFile, INFINITE);// Wait until all data is in the buffer
```

The *WaitForSingleObject* function, when you call it with the handle to an asynchronous I/O device, will wait until the operating system has completed the appropriate processing for the device before it releases and lets the thread continue.

UNDERSTANDING WORKING-SET SIZE QUOTAS 1489

When your programs perform asynchronous I/O, the operating system will maintain a list of pending I/O requests. The operating system fixes the size of the list at system start-up. Occasionally, an asynchronous I/O request may fail because the list of pending I/O requests is already full. If the list is full when you issue another request, *ReadFile* and *WriteFile* will return *False,* and *GetLastError* will return either *ERROR_INVALID_USER_BUFFER* or *ERROR_NOT_ENOUGH_MEMORY.* Moreover, when you issue an I/O request, the system must "page lock" your program's data buffer. The data buffer is part of the process's working set, and each process has a maximum working set. A process's *working set* is the set of memory pages currently visible to the process in physical RAM memory. If you do not have enough room in your process's working set, issuing the I/O request will cause it to fail and *GetLastError* will return *ERROR_NOT_ENOUGH_QUOTA.* You can increase your process's working set size by calling *SetProcessWorkingSetSize,* which Tip 1490 explains in detail.

1490 SETTING HIGHER OR LOWER QUOTAS

As you learned in Tip 1489, your programs can increase their working set sizes (the set of memory pages currently visible to the process in physical RAM memory), should the programs require additional space within their working set. These pages are resident and available for an application to use without triggering a page fault. The size of a process's working set is specified in bytes. The minimum and maximum working set sizes affect the virtual memory paging behavior of a process. The *SetProcessWorkingSetSize* function sets the minimum and maximum working set sizes for a process you specify. You will use the *SetProcessWorkingSetSize* function within your programs as shown in the following prototype:

```
BOOL SetProcessWorkingSetSize(
    HANDLE hProcess,               // open handle to the process of interest
    DWORD dwMinimumWorkingSetSize,// specifies minimum working set size
    DWORD dwMaximumWorkingSetSize // specifies maximum working set size
);
```

The *hProcess* parameter is an open handle to the process whose working set sizes you will set. Under Windows NT, the handle must have the *PROCESS_SET_QUOTA* access right. The *dwMinimumWorkingSetSize* parameter specifies a minimum working set size for the process. The virtual memory manager tries to keep at least this much memory resident in the process whenever the process is active. The *dwMaximumWorkingSetSize* parameter specifies a maximum working set size for the process. The virtual memory manager tries to keep no more than this much memory resident in the process whenever the process is active and memory is in short supply. If both *dwMinimumWorkingSetSize* and *dwMaximumWorkingSetSize* have the value *0xFFFFFFFF,* the function temporarily trims the working set of the process you specify to zero. This essentially swaps the process out of physical RAM memory. If the function succeeds, it returns non-zero. If the function fails, it returns zero. Call *GetLastError* to obtain extended error information. You can empty the working set of the specified process by specifying the value *0xFFFFFFFF* for both the minimum and maximum working set sizes. If the values of either *dwMinimumWorkingSetSize* or *dwMaximumWorkingSetSize* are greater than the process's current working set sizes, the specified process must have the *SE_INC_BASE_PRIORITY_NAME* privilege. The operating system allocates working set sizes on a first-come, first-served basis. For example, if an application successfully sets 40Mb as its minimum working set size on a 64Mb system, and a second application requests a 40Mb working set size, the operating system denies the second application's request.

Using the *SetProcessWorkingSetSize* function to set an application's minimum and maximum working set sizes does not guarantee that the operating system will reserve the requested memory, or that the memory will remain resident at all times. When the application is idle, or a low-memory situation causes a demand for memory, the operating system can reduce the application's working set. An application can use the *VirtualLock* function to lock ranges of the application's virtual address space in memory; however, that can potentially degrade the system's performance.

When you increase the working set size of an application, you are taking away physical memory from the rest of the system. This can degrade the performance of other applications and the system as a whole. Taking away physical memory can also lead to failures in operations that require physical memory to be present; for example, creating processes, threads, and a kernel pool. Therefore, you must use the *SetProcessWorkingSetSize* function carefully. You must always consider the performance of the whole system when you are designing an application.

1491 UNDERSTANDING THE GETLASTERROR FUNCTION

As you have learned throughout the last 240 Tips, you can often obtain more information about a program error. Generally, your programs should call the *GetLastError* function to obtain more information. The *GetLastError* function returns the calling thread's last-error code value. The operating system maintains the last-error code on a per-thread basis. Multiple threads do not overwrite each other's last-error code. You will use the *GetLastError* function within your programs as shown here:

```
DWORD GetLastError(void);
```

The function returns the calling thread's last-error code value. Functions call the *SetLastError* function to set the last-error code value. You should call the *GetLastError* function immediately when a function's return value indicates that such a call will return useful data (such as further extended error information) because some functions will call *SetLastError(0)* when they succeed, wiping out the error code the most recently failed function previously set.

Most functions in the Win32 API that set the thread's last error code value set it when they fail; a few functions set it when they succeed. A return value error code, such as *False, NULL, 0xFFFFFFFF,* or –1, typically indicates function failure. Error codes are 32-bit values (bit 31 is the most significant bit). Bit 29 is reserved for application-defined error codes; no system error code has this bit set. If you are defining an error code for your application, set bit 29 to one. Setting bit 29 to one indicates that an application has defined the error code, and ensures that your error code does not conflict with any error codes the operating system defines. You can use the *FormatMessage* function to format output from the result of a call to *GetLastError*. Tip 1492 explains the *FormatMessage* function.

FORMATTING ERROR MESSAGES WITH FORMATMESSAGE 1492

As you learned in Tip 1491, your programs can use the *GetLastError* function to obtain a numeric representation of a thread's last error. However, you will more commonly want to see the string representation of the thread's last error. The *FormatMessage* function formats a message string. The function requires a message definition as input. The message definition can come from a buffer passed into the function. It can come from a message table resource in an already-loaded module. In addition, the calling process can ask the function to search the system's message table resource(s) for the message definition. The function finds the message definition in a message table resource based on a message identifier and a language identifier. The function copies the formatted message text to an output buffer, processing any embedded insert sequences, if requested. You will use the *FormatMessage* function within your programs as shown in the following prototype:

```
DWORD FormatMessage(
    DWORD dwFlags,           // source and processing options
    LPCVOID lpSource,        // pointer to message source
    DWORD dwMessageId,       // requested message identifier
    DWORD dwLanguageId,      // language identifier for requested message
    LPTSTR lpBuffer,         // pointer to message buffer
    DWORD nSize,             // maximum size of message buffer
    va_list *Arguments       // address of array of message inserts
);
```

Table 1492.1 lists the parameters for the *FormatMessage* function.

Parameter	Description
dwFlags	Contains a set of bit flags that specify aspects of the formatting process and how to interpret the *lpSource* parameter. The low-order byte of *dwFlags* specifies how the function handles line breaks in the output buffer. The low-order byte can also specify the maximum width of a formatted output line. You can specify a combination of the bit flags Table 1491.2 details.
lpSource	Specifies the message definition's location. The type of this parameter depends upon the settings in the *dwFlags* parameter. If you set *dwFlags* to *FORMAT_MESSAGE_FROM_HMODULE*, *lpSource* is an *hModule* of the module that contains the message table to search. On the other hand, if you set *dwFlags* to *FORMAT_MESSAGE_FROM_STRING*, *lpSource* is an *LPTSTR* that points to unformatted message text. If you do not set either of these flags in *dwFlags*, then the function ignores *lpSource*.

***Table 1492.1** The parameters for the **FormatMessage** function.* *(continued on following page)*

Parameter	Descritption
dwMessageId	Specifies the 32-bit message identifier for the requested message. The function ignores this parameter if *dwFlags* includes *FORMAT_MESSAGE_FROM_STRING*.
dwLanguageId	Specifies the 32-bit language identifier for the requested message. The function ignores this parameter if *dwFlags* includes *FORMAT_MESSAGE_FROM_STRING*. If you pass a specific *LANGID* in this parameter, *FormatMessage* will return a message for that *LANGID* only. If the function cannot find a message for that *LANGID*, it returns *ERROR_RESOURCE_LANG_NOT_FOUND*.
lpBuffer	Points to a buffer for the formatted (and *NULL*-terminated) message. If *dwFlags* includes *FORMAT_MESSAGE_ALLOCATE_BUFFER*, the function allocates a buffer using the *LocalAlloc* function, and places the address of the buffer at the address specified in *lpBuffer*.
nSize	If you do not set the *FORMAT_MESSAGE_ALLOCATE_BUFFER* flag, this parameter specifies the maximum number of bytes (ANSI version) or characters (Unicode version) that the program can store in the output buffer. If you set *FORMAT_MESSAGE_ALLOCATE_BUFFER*, this parameter specifies the minimum number of bytes or characters to allocate for an output buffer.
Arguments	Points to an array of 32-bit values that are used as insert values in the formatted message. %1 in the format string indicates the first value in the *Arguments* array; %2 indicates the second argument; and so on.

Table 1492.1 *The parameters for the* ***FormatMessage*** *function. (continued from previous page)*

The function's interpretation of each 32-bit value depends on the formatting information contained in the *dwFlags* parameter and the location of the actual message definition. The default is to treat each value as a pointer to a *NULL*-terminated string. By default, the *Arguments* parameter is of type *va_list**, which is a language- and implementation-specific data type for describing a variable number of arguments. If you do not have a pointer of type *va_list**, then specify the *FORMAT_MESSAGE_ARGUMENT_ARRAY* flag and pass a pointer to an array of 32-bit values; those values are input to the message formatted as the insert values. Each insert must have a corresponding element in the array. Table 1492.2 details the format flags for the *dwFlags* parameter.

Value	Meaning
FORMAT_MESSAGE_ALLOCATE_BUFFER	Specifies that the *lpBuffer* parameter is a pointer to a *PVOID* pointer, and that the *nSize* parameter specifies the minimum number of bytes (ANSI version) or characters (Unicode version) to allocate for an output message buffer. The function allocates a buffer large enough to hold the formatted message, and places a pointer to the allocated buffer at the address *lpBuffer* specifies.
FORMAT_MESSAGE_IGNORE_INSERTS	Specifies that the function must ignore insert sequences in the message definition and pass them through to the output buffer unchanged. This flag is useful for fetching a message for later formatting. If you set this flag, the function ignores the *Arguments* parameter.

Table 1492.2 *The possible values for the* ***dwFlags*** *parameter. (continued on following page)*

Value	Meaning
FORMAT_MESSAGE_FROM_STRING	Specifies that *lpSource* is a pointer to a *NULL*-terminated message definition. The message definition may contain insert sequences, just as the message text in a message-table resource may. You cannot use this format flag with *FORMAT_MESSAGE_FROM_HMODULE* or *FORMAT_MESSAGE_FROM_SYSTEM*.
FORMAT_MESSAGE_FROM_HMODULE	Specifies that *lpSource* is a module handle containing the message-table resource(s) to search. If this *lpSource* handle is *NULL*, the function will search the current process' application image file. You cannot use this format flag with *FORMAT_MESSAGE_FROM_STRING*.
FORMAT_MESSAGE_FROM_SYSTEM	Specifies that the function should search the system message-table resource(s) for the requested message. If you specify this flag with *FORMAT_MESSAGE_FROM_HMODULE*, the function searches the system message table if it does not find the message in the module *lpSource* specifies. You cannot use this format flag with *FORMAT_MESSAGE_FROM_STRING*.
FORMAT_MESSAGE_ARGUMENT_ARRAY	Specifies that the *Arguments* parameter is *not* a *va_list** structure, but instead is just a pointer to an array of 32-bit values that represent the arguments.

Table 1492.2 The possible values for the ***dwFlags*** *parameter. (continued from previous page)*

If the function succeeds, it returns the number of bytes (ANSI version) or characters (Unicode version) stored in the output buffer, excluding the terminating *NULL* character. If the function fails, it returns zero. To get extended error information, call *GetLastError*.

Within the message text, the function supports several escape sequences for dynamic message formatting. Table 1492.3 shows these escape sequences and their meanings. All escape sequences start with the percent character (%).

Escape Sequence	Meaning
%0	Terminates a message text line without a trailing newline character. You can use this escape sequence to build up long lines or to terminate the message itself without a trailing newline character. It is useful for prompt messages.
%n!printf format string!	Identifies an insert. The value of *n* can be in the range 1 through 99. The *printf* format string (which must have exclamation marks bracketing it) is optional and defaults to *!s!* if you do not specify it. The *printf* format string can contain the * specifier for either the precision or the width component. If you specify * for one component, the *FormatMessage* function uses insert %*n*+1; it uses %*n*+2 if you specify * for both components. The function does not support floating-point *printf* format specifiers *e, E, f,* and *g*. The alternative is to use the *sprintf* function to format the floating-point number into a temporary buffer, then use that buffer as the insert string.

Table 1492.3 The possible escape sequences of the %-prefaced formatting characters.

The function formats any other non-digit character following a percent sign in the output message as simply the character without the percent sign. Table 1492.4 lists some examples of non-formatting character output.

Format string	Resulting output
%%	A single percent sign in the formatted message text.
%n	A hard line break when the format string occurs at the end of a line. This format string is useful when *FormatMessage* is supplying regular line breaks so the message fits in a certain width.
%space	A space in the formatted message text. You can use this format string to ensure the appropriate number of trailing spaces in a message text line.
%.	A single period in the formatted message text. You can use this format string to include a single period at the beginning of a line without terminating the message text definition.
%!	A single exclamation point in the formatted message text. You can use this format string to include an exclamation point immediately after an insert without its being mistaken for the beginning of a *printf* format string.

***Table 1492.4** Sample output of non-formatting characters.*

The *GenerateError.cpp* program Tip 1491 introduced uses the *FormatMessage* function extensively.

1493 ASYNCHRONOUS I/O WITH AN EVENT KERNEL OBJECT

In Tip 1490, you used the *WaitForSingleObject* function together with a device handle to an asynchronous device to perform asynchronous I/O. While working with a device kernel object, as Tip 1490 detailed, is relatively simple and straightforward, is not particularly useful when handling multiple I/O requests simultaneously. If, for example, you are trying to perform multiple asynchronous I/O requests against a single file simultaneously, waiting for the file handle will not help your processing because it will become signaled when the first event completes, and you must wait again for it to release—which could result in waiting forever.

You can also use the *CreateEvent* function to create an event kernel object. You can then identify that object within the *hEvent* member of the *OVERLAPPED* structure you pass to your asynchronous I/O request function (either *ReadFile* or *WriteFile*). When you pass an event in such a manner, the operating system will automatically set the event to signaled when the I/O operation completes. However, because your program can set a different event for each I/O operation, your programs can appropriately respond to the completion of one I/O operation and not the other.

Each time you perform an asynchronous I/O operation, your program should create a new event for that operation. That way, each time the operating system completes its processing, it will set the event for the calling operation to a signaled state. As Tip 1494 describes, you can then wait for the events you want to complete.

1494 USING WAITFORMULTIPLEOBJECTS WITH ASYNCHRONOUS I/O

As you have learned, you can use *WaitForMultipleObjects* to wait for one of many events to occur, or to wait for some subset of many events to occur. When you perform asynchronous I/O, you should use *WaitForMultipleObjects* to synchronize your threads to a certain set of events. Your program should call *WaitForMultipleObjects* with the handles of all the events that the operating system must complete before your program can continue its processing, and then wait for those events to become a signaled state. Generally, you will perform such processing using code similar to the following code fragment:

```
Event[1] = HANDLE CreateEvent(LPSECURITY_ATTRIBUTES lpEventAttributes,
               BOOL bManualReset, BOOL bInitialState, LPCTSTR Event1);
Overlapped1.hEvent = Event[1];
ReadFile(hFile, bBuffer, sizeof(bBuffer_, &dwNumBytesRead, &Overlapped1);
Event[2] = HANDLE CreateEvent(LPSECURITY_ATTRIBUTES lpEventAttributes,
               BOOL bManualReset, BOOL bInitialState, LPCTSTR Event2);
```

```
Overlapped2.hEvent = Event[2];
ReadFile(hFile, bBuffer, sizeof(bBuffer_, &dwNumBytesRead, &Overlapped2);
// Additional processing here
DWORD WaitForMultipleObjects(2, CONST HANDLE *Event, BOOL bWaitAll, INFINITE);
```

The code fragment creates an event and passes that event to the first read action. Next, the code fragment creates a second event and passes it to the second read action. Finally, the code waits for both events to return before it continues its processing.

Event kernel objects are very useful for managing asynchronous I/O. The danger with using event kernel objects is if you set an event kernel object to an auto-reset event, because it is possible that a thread will hang forever while it waits for the auto-reset event to reset, even though the function previously completed the file I/O operation. If you call *GetOverlappedResult* to determine how many bytes the I/O operation transferred successfully, it will reset the event to the unsignaled state. In short, closely observe the sequence of functions you execute when using event kernel objects to manage asynchronous I/O.

INTRODUCING I/O COMPLETION PORTS 1495

The fourth technique that your programs can use to perform asynchronous I/O is the use of I/O completion ports. You will generally use I/O completion ports when you design a program that will service hundreds or even thousands of users (such as a Web server). I/O completion ports are extremely secure and robust and can safely handle such high volumes of communication activities. When you create a service application, you will generally do so in one of two ways:

- In the *serial model*, a single thread waits for a client to make a request (usually over the network). When the request comes in, the thread wakes and handles the client's request.
- In the *concurrent model*, a single thread waits for a client request and then creates a new thread to handle the request. While the new thread handles the client's request, the original thread loops back around and waits for another client request. When the thread handling the client's request completes its processing, the thread dies.

The serial model is a very limited model, in that it does not handle multiple, simultaneous requests well (because only a single thread handles requests). In contrast, the concurrent model is capable of handling an extremely large number of requests simultaneously, because each request receives its own thread. When you design Windows NT services, your programs will generally use the concurrent model. You will only use I/O completion ports in conjunction with applications that use the concurrent model.

Clearly, the creation of a concurrent model service is well beyond the scope of this book. It is sufficient for you to understand the difference between serial model services and concurrent model services as you continue to develop your programs.

Note: *You can use I/O completion ports only under Windows NT. Windows 95 does not have the functionality necessary to implement I/O completion ports.*

USING ALERTABLE I/O FOR ASYNCHRONOUS PROCESSING 1496

As you have learned, whenever a function creates a thread, the system also creates a message queue for that thread and associates it with the thread. The operating system also creates another queue for that thread, known as the *Asynchronous Procedure Call* (APC) queue. The operating system uses the low-level functions within the kernel to create and maintain the APC queue. Because the operating system uses low-level kernel functions to maintain the Asynchronous Procedure Call queue, the APC queue is a very fast and efficient method of managing asynchronous I/O.

It is possible for your programs to make I/O requests and have functions send the results of the I/O requests directly to the calling thread's Asynchronous Procedure Call queue. To send completed I/O requests to your thread's Asynchronous Procedure Call queue, you will use the *ReadFileEx* and *WriteFileEx* functions, as shown here:

```
BOOL ReadFileEx(HANDLE hFile, LPVOID lpBuffer, DWORD nNumberOfBytesToRead,
     LPOVERLAPPED lpOverlapped, LPOVERLAPPED_COMPLETION_ROUTINE
     lpCompletionRoutine );

BOOL WriteFileEx(HANDLE hFile, LPCVOID lpBuffer, DWORD nNumberOfBytesToWrite,
     LPOVERLAPPED lpOverlapped, LPOVERLAPPED_COMPLETION_ROUTINE
     lpCompletionRoutine );
```

As you can see, both functions accept, as their last parameter, the address of a completion routine to execute when they complete their processing. Tip 1498 explains both *ReadFileEx* and *WriteFileEx* in more detail. You must use the following prototype for the completion routine that both functions use:

```
VOID WINAPI FileIOCompletionRoutine(
   DWORD dwErrorCode,                  // completion code
   DWORD dwNumberOfBytesTransfered,    // number of bytes transferred
   LPOVERLAPPED lpOverlapped           // pointer to structure with I/O information
);
```

The *dwErrorCode* parameter specifies the I/O completion status. The *dwErrorCode* parameter may be one of the values Table 1496 shows.

Value	Meaning
0	The I/O was successful
ERROR_HANDLE_EOF	The function tried to read past the end of the file

***Table 1496** The possible values for the **dwErrorCode** parameter.*

The *dwNumberOfBytesTransfered* parameter specifies the number of bytes transferred. If an error occurs, this parameter is zero. The *lpOverlapped* parameter points to the *OVERLAPPED* structure the asynchronous I/O function specifies. Windows does not use the *hEvent* member of the *OVERLAPPED* structure; the calling application may use this member to pass information to the completion routine. Windows does not use the *OVERLAPPED* structure after the program calls the completion routine, so the completion routine can deallocate the memory the *OVERLAPPED* structure uses.

The *FileIOCompletionRoutine* function is a placeholder for an application-defined or library-defined function name. Returning from the *FileIOCompletionRoutine* function lets Windows call another pending I/O completion routine. All waiting completion routines are called before the alertable thread's wait is satisfied with a return code of *WAIT_IO_COMPLETION*. Windows may call the waiting completion routines in any order. Windows may or may not call the routines in the order the program completes the I/O functions. Each time Windows calls a completion routine, it uses some of the application's stack. If the completion routine does additional asynchronous I/O and alertable waits, the stack may grow.

1497 Alertable I/O Only Works on Windows NT

As you have learned, alertable I/O is an advanced technique for handling asynchronous I/O that uses the I/O message queue and one or more callback functions. Because alertable I/O uses the extended versions of the *ReadFile* and *WriteFile* functions, your programs can only use alertable I/O if you are sure that they will only run on Windows NT systems. If you try to use *ReadFileEx* or *WriteFileEx* on a Windows 95 or Win32 system, either function will return *False* and not perform any processing. A call to *GetLastError* will return *ERROR_CALL_NOT_IMPLEMENTED*. Do not try to use alertable I/O on a Windows 95 machine, as it will result in unpredictable effects.

USING READFILEEX AND WRITEFILEEX 1498

As you learned in Tip 1494, your programs can use the *ReadFileEx* and *WriteFileEx* functions to perform asynchronous file I/O under Windows NT. The *ReadFileEx* function reads data from a file asynchronously. Programmers designed the *ReadFileEx* function solely for asynchronous operation, unlike the *ReadFile* function, which programmers designed for both synchronous and asynchronous operation. *ReadFileEx* lets an application perform other processing during a file read operation. The *ReadFileEx* function reports its completion status asynchronously, calling a completion routine you specify when it completes reading and the calling thread is in an alertable wait state. You will use the *ReadFileEx* function within your programs as shown in the following prototype:

```
BOOL ReadFileEx(
   HANDLE hFile,                       // handle of file to read
   LPVOID lpBuffer,                    // address of buffer
   DWORD nNumberOfBytesToRead,         // number of bytes to read
   LPOVERLAPPED lpOverlapped,          // address of offset
   LPOVERLAPPED_COMPLETION_ROUTINE
      lpCompletionRoutine              // address of completion routine
);
```

The *ReadFileEx* function accepts the parameters Table 1498 details.

Parameter	Description
hFile	An open handle that specifies the file entity from which to read. You must create this file handle with the *FILE_FLAG_OVERLAPPED* flag and must have *GENERIC_READ* file access. The *hFile* parameter can be any handle the *CreateFile* function opened with the *FILE_FLAG_OVERLAPPED* flag.
lpBuffer	Points to a buffer that receives the data the function reads from the file. The application should not use this buffer until the function completes the read operation.
nNumberOfBytesToRead	Specifies the number of bytes for the function to read from the file.
lpOverlapped	Points to an *OVERLAPPED* data structure that supplies data for the function to use during the asynchronous (overlapped) file read operation. If the file *hFile* specified supports byte offsets, the calling process of *ReadFileEx* must specify a byte offset within the file at which reading should begin. The calling process specifies the byte offset by setting the *OVERLAPPED* structure's *Offset* and *OffsetHigh* members. If the file entity *hFile* specified does not support byte offsets (for example, if it is a named pipe), the calling process must set the *Offset* and *OffsetHigh* members to zero, or *ReadFileEx* fails. The *ReadFileEx* function ignores the *OVERLAPPED* structure's *hEvent* member. *ReadFileEx* signals completion of its read operation by calling, or queueing a call to, the completion routine *lpCompletionRoutine* points to, so it does not need an event handle. The *ReadFileEx* function does use the *OVERLAPPED* structure's *Internal* and *InternalHigh* members. An application should not set these members. The *OVERLAPPED* data structure *lpOverlapped* points to must remain valid for the duration of the read operation.
lpCompletionRoutine	Points to the completion routine for Windows to call when the read operation is complete and the calling thread is in an alertable wait state.

***Table 1498** The parameters for the **ReadFileEx** function.*

If the *ReadFileEx* function succeeds, it returns non-zero. If the function fails, it returns zero. To get extended error information, call *GetLastError*. If the function succeeds, the calling thread has an asynchronous I/O operation pending: the overlapped read operation from the file.

When the overlapped read operation completes, and the system blocks the calling thread in an alertable wait state, the system calls the function that *lpCompletionRoutine* points to, and the wait state completes with a return code of *WAIT_IO_COMPLETION*.

If the function succeeds and the file reading operation completes, but the calling thread is not in an alertable wait state, the system queues the completion routine call, holding the call until the calling thread enters an alertable wait state. If *ReadFileEx* tries to read past the end of the file, the function returns zero, and *GetLastError* returns *ERROR_HANDLE_EOF*.

If another process locks a portion of the file *hFile* specifies, and the read operation specified in a call to *ReadFileEx* overlaps the locked portion, the call to *ReadFileEx* fails. If *ReadFileEx* tries to read data from a mailslot whose buffer is too small, the function returns *False*, and *GetLastError* returns *ERROR_INSUFFICIENT_BUFFER*.

Applications must not read from nor write to the input buffer that a read operation is using until read operation completes. A premature access to the input buffer may lead to corruption of the data the function reads into that buffer. The *ReadFileEx* function may fail if there are too many outstanding asynchronous I/O requests. In the event of such a failure, *GetLastError* can return *ERROR_INVALID_USER_BUFFER* or *ERROR_NOT_ENOUGH_MEMORY* (as you learned in Tip 1488).

If you try to read from a floppy drive that does not have a floppy disk, the system will display a message box prompting you to retry the operation. To prevent the system from displaying this message box, call the *SetErrorMode* function with *SEM_NOOPENFILEERRORBOX*. If *hFile* is a handle to a named pipe or other file entity that does not support the byte-offset concept, the *Offset* and *OffsetHigh* members of the *OVERLAPPED* structure *lpOverlapped* points to must be zero, or *ReadFileEx* fails.

1499 USING A CALLBACK COMPLETION ROUTINE

As you learned in Tip 1497, your programs must use a callback completion routine with both *ReadFileEx* and *WriteFileEx*. You must use the following prototype for the completion routine that both functions call:

```
VOID WINAPI FileIOCompletionRoutine(
   DWORD dwErrorCode,                  // completion code
   DWORD dwNumberOfBytesTransfered,    // number of bytes transferred
   LPOVERLAPPED lpOverlapped           // pointer to structure with I/O information
);
```

When you call one of the *Wait* objects and place the thread into an alertable state, the operating system first checks your thread's Asynchronous Procedure Call (APC) queue. If at least one entry is in the queue, the system does not put the thread to sleep; instead, the system pulls the entry from the Asynchronous Procedure Call queue and your thread calls the callback routine, passing it the completed I/O request's error code, number of bytes transferred, and the address of the *OVERLAPPED* structure the thread originally passed in the I/O request. After the callback routine performs its processing, the system again checks for more entries in the Asynchronous Procedure Call queue. If more entries exist, the system will pass them to the callback routine in order. If no more entries exist, the alertable function returns, and the thread continues its processing without ever sleeping. Therefore, the only time your thread sleeps is when there are no entries in the Asynchronous Procedure Call queue.

As you have learned, your programs can use the powerful alertable I/O techniques to perform robust processing of asynchronous file input and output. The CD-ROM that accompanies this book includes the program *Alertable_IO.cpp*, which uses the concepts of alertable I/O to perform a simple copy task. When you compile and execute the program, it will use alertable I/O to prepare and copy a file, and alert you to its processing and completion.

When the program begins, it creates a set of I/O requests. To do so, it initializes a set of *MAX_PENDING_IO_REQS* structures, which it will use to inform the operating system of the largest possible number of simultaneous I/O requests. Each structure contains an *OVERLAPPED* structure, though none contains a pointer within the *hEvent* member. In addition to the *OVERLAPPED* structure each I/O request requires, each request requires a memory buffer, which the program maintains within the *IO_REQS* structure as well.

After the program initializes an *IO_REQS* structure, it calls the *ReadFileEx* function to issue the request for the read file from the operating system. At that point, the program begins to use alertable I/O. The process immediately returns to the dialog box prompt, and the user can immediately enter another file to copy. However, in the background, *ReadFileEx* is finding and reading the file. When it completes, it alerts the process (through the callback routine), which uses the information that it previously read to write the file copy. Although the program is too long to list in its entirety here, the two callback functions are worthwhile to analyze:

```
void WINAPI WriteCompletionRoutine(DWORD dwErrorCode,
   DWORD dwNumberOfBytesTransferred, LPOVERLAPPED pOverlapped);

void WINAPI ReadCompletionRoutine(DWORD dwErrorCode,
   DWORD dwNumberOfBytesTransferred, LPOVERLAPPED pOverlapped);
 {
   PIOREQ pIOReq = (PIOREQ) pOverlapped;
   chASSERT(dwErrorCode == NO_ERROR);
   g_cs.nReadsInProgress--;

   // Round up the number of bytes to write to a sector boundary

   dwNumberofBytesTransferred = (dwNumberOfBytesTransferred + g_cs.dwPageSize-1) &
      ~(g_cs.dwPageSize-1);
   chVERIFY(WriteFileEx(g_cs.hFileDst, pIOReq->pbData, dwNumberofBytesTransferred,
           pOverlapped, WriteCompletionRoutine));
   g_cs.nWritesInProgress++;
 }
void WINAPI WriteCompletionRoutine(DWORD dwErrorCode,
   DWORD dwNumberOfBytesTransferred, LPOVERLAPPED pOverlapped);
 {
    PIOREQ pIOReq = (PIOREQ) pOverlapped;
    chASSERT(dwErrorCode == NO_ERROR);
    g_cs.nWritesInProgress--;

    if (g_cs.ulNextReadOffset.Quadpart < g_cs.ulFileSize.Quadpart)
    {
      // the function hasn't read past the end of file yet
      // so, read the next chunk of data

      pOverlapped->Offset = g_cs.ulNextReadOffset.LowPart;
      pOverlapped->OffsetHigh = g_cs.ulNextReadOffset.HighPart;
      chVERIFY(ReadFileEx(g_cs.hFileSrc, pIOReq->pbData, BUFFSIZE,
              pOverlapped, ReadCompletionRoutine));
      g_cs.nReadsInProgress++;
      g_cs.ulNextReadOffset.Quadpart += BUFFSIZE;
    }
 }
```

As you can see, the *ReadFileCompletionRoutine* callback function calls the *WriteFileEx* function with the information *ReadFileEx* returns and the address of the *WriteFileCompletionRoutine* callback function. The *WriteCompletionRoutine* checks the current size of the copy each time *WriteFileEx* returns; if the copy is not complete, *WriteCompletionRoutine* changes the offset location within the file and calls the *ReadFileEx* function again. If the copy is complete, the function exits, and the program handles file cleanup elsewhere. As you can see, implementing alertable I/O within your programs is a relatively simple, and highly useful, way to perform asynchronous I/O.

1500 UNDERSTANDING THE .NET ENVIRONMENT

Over the past 10 years, the computing model has moved from the mainframe to a highly distributed client/server environment. Today, developers make extensive use of Active Server Pages, machine independent code (such as Java applets), as well as Perl- and JavaScript-based solutions. Microsoft's .NET architecture is an attempt to provide programmers with a set of tools for creating distributable solutions that are platform (operating system, hardware, and programming language) independent.

The .NET environment provides a collection of building blocks developers can employ to simplify application development. C# is simply one tool in the .NET toolbox. In the future, programmers will build .NET applications using C#, Visual Basic .NET, and other programming languages. Even more importantly, programmers will readily share code created in any language.

What then, is .NET? In general, .NET is a collection of tools that create an infrastructure for building distributed applications. The .NET architecture requires client support (within the browser), server support, and tools programmers can use to build .NET applications.

1501 UNDERSTANDING .NET APPLICATIONS

As you just learned, the .NET environment provides tools that create an infrastructure that programmers can use to build distributed applications. In general, the .NET environment is well suited for any application you want to expose (make available) to users as a Web service, for example, company inventory data for the sales staff, available seating for ticket sales, stock trading information, product shipping information, or company locator information.

You may be thinking that you already know how to implement many of these applications using existing languages and tools. The .NET environment leverages existing software capabilities by providing a more secure computing environment, increasing code reuse (code written in another language, such as Visual Basic .NET, is readily usable within C#, for example), and reduces programming errors by moving tasks, such as memory management, to the operating system, as opposed to the application.

1502 OTHER ADVANTAGES OF THE .NET ENVIRONMENT

The .NET environment provides an infrastructure that promotes code reuse, which, in turn, reduces the amount of code developers must write for new applications. In the past, programmers often reused code written in the same programming language. In other words, a programmer who was creating a new application in C++ could integrate classes created for another application using C++. Within the .NET environment, the C++ programmer can integrate code written in Visual Basic .NET into an application.

The .NET environment provides programmers with the ability to create platform independent code, much like Java virtual-machine code, which the browser compiles to a specific platform using just-in-time (JIT) compilation.

The .NET environment standardizes run-time libraries across programming languages by implementing a large base class library, whose objects and methods are readily available to applications written in any programming language.

The .NET environment provides a single integrated development environment (IDE), the Visual Studio, shown in Figure 1502. Using Visual Studio, programmers can use the same development tools to create, test, and debug applications they write using a variety of programming languages.

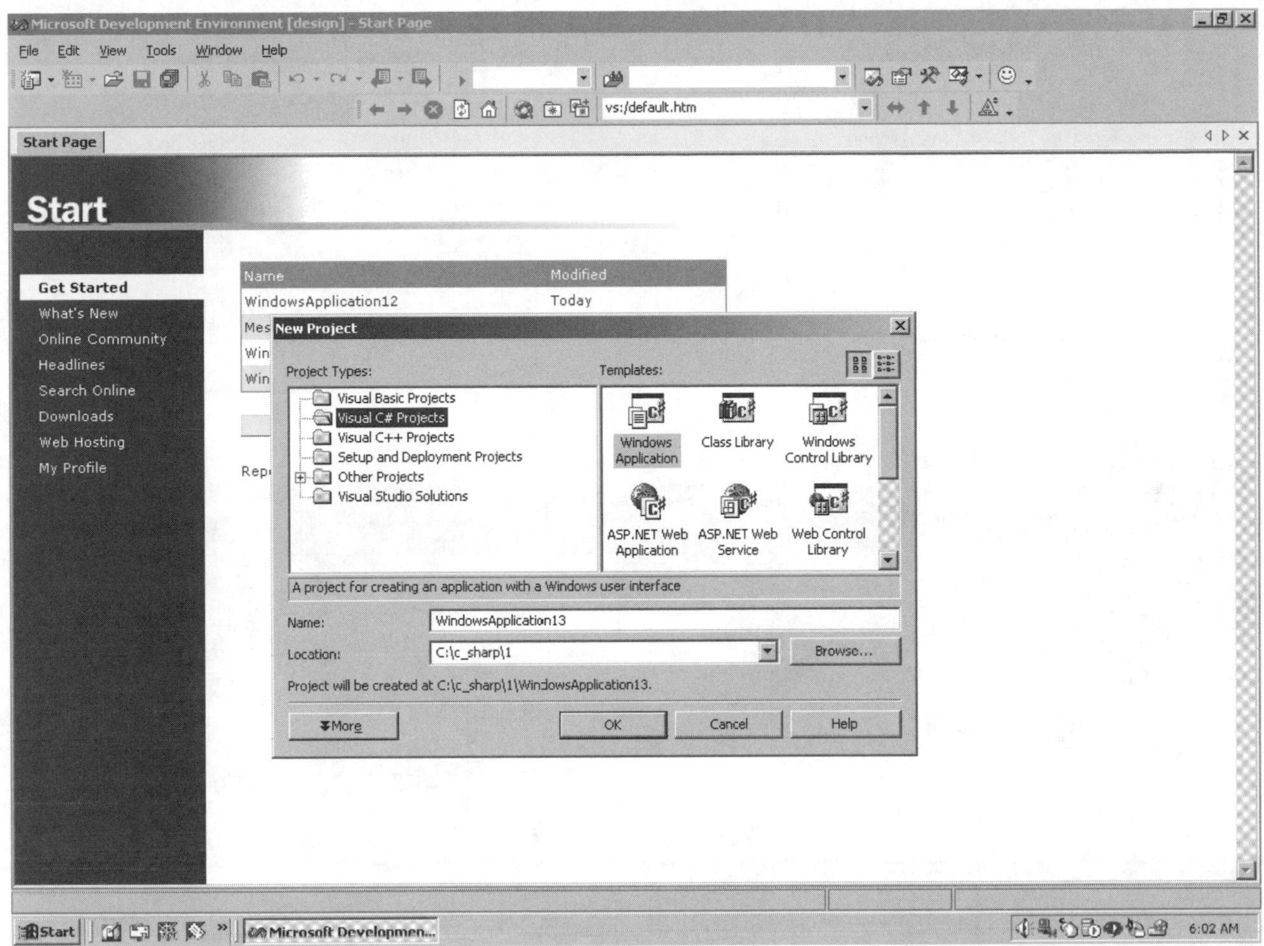

Figure 1502 Using the Visual Studio, programmers can create .NET applications using a range of programming languages.

UNDERSTANDING THE VISUAL STUDIO 1503

Microsoft provides the Visual Studio as the integrated development environment most programmers will use to create .NET applications. Admittedly, programmers can create .NET applications using a simple text editor and by performing command-line compilations. In fact, that is exactly how you will create your first C# programs. However, as shown in Figure 1503, the Visual Studio provides multiple windows within which you can display various source files, object class information, program output, debug information, and more. Further, the Visual Studio provides a set of Wizards you can use to create projects and simplify your application deployment. Several of this book's tips look at specific development features within the Visual Studio.

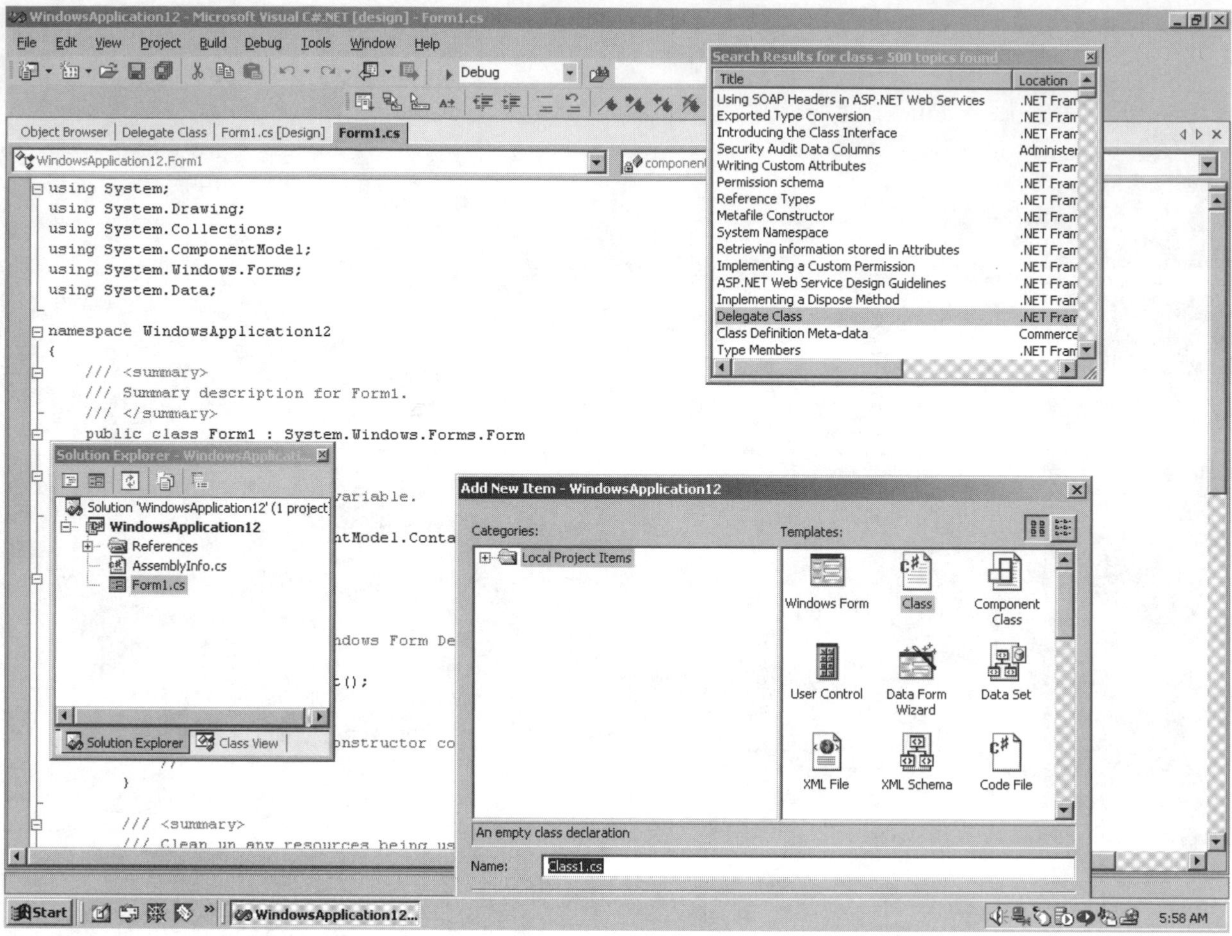

Figure 1503 Taking a closer look at the Visual Studio's integrated development environment.

1504 UNDERSTANDING THE COMMON LANGUAGE RUNTIME (CLR)

For years, programmers have used functions built into a programming language's run-time library to perform specific tasks. The C programming language, for example, provides hundreds of run-time library routines that programmers can use to perform file output, string manipulation, arithmetic operations, and more. Normally, C programs link to the run-time library functions, which places the library code within the application's executable file. Programmers refer to the process of linking library code to an application as *early-stage binding*.

Windows programs, in contrast, make use of dynamic link libraries (DLLs) to implement *late-stage binding*. When a program uses a DLL function, the program does not actually bind to the function until the program is actually running. In this sense, DLLs more closely correspond to the concept of a "run-time" library. As the program executes (during run time) the program will locate and bind to the functions it needs.

The .NET environment provides the Common Language Runtime (CLR) which provides a wide range of functions that programs can access during run time. In the future, the CLR will replace the code that exists today within many

DLLs. By eliminating such DLLs, the Common Language Runtime will eliminate many bugs that occur as a result of users having the wrong version of a DLL.

The Common Language Runtime also provides a collection of objects that applications can use to perform specific tasks. Further the Common Language Runtime performs memory management operations beyond simple "garbage collection," which simplifies the steps programmers must perform to allocate and use dynamic memory or to create and use object instances.

A key function of the Common Language Runtime is that only one CLR exists. That means the Common Language Runtime routines and objects you access within a C# program are identical to the ones you might use within an application you write using Visual Basic .NET. Microsoft's goal is to develop a Common Language Runtime capable of running on a myriad of devices, from Windows-based PCs to small, handheld devices.

UNDERSTANDING METADATA (DATA ABOUT DATA) 1505

If you have created HTML-based Web sites, you may have encountered the <meta> tag, which you can use to provide the browser with additional information about the Web page. Using the <meta> tag, for example, users can control a Web page's refresh operations, redirect users to a different page, provide information Web robots (programs that scour the Web for search engines) can use to categorize sites, and much more.

In short, metadata is data about data. Assume for example that you have a video-clip object (the data). Using metadata (data about the data), you might inform an application about the video clip's format, the number of frames per second within the clip, the clip's resolution, and so on.

Behind the scenes, the .NET Common Language Runtime makes extensive use of metadata to describe objects. In the case of a class object, for example, the metadata might provide specifics about *public* data member and method. The metadata may also specify a version number, security information, modification dates and times, and so on.

Within the .NET environment, you will encounter the term metadata on a regular basis. Simply understand that metadata lets one application provide information to another application about its data. Often, applications format metadata using XML (the extensible markup language).

UNDERSTANDING INTERMEDIATE LANGUAGES (ILS) 1506

As you may know, Java applets are unique because they compile to a virtual-machine code that is not specific to any one computer. As such, users can run the same Java applet on a Windows-based PC, a Mac, or a Linux-based workstation. Within a browser is special code that programmers refer to as the Java Virtual Machine. When a user opens a Web page that contains a Java applet, the browser loads the applet into its Java Virtual Machine, which, in turn, converts the "generic" Java virtual-machine code into the executable format of the user's machine.

The .NET environment supports similar intermediate languages, which the Microsoft documentation simply refers to as ILs. When you create a .NET application, your program code will exist in the form of an intermediate language (an IL). When the user runs your application, the .NET environment, in turn, will use just-in-time compilers to compile your code to a platform specific executable format.

1507 WHAT IT TAKES TO USE .NET APPLICATIONS

To run .NET applications, you need a server that supports .NET as well as a browser that provides client-side support for .NET. In general, today, the client-side browser for .NET applications is Internet Explorer (version 5.5 or later). Most .NET applications are best suited for Windows 2000. You can run many .NET applications on Windows 98; however, to fully support server-based applications, you must run the Internet Information Services (IIS) on Windows 2000.

In the future, most server-based programs will evolve to provide .NET support. Today, Microsoft is building .NET support into its Enterprise servers, which include such servers as Application Center 2000, BizTalk Server 2000, SQL Server 2000, Commerce Server 2000, Exchange Server 2000, and others.

Programmers can create .NET applications using a number of programming languages, ranging from C# and Visual Basic .NET to Cobol. In fact, programmers can share solutions they develop in a language-independent fashion.

1508 UNDERSTANDING THE .NET COMMON TYPES

As you have learned, a variable's type specifies a range of values the variable can store and a set of operations programs can perform on the variable. For example, an integer value can store a range of positive and negative counting numbers and programs can use integers within operations that add, subtract, multiply, divide and so on. In contrast, you could not divide, or multiply two character strings.

Historically, different programming languages have used different data types. Further, depending on the platform, the same programming language (C for example), might implement the same data type differently (one system might, for example, implement the type *int* using 16 bits, while a second system might use 32 bits).

The .NET environment defines a common set of data types which each .NET language uses. By standardizing data types in this way, the .NET environment simplifies the process of sharing code across different languages and platforms.

Table 1508 lists the .NET standard data types.

Type	Contents
Boolean	Represents a true or false value.
Byte	Represents a value in the range 0 to 255.
Char	Represents a 16-bit Unicode character.
DateTime	Represents a date and time value.
Decimal	Represents a value with 28 significant digits in the range –79,228,162,514,264,337,593,543,950,336 to 79,228,162,514,264,337,593,543,950,335.
Double	Represents a floating-point value using 64 bits.
GUID	Represents a Globally Unique Identifier using 128 bits.
Int16	Represents a value in the range –32,678 to 32,677.
Int32	Represents a value in the range –2,147,483,648 to 2,147,483,647.

Table 1508 The .NET standard data types. *(continued on following page)*

Type	Contents
Int64	Represents a value in the range–9,223,372,036,854,775,808 through 9,223,372,036,854,775,807.
Sbyte	Represents a value in the range –128 to 127.
Single	Represents a floating-point value using 32 bits.
TimeSpan	Represents a positive or negative time interval.

Table 1508 The .NET standard data types. (continued from previous page)

UNDERSTANDING ASSEMBLIES 1509

Programmers refer to the process of releasing a .NET application for use as *deployment*. Programmers deploy .NET applications using one or more *assemblies*.

Within the .NET environment, an assembly consists of one or more files. Each assembly is "self-describing" in that the assembly uses metadata to describe its contents and the resources that must exist for the assembly to run. Programmers refer to the assembly's metadata description as the assembly's *manifest*.

A .NET application consists of one or more assemblies. Each assembly has a public key (like a public key that you would use for a digital signature) that ensures the assembly's uniqueness. In later tips, you will build and deploy your own assemblies. Figure 1509 illustrates a .NET assembly.

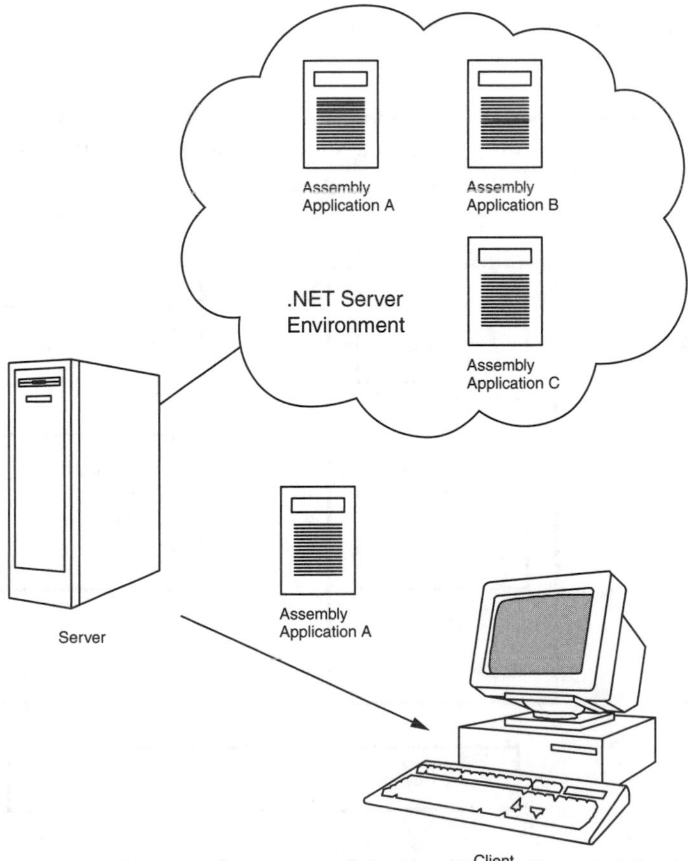

Figure 1509 *Programmers deploy .NET applications using assemblies.*

1510 METADATA IS NOT NEW TO .NET

The concept of using metadata or data to describe other data is not new to the .NET environment. Within HTML and WML, developers use the <meta> tag to provide browsers (or other agent programs) with additional information regarding the current page. For example, a Web page may use metadata to provide a search engine with a summary of the page's keywords. Or, a Web page may use metadata to direct the browser to perform specific refresh operations. Likewise, the Common Object Model (COM) uses metadata to expose objects within a type library.

Unlike COM and COM+, the .NET environment always stores metadata with the component itself (the metadata creates the assembly's manifest). In this way, the metadata is always current.

The .NET environment makes extensive use of XML to represent metadata. In fact, the .NET environment even provides objects applications can use to read and write XML-based data.

1511 UNDERSTANDING .NET APPLICATIONS

A .NET application is a program built specifically for the .NET environment. A .NET application consists of one or more *assemblies*. In general, an assembly is a package that contains one or more modules. A module is an individual file which corresponds to either a dynamic link library (DLL) or an executable. Each module uses metadata to describe its contents.

The module's code is in an intermediate language—as opposed to a platform specific language. Each object within a module is defined by (or is encapsulated within) a type. Within the .NET environment, a type defines the object's methods, properties, and fields.

A type's methods correspond to its functions. The type's fields correspond to the object's data values. Properties are similar to fields but normally have programming code that validates the values applications can assign to them. Figure 1511 illustrates the contents of a .NET application.

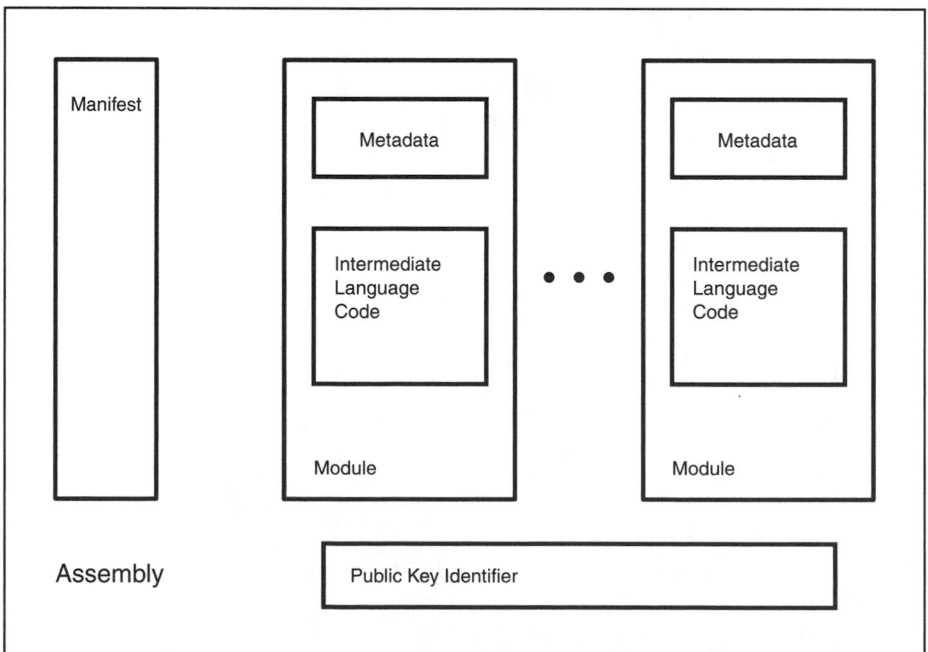

***Figure 1511** The make up of a .NET application.*

UNDERSTANDING THE SYSTEM OBJECT 1512

Within the .NET Common Language Runtime (CLR), all types are derived from a special system object named *System.Object*. Every object, therefore, supports four key methods, which Table 1512 defines. As you read through C# articles and books, you will encounter discussions where the articles refer to an object as a "true .NET object." In general, to be a "true .NET object," the object must have been derived from *System.Object*.

Method	Purpose
Equals	Tests whether one object equals another. Two reference variables are equal if they reference the same object.
GetHashCode	Returns an object's unique hash code. Two objects are equal if they have the same hash code.
GetType	Returns a Type object an application can use to retrieve an object's metadata.
ToString	A base class function (which classes can override) that returns a character string that represents the object's value.

***Table 1512** Methods all objects derive from System.Object.*

UNDERSTANDING REFERENCE AND VALUE TYPES 1513

Within the .NET Common Language Runtime, types are either value or reference. The distinction between the two types relates to where the object will reside within memory.

Reference type objects will always reside in a special memory region called the *heap*. If you have allocated memory dynamically using the C/C++ malloc function, for example, the memory you allocated came from the heap.

In contrast, value type objects reside on the stack. When you pass parameters to functions, programming languages normally place a copy of the parameter's value on the stack.

Within the .NET environment, the heap is unique in that the Common Language Runtime manages the heap memory. Unlike the C/C++ environments that require programmers to release (free) resources they allocate from the heap when the program no longer needs the resource, the Common Language Runtime will perform such operations automatically, freeing the programmer of the responsibility. In other words, the Common Language Runtime can detect when the object is no longer in use and will automatically release the object's memory back to the heap.

To manage memory in this way, the Common Language Runtime periodically runs a special program called the *Garbage Collector*, whose job is to determine which objects are no longer in use and to release the object's memory back to the heap.

In later tips, you will learn how to create reference and value variables within C# programs.

UNDERSTANDING REFLECTION 1514

As you have learned, with the .NET environment, objects use metadata to describe themselves to applications. Programmers refer to an object's ability to describe itself as *reflection*. Using the .NET reflection capabilities, programs can ask an object to specify its methods, fields, and properties. Figure 1514 illustrates how programs use reflection to query an object.

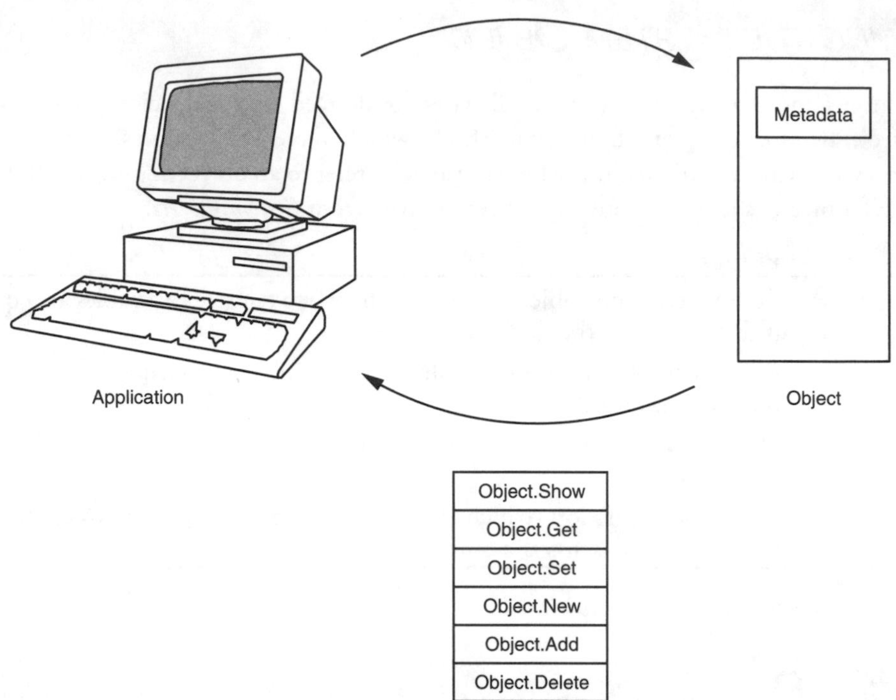

Figure 1514 Using reflection, a program can query an object regarding its data and method members.

As you will learn, applications can use the reflection API to query an object. Table 1514 briefly describes three commonly used methods that return information about an object. In the following tip, you will use the methods Table 1514 presents to query the .NET Garbage Collector object.

Method	Purpose
Type.GetFields	Returns a list of the object's data fields.
Type.GetMethods	Returns a list of the object's member functions.
Type.GetProperties	Returns a list of the object's properties.

Table 1514 Methods that return information about an object.

1515 Interacting with Managed Data

Within the .NET environment, most objects will be reference type objects that reside in the heap. As discussed, the Common Language Runtime manages the heap and the objects it contains. By managing the heap in this way, the Common Language Runtime eliminates the programmer's need to release memory that corresponds to objects that are no longer in use.

To perform memory management operations, the Common Language Runtime relies on special programs to perform garbage collection. In general, the Garbage Collector, which is built into the Common Language Runtime, will run at fixed intervals to release objects that are no longer in use and to move objects in memory, as necessary, to reduce memory fragmentation.

Depending on the processing your program performs, there may be times when you will want your program to initiate the Garbage Collector's processing. For example, a program that just released several very large objects or a program that is about to allocate several large objects may want the Garbage Collector to first clean up the heap. In such cases, your programs can initiate the Garbage Collector's processing by calling the GC.collect function. The following program, *GarbageInfo.cs*, uses reflection (a program's ability to query an object) to display information regarding the Garbage Collection object:

```
using System;
using System.Reflection;

class clsGarbageInfo
{
   static void Main()
   {
          Type t = typeof(System.GC);
          MethodInfo [] Methods = t.GetMethods();

          foreach (MethodInfo Method in Methods)
                Console.WriteLine("{0}", Method.Name);
   }
}
```

If you compile and run the program from the command line, the program will display output similar to the following:

```
C:\Subdir> GarbageInfo  <Enter>
GetHashCode
Equals
ToString
GetGeneration
Collect
Collect
KeepAlive
GetGeneration
get_MaxGeneration
WaitForPendingFinalizers
SuppressFinalize
ReRegisterForFinalize
GetTotalMemory
GetType
```

UNDERSTANDING THE .NET CLASS FRAMEWORK 1516

As previously discussed, the .NET environment moves objects and run-time library routines away from a specific programming language and into an infrastructure that all languages can access. The .NET class framework organizes classes, interfaces, and structures using namespaces, the most common of which is the *System* namespace.

Within the *System* namespace, you will find additional, second-level names, such as *IO* (*System.IO*), which programs can use to perform stream-based operations; *Reflection* (*System.Reflection*), which programs can use to query an object about its data, properties, and methods; and *Exception* (*System.Exception*) which defines default exception handling.

Normally, at the start of a C# program, you will find one or more using clauses, which specify a namespace that defines the routines the program will access, as shown here:

```
using System;
using System.IO;
using System.Reflection;
using System.Exception;
```

Table 1516 briefly summarizes several of the key *System* sub-level namespaces.

Namespace	Purpose
System.CodeDOM	Classes applications can use to present a source code document.
System.Collections	Classes that describe a variety of collection-based data structures, such as arrays, linked lists, and queues.
System.Configuration	Classes applications can use to locate system configuration information.
System.Data	Classes that support data input from sources such as SQL.
System.Diagnostics	Classes you can use to debug a .NET application.
System.DirectoryServices	Classes that provide applications with an interface to the Active Directory.
System.Drawing	Classes applications can use to create GDI-based images.
System.Globalization	Classes that support international features, such as date, time, and currency formats.
System.IO	Classes that support stream-based I/O operations.
System.Management	Classes that interface with the Windows Management Instrumentation services.
System.Messaging	Classes applications can use to send and receive messages across message queues.
System.Net	Classes applications can use to perform network operations.
System.Reflection	Classes applications can use to query an object about the object's capabilities.
System.Runtime	Classes applications can call to interact with assemblies, COM objects, and remote objects.
System.Security	Classes applications can call for encryption, access control, and more.
System.Text	Classes applications can call to manipulate ASCII and Unicode text.
System.Timers	Classes that provide timer-event support.
System.Web	Classes applications can call to interact with a browser.
System.Windows.Forms	Classes applications can call to create Windows-based forms.
System.XML	Classes that support XML-based operations, such as reading and writing XML-based content.

Table 1516 Sub-level namespaces within the System namespace.

1517 Building and Compiling Your First C# Program

One of the easiest ways to get up and running with C# is simply to create a few simple programs using a text editor that you then compile and run from the command line. Normally, to access the command line, you select the Start menu Run option and type command. Windows, in turn, will open an MS-DOS window. To compile programs within the MS-DOS window, you normally must then set one or more environment settings the compiler will use to locate library files and other executables. To simplify this process, Visual Studio provides a command window within which you should develop programs from the command line (as opposed to using the Windows command prompt). That's because the Visual Studio command window sets up the environment variables for the C# compiler. To open the Visual Studio command window, perform these steps:

1. Select the Start menu Programs options. Windows will display the Programs submenu.

2. Within the Programs submenu, select the Microsoft Visual Studio.NET 7.0 item and then choose the Visual Studio.NET Command Prompt.

When you create programs that run from the command and display their output within an MS-DOS window, programmers refer to the program as a *console program*. To create a C# program, you can use any text editor, such the Windows Notepad. When you create a source file using a text editor, you use the *.cs* file extension. For example, using Notepad, create the following source file, *Book.cs*:

```
using System;

class clsBooks
{
    static void Main()
    {
        Console.WriteLine("Jamsa's C/C++/C# Programmer's Bible");
    }
}
```

To compile the program from the command line, issue the following csc command:

```
C:\subdir> csc book.cs  <Enter>
```

If your source file does not contain any syntax errors, the C# compiler will create an executable file named Book.exe that you can execute from the command line as follows:

```
C:\subdir> book  <Enter>
Jamsa's C/C++/C# Programmer's Bible
```

UNDERSTANDING THE CONSOLE.WRITELINE FUNCTION 1518

To display output from a command line program in C, you made extensive use of the *printf* function. Likewise, within C++, you displayed output using the *cout* stream. Within the C# environment, programmers refer to programs that run within an MS-DOS-based window as console-based programs. To display output to a console window using C# program, you can use the *Console.WriteLine* (note the uppercase L) function. As you will learn, the *WriteLine* function is very much like a *printf* function that includes a newline (\n) character to advance the cursor to the start of the following line after the output has been displayed. The following C, C++, and C# statements, for example, each will display the message Hello, World! within the console window and then advance the cursor to the start of the new line:

```
printf("Hello, World!\n");                // C

cout << "Hello, World" << endl;           // C++

Console.WriteLine("Hello, World");        // C#
```

The *WriteLine* function resides in the *Console* class, which, in turn, resides within the *System* namespace. To use the *WriteLine* function, you can specify the complete name, *System.Console.WriteLine*, or you can use the *using* keyword to include the *System* namespace and then simply refer to the function as *Console.WriteLine*.

Note, however, that because Console is a class, as opposed to a namespace sublevel, you cannot use the *using* keyword with *System.Console* and then refer to the function simply as *WriteLine*. Should you try to include *System.Console* within a *using* clause, the C# compiler will generate a syntax error.

1519 Using the Console.Write Function

As you have learned, the *Console.WriteLine* function displays output to the console window and then advances the cursor to the start of the next line. In a similar way, the *Console.Write* function displays output, but unlike *WriteLine*, the Write function does output a carriage return and linefeed to advance the cursor. The following program, *WriteDemo.cs*, uses the function to display the words Curly, Moe, and Larry on the same line:

```
using System;

class clsWriteDemo
 {
   static void Main()
    {
        Console.Write("Curly ");
        Console.Write("Larry ");
        Console.Write("Moe");
    }
 }
```

After you compile and execute the program, your screen will display the following output:

```
C:\Subdir> WriteDemo  <Enter>
Curly Larry Moe
```

1520 Using Escape Sequences with Console.WriteLine

As briefly discussed in the preceding tips, the *Console.WriteLine* and *Console.Write* functions behave very much like the C *printf* function and the C++ *cout* stream. Within a character string that you output using *Console.WriteLine* or *Console.Write*, you can place one or more of the escape sequences listed in Table 1520, such as the newline character (\n) which will direct the function to advance the cursor to the start of the next line.

In general, like C/C++ the C# compiler recognizes the backslash character (\) as an *escape* character. When the compiler encounters the backslash, the compiler will use the character that follows the backslash to select a predefined operation.

Escape Sequence	Purpose
\a	Sounds the computer's built-in bell
\b	Moves the cursor back one space
\f	Generates a formfeed that ejects the current page of printed output
\n	Generates a carriage return and linefeed that advances the cursor to the start of the next line
\r	Generates a carriage return only that moves the cursor to the start of the current line
\t	Advances the cursor to the next tab stop
\v	Advances the cursor to the next vertical tab
\"	Displays the double quote character
\\	Displays the backslash character

Table 1520 C# escape sequences.

The following program, *Escape.cs*, illustrates the use of several of the escape sequences:

```
using System;

class clsEscape
 {
   static void Main()
    {
        Console.WriteLine("\aBell\aBell\aBell");
        Console.Write("\tTab");
        Console.Write("\tTab");
        Console.Write("\tTab");
        Console.Write("\n");
        Console.WriteLine("\"Hello, World!\"");
    }
 }
```

After you compile and execute this program, your screen will display the following output:

```
C:\Subdir> Escape  <Enter>
BellBellBell
   Tab    Tab    Tab
"Hello, World"
```

HANDLING THE BACKSLASH CHARACTER IN C# STRINGS 1521

As you learned in Tip 1520, the C# compiler, like C and C++, treats the backslash character as an escape character. When your program's hard code directory names include backslashes, you must remember to place two backslash characters for each backslash within the directory name. For example, to open the file named *Sample.Dat* that resides in the folder *C:\Demos\Tips*, you must specify the pathname *C:\Demos\Tips\Sample.Dat*. To represent each backslash character within a character string that contains the pathname, you must place two backslash characters. As such, the character string would contain the following:

```
string path = "C:\\Demos\\Tips\\Sample.Dat"
```

In addition to letting you use the double backslashes in this way, C# also lets you precede a character string with the at sign (@) character that directs the compiler not to perform escape-sequence processing. In this way, you would represent the string as follows:

```
string path = @"C:\Demos\Tips\Sample.Dat";
```

The following program, *NoEscape.cs*, illustrates the use of the at sign character to disable escape-sequence processing:

```
using System;

class clsNoEscape
 {
   static void Main()
    {
      string pathname = @"C:\Tip-1600\Tip1521\NoEscape.cs";

      Console.WriteLine(pathname);
    }
 }
```

After you compile and execute this program, your screen will display the following output:

```
C:\Subdir> NoEscape  <Enter>
C:\Tip-1600\Tip1521\NoEscape.cs
```

1522 Using the Plus Sign (+) to Concatenate Strings Within Console.WriteLine and Console.Write

As you will learn, using the plus sign, your program can concatenate (combine) two string values. For example, the following operation concatenates a filename to a folder name to create a complete pathname:

```
string Filename = "Sample.Dat";

string Folder = "C:\\Demos\\Tips\\" + "Sample.Dat";

string path = Folder + Filename;
```

When your programs display output using the *Console.WriteLine* and *Console.Write* functions, there may be times when the character string output becomes quite long. To better format your text, you may want to take advantage of string concatenation to pass the function a long string, as shown here:

```
Console.WriteLine("Using the concatenation operator (+) " +
                  "you can break a long string value " +
                  "into several strings to better " +
                  "your output. ");
```

1523 Using Console.WriteLine to Display Values

In the previous tips, you used the *Console.WriteLine* and *Write* functions to display messages to the console window. In each example, your application simply wrote character strings:

```
Console.WriteLine("Hello, world!");
```

Within C programs, you made extensive use of the *printf* function to display specific values:

```
printf("The number is %d", 3+7);
```

Likewise, within C++, you used the *cout* stream to display similar output:

```
cout  <<  "The number is " << 3 + 7;
```

Using the *Console.WriteLine* and *Write* functions, you can display values by using placeholders in the form {0}, {1}, and so on. The following statement, for example, uses placeholders within the *Console.WriteLine* function:

```
Console.WriteLine("The number is {0}", 3 + 7);
```

In this case, the WriteLine function will substitute the value 10 for the placeholder {0}. The following statement uses three placeholders:

```
Console.WriteLine("The result of {0} + {1} = {2}",
   3, 7, 3+7);
```

In this case, the function will substitute the value 3 for the placeholder {0}, the value 7 for the placeholder {1}, and the value 10 for the placeholder {2}.

The following program, *Placeholders.cs*, illustrates the use of several different placeholder and value substitutions:

```
using System;

class clsPlaceholders
 {
   static void Main()
    {
```

```
      Console.WriteLine("Line {0} output", 1);
      Console.WriteLine("{0} {1} {2}", 1, 2, 3);
      Console.WriteLine("{0} {0} {0}", 1);
      Console.WriteLine("Hello {0}", "World!");
      Console.WriteLine("{0} {1} {2}", 1, "Big", 3.44);
   }
}
```

As you can see in the third WriteLine function call, C# lets you use the same placeholder value more than one time. When you compile and execute this program, your console window will display the following output:

```
C:\Subdir> Placeholders  <Enter>
Line 1 output
1 2 3
1 1 1
Hello World!
1 Big 3.44
```

YOU MUST SPECIFY A VALUE FOR EACH PLACEHOLDER

In the previous tip, you learned that by using placeholders such as {0} and {1}, you can display values and variables using the *Console.WriteLine* and *Write* functions. When you specify placeholders, you must make sure that you include values the function is to substitute for each placeholder. If you specify the placeholders {0} and {1}, for example, and only provide one value, the function will generate an exception. If your program does not handle the exception, your program will immediately end. The following program, *NoValueError.cs*, specifies two placeholders, but provides only one value:

```
using System;

class clsNoValueError
 {
   static void Main()
    {
       Console.WriteLine("First value {0} second {1}", 1);
    }
 }
```

When you run this program, the *WriteLine* function will generate an exception that ends your program and the console window will display an error message that briefly describes the exception. In later tips, you will learn how to detect and respond to such exceptions.

FORMATTING OUTPUT WITH PLACEHOLDERS

When C programs display output using the *printf* function, the programs can format the output by including format specifiers, such as %3d or %7.2f. When your programs use the *Console.WriteLine* and *Write* functions to display output, your programs can place a format specifier after the placeholder number, such as {1, d} or {2, 7:f}. The format specifier can include an optional width value, followed by a colon and a character that specifies the value's type. Table 1525 briefly describes the type specifiers.

Specifier	Value Type
C or c	Local currency format.
D or d	Integer value.
E or e	Scientific notation.
F or f	Floating point.
G or g	Selects scientific or floating point depending on which is most compact.
N or n	Numeric formats which includes commas for large values.
X or x	Hexadecimal formats

Table 1525 Format specifiers for use with Console.WriteLine and Console.Write.

The following program, *FormatValues.cs*, illustrates the use of the format and width specifiers:

```
using System;

class clsFormatValues
 {
   static void Main()
    {
      Console.WriteLine("See spacing {0, 1:D}", 1);
      Console.WriteLine("See spacing {0, 2:D}", 2);
      Console.WriteLine("See spacing {0, 3:D}", 3);

      Console.WriteLine("See decimals {0, 7:f1}", 3.1415);
      Console.WriteLine("See decimals {0, 7:f2}", 3.1415);
      Console.WriteLine("See decimals {0, 7:f3}", 3.1415);
      Console.WriteLine("See decimals {0, 7:f4}", 3.1415);
      Console.WriteLine("Decimal {0, 3:D} Hex {0, 3:X}", 10);

     Console.WriteLine("See commas {0, 0:N}", 123456789);
    }
 }
```

When you compile and run this program, the console window will display the following output:

```
C:\Subdir> FormatValues <ENTER>
See spacing  1
See spacing   2
See spacing    3
See decimals      3.1
See decimals     3.14
See decimals    3.142
See decimals   3.1415
Decimal 10 Hex A
See commas 123,456,789.00
```

1526 Specifying Precision Within an Output Format

In the previous tip, you learned to use several different format specifiers within the *Console.WriteLine* and *Console.Write* functions. When your program displays floating-point values, there will be times, such as when the value represents currency, when you will want to specify the number of digits the functions display to the right of the decimal point. For example, assume your program must display the variable *Amount*, which contains the value 0.123456790. To control the number of digits the *WriteLine* and *Write* functions display, you specify the placeholder, width, format specifier, and number of digits, as shown here:

```
Console.WriteLine("See decimals {0, 12:f1}",
   0.123456789);  // 0.1
Console.WriteLine("See decimals {0, 12:f9}",
   0.123456789);  // 0.123456789
```

The following program, *ShowPrecision.cs*, illustrates the use of the *Console.WriteLine* function with several different E, F, and G format specifications:

```
using System;

class clsShowPrecision
 {
   static void Main()
    {
       Console.WriteLine("See decimals {0, 12:f1}",
          0.123456789);
       Console.WriteLine("See decimals {0, 12:f2}",
         0.123456789);
       Console.WriteLine("See decimals {0, 12:f3}",
         0.123456789);
       Console.WriteLine("See decimals {0, 12:f4}",
         0.123456789);
       Console.WriteLine("See decimals {0, 12:f5}",
         0.123456789);
       Console.WriteLine("See decimals {0, 12:f6}",
         0.123456789);
       Console.WriteLine("See decimals {0, 12:f7}",
         0.123456789);
       Console.WriteLine("See decimals {0, 12:f8}",
         0.123456789);
       Console.WriteLine("See decimals {0, 12:f9}",
         0.123456789);

       Console.WriteLine("See decimals {0, 12:e1}",
         0.123456789);
       Console.WriteLine("See decimals {0, 12:e2}",
         0.123456789);
       Console.WriteLine("See decimals {0, 12:e3}",
         0.123456789);
       Console.WriteLine("See decimals {0, 12:e4}",
         0.123456789);
       Console.WriteLine("See decimals {0, 12:e5}",
         0.123456789);
       Console.WriteLine("See decimals {0, 12:e6}",
         0.123456789);
       Console.WriteLine("See decimals {0, 12:e7}",
         0.123456789);
       Console.WriteLine("See decimals {0, 12:e8}",
         0.123456789);
       Console.WriteLine("See decimals {0, 12:e9}",
         0.123456789);

       Console.WriteLine("See decimals {0, 12:g1}",
         0.123456789);
       Console.WriteLine("See decimals {0, 12:g2}",
         0.123456789);
       Console.WriteLine("See decimals {0, 12:g3}",
         0.123456789);
       Console.WriteLine("See decimals {0, 12:g4}",
         0.123456789);
```

```
        Console.WriteLine("See decimals {0, 12:g5}",
          0.123456789);
        Console.WriteLine("See decimals {0, 12:g6}",
          0.123456789);
        Console.WriteLine("See decimals {0, 12:g7}",
          0.123456789);
        Console.WriteLine("See decimals {0, 12:g8}",
          0.123456789);
        Console.WriteLine("See decimals {0, 12:g9}",
          0.123456789);
    }
}
```

After you compile and run this program, your screen will display the following output:

```
C:\Subdir> ShowPrecision  <Enter>
See decimals          0.1
See decimals         0.12
See decimals        0.123
See decimals       0.1235
See decimals      0.12346
See decimals     0.123457
See decimals    0.1234568
See decimals   0.12345679
See decimals  0.123456789
See decimals      1.2e-001
See decimals     1.23e-001
See decimals    1.235e-001
See decimals   1.2346e-001
See decimals 1.23457e-001
See decimals 1.234568e-001
See decimals 1.2345679e-001
See decimals 1.23456789e-001
See decimals 1.234567890e-00
See decimals          0.1
See decimals         0.12
See decimals        0.123
See decimals       0.1235
See decimals      0.12346
See decimals     0.123457
See decimals    0.1234568
See decimals   0.12345679
See decimals  0.123456789
```

1527 FORMATTING OUTPUT VALUES USING ### AND 000

In the previous Tip, you learned how to use width and format specifiers within the *Console.WriteLine* and *Write* functions to format your output. In addition to the width and format specifiers, the functions also let you use the pound sign (#) to format your data. For example, the following statement directs *Console.WriteLine* to display a floating-point value with two digits to the right of the decimal point:

```
Console.WriteLine("The value is {0 0:###.##}", Value);
```

When you use the pound sign character to specify an output format, the WriteLine function will not display leading zeros. In other words, the function would output the value 0.123 as simply .123. When you want to display leading zeros, you can replace the pound sign with a zero, as shown here:

```
Console.WriteLine("The value is {0 0:000.00}", Value);
```

The following program, *PoundSignFormat.cs*, illustrates the use of the pound sign to format a variety of numeric values:

```
using System;

class clsPoundSignFormat
{
   static void Main()
   {
           Console.WriteLine("{0, 0:#.#}", 0.123456789);
           Console.WriteLine("{0, 0:#.###}", 0.123456789);
           Console.WriteLine("{0, 0:#.######}", 0.123456789);

           Console.WriteLine("{0, 0:0.0}", 0.123456789);
           Console.WriteLine("{0, 0:0.000}", 0.123456789);
           Console.WriteLine("{0, 0:0.000000}", 0.123456789);

           Console.WriteLine("{0, 0:#,###,###}", 123);
           Console.WriteLine("{0, 0:#,###,###}", 1234);
           Console.WriteLine("{0, 0:#,###,###}", 123456789);
   }
}
```

When you compile and execute this program, the console window will display the following output:

```
C:\Subdir> PoundSignFormat  <Enter>
.1
.123
.123457
0.1
0.123
0.123457
123
1234
123,456,789
```

UNDERSTANDING THE C# MAIN() FUNCTION 1528

Within C and C++, the first function your programs execute is the function *main*. Within C# programs, the first function that executes is also named *Main*. However, in C#, the function's name begins with an uppercase letter. The following program, *FirstMain.cs*, creates a C# application that performs no processing. The program will run, call the *Main* function that contains no statements, and will end:

```
using System;

class clsFirstMain
{
    static void Main()
      {
          // Do nothing
      }
}
```

Within C#, all functions (and that includes *Main*) must be a member of a class. Normally, however, your code cannot call a class member function until you create a class object. You may have already guessed your dilemma. To create an object, you must run *Main* function. However, because *Main* is a class member function, it would appear that you must first create an object before you can run *Main*. This is where the *static* keyword comes into play. As you learned when you examined C++ classes, if you precede a class member function name with the *static* keyword, the compiler will let you call the member function even if no instances of the object exist. The same use of the *static* keyword holds true in C#, which explains why you must declare the Main function as *static*.

1529 Using the void Keyword with the Main Function

As you learned when you examined C and C++ functions, when a function does not return a value, you specify the function's type as *void*. Like all functions, *Main* can return a value. In the case of *Main*, the function returns the value to the operating system. If your program's *Main* function does not return a value, you should declare the function's type as *void*, as shown here:

```
static void Main()
{
    //  Statements
}
```

If your program's *Main* returns an exit status value to the operating system, you should define *Main* as follows:

```
static void Main()
{
    //  Statements

    return(1);
}
```

In C and C++ programs, you can pass command-line parameters to the *main* function which your programs can access using the *argv* and *argc* variables:

```
void main(char *argv[], int argc)
```

In C#, your programs can access command-line arguments using an array of character strings:

```
static void Main(string[] args)
```

Within C and C++ programs, you normally use the *void* keyword within the function *main's* parameter list when the function does not use command-line arguments. C#, however, does not support the use of *void* to indicate no parameters. If you place the *void* keyword within *Main's* parameter list, as follows, the C# compiler will generate syntax error:

```
static void Main(void)  // void in parameter list
                        // creates an error in C#
```

1530 Accessing Command-Line Arguments

Within C#, you can access the command-line arguments a user specifies when he or she runs a program by using a character string array that the operating system passes to *Main*. To declare *Main* so it is capable of receiving command-line parameters, you must specify a parameter as follows:

```
static void Main(string[] args)
```

To determine the number of parameters the user specified on the command line, you can use the *String* type's *Length* member. The following program, *CmdCount.cs*, uses the *Length* member to specify the number of command-line arguments the program receives:

```
using System;

class clsCmdCount
{
    static void Main(string[] args)
      {
         Console.WriteLine("Number of parameters is {0}",
           args.Length);
      }
}
```

After you compile the program, run the program by passing several command-line parameters to the program as shown here:

```
C:\Subdir> CmdCount  AAA BBB CCC DDD  <Enter>
```

The program, in turn, will display a count of the number of parameters it receives:

```
Number of parameters is 4
```

In a similar way, the following, *ShowCmdLine.cs*, uses a *for* loop to display each command-line parameter on its own line. To display the command-line parameter, the program uses the *Console.WriteLine* function:

```
using System;

class clsShowCmdLine
{
    static void Main(string[] args)
      {
          int i;

          for (i = 0; i < args.Length; i++)
             Console.WriteLine("Parameter {0} is {1}",
               i, args[i]);
      }
}
```

After you compile this program's source code, run the program with a command line similar to the following:

```
C:> ShowCmdLine AAAA BBBB CCCC DDDD  <Enter>
```

The program, in turn, will display each command-line parameter on its own line as follows:

```
Parameter 0 is AAAA
Parameter 1 is BBBB
Parameter 2 is CCCC
Parameter 3 is DDDD
```

DEFINING MULTIPLE MAINS 1531

As you have learned, within a C# program, *Main* is the first function the application executes. As you create large applications, there may be times, possibly for testing purposes, when you want to place a *Main* function within two or more classes. Then, should you later need to test the class, you can recompile the application and direct the compiler to call that class's *Main* function when the program starts, which may in turn, perform specific test-related operations. The following C# program, *TwoMains.cs*, defines two classes, each of which contains a *Main* function:

```
using System;

class clsOne
{
```

```
    static void Main()
    {
        Console.WriteLine("Hello, World – Class One");
    }
}

class clsTwo
{
    static void Main()
    {
        Console.WriteLine("Goodbye, World – Class Two");
    }
}
```

When you compile the program, you must tell the C# compiler which *Main* function you want to use as the program's entry point by using the */main* switch. The following command line, for example, directs the compiler to use the *Main* that resides in *clsTwo*:

```
C:\Subdir> csc /main:clsTwo TwoMains.cs <Enter>
```

If you later want to use the *Main* function that resides within *clsOne* as the entry point, simply recompile the program using the following command line:

```
C:\Subdir> csc /main:clsOne TwoMains.cs <Enter>
```

1532 UNDERSTANDING C# KEYWORDS

As you have learned, keywords have special meaning to the compiler. When you declare a variable or define a function, you cannot use a keyword for an identifier. If you use a keyword for a variable or function name, the C# compiler will generate a syntax error and your program will not compile. Table 1532 specifies the C# keywords.

abstract	as	base	bool	break	byte	case
catch	char	checked	class	const	continue	decimal
default	delegate	do	double	else	enum	event
explicit	extern	false	finally	fixed	float	for
foreach	goto	if	implicit	in	int	interface
internal	is	lock	long	namespace	stackalloc	new
null	object	operator	out	override	params	private
protected	public	readonly	ref	return	sbyte	sealed
sizeof	static	string	struct	switch	this	throw
true	try	typeof	uint	ulong	unchecked	unsafe
ushort	using	virtual	short	void	while	

Table 1532 C# keywords.

As you examine C# programs, there may be times when you see a keyword preceded by an at sign (@), such as *@goto*. C# considers a keyword preceded by an @ sign as a valid identifier. In general, to avoid confusion, you should not use the at sign and reserved word in this way. However, you may encounter such use within another program. The following program, *AtReserved.cs*, illustrates the use of the at sign to make a reserved word a valid identifier:

```
using System;

class clsAtReserved
{
   static void Main()
    {
        int @goto = 1700;
        string @for = "Jamsa's C/C++/C# Programmer's Bible";

        Console.WriteLine("{0}  {1}", @for, @goto);
    }
}
```

Because the code precedes the *goto* keyword with an at sign (@), the program can use the combination as a variable. If you remove the at sign and compile the application, the C# compiler will generate a syntax error and the program will not compile.

TAKING A CLOSER LOOK AT C# TYPES 1533

As you have learned, a variable's type specifies the range of values the variable can store and the set of operations the program can perform on the variable. To declare a variable within C#, you must specify the variable's type, followed by the variable's name. Table 1533 briefly describes the C# data types.

Type	Sizeof(Type)	Description
bool	2	Boolean value (true or false)
byte	1	Unsigned byte 0 to 255
char	2	Unicode character 0 to 65,535
decimal	16	Fixed precision floating-point value well suited for currency calculations -79,228,162,514,264,337,593,543,950,336 through 79,228,162,514,264,337,593,543,950,335
double	8	Double-precision floating point $5.0x10^{-324}$ to $1.7x10^{+308}$
float	4	Floating-point number $1.5x10^{-45}$ to $3.4x10^{+38}$
int	4	Signed integer -2,147,483,648 to 2,147,483,647
long	8	Signed integer -9,223,372,036,854,775,808 to 9,223,372,036,854,775,807
sbyte	1	Signed byte -128 to 127
string	varies	Unicode string
short	2	Signed short integer -32,768 to 32,767
uint	4	Unsigned integer 0 to 4,294,967,295
ulong	8	Unsigned integer 0 to 18,446,744,073,709,551,615
ushort	2	Unsigned short integer 0 to 65,535

Table 1533 The C# data types.

CREATING VALUE-TYPE VARIABLES 1534

As you have learned, within the .NET environment, the Common Language Runtime (CLR) manages the heap memory. Variables that reside on the heap are *reference variables*. In contrast, variables that reside on stack are *value variables*. When you perform a standard declaration, such as the following, C# will create the variables on stack, meaning C# will create value variables:

```
float Salary;
int age;
```

For now, simply understand that unless you use the *new* operator, C# will likely create a value variable that resides on the stack. In the following Tip, you will learn how to create reference variables that reside within the heap.

1535 Creating Reference-Type Variables

As you learned in the previous tip, when you create a reference variable, C# will place the variable within the heap. Within the .NET environment, the Common Language Runtime provides code that manages the heap's contents. Earlier in this book, you used the *malloc* C run-time library to allocate memory dynamically from the heap. Later, when your program was done with the memory, your programs released the memory back to the heap using the *free* function. In a similar way, within C++ programs, your code used the *new* operator to allocate memory and the *dispose* operator to release it.

Because the Common Language Runtime (CLR) manages the heap, programmers do not have to worry about releasing the memory they allocate. Instead, the Common Language Runtime will determine when the memory is no longer in use and it will automatically release memory.

Normally, you will use reference-type variables to hold class instances. However, using the *new* operator, you can create a reference variable for any data type. The following statement, for example, creates a reference variable of type *int*:

```
int Age = new int();
```

Normally, you would not use reference variables for simple data types. Instead, as shown later in this book, you will use reference variables on a regular basis to create class data.

1536 C# Forces You to Assign Values to Variables

To reduce errors that result from the use of variables to which your code has not yet assigned a value, C# will not let your program use a variable until your code assigns a value to the variable. When your code tries to use a variable for which the compiler has not encountered a previous value assignment, the compiler will generate a syntax error and your program will not compile. The following program, *NoValueError.cs*, for example, tries to use the variable *UserNumber* before the variable has been assigned a value:

```
using System;

class clsNoValueError
{
    static void Main()
      {
           int UserNumber;

           Console.WriteLine("{0}", UserNumber);
      }
}
```

When you compile the *NoValueError.cs* program, the C# compiler will generate a syntax error that tells you that you must initialize a variable before you use it.

USING UNINITIALIZED VARIABLES AS PARAMETERS 1537

In the previous tip you learned that to reduce possible errors due to the use of uninitialized variables, C# will not let your code use a variable until you have assigned a value to the variable. The following program, *ParameterError.cs*, for example, tries to pass the variable *UserNumber* to the function *ShowVariable*:

```
using System;

class clsParameterError
{
    static void ShowVariable(int x)
       {
           Console.WriteLine("{0}", x);
       }

    static void Main()
      {
          int UserNumber;

          ShowVariable(UserNumber);
      }
}
```

Again, because you have not yet assigned a value to the variable, the C# compiler will generate a syntax error and will not compile your program.

Often, however, programs will use functions to initialize variables. If the C# compiler would not let your programs pass an uninitialized variable to a function, C# would defeat a common programming practice. As it turns out, in cases where a function initializes a variable, you simply must precede the variable with the *out* keyword when you call the function and within the function definition:

```
SomeFunction(out VariableName);    // Function call

// Function definition
void SomeFunction(out int VariableToInitialize)
  {
      VariableToInitialize = 7;
  }
```

The following program, *InitVariable.cs*, illustrates the use of the *out* keyword to initialize a variable:

```
using System;

class clsInitVariable
{
    static void InitializeVariable(out int x)
     {
          x = 1700;
     }

    static void Main()
     {
          int UserNumber;

          InitializeVariable(out UserNumber);

          Console.WriteLine("{0}", UserNumber);
     }
}
```

As you can see, the program places the *out* keyword within the function's parameter list and before the variable name when the program passes the variable to the function.

1538 C# Does Not Support the #define and #include Directives

Within C and C++ programs, you have made extensive use of the #define preprocessor directive to create constants and macros and the #include directive to insert a header or source file within your program. C# does not support the #define or #include directives.

To create a constant within a C# program, you can use the *const* keyword or you can create a read-only variable to which C# lets your programs assign a value, one time, at run time. C# does support function prototypes, because C# does not support global functions (all functions must be declared within a class). Thus, because you cannot use #define to create constants and you do not need function prototypes, C# does not require include files, and hence does not support the #include directive.

1539 Defining a Variable as ReadOnly

In earlier tips, you learned how to use the *const* keyword to create a constant within C++ programs:

```
const int TipCount = 1700;
```

If you try to change the value of a variable that uses the *const* attribute, the C# compiler will generate a syntax error and your program will not compile. Although the *const* keyword lets you create a constant, you must know the constant's value when you compile the program. Should the constant change, you must edit your program, make the change, and then recompile.

Within C#, you can use the *readonly* keyword to create a constant to which the compiler lets you assign a value on time. After you assign a value to *readonly* class-member variable, you cannot change the variable's value. To assign a value to a *readonly* variable, you must either assign the value when you declare the variable or you must assign the value from within a constructor function (which might, for example, read the variable's value from the user or from a file). If you assign the value to a *readonly* variable when you declare the variable, the readonly attribute is essentially the same as *const*. The following program, *AssignReadonly.cs*, uses a constructor function to assign a value to a *readonly* variable:

```
using System;

class clsAssignReadonly
{
   public static readonly int UserNumber = 1700;

   static void Main()
   {
         Console.WriteLine("{0}", UserNumber);
   }
}
```

In this case, the program initializes the *readonly* class member at declaration. However, the program could have used a constructor function to initialize the variable.

CREATING ENUMERATIONS WITHIN C# 1540

As you have learned, an enumeration lets you assign constant values to a list of related identifiers, which your programs then later use by name. For example, the following enumeration creates an enumerated list named DaysOfWeek:

```
enum DaysOfWeek
  {
     Sunday, Monday, Tuesday, Wednesday, Thursday,
     Friday, Saturday
  }
```

When you create an enumerated list within a program, the compiler does not actually create any programming statements for the list. Instead, the compiler uses the list to replace the list's identifiers that appear throughout your code with the identifier's corresponding value.

When you create an enumerated list within a C# program, you can place the list within or outside of a class definition. If you place the list within a class definition, only that class is aware of the list. If you try to use one of the list's identifiers outside of the class, the C# compiler will generate an error message for an unknown identifier.

USING CHECKED VARIABLES TO DETECT OVERFLOW ERROR 1541

As you have learned, an object's type defines a set of values the object can store and a set of operations your programs can perform on an object. For example, a variable of type *int* can store values in the range -2,147,483,648 to 2,147,483,647. When you exceed the range of values a variable can store, an overflow error occurs. For example, the following program, *ShowOver.cs*, assigns an *int* variable the value 2,147,483,647. The program then increments the variable's value and displays the result. Because of the overflow error, the variable does not contain the value 2,147,483,648, but rather, it contains –2,147,483,468:

```
using System;

class clsShowOver
  {
    static void Main()
      {
        int Value = 2147483647;

        Value++;

        Console.WriteLine("{0}", Value);
      }
  }
```

To compile this program from the command line, issue the following command:

```
C\Subdir> csc  ShowOver.cs   <Enter>
```

When you execute the program, the program will display the following output:

```
C:\Subdir> ShowOver  <Enter>
-2,147,483,468
```

Within C#, you can direct your programs to detect overflow errors by adding the /checked switch to your command-line compilation. When you direct C# to detect overflow errors, C# will generate (throw) an exception when the overflow occurs. The following command line uses the /checked switch to compile the previous program:

```
C\ > csc  /checked  ShowOver.cs   <Enter>
```

In this case, when you execute the program, it will display an error message (due to the exception) that tells you an overflow occurred.

In addition to using the /checked word to change C#'s default overflow checking, you can use the *checked* and *unchecked* keywords within your code to enable or disable checking. For example, the following code uses the *checked* keyword to enable overflow checking for the previous operation:

```
using System;

class clsShowOver
 {
   static void Main()
    {
      int Value = 2147483647;

      checked
        {
           Value++;
        }

       Console.WriteLine("{0}", Value);
    }
 }
```

1542 UNDERSTANDING TYPE CONVERSIONS

As is the case with C and C++, when you assign a value of one type to a variable of a different type with C#, your application can perform an implicit type conversion or you may have to explicitly cast the conversion. C# will implicitly cast a lower-level type, such as a *float*, to a higher-level type, such as a *double*, because no data is lost during such an assignment. However, when you assign the value of type *double* to a variable of type *float*, you will lose precision. In such cases, you must force the conversion by performing a cast. If you do not explicitly cast a higher-type value to a lower-type variable, the C# compiler will display a syntax error and your code will not compile. The following C# program, *ConvertDemo.cs*, illustrates an implicit and explicit type conversion:

```
class clsConvertDemo
  {
   static void Main()
   {
         float fltValue;
         double dblValue = 0.1234567890;

         int intValue = 77;
         long lngValue;

         lngValue = intValue;           // Implicit

         fltValue = (float) dblValue;   // Explicit

         Console.WriteLine("{0} {1}", lngValue, fltValue);
   }
}
```

USING CONVERSION FUNCTIONS 1543

In Tip 1542, you learned how to explicitly cast a value of one type to another. C# also provides your applications a set of conversion routines you can call to convert a value from one type to another. Within C#, everything is an object. Even numeric values such as 1543 are objects. Each C# object, as you have learned, inherits methods from the Object class. One such method is *toString*, which converts the object's value into a character-string format. The following C# program, *ToStringDemo.cs*, illustrates the use of the *toString* method:

```
using System;

class clsToStringDemo
{
   static void Main()
   {
         int Count = 1700;
         double Temp = 98.6;

         Console.WriteLine("{0}", Count.ToString());
         Console.WriteLine("{0}", Temp.ToString());
         Console.WriteLine("{0}", 2002.ToString());
   }
}
```

Note that even the constant 2002 (which is an object) can call the *ToString* method. After you compile and execute this program, your screen will display the following output:

```
C:\Subdir> ToStringDemo  <Enter>
1700
98.6
2002
```

Depending on an object's type, the object may have several other conversion methods, such as *ToChar*, *ToInt*, *ToDouble*, and so on. For more information on C# data-type conversions, see the C# *Convert* class.

REVIEWING C# OPERATORS 1544

Like C and C++, C# provides the standard operators, such as +, -, /, and * for addition, subtraction, division, and multiplication. Table 1544.1 briefly describes the C# unary operators, which perform their operation on a single operand, such as incrementing or decrementing a variable's value.

Operator	Description	Purpose
~	One's complement	Inverts each bit within a value, changing bits with the value 1 to 0, and vice versa.
!	Negation	Negates an expression, returning *true* if an expression is *false* or *false* if the expression is *true*.
-	Unary minus	Sets an expression's numeric sign to negative, making a positive expression negative and a negative expression positive.
+	Unary plus	Sets an expression's numeric sign to positive.
++	Increment	Adds one to the value of a variable.
--	Decrement	Subtracts one from the value of a variable.

Table 1544.1 C# unary operators. *(continued on following page)*

Operator	Description	Purpose
new	Memory allocate	Allocates a memory location within the heap to store a reference type object.
sizeof	Size of	Returns an object's size (the number of bytes needed to store the object) in bytes.

Table 1544.1 C# unary operators. (continued from previous page)

Table 1544.2 lists the C# binary operators, which perform their operation on two operands.

<table>
<tr><th>Operator</th><th>Description</th><th>Purpose</th></tr>
<tr><td>+, -, *, /, %</td><td>Standard arithmetic</td><td>Performs addition, subtraction, multiplication, division, and modulo division.</td></tr>
<tr><td><<, >></td><td>Shift</td><td>Moves the bits in the left operand to the left or right (depending on << or >>) the number of bit positions specified by the right operand.</td></tr>
<tr><td><. >, <=, >=, ==, !=, is</td><td>Relational</td><td>Compares two expressions, returning a true or false value.</td></tr>
<tr><td>&, |, ^</td><td>Bitwise</td><td>Performs bitwise AND, OR, and Exclusive OR operations on bits in the left operand and right operand.</td></tr>
<tr><td>&&, ||</td><td>Logical</td><td>Performs a logical AND or a logical OR operation on two Boolean (true or false) values.</td></tr>
<tr><td>=</td><td>Assignment</td><td>Assigns the value of the left operand to the value of the right operand.</td></tr>
</table>

Table 1544.2 C# binary operators.

Table 1544.3 summarizes the relational operators used in C#.

Operator	Description	Result
<	Less than	Returns *true* if first operand is less than the second, or *false* otherwise.
>	Greater than	Returns *true* if the first operand is greater than the second, or *false* otherwise.
<=	Less than or Equal to	Returns *true* if the first operand is less than or equal to the second or *false* otherwise.
>=	Greater than or Equal to	Returns *true* if the first operand is greater than or equal to the second, or *false* otherwise.
==	Equal to	Returns *true* if the first operand is equal to the second, or *false* otherwise.
! =	Not Equal to	Returns *true* if the first operand is not equal to the second or *false* otherwise.

Table 1544.3 C++ relational operators.

EXAMINING C# OPERATOR PRECEDENCE 1545

To ensure that programs evaluate expressions consistently, C#, like C and C++, provides an operator precedence. Table 1545 organizes the C# operators by precedence. To override the operator precedence, your programs can group the expressions you want C# to evaluate first within parentheses.

Operator	Meaning	Example
.	Member of	object.member_name
[]	Subscript	array[element]
()	Function call	FunctionName (arguments)
()	Subexpression	(Expression)
++	Postfix increment	variable++
--	Postfix decrement	variable--
new	Allocate memory	new type
typeof	Type retrieval	typeof(object type)
sizeof	size of type	sizeof(object type)
checked	range checking	checked (expression)
unchecked	range checking	unchecked (expression)
!	Logical not	! expression
++	Prefix increment	++variable
--	Prefix decrement	--variable
~	Ones complement	~variable
+	Unary plus	+ expression
-	Unary minus	- expression
()	Cast	(type) expression
*	Multiply	expression * expression
/	Divide	expression / expression
%	Modulo	expression % expression
+	Addition	expression + expression
-	Subtraction	expression – expression
<<	Shift left	expression << expression
>>	Shift right	expression >> expression
<	Less than	expression < expression
>	Greater than	expression > expression
<=	Less than or equals	expression <= expression
>=	Greater than or equals	expression >= expression
is	Compatibility	expression *is* object type
==	Equality	expression == expression

Table 1545 C# operator precedence. *(continued on following page)*

!=	Not Equal	expression != expression
&	Bitwise AND	expression & expression
|	Bitwise OR	expression | expression
^	Bitwise XOR	expression ^ expression
&&	Conditional AND	expression && expression
||	Conditional OR	expression || expression
?:	Conditional	x = expression ? expression : expression
	Assignment	= *= /= %= += -= <<= >>= &= ^= |=

Table 1545 C# operator precedence. (continued from previous page)

1546 AVOID SPACES WITHIN INCREMENT AND DECREMENT OPERATORS

As you know, the increment (++) and decrement (--) operators let your programs add or subtract 1 from a variable's current value. When you use these operators within C#, you must watch for errant spaces between the double-plus or double-minus signs. For example, the following statement uses the increment operator to increment the variable *Counter*, by one:

```
++Counter;
```

Assume, for example, that when you typed the previous statement, you errantly typed an extra space between the two plus signs, as follows:

```
+ +Counter;
```

Unlike C and C++ (which will ignore the space between the plus signs), the C# compiler will treat the plus signs as two unary plus operators, and as such, will not increment the variable's value—nor will the compiler generate an error message. The following program, *ErrantSpace.cs*, places spaces between the plus signs and minus signs in an increment and decrement operation:

```
using System;

class clsErrantSpaces
 {
    static void Main()
      {
           int x = 100;
           int y = 100;
           int a, b;

           a = + +x;
           b = - -y;
           Console.WriteLine("{0} {1}", a, b);
      }
 }
```

When you compile and execute this program, your screen will display the following output:

```
C:\Subdir> ErrantSpaces  <Enter>
100 100
```

Because the space between the plus and minus signs causes the C# compiler to treat the increment and decrement operators as unary operators, the program does not increment or decrement the corresponding variables.

UNDERSTANDING UNSAFE CODE 1547

Unlike the C and C++ programming languages, C# applications do not make extensive use of pointers. Because programmers often misuse pointers, which leads to errors, C# considers the use of pointers an "unsafe" practice. If you must use pointers within a function, for example, you can use the *unsafe* keyword within the function header, as shown here:

```
public unsafe void SomeFunction(int x, int y, int z)
{
   int *temp;  // Pointer variable
}
```

C# does not require that you declare an entire function as unsafe. Instead, you can place the unsafe keyword within a specific statement, or you can specify a block of statements is unsafe, as shown here:

```
unsafe
  {
     // Statements
  }
```

As you have learned, within C#, the Common Language Runtime manages objects you create on the heap. When you declare code as "unsafe" the Common Language Runtime does not manage the code's variables. Within code that you declare as *unsafe* you may declare and use pointers. To compile, from the command line, a program that uses unsafe code, you must include the */unsafe* switch when you invoke the compiler, as shown here:

```
C:\Subdir> csc  /unsafe  SomeProgram.cs   <Enter>
```

The following program, *UnsafeDemo.cs*, uses pointers to swap the value of two variables within the function *swap*. Because of the pointer use, the program must use the C# *unsafe* keyword:

```
using System;

class clsUnsafeDemo
{
  unsafe static void Swap(int *x, int *y)
   {
     int z;

     z = *x;
     *x = *y;
     *y = z;
   }

  static void Main()
   {
     int Big = 1000;
     int Small = 0;

     unsafe
      {
        Swap(&Big, &Small);
      }

     Console.WriteLine("Big {0}  Small {1}", Big, Small);
   }
}
```

1548 C# CONSIDERS THE SIZEOF OPERATOR UNSAFE

In the previous tip, you learned that if your programs use pointers, you must declare the corresponding code "unsafe." In a similar way, if your program uses the *sizeof* operator, which returns the number of bytes C# needs to store an object, you must declare the corresponding code as unsafe. The following program, *ShowSizes.cs*, uses the sizeof operator to display the size (in bytes) of several of the standard C# types:

```
using System;

class clsShowSizes
 {
   static void Main()
    {
      unsafe
       {
          Console.WriteLine("int is {0} bytes",
              sizeof(int));
          Console.WriteLine("float is {0} bytes",
              sizeof(float));
          Console.WriteLine("decimal is {0} bytes",
              sizeof(decimal));
          Console.WriteLine("byte is {0} bytes",
              sizeof(byte));
       }
    }
 }
```

To compile this program from the command line, you must include the /unsafe switch as shown here:

```
C:\Subdir> csc  /unsafe  ShowSizes.cs  <Enter>
```

After you run the program, your screen will display the following output:

```
C:\Subdir> ShowSizes  <Enter>
int is 4 bytes
float is 4 bytes
decimal is 16 byte
byte is 1 bytes
```

1549 MAKING DECISIONS WITHIN A C# PROGRAM

As you have learned, programs use conditional operations, such as *if* and *if-else* statements to make decisions. C# fully supports the C/C++ *if* and *if-else* statements. In addition, C# supports the *switch* statement. However, C# changes the *switch* statement to eliminate the fall through processing. As you know, within C and C++, after a *switch* statement finds a matching case, the *switch* statement considers the remaining cases valid as well, and unless you provide a *break* statement, the *switch* statement will execute the corresponding statements. The following C program, *CountVowels.c*, for example, loops through the letters of the alphabet, using a *switch* statement to count the number of vowels:

```
#include <stdio.h>

void main(void)
 {
   int letter, Vowel_Count = 0;

   for (letter = 'A'; letter <= 'Z'; letter++)
     switch (letter) {
       case 'A':
```

```
        case 'E':
        case 'I':
        case 'O':
        case 'U': Vowel_Count++;
      }

    printf("Number of vowels is %d\n", Vowel_Count);
  }
```

As you can see, by taking advantage of the "fall through," the *switch* statement's code is quite simple. The C# developers, however, wanted to reduce the chance of errors due to a programmer forgetting to insert a *break* statement to prevent fall through. If you cut and paste the *switch* statement from the previous program into a C# program, and then compile the code using C#, the compiler will generate a syntax error due to the *case* statement fall through.

Testing a Variable Assignment in C# 1550

Within C and C++ programs, programmers make extensive use of the *if* statement to determine if a variable contains a zero or non-zero value. As you may recall, C and C++ treat the value 0 as false and any non-zero value as true. The following code fragment illustrates the use of an *if* statement in C/C++ to test a variable for a non-zero value:

```
if (SomeVariable)
  printf("Varible is non-zero, true\n");
else
  printf("Varible is zero, false\n");
```

Within C#, programs cannot test a variable for a zero or non-zero value in this way. Within C#, an if statement must evaluate a true or false value, as opposed to a zero or non-zero value. Thus, within C#, you must specify a comparison within an if statement that generates a true or false result, as shown here:

```
if (SomeVariable != 0)
  Console.WriteLine("Varible is non-zero");
else
  Console.WriteLine("Varible is zero");
```

In a similar way, many C/C++ programs loop until a zero or *NULL* value occurs. Again, within C#, within iterative constructs such as a *while* loop, you must specify a comparison that evaluates to a Boolean result. If you specify a condition, such as the following that does produce a Boolean result, the C# compiler will generate a syntax error:

```
for (int x = 10; x; x--)
  {
    // Loop statements
  }
```

Repeating Statements Within a C# Program 1551

As you have learned, programs use iterative (or looping) constructs to repeat one or more statements within a program. Like C and C++, C# fully supports the *for*, *while*, and *do-while* loops. C# also supports the use of the *goto* statement. As you will learn in the following tip, C# provides the *foreach* construct which lets applications loop through each element of an array.

Within C#, the *for*, *while*, and *do-while* loops are identical to their C and C++ counterparts. The *goto* statement within C# can branch to a label, a case statement, or the default case statement. The following program, *GotoDemo.cs*, illustrates the use of the *goto* statement within a C# application that implements a loop that displays the numbers 1 through 10:

```
using System;

class clsGotoDemo
  {
   static void Main()
    {
      int Count = 0;

      Repeat:
        Console.WriteLine("{0}", Count);
        if (Count++ < 10)
          goto Repeat;
    }
  }
```

1552 LOOPING THROUGH ARRAY ELEMENTS USING FOREACH

Throughout this book, you have used *for* loops to display the contents of an array. As you know, when you use a *for* loop, you must initialize, test, and later increment the loop's control variable. Should you omit one of the steps, or use the wrong variable within one of the steps, an error will occur that may create an infinite or unending loop. To simplify the process of looping through array elements, C# provides the *foreach* statement, the format of which is as follows:

```
foreach (int index in ArrayVariable)
 {
    // Statements
 }
```

The *foreach* loop does not assign index values to the loop's control variable, but rather, the loop assigns the array's actual values. Assume, for example, that you have an array named *Employees* that contains the names of each of your company's employees. Using a *foreach* statement, you can display the employee names as follows:

```
foreach (int Emp in Employees)
 {
    System.Console.WriteLine(Emp);
 }
```

The following program, *ForEachDemo.cs*, uses the *foreach* statement to loop through several different arrays:

```
using System;

class clsForEachDemo
{
  static void Main()
   {
     char [] Alphabet = new char[26];

     int [] One_Ten = new int[10];

     // use a for loop to initialize the character array
     for (int i = 0; i < 26; i++)
       Alphabet[i] = (char) ((char) i + 'A');

     for (int i = 0; i < 10; i++)
       One_Ten[i] = i;

     // display the arrays
```

```
      foreach (int Value in One_Ten)
        Console.Write("{0}", Value);

      Console.WriteLine();

      foreach (int letter in Alphabet)
        Console.Write("{0}", Convert.ToChar(letter));
    }
  }
```

Because the *foreach* loop does not provide index values, the previous program used *for* loops to initialize each array.

USING FOREACH TO DISPLAY COMMAND-LINE ARGUMENTS 1553

As you learned in the previous tip, the C# *foreach* statement provides a convenient way for programs to loop through the elements in an array. Further, as you have learned, C# places a program's command-line parameters within an array of character strings. The following program, *ForEachCmd.cs*, uses a *foreach* statement to display a program's command-line arguments:

```
using System;

class clsForEachCmd
 {
  static void Main(string [] args)
   {
     foreach (string Argument in args)
       Console.WriteLine("{0}", Argument);
   }
}
```

After you compile this program, invoke the program using the command line shown here:

```
C:\Subdir> ForEachCmd  AAA  BBB  CCC  DDD   <Enter>
```

The program, in turn, will display the following output:

```
C:\Subdir> ForEachCmd  AAA BBB  CCC DDD  <Enter>
AAA
BBB
CCC
DDD
```

USING CONSOLE.READLINE TO READ KEYBOARD INPUT 1554

In several of the previous tips, the programs have used the *Console.WriteLine* function to display output to the screen display. In a similar way, the *Console.ReadLine* function reads a line of text the user enters from the keyboard and assigns the text to a variable. For example, to read an integer number from the keyboard, you would assign the value the *Console.ReadLine* function returns to a variable of type *string*, as shown here:

```
String UserName;
UserName = Console.ReadLine();
```

To assign the user input to an *int* variable, you can use the Convert class to convert the string the ReadLine function returns to an integer value, as shown here:

```
int Result;
Result = Convert.Int16(Console.ReadLine());
```

The following program, *ReadLineDemo.cs*, uses the *Console.ReadLine* function to assign user input to two different variables:

```
using System;

class clsReadLineDemo
{
  static void Main(string [] args)
   {
     string Name;
     int Age;

     Console.Write("Enter your name: ");
     Name = Console.ReadLine();

     Console.Write("Enter your age: ");
     Age = Convert.ToInt16(Console.ReadLine());

     Console.WriteLine("Hello {0}, it's great to be {1}",
       Name, Age);
   }
}
```

When programs use the *Console.ReadLine* function to read user input from the keyboard, the function will read all the characters the user inputs including the carriage return and linefeed that the Enter key generates. The function, however, will not return the carriage return and linefeed characters within the input. Often programs will want to read keyboard input one character at a time. In such cases, the program can use the *Console.Read* function.

1555 Using Console.Read to Read One Character at a Time

In tip 1554, you used the *Console.ReadLine* function to read values from the keyboard one line of input at a time. Depending on the processing your application performs, there may be times when you only want to read one character at a time. In such cases, your applications can use the *Console.Read* function. The following program, *MenuChoice.cs*, displays a list of menu options to the user and then uses the *Console.Read* function to read the user's selection:

```
using System;

class clsMenuChoices
{
  static void Main()
   {
     char menuChoice;

     bool done = false;

     Console.WriteLine("Main Menu\n ");
     Console.WriteLine("A   First Option");
     Console.WriteLine("B   Second Option");
     Console.WriteLine("C   Third Option");
     Console.WriteLine("D   Fourth Option");
     Console.Write("Choice: ");

     do {
      menuChoice = (char) Console.Read();
      if ((menuChoice == 'A') || (menuChoice == 'a'))
        done = true;
      else if ((menuChoice == 'B') || (menuChoice == 'b'))
```

```
        done = true;
      else if ((menuChoice == 'C') || (menuChoice == 'c'))
        done = true;
      else if ((menuChoice == 'D') || (menuChoice == 'd'))
        done = true;
    } while (! done);

    Console.WriteLine("You selected {0}", menuChoice);
  }
}
```

Although the *Console.ReadLine* function returns a character at a time, the function will not return the character until the user presses the Enter key. Take time to experiment with the previous program. First run the program and enter one of the menu choices, such as A. The program will display your selection and end. Then run the program and enter a value such as Q and press Enter. When you make your selection, the Enter key will advance the cursor to the start of the line that follows the Choice: prompt. Next, try entering the letters ZXYA and then press Enter. Within the program's do loop, the code will simply read and ignore the letters ZXY and will then process the letter A as your selection.

C# FUNCTIONS MUST BELONG TO A CLASS 1556

When you create programs within C++, some functions may reside within a class as a class method and some may reside outside of a class. Within C#, in contrast, all functions must reside within a class. You cannot define a function outside of a class.

Often, you may have a "generic" function that applies to many classes. In such cases, you might create a class called *Utilities* and define the function as *static* within the class. By defining the function as *static*, programs can call the function although no instance of the *utilities* class may exist.

The following program, *StaticFunctions.cs*, creates two *static* functions within the *Utilities* class, and a third function which is a member of class *clsStaticFunctions*. The program then calls each of the functions:

```
using System;

class Utilities
{
  static public void Hello()
   {
     Console.WriteLine("Hello, C# World");
  }

  static public double Average(int a, int b)
   {
     return((a + b) / 2.0);
   }
}

class clsStaticFunctions
{
  static void LocalHello()
   {
     Console.WriteLine("Hello, from local function");
  }

  static void Main()
   {
     Utilities.Hello();
```

```
      Console.WriteLine("The average of {0} and {1} is {2}",
         5, 15, Utilities.Average(5, 15));
      LocalHello();
    }
}
```

Note that to make the *Hello* and *Average* functions callable from outside of the *Utilities* class, the program defines the two functions as *public*.

1557 USING IN AND OUT PARAMETERS

Within C and C++ programs, to change a variable's value within a function, you must pass the function a pointer to the variable. C#, however, considers the use of pointers "unsafe" code. To change a variable's value within a C# function, you must precede the variable name with the *ref* keyword when you call the function:

```
ChangeFirst(ref c, d);
```

Further you must precede the parameter name within the function definition with *ref* keyword as well:

```
static void ChangeFirst(ref int a, int b)
  {
     a = 1001;
     b = 2002;
  }
```

The following program, *InOut.cs*, calls four different functions. The first function does not use the *ref* keyword for either parameter. When the function's execution completes, changes the function has made to the parameter values do not remain in effect. The second function uses the *ref* keyword to change the first parameter's value. Likewise, the third function changes the second parameter's value, and the fourth function changes both parameters:

```
using System;

class InOut
{
  static void NoChange(int a, int b)
   {
     a = 1001;
     b = 2002;
   }

  static void ChangeFirst(ref int a, int b)
   {
     a = 1001;
     b = 2002;
   }

  static void ChangeSecond(int a, ref int b)
   {
     a = 1001;
     b = 2002;
   }

  static void ChangeBoth(ref int a, ref int b)
   {
     a = 1001;
     b = 2002;
   }

  static void Main()
   {
```

```
      int a = 0, b = 0, c = 0, d = 0;
      int e = 0, f = 0, g = 0, h = 0;

      NoChange(a, b);
      Console.WriteLine("{0} {1}", a, b);

      ChangeFirst(ref c, d);
      Console.WriteLine("{0} {1}", c, d);

      ChangeSecond(e, ref f);
      Console.WriteLine("{0} {1}", e, f);

      ChangeBoth(ref g, ref h);
      Console.WriteLine("{0} {1}", g, h);
    }
}
```

After you compile and execute this program, your screen will display the following output:

```
C:\Subdir> InOut  <Enter>
0 0
1001 0
0 2002
1001 2002
```

HANDLING A VARIABLE NUMBER OF PARAMETERS 1558

As you create functions within C#, there may be times when you do not know the number of parameters you will pass to a function. For example, throughout this book's C# tips, you have made extensive use of the *Console.WriteLine* function to display output to the console window. When your programs call the *Console.WriteLine* function, you can pass the function a variable number of parameters, as shown here:

```
Console.WriteLine("{0}", 1);
Console.WriteLine("{0} {1}", 1, 2);
Console.WriteLine("{0} {1} {2}", 1, 2, 3);
```

Within C#, you can create a function that supports a variable number of parameters by using the *params* keyword within the function definition to specify an array that will contain each of the values.

In the tips that follow, you will examine C# arrays in detail. The following program, *VarParams.cs*, uses arrays to handle a variable number of function parameters:

```
using System;

class VarParams
{
  static void ShowAllStrings(params string [] args)
   {
     foreach(string Parameter in args)
       Console.Write("{0} ", Parameter);

       Console.WriteLine();
   }

  static void ShowAllInts(params int [] args)
   {
     foreach(int Parameter in args)
        Console.Write("{0} ", Parameter);
     Console.WriteLine();
   }
```

```
  static void Main()
   {
     ShowAllStrings("A", "B", "C");
     ShowAllStrings("A", "B");

     ShowAllInts(1);
     ShowAllInts(1, 2);
     ShowAllInts(1, 2, 3);
   }
}
```

After you compile and execute this program, your screen will display the following output:

```
C:\Subdir> VarParams  <Enter>
A B C
A B
1
1 2
1 2 3
```

1559 Using Arrays Within C#

As you know, an array is a data structure that lets you store multiple values of the same type within a single variable. Within C#, arrays are reference-type variables derived from the *System.Array* class. As you will learn in the tips that follow, the array class provides several powerful methods you can use to sort, search, and even reverse the order of elements within an array.

To create an array within C#, you use the *new* operator. As you have learned, the *new* operator directs C# to create a reference variable, which resides in the heap. To begin, you must specify the array's type, followed by right and left facing brackets ([]) and the array variable's name. The following statement, for example, defines, but does not allocate space for, an array of integer values:

```
int [] ClassScores;
```

Next, to allocate space for the array, you must assign the result of the new operator to the array variable. When you invoke new, you must specify the array's type and the number of elements you desire. The following statement, for example, allocates space for 100 integer values:

```
ClassScores = new int [100];
```

Normally, you will declare the array variable and allocate space in the same statement, as shown here:

```
int [] ClassScores = new int [100];
```

Like C and C++, the C# compiler uses zero-based indexing to access array elements. To access elements within the *ClassScores* array, for example, your program would use elements in the range *ClassScores[0]* to *ClassScores[99]*.

The following program, *ArrayDemo.cs*, creates an array named *Alphabet* of type byte, and then assigns to the array the ASCII values that correspond to the letters A through Z:

```
using System;

class clsArrayDemo
{
  static void Main()
   {
     char [] Alphabet = new char[26];
     int index;
```

```
      for (index = 0; index < 26; index++)
        Alphabet[index] = Convert.ToChar(65+index);

      for (index = 0; index < 26; index++)
        Console.Write(Alphabet[index]);
    }
}
```

To assign the array elements, the program calls the *Convert.ToChar* function, which converts the numeric values to a Unicode character. The value 65 corresponds to the ASCII character "A".

INITIALIZING AN ARRAY WITHIN C# 1560

In the previous tip, you learned how to allocate memory to store an array of values. Often when you create an array, your code may want to initialize the array entries. In such cases, your code can place the array values within right and left facing braces ({}), separating the values with commas. The following statement, for example, allocates an array of integer values and then initializes the array elements:

```
int [] Prices = new int [5] {100, 500, 250, 300, 400};
```

When programs initialize arrays during the array's declaration, the programs often do not specify the number of elements in the array. Instead, the programs let the compiler determine the number of elements by counting the number of initial values the program specifies. The following statement allocates memory for and initializes an array of three floating-point values:

```
float [] Prices = new float [] {0.10, 0.50, 1.0};
```

The following program, *InitializeAlphabet.cs,* changes the program you examined in Tip 1559 to initialize the array Alphabet at declaration:

```
using System;

class clsInitializeAlphabet
{
  static void Main()
   {
     char [] Alphabet = new char []  {
       'A', 'B', 'C', 'D', 'E', 'F', 'G', 'H', 'I',
       'J', 'K', 'L', 'M', 'N', 'O', 'P', 'Q', 'R', 'S',
       'T', 'U', 'V', 'W', 'X', 'Y', 'Z'
      };

     int index;

     for (index = 0; index < 26; index++)
       Console.Write(Alphabet[index]);
   }
}
```

DETERMINING AN ARRAY'S LENGTH 1561

Within C and C++, programs that use arrays have to track the number of elements the array contains. As discussed, within C#, all arrays are derived from the *Array* class. The *Array* class provides the *Length* member that tracks the array size. Within your code, for example, you might use the *Length* member within a *for* loop that processes the array elements:

```
for (Index = 0; Index < SomeArray.Length; Index++)
  {
    // Statements
  }
```

The following *UseLength.cs* program changes the *InitializeAlphabet.cs* program you examined in Tip 1560 to use the *Length* member to display the contents of the Alphabet array:

```
using System;

class clsUseLength
{
  static void Main()
   {
     char [] Alphabet = new char []  {
        'A', 'B', 'C', 'D', 'E', 'F', 'G', 'H', 'I',
        'J', 'K', 'L', 'M', 'N', 'O', 'P', 'Q', 'R', 'S',
        'T', 'U', 'V', 'W', 'X', 'Y', 'Z'
      };

     int index;

     for (index = 0; index < Alphabet.Length; index++)
       Console.Write(Alphabet[index]);
   }
}
```

In a similar way, the following *ArraySizes.cs* program uses the *Length* member to display the number of elements in several different arrays:

```
using System;

class clsUseLength
{
  static void Main()
   {
     int [] Big = new int[1000];
     double [] Currency = new float [15];
     char [] Letters = new char [1500];


      Console.WriteLine("Number of elements is Big is {0}",
         Big.Length);
      Console.WriteLine("Number of elements is Currency is {0}", Currency.Length);
      Console.WriteLine("Number of elements is Letters is {0}", Letters.Length);
   }
}
```

After you compile and execute this program, your screen will display the following output:

```
C:\Subdir> ArraySizes   <Enter>
Number of elements is Big is 1000
Number of elements is Currency is 15
Number of elements is Letters is 1500
```

1562 C# ENFORCES ARRAY BOUNDARIES

When you create an array within C, C++, or C#, the compiler will allocate memory to hold the number of elements you specify. The first array element is always element 0 (*arrayName[0]*). The final element corresponds to the number of entries within the array, minus one.

To prevent programs from exceeding an array's boundaries, C# (unlike C and C++) examines each array element your program uses to ensure the element falls within the array's bounds. If your program either intentionally or errantly tries to access an element outside of the array boundaries, C# will generate an exception, which, by default, will end the application.

The following program, *InvalidArrayBoundary.cs*, tries to assign a value to element 5 of an array that supports elements 0 through 4. When the program tries the assignment, C# will generate an exception that will terminate the program:

```
using System;

class clsInvalidArrayBoundary
{
  static void Main()
  {
    int [] small = new int [5];

    small[5] = 1700;    // Invalid array boundary
  }
}
```

CATCHING AN INVALID ARRAY BOUNDS EXCEPTION 1563

In Tip 1562 you learned if you assign a value to an array index that falls outside of the array's boundary, C# will generate an exception. Specifically, when C# detects the invalid array element, C# will generate the *IndexOutOfRangeException* exception which your programs can catch. The following program, *CatchArrayException.cs*, again tries to assign a value that is outside of the array bounds. Rather than terminate when the exception occurs, the code catches the exception and displays an error message to the user before ending the program:

```
using System;

class clsCatchArrayException
{
   static void Main()
   {
      int [] small = new int [5];

      try
      {
         small[5] = 1700;    // Invalid array boundary
      }
      catch (IndexOutOfRangeException e)
      {
        Console.WriteLine ("Program tried to access " +
           "an invalid array boundary");
      }
   }
}
```

After you compile and run this program, your screen will display the following output:

```
C:\Subdir> CatchArrayException  <Enter>
Program tried to access an invalid array boundary
```

1564 RESIZING AN ARRAY

Within C#, after you declare an array and allocate memory, the array's size is fixed. However, an application can use the *new* operator to allocate a larger or small memory region, the location of which you can assign to the array. If you use the *new* operator to resize an array in this way, you will lose the original contents of the array. Your code can, however, copy the array's current contents to a temporary location, allocate the memory, and then copy the array contents from the temporary location into the new array. Because the Common Language Runtime manages the heap space, your program does not have to worry about freeing the array's old memory location. The following program, *Resize.cs*, allocates an array of five integer values. The program then displays the values and then resizes the array using the technique just discussed:

```
using System;

class Resize
{
  static void Main()
   {
     int [] Values = new int [5];
     int i;

     for (i = 0; i < Values.Length; i++)
       Values[i] = i;

     for (i = 0; i < Values.Length; i++)
       Console.Write("{0}", Values[i]);

     Console.WriteLine();

     int [] Temp = new int[5];

     for (i = 0; i < Values.Length; i++)
         Temp[i] = Values[i];

      Values = new int[10];

      for (i = 0; i < Temp.Length; i++)
         Values[i] = Temp[i];

      for (i = 0; i < Values.Length; i++)
          Console.Write("{0}", Values[i]);
   }
}
```

In this case, the program copied the array's content using a *for* loop. The program could have, however, simply used the *Array* class *Copy* method, as shown here:

```
Array.Copy(Values, 0, Temp, 0, Values.Length);
```

The *Copy* method lets you specify the first element in the array that you want to copy and the target element in the destination array, as well as the number of elements you want to copy.

After you compile and execute the *Resize* program, your screen will display the following output:

```
C:\Subdir> Resize  <Enter>
01234
0123400000
```

TAKING A CLOSER LOOK AT THE ARRAY TYPE 1565

As you have learned, with C#, all arrays are derived from the *System.Array* class. As such, the arrays you create can use the *Array* class members. In previous tips, for example, programs have used the *Length* member to determine the number of elements in an array and the *Copy* function to copy one array's elements to a second array. Table 1565 lists *Array* class methods you may want to call from within your applications to manipulate an array.

Method	Purpose
BinarySearch	Searches a sorted array for a specific value.
Clear	Assigns the value zero to a range of array elements.
Copy	Copies values from one array to another.
IndexOf	Searches an array and returns the index of the *first* occurrence of a specific value.
LastIndexOf	Searches an array and returns the index of the *last* occurrence of a specific value.
Reverse	Reverses the order of array elements.
Sort	Sorts the elements in an array in ascending order.

Table 1565 Array class methods.

SORTING AND SEARCHING AN ARRAY 1566

As you just learned, the *System.Array* class, from which all C# arrays are derived, provides methods your programs can use to sort and search an array. Sorting and searching are common array operations. The following program, *SortSearch.cs,* uses the I class *Sort* function to sort an array and then uses *BinarySearch* and *IndexOf* functions to search the array for specific values. The *BinarySearch* function requires that the array's elements are in order, whereas the *IndexOf* function does not:

```
using System;

class SortSearch
{
  static void Main()
   {
    int [] Values = new int [] {9, 1, 8, 2, 7, 3, 6, 4, 5};

    foreach (int i in Values)
      Console.Write(i);

    Console.WriteLine();

    Array.Reverse(Values);

    foreach (int i in Values)
      Console.Write(i);

    Console.WriteLine();

    Array.Sort(Values);
    foreach (int i in Values)
```

```
      Console.Write(i);

    Console.WriteLine("\nThe value 5 is at offset {0}",
      Array.BinarySearch(Values,0, Values.Length, 5));

    Console.WriteLine("The value 3 is at offset {0}",
      Array.IndexOf(Values, 3));
  }
}
```

After you compile and execute the program, your screen will display the following output:

```
C:\Subdir> SortSearch.cs  <Enter>
918273645
546372819
123456789
The value 5 is at offset 4
The value 3 is at offset 2
```

1567 Declaring Multidimensional Arrays Within C#

As you have learned, using arrays, your programs can store multiple values of the same type within one variable. C# arrays differ from arrays in C and C++ in that they are based on the *Array* type, which provides several methods your programs can call on to manipulate the array as well as built-in boundary checking.

As the information your applications store becomes more complex, there will be times when you must use multidimensional arrays to store information. You might, for example, use a 2D array to represent the positions within a tic-tac-toe board, or, you might use a 2D array to represent your company's sales, by day, for each month of the year.

To create a 3x3 array capable of representing the tic-tac-toe board, your program will use the following statement:

```
int [][] Board = new [3][];
```

When you declare a 2D array within C#, you do not specify the second dimension. Instead, you must later use the new operator to allocate the next dimension individually, as shown here:

```
Board[0] = new int [3];
Board[1] = new int [3];
Board[2] = new int [3];
```

After you allocate your array in this way, you can assign values to the array as follows:

```
Board[0][0] = 0;
Board[0][1] = 0;
Board[0][2] = 0;
Board[1][0] = 0;
Board[1][1] = 0;
Board[1][2] = 0;
Board[2][0] = 0;
Board[2][1] = 0;
Board[2][2] = 0;
```

The following program, *TicTacToe.cs*, assigns values to an array that represents a tic-tac-toe board and then displays the array's contents:

```
using System;

class TicTacToe
{
  static void Main()
   {
```

```
      int [][] Board = new int[3][];

      for (int i = 0; i < Board.Length; i++)
        Board[i] = new int[3];

      Board[0][0] = 1;   // x
      Board[0][1] = 0;   // space
      Board[0][2] = 2;   // y
      Board[1][0] = 0;   // space
      Board[1][1] = 1;   // x
      Board[1][2] = 2;   // y
      Board[2][0] = 2;   // 2
      Board[2][1] = 0;   // space
      Board[2][2] = 1;   // x

      for (int i = 0; i < 3; i++)
        {
            for (int j = 0; j < 3; j++)
              if (Board[i][j] == 0)
                  Console.Write("  ");
              else if (Board[i][j] == 1)
                  Console.Write("X ");
              else if (Board[i][j] == 2)
                  Console.Write("Y ");

            Console.WriteLine();
        }
    }
}
```

After you compile and execute this program, your screen will display the following output:

```
C:\Subdir> TicTacToe  <Enter>

X   Y
  X Y
Y   X
```

UNDERSTANDING "JAGGED" ARRAYS 1568

In Tip 1567, you learned how to create a 2D array within C#. As you learned, creating the array was a two-step process. First, you allocated the array's primary dimension. Then, you individually allocated the second dimension for each of the existing array elements. The array that you created in Tip 1567 was a 3x3 array. C#, however, does not require that each dimension of your arrays is the same. Assume, for example, that you must create an array to store values for each day of the year. To begin, you might create a 2D array named *Months*, as shown here:

```
int [][] Months = new int [12][];
```

As you can see, the array has 12 elements, one for each month of the year. Then, you might allocate the elements to hold the days for each month as follows:

```
Months[0]  = new int[31]; // January
Months[1]  = new int[29]; // Assume leap year
Months[2]  = new int[31]; // March
Months[3]  = new int[30]; // April
Months[4]  = new int[31]; // May
: :
Months[11]  = new int[31]; // December
```

When you create a multidimensional array, the number of elements in a dimension can vary. C# programmers refer to the array as a jagged array. The following program, *JaggedArray.cs*, creates a jagged array based on the number of days in each month. Then the program uses the *GetLength* method to determine the number of days of storage for each month:

```
using System;

class JaggedArray
{
   static void Main()
   {
      int [][] Months = new int [12][];

      Months[0] = new int [31];
      Months[1] = new int [29];
      Months[2] = new int [31];
      Months[3] = new int [30];
      Months[4] = new int [31];
      Months[5] = new int [30];
      Months[6] = new int [31];
      Months[7] = new int [31];
      Months[8] = new int [30];
      Months[9] = new int [31];
      Months[10] = new int [30];
      Months[11] = new int [31];

      for (int i = 0; i < 12; i++)
        Console.WriteLine("Month {0} has {1} days",
           i, Months[i].GetLength(0));
   }
}
```

After you compile and execute this program, your screen will display the following output:

```
C:\Subdir> JaggedArray  <Enter>
Month 0 has 31 days
Month 1 has 29 days
Month 2 has 31 days
Month 3 has 30 days
Month 4 has 31 days
Month 5 has 30 days
Month 6 has 31 days
Month 7 has 31 days
Month 8 has 30 days
Month 9 has 31 days
Month 10 has 30 days
Month 11 has 31 days
```

1569 CREATING A 3D OR 4D ARRAY

In the previous two tips, you created two different two-dimensional arrays. Within C#, you can create arrays of many dimensions. The following statement, for example, creates a three-dimensional array:

```
int [][][] xyz = new int [2][][];
```

In a similar way, the following statement would create a four-dimensional array:

```
int [][][][] xyz_time = new int [2][][][];
```

As before, after you create the array, you must initialize each dimension separately. As you create each dimension, you can create jagged arrays, as discussed in Tip 1568.

Assume, for example, that you create a program that tracks the number of hours you work on a project each day over the course of a year. You might, for example, expand the *Months* array to track hours for each day, as follows:

```
int [][][] Months = new int [12][][];

Months[0]  = new int [31][];
Months[1]  = new int [29][];
Months[2]  = new int [31][];
Months[3]  = new int [30][];
Months[4]  = new int [31][];
Months[5]  = new int [30][];
Months[6]  = new int [31][];
Months[7]  = new int [31][];
Months[8]  = new int [30][];
Months[9]  = new int [31][];
Months[10]  = new int [30][];
Months[11]  = new int [31][];
```

Because each day will have 24 possible hours to which you can assign a specific project, you can easily allocate the third dimension within a *for* loop, as follows:

```
for (int i = 0; i < 12; i++)
  for (int j = 0; j < Months[i].GetLength(0); j++)
    Months[i][j] = new int[24];
```

USING STRINGS IN C# 1570

Within C and C++, programs use NULL-terminated character arrays to store strings. C#, in contrast, provides a *string* type which can store 16-bit Unicode characters. The C# *string* type is actually an alias for the *System.String* class. Thc following statements create and initialize several *string* objects:

```
string Title = "Jamsa's C/C++/C# Programmer's Bible";
string Topic = "Examines C, C++, and C# programming";
string Section = "Using C# strings"
```

To determine the number of characters within a *string* object, your programs do not have to loop through the characters in search of the NULL. In fact, C# strings do not use the NULL character. Instead, your programs can use the *Length* property.

The following program, *StringDemo.cs*, illustrates the use C# strings:

```
using System;

class StringDemo
{
  static public string AssignIt(string Str)
   {
     return (Str);
   }

  static public void ChangeIt(ref string Str)
   {
     Str = "New string";
   }

  static public void NoChange(string Str)
   {
```

```
      Str = "New string";
    }

   static void Main()
    {
      string Title = "Jamsa's C/C++/C# Programmer's Bible";
      string Topic;
      string Day = "Tuesday";

      Console.WriteLine(Title);
      Console.WriteLine(Day);

      Topic = AssignIt("C# strings");
      Console.WriteLine(Topic);

      NoChange(Title);
      Console.WriteLine(Title);

      ChangeIt(ref Title);
      Console.WriteLine(Title);
    }
}
```

After you compile and execute this program, your screen will display the following output:

```
C:\Subdir> StringDemo  <Enter>
Jamsa's C/C++/C# Programmer's Bible
Tuesday
C# strings
Jamsa's C/C++/C# Programmer's Bible
New string
```

Note that the program assigns values to the strings at declaration using a function that returns a value of type *string*, and within a function that receives a reference parameter. When you create *string*, C# creates the *string* as a value object. To change the string's contents within a function, you must pass the *string* to the function by reference. If you examine the program's output, you will find that the function *NoChange*, which receives a value parameter, does not change the string's contents, whereas the function *ChangeIt*, which receives a reference, successfully assigns a new value to the *string*.

1571 COMMON C# STRING OPERATIONS

In Tip 1570, you learned how to create *string* objects within a C# program. When your programs work with strings, there are several operations you will perform on a regular basis, such as comparing strings, appending (concatenating) one string to another, and determining a string's length.

To determine a string's length, you do not search the string for the NULL character, as you would in C or C++. Instead, you use the *Length* property. To compare strings, you can use the double equal signs, much like you would compare two numbers. Finally to concatenate one string to another, you use the plus sign (+). The following program, *StringOperations.cs*, illustrates the use of several common *string* operations:

```
using System;

class StringOperations
{

  static void Main()
   {
```

```
      string WeekDay = "Tuesday";
      string Day = "1st";
      string Month = "January";
      string Year = "2002";

      string Today;

      Console.WriteLine("Today is {0} the {1} of {2} {3}",
           WeekDay, Day, Month, Year);

      string Date = Month + " " + Day + " " + Year;
      Console.WriteLine("Date: {0}", Date);

      if (WeekDay == "Tuesday")
        Console.WriteLine("It's Tuesday!");

   }
}
```

After you compile and execute this program, your screen will display the following output:

```
C:\Subdir> StringOperations  <Enter>
Today is Tuesday the 1st of January 2002
Date: January 1st 2002
It's Tuesday!
```

USING THE FORMAT SPECIFIERS TO FORMAT STRING DATA 1572

Within C and C++ programs, programmers use the *sprintf* function to format data within a character string:

```
sprintf(SomeString, "%7.2", SomeValue);
```

Within C# programs, you can use the *String.Format* function and the width and format specifiers you used with *Console.WriteLine* to format a character string's contents, as shown here:

```
StrFmt = String.Format("{0, 7:f4}", 0.123456789);
```

The following program, *StringFormat.cs*, illustrates the use of the *Format* function to format data within a character string:

```
using System;

class clsStringFormat
{
  static void Main()
   {
     string StrFmt;

     StrFmt = String.Format("{0, 7:f4}", 0.123456789);
     Console.WriteLine(StrFmt);

     StrFmt = String.Format("{0, 0:0.000}",
           0.123456789);
     Console.WriteLine(StrFmt);

     StrFmt = String.Format("{0, 7:d}", 1700);
     Console.WriteLine(StrFmt);
   }
}
```

After you compile and execute this program, the console window will display the following output:

```
C:\Subdir> StringFormat  <Enter>
0.1235
0.123
   1700
```

1573 CONVERTING A STRING TO UPPER- OR LOWERCASE

In the previous tip, you used the *String* class *Format* function to assign a formatted string to a *string* variable. The *String* class has several other methods your programs can use to perform common operations. For example, the following program, *ToUpperLower.cs*, uses the *String* class *ToUpper* and *ToLower* functions to convert a string's contents to upper- and lowercase:

```
using System;

class clsToUpperLower
{
  static void Main()
   {
     string Title = "Jamsa's C/C++/C# Programmer's Bible";

     Console.WriteLine(Title.ToLower());
     Console.WriteLine(Title.ToUpper());
     Console.WriteLine(Title);
   }
}
```

After you compile and execute this program, your screen will display the following output:

```
C:\Subdir> ToUpperLower  <Enter>
jamsa's c/c++/c# programmer's bible
JAMSA'S C/C++/C# PROGRAMMER'S BIBLE
Jamsa's C/C++/C# Programmer's Bible
```

1574 FINDING A CHARACTER WITHIN A STRING

When your programs work with strings, you may often need to search a string for the first or last occurrence of a specific character. Within C and C++, your programs will use functions such as *strchr* to perform such operations. Within C#, your programs can take advantage of the *String* class *IndexOf* and *LastIndexOf* methods. The following program, *FindCharacter.cs*, uses the *String* class methods to locate the first and last occurrence of the character C in the string "Jamsa's C/C++/C# Programmer's Bible:"

```
using System;

class clsFindCharacter
{
  static void Main()
   {
     string Title = "Jamsa's C/C++/C# Programmer's Bible";

     Console.WriteLine(Title);
     Console.WriteLine("First C is at {0}",
        Title.IndexOf('C'));
     Console.WriteLine("Last C is at {0}",
        Title.LastIndexOf('C'));
   }
}
```

After you compile and execute this program, your screen will display the following output:

```
C:\Subdir> FindCharacters  <Enter>
Jamsa's C/C++/C# Programmer's Bible
First C is at 8
Last C is at 14
```

INSERTING AND REMOVING SUBSTRINGS 1575

When you work with strings, there will be many times when you must insert or remove a substring into or from a string. Using the *String* class *Insert* method, you can insert a substring at the index location you specify. Likewise, using the *String* class *Remove* function lets you remove characters from a string. The following program, *SubString.cs*, uses the *String* class *Insert* and *Remove* methods to manipulate a character string:

```
using System;

class clsSubString
{
  static void Main()
   {
     string Title = "Jamsa's Programmer's Bible";

     Title = Title.Insert(8, "C/C++/C# ");

     Console.WriteLine(Title);

     Title = Title.Remove(0, 8);

     Console.WriteLine(Title);
   }
}
```

After you compile and execute this program, your screen will display the following output:

```
C:\Subdir>  SubString  <Enter>
Jamsa's C/C++/C# Programmer's Bible
C/C++/C# Programmer's Bible
```

DETERMINING HOW A STRING STARTS AND ENDS 1576

When your programs must parse character strings for substrings, there will often be times when you must know if a string starts with a specific substring. For example, assume that you have a list of cities, states, and zip codes, and you need to locate each occurrence of the city Houston. Using the *String* class *StartsWith* method, you can determine if the string starts with the substring "Houston" as shown here:

```
if (CityStateZip.StartsWith("Houston"))
  // Statements
```

In a similar way, you can use the *String* class *EndsWith* function to determine if a string ends with a specific substring. Using the *EndsWith* function, for example, you can determine if a *CityStateZip* string ends with the zip code 77469 as follows:

```
if (CityStateZip.StartsWith("Houston"))
  // Statements
```

The following program, *StartsEnds.cs*, illustrates the use of the *StartsWith* and *EndsWith* functions:

```
using System;

class clsStartsEnds
{
  static void Main()
   {
     string Office1 = "Houston, Texas 77479";
     string Office2 = "Las Vegas, Nevada 85023";

     if (Office1.StartsWith("Houston"))
       Console.WriteLine("{0} starts with Houston",
         Office1);
     else
       Console.WriteLine("{0} does not start with Houston",
         Office1);

     if (Office2.StartsWith("Houston"))
       Console.WriteLine("{0} starts with Houston",
         Office2);
     else
       Console.WriteLine("{0} does not start with Houston",
         Office2);

     if (Office1.EndsWith("85023"))
       Console.WriteLine("{0} ends with 85023", Office1);
     else
       Console.WriteLine("{0} does not end with 85023",
         Office1);

     if (Office2.EndsWith("85023"))
       Console.WriteLine("{0} ends with 85023", Office2);
     else
       Console.WriteLine("{0} does not end with 85023",
         Office2);
   }
}
```

After you compile and execute this program, your screen will display the following output:

```
C:\Subdir> StartsEnds  <Enter>
Houston, Texas 77479 starts with Houston
Las Vegas, Nevada 85023 does not start with Houston
Houston, Texas 77479 does not end with 85023
Las Vegas, Nevada 85023 ends with 85023
```

1577 Splitting a Delimited String

In several of the past few tips, you have learned to determine if a character string contains a substring and how to remove the substring. Often, programs must parse delimited strings that contain substrings that are separated by a specific character. For example, the following string separates a city, state, and zip code using a forward slash (/) character: "Houston/Texas/77469". Using the *String* class *Split* function, you can separate a delimited character string into an array of strings that contain each of the substrings. The following program, *ParseString.cs*, illustrates the use of the *Split* function:

```
using System;

class clsParseString
{
```

```
  static void Main()
   {
     string Office = "Houston/Texas/77479";

     string [] AddressParts;
     char [] Delimiter = new char[] {'/'};

     AddressParts = Office.Split(Delimiter, 3);

     foreach (string Piece in AddressParts)
       Console.WriteLine(Piece);
   }
}
```

After you compile and execute this program, your screen will display the following output:

```
C:\Subdir> ParseString  <Enter>
Houston
Texas
77479
```

CHANGING STRINGS AND ARRAYS WITHIN FUNCTIONS 1578

In Tip 1557, you learned that to change a variable within a function, you must pass the variable to the function by reference, using the C# *ref* keyword. By default, when your programs pass an array to a function, C# passes the array as a reference variable. As such, even if you do not specify the *ref* keywords, the function can change the array's contents. In contrast, when your programs pass a *string* variable to a function, the C# passes the variable by value, which means that if you want to change the variable's value, you must precede the variable name with *ref* keyword.

The following program, *StrArrayParam.cs*, passes an array to the function *ChangeArray* which reverses the array's contents. The program then passes a *string* variable (by reference) to the function *ChangeString*, which converts the string's characters to uppercase:

```
using System;

class StrArrayParam
{
  static void ChangeArray(int [] Values)
   {
     Array.Reverse(Values);
   }

  static void ChangeString(ref string SomeString)
   {
     SomeString = SomeString.ToUpper();
   }

  static void Main()
   {
     int [] Values = new int[] {1, 2, 3, 4, 5};
     string Title = "Jamsa's C/C++/C# Programmer's Bible";

     foreach (int Value in Values)
       Console.Write("{0} ", Value);

     Console.WriteLine("\n{0}", Title);

     ChangeArray(Values);
```

```
      ChangeString(ref Title);

      foreach (int Value in Values)
        Console.Write("{0} ", Value);

      Console.WriteLine("\n{0}", Title);
    }
  }
```

After you compile and execute this program, your screen will display the following output:

```
C:\Subdir> StrArrayParams  <Enter>
1 2 3 4 5
Jamsa's C/C++/C# Programmer's Bible
5 4 3 2 1
JAMSA'S C/C++/C# PROGRAMMER'S BIBLE
```

Take time to experiment with this program by removing, for example, the keyword *ref* that precedes the *string* variable within the function call and function definition. You will learn that if you do not pass the *string* variable by reference, the function cannot change the variable's contents.

1579 Revisiting C# Namespaces

Earlier in this book you used namespaces within C++ programs to better organize the scope of your variables, classes, and functions. C# and the .NET environment make extensive use of namespaces. Throughout the C# tips, you will make extensive use of the Common Language Runtime routines, which will require that your programs must include the namespaces via the *using* statement. Through the previous tips, for example, your programs have made extensive use of the *System* namespace.

As your programs become more complex, you will create your own namespaces by placing the statements within namespace block statements. The following application, for example, creates the MyStuff namespace:

```
namespace MyStuff
{
    //  Add object definitions here
}
```

To access an object that resides within a namespace (other than the current namespace), you simply place the object's name after a period and the namespace name, such as *MyStuff.SomeFunction()*.

As you have learned, the .NET environment encourages code sharing. As you begin to create your own libraries for use by other programs, you can avoid class name conflicts (which would occur when a class name you use within your library matches a class name within the application program). By placing your library's class definitions within a namespace, you reduce such conflicts.

1580 Creating Your Own Namespace Within C#

As you have learned, when you place the classes you create within a unique namespace, you reduce potential class name conflicts. To create a namespace within a C# application or library, you can use a namespace block statement, as shown here:

```
namespace SomeName
  {
     // Statements
  }
```

The following application, *NamespaceDemo.cs*, creates a namespace named Utility. Within the namespace, the code creates a class that contains several static functions that the program can later call without having to create an instance of the class. To use the class, the program precedes the class name with the *Utility* name statement at the start of the code:

```
using System;

namespace Utility
{
  class Messages
   {
      static public void SayHello()
       {
         Console.WriteLine("Hello, C# World!");
       }

      static public void SayGoodbye()
       {
         Console.WriteLine("C ya later!");
       }

      static public void SayMessage(string Message)
       {
         Console.WriteLine(Message);
       }
   };
}

  class NamespaceDemo
   {
      static void Main()
       {
         Utility.Messages.SayHello();
         Utility.Messages.SayMessage("Namespaces are easy");
         Utility.Messages.SayGoodbye();
   }
}
```

To eliminate the need to precede the Messages class with the *Utility* namespace, the program can use the *using Utility* statement as shown here:

```
using Utility;

class NamespaceDemo
{
  static void Main()
   {
         Messages.SayHello();
         Messages.SayMessage("Namespaces are easy");
         Messages.SayGoodbye();
   }
}
```

The program cannot use the statement using *Utility.Messages* because *Messages* is the name of a class, and not a sub-level namespace.

RESOLVING NAME AMBIGUITY 1581

As you have learned, namespaces help you reduce name conflicts. That said, there may be times when you use different namespaces when the C# compiler needs you to specify more information before it can determine to which class your code is referring. For example, the following program, *TwoMessageClasses.cs*, creates two namespaces, one named Hello, and one named Goodbye. Both namespaces contain a class named Message. However, because the classes reside in separate namespaces, the class names do not conflict when you precede the class names with the corresponding namespace. However, if the program specifies only the class name, as shown here, the compiler cannot determine the class you desire and will generate a syntax error:

```
using System;
using Hello;
using Goodbye;

class NamespaceDemo
{
  static void Main()
   {
     Messages.SayHello();

     Messages.SayMessage("Namespaces are easy");

     Messages.SayGoodbye();
   }
}

namespace Hello
{
  class Messages
   {
     static public void SayHello()
      {
         Console.WriteLine("Hello, C# World!");
      }

     static public void SayMessage(string Message)
      {
         Console.WriteLine("In Hello: {0}", Message);
      }
   };
}

namespace Goodbye
{
  class Messages
   {
      static public void SayGoodbye()
      {
          Console.WriteLine("C ya later!");
      }

     static public void SayMessage(string Message)
      {
        Console.WriteLine("In Goodbye: {0}", Message);
      }
   };
}
```

To resolve the syntax error, precede each class with the corresponding namespace, as shown here:

```
class NamespaceDemo
{
  static void Main()
   {
     Hello.Messages.SayHello();
```

```
      Hello.Messages.SayMessage("Namespaces are easy");

      Goodbye.Messages.SayGoodbye();
    }
}
```

NESTING ONE NAMESPACE WITHIN ANOTHER 1582

As you examine C# applications and class libraries, you may often encounter cases where a programmer has nested one namespace within another. Namespaces exist to help you resolve name conflicts. As the size and complexity of your applications or class libraries increase, you may find it convenient to nest one namespace within another. For example, the following application, *NestNamespace.cs*, creates a namespace named *Utility*. Within that namespace, the code defines the *Disk*, *File*, *Keyboard* namespaces:

```
class NamespaceDemo
{
  static void Main()
  {
    Utility.Disk.HardDisk.ShowDriveType();

    Utility.File.FileInfo.ShowFilename();

    Utility.Keyboard.Kbd.ShowStatus();
  }
}

namespace Utility
{
  namespace Disk
  {
    class HardDisk
    {
      static public void ShowDriveType()
      {
        System.Console.WriteLine("100Gb hard disk, "
              +  " 12ms access time");
      }
    };
  }

  namespace File
  {
    class FileInfo
    {
          static public void ShowFilename()
        {
          System.Console.WriteLine("Filename is" +
               " NestNamespaces.cs");
        }
      };
  }

  namespace Keyboard
  {
    class Kbd
    {
        static public void ShowStatus()
        {
          System.Console.WriteLine("Caps Lock Set");
```

```
            }
        };
    }
}
```

To call the functions, the code specifies the namespace, sub-level namespace, class name, and method name. The code could, in turn, specify using statements and then simply specify the class names, as shown here:

```
using Utility.Disk;
using Utility.File;
using Utility.Keyboard;

class NamespaceDemo
{
  static void Main()
   {
     HardDisk.ShowDriveType();

     FileInfo.ShowFilename();

     Kbd.ShowStatus();
  }
}
```

1583 Using Multiple Source Code Files

As your programs increase in size and complexity, there may be times when you will want to organize your program's source within two or more source files. As you have learned, C# does not support the *#include* preprocessor directive, which you might use within C or C++ to include an additional source file.

If you compile programs from the command line, the C# compiler will let you place each source code file's name within the compiler's command line:

```
C:\Subdir> csc  File1.cs File2.cs File3.cs  <Enter>
```

In Tip 1582, you created the *Utility* namespace which contained the *Disk*, *File*, and *Keyboard* nested namespaces. The following file, *Utility.cs*, contains the *Utility* namespace:

```
namespace Utility
{
  namespace Disk
   {
     class HardDisk
       {
         static public void ShowDriveType()
          {
               System.Console.WriteLine("100Gb hard disk, "
                   +  " 12ms access time");
          }
       };
   }

   namespace File
   {
     class FileInfo
       {
          static public void ShowFilename()
           {
             System.Console.WriteLine("Filename is" +
                  " NestNamespaces.cs");
```

```
            }
        };
    }

    namespace Keyboard
    {
        class Kbd
        {
            static public void ShowStatus()
            {
                System.Console.WriteLine("Caps Lock Set");
            }
        };
    }
}
```

Next, the following file, *UseUtil.cs*, contains the main program code that uses the *Utility* namespace:

```
using Utility.Disk;
using Utility.File;
using Utility.Keyboard;

class NamespaceDemo
{
  static void Main()
  {
    HardDisk.ShowDriveType();

    FileInfo.ShowFilename();

     Kbd.ShowStatus();
  }
}
```

To compile the program, you simply include both filenames within the compiler's command line, as shown here:

```
C:\Subdir> csc  UseUtil.cs  Utility.cs  <Enter>
```

REVISITING C# COMMENTS 1584

Like C and C++, C# supports the two types of comments. To place a comment at the end of a line, you can place the two slashes (//), as shown here:

```
int Employee_id; // Uniquely identifies each employee

double Employee_Salary;  // Monthly salary
```

To comment a section of code, you can use a block comment that starts with /* and later ends with */ as shown here:

```
/*
 * Utility namespace
 *
 * Written by: K. Jamsa 3/01/01.
 *
 * Provides a collection of disk and file utilities.
 *
 */
```

Within the Visual Studio, you can create a Web page that provides documentation for your program. Within the Web page, you view information on your classes, namespaces, and more, as shown in Figure 1584.

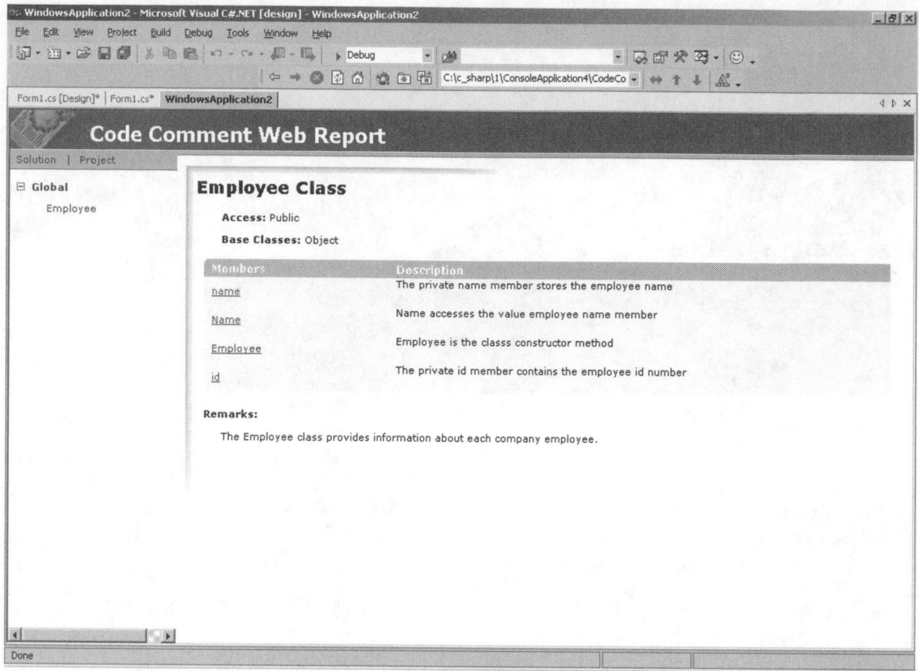

Figure 1584 *Viewing a program's Web documentation created by the Visual Studio.*

To direct the Visual Studio to use a program's documentation to create a Web page, select the Tools menu Build comment Web Pages option. The Visual Studio, in turn, will automatically create a Web page you can use to view your program's object, and documentation.

1585 C# Does Not Support Function Prototypes

In the C++ programs this book presents, the code makes extensive use of function prototypes that define a function's return and parameter types. Using function prototypes, an application can inform the C++ compiler of a function's use when the function is declared outside of the source file. Within C++, programs make extensive use of header files (which programmers sometimes refer to as include files) that define function prototypes. The header files contain the function prototypes for functions defined within run-time libraries.

C# does not support function prototypes. If you specify a function prototype within your code, the C# compiler will generate a syntax error. Within C#, every function must reside within a class. There is, therefore, no such thing as a "global" function. To use a function, you must specify the class name. If the C# compiler cannot locate the corresponding class, the compiler will generate a syntax error.

1586 C# Eliminates the Use of void in a Parameter List

Within C and C++ programs, you normally use the *void* keyword within a parameter list to inform the compiler (and other programmers who are reading your code) that the function does not receive parameters. For example, the following statements illustrate a common definition of the *main* function within C and C++:

```
void main(void) // void is not valid in a C# parameter list
  {
     // Statements
  }
```

Within C# programs, you only use the *void* keyword to specify that a function does not return a value. If you place the *void* keyword within a parameter list, the C# compiler will generate a syntax error. The following program, *MainVoid.cs*, defines the *Main* function with empty parentheses, to indicate the function does not support parameters. Experiment with the function code by placing the *void* keyword between the parentheses. When you later compile the program, the C# compiler will generate a syntax error:

```
using System;

class clsMainVoid
{
  static void Main()
   {
     Console.WriteLine("Hello, C#");
   }
}
```

OVERLOADING FUNCTIONS IN C# 1587

When you create functions, there will be many times when you find that you are creating the same functions for various data types. For example, you might create the *maximum* and *minimum* functions that return the larger and smaller of two values of type *int*, as shown here:

```
static public int Maximum(int a, int b)
 {
   return ((a > b) ? a: b);
 }
```

Later, you may decide that you need similar functions for values of type *double*:

```
static public double Maximum(double a, double b)
 {
   return ((a > b) ? a: b);
 }
```

Normally, functions must have unique names. However, in this case, the function names suit the operations the code is performing. Rather than forcing you to create a series of functions named *int_maximim*, *double_maximum*, *long_maximum*, and so on, C# lets you override the function names, using the same function names for the different data types. When the C# compiler compiles your code, it uses the parameters to functions to determine which function you want to use. The following program, *FunctionOverride.cs*, illustrates the use of function overloading:

```
using System;

class clsFunctionOverride
{
  static public int Maximum(int a, int b)
   {
     return ((a > b) ? a: b);
   }

  static public long Maximum(long a, long b)
   {
     return ((a > b) ? a: b);
   }

  static public double Maximum(double a, double b)
   {
     return ((a > b) ? a: b);
   }
```

```
    static void Main()
    {
      int a = 1001, b = 2002;
      double c = 10.0, d = 10.2;
      long e = 100000, f = 200000;

      Console.WriteLine("Max of  {0} and {1} is {2}",
           a, b, Maximum(a, b));
      Console.WriteLine("Max of  {0} and {1} is {2}",
           c, d, Maximum(c, d));
      Console.WriteLine("Max of  {0} and {1} is {2}",
           e, f, Maximum(e, f));
    }
}
```

After you compile and execute this program, your screen will display the following output:

```
C:\Subdir> FunctionOverride  <Enter>
Max of  1001 and 2002 is 2002
Max of  10 and 10.2 is 10.2
Max of  100000 and 200000 is 200000
```

When you override functions, the C# compiler creates a *signature* for each function that combines the function name (including the namespace and class names) with the parameter list. In the previous program, because the types within the parameter lists varied, the functions each had a unique signature. Should the C# compiler encounter two functions that have the same signature, the compiler will generate a syntax error.

1588 C# Does Not Support Bit Fields

Earlier in this book, you learned that the C and C++ programming languages support bit fields that let you use members of a structure to represent bit values of a specific length. For example, the following Date structure uses bit fields to represent month, day, and year:

```
struct Date                  // Not valid in C#
  {
     unsigned month:4;
     unsigned day:5;
     unsigned year: 7;
  };
```

Using bit fields lets programs reduce their storage requirements by representing information compactly. Unfortunately, C# does not support bit fields. If you try to use the bit-field syntax within a C# application, the compiler will generate a syntax error.

1589 In C# Structures Differ from Classes

Within this book's C++ tips, you learned that structures and classes were quite similar. In C#, you create a structure just as you do within C++ by defining the structure members, which can be variables or methods. The following program, *StructureDemo.cs*, creates a simple structure that contains variable and method members:

```
using System;

class clsStructureDemo
{
  struct Employee
   {
```

```
      public string Name;
      public int Age;
      public double Salary;

      public Employee(string name, int age, double salary)
       {
           Name = name;
           Age = age;
           Salary = salary;
       }

      public void ShowEmployee()
       {
           Console.WriteLine("Name {0}", Name);
           Console.WriteLine("Age {0}", Age);
           Console.WriteLine("Salary {0}", Salary);
       }
 };

      static void DoesntChangeIt(Employee E)
       {
           E.Salary = 100000;
       }

      static void ChangeIt(ref Employee E)
       {
           E.Salary = 100000;
       }

      static void Main()
       {
         Employee Boss = new Employee("Johnson", 60, 85000);

         Boss.ShowEmployee();

         DoesntChangeIt(Boss);
         Boss.ShowEmployee();

         ChangeIt(ref Boss);
         Boss.ShowEmployee();
       }
 }
```

Note that the program initializes the structure using a constructor method, much like you would initialize a class instance. When you use constructor methods with C# structures, the method must receive parameters. You cannot try to compile a *struct* constructor method that has an empty parameter list or the C# compiler will generate a syntax error.

When you create a class, C# places the object instance within the heap, which is managed by the Common Language Runtime. When you create a structure, in contrast, C# places the variable in the stack. In other words, a class creates a reference object and a structure creates a value object. Note that the program calls two functions, each of which assigns the value 100,000 to the *Salary* field. The first function, *DoesNotChangeIt*, receives a value parameter from the program (remember, a structure variable, unlike a class, is a value object). As such, the change the function makes to the field does not remain in effect after the function ends. The second function, *ChangeIt*, receives a reference parameter, which lets the function successfully change the structure field.

After you compile and execute this program, your screen will display the following output:

```
C:\Subdir> StructureDemo  <Enter>
Name Johnson
Age 60
Salary 85000
Name Johnson
Age 60
Salary 85000
Name Johnson
Age 60
Salary 100000
```

1590 REVISITING C# CLASSES

Throughout the C# tips you have examined thus far, you have placed the Main program within a variety of classes. You have also used several C# classes, such as *Array*, *String*, and *Console*. In the tips that follow, you will create and instantiate your own classes within C#. As you will learn, C# classes are quite similar to those you examined within C++. Using C# classes, you can take advantage of inheritance, polymorphism, and more.

C# is an object-oriented program. In fact, within C#, everything is an object—derived from the base *System.Object* class. As you have learned, unlike C++, you cannot create a C# program that does define at least one class.

If you are familiar with C++ classes, you may find many of the tips that follow present information that is familiar. However, there are some differences between C++ and C# classes you must note. C#, for example, does not support multiple inheritance.

Within C#, each class you create is a *reference* object that resides on a heap. As discussed, the Common Language Runtime (CLR) manages objects on the heap and will automatically discard the object when you are no longer using it.

1591 DEFINING A C# CLASS

As you have learned, to create a class within C#, you use a class definition that defines the class members:

```
class ClassName
  {
     // Class Methods and variables
  };
```

When you name your classes, you may want to precede the class name with the letters *cls* to indicate to other programmers that the identifier corresponds to a class, such as *clsEmployee* or *clsFiletype*.

Every C# program must have at least one class. The following program, *FirstClass.cs*, creates a simple class named *clsMessage*, which defines a constructor method, the *ShowMessage* method, and one variable member:

```
class clsMessage
{
  public clsMessage(string Msg)
   {
     Message = Msg;
   }

  public void ShowMessage()
   {
     System.Console.WriteLine(Message);
   }

  public string Message;
}
```

```
class clsFirstClass
{
  static void Main()
   {
     clsMessage Title = new clsMessage("Jamsa's C/C++/C#" +
                                       " Programmer's Bible");
     Title.ShowMessage();
   }
}
```

As you can see, the class defines a constructor method (that has the same name as the class) that initializes the class *Message* member. To create an instance of the class, the code uses the *new* operator.

USING NEW TO CREATE A CLASS INSTANCE 1592

As you learned within this book's C++ tips, creating a class instance is a two-step process. First, you must define the class. Then, you use the *new* operator to create the instance. The *new* operator, in turn, will allocate space on the heap that holds the object instance. Within the C# environment, the Common Language Runtime manages objects on the heap. Within C#, therefore, after you allocate space for an object instance using *new*, you do not later have to dispose of the space, as you must do within C++. Instead, the Common Language Runtime will determine when you are no longer using the object and will dispose of the object's memory for you automatically.

To create an object instance using the new operator, you specify the class type, class name, followed by an assignment operator and *new*, and then the class constructor, as shown here:

```
clsClassTypeName ClassVariable = new clsClassTypeName();
```

The following program, *CreateClassInstance.cs*, defines a simple class named *DemoClass* and then uses the *new* operator to create an instance of the class. Then the program uses one of the class methods to display the class member's value:

```
using System;

class DemoClass
{
  public DemoClass()
   {
     Value = 1700;
   }

  public void ShowValue()
   {
     System.Console.WriteLine("{0}", Value);
   }

  public int Value;
};

class clsCreateClassInstance
{
  static void Main()
   {
     DemoClass SomeVariable = new DemoClass();

     SomeVariable.ShowValue();
  }
}
```

The following program, *LimitMessage.cs*, for example, requires the user to change the *Message* variable either via the constructor method or the *AssignMessage* method. Within both methods, the code limits the message to 128 characters. If the message length exceeds 128 characters, the functions assign an empty string to the message:

```
class clsMessage
{
  public clsMessage(string Msg)
   {
      if (Msg.Length <= 128)
         Message = Msg;
       else
          Message = "";
   }

  public void AssignMessage(string Msg)
   {
      if (Msg.Length <= 128)
         Message = Msg;
       else
          Message = "";
   }

  public void ShowMessage()
   {
     System.Console.WriteLine(Message);
   }

  private string Message;
}

class clsLimitMessage
{
  static void Main()
   {
     clsMessage Title = new clsMessage("Jamsa's C/C++/C#" +
          " Programmer's Bible");
     Title.ShowMessage();

     Title.AssignMessage("Using Private Class Members");
     Title.ShowMessage();
   }
}
```

1597 Controlling Class Access Using public and internal

When you define a class, C# lets you specify an access level for the class itself. By default, if you do not specify an access control, C# uses *public* level access. There may be times, however, when you want to limit the code that can access the class. In such cases, you define the class as *internal*. When you use the *internal* keyword, C# will treat the class as public within the current module, but code outside of the module cannot access the class. The following file, *PublicInternal.cs*, defines two classes, one public and one internal. Within the module, C# treats the *internal* class as *public*, which means the other class can freely use the class. Outside of the module, however, code can use the class:

```
internal class clsSecretMessage
{
  public clsSecretMessage(string Txt)
   {
     Text = Txt;
   }
```

```
  public void AssignText(string Txt)
   {
     Text = Txt;
   }

  public void ShowText()
   {
     System.Console.WriteLine(Text);
   }

  private string Text;
}


public class clsMessage
{
  clsSecretMessage MessageHolder;

  public clsMessage(string Msg)
   {
     MessageHolder = new clsSecretMessage(Msg);
   }

  public void AssignMessage(string Msg)
   {
     MessageHolder.AssignText(Msg);
   }

  public void ShowMessage()
   {
     MessageHolder.ShowText();
   }
}
```

USING THE INTERNAL ACCESS CONTROL SPECIFIER 1598

In the previous Tip, you used the *internal* access control specifier to limit access to a class. Within a class, you can use the *internal* access control to limit access to class members as well, either methods or class variables. When you use the *internal* access control specifier with a class member, the member will appear as *public* to other classes within the module, but the member will not be accessible to code outside of the module. The following program, *InternalMembers.cs*, uses the *internal* access control for several class members, including a member variable. Note that because the *internal* member appears as *public* to code within the current module, the code can directly access the member variable using the dot operator:

```
class clsMessage
{
  internal clsMessage(string Msg)
   {
     Message = Msg;
   }

  internal void AssignMessage(string Msg)
   {
     Message = Msg;
   }

  internal void ShowMessage()
   {
```

```
      System.Console.WriteLine(Message);
    }

  internal string Message;
}


class clsLimitMessage
{
  static void Main()
   {
     clsMessage Title = new clsMessage("Jamsa's C/C++/C#" +
          " Programmer's Bible");
     Title.ShowMessage();

     Title.AssignMessage("Using Private Class Members");
     Title.ShowMessage();

     Title.Message = "Internal members are public";
     Title.ShowMessage();
   }
}
```

Note the last two lines of the program code. Because the *internal* variables are *public* within the current module, the code can directly access the variable's value using the dot operator. To prevent such access, you must define the member variable as *protected.*

1599 SUMMARIZING C# ACCESS CONTROL

In several of the previous tips, you have used the C# access control specifiers to limit access to classes and class members. In later tips, when you examine inheritance, you will examine the protected access control and the protected internal control. For now, Table 1599 briefly summarizes the C# access control specifiers.

Access Control	Restriction
internal	Restricts access to code within the current module.
private	Restricts access to code within the current class.
protected	Restricts access to code within the current class or classes derived from that class.
protected internal	Restricts access to the current class and classes derived from the class that reside in the current module.
public	Provides access to any class.

Table 1599 The C# access control specifers.

1600 INITIALIZING MEMBERS USING CONSTRUCTOR FUNCTIONS

As you learned in the C++ section of this book, to simplify the steps you must perform to initialize class member variables, object-oriented programming languages use constructor functions. As you will learn, C#'s use of constructor functions is similar to C++: the constructor name must be the same as the class name, you normally declare the constructor function as *public,* the constructor does not return a value, and so on.

The following program, *ConstructorDemo.cs,* creates a simple class named *clsBook* and uses a constructor function of the same name to initialize the class member variables:

```
using System;

class clsBook
{
  public clsBook(string title, string author, string topic)
   {
     Title = title;
     Author = author;
     Topic = topic;
   }

  public void ShowBook()
   {
     System.Console.WriteLine(Title);
     System.Console.WriteLine(Author);
     System.Console.WriteLine(Topic);
   }

  public string Title;
  public string Author;
  public string Topic;
};

class clsCreateClassInstance
{
  static void Main()
   {
     clsBook Book = new clsBook("Jamsa's C/C++/C#" +
           " Programmer's Bible", "Jamsa",
           "C, C++, and C# programming");
     Book.ShowBook();
   }
}
```

USING THE THIS KEYWORD 1601

As you learned in the C++ section of this book, when you use a constructor method to initialize class member variables, there may be times when the names of parameters you pass to the constructor are quite similar to the class member variable names. For example, the following constructor initializes a class name *clsBook* with three parameters whose names differ from the class members only by case:

```
class clsBook
{
  public clsBook(string title, string author, string topic)
   {
     Title = title;
     Author = author;
     Topic = topic;
   }

  // other class statements
}
```

To avoid confusion between the parameter names and the class member names, you may want to precede the class member names with the *this* keyword which tells the C# compiler (and other programmers who are reading your code) that you are referring to the current object instance. The following program, *UseThis.cs*, uses the *this* keyword within a constructor method:

```
using System;

class clsBook
{
  public clsBook(string Title, string Author, string Topic)
   {
     this.Title = Title;
     this.Author = Author;
     this.Topic = Topic;
   }

  public void ShowBook()
   {
     System.Console.WriteLine(Title);
     System.Console.WriteLine(Author);
     System.Console.WriteLine(Topic);
   }

  public string Title;
  public string Author;
  public string Topic;
};

class clsCreateClassInstance
{
  static void Main()
   {
     clsBook Book = new clsBook("Jamsa's C/C++/C#" +
           " Programmer's Bible", "Jamsa",
           "C, C++, and C# programming");
     Book.ShowBook();
   }
}
```

Note that in this case, the parameter names to the constructor method are identical, even in case, to the class member names. However, because the constructor precedes the class member names with the *this* keyword, the C# compiler (and programmers reviewing the code) knows which names correspond to the parameters and which to the class members.

1602 C# Supports Static Class Members

As does the C++ programming language, C# lets you define class methods and members as *static*. When you create a static class member variable, that variable's value is shared by each class instance you create. Further, you can use the variable's value even if the program does not create an object instance. Likewise, when you define a class method as *static*, C# lets you call the method even if you have not created a class instance. Throughout the previous tips, you have made extensive use of the *static* class method *Main*.

The following program, *StaticMember.cs*, defines a class that contains *static* member variable *Title* which contains the title of this book. The class also defines the static method *ShowTitle* which displays the variable's value. The program uses the class member variable and method without creating an instance of the class:

```
using System;

class clsBook
{
  public static void ShowTitle()
   {
     System.Console.WriteLine(Title);
```

```
    }

   public static string Title = "Jamsa's C/C++/C# +
      "Programmer's Bible";
};

class clsStaticMember
{
   static void Main()
    {
      clsBook.ShowTitle();
    }
}
```

CREATING A PRIVATE CONSTRUCTOR 1603

Normally, when you create a class constructor function, you use the *public* access specifier so a program can call the method to initialize class member variables. Sometimes, however, you may create a class simply to hold static class members or static methods. In such cases, you will not want the program to create a class instance. To prevent a program from creating an instance of a class, you can simply define the constructor function as private.

The following program, *PrivateConstructor.cs*, defines the class that contains static members and methods. The code class defines its constructor as *private* to prevent an instance of the class:

```
using System;

class clsBook
{
   private clsBook()
    {
    }

   public static void ShowBook()
    {
      System.Console.WriteLine(Title);
      System.Console.WriteLine(Author);
      System.Console.WriteLine(Topic);
    }

   public static string Title = "Jamsa's C/C++/C#" +
         " Programmer's Bible";
   public static string Author = "Jamsa";
   public static string Topic = "C, C++, and C#" +
         " programming";
};

class clsCreateClassInstance
{
   static void Main()
    {
      clsBook.ShowBook();
    }
}
```

USING A STATIC CONSTRUCTOR METHOD 1604

Within C#, unlike the C++ programming language, a class can declare a constructor method as *static*. A *static* constructor method can only access *static* properties, fields, and methods in the class.

When you use the *static* keyword for a constructor method, you cannot specify an access level such as *public* or *private*. Further, you cannot call the *static* constructor. Instead, the Common Language Runtime will call the function for you before you create an instance of the class and before you access a *static* class member, should you not create a class instance.

The following program, *StaticConstructor.cs*, illustrates the use of a *static* constructor method:

```
using System;

class clsBook
{
  static clsBook()
   {
     Title = "Jamsa's C/C++/C# Programmer's Bible";
   }

  public static string Title;
};

class clsStaticMember
{
  static void Main()
   {
     Console.WriteLine(clsBook.Title);
   }
}
```

1605 OVERLOADING CLASS CONSTRUCTOR FUNCTIONS

As the classes you create become more complex, you may want to provide programmers with different ways to call the constructor method in order to initialize different class members. For example, assume that you have a Message class that contains member variables that contain the message text, the font you want the program to use to display the message, and a text color. To simplify use of the class, you might create three different constructors, one that only requires a message, one that requires a message and a font, and one that requires a message, font, and color.

Within C#, you can overload the constructor function by defining multiple constructors. The following program, *OverloadConstructor.cs*, creates three different versions of the constructor method:

```
using System;

class clsMessage
{
   public clsMessage(string message)
   {
          Message = message;
          Font = "Arial";
          Color = "Black";
          }

   public clsMessage(string message, string font)
   {
          Message = message;
          Font = font;
          Color = "Black";
   }

   public clsMessage(string message, string font, string color)
```

```
    {
            Message = message;
            Font = font;
            Color = color;
    }

    public void ShowMessage()
    {
            System.Console.WriteLine(Message);
            System.Console.WriteLine(Font);
            System.Console.WriteLine(Color);
    }

    public string Message;
    public string Font;
    public string Color;
};

class clsStaticMember
{
    static void Main()
    {
            clsMessage Msg_A = new clsMessage("Hello");
            clsMessage Msg_B = new clsMessage("Hola", "Times");
            clsMessage Msg_C = new clsMessage("Good bye",
              "Helvetica", "Blue");

            Msg_A.ShowMessage();
            Msg_B.ShowMessage();
            Msg_C.ShowMessage();
    }
}
```

USING A DESTRUCTOR FUNCTION TO CLEAN UP AN OBJECT 1606

Just as C# calls a class constructor function when you create a class instance, C# calls a destructor function when the Common Language Runtime later discards the object. Depending on the processing your class performs, you may want to use the destructor function to update a log file or even to store the class information within a file. To create a destructor function, you create a method that contains the class name that you precede with the tilde character (~), as shown here:

```
~clsClassName()
{
    //  Statements here
}
```

When you create a destructor function, you do not specify a return type or an access specifier, such as *public* or *private*. Further, the constructor cannot receive parameter values. The following program, *DestructorDemo.cs*, illustrates the use of a destructor function. In this case, the destructor method simply displays a message to the console letting you know the function has been called:

```
class clsSample
{
  public clsSample(string name)
   {
     Name = name;
   }

  ~clsSample()
```

```
   {
     System.Console.WriteLine("Calling destructor for {0}",
         Name);
   }

  public string Name;
};

class clsStaticMember
{
  static void Main()
   {
     clsSample First = new clsSample("First");
     clsSample Second = new clsSample("Second");
     clsSample Third = new clsSample("Third");
   }
}
```

1607 Understanding the Finalize Method

In the previous tip, you learned how to create a destructor function that automatically executes before the Common Language Runtime discards an object. As you examine destructors, you may encounter discussions of the *Finalize* method. As it turns out, the *System.Object* class, from which all C# objects are defined, contains a *Finalize* method that exists to let the object perform clean-up processing before the Common Language Runtime discards the object. By default, the *Finalize* method does nothing.

Within some programming languages, you can override the *Object.Finalize* method. Within C#, however, you should not explicitly override the function. Instead, when you create a destructor function using the syntax shown in the previous tip, the C# compiler will override the function for you using your destructor's code. For example, assume that your class uses the following destructor code:

```
  ~clsSample()
   {
     System.Console.WriteLine("Calling destructor for {0}",
         Name);
   }
```

The C# compiler will use the code to override the *Finalize* method as follows:

```
protected override void Finalize()
{
   try
   {
     System.Console.WriteLine("Calling destructor for {0}",
         Name);
   }
   finally
   {
      base.Finalize();
   }
}
```

1608 Determining the Space Allocated on the Heap

When your programs create reference variables, the variables will reside within the heap, which the Common Language Runtime manages. As you allocate space on the heap, there may be times when you are interested in the amount of space that has been allocated. For example, assume that your program allocates 1Mb of heap space to hold a large object. Later, your program releases the object. By monitoring the available heap space, you may be able to determine when Garbage Collector has truly discarded the object.

To determine the amount of space allocated on the heap, your programs can call the *GC.GetTotalMemory* method. The following program, *HeapSpaceAllocated.cs*, uses the *GC.GetTotalMemory* method to display the heap's use as the program allocates and discards various objects:

```
using System;

class clsHeapSpaceAllocated
{
  static void Main()
   {
     Console.WriteLine("Heap allocated {0}",
        GC.GetTotalMemory(true));

     byte [] BigArray = new byte[1000000];
     Console.WriteLine("Heap allocated {0}",
        GC.GetTotalMemory(true));

     double [] DblArray = new double[1000000];
     Console.WriteLine("Heap allocated {0}",
        GC.GetTotalMemory(true));

     BigArray = new byte[1];
     Console.WriteLine("Heap allocated {0}",
        GC.GetTotalMemory(true));

     DblArray = new double[1];
     Console.WriteLine("Heap allocated {0}",
        GC.GetTotalMemory(true));
   }
}
```

The program passes the value *true* to the function to direct it to collect any available space before it returns the heap's use. After you compile and execute this program, your screen will display the following output:

```
C:\Subdir> HeapSpaceAllocated  <Enter>
Heap allocated 4120
Heap allocated 1012228
Heap allocated 9012240
Heap allocated 8012244
Heap allocated 12252
```

FORCING THE GARBAGE COLLECTOR TO COLLECT 1609

As you have learned, within the Common Language Runtime, a special program, the Garbage Collector, manages the heap, periodically discarding objects that are no longer in use. Before the Garbage Collector discards an object, the Garbage Collector will call the object's destructor method, which gives the object a chance to perform any "clean-up" processing it requires.

Depending on your program's processing, there may be times when you will want the Garbage Collector to run. Say, for example, your code has discarded several objects that consume a large amount of memory. By forcing the Garbage Collector to discard the objects, you can immediately free up memory. To force the Garbage Collector to run, your programs can call the *GC.Collect* method.

The following program creates three instances of an object. However, the program assigns each instance to the same variable, which means, after an assignment, the previous object is no longer in use, which means it is suitable for garbage collection. The program calls the *GC.Collect* function to force a collection:

```
using System;

class clsSample
{
  public clsSample(string name)
   {
     Name = name;
   }

  ~clsSample()
   {
     Console.WriteLine("Calling the destructor for {0}",
        Name);
   }

  public string Name;
};

class clsForceCollection
{
  static void Main()
   {
     Console.WriteLine("Heap allocated {0}",
        GC.GetTotalMemory(true));
     clsSample Object = new clsSample("First");
     Object = new clsSample("Second");
     GC.Collect();
     Object = new clsSample("Third");
     Console.WriteLine("Heap allocated {0}",
        GC.GetTotalMemory(true));
   }
}
```

After you compile and execute this program, your screen will display output similar to the following:

```
C:\Subdir> ForceCollection  <Enter>
Heap allocated 4208
Calling the destructor for First
Calling the destructor for Second
Heap allocated 12396
Calling the destructor for Third
```

When your program calls the *GC.Collect* function you may accelerate the timing of when the Garbage Collector will perform its task. However, the process will occur asynchronously to your program's execution, meaning, when your program calls the *GC.Collect* method, your program won't stop until the garbage collection is complete. Instead, your program will simply put the collection process into motion.

1610 UNDERSTANDING GARBAGE COLLECTION GENERATIONS

As the Garbage Collector within the Common Language Runtime performs its operations, it keeps track of objects by generations. Meaning, the Garbage Collector assigns a generation value to each object. The newest objects, for example, reside in generation 0 and the older objects in other generations. The following program, *GarbageGenerations.cs*, uses the *GC.GetGeneration* method to display the generation for various objects as the system executes. The program also uses the *GC.MaxGenerations* property to display the oldest generation the system will track:

```
using System;

class clsGarbageGenerations
{
  static void Main()
   {
     Console.WriteLine("Max Generations supported {0}",
          GC.MaxGeneration);

    byte [] FirstArray = new byte[1000000];
    Console.WriteLine("FirstArray is generation {0}",
          GC.GetGeneration(FirstArray));

     Console.WriteLine("Heap allocated {0}",
          GC.GetTotalMemory(true));

    double [] SecondArray = new double[1000000];
    Console.WriteLine("\nFirstArray is generation {0}",
          GC.GetGeneration(FirstArray));
    Console.WriteLine("SecondArray is generation {0}",
          GC.GetGeneration(SecondArray));
    Console.WriteLine("Heap allocated {0}",
          GC.GetTotalMemory(true));

    byte [] ThirdArray = new byte[1];
    Console.WriteLine("\nFirstArray is generation {0}",
          GC.GetGeneration(FirstArray));
    Console.WriteLine("SecondArray is generation {0}",
          GC.GetGeneration(SecondArray));
    Console.WriteLine("ThirdArray is generation {0}",
          GC.GetGeneration(ThirdArray));
    Console.WriteLine("Heap allocated {0}",
          GC.GetTotalMemory(true));

    double [] FourthArray = new double[1];
    Console.WriteLine("\nFirstArray is generation {0}",
          GC.GetGeneration(FirstArray));
    Console.WriteLine("SecondArray is generation {0}",
          GC.GetGeneration(SecondArray));
    Console.WriteLine("ThirdArray is generation {0}",
          GC.GetGeneration(ThirdArray));
    Console.WriteLine("FourthArray is generation {0}",
          GC.GetGeneration(FourthArray));
    Console.WriteLine("Heap allocated {0}",
          GC.GetTotalMemory(true));
   }
}
```

After you compile and execute this program, your screen will display the following output:

```
C:\Subdir> GarbageGenerations  <Enter>
Max Generations system supports 2
FirstArray is generation 2
Heap allocated 1012732

FirstArray is generation 2
SecondArray is generation 2
Heap allocated 9012732

FirstArray is generation 2
SecondArray is generation 2
```

```
ThirdArray is generation 0
Heap allocated 9012748

FirstArray is generation 2
SecondArray is generation 2
ThirdArray is generation 1
FourthArray is generation 0
Heap allocated 9012768
```

1611 Revisiting Class Inheritance

When you examined object-oriented programming within this book's tips on C++, you learned that by using *inheritance* you can build (derive) a new class based upon the members of an existing class (which programmers refer to as the *base* class). The new class, in turn, inherits access to the base class members.

C# supports class inheritance. To derive a new class from an existing class, within C#, you use the same format you used within C++, meaning, you specify the new class name, followed by a colon and the base class name, as shown here:

```
class clsNewClassName : clsBaseClassName
{
  // New class members beyond those in the base class
}
```

Normally, when you derive one class from another in this way, you add new members (methods, variables, or both) to the new class, which builds up on the base class. The following program, *PhoneNumber.cs*, creates a base class named *PhoneNumber*. The program then derives the *LongDistance* class from the base by adding support for an area code:

```
class clsPhoneNumber
{
  public clsPhoneNumber()
   {
     // Do nothing
   }

  public void ShowPhoneNumber()
   {
     System.Console.WriteLine(PhoneNumber);
   }

  protected string PhoneNumber;
};


class clsLongDistance : clsPhoneNumber
{
  public clsLongDistance(string AreaCodeDigits,
            string Digits)
   {
     AreaCode = AreaCodeDigits;
     base.PhoneNumber = Digits;
   }


  public void ShowAreacodeAndNumber()
   {
     System.Console.Write(AreaCode + "-");
     base.ShowPhoneNumber();
   }
```

```
  private string AreaCode;
};


class clsPhoneDemo
{
  static void Main()
   {
     clsLongDistance Information =
        new clsLongDistance(“800”, “555-1212”);

     Information.ShowAreacodeAndNumber();
   }
}
```

Note that this code does not use the base class constructor to initialize the base class, but rather, simply assigns the phone number to the base class member variable. In the next tip, you will learn how to call the base class constructor from the derived class constructor to simplify the member variable initialization.

To access the base class members (variable and methods) the derived class uses the C# *base* keyword. For example, to assign a value to the base class PhoneNumber variable, the derived class uses the following statement:

```
base.PhoneNumber = Digits;
```

Likewise, to call the base class ShowPhoneNumber method, the derived class uses the following:

```
base.ShowPhoneNumber();
```

In the following tip, you will learn how to call the base class constructor from the derived class to initialize base class members.

CALLING THE BASE CLASS CONSTRUCTOR FUNCTION 1612

As you know, constructor functions exist to simplify the process of initializing class member variables. When you derive a class from a base class, you will likely need to initialize one or more base class member variables. The easiest way to do so is to call the base class constructor function. To call the base class constructor within C#, you must use the *base* keyword as shown here:

```
public clsDerivedClass(int a, int b), : base(b)
  {
     SomeMember = a;
  }
```

In this case, the derived class constructor function receives two parameters, *a* and I. The derived class function passes the *b* parameter on to the base class constructor by invoking the special *base* method.

The following program, *DeriveMultiprocessor.cs*, derives a multiprocessor class from the base PC class:

```
class clsPC
{
  public clsPC(double speed, int RAM)
   {
     CPU_speed = speed;
     System_RAM = RAM;
   }

  public void ShowPC()
   {
     System.Console.WriteLine(“CPU {0}”, CPU_speed);
```

```
      System.Console.WriteLine("RAM {0}", System_RAM);
    }

   protected double CPU_speed;
   protected int System_RAM;
};


class clsMultiprocessor : clsPC
{
   public clsMultiprocessor(double speed1,
      double speed2, int RAM) : base(speed2, RAM)
    {
      CPU_A_speed = speed1;
    }

   public void ShowMultiprocessor()
    {
      System.Console.WriteLine("First CPU {0}",
         CPU_A_speed);
      base.ShowPC();
    }

   private double CPU_A_speed;
};


class clsDeriveMultiprocessor
{
   static void Main()
    {
      clsMultiprocessor FastPC = new clsMultiprocessor(1.4,
        1.4, 512);

      FastPC.ShowMultiprocessor();
    }
}
```

1613 Unlike C++, C# Does Not Allow Multiple Inheritance

Earlier in this book's C++ tips you used multiple inheritance to derive a class from two or more base classes. C#, however, does not support multiple inheritance. As you will learn later in these tips, using interfaces, you can achieve a form of multiple inheritance.

Multiple inheritance occurs when a class inherits properties from two or more classes at the same time. This is different than deriving a class from a base class, which itself is derived from another base. As you have learned, within C#, all classes are derived from the *System.Object* class. As such, when you use inheritance to derive an object, you create an inheritance chain that goes back to *System.Object*. The inheritance chain is not multiple inheritance, which happens when a class inherits properties from two or more classes at the same time. C++ supports multiple inheritance, whereas C# does not.

1614 A Closer Look at Constructor and Destructor Execution

As you have learned, when you derive a class from a base class, the derived class can call the base class constructor method to initialize base class members. Later, when you are done with the derived class and the Common Language Runtime is ready to discard the object, the Common Language Runtime function will call the object's destructor function.

A C# program calls destructors in the reverse order it calls constructors. First, the derived class destructor executes, then the base class. If there is more than one generation in the hierarchy, the most distant ancestor executes last.

The following program, *MonitorConstructorsAndDestructors.cs*, derives a class from a base class. Within each of the constructor and destructor methods for each class, the code simply displays a message to let you know the function has been called:

```
using System;

class clsBase
{
   public clsBase()
   {
         Console.WriteLine("In base constructor");
   }

   ~clsBase()
   {
         Console.WriteLine("In base destructor");
   }
};

class clsDerived : clsBase
{
   public clsDerived() : base()
   {
         Console.WriteLine("In derived class constructor");
   }

   ~clsDerived()
   {
         Console.WriteLine("In derived class destructor");
   }
};

class clsMonitorConstructorsAndDestructors
{
   static void Main()
   {
         clsDerived SomeObject = new clsDerived();
   }
}
```

After you compile and execute this program, your screen will display the following output:

```
C:\Subdir> MonitorConstructorsAndDestructors  <Enter>
In base constructor
In derived class constructor
In derived class destructor
In base destructor
```

As you can see, for constructor functions, C# will call the base class constructor methods first. Likewise for destructor methods, C# will call the base class destructor method last.

USING A SEALED CLASS TO PREVENT INHERITANCE 1615

As you create classes, there may be times when you do not want other classes to derive a new class from a specific class. In such cases, you can use the *sealed* keyword to prevent a class from being used as a base class. The *sealed* keyword marks the end of an inheritance chain:

```
sealed class clsBook
{
  public clsBook(string Title, string Author, string Topic)
   {
     this.Title = Title;
     this.Author = Author;
     this.Topic = Topic;
   }

  public void ShowBook()
   {
     System.Console.WriteLine(Title);
     System.Console.WriteLine(Author);
     System.Console.WriteLine(Topic);
  }

  public string Title;
  public string Author;
  public string Topic;
};
```

In this case, if you try to derive a new class from the *clsBook* class, the C# compiler will generate syntax errors and your code will not compile.

1616 USING VIRTUAL FUNCTIONS

When you use inheritance to derive one class from another, the derived class will inherit the base class methods and member variables. There may be times, however, when you may want the derived class to override a base class method. In such cases, you can declare the method as *virtual*. If the derived class implements the corresponding function, C# will use the derived class method. If the derived class does not implement the function, C# will use the base class method. As you will learn, virtual functions play a key role in polymorphic objects, which may change forms as a program executes. The following program, *UseVirtual.cs*, illustrates the use of *virtual* functions. The program will derive two classes from a base class. The first class will override the base class *virtual* function, whereas the second will not:

```
class clsTheBaseClass
{
  public clsTheBaseClass(string Info)
   {
     Text = Info;
   }

  public virtual void ShowInfo()
   {
     System.Console.WriteLine(Text);
   }

  protected string Text;
};

class clsDerivedA : clsTheBaseClass
{
  public clsDerivedA(string data) : base(data)
   {
   }

   public override void ShowInfo()
   {
```

```
      System.Console.WriteLine("In Overridden function: {0}",
          Text);
    }
};

class clsDerivedB : clsTheBaseClass
{
  public clsDerivedB(string data) : base(data)
    {
    }

};

class clsUseVirtual
{
  static void Main()
    {
      clsDerivedA A = new clsDerivedA("Class A");
      clsDerivedB B = new clsDerivedB("Class B");

      A.ShowInfo();
      B.ShowInfo();
    }
}
```

Within the class *clsDerivedA*, the class defines a ShowInfo method that overrides the base class virtual function. Note that the code places the *override* before the function name to inform the compiler that it wants to override the *virtual* function. After you compile and execute this program, your screen will display the following output:

```
C:\Subdir> UseVirtual  <Enter>
In Overridden function: Class A
Class B
```

As you can see, the *DerivedA* class uses its own method to display the output, whereas *DerivedB* uses the base class method.

REVISITING OBJECT POLYMORPHISM 1617

As you learned in the C++ section of this book, a polymorphic object has the ability to change forms as the program executes. C# supports polymorphic objects. The following program, *PolyPhone.cs*, creates a polymorphic *Phone* object that can change forms to become either a rotary phone or a tone phone:

```
class clsPhone
{
  public clsPhone(string number)
   {
     Number = number;
   }

  public virtual void Dial(string NumberToCall)
   {
     System.Console.WriteLine("Beep, Beep: Tone Dial");

         System.Console.WriteLine("Calling {0}",
           NumberToCall);
   }

  protected string Number;
};
```

```
class clsRotaryPhone : clsPhone
{
  public clsRotaryPhone(string number) : base(number)
   {
   }

  public override void Dial(string NumberToCall)
   {
    System.Console.WriteLine("In Rotary Dial calling: {0}",
       NumberToCall);
   }
};

class clsTonePhone : clsPhone
{
    public clsTonePhone(string number) : base(number)
    {
    }
};

class clsUseVirtual
{
  static void Main()
   {
     clsTonePhone Cellular = new clsTonePhone("555-1212");
     clsRotaryPhone GrandmasPhone  = new
         clsRotaryPhone("555-5555");

     Cellular.Dial("800-555-1212");
     GrandmasPhone.Dial("800-555-1212");

     System.Console.WriteLine("\n\nAbout to Create a " +
          "Polymorphic phone");

     clsPhone Phone = Cellular;
     Phone.Dial("800-555-1212");

     Phone = GrandmasPhone;
     Phone.Dial("800-555-1212");
   }
}
```

As you can see, the code defines a *clsPhone* object to which it assigns a tone-dial phone object to place a call, and then a rotary-dial phone object to place a call. In other words, during the program's execution the *Phone* object changes forms, from a tone-dial phone to a rotary phone. After you compile and execute this program, your screen will display the following output:

```
C:\Subdir> PolyPhone  <Enter>
Beep, Beep: Tone Dial
Calling 800-555-1212
In Rotary Dial, calling: 800-555-1212


About to Create a Polymorphic phone
Beep, Beep: Tone Dial
Calling 800-555-1212
In Rotary Dial, calling: 800-555-1212
```

USING ABSTRACT CLASS METHODS IN C# 1618

In the previous tips, you used *virtual* functions and the *override* keyword to let a derived class override a base class method. Just as there may be times when you will want to provide a derived class with the ability to override a method, there will also be times when you will want to force a derived class to implement a base class method. In such cases, your applications must define the base class and the method as an *abstract*, which tells the compiler that a derived class must implement the method and which prevents an application from creating an instance of the base class. The following program, *UseAbstract.cs*, illustrates the use of an *abstract* method within C#:

```
abstract class clsBase
{
   public clsBase(string text)
   {
         Content = text;
   }

   public abstract void ShowInfo();

   protected string Content;
};

class clsDerived : clsBase
{
   public clsDerived(string text) : base(text)
   {
   }

   public override void ShowInfo()
   {
         System.Console.WriteLine(Content);
   }
};

class clsUseAbstract
{
   static void Main()
    {
     clsDerived A = new clsDerived("Implements the" +
        " abstract method");

     A.ShowInfo();
   }
}
```

C# DOES NOT SUPPORT CLASS FRIENDS 1619

Within C++ programs, class can define another class (or a method within another class) as a *friend*, which it allows to access its own private class members (or specific members). The following C++ class definition, for example, defines a class named Laurel and within that class defines a friend named Hardy:

```
class Laurel {
  public:
    friend Hardy;// Not valid in C#
    // Other class members
  private:
    // private members here
};
```

C# does not support class friends. In fact, C# has done away with the *friend* reserved word. If you try to declare a *friend* class within a C# application, the compiler will generate a syntax error.

1620 USING REFLECTION TO QUERY A CLASS

As you have learned, C# provides the ability for a program to query a class to determine the members and methods the class exposes for use. Programmers refer to the ability of a class to handle such queries as *reflection*. When you create a class within C#, your class automatically receives the ability to respond to such queries. For example, the following program, *QueryMyClass.cs*, creates a class named *Demo* that contains several member variables and methods. The program uses reflection to display the names of the class members and methods:

```
using System;
using System.Reflection;

class Demo
{
   public int MemberOne;
   public int MemberTwo;
   public int MemberThree;
   private int MemberFour;
   private int MemberFive;

   public void MethodOne()
   {
   }

   public void MethodTwo()
   {
   }

   private void MethodThree()
   {
   }
};

class clsQueryMyClass
{
   static void Main()
   {
          Type t = typeof(Demo);

          MethodInfo [] Methods = t.GetMethods();

          Console.WriteLine("Methods");

          foreach (MethodInfo Method in Methods)
                 Console.WriteLine("{0}", Method.Name);

          Console.WriteLine("\n\nMembers");
          MemberInfo [] Members = t.GetMembers();

          foreach (MemberInfo Member in Members)
          Console.WriteLine("{0}", Member.Name);
   }
}
```

After you compile and execute this program, your screen will display the following output:

```
C:\Subdir> QueryMyClass  <Enter>
```

```
Methods
GetHashCode
Equals
ToString
MethodOne
MethodTwo
GetType

Members
MemberOne
MemberTwo
MemberThree
GetHashCode
Equals
ToString
MethodOne
MethodTwo
GetType
```

As you can see, the program calls the *GetMethods* and *GetMembers* methods. Note that both functions return only the *public* methods. The methods do not return the member variables or methods the class defines as *private*. Also note that the *GetMembers* method returns member variables and methods. In the next tip, you will learn how to use the *GetFields* method to return only member variables. Also note that the functions return more methods and member variables than the class defines. That is because every class you declare in C# inherits members from the *System.Object* class, from which all classes are derived.

USING REFLECTION TO RECEIVE CLASS FIELDS 1621

In Tip 1620, you used reflection to query a program-defined class about its members. As you learned, when your programs call the *GetMembers* method, C# returns all public member variable and methods. Depending on your program's processing, there will be times when you will want only information about member variables. In such cases, you can direct the *GetFields* function to return only members of a specific type, or you can use query to further query each member. The following program, *GetClassFields.cs*, uses the *GetFields* method to retrieve the class member variables:

```
using System;
using System.Reflection;

class Demo
{
   public int MemberOne;
   public int MemberTwo;
   public int MemberThree;
   private int MemberFour;
   private int MemberFive;

   public void MethodOne()
   {
   }

   public void MethodTwo()
   {
   }

   private void MethodThree()
   {
   }
};
```

```
class clsGetClassFields
{
   static void Main()
   {
         Type t = typeof(Demo);

         FieldInfo [] Fields = t.GetFields();

         Console.WriteLine("Fields");

         foreach (FieldInfo Field in Fields)
         Console.WriteLine("{0}", Field.Name);
   }
}
```

After you compile and execute this program, your screen will display the following output:

```
C:\Subdir> GetClassFields  <Enter>
Fields
MemberOne
MemberTwo
MemberThree
```

1622 OVERLOADING OPERATORS WITHIN C#

In earlier tips, you learned how to overload functions within a C# application. In a similar way, there may be times when you create a class when you want to override a C# operator to change the operation the operator performs. In such cases, you can define a class method that implements the operation you desire. The following statement, for example, illustrates how you would overload the plus (+) operator:

```
Static public ReturnType operator +(clsSomeName x,
                                    clsSomeName y)
```

When you override an operator, you will normally use the *public* access specifier to make an operator accessible outside of the class definition. The number of arguments to method should match the number of operands the operator uses. Unlike C++, which lets you overload almost every operator, C# only lets you overload only the unary, binary, and comparison operators. C# does not let you overload the assignment operator.

The following program, *OperatorOverload.cs*, overloads the plus and minus operators for a class:

```
class clsItem
{
   public clsItem(string text, double price)
   {
         Item = text;
         Price = price;
   }

   static public double operator +(clsItem x, clsItem y)
   {
         return(x.Price + y.Price);
   }

   public void ShowItem()
   {
         System.Console.WriteLine("{0} ${1}", Item, Price);
   }

   private string Item;
```

```
   private double Price;
};

class clsOperatorOverload
{
   static void Main()
   {
         clsItem Purse = new clsItem("Hand bag", 25.00);
         clsItem Shoes = new clsItem("Loafers", 75.00);

         Purse.ShowItem();
         Shoes.ShowItem();
         System.Console.WriteLine("Total cost ${0}",
               Purse + Shoes);
   }
}
```

After you compile and execute this program, your screen will display the following output:

```
C:\Subdir> OperatorOverload  <Enter>
Hand bag $25
Loafers $75
Total cost $100
```

1623 DECLARING A CLASS ARRAY

In previous tips, you have created single and multidimensional arrays of various types. Within C#, just as you can create an array of *int* values, you can create an array of user-defined classes as well. The following statement, for example, creates an array capable of holding ten elements of type *DemoClass*:

```
DemoClass [] ArrayName = new DemoClass[10];
```

The following program, *ClassArray.cs*, creates an array that contains several class members and then uses array elements to access the various objects:

```
using System;

class DemoClass
{
  public DemoClass(int value)
   {
     Value = value;
   }

  public void ShowValue()
   {
     Console.WriteLine("{0}", Value);
   }

  int Value;
}

class clsClassArray
{
  public static void Main ()
   {
     DemoClass [] ClassArray = new DemoClass[10];

     for (int i = 0; i < ClassArray.Length; i++)
       ClassArray[i] = new DemoClass(i);
```

```
      foreach (DemoClass element in ClassArray)
        element.ShowValue();
    }
}
```

As you can see, the program declares the array *ClassArray* and then uses a *for* loop to create 10 instances of *DemoClass* objects. Next, the code uses a *foreach* loop to move through the array of class objects, one object at a time. Within the *foreach* loop, the code uses the *ShowValue* method to display the current object's value.

1624 BOXING A C# VARIABLE

As you have learned, C# stores reference variables within the heap, which the Common Language Runtime manages. Value variables, in contrast, reside in the stack. Within C#, everything is an object. To simplify operations, C# assigns each value type with a hidden (or behind-the-scenes) reference type, which C# automatically creates when a program casts the value-type variable to a reference type. C# refers to the process of converting a value-type variable to a reference variable as *boxing* the variable.

To box a variable, you simply assign the variable to a variable of type *object*, as shown here:

```
int SomeVariable = 1701;

object ReferenceValue = SomeVariable;  // Box the variable
```

The following program, *BoxVariable.cs*, boxes a variable of type *int* and a variable of type *float* and then passes the reference variables to a function that displays the object's value:

```
using System;

class clsBoxVariable
{
  public static void ShowValue(object ReferenceValue)
   {
     Console.WriteLine(ReferenceValue.ToString());
   }

  public static void Main ()
   {
     int iValue = 1700;
     double fValue = 1.1;

     object a, b;

     a = iValue;  // Box the variable
     b = fValue;  // Box the variable

     ShowValue(a);
     ShowValue(b);
   }
}
```

1625 UNBOXING A C# VARIABLE

In Tip 1624, you learned how to box to a variable by converting a value-type variable to a reference variable. In a similar manner, C# refers to the process of converting a reference-type variable back to a value-type variable as *unboxing*. The following program, *UnboxDemo.cs*, boxes a variable of type *int* that the code passes to the function

ChangeValues. Within the function, the code changes the reference variable's value. The code then unboxes the reference variable by converting it back to a value-type variable:

```
using System;

class clsBoxVariable
{
  public static void ChangeValue(ref object ReferenceValue)
   {
     ReferenceValue = 2002;
   }

  public static void Main ()
   {
     int iValue = 1700;
     object a;

     a = iValue;  // Box the variable

     ChangeValue(ref a);

     iValue = (int) a;

     Console.WriteLine("{0}", iValue);
   }
}
```

After you compile and execute this program, your screen will display the following output:

```
C:\Subdir> UnboxDemo  <Enter>
2002
```

UNDERSTANDING STANDARD CLASS CONSTRUCTORS 1626

In several of the previous examples, you have made extensive use of class constructor methods to initialize class member variables. Within C#, as you have learned, everything is an object, including the standard types, such as *int* and *double*. Within C#, therefore, the standard types, like other classes, support class constructor methods. The standard constructor functions assign the value 0 to the variables you create. Unfortunately, the default standard constructors are parameterless functions. As such, you cannot use the constructor function to initialize a standard-type variable when you create them:

```
int Count = new int(1700);     // C# does not let you use a
                               // default standard-type
                               // constructor method to
                               // initialize a variable
```

USING THE STACK CLASS 1627

In earlier tips, you used the C# *Array* class to store two or more values of the same type. As you have learned, after you declare an array, you cannot resize the *Array* object. For cases when you need a dynamic data structure whose size can change as your program executes, your programs can use a *Stack* object. The *Stack* object type stores values of type *Object* (the base object type from which other C# objects are derived). To declare a *Stack* object, you use a declaration similar to the following:

```
Stack Values = new Stack();
```

Depending on your application's storage needs, you can increase or decrease the stack object's size by passing a size to the Stack constructor method:

```
Stack Values = new Stack(1000);
```

Within a *Stack* object, you can store values of various types, meaning you can place values of type *int*, *double*, *String*, as well as other object types. To place a value into the stack, you use the *Push* method. To remove the top element from the stack, you use the *Pop* method. When you place various value types within a stack, you may need to later convert the value back to the original type when you perform a pop operation. For example, the following statement pops a value of type *double* from a stack object:

```
double Value;

Value = Convert.ToDouble(SomeStack.Pop());
```

The following program, *StackDemo.cs*, illustrates the use of *Stack* objects. The program uses the *Stack* object Count method to track the number of items in the stack. Also to use the *Stack* object, the code must use the *System.Collections* namespace:

```
using System;
using System.Collections;

public class StackDemo
{
   public static void Main()
   {
          // Create and initialize a new Stack
          Stack Values = new Stack();

          Values.Push(100);
          Values.Push(200);
          Values.Push(300);
          Values.Push(400);
          Values.Push(500);

          int Count = Values.Count;  // Get the number of
                                     // items in the stack

       for (int i = 0; i < Count; i++)
            Console.WriteLine("{0}", Values.Pop());
   }
}
```

A *Stack* object creates a last in, first out data structure, meaning the last value that you place into the stack is the first value that you remove. After you compile and execute this program, your screen will display the following output:

```
C:\Subdir> StackDemo  <Enter>
500
400
300
200
100
```

1628 CREATING A NAMESPACE ALIAS

Throughout this book's C# programs, you have made extensive use of namespaces. Using namespaces, you can better organize the identifiers you use within your code. As you have learned, C# lets you create multiple namespace levels, as shown here:

```
// Examples of custom namespaces

using Utilities.Disk.File.BinaryOperations;
using Utilities.Printer.Fonts.Truetype;
using Networks.Administrator.Toolkit;
```

Within your code, you can take advantage of the *using* keyword as just shown, or you can specify the complete namespace to an object as shown here:

```
Utilities.Disk.File.BinaryOperations.Stream.Open("File");
```

Rather than using a complete namespace name in this way, C# lets you create a namespace alias—an abbreviated name—you can use to reference the namespace. To create a namespace alias, you use the *using* keyword as shown here:

```
using FileOps = Utilities.Disk.File.Binary.Operations;
```

After you create the namespace alias, you can use the alias within your code in place of the complete namespace, as shown here:

```
FileOps.Stream.Open("File");
```

YET ANOTHER MULTIPLE ARRAY FORMAT 1629

Earlier in this book, you learned how to create multidimensional arrays in C#. For example, the following statements create a 3x3 array of integer values:

```
int [][] Board = new int [3][];

Board[0] = new int [3];
Board[1] = new int [3];
Board[2] = new int [3];
```

In addition to using this format to define multidimensional arrays, C# lets you use the following format to create a two-dimensional array:

```
int [,] Board = new int [3,3];
```

The following program, *Fill2Darray.cs*, creates a two-dimensional array using the format just shown and then assigns values to each array element:

```
class clsFill2Darray
{
   static void Main()
   {
         int [,] Board = new int [3,3];
       int x = 0;

         for (int i = 0; i <= 2; i++)
                for (int j = 0; j <= 2; j++)
                   Board[i,j] = x++;

         for (int i = 0; i <= 2; i++)
                for (int j = 0; j <= 2; j++)
                      System.Console.WriteLine(Board[i,j]);
   }
}
```

After you compile and execute this program, your screen will display the following output:

```
C:\Subdir> Fill2Darray  <Enter>
0
1
2
3
4
5
6
7
8
```

1630 Handling Stack Object Exceptions

As you learned in Tip 1629, you store values within a *Stack* object using the *Push* method. Later, when your program must retrieve values from the *Stack*, your programs use the *Pop* method. Using the *Count* method, programs can determine the number of objects that reside in the *Stack*. If a program tries to pop a value from an empty stack, C# will generate an exception. The following program, *PopException.cs*, places three objects in the stack and then later tries to pop four values, which causes an exception. In this case, however, the program detects and handles the exception:

```
using System;
using System.Collections;

public class StackDemo
 {
   public static void Main()
    {
      // Create and initialize a new Stack
      Stack Values = new Stack();

      Values.Push("Jamsa's C/C++/C#");
      Values.Push("Programmer's");
      Values.Push("Bible");

      try
       {
         for (int i = 0; i < 4; i++)
           Console.WriteLine("{0}", Values.Pop());
       }
      catch (Exception e)
       {
         Console.WriteLine("Exception in Pop operation");
       }
    }
 }
```

After you compile and execute this program, your screen will display the following output:

```
C:\Subdir> PopException  <Enter>
Bible
Programmer's
Jamsa's C/C++/C#
Exception in Pop operation
```

TAKING A CLOSER LOOK AT C# PREPROCESSING 1631

As you have learned, the C# compiler does not support C and C++ directives such as *#include*. As it turns out, the C# compiler does support some "preprocessor-like" directives. However, the C# does not have an actual "preprocessor" step.

For example, within C#, you cannot use the *#define* directive to create a constant or macro definition, as you might in C or C++. You can, however, use the *#define* directive to define an identifier you can use to control C# preprocessing. Assume, for example, you want to define a directive named *DEBUGGING* that you use to enable or disable the inclusion of debug error messages. To do so, you would use the *#define* directive as follows:

```
#define DEBUGGING
```

Later, within your source code, you can use the *#if* directive to perform specific processing if the *DEBUGGING* identifier is defined, as shown here:

```
#if DEBUGGING
  System.Console.WriteLine("Some message information");
#endif
```

Within C#, you can use the *#define*, *#undef*, *#if*, *#elif*, *#else*, and *#endif* directives.

UNDERSTANDING PREPROCESSOR REGIONS 1632

As you examine C# programs, there may be times when you encounter *#region* and *#endregion* preprocessor directives. In general, these two directives let programmers mark specific sections of code:

```
#region Optional Text
  // Program statements here
#endregion Optional Text
```

C#, itself, makes no use of programmer-defined regions. Instead, the directives exist to support other programming tools, such as a performance profiler or automated comment generator.

USING THE IS OPERATOR 1633

When you write code to work with a variety of objects, there may be times when you must know if an object is a specific object type or if the object is derived from a specific type. In such cases, your programs can use the C# *is* operator. For example, the following statement tests if an object is a *ToneDial* object (or has been derived from the *ToneDial* class):

```
if (PhoneObject is ToneDial)
  {
    // Statements
  }
```

The following program, *IsOperator.cs*, uses the *is* operator to examine a polymorphic object to determine the object's current type:

```
class clsPhone
{
  public clsPhone(string number)
   {
     Number = number;
   }

  public virtual void Dial(string NumberToCall)
```

```
    {
      System.Console.WriteLine("Beep, Beep: Tone Dial");

      System.Console.WriteLine("Calling {0}", NumberToCall);
    }

  protected string Number;
};

class clsRotaryPhone : clsPhone
{
  public clsRotaryPhone(string number) : base(number)
    {
    }

  public override void Dial(string NumberToCall)
    {
      System.Console.WriteLine("Rotary Dial, calling: {0}",
                 NumberToCall);
    }
};

class clsTonePhone : clsPhone
{
  public clsTonePhone(string number) : base(number)
    {
    }
};

class clsIsOperator
{
  static void Main()
    {
      clsTonePhone Cellular = new clsTonePhone("555-1212");
      clsRotaryPhone GrandmasPhone =
         new clsRotaryPhone("555-5555");

      Cellular.Dial("800-555-1212");
      GrandmasPhone.Dial("800-555-1212");

      System.Console.WriteLine("\n\nAbout to Create a" +
          " Polymorphic phone");

      clsPhone Phone = Cellular;
      Phone.Dial("800-555-1212");

      if (Phone is clsTonePhone)
       System.Console.WriteLine("The object is a" +
          " tone phone");

      Phone = GrandmasPhone;
      Phone.Dial("800-555-1212");

      if (Phone is clsRotaryPhone)
        System.Console.WriteLine("The object is a" +
          " rotary phone");
    }
}
```

After you compile and execute this program, your screen will display the following output:

```
C:\Subdir> IsOperator  <Enter>
Beep, Beep: Tone Dial
Calling 800-555-1212
In Rotary Dial, calling: 800-555-1212

About to Create a Polymorphic phone
Beep, Beep: Tone Dial
Calling 800-555-1212
The object is a tone phone
In Rotary Dial, calling: 800-555-1212
The object is a rotary phone
```

UNDERSTANDING CLASS CONVERSION AND THE AS OPERATOR 1634

As you know, when a C# program assigns a value of one type to a variable of a different type, C# performs a cast operation. Normally, C# will perform implicit downcast operations, meaning, C# does not require the program code to perform any type of conversion when the code assigns a value of a "lower" type to a "higher" type. For example, C# will implicitly convert a value of type *int* to a value of type *double*. C#, however, will not implicitly cast a value from a higher type to a lower type. To perform such conversions, your code must perform an explicit conversion.

In a similar way, C# lets you assign a value of one class type to another, provided the classes relate in some way through inheritance. For example, C# will perform an implicit *upcast* operation to convert a derived class to its base class. In contrast, a C# program must explicitly *downcast* a base class to a class derived from the base.

When a downcast operation fails, C# will generate an *InvalidCastException* exception. Using the C# *as* operator, a program can direct C# to assign a class the *null* value when a downcast operation fails, as opposed to generating an exception:

```
BaseClassVariable = DerivedClassVariable as DerivedClassName;
```

The following program, *AsOperator.cs*, illustrates the use of the *as* operator:

```
using System;

class clsPhone
{
  public clsPhone(string number)
    {
      Number = number;
    }

  public virtual void Dial(string NumberToCall)
    {
      System.Console.WriteLine("Beep, Beep: Tone Dial");

      System.Console.WriteLine("Calling {0}", NumberToCall);
    }

  protected string Number;
};

class clsRotaryPhone : clsPhone
{
  public clsRotaryPhone(string number) : base(number)
    {
    }

  public override void Dial(string NumberToCall)
    {
```

```
      System.Console.WriteLine("Rotary Dial, calling: {0}",
           NumberToCall);
    }
};

class clsAsOperator
{
  static void Main()
    {
      clsPhone BasePhone = new clsPhone("555-1212");
      clsRotaryPhone DerivedPhone  = new
         clsRotaryPhone("555-5555");

      // Try downcast
      DerivedPhone = BasePhone as clsRotaryPhone;

      if (DerivedPhone == null)
        System.Console.WriteLine("DerivedPhone is NULL");
      else
        System.Console.WriteLine("BasePhone is downcast");

      // Try downcast
      try
       {
         DerivedPhone = (clsRotaryPhone) BasePhone;
       }
     catch (Exception e)
       {
         System.Console.WriteLine("Invalid cast" +
             " exception caught");
       }

     DerivedPhone  = new clsRotaryPhone("555-5555");

     // Try upcast
     BasePhone = DerivedPhone as clsPhone;

     if (BasePhone == null)
       System.Console.WriteLine("BasePhone is NULL");
     else
       System.Console.WriteLine("DerivedPhone has" +
           " been upcast");

     try
      {
        BasePhone = (clsPhone) DerivedPhone;
        System.Console.WriteLine("DerivedPhone again" +
            " upcast");
      }
    catch (Exception e)
      {
        System.Console.WriteLine("Invalid cast" +
           " exception caught");
      }
    }
}
```

After you compile and execute this program, your screen will display the following output:

```
C:\Subdir> AsOperator  <Enter>
DerivedPhone is NULL
Invalid cast exception caught
```

```
BasePhone is upcast
BasePhone is again upcast
```

C# DOES NOT SUPPORT DEFAULT PARAMETER VALUES 1635

Within this book's C++ tips, you took advantage of default parameters for functions. For example, the following C++ function, *InitializeParameters*, assigns default values to its three parameters:

```
// C# does not support default parameter values

void InitializeParameters(int a=0, int b = 1, int c = 2)
 {
     cout << a << endl;
     cout << a << endl;
     cout << a << endl;
 }
```

Within a C++ program, an application can invoke the function with 1, 2, or 3 parameter values, as shown here:

```
// C# does not support default parameter values

InitializeParameters(100, 200, 300);
InitializeParameters(100, 200);
InitializeParameters(100);
```

If you try to use default parameters within a C# program, the compiler will generate syntax errors and will not compile the code.

UNDERSTANDING THE FIXED KEYWORD 1636

As you have learned, C# does not promote the use of pointers. In fact, to use pointers within a C# program, you must declare your program as *unsafe*. As you examine C# programs that use pointers, there may be times when you encounter the *fixed* keyword at the start of a program statement. When C# encounters the fixed keyword, C# will freeze the memory locations of reference variables (which reside in the heap and which the Common Language Runtime can move to manage memory as it requires) for the duration of the statement. Using the *fixed* keyword, a program can, for example, use a pointer to an array to move through the array elements without the concern that during the operation, the Common Language Runtime may move the array to a new location on the heap. The following program, *FixedDemo.cs*, uses the *fixed* keyword to freeze an array for which the code passes a pointer to a function that uses the pointer to display the array's contents:

```
using System;

unsafe class clsFixedDemo
{
  public static void ShowArray(int *p, int elements)
   {
     int i = 0;

     while (i++ < elements)
       Console.WriteLine("{0}", *p++);
   }

  public static void Main ()
   {
     int *p;
```

```
      int [] Values = new int [] {1, 2, 3, 4, 5};

      fixed
       {
         p = &Values;
         ShowArray(p, Values.Length);
       }
    }
 }
```

1637 UNDERSTANDING DELEGATES

As your programs execute, there may be times when the Common Language Runtime must call a function that resides within your code, such as when a control generates an *event* for which your code defines an event handler. Within Windows-based applications, programmers often define *callback* functions, which Windows can use to access your code.

Within C or C++, your program would pass a pointer to callback function, which Windows would later use to call the function. C#, however, does not promote the use of pointers. Instead, C# lets you define a *delegate*.

When you create a delegate, you essentially define a function signature, but you do not implement the corresponding function. For example, the following statement creates a delegate for a function named *ShowMessage*:

```
public delegate void ShowMessage(string Message);
```

After you declare the delegate, you must create a variable that stores the delegate. To do so, you declare a variable of the type that corresponds to the delegate name, which, in this case, is *ShowMessage*:

```
public static ShowMessage DelegateHolder = null;
```

By initializing the variable with the *null* value, your code can later test the variable to insure that it has been assigned the address of a function. The following program, *DelegateDemo.cs*, creates a delegate and then assigns different functions to the value that the program uses the delegate to call:

```
using System;

class clsDelegateDemo
{
  public delegate void ShowMessage(string Message);
  public static ShowMessage DelegateHolder = null;

  public static void Main ()
   {
     DelegateHolder = new ShowMessage(UpperMessage);

     DelegateHolder("Jamsa's C/C++/C# Programmer's Bible");

     DelegateHolder = new ShowMessage(LowerMessage);

     DelegateHolder("Jamsa's C/C++/C# Programmer's Bible");
   }

  private static void UpperMessage(string Message)
   {
     Console.WriteLine(Message.ToUpper());
   }

  private static void LowerMessage(string Message)
   {
```

```
      Console.WriteLine(Message.ToLower());
    }
}
```

Note that the functions the program assigns to the delegate have the same return type and parameter types. After you compile and execute this program, your screen will display the following output:

```
C:\Subdir> DelegateDemo  <Enter>
JAMSA'S C/C++/C# PROGRAMMER'S BIBLE
jamsa's c/c++/c# programmer's bible
```

COMPOSING A DELEGATE 1638

In the previous tip, your program assigned one method to a delegate. Depending on the processing your program performs, there may be times when you want to assign two or more methods to a delegate—creating a *multicast delegate*. Programmers refer to the process of assigning methods to a delegate as composing to the delegate.

When you assign two or more methods to a delegate, C# will automatically call each method, in the order you subscribed them. To compose a delegate, you use the += operator, as shown here:

```
DelegateHolder += new ShowMessage(UpperMessage);
```

To remove a method from the delegate's method list, you use the -= operator:

```
DelegateHolder -= new ShowMessage(UpperMessage);
```

The following program, *ComposeDelegate.cs*, first assigns a single method to a delegate, which the program then calls. The code then assigns a second method to the delegate and calls the delegate. Finally, the code removes the second method from the delegate's method list and calls the delegate for a final time:

```
using System;

class clsComposeDelegate
{
  public delegate void ShowMessage(string Message);
  public static ShowMessage DelegateHolder = null;

  public static void Main ()
    {
      DelegateHolder += new ShowMessage(UpperMessage);
      DelegateHolder("Jamsa's C/C++/C# Programmer's Bible");
      Console.WriteLine();

      DelegateHolder += new ShowMessage(LowerMessage);
      DelegateHolder("Jamsa's C/C++/C# Programmer's Bible");
      Console.WriteLine();

      DelegateHolder -= new ShowMessage(LowerMessage);
      DelegateHolder("Jamsa's C/C++/C# Programmer's Bible");
    }

  private static void UpperMessage(string Message)
    {
      Console.WriteLine(Message.ToUpper());
    }

  private static void LowerMessage(string Message)
```

```
    {
      Console.WriteLine(Message.ToLower());
    }
}
```

After you compile and execute this program, your screen will display the following output:

```
C:\Subdir> ComposeDelegate  <Enter>
JAMSA'S C/C++/C# PROGRAMMER'S BIBLE

JAMSA'S C/C++/C# PROGRAMMER'S BIBLE
jamsa's c/c++/c# programmer's bible

JAMSA'S C/C++/C# PROGRAMMER'S BIBLE
```

1639 Handling Multicast Delegate Return Values

In Tip 1638, you created a multicast delegate that called two different methods. The methods in Tip 1638 each were declared as type *void* and thus, did not return a value. When the methods for a multicast delegate return a value, the delegate will return to the caller only the return value of the last method called. The following program, *DelegateReturnValue.cs*, assigns two methods, each that return values of the type *int* to a multicast delegate. When the program calls the delegate, the delegate invokes both methods. However, the delegate returns only the value of the last method called:

```
using System;

class clsComposeDelegate
{
  public delegate int ShowMessage(string Message);
  public static ShowMessage DelegateHolder = null;

  public static void Main ()
    {
      int result;

      DelegateHolder += new ShowMessage(UpperMessage);
      result = DelegateHolder("AAAAA");
      Console.WriteLine("Result: {0}", result);

      DelegateHolder += new ShowMessage(LowerMessage);
      result = DelegateHolder("BBB");
      Console.WriteLine("Result: {0}", result);
    }

  private static int UpperMessage(string Message)
    {
      Console.WriteLine(Message.ToUpper());
      return(1);
    }

  private static int LowerMessage(string Message)
    {
      Console.WriteLine(Message.ToLower());
      return(2);
    }
}
```

After you compile and execute this program, your screen will display the following output:

```
C:\Subdir> DelegateReturnValue  <Enter>
AAAAA
Result: 1
BBB
bbb
Result: 2
```

UNDERSTANDING CLASS PROPERTIES 1640

In previous tips, you used C# methods to validate the values a program assigns to class member variables. For example, the following program, *ValidateWeight.cs*, uses a *setValue* function that restricts the program to assigning a weight value in the range 25 to 50 to the *Weight* class member. If the program passes an invalid weight to the function, the code assigns the weight value -1:

```
using System;

class Package
{
  public Package(int Weight)
   {
     setValue(Weight);
   }

  public void setValue(int Weight)
   {
     if ((Weight >= 25) && (Weight <= 50))
       this.Weight = Weight;
     else
       this.Weight = -1;
   }

  public void showPackage()
   {
     Console.WriteLine("Weight: {0}", Weight);
   }

  private int Weight;
}

class clsValidateWeight
{
  public static void Main ()
   {
     Package Box = new Package(50);
     Package Envelope = new Package(35);
     Package Crate = new Package(500);

     Box.showPackage();
     Envelope.showPackage();
     Crate.showPackage();
   }
}
```

After you compile and execute this program, your screen will display the following output:

```
C:\Subdir> ValidateWeight  <Enter>
Weight: 50
Weight: 35
Weight: -1
```

To simplify the process of assigning, yet validating a class member's value, C# provides properties. In general, a property is a hidden class member (*protected* variable) that uses special *get* and *set* methods to return and assign the class member variable. After you define a property, the program code can use the assignment operator to assign a value to or to retrieve a value from the property. The following program, *UseProperties.cs*, provides processing similar to that in the previous program. However, as you can see, rather than call class methods to assign and retrieve the values, the program simply uses the assignment operator:

```
using System;

class Package
{
  public Package(int W)
   {
     this.Weight = W;
   }

  private int weight;

  public int Weight
   {
     set
      {
        if ((value >= 25) && (value <= 50))
          weight = value;
        else
          weight = -1;
      }

     get
      {
        return weight;
      }
   }

  public void showPackage()
   {
     Console.WriteLine("Weight: {0}", Weight);
   }
}


class clsValidateWeight
{
  public static void Main ()
   {
     Package Box = new Package(50);
     Package Envelope = new Package(35);
     Package Crate = new Package(500);

     Box.showPackage();
     Envelope.showPackage();
     Crate.showPackage();

     Crate.Weight = 25;
     Box.Weight = 300;
     Envelope.Weight = 250;

     Box.showPackage();
     Envelope.showPackage();
     Crate.showPackage();
   }
}
```

As you can see, when you create a *set* accessor, your code uses the special *value* keyword to which C# assigns the value the program is assigning to the property. After you compile and execute this program, your screen will display the following output:

```
C:\> UseProperties  <Enter>
Weight: 50
Weight: 35
Weight: -1
Weight: -1
Weight: -1
Weight: 25
```

INHERITING PROPERTIES 1641

In Tip 1640, you learned that C# provides class properties that you can use to assign values to and to receive values from class member variables. When a base class defines one or more properties, a class that derives itself from the base class inherits those properties. The following program, *InheritProperties.cs*, provides an example.

```
using System;

class Package
{
  public Package(int W)
   {
     this.Weight = W;
   }

  private int weight;

  public int Weight
   {
     set
      {
        if ((value >= 25) && (value <= 50))
          weight = value;
        else
          weight = -1;
      }

     get
      {
        return weight;
      }
   }

  public void showPackage()
   {
     Console.WriteLine("Weight: {0}", Weight);
   }
}

class CardboardBox : Package
{
  public CardboardBox(int weight, int paperweight) :
         base(weight)
   {
     PaperWeight = paperweight;  // Strength of cardboard
   }
```

```
   public void ShowBox()
    {
      Console.WriteLine("Box supports {0} pounds",
        BoxWeight);
      base.showPackage();
    }

   private int BoxWeight;

   public int PaperWeight
    {
      set
       {
         if ((value > 0) && (value <= 100))
           BoxWeight = value;
         else
           BoxWeight = -1;
       }

      get
       {
         return BoxWeight;
       }
    }
}

class clsInheritProperties
{
  public static void Main ()
   {
     CardboardBox Box = new CardboardBox(50, 75);
     Box.ShowBox();

     Box.PaperWeight = 200;
     Box.ShowBox();

     Box.PaperWeight = 80;
     Box.Weight = 33;
     Box.ShowBox();

     Box.PaperWeight = 50;
     Box.Weight = 200;
     Box.ShowBox();
   }
}
```

As you can see, the *CardboardBox* class inherits properties defined within the *Package* class and can access the *Package* class properties directly. After you compile and execute this program, your screen will display the following output:

```
C:\Subdir> InheritProperties  <Enter>
Box supports 75 pounds
Weight: 50
Box supports -1 pounds
Weight: 50
Box supports 80 pounds
Weight: 33
Box supports 50 pounds
Weight: -1
```

USING MULTIPLE PROPERTIES 1642

In Tip 1640, you created a class that used *get* and *set* accessors to assign values to class properties. Within a class, you can define multiple properties, for each of which you must define corresponding *get* and *set* methods. The following program, *MultipleProperties.cs*, creates the *Package* class that defines two separate properties. To implement the properties, the class defines *get* and *set* accessors for each:

```
using System;

class Package
{
  public Package(int W, string Contents)
   {
     this.Weight = W;
     this.Contents = Contents;
   }

  private int weight;

  public int Weight
   {
     set
      {
        if ((value >= 25) && (value <= 50))
          weight = value;
        else
          weight = -1;
      }

     get
      {
        return weight;
      }
   }

  private string contents;

  public string Contents
   {
     set
      {
        if (value.Length < 20)
          contents = value;
        else
          contents = "Invalid contents";
      }

     get
      {
        return contents;
      }
   }

  public void showPackage()
   {
     Console.WriteLine("Weight: {0}", Weight);
     Console.WriteLine("Content: {0}", Contents);
   }
}
```

```
class clsMultipleProperties
{
  public static void Main ()
   {
     Package Box = new Package(50, "Books");
     Package Envelope = new Package(35, "Contracts");
     Package Crate = new Package(500,
        "Very long description of contents");

     Box.showPackage();
     Envelope.showPackage();
     Crate.showPackage();
   }
}
```

After you compile and execute this program, your screen will display the following output:

```
C:\Subdir> MultipleProperties  <Enter>
Weight: 50
Content: Books
Weight: 35
Content: Contracts
Weight: -1
Content: Invalid contents
```

1643 CREATING A READONLY PROPERTY

As you have learned in the previous tips, a *property* is a class field whose value a program accesses through the use of *get* and *set* accessor methods. Depending on the property's purpose, there may be times when you will want the property to be readonly—meaning, you do not want the program to be able to change the property's value. In such cases, you simply define the property's *get* accessor method, but you do not define a *set* accessor. The following program, *ReadonlyProperty.cs*, for example, illustrates how you create a readonly property.

```
using System;

class Package
{
  public Package(int W, string Contents)
   {
     this.Weight = W;
     this.Contents = Contents;
   }

  private int weight;

  public int Weight
   {
     set
      {
        if ((value >= 25) && (value <= 50))
          weight = value;
        else
          weight = -1;
      }

     get
      {
        return weight;
      }
   }
```

```
  private string contents;

  public string Contents
   {
     set
      {
        if (value.Length < 20)
          contents = value;
        else
          contents = "Invalid contents";
      }

     get
      {
        return contents;
      }
   }

  private string companyName = "ACE Packaging Company";

  public string CompanyName
   {
     get
      {
        return companyName;
      }
   }

  public void showPackage()
   {
     Console.WriteLine("Weight: {0}", Weight);
     Console.WriteLine("Content: {0}", Contents);
     Console.WriteLine("Company: {0}", CompanyName);
   }
}

class clsMultipleProperties
{
  public static void Main ()
   {
     Package Box = new Package(50, "Books");
     Package Envelope = new Package(35, "Contracts");
     Package Crate = new Package(500,
        "Very long description of contents");

     Box.showPackage();
     Envelope.showPackage();
     Crate.showPackage();
   }
}
```

As you can see, the *Package* class provides a *get* accessor for the *companyName* property, but does not provide a *set* accessor, which prevents the program from changing the property's value. After you compile and execute this program, your screen will display the following output:

```
C:\Subdir> ReadonlyProperty  <Enter>
Weight: 50
Content: Books
Company: ACE Packaging Company
Weight: 35
Content: Contracts
Company: ACE Packaging Company
```

```
Weight: -1
Content: Invalid contents
Company: ACE Packaging Company
```

1644 UNDERSTANDING C# INDEXERS

In earlier tips, you have learned how to create and access values stored within C# arrays. As you have learned, to access an array element, you must specify an index into the array:

```
Value = SomeArray[index];
```

Within C#, an *indexer* is a class property that lets you access an object as if the object was an array. For example, assume that you have defined a *File* class. Within the class, you can create an indexer that you might use to retrieve the tenth byte in the file as follows:

```
ByteValue = MyFile[9];  // Element 9 is the tenth value
                        // in a zero-based array
```

Likewise, assume you have created a database class. Again using an indexer within the class, you might retrieve the fifth record in the database as follows:

```
Employee = EmployeeDatabase[4];
```

To create an indexer, you define *get* and *set* accessor methods within the class, much like you would for a property. However, for an indexer, you specify an argument that serves as an index value. The following program, *IndexerDemo.cs*, for example, creates a *Titles* class that uses an indexer to present the object's contents as an array:

```
using System;

class Titles
{
  public Titles()
   {
     Contents = new String[25];

     for (int i = 0; i < Contents.Length; i++)
        Contents[i] = null;

     length = 25;
   }

  public object this[int index]
   {
     set
      {
        if ((index >= 0) && (index < Contents.Length))
          Contents[index] = value.ToString();
         else
          Contents[index] = null;
      }

     get
      {
        if ((index >= 0) && (index < Contents.Length))
          return Contents[index];
                 else
                        return(null);
          }
   }
```

```
    private int length;

    public int Length
    {
           get
           {
                  return(length);
           }
    }

    private string [] Contents;
}

class clsIndexerDemo
{
    public static void Main ()
    {
           Titles Books = new Titles();

           Books[0] = "Rescued by C#";
           Books[3] = "Jamsa's C/C++/C# Programmer's Bible";
           Books[10] = "Advanced C#";

           for (int i = 0; i < Books.Length; i++)
             if (Books[i] != null)
             Console.WriteLine(Books[i]);
    }
}
```

After you compile and execute this program, your screen will display the following output:

```
C:\Subdir> IndexerDemo  <Enter>
Rescued by C#
Jamsa's C/C++/C# Programmer's Bible
Advanced C#
```

REVISITING C# EXCEPTIONS 1645

Within this book's C++ tips, you first learned that exceptions are error conditions which are beyond the type of errors a programmer should expect to occur during an application's normal course of operations. For example, when a program opens a file, the programmer should test to ensure the program successfully opened a file. Likewise, when a program establishes a network connection, the programmer should test to verify the connection was properly established. An exception, in these cases, might be a disk running out of space during a file operation or the network connection being dropped during the course of a transmission.

When such error conditions occur, a program generates or "throws" an exception. Many of the methods within the Common Language Runtime, for example, will throw exceptions for various conditions.

By default, if a C# program does not detect and respond to an exception, the program will display an error message regarding the exception and the program will end. To detect an exception, the program must perform the operation that is capable of throwing the exception within a *try* statement:

```
try
  {
     // Some operation
  }
```

Following the *try* statement, the code must place one or more *catch* statements that test for the occurrence of a specific exception. The code can also include a *finally* statement that handles any "uncaught" exceptions.

The following program, *ExceptionDemo.cs*, detects an exception that C# raises when the code tries to assign a value to an invalid index within an array:

```
using System;
using System.Collections;

public class ExceptionDemo
{
  public static void Main()
   {
     int [] Values = new int[100];
     int index = -1;

     try
      {
         Values[index] = 1700;
      }
     catch (System.Exception e)
      {
         Console.WriteLine("Caught exception");
         Console.WriteLine("{0}", e.ToString());
      }
   }
}
```

After you compile and execute this program, your screen will display the following output:

```
C:\Subdir> ExceptionDemo  <Enter>
Caught exception
System.IndexOutOfRangeException: Exception of type
System.IndexOutOfRangeException was thrown.
   at ExceptionDemo.Main() in c:\c_sharp\1\consoleapplication7\class1.cs:line 13
```

1646 UNDERSTANDING THE SYSTEM.EXCEPTION OBJECT

Within C#, all exceptions must be of the type *System.Exception* or derived from the type. When your application throws an exception, your application actually creates an instance of an *Exception* object:

```
throw new Exception();
```

When your programs create an exception, your code can use one of four constructor methods:

```
Exception();              // Default parameters

Exception(string);        // Assigns the error message specified

Exception(SerializationInfo, StreamingContext);
                          // Uses serialized data

Exception(string, Exception);
                          // Specifies an error message and a
                          // reference to an inner exception that
                          // is actually generating the exception
```

1647 DERIVING A CUSTOM EXCEPTION

Within C#, all exceptions are based on the *System.Exception* class. Depending on your program's needs, there may be times when you will want to create a custom exception. For example, you may want to add a date and time field to the exception that your code can use to monitor exactly when the exception occurred. In such cases, your programs can derive a class from the *System.Exception* class. The following program, *CustomException.cs*, derives the *BookException* class from the *SystemException* class to add a field that tracks the date and time the exception occurred:

```
using System;
using System.Collections;

class BookException : System.Exception
{
  public BookException() : base()
   {
     ExceptionDateTime = DateTime.Now;
   }

  public DateTime ExceptionDateTime;
}

class Book
{
  public Book(string title, int pages)
   {
     this.title = title;

     if (pages > 500)
       throw (new BookException());
     else
       this.pages = pages;
   }

  private string title;
  private int pages;
}

public class CustomException
{
  public static void Main()
   {
     try
      {
        Book MyNovel = new Book("American Dream", 600);
      }
     catch (BookException e)
      {
        Console.WriteLine("Caught exception");
        Console.WriteLine("{0}",
           e.ExceptionDateTime.ToString());
      }
   }
}
```

After you compile and execute this program, your screen will display the following output:

```
C:\Subdir> CustomException  <Enter>
Caught exception
2/17/2002 12:53:22 PM
```

USING THE SYSTEM.EXCEPTION CLASS STACKTRACE METHOD 1648

As you debug programs that are encountering exceptions, or possibly when your working code responds to an exception, there may be times when you will want to know which functions the code called immediately before the exception occurred. In such cases, your code can use the *Exception* class *StackTrace* member that contains the names of called functions that currently reside within the stack. In other words, the *StackTrace* member contains the names of functions that the program has called that have not yet completed their processing. The following program, *StackTraceDemo.cs*, calls a function named *One,* which in turn calls the function named *Two,* which in turn calls the function named *Three.* Within function Three, the code generates a divide by zero exception which the code catches. Within the exception handler the code displays the value of the *StackTrace* method, which contains the function names:

```
using System;

class clsMain
{
   public static void One()
    {
      Console.WriteLine("In function One");
      Two();
    }

   public static void Two()
    {
      Console.WriteLine("In function Two");
      Three();
    }

   public static void Three()
    {
      int result, zero = 0;

      Console.WriteLine("In function Three");
      result = 1 / zero;
    }

   static public void Main ()
    {
      try
       {
          One();
       }
      catch (System.Exception e)
       {
          Console.WriteLine("Exception occurred");
          Console.WriteLine(e.StackTrace);
       }
    }
}
```

After you compile and execute this program, your screen will display the following output:

```
C:\Subdir> StackTraceDemo  <Enter>
In function One
In function Two
In function Three
Exception occurred
   at clsMain.Three() in c:\Subdir\StackTraceDemo:line 20
   at clsMain.Two() in c:\Subdir\StackTraceDemo.cs:line 13
   at clsMain.One() in c:\Subdir\StackTraceDemo.cs:line 7
   at clsMain.Main() in c:\Subdir\StackTraceDemo.cs:line 27
```

1649 UNDERSTANDING C# INTERFACES

Throughout several of this book's C# tips, you have used inheritance to derive a class that uses the methods, fields, and properties of a base class. As the complexity of your classes increase, there may be times when you will want a class (or several classes) to use a similar set of members that, unfortunately, do not reside within one base class. In such cases, you can use a C# interface.

To begin, you can think of an interface as an abstract class that contains only pure-virtual methods, which means, a class that implements an interface must provide definitions for each of the methods the interface defines.

Syntactically, a C# interface looks similar to a class. The interface can specify methods, events, properties, and indexers (but not variables). However, unlike a class definition, the interface does not implement any of the methods.

To define an interface, you replace the *class* keyword with *interface* and then you specify methods, events, properties, and indexers within the braces that follow:

```
interface InterfaceName
{
   void MethodName_1(int a, int b):
   int MethodName_2():
   double MethodName_3(string b):
   void MethodName_4(double b):
}
```

Note that the interface does not precede the method names with the abstract keyword. The fact that a class that implements the interface must define the corresponding methods is implied within in the interface definition. An interface does not create a data structure. Instead, it simply creates a template for an abstract class for which classes that implement the interface must specify the interface's method definitions.

IMPLEMENTING AN INTERFACE 1650

After you define an interface within your code, another code must implement the interface by providing definitions for each of the methods the interface specifies. To implement an interface, a class uses a syntactic format similar to what you would use to derive one class from another, as shown here:

```
class NewClassName: InterfaceName
```

The following program, *ImplementInterface.cs*, creates an interface named *Book* that specifies several interface methods. Next, the program implements a class named *NewBook* that implements the interface:

```
using System;
using System.Collections;

interface Book
{
  void ShowBook();
  string GetTitle();
  int GetPages();
  void SetPages(int pages);
}

class NewBook: Book
{
  public string title;
  public int pages;
  public string author;

  public NewBook(string title, string author, int pages)
   {
     this.title = title;
     this.pages = pages;
     this.author = author;
```

```
  }

  public string GetTitle()
   {
     return(title);
   }

  public int GetPages()
   {
     return(pages);
   }

  public void SetPages(int pages)
   {
     this.pages = pages;
   }

  public void ShowBook()
   {
     Console.WriteLine("Title: {0}", title);
     Console.WriteLine("Author: {0}", author);
     Console.WriteLine("Pages: {0}", pages);
   }
}

public class ImplementInterface
{
  public static void Main()
   {
     NewBook MyNovel = new
       NewBook("American Dream", "Smith", 600);
     MyNovel.ShowBook();
   }
}
```

After you compile and execute this program, your screen will display the following output:

```
C:\Subdir> ImplementInterface  <Enter>
Title: American Dream
Author: Smith
Pages: 600
```

1651 Implementing Multiple Interfaces

As you have learned, C# does not support multiple inheritance, which means a C# class cannot use members from two or more base classes. A class, however, can implement multiple interfaces. When a class implements multiple interfaces in this way, the class must define each interface's methods. For example, the following program, *MultipleInterfaces.cs*, creates two interfaces, one named *Person* and one *Office*. The program then uses the interfaces to implement an *Employee* interface:

```
using System;
using System.Collections;

interface Person
 {
   void ShowName();
   void ShowPhone();
   void ShowAddress();
 }
```

```
interface Office
 {
   void ShowOffice();
   void ShowOfficePhone();
 }

class Employee: Person, Office
{
  private string Name;
  private string Phone;
  private string Address;
  private string OfficeNumber;
  private string OfficePhone;

  public Employee(string name, string phone,
     string address, string officenumber,
     string officephone)
   {
     this.Name = name;
     this.Phone = phone;
     this.Address = address;
     this.OfficeNumber = officenumber;
     this.OfficePhone = officephone;
   }

  public void ShowName()
   {
     Console.WriteLine("Name: {0}", Name);
   }

  public void ShowPhone()
   {
     Console.WriteLine("Phone: {0}", Phone);
   }

  public void ShowAddress()
   {
     Console.WriteLine("Address: {0}", Address);
   }

  public void ShowOffice()
   {
     Console.WriteLine("Office: {0}", OfficeNumber);
   }

  public void ShowOfficePhone()
   {
     Console.WriteLine("Office Phone: {0}", OfficePhone);
   }

  public void ShowEmployee()
   {
     ShowName();
     ShowPhone();
     ShowAddress();
     ShowOffice();
     ShowOfficePhone();
   }
 }

 public class MultipleInterfaces
  {
```

```
  public static void Main()
   {
     Employee Worker = new Employee("Jones", "555-1212",
        "123 Main Street", "3A", "555-1213");

     Worker.ShowEmployee();
   }
}
```

After you compile and execute this program, your screen will display the following output:

```
C:\Subdir> MultipleInterfaces  <Enter>
Name: Jones
Phone: 555-1212
Address: 123 Main Street
Office: 3A
Office Phone: 555-1213
```

1652 REIMPLEMENTING AN INTERFACE

When a class inherits a method from a base class that the base class implemented from an interface, the derived class can reimplement the interface methods. When a derived class reimplements an interface method in this way, the class must precede the method type and name with the new keyword, as shown here:

```
public new void ShowBook()
 {
   Console.WriteLine("Book title: {0}", title);
   Console.WriteLine("Book author: {0}", author);
 }
```

The following program, *ReimplementInterface.cs*, creates a *BaseBook* class that implements the methods defined in the *Book* interface. Next, the program derives the class *DerivedBook* from the *BaseBook* class, which, in turn, reimplements two interface class methods:

```
using System;
using System.Collections;

interface Book
{
  void ShowBook();
  string GetTitle();
  int GetPages();
  void SetPages(int pages);
}

class BaseBook: Book
{
  public string title;
  public int pages;
  public string author;

  public BaseBook(string title, string author, int pages)
   {
     this.title = title;
     this.pages = pages;
     this.author = author;
   }

  public string GetTitle()
   {
```

```
      return(title);
    }

  public int GetPages()
    {
      return(pages);
    }

  public void SetPages(int pages)
    {
      this.pages = pages;
    }

  public void ShowBook()
    {
      Console.WriteLine("Title: {0}", title);
      Console.WriteLine("Author: {0}", author);
      Console.WriteLine("Pages: {0}", pages);
    }
}

class DerivedBook: BaseBook, Book
{
  public DerivedBook(string title, string author,
       int pages) : base(title, author, pages)
    {
    }

  public new string GetTitle()
    {
      return("Books title is " + title);
    }

  public new void ShowBook()
    {
      Console.WriteLine("Book title: {0}", title);
      Console.WriteLine("Book author: {0}", author);
    }
}

public class ReimplementInterface
{
  public static void Main()
    {
      DerivedBook Novel = new
        DerivedBook("American Dream", "Jones", 555);

      Novel.ShowBook();
    }
}
```

After you compile and execute this program, your screen will display the following output:

```
C:\Subdir> ReimplementInterface  <Enter>
Book title: American Dream
Book author: Jones
```

USING REFLECTION AND C# INTERFACES 1653

As you have learned, reflection is the ability to query a class to determine class member information. In the previous tips, you have created several classes that implement interfaces. When a class implements an interface, the interface methods, properties, and events become part of the new class. As such, if you perform a reflection query against the new class, the results will include the interface members. The following program, *QueryInterfaceImplementation.cs*, for example, uses reflection to query the *Employee* class you created in Tip 1651, which implements two interfaces (the *Person* and *Office* interface):

```
using System;
using System.Collections;
using System.Reflection;

interface Person
 {
   void ShowName();
   void ShowPhone();
   void ShowAddress();
 }

interface Office
 {
   void ShowOffice();
   void ShowOfficePhone();
 }

class Employee: Person, Office
{
  private string Name;
  private string Phone;
  private string Address;
  private string OfficeNumber;
  private string OfficePhone;

  public Employee(string name, string phone,
     string address, string officenumber,
     string officephone)
   {
     this.Name = name;
     this.Phone = phone;
     this.Address = address;
     this.OfficeNumber = officenumber;
     this.OfficePhone = officephone;
   }

  public void ShowName()
   {
     Console.WriteLine("Name: {0}", Name);
   }

  public void ShowPhone()
   {
     Console.WriteLine("Phone: {0}", Phone);
   }

  public void ShowAddress()
   {
     Console.WriteLine("Address: {0}", Address);
   }

  public void ShowOffice()
   {
     Console.WriteLine("Office: {0}", OfficeNumber);
   }
```

```
  public void ShowOfficePhone()
   {
     Console.WriteLine("Office Phone: {0}", OfficePhone);
   }

  public void ShowEmployee()
   {
     ShowName();
     ShowPhone();
     ShowAddress();
     ShowOffice();
     ShowOfficePhone();
   }
}

public class MultipleInterfaces
 {
   public static void Main()
    {
      Employee Worker = new Employee("Jones", "555-1212",
         "123 Main Street", "3A", "555-1213");

      Type t = typeof(Employee);
      MethodInfo [] Methods = t.GetMethods();

      foreach (MethodInfo Method in Methods)
        Console.WriteLine("{0}", Method.Name);
    }
 }
```

After you compile and execute this program, your screen will display the following output:

```
C:\Subdir> QueryInterfaceImplementation  <Enter>
ShowOfficePhone
ShowOffice
ShowAddress
ShowPhone
ShowName
GetHashCode
Equals
ToString
ShowEmployee
GetType
```

KEY INTERFACES DEFINED WITHIN THE .NET ENVIRONMENT 1654

Just as the .NET environment provides standard types and common runtime library routines, the .NET environment also defines three commonly used interfaces: *ICloneable*, *IComparable*, and *IFormattable*.

The *ICloneable* interface provides the *Clone* method that lets an object create a copy of itself. Some classes that implement the *ICloneable* interface include each of the container classes, such *Array*, *ArrayList*, and so on.

The *IComparable* interface provides a *CompareTo* method that programs can use in searching and sorting functions to compare two values of the type that implements the interface. The *CompareTo* method returns one of the following values:

Less than zero	The first object is less than the second
Zero	The objects are equal
Greater than zero	The first object is greater than the second

Finally, the *IFormattable* interface provides the *ToString* method which applications can use to display a string representation of the object's value.

In Tip 1655, you will learn how your applications can determine if an object implements one of these key interfaces as your program executes, which will let your program decide during run time if it can take advantage of the method.

1655 DETERMINING IF A CLASS IMPLEMENTS AN INTERFACE

When your programs use classes defined outside of your code, there may be times when you must determine if a class implements a specific interface before you try to use the class to call an interface method. For example, in Tip 1654 you examined the three common interfaces, *ICloneable*, *IComparable*, and *IFormattable*. If your code knows that an object implements one of these interfaces, your code can then call the corresponding method.

To determine if a class truly implements an interface, your code can use the C# *is* operator. As you have learned, the C# *is* operator returns the value *true* if a class is a specific class type or is derived from the class. In addition, the operator will return *true* if a class implements the interface you specify.

The following program, *TestInterface.cs*, uses the *is* operator to determine if various classes implement different interfaces:

```
using System;
using System.Collections;

class SampleClass
{
  public SampleClass(int value)
   {
     this.value = value;
   }

  private int value;
}

public class ArrayClone
{
  public static void Main()
   {
     int [] A = new int [] {1, 2, 3, 4, 5};

     SampleClass Obj = new SampleClass(1700);

     if (A is System.ICloneable)
       Console.WriteLine("A is ICloneable");
     else
       Console.WriteLine("A is not ICloneable");

     if (A is System.IFormattable)
       Console.WriteLine("A is IFormattable");
     else
       Console.WriteLine("A is not IFormattable");

     if (Obj is System.ICloneable)
       Console.WriteLine("Obj is ICloneable");
     else
       Console.WriteLine("Obj is not ICloneable");

     if (Obj is System.IFormattable)
       Console.WriteLine("Obj is IFormattable");
```

```
      else
        Console.WriteLine(“Obj is not IFormattable”);
    }
}
```

After you compile and execute this program, your screen will display the following output:

```
C:\Subdir> TestInterface  <Enter>
A is ICloneable
A is not IFormattable
Obj is not ICloneable
Obj is not IFormattable
```

RESOLVING INTERFACE NAME CONFLICTS 1656

When a class implements two or more interfaces, there may be times when a method name in one interface conflicts within a method name within another. For example, the following interfaces both implement a method named *ShowMessage*:

```
interface Wireless
 {
   void ShowMessage();
 }

interface Network
 {
   void ShowMessage();
 }
```

When such method name conflicts occur, the class that implements the interfaces should precede the method names with the interface name, followed by a dot operator, as shown here:

```
  void Wireless.ShowMessage()
   {
     Console.WriteLine(“{0}”, WirelessMsg);
   }
```

Later to call one of the method names, the code must specifically cast the class to the corresponding interface type. The following statement, for example, would direct the code to call the *Wireless* interface *ShowMessage* implementation:

```
((Wireless)Data).ShowMessage();
```

The following program, *InterfaceNameConflicts.cs*, uses complete interface names and dot operators to resolve interface method name conflicts:

```
using System;

interface Wireless
 {
   void ShowMessage();
 }

interface Network
 {
   void ShowMessage();
 }

class Packet: Wireless, Network
```

```
{
  void Wireless.ShowMessage()
   {
     Console.WriteLine("{0}", WirelessMsg);
   }

  void Network.ShowMessage()
   {
     Console.WriteLine("{0}", NetworkMsg);
   }

  public Packet(string WMsg, string NMsg)
   {
     WirelessMsg = WMsg;
     NetworkMsg = NMsg;
   }

  private string WirelessMsg;
  private string NetworkMsg;
}

public class InterfaceNameConflicts
 {
   public static void Main()
    {
       Packet Data = new Packet("Hello Wireless",
         "Hello Wired");

       ((Wireless)Data).ShowMessage();
       ((Network)Data).ShowMessage();
    }
}
```

1657 Understanding C# Events

As you learned within this book's Windows tips, Windows often notifies applications that a specific action has occurred, such as a mouse click, keyboard key press, or a window resizing, by generating an event. A program, in turn, can provide special code, which programmers refer to as an *event handler*, that the program runs each time the event occurs.

Within C#, a class can contain one or more event members that correspond to the events the class can generate. An *Inventory* class, for example, might raise events when inventory levels become too high or too low. Likewise, a *NuclearReactor* class might generate events as the reactor's temperatures change.

A class that defines an event normally also defines a delegate that defines the corresponding event handler. By using a delegate to define the event handler, other classes can subscribe to the delegate a method the class wants C# to call when the event occurs. To raise an event, the class simply creates an instance of the event type. C#, in turn, will then call the handlers that have subscribed for event notification.

1658 Defining an Event and an Event Delegate

The following *NuclearReactorMonitor* class generates the *HighTemp* event, which the class throws when the reactor's temperature is above 200 degrees. Before the class can define the event, the class creates a delegate (a signature holder) that will store the method the program invokes to handle the event. The class then declares the event as the delegate type:

The delegate for the event handler must receive two parameters. The first specifies the object that is generating the event and the second parameter contains arguments specific to the event. The event arguments must be a class type derived from the *System.EventArgs* class, as shown here:

```
public class ReactorEventArgs : EventArgs
{
   public ReactorEventArgs(int temp)
   {
         AlarmTemp = temp;
   }

   public int AlarmTemp;
}
```

The following statements define the *NuclearReactorMonitor* class, which first defines the event and later (within the *SetTemp* method) generates the event:

```
class NuclearReactorMonitor
{
  public delegate void TempHandler(object source,
    ReactorEventArgs e);

  public event TempHandler HighTemp;

  public NuclearReactorMonitor()
   {
     this.temp = 0;
   }

  public void SetTemp(int temp)
   {
     if (temp > 200)
       {
          if (HighTemp != null)
            HighTemp(this, new ReactorEventArgs(temp));
       }
     else
      this.temp = temp;
   }

   public int temp;
}
```

To generate the event, the class first verifies that the handler is defined and then calls the handler method, passing to the method the object that is generating the event (*this*) and a new instance of an event arguments:

```
HighTemp(this, new ReactorEventArgs(temp));
```

HANDLING AN EVENT 1659

To handle an event generated by a C# class, a second class must define a method that will serve as an event handler and then subscribe the method to the delegate. The following program, *HandleEvent.cs*, creates *Generator* class, which, in turn, creates an instance of the *NuclearReactorMonitor* class. The *Generator* class, in turn, creates the *TemperatureHandler* method which it subscribes to the delegate to handle the event:

```
using System;

public class ReactorEventArgs : EventArgs
{
  public ReactorEventArgs(int temp)
   {
     AlarmTemp = temp;
   }

  public int AlarmTemp;
}

class NuclearReactorMonitor
{
  public delegate void TempHandler(object source,
     ReactorEventArgs e);

  public event TempHandler HighTemp;

  public NuclearReactorMonitor()
   {
     this.temp = 0;
   }

  public void SetTemp(int temp)
   {
     if (temp > 200)
      {
        if (HighTemp != null)
         HighTemp(this, new ReactorEventArgs(temp));
      }
    else
      this.temp = temp;
   }

  public int temp;
}

class Generator
{
  public Generator()
   {
     GeneratorMonitor = new NuclearReactorMonitor();

       GeneratorMonitor.HighTemp += new
     NuclearReactorMonitor.TempHandler(TemperatureHandler);
   }

  public void TemperatureHandler(object source,
     ReactorEventArgs e)
   {
     Console.WriteLine("Handling event - Temp: {0}",
       e.AlarmTemp);
   }

  public void AssignTemp(int temp)
   {
     GeneratorMonitor.SetTemp(temp);
   }
```

```
    NuclearReactorMonitor GeneratorMonitor;
  }

  public class ClassEvent
  {
    public static void Main()
     {
       Generator Site = new Generator();

       Site.AssignTemp(250);
     }
  }
```

After you compile and execute this program, your screen will display the following output:

```
C:\Subdir> HandleEvents  <Enter>
Handling event – Temp: 250
```

CREATING MULTITHREADED C# PROGRAMS 1660

Throughout the previous C# tips, each application has used a single thread of execution, meaning, each application had a single stream of instructions the CPU ran to execute the program. In contrast, a multithreaded application contains two or more threads of execution. A word processing program, for example, might use one thread of execution to spell check a document, and a second thread of execution that lets the word processor simultaneously print a different document.

When an application uses multiple threads of execution, the application's threads do not actually run at the exact same time. Instead, behind the scenes, thread management software quickly switches CPU control among the threads, meaning, each thread within the application will run for a slice of time, followed by a different thread, which in turn, is followed by yet another thread. Because the thread management software switches so quickly between the threads, the threads appear to run simultaneously.

To use multiple threads within a C# application, you must first specify that the application will use the *System.Threading* namespace:

```
using System.Threading;
```

Next, your code must specify a method that implements the *Thread* class *ThreadStart* delegate, which corresponds to the first method a newly created thread will execute. Finally, your code must create a *Thread* class object, passing to the object's constructor the method you want the thread to first execute. Then, your code starts the thread's execution by calling the *Thread* class *Start* method.

CREATING A SIMPLE MULTITHREADED EXAMPLE 1661

In Tip 1660, you learned the steps your code must perform to create a multithreaded application. The following program, *ABCThreads.cs*, created three threads, named *A*, *B*, and *C*, each of which simply displays the letter A, B, or C to the console window 250 times:

```
using System;
using System.Threading;

public class ABCThreads
{
  public static void ShowA()
   {
```

```
      for (int i = 0; i < 250; i++)
        Console.Write("A");
    }

  public static void ShowB()
    {
      for (int i = 0; i < 250; i++)
        Console.Write("B");
    }

  public static void ShowC()
    {
      for (int i = 0; i < 250; i++)
        Console.Write("C");
    }

  public static void Main()
    {
      ThreadStart ThreadA = new ThreadStart(ShowA);
      ThreadStart ThreadB = new ThreadStart(ShowB);
      ThreadStart ThreadC = new ThreadStart(ShowC);

      Thread A = new Thread(ThreadA);
      Thread B = new Thread(ThreadB);
      Thread C = new Thread(ThreadC);

      Console.WriteLine("About to start threads");

      A.Start();
      B.Start();
      C.Start();

      Console.WriteLine("Started threads");
      Console.ReadLine();
    }
}
```

In this case, the program defines three different methods that the different threads will execute. Then, the code creates three *ThreadStart* objects that correspond to each method the threads will execute. Then, the code creates three thread objects, which the code then starts using the *Thread* class Start method. After you compile and execute this application, your screen will display the following output:

```
C:\Subdir> ABCThreads  <Enter>
About to start threads
Started threads
AAAAAAAAAAAAAAAAAAAAAAAAAAAAAAAAA AAAAA AAAAAA AAAAAAA AAAAA AAA AAAAAAA AAAAA AAA
AAAAAAAAAAAAAAAAAAAAAAAAAAAAAAAAA AAAAA AAAAAA AAAAAAA AAAAA AAA AAAAAAA AAAAA AAA
AAAAAAAAAAAAAAAAAAAAAAAAAAAAAAAAA AAAAA AAAAAA AAAAAAA AAAAA AAA AAAAAAA AAAAA AAA
AAAAAA AAAAAAACCC CCCCC CCCCCC CCCCBB BBBBB BBBBB BBBBBB BBBBBBB BBBBB BBB BBBBBBB BBBBB BBB
BBBBBB BBBBB BBBBB BBBBB BBBBBB BBBBBB BBBBB BBBBB BBBBBB BBBBBBB BBBBB BBB BBBBBBB BBBBB BBB
BBBBBB BBBBB BBBBB BBBBB BBBBBB BBBBBB BBBBB BBBBB BBBBBB BBCCCCC CCCCC CCC CCCCCCC CCCCC CCC
CCCCCC CCCCC CCCCC CCCCC CCCCCC CCCCCC CCCCC CCCCC CCCCCC CCCCCCC CCCCC CCC CCCCCCC CCCCC CCC
CCCCCC CCCCC CCCCC CCCCC CCCCCC CCCCCC CCCCC CCCCC CCCCCC CCCCCCC CCCCC CCC CCCCCCC CCCCC CCC
CCCCCC CCCCC CCCCC CCCCC CCCCCC CCCCCC CCCCC CCCCC CCCBBB BBBBBBB BBBBB BBB BBBBBBB BBBBB BBB
BBBBBB BBBBB BBBBB BBBBB BBBBBB BBBBBB BBBBBB B
```

As you can see, the thread management software runs the *A* thread for a period of time, followed, in this case, by the *C* thread, and the *B* thread. Then, the code runs the *C* thread again for a period of time, followed by the *B* thread.

PUTTING A THREAD TO SLEEP 1662

Depending on the processing threads perform within your application, there may be times when you will want to pause a thread for a specific period of time. In such cases, your programs can call the *Thread* class *Sleep* function, passing to the function the number of milliseconds for which you want the thread to sleep. The following statement, for example, directs thread *A* to sleep for 3 seconds (3000 milliseconds):

```
A.Sleep(3000);
```

The following program, *SleepThreads.cs*, modifies the program you examined in Tip 1661 slightly by changing function *ShowA* so that it sleeps for 3 seconds, function *ShowB* so it sleeps for 2 seconds, and *ShowC* so that it sleeps for 1 second:

```
using System;
using System.Threading;

public class ABCThreads
{
  public static void ShowA()
   {
     Thread.Sleep(3000);

     for (int i = 0; i < 250; i++)
       Console.Write("A");
   }

  public static void ShowB()
   {
     Thread.Sleep(2000);
     for (int i = 0; i < 250; i++)
       Console.Write("B");
   }

  public static void ShowC()
   {
     Thread.Sleep(1000);
     for (int i = 0; i < 250; i++)
       Console.Write("C");
   }

  public static void Main()
   {
     ThreadStart ThreadA = new ThreadStart(ShowA);
     ThreadStart ThreadB = new ThreadStart(ShowB);
     ThreadStart ThreadC = new ThreadStart(ShowC);

     Thread A = new Thread(ThreadA);
     Thread B = new Thread(ThreadB);
     Thread C = new Thread(ThreadC);

     Console.WriteLine("About to start threads");

     A.Start();
     B.Start();
     C.Start();
```

```
        Console.WriteLine("Started threads");
        Console.ReadLine();
      }
  }
```

After you compile and execute this program, your screen will display the following output:

```
C:\Subdir> SleepThreads  <Enter>
About to start threads
Started threads
CCCCCCCCCCCCCCCCCCCCCCCCCCCCCCCCCCCCCCCCCCCCCCCCCCCCCCCCCCCCCCCCCCCCCCCCCCCCCC
CCCCCCCCCCCCCCCCCCCCCCCCCCCCCCCCCCCCCCCCCCCCCCCCCCCCCCCCCCCCCCCCCCCCCCCCCCCCCC
CCCCCCCCCCCCCCCCCCCCCCCCCCCCCCCCCCCCCCCCCCCCCCCCCCCCCCCCCCCCCCCCCCCCCCCCCCCCCC
CCCCCCCCCCCCBBBBBBBBBBBBBBBBBBBBBBBBBBBBBBBBBBBBBBBBBBBBBBBBBBBBBBBBBBBBBBBBBB
BBBBBBBBBBBBBBBBBBBBBBBBBBBBBBBBBBBBBBBBBBBBBBBBBBBBBBBBBBBBBBBBBBBBBBBBBBBBBB
BBBBBBBBBBBBBBBBBBBBBBBBBBBBBBBBBBBBBBBBBBBBBBBBBBBBBBBBBBBBBBBBBBBBBBBBBBBBBB
BBBBBBBBBBBBBBBBBBBBBBBBBBAAAAAAAAAAAAAAAAAAAAAAAAAAAAAAAAAAAAAAAAAAAAAAAAAAAA
AAAAAAAAAAAAAAAAAAAAAAAAAAAAAAAAAAAAAAAAAAAAAAAAAAAAAAAAAAAAAAAAAAAAAAAAAAAAAA
AAAAAAAAAAAAAAAAAAAAAAAAAAAAAAAAAAAAAAAAAAAAAAAAAAAAAAAAAAAAAAAAAAAAAAAAAAAAAA
AAAAAAAAAAAAAAAAAAAAAAAAAAAAAAAAAAAAAAAAAA
```

As you expect, thread *C*, which sleeps the least amount of time, completes its processing first, followed by thread *B* and then thread *A*.

1663 SUSPENDING, RESUMING, AND TERMINATING THREAD EXECUTION

In Tip 1662, you learned how to use the *Thread* class *Sleep* function to suspend a thread's execution for a specific number of milliseconds. Depending on the processing your application performs, there may be times when you want to suspend a thread for an undetermined amount of time. In such cases, your programs can use the *Thread* class *Suspend* method. For example, the following statement suspends thread *A's* execution:

```
A.Suspend();
```

When you suspend a thread's execution in this way, the thread management software will not execute the corresponding thread until the application later resumes the thread's processing by calling the *Thread* class *Resume* method:

```
A.Resume();
```

In a similar way, there may be times when your application must end a thread's execution. In such cases, your programs can call the *Thread* class *Abort* method:

```
A.Abort();
```

The following program, *AbortThreads.cs*, starts three threads, *A*, *B*, and *C*. The code then uses the *Thread* class *Abort* method to terminate threads *A* and *B*.

```
using System;
using System.Threading;

public class ABCThreads
{
  public static void ShowA()
   {
     for (int i = 0; i < 250; i++)
       Console.Write("A");
   }
```

```
  public static void ShowB()
   {
     for (int i = 0; i < 250; i++)
       Console.Write("B");
   }

  public static void ShowC()
   {
     for (int i = 0; i < 250; i++)
       Console.Write("C");
   }

  public static void Main()
   {
     ThreadStart ThreadA = new ThreadStart(ShowA);
     ThreadStart ThreadB = new ThreadStart(ShowB);
     ThreadStart ThreadC = new ThreadStart(ShowC);

     Thread A = new Thread(ThreadA);
     Thread B = new Thread(ThreadB);
     Thread C = new Thread(ThreadC);

     Console.WriteLine("About to start threads");

     A.Start();
     A.Abort();

     B.Start();
     B.Abort();

     C.Start();

     Console.WriteLine("Started threads");
     Console.ReadLine();
  }
}
```

After you compile and execute this program, your screen will display the following output:

```
C:\Subdir> AbortThreads  <Enter>
About to start threads
AAAAAAAAAAAAAAAAAAAAAAAAAAAAAAAAAA AAAAA AAAAA AAAAAA AAAAAAAAAAAAAAAAAAAAAAA AAAAA AAAAAAA
AAAAAAAAAAAAAAAAAAAAAAAAAAAAAAAAAA AAAAA AAAAA AAAAAA AAAAAAAAAAAAAAAAAAAAAAAStarted
threadsCCCC CCCCC CCCCC CCCCCC CCCCCC CCCCC CCCCC CCCCCC CCCCCCC CCCCC CCC CCCCCCC CCCCC CCCCC CC
CCCCCC CCCCC CCCCC CCCCC CCCCCC CCCCCC CCCCC CCCCC CCCCCC CCCCCCC CCCCC CCC CCCCCCC CCCCC CCCCC CC
CCCCCC CCCCC CCCCC CCCCC CCCCCC CCCCCC CCCCC CCCCC CCCCCC CCCCCCC CCCCC CCC CCCCCCC CCCCC CCCCC CC
```

As you can see, thread *A* starts its execution before the program aborts it. In this case, thread *B* does not begin. Thread *C*, in contrast, runs to completion.

Understanding Thread Race Conditions 1664

When applications use multiple threads, there are times when the threads can actually interfere with one another because the threads, which may have explicit timing dependencies, get out of sync due to CPU switching. For example, assume that one thread must produce an item before another thread can consume the item. Further, assume that the producer thread cannot create an item until the consumer thread consumes an existing item, so there is room within a buffer, for example, for a new item. The first thread, named *Producer*, executes the following code:

```
  public static void Producer()
   {
     int value = 0;

     do
      {
        if (buffer.empty)
          {
            buffer.empty = false;
            buffer.item = value;
            value = (value == 1) ? 0: 1;
            Console.WriteLine("P: {0}", buffer.item);
          }
      } while (true);
   }
```

Likewise, the *Consumer* thread executes the following code:

```
  public static void Consumer()
   {
     int value;

     do
      {
        if (! buffer.empty)
          {
            buffer.empty = true;
            Thread.Sleep(1000);
            value = buffer.item;
            Console.WriteLine("C: {0}", value);
          }
      } while (true);
   }
```

As you can see, the Producer thread assigns, in order, the value 0, then the value 1, then the value 0, then the value 1, and so on, to the variable *buffer.item*. The following program, *ProducerConsumer.cs*, uses threads to run the *Producer* and *Consumer* methods:

```
using System;
using System.Threading;

public class ProducerConsumer
{
  class buffer
   {
     public static int item = 0;
     public static bool empty = true;
   }

  public static void Producer()
   {
     int value = 0;
     do
      {
        if (buffer.empty)
          {
            buffer.empty = false;
            buffer.item = value;
            value = (value == 1) ? 0: 1;
            Console.WriteLine("P: {0}", buffer.item);
          }
      } while (true);
```

```
    }

    public static void Consumer()
     {
       int value;

       do
        {
          if (! buffer.empty)
            {
              buffer.empty = true;
              Thread.Sleep(1000);
              value = buffer.item;
              Console.WriteLine("C: {0}", value);
            }
        } while (true);
    }

    public static void Main()
     {
       ThreadStart produce = new ThreadStart(Producer);
       ThreadStart consume = new ThreadStart(Consumer);

       Thread ProducerThread = new Thread(produce);
       Thread ConsumerThread = new Thread(consume);

       ProducerThread.Start();
       ConsumerThread.Start();
    }
}
```

After you compile and execute this program, the Producer thread will produce an item and the Consumer will consume an item, in sequence, as shown here:

```
C:\Subdir> ProducerConsumer  <Enter>
P: 0
C: 0
P: 1
C: 1
P: 0
C: 0
P: 1
C: 1
P: 0
C: 0
P: 1
C: 1
```

Although the program appears to display the output that you would expect, the code is intentionally written so that a potential bug can arise. If you examine the *Consumer* method, for example, note that when the if statement is true (meaning the buffer is empty), the code immediately sets the buffer empty flag to true (before the code actually consumes the item):

```
          if (! buffer.empty)
            {
              buffer.empty = true;
```

Next, to slow the output, the thread sleeps for 1 second. Assume that while the thread is sleeping, the thread management software switches control of the CPU to the *Producer* thread. The fact that the consumer has set the buffer empty flag to true lets the *Producer* thread place a new item in the buffer before the *Consumer* consumed the previous item! The error occurs because the *Producer* and *Consumer* provide code that synchronizes the other's operation.

However, the threads do not have control over when the thread management software may switch control from one thread to the next. In the tips that follow, you will learn how to eliminate the possibility of such errors.

1665 USING THE C# LOCK STATEMENT TO PROTECT A RESOURCE

As you learned in Tip 1664, when applications use multiple threads, there are times when the applications encounter errors due to "race conditions" or synchronization errors among the threads. To help applications better coordinate thread activities, C# applications can use the *lock* statement. In general, the *lock* statement lets an application secure a resource (such as a data buffer) in such a way that threads cannot access the object after another thread has locked it. Should a thread try to access locked resources, the thread management software will suspend the thread until the thread holding the lock releases it. The format of the lock statement is as follows:

```
lock (referenceVariableName)
 {
    // Statements that use the resource
 }
```

To use the *lock* statement, you must specify a reference variable upon which you want to apply the lock. After the thread that issues the lock statement completes the last statement the lock contains, the thread management software unlocks the resources and resumes the execution of a thread that may have been suspended waiting for the lock to be released. The following program, *LockDemo.cs*, changes the *ProducerConsumer.cs* program presented in Tip 1664 to use the lock statement to protect the buffer resource:

```
using System;
using System.Threading;

public class LockDemo
{
  class buffer
   {
     public static int item = 0;
     public static bool empty = true;
     public static object BufferLock = new Object();
   }

  public static void Producer()
   {
     int value = 0;

     do
      {
         lock(buffer.BufferLock)
          {
            if (buffer.empty)
              {
                buffer.empty = false;
                buffer.item = value;
                value = (value == 1) ? 0: 1;
                Console.WriteLine("P: {0}", buffer.item);
              }
          }
      } while (true);
   }

  public static void Consumer()
   {
     int value;
```

```
      do
       {
          lock (buffer.BufferLock)
           {
             if (! buffer.empty)
               {
                 buffer.empty = true;
                 Thread.Sleep(500);
                 value = buffer.item;
                 Console.WriteLine("C: {0}", value);
               }
           }
        } while (true);
    }

   public static void Main()
    {
      ThreadStart produce = new ThreadStart(Producer);
      ThreadStart consume = new ThreadStart(Consumer);

      Thread ProducerThread = new Thread(produce);
      Thread ConsumerThread = new Thread(consume);

      ProducerThread.Start();
      ConsumerThread.Start();
    }
}
```

USING THE MONITOR CLASS TO SYNCHRONIZE THREADS 1666

In Tip 1665, you learned how to use the C# *lock* statement to synchronize thread operations. As you examine C# applications that use multiple threads, you may encounter use of the *Monitor* class. Like the C# *lock* statement, the *Monitor* class places a lock on an object. To place a lock on an object, a thread executes the *Monitor.Enter* method, as shown here:

```
Monitor.Enter(buffer.BufferLock);
```

When a thread invokes the *Montior.Enter* class, the thread management software will grant the thread access to the resource, provided another thread has not previously locked the resource by using *Monitor.Enter* method. If another thread has locked the resource, the second thread will suspend until the resource becomes available. To release the lock, the thread executes the *Monitor.Exit* method:

```
Monitor.Exit(buffer.BufferLock);
```

The advantage of using the *Monitor* class over the *lock* statement is that the *Monitor* class provides a set of methods that threads can call to determine if an object is currently locked. Using these methods, a thread can decide if it wants to suspend itself waiting for the object, or whether it should perform other processing. The disadvantage of using the *Monitor* class is that you must make sure that each time you call a *Monitor.Enter* method, you later call *Monitor.Exit*. Otherwise, the object will remain locked indefinitely.

The following program, *UseMonitor.cs*, uses the *Monitor* class to lock and later unlock the *buffer.BufferLock* object.

```
using System;
using System.Threading;

public class LockDemo
{
  class buffer
```

```
   {
     public static int item = 0;
     public static bool empty = true;
     public static object BufferLock = new Object();
   }

   public static void Producer()
   {
     int value = 0;

     do
      {
         Monitor.Enter(buffer.BufferLock);

         if (buffer.empty)
           {
             buffer.empty = false;
             buffer.item = value;
             value = (value == 1) ? 0: 1;
             Console.WriteLine("P: {0}", buffer.item);
           }

          Monitor.Exit(buffer.BufferLock);
      } while (true);
   }

  public static void Consumer()
   {
     int value;

     do
      {
         Monitor.Enter(buffer.BufferLock);

         if (! buffer.empty)
           {
              buffer.empty = true;
              Thread.Sleep(500);
              value = buffer.item;
              Console.WriteLine("C: {0}", value);
           }

         Monitor.Exit(buffer.BufferLock);

      } while (true);
   }

  public static void Main()
   {
     ThreadStart produce = new ThreadStart(Producer);
     ThreadStart consume = new ThreadStart(Consumer);

     Thread ProducerThread = new Thread(produce);
     Thread ConsumerThread = new Thread(consume);

     ProducerThread.Start();
     ConsumerThread.Start();
   }
}
```

USING MONITOR.TRYENTER TO PREVENT THREAD BLOCKING 1667

In Tip 1666, you learned how to use the *Monitor* class to lock a resource shared by multiple threads. As you learned, to lock a resource, a thread executes the *Monitor.Enter* method. If a second thread calls the *Monitor.Enter* method for a locked object, that thread will suspend its execution until the resource becomes available. Depending on the processing a thread performs, you may prefer that your thread perform other processing (as opposed to suspending) until the resource becomes available. In such cases, the thread can call the *Monitor.TryEnter* method, which either grants the thread with access to a resource that is not locked, or lets the thread continue by performing processing unrelated to the resource.

The following program, *TryEnter.cs*, changes the *UseMonitor.cs* program to use the *Monitor.TryEnter* method. When either the *Producer* or *Consumer* threads gain access to the resource, the threads will perform their normal processing. Otherwise, the *Producer* thread will repeatedly display the letter P while it waits for the lock. Likewise, the *Consumer* thread will repeatedly display the letter C:

```
using System;
using System.Threading;

public class LockDemo
{
  class buffer
   {
     public static int item = 0;
     public static bool empty = true;
     public static object BufferLock = new Object();
   }

  public static void Producer()
   {
     int value = 0;

     do
      {
        if (Monitor.TryEnter(buffer.BufferLock))
          {
            if (buffer.empty)
              {
                buffer.empty = false;
                buffer.item = value;
                value = (value == 1) ? 0: 1;
                Console.WriteLine("P: {0}", buffer.item);
              }
            Monitor.Exit(buffer.BufferLock);
          }
        else
          Console.Write("P");
      } while (true);
   }

  public static void Consumer()
   {
     int value;

     do
      {
        if (Monitor.TryEnter(buffer.BufferLock))
          {
            if (! buffer.empty)
              {
```

```
                    buffer.empty = true;
                    Thread.Sleep(500);
                    value = buffer.item;
                    Console.WriteLine("C: {0}", value);
                 }
               Monitor.Exit(buffer.BufferLock);
            }
          else
            Console.Write("C");
        } while (true);
    }

   public static void Main()
    {
      ThreadStart produce = new ThreadStart(Producer);
      ThreadStart consume = new ThreadStart(Consumer);

      Thread ProducerThread = new Thread(produce);
      Thread ConsumerThread = new Thread(consume);

      ProducerThread.Start();
      ConsumerThread.Start();
    }
 }
```

After you compile and execute this program, you will see how the half-second delay within the Consumer thread causes the Producer thread to experience numerous delays when it tries to allocate the buffer:

```
C:\Subdir> TryEnter  <Enter>
P: 0
CCCCCCCCCCCCCCCCCCCCCCCCCCCCCCCCCCPPPPPPPPPPPPPPPPPPPPPPPPPPPPPPPPPPPPPPPPPPPPPPPPPPPPPPPPPPPP
PPPPPPPPPPPPPPPPPPPPPPPPPPPPPPPPPPPPPPPPPPPPPPPPPPPPPPPPPPPPPPPPPPPPPPPPPPPPPPPPPPPPPPPPPPPPPPP
PPPPPPPPPPPPPPPPPPPPPPPPPPPPPPPPPPPPPPPPPPPPPPPPPPPPPPPPPPPPPPPPPPPPPPPPPPPPPPPPPPPPPPPPPPPPPPP
PPPPPPPPPPPPPPPPPPPPPPPPPPPPPPPPPPPPPPPPPPPPPPPPPPPPPPPPPPPPPPPPPPPPPPPPPPPPPPPPPPPPPPPPPPPPPPP
PPPPPPPPPPPPPPPPPPPPPPPPPPPPPPPPPPPPPPPPPPPPPPPPPPPPPPPPPPPPPPPPPPPPPPPPPPPPPPPPPPPPPPPPPPPPPPP
PPPPPPPPPPPPPPPPPPPPPPPPPPPPPPPPPPPPPPPPPPPPPPPPPPPPPPPPPPPPPPPPPPPPPPPPPPPPPPPPPPPPPPPPPPPPPPP
PPPPPPPPPPPPPPPPPC: 0
PPPPPPPPPPPPPPPPPPPPPPPPPPPPPPPPPPPPP: 1
CCCCCCCCCCCCCCCCCCCCCCCCCCCCCCCCCCCCCCCCCCCCCCCCPPPPPPPPPPPPPPPPPPPPPPPPPPPPPPPP
PPPPPPPPPPPPPPPPPPPPPPPPPPPPPPPPPPPPPPPPPPPPPPPPPPPPPPPPPPPPPPPPPPPPPPPPPPPPPPPP
PPPPPPPPPPPPPPPPPPPPPPPPPPPPPPPPPPPPPPPPPPPPPPPPPPPPPPPPPPPPPPPPPPPPPPPPPPPPPPPP
PPPPPPPPPPPPPPPPPPPPPPPPPPPPPPPPPPPPPPPPPPPPPPPPPPPPPPPPPPPPPPPPPPPPPPPPPPPPPPPP
PPPP
```

1668 CONTROLLING THREAD PRIORITY

As you have learned, when an application uses multiple threads, thread management software will assign each thread a small slice of CPU processing time (which programmers refer to as a *timeslice*). The thread management software will normally distribute the timeslices across the threads in a uniform fashion, meaning, if two threads are running, each would get 50% of the program's CPU timeslice (remember, other programs are likely running within the operating system, each of which is getting its own timeslice).

Depending on your program's processing, you may have some threads that perform more important processing than others. In such cases, your programs can increase or decrease a thread's priority. The higher the thread's priority, the larger the amount of CPU processing time the thread will receive.

To change a thread's priority, your application can assign a priority value to the *Thread* class *Priority* member, using an enumerated value that corresponds to the priority level you desire: *Normal, Above Normal, Below Normal, Highest, Lowest.* The following program, *ThreePriorities.cs*, creates three threads, *A, B, C.* The code then assigns thread *A* the *AboveNormal* priority, thread *B* the *Normal* priority, and thread *C BelowNormal*:

```
using System;
using System.Threading;

public class ThreePriorities
{
  public static void ShowA()
   {
     for (int i = 0; i < 250; i++)
       Console.Write("A");
   }

  public static void ShowB()
   {
     for (int i = 0; i < 250; i++)
       Console.Write("B");
   }

  public static void ShowC()
   {
     for (int i = 0; i < 250; i++)
       Console.Write("C");
   }

  public static void Main()
   {
     ThreadStart ThreadA = new ThreadStart(ShowA);
     ThreadStart ThreadB = new ThreadStart(ShowB);
     ThreadStart ThreadC = new ThreadStart(ShowC);

     Thread A = new Thread(ThreadA);
     Thread B = new Thread(ThreadB);
     Thread C = new Thread(ThreadC);

     A.Priority = ThreadPriority.AboveNormal;
     B.Priority = ThreadPriority.Normal;
     C.Priority = ThreadPriority.BelowNormal;

     Console.WriteLine("About to start threads");

     A.Start();
     B.Start();
     C.Start();

     Console.WriteLine("Started threads");
     Console.ReadLine();
   }
}
```

The *ThreadPriority* class contains the priority values you can use to assign priority to a thread. After you compile and execute this program, your screen will display the following output:

```
C:\Subdir> ThreePriorities  <Enter>
About to start threads
AAAAAAAAAAAAAAAAAAAAAAAAAAAAAAAAAAAAAAAAAAAAAAAAAAAAAAAAAAAAAAAAAAAAAAAAAAAAAAAAAAAAAA
AAAAAAAAAAAAAAAAAAAAAAAAAAAAAAAAAAAAAAAAAAAAAAAAAAAAAAAAAAAAAAAAAAAAAAAAAAAAAAAAAAAAAA
AAAAAAAAAAAAAAAAAAAAAAAAAAAAAAAAAAAAAAAAAAAAAAAAAAAAAAAAAAAAAAAAAAAAAAAAAAAAAAAAAAAAAA
AAAAAAAAAAAAAAAAAAAAAAAAAAAAAAAAAAAAAAAAAAAAAAAAAAAAAAAAAAAAStarted threads
BBBBBBBBBBBBBBBBBBBBBBBBBBBBBBBBBBBBBBBBBBBBBBBBBBBBBBBBBBBBBBBBBBBBBBBBBBBBBBBBBBBBBB
BBBBBBBBBBBBBBBBBBBBBBBBBBBBBBBBBBBBBBBBBBBBBBBBBBBBBBBBBBBBBBBBBBBBBBBBBBBBBBBBBBBBBB
BBBBBBBBBBBBBBBBBBBBBBBBBBBBBBBBBBBBBBBBBBBBBBBBBBBBBBBBBBBBBBBBBBBBBBBBBBBBBBBBBBBBBB
BBBBBBBBBBBBBBBBBBBBBBBBBBBBBBBBBBBBBBBBBBBBCCCCCCCCCCCCCCCCCCCCCCCCCCCCCCCCCCCCCCCCCC
CCCCCCCCCCCCCCCCCCCCCCCCCCCCCCCCCCCCCCCCCCCCCCCCCCCCCCCCCCCCCCCCCCCCCCCCCCCCCCCCCCCCCC
CCCCCCCCCCCCCCCCCCCCCCCCCCCCCCCCCCCCCCCCCCCCCCCCCCCCCCCCCCCCCCCCCCCCCCCCCCCCCCCCCCCCCC
CCC
```

1669 UNDERSTANDING C# ATTRIBUTES

As you have learned, the .NET environment makes extensive use of metadata, which applications can query to retrieve information (data) about data. Using the C# *Reflection* class, programs can query classes about the capabilities the class provides.

To extend the amount of information an application can store about its capabilities, C# provides attributes. In general, an attribute is a characteristic. A class can have attributes, methods can have attributes (such as the parameters the method supports), fields can have attributes, and so on.

For example, assume that an application stores session information within a file named *SESSION_ID.txt* that resides within the directory *USER_SESSION*. The problem of using a file to store the value in this way is that the program contains information and the file contains information, but there is nothing other than the possible comments within the source code that tie the two together. A better solution would be for the program to use an attribute to store information within the program itself that another program can query to learn about the file data.

To specify an attribute within your code, you place the attribute within left and right brackets, as shown here:

```
class SomeClass
{
   [FileandDirectory("USER_SESSION", "SESSION_ID.txt")]
   public string SessionFile;
}
```

In Tip 1670, you will learn that to create an attribute you define a class that you derive from the *System.Attribute* class.

1670 CREATING AN ATTRIBUTE

To create an attribute within a C# program, you will create a class variable that will store the attribute information. You will derive the class variable from the *System.Attribute* class. The following *FileandDirectoryAttribute* class creates an attribute based on the Attribute class:

```
class FileandDirectoryAttribute: Attribute
{
   public FileandDirectoryAttribute(string directory,
      string file)
   {
         this.directory = directory;
         this.file = file;
```

```
    }

    public string directory;
    public string file;
}
```

After you define the attribute, you can use it by placing the corresponding name (minus the word attribute) within brackets before the item to which you want to assign the attribute, as shown here:

```
class SomeClass
{
    public SomeClass(string value)
    {
            SessionFile = value;
    }

    [FileandDirectory("USER_SESSION", "SESSION_ID.txt")]
    public string SessionFile;
}
```

QUERYING CLASS ATTRIBUTE INFORMATION 1671

As you have learned, C# attributes let you associate information within programs, classes, and class members. If a program defines attribute information, a program can query that information using the *System.Type* class methods. The following program, *ShowAttributes.cs*, for example, queries a field about its attributes:

```
using System;
using System.Reflection;

class FileandDirectoryAttribute: Attribute
{
  public FileandDirectoryAttribute(string directory,
     string file)
   {
     this.directory = directory;
     this.file = file;
   }

  public string directory;
  public string file;
}

class SomeClass
{
  public SomeClass(string value)
   {
     SessionFile = value;
   }

   [FileandDirectory("USER_SESSION", "SESSION_ID.txt")]
  public string SessionFile;
}

public class ShowAttribute
 {
  public static void Main()
   {
     SomeClass Demo = new
         SomeClass("C:\\USER_SESSION\\SESSION_ID.txt");
```

```
      Type type = Type.GetType("SomeClass");

      foreach(FieldInfo field in type.GetFields())
        foreach (Attribute attr in
          field.GetCustomAttributes(true))
         {
           FileandDirectoryAttribute fileattr =
              attr as FileandDirectoryAttribute;
           Console.WriteLine("{0}", fileattr.directory);
           Console.WriteLine("{0}", fileattr.file);
         }
    }
}
```

To start, the program uses the *Type.GetType* method to get information about the class *SomeClass*. Next, the program uses the *GetFields* method to retrieve the class methods. Then, for each field, the code uses the *GetCustomAttributes* method to retrieve the field's attribute information.

After you compile and execute this program, your screen will display the following output:

```
C:\Subdir> ShowAttributes  <Enter>
USER_SESSION
SESSION_ID.txt
```

1672 WORKING WITH MULTIPLE ATTRIBUTES

As you have learned, C# lets programs assign attributes to classes, methods, and fields. The following program, *MultipleAttributes.cs*, assigns attributes to a variety of elements. The code then retrieves and displays the attributes:

```
using System;
using System.Reflection;

class FileandDirectoryAttribute: Attribute
{
  public FileandDirectoryAttribute(string directory,
      string file)
   {
     this.Directory = directory;
     this.File = file;
   }

  protected string directory;
  public String Directory
   {
     get { return(directory); }
     set { directory = value; }
   }

  protected string file;
  public String File
   {
     get { return(file); }
     set { file = value; }
   }
}


class VersionInformationAttribute: Attribute
{
  public VersionInformationAttribute(string version)
   {
```

```
      this.version = version;
    }

  public string version;
}

class ProgrammerInformationAttribute: Attribute
{
  public ProgrammerInformationAttribute(string name,
      string date)
    {
      this.name = name;
      this.date = date;
    }

  public string name;
  public string date;
}

[VersionInformation("1.0A")]
[ProgrammerInformation("Jamsa", "9/30/02")]
class SomeClass
{
  public SomeClass(string value)
    {
      SessionFile = value;
    }

  [FileandDirectory("USER_SESSION", "SESSION_ID.txt")]
  public string SessionFile;
}

public class MultipleAttribute
 {
  public static void Main()
   {
     SomeClass Demo = new
       SomeClass("C:\\USER_SESSION\\SESSION_ID.txt");

     Type type = typeof(SomeClass);

     foreach (Attribute attr in
        type.GetCustomAttributes(true))
      {
         if (attr is ProgrammerInformationAttribute)
          {
            Console.WriteLine("{0} {1}", attr,
              ((ProgrammerInformationAttribute)attr).name);

            Console.WriteLine("{0} {1}", attr,
              ((ProgrammerInformationAttribute)attr).date);
          }
        else if (attr is VersionInformationAttribute)
          Console.WriteLine("{0} {1}", attr,
            ((VersionInformationAttribute)attr).version);
      }

     foreach(FieldInfo field in type.GetFields())
      {
        foreach (Attribute attr in
          field.GetCustomAttributes(true))
          {
```

```
            if (attr is FileandDirectoryAttribute)
            {
              Console.WriteLine("{0} {1}", attr,
              ((FileandDirectoryAttribute)attr).Directory);

              Console.WriteLine("{0} {1}", attr,
               ((FileandDirectoryAttribute)attr).File);
            }
          }
        }
      }
    }
```

After you compile and execute this program, your screen will display the following output:

```
C:\Subdir> MultipleAttributes  <Enter>
ProgrammerInformationAttribute Jamsa
ProgrammerInformationAttribute 9/30/02
VersionInformationAttribute 1.0A
FileandDirectoryAttribute USER_SESSION
FileandDirectoryAttribute SESSION_ID.txt
```

1673 Using the C# Math Class

In this book's C and C++ tips, your programs used run-time library functions to perform such operations as calculating a value's square root or using the trigonometric functions to calculate a sine and cosine.

Within C#, programs can use the *Math* class to perform such operations. The *Math* class provides the set of static methods listed in Table 1673. Because the methods are static, programs can call the methods without having to create an instance of the *Math* class.

Method	Purpose
Abs	Returns the absolute value
Acos	Returns the arc-cosine
Asin	Returns the arc-sine
Atan	Returns the arc-tangent
Atan2	Returns the arc-tangent of a value that is the quotient of two numbers
Ceiling	Returns the smallest whole number greater than or equal to a specified value
Cos	Returns the cosine
Cosh	Returns the hyperbolic cosine
Exp	Returns e raised to the specified power.
Floor	Returns the largest whole number less than or equal to a specified value
IEEERemainder	Returns the remainder of a division
Log	Returns the logarithm
Log10	Returns the base 10 logarithm
Max	Returns the larger of two values
Min	Returns the smaller of two values
Pow	Returns a value raised to the specified power
Round	Returns the number nearest the specified value

Table 1673 The C# Math class methods. *(continued on following page)*

Method	Purpose
Sign	Returns a value indicating the sign of a number
Sin	Returns the sine
Sinh	Returns the hyperbolic sine
Sqrt	Returns the square root
Tan	Returns the tangent
Tanh	Returns the hyperbolic tangent

Table 1673 The C# Math class methods. (continued from previous page)

The following program, *MathClass.cs*, illustrates the use of several *Math* class methods:

```
using System;

class clsUseThis
{
  static void Main()
   {
     Console.WriteLine(“Sqrt of {0} is {1}”,
        25, Math.Sqrt(25));
     Console.WriteLine(“Value of 10 raised to 5 is {0}”,
        Math.Pow(10, 5));
     Console.WriteLine(“Log10 of 100 is {0}”,
        Math.Log10(100));
     Console.WriteLine(“Max of 10 and 25 is {0}”,
        Math.Max(10, 25));
   }
}
```

After you compile and execute this program, your screen will display the following output:

```
C:\Subdir> MathClass  <Enter>
Sqrt of 25 is 5
Value of 10 raised to 5 is 100000
Log10 of 100 is 2
Max of 10 and 25 is 25
```

REVISITING C# COLLECTIONS 1674

In several of the preceding tips, you have used the *Array* and *Stack* classes to store multiple values within a single variable. C# programmers refer to data structures that store multiple values in this way as *collections*. In addition to the *Array* and *Stack* classes, C# provides a standard set of *collections* classes that Table 1674 describes.

Method	Purpose
Array	Groups items of the same type. Provides routines to search, sort, and manipulate elements
ArrayList	A dynamically-sized array that provides array-manipulation routines
BitArray	A dynamically-sized array of Boolean values that use 1 bit to store each element
Hashtable	A data structure that creates an index (a key) for each element based on the element's value

Table 1674 The C# Math class methods. (continued on following page)

Method	Purpose
Queue	A first-in first-out data structure
SortedList	A data structure stores sorted data that is accessible by a key value or index
Stack	A last-in first-out data structure
StringCollection	A data structure that provides methods to manipulate a group of strings

Table 1674 C# collection classes and interfaces. (continued from previous page)

1675 Using XML-Based Comments

In earlier tips, you learned that Visual Studio lets you create Web pages that contain documentation regarding your application. In addition, you have learned that C# supports the C and C++ comment forms shown here:

```
// Comment

/* Comment */
```

In addition, C# supports a third comment type, called a *document comment*, that uses three slashes (///). Within a document comment, you can embed XML tags that group related comments. By enclosing your comments within predefined XML tags, programs such as Visual Studio that let you automate documentation can better parse and display the comments that appear in your source code.

The C# programming environment predefines several XML tags you can use to comment your code. Table 1675 briefly describes the predefined XML tags you can use within C# document contents. Tip 1676 illustrates the use of the several of the XML-based comment tags.

XML Tag	Purpose
<c>	Indicates a code reference within a text description
<code>	Indicates multiple lines of code
<example>	Illustrates the use of a method or member
<exception>	Documents an exception class
<include>	Lets you include comments that reside in a different file
<list>	Creates a list, such as a numbered or bulleted list
<para>	Lets you structure the text within a comment to create a paragraph
<param>	Describes a method parameter
<paramref>	Lets you highlight a word that corresponds to a parameter
<permission>	Describes a member's security access
<remarks>	Provides overview information
<returns>	Specifies a method's return value
<see>	Lets you create a link to another reference
<seealso>	Lets you specify a link to related topics
<summary>	Describes a member
<value>	Describes a property

Table 1675 Predefined XML tags for use in C# document comments.

EXTENDING YOUR PROGRAM'S COMMENTS USING XML TAGS 1676

As you learned in the previous tip, C# supports documentation comments within which you can place XML-based tags to better organize your documentation. The following program, XMLComments.cs, illustrates the use of XML-based documentation:

```
using System;

/// <remarks>
/// The Employee class provides information about
/// each company employee.
/// </remarks>
public class Employee
{
    /// <summary>
    /// The private name member stores the employee name
    /// </summary>
    /// <value>name is the private class member that stores
    /// the employee name</value>
    private string name;

    /// <summary>Name accesses the value employee name
    /// member</summary>
    /// <value>Name accesses the value employee name
    /// member</value>
    public string Name
    {
           get
           {
                  return name;
           }

           set
           {
                  name = value;
           }
    }

    /// <summary>Employee is the class constructor method
    /// </summary>
    /// <param name="name">Contains the employee's
    /// name</param>
    /// <param name="id">Contains the employee's id
    /// number</param>
    public Employee(string name, int id)
    {
           this.Name = name;
           this.id = id;
    }

    /// <summary>The private id member contains the employee
    /// id number</summary>
    /// <value>id contains the employee id number</value>
    private int id;
}
```

If, for example, you use the Visual Studio to build a documentation Web page for the application, the Visual Studio will use the XML comments to create the documentation shown in Figure 1676.

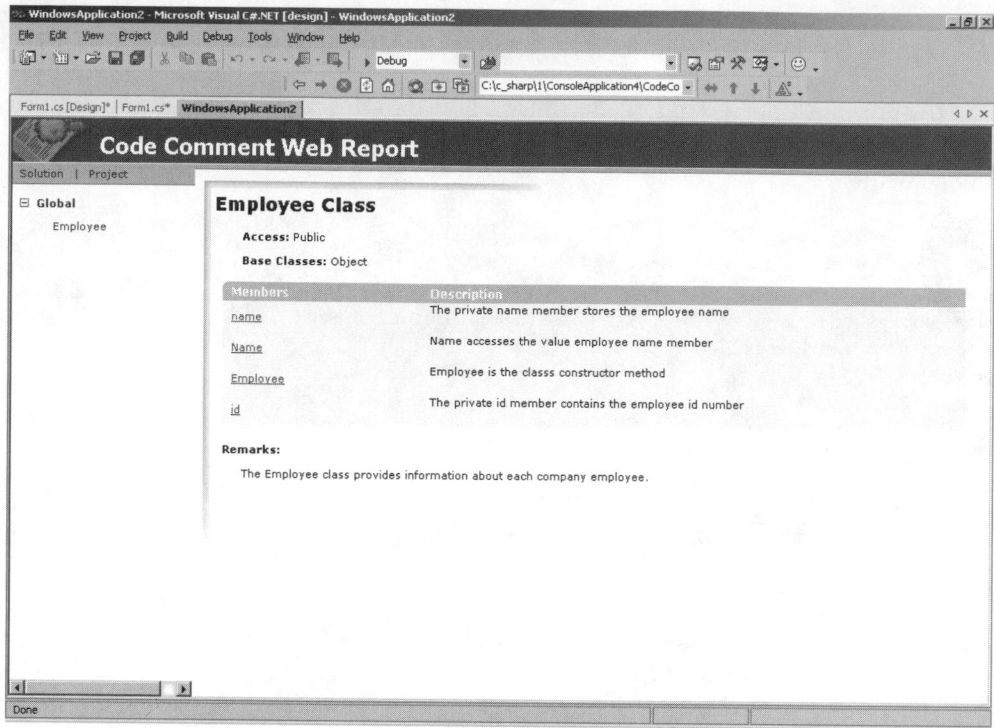

Figure 1676 Displaying XML-based comments within a program's documentation Web pages.

1677 CREATING A SIMPLE WINDOWS-BASED PROGRAM

The Visual Studio exists to simplify the process of creating Windows-based programs. That said, you can use the standard text editor and then compile a Windows-based program from the command line—for some programmers, old habits die hard. To display output within a Windows-based program, you will no longer use the *Console* routines, such as *Console.WriteLine*, but rather, you may (initially as you are getting started) display messages using a message box, such as that shown in Figure 1677.

Figure 1677 Using a message box to display Windows-based output.

To display text within a message box, you can call the *MessageBox.Show* method, as shown here:

```
MessageBox.Show("Hello, Windows World!", "Greeting");
```

The *MessageBox.Show* function resides within the *System.Windows.Forms* namespace. To use the function, you will normally place a using clause near the start of your program as follows:

```
using System.Windows.Forms;
```

The following program, *MessageBox.cs*, uses the *MessageBox.Show* function to create the message box previously shown in Figure 1676:

```
using System;
using System.Windows.Forms;

public class clsMessageBox
{
    public static void Main()
    {
      MessageBox.Show ("Hello, Windows world!",
        "Greeting");
    }
}
```

When you compile programs within the Visual Studio, the Visual Studio will normally keep track of the modules (such as dynamic link libraries) it must include within your program to support the functions you call. C# programmers refer to these modules as *references* (do not confuse the module term with reference variables that reside in the heap). To include the references within the command line that compiles your program, you must use the */r* switch to specify each reference's filename. To compile the *MessageBox.cs* program, for example, you must include three references, as shown here:

```
C\Subdir> csc /r:System.dll /r:System.Windows.Forms.dll /r:Microsoft.Win32.Interop.dll
MessageBox.cs  <Enter>
```

After you compile your program, you can run the program from either the command, by simply typing the program's name, or from within the Windows environment, by double-clicking your mouse on the program file:

```
C\Subdir> MessageBox  <Enter>
```

USING A FILE TO STORE REFERENCE MODULE NAMES 1678

In the previous Tip, you used the /r switch to specify reference modules when you compiled a Windows-based program from the command line. Although most programmers will normally create Windows-based programs from within the Visual Studio, some programmers will insist on working from the command line.

To simplify the process of insuring the compiler has the correct names for each reference module, you can create an ASCII file that contains the /r switches and file names. For example, to compile the *MessageBox.cs* program, you must specify the *System.dll*, *Forms.dll*, and *Interop.dll* files. To do so, you might create a file named *MessageBoxRefs.txt* that contains the following entries:

```
/r:System.dll
/r:System.Windows.Forms.dll
/r:Microsoft.Win32.Interop.dll
```

Then, when you compile your program, you can use the filename within the command line as shown here:

```
C:\Subdir> csc @MessageBoxRefs.txt  MessageBox.cs  <Enter>
```

Within the command line, the at sign (@) directs the C# compiler to examine the file *MessageBoxRefs.txt* for additional command-line options.

1679 Using Visual Studio to Build a Simple Windows Program

Throughout the previous tips, you have used an editor such as Notepad to create your program source files and then you compiled and ran your programs from the command line. The Microsoft Visual Studio, however, exists to help you create Windows-based programs. Within the Visual Studio, you can create and edit your program source code in an environment that highlights the C# syntax, compile and debug your applications, take advantage of programming Wizards that help you perform a variety of common tasks, and access an extensive online help. In the Tips that follow, you will make extensive use of the Visual Studio to create a range of programs. To begin, you will use the Visual Studio to create the previous *MessageBox.cs* that you created in Tip 1676:

```
using System;
using System.Windows.Forms;

public class clsMessageBox
{
    public static void Main()
    {
        MessageBox.Show ("Hello, Windows world!",
          "Greeting");
    }
}
```

To start the Microsoft Visual Studio, perform these steps:

1. To run the Visual Studio, select the Start menu Programs option. Windows will display the Programs menu.
2. Within the Programs menu, select Microsoft Visual Studio.NET (the option may be followed by a version number).
3. Windows will display a submenu. Within the submenu, again choose Microsoft Visual Studio.Net. Windows will display the Visual Studio, as shown in Figure 1679.1.

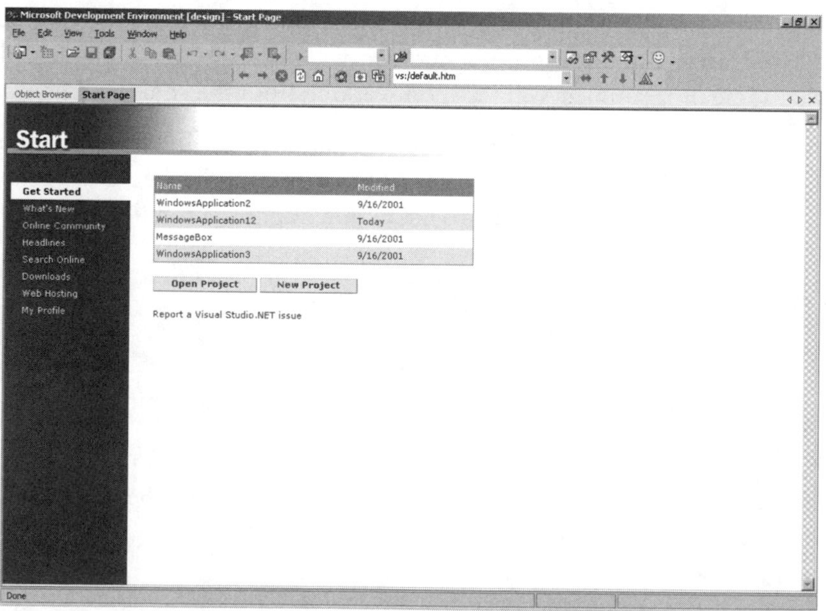

Figure 1679.1 The Microsoft Visual Studio integrated development environment (IDE).

To create the *MessageBox.cs* program, perform these steps:

1. Select the File menu New option and choose Project. The Visual Studio, in turn, will display the New Project dialog box, as shown in Figure 1679.2.

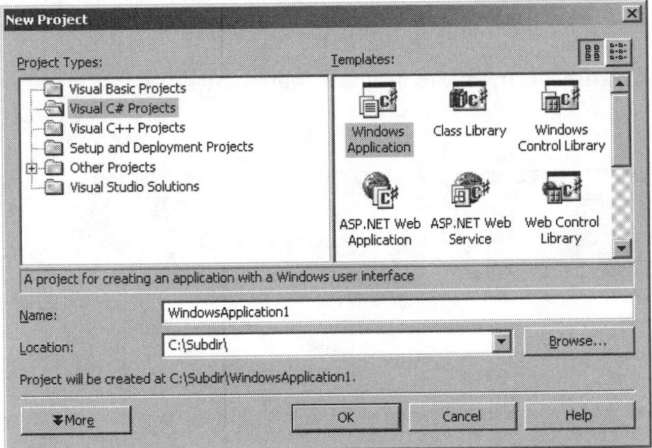

Figure 1679.2 The New Project dialog box.

2. Within the New Project dialog box name field, type in the project name MessageBox. Within the Location field, select the folder within which you want to place the project files.

3. Within the New Project dialog box, double-click your mouse on the Windows Application icon. The Visual Studio will create a new Windows option, displaying the form within which you can place application controls, such as fields, checkboxes, buttons, and so on, as shown in Figure 1679.3. You will not use the form to create the *MessageBox.cs* application.

Figure 1679.3 A new Windows application begins with a form.

4. When you create a new Windows application, the Visual Studio provides you with the program statements necessary to create and display an initial form. In this case, you will not use the program statements. To delete the statements, select the View menu Code option to display the source-code statements. Then, select the Edit menu Select All option and choose Delete. The Visual Studio will remove the program statements, displaying an empty code window.
5. Within the code window, type in the program statements previously shown for *MessageBox.cs*, as shown in Figure 1679.4.

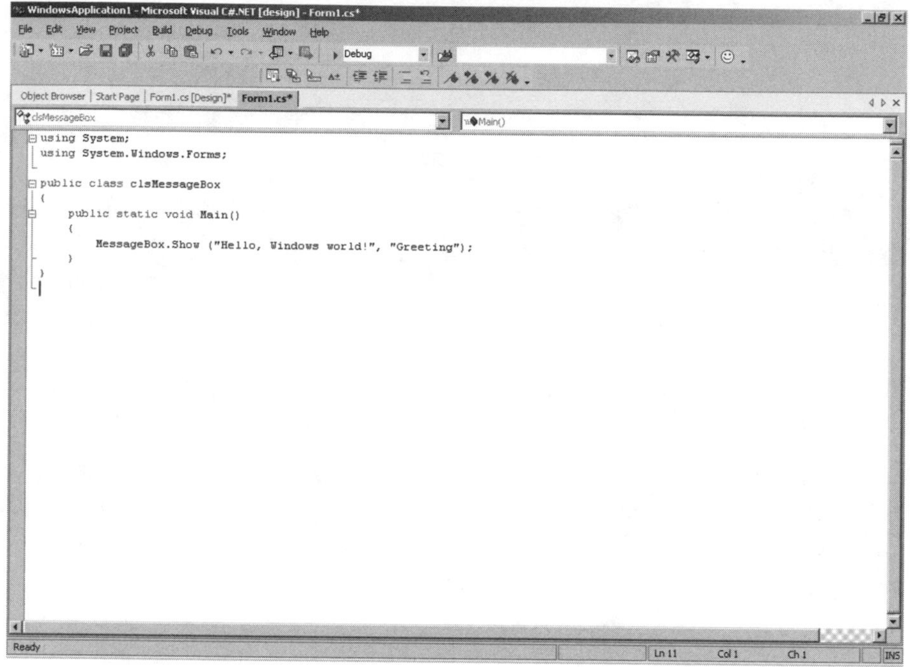

Figure 1679.4 Placing program statements within the Visual Studio code window.

6. To save your program code, select the File menu Save Form1.cs As menu option. The Visual Studio will display the Save As dialog box.
7. Within the Save As dialog box, type MessageBox.cs and choose OK.
8. To compile the program, select the Build menu Build option.
9. To run the program, select the Debug menu Start option. The Visual Studio will display the program's message box.

After you build your project, the Visual Studio will place an executable program file with the name *MessageBox.exe* in a folder on your disk within your *MessageBox* project files. To run your program from within Windows, you can simply double-click your mouse on the program file.

1680 CREATING A PROJECT WITHIN VISUAL STUDIO

In the previous Tip, you used the Visual Studio File menu New option and chose Project to create a Windows application. The Visual Studio, in turn, displayed the New Project dialog box, as shown in Figure 1679, within which you can select the project type you desire.

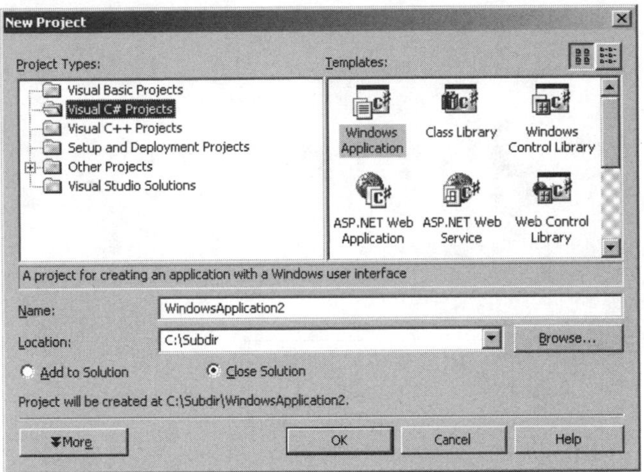

Figure 1680 Using the Visual Studio New Project dialog box to select a project type.

Using the Visual Studio, you can create a wide range of application types, from C# programs to Visual C++ programs and even programs in Visual Basic. Further, the Visual Studio lets you specify the specific type of application you need to build, from a console application, to a Windows program, to a class library, to a variety of services.

When you select a project type within the New Project dialog box, the Visual Studio uses a related template that provides you with code you can use to develop your application quickly. Table 1680 briefly discusses the various project types. Depending on the project type you select, the Visual Studio will open an appropriate template you can use to get started.

Project Type	Purpose
ASP.NET Web Application	An application that provides Web pages for a user interface.
ASP.NET Web Service	A Web-based service callable from other applications.
Class Library	Places one or more classes within a library for use by other applications.
Console Application	An application that executes from the command line.
Empty Project	An empty solution file within which you can place a local application.
Empty Web Project	An empty solution file within which you can place a Web application.
Web Control Library	A library of controls callable from Web-based applications.
Windows Application	A Windows application that provides forms for a user interface.
Windows Control Library	A library of controls for use in other Windows-based applications.
Windows Service	A Windows-based service callable from other applications.

Table 1680 Project templates for C# projects.

RAPIDLY BUILDING A FORM-BASED WINDOWS APPLICATION 1681

As you learned in Tip 1679, when you use the Visual Studio File menu New Project option to create a Windows application, the Visual Studio will display a blank form upon which your program can place various controls, as shown in Figure 1681.1.

Figure 1681.1 The Visual Studio displays a blank form when you create a new Windows application.

If you have programmed within the Visual Basic environment, you should be quite familiar with the use of forms and controls, such as text boxes, radio buttons, lists, and so on which you place onto a form to build an application's user interface. In this tip, you will get a feel for just how much power the Visual Studio provides to you as you build programs. You will use the Visual Studio to automatically generate the application *HelloWindows.cs*, that displays the form shown in Figure 1681.2.

Figure 1681.2 The output of the HelloWindows application.

To create and build the *HelloWindows* application using the Visual Studio, perform these steps:

1. Within the Visual Studio, select the File menu New option and choose Project. The Visual Studio, in turn, will display the New Project dialog box.
2. Within the New Project dialog box, type in the project name *HelloWindows* and select a directory location within which you want Windows to place your project files. Next, double-click your mouse on the Windows Application icon. Windows will display a blank form as previously shown in Figure 1681.1.
3. Select the Visual Studio View menu Toolbox option to display a window of controls you can place on a form, as shown in Figure 1681.3.

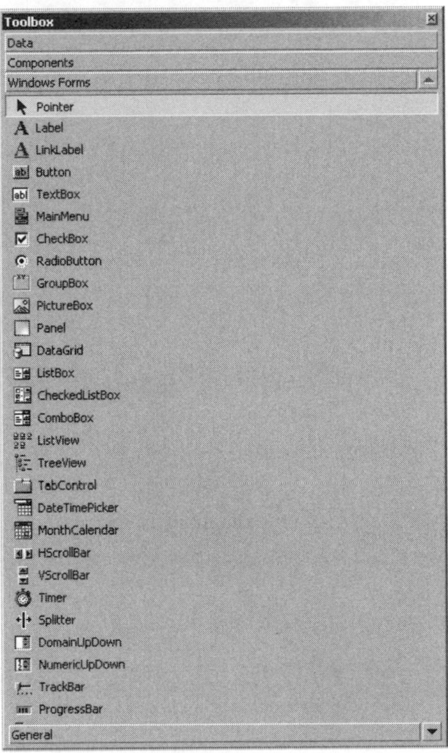

Figure 1681.3 The Visual Studio Toolbox contains controls you can place onto a form.

4. Within the Toolbox, drag the Label control onto the form. The Visual Studio will display a control box on the form for the label and a Properties dialog box you can use to assign attribute values to the control, as shown in Figure 1681.4.

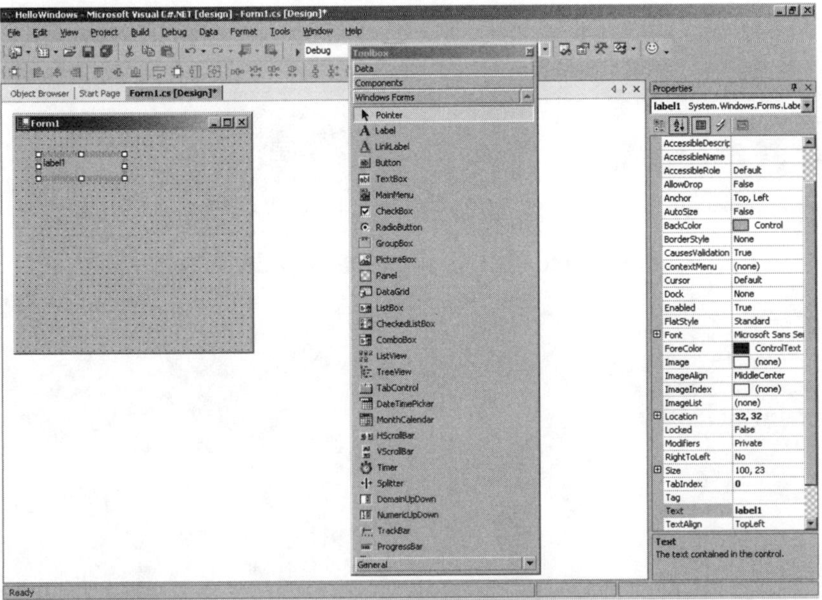

Figure 1681.4 The Visual Studio places a box around the current control and displays a Properties dialog box you can use to assign attributes to the control.

5. Within the Properties dialog box, use the slider to scroll down through the control's attributes and locate the Text attribute. Click your mouse within the Text property box, and then replace the text *Label1* with the text *Hello, Windows!*
6. Select the Build menu Build option to compile your program.
7. Select the Debug menu Start option to run your program.

Without writing any code, you have created a Windows-based program within the Visual Studio. In Tip 1682, you will take a look at the code the Visual Studio created for you to build your program.

1682 Viewing Your Source Code Within the Visual Studio

In the previous tip, you let Visual Studio automatically create a program for you that displayed a label within a form. To create the program, the Visual Studio generated the program's C# source code for you. To view the program source, select the View menu Code option. Visual Studio, in turn, will display the source code as shown in Figure 1682.

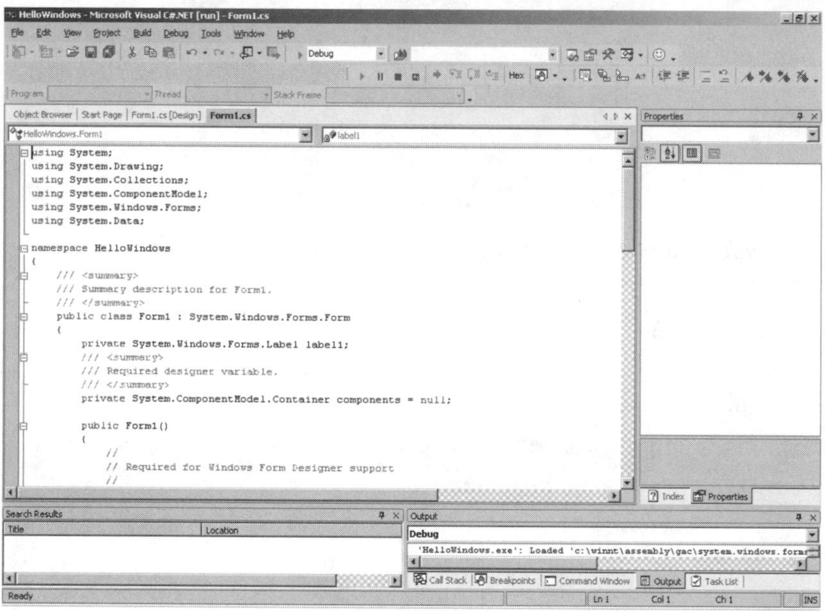

Figure 1682 Displaying source code within the Visual Studio.

To display the form, the Visual Studio created the following source code:

```
using System;
using System.Drawing;
using System.Collections;
using System.ComponentModel;
using System.Windows.Forms;
using System.Data;

namespace WindowsApplication
{
  /// <summary>
  /// Summary description for Form1.
  /// </summary>
  public class Form1 : System.Windows.Forms.Form
   {
```

```
private System.Windows.Forms.Label label1;
/// <summary>
/// Required designer variable.
/// </summary>
private System.ComponentModel.Container
  components = null;

public Form1()
 {
    //
    // Required for Windows Form Designer support
    //
    InitializeComponent();

    //
    // TODO: Add any constructor code after
    // InitializeComponent call
      //
 }

    /// <summary>
    /// Clean up any resources being used.
    /// </summary>
 protected override void Dispose( bool disposing )
  {
    if( disposing )
     {
       if (components != null)
          {
            components.Dispose();
          }
     }
    base.Dispose( disposing );
 }

 #region Windows Form Designer generated code
 /// <summary>
 /// Required method for Designer support-do not modify
 /// the contents of this method with the code editor.
 /// </summary>
 private void InitializeComponent()
  {
    this.label1 = new System.Windows.Forms.Label();
    this.SuspendLayout();
    //
    // label1
    //
    this.label1.Location = new
        System.Drawing.Point(64, 16);
    this.label1.Name = "label1";
    this.label1.TabIndex = 0;
    this.label1.Text = "Hello, Windows!";
    //
    // Form1
    //
    this.AutoScaleBaseSize =
       new System.Drawing.Size(5, 13);
    this.ClientSize = new System.Drawing.Size(292, 273);
    this.Controls.AddRange(new
       System.Windows.Forms.Control[] {this.label1});
    this.Name = "Form1";
    this.Text = "Form1";
```

```
          this.ResumeLayout(false);
        }
      #endregion

      /// <summary>
      /// The main entry point for the application.
      /// </summary>
      [STAThread]
      static void Main()
       {
          Application.Run(new Form1());
       }
   }
}
```

As you can see, the code that the Visual Studio creates contains XML-based comments, attributes, and inheritance.

1683 Using the Visual Studio's Common Controls

Within C# applications, programmers use forms to build the user interface. When you begin a program, you start with an empty form, which you can view as a clean slate onto which you drag controls, such as labels, text fields, radio buttons, checkboxes, and more. The Visual Studio provides such controls for you on the Toolbox, which you can display by selecting the View menu Toolbox option. By taking advantage of common controls the Visual Studio provides, you greatly simplify the amount of code and text you must write as you design your application's user interface. For years, Visual Basic programmers have made extensive use of controls to design an application's user interface.

In general, using the Visual Studio Toolbox, you can drag and drop a control onto a form. Behind the scenes, the Visual Studio will create the C# programming code that creates and displays the control. The "common controls" are so named because they provide users with a consistent user interface that is "common" across Windows-based applications. Figure 1683 illustrates a form onto which a programmer has dropped several common controls.

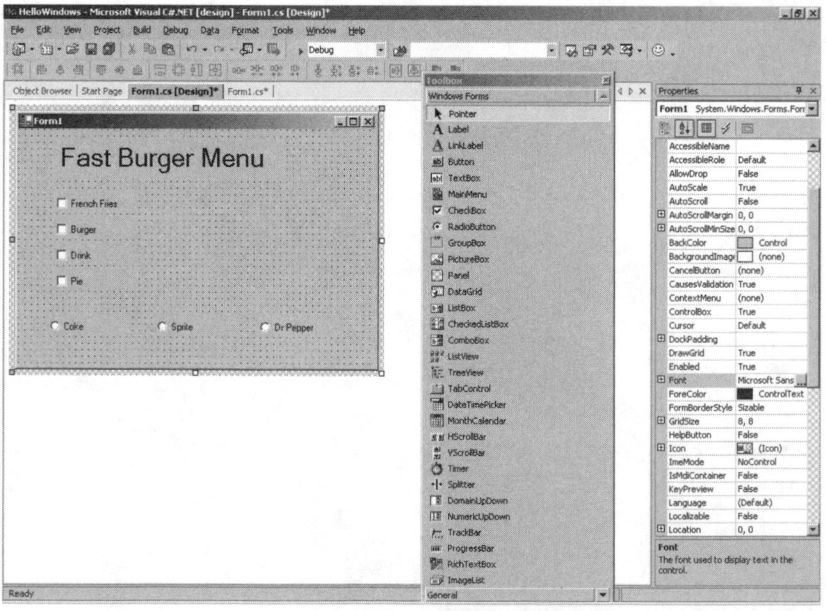

Figure 1683 Using the Visual Studio to drag and drop controls that build a user interface.

Table 1683 briefly describes the common controls you can use within your C# applications.

Control	Description
Button	Creates buttons such as Cancel.
Checkbox	Creates a selectable checkbox. Often used with other Checkbox controls from which users can select multiple boxes.
Combo Box	Combines an Edit and List box.
Date/Time Picker	Displays a dialog box that lets a user enter date and time information.
Edit Box	A box for entering or editing text.
Group Box	Group buttons and other controls.
Horizontal Scroll Bar	Creates a horizontal scroll bar.
Label	Contains a static text message.
List	Contains a list of text with icons.
List Box	Contains a list of strings.
Progress	A percentage bar that indicates the progress of an operation.
Radio Button	Creates a selectable button. Normally used with other buttons of which the user can select only one option.
Rich Edit	An Edit control supporting many text-formatting characters.
Slider	A control that lets the user select a value from a range of values by sliding a control across a bar.
Spin Button	Arrow buttons upon which the user can click his or her mouse to increment or decrement a value.
Static Text	Unchanging text, normally used for a control label.
Status Bar	A control to which applications can display program status messages.
Tab	A control that provides a tab upon which the user can select to chose a control, much like a notebook tab.
Toolbar	A control that contain icons (buttons) to initiate specific operations.
ToolTip	Creates a pop-up (balloon-like) window that contains text that describes a tool's use.
Tree	Displays a hierarchical list, much like a Windows Explorer directory list.
Vertical Scroll Bar	Creates a vertical scroll bar.

Table 1683 *The basic Windows common controls and their C# names.*

CUSTOMIZING COMMON CONTROLS 1684

In Tip 1682, you examined the Visual Studio toolbox and many of the common controls that you can drag and drop onto a form. When you place a control onto a form, the Visual Studio lets you customize a variety of settings, some of which apply to every control (such as the form's coordinates), and some of which are specific to the control. To assign properties to a specific control that appears on a form, click your mouse on the form. The Visual Studio, in turn, will display a box-like frame around the control to indicate that it is the current control. Next, the Visual Studio will display a Properties window that you can use to assign the attribute settings. To delete a control from a form, simply click your mouse on the form to select the control and then press the Delete key. To size a control on the form, use your mouse to drag the corner of the control's frame in or out to increase the size of the frame as you desire.

Figure 1683.1, for example, illustrates a Date/Time picker control on a form. As you can see, using the control's property settings, you can change the control's size, default date and time, and so on.

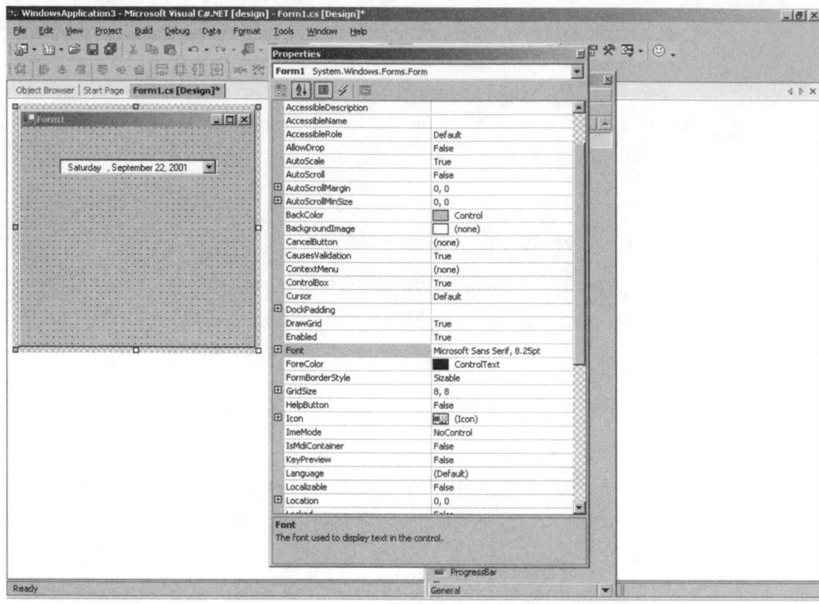

Figure 1684.1 Using a properties window to assign attributes to the selected control.

As you use the property settings to change a control's attributes, the Visual Studio, in turn, will automatically update the control's corresponding source code within your application to reflect the change. For example, Figure 1684.2 illustrates a form that contains a Monthly Calendar control.

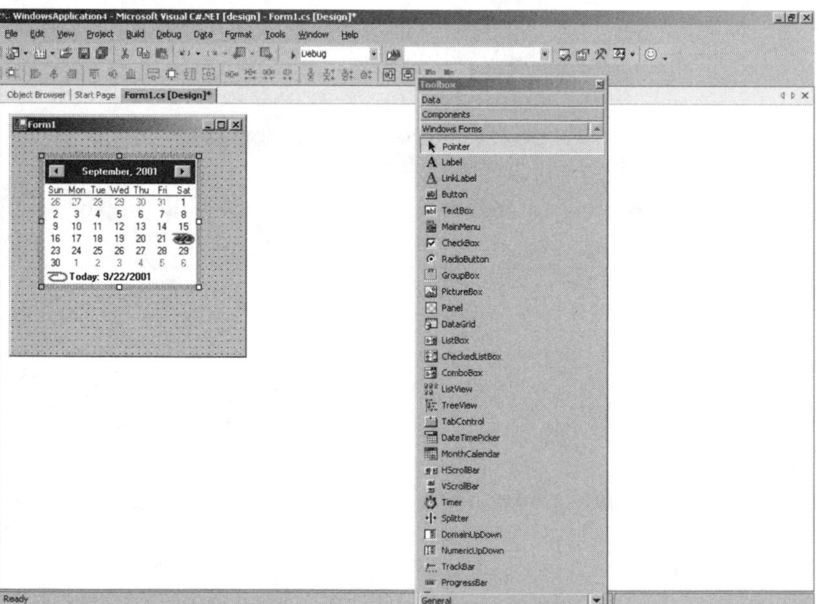

Figure 1684.2 A Monthly Calendar control on a form.

By default, when you place a Monthly Calendar control on a form, the Visual Studio will place the following program statements within your code:

```
this.monthCalendar1 = new
    System.Windows.Forms.MonthCalendar();
this.SuspendLayout();

//
// monthCalendar1
//

this.monthCalendar1.Location = new
   System.Drawing.Point(40, 16);
this.monthCalendar1.Name = "monthCalendar1";
this.monthCalendar1.TabIndex = 0;
```

If you use the Visual Studio to change the Monthly Calendar's attributes, the Visual Studio will change code accordingly, as shown here:

```
this.monthCalendar1 = new
    System.Windows.Forms.MonthCalendar();
this.SuspendLayout();

//
// monthCalendar1
//

this.monthCalendar1.Location = new
    System.Drawing.Point(40, 16);
this.monthCalendar1.Name = "monthCalendar1";
this.monthCalendar1.ShowToday = false;
this.monthCalendar1.ShowTodayCircle = false;
this.monthCalendar1.TabIndex = 0;
this.monthCalendar1.TitleBackColor =
    System.Drawing.Color.FromArgb(((System.Byte)(255)),
      ((System.Byte)(128)), ((System.Byte)(128)));

this.monthCalendar1.TodayDate = new
    System.DateTime(2001, 9, 27, 0, 0, 0, 0);
```

Likewise, as you change your program's source code, the Visual Studio will reflect the changes within the control's property settings.

INTERACTING WITH COMMON CONTROLS 1685

The Visual Studio provides many powerful controls your applications can use by simply dragging and dropping the control onto a form. Using the Visual Studio, drag and drop the following *MonthCalendar* and *DateTimePicker* controls onto a form (to fit the controls onto a form, you may have to increase the form's size). Your form, in turn, should appear as shown in Figure 1684.

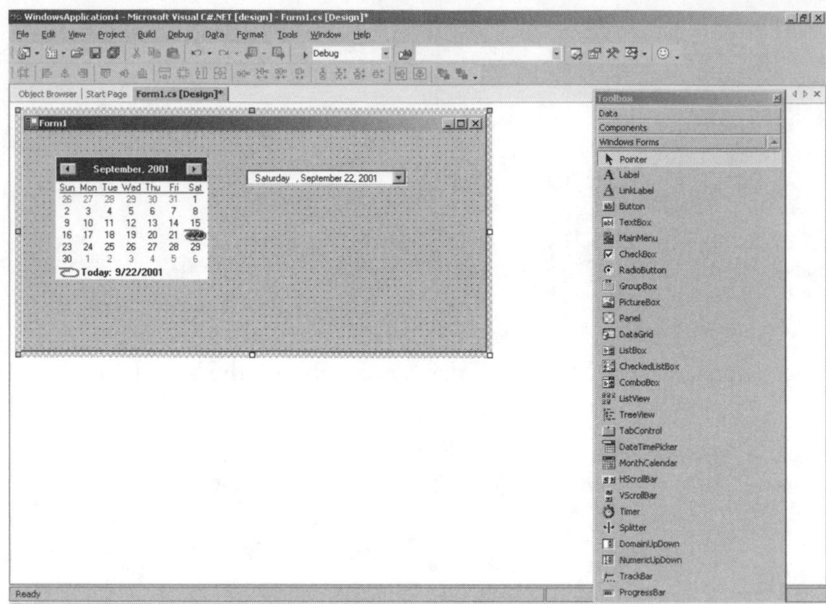

Figure 1685 Placing common controls on a form.

Depending on the control you are using, how your application will interact with the control will differ. In some cases your code will call specific methods to get a specific control value; in other cases, the control's method will return several values within a string that you will parse; and, as you will learn later, there will be times when your code must respond to handle events a control on the form generates.

For example, the following program, *GetControlValues.cs*, displays the values the user selectsfor the controls shown in Figure 1684. The majority of the program code is created automatically by the Visual Studio. Within the *Main* method, however, place the following statements:

```
Form1 DemoForm = new Form1();
Application.Run(DemoForm);

MessageBox.Show(DemoForm.monthCalendar1.ToString(),
   "Date");
MessageBox.Show(DemoForm.dateTimePicker1.ToString(),
   "Date/Time");
```

The following statements illustrate the program's complete source code:

```
using System;
using System.Drawing;
using System.Collections;
using System.ComponentModel;
using System.Windows.Forms;
using System.Data;

namespace GetControlValues
{
  /// <summary>
  /// Summary description for Form1.
  /// </summary>
  public class Form1 : System.Windows.Forms.Form
  {
    private System.Windows.Forms.MonthCalendar
```

```
        monthCalendar1;
    private System.Windows.Forms.DateTimePicker
        dateTimePicker1;

    /// <summary>
    /// Required designer variable.
    /// </summary>
    private System.ComponentModel.Container
      components = null;

    public Form1()
     {
      //
      // Required for Windows Form Designer support
      //
      InitializeComponent();

      //
      // TODO: Add any constructor code after
      // InitializeComponent call
      //
     }

    /// <summary>
    /// Clean up any resources being used.
    /// </summary>
    protected override void Dispose( bool disposing )
     {
       if( disposing )
         {
           if (components != null)
            {
              components.Dispose();
            }
         }
       base.Dispose( disposing );
     }

     #region Windows Form Designer generated code
     /// <summary>
     /// Required method for Designer support-do not modify
     /// the contents of this method with the code editor.
     /// </summary>
     private void InitializeComponent()
      {
        this.monthCalendar1 =
          new System.Windows.Forms.MonthCalendar();
        this.dateTimePicker1 =
          new System.Windows.Forms.DateTimePicker();
        this.SuspendLayout();
        //
        // monthCalendar1
        //
        this.monthCalendar1.Location =
          new System.Drawing.Point(32, 16);
        this.monthCalendar1.Name = "monthCalendar1";
        this.monthCalendar1.TabIndex = 0;
        //
        // dateTimePicker1
        //
        this.dateTimePicker1.Location =
          new System.Drawing.Point(272, 40);
```

```
            this.dateTimePicker1.Name = "dateTimePicker1";
            this.dateTimePicker1.TabIndex = 1;
            //
            // Form1
            //
            this.AutoScaleBaseSize =
              new System.Drawing.Size(5, 13);
            this.ClientSize =
              new System.Drawing.Size(520, 309);
            this.Controls.AddRange(new
              System.Windows.Forms.Control[] {
                this.dateTimePicker1,
                this.monthCalendar1});
                this.Name = "Form1";
                this.Text = "Form1";
                this.ResumeLayout(false);
        }
      #endregion

      /// <summary>
      /// The main entry point for the application.
      /// </summary>
      [STAThread]
      static void Main()
        {
          Form1 DemoForm = new Form1();
          Application.Run(DemoForm);

          MessageBox.Show(DemoForm.monthCalendar1.ToString(),
            "Date");
          MessageBox.Show(DemoForm.dateTimePicker1.ToString(),
            "Date/Time");
        }
    }
}
```

In this case, the program retrieves the *monthCalendar1* and *dateTimePicker1* control values by simply accessing the control's *ToString* method. When you run this program, a form will appear on your screen that contains the two controls. After you select the values you desire, close the form (in later tips you will learn how to use events to respond to button operations that would let you place, for example, an OK or Close button on the form). The program will then display two message boxes that contain the values you selected.

1686 Using the Common Dialogs

In several of the previous tips, you have used the C# common controls within your programs to build a common user interface. In addition to the common controls, the C# programming environment provides a collection of common dialog boxes that programs can use to open files, save files, print documents, and so on. Table 1685 briefly describes each of the common dialogs. Figure 1686, for example, illustrates the Font dialog box, which applications can display using a *FontDialog* object.

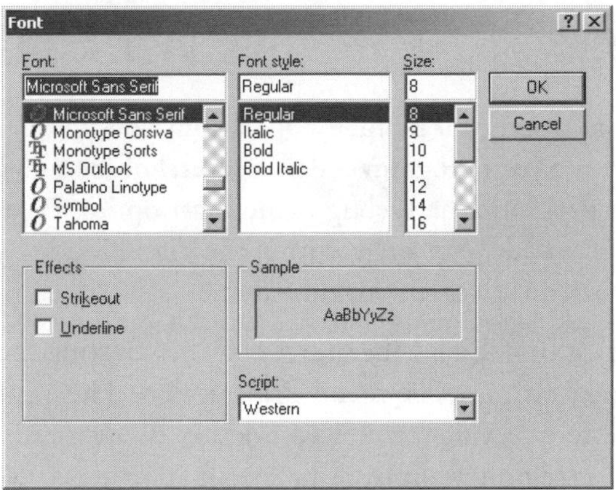

Figure 1686 The Font dialog box.

Dialog	Purpose
ColorDialog	Lets the user select an existing color or define and select a custom color.
FontDialog	Lets a user select an existing font.
OpenFileDialog	Lets a user open a file.
PageSetupDialog	Lets a user customize how an application will print information.
PrintDialog	Lets a user select a printer and choose the portion of a document to print.
SaveFileDialog	Lets a user save a file.

Table 1686 The common dialog boxes available to C# applications.

The following program, *UseCommonDialogs.cs*, illustrates the use of the several of the common dialogs within a C# program:

```
using System;
using System.Drawing;
using System.Windows.Forms;

class clsMain
{
   static public void Main ()
   {
         ColorDialog color = new ColorDialog();
         color.ShowDialog ();
         MessageBox.Show(color.ToString(), "Color");

         OpenFileDialog openfile = new OpenFileDialog();
         openfile.ShowDialog ();
         MessageBox.Show(openfile.FileName, "File");

         SaveFileDialog savefile = new SaveFileDialog();
         savefile.ShowDialog ();
         MessageBox.Show(savefile.FileName, "File");

         FontDialog font = new FontDialog();
         font.ShowDialog ();
         MessageBox.Show(font.ToString());
   }
}
```

1687 Performing Common Operations Within the Visual Studio

As you work within the Visual Studio, there will be many operations you will perform on a regular basis, such as compiling, running, saving, and printing your program code. In several of the previous tips, you have used the Build menu Build option to compile your program, the Debug menu Start option to run your application, and the File menu Save As option to save your source code. Normally, within the Visual Studio, you will want to save each of your project files, and not just your source code. To save your project files, select the File menu Save All option.

To print your program source, you must first display the source code by selecting the View menu Code option. Then, you can print your source code by selecting the File menu Print option. The Visual Studio, in turn, will display a print dialog box, which you can use to select the printer you desire. By default, the Visual Studio will print your entire source file. In some cases, you may only want to print specific portions of your code. In such cases, use your mouse or keyboard arrow keys (while holding down the Shift key) to select the code you want to print and then select the File menu Print option. Within the Print dialog box, choose the Selection option and choose OK.

The Visual Studio provides extensive online help. To access the help within Visual Studio, select the Help menu. Figure 1687, for example, displays the Visual Studio help text for the *MonthlyCalendar* class that you used in Tip 1684. Within the Visual Studio online help, you can view information regarding Common Language Runtime routines, standard types, as well as details regarding class members, methods, properties, and more.

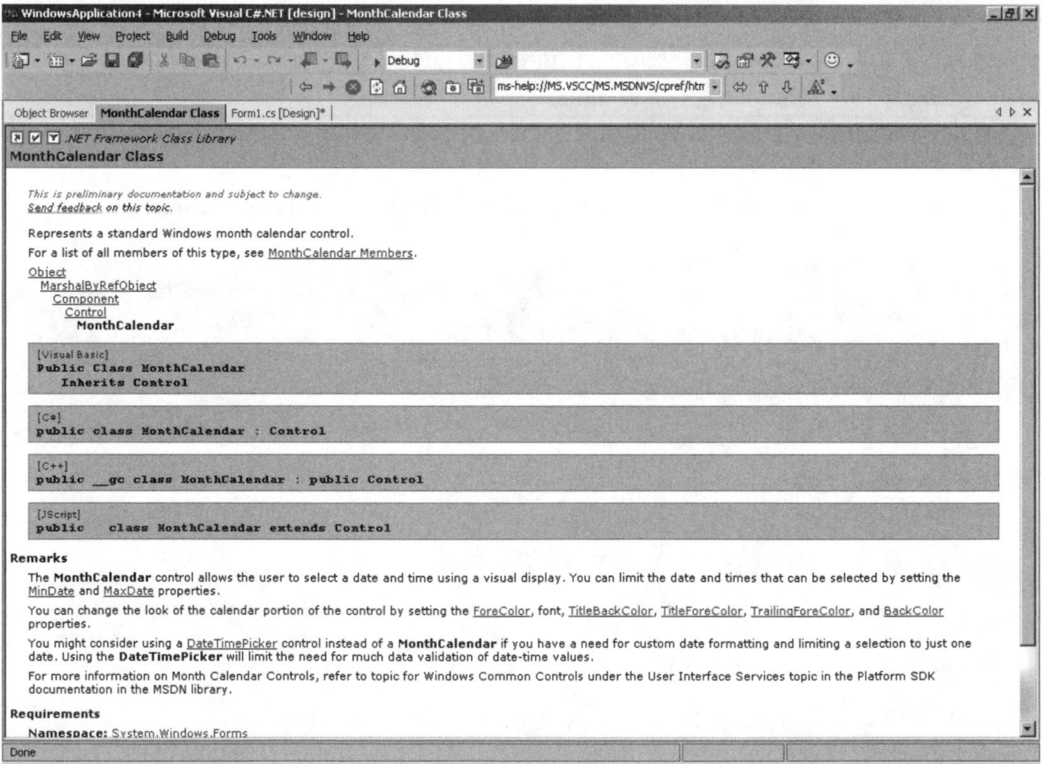

Figure 1687 Using the Visual Studio online help facility.

As the complexity of your programs increases, finding errors (debugging) your applications can become more difficult. To help you locate errors within your code, the Visual Studio provides debugging software you can use to execute your program one statement at a time, or until the code reaches a specific statement you select, called a breakpoint. To access the Visual Studio debugging tools, select the Debug menu. As you debug programs, there will be many times when you will want your program to run until it reaches a specific statement. To place such a debug breakpoint within your code, right-click your mouse on the statement you desire. The Visual Studio, in turn, will display a pop-up menu. Within the menu, select the Insert Breakpoint option. The Visual Studio, in turn, will highlight the statement to indicate the breakpoint. When you later select the Debug menu Start option, the Visual Studio will run the program statements that appear prior to your breakpoint.

HANDLING CONTROL-GENERATED EVENTS 1688

In previous tips, you learned how to use delegates to create event handlers, methods which automatically execute when the event occurs. Many of the controls you can place on a form generate events that your program can handle to determine if the user has selected a specific control, change the contents of a control, and so on. The Visual Studio makes it very easy for you to define handlers for such events. To better understand the event handling process, create a new Windows project within the Visual Studio. Next, using the Toolbox, drag an *ab* button onto the form. Using the control's Properties window, change the control's text from *Button1* to OK. Your form will appear as shown in Figure 1688.1.

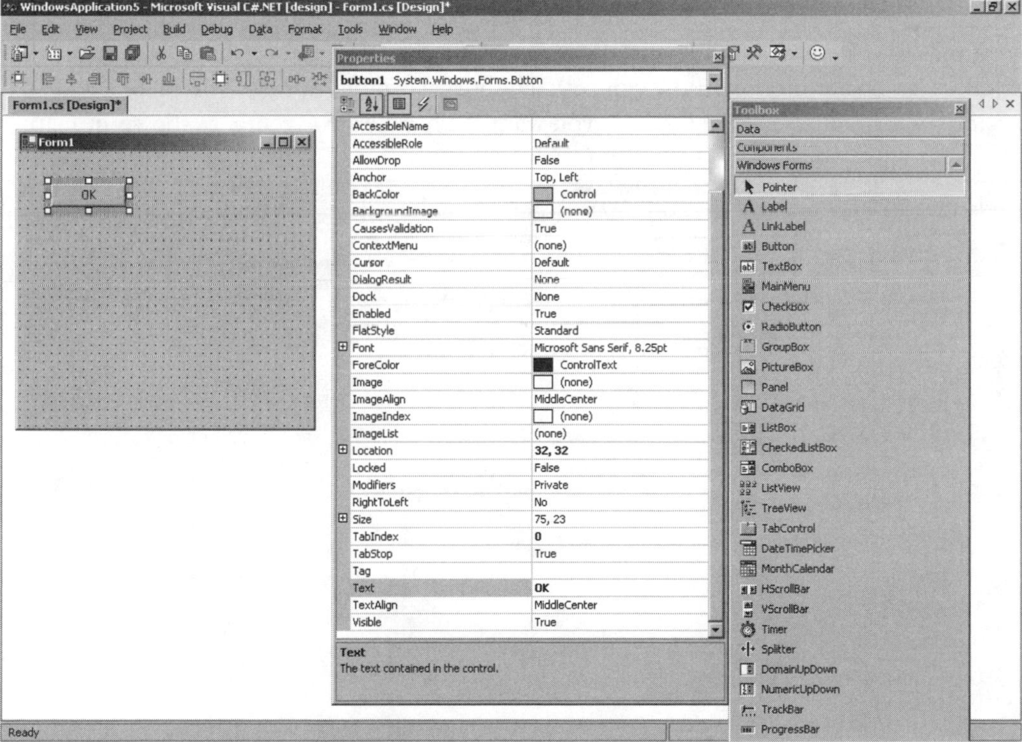

Figure 1688.1 Placing an OK button on a form.

At the Properties window, you should see a small icon that contains a lightning bolt, which corresponds to the control's events. Click your mouse on the lightning bolt icon. The Properties window, in turn, will display a list of the control's events, as shown in Figure 1688.2.

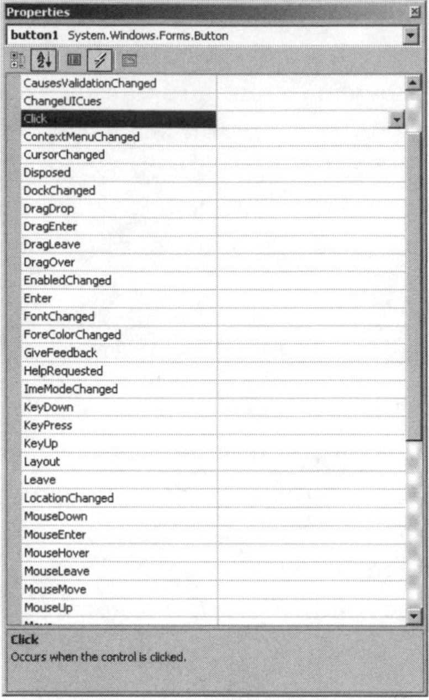

Figure 1688.2 Displaying a control's events.

Within the control's list of events, double-click your mouse on the Click event (that the control will generate each time the user clicks his or her mouse on the control). The Visual Studio, in turn, will place a code template within your program into which you can write the code you want the handler to execute, as shown in Figure 1688.3.

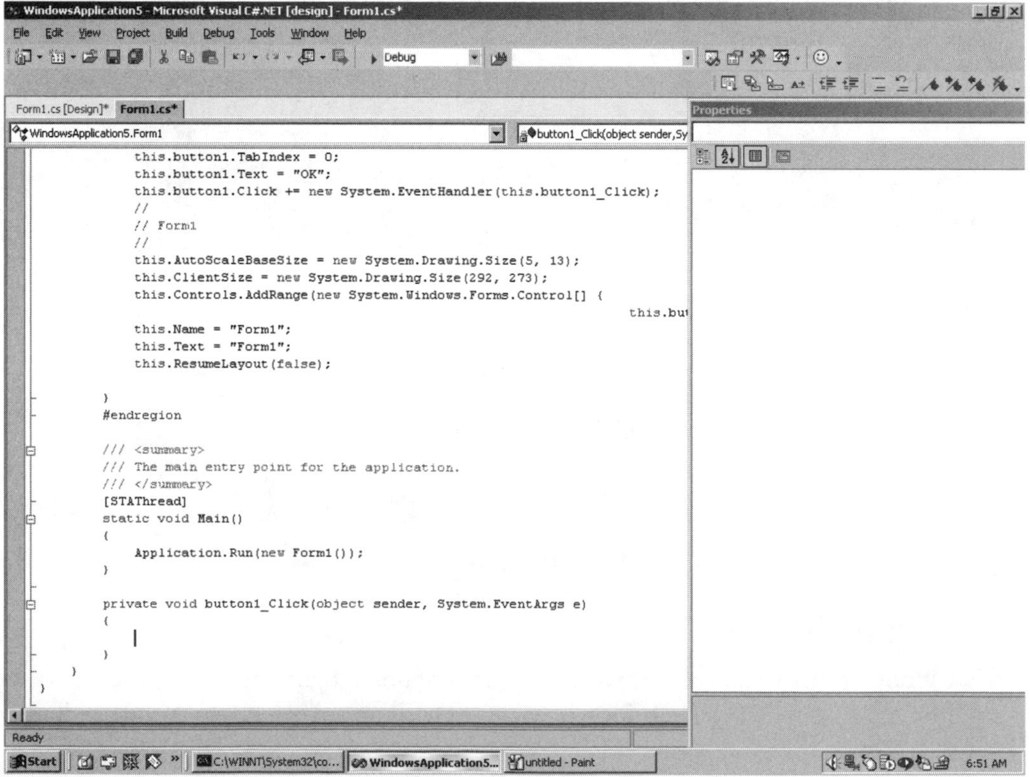

Figure 1688.3 Using event-handler code generated by the Visual Studio.

Note how the event-handler code resembles the code you created in previous tips. In this case, within the event handler, simply place the code:

```
MessageBox.Show("Mouse Click Caught",
   "Event Handler Demo");
```

Each time you click your mouse on the control, the event handler, in turn, will display the message box. Thus, creating an event handler for a control within the Visual Studio is as easy as double-clicking your mouse on the control event you desire and then typing in the code you want the handler to execute.

HIDING, REDISPLAYING, AND CLOSING FORMS 1689

In Tip 1684, you created the *UseControl.cs* application that displayed a form that contains a *MonthlyCalendar* control and a *DateTimePicker* control. At that time, the form did not have any buttons you could use to control the form's processing. Within the Visual Studio, open the UseControl project and then add to the form the OK button, as shown in Figure 1689.

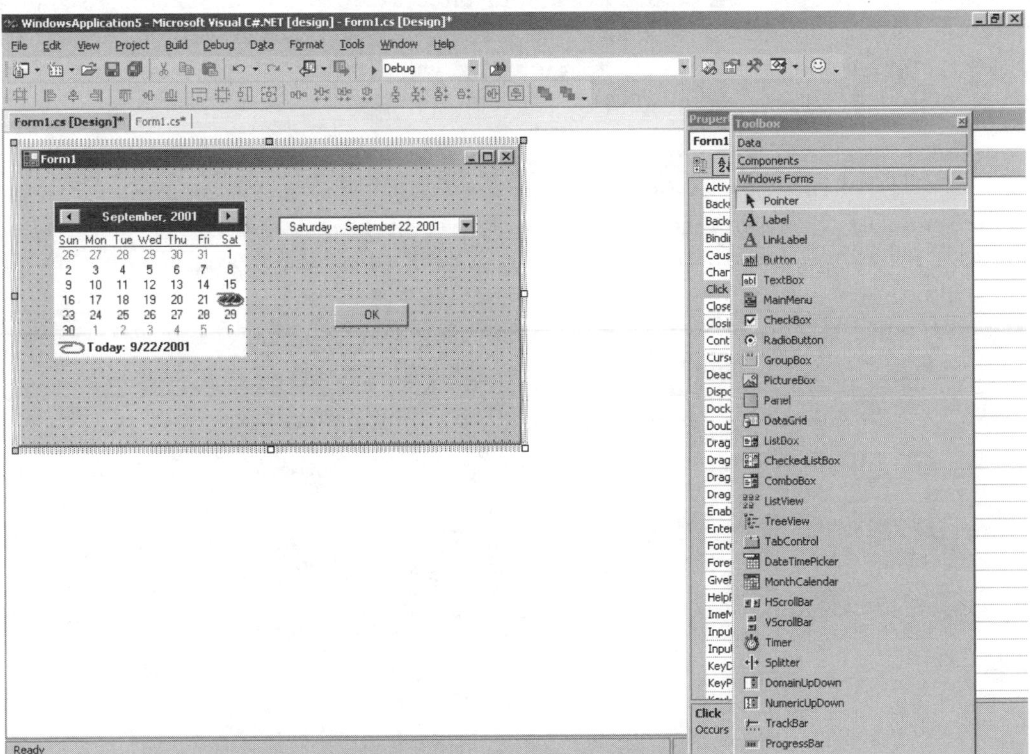

Figure 1689 Adding an OK button to a form.

Next, use the Properties window to display the OK button's events and double-click your mouse on the Click event. Within the event handler, place the following statements:

```
MessageBox.Show("About to hide the form", "Form Message");

this.FindForm().Hide();
MessageBox.Show("About to redisplay the form",
   "Leaving Event Handler");
this.FindForm().Show();
```

Note the event handler uses the code *this.FindForm()* to get the control's corresponding form. Using the *Hide* method, the code can hide (but not destroy) the form. Later using the *Show* method, the code can redisplay the form.

You could also, for example, move the code from the *Main* method that displays the user's selections into the event handler, and possibly then not redisplay the form, as shown here:

```
this.FindForm().Hide();
MessageBox.Show(this.FindForm().monthCalendar1.ToString(),
     "Date");
 MessageBox.Show(this.FindForm().dateTimePicker1.ToString(),
     "Date/Time");

this.FindForm.Close();
```

In this case, when the code executes the *Close* method the code will destroy the code form.

1690 Adding Forms to a C# Project

In several of the previous tips, you have created a Windows project that used a single form. As your applications become more complex, you may use multiple forms to create your user interface. Fortunately, the Visual Studio makes it very easy for you to add forms to your application and will automatically generate much of the code your application needs to display and manage the form.

To better understand the process of using multiple forms within an application, create a new Windows project that you name *MultipleForms*. When Windows displays the default *Form1*, select the Project menu Add Windows Form option two times to add *Form2* and *Form3* to the project, as shown in Figure 1690.1. Each time you select the Project menu Add Windows Form option, the Visual Studio will display the Add New Item dialog box. Within the dialog box, double-click your mouse on the Windows Form option.

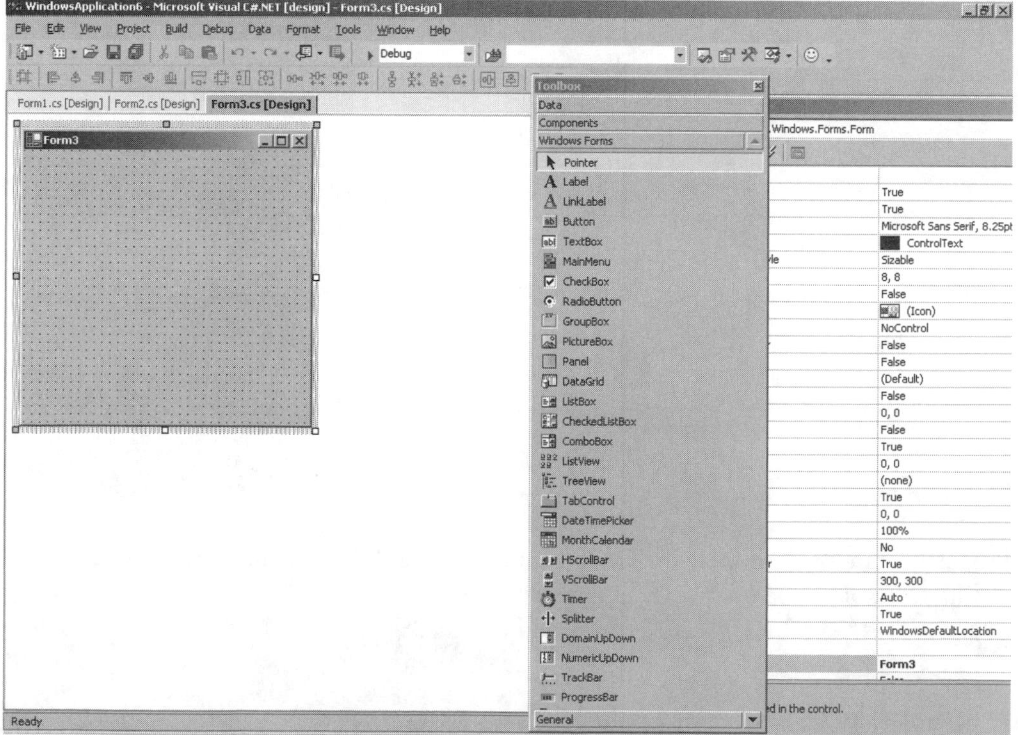

Figure 1690.1 Using multiple forms within a Windows-based project.

When you place multiple forms within a project, the Visual Studio will separate each form's source within its own file. To place a control onto a specific form, you simply select the design view for the form you desire and then drag and drop the controls you need from the Visual Studio Toolbox.

As discussed, the Visual Studio will generate automatically many of the programming statements you will need to manage each form and the controls it contains. The Visual Studio will place the code for your program's *Main* method within *Form1*.

Within the *Main* method within *Form1*, place the following statements that create and show form objects for *Form2* and *Form3,* and then start *Form1* using the *Application.Run* method:

```
static void Main()
 {
   Form2 F2 = new Form2();
   Form3 F3 = new Form3();
   F2.Show();
   F3.Show();
   Application.Run(new Form1());
 }
```

After you compile and execute this application, the program will display the three forms as shown in Figure 1690.2. Using your mouse, you can click on a form to bring the form to the forefront, as the active form. You can also individually close forms. If you close Form1, however, the program will close all of the forms and will end. In Tip 1691, you will learn how to better control form operations.

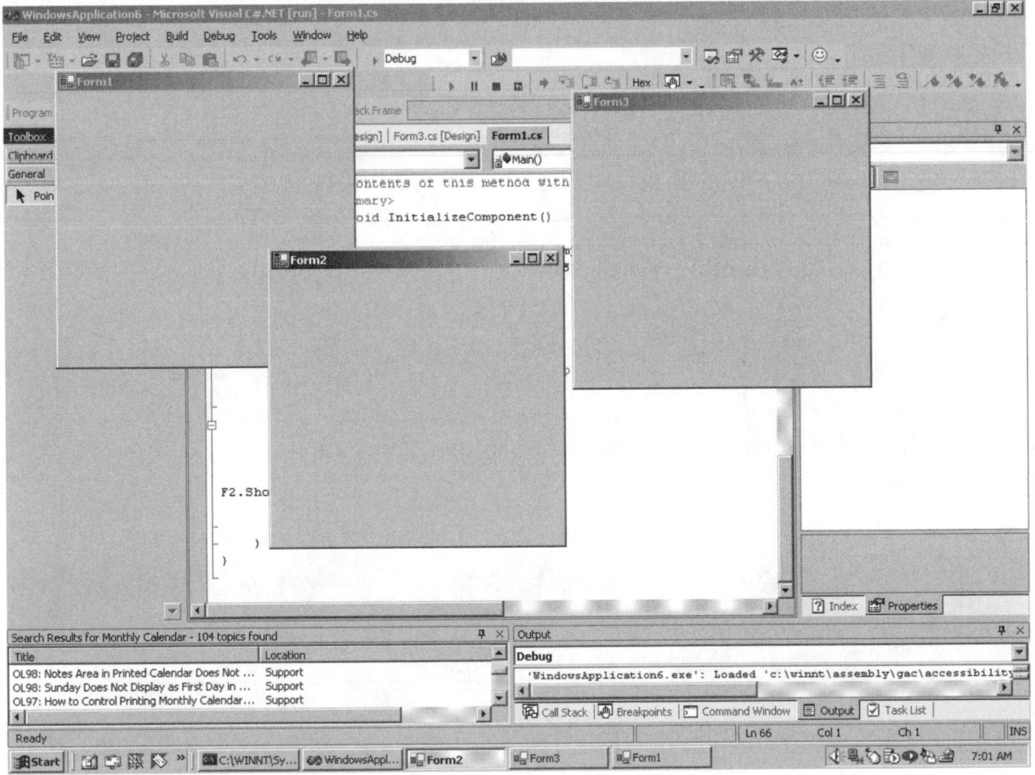

Figure 1690.2 Displaying multiple forms at one time.

1691 DISPLAYING AND HIDING A FORM

In Tip 1690, you created an application that opened and displayed three forms on the user's screen. When your applications use multiple forms, you will normally want your applications to maintain control of when each form appears and how the users traverses the forms. The following code changes the previous program slightly to provide buttons the user can click to move from one form to the next, or to the previous form. If the user tries to close *Form2*, the application will redisplay *Form1*. Likewise, if the user closes *Form3*, the application will redisplay *Form2*.

Within *Form1*, place the following code within the Click event handler for the Next button:

```
private void button1_Click(object sender, System.EventArgs e)
 {
   Form2 f2 = new Form2();
   this.Hide();
   f2.ShowDialog();
   this.Show();
 }
```

In this case, the code creates a new instance of the *Form2* class, hides itself, and then displays the new *Form2* class as a dialog box. When the *Form2* class completes its processing, this form will redisplay itself and resume its processing.

Within *Form2*, place the following code within the Click event handler for the Next button:

```
private void button1_Click(object sender, System.EventArgs e)
 {
   Form3 f3 = new Form3();
   this.Hide();
   f3.ShowDialog();
   this.Show();
 }
```

As you can see, the processing the *Form2* class performs when the user clicks on the Next button is nearly identical to that performed by *Form1*, with the exception that it displays a new instance of *Form3*.

Then, place the following code within the event handler for the Previous button:

```
private void button2_Click(object sender, System.EventArgs e)
 {
   this.FindForm().Close();
 }
```

When the user clicks his or her mouse on the Previous button, the form simply closes itself, which returns control to the caller (which in this case is *Form1*).

Finally, within *Form3*, place the following code within the event handler for the Previous button:

```
private void button1_Click(object sender, System.EventArgs e)
 {
   this.FindForm().Close();
 }
```

1692 USING THE VISUAL STUDIO TO BUILD CONSOLE APPLICATIONS

Throughout this book, you have created numerous Console-based applications that display their output using the *Console.WriteLine* method. Often, when you create simple Console-based applications, you will use the command line C# compiler (csc) to compile your source code. However, the Visual Studio makes it easy for you to build, run, and debug Console-based applications.

To create a Console-based application within the Visual Studio, select the File menu New option and choose Project. The Visual Studio, in turn, will display the New Project Type dialog box. Within the dialog box, type in the name you want to use for your program, and then double-click your mouse on the Console Application icon. The Visual Studio, in turn, will open a code window, within which is provided some template source code to get you started faster, as shown in Figure 1692.

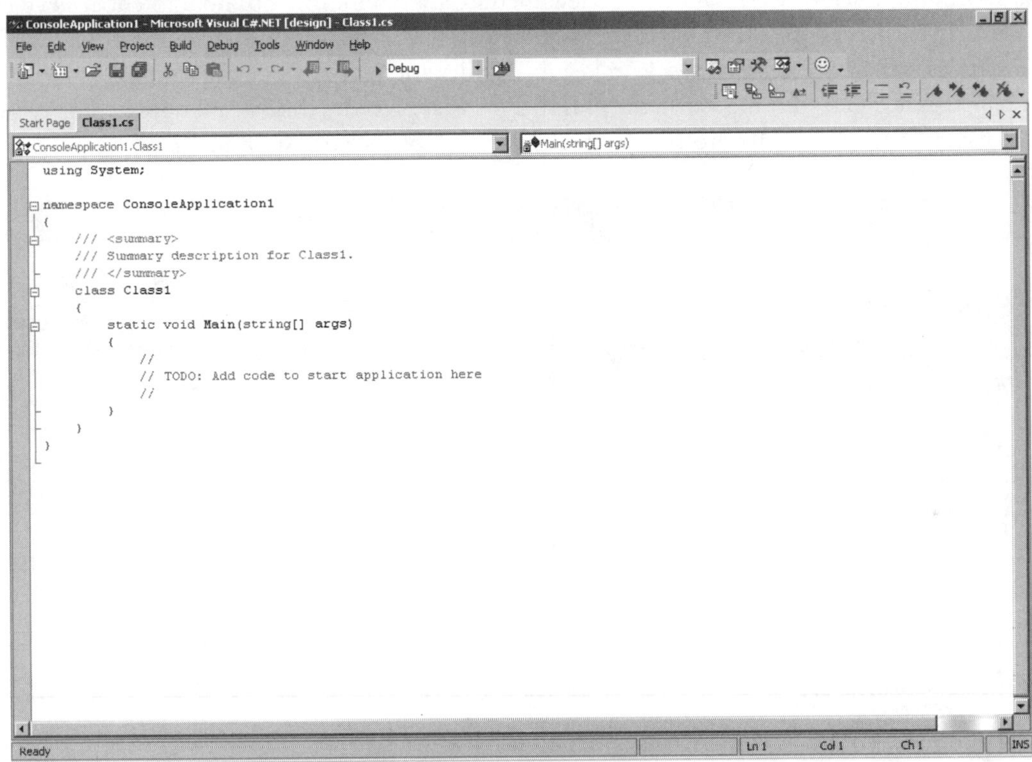

Figure 1692 Using the Visual Studio to create a Console-based application.

Assume, for example, that your Console-based application contains the following statements:

```
using System;

class ConsoleDemo
{
  static void Main()
   {
      Console.WriteLine("Hello Kris");
   }
}
```

To build your program within the Visual Studio, you select the Build menu Build option. Next, to run your program, you select the Debug menu Start option. Unfortunately, your program will normally quickly open a console window, display its output, and then close the window before you have a chance to see the output. One way to prevent the Visual Studio from closing your program's output window so quickly is to place a breakpoint after your program's last statement. You can insert a breakpoint within your code by right-clicking your mouse on the program statement you desire and then choosing the Insert Breakpoint option from the pop-up menu that appears. When the Visual Studio debugger runs your program, it will stop the program's execution when the debugger reaches the breakpoint, which will give you an opportunity to click your mouse on the Console window to display the Console-based output.

1693 Using the Object Browser to Research Classes

As you write programs, there will be many times when you will need to look up specific class methods and members. In such cases, the Visual Studio Object Browser makes researching class information very quick and easy. To open the Object Browser window, select the Visual Studio View menu Other Windows option and choose Object Browser. The Visual Studio, in turn, will open the Object Browser. By default, it will let you view objects that reside within the current application as well as objects that reside in key .NET files and namespaces. For example, to view the objects in the System namespace, click your mouse on the plus sign (+) that precedes the System option. The Object Browser will expand its display, as shown in Figure 1693.1, to display each of the corresponding objects.

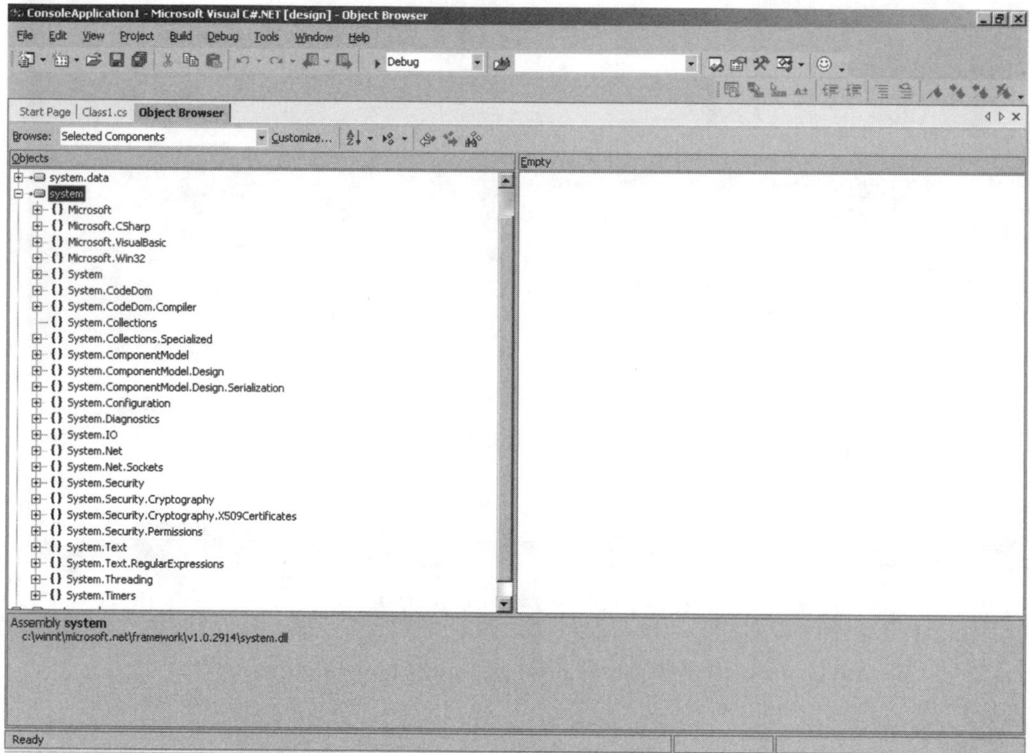

Figure 1693.1 Using the Object Browser to view System objects.

If you continue to click on the plus sign that precedes entries within the Object Browser tree, you can eventually display specifics about an object's methods, properties, fields, and interfaces, as shown in Figure 1693.2.

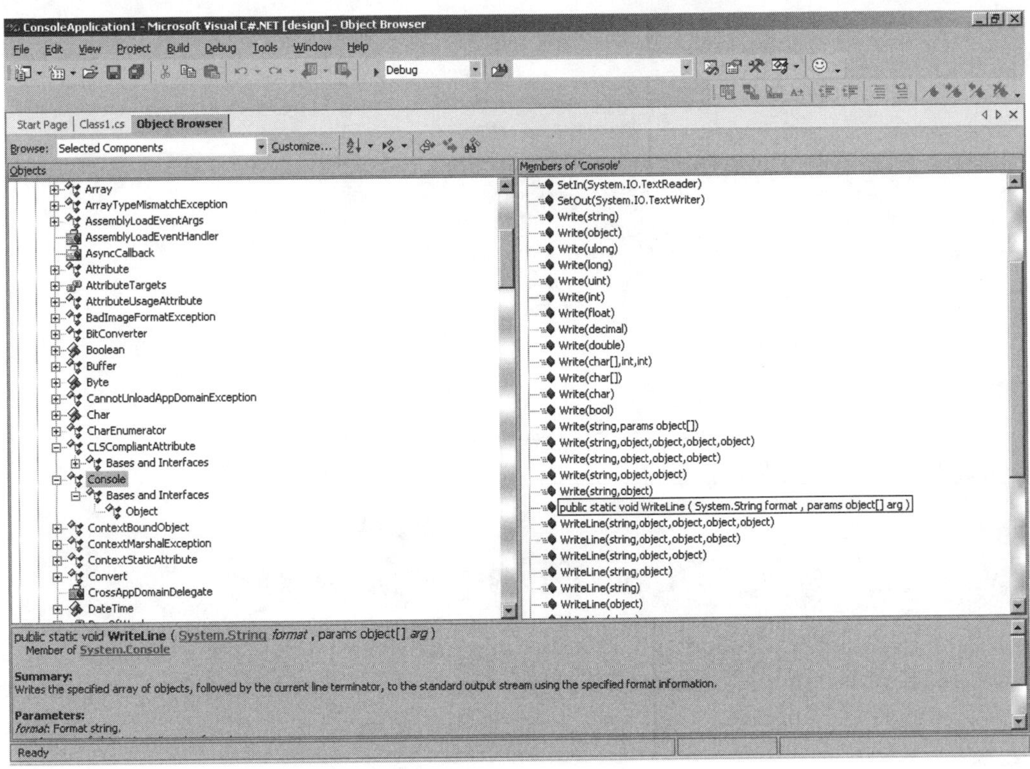

Figure 1693.2 Using the Object Browser to display class specifics.

USING THE SOLUTION EXPLORER TO MANAGE YOUR PROJECT 1694

Throughout this book's tips, you have been able to implement most of your programs using a single C# source file. As your C# applications become more complex, you may use multiple source files that contain individual classes, library files, and other resources (such as graphics). When you create a project within the Visual Studio, the project organizes the various files for you. If you take time to examine the folders within which you have stored the Windows-based projects you have created during recent tips, you will find files within a variety of extensions.

You will normally encounter a file with the *csproj* extension that contains your C# project file. Next, you may find a file with the *sln* extension that programmers refer to as a *solution file*. In general, the solution file stores information about your project (such as the locations of files on your disk). Finally, you may encounter a file with the *suo* extension, which contains the options (suo stands for solution user options) you can select within the Visual Studio, and which lets you customize the integrated development environment on aproject basis.

As you work with multiple files within the Visual Studio, you should take advantage of the Visual Studio's Solution Explorer, shown in Figure 1694. Using the Solution Manager, you can move quickly between each of your project files or manage your files by performing copy, move, and delete operations. To open the Solution Explorer, select the View menu Solution Explorer option.

Figure 1694 Using the Solution Explorer to manage a project.

1695 USING THE VISUAL STUDIO CLASS VIEW

As your classes become more complex and as the number of classes your programs use increases, you should take advantage of the Visual Studio Class View to quickly navigate your classes. Using the Class View, you can view class methods, fields, properties, indexers, base classes, and interfaces. For example, the following program, *InhertitProperties.cs*, derives a class from a base class that implements an interface:

```
using System;

class Package
{
  public Package(int W)
   {
     this.Weight = W;
   }

  private int weight;

  public int Weight
   {
     set
      {
        if ((value >= 25) && (value <= 50))
          weight = value;
        else
          weight = -1;
      }

     get
      {
        return weight;
      }
   }
```

```
  public void showPackage()
   {
     Console.WriteLine("Weight: {0}", Weight);
   }
}

class CardboardBox : Package
{
  public CardboardBox(int weight, int paperweight) :
         base(weight)
   {
     PaperWeight = paperweight;  // Strength of cardboard
   }

  public void ShowBox()
   {
     Console.WriteLine("Box supports {0} pounds",
       BoxWeight);
     base.showPackage();
   }

  private int BoxWeight;

  public int PaperWeight
   {
     set
      {
        if ((value > 0) && (value <= 100))
          BoxWeight = value;
        else
          BoxWeight = -1;
      }

     get
      {
        return BoxWeight;
      }
   }
}

class clsInheritProperties
{
  public static void Main ()
   {
     CardboardBox Box = new CardboardBox(50, 75);
     Box.ShowBox();

     Box.PaperWeight = 200;
     Box.ShowBox();

     Box.PaperWeight = 80;
     Box.Weight = 33;
     Box.ShowBox();

     Box.PaperWeight = 50;
     Box.Weight = 200;
     Box.ShowBox();
   }
}
```

Using the Class View, you can traverse your program's classes as shown in Figure 1695. To display the Class View Window, select the View menu Class View option. The Class View window, in turn, will display your program's class hierarchy. By clicking your mouse on an entry within the Class View window, you can quickly jump to the corresponding source code within your application.

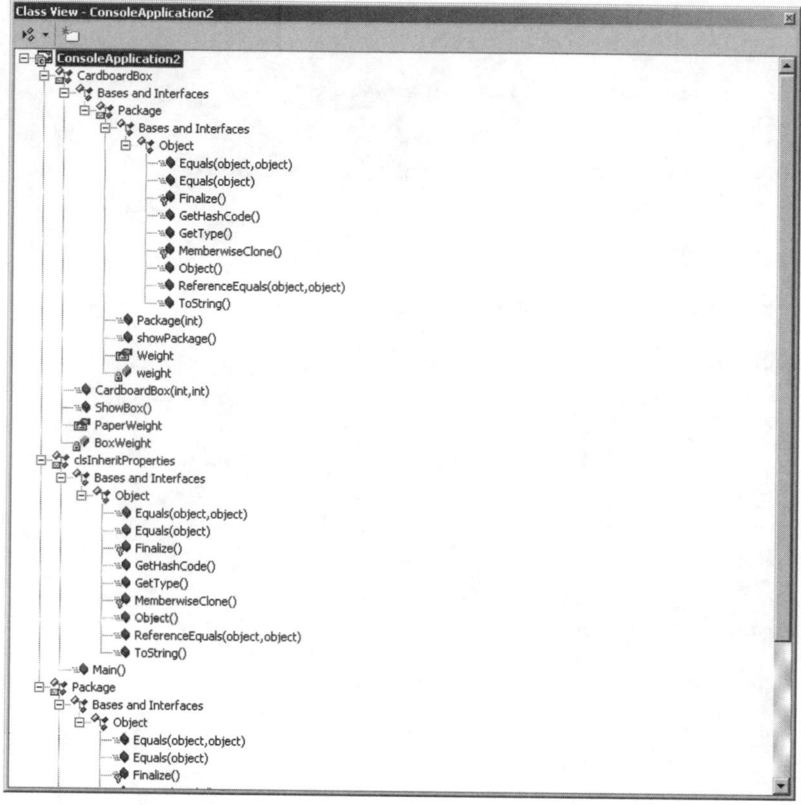

Figure 1695 Using the Visual Studio Class View to traverse a class.

1696 Using the Visual Studio Class Wizard

In several of the previous tips, you have used the Visual Studio to drag and drop forms and controls into your application. As you use the Visual Studio's visual environment to move such objects, Visual Studio, behind the scenes, automatically generates the corresponding source code your program must execute to interact with the object.

In a similar way, the Visual Studio provides a tool called the Class Wizard that you can use to simplify the steps you must perform to add a class to your code. For example, assume that you want to add the following class to your program:

```
class Employee
{
    public Employee(string name, int id)
     {
        this.name = name;
        this.id = id;
     }
    void ShowEmployee()
     {
```

```
        Console.WriteLine("Name: {0} id: {1}", name, id);
    }

    public string name;
    public int id;
}
```

Obviously, you can simply type the class statements into your program's source code. However, the Visual Studio Class Wizard will walk you through much of the class-creation process using a graphical user interface. To start the Class Wizard, perform these steps:

1. Within the Visual Studio, open the Class View window by selecting the View menu Class View option, as discussed in Tip 1695.
2. With the Class View window open, select the Project menu Add Class option. The Visual Studio, in turn, will display the C# Class Wizard dialog box, as shown in Figure 1696.1.

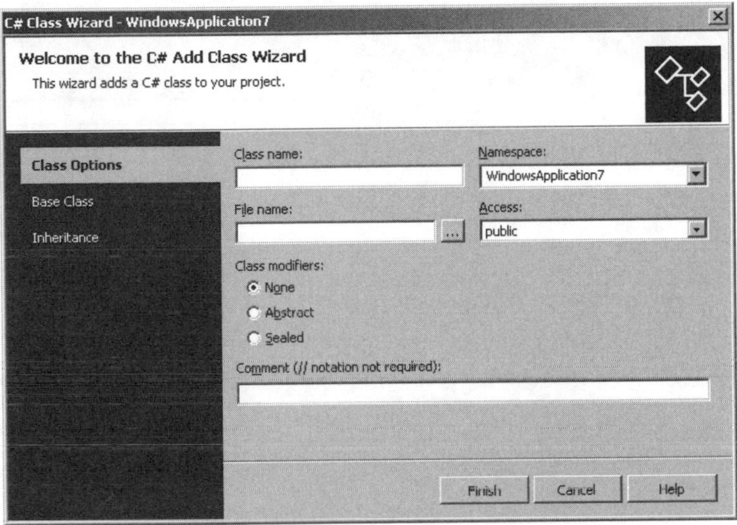

Figure 1996.1 The C# Class Wizard dialog box.

3. Within the C# Class Wizard dialog box, type in the class name Employee and then click your mouse on the Finish button. The Visual Studio, in turn, will create a class template that contains the following code, within which you can complete the class definition:

```
using System;

namespace ApplicationNamespace
{
   /// <summary>
   ///
   /// </summary>
   public class Employee
   {
         public Employee()
         {
               //
               // TODO: Add constructor logic here
               //
         }
   }
}
```

Take time to experiment with the Class Wizard. As you will find, the Class Wizard supports inheritance and can quickly generate code that ties a new class to a base class.

1697 DEFINING GENERAL ASSEMBLY INFORMATION

As you have learned, the .NET environment deploys applications as an assembly. Throughout this book, you have learned that .NET applications can rely on metadata to query applications, classes, and other objects about their capabilities. The same is true for an assembly. At the start of each assembly is a manifest that contains information about the assembly's purpose and capabilities. Using the Visual Studio Solution Explorer, you can quickly access and assign your application's general assembly fields, shown in Figure 1697. Using the general fields, you can specify your application's copyright information, version number, company name, and more. The assembly stores the information using C# attributes. In Tip 1698, you will learn how to query the assembly settings.

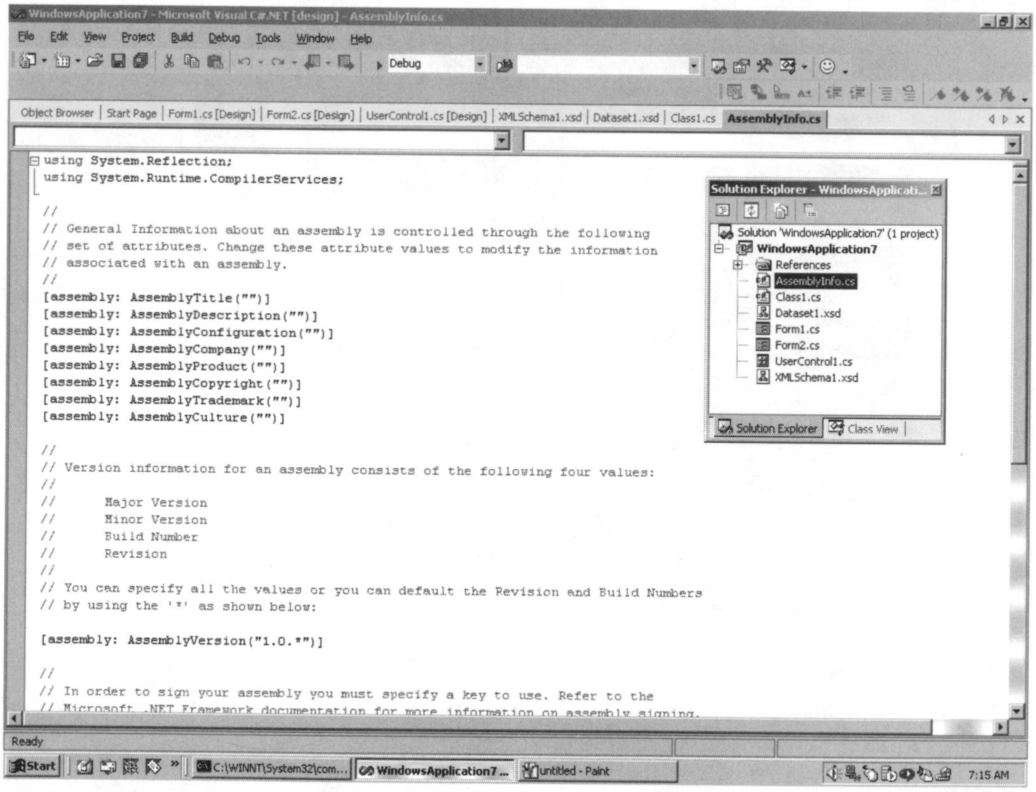

Figure 1697 Using the Solution Explorer to assign general assembly information.

1698 VIEWING AN ASSEMBLY'S GENERAL CONTENTS

In Tip 1697, you learned how to use the Visual Studio Solution Exporer to assign values to an application's general assembly fields. Using the C# Reflection class, an application can query the assembly's contents. The following program, *ShowGeneralAssembly.cs*, assumes you have used the Solution Explorer to assign values such as the following to the general Assembly entries for the Title, Description, Version, and Copyright attributes:

```
[assembly: AssemblyTitle("Show General Assembly")]
[assembly: AssemblyDescription("Access assembly data")]
```

```
[assembly: AssemblyCompany("Jamsa Press")]
[assembly: AssemblyCopyright("Copyright (C) JMG 2002")]
```

The program then uses the *Assembly* class *GetExecutingAssembly* to retrieve the assembly for the current program. The program then loops through the assembly attributes, displaying the values for the above attributes:

```
using System;
using System.Reflection;

public class ShowGeneralAssembly
{
  public static void Main()
   {
     Assembly Info = Assembly.GetExecutingAssembly();

     foreach (Attribute attr in
        Info.GetCustomAttributes(true))
      {
        if (attr is AssemblyCompanyAttribute)
          Console.WriteLine("{0}",
            ((AssemblyCompanyAttribute)attr).Company);
        else if (attr is AssemblyCopyrightAttribute)
          Console.WriteLine("{0}",
            ((AssemblyCopyrightAttribute)attr).Copyright);
        else if (attr is AssemblyDescriptionAttribute)
          Console.WriteLine("{0}",
         ((AssemblyDescriptionAttribute)attr).Description);
        else if (attr is AssemblyTitleAttribute)

             Console.WriteLine("{0}",
              ((AssemblyTitleAttribute)attr).Title);
      }
   }
}
```

After you compile and execute this program (remember you must first use the Solution Explorer to assign values to the assembly attributes), your screen will display the following output:

```
C:\Subdir> ShowGeneralAssembly   <Enter>
Copyright (C) JMG 2002
Jamsa Press
Access assembly data
Show General Assembly
```

REVISITING VERSIONING 1699

As discussed, a key feature of the .NET environment is that the applications can support versioning, which reduces the pain and suffering users experience when new software releases occur. As you have likely experienced, in the past, a user might install a new software program which updated a dynamic link library file (a DLL) that caused other programs dependent on the older DLL to fail. Troubleshooting such DLL conflicts is frustrating and quite time consuming.

Within a .NET assembly, the manifest metadata can provide applications with extensive information about the application (or library) and its classes. In fact, by using C# attributes, an application could place multiple versions of software within the same application. Then by querying the metadata, an application can determine which routine it should invoke based on the version support it requires.

The following program somewhat demonstrates how such versioning might work. In this case, for simplicity, all the code resides within one program. In a real-world example, the code might combine one application that must use a class that resides in the library file. Within the library, assume that two versions of the code exist. The application passes a value to the library that specifies the version it requires. The library, in turn, examines a "go-between" class for a class that corresponds to the version. To determine which version each class supports, the code queries each class's attribute information. When the code finds a class whose version matches the one that the application has selected, the code calls the corresponding class.

In this case, the application lets you select to run the Old version or the New version. When the application starts, it displays a form similar to that shown in Figure 1699.

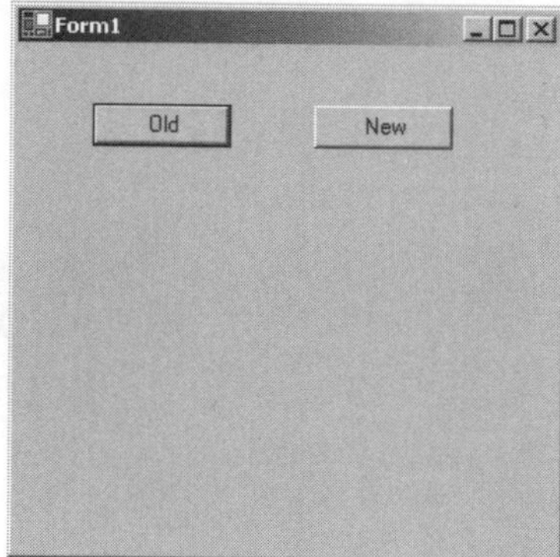

Figure 1699 Letting the user select the version desired.

Next, the code uses C# reflection capabilities to query the go-between class to determine the classes it contains. For each class, the code examines the class attributes to determine which version the class implements. If the class version matches the version the user selected, the code calls that class:

```
using System;
using System.Reflection;
using System.Drawing;
using System.Collections;
using System.ComponentModel;
using System.Windows.Forms;

public class Form1 : System.Windows.Forms.Form
 {
   private System.Windows.Forms.Button button1;
   private System.Windows.Forms.Button button2;
   private System.ComponentModel.Container
     components = null;

   public Form1()
    {
      InitializeComponent();
    }

   protected override void Dispose( bool disposing )
    {
      if( disposing )
        {
          if(components != null)
            {
              components.Dispose();
            }
```

```
        }
      base.Dispose( disposing );
    }

    #region Windows Form Designer generated code
    private void InitializeComponent()
     {
       this.button1 = new System.Windows.Forms.Button();
       this.button2 = new System.Windows.Forms.Button();
       this.SuspendLayout();
       //
       // button1
       //
       this.button1.Location = new
          System.Drawing.Point(40, 32);
       this.button1.Name = "button1";
       this.button1.TabIndex = 0;
       this.button1.Text = "Old";
       this.button1.Click += new
          System.EventHandler(this.button1_Click);
       //
       // button2
       //
       this.button2.Location = new
          System.Drawing.Point(160, 32);
       this.button2.Name = "button2";
       this.button2.TabIndex = 1;
       this.button2.Text = "New";
       this.button2.Click += new
         System.EventHandler(this.button2_Click);
       //
       // Form1
       //
       this.AutoScaleBaseSize = new
         System.Drawing.Size(5, 13);
       this.ClientSize = new System.Drawing.Size(292, 273);
       this.Controls.AddRange(new
         System.Windows.Forms.Control[] {
           this.button2,
           this.button1});
       this.Name = "Form1";
       this.Text = "Form1";
       this.ResumeLayout(false);
     }
     #endregion

    private void button1_Click(object sender,
       System.EventArgs e)
     {
       InterfaceClass.Choice = "Old";
       this.FindForm().Close();
     }

     private void button2_Click(object sender,
       System.EventArgs e)
       {
        InterfaceClass.Choice = "New";
        this.FindForm().Close();
       }
  }

class VersionInformationAttribute: Attribute
```

```
{
  public VersionInformationAttribute(string version)
   {
     this.version = version;
   }

  public string version;
}

class InterfaceClass
{
  public static string Choice;

  public static void CallRoutine(string choice)
   {
     if (choice == "New")
       new InterfaceClass.NewClass();
     else if (choice == "Old")
       new InterfaceClass.OldClass();
   }

  [VersionInformation("Old")]
  public class OldClass
   {
     public OldClass()
      {
        MessageBox.Show("Calling Old Version",
          "Old Version");
      }
   }

   [VersionInformation("New")]
  public class NewClass
   {
     public NewClass()
      {
        MessageBox.Show("Calling New Version",
          "New Version");
      }
   }
}

public class TwoVersions
 {
  public static void Main()
   {
     Application.Run(new Form1());

     Type type = typeof(InterfaceClass);

     foreach (MemberInfo Member in type.GetMembers())
       {
         foreach (Attribute attr in
             Member.GetCustomAttributes(true))
          {
            if (attr is VersionInformationAttribute)
             if
              (((VersionInformationAttribute)attr).version
               == InterfaceClass.Choice)
              {
            InterfaceClass.CallRoutine(
                    InterfaceClass.Choice);
```

```
            }
          }
        }
    }
}
```

REVISITING C# REFLECTION 1700

Throughout this book you have used the Reflection class to query a wide range of objects. As you have progressed within your knowledge of C#, you have examined classes, fields, properties, indexers, methods, interfaces, delegates, attributes, and much more. This book's final program, RevisitReflection.cs, presents a relatively complex class that uses many different class capabilities. The program queries the class using reflection to show just how much you can learn from a class about itself:

```
using System;
using System.Reflection;

class VersionInformationAttribute: Attribute
{
  public VersionInformationAttribute(string version)
   {
     this.version = version;
   }

  public string version;
}

interface SomeInterface
 {
   void SayHello();
 }

class SomeBase : SomeInterface
{
  public SomeBase()
   { // Do nothing
   }

  public void SayHello()
   {
     Console.WriteLine("Hello, C# World!");
   }
}

[VersionInformation("1.0A")]
class SomeClass : SomeBase
{
  public SomeClass(string value): base()
   {
     SomeString = value;
   }

  public string SomeString;

  private int somefield;

  public int Somefield
   {
     set
```

```
      {
        if (value <= 100)
          somefield = value;
        else
          somefield = -1;
      }
      get
      {
         return somefield;
      }
   }
}

public class RevisitReflection
{
  public static void Main()
   {
     Type type = typeof(SomeClass);
     Console.WriteLine("Class Attributes");

     foreach (Attribute attr in
       type.GetCustomAttributes(true))
      {
        Console.WriteLine(attr);
      }

     Console.WriteLine("\nClass Fields");
     foreach(FieldInfo field in type.GetFields())
      {
        Console.WriteLine(field);
      }

      Console.WriteLine("\nClass Methods");
      foreach(MethodInfo method in type.GetMethods())
       {
         Console.WriteLine(method);
       }

      Console.WriteLine("\nClass Properties");
      foreach(PropertyInfo property in
         type.GetProperties())
       {
         Console.WriteLine(property);
       }

      Console.WriteLine("\nClass Members");
      foreach(MemberInfo member in type.GetMembers())
       {
         Console.WriteLine(member);
       }

       Console.WriteLine("\nClass base type: {0}",
         type.BaseType);

       Console.WriteLine("\nClass Interface");
       foreach(object inter in type.GetInterfaces())
        {
          Console.WriteLine(inter);
        }
   }
}
```

After you compile and execute this program, your screen will display the following output:

```
C:\Subdir> RevisitReflection.cs  <Enter>
Class Attributes
VersionInformationAttribute

Class Fields
System.String SomeString

Class Methods
Void SayHello()
Int32 GetHashCode()
Boolean Equals(System.Object)
System.String ToString()
Void set_Somefield(Int32)
Int32 get_Somefield()
System.Type GetType()

Class Properties
Int32 Somefield

Class Members
System.String SomeString
Void SayHello()
Int32 GetHashCode()
Boolean Equals(System.Object)
System.String ToString()
Void set_Somefield(Int32)
Int32 get_Somefield()
System.Type GetType()
Void .ctor(System.String)
Int32 Somefield

Class base type: SomeBase

Class Interface
SomeInterface
```

Index